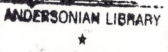

RAPIDLY
SOLIDIFIED
ALLOYS

MATERIALS ENGINEERING

1. Modern Ceramic Engineering: Properties, Processing, and Use in Design. Second Edition, Revised and Expanded, *David W. Richerson*
2. Introduction to Engineering Materials: Behavior, Properties, and Selection, *G. T. Murray*
3. Rapidly Solidified Alloys: Processes · Structures · Applications, *edited by Howard H. Liebermann*
4. Fiber and Whisker Reinforced Ceramics for Structural Applications, *David Belitskus*

Additional Volumes in Preparation

Thermal Analysis of Materials, *Robert F. Speyer*

Friction and Wear of Ceramics, *edited by Said Jahanmir*

Mechanical Properties of Metallic Composites, *edited by Shojiro Ochiai*

RAPIDLY SOLIDIFIED ALLOYS

Processes
Structures
Properties
Applications

edited by

Howard H. Liebermann

AlliedSignal Inc.
Parsippany, New Jersey

Marcel Dekker, Inc. New York • Basel • Hong Kong

Library of Congress Cataloging-in-Publication Data

Rapidly solidified alloys: processes, structures, properties, applications / edited by Howard H. Liebermann.
p. cm. -- (Materials engineering; 3)
Includes bibliographical references and index.
ISBN 0-8247-8951-2
1. Alloys--Rapid solidification processing. I. Liebermann, Howard H. II. Series.
TS247.R368 1993
671.3--dc20 93-7516
CIP

The publisher offers discounts on this book when ordered in bulk quantities. For more information, write to Special Sales/Professional Marketing at the address below.

This book is printed on acid-free paper.

MARCEL DEKKER, INC.
270 Madison Avenue, New York, New York 10016

Current printing (last digit):
10 9 8 7 6 5 4 3 2 1

PRINTED IN THE UNITED STATES OF AMERICA

Preface

During the last thirty years, the rapid solidification of metals and alloys has evolved from laboratory-scale research experiments to commercial-scale manufacturing operations. Thousands of technical papers and patents and many books and conference proceedings on the subject have been published. The purpose of this book is to serve as a resource for both students and researchers who require a timely, comprehensive treatment of the subject as provided by contributors who are world-recognized experts in their respective fields.

The organization of this book progresses logically from processes to structures and properties, and finally to applications. I have attempted to make each of these three sections represent a broad, yet focused diversity of topics. Extensive referencing of each chapter will facilitate the search for additional information on very specific topics and a materials index is also included. Great care has been taken to provide as much coherency and uniformity as possible in a work with so many different contributors, while still retaining the individual style of each author.

I sincerely thank each of the contributing authors for their dedication, diligence, and patience in working with me in the venture that resulted in this book. I thank the staff of the publisher, Marcel Dekker, Inc., for their high level of cooperation and interest in the production of this work, my colleagues working at AlliedSignal Advanced Materials, Amorphous Metals business unit for their advice, and Amorphous Metals for providing the secretarial services required to handle the many communications during the writing of this book. Finally, I am especially grateful to my wife Lynda and children Daniel and Amanda for their patience with me during the course of this book's production.

Howard H. Liebermann

Contents

Preface iii
Contributors ix

Part I Processes

1 Background to Rapid Solidification Processing 1
R. W. Cahn

2 Fundamentals of Solidification at High Rates 17
W. J. Boettinger and J. H. Perepezko

3 Undercooling and Solidification 79
D. M. Herlach and R. Willnecker

4 Processing Principles in Rapid Solidification 103
Paul Hideo Shingu and Keiichi N. Ishihara

5 Rapid Solidification Processing of Composites 119
E. R. Murray and D. G. Ast

6 Production Techniques of Alloy Wires by Rapid Solidification 139
M. Hagiwara and A. Inoue

7 Synthesis and Consolidation of Amorphous Metallic Powders 157
R. B. Schwarz

8 Rapidly Solidified Surface Layers by Laser Melting 195
E. Schubert and H. W. Bergmann

Part II Structures and Properties

9 Atomic Structure of Rapidly Solidified Alloys 231
T. Egami

10 Structural Relaxation and Atomic Transport in Amorphous Alloys 269
A. L. Greer

11 Phase Transformations in Rapidly Solidified Alloys 303
Uwe Köster and Uwe Schünemann

12 Microstructure-Property Relations in Rapidly Solidified Crystalline Alloys 339
S. K. Das and F. H. Froes

13 Recent Studies of the Mechanical Properties of Amorphous Alloys 379
J. C. M. Li

14 Electronic Transport Properties 431
T. Richmond and H.-J. Güntherodt

15 Low-Temperature Properties of Rapidly Solidified Alloys 461
H. v. Löhneysen

16 Intrinsic Magnetic Properties of Amorphous Alloys 505
T. Mizoguchi

17 Engineering Magnetic Properties 553
A. Hernando and M. Vázquez

18 Chemical Properties of Rapidly Solidified Alloys 591
K. Hashimoto

Part III Applications

19 Applications of Rapidly Solidified Soft Magnetic Alloys 617
Carl H. Smith

20 Rapidly Solidified Nd-Fe-B Alloy for Permanent Magnet Applications 665
M. S. Guthrie

21 Brazing and Soldering with Rapidly Solidified Alloys 691
A. Rabinkin and Howard H. Liebermann

22 Mechanical, Chemical, and Electrical Applications of Rapidly Solidified Alloys 737
C. Suryanarayana and F. H. Froes

23 Rapid Solidification: Origins, Present, and Future 755
H. Jones

Materials Index 765
Subject Index 777

Contributors

D. G. Ast, Ph.D. Professor, Department of Materials Science and Engineering, Cornell University, Ithaca, New York

H. W. Bergmann, Ph.D. Professor, Division of Metals Technology, Department of Material Science, University of Erlangen Nürnberg, Erlangen, Germany

W. J. Boettinger, Ph.D. Metallurgist, Metallurgy Division, National Institute of Standards and Technology, Gaithersburg, Maryland

R. W. Cahn, F.R.S. Professor, Department of Materials Science and Metallurgy, University of Cambridge, Cambridge, England

S. K. Das, Ph.D. Senior Manager, Metals Laboratory, AlliedSignal Corporate Research and Technology, Morristown, New Jersey

T. Egami, Ph.D. Professor, Department of Materials Science and Engineering, University of Pennsylvania, Philadelphia, Pennsylvania

F. H. Froes, M.S., Ph.D. Director and Professor, Institute for Materials and Advanced Processes, University of Idaho, Moscow, Idaho

A. L. Greer, M.A., Ph.D. University Lecturer, Department of Materials Science and Metallurgy, University of Cambridge, Cambridge, England

H.-J. Güntherodt, Dr. Sc.nat., Professor, Department of Physics, University of Basel, Basel, Switzerland

M. S. Guthrie* Delco Remy Magnequench, General Motors Corporation, Anderson, Indiana

M. Hagiwara Unitika Research Center, Unitika Ltd., Uji, Japan

K. Hashimoto, D.Sc. Professor, Institute for Materials Research, Tohoku University, Sendai, Japan

D. M. Herlach Institut für Raumsimulation, Deutsche Forschungsanstalt für Luft-und Raumfahrt, Köln, Germany

A. Hernando Instituto de Magnetismo Aplicado, Madrid, Spain

A. Inoue Institute for Materials Research, Tohoku University, Sendai, Japan

Keiichi N. Ishihara Assistant Professor, Department of Metal Science and Technology, Kyoto University, Kyoto, Japan

H. Jones, Ph.D., F.I.M. Professor, Department of Materials Engineering, University of Sheffield, Sheffield, England

Uwe Köster Professor, Department of Chemical Engineering, University of Dortmund, Dortmund, Germany

Howard H. Liebermann, Ph.D. Manager, Research and Development, Amorphous Metals, AlliedSignal Advanced Materials, Parsippany, New Jersey

J. C. M. Li, Ph.D. Professor, Department of Mechanical Engineering, University of Rochester, Rochester, New York

H. v. Löhneysen Professor, Institute of Physics, University of Karlsruhe, Karlsruhe, Germany

T. Mizoguchi, Ph.D. Professor, Department of Physics, Faculty of Science, Gakushuin University, Tokyo, Japan

---

* *Present affiliation:* Regional Applications Engineering Manager, Magnet Division, Stackpole Carbon Company, Kane, Pennsylvania

E. R. Murray, M.S., Ph.D.* Cornell University, Ithaca, New York

J. H. Perepezko Professor, Department of Materials Science and Engineering, University of Wisconsin-Madison, Madison, Wisconsin

A. Rabinkin, Ph.D. Group Leader, Materials Development, Amorphous Metals, AlliedSignal Advanced Materials, Parsippany, New Jersey

T. Richmond, B.Sc., Dr. Department of Physics, University of Basel, Basel, Switzerland

E. Schubert Department Leader, Department of New Materials and New Technologies, ATZ-EVUS, Vilseck, Germany

Paul Hideo Shingu, Ph.D. Professor, Department of Metal Science and Technology, Kyoto University, Kyoto, Japan

Uwe Schünemann Flachglas AG, Gelsenkirchen, Germany

R. B. Schwarz, Ph.D. Center for Materials Science, Los Alamos National Laboratory, Los Alamos, New Mexico

Carl H. Smith, Ph.D. Supervisor of Magnetic Alloys Research, Metals Laboratory, AlliedSignal Research and Technology, Morristown, New Jersey

C. Suryanarayana, M.Sc. (Eng.), Ph.D. Professor, Institute for Materials and Advanced Processes, University of Idaho, Moscow, Idaho

M. Vázquez Instituto de Magnetismo Aplicado, Madrid, Spain

R. Willnecker Institut für Raumsimulation, Deutsche Forschungsanstalt für Luft-und Raumfahrt, Köln, Germany

---

* *Present affiliation:* Instructor and Department Assistant, Department of Physics, Haverford College, Haverford, Pennsylvania

1

Background to Rapid Solidification Processing

R. W. Cahn
University of Cambridge, Cambridge, England

I. QUENCHING IN THE SOLID STATE: METHODS AND LIMITATIONS

Hot steel has been quenched into water or oil, to harden it, for a millenium at least, and the technique has acquired many subtle variants. Numerous solutes, some of the ancient ones with magical overtones, have been added to quenching water to improve swords and armor; in recent years, polymer additions have enhanced control over quenching rates in heat treatment shops. In this process, 1000 K/s would be reckoned an exceptionally high rate of cooling, dangerous because apt to cause cracking.

Attainable cooling rates depend primarily on the dimensions of the workpiece, and high cooling rates have been attained for research purposes by quenching thin foils by means of a fast-moving stream of a gas possessing high thermal conductivity. According to a 1951 paper by Duwez [1], with a metallic foil 50 μm thick quenched in a helium stream moving at several hundred feet per second, cooling rates of 10,000–20,000 K/s can be achieved. That is probably close to the practicable limit for solid-state quenching. Continuous heating followed by immediate quenching of thin sheet just after cold-rolling—the so-called continuous annealing process for steel sheet—is now normal industrial practice in Japan [2].

Quenching of hot steel is of course only one of many industrial applications of solid-state quenching, all of them limited to modest cooling rates. All such applications have as their aim to preserve at room temperature a *metastable* phase structure, either for use "as is" or for subsequent modification in the direction of thermodynamic equilibrium without necessarily reaching it, as in the tempering of hardened steel. It is no exaggeration to say that the artful control of metastability is the metallurgist's central skill.

Often, a solid-state quench preserves metastably the structure stable at the temperature from which the alloy is quenched, though under appropriate circumstances the alloy can transform to another nonequilibrium structure during the quench: this of course happens during the quenching of austenite to generate martensite. It is natural for an inquisitive metallurgist to wonder what would happen if an alloy is quenched from the molten state: the preservation, metastably, of the structure stable at the starting temperature would yield a congealed liquid, that is to say, a glass. Alternatively, from analogy with the generation of martensite, one might expect the creation of a metastable crystalline structure during the quench. Such inquisitiveness on the part of an unusually adventurous metallurgist, Pol Duwez, gave rise to the subject matter of this book.

Since vapor is the third state of matter, it should also be possible to quench from that state. Before turning to the proper topic of this book, then, I shall briefly look at this alternative approach. I shall use the term "rapid quenching" to denote, collectively, quenching from either the liquid or the vapor state, because both methods permit much faster quenching rates than quenching from the solid starting point.

II. EARLY EXPERIMENTS ON QUENCHING FROM THE VAPOR PHASE

Evaporation in vacuo has long been used to create thin films for research purposes, and whenever the substrate is cooler than the evaporation source, the vapor is necessarily quenched on deposition. Possibly the first record of a clearly anomalous structure being generated in this way is by the German physicist Kramer [3], who claimed to have produced amorphous metals (for example, Sb) by this method. However, this work had no follow-up for many years, and the first application of this technique to excite widespread attention was by two physicists in Göttingen, Buckel and Hilsch [4–7]. Their interest was in superconductivity and vapor deposition for them was merely a means of preparing thin specimens of various metals and alloys; since they were about to measure electrical conduction at temperatures maintained by the use of liquid helium, they reckoned to save time by depositing their films direct on to substrates held at 4 K, and in this way they incidentally assured a very drastic quench from the vapor phase.

Buckel and Hilsch found surprising anomalies in the superconducting behavior of these films when compared with specimens cut from "massive" solids made by casting in the usual way, and accordingly they examined the structure of their anomalous films of Bi, Ga, Sn, and Sn-Cu alloys, including an early use of electron diffraction [6]. They concluded that the films were *feinkristallin* or even *feinstkristallin*; the former term was asserted to imply a grain size > 100 Å, the latter, a grain size ≈ 40 Å. A little later it had become clear that some of these films, at least, were in fact amorphous: this was true of Bi, Ga, and Sn-Cu. In 1956 Buckel and Hilsch [7] also established that the normal solid solubility of one metal in another could be greatly exceeded by films prepared by vapor quenching (the term was not yet in use): thus, the normally very slight solubility of Cu in Sn could be stretched beyond 20 at% and that of Bi in Sn to 45 at%. When this concentration of Bi was exceeded, both components were deposited in the *feinstkristallin* state.

Thus, this classic early study established two key features of rapid quenching: the extension of solid solubility and the creation of vitrified, i.e., amorphous (or glassy) metals or alloys. In fact, through the accident of the authors' preferred metals, they chanced at this early date to find two of the few metals known to be vitrifiable in the pure, unalloyed state.

However, there is still disagreement whether the Bi films made by Buckel and Hilsch's method are in fact amorphous, or only microcrystalline. Thus, Bergmann [8], in a comprehensive review of amorphous superconductivity, cites a study by Comberg et al. [9] which claims to show, on the basis of Hall effect measurements in particular, that vapor-quenched Bi is in fact microcrystalline. Ever since Buckel and Hilsch's pioneering study, arguments have raged between those who identify a particular product as being amorphous and those who believe it to be microcrystalline. Only very recently have Chen and Spaepen [10] invented a method, the use of a differential scanning microcalorimeter to establish the form of the isothermal heat release during annealing of a disputed phase, which seems to be capable of making a decisive distinction between amorphous and microcrystalline materials (because in the one case nucleation and growth of grains is involved, while in the other, preexisting grains grow, and the two situations have different heat-release fingerprints). Somewhat later, it was established that other metals such as Pb, Sn, In, Tl, etc., could also be forced into the amorphous state by codepositing them with 10–20 at% of a second component (for details, see Bergmann [8]).

In recent years, vapor quenching has from time to time been used to make products of substantial size, for industrial use as well as research, by means of a "high-rate" approach, using either sputtering [11] or high-power thermal evaporation [12,13] to make the vapor, but it is fair to say that (at least as far as can be judged from the open literature) the method has never really "taken off" as an industrial production method when compared with quenching from the liquid state. No doubt the great expense of vapor quenching, even the high-rate variety, is the reason for this hesitation to apply the approach.

III. AMORPHOUS ALLOYS BY ELECTRODEPOSITION

In the pre-Duwez era of which I have been writing, there is only one other record of an amorphous alloy being made, and this in fact antedated Buckel and Hilsch. Brenner, the leader of the distinguished electrodeposition group at the National Bureau of Standards which specialized in the difficult art of electrodepositing alloys as distinct from pure metals, had reported that Ni-P alloys can be deposited in amorphous form [14]. Subsequently, much further work was done on these alloys (which are widely applied in industry as wear-resistant coatings) [15] and there is no doubt as to their amorphous nature when the composition is just right.

IV. THE ORIGINS OF QUENCHING FROM THE LIQUID STATE

Pol Duwez, an imaginative metallurgist working at the California Institute of Technology, is universally regarded as the father of rapid quenching from the liquid state. While he is certainly the father, he is not, in Shakespeare's phrase, the "onlie begetter." This is because others before him made wires and ribbons by rapid freezing, but they appear to have been concerned primarily if not exclusively with inventing a production technique for cheaply making the relevant shapes, whereas Duwez was explicitly interested in the metallurgical consequences of the rapid cooling—that is, he was looking at the technique from a researcher's viewpoint while his predecessors were looking at it from a production engineer's perspective. The considerations that led him to his innovations are very clearly set out in his own accounts of his early researches [16,17]. He had a general, and a particular, objective in mind. The general objective was to see whether quenching from the liquid might metastably retain at room temperature a congealed liquid, that is, a glass, or alternatively create previously unknown metastable structures during the quench itself. In other words, he wished to discover whether quenching from the liquid would recreate the behavior, mutatis mutandis, known to be characteristic of quenching from the solid state. Duwez's particular objective was to solve a puzzle from alloy theory: according to Hume-Rothery's celebrated rules applicable to solid solutions based on copper and noble metals, gold, silver, and copper should all be entirely intersoluble in pairs because of their similar atomic radii, whereas in fact, while Cu-Au and Ag-Au alloys behave in this way, Cu-Ag alloys do not, and form a eutectic system, admittedly with extensive solid solubility in the terminal phases. Duwez realized that solid quenching could be of no use in resolving this puzzle, since quenching a two-phase solid structure could scarcely be expected to convert it into a metastable single-phase structure. However, quenching from a single-phase *liquid* could in principle convert this into a metastable single-phase crystalline solid. Duwez wrote explicitly in 1967 that "the possibility of removing this rather exceptional case from the list of binary alloys which did not follow the Hume-Rothery criteria was the main incentive for finding an experimental technique capable of achieving extreme rates of cooling from the liquid state" [17]. In one of the very first publications in the field, Duwez et al. [18] were able to show that melt-quenched Cu-Ag alloys in fact showed complete metastable intersolubility.

Duwez had some (American) precursors and there was also another American who made his inventions around the same time as Duwez, on the other side of the country. The precursors include Lang, who in 1871 patented a simple form of melt-spinning wires of low-melting alloys, Strange and Pim, who in 1908 again patented [19] a form of *melt spinning* (formation of a quenched ribbon from liquid alloy, using a metal wheel to abstract heat), and Pond, who in patents of 1958, 1959 and 1961 which later became influential, both invented free-flight melt spinning (in which a stable liquid jet issuing from a nozzle is quenched by the surrounding gas while still in flight) [20] and reinvented the 1908 technique of chill-block melt spinning [21]. (Melt spinning and its derivatives, which as we have seen have been repeatedly reinvented, have become by far the most common methods of melt quenching, even though Duwez's early work did not depend at all on this method). Apart from these investigators, Duwez in 1967 himself cited the work of the Germans Falkenhagen and Hofmann [22], who by casting alloy melts into copper molds refrigerated with liquid nitrogen were able to extend solid solubilities in several systems. (He might have added a 1937 initiative by the Germans Gröber and Hanemann [23] who invented a "quench-injection" method of making wire by sucking liquid alloy into a cylindrical cavity in a chilled copper mold.) He also cited Olsen and Hultgren [24] (like himself in California), who in 1950 (shortly before Duwez's early work on gas quenching of solid foils) injected tiny liquid alloy (Ni-Cu) globules into iced brine to study segregation and estimated a (calculated) cooling rate of nearly 100,000 K/s. It is to be presumed that this estimate of cooling rate, substantially greater than anything achieved in solid-state quenching, must have encouraged Duwez in his pioneering approach to melt quenching.

Duwez decided in 1959 that the most favorable conditions for achieving rapid cooling from the melt required the rapid creation of a *thin* layer of melt in intimate and effective contact with a solid of good thermal conductivity. (He had tried Olsen and Hultgren's approach with a concentrated Cu-Ag alloy and did not succeed in creating a metastable solid solution, and so he decided to eschew quenching into a fluid medium.) He was therefore obliged to try the only alternative, the use of a solid substrate (or, as we say nowadays, a chill block). He tried two versions, the gun and the piston-and-anvil apparatus, shown in Figs. 1 and 2, respectively. In the gun, a high-pressure reservoir of inert gas is built up above a Mylar diaphragm, and when the compressed gas finally ruptures, a shock wave forms and breaks up the small melt drop into minute droplets which are propelled rapidly onto a polished copper substrate strip. The tiny flakes thus generated are ideal for x-ray diffraction and (as was found a little later) for TEM. In the piston-and-anvil device, later modified into a two-piston variant, a falling melt drop breaks a light beam and this releases a pneumatically operated piston that squashes the falling drop against the anvil piston, between two copper sheets (Fig. 2). This generates larger and thicker foils, useful when physical properties such as resistivity are to be measured.

It should be mentioned here that, at about the same time as Duwez's first papers in the field, two Russians, Miroshnichenko and Salli, quite independently published

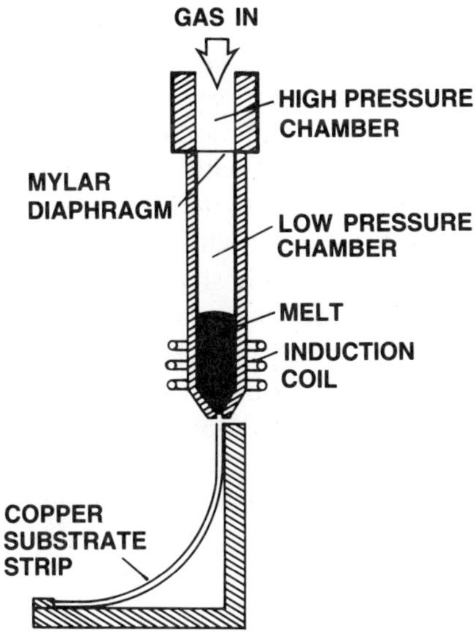

Figure 1 Gun technique of Duwez for rapid solidification of melts.

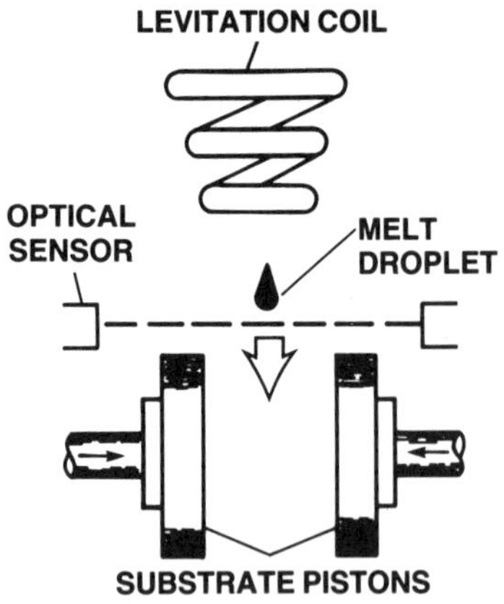

Figure 2 Drop smasher, using levitation melting by induction (the pistons are pneumatically or electromagnetically accelerated).

an account of a variant of the Duwez gun [25]. However, there has been very little follow-up from the Soviet Union since then.

Not only were Duwez and his co-workers able, by means of these instrumental innovations, to answer his own question about Cu-Ag alloys, but his colleague Klement [26] was able to show an extension of solid solubility (as in Buckel and Hilsch's work) in Ag-Ge alloys, where the solubility of Ge in Ag was extended, an observation judged to be in accordance with Hume-Rothery's predictions; in the same study, a new metastable intermetallic compound, not found in equilibrium, was observed in the Ag-Ge system. (A little later, an anomalous simple cubic phase was discovered in the Te-Au system [27].)

Viewed from the perspective of scientific fruitfulness, undoubtedly the most pregnant of the discoveries of the Duwez group was the finding of a glass in the Au-Si system [28]. Hardly had this discovery been published (*Nature* can publish very rapidly), when Cohen and Turnbull [29] published another paper in the same journal in which they made the suggestion that the ready formation of a metallic glass in the Au-Si system at around 25 at% Si was connected with the existence of a deep eutectic in equilibrium near this composition. They pointed out that this gave the melt the chance to cool stably to a low temperature at which its viscosity was quite high and therefore diffusion in the melt was relatively sluggish; crystallization, which requires diffusion, would therefore be sluggish enough to enable the melt to congeal during a rapid quench. This at once provided a systematic basis for finding other glass-forming systems, as Duwez [17] has himself recorded, and this led quickly, inter alia, to the discovery of the important Pd-Si glass-forming alloys [30]. Soon after, Duwez used the electron microscope both to confirm the amorphous nature of the quenched Pd-Si alloy and to study its crystallization in situ in the microscope. He may well have been the first to apply TEM to a melt-quenched alloy.

Morrel Cohen [31] has recently given a personal account of the chance circumstances (the sharing of a room by the two authors at the Cavendish Laboratory) which led to the writing of their influential joint paper; since that time, Turnbull and Cohen have written a long series of joint papers on glass formation, in both metallic and nonmetallic systems, and both have been extremely influential in this field. Turnbull and Cohen's analysis was so important because it offered a reasoned basis for choosing promising alloys to turn into metallic glasses. Its role was similar to Merica's recognition in the early 1920s that a terminal solid solubility decreasing sharply with falling temperature is a necessary precondition of age hardening. Until that recognition, only one age-hardening alloy was known; thereafter, their number increased rapdily.

Once the paper on Pd-Si glasses had been published in 1965, the early history of melt quenching was essentially complete. The scientific topics had been mapped out and a very fruitful period of research was under way. The technological promise was slower in delivery, as we shall see.

A note on nomenclature. Duwez was in the habit of speaking and writing cautiously of "rapid quenching from the liquid state"; this is something of a mouthful, and the shorter form, "melt quenching," soon established itself. The onomatopoeic term "splat quenching" then surfaced; Duwez himself found it distasteful and avoided it, and its usage appears to have come and gone. As melt quenching increasingly entered industrial practice, the term "rapid solidification processing" appeared and finds increasing favor. It is used in the title of this chapter.

The piston-and-anvil method in due course was named "drop smashing," and this is still sometimes used. Specific terms that apply to particular quenching methods, such as melt spinning, melt extraction, planar flow casting, centrifugal atomization, and the like, are in stable use. "Vapor quenching" is also an established term, as are more recent coinages such as "solid-state amorphization reaction." An attempt was made some years ago to introduce the term "liquisol quenching" for quenching from the liquid to the solid state, but it did not find favor.

The international conferences on melt quenching which have been held at intervals since 1970 have mostly had the title Rapid Quenching from the Melt. A purist might object that "rapid quenching" is a tautology, but when one recalls that quenching from the melt allows much faster cooling than quenching from the solid state, perhaps the apparent tautology may be accepted as making a valid distinction.

V. MELT SPINNING AND ITS DERIVATIVES AND ANALOGUES

It cannot sufficiently be emphasized that the driving force of the innovations due to Duwez and his collaborators was scientific curiosity: melt quenching was initially a means to a purely scientific end. By contrast, the early studies and patents concerned with making wires and ribbons directly from the melt by free-flight melt spinning, chill-block melt spinning and other techniques such as quench injection appear to have been aimed at improving industrial production techniques, and the fast quench was, if anything, an incidental nuisance (it certainly was troublesome in the development of quench injection). Again, the "Taylor wire method" of 1924 for making extremely fine wires by drawing down a metallic rod encased by an oxide glass tube [32] was introduced as just that, a way of achieving the difficult objective of producing very fine wires. After a 50-year gap this method was resuscitated and very recently a comprehensive survey of its potential has been

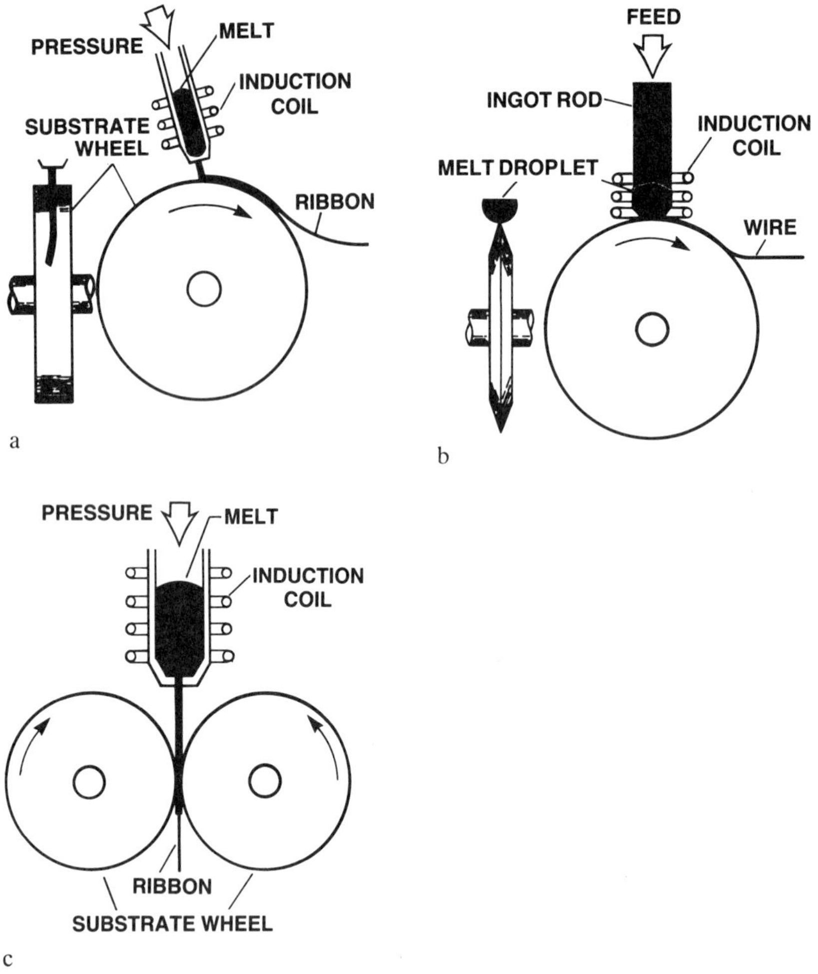

Figure 3 Principal methods of rapid quenching of alloy ribbon or fiber from the melt: (a) chill-block melt spinning; (b) pendant-drop melt extractions; (c) twin-roller quenching; (d) planar flow casting (PFC); (e) in-rotating water spinning (INROWASP) of wire.

published [33]; the emphasis is still more on the method's production benefits than on its rapid quenching characteristics. Likewise, free-flight melt spinning as patented by Pond in 1959 came to be better understood (the central problem was recognized to be the avoidance of Rayleigh instabilities which break up a stable jet into drops, and this can be achieved by alloying the metal with an oxide-forming constituent to create a rapidly formed and stable surface oxide film) and the technique was tried out, unsuccessfully, as a possible way of making tire cord on a large scale [34]. Nevertheless, the method has found uses, some of which at least exploit its rapid quenching characteristics [35].

(Single-roller) chill-block melt spinning (CBMS) is another matter altogether: initially developed as a production method, it was then used for research purposes by Pond and Maddin (in a variant which involved an outward facing jet impinging

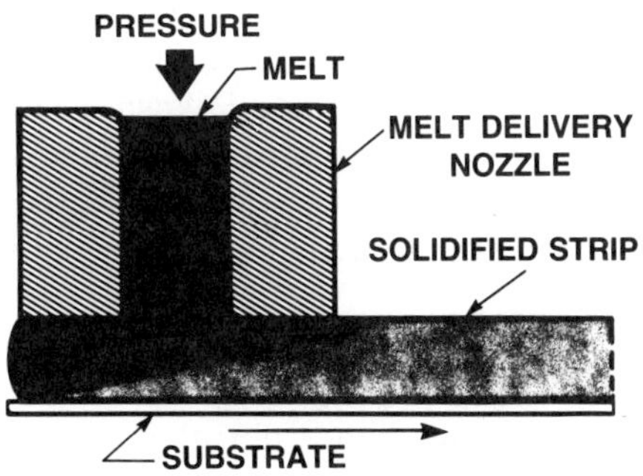

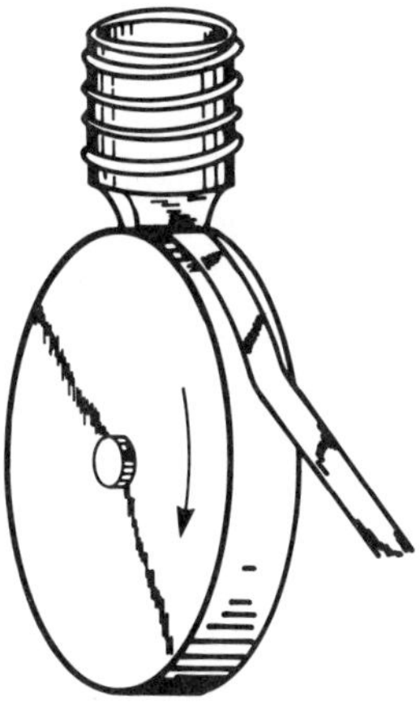

d

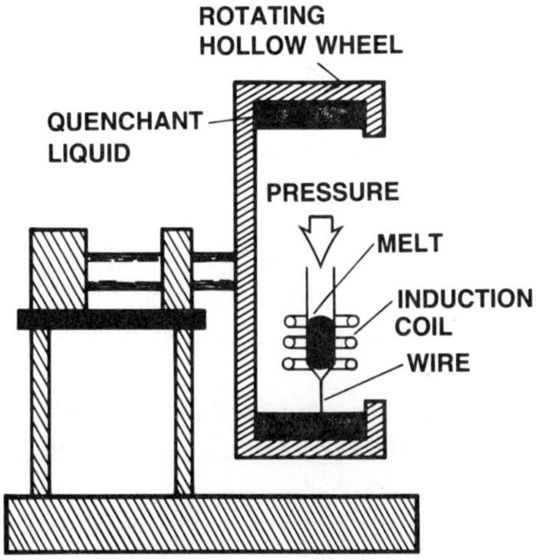

e

on the *inner* surface of a rotating ring) [36] and soon afterwards applied by Masumoto and Maddin to the production of ribbon substantial enough to make tensile test pieces [37]. Masumoto took the technique back to Japan and there it has been used in a long and distinguished series of researches on both metallic glasses and melt-quenched crystalline alloys during the past 20 years. The focus in all this work has been on investigation and exploitation of rapid quenching. Most commonly used melt-spinning devices today operate by projecting a melt jet on the *external* surface of a rotating wheel. The origin of the modern design of this deceptively simple device (deceptive because of the many hidden variables which govern the quality of the resultant ribbon) can be traced to a patent by Bedell [38].

The more recent development of chill-block melt spinning in its most important form, that of the single-roller device, is beyond the scope of this introductory survey. An excellent and detailed account of the development of this and all the other production techniques of rapid quenching, together with an unusually full bibliography, can be found in a recent book by Anantharaman and Suryanarayana [39].

Another variant of single-roller CBMS produces melt-quenched wires, typically 100–200 mm in diameter. This method was invented and perfected in Japan (again, it is much more difficult to get right than appears at first sight) and has largely been exploited there for both research and production. Ohnaka and Fukusako [40] first reported on the technique of directing a jet *outward* into a rotating annulus of water: the method can generate either powder or a wire, according to whether the jet is stable or breaks up. A more accessible account in English has recently been published [41].

There is no need to dwell here on the double-roller form of CBMS [42], on which a good deal of effort has been expended though its quenching performance is not as good as that of single-roller CBMS. The subtleties concerning this method are well outlined by Anantharaman and Suryanarayana [39].

The key development that permitted single-roller CBMS to become a major industrial production technique was its conversion into the process which has come to be known as "planar flow casting." The Pond-Maddin-Masumoto version of single-roller CBMS did not permit wide sheets to be made; the width limit was only a few millimeters. The new process, patented by Narasimhan [43], involves a jet orifice held very close to the rotating wheel so that the resultant melt puddle is physically constrained between the wheel and the lips of the nozzle. With this geometry (the details of which have to be precisely right) it is possible to use nozzles in the form of long slits and hence to make wide sheet, more than simply a few centimeters across, of high uniformity in dimensions and properties. The technique is now commercially exploited to make soft magnetic foil for transformers and foils for other uses, such as brazing and soldering. Another variant of the process entails the production of a variety of crystalline alloys which are immediately comminuted and consolidated into various massive shapes.

What is publicly known or can be deduced about the planar flow casting process (its details are carefully protected) can be found in Anantharaman and Suryanarayana's book.

A close relative of single-roller CBMS is the melt-extraction process—actually a family of processes—in which, essentially, a rotating chill block having a circumferential knife-edge dips into a melt and extracts a filament of D-shaped cross section. The great benefit of this method, in at least one of its variants (the pendant-drop version) is that problems with crucibles are bypassed. The technique goes back to Pond's 1959 patent [20] and was largely developed in Battelle's Columbus Laboratories; some details can be found in previously cited reviews [35,39]. The technique is in commercial use to make stainless steel reinforcement for refractories and concrete.

Fig. 3 shows in diagrammatic form a number of the techniques which have been outlined in this section.

VI. OTHER MELT-QUENCHING AND RELATED TECHNIQUES

The most important industrial family of techniques, other than the CBMS methods, undoubtedly is the range of atomization techniques for making melt-quenched powder. When powder metallurgy began, between the World Wars I and II, atomization was the simplest and most convenient method of making powders for sintering, and rapid quenching was not considered as a benefit. Grant, at M.I.T., was probably the first to investigate the potential of alloys (his early interest was largely focused on aluminum) made by generating powders quenched as rapidly as possible and then consolidated; his early experiments were reported at the First International Conference on Rapidly Quenched Metals [44] (it was then entitled a "Conference on Metastable Metallic Alloys"). Numerous improvements in the techniques of making melt-quenched powders came in the 1970s and early 1980s; they included notably the use of ultrasonic atomization, introduced in the late 1970s [45] and largely developed by Grant's group. Other variants include centrifugal atomization (itself developed in several variants), twin-roll atomization, soluble-gas atomization, electrohydrodynamic atomization and, most important, the numerous spray deposition methods originated in the U.K. [46–48] and elsewhere. Some of these processes produce loose powder which must then be warm-consolidated in such a way as not to lose the properties induced by melt quenching, others produce a solid preform immediately. Perhaps the most influential of these processes has been the British "Osprey" process, a form of "spray forging" on to a moving substrate, so-called because the technique permits such a high density ($<1\%$ porosity) that the preform can be subsequently forged. These industrially vital techniques are well surveyed by Anantharaman and Suryanarayana [39].

A somewhat related technique is spark erosion (also called electrical discharge machining). The method has long been used to shape hard metals which are difficult or impossible to shape by cutting, but recently it has been used to make melt-quenched powder in relatively small quantities. For the purpose of making powder, a series of electrical discharges is passed between an electrode and a bed of pieces of the same alloy immersed in a dielectric liquid. Tiny pieces of alloy are torn off, melted and then converted into a plasma bubble, and these pieces then recondense and solidify very fast. The technique has largely been developed by Walter at G.E. in America [49]. Very refractory alloys can be melt-quenched to powder by this means, without difficulty, but there is the problem of chemical interaction between the molten droplets and the dielectric liquid, which can be serious for high-melting metals [50].

Melt-quenching can be achieved locally, at surfaces, by laser treatment. This was originally regarded as a research tool; early studies [51] with pulsed lasers led to the demonstration of solubility extension [52] (in fact, this study was devoted to the same Cu-Ag alloys which had sparked off the whole field 14 years earlier), while another study demonstrated metastable phase formation in Ag-Ge alloys [53]. Later, laser treatment (both pulsed and steady) became a widespread industrial surface-hardening procedure; its importance is attested by a long series of conference reports and survey volumes, the most recent of which has just appeared [54]. Laser treatment is now mainly applied as an industrial process, but is still occasionally used as a research tool in metallurgy. For instance, a pulsed laser has been used to convert pure gallium held at low temperatures into a glass [55]; the vitrification of this metal has fascinated researchers ever since Buckel and Hilsch's pioneering studies of 1954 to 1956.

Finally, there is a range of techniques that achieve the same end result as melt quenching, but by radically different means (not involving the formation of a liquid phase at any stage). These include the various ion processes (ion implantation and ion mixing), irradiation amorphization, mechanical alloying and the so-call solid-state amorphization reaction. These are of only marginal relevance in the context of the present book, so the bare mention must suffice. It is perhaps worth adding that solid-state amorphization (in which the interdiffusion of two metallic thin films generates an amorphous interlayer), was only discovered in 1983 [56] but has already justified the convocation of a specialised conference, held in 1987 [57].

VII. CALIBRATION OF COOLING RATES IN MELT-QUENCHING PROCESSES

When the early experiments were being done by Duwez and his immediate successors, little interest was expressed in the determination of the actual cooling rates achieved by the various melt-quenching techniques: the proof of the pudding was in the eating; that is, the metallurgical consequences were what mattered. A little later, when interest began to develop in the factors governing critical cooling rates needed for various metallic melts to vitrify, it became desirable to measure cooling rates with reasonable accuracy, for comparison with theoretical predictions. The situation arose in 1976 to 1977, when Davies [58] published a formal analysis leading to the prediction of a critical cooling rate in terms of the equilibrium freezing temperature and the glass transition temperature, and applied it to particular cases [59]. (This episode has been summarized by Cahn [60].)

The methods used to estimate cooling rates fall into two main classes: direct and indirect. The indirect were first to be extensively examined: the mean cooling rate has been related to secondary dendrite spacings or alternatively to eutectic lamella spacings. Most of this was done in 1979 to 1982, by Jones in England, and very well defined straight-line relations were determined between the logarithms of the spacing and the cooling rates; such plots could then be extrapolated into the domain of very fast cooling, which was not then susceptible to direct measurement. For cooling rates of powder particles during atomization, the only recourse has been to theoretical calculations, which have become very involved. Rather than going into details here, I refer readers to a survey [61].

The direct methods involve measurement of temperature/time relationships by means of microthermocouples or optical pyrometry. This type of approach began extensively in the 1980s: thus Duflos and Cantor [62,63] used microthermocouples to measure cooling rates in a drop smasher, and established (an important point that has often been ignored) that the cooling rate is a function of the temperature and decreases very sharply as the specimen cools in the solid state. Subsequently, Cantor, working in Oxford with his collaborators, has used pyrometry ingeniously to determine the cooling rate history of a ribbon in single-roller CBMS [64,65]. This kind of direct approach is of course essential when glasses are being made, since there are no microstructural features such as dendrites which can be exploited for indirect estimation of mean cooling rates; furthermore, even where indirect techniques can be used, they have the drawback that only a single *mean* cooling rate can be estimated. Cantor and Duflos [63] have established that with sufficiently fast moving pistons, the drop smasher is a more efficient quenching device than the melt spinner.

Cooling rates in laser surface treatment of alloys have hitherto only been estimated by theoretical analysis. Cline and Anthony [66] and Mehrabian [67] were among the first to undertake this, and the activity still continues [68]. Perhaps in due course optical pyrometry will be successfully applied to this problem. Lasers are very extensively used in the semiconductor industry for the process of "laser annealing" of silicon microcircuits after ion implantation of dopants, which damages the silicon lattice. A very sophisticated body of experimental procedure and analysis has developed, which allows processes such as freezing velocities of (momentarily) liquefied silicon to be accurately assessed, and temperature histories to be indirectly determined. One good survey of this body of knowledge is by Sedgwick [69]. Unfortunately there has been almost no interaction between the two communities, the laser-treating metallurgists and the laser-annealing semiconductor specialists. This is a pity, since the two approaches have something to teach each other's exponents.

VIII. APPLICATIONS OF MELT QUENCHING

This chapter is devoted to the *background* of melt quenching, alias rapid solidification processing or RSP, and thus it is not really appropriate to discuss here the gradual introduction of RSP into industrial practice. Most of this has happened during the past decade, and it certainly has much further to go, whereas the scientific underpinning of the whole enterprise, with which this chapter is concerned, has probably reached and passed its apogee.

It must suffice here to make just two points. First, the most important application of RSP in terms of value and visibility is undoubtedly the use of wide ferrous glass ribbons made by planar flow casting, pioneered by the Allied-Signal Corporation in America, as transformer windings for use at power frequency, a subject fully treated in Chapter 19. Second, the *earliest* large-scale application of RSP to industrial practice was to tool steels. The American firm, Crucible Speciality Metals, in 1970 began to market steels made by its crucible particle metallurgy (CPM) process—it is still marketing them today—while Aurora Steel of Sheffield, England, for a while sold a comparable product at about the same time. The CPM process involves gas-atomizing molten steel, sieving and canning the powder, and hot-pressing it isostatically. This must have been a very early application of HIP. The product has a much finer and more uniform distribution of carbides, at a higher volume fraction, than conventionally processed tool steels. Higher total alloying contents, with consequently greater hardness, than in conventionally processed steels are feasible without unacceptable brittleness. This pioneering application pinpoints a vital fact which took a long time to be generally accepted: to get the maximum benefits out of RSP, it is necessary to redesign the alloy composition, generally in the direction of higher alloying contents. There is only limited benefit to be got out of RSP applied to alloys of traditional compositions.

REFERENCES

1. P. Duwez, *Trans. AIME*, *191*, 765 (1951).
2. N. Ohashi, in *Suppl. Vol. 1 of the Encyclopedia of Materials Science and Engineering*, edited by R. W. Cahn (Pergamon, Oxford, 1988), p. 85.
3. J. Kramer, *Ann. Phys. 19*, 37 (1934); *Z. Phys.*, *106*, 675 (1937).
4. W. Buckel and R. Hilsch, *Z. Phys.*, *131*, 420 (1952).
5. W. Buckel and R. Hilsch, *Z. Phys.*, *138*, 109 (1954).

6. W. Buckel, *Z. Phys.*, *138*, 136 (1954).
7. W. Buckel and R. Hilsch, *Z. Phys.*, *146*, 27 (1956).
8. G. Bergmann, *Phys. Rep.*, *27*, 159 (1976).
9. A. Comberg, S. Ewert, and H. Wühl, *Z. Phys. B*, *20*, 165 (1975).
10. L. C. Chen and F. Spaepen, *Nature*, *336*, 366 (1988); idem, *J. Appl. Phys.*, *69*, 679 (1991).
11. S. D. Dahlgren, in *Rapidly Quenched Metals III*, edited by B. Cantor (Metals Society, London, 1978), p. 36.
12. R. L. Bickerdike, D. Clark, J. N. Eastabrook, G. Hughes, W. N. Mair, P. G. Partridge, and H. C. Ranson, *Int. J. Rapid Solidification*, *2*, 1 (1986).
13. R. W. Gardiner and M. C. McConnell, *Metals Mater.*, *3*, 254 (1987).
14. A. Brenner, D. E. Couch, and E. K. Williams, *J. Res. Nat. Bur. Stand.*, *44*, 109 (1950).
15. A. Brenner, *Electrodeposition of Alloys: Principles and Practice*, 2nd ed. (Academic Press, New York, 1963).
16. P. Duwez, *Trans. ASM*, *60*, 607 (1967).
17. P. Duwez, in *Progress in Solid-State Chemistry*, Vol. 3, edited by H. Reiss (Pergamon, Oxford, 1967), p. 377.
18. P. Duwez, R. H. Willens, and W. Klement, Jr., *J. Appl. Phys.*, *31*, 36 (1960).
19. E. A. Strange and C. H. Pim, U.S. Patent No. 905,758 (1908).
20. R. B. Pond, U.S. Patent No. 2,879,566 (March 1959); U.S. Patent No. 2,976,590 (March 1961).
21. R. B. Pond, U.S. Patent No. 2,825,108 (March 1958).
22. G. Falkenhagen and W. Hofmann, *Z. Metallkde.*, *43*, 69 (1952).
23. H. Gröber and H. Hanemann, *Arch. Eisenhüttenw.*, *11*, 199 (1937).
24. W. T. Olsen, Jr., and R. Hultgren, *Trans. AIME*, *188*, 1223 (1950).
25. I. S. Miroshnichenko and I. V. Salli, *Indust. Lab.*, *25*, 1463 (1959) (English translation of paper in *Zav. Lab.*, *25*, 1398 (1959).)
26. W. Klement, Jr., *J. Inst. Metals*, *90*, 27 (1961).
27. H. L. Luo and W. Klement, Jr., *J. Chem. Phys.*, *36*, 1870 (1962).
28. W. Klement, Jr., R. H. Willens, and P. Duwez, *Nature*, *187*, 869 (1960).
29. M. H. Cohen and D. Turnbull, *Nature*, *189*, 131 (1961).
30. P. Duwez, R. H. Willens, and R. C. Crewdson, *J. Appl. Phys.*, *36*, 2267 (1965).
31. M. H. Cohen, in *Phase Transitions in Condensed Systems—Experiments and Theory, a Festschrift in Honor of D. Turnbull*, edited by G. S. Cargill III, F. Spaepen, and K. N. Tu, Symp. Proc., Vol. 57 (Materials Research Society, Pittsburgh, 1987), p. 31.
32. G. F. Taylor, *Phys. Rev.*, *23*, 655 (1924).
33. I. W. Donald, *J. Mater. Sci.*, *22*, 2661 (1987).
34. J. W. Mottern and W. J. Privott, editors, *Spinning Wires from Molten Metal*, A. I. Ch. E. Symp. Ser., Vol. 74 (A. I. Ch. E., New York, 1978).
35. R. B. Pond, R. E. Maringer, and C. E. Mobley, in *New Trends in Materials Processing* (A.S.M., Metals Park, OH, 1976), p. 128.
36. R. B. Pond and R. Maddin, *Trans. Met. Soc. AIME*, *245*, 2475 (1969).
37. T. Masumoto and R. Maddin, *Acta Metall.*, *19*, 725 (1971).
38. J. R. Bedell, U.S. Patent No. 3,862,658, assigned to Allied Chemical Corp. (January 1975).
39. T. R. Anantharaman and C. Suryanarayana, *Rapidly Solidified Metals* (Trans Tech Pub., Switzerland, 1987), p. 25. See also: C. Suryanarayana, in *Processing of Metals and Alloys*, edited by R. W. Cahn (VCH, Weinheim, Germany, 1991), p. 57.
40. I. Ohnaka and T. Fukusako, *J. Japan Inst. Metals*, *42*, 415 (1978).
41. I. Ohnaka, in *Suppl. Vol. I to the Encyclopedia of Materials Science and Engineering*, edited by R. W. Cahn (Pergamon, Oxford, 1988), p. 584.
42. H. S. Chen and C. E. Miller, *Rev. Sci. Instr.*, *41*, 1237 (1970).

43. M. C. Narasimhan, U.S. Patent 4,142,571, assigned to Allied Chemical Corp. (March 1979).
44. N. J. Grant, in *Proc. (First) Int. Conf. Metastable Metallic Alloys, Brela, 1970* (Fizika, Vol. II, Supp. 2, 1970), paper 16.1.
45. V. Anand, A. J. Kaufman, and N. J. Grant, in *Rapid Solidification Processing: Principles and Technology II*, edited by R. Mehrabian, B. H. Kear, and M. Cohen (Claitor's, Baton Rouge, LA, 1980), p. 273.
46. A. R. E. Singer, British Patent No. 1,262,471 (1968).
47. A. R. E. Singer, *Metals Mater.*, *4*, 246 (1970).
48. R. G. Brooks, A. G. Leatham, J. S. Coombs, and C. Moore, *Metallurg. Metal Forming*, *9* (no. 4), 1 (1977).
49. A. E. Berkowitz and J. L. Walter, *J. Mater. Res*, *2*, 275 (1987).
50. R. W. Cahn, J. L. Walter, and D. W. Marsh, in *Proc. Sixth. Int. Conf. Rapidly Quenched Metals* (Montréal, 1987), *Mat. Sci. Eng.*, *98*, 33 (1988).
51. W. A. Elliott, F. P. Gagliano, and G. Krauss, *Appl. Phys. Lett.*, *21*, 23 (1972).
52. W. A. Elliott, F. P. Gagliano, and G. Krauss, *Metall. Trans.*, *4*, 2031 (1973).
53. M. Laridjani, P. Ramachandrarao, and R. W. Cahn, *J. Mater. Sci.*, *7*, 627 (1972).
54. B. L. Mordike, editor, *Laser Treatment of Materials* (DGM Informationsgesellschaft Verlag, Germany, 1988). See also: B. L. Mordike, in *Processing of Metals and Alloys*, edited by R. W. Cahn (VCH, Weinheim, Germany, 1991), p. 111.
55. J. Fröhlingsdorf and B. Stritzker, in Ref. 54, p. 63.
56. X. L. Yeh, K. Samwer, and W. L. Johnson, *Appl. Phys. Lett.*, *42*, 242 (1983).
57. Proc. Conf. on Solid State Amorphizing Transformations, Los Alamos, August 1987, *J. Less-Common Metals*, *140*, 1 (1988).
58. H. A. Davies, *Phys. Chem. Glasses*, *17*, 159 (1976).
59. B. G. Lewis and H. A. Davies, in *The Structure of Non-Crystalline Materials*, edited by P. H. Gaskell (Taylor and Francis, London, 1977), p. 89.
60. R. W. Cahn, *Contemp. Phys.*, *21*, 43 (1980).
61. R. W. Cahn, in *Physical Metallurgy*, 3rd ed., edited by R. W. Cahn and P. Haasen (North-Holland, Amsterdam, 1983), p. 1785.
62. B. Duflos and B. Cantor, *Acta Metall.*, *30*, 323 (1982).
63. F. Duflos and B. Cantor, *J. Mater. Sci.*, *22*, 3765 (1987).
64. C. Hayzelden, J. J. Rayment, and B. Cantor, *Acta Metall.*, *31*, 379 (1983).
65. A. G. Gillen and B. Cantor, *Acta Metall.*, *33*, 1813 (1985).
66. H. E. Cline and T. R. Anthony, *J. Appl. Phys.*, *48*, 3895 (1977).
67. R. Mehrabian, *Int. Metall. Rev.*, *27*, 185 (1982).
68. E. Geissler and H. W. Bergmann, in Ref. 54, p. 101.
69. T. D. Sedgwick, *J. Electrochem. Soc.*, *130*, 484 (1983).

2

Fundamentals of Solidification at High Rates

W. J. Boettinger
National Institute of Standards and Technology, Gaithersburg, Maryland

J. H. Perepezko
University of Wisconsin-Madison, Madison, Wisconsin

I. INTRODUCTION

Throughout the history of metallurgical processing, quenching treatments have had an important role in the production and control of alloy microstructures. Indeed the early development of rapid solidification originated with the attempts by Duwez and co-workers [1] to modify an equilibrium solidification structure by a rapid quenching of the liquid alloy. Duwez based this approach on some of the earlier work by Falkenhagen and Hoffmann [2] where initial nucleation undercoolings and cooling rates up to 10^5 K/s were determined for a number of metals and alloys. Building on these initial successes, subsequent studies established a survey of a wide range of alloy systems and interesting metallurgical structures as reviewed by Cahn [3] and Jones [4]. In addition, a good deal of early work in the field focused on equipment invention and evaluation, most notably the development of melt spinning by Pond [5].

There are serious practical difficulties with the accurate measurement of temperature in most methods of rapid liquid quenching. Besides the problems of measurement sensitivity and response, the relatively large latent heat of crystallization causes recalescence and an uncertain relationship between liquid cooling rate and temperature during solidification. As a result, many of the reports of liquid cooling rate evaluation are based on heat flow calcuations. Alternatively, the size scale of microstructural features such as dendrites or cells are often related to the cooling rate or local solidification time. Although such relationships may be valid for secondary arm spacings in fully developed dendritic alloys, the extrapolation of size scale relationships determined at slow cooling rates to estimate high cooling-rate values is of uncertain validity.

Much of the early rapid solidification (RS) processing work was focused on cooling rate because of its common usage as a variable in solid-state transformations, where heat evolution is usually negligible. At the same time, the application of time-temperature-transformation (TTT) diagrams representing the kinetics of crystallization at different melt undercoolings was developed to estimate the critical cooling rates required to avoid crystallization and thus cause glass formation [6]. Because no heat is evolved when crystallization is avoided, the specification of a cooling rate is meaningful for vitrification. It is not, however, adequate for a complete analysis when crystallization occurs. The importance of the role of undercooling becomes more apparent with the realization that metastable phases can form during slow cooling. Initially undercooled droplets [7,8] as well as controlled directional solidification studies [9,10] emphasize undercooling controlled kinetics. In the former case, the initial melt undercooling can control the nucleation phase selection and generate high solidification rate. In the latter case, the imposed solidification velocity controls the operating interfacial undercooling. In both cases subsequent competitive growth of phases also contributes to phase selection and the evolution of the dominant microstructure. From this viewpoint, it becomes clear that rapid solidification should not always be equated to rapid quenching.

The present discussion focuses primarily on the principles underlying the general methods of importance to melt processing: powders with cooling mainly by convection and melt streams with conductive cooling to a quenching substrate. In spite of the appearance of seemingly different sample configurations and conditions, there is justification for treating many of the solidification characteristics involved in these processes on a common basis. Such a normalization of behavior is possible if the solidification kinetics involved in the various processing methods are viewed in terms of interface undercooling and solidification velocity. From this point of view general principles governing the liquid to solid transformation at high rates can provide a unified framework to understand the variety of processing approaches used and the microstructures obtained in rapid solidification. The central principles cover four areas: heat flow, thermodynamics, nucleation kinetics, and growth kinetics. To a certain extent, heat flow determines the relationship between externally controllable processing parameters and the internal fundamental solidification parameters. The thermodynamics of metastable and nonequilibrium solidification sets the possible range of solidification product phases while nucleation and growth kinetics determine the detailed microstructural evolution. Within these four areas, specific features are examined including the role of melt subdivision, undercooling and liquid cooling rate on nucleation kinetics and the role of solute redistribution on growth kinetics.

II. CONTROLLING PROCESSING VARIABLES

The techniques usually considered to produce rapidly solidified alloys are melt spinning, planar flow casting, or melt extraction, which produce thin (~20–100 μm) ribbon, tape, sheet, or fiber; atomization, which produces powder (~10–200 μm); and surface melting and resolidification to produce thin surface layers (0.1–200 μm). All of these techniques may be thought of as casting where at least one physical dimension is small.

Rapid solidification of crystalline alloys clearly involves the rapid removal of the latent heat of fusion by an appropriate heat sink. Two types of sinks are important: external and internal. External heat sinks are massive (relative to the casting size) metal substrates such as the wheel in melt spinning, the underlying unmelted solid in surface melting, or large volumes of high-velocity-cooling gas such as found in most atomization processes. The effectiveness of these heat sinks is characterized by a heat transfer coefficient, h. For surface melting infinite values of h are appropriate because the melt is in contact with its own solid; that is, the temperature is continuous across a melt-crystal interface. For atomization the heat transfer coefficient is often dominated by convective/conductive cooling which increases dramatically with decreasing powder size. Jones [4] gives an expression for h which also depends on the properties of the quenching gas and the relative velocity between the droplet and the gas. Values for h typically range from 10^{-3} to 10^2 W/cm^2-K. For quenching against a substrate h values are difficult to measure and depend on the details of the surface of the quenching substrate. Estimates, often using dendrite arm spacings, vary widely (10^{-1} to 10^2 W/cm^2-K). Clearly such large uncertainties make detailed predictions difficult for substrate quenching.

A second type of heat sink is internal in the sense that the casting is its own heat sink. This occurs when the alloy is undercooled below its liquidus temperature prior to the initiation of solidification. The release of latent heat can then be absorbed by the sample itself. When the undercooling of the liquid prior to solidification, ΔT, is spatially uniform, the effectiveness of this heat sink is expressed by the dimensionless undercooling, $\Delta\theta$, given by

$$\Delta\theta = \frac{\Delta T}{L/C} \qquad (1)$$

where C is the heat capacity per unit volume of liquid and L is the latent heat per unit volume (L/C for Al is 364 K). Hypercooling is defined by the condition where $\Delta\theta \geq 1$ and complete isenthalpic solidification is possible [11]. If $\Delta\theta > 1$, all of the latent heat can be absorbed by the casting without reheating to the liquidus temperature, T_L, and without need for any external heat extraction during the actual solidification process. This type of heat sink is very effective in causing high solidification velocities. Unfortunately, an a priori knowledge of the nucleation temperature of an alloy is as difficult to determine as the heat transfer coefficient. One important general trend, however, is the increase in initial undercooling with decreasing particle size [8,12].

A. Control of Liquid Undercooling

The physical shape of rapidly solidified samples is clearly of importance in evaluating the overall thermal conditions for the different processing methods and sample configurations. However, it is of perhaps equal importance to consider the effect of sample geometry on the extent of melt undercooling at the onset temperature of nucleation, T_N, and rapid growth. For example, with powders the production method and sample shape are known to have an important role in allowing for the isolation of the most active internal nucleation sites to a small fraction of the powder population [8,12]. In this case, regardless of the imposed cooling rate, the powder configuration and associated melt subdivision are conducive to high melt undercooling.

For a melt stream or surface melt layer configuration the geometric shape does not appear to have a direct influence on a physical nucleant isolation by melt subdivision. However, the thermal transport and solute redistribution conditions,

Table 1 Undercooled Melt Processing: Powders

Parameter	Undercooling response	Remarks
Powder size	Increased ΔT with size refinement at constant $\dot{T}$	Nucleant isolation follows Poisson statistics
Powder coating	Function of coating structure and chemistry; major effect in limiting undercooling of the finest powders	Most effective coating is catalytically inert; ΔT in nucleation is usually heterogeneous
Cooling rate	ΔT generally increases with increasing $\dot{T}$	Changing $\dot{T}$ can alter the operative nucleation kinetics
Melt superheat	System specific	Appears to be related to coating catalysis
Alloy composition	T_N follows trend of T_L	Melt purity not usually critical; near glass transition, T_N decreases rapidly
Pressure	T_N parallels melting curves trend	Change in response can signal alternate phase formation

accompanying the rapid progression of a solidification front under a high rate of heat extraction to the underlying substrate, may be shown to lead to a localization of nucleant influence, so that melt undercooling is obtained for a sufficient time period to allow for the competitive development of structural modifications.

Based upon experimental experience an effective approach that may be applied to obtain a large melt undercooling involves the cooling of a collection of fine liquid metal droplets [13]. By dispersing a liquid sample into a large number of small, independent drops with sizes from 5–40 μm only a small fraction of these drops will contain potent nucleants so that the majority of the drops can display a large undercooling. Past experience with the droplet method has identified a number of processing parameters that govern the optimization of undercooling in powder samples. These processing variables include droplet size refinement, droplet surface coating catalysis, uniformity of coating, melt superheat, cooling rate [8,13–16], alloy composition [13], and applied pressure [17]. The key responses to changes in these processing variables are highlighted in Table 1. Even when these processing conditions are arranged to produce maximum undercooling, most experience suggests that solidification is still initiated by a heterogeneous nucleation site associated with the sample surface [18]. Therefore, it appears that a close attention to the nature of the powder surface coating is of prime importance in achieving reproducible, large undercooling values in fine powders.

B. Analysis of Heat Flow During Atomization and Melt Spinning

The interplay of initial undercooling and external heat transfer is clearly shown in the schematic representations of the two melt-spinning conditions given in Fig. 1. In (a), the liquid-solid interface is shown to begin very close to the point where the liquid first comes into contact with the wheel. In (b), liquid exists for some distance (time) in contact with the wheel without nucleation. During this time the

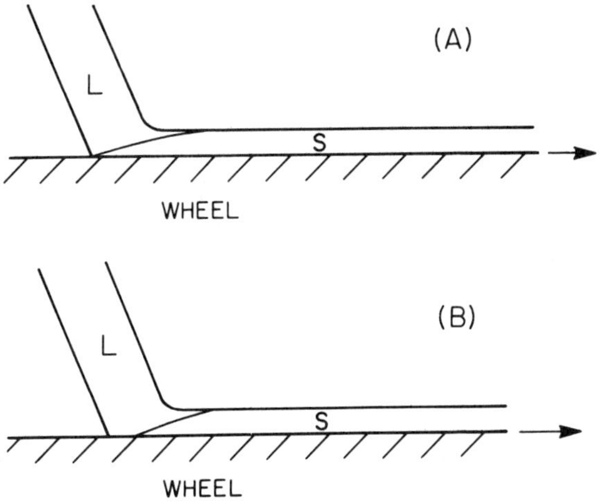

Figure 1 Schematic representation of the liquid-solid interface position during melt spinning if nucleation occurs near the wheel surface. In (A) nucleation occurs at a temperature near the alloy liquidus. In (B) nucleation occurs at large undercooling (that is, solid-liquid interface is displaced downstream). Vertical distances are exaggerated for clarity.

liquid can become undercooled and the liquid-solid interface, once nucleated at the wheel surface, will move at a higher rate than would the interface shown in (a). That is, a combination of high initial undercooling and high heat transfer coefficient promotes the "most" rapid solidification [19,20]. Some authors [21] argue that when the ribbon thickness depends inversely on the square root of the wheel speed, the cooling is nearly ideal ($h \rightarrow \infty$). However, several other mechanisms which can contribute to ribbon formation, including growth into an undercooled melt or growth of a momentum boundary layer, would also result in the same square root relation between ribbon thickness and wheel speed.

Theoretical details of the relative roles of the magnitudes of the heat transfer coefficient and of the initial undercooling have been examined for powders by Levi and Mehrabian [22] and for quenching against a substrate by Clyne [23]. A major complication occurs for heat flow analysis even for pure materials when high solidification velocities are involved. The temperature of the interface, T_I, cannot be treated as a constant, equal to the melting point, T_M, but rather is a function of the interface velocity. As a result, the heat flow analysis depends on the details of this function which can be very complicated, especially for dendritic growth and/or for alloys. Levi and Mehrabian as well as Clyne treat the case of a pure metal freezing with a smooth (nondendritic) liquid-solid interface governed by a kinetic law for the interface velocity V given by $V = \mu(T_M - T_I)$, where μ is the linear interface attachment coefficient.

Even for pure metals, the value of the coefficient is not certain. For aluminum, Levi and Mehrabian use a range of values for μ between 2 and 50 cm·s^{-1}·K^{-1} while Clyne uses 4 cm·s^{-1}·K^{-1}. The model of collision-limited growth [24] would suggest a value for μ near 300 cm·s^{-1}·K^{-1} for Al. Despite this uncertainty, these calculations show the importance of the initial undercooling, ΔT, on the development of high solidification velocities. Fig. 2, adapted from Levi and Mehrabian,

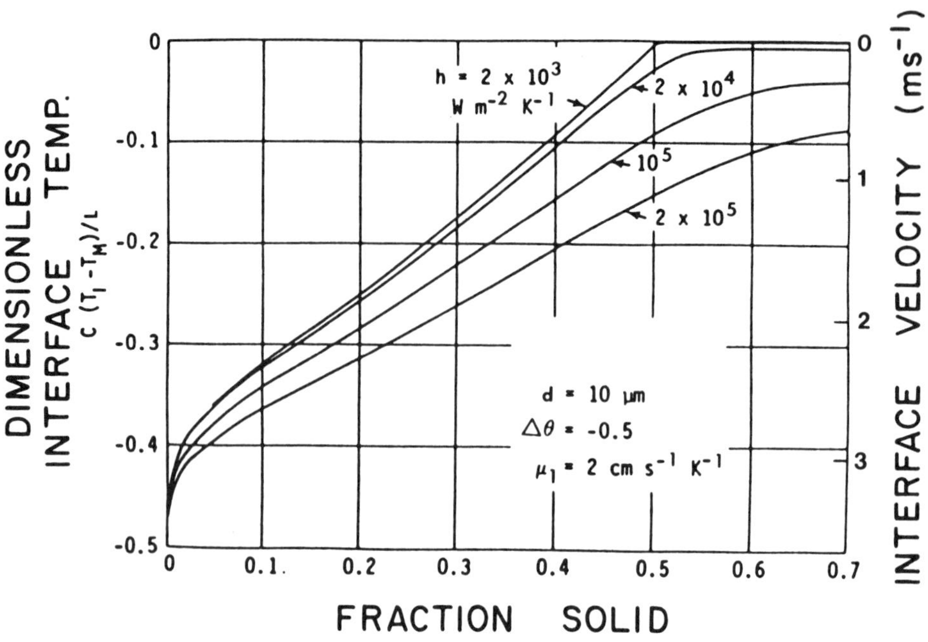

Figure 2 Calculated interface temperature [22] and interface velocity (added by the present authors) for the solidification of a powder particle initially undercooled by 1/2(L/C). The temperature rises and the velocity falls as growth proceeds from the point of nucleation on the powder surface across the powder particle. The effect of various values of the heat transfer coefficient h is also shown.

shows the interface temperature and interface velocity as solidification proceeds from one side of a powder particle to the other (increasing fraction solid). The curves show the case of an initial dimensionless undercooling of 0.5 (~182 K for Al) for various values of the heat transfer coefficient h. The velocity starts at a high value (>3 m/s) and slows as the interface crosses the particle. This slowing is due to the evolution of the latent heat at the liquid-solid interface and the resultant reduction in the interface undercooling. The effect of changing the heat transfer coefficient by two orders of magnitude affects primarily the velocity after the fraction solid exceeds the dimensionless initial undercooling (0.5 in this case). Growth for small fraction solid is controlled by internal heat flow, while growth for large fraction solid is controlled by external heat flow. If no initial undercooling occurred, the growth velocities for the entire particle would be near those seen at high fraction solid (Fig. 2) which are typically around 10 cm/s.

Similar effects of initial undercooling are shown by Clyne [23] for quenching against a substrate. Figure 3 (from Clyne) shows temperature-time plots at two positions inside a 50-μm-thick layer of an Al melt in contact with a substrate with $h = 10^2\ W{\cdot}cm^{-2}{\cdot}K^{-1}$. For example, at a position 5 μm from the substrate, the temperature in the liquid drops until nucleation occurs at the substrate at $\Delta\theta = 0.38$, which is calculated from homogeneous nucleation theory. The temperature at this position then rises rapidly as the liquid-solid interface proceeds from the substrate to the 5-μm position. Thereafter, the temperature at this position falls

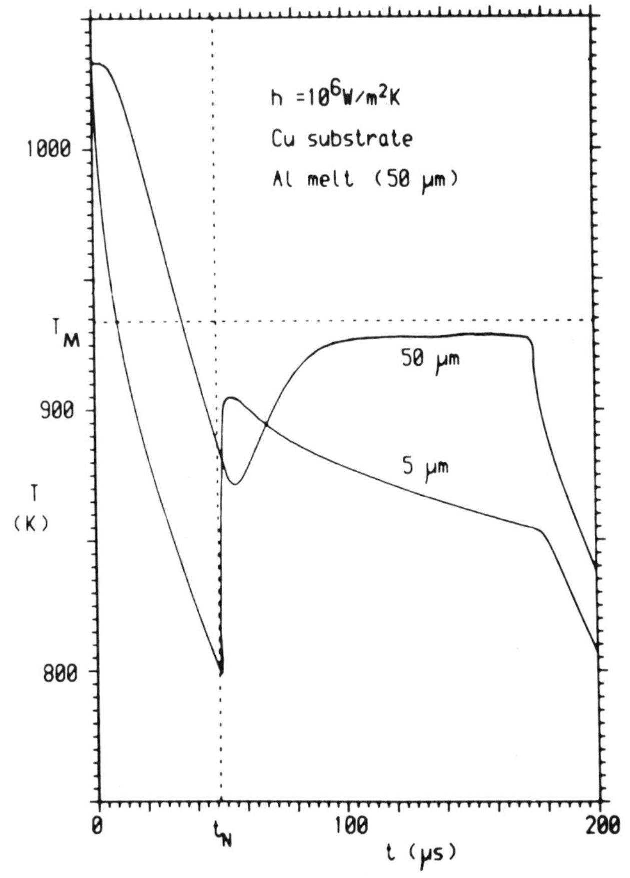

Figure 3 Calculated temperature-time histories [23] for two positions within a liquid layer 5 and 50 μm from a chilling substrate. The nucleation is assumed to occur at the substrate surface and is estimated to occur at an undercooling of ~0.4 L/C. The recalescence after the passing of the liquid-solid interface is evident at both positions.

slowly until the interface reaches the 50-μm position (end of solidification) and then falls more rapidly. A similar plot for the 50-μm position shows a small initial drop in temperature followed by a rise as the interface approaches. The higher temperature when the interface passes the 50-μm position compared to the 5-μm position shows the slow down of the liquid-solid interface across the sample thickness. As described below for concentrated alloys, much slower solidification rates occur in samples initially undercooled by the same amount.

III. THERMODYNAMICS

Although rapid solidification is widely considered as a "nonequilibrium" process, it is clear that different degrees of departure from full equilibrium occur and constitute a hierarchy which is followed with increasing solidification rate. This hierarchy is shown in Table 2.

Table 2 Hierarchy of Equilibrium

I. *Full Diffusional (Global) Equilibrium*
  a. No chemical potential gradients (compositions of phases are uniform)
  b. No temperature gradients
  c. Lever rule applicable
II. *Local Interfacial Equilibrium*
  a. Chemical potential continuous across the interface
  b. Phase diagram gives compositions and temperatures only at liquid solid interface
  c. Corrections made for interface curvature (Gibbs-Thomson effect)
III. *Metastable Local Interfacial Equilibrium*
  a. Important when stable phase cannot nucleate or grow sufficiently rapidly
  b. Metastable phase diagram (a true thermodynamic phase diagram missing the stable phase or phases) gives the interface conditions
IV. *Interfacial Nonequilibrium*
  a. Phase diagram fails to give temperature and compositions at interface
  b. Chemical potentials are not equal at interface
  c. Free-energy functions of phases still lead to criteria for the impossible.

The conditions required for global equilibrium (I) are usually obtained only after long-term annealing. Chemical potentials and temperature are uniform throughout the system. Under such conditions no changes occur with time.

During solidification or other phase transitions, gradients of temperature and composition must exist within the phases. However, one may often accurately describe the overall kinetics using diffusion equations to describe temperature and composition within the phases and using the phase diagram to give the possible temperature and compositions for boundaries between the phases. The Gibbs-Thomson effect is included to determine shifts in equilibrium due to boundary curvature. This is called local equilibrium (II). The approach is never strictly valid, but it is based on the notion that interfaces will equilibrate much more quickly than will bulk phases. The conditions present in (II) are widely used to model the majority of solidification and phase transformations excluding massive (partitionless) and martensitic transformations. For example, under the assumptions of fast diffusion in the liquid phase, very slow diffusion in the solid phase, and local equilibrium at the interface, the Scheil equation describes quite accurately the "nonequilibrium" coring or microsegregation in conventional castings [25].

Metastable equilibrium (III) can also be used locally at interfaces and is important in ordinary metallurgical practice. The change of cast iron from the stable gray form (austenite and graphite) to the metastable white form (austenite and cementite) with increasing solidification rate (and interface undercooling) is a familiar example [26]. The eutectic temperature and composition for white cast iron is a measurable thermodynamic transformation temperature as is any stable eutectic temperature. Metastable equilibrium occurs when the free energy is a minimum, but not an absolute minimum. It is often thought of as a "constrained" equilibrium. The constraint for the existence of this metastable equilibrium between liquid, austenite, and cementite is the absence of graphite. Graphite can be absent because its nucleation or growth is difficult. Subject to this constraint, the free energy is minimized and a large fluctuation (nucleation) is necessary to reach the stable equilibrium. The growth kinetics for white cast iron is treated using metastable local equilibrium at the solidification front. When solidification is complete, a two-phase mixture of austenite and cementite can exist in a global metastable equilibrium. In general, local metastable equilibrium is important during rapid solidification because many equilibrium phases, especially those with complex crystal structures, seem to have difficulties with nucleation and/or growth. Hence, metastable phase diagrams are important in describing interface conditions for many rapid solidification processes [27].

Significant loss of interfacial equilibrium (IV in Table 2), whether for a stable or a metastable phase, is thought to become important for simple crystalline phases when the crystal growth rate exceeds the diffusive speed of solute atoms exchanging between the liquid and solid, V_D. An *upper bound* on this diffusive speed is D/a_0, where D is the liquid diffusion coefficient and a_0 is the interatomic dimension. Experiments on doped Si [28] and on metallic alloys [29] have shown that for crystal growth rates >1 m/s, significant interfacial nonequilibrium effects exist and solute is trapped into the solid at levels exceeding the equilibrium solubility. Despite the loss of interface equilibrium, the free-energy functions of the solid and liquid phases can be used to restrict the range of compositions that can exist at the interface at various temperatures, as shown by Baker and Cahn [30]. This restriction is obtained by the requirement that only those processes that reduce the free energy of the system are possible. Figure 4 shows the region of allowable solid compositions at

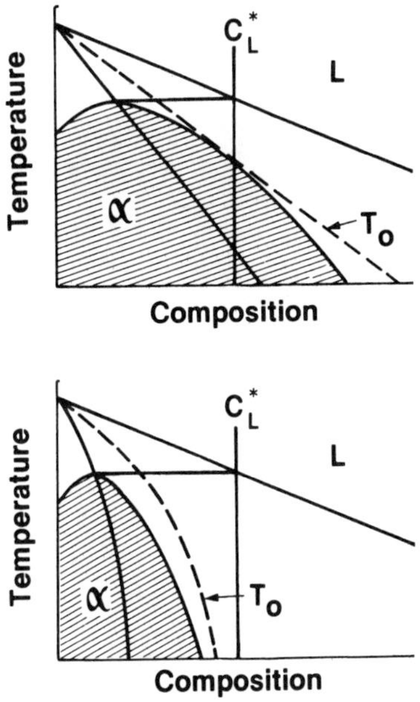

Figure 4 Regions of thermodynamically allowed solid composition that may be formed from liquid of composition C_L^* at various temperatures. The value of T_0 is the highest temperature at which partitionless solidification can occur. In (b) the T_0 temperature plunges and partitionless solidification ($C_L^* = C_S^*$) is impossible for liquid of composition C_L^* [33].

the interface for a fixed liquid composition at the interface for various interface temperatures. While one might consider a further level in the hierarchy where the free-energy functions for the phases must be abandoned, this will not be discussed in the present work.

A. Metastable Phases Diagrams

Whether a rapidly solidified microstructure is composed of stable equilibrium phases or metastable phases depends on the nucleation and growth kinetics of the competing product phases. A schematic representation of the role of nucleation and growth kinetics for phase selection as a function of processing conditions is shown in Fig. 5. The thermodynamic relationships for the molar free energy of a pure material as a liquid, a stable phase α, and a metastable phase β are shown in Fig. 5a. The melting point of the metastable β phase is a well-defined thermodynamic quantity given by T_M^β in Fig. 5a. In Figs. 5b, c, a possible pair of functions are depicted to illustrate the role of kinetics in phase selection under conditions that favor the nucleation and growth of the metastable β phase.

For the *nucleation* of α or β from the liquid, the dominant product phase is determined to a large extent by the lowest value of the activation energy barrier

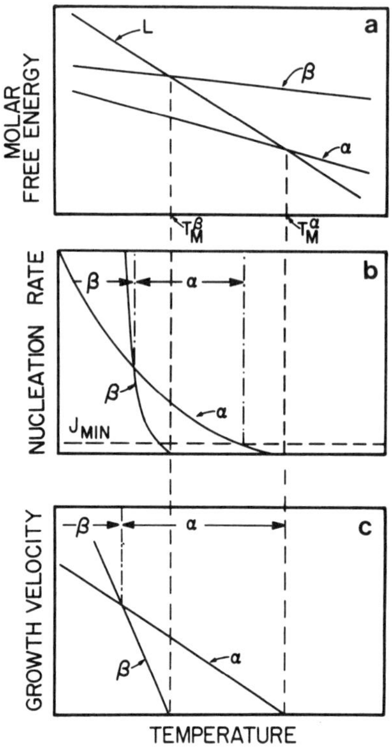

Figure 5 Schematic representation of the operation of competitive phase selection kinetics which favors the formation of a metastable phase β from the liquid L at low temperature in spite of (a) the thermodynamic stability of α. (b) Shows the temperature range for faster nucleation of β phase while (c) shows the temperature range for faster growth of the β phase.

for nucleation, ΔG^*, of α or β. The value of ΔG^* for each solid phase is a function of the amount of undercooling below the respective melting points, the liquid-crystal surface energy and the potency of any catalytic surfaces present which act as heterogeneous nucleation sites for α or β. For the nucleation rate relationships presented in Fig. 5b, nucleation of α dominates at low undercoolings below the stable melting point, whereas nucleation of β can dominate at lower temperatures. Several examples of a nucleation-controlled transition in phase selection with increasing undercooling have been reported [7,8,31]. This behavior is controlled by interfacial energies associated with the nucleus, liquid, and active catalytic sites and occurs even though the bulk thermodynamic driving force is usually greater for the formation of the stable phase.

The *growth* of a stable phase can also be difficult. Following nucleation of two phases from the melt, one phase may grow so much more quickly that it will dominate the microstructure of an RS alloy. A simple example is illustrated in Fig. 5c, which shows schematically the growth rate of the α and β phases from the liquid as a function of interface temperature in a situation favoring metastable phase formation. At large undercooling, the metastable β phase grows faster than the stable phase α. Such a situation could exist for pure materials when the α phase grows with a faceted interface and growth is sluggish due to the difficulties of interface attachment. A simple criterion for the development of a faceted growth interface was developed by Jackson [32]. When a dimensionless alpha factor, which is the ratio of the entropy of fusion (latent heat per mole/absolute melting temperature) to the universal gas constant, is greater than about four, faceted growth is expected for close-packed interfaces. Then one usually obtains a steep velocity-undercooling curve as shown for the α phase in Fig. 5c. For alloys, the analysis of growth competition for various phases also requires including the difficulty of solute redistribution in the different growth morphologies present (for example, dendrites, eutectics). The nucleation and growth curves presented in Fig. 5 pertain to relatively low undercoolings. For thermally activated kinetics, both rates exhibit a maximum at lower temperatures. For example, such a maximum occurs for eutectic growth due to the temperature dependence of the diffusion coefficient [33].

While the plots shown in Fig. 5 are schematic, they can be constructed in a quantitative manner for specific systems and metastable phases of interest. The first level of information needed for such analysis is the melting temperature of a metastable phase, which identifies its range of possible formation from the melt. Most often the melting temperature of a metastable phase is calculated from the analysis of a thermodynamic cycle [34], but there are examples of direct experimental measurement under favorable kinetic conditions [35–37].

For a pure component that exhibits an allotropic transition between a low-temperature β phase and a high-temperature α phase, the melting of β at T_m^β lies between the melting of α at T_m^α and the allotropic transition at $T_{\alpha/\beta}$. When specific heat corrections are neglected the value of T_m^β is given as

$$T_m^\beta = \frac{[\Delta S_{\alpha/\beta} T_{\alpha/\beta} + \Delta S_m^\alpha T_m^\alpha]}{\Delta S_{\alpha/\beta} + \Delta S_m^\alpha} \tag{2}$$

where $\Delta S_{\alpha/\beta}$ is the entropy change for the allotropic transition and ΔS_m^α is the entropy of melting for α phase.

While allotropic transitions are not observed for many pure components other crystal structures are stabilized by high pressure or with the addition of alloying elements. In these cases a thermodynamic analysis of the phase equilibria permits estimates of the so-called lattice stability or the relative stability of different crystal structures of a given component, as compared to the equilibrium crystal lattice structure. An example is provided in Fig. 6, which shows the temperature-pressure phase diagram for pure Bi [38]. The Bi(II) phase becomes the stable solidification product as pressure increases above about 1.7 GPa. An extrapolation of the melting curve for Bi(II) to 1 atm. (based on the Clausius-Clapeyron relation) yields a melting point of 174°C. This melting point identifies the minimum undercooling below the Bi(I) melting point required to form Bi(II) from the melt at ambient pressure and is the basis for developing the lattice stability for the Bi(II) structure. Clearly, such lattice stability estimates are sensitive to the accuracy of the alloy phase equilibria or elevated pressure measurements that are used for analysis. For the purpose of comparison, a partial listing of the melting points of the common BCC, FCC, and HCP structures for a number of metals of interest in RS that are obtained from lattice stability expressions is provided in Table 3. Lattice stability values may be obtained for other crystal structure types including intermediate alloy phase structures. All of the values listed in Table 3 are based on the extensive work of Kaufman [39], except for the case of Fe [27,40,41].

With this first level assessment it is possible to identify candidate systems in which alternate metastable crystal structures may be produced during RS. For example, if an undercooling level ($\Delta T/T_m$) of about 0.3 below the stable phase melting point or alloy liquidus is used as a basis for evaluation, then alloys based

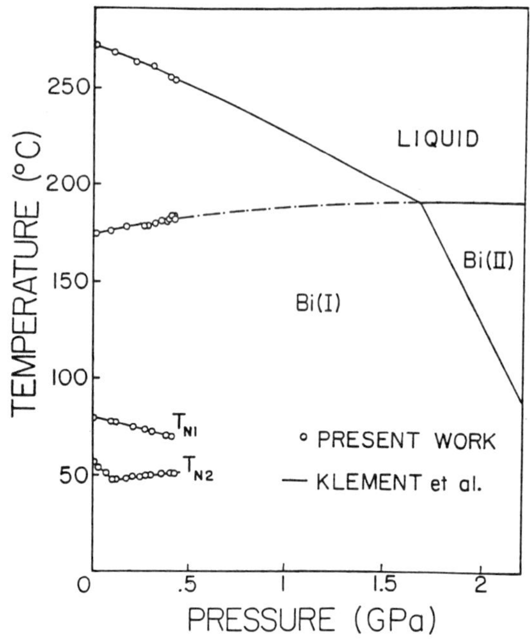

Figure 6 The pressure-temperature phase diagram for Bi with experimental data for droplet nucleation and melting behavior. The T_{N1} and T_{N2} trends refer to the pressure dependence of the nucleation temperature at different undercooling levels. The extrapolated melting curve for the Bi(II) phase is given by the dot-dash curve. From Ref. [38].

Table 3 Metastable Phase Melting Points of Pare Elements

Metal	Melting point (K)[a]	Undercooling (K)[b]	
		ΔT	$\Delta T/T_m$
Al	$T_m^{\alpha} = 931$	—	—
	$T_m^{\epsilon} = 539$	392	.42
	$T_m^{\beta} = 94$	837	.89
Be	$T_m^{\beta} = 1556$	—	—
	$T_m^{\epsilon} = 1547$	9	.05
	$T_m^{\alpha} = 1136$	420	.27
Co	$T_m^{\alpha} = 1768$	—	—
	$T_m^{\epsilon} = 1705$	63	.035
	$T_m^{\beta} = 805$	963	.54
Cr	$T_m^{\beta} = 2175$	—	—
	$T_m^{\epsilon} = 1175$	1000	.46
	$T_m^{\alpha} = 860$	1315	.60
Cu	$T_m^{\alpha} = 1357$	—	—
	$T_m^{\epsilon} = 1142$	215	.158
	$T_m^{\beta} = 1081$	276	.20
Fe	$T_m^{\beta} = 1536$	—	—
	$T_m^{\alpha} = 1531$	5	.003
Mg	$T_m^{\epsilon} = 922$	—	—
	$T_m^{\beta} = 650$	272	.228
	$T_m^{\alpha} = 610$	312	.295
Nb	$T_m^{\beta} = 2740$	—	—
	$T_m^{\epsilon} = 1420$	1320	.48
	$T_m^{\alpha} = 1170$	1570	.57
Ni	$T_m^{\alpha} = 1725$	—	—
	$T_m^{\epsilon} = 1445$	280	.16
	$T_m^{\beta} = 885$	840	.486
Ti	$T_m^{\beta} = 1940$	—	—
	$T_m^{\epsilon} = 1697$	243	.125
	$T_m^{\alpha} = 1421$	519	.267
Zn	$T_m^{\epsilon} = 692$	—	—
	$T_m^{\alpha} = 616$	76	.11
	$T_m^{\beta} = 551$	141	.20
Zr	$T_m^{\beta} = 2125$	—	—
	$T_m^{\epsilon} = 1820$	305	.14
	$T_m^{\alpha} = 1170$	955	.45

[a]Phase identity: $\alpha \equiv$ FCC, $\epsilon \equiv =$ HCP, $\beta \equiv$ BCC.
[b]ΔT is referenced to the stable high-temperature-phase melting point, T_m.

on Be, Co, Cu, Fe, Mg, Ni, Ti, Zn, and Zr are likely candidates for the formation of one or more alternate crystal structures. However, Al, Cr, and Nb base alloys will not be good candidates. Beyond the first level of comparison in Table 3, the elemental lattice stabilities together with thermodynamic models of alloy solution behavior serve as the basis for the computational analysis of binary and higher-order phase equilibria.

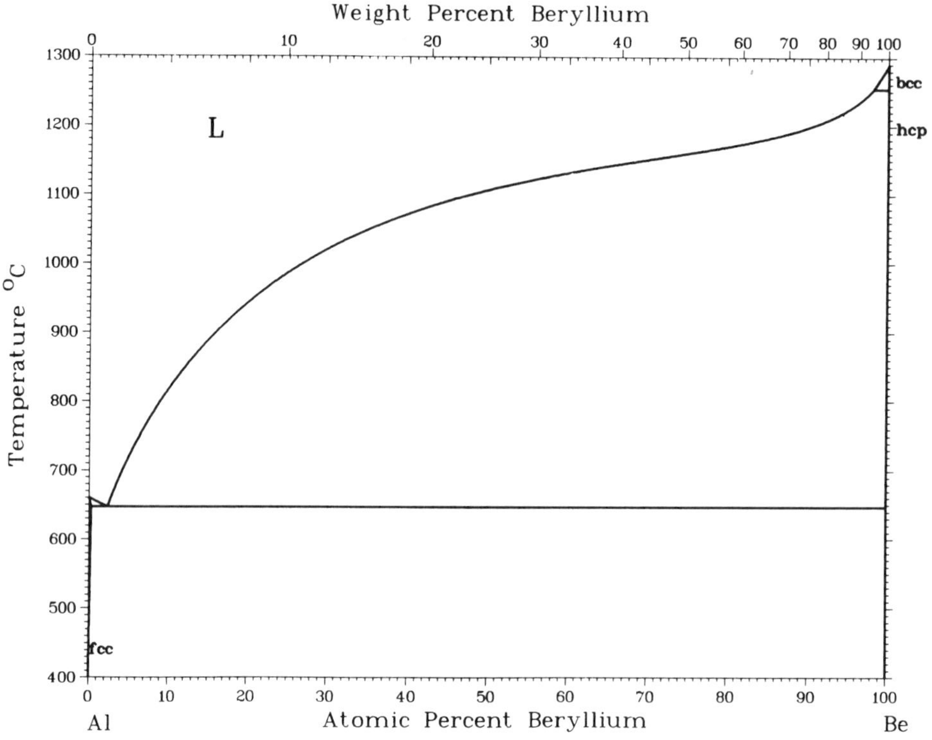

A

Figure 7 Al-Be stable and metastable phase diagrams [42]: (A) stable phase diagram (B) superposition of several separate metastable diagrams involving the liquid, FCC Al, FCC Be, HCP Be, and BCC Be.

An example of a series of calculated metastable binary phase diagrams is shown in Figs. 7 and 8 for the Al-Be system [42]. These diagrams were prepared by determining free-energy functions which fit measured phase boundaries, combined with a calculated lattice stability value for pure FCC Be. Changes in this value can alter the diagrams significantly, however the range of possibilities is still illuminating. Figure 7a shows the stable equilibrium diagram without any metastable phase boundaries—a "simple" eutectic system. Figure 7b shows a metastable diagram as usually presented, including the FCC T_0 curve and liquid spinodal. It is a superposition of several separate metastable diagrams which must not be used in a mixed manner. A statement of the missing phase or phases uniquely defines a single metastable diagram. Figure 8a shows the case where both the HCP-Be phase and the BCC-Be phase are missing. In this case the liquid contains a miscibility gap with a critical point at ~70 at% Be and ~1100°C. A short segment of the liquidus curve for FCC-Be develops along with a monotectic reaction $L_2 \rightarrow L_1$ + FCC-Be at ~96 at% Be and ~845°C and a eutectic reaction $L_1 \rightarrow \alpha$-Al + FCC-Be at ~41 at% Be and ~636°C. Figure 8b is an enlargement of Fig. 7b for Al-rich alloys.

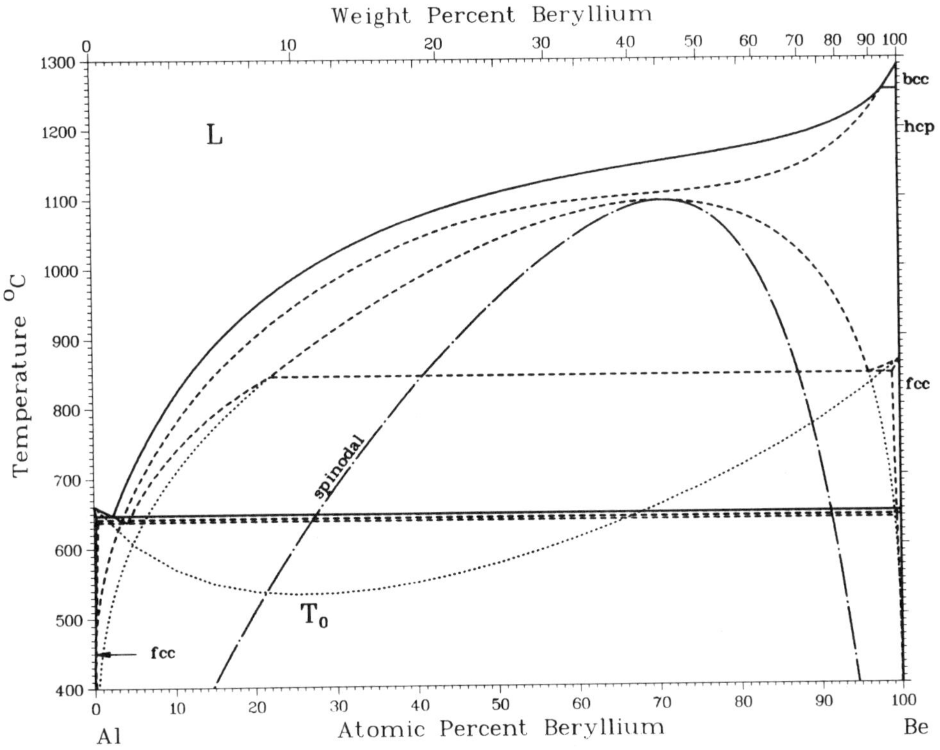

B

The possibility of the formation of several crystalline forms of Be in combination with α-Al is evident due to the close proximity of their liquidus curves and associated eutectics, and their modest undercoolings below the stable eutectic. Hence, in principle, for a given alloy composition, various solidification paths and microstructures can develop in different processing treatments depending on the difficulty of nucleation or growth of the various stable or metastable equilibrium phases.

The calculation of phase diagrams is an ongoing area of research performed by groups worldwide and often published in the journal *CALPHAD*. The results of this work are most useful when good phase boundary and calorimetric data exist so that the accuracy of thermodynamic modeling can be tested and the calculations refined. However, most often, phase equilibra data are available only over limited ranges of composition and temperature for the phases of interest.

In order to illustrate this point the results shown in Fig. 9 provide a summary of actual measurements of metastable phase boundaries in the Pb-Sn system and a comparison of the measurements with two thermodynamic model calculations of the phase equilibria [43,44]. Within the region of stable equilibria the quasi-regular and quasi-subregular solution models provide essentially equivalent fits within the uncertainty of data. However, within the metastable region there are significant differences between the two models and only the quasi-subregular model provides a satisfactory fit to the measurements.

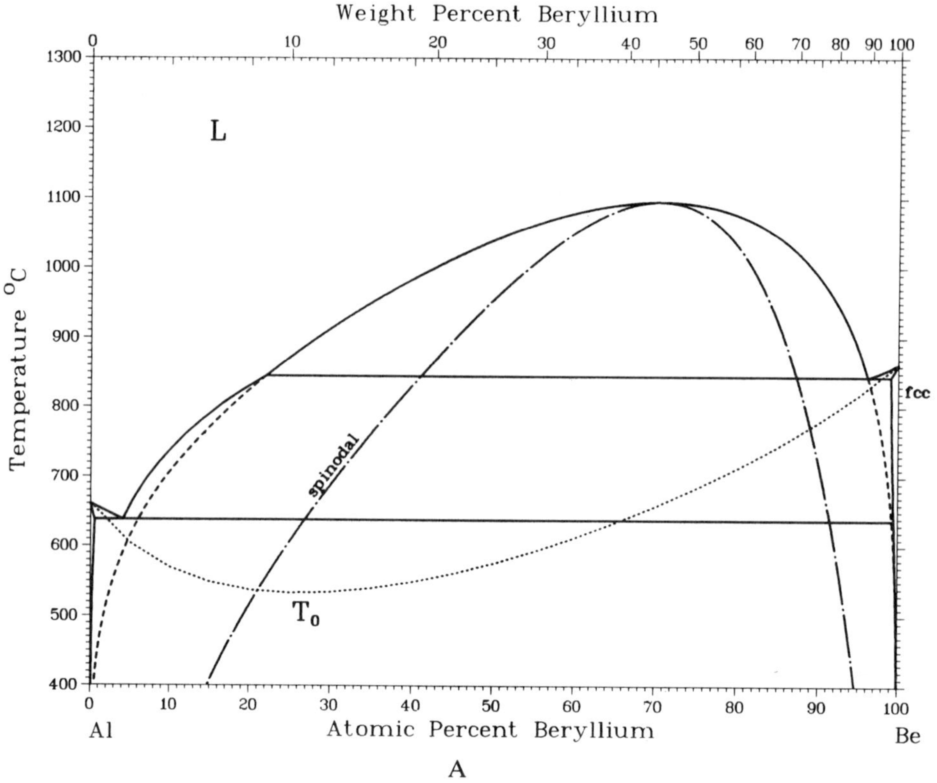

Figure 8 Al-Be metastable phase diagrams [42]: (A) single metastable diagram when BCC and HCP Be are absent, (B) enlargement of Fig. 7b in the Al-rich range of compositions.

B. T_0 Curves

Metastable equilibrium is a true thermodynamic equilibrium in the sense that the free energy is minimized [30,45]. Chemical potentials of the components for each phase involved in the metastable equilibrium are equal. In Table 2, however, another situation is described in (iv), and relates to a situation where chemical potentials are not equal across an interface growing at a high rate and large undercooling. These rapid growth rates can trap the solute into the freezing solid at levels exceeding the equilibrium value for the corresponding liquid composition present at the interface. The increase in chemical potential of the solute across the interface must be balanced by the decrease in chemical potential of the solvent in order for crystallization to occur; that is, to yield a net decrease in free energy [30]. To achieve this the interface temperature must lie significantly below the liquidus temperature, whether stable or metastable.

For any pair of liquid and solid compositions, a thermodynamic temperature exists which is the highest interface temperature at which crystallization can occur. This was shown in Fig. 4. One often considers a limiting case called partitionless solidification, which intuitively is favored at very high solidification rate, where the composition of the solid equals the composition of the liquid at the interface. This thermodynamic temperature is called the T_0 temperature and is the temperature at which the molar free energies of the liquid and solid phases are equal for the

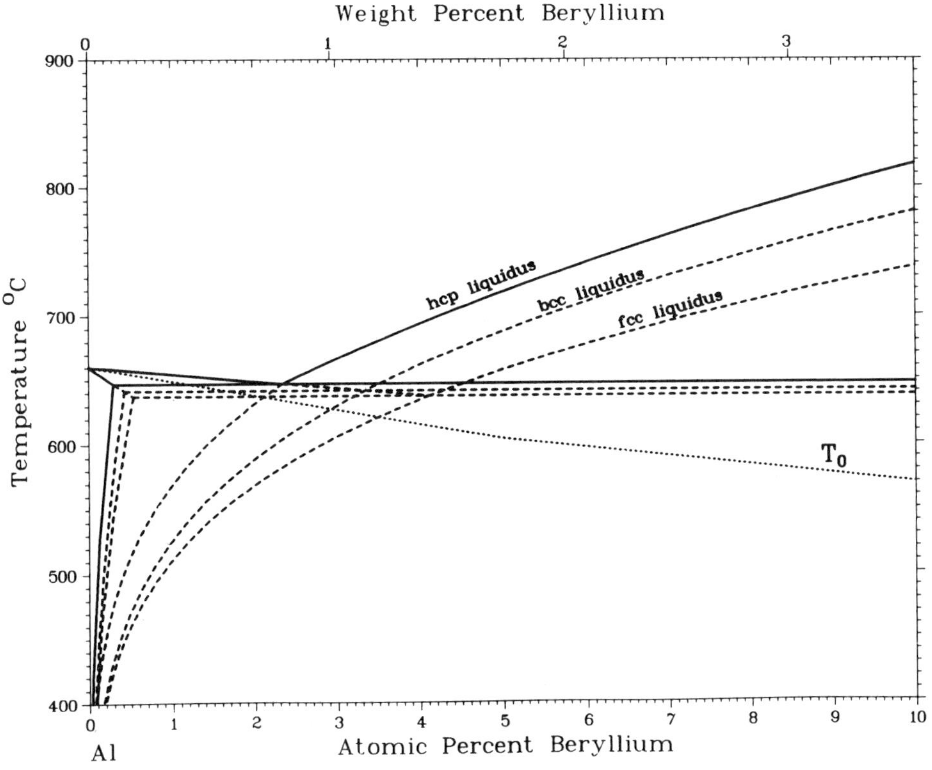

B

composition of interest. The locus of T_0 over a range of composition constitutes a T_0 curve and represents an important piece of thermodynamic information which should be included whenever possible in phase diagram calculations. As illustrated in Fig. 4, T_0 curves represent only part of the thermodynamic information available, restricting solidification without local equilibrium.

The geometrical structure of T_0 curves can be considered by illustration on a few simple phase diagrams with reference to RS phenomena. The T_0 curves exist, of course, for the liquid with both stable or metastable phases, and lie between the liquidus and solidus for those phases. Figure 10 shows schematically, possible T_0 curves for two eutectic phase diagrams [33]. One important use of these curves is to determine whether a bound exists for the extension of solubility by rapid melt quenching. If the T_0 curves plunge to very low temperatures as in Fig. 10a, single phase α or β crystals with composition beyond their respective T_0 curves cannot be formed from the melt. In fact, for phases with a retrograde solidus, the T_0 curve plunges to absolute zero at a composition equal to the liquidus composition at the retrograde temperature, thus placing a bound on solubility extension [46]. Experiments on laser-melted Si seem to confirm this bound [28]. Eutectic systems with plunging T_0 curves are good candidates for easy metallic glass formation [47]. An alloy in the center of such a phase diagram can only crystallize into a mixture of solid phases with different compositions regardless of the departure from equilibrium. The diffusional kinetics of this separation from the liquid phase frequently depresses the solidification temperature to near the glass transition, T_g, where an increased liquid viscosity effectively halts crystallization.

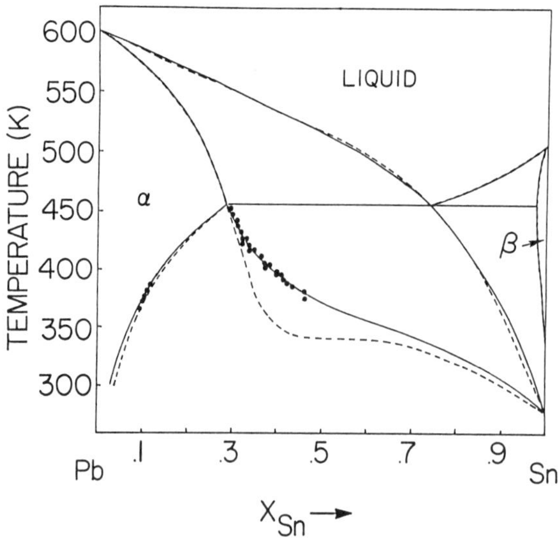

Figure 9 Calculated Pb-Sn phase diagram including metastable ($\alpha + L$) extensions determined from a quasi-regular (dashed line) or quasi-subregular solution model (solid line) along with experimental data points.

In contrast, alloys with T_0 curves which are only slightly depressed below the stable liquidus curves, as in Fig. 10b, make good candidates for solubility extension and unlikely ones for glass formation. Regions of composition are shown where partitionless solidification is thermodynamically possible. In fact for compositions below both T_0 curves, a mixture of α and β phase could form wherein each phase has the same composition as the liquid.

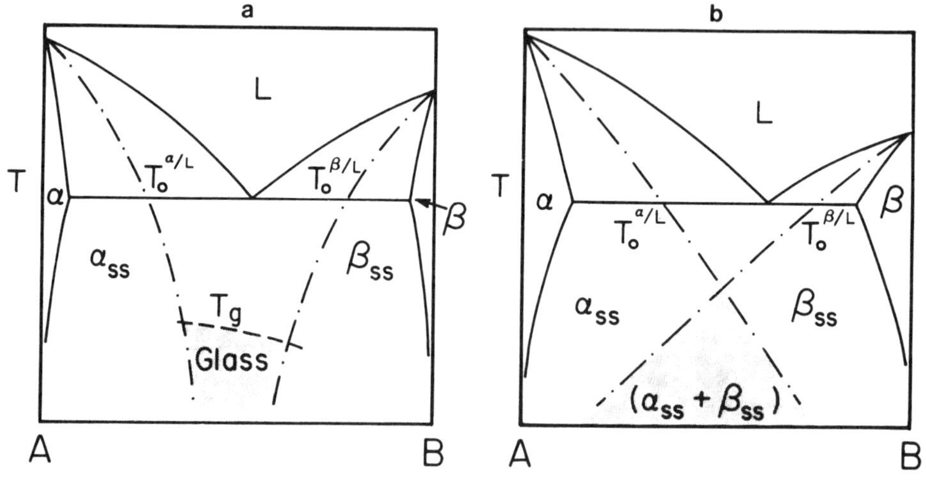

Figure 10 Schematic representation of T_0 curves for the liquid to crystal transformation in two types of eutectic systems [19].

IV. NUCLEATION KINETICS

For high levels of liquid undercooling the range of possible product structures and variety of solidification reaction paths indicated by a consideration of the hierarchy of metastable phase equilibria, demonstrates that even so-called simple eutectic alloy systems offer several opportunities for microstructure modification. The selection of a given pathway is dependent on the operational nucleation kinetics and competitive growth kinetics. A general analysis of nucleation rates in order to develop specific predictions is sensitive to a number of uncertain parameters including the effectiveness and density of heterogeneous nucleation sites, interfacial energies, and driving free energy. However, in the context of rapid quenching treatments, there are a number of useful features and relations that are possible to develop by considering limiting conditions of crystal nucleation.

A. Nucleation Rate

At the onset it is important to note that the process of diffusional nucleation is sporadic both temporally and spatially among the active sites involved. Nucleation of a crystalline phase during solidification may be classified into two mechanistic types. When nucleation occurs throughout the volume of liquid without an association with any catalytic nucleation sites, it represents homogeneous nucleation. For the majority of processing conditions, nucleation is initiated at some catalytic site as heterogeneous nucleation. These sites may be distributed on the exterior melt surface or within the bulk volume of the liquid sample. In either case the nucleation rate is described per unit area of catalytic surface exposed to the melt. To illustrate the main aspects of nucleation, simple classical rate theory is used. Based on classical theory [6,48,49] a general expression for the steady state nucleation rate, J_i, can be represented as

$$J_i = \Omega_i \exp\left[-\frac{\Delta G^* f(\theta)}{kT} \right] \tag{3}$$

where J_i relates to either heterogeneous surface nucleation, J_a, homogeneous nucleation, J_v, or heterogeneous nucleation on volume-distributed catalytic sites, J_s. Respective values for the prefactor, Ω_i, activation barrier, ΔG^*, and contact angle function, $f(\theta)$, are used in Eq. (3) and kT is the thermal energy. The expressions for Ω_i involve a product of a nucleation site density on a sample surface or volume basis, the number of atoms on a nucleus surface and a liquid jump frequency. For most cases, $\Omega_v = 10^{30}/\eta$ cm^{-3}·s^{-1} and $\Omega_a = \phi 10^{22}/\eta$ cm^{-2}·s^{-1} with η, the liquid shear viscosity (in poise), given by [4] as

$$\eta = 10^{-3.3} \exp\left[\frac{3.34 T_L}{T - T_g}\right] \tag{4}$$

in terms of liquidus temperatures T_L and ϕ, the fraction of sample surface sites active. For J_v and J_a, ΔG^* is given by

$$\Delta G^* = \frac{b\sigma^3}{\Delta G_v^2} \tag{5}$$

where σ is the liquid-solid interfacial energy, ΔG^* is the driving free energy for nucleation of a unit volume of product phase, and $b = 16\pi/3$ for spherical nuclei.

With planar catalytic sites and spherical nuclei $f(\theta) = [2 - 3\cos\theta + \cos^3\theta]/4$. For heterogeneous volume nucleation, the value of Ω_s depends on the specific catalyst concentration, size, and the form of ΔG^*, which is related to the details of the catalyst-nucleus interaction [50].

The classical nucleation model is based upon the formation of critical clusters by the addition or dissolution of individual atoms or molecules. At steady state, the distribution of cluster sizes is stationary and the isothermal nucleation rate is a constant, as represented by Eq. (3). During rapid melt quenching, and especially at high undercooling near T_g, atomic transport may not be adequate to maintain cluster equilibrium. Under conditions of relatively sluggish atomic motion, the initial cluster distribution can be different from the steady-state value and the time interval required to reach steady state represents an effective delay time. During this transient period the time-dependent nucleation rate, $J_i(t)$, increases continuously and once the steady-state value is reached, the number of nucleated crystals, N, on a volume or area basis can be represented as $N = J_i(t - \tau)$, where τ is an effective delay time [51]. Transient nucleation has been reported by Kelton and Greer [51–53] to affect significantly the number of quenched-in nuclei and the relative ease of glass formation. For example, in a Au–19 at% Si alloy the calculated critical cooling rate for glass formation decreased from 10^7 K/s for steady-state conditions to 10^5 K/s for the transient conditions [52]. In addition, at the transition temperature below which transient effects become important, the delay time, τ, and cooling rate, $\dot{T}$, are inversely proportional. For several materials the relationship $\tau\dot{T} \sim 1$ K holds and may be used to estimate the relative importance of transient effects during a given RS processing treatment [53]. Current experience indicates that non-steady-state effects are likely to be most important for glass formation during rapid quenching [52] or during devitrification upon heating [54,55].

B. Nucleation Rate Parameters

Nucleation rate is a relatively steep function of temperature with a magnitude determined principally by the exponential term involving ΔG^* at the nucleation temperature T_N, and to a lesser extent by the prefactor term. In evaluating the temperature dependence of J_i a constant value based upon theory is usually taken for Ω_v; but for Ω_a and Ω_s it is necessary to consider that the catalytic site density may vary for different conditions. Similarly, little information is available to judge the catalyst potency, so that a range of values for $f(\theta)$ is normally used in calculations. However, the most important parameters in determining J_i and hence the *maximum* undercooling level are ΔG_v and σ, which have received continued experimental and theoretical study.

In pure metals the value of ΔG_v can be calculated directly from

$$\Delta G_v = \frac{L\,\Delta T}{T_M} - \int_{T_N}^{T_M} \Delta C\,dT + T\int_{T_N}^{T_M} \frac{\Delta C}{T}\,dT \tag{6}$$

where ΔC is the heat capacity difference between undercooled liquid and crystal. There have been only relatively few measurements of liquid heat capacity C for the undercooled state. Although the magnitude of the effect varies between the metals shown in Fig. 11, there is a continuous rise in C with decreasing temperature

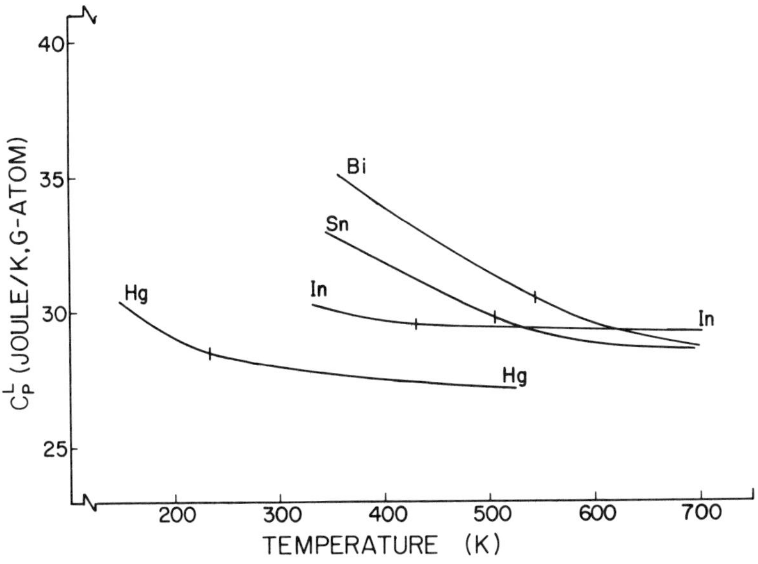

Figure 11 Heat capacity of several liquid metals as a function of temperature extending into the undercooled regime.

and a concomitant growing divergence in ΔC with increased undercooling. Most often measurements are not available and approximations for ΔC are used in Eq. (6). For pure metals, measurements [56] indicate that ΔG_v is represented to within a few percent by the relation proposed by Turnbull [48], based upon $\Delta C = 0$, giving $\Delta G_v = L(1 - T_r)$ with $T_r = T/T_M$. In the case of alloys there are also only limited measurements of the enthalpy change during solidification and only a few measurements of C_p^l for the undercooled liquid state. Often the latent heat for an alloy is taken as the weighted average of the latent heat values of the components and ΔC corrections are neglected. However, as measurements for the Pb-Bi eutectic alloy given in Fig. 12 have indicated, such approximations can lead to significant errors [8]. The enthalpy results shown in Fig. 12 define the undercooling for the onset of hypercooling for solidification of a metastable single phase and for solidification of the equilibrium eutectic mixture. If a weighted average of the components values is used to evaluate the dimensionless undercooling from Eq. (1) a value of 292°C is obtained, which is significantly different from the experimental value of 132°C. Moreover, it is useful to note that the latent heat of the metastable phase is about two-thirds the value for the eutectic.

For glass forming alloys where very deep undercoolings are achieved, the rising trend of C_p^l with undercooling becomes especially significant. As revealed in Fig. 13, the increase of C_p^l can reach values about twice that of stable liquid [57]. In fact, the termination of the rise in C_p^l represents the thermal manifestation of the glass transition, T_g. In this case, ΔC corrections are important in obtaining ΔG_v values that can be used to evaluate the competition between crystal nucleation and glass formation. In the absence of experimental measurements (which are pre-

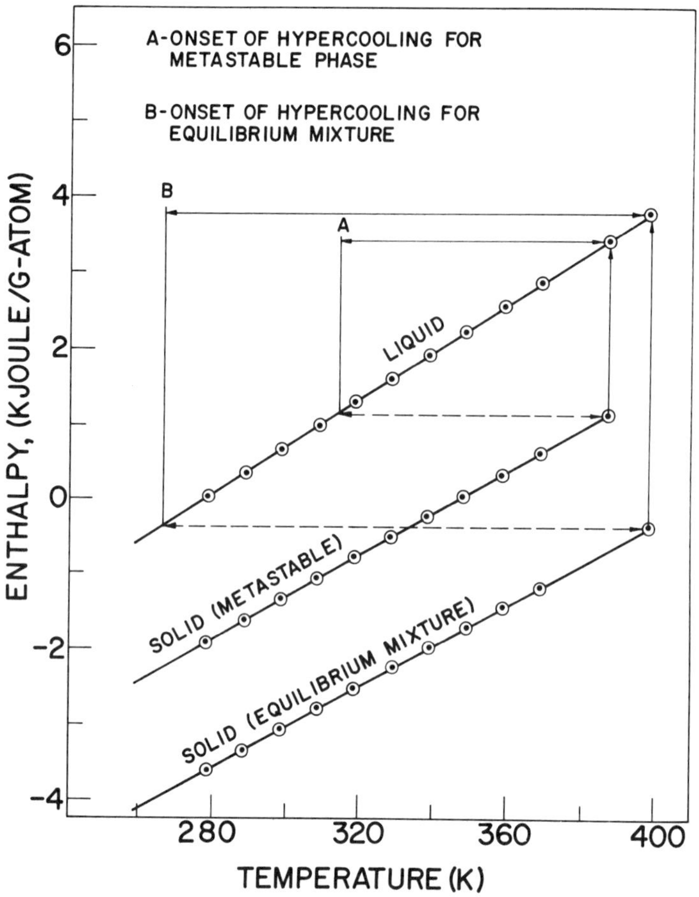

Figure 12 Temperature dependence of the enthalpy for a Pb-Bi eutectic alloy including a metastable solid and stable solid phases.

ferred) it appears that the approximation model suggested by Thompson and Spaepen [58] for ΔG_v as

$$\Delta G_v = L\,\Delta T \frac{2T}{T_M(T_M + T)} \tag{7}$$

provides a useful estimate when the suppression of partitionless crystallization is being evaluated as the competitor to glass formation.

For the more common case of nucleation of a solid of composition, C_s, that is different from the undercooled liquid alloy composition, the free-energy change is given by

$$\Delta G_v = \frac{(\mu_s^A - \mu_L^A)(1 - C_s) + (\mu_s^B - \mu_L^B)C_s}{V_m} \tag{8}$$

where μ_s^A and μ_s^B are the chemical potentials for species A and B in the solid, μ_L^A and μ_L^B are the component chemical potentials for the liquid and V_m is the molar

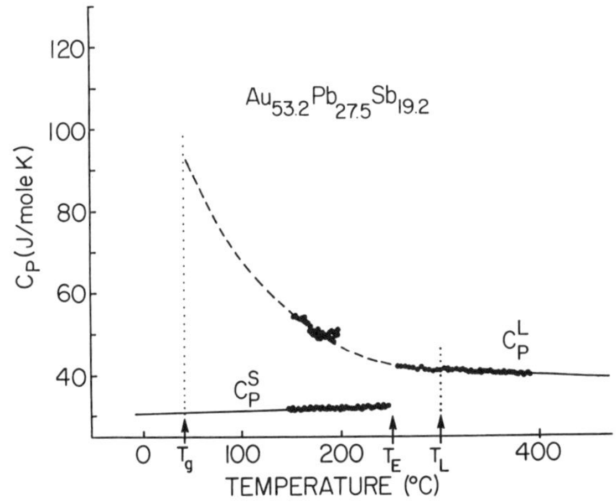

Figure 13 Heat capacity of solid and stable and metastable undercooled liquid in a $Au_{53.2}Pb_{27.5}Sb_{19.2}$ alloy.

volume. In *alloys* the free energy for nucleus formation is a function of cluster size and cluster composition so that ΔG^* in the classical model refers to critical values of nucleus composition and size. It is apparent from Eq. (8) that ΔG_v is maximized for a composition where $\mu_s^A - \mu_L^A = \mu_s^B - \mu_L^B$; that is, by the parallel tangent construction. The maximum driving-force condition has been proposed [59,60] to find the favored nucleus composition but the concurrence of ΔG^* and the maximum ΔG_v in composition also requires that σ and $f(\theta)$ have no composition dependence.

Experience with alloy undercooling indicates that the composition dependence of T_N reflects the composition dependence of T_L. For example, in the Pb-Sb system [61] the undercooling results in Fig. 14 reveal that T_N follows a similar trend to T_L even for different T_N levels resulting from catalytic sites of different potency (that is, different surface coatings). The maximum ΔG_v condition for nucleus composition has been used to interpret the composition dependence of T_N [60]. However, the generality of the interpretation requires further examination; especially in light of the fact that the many cases of metastable phase formation during RS do not satisfy a maximum driving-force condition and the more complex nature of nucleation in two component systems [62].

In the study of nucleation one of the most important parameters is σ, since the nucleation rate is proportional to $\exp(-\sigma^3)$ and is thus very sensitive to even slight uncertainty in the σ value. Quantitative growth rate calculations also depend on the value for σ through the Gibbs-Thomson effect which changes the equilibrium temperature of curved interfaces (Table 2). A direct equilibrium measurement of σ even with a large uncertainty has been very difficult. Usually values of σ are calculated from classical nucleation theory using the experimentally obtained maximum undercooling values which are assumed to be associated with homogeneous nucleation. Unless the maximum undercooling value is proven to refer to homogeneous nucleation the calculated σ value is expected to underestimate the true value. Furthermore the nucleation-derived values for σ pertain to the undercooling temperature, T_N, and not the melting temperature, for which an equilibrium mea-

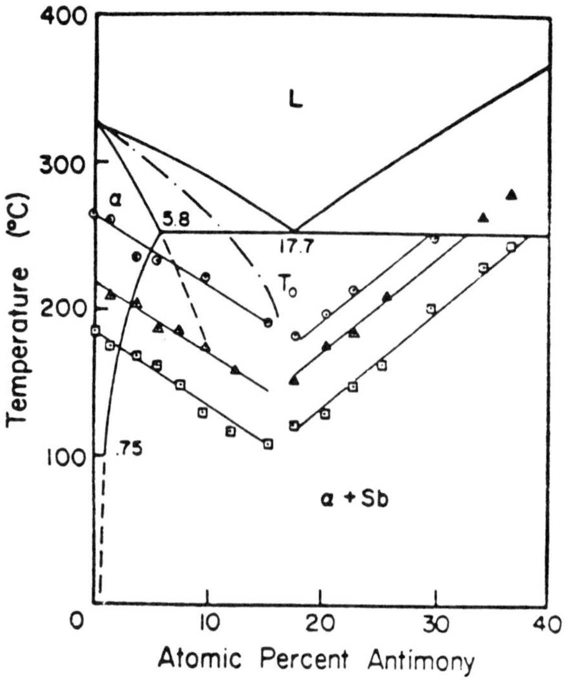

Figure 14 Summary of undercooling behavior in Pb-Sb alloys [61]. Undercooling trends at different levels are produced by different droplet surface coating treatments.

surement is possible. This raises an important question relating to the temperature dependence of σ. There have been a number of theoretical approaches to study the structural properties of the solid-liquid interface on the basis of various model constructions [63–67]. The models range from limiting cases in which σ is calculated for enthalpic bond breaking arguments or derived solely from a basis of entropic considerations to intermediate approaches where a weighted contribution of enthalpic and entropic terms is incorporated into the evaluation. For example, an entropic model would suggest that a bcc phase would have a lower σ value than an fcc phase; thus bcc phases would be favored during competitive nucleation. While cases of metastable bcc phases formation during RS have been reported [68,69] there are also cases where exceptions have been observed [34,70]. Other approaches based upon enthalpic models for σ have been applied to calculate the nucleation conditions for suppression of intermetallic phase formation in Al alloys [71] and the partitionless formation of crystals in highly undercooled alloy liquids [72,73]. Certainly, an improvement in the independent experimental determination of σ for metals would be invaluable in allowing a clearer evaluation of theoretical interface models and more reliable nucleation and growth rate calculations.

C. Nucleation Kinetics During Continuous Cooling

In general, the critical condition required to observe the onset of nucleation experimentally (usually one nucleus per sample) during isothermal holding over a time, t, in a sample liquid volume, v, or catalytic surface area, a or s, is given by

$J_v vt = 1$ for volume-dependent nucleation and $J_a at = 1$ or $J_s st = 1$ for heterogeneous nucleation. For continuous cooling this conditions requires some modification [74] to account for the total number of nuclei, N, formed on a volume or area basis during cooling from T_M to T_N at a constant rate $\dot{T}$ and is represented by

$$N = \frac{1}{\dot{T}} \int_{T_N}^{T_M} J_i(T)\, dT \tag{9}$$

While this expression may be evaluated numerically, a useful estimate is possible by expanding $J_i(T)$ near T_N to first order and changing the range of integration to include only positive values of this approximation to $J_i(T)$. This yields for $N = 1$ and surface nucleation

$$J_a(T_N)a \frac{\Delta T}{\dot{T}} = 2 \frac{\Delta G^*}{k} f(\theta)\left[\frac{3T_N - T_M}{T_N^2}\right] = K(T_N) \tag{10}$$

For $0.6 \leq T_N/T_M \leq 0.7$, $K(T_N)$ will typically range from 100 to 200. Letting $t = \Delta T/\dot{T}$, $J_a at = K(T_N)$. From this basis the heterogeneous nucleation time for the appearance of crystals during continuous cooling can be represented as a function of reduced temperature, $T_r = T/T_N$, [18] as

$$\ln t = -\ln[a\Omega_a K(T_N)] + \frac{16\pi\sigma^3 f(\theta)}{3kL^2 T_M (1 - T_r)^2 T_r} \tag{11}$$

The cooling-rate dependence of the nucleation temperature associated with heterogeneous nucleation, which is reflected in the shape of the transformation diagram (breadth of the C curve), is represented in general for fixed $K(T_N)$ as

$$\frac{dT_r}{d(\ln t)} = \frac{3L^2 k T_M}{16\pi\sigma^3 f(\theta)} \left[\frac{(1 - T_r)^3 T_r^2}{3T_r - 1}\right] \tag{12}$$

where the possible temperature dependence of Ω_a, L, and $f(\theta)$ has been neglected to simplify illustration. For homogeneous nucleation $f(\theta) = 1$, but for heterogeneous nucleation $0 \leq f(\theta) < 1$. Transformation diagrams for different types of active heterogeneous sites at different sample locations will use separate values of $f(\theta)$ and Ω_a and have different breadths representing different temperature dependencies of the nucleation rate. In Eq. (12), $dT_r/d(\ln t)$ becomes infinite at $T_r = 1/3$ as the nucleation time is minimized during cooling. Therefore, $T_r = 1/3$ is believed to estimate a lower bound to the onset of sensible nucleation of a crystalline phase during cooling for a pure metal provided that Ω_s, σ, $f(\theta)$, and L are not strongly temperature dependent.

A similar maximum undercooling limit may be expected in alloys although the simple expression presented in Eq. (12) cannot be expected to hold without modification. For example, within glass forming composition ranges near T_g, the diffusive attachment frequency in the liquid changes markedly with undercooling and would affect an undercooling limit and the T_r vs. $\dot{T}$ relationship significantly.

D. Powder Size Distribution Statistics for Nucleation

During cooling the controlling kinetics will be influenced strongly by the active catalyst (heterogeneity) distribution. A striking example is illustrated in Fig. 15,

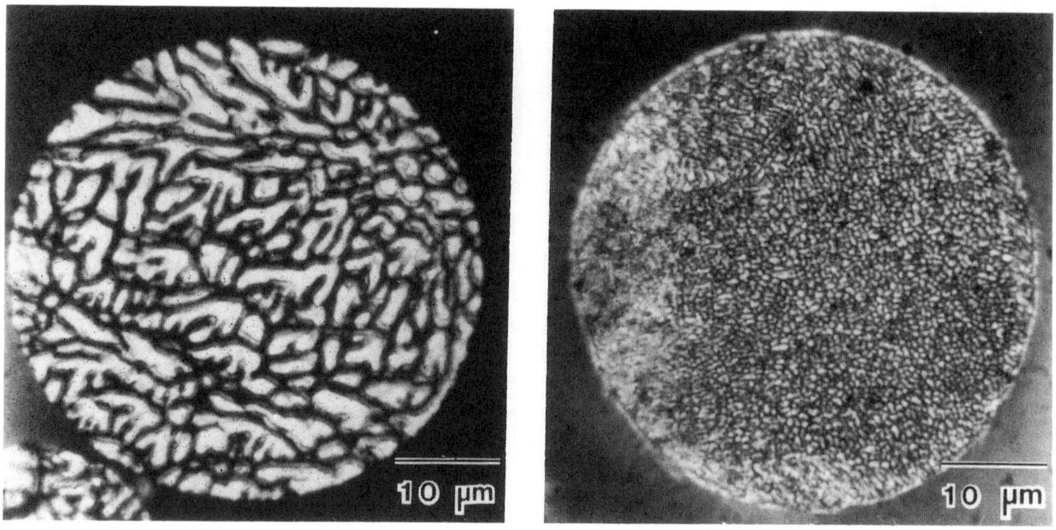

(a) (b)

Figure 15 Optical micrographs of Al–4.5 wt% Cu powder particles of the same size from a single atomizing run with dramatically different degrees of structural refinement. Most probably (a) solidified with little undercooling prior to nucleation while (b) was free of potent catalytic nucleation sites and achieved significant undercooling.

which shows different microstructures in two Al-4.5 wt% Cu powders of the same diameter from a single atomization run. In Fig. 15a a coarse dendritic structure is evident, while in Fig. 15b a fine-scale cellular morphology is present. Although it is conceivable that the two powder particles underwent very different cooling rates, it is more likely, considering the frequency of such events, that the different morphologies evolved because of solidification initiated at different melt undercoolings due to the presence or absence of a heterogeneity. In order to understand this behavior an analysis of nucleant isolation and distribution within undercooled volumes is necessary.

As long as there is a single type of nucleation kinetics with a steep temperature dependence, the effect of the amount of undercooled volume on the nucleation temperature is relatively small. In fact, for isokinetic behavior, a narrow range of nucleation temperatures (<10°C) is expected even with a sample of a broad size distribution of undercooled powders. However in the case of a number of pure metals and alloys, nucleation in a collection of powders is often observed to occur over a wide temperature range (>20°C). In this case the operation of a single type of nucleation kinetics in all of the powders is unlikely, and the broad nucleation range is most likely associated with the superposition of several nucleation kinetics from different heterogeneities that are activated in different powder sizes at different temperatures during the cooling process. As a result, nucleation at different levels of undercooling can result in the formation of different structures in different isolated undercooled volumes.

If the nucleating sites contained within a given liquid volume are distributed randomly, the arrangement of nucleants among the droplets for a high degree of dispersal may be described by a Poisson distribution. For this case the nucleant free droplet fraction, X, among droplets is represented by $X = \exp(-mv)$ where m is the average number of nucleants per volume. Based on the reported experience with powder samples, it appears that undercooling effects can become apparent for size refinement below about 100 μm and can become appreciable for powder sizes less than about 10 μm. This suggests that typical values for nucleant densities within the volume of a melt should range from about 10^6–10^9 cm^{-3}. Although these nucleant densities are inferred from powder sample behavior, it is reasonable to accept the values for molten layers or streams. The relatively sharp selection of a given X value (for example, $X = 0.9$) with isolation volume is illustrated in Fig. 16 for m values from 2×10^6cm^{-3} to 2×10^9cm^{-3}. This not only indicates the important role of size refinement in achieving large undercooling, but also that very small active nucleant concentrations can influence the undercooling strongly. Indeed, there is some evidence that impurities in specific cases can play a role in limiting undercooling behavior. Similar relationships can be developed for surface nucleant distributions.

Beyond the effect of size distribution statistics on nucleant isolation to allow for the operation of a kinetic transition in product phase selection, it is important to learn whether or not size distribution effects can also make it possible to observe a transition in the nature of a nucleation process. The transition from heterogeneous surface to homogeneous nucleation is of interest. One way of examining such a case involves the consideration of conditions needed for the number of nuclei produced by the two mechanisms, $J_v vt$ and $J_a at$, to be of comparable magnitude in a sample. This occurs if the ratio of the nucleation rates, $J_a/J_v = v/a$; that is, the ratio of sample volume to catalytic area. The ratio of the two nucleation rates from Eq. (3) is given by

$$\frac{J_a}{J_v} = \frac{\Omega_a}{\Omega_v} \exp\left[\frac{\Delta G^*}{kT}(1 - f(\theta))\right] \tag{13}$$

The onset of sensible nucleation is usually represented by a limiting value of $\Delta G^* \simeq 60\,kT$. Figure 17 shows a plot of J_a/J_v against θ according to Eq. (13). For

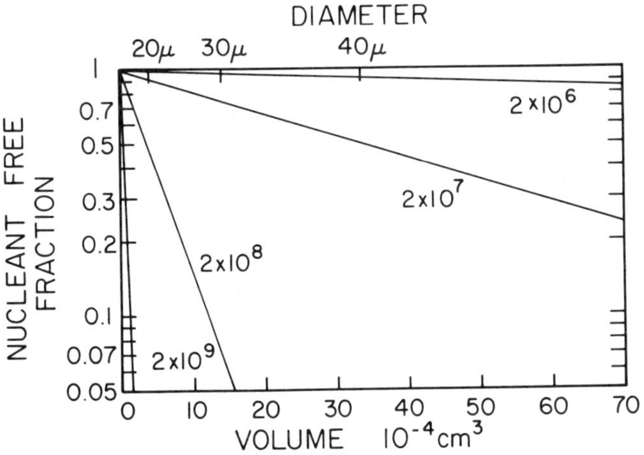

Figure 16 Nucleant-free fraction in different powder sizes for several nucleant densities.

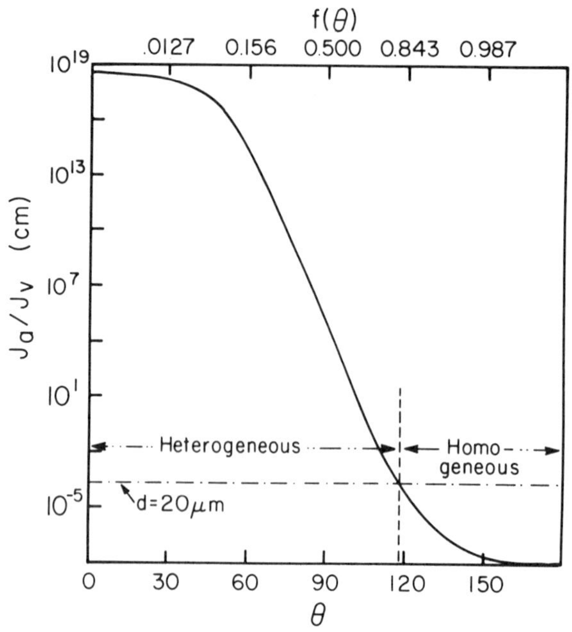

Figure 17 Relative magnitudes of heterogeneous surface to homogeneous nucleation in a 20-μm drop as a function of contact angle.

a 20 μm diameter powder in which the entire surface is assumed to be an active catalytic surface, $v/a = 3.3 \times 10^{-4}$ cm, and the critical contact angle, θ_t, for the transition is 118° as shown in Fig. 17. At $\theta = 118°$, a 20-μm-diameter powder has an equal probability to initiate solidification by a heterogeneous surface or homogeneous nucleation process. Homogeneous nucleation is favored to dominate against a background of heterogeneous nucleation for nucleants described by contact angles in excess of 118°. Decreasing the relevant values of ϕ (which changes Ω_a), and ΔG^* below the values used in Fig. 17 will also result in some reduction in θ_t from 118°. The main origin of this kinetic transition is related to the larger magnitude of the prefactor for homogeneous nucleation. Consequently, it is not necessary for a given undercooled volume to be free of *all* nucleants; it is only necessary for the isolation induced by subdivision or other processes to confine the influence of the most potent nucleants associated with $\theta < \theta_t$ in order to possibly observe the homogeneous nucleation undercooling. However, experimental access to the regime of homogeneous nucleation remains quite difficult.

E. Competitive Nucleation Kinetics During Cooling

In order to understand the kinetic competition between different phases during nucleation, the comparison of thermodynamic stability is frequently not sufficient. For nucleation kinetics, the driving free energy is not the only parameter of importance; the nucleation site density and the magnitude of ΔG^* associated with a given phase are also most important. As an illustration of the application of nucleation kinetics to reveal the relationships between cooling rate, undercooling, catalyst distribution and phase selection, it is useful to consider the competition between two different nucleation kinetics during continuous cooling in several different situations that are illustrated in Fig. 18. As an initial example, when two catalysts of different population density and potency are in the same volume (B and C), the undercooling limit at T_{r1} is determined by catalyst C at cooling rate $\dot{T}_1$. At cooling rate $\dot{T}_2$, however, the undercooling increases to T_{r2} by circumventing the catalytic effect of C. Thus, at a continually increasing cooling rate, there can be a significant improvement of undercooling as the catalytic effects of highly potent nucleants are circumvented in a given volume. When two different phases are competing with each other at the same catalytic surface present in an undercooled liquid, the magnitude of the activation free-energy barrier becomes the most dominant factor, according to nucleation theory. Therefore, in order to favor nucleation of a metastable phase at a given cooling rate, the activation free-energy barrier for the metastable phase must be lower than that for the equilibrium phase. In terms of Eqs. (3) and (5) this condition may be expressed as

$$\frac{\sigma_m^3 f(\theta_m)}{(L_m)^2(1 - T_r')^2} < \frac{\sigma^3 f(\theta)}{L^2(1 - T_r)^2} \tag{14}$$

where σ_m and θ_m are the metastable phase-liquid interfacial energy and the metastable phase-catalytic surface contact angle respectively, L_m is the heat of fusion per unit volume of the metastable phase, and T_r' is the reduced nucleation temperature of the metastable phase. Since $T_r' > T_r$ and usually $L_m < L$, Eq. (14) implies that $\sigma_m^3 f(\theta_m) < \sigma^3 f(\theta)$ for favorable nucleation of the metastable phase. This result again focuses on nucleation site potency and interfacial energy as critical factors in phase selection.

It has been shown in Eq. (12) that the breadth of the transformation kinetics diagram associated with different nucleation sites is proportional to $L^2 T_M/\sigma^3 f(\theta)$. In this case, the breadth of the transformation diagram for the nucleation of the metastable phase (A) is greater than that for the equilibrium (B), but the noses of the two diagrams are placed at the same position, as illustrated in Fig. 18. Unless

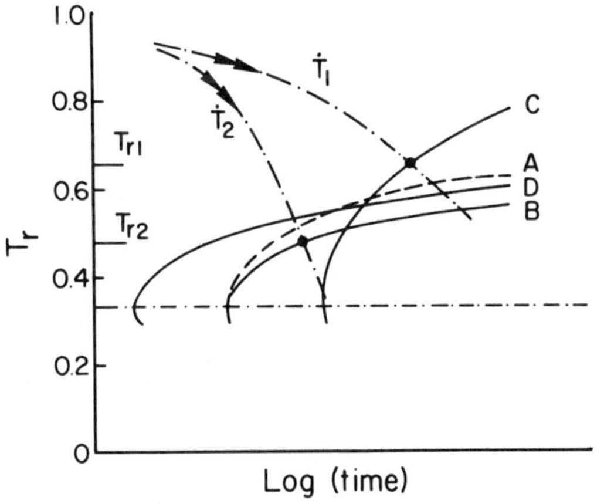

Figure 18 Time-temperature-transformation diagram representing different nucleation kinetics that may occur during continuous cooling of undercooled liquids [18].

there are nucleants of higher catalytic potency for nucleation of the equilibrium phase, nucleation of the metastable phase is dominant regardless of the cooling rate. Alternately, if the nucleation barrier for A is higher than that for B, the equilibrium phase will nucleate at all cooling rates. In addition, it is also possible for the nucleation of the equilibrium phase to be favored by an increase of the cooling rate if the nucleation site density for the equilibrium phase (curve D) is greater than that for the metastable phase. For these conditions an increase in cooling rate can result in a decreasing yield of a metastable product. In fact, a transition in nucleation from surface-dependent to volume-dependent kinetics with cooling (such as C to D in Fig. 18) can yield a large increase in nucleation rate and may contribute to a propensity for multiple nucleation, as reported in quenched Al-Si alloy droplets [22]. Other factors involving the specific thermal history of droplets can also contribute to the development of multiple nucleation, as described in Section IVG. Furthermore, the combination of kinetic transitions related to size distribution statistics and cooling rate effects can provide a useful experimental probe of nucleation behavior and also an effective processing tool for optimizing or controlling the product phase yield during RS.

F. Examples of Nucleation Kinetics Analysis

1. *Metastable Phase Formation in a* Pb-Bi *Eutectic Alloy*

The approach to the continuous cooling transformation kinetics outlined above has been applied to analyze metastable *x*-phase formation in a Pb-56 at% Bi alloy [75]. In this alloy, extensive undercooling ($T_r = 0.7$) occurs yielding nucleation of a metastable single phase product with a well-defined crystallization onset, and thermal analysis measurements indicate that T_N decreases from 265 K to 255 K over a cooling rate range of 10 to 320 K/min. The continuous cooling nucleation results can be fitted to Eq. (11) in terms of a single nucleation frequency with a slope of 1.74×10^7 K·J^2·cm^{-6} and a value for $\Omega_a aK(T_n)$ of 3.3×10^{22} s^{-1}. Instead of the usual approximation for $\Delta G_v = L(1 - T_r)$, the ΔG_v values used were derived from direct calorimetric measurement of the latent heat of the metastable *x*-phase, the temperature dependence of ΔC and reported molar volumes. The prefactor term, Ω_a is in reasonable agreement with the classical theory result if only a portion of the surface area represents an active catalytic site for metastable phase nucleation (i.e., $\phi \simeq 0.01$).

If Ω_a, σ, and $f(\theta)$ are treated as constant, evaluation of Eq. (11) can be used to develop the transformation diagram based on $J_a at = 1$ that is shown in Fig. 19. Although the calculation should be treated as approximate since estimates were used for several of the parameters, the results do illustrate some interesting features regarding the transformation kinetics. For example, if the linear approximation for ΔG_v had been applied instead of the measured values, the nose of the transformation diagram would be located at $T_r = 1/3$ or 128 K rather than 187 K. Also, the cooling rates required to reach the nose of the curve differ by a factor of 10^3 for the different ΔG_v values. This observation reveals the importance of using measured values for the nucleation parameters even if they represent a limited temperature range. It is also of interest to compare the droplet kinetics behavior with rapid quenching results. In this regard the calculated transformation diagram is consistent with the observation that suppression of the metastable phase by splat

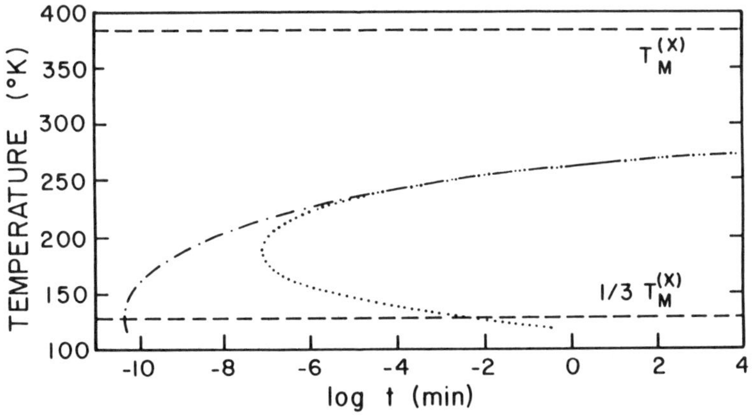

Figure 19 Calculated heterogeneous nucleation kinetics for a metastable x-phase in Pb–56 at.% Bi eutectic alloy droplets. The dotted curve is based on measured ΔG_v values and the dot-dash curve is derived from the linear ΔG_v approximation.

quenching is difficult [76]. According to Fig. 19 a critical cooling rate in excess of 5×10^7 K/s is required to bypass x-phase nucleation.

2. *Crystallization Kinetics During Drop Tube Processing of a* Ti-*40* at.%Al *Alloy*

A technique which provides conditions favorable for attaining large liquid undercoolings in high-melting-point materials is drop tube processing [77], where powder samples undergo melting and solidification during free fall through a vertical cylindrical chamber. Processing parameters such as powder size, powder surface coating, cooling rate, and melt superheat can be varied to analyze the effect of these parameters on the undercooling level [78]. An important advantage of the drop tube method is the capability to process high-temperature and reactive materials in a containerless manner.

This experimental technique has been recently applied to investigate the kinetic competition between α (hcp) and β (bcc) crystallization in a Ti-40 at% Al alloy [79]. An observed change in phase selection from an equilibrium β solidification product in coarse powder to α formation from the melt in fine powder is consistent with an increase in undercooling with decreasing droplet size. Using image analysis and SEM to quantitatively assess the microstructural abundance of α and β as a function of droplet diameter, a comparison can be made between powders processed in He gas and those processed in Ar to examine the effect of the higher conductivity He on the phase selection (Fig. 20). To provide a basis for the comparison, an expression for the cooling rate of a falling particle as a function of droplet size and gas environment can be written as

$$\frac{dT}{dt} = -\frac{\epsilon A \sigma_{sb}}{mC}(T^4 - T_\infty^4) - \frac{ha}{mC}(T - T_{gas}) \tag{15}$$

where ϵ is the emissivity of the liquid, a is the surface area of the droplet, σ_{sb} is the Stefan-Boltzmann constant, m is the mass of the droplet, C is the specific heat, h is the heat transfer coefficient, t is time, T_∞ is room temperature, and T_{gas} is the

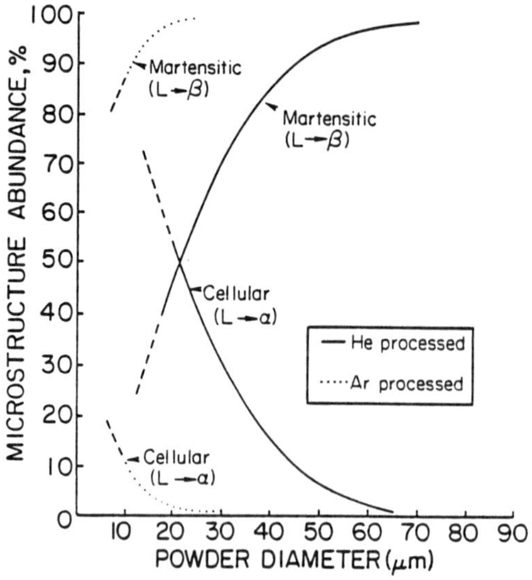

Figure 20 Summary of microstructural analysis for drop tube processed Ti–40 at.% Al powder showing the relative yield of the α and β phases as a function of powder diameter for He and Ar atmospheres.

gas temperature [80]. The first and second terms in Eq. (15) represent the contribution of radiative and convective cooling, respectively, to the overall cooling rate. Assuming Newtonian cooling conditions during free fall and no change in the controlling nucleation kinetics, an observed shift in the yield curves of α and β with a change in gas atmosphere (Fig. 20) may be described by a simple scaling of

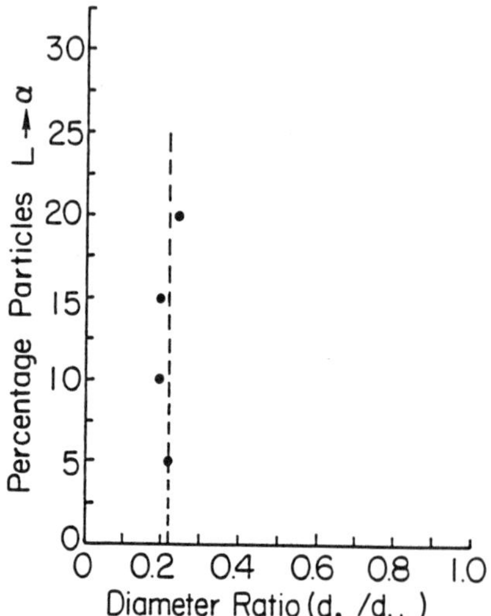

Figure 21 The rescaling of the results in Fig. 20 in terms of the diameter ratio for Ar and He processed powder to indicate isokinetic behavior for metastable α-phase nucleation.

cooling rate [81]. In other words, the droplet sizes required to produce an equivalent volume fraction of metastable product under Ar and He atmospheres are approximately equal to those required to produce equivalent cooling rates during free fall. One method of expressing the equivalence in thermal conditions for drop tube processed samples is through a simple comparison of the droplet diameters required to produce a given percentage of the population solidifying as α under He and Ar environments, as summarized in Fig. 21.

The observed scaling of phase selection with cooling rate is consistent with a single type of active heterogeneous nucleation site operating at the sample surface for the formation of the metastable α-phase. If nucleation of equilibrium β-phase also occurs on another site, that fraction of the droplet population which does not contain this site will then undercool to form α. Therefore, based upon heterogeneous surface nucleation, Poisson statistics [18] can be used to interpret the microstructural abundance of α as a function of droplet size for the two gas environments [82].

G. Multiple Nucleation During Continuous Cooling

While a consideration of competitive phase selection kinetics is necessary, it is incomplete since the thermal conditions immediately following the *initial* nucleation event have an important influence on the possibility of further nucleation events. The thermal history reflects the balance between the imposed external cooling and the recalescence due to the latent heat emitted by a rapidly advancing solidification front. The thermal path following the initial nucleation appears to have the greatest effect under high-cooling-rate conditions. For example, the solidification of undercooled powders during slow cooling invariably yields single-grain structures, but during rapid quenching of undercooled powders multigrain structures have been reported for a number of alloys.

The problem of quantitatively describing the relationship between the thermal history and solidification behavior is not simple. Several detailed numerical calculations have been developed to treat various aspects of the thermal history after the initial nucleation event [22,23]. However, useful insight into the basic features of this thermal history can be developed from an approximate model. This approach is limited and clearly not rigorous, but it does serve to illustrate the essential features of behavior for different processing methods.

During continuous cooling at a rate $\dot{T}$ the temperature in a volume ν_s, assumed spatially isothermal, following a single nucleation event at time $t = 0$ at a temperature T_N can be estimated by

$$T = T_N - \dot{T}t + \frac{4\pi V^3}{3\nu_s}\left(\frac{L}{C}\right)t^3 \tag{16}$$

where V is the radial growth rate of a spherical crystal assumed constant near T_N. For a powder sample the volume ν_s may be thought of as the powder volume. For a sample not limited in size in some dimensions, such as melt-spun ribbon, the volume ν_s may be thought of as the reciprocal of the nucleant density. Clearly the temperature given by Eq. (16) is an average within the volume, ν_s: some regions near the growing crystal are above this temperature while regions away from the growing crystal are below this temperature.

From Eq. (16) the temperature can be shown to continue to drop for a time, t^*, to a temperature, T_N^*, prior to recalescence, as given by

$$t^* = \left\{ \frac{v_s C \dot{T}}{4\pi V^3 L} \right\}^{1/2} \tag{17}$$

$$(T_N - T_N^*)^2 = \frac{1}{9\pi} \left(\frac{C}{L} \right) v_s \left(\frac{\dot{T}}{V} \right)^3 \tag{18}$$

Following time t^*, recalescence causes the temperature to rise, returning to T_N in a time $t^*\sqrt{3}$ and exceeding T_N to some maximum temperature at longer time before final solidification. Clearly a second nucleation event in v_s will alter the thermal history. The thermal history profile described by Eq. (16) is illustrated schematically in Fig. 22.

Conditions that favor *large* values of $T_N - T_N^*$ and t^* are of interest. Here the importance of high values of $\dot{T}$ and v_s is evident, but also the effect of relatively low values of V can be appreciated. Solidification structures with low values of V usually involve growth with significant solute redistribution due to the thermodynamic impossibility of partitionless solidification. Maximum velocities in this case are typically estimated to range up to about 10 cm/s. Although Eq. (16) is based on a constant V and will tend to overestimate $T_N - T_N^*$, a fair upper bound can be obtained. With $V \simeq 10$ cm/s it is possible that $T_N - T_N^*$ will range from 10 to 100 K for cooling rates from 10^5 to 10^6 K/s for $v_s = 10^6/\text{cm}^3$ and $L/C \sim 300$ K. Much smaller $T_N - T_N^*$ values occur for collision-limited growth. In all cases it should be noted that Eqs. (16)–(18) and the model illustrated in Fig. 22 are intended to apply approximately to relatively low values of $T_N - T_N^*$ (order of 10 K).

Two interesting features can be noted from this simple model. Even with a processing method involving cooling by conduction to an underlying substrate where potent nucleation sites may be present, it is possible to develop bulk melt undercooling levels which are sufficient to significantly alter the solidification velocity and allow for metastable phase formation. Similarly, due to the strong tem-

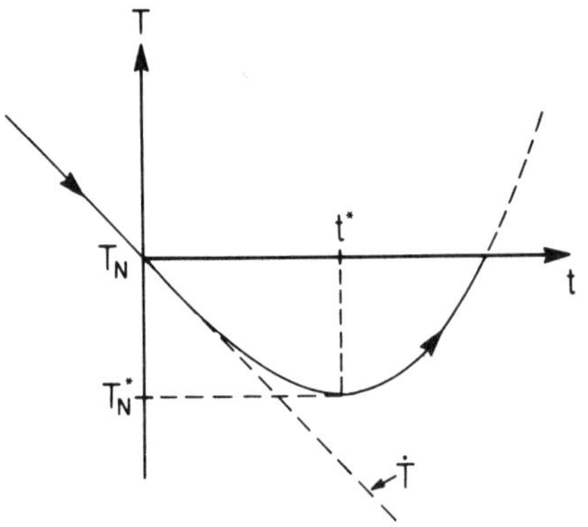

Figure 22 Approximate thermal history profile following nucleation of an undercooled liquid.

perature dependence of the nucleation rate, the additional undercooling $T_N - T_N^*$ can allow for the development of multiple nucleation events which can yield fine-grain, microcrystalline structures. Using this approach, the incidence of extra grains has been described by Turnbull [83] and calculations of grain size for pure metals have been performed [22,84–87]. The calculations are subject to the same uncertainty about nucleation rate parameters as described above. Although such a mechanism can account for the evolution of a microcrystalline structure, there are other possible origins including heterogeneous volume nucleation on a high density of internal sites. Furthermore, severe reduction in V can be realized as T_N^* approaches T_g due to a reduction in liquid diffusivity. While these conditions imply that transient nucleation effects also become important, a large increase in grain density is expected as well.

The cross-sectional TEM micrograph of a melt spun Nb-25 at% Si alloy ribbon in Fig. 23 provides a striking illustration of these effects [88]. Away from the wheel side of the ribbon a mixture of glass and microcrystallites is observed with grain sizes of about 15 nm. Toward the ribbon center grain sizes for a mixture of α-Nb and Nb_5Si_3 of about 0.1 μm are typical. Clearly the growth rate of the two-phase mixture is slow compared to a single-phase material and contributes to the abundant nucleation in this material after the initial crystallization event. Also, the observation of a glass-crystal transition region and a grain size gradient with distance from the wheel supports the importance of both cooling rate and undercooling during processing.

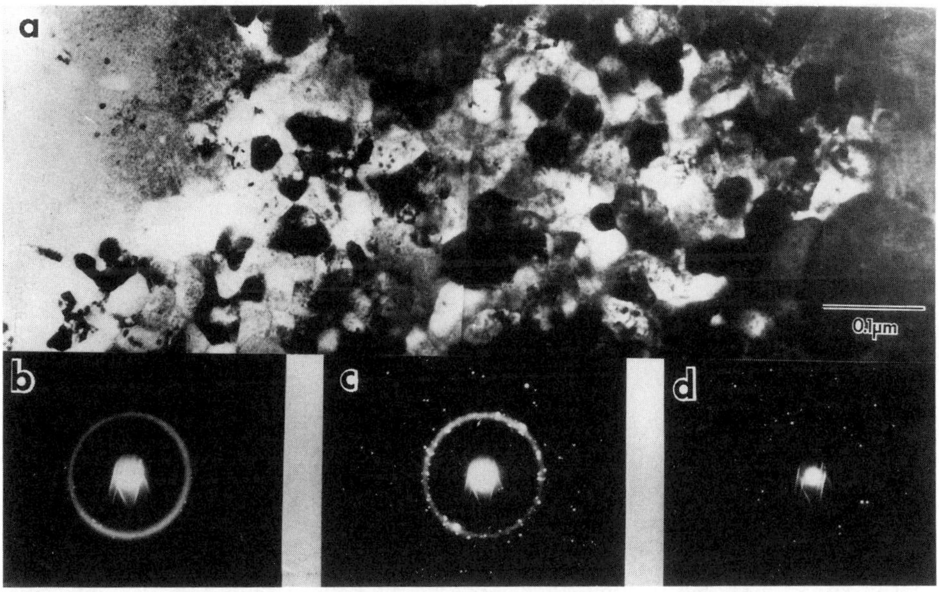

Figure 23 (a) Taper section thin foil TEM micrograph of a melt-spun Nb–25 at% Si alloy which shows transition from glassy structure (left) to progressively larger grain microcrystalline structure of α-Nb and Nb_5Si_3 as a function of distance from wheel surface. (b, c, d) the corresponding SADP for the transition [88].

V. GROWTH KINETICS

The analysis of growth kinetics during rapid solidification requires the solution to a complex moving boundary problem for the solid-liquid interface. Both solute and heat diffusion equations must be solved in the liquid and solid subject to conditions on external boundaries and at the liquid-solid interface. The goal of this analysis is to predict the phase or phases that grow the fastest and dominate the microstructure and to predict the growth morphology or interface shape and hence the segregation pattern in the frozen solid. This problem poses severe mathematical difficulties even for slow-speed solidification; for example, the shape of a growing dendrite with side branches. In the following, attention is given only to those aspects particular to rapid solidification. Consideration is first given to the deviation of liquid-solid interface conditions from local equilibrium and second to changes in solutions to the diffusion equations that are necessary for high-rate solidification.

A. Nonequilibrium Liquid-Solid Interface Condition

As indicated in (IV) of Table 2, liquid-solid interface conditions can deviate significantly from the predictions of the phase diagram at high solidification rate. In what follows a model to describe this deviation is described. The predictions of the model span the range from slow velocity, where local equilibrium is usually valid, to high velocities where partitionless solidification occurs.

Baker and Cahn [30] described the interface conditions for solidification of a binary alloy in terms of two response functions. One choice for the response functions describes the interface temperature T_I and the composition C_s^* of the solid at the interface [89]. These response functions can be written as follows:

$$T_I = T^*(V, C_L^*) - T_M \Gamma K_m \tag{19a}$$

$$C_s^* = C_L^* k^*(V, C_L^*) \tag{19b}$$

where V is the local interface velocity, C_L^* is the composition of the liquid at the interface, Γ is the capillarity constant (σ/L), and K_m is the mean curvature of the solid-liquid interface. The functions T^* and k^* must be determined by a detailed kinetic model for the interface. At zero velocity they are very simply related to the phase diagram: $T^*(O, C_L^*)$ is the equation for the phase diagram liquidus and $k^*(O, C_L^*)$ is the equation for the equilibrium partition coefficient k_E, which can depend on composition. The dependence of k^* on curvature is thought to be negligible [25].

During solidification, the temperature and the compositions at the interface are constrained by thermodynamics. There must be a net decrease in the free energy per mole, ΔG, to form an infinitesimal amount of solid of composition C_s^* from liquid of composition C_L^* [30]. The free-energy change on a volume basis is given by Eq. (8) with C_s^* substituted for C_s. Under the usual assumption of local equilibrium at the solid-liquid interface, which is useful at slow rates of solidification, the chemical potential of the solid, μ_s^i, and the chemical potential of the liquid, μ_L^i, must be equal for each component i ($=A, B$). During rapid solidification, this is not necessarily the case and nonequilibrium conditions at the interface can exist. The term "solute trapping" is commonly used when the partition coefficient deviates from the equilibrium value but, as originally defined, was restricted to the case in which the chemical potential of the solute increases during solidification. The T_0 curve which was described previously is the locus of compositions and temperatures for which $\Delta G = 0$ and the liquid and solid phase compositions are equal.

Several models for the dependence of the partition coefficient on velocity, Eq. (19b), have been formulated. The model formulated by Baker [90] is quite general. Other theories [91,92] predict that the partition coefficient changes monotonically from its equilibrium value to unity as the growth velocity increases. In these models the interface partition coefficient is a function of a dimensionless velocity V/V_D where V_D is the diffusive speed for atom exchange between the crystal and the liquid. At the present time, one probably should view V_D as a materials parameter to be determined by experiment. It is a ratio of the diffusion coefficient, D_i, for that exchange to the interatomic distance, a_0. This diffusion coefficient is bounded by that of the liquid and the solid, but appears to be closer to that of the liquid. The functional form of the model proposed by Aziz [91] and by Jackson et al. [92] for nonfaceted growth is given as

$$k^*(V) = \frac{k_E + V/V_D}{1 + V/V_D} \tag{20}$$

which is valid only for dilute alloys where the composition dependence of k_E is neglected. At a velocity of V_D, the partition coefficient is the mean of the equilibrium partition coefficient and unity. Because Eq. (20) has no dependence on composition, it cannot treat the situation shown in Fig. 10, in which partitionless solidification is impossible for some compositions. The Aziz model has been generalized to treat nondilute alloys [93].

An analytical expression for the interface temperature, Eq. (19a) can be obtained from the quantity ΔG evaluated for dilute solutions [106,122]. For liquid of composition C_L^* which is transformed to solid of composition C_s^* at a temperature T_I, Eq. (8), for dilute solutions is equivalent to

$$\frac{\Delta G}{R_g T_I} = -\frac{L_m}{R_g T_M^2}(T_M + m_L C_L^* - T_I) + C_L^*\left(k_E - k^*\left(1 - \ln\frac{k^*}{k_E}\right)\right) \tag{21}$$

In this expression m_L is the liquidus slope, T_M is the pure metal (solvent) melting point, L_m is the latent heat per mole and R_g is the gas constant. Using the expression for ΔG given above and the assumption of a linear kinetic law for the interface velocity, $V = -V_0(\Delta G/R_g T_I)$ where V_0 is a constant related to the ease of crystallization, one obtains the two response functions for a flat interface as follows:

$$T_I = T_M + m_L C_L^*\left[1 + \frac{1}{1 - k_E}\left(k_E - k^*\left(1 - \ln\frac{k^*}{k_E}\right)\right)\right] - \frac{R_g T_M^2}{L_m}\frac{V}{V_0} \tag{22a}$$

$$C_s^* = k^* C_L^* \tag{22b}$$

where k^* is given by Eq. (20). For a curved interface, Eq. (22a) would contain an additional term for the Gibbs-Thomson effect, as given in Eq. (19). Note that at $V = 0$,

$$T_I = T_M + m_L C_L^* \tag{23a}$$

$$C_s^* = k_E C_L^* \tag{23b}$$

which are the conditions for local interface equilibrium. As k^* goes to unity, the term in brackets in Eq. (22a) changes the effective liquidus slope from the equilibrium value of m_L to the slope of the T_0 curve, which is $m_L[(\ln k_E)/(1 - k_E)]$. The last term in Eq. (22a) can be identified as a interface kinetic undercooling which is proportional to V.

The choice of the parameters V_0 and V_D as well as other assumptions in the above model clearly require experimental measurements. For a liquid diffusion coefficient typical of metals (2.5×10^{-5} cm^2/s) and a length scale of 0.5 nm, V_D should be less than 5 m/s. Experiments on doped Si [28] and on Al-Sn alloys [29] suggest that V_D is between 2 and 10 m/s. The value for V_0 has not been measured although some factors regarding its choice can be discussed. For pure metals, V_0 is proportional to the usual linear interface attachment kinetic coefficient, μ, namely, $\mu = L_m V_0 / R_g T_M^2$. Hence faceted crystals should have small values of V_0 while unfaceted metallic-like crystals should have large values. For the latter case V_0 may approach the velocity of sound in the liquid metal. This is called collision-limited growth [24] and most likely applies for disordered close-packed crystals. For non-close-packed crystals, especially those which are chemically ordered, crystallization kinetics are at least partially diffusion-limited and V_0 would be related to the ratio D/a_0, which is usually less than the speed of sound. This latter point has been used to explain regions of glass formation which are slightly larger than that shown in Fig. 10 for some alloys [94].

Figure 24 shows a composite plot of the two response functions obtained using Eqs. (22) superimposed on a phase diagram including the liquidus, solidus, and T_0 curves. The composition of the solid at the interface and the interface temperature are plotted along a curve parameterized by interface velocity for a given fixed concentration in the liquid at the interface. The figure is based on a phase diagram for Ag-Cu with $k_E = 0.44$ and $m_L = -5.6$ K/at%, a V_D value of 5 m/s and two values of V_0 (infinity and 2×10^3 m/s). At zero velocity the composition of the solid lies on the solidus curve. At intermediate velocities (about 10 cm/s) the composition of the solid moves toward the composition of the liquid with a small increase in undercooling. At high velocities the solid composition is near the liquid composition and the interface temperature is near the T_0 curve. The broken curve shows the case where V_0 is infinite. This curve is the thermodynamic bound on the solid composition at the interface for a given liquid composition at the interface and various interface temperatures. The broken curve can also be obtained by letting $\Delta G = 0$ in Eq. (21). The full curve, for a finite value of V_0, shows the interface temperature plunging quite rapidly. Alternately, one could plot C_L^* and T_I for fixed C_s^*. This would correspond to the physical situation of steady-state solidification of an alloy of bulk composition C_s^* at different velocities.

An alternate way to plot the interface conditions is to use Eqs. (22) to draw "interface condition diagrams" as suggested by Aziz and Kaplan [93]. At any

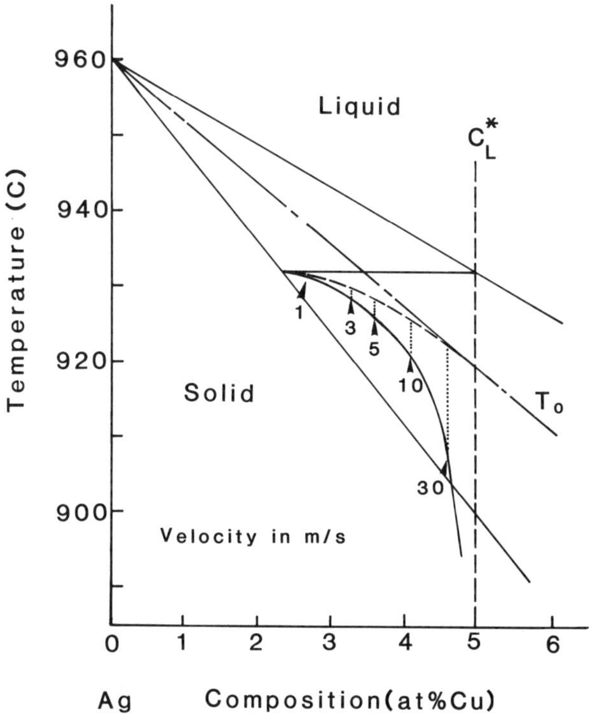

Figure 24 Interface temperature and solid compositions for solid that forms at the indicated velocities (in m/s) from a liquid of fixed composition C_L^* at the interface. Figure shown for Ag-Cu with $m_L = -5.6$ K/at%, $k_E = 0.44$, $C_L^* = 5$ at% Cu, $V_D = 5$ m/s. The dashed curve shows the case for $V_0 = \infty$ while the solid curve shows the case for $V_0 = 2 \times 10^3$ m/s [122].

velocity of interest, a pair of straight lines (for dilute alloys) will represent liquid and solid compositions which can be present at the interface for any interface temperature. These diagrams will be a function of the two kinetic parameters V_D and V_0. Figure 25 shows a plot of the interface condition diagrams for Ag-Cu alloys for various velocities and for $V_0 = 2 \times 10^5$ cm/s (for collision limited growth) and $V_D = 2 \times 10^2$ cm/s. As the velocity is increased, the "liquidus" and "solidus" will converge to a single line near but below the T_0 curve.

Whether or not partitionless solidification can occur at temperatures above the solidus is a frequent topic of discussion. Using the model above, the answer to this question depends on the values of the kinetic parameters, V_0 and V_D. Generally, a high V_0 and a low V_D will lead to partition coefficients very close to one at temperatures above the solidus. On the other hand, a low V_0 and a high V_D lead to partition coefficients very close to one only at temperatures far below the solidus.

This analysis provides a pair of thermodynamically consistent response functions for dilute alloys for the conditions *only* at the liquid-solid interface. For concentrated alloys no simple expressions can be written because k_E depends on composition. However, given a thermodynamic description of the liquid and solid phases and values for the two kinetic parameters V_D and V_0, the response functions

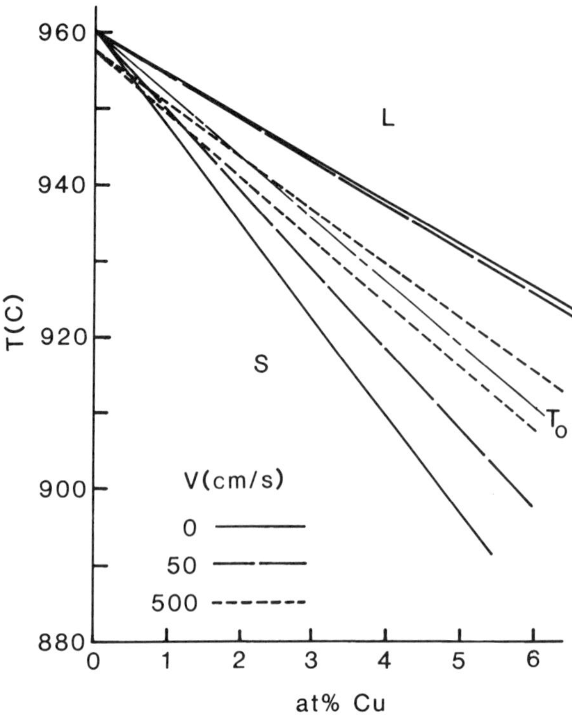

Figure 25 Interface condition diagrams for dilute Ag-Cu alloys showing pairs of "liquidus" and "solidus" lines for interface velocities of 0, 50, and 500 cm/s. $V_0 = 2 \times 10^5$ cm/s, $V_D = 2 \times 10^2$ cm/s [106].

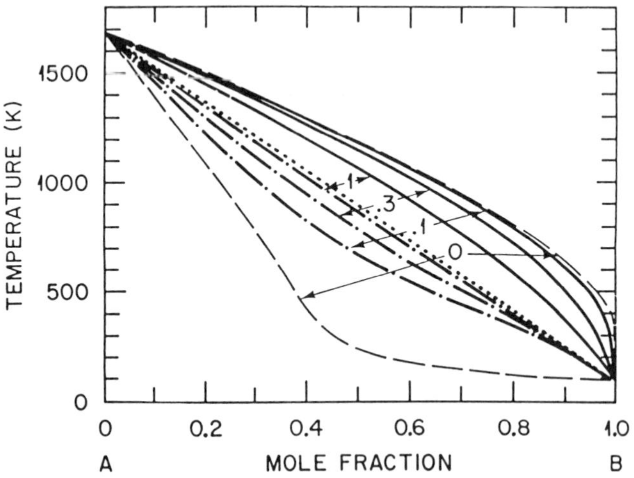

Figure 26 Kinetic interface condition diagram [93] for ideal liquid and solid with melting points as shown, entropy of fusion is R, and $V_0/V_D = 100$. Dashed lines: equilibrium liquidus and solidus ($V = 0$), dash-dot lines: kinetic liquidus and solidus at the indicated velocities (V/V_D), dotted line: T_0 curve.

can be calculated numerically [93]. Figure 26 shows the result for all compositions of an ideal binary alloy. Effective liquidus and solidus curves are shown for each interface velocity. These curves give the possible combinations of C_L^*, C_S^*, and T_I as interface conditions at the indicated velocities. It must be emphasized that these conditions must be combined with solute redistribution and heat flow analyses in the bulk liquid and solid phases to permit the full analysis of rapid solidification problems.

Recent theoretical work [95] has extended these ideas to the solidification of ordered intermetallic compounds. Experiments indicate that some compounds, which are normally ordered at the solidus, can be forced to solidify as the disordered form of the crystal structure [96–100]. The theory treats this by a consideration of solute trapping on each sublattice of the ordered phase. At high rates, there is insufficient time to proportion the solute onto each sublattice and a disordered solid can be formed. Often, however, this disordered phase reverts to the equilibrium ordered phase during solid-state cooling with a resultant microstructure consisting of a high density of antiphase domains.

B. Solidification Controlled by Solute Diffusion

The solidification of alloys requires redistribution of heat and solute in the bulk phases as long as there are discontinuities in enthalpy (the latent heat) and composition at the interface, respectively. Because the solute diffusivities are many orders of magnitude smaller than the thermal diffusivities, details of microstructure tend to be determined by solute diffusion. As discussed in the previous section, only at growth rates on the order of 1 m/s, where k^* may approach unity, can one ignore the solute redistribution problem. This condition is not achieved in most commercial rapid solidification methods.

Analysis of solute redistribution in the liquid phase involves the dimensionless solute Peclet number, $P_c = Vl/2D$, where V is the interface velocity, l is a characteristic length scale of the microstructure being analyzed, and D is the liquid diffusion coefficient. Typical values of V and l which occur for slow solidification rates indicate that $P_c \ll 1$ and approximate solutions to the diffusion equation valid at low Peclet numbers are employed to treat conventional solidification problems. However, many rapid solidification problems involve situations where $P_c \sim 1$ and where $k^* \neq 1$. As summarized in Table 4, at least four topics in solidification must be viewed differently at low and high values of P_c: interface stability, microsegregation, dendritic growth, and eutectic growth. It is also useful to add the topic of solute trapping to Table 4. However, the diffusion models used for this analysis are typically atomistic in nature, rather than based on the continuum approach used to analyze diffusion for the other four topics, and the D value used is not that for the liquid.

1. *Interface Stability*

In some cases, absolute stability provides a mechanism whereby an alloy can solidify at high rate with a planar liquid-solid interface and hence without microsegregation in the solid even though a condition of $k^* = 1$ has not been achieved. Results on Ag-1 Cu and Ag-5 wt% Cu alloys solidified using electron beam surface melting and resolidification at known velocities [101] show a transition from cellular to cell-

Table 4 Effect of Solute Peclet Number ($Vl/2D$) on Various Solidification Topics

Subject	l	$P_c \ll 1$	$P_c \sim 1$
Interface shape stability	Interface perturbation wavelength	Constitutional supercooling	Absolute stability
Microsegregation	Primary dendrite or cell spacing	Scheil-type analysis	Flat solute profiles
Dendritic growth	Dendritic tip radius	Trivedi, Lipton-Glicksman-Kurz	Boettinger-Coriell, Lipton-Kurz-Trivedi
Eutectic growth	Eutectic spacing	Jackson-Hunt analysis	$\lambda^2 V \neq$ constant or no eutectic structures
Solute trapping	Interatomic dimensions	Local equilibrium	Partitionless solidification

free structures as the growth rate is increased. This kind of microstructural transition can be understood by including the effects of liquid-solid interfacial energy into constitutional supercooling analysis using the full morphological stability theory [102,103]. The range of solidification rates where surface energy is important in promoting a planar interface corresponds to high Peclet numbers. From this theory, a planar interface is stable for an alloy of composition, C_0, if the growth rate, V, exceeds a critical value given by

$$V = \frac{m_L D(1 - k_E)C_0}{k_E^2 T_M \Gamma} \tag{24}$$

and if the net heat flow is into the solid. Equation (24) does not apply when growth occurs into an undercooled melt. For many practical alloys, the velocity requirement of Eq. (24) is so high that the assumption used in the theory of local equilibrium is invalid. For this purpose the original Mullins and Sekerka theory has been extended to include the case where the partition coefficient depends on velocity [104]. This modification tends to lower the critical velocity for cell-free growth.

The theory has also been examined for negative temperature gradients [103,105]. Negative temperature gradients exist at the liquid-solid interface when initial undercooling of the melt occurs. In this case, planar growth is always unstable and cellular or dendritic interfaces should ensue. However for any growth rate, very small wavelength perturbations of the interface shape are stabilized by surface energy and cannot develop into a cellular interface. For example, in the solidification of undercooled submicron powders, a cellular structure is often absent in the first region to solidify [22]. This effect may be explained by the fact that the powder is smaller than the smallest unstable wavelength [106,107].

2. *Microsegregation*

Even though microsegration-free crystalline microstructures are obtained for some alloys under certain conditions, many commercially important alloys exhibit very fine cellular structures. Geometrically, the cellular structures that are obtained at high rate differ little from cells observed at slow rates except for the scale. However, the details of the microsegregation (composition) profile within cells, the volume fraction of intercellular material and/or the actual identity of phases found in intercellular regions may differ from those found in more slowly solidified alloys.

The amount of solute incorporated into the cell interiors has a direct influence on the volume fraction of intercellular phases and the degree of precipitation possible in the cell interiors during subsequent thermomechanical treatment. Several measurements have been reported for cellular solute profiles in rapidly solidified alloys produced with unknown or calculated growth rates [69,108–110]. Figure 27 shows solute profiles [111,112] measured from samples where the growth rate was determined experimentally using the electron beam melting and resolidification technique. One notes a general increase in the Cu content of the cells with increasing solidification rate. A reduction in the amount of second phase at the cell walls also occurs. This trend is expected from various theories for dendritic growth (described below) even with the assumption of local interfacial equilibrium.

Except for the sample solidified at 0.1 cm/s, the solute profiles in Fig. 27 are relatively uniform and do not show the characteristic U-shaped profile expected from Scheil-type analysis, where the assumption of no concentration gradients in the liquid between solidifying cells is usually made. This assumption seems only valid when the cell spacing is much less than D/V, that is, $P_c \ll 1$. Such is not the case in these experiments where the solute Peclet numbers using the measured cell spacings and growth rates range from 0.5 for the U-shaped profile to 25 for the sample solidified at 18 cm/s. Theoretical analysis of cell shapes and corresponding solute profiles is an active area of research [113].

Another interesting aspect sometimes related to cellular solidification involves the formation of small precipitates and dispersoids [114] which are very important for strengthening. In some alloys, precipitates grow from supersaturated solid solution in a rather normal fashion during cooling to room temperature. In several alloys the dispersoids are thought to form directly from the liquid phase with spacings related to the cell spacing. Rapidly solidified alloys with extended solubility can be heat-treated only once (often during consolidation). Thus any segregation spacing inherited from the solidification process can greatly affect the spacing of precipitates or other second phase particles. For example, Al–3.7 wt% Ni–1.5 wt%

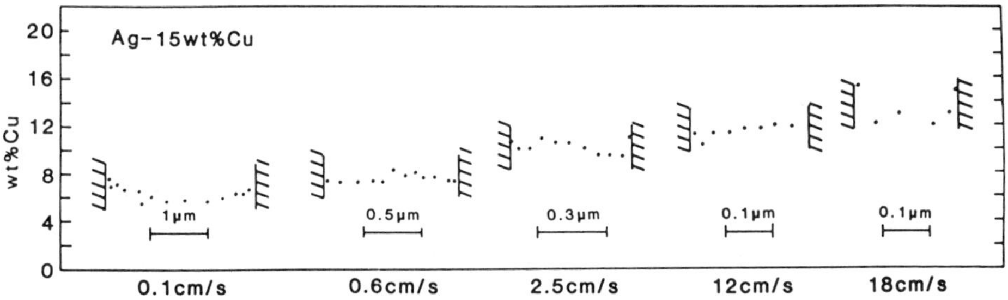

Figure 27 Microsegregation profiles measured by STEM across cells in Ag–15 wt% Cu alloys solidified at the indicated rates. The intercellular regions (crosshatched) contain eutectic ($V < 2.5$ cm/s) or only Cu-phase ($V > 2.5$ cm/s).

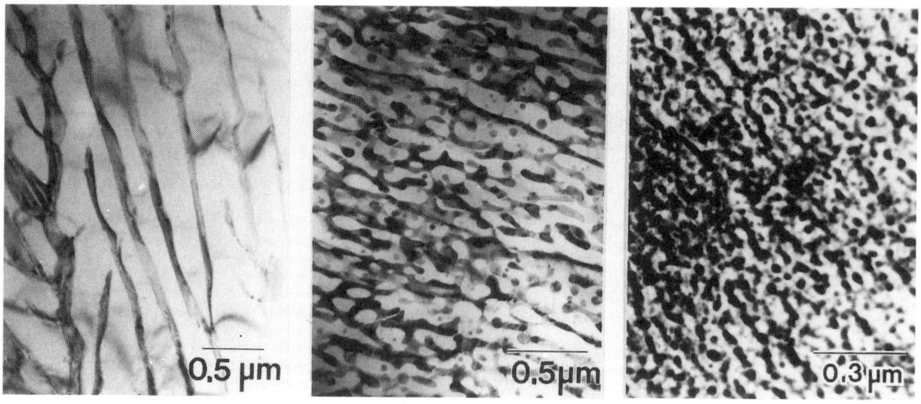

(a) (b) (c)

Figure 28 Transmission electron micrographs of cellular structures of α-Al with intercellular Al_9FeNi second phase in an Al–3.7 wt% Ni–1.5 wt% Fe alloy following moving-spot electron-beam surface melting. An increase in solidification rate results in a transition in the morphology of the intercellular phase, Al_9FeNi, from (a) continuous layers at 2.5 cm/s to (b) semicontinuous layers to at 100 cm/s to (c) discrete dispersoids at 500 cm/s. The growth direction is in the plane of the micrograph [115].

Fe alloy solidifies with cells of FCC aluminum and intercellular intermetallic Al_9FeNi (Fig. 28). As the growth rate increases one observes a transition in the morphology of the intercellular phase from continuous (in the growth direction) to discrete particles. At high rates the cellular interface pinches off isolated droplets of liquid which subsequently freeze to form the structure in Fig. 28c [115]. Calculations of cell shapes under conditions far from those giving planar interfaces [116] show a deepening of the grooves at the cell boundaries and subsequent collapse of the side walls to pinch off liquid droplets.

3. *Eutectic Growth*

For eutectic growth in conventionally solidified alloys the solute Peclet number based on the eutectic spacing is normally much less than one ($\sim 10^{-3}$). Based on this fact, the Jackson-Hunt analysis of eutectic solidification, which explicitly assumes that $P_c \ll 1$, predicts the eutectic spacing to decrease as $V^{-0.5}$ [117]. However, using controlled rapid solidification techniques, it has been shown that a maximum growth rate exists for eutectic structures in several alloys with eutectic spacings no smaller than 20 nm, namely, Pd-Cu-Si (at 2 mm/s), Ag-Cu (at 2.5 cm/s) and NiAl-Cr (at 2.5 cm/s) eutectics [47,101,118]. Attempts to force growth above these rates causes the development of different morphologies. The Pd-Cu-Si alloy takes on a glassy structure, the Ag-Cu alloy (at a somewhat higher rate) becomes a supersaturated single FCC phase, while the NiAl-Cr alloy becomes a supersaturated B2 NiAl(Cr) phase. The particular phase selection in the different alloys is determined to a large extent by the thermodynamic options available to the particular alloy system, as described in Fig. 10. Nonetheless, the growth rate maximum for the eutectic structure suggest that modification of eutectic theory is necessary.

The simplest modification is the inclusion of the temperature dependence of the liquid diffusion coefficient directly into the Jackson-Hunt theory [33,101]. Normally this is neglected in solidification theory but is commonly used in solid-state eutectoid growth to describe the "nose" of TTT curves. This modification is important for rapid solidification because the growth front may be 100–200 K below the eutectic temperature, compared to only a few K for slow solidification. The inclusion of this temperature dependence causes the existence of a maximum growth rate, typically in the cm/s range, and a minimum spacing, typically in the 20-nm range. The actual values depend on the phase diagram shape and the rate at which the diffusion coefficient decreases below the eutectic temperature (particularly for glass-forming systems). The Peclet number based on this minimum eutectic spacing and maximum growth rate is ~0.1.

By including the effect of a temperature-dependent diffusion coefficient and altering the diffusion solution, Trivedi, Magnin, and Kurz [119] have recently extended the Jackson-Hunt diffusion analysis to high Peclet numbers. If the partition coefficients of the two solid phases (assumed independent of velocity in the theory) are low (wide composition separation between solid phases), the growth rate maximum exists as described above and the constant $(\text{spacing})^2$ (velocity) relation is lost. If the partition coefficients are close to unity, a transition to single-phase growth of one of the solid phases is predicted. In either case, eutectic structures finer than a few tens of nanometers will not occur.

When eutectic microstructures occur in rapidly solidified samples, they often appear at compositions other than that of the thermodynamic eutectic. A competitive analysis, using the growth rates of dendrites of each of the two phases in the eutectic and the eutectic itself at various temperatures, can lead to a shifting of the range of compositions where two-phase eutectic growth occurs (the coupled zone) as the undercooling or rate increases. When the eutectic involves a facetted (for example, intermetallic) and a nonfaceted phase (for example, a terminal solid solution), the coupled zone shifts in the direction of the intermetallic compound. The composition of the eutectic structure does not correspond to the thermodynamic eutectic composition. When neither phase in the eutectic is facetted or when both phases are facetted, the coupled zone does not usually shift. In both cases, the composition range of the coupled zone broadens with increasing growth rate.

This competitive growth effect also leads to a change in the identity of the primary or first phase to form during rapid solidification. There are many situations where the primary solidification phase formed during rapid solidification is different from that observed for slow solidification. Many examples are found in aluminum alloys containing transition elements such as iron, manganese, or chromium, independent of the processing method. If the alloy is hypereutectic, slowly cooled castings will contain intermetallics such as Al_9FeNi, Al_3Fe, or Al_6Mn as the primary (or first) phase to solidify. However, under rapid solidification conditions the primary phase in these alloys is the aluminum solid solution usually found as cells with an intercellular intermetallic phase. The transition of the identity of the primary phase from the intermetallic to aluminum solid solution results from the same velocity dependence of the competition between the growth of primary (dendritic) aluminum, intermetallic, and eutectic that causes the shifting of the coupled zone described above. In some cases, an intermetallic that is not on the equilibrium phase diagram may compete with aluminum; for example, metastable Al_6Fe in Al-Fe alloys.

4. *Dendritic Growth*

Solidification of primary phases into undercooled melts is usually thought to be fully dendritic (with side branches). Although there is mounting experimental evidence that cellular structures can also be formed in this situation, solidification theory is presently not able to explain this transition. Nevertheless, it is instructive to examine recent developments in dendritic growth theory for the insight it provides into rapidly solidified microstructures.

In recent years considerable advances have occurred in the theory of dendritic growth, at least with regard to the tip region. The first advance was the application of marginal stability to determine the tip radius [120] for pure materials. This replaced the maximum-growth-rate hypothesis which underestimated measured tip radii and overestimated the growth rates. Marginal stability is a concept in which the operating tip radius under a given set of growth conditions will be the largest stable value; i.e., any further increase would lead to tip splitting. Commonly the Mullins-Sekerka wavelength for the temperature and composition gradients at the tip are set equal to the tip radius.

The second advance, which was particularly important practically, was the treatment of alloy dendrites by Lipton, Glicksman, and Kurz (LGK) [121]. The theory predicts the growth rate, V, tip radius, R, and composition of an isolated dendrite tip as a function of bath undercooling, ΔT, and bulk alloy composition, C_0. The theory includes the effects of solute redistribution, latent heat removal, and interface curvature, and assumes local interface equilibrium. Here the bath undercooling is the difference between the liquidus temperature for the bulk alloy composition, T_L, and the temperature of the liquid far from the dendrite tip, T_b (the bath temperature which is the same as the nucleation temperature). The growth rate is calculated by decomposing the (bath) undercooling into three parts: the thermal undercooling, ΔT_t, the solute undercooling, ΔT_s, and the curvature undercooling, ΔT_c, according to $\Delta T = \Delta T_t + \Delta T_s + \Delta T_c$.

An important result of the LGK theory is how strongly composition and hence solute redistribution affects the growth rate. Figure 29 shows that for fixed bath undercooling below the liquidus, the calculated dendritic growth rate exhibits a maximum for dilute compositions and then falls by several orders of magnitude for concentrated alloys. This slowing of growth is caused by the difficulty of solute redistribution relative to latent heat redistribution. Because nucleation temperatures often parallel the liquidus, the use of fixed undercooling is a reasonable basis for comparison of alloys of different composition. Results are shown in dimensionless form. The bath undercooling is expressed as a fraction of L/C, the composition is expressed as the ratio of the primary phase liquidus-solidus temperature difference, ΔT_0, to L/C; the growth rate is expressed as a fraction of the velocity $2\alpha(L/C)/T_M\Gamma$, where α is the thermal diffusivity of the liquid. For aluminum alloys a dimensionless bath undercooling of 0.1 corresponds to 36.4 K, a dimensionless composition of 0.1 corresponds to a composition having a freezing range of 36.4 K and a dimensionless growth rate of 10^{-5} corresponds to 248 cm/s.

In Fig. 30 the effect of composition on the thermal, solute, and curvature undercooling for a dimensionless bath undercooling of 0.1 is superimposed on a phase diagram for $k_E = 0.1$. Note that when the line for the bath (or nucleation) temperature crosses the solidus curve, the distribution of undercooling changes sig-

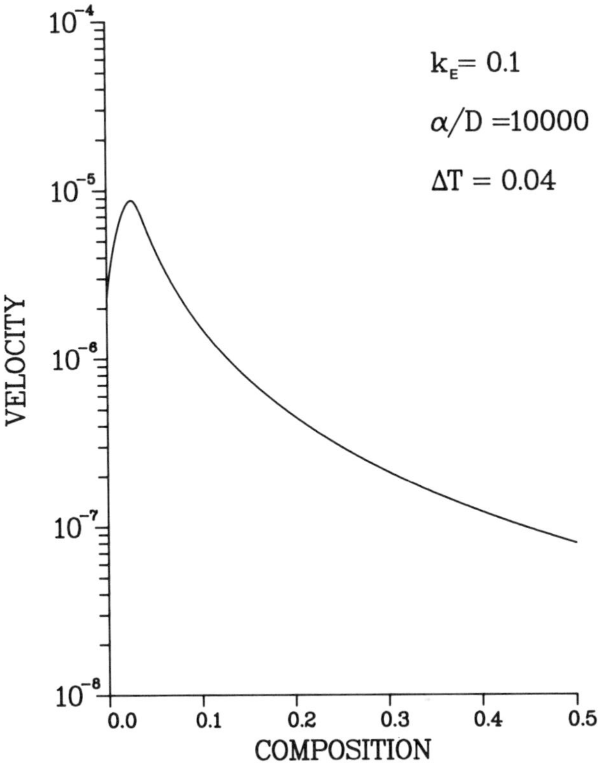

Figure 29 Calculated dimensionless dendritic growth velocity versus dimensionless composition for $k_E = 0.1$ and a dimensionless bath undercooling of 0.04 (0.04 L/C) [122].

nificantly. On the left (dilute alloys) where $\Delta T > \Delta T_0$, ΔT_t is large, suggesting the dominance of latent heat redistribution. On the right (more concentrated alloys) where $\Delta T < \Delta T_0$, ΔT_t is small, suggesting the dominance of solute redistribution. Similar plots are found for different values of k_E.

The predictions of the LGK theory can usually be fit by an equation of the form

$$V = \mu_n \, \Delta T^n \tag{25}$$

where $\Delta T = T_L - T_b$. The parameter μ_n depends on alloy composition and $n \simeq 2.84$. When ΔT_t is small (concentrated alloys) the temperature of the dendrite tip, T^*, is very close to that of the undercooled melt far from the tip, T_b. Thus, T_b and T^* can be used interchangeably to calculate ΔT for use in Eq. (25). In other words, when $\Delta T < \Delta T_0$ solute effects dominant the dendritic growth problem, and the bath undercooling is almost identical to the tip undercooling. Equivalently, the temperature gradient (even the sign of the gradient) has little effect on the growth rate for concentrated alloys ($\Delta T < \Delta T_0$).

The third advance, which is particularly important for large bath undercoolings, involved modifications to treat the important cases of large P_c [106,122,123] and large thermal Peclet number, P_t [105,123] where $P_t = VR/2\alpha$ with α being the

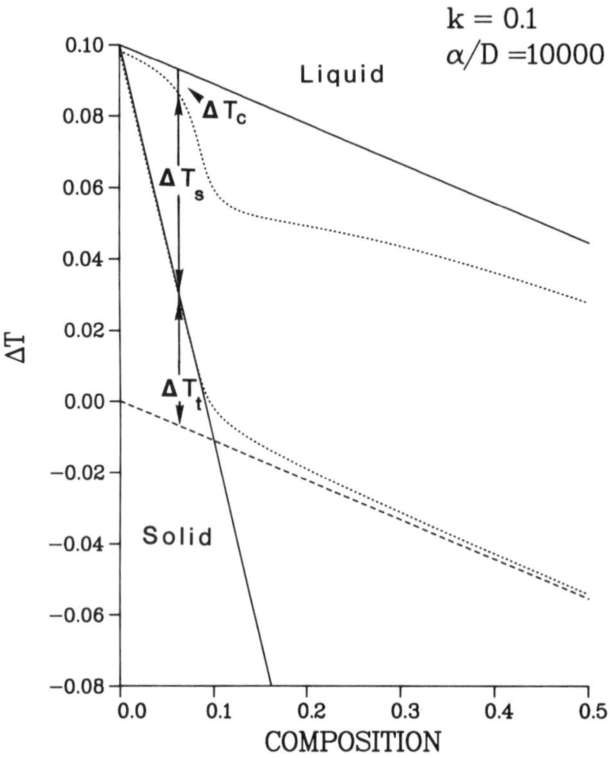

Figure 30 Relative size of the curvature undercooling ΔT_c solute undercooling, ΔT_s, and thermal undercooling, ΔT_t, for dendritic growth with a bath undercooling of 0.1 superimposed on a dimensionless phase diagram. For compositions where the bath undercooling exceeds the alloy freezing range, ΔT_t is not small. For compositions where the bath undercooling is less than the alloy freezing range, ΔT_t is small [122].

liquid thermal diffusivity; this third advance also includes nonequilibrium interface conditions [106,124]. The details of these modifications are beyond the scope of this chapter; however, we present some calculated results in Fig. 31 for the growth rate as a function of bath (nucleation) temperature for dendrites of FCC austenite and metastable BCC ferrite of an Fe–15 at.% Ni alloy. This is of interest because the BCC phase has long been known to form preferentially in rapidly solidified Fe-Ni alloys [125]. The liquidus temperatures for FCC and BCC (1770 and 1742 K, respectively) as well as k_E values of 0.87 and 0.71, respectively, are obtained from a calculated phase diagram [126]. As seen in the figure, the growth rate of the FCC phase is always greater than for the BCC phase, regardless of the nucleation temperature. Thus in this case, phase selection must be caused by preferential nucleation of the BCC phase, as had been developed by others [127]. The bump in the curve for the BCC phase is a real effect and corresponds to high P_c effects on the operating tip radius.

In summary, the relationship between initial undercooling and growth rate depends strongly on alloy composition. More fundamentally, it also depends on the morphology of the liquid-solid interface. Often, heat flow calculations which at-

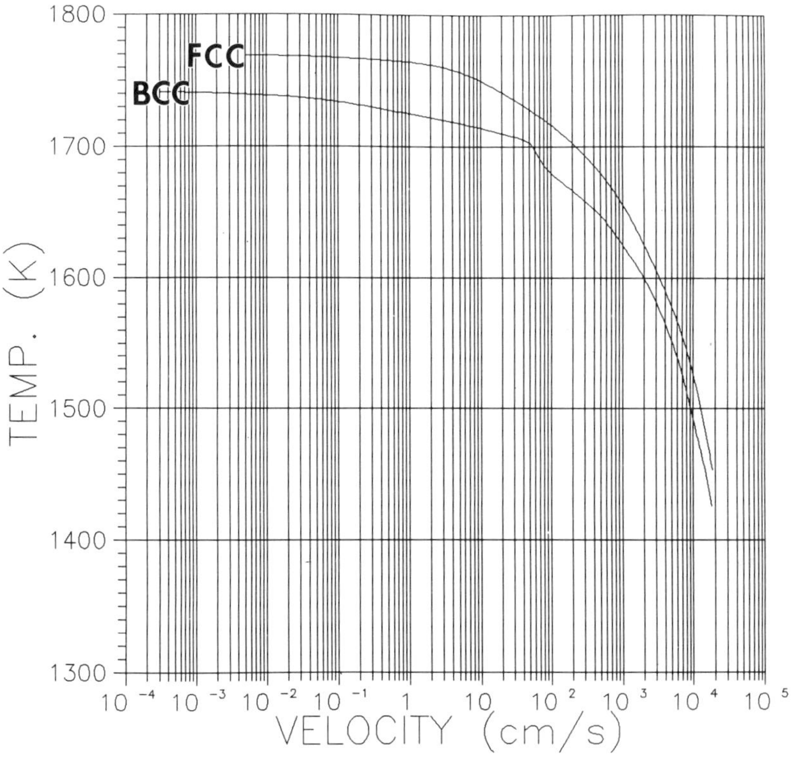

Figure 31 Calculated dendrite growth velocity at different (bath) temperatures for FCC and BCC phases in an Fe–15 at.%Ni alloy using modified theory [106] correct for high undercooling.

tempt to determine the relationship between initial undercooling and growth rate assume a smooth interface and a linear kinetic relationship such as is used for interface attachment kinetics. Such an approach is clearly incorrect for alloys which solidify with a dendritic, cellular or eutectic interface.

VI. AN EXAMPLE—ATOMIZED AL–8 WT% FE POWDER

In the following example many of the concepts just developed are used to understand the results of a study of atomized Al–8 wt% Fe powder [128]. Four distinct microstructures were observed which were a function of powder diameter, as shown in Fig. 32. For powders with diameters larger than 40 μm, primary intermetallic Al_3Fe was generally observed. This is the primary phase expected from the stable phase diagram. For powders 30 μm in diameter, eutectic structure of α-Al + Al_6Fe was observed most commonly (Fig. 33). This metastable eutectic forms only when the Al_3Fe phase is absent in the microstructure. As judged by the eutectic spacing, the solidification rate across the powder is relatively uniform. For powders between 5 and 20 μm, a two-zone structure was observed which is composed of extremely fine (~50 nm) cells of α-Al (termed microcellular) and coarser (~0.5 μm) cells of

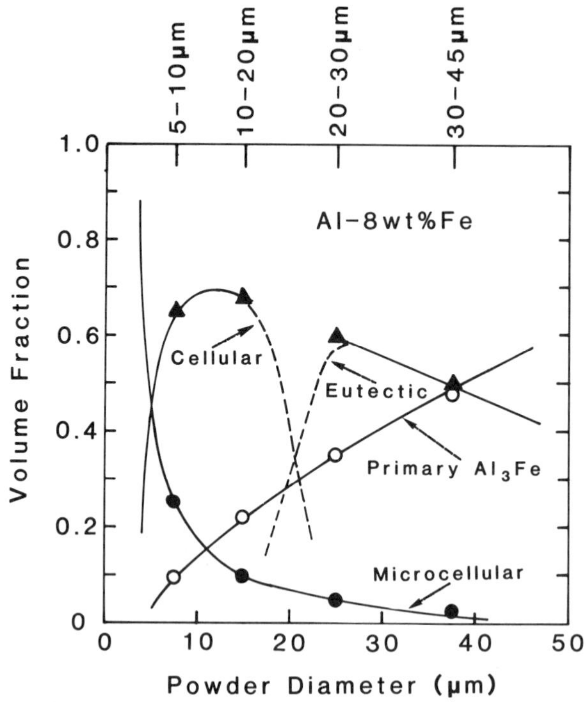

Figure 32 Volume fraction of the microcellular, cellular, eutectic, and primary intermetallic structures observed in Al–8 wt% Fe powders of different diameters [128].

α-Al (Fig. 34). The structure is seen to emanate from a point on the powder surface at the left where nucleation has occurred. A dramatic change in cell spacing is observed and suggests a highly nonuniform solidification rate across the powder.

Each of these microstructures is well known in Al-Fe alloys. Indeed the transition of microstructure for a single alloy composition as a function of powder size is a dramatic example of competitive growth kinetics between stable and metastable phases with cellular and eutectic morphologies and the development of initial undercooling in powders.

A starting point for the analysis of these results is a series of metastable phase diagrams, shown in Fig. 35 [42]. The presence of the eutectic of α-Al and Al_6Fe when Al_3Fe is absent suggests examining a metastable phase diagram in which the Al_3Fe phase has been omitted. Murray has calculated this diagram which exposes the Al_6Fe liquidus and the metastable eutectic at ~650°C and ~3.2 wt% Fe, as shown in Fig. 35. Also shown is the extension of the α-aluminum liquidus if the Al_6Fe phase were also absent. In contrast with the thermodynamic metastable eutectic composition, the observed metastable eutectic microstructure has an average composition of 8 wt% Fe, suggesting that the coupled zone for the metastable eutectic has shifted. The metastable diagram provides the necessary interface conditions to perform the calculation of the coupled zone. The diagram also shows that at 8 wt% Fe, the liquid must be undercooled below the stable Al_3Fe liquidus

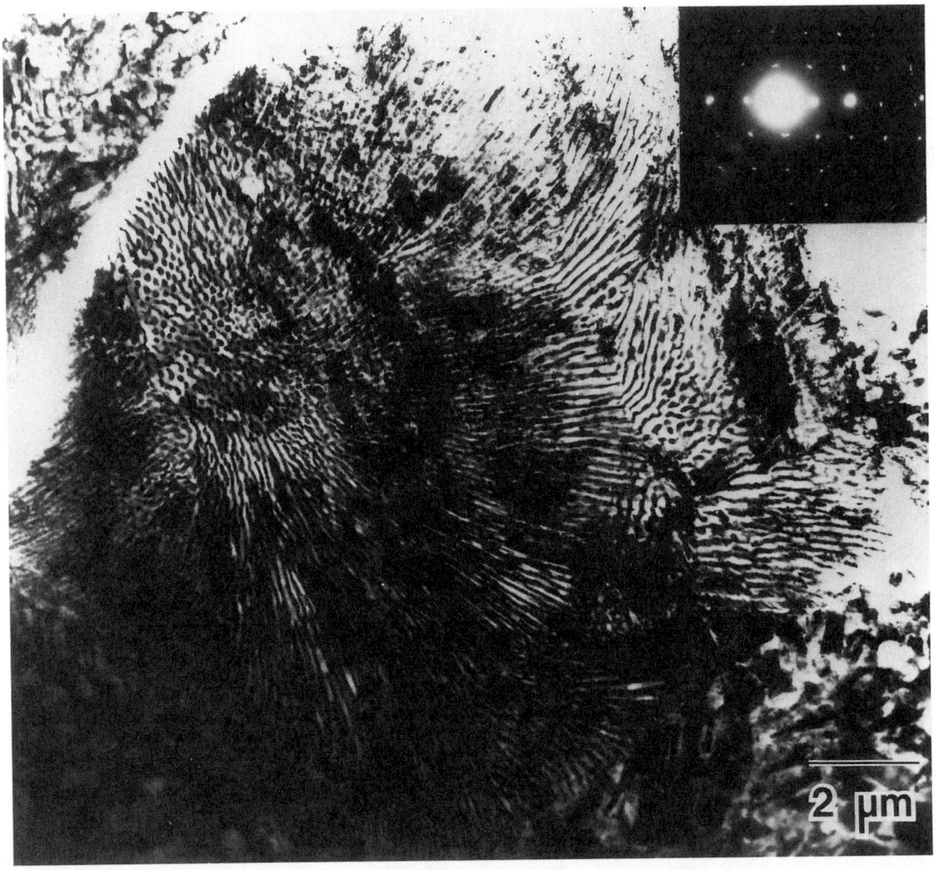

Figure 33 TEM micrograph of a fully eutectic structure typical of that seen in 20–30 μm diameter powder particles of Al–8% Fe. The SADP identifies the metastable phase Al_6Fe which occurs as rods in a matrix of α-Al to form the eutectic structure [128].

by ~200 K to get formation of α-Al and Al_6Fe eutectic and by ~220 K to get formation of cells of α-Al.

The small powders exhibit a two-zone structure of α-Al cells typical of a sample which undergoes recalescence. Thus the amount of undercooling below the α-Al metastable liquidus prior to nucleation significantly effects the heat flow analysis. On the other hand, in the larger diameter powders, the absence of a two-zone structure indicates that a consideration of recalescence is not necessary and that the degree of undercooling prior to solidification *does not* contribute significantly to the heat flow analysis.

This difference is partly due to the presence of a concentration of potent catalytic sites for nucleation within the bulk alloy which do not become isolated into a minor fraction of the powder particles unless the particle size is less than ~10 μm. This agrees quite well with the experimental observation of a large increase in the undercooling capacity of Al-Fe alloys in the 10-μm size range [14]. For the small

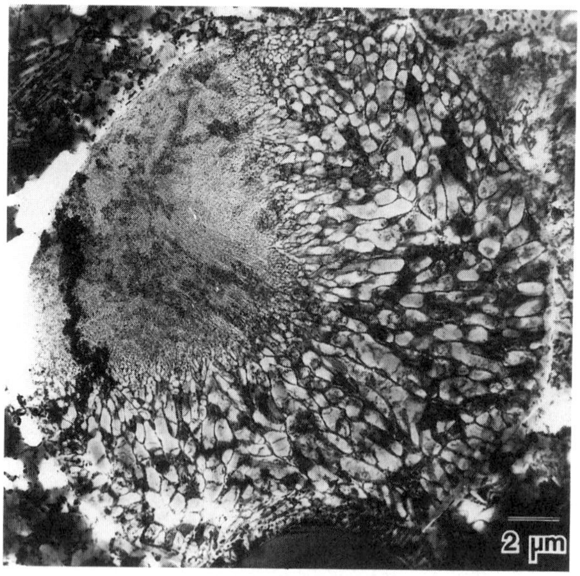

Figure 34 TEM micrograph of a Al–8% Fe powder particle. A single nucleation site (left) initiates solidification at a high interface rate to produce a fine cellular structure. Recalescence slows the interface rate to produce the coarser cellular structure at the right [128].

powder particles, nucleation occurs at the surface and is therefore also heterogeneous albeit on much less potent catalytic sites. The difference is also due to the sluggish kinetics of intermetallic and eutectic growth which retards the recalescence effect.

This discussion will therefore be divided into two sections: the first deals with the large particles, in which solidification velocity is controlled primarily by extraction of heat to the powder exterior and where the microstructures formed are cellular, eutectic, or primary intermetallic. The second section deals with the small particles, in which the solidification velocity is primarily a function of the initial undercooling and the two-zone microcellular-cellular structures are observed.

A. A Simple Model for Powder Solidification with No Initial Undercooling

In order to derive a simple expression for the dependence of solidification velocity on powder size, a spherical powder particle of diameter, d, will be approximated by a cylinder of diameter, d, and height, d. From a thermal viewpoint, solidification will occur in an isothermal sample with a planar interface moving at velocity, V_e, from one cylinder face to the other. Prior to nucleation, the temperature difference within a powder particle is small if the Biot number, $B_i = hd/k_L \ll 1$, where k_L is the thermal conductivity of the liquid alloy. For a 20-μm-diameter powder particle this condition is satisfied if $h \ll 4.5 \times 10^2$ W/cm^2·K. After nucleation, release of latent heat of fusion will be effectively spread throughout the entire powder particle if $Vd/\alpha \ll 1$. For a 20-μm-diameter particle this condition is satisfied if $V \ll 170$ cm/s.

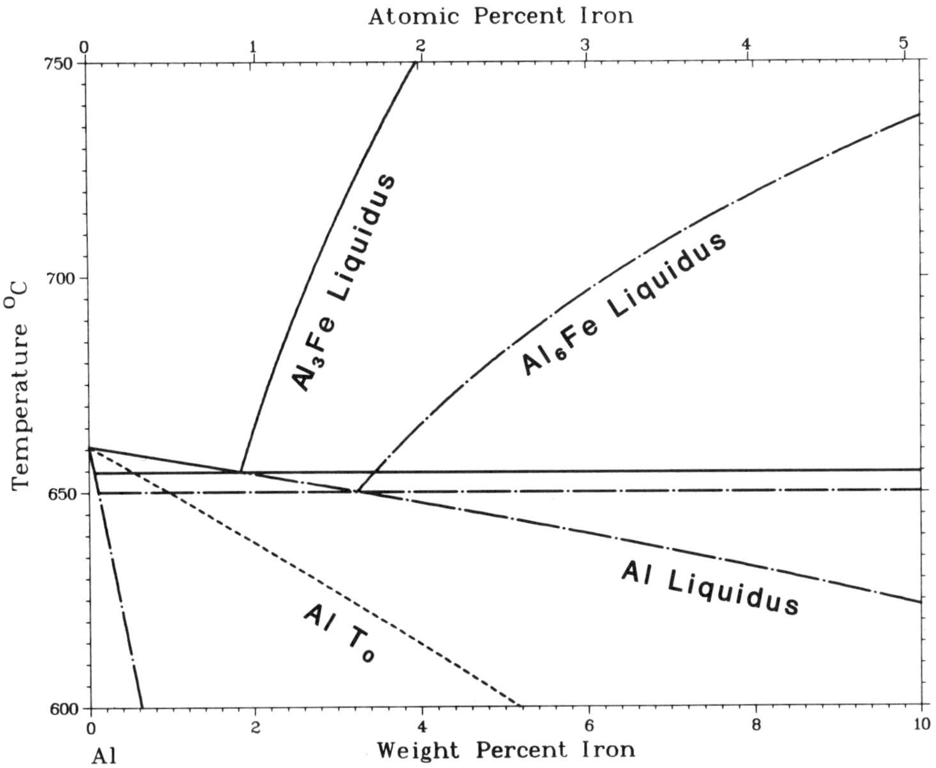

Figure 35 Al-rich portion of the Al-Fe phase diagram [42] showing the stable (Al-Al_3Fe) and the metastable (Al-Al_6Fe) diagram. The extended α-Al liquidus and solidus as well as the T_0 curve for partitionless solidification of α-Al are also shown. The α-Al solvus curves are omitted for clarity.

Assuming the powder is isothermal at any instant, a simple heat balance during solidification gives the solidification velocity, V_e, controlled by external heat extraction as

$$V_e = \frac{6h\,\Delta T_{\text{gas}}}{L} \tag{26}$$

where ΔT_{gas} is the temperature difference between the sample and the cooling gas. The predictions of this expression agree reasonably well with numerical calculations of "minimum velocities" given by Levi and Mehrabian [22] in their Table IV using spherical geometry and various heat transfer coefficients. The cooling rate, $\dot{T}$, prior to nucleation using the same assumptions is given by

$$\dot{T} = \frac{6h\,\Delta T_{\text{gas}}}{Cd} \tag{27}$$

For the present case, cooling was considered to be primarily by conduction to the argon gas remaining in the atomization chamber [128] and hence $h = k_{\text{gas}}/d$, where k_{gas} is the thermal conductivity of argon at the pressure in the atomizing

Table 5 Estimated Heat Transfer Coefficients and Growth Rate for Powders with No Initial Undercooling.

d (μm)	h (W/cm^2·K)	V_e (cm/s)	$\dot{T}$ (K/s)
5	0.54	1.99	1.5×10^6
10	0.27	0.99	3.6×10^5
20	0.13	0.50	9.1×10^4
30	0.09	0.33	4.0×10^4
40	0.07	0.25	2.2×10^4
50	0.05	0.20	1.5×10^4
100	0.03	0.10	3.6×10^3

chamber. Table 5 gives values of h, V_e, and $\dot{T}$ for various diameter powder particles using $k_{gas} = 1.34 \times 10^{-4}$ W/cm·K, and $\Delta T_{gas} = 600$ K.

The observed microstructures in the large diameter powders are consistent with the effect of particle size on the calculated values of V_e. From the directional solidification experiments of Hughes and Jones [10] shown in Fig. 36, the microstructure of Al–6 wt% Fe changes from a primary intermetallic structure of Al_3Fe to a fully eutectic structure of α-Al + Al_6Fe and finally to a cellular structure of α-Al over the range of growth velocity from 0.1 to 1 cm/s. Slightly higher transition velocities would be expected for Al–8 wt% Fe. An identical change in microstructure is observed for Al–8 wt% Fe powders with diameters between 50 and 10 μm where the estimated growth rate based solely on external heat extraction also changes from ~0.2 to 1 cm/s (see Table 5). A comparison of the microstructural

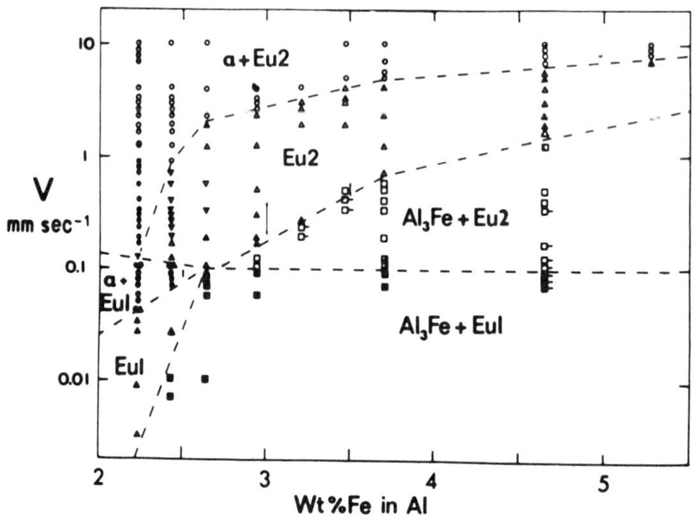

Figure 36 From Hughes and Jones [10]. Dominant growth morphologies as a function of growth velocity and alloy concentration in Al-Fe alloys solidified with temperature gradient of 20 K/mm. EU1 = Al + Al_3Fe eutectic, EU2 = Al + Al_6Fe eutectic, α = dendritic α-Al solid solution.

transitions obtained by directional solidification and by atomization at the same growth velocity is possible without regard to temperature gradient because solute redistribution dominates the kinetic relationships between growth rate and undercooling for these Al-Fe alloys.

Hughes and Jones [10] and more recently Kurz and Trivedi [129] have analyzed the competitive growth kinetics of Al_3Fe, α-Al + Al_6Fe eutectic, and α-Al theoretically to quantify the microstructural transitions in the directionally solidified Al-Fe alloys. They detail the reasons for the shift in the coupled zone for the α-Al + Al_6Fe out to 8 wt% Fe. This is an example of a shift of the compositions of the coupled zone away from a nonfaceted phase (α-Al).

The possibility of *initial* undercooling was neglected in the heat flow analysis to determine the solidification velocity because no two-zone microstructures which typically accompany recalescence were observed in the large powders. This does not imply that these large particles are not undercooled during solidification. The growth of the intermetallics or the metastable eutectic at the above growth rates requires very large interface undercoolings which is would be nearly equal to the bulk or bath undercooling in the present case. This is precisely the reason why they can compete for dominance of the microstructure despite very different liquidus temperatures.

B. A Simple Model for Solidification with Initial Undercooling

Using the same geometry and assumptions to model powder solidification as described above, the velocity, V_r, during recalescence of an initially undercooled powder with no external heat extraction during solidification can be obtained simply. This is necessary to describe the two-zone microstructures seen in the small powders. In this case, however, one needs a specific model of how the growth rate depends on interface undercooling. A heat balance will give the position of the interface, x, measured from the nucleating cylinder face as a function of the particle temperature, T, and nucleation temperature, T_N, span as

$$\frac{x}{d} = \frac{T - T_N}{L'/C} \tag{28}$$

The parameter L' is a fraction of the normal latent heat given by the volume fraction of solid within the microcellular structure. This fraction is estimated from micrographs as 0.7 in the present case. The nucleation temperature is not known a priori due to the unknown nature of the catalytic sites. If the interface velocity can be expressed by a kinetic law of the form

$$V = \mu_n(T_L - T)^n \tag{29}$$

where T_L is the alloy liquidus temperature, the growth rate V_r can be written using Eq. (28) as

$$V_r = V_N\left(1 - \frac{1}{\Delta\theta'}\frac{x}{d}\right)^n \tag{30}$$

where $\Delta\theta' = (T_L - T_N)/(L'/C)$, and V_N is the initial growth rate given by

$$V_N = \mu_n(T_L - T_N)^n \tag{31}$$

For a pure metal, $n = 1$ and μ_1 is the linear kinetic coefficient. However in the present case of cellular growth of Al–8 wt% Fe, n and μ_n will be obtained using LGK dendritic theory. Because the freezing range ΔT_0 is very large compared to L/C and is estimated as 740 K for this alloy [42], the growth is dominated by solute redistribution, the temperature gradient is relatively unimportant and in addition the bath and dendritic tip undercoolings can be used interchangeably. For Al–8 wt% Fe solidifying as dendrites of α-Al, the Lipton, Glicksman, Kurz theory gives $\mu_n = 1.49 \times 10^{-4}$ cm/s and $n = 2.84$. When $Vd/\alpha > 1$ the assumption of an isothermal sample is invalid and the interface will be subject to a local recalescence which will slow growth more quickly as shown by Levi and Mehrabian [22]. For a 10-μm particle, this effect is only important when $V > 340$ cm/s and will be neglected here.

Figure 37 shows the predicted velocity vs. distance solidified using Eqs. (26) and (30) for a 10-μm-diameter powder particular of Al–8 wt% Fe freezing as α-Al. The sum of these velocities is thought to represent the situation which occurred in the powder shown in Fig. 34. Note that the volume fraction of microcellular struc-

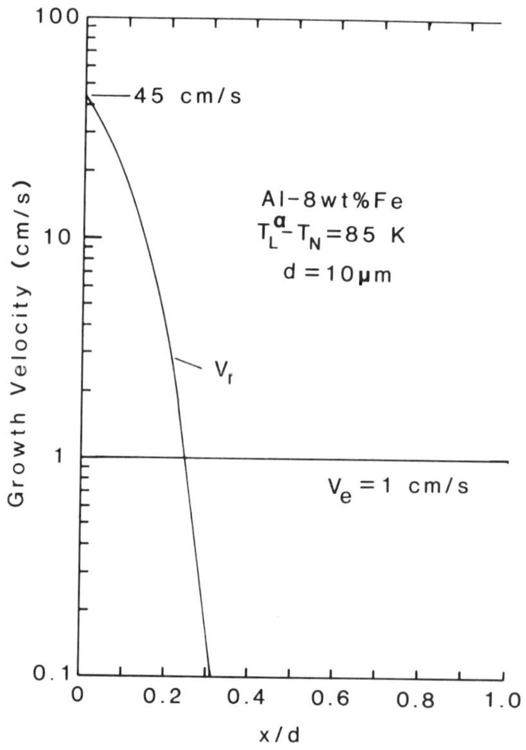

Figure 37 Estimated growth velocity versus fractional distance across an initially undercooled powder particle for solidification of α-Al cellular structures. The curve V_r is the velocity during recalescence and V_e is the velocity controlled by the external heat extraction. The estimates are for an initial undercooling ~85 K below the α-Al liquidus which produces an initial growth rate of 45 cm/s. The heat transfer coefficient is 0.27 J/cm²·s·K and produces a growth rate of 1 cm/s after recalescence is complete.

ture in the powder is ~1/3 and corresponds to the choice of $\Delta\theta' = 1/3$ or a nucleation temperature ~$1/3(L'/C)$ or 85 K below the metastable α-Al liquidus. Also note that $V_N = 45$ cm/s and $V_e = 1$ cm/s from Eq. (31) and Table 5 respectively. Due to the large decrease in the growth rate, the cell spacing increases dramatically across the powder.

Because the choice of μ_n and n are determined by solute redistribution in the present case, an initial undercooling of 85 K does not produce as high an initial growth rate as would be expected for a kinetic relation typical of pure metals. This also explains why, for Al–8% Fe, with its large value of ΔT_0, the zone of microstructure formed during recalescence exhibits microsegregation. Partitionless solidification requires growth rates in the range of m/s as described earlier. The observations of similar two zone microstructures in melt-spun ribbons of Al–8% Fe is a further illustration of the importance of undercooling in understanding solidification during all RS processing methods.

VII. SUMMARY

One of the most attractive features of RS is the demonstrated capability of yielding microstructural transitions over a range of processing conditions in a variety of major alloy systems. Whether these microstructural transitions appear as modifications of dispersoid or second-phase particle size and density, or as alterations in the morphological appearance of solidification products, or as more drastic changes of the product crystal structure or crystallinity, they represent a powerful method of structural control. Often, RS is viewed as a nonequilibrium process. For the most part this view is incorrect, for the hierarchy of equilibrium incorporates metastable states and local interfacial equilibrium which pertain to much of RS treatment and structures. Indeed, the suppression of dendritic segregation and the associated segregation by RS is a good example of the use of rapid solidification to achieve a final state closer to equilibrium than can be obtained by conventional methods.

In general, the a priori prediction of RS structure for a given alloy based upon solidification kinetics analysis is not yet possible, but metastable phase diagrams can reveal the possible range of product phases. The interpretation of nucleation kinetics in terms of undercooling offers promise for further analysis development. There has been good progress in the analysis of competitive growth rates so that it is now possible to evaluate, at least qualitatively, kinetic trends.

It is clear that the growing experience in RS has required a reevaluation and further development of solidification kinetic theory to interpret the new structures and behavior. At the same time it has also become apparent that an optimization of RS structure and potential benefits may not be achieved with conventional alloy compositions. In fact, the application of fundamental solidification kinetics and metastable phase diagram analysis offers a basis for an effective RS alloy design. Indeed, the theoretical and experimental examination RS continues to provide new advances in understanding and potential microstructural options.

The contents of this chapter are based upon the treatment presented at the symposium, *Rapidly Solidified Crystalline Alloys*, edited by S. K. Das, B. H. Kear, and C. M. Adam, TMS-AIME, Warrendale, PA, (1985) pp. 21–58, which has been expanded and updated as of January, 1990.

ACKNOWLEDGMENT

The authors would like to acknowledge support from ARO (DAAL 03-90-G-0042), NASA (NAG-8-771) for JHP, and DARPA (Order # 6065) for WJB. Also discussions with W. Allen, L. Bendersky, J. W. Cahn, S. R. Coriell, K. Ohsaka, R. J. Schaefer, and D. Schechtman are greatly appreciated.

REFERENCES

1. P. Duwez, R. H. Willens, and W. Klement, *J. Appl. Phys.*, *31*, 1136, 1137 (1960).
2. G. Falkenhagen and W. Hoffmann, *Z. Metallkde*, *43*, 69 (1952).
3. R. W. Cahn, in *Physical Metallurgy*, Part I, edited by R. W. Cahn and P. Haasen (North-Holland, Amsterdam, 1983), p. 1780.
4. H. Jones, *Rapid Solidification of Metals and Alloys,* Monograph #8, (Institution of Metallurgists, London, 1982).
5. R. B. Pond, U.S. Patent 2,825,108 (1958).
6. D. Turnbull, *Contemp. Phys.*, *10*, 473 (1969).
7. J. H. Perepezko and I. E. Anderson, in *Synthesis and Properties of Metastable Phases*, edited by E. S. Machlin and T. J. Rowland (TMS-AIME, Warrendale, PA, 1980), p. 31.
8. J. H. Perepezko, in *Rapid Solidification Processing: Principles and Technologies II*, edited by R. Mehrabian, B. H. Kear, and M. Cohen (Claitor's, Baton Rouge, LA, 1980), p. 56.
9. C. M. Adam and L. M. Hogan, *J. Austral. Inst. Metal*, *17*, 81 (1972).
10. I. R. Hughes and H. Jones, *J. Mater. Sci.*, *11*, 1781 (1976).
11. M. E. Glicksman and R. J. Schaefer, *J. Cryst. Growth*, *1*, 297 (1967).
12. D. Turnbull, *J. Appl. Phys.*, *21*, 1022 (1950).
13. J. H. Perepezko and J. S. Paik, in *Rapidly Solidified Amorphous and Crystalline Alloys,* edited by B. H. Kear, B. C. Giessen, and M. Cohen (North-Holland, Amsterdam, 1982), p. 49.
14. S. E. Le Beau, J. H. Perepezko, B. A. Mueller, and G. J. Hildeman, in *Rapidly Solidified Powder Aluminum Alloys*, *ASTM STP 890*, edited by M. E. Fine and E. A. Starke, Jr. (ASTM, Philadelphia, PA, 1986), p. 118.
15. B. A. Mueller and J. H. Perepezko, in *Aluminum Alloys—Their Physical and Mechanical Properties*, edited by E. A. Starke, Jr., and T. H. Sanders, Jr. (EMAS, West Midlands UK, 1986), p. 201.
16. B. A. Mueller and J. H. Perepezko, *Metall. Trans.*, *18a*, 1143 (1987).
17. W. Yoon, J. S. Paik, D. La Court, and J. H. Perepezko, *J. Appl. Phys.*, *60*, 3489 (1986).
18. J. H. Perepezko, B. A. Mueller, and K. Ohsaka, in *Undercooled Alloy Phases*, edited by E. W. Collings and C. C. Koch, (TMS, Warrendale, PA, 1987), p. 289; J. H. Perepezko, *Mater. Sci. Eng.*, *65*, 125 (1984).
19. J. H. Perepezko and W. J. Boettinger, in *Surface Alloying by Ion, Electron and Laser Beams* (ASM, Metals Park, OH, 1987), p. 51.
20. H. Jones, *Mater. Sci. Eng.*, *51* (1969).
21. S. C. Huang and R. P. LaForce, in *Rapidly Solidified Metastable Materials*, edited by B. H. Kear and B. C. Giessen (Elsevier, New York 1984), p. 125.
22. C. G. Levi and R. Mehrabian, *Metall. Trans.*, *13A*, 13 and 221 (1982).
23. T. W. Clyne, *Metall. Trans.*, *15B*, 369 (1984).
24. D. Turnbull and B. G. Bagley, in *Treatise on Solid State Chemistry*, Vol. V, edited by N. B. Hannay (Plenum, New York, 1975), p. 513.
25. M. C. Flemings, *Solidification Processing* (McGraw-Hill, New York, 1974).

26. H. Jones and W. Kurz, *Metall. Trans.*, *11A*, 1265 (1980).
27. J. H. Perepezko and W. J. Boettinger, in *Alloy Phase Diagrams*, edited by L. H. Bennett, T. B. Massalski., and B. C. Giessen (Elsevier, New York, 1983), p. 223.
28. C. W. White, D. M. Zehner, S. U. Campisano, and A. G. Cullis, in *Surface Modification and Alloying by Lasers, Ion, and Electron Beams*, edited by J. M. Poate, G. Foti, and D. C. Jacobson (Plenum, New York, 1983), p. 81.
29. S. T. Picraux and D. M. Follstaedt, in Ref. 28, p. 2, J. M. Poate, G. Foti, and D. C. Jacobson, editors (Plenum, New York, 1983), p. 287.
30. J. C. Baker and J. W. Cahn, in *Solidification* (ASM, Metals Park, OH, 1971), p. 23.
31. L. Bosio, A. Defrain, and I. Epelboin, *J. Phys.*, *27*, 61 (1966).
32. K. A. Jackson, in *Growth and Perfection of Crystals*, edited by R. H. Doremus et al. (Wiley, New York, 1958), p. 319.
33. W. J. Boettinger, in *Rapidly Solidified Amorphous and Crystalline Alloys*, edited by B. H. Kear, B. C. Giessen, and M. Cohen (North-Holland, Amsterdam, 1982), p. 15.
34. D. M. Follstaedt, P. S. Peercy, and J. H. Perepezko, *Appl. Phys. Lett.*, *48*, 338 (1986).
35. J. A. Knapp and D. M. Follstaedt, *Phys. Rev. Lett.*, *58*, 2454 (1987).
36. D. M. Follstaedt and J. A. Knapp, in *Selected Topics in Electronic Materials*, edited by B. R. Appleton, D. K. Biegelsen, W. L. Brown, and J. A. Knapp (Materials Research Society, Pittsburgh, PA, 1988), p. 263.
37. W. P. Allen, H. J. Fecht, and J. H. Perepezko, *Scripta Metall.*, *23*, 643 (1989).
38. W. Klement, A. Jayaraman, and G. C. Kennedy, *Phys. Rev.*, *131*, 632 (1963).
39. L. Kaufman and H. Bernstein, *Computer Calculation of Phase Diagrams* (Academic Press, New York 1970).
40. J. Ågren, *Metall. Trans.*, *10A*, 1847 (1979).
41. Y.-Y. Chuang, Y. A. Chang, and R. Schmid, *Metall. Trans.*, *17A*, 1361 (1986).
42. J. L. Murray, unpublished research, NBS, 1985.
43. H. J. Fecht and J. H. Perepezko, *Metall. Trans.*, *20A*, 785 (1989).
44. H. J. Fecht, M.-X. Zhang, Y. A. Chang, and J. H. Perepezko, *Metall. Trans.*, *20A*, 795 (1989).
45. J. W. Cahn, in *Rapid Solidification Processing: Principles and Technologies II*, edited by R. Mehrabian, B. H. Kear, and M. Cohen (Claitor's, Baton Rouge, LA, 1980), p. 24.
46. J. W. Cahn, S. R. Coriell, and W. J. Boettinger, in *Laser and Electron Beam Processing of Materials*, edited by C. W. White and P. S. Peercy (Academic Press, New York, 1980), p. 89.
47. W. J. Boettinger, in *Proc. Fourth Int. Conf. on Rapidly Quenched Metals*, edited by T. Masumoto and K. Suzuki (Japan Institute of Metals, Sendai, 1982), p. 99.
48. D. Turnbull, *J. Chem. Phys.*, *20*, 411 (1952).
49. D. Turnbull, in *Solid State Phys.*, Vol. III, edited by F. Seitz and D. Turnbull (Academic Press, New York, 1956), p. 225.
50. D. Turnbull, *Prog. Mater. Sci—Chalmers Ann.* Vol. edited by J. W. Christian, P. Haasen, and T. B. Massalski (Pergamon, London, England, 1981), p. 269.
51. K. F. Kelton, A. L. Greer, and C. V. Thompson, *J. Chem. Phys.*, *79*, 6261 (1983).
52. K. F. Kelton and A. L. Greer, in *Rapidly Quenched Metals*, Vol. I, edited by S. Steeb and H. Warlimont, (Elsevier, New York, 1985), p. 223.
53. K. F. Kelton and A. L. Greer, *J. Non-Cryst. Solids*, *79*, 295 (1986).
54. C. V. Thompson, A. L. Greer, and F. Spaepen, *Acta Metall.*, *31*, 1883 (1983).
55. U. Köster and U. Herold, in Ref. 437, p. 717.
56. J. H. Perepezko and J. S. Paik, *J. Non-Cryst. Solids*, *61*, 113 (1984).
57. M. C. Lee, H. J. Fecht, J. L. Allen, J. H. Perepezko, K. Ohsaka, and W. L. Johnson, *Mater. Sci. Eng.*, *97*, 301 (1988).
58. C. V. Thompson and F. Spaepen, *Acta Metall.*, *27*, 1855 (1979).

59. M. Hillert, *Acta Metall.*, *1*, 764 (1953).
60. C. V. Thompson and F. Spaepen, *Acta Metall.*, *31*, 202 (1983).
61. J. J. Richmond, J. H. Perepezko, S. E. LeBeau, and K. P. Cooper, in *Rapid Solidification Processing: Principles and Technologies III*, edited by R. Mehrabian (NBS, Washington, DC, 1983), p. 90.
62. K. Binder and D. Stauffer, *Adv. Phys.*, *25*, 343 (1976).
63. A. S. Skapski, *Acta Metall.*, *4*, 576 (1956).
64. F. Spaepen, *Acta Metall.*, *23*, 729 (1975).
65. R. W. Ewing, *Phil. Mag.*, *25*, 778 (1972).
66. A. Bonnissent, J. L. Finney, and B. Mutaftschiev, *Phil. Mag.*, *8*, 42 (1980).
67. D. W. Oxtoby and A. D. J. Haymet, *J. Chem. Phys.*, *76*, 6282 (1982).
68. R. E. Cech, *J. Metals*, *206*, 585 (1956).
69. T. F. Kelly, G. B. Olson, and J. B. Vander Sande, in Ref. 13, p. 343.
70. C. G. Woychik, D. H. Lowndes, and T. B. Massalski, *Acta Metall.*, *33*, 1861 (1985).
71. N. Saunders and P. Tsakiropoulos, *Mater. Sci. Tech.*, *4*, 157 (1988).
72. K. N. Ishikhara, M. Maeda, and P. H. Shinyu, *Acta Metall.*, *33*, 2133 (1985).
73. P. Gressin, N. Eustathopoulos, and P. Desre, *Scripta Metall.*, *17*, 711 (1983).
74. J. P. Hirth, *Met. Trans.*, *9A*, 401 (1978).
75. J. H. Perepezko, B. A. Mueller, J. J. Richmond, and K. P. Cooper, in *Rapidly Quenched Metals*, edited by S. Steeb and H. Warlimont (Elsevier, New York, 1985), p. 43.
76. C. Barromee-Gautier, B. C. Giessen, and N. J. Grant, *J. Chem. Phys.*, *48*, 1905 (1968).
77. D. S. Shong, J. A. Graves, Y. Ujiie, and J. H. Perepezko, in *Materials Processing in the Reduced Gravity Environment of Space*, edited by R. H. Doremus and P. C. Nordine (Materials Research Society, Pittsburgh, PA, 1987), p. 17.
78. J. H. Perepezko, J. A. Graves, and B. A. Mueller, In *Processing of Structural Metals by Rapid Solidification*, edited by F. H. Froes and S. J. Savage (ASM, Metals Park, OH, 1987), p. 13.
79. J. A. Graves, Ph.D. thesis, University of Wisconsin, (1987).
80. G. H. Geiger and D. R. Poirier, *Transport Phenomena in Metallurgy* (Addison-Wesley, Reading, MA, 1973).
81. T. W. Clyne, R. A. Ricks, and P. J. Goodhew, *Int. J. Rapid Solidification*, *1*, 59 (1985).
82. D. J. Thoma, W. P. Allen, and J. H. Perepezko, University of Wisconsin, 1989 (unpublished).
83. D. Turnbull, in *Proc. of 2nd Israeli Mater. Eng. Conf.*, edited by E. Gill and S. I. Rokhlin (Beersheva, Israel, 1984), p. 1.
84. I. Maxwell and A. Hellawell, *Acta Metall.*, *23*, 229 (1975).
85. P. G. Boswel and G. A. Chadwick, *Scripta Metall.*, *11*, 459 (1977).
86. C. G. Levi, *Met. Trans.*, *19A*, 699 (1988).
87. R. D. Field, Ph.D. thesis, University of Illinois—Champaign, 1982.
88. L. A. Bendersky, F. S. Biancaniello, W. J. Boettinger, and J. H. Perepezko, *Mater. Sci. Eng.*, *89*, 151 (1987).
89. W. J. Boettinger, S. R. Coriell, and R. F. Sekerka, *Mater. Sci. Eng.*, *65*, 27 (1984).
90. J. C. Baker, Interfacial Partitioning During Solidification, Ph.D. thesis, Massachusetts Institute of Technology, 1970, Chap. V (see also Ref. 46).
91. M. J. Aziz, *J. Appl. Phys.*, *53*, 1158 (1982).
92. K. A. Jackson, G. H. Gilmer, and H. J. Leamy, in *Laser and Electron Beam Processing of Materials*, edited by C. W. White and P. S. Peercy, (Academic Press, New York, 1980), p. 104.
93. M. J. Aziz and T. Kaplan, *Acta Metall.*, *36*, 2335 (1988).

94. C. J. Lin, F. Spaepen, D. Turnbull, *J. Non-Cryst. Solids*, *61–62*, 767 (1984).
95. W. J. Boettinger and M. J. Aziz, *Acta Metall.*, *37*, 3379 (1989).
96. C. C. Koch, J. A. Horton, C. T. Liu, O. B. Cavin, and J. O. Scarborough, in Ref. 61, p. 264.
97. A. Inoue, T. Masumoto, H. Tomioka, and N. Yano, *Int. J. Rapid Solidification*, *1*, 115 (1984).
98. S. C. Huang, E. L. Hall, K. M. Chang, and R. P. Laforce, *Metall. Trans.*, *A17*, 1685 (1986).
99. W. J. Boettinger, L. A. Bendersky, F. S. Biancaniello, and J. W. Cahn, *Mater. Sci. Eng.*, *8*, 273 (1988).
100. S. C. Huang and E. L. Hall, *Mater. Res. Soc. Symp. Proc.*, *Symp.*, *133*, 373 (1989).
101. W. J. Boettinger, D. Shechtman, R. J. Schaefer, and F. S. Biancaniello, *Metall. Trans.*, *15A*, 55 (1984).
102. W. W. Mullins and R. F. Sekerka, *J. Appl. Phys.*, *35*, 444 (1964).
103. S. R. Coriell and R. F. Sekerka, in Ref. 8, p. 35.
104. S. R. Coriell and R. F. Sekerka, *J. Cryst. Growth*, *61*, 499 (1983).
105. R. Trivedi and W. Kurz, in *Science and Technology of the Undercooled Melt*, NATO ASI Series E-N0114, edited by P. R. Sahm, H. Jones, and C. M. Adam (Martinus-Nijhoff, Dordrecht, 1986), p. 260.
106. W. J. Boettinger, S. R. Coriell, and R. Trivedi, in *Rapid Solidification Processing: Principles and Technologies IV*, edited by R. Mehrabian and P. A. Parrish (Claitor's, Baton Route, LA, 1988), p. 13.
107. O. Salas and C. G. Levi, *Int. J. Rapid Solidification*, *4*, 1 (1988).
108. L. J. Mazur and M. C. Flemings, in Ref. 47, Vol. II, p. 1557.
109. T. Z. Kattamis and R. Mehrabian, *J. Mater. Sci.*, *9*, 1446 (1974).
110. H. Palacio, M. Solari, and H. Biloni, in *Physical Metallurgy*, Part I, edited by R. W. Cahn and P. Haasen (North-Holland, Amsterdam, 1983), p. 1780.
111. L. A. Bendersky and W. J. Boettinger, in *Rapidly Quenched Metals*, edited by S. Steeb and H. Warlimont (Elsevier, New York, 1985), p. 887.
112. W. J. Boettinger, L. A. Bendersky, S. R. Coriell, R. J. Schaefer, and F. S. Biancaniello, *J. Cryst. Growth*, *80*, 17 (1987).
113. G. B. McFadden, R. F. Boisvert, and S. R. Coriell, *J. Cryst. Growth*, *84*, 371 (1987); N. Ramprasad, M. J. Bennett, and R. A. Brown, *Phys. Rev. B*, *38*, 583 (1988).
114. D. C. Van Aken and H. L. Fraser, *Acta Metall.*, *33*, 963 (1985).
115. W. J. Boettinger, L. A. Bendersky, R. J. Schaefer, and F. S. Biancaniello, *Metall. Trans.*, *19A*, 1101 (1988).
116. L. H. Ungar and R. A. Brown, *Phys. Rev. B*, *31*, 5931 (1985).
117. K. A. Jackson and J. D. Hunt, *Trans. TMS-AIME*, *236*, 1129 (1966).
118. W. J. Boettinger, D. Shechtman, T. Z. Kattamis, and R. J. Schaefer, in *Rapidly Quenched Metals*, edited by S. Steeb and H. Warlimont (Elsevier, New York, 1985), p. 871.
119. R. Trivedi, P. Magnin, and W. Kurz, *Acta Metall.*, *35*, 971 (1987).
120. J. S. Langer and H. Mueller-Krumbaar, *Acta Metall.*, *26*, 1861 (1978); M. E. Glicksman, in *Crystal Growth of Electronic Materials*, edited by E. Kaldis (Elsevier, New York, 1984), Chapter 2.
121. J. Lipton, M. E. Glicksman, and W. Kurz, *Mater. Sci. Eng.*, *65*, 57 (1984).
122. W. J. Boettinger and S. R. Coriell, in Science and Technology of the Undercooled Melt, NATO ASI Series E-N0114, edited by P. R. Sahm, H. Jones, and C. M. Adam (Martinus-Nijhoff, Dordrecht, 1986), p. 81.
123. J. Lipton, W. Kurz, and R. Trivedi, *Acta Metall.*, *35*, 957 (1987).
124. R. Trivedi, J. Lipton, and W. Kurz, *Acta Metall.*, *35*, 965 (1987).
125. R. E. Cech, *Trans. AIME*, *206*, 585 (1956).

126. Y.-Y. Chuang, K.-C. Hsieh, Y. A. Chang, *Metall. Trans.*, *17A*, 1377 (1986).
127. T. F. Kelly and J. B. VanderSande, *Int. J. Rapid Solidification*, *3*, 51 (1987).
128. W. J. Boettinger, L. A. Bendersky, and J. G. Early, *Metall. Trans.*, *17A*, 781 (1986).
129. R. Trivedi and W. Kurz, in *Solidification Processing of Eutectic Alloys*, edited by D. M. Stefanescu, G. J. Abbaschian, and R. J. Bayuzick (TMSAIME, Warrendale, PA, 1988), p. 3.

3

Undercooling and Solidification

D. M. Herlach and R. Willnecker
Institut für Raumsimulation, Deutsche Forschungsanstalt für Luft- und Raumfahrt, Köln, Germany

I. INTRODUCTION

Interest in solidification phenomena progressively increased during the past two or three decades since novel structures and phases of materials have been made possible by advances in sample preparation techniques. Rapid quenching, in particular, has been employed successfully for the production of metastable solid phases directly from the melt.

If a melt is cooled at a sufficiently high rate it can solidify directly into a metastable state by bypassing nucleation and growth of the stable phase. The term "metastability" refers to an excess energy frozen-in during rapid solidification, implying that a metastable state is out of thermal equilibrium. In general, a broad spectrum of metastable structures can be classified according to their energy in excess of that of the stable structure, measured in units of RT_m, where T_m is the melting temperature of the respective material [1]. The morphological metastability of microcrystalline materials, compositionally modulated films, and interphase dispersions arises from the interfacial energy contributions ($\sim 0.1RT_m$). The structural metastability of intermetallic compounds and amorphous solids is due to a higher Gibbs free energy of the metastable structure in comparison with that of the stable state ($\sim 0.5RT_m$). Finally, compositional metastability exists in supersaturated solid solutions with a maximum excess energy up to RT_m.

The high cooling rates associated with rapid quenching have often led to its being treated on the basis of kinetic aspects alone. However, from a more generalized point of view, the requirements for the occurence and persistance of metastability are threefold [2]:

1. Thermodynamic exposure of the structure
2. Kinetic preference of the metastable state
3. Kinetic entrapment of the solidifying structure

In particular, conditions 1 and 2 are associated with the degree of undercooling attained prior to solidification, while condition 3 depends on external constraints on the system, e.g., the rate with which the sample is cooled.

From these considerations it follows directly that undercooling is a necessary precondition for the nonequilibrium solidification of the melt into metastable phases. The first step which initiates crystallization and consequently limits melt undercooling is nucleation, which selects the starting phase for solidification, stable or metastable, depending on its thermodynamic and kinetic characteristics. The activation energy to form critical nuclei of the respective solid phases depends on the interfacial energy, the thermodynamic driving force and, in presence of heterogeneities, on their catalytic potency. Despite the fact that the driving force is largest for a stable phase, the activation energy for nucleation can be lower for metastable phases because of their different interfacial energies and catalytic potency factors [3].

Following nucleation, growth completes the crystallization of the undercooled melt. Rapid solidification requires a large temperature gradient at the crystallization front which can be achieved either by spreading the melt quickly and thinly over a cool substrate or by undercooling the melt by a substantial amount prior to nucleation. The analysis and modeling of rapid solidification with respect to the development of metastable phases and refined microstructures requires a detailed knowledge of crystal growth behavior. However, rapid-quenching technologies do not allow direct investigations of crystal growth, since sample solidification always starts at the interface with the cooling substrate. The solid-liquid interface is, therefore, concealed by the melt puddle and, consequently, is inaccessible for direct measurements. However, as recently demonstrated, rapid growth can be directly studied in levitation-melted and undercooled samples by measuring the fast temperature rise during recalescence as the latent heat of fusion is released during solidification [4,5].

In the present work, a survey of nucleation studies in undercooled liquids is presented and the subsequent growth behavior discussed preponderantly in terms of rapid dendritic growth. The consequences of rapid solidification are evaluated with respect to the formation of metastable grain-refined microstructures.

II. UNDERCOOLING AND NUCLEATION

The ease with which a melt may be undercooled depends on both the catalytic potency of the heterogeneous nucleants present and on the dynamics of the nucleation process. Consequently, there are two methods of achieving high undercooling: removal of heterogeneous nucleants of high potency from the melt and rapid quenching to bypass nucleation on these heterogeneities. Rapid-quenching technologies such as chill-block melt spinning are state-of-the-art practice for the production of metastable thin ribbons on an industrial scale. However, the suitability of such processes for conducting detailed investigations of crystal nucleation and growth in the solidification process remains limited, since large temperature inhomogeneities exist along the melt puddle. Moreover, depending on the catalytic potency of the substrate, enhanced nucleation rate at the melt/substrate interface may occur.

An alternative approach to achieve large undercoolings consists of the passivation and removal of the most potent heterogeneous nucleants; in this way, such undercoolings may be obtained at slow cooling rates. This procedure is particularly suited for direct investigations of nucleation and growth phenomena. This chapter will therefore concentrate on these methods with particular emphasis on the discussion of the undercooling behaviour and its consequences on the crystallization process.

A. Experimental Approaches

Heterogeneous nucleation may take place on surface or interfacial nucleants or on volume nucleants. Isolation of volume heterogeneities has been assumed to be possible by subdividing a macroscopic sample into a large number of small particles [6]. If the heterogeneities are distributed randomly within the volume, one should expect a steep rise in the probability of finding a nucleant-free segment with increasing particle refinement. A large variety of pure metals has been investigated by this technique, resulting in undercoolings up to $0.20T_m$. This technique was improved by Turnbull [7] and later by Rasmussen and Perepezko [8,9] by preparing extremely fine dispersions of metallic droplets into an emulsion liquid. This led to a great extension of undercooling, especially of low-melting materials such as Bi, In, Hg, Pb and Ga, up to a value of $\Delta T < 0.58T_m$ in case of Ga [10]. An analysis shows that even these undercooling results can be related to surface or interface nucleation rather than to volume nucleation. An exception is an undercooling of $\Delta T \sim 0.33T_m$ in Hg, which appears to correspond to the onset of homogeneous nucleation [7].

Assuming that heterogeneous nucleation at the interface with the container wall is the dominating process initiating solidification, one may try to deactivate this catalytic nucleation site either by using inert walls or by avoiding any contact with a solid or liquid medium by containerless processing. Bardenheuer and Bleckmann [11] were the first to demonstrate that Fe and Ni bulk melts undercool by an amount of $\Delta T \sim 0.17T_m$ if the samples are enclosed into a glassy slag. This has been confirmed by later experiments applying this method to other metals and alloys [12]. More recently, experiments on bulk Pd-Ni-P melts embedded within liquid B_2O_3 revealed undercoolings up to $\Delta T = T_m - T_g$ (T_g = glass transition temperature) succeeding in the preparation of fully amorphous bulk samples at extremely low cooling rates of approximately 1 K/s [13].

Flemings and his co-workers [4] combined the methods of melt fluxing and electromagnetic levitation for undercooling experiments on multigram samples of transition metal alloys. In this technique, the melt is embedded in an inorganic glass and the whole arrangement is placed within an rf coil, allowing simultaneous levitation and inductive heating of the sample. Such experiments enable the solidification rate and microstructural development to be studied as a function of initial undercooling.

These methods have been extended to containerless processing in pure or even ultrapure environments, which provide optimum cleanliness and accessibility. The electromagnetic levitation technique was first employed to measure the mass density of levitated liquid drops both above and below the melting temperature [14]. More recently, this method has been employed for undercooling and nucleation studies

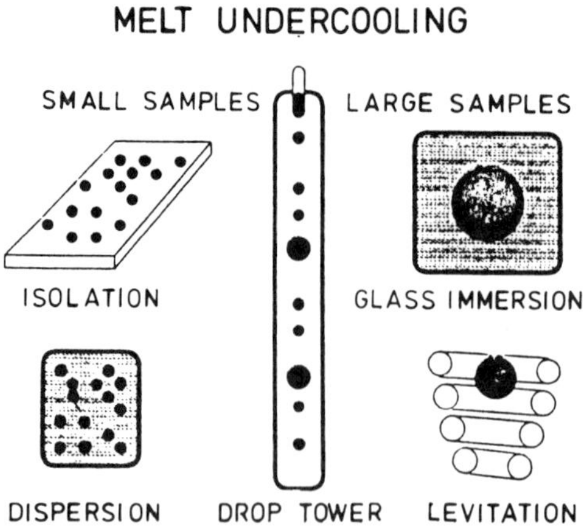

Figure 1 Experimental methods for undercooling molten metals and alloys.

[15]. Containerless processing of bulk melts resulted in undercoolings comparable to those obtained by the droplet dispersion technique. This indicates that even bulk melts can be undercooled substantially if container-wall-induced nucleation is completely eliminated and surface nucleants consisting of metal oxides are reduced. The limitation of this technique is a material-dependent lower temperature limit imposed by gravity, which might be overcome in future space experiments [16].

A simpler technique to perform containerless undercooling experiments is the use of drop towers or drop tubes. This approach has been increasingly used in recent years for nucleation studies [17,18], investigations of the development of refined microstructures [19] and metastable phase formation [19–21] from an undercooled melt. However, such experiments, in which a melt is dispersed into small droplets solidifying during free fall, do not allow direct diagnostics of the solidification process, particularly in terms of temperature measurements. On the other hand, drop tube experiments enable containerless solidification to be combined with the capability of rapid cooling. This is essential in the analysis of ripening and coarsening effects on microstructure development during and after solidification of deeply undercooled melts [22]. It is apparent that complementary information is obtained from drop tube and levitation experiments. The latter measures the thermal history of an individual sample, whereas the former provides a statistical analysis of the nucleation and growth processes as a function of droplet size and cooling rate. Figure 1 shows an overview of the experimental techniques used for undercooling of metallic melts. Deep undercooling of metals by slow cooling methods was first achieved by experiments that divided the melt into many small particles. Large samples can be undercooled by immersion in an inert, amorphous slag, or levitated in an electromagnetic field. Drop tubes can undercool small samples for statistical analysis [3].

B. Nucleation Phenomena

The first attempts to describe nucleation in a supersaturated vapor came from a theoretical approach by Volmer and Weber [23]. This theory was later improved by Becker and Döring [24], who developed a kinetic theory of nucleation. An extension to the liquid-solid phase transition was undertaken by Turnbull and Fisher [25]. In the following, we will develop the basic concept of this theory, first concentrating on homogeneous nucleation and then heterogeneous nucleation due to foreign phases within the melt or at interfaces. Special emphasis is also focused on transient effects which become important when liquids are rapidly cooled [26].

Atomic fluctuations within a liquid will give rise to instantaneous formation of solid-like clusters. Above the melting temperature, these clusters will decay spontaneously, since the Gibbs free energy of the liquid is lower than that of the solid (Fig. 2). However, as we undercool the liquid by an amount $\Delta T = T_E - T > 0$ the free energy of the solid becomes less than that of the liquid by the difference $\Delta G_v = G^s - G^L < 0$. Consequently, we expect a spontaneous growth of the clusters to form a crystal nucleus. On the other hand, a cluster implies a solid-liquid interface, with the energy σ which is always positive and therefore acts as a barrier against crystallization. The free energy of formation of a cluster of n atoms ΔG_n is given by

$$\Delta G_n = n \cdot \Delta G_v + A_n \cdot \sigma \tag{1}$$

with A_n the surface area of the cluster. As shown schematically in Fig. 3, ΔG_n is initially positive and goes through a maximum at the critical nucleus size, n^*. Clusters or embryos smaller than n^* tend to shrink whereas clusters or nuclei of size larger than n^* tend to grow.

The nucleation rate is obtained by considering the reaction kinetics of the formation and decay, respectively of clusters. Clusters are assumed to arise by reactions of the type

$$\begin{aligned} E_{n-1} + E_1 &\underset{k_n^-}{\overset{k_{n-1}^+}{\rightleftarrows}} E_n \\ E_n + E_1 &\underset{k_{n+1}^-}{\overset{k_n^+}{\rightleftarrows}} E_{n+1} \end{aligned} \tag{2}$$

where E_n denotes a cluster consisting of n atoms, E_1 is a single atom, k_{n-1}^+ and k_n^- are the rate of addition of atoms to a cluster E_{n-1} and the rate of loss of atoms from E_n, respectively.

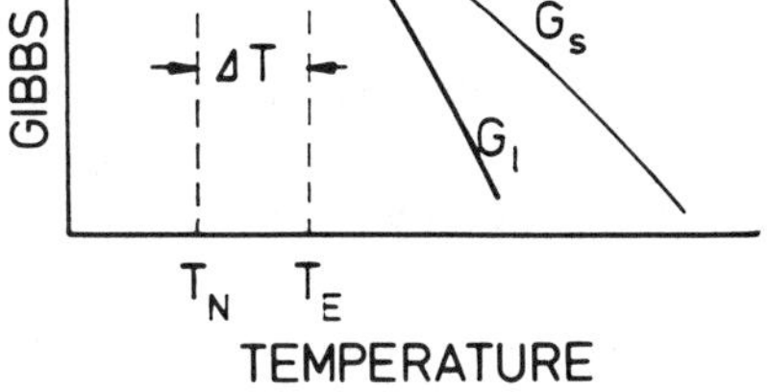

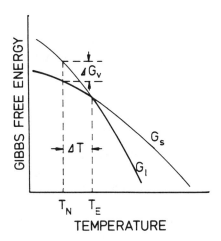

Figure 2 Temperature dependence of the free energy of solid and liquid phases.

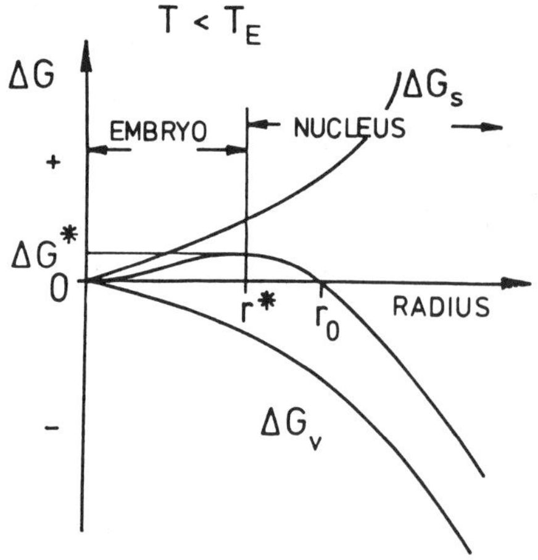

Figure 3 Free energy change during the formation of a nucleus as a function of its radius.

Under the assumption of a dynamical equilibrium between cluster formation and decay, the cluster population N_n^e will be given by

$$N_n^e = N \exp\left(-\frac{\Delta G_n}{kT}\right) \tag{3}$$

with N the total numbers of atoms within the liquid and k Boltzmann's constant. Under steady-state conditions, the forward reaction according to Eq. (2) is independent of both time, t, and cluster size, n; steady-state nucleation rate, I^s, is then given by

$$I^s = N_n^s k_n^+ - N_{n+1}^s k_{n+1}^- \tag{4}$$

The reaction coefficients, k, are proportional to the number of atomic sites on the surface of the cluster and to the atomic jump rate between liquid and cluster; it is assumed that the latter can be described on the basis of atomic diffusion across the interface:

$$D = D_0 \exp\left(-\frac{\Delta G_D}{kT}\right) \tag{5}$$

with ΔG_D the activation energy for atomic diffusion. Using Eqs. (3)–(5) and taking into account the number of critical clusters N_n^e under steady state conditions yields an expression for the nucleation rate:

$$I^s = C \exp\left(-\frac{\Delta G_D}{kT}\right) \exp\left(-\frac{\Delta G_n^*}{kT}\right) \tag{6}$$

with C a prefactor containing the number of potential nucleation sites and the atomic vibration frequency. Under the assumption of the validity of the Stokes-Einstein relation, the first exponential term can be expressed by the temperature dependence of the viscosity [27].

The above considerations neglect time dependence in the evolution of the steady-state cluster population. However, the establishment of the steady-state distribution of clusters needs a finite time, characterized by the transient τ. Several attempts have been made to describe the time dependence of the isothermal nucleation rate. An expression yielding a fair description of experimental results obtained especially on silicate systems is due to Kashchiev [28]:

$$I_t = I^S\left[1 + 2\sum_{\infty}^{m=1} (-1)^m \exp\left(-\frac{m^2 t}{\tau}\right)\right] \tag{7}$$

where m is a cluster size smaller than n^*.

The above treatment pertains to homogeneous nucleation only. In practice, however, heterogeneous nucleation plays the more dominant role. For this case, the number of potential nucleation sites is much smaller, these being wall sites or sites on catalytic particles within the melt. Heterogeneous nucleants will be active if they reduce the activation energy barrier, ΔG_n^*, for the formation of critically sized nuclei. Since this barrier arises from the influence of the interface energy, any change in interface conditions introduced through the introduction of an additional surface will have a considerable influence.

Consider a spherical cap-shaped cluster growing on an interface. Having the critical radius r^*, the cap would be stable against decay, except for its contact circle, which has a radius $r^* \sin\vartheta < r^*$. The cap can, however, be stabilized by surface forces under the condition

$$\sigma_{l,s} = \sigma_{c,s} + \sigma_{l,c}\cos\vartheta \tag{8}$$

where the subscripts, c, l, and s stand for cap, liquid and substrate, respectively. This reduces the total volume, and consequently, the activation energy, ΔG^*, to form a critical nucleus by the factor

$$f(\vartheta) = 0.25(2 - 3\cos\vartheta + \cos^3\vartheta) \tag{9}$$

which is smaller than unity for any contact angle below $\vartheta = 180°$. For this reason, the onset of heterogeneous nucleation will limit the undercooling of a melt long before homogeneous nucleation can be active.

Figure 4 gives a schematic overview of the isothermal evolution of homogeneous and heterogeneous nuclei both with and without transient effects. Throughout the discussion above, fluctuations have been considered in the size of the nucleus but not in its composition, so the results are only applicable to pure metals. When concentration is introduced as a further thermodynamic variable, the size and composition of the most likely nucleus is obtained by looking for the smallest activation barrier, $\Delta G^*(n, x)$ with n the size and x the concentration of the nucleus, respectively. According to Eq. (1), one gets the minimum activation energy from the extrema obtained from $\partial/\partial n(\Delta G_n) = 0$ and $\partial/\partial x(\Delta G_n) = 0$. Thompson and Spaepen [29] have shown that in the simple case of a regular solution model, the concentration of the critical nucleus is given by

$$x_N^A = \left((1 - x_l^A)\frac{\exp(y)}{x_l^A} + 1\right)^{-1} \tag{10}$$

with

$$y = \frac{T_2}{T}\ln\left(\frac{x_l^A(1 - x_s^A)}{x_s^A(1 - x_l^A)} - \frac{T_l - T}{RT}(\Delta S_m^A - \Delta S_m^B)\right) \tag{11}$$

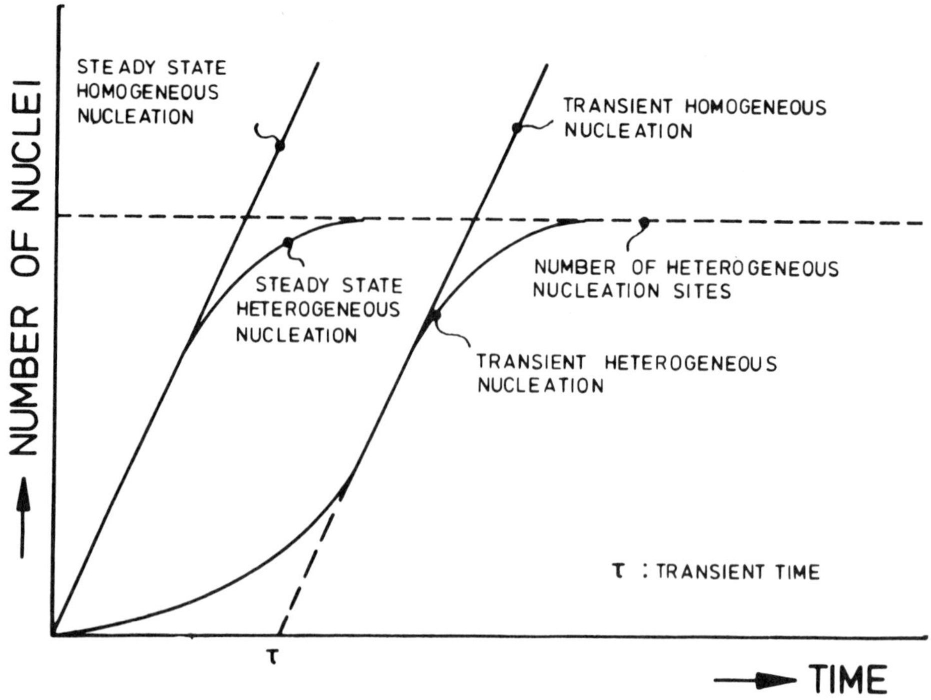

Figure 4 Isothermal evolution of the number of nuclei as a function of time, for homogeneous and heterogeneous nucleation, neglecting and taking into account transient effects.

x_l^A and x_s^A denote the concentration of component A in liquid and solid state, respectively, according to the phase diagram of a binary A-B alloy. ΔS_m^A and ΔS_m^B are the entropies of fusion of the pure elements A and B, respectively. Thus, $\Delta G^*(n, x)$ is determined from the concentration-dependence of the interface energy, σ, and the free energy difference, ΔG_v. Using the negentropic model as developed by Spaepen et al. [30] to estimate σ, its concentration dependence can be expressed by evaluating $V(x)$ and $\Delta S(x)$, with $V(x)$ the molar volume and $\Delta S(x)$ the entropy of fusion of the alloy having a composition x.

C. Results on Undercooling and Nucleation

Nucleation theory provides the concept of homogeneous nucleation and raises the question of whether such nucleation behavior can be detected experimentally. From the fact that homogeneous nucleation requires the maximum undercooling under fixed processing conditions, because of its highest activation energy to form critical nuclei, it is evident that the undercoolings at the onset of measurable homogeneous nucleation must equal or exceed the maximum reported experimental values. This is a necessary but not a sufficient condition, since even in case of maximum experimental undercooling, a further extension of ΔT in future experiments cannot be ruled out. Other criteria are needed in order to test the question of homogeneous nucleation. In the case of homogeneous nucleation one would expect a characteristic behavior in

- The dependence of the nucleation frequency on the size of the sample, scaling with the volume
- High nucleant densities, scaling with the number of atoms within the melt, since every atom can act as a nucleus site
- A dependence of the number of nuclei on cooling rate [17]

The literature contains only a few examples of nucleation measurements where only one, let alone all, of the above criteria are fulfilled. As reported by Turnbull, the nucleation frequency of 2–8-micron-diameter mercury droplets coated by a mercury laurate film increases systematically with the volume of the droplets. This has been interpreted as convincing evidence of homogeneous nucleation [7]. By studying the crystallization behavior of amorphous Fe-B alloys, Greer [31] found a very high density of crystallites ($\sim 10^{12}/cm^3$) randomly situated and oriented. From this result it was concluded that copious homogeneous nucleation had occurred in the molten alloy during the quenching process. More recently, the nucleation behavior of the Al-Mn icosahedral phase has been investigated in micron- and submicron-sized droplets of $Al_{86}Mn_{14}$ produced by atomization [32]. An extremely high density of nucleation sites ($\sim 10^{18}/cm^3$), was found and this was attributed to homogeneous nucleation; the similar short-range order in solid icosahedral phases [33] and simple metallic liquids [34] should facilitate the formation of homogeneous nuclei of icosahedral structure.

Perepezko et al. [35] reported results of the maximum undercooling observed in Sn-Bi alloys of various concentrations, by the droplet emulsion technique. A reanalysis of these undercooling values in terms of homogeneous nucleation theory of binary alloys leads to a sufficient description of the data [29]. However, proof of homogeneous nucleation in Sn-Bi would require additional investigations of nucleation frequency according to the criteria mentioned above. Figure 5 gives the phase diagram for Cu-Ni together with undercooling results obtained by Cech and Turnbull [36] on small particles (open dots). In this case, the experimental results can only be described assuming heterogeneous nucleation with a catalytic potency factor $f(\vartheta) = 0.16$ [37]. Results of levitation undercooling experiments on samples of the same system, although of much larger size (diameter $\sim$ 6 mm), are shown for comparison (full dots) [38]; they also indicate heterogeneous nucleation with a similar but slightly enhanced catalytic potency factor $f(\vartheta) = 0.19$ for all alloys investigated. The large undercoolings obtained for the bulk samples, comparable to or even greater than those of the small particles, suggest that surface-induced heterogeneous nucleation limits the undercooling behavior, rather than heterogeneous nucleation throughout the volume. This is confirmed by levitation experiments on bulk Fe-Ni alloys [15] where the maximum undercoolings can be described by the same value of $f(\vartheta)$ for the Ni-rich samples. This surface nucleation has been associated with the formation of NiO on the surface of the sample. NiO is stable up to a temperature of 2240 K, hence a thin solid NiO layer at the surface will be present even when the bulk of the sample is molten.

The dominant role of surface nucleation is also evidenced by drop tube experiments on the glass-forming alloy $Pd_{82}Si_{18}$ [17]. Figure 6 shows a diagram where the fraction of droplets, x, solidified in glassy state during free fall is plotted as a

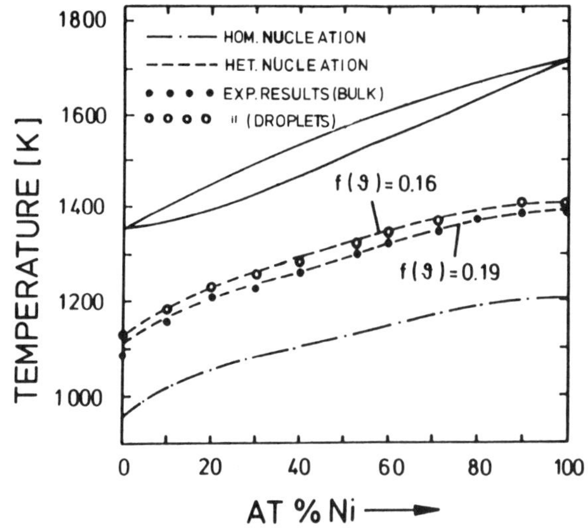

Figure 5 Phase diagram for Cu-Ni, showing the undercooling results of Cech and Turnbull (open dots) [36] on small particles and those of Willnecker et al. (closed dots) [38] obtained on levitated bulk samples. The lines through the data are the predictions of nucleation theory, using a nucleation reduction factor $f(\vartheta) = 0.16$ for the small particles [37] and $f(\vartheta) = 0.19$ for the bulk samples. For comparison, the dashed line gives the undercooling as expected in the limiting case of homogeneous nucleation.

function of droplet diameter, d (open dots). Assuming that the formation of amorphous phase requires the avoidance of any nucleation event and, further, that the droplets solidify either completely glassy or completely crystalline, the measurements could be analyzed with respect to different nucleation processes. As can be seen from Fig. 6, the results suggest that heterogeneous surface nucleation is dominant. This is probably due to the formation of SiO_2 by an oxidation reaction at the droplet surface.

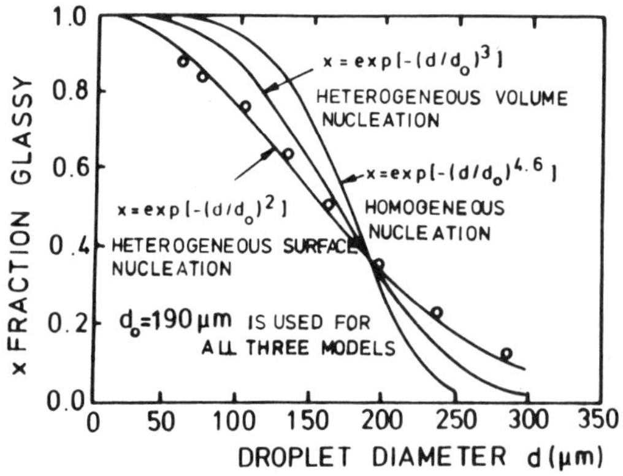

Figure 6 A comparison of the drop tube results for $Pd_{82}Si_{18}$ with the predictions of three nucleation models [17].

On the other hand, drop tube experiments on the similar system $Pd_{77.5}Cu_6Si_{16.5}$ indicate heterogeneous nucleation both within the volume of the droplets and on their surface [39]. Volume heterogeneities arise from the preformation of the Pd_3Si intermetallic phase prior to quenching; this process can be hindered by superheating the melt to a temperature above the liquidus temperature of Pd_3Si, leading to a marked improvement in the glass-forming ability. Under such circumstances, fully glassy spheres up to 1.5 mm in diameter are obtained [40]. These results are in agreement with investigations of the undercooling behavior of the same system and its analysis by nucleation theory [41].

A strong influence of the melt superheat on glass-forming ability has been confirmed by drop tube experiments performed on Cu-Zr alloys [18]. Heating the melt above the liquidus temperature of the intermetallic Cu_3Zr and ensuring a high purity of the gas atmosphere reduces heterogeneous nucleation both within the volume and on the surface of the droplets with the consequence that fully glassy solidification may be obtained in particles up to 250 μm in diameter. Relating droplet diameter to cooling rate yields a critical cooling rate for fully amorphous solidification, in the containerlessly processed droplets, which is lower by a factor of up to 20 when compared with that of rapid-quenching experiments [42].

The above considerations concern nucleation processes within undercooled melts. Once nucleation is completed, subsequent growth leads to full crystallization of the melt. Thereby, besides nucleation, the mechanisms of growth and its kinetics selects the phase and microstructure of the final solidification product. The following section, therefore, will concentrate on growth phenomena with special emphasis on rapid dendritic growth into undercooled melts, its description and consequence on microstructure evolution.

III. SOLIDIFICATION

A. General Remarks

The basic principle of rapid solidification is to achieve high cooling rates from the liquid state via large thermal gradients. Rapid-quenching techniques are the most commonly used methods, providing rapid external heat extraction through a solid phase. For the preparation of materials with reproducible properties, a detailed insight into the phenomena of the rapid solidification process is a mandatory precondition. A poorly developed theoretical description of the formation of rapidly quenched materials has led to technological development resulting largely from empirical methodology. In fact, while rapid quenching is a widely applied technique for the preparation of materials, it offers little potential for establishing a conclusive picture of the kinetics of the solidification process.

In the past, considerable interest has been focused on in situ measurements of the melt-spinning process [43–47]. The interpretation of the experimental results was complicated by the fact that high-speed thermocameras or photodetectors can only measure surface temperatures. The temperature gradient inside the melt puddle is experimentally inaccessible to measurement. Therefore, such treatments are of limited help in understanding of microstructural development.

During recent years, experiments have been undertaken to provide more insight into a material's evolution during rapid solidification processing. The techniques

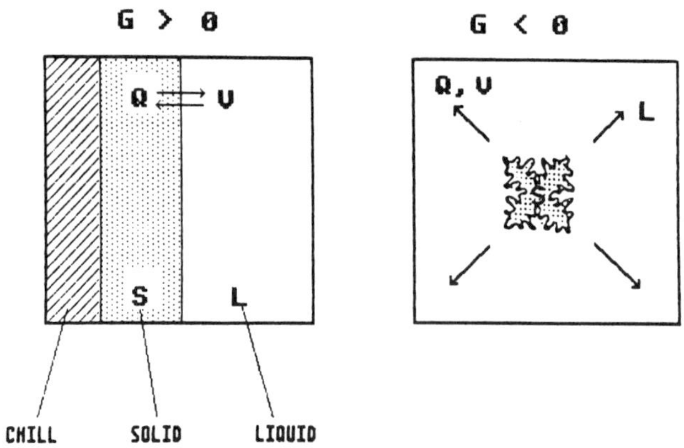

Figure 7 A comparison of heat flow direction in constrained growth (left-hand side) and free dendritic growth (right-hand side). G is the thermal gradient ahead of the solid-liquid interface.

developed make use of an alternative method to achieve high interfacial velocities with the remarkable result that the early stages of rapid solidification may be studied directly. As already stated, a large heat extraction rate is a necessary condition in rapid solidification processes; this is commonly realized by quenching the liquid on a cold substrate, which ensures a large driving force for crystallization. An alternative method is based on the same principle, but uses the undercooled liquid as the required heat sink. The essential difference between this and rapid-quenching techniques is the inverted direction of the thermal gradient in the liquid. Figure 7 illustrates the differences in heat flow. In both cases the thermal undercooling of the solid-liquid interface ΔT_l determines the resulting growth kinetics. Differences occur regarding the growth morphologies because the advancing interface, in case of a negative temperature gradient, will always be unstable, leading to dendritic growth. In the following, a review will be given of the experimental work on rapid solidification in undercooled liquids and the concomitant theoretical progress in this field, with emphasis on the steady-state aspects of dendritic growth.

B. Modeling of Dendritic Growth

Theoretically, the growth of a solid into its undercooled melt is termed unconstrained or free dendritic growth. This is quite dissimilar to the constrained growth occurring in all directional solidification experiments, in which heat flow is opposite to the growth direction and the growth rate as well as the temperature gradient is fixed and controlled. In the case of unconstrained growth the dominant and most important morphology is the dendrite.

The first qualitative description of dendritic growth behavior comes from the work of Papapetrou [48]. A semiquantitative approach for this kind of solidification process was given by Fisher [49] and Zener [50]. In 1947, Ivantsov showed that the main aspects of dendritic growth could be described quantitatively by macroscopic heat flow theory [51]. In a more general way, Horvay and Cahn found that an elliptical paraboloid was an exact solution of the diffusion equation [52]. To predict steady-state conditions, the assumptions that the crystalline surface is isothermal and growth proceeds in a shape-preserving manner have been made. Nonlinear boundary conditions have been ignored for simplification. This leads to the formula presented by Ivantsov [51]:

$$\Delta T_t = \frac{\Delta H_m}{c_p^l} P_t e^{P_t} E_1(P_t) \equiv \frac{\Delta H_m}{c_p^l} \mathrm{Iv}(P_t) \tag{11}$$

Here ΔT_t is the thermal bath undercooling, ΔH_m the latent heat of fusion, c_p^l the specific heat of the liquid, and P_t is the thermal Péclet number ($VR/2a$, with V the growth rate, R the dendrite tip radius, and a the thermal diffusivity of the liquid). $E_1(P_t)$ defines the exponential integral function $\int_{P_t}^{\infty}(e^{-z}/z)\,dz$ and $\mathrm{Iv}(P_t)$ is the Ivantsov function.

Equation (11) describes the steady-state solution for the thermal field around a dendrite tip in the form of a paraboloid of revolution. This corresponds to crystal growth in a pure undercooled melt, wherein the major transport process is heat conduction. From (11), the velocity of growth as a function of undercooling could not be predicted uniquely because the dendritic tip radius remains unknown. The previous theoretical approach was modified to include nonisothermal boundary conditions [53]. A term ΔT_R can be added to Eq. (11), reflecting the influence of the capillarity (Gibbs-Thomson) effect in lowering the equilibrium temperature of the interface. This prevents dendrites from growing with an infinite velocity as $R \rightarrow 0$ and leads to a possible maximum growth rate, $V_{\max}$. However, experiments by Glicksman on growth in undercooled transparent model systems gave reliable evidence that $V_{\max}$ is not reached by real dendrites growing at a given undercooling [54]. The "working radius" of the dendrite tip was measured and found to be much larger than predicted by this "extremum hypothesis"; the corresponding growth velocity deviated from $V_{\max}$ by a factor of 6.5.

To date, the stability criterion is widely used to determine the tip radius and consequently to predict the resulting growth speed. This marginal stability hypothesis was a result of an analysis of the morphological stability of the growing dendrite tip. It was especially the work of Langer and Müller-Krumbhaar [55] which showed that the dendrite tip becomes unstable against perturbations when a critical tip size is reached. From their sophisticated numerical analysis, it was concluded that the operating point of the tip corresponds to the slowest stable growth mode. This result is compatible with experimental observations [54,56]. In the limit of small Péclet numbers, the tip radius for a pure melt was calculated to be [55]

$$R^2 = \frac{2ad_0}{\sigma^* V} \tag{12}$$

with σ^* the stability constant and d_0 the capillarity length defined as $\Gamma c_p^l/\Delta H_m$ (Γ is the Gibbs-Thompson coefficient). Equation (12) in connection with (11) provides a unique relationship for ΔT dependence of V and R. A detailed review of various stability theories and a comparison of their predictions with experimental results is given by Huang and Glicksman [56].

From a technological viewpoint, the more interesting growth phenomena are those connected with rapid solidification processes in alloy systems. Recently, the theories of free dendritic growth have been extended to large Péclet numbers and binary alloys, where the modeling takes into account the coupled processes of heat and mass transport. Lipton, Glicksman, and Kurz developed a first approach using the stability concept to predict crystal growth rate in undercooled binary alloy melts [57]. Boettinger and Coriell presented an extension of this model to large chemical Péclet numbers [58]. A more general solution was obtained by Lipton, Kurz, and Trivedi (LKT) for the case of large thermal and solutal Péclet numbers [59]. From this theory, the tip radius can be expressed as

$$R = \frac{\Gamma/\sigma^*}{\dfrac{P_t\,\Delta H_m}{c_p^l}(1 - n) + 2m_l P_c C_0 \dfrac{k + 1}{1 - (1 - k)\mathrm{Iv}(P_c)}(1 + g)}$$

$$n = \frac{1}{\sqrt{1 + 1/\sigma^* P_t^2}} \qquad g = \frac{2k}{1 - 2k - \sqrt{1 + 1/\sigma^* P_c^2}} \tag{13}$$

with m_l the slope of the liquidus, P_c the chemical Péclet number ($VR/2D$, where D is the diffusion coefficient of the solute in the liquid alloy), C_0 the alloy composition and k the equilibrium partition coefficient.

Following LKT theory, the bulk undercooling can be divided as

$$\Delta T = \Delta T_t + \Delta T_c + \Delta T_R \tag{14}$$

where ΔT_c represents the constitutional undercooling, ΔT_t the thermal undercooling defined by Eq. (11), and ΔT_R the curvature undercooling. Equation (14) in connection with Eq. (13) allows a unique prediction of the tip radius and the growth velocity at a given undercooling for alloy systems.

The LKT theory has been derived with the assumption that no kinetic undercooling limits the solidification process and that the equilibrium partition coefficient can be used. At large solidification velocities, these approximations are no longer valid. To introduce kinetic effects, Eq. (14) has to be modified by a term ΔT_k, first introduced by Trivedi et al. [60], reflecting the possible maximum atomic mobility at the solid-liquid interface. The term ΔT_k could be derived from the collision limited growth model of Turnbull [61].

Large solidification velocities lead to partitionless growth. A necessary but not sufficient condition for partitionless solidification is a solid-liquid interfacial temperature below T_0 in order that the solid phase of the melt composition will have a lower free energy than that of the equilibrium solid composition. The second requirement is an interface speed of the order of the diffusive velocity, which is defined as D/λ (where λ is the diffusive jump distance). Models for a growth velocity-dependent partition coefficient $k(V)$ have been presented by different authors [62,63]. Aziz has formulated a model for continuous growth such that [62]

$$k_v = \frac{k + (a_0 \cdot V/D)}{1 + (a_0 \cdot V/D)} \tag{15}$$

where a_0 is a length scale in the order of the interatomic distance. The predicted relationship has been verified (at least in the regime of interface speeds up to 10 m/s) with pulsed laser experiments on doped semiconductors [64]. The effect of a velocity-dependent partition coefficient on the dendritic growth velocity was recently analyzed by Trivedi et al. [60]. A significant effect on the growth velocity was found, resulting in a sharp discontinuous increase of the solidification velocity when $k \rightarrow 1$ ($\Delta T_c \rightarrow 0$), equivalent to the case of thermally controlled growth. The growth behavior in a deeply undercooled alloy approaches that of a single component material, leading to segregation-free dendritic structures.

In view of these summarized theoretical models and predictions, experimental results play an important role for their justification and their applicability to metallic systems. In the following section, we will proceed to a description of experiments in this field with emphasis on quantitative measurements of large solidification rates.

C. Measurements of Dendritic Solidification in Undercooled Metals

We will not try to give a complete review of existing experimental studies on dendritic growth. Instead we will place emphasis on those experiments which further our understanding of free dendritic growth phenomena in metallic systems. A review of the measured growth rates of crystals of pure materials from the poorly undercooled melt is given by Kobayashi et al. [65]. The results can be described by the equation $v = a(\Delta T)^b$, where a and b are constants for a particular system. The results do not lend themselves to extensive theoretical analysis due to the small range of undercooling studied. The first quantitative measurements of large solidification velocities in undercooled pure metallic melts were performed by Walker [66]. The technique of glass encasement was employed to achieve undercoolings of up to 250 K in Ni and Co samples contained in vitreous silica boats. Dendritic growth rates were measured by means of two photodiodes which were used to determine the time lag of recalescence between two silica tubes immersed in the melt. The velocity data for Ni could be fitted by $V \sim \Delta T^2$ up to an undercooling of about 175 K, at which the velocity was about 40 m/s. Beyond such an undercooling, growth rates deviate markedly from the relationship established for lower velocities and show large scatter. Walker put forward the idea of cavitation at the solid-liquid interface to explain this behavior. We shall return to this point at the end of this chapter.

Colligan and Bayles reported measurements of $V(\Delta T)$ on undercooled Ni samples, as determined by direct optical measurements as well as by high-speed cinematographic technique [67]. The propagating crystallization front was observed by detecting changes in sample surface brightness due to recalescence. Despite the large scatter in the data, an analysis of the results yielded the relationship $V \sim \Delta T^{1.8}$, in rough accordance with the previous data of Walker.

A first theoretical consideration of the experimental results of Walker and of Colligan and Bayles was carried out by Coriell and Turnbull [68]. A qualitative fit of the growth data could be made by using the previous theoretical treatments of Ivantsov [51] and Langer and Müller-Krumbhaar [55], as well as by regarding the interface kinetics.

The same technique applied by Walker was also used by Suzuki et al. to determine growth velocities of undercooled copper and silver melts [69]. From the discussion of their findings, some insufficiencies of the experimental method become obvious. Fig. 8 shows the results obtained for pure copper. The large scatter of

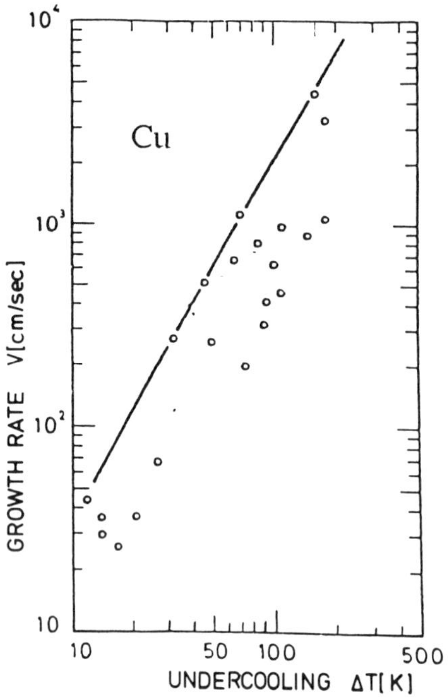

Figure 8 Measured growth rates as a function of undercooling for copper [69].

growth rates is a characteristic feature of this kind of measurement. The authors could attribute this behavior to the interaction of the growing dendrites with the mold wall or with other dendrites. Therefore, the solidified structure has to be examined for each experiment to determine the crystal growth direction and to ensure the correct measurement of the propagation velocity of primary dendrite stems. Consequently, the straight line in Fig. 8, connecting the measured maximum growth velocities, represents the growth rates of free dendrites as a function of undercooling.

Figure 9 shows a comparison of the results with actual dendrite growth theory. The solid line is obtained following the formalism of LKT to predict $V(\Delta T)$, but ignoring the kinetic resistance due to the configurational rearrangement at the interface. The experimental results differ markedly from the calculated values, showing that kinetic effects cannot be ignored at large solidification rates, even in pure metals. This is confirmed by recent measurements of undercooled Ni samples using the electromagnetic levitation technique, as discussed later [3,5,70]. The dashed line in Fig. 9 represents the calculation, taking into account the kinetic term ΔT_k. The results are in good agreement with the experimental values, demonstrating the important role of atomic attachment kinetics at a rapidly propagating solidification front.

Further extensive measurements of dendrite growth velocities have been presented for Ni and Ni-Sn alloys by Flemings et al. [71–73]. These authors observed the solidification of inductively melted metallic drops, of about 10 mm diameter, encased in glass. Both cinematography and high-speed optical temperature mea-

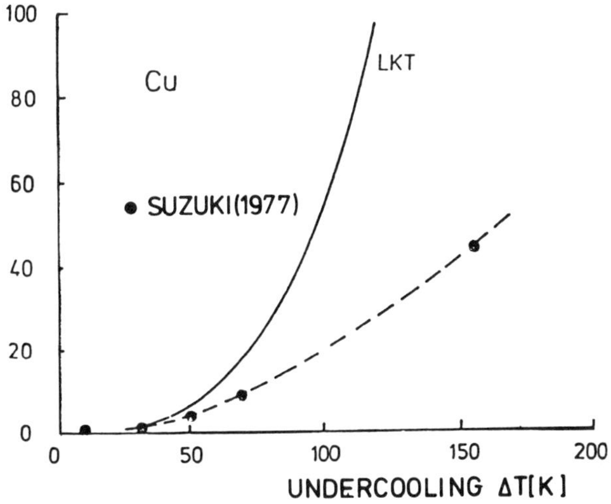

Figure 9 Comparison of measured free dendritic growth rates for copper (Fig. 8) with computations in the framework of growth theories. The solid curve was calculated using the theory of LKT. The dashed line was calculated following the formula of LKT modified with an interface kinetic term.

surements were employed to study dendrite growth velocities. For the global optical measurements a fast two-color pyrometer was used to resolve the recalescence time of the solidifying sample. Recalescence times were divided by the sample dimension to calculate the dendritic growth rates. Measured velocities ranged from about 0.07 to 100 m/s (for the pure system) at the largest undercooling. Figure 10 shows the results for Ni and two Ni-Sn alloys [73]. The data were analyzed in the framework of recent dendrite growth theory [59]. While the experimental values can be fitted assuming local equilibrium at the solid-liquid interface, the authors state that further experimental tests are needed, particularly at large undercoolings where the assumption of local equilibrium at the interface is unjustified and solute trapping is expected. A disadvantage of the measurements of Flemings et al. is the undefined nucleation point and the consequently unknown dendrite pathway. This may lead to an error which cannot be estimated quantitatively. In the regime of large undercoolings, the data are not sufficiently accurate to resolve the effects of partionless solidification which are theoretically predicted.

An alternative technique to measure accurate free dendrite growth velocities was recently demonstrated by Schleip et al. [5]. Figure 11 shows a schematic illustration of the experimental method. Measurements were performed on freely levitated molten alloy spheres, where high undercoolings were attained by the elimination of the heterogeneous nucleation effects of the container walls [15]. Recalescence time of the solidifying sample was detected via a high-speed photosensing device consisting of two adjacent fast Si photodiodes (response time less than 1 μs). The projection of the measuring areas on the sample surface is shown in Fig. 11a. Accurate values of the growth rates can only be determined if the measured recalescence time can be associated with the propagation path of the dendrites in a definite way. For this reason, the undercooled samples were triggered

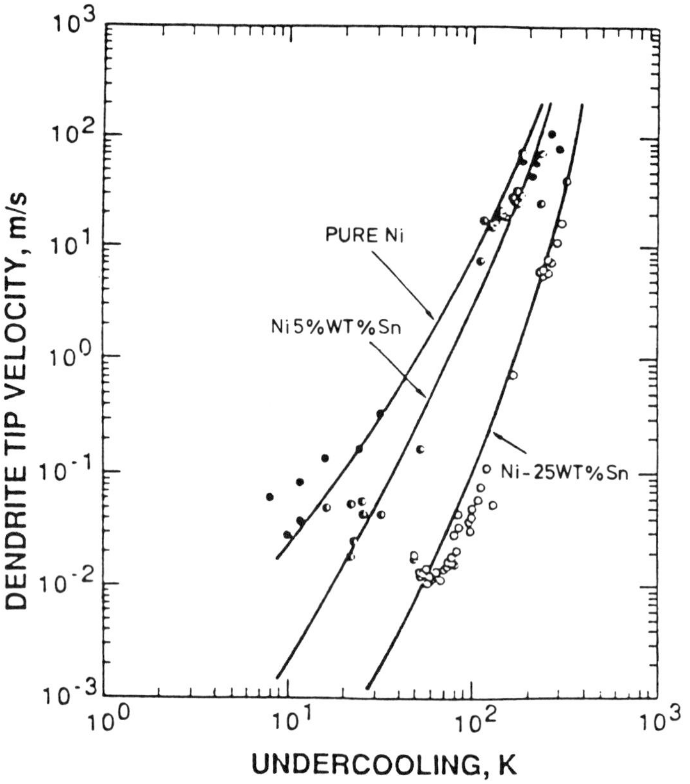

Figure 10 Measured dendritic growth velocity as a function of undercooling for Ni and Ni-Sn alloys [73]. Solid lines indicate the theoretical predictions [59].

to initiate solidification at a well-defined point in a plane normal to the viewing direction. Dendrites proceed isotropically from the trigger point throughout the sample volume, as shown schematically in Fig. 11b and confirmed by metallographic investigation of the solidified spheres. The left side of Fig. 11 shows typical signal-time profiles of the two detectors during the solidification process. Analysis of the two signals results in a constant growth velocity. This confirms previous experimental results on the solidification of Bi samples in moderately undercooled melts by Glicksman [74]. It also demonstrates that free dendritic growth proceeds under steady-state conditions.

The technique described above was applied to measurements of growth velocities in various metallic systems [70]. The most important results were obtained on a copper-nickel alloy as shown in Fig. 12 [75]. The data show a power law behavior, $V \sim \Delta T^b$, for undercoolings below a critical value ΔT^* of about 190 K. In the vicinity of ΔT^*, growth velocity increases sharply and shows a nearly linear behavior with ΔT at large undercoolings. The data do not agree with the approximation of complete solute redistribution at the interface, represented by the dashed line in Fig. 12 ($k = 0.81$ is the equilibrium partition coefficient of the alloy). The observed values up to ΔT^* can be fitted by assuming deviations from local equilibrium at the interface as theoretically predicted in the case of large growth rates [60]. How-

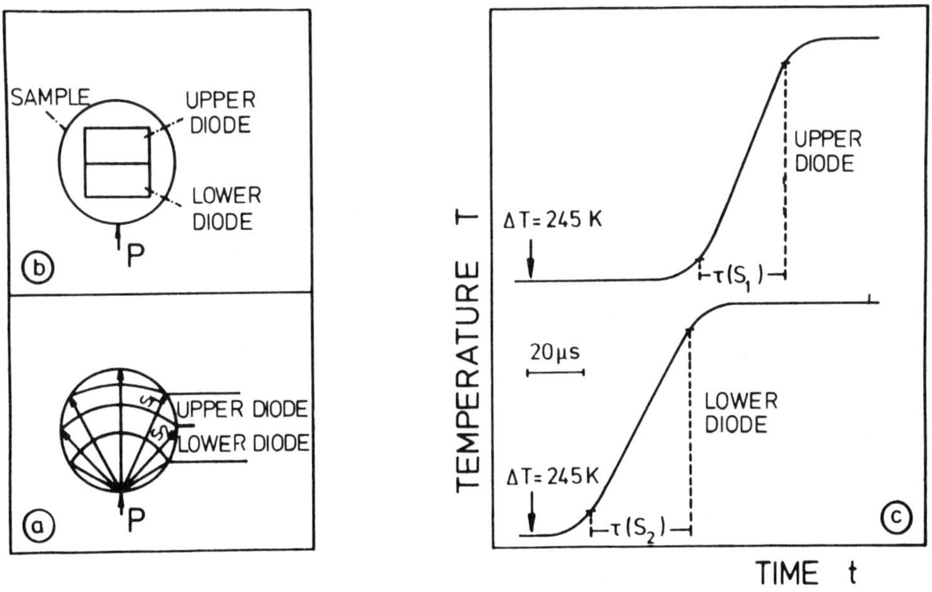

Figure 11 Solidification velocity measurement on a levitated metal sample via a two-element photodetector [5]: (a) projection of the measuring areas on the samples surface; (b) cross section through the sample, showing the different path lengths in the volume corresponding to the upper and lower field of view; (c) simultaneous recording of the signals during recalescence.

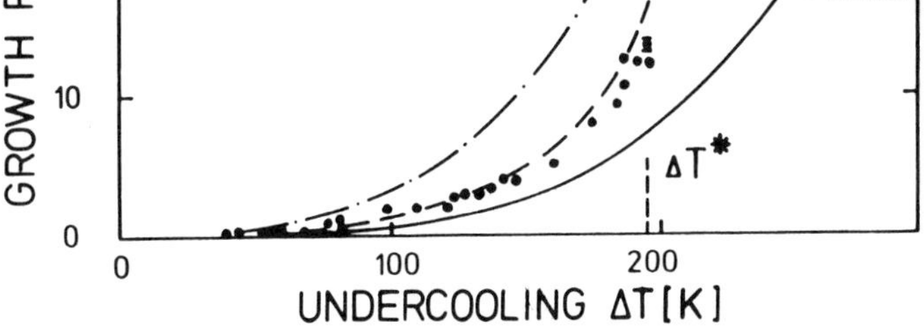

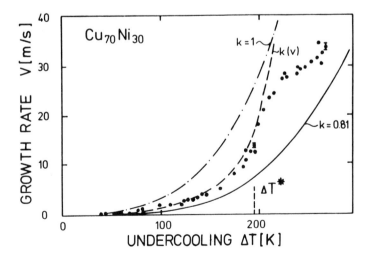

Figure 12 Measured growth rates as a function of undercooling for a $Cu_{70}Ni_{30}$ alloy. The curves represent calculations following free dendritic growth theories [59,60] assuming diffusional equilibrium ($k = .81$), partitionless solidification ($k = 1$) and a velocity-dependent partition coefficient ($k(v)$).

ever, the extreme limit of polymorphous solidification (curve $k = 1$ in Fig. 12) is not achieved. The measurements suggest an increasing partition coefficient with increasing undercooling and growth velocity. As shown by the solid line in Fig. 12, quantitative agreement of the results with theoretical calculation can be attained using the approach of Trivedi et al. [60], which includes a growth velocity-dependent partition coefficient according to the model of Aziz (Eq. (15).

As outlined earlier in this chapter, partitionless solidification requires an interface temperature below the T_0-line in the phase diagram. The detailed analysis of the data for $Cu_{70}Ni_{30}$ alloy within the framework of dendritic growth theory reveals an interfacial temperature of 1466 K in the vicinity of ΔT^*. This is in excellent agreement with the T_0-value of the particular alloy composition and confirms the onset of solute trapping.

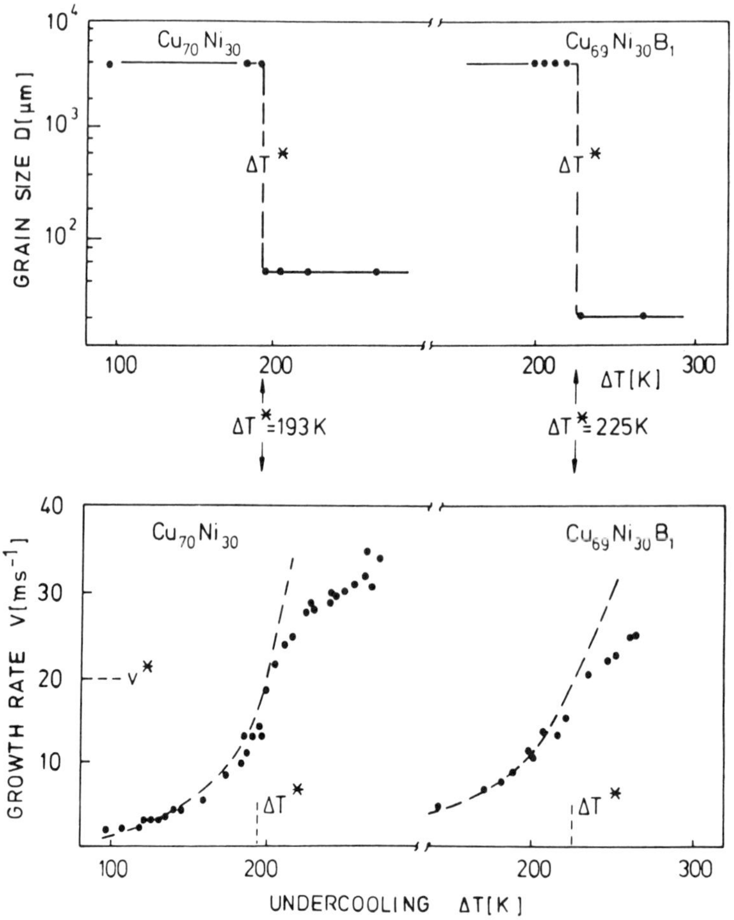

Figure 13 Grain size as a function of undercooling for $Cu_{70}Ni_{30}$ and $Cu_{69}Ni_{30}B_1$ (upper part). Measured growth rates as a function of undercooling for the same alloys (lower part). The observed critical undercooling for grain refinement is strongly correlated to the value of ΔT^*.

Interestingly, at undercoolings in excess of ΔT^*, the experimental observations do not follow the predicted curve; it is in this region where the most interesting microstructural effects, namely complete solute trapping and grain refinement, are known to occur. The experimentally determined linear relationship between growth velocity and undercooling cannot be explained by existing theories. It had originally been suggested by Walker that a cavitation mechanism, resulting from the rapid liquid flow during solidification, affected both the growth behavior and the subsequent morphological development [66]. In this context, Willnecker was recently able to demonstrate that the critical temperature ΔT^* can be related to the onset of the grain refinement processes [70]. Figure 13 shows the dependence of growth velocity and grain diameter on initial of undercooling for Cu-Ni alloys. A small addition of boron increased the critical undercooling for grain refinement by 32 K, the same increase is observed for ΔT^*. The origin of the sudden change in microstructure and growth behavior at large undercoolings has not been clarified unambiguously, despite much effort in this area [76]. Further measurements are needed to decide whether one or any of the many theories put forward satisfactorily describe these effects.

IV. CONCLUSION

We want to close this chapter with some concluding remarks about the central position of undercooling experiments between the various rapid solidification techniques.

Undercooling and subsequent crystallization are important aspects for understanding rapid solidification processes. It has been recognized in recent years that the concept of solidification rate has to be shifted from emphasizing cooling rate to now considering solid-liquid interface velocity. Interfacial undercooling together with the resulting interface kinetics constitutes a key aspect of rapid solidification.

Undercooling investigations and measurements of recalescence behavior provide a direct insight into the first stages of transformation in rapid solidification processes. The basic studies summarized here reveal the capability to resolve the crystallization sequences in detail. Experimental results in conjunction with the impressive theoretical work on dendritic growth phenomena enable us to establish a unique relationship between interfacial undercooling and crystal growth velocity.

In the future, it will be instructive to apply the fundamental results of undercooling experiments and solidification kinetics to commonly used rapid solidification techniques. The most suitable technique for this purpose is, of course, melt atomization. Up to now, little attention has been paid to the comparison of solidification in the melt-spinning process and solidification from bulk undercooled melts: a first attempt has been made by Caesar et al. for Cu-Ni alloys [77] and by Liebermann et al. for Al-Fe-V alloys [78]. Further research aimed at the comprehensive understanding of microstructure evolution and compositional distribution during solidification will improve our ability to achieve a desired microstructure from a particular rapid solidification process. From this viewpoint we expect increasing attention to be paid to fundamental studies on undercooling and solidification.

ACKNOWLEDGEMENTS

The authors want to thank Dr. R. F. Cochrane for a careful reading of the manuscript and valuable discussion, and Prof. Dr. B. Feuerbacher for continuous support of the present work.

REFERENCES

1. D. Turnbull, *Metall. Trans.*, *12A*, 695 (1981).
2. D. Turnbull, in *Undercooled Alloy Phases*, edited by E. W. Collings and C. C. Koch (TMS-AIME, 1986), p. 3.
3. B. Feuerbacher, Materials Science Reports, *4*, 1 (1989).
4. Y. Wu, T. J. Piccone, Y. Shiohara, and M. C. Flemings, *Metall. Trans.*, *18A*, 915 (1987).
5. E. Schleip, R. Willnecker, D. M. Herlach, and G. P. Görler, *Mater. Sci. Eng.*, *98*, 39 (1988).
6. D. Turnbull and R. E. Cech, *J. Appl. Phys.*, *21*, 804 (1950).
7. D. Turnbull, *J. Chem. Phys.*, *20*, 411 (1952).
8. D. H. Rasmussen and C. R. Loper, *Acta Metall.*, *23*, 1215 (1975).
9. J. H. Perepezko, D. H. Rasmussen, I. E. Andersen, and C. R. Loper, in *Proc. Int. Conf. Solidification and Casting of Metals* (Metals Soc., London, 1979), p. 169.
10. J. H. Perepezko, *Mater. Sci. Eng.*, *65*, 125 (1984).
11. P. Bardenheuer and R. Bleckmann, *Mitteilg. KWI f. Eisenforschung*, *21*, 201 (1939).
12. T. Z. Kattamis, *Z. Metallkde.*, *61*, 856 (1970).
13. H. W. Kui, A. L. Greer, and D. Turnbull, *Appl. Phys. Lett.*, *45*, 615 (1985).
14. S. Y. Shiraishi and R. G. Ward, *Can. Met. Quart.*, *3*, 117 (1964).
15. R. Willnecker, D. M. Herlach, and B. Feuerbacher, *Appl. Phys. Lett.*, *49*, 1339 (1986).
16. J. Piller, R. Knauf, P. Preu, G. Lohöfer, and D. M. Herlach, In *Proc. Sixth Eur. Symp. Material Sciences under Microgravity Conditions* (ESA SP-256, 1987), p. 437.
17. A. J. Drehman and D. Turnbull, *Scripta Metall.*, *15*, 543 (1981).
18. F. Gillessen and D. M. Herlach, *Mater. Sci. Eng.*, *97*, 147 (1988).
19. R. F. Cochrane, P. V. Evans, and A. L. Greer, *Mater. Sci. Eng.*, *98*, 99 (1988).
20. L. L. Lacy, M. B. Robinson, and T. J. Rathz, *J. Cryst. Growth*, *51*, 47 (1981).
21. W. H. Hofmeister, M. B. Robinson, and R. J. Bayuzick, *Appl. Phys. Lett.*, *49*, 1342 (1986).
22. R. F. Cochrane, D. M. Herlach, and R. Willnecker, in *INDO-US Workshop on Metastable Microstructures*, to appear.
23. H. V. Volmer and A. Weber, *Z. Phys. Chem.*, *24*, 719 (1935).
24. R. Becker and W. Döring, *Ann. Phys.*, *24*, 719 (1935).
25. D. Turnbull and J. C. Fisher, *J. Chem. Phys.*, *17*, 71 (1949).
26. K. F. Kelton, A. L. Greer, and C. V. Thompson, *J. Chem. Phys.*, *79*, 6261 (1983).
27. D. Turnbull, *Contemp. Phys.*, *10*, 473 (1969).
28. D. Kashchiev, *Surf. Sci.*, *14*, 209 (1969).
29. C. V. Thompson and F. Spaepen, *Acta Metall.*, *31*, 2021 (1983).
30. F. Spaepen and R. B. Meyer, *Scr. Metall.*, *10*, 257 (1976).
31. A. L. Greer, *Acta Metall.*, *30*, 171 (1982).
32. L. A. Bendersky and S. D. Ridder, *J. Mater. Res.*, *1*, 405 (1986).
33. D. Shechtman, I. Blech, D. Gratias, and J. W. Cahn, *Phys. Rev. Lett.*, *53*, 1951 (1984).
34. F. C. Frank, *Proc. R. Soc. London Ser.*, *A 215*, 43 (1952).
35. J. H. Perepezko, in *Rapid Solidification Processing, Principles and Technologies, II*, edited by R. Mehrabian, B. H. Kear, and M. Cohen (Claitor's, Baton Rouge, LA, 1980), p. 56.

36. R. E. Cech and D. Turnbull, *J. Metals*, *191*, 242 (1951).
37. C. V. Thompson and F. Spaepen, in *Materials Processing in the Reduced Gravity Environment on Space*, edited by G. E. Rindone (North-Holland, Amsterdam, 1982), p. 603.
38. R. Willnecker, D. M. Herlach, and B. Feuerbacher, *Mater. Sci. Eng.*, *98*, 85 (1988).
39. F. Gillessen, D. M. Herlach, and B. Feuerbacher, *J. Less-Common Metals*, *145* (1988).
40. J. Steinberg, A. E. Lord, Jr., L. L. Lacy, and J. Johnson, *Appl. Phys. Lett.*, *38*, 135 (1981).
41. D. M. Herlach, and F. Gillessen, *J. Phys. F*, *17*, 1635 (1987).
42. Y. Nishi, T. Morohoshi, M. Kawakami, K. Suzuki, and T. Masumoto, in *Proc. Fourth Int. Conf. Rapidly Quenched Metals*, edited by T. Masumoto and K. Suzuki (Sendai, 1982), p. 111.
43. S. C. Huang, H. C. Fiedler, *Mater. Sci. Eng.*, *51*, 39 (1981).
44. M. J. Tenwick and H. A. Davies, in *Rapidly Quenched Metals*, Vol. I, edited by S. Steeb and H. Warlimont, (North-Holland, Amsterdam 1985), p. 67.
45. A. G. Gillen and B. Cantor, *Acta Metall.*, *33*, 1813 (1985).
46. G. Stephani, H. Mühlbach, H. Fiedler, and G. Richter, *Mater. Sci. Eng.*, *98*, 29 (1988).
47. P. Cremer and J. Bigot, *Mater. Sci. Eng.*, *98*, 95 (1988).
48. A. Papapetrou, *Z. Kristallogr.*, *92*, 11 (1936).
49. J. C. Fisher, in *Principles of Solidification*, edited by B. Chalmers (Wiley, New York, 1964), p. 105
50. C. Zener, *Trans. AIME*, *167*, 550 (1946).
51. G. P. Ivantsov, *Dokl. Akad. Nauk. SSR*, *58*, 569 (1947).
52. G. Horvay and J. W. Cahn, *Acta Metall.*, *9*, 651 (1961).
53. M. E. Glicksman and R. J. Schäfer, *J. Cryst. Growth*, *1*, 297 (1967); *2*, 239 (1968).
54. M. E. Glicksman, R. J. Schäfer, and J. A. Ayers, *Metall. Trans*, *7A*, 1747 (1976).
55. J. S. Langer and H. Müller-Krumbhaar, *Acta Metall*, *26*, 1681 (1978).
56. S. C. Huang and M. E. Glicksman, *Acta Metall.*, *29*, 701 (1981).
57. J. Lipton, M. E. Glicksman, and W. Kurz, *Mater. Sci. Eng.*, *65*, 57 (1984).
58. W. J. Boettinger and S. R. Coriell, in *Science and Technology of the Undercooled Melt*, edited by P. R. Sahm, H. Jones, and C. M. Adam (Martinus-Nijhoff, Dordrecht, 1986). p. 81.
59. J. Lipton, W. Kurz, and R. Trivedi, *Acta Metall.*, *35*, 957 (1987).
60. R. Trivedi, J. Lipton, and W. Kurz, *Acta Metall.*, *34*, 1663 (1986).
61. D. Turnbull, *Metall. Trans.*, *12A*, 693 (1981).
62. M. J. Aziz, *J. Appl. Phys.*, *53*, 1158 (1982).
63. H. A. Jackson, G. H. Gilmer, and H. J. Leamy, in *Laser and Electron Beam Processing of Materials*, edited by C. W. White and P. S. Peercy (Academic Press, New York, 1980), p. 104.
64. M. J. Aziz, in Ref. 2, p. 375.
65. H. F. Kobayashi, M. Kumikawa, and P. H. Shingu, *J. Cryst. Growth*, *67*, 85 (1984).
66. J. L. Walker, in Ref. 49, p. 114.
67. G. A. Colligan and B. J. Bayles, *Acta Metall.*, *10*, 895 (1962).
68. S. R. Coriell and D. Turnbull, *Acta Metall.*, *30*, 2135 (1982).
69. T. Suzuki, S. Toyoda, T. Umeda, and Y. Kimura, *J. Cryst. Growth*, *38*, 123 (1977).
70. R. Willnecker, thesis, University of Bochum, Bochum, Germany, 1988.
71. Y. Wu, T. J. Piccone, Y. Shiohara, and M. C. Flemings, *Metall. Trans.*, *18A*, 915 (1987).
72. T. J. Piccone, Y. Wu, Y. Shiohara, and M. C. Flemings, *Metall. Trans.*, *18A*, 925 (1987).
73. T. J. Piccone, Y. Wu, Y. Shiohara, and M. C. Flemings, in *Solidification Processing*, edited by J. Beech and H. Jones (Inst. Metals, London, 1988), p. 268.

74. M. E. Glicksman, *Acta Metall.*, *13*, 1231 (1965).
75. R. Willnecker, D. M. Herlach, and B. Feuerbacher, *Phys. Rev. Lett.*, *62*, 2707 (1989).
76. Recent review: A. Munitz and G. J. Abbaschian, in Ref. 2, p. 23.
77. C. Caesar, U. Köster, R. Willnecker, and D. M. Herlach, *Mater. Sci. Eng.*, *98*, 339 (1988).
78. H. H. Liebermann and R. L. Bye, in *Rapidly Solidified Crystalline Alloys*, edited by S. K. Das, B. H. Kear, and C. M. Adam (TMS-AIME, 1985), p. 61.

4

Processing Principles in Rapid Solidification

Paul Hideo Shingu and Keiichi N. Ishihara
Kyoto University, Kyoto, Japan

I. INTRODUCTION

Aside from the purpose of rapid quenching, the formation of ribbon or wire directly from molten metals and alloys has long been an aim of engineers. Such processes can eliminate the elaborate mechanical reduction of cast ingots to small final dimensions. The essence of the technology lies in the formation of the material final shape while still in the liquid state and the conservation of that shape on transformation to the solid state. In such processes difficulty always exists, at the point of contact between the melt stream to the cooling media. Because of the high fluidity of melt, the stream shape can easily be distorted or even disrupted so that the maintenance of the original orifice or slot cross-sectional shape by the melt is always quite difficult. Nevertheless, some successes have been achieved; examples of successful processes include planar-flow casting for ribbon formation and free-jet casting for wire formation. An illustrative introduction of various rapid quenching techniques is given by Jones [1]. In the case of ribbon formation, a breakthrough happened when a process was invented in which, instead of trying to have the melt maintain the cross-sectional shape of the slot as it was ejected into the cooling medium, the ribbon is formed out of a melt "puddle" on a fast-moving solid surface [2,3]. The principle of this process, which will be described in some detail in the next section, is based on the dragging out of a liquid boundary layer, which naturally takes the shape of a ribbon, out of the melt puddle. In the case of the direct wire formation from the melt, the conservation of the circular cross-sectional shape of the melt formed by the orifice is essential. The break down of the shape is inevitable when solid coolant is used so that only semicircular or oblique cross-sectioned wire can be made in such a case. The melt extraction method [4] is one of such technique. Spinning of molten alloy while surrounded by semimolten glass to keep the shape of liquid metal via the high viscosity of glass is called the Taylor process [5]. While this process may be used for some alloys, the reaction between the melt and glass which takes place in many cases limits the choice of metal and glass combinations to be spun. Also, the cooling rate of this wire casting process is reduced by the existence of low heat conductive glass around the metal wire during cooling. An innovative method by Ohnaka [6] which utilizes the nearly perfect laminar flow of liquid coolant inside a rotating drum as cooling medium allows preservation of the circular cross section of the orifice in the wire. This process is called the in-rotating liquid-spinning process (INROLISP).

In the following sections, the principle of ribbon formation by the so-called single-roller method, which is the most commonly used method of ribbon formation from the melt, will be given. Some remarks on twin roller method will also be given. The principle of the INROLISP will be also mentioned. The difference between the amorphous and crystalline materials in forming the ribbon or wire from the melt will be stressed based on some experimental results. Finally, the principle of compaction of amorphous alloys will be mentioned. Amorphous alloys are usually produced in the form of thin ribbon, wire, or powder due to the necessity of rapid heat extraction in the production process. In order to obtain bulk shapes of amorphous alloys some way of compaction must be used. Sintering at high temperatures cannot be applied to amorphous alloy because of the occurrence of crystallization. However, the fluidlike low-viscosity state, which is characteristic of amorphous metals at temperatures near T_g, the glass temperature, can be utilized for the compaction. Such a process may be regarded as mechanical forming of the metastable liquid state of alloys.

II. RIBBON FORMATION BY THE SINGLE-ROLLER METHOD

A. Chill-Block Casting

The single-roller method is most commonly used for ribbon formation directly from the molten metal. One of the techniques falling into this category is "chill-block casting," in which molten metal is flowed onto a moving chill substrate and ribbon is formed. The technique of the chill block casting has a long history which dates back to the 19th century [7].

Kavesh [8] first pointed out the importance of the formation of a stable "puddle" on the moving chill substrate in order to produce geometrically uniform ribbon. He applied the concept of thermal and momentum transport modes for the analysis of the engineering parameters of ribbon formation out of the puddle. In the case of the thermal transport-controlled model, ribbon formation occurs by solidification of liquid due to the heat flow into the chill block. The solidified ribbon is extracted from the puddle, hence, the thickness of a ribbon corresponds to that of the solidified layer. On the other hand, in the case of the momentum transport-controlled model, the liquid film is dragged out from the puddle to solidify farther downstream. It was thought, in early stages of the study of chill-block casting, that the thermal transport mechanism dominates the ribbon formation process. This was substantiated by Kavesh's calculation based on the Prandtl number of the melt, which indicated that thermal propagation was roughly three to nine times faster than momentum propagation. However, much experimental evidence of single-roller experiments revealed the existence of unsolidified liquid downstream of the puddle, generally supporting the dominance of momentum boundary layer propagation [9]. Nevertheless, in very special cases where the puddle thickness is forced to become very small, or the thermal contact between the liquid and substrate is extremely good, thermal transport may become important. In the following section, the momentum transport process will be reviewed in some detail and a possible modification of ribbon formation theory, due to the cooperative contribution of thermal transport, will be discussed. Differences in the casting of crystalline and amorphous alloys will be then described.

B. Boundary Layer Formation

When molten alloy is supplied onto a rapidly moving substrate, it will be repelled by the substrate as liquid globules unless wetting between the melt and the substrate takes place. Such globulization of the melt tends to occur when the velocity of the substrate is insufficient. Increased substrate velocity induces molecular contact between the melt and substrate to enhance wetting. When sufficient wetting is attained a puddle is formed from which a film of liquid is drawn. The formation of such a liquid film can readily be confirmed for the single-roller technique when stream of water is directed onto the roller. When the roller revolution rate is insufficient, water droplets are formed on impact while at higher roller revolution rate, a persistent water film can be generated along the entire circumference of the roller.

The liquid film is formed within the liquid puddle as a momentum (or velocity) boundary layer. This boundary layer is usually defined as the contour of the position within the liquid wherein the local velocity has achieved a certain given fraction of the velocity of the moving solid substrate. Thus, the 1% boundary layer means that the fluid flow velocity is less than 1% of the substrate velocity outside this layer. The analysis of the velocity boundary layer on the rotating wheel may be treated most simply as a modification of the Blasius problem introduced in Schlichting [10]. Referring to Fig. 1 for the notations of coordinates and velocity directions, the basic equations for the flow of liquid on the moving substrate are

$$u \frac{\partial u}{\partial x} + v \frac{\partial u}{\partial y} = \nu \frac{\partial^2 u}{\partial y^2} \tag{1}$$

$$\frac{\partial u}{\partial x} + \frac{\partial v}{\partial y} = 0 \tag{2}$$

The boundary conditions are

$$u = U_w, \quad v = 0 \qquad \text{at } y = 0 \tag{3}$$

$$u = 0 \qquad \text{at } y = \infty \tag{4}$$

where ν is the kinematic viscosity of liquid and U_w is the velocity of the surface of a roller. The simultaneous partial differential equations and boundary conditions

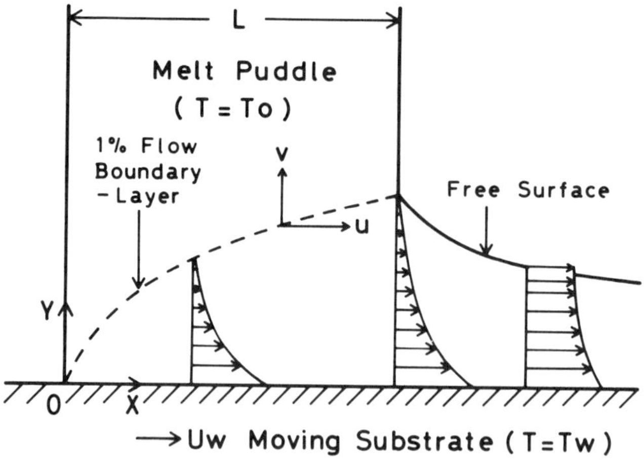

Figure 1 Schematic drawing showing the formation of flow boundary layer on the moving substrate.

(1)–(4) can be transformed to an ordinary differential equation and boundary conditions by the use of a similarity variable $\eta = y/\sqrt{\nu x/U_w}$ and the flow function $\phi = \sqrt{\nu U_w x}\, f(\eta)$,

$$2f''' + f''f = 0 \qquad (1)'$$

with boundary conditions

$$f = 0, \quad f = 1 \qquad \text{at } \eta = 0 \qquad (2)'$$

$$f' = 0 \qquad \text{at } \eta = \infty \qquad (3)'$$

The prime denotes differentiation with respect to η. This two-point boundary value problem was solved numerically using the physical properties of the Fe-Ni-P-B alloy, which is a typical transition metal–metalloid type glass-forming alloys [11]. Figure 2 shows the calculated result of 1% and 10% boundary layer thicknesses. In this treatment, the length of puddle was obtained from experimental results since the analytical estimation of puddle length is quite difficult. Difficulties exist in the mathematical treatment of the free-liquid surface. Recent developments in the numerical method of calculating transient fluid flow with free boundaries may be applied for this problem. Figure 3 shows examples of the results of such calculations, indicating the decrease in puddle size with the increase in substrate velocity [12]. This tendency is in qualitative agreement with the observed results. When the thermal contact between substrate and molten alloy in the puddle becomes extremely good the assumption of momentum transport domination cannot be asserted. This thermal contact may well be expressed in terms of h_i, the interfacial heat transfer coefficient between substrate and melt. For good thermal contact (large h_i) sufficient reduction of temperature in the puddle takes place to cause solidification of liquid. In the case of glass-forming melts, the effect of temperature

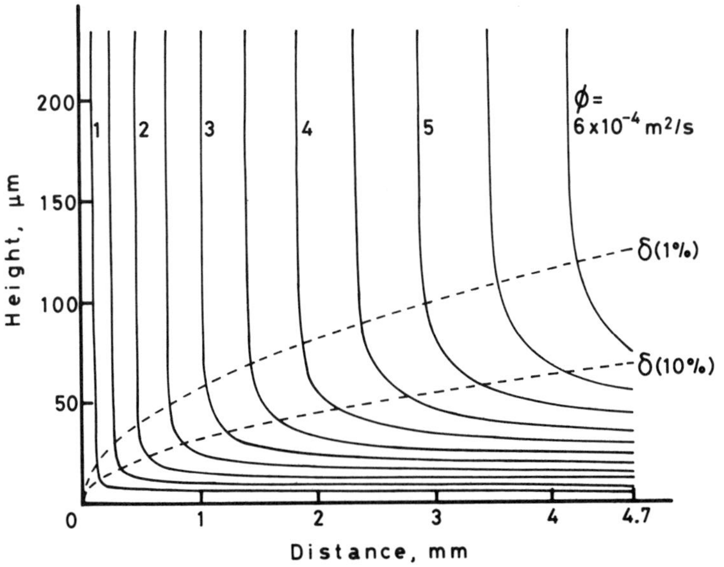

Figure 2 Fluid flow lines within the puddle. Boundary layers (1%) and (10%) indicating, respectively, the positions of 1% and 10% of the substrate velocity are shown by dashed lines.

reduction of liquid in the puddle is associated with an increase in viscosity. In this case, the partial differential equations to be solved are

$$\frac{\partial u}{\partial t} + u\frac{\partial u}{\partial x} + v\frac{\partial u}{\partial y} = \frac{\partial}{\partial y}\left(\nu\frac{\partial u}{\partial y}\right) \tag{5}$$

$$\frac{\partial u}{\partial x} + \frac{\partial v}{\partial y} = 0 \tag{6}$$

$$\frac{\partial T}{\partial t} + u\frac{\partial T}{\partial x} + v\frac{\partial T}{\partial y} = \alpha\frac{\partial^2 T}{\partial y^2} \tag{7}$$

where T and α refer, respectively, to temperature and thermal diffusivity. Initial and boundary conditions are

$$u = 0, \quad \frac{\partial v}{\partial x} = 0, \quad \frac{\partial T}{\partial x} = 0 \qquad \text{at } x = 0 \tag{8}$$

$$u = U_w, \quad v = 0, \quad K\frac{\partial T}{\partial y} = h_i(T - T_w) \qquad \text{at } y = 0 \tag{9}$$

$$u = 0, \quad T = T_0 \qquad \text{at } y = \infty \tag{10}$$

$$u = 0, \quad v = 0, \quad T = T_0 \qquad \text{at } t = 0 \tag{11}$$

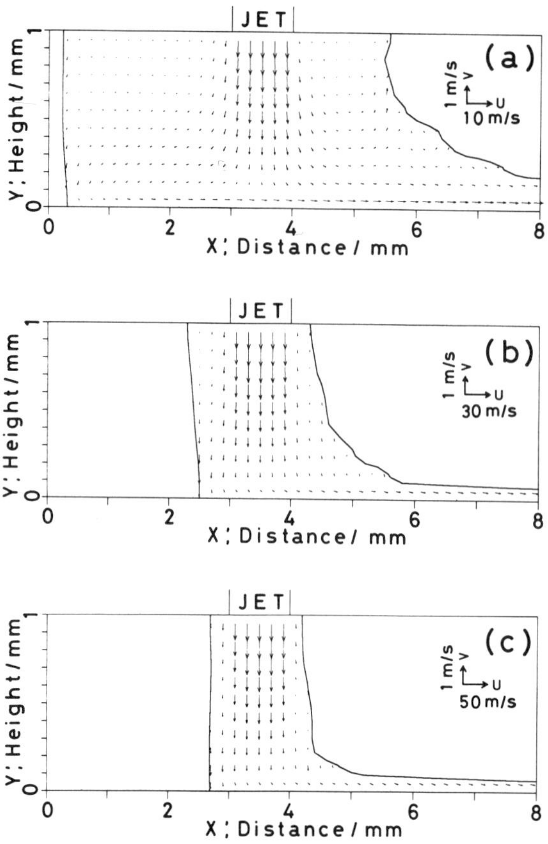

Figure 3 Calculated fluid flow velocities in the puddle for three different substrate velocities. (a) $U_w = 10$ m s^{-1}, (b) $U_w = 30$ m s^{-1}, (c) $U_w = 50$ m s^{-1}. The orifice length is 1 mm for all three cases.

where k is the thermal conductivity, T_w and T_0 respectively refer to the roller surface and the initial melt temperatures. The change in kinematic viscosity of undercooled liquid may be given by the Vogel-Fulcher relation [13,14],

$$\nu = \frac{A}{\rho} \exp\left(\frac{B}{T - C}\right) \tag{12}$$

where A, B, and C are numerical constants and ρ is the density of melt. The result of calculations indicates that, for h_i value of as high as 4.2×10^5 W/m^{2}K, the process is still dominated by momentum transport, as shown in Fig. 4.

After a liquid boundary layer is dragged out of the puddle, the final thickness of a ribbon will be reached when the flow velocity of ribbon free surface reaches to that of the ribbon roller surface. Figure 5 shows the calculated final ribbon thickness, assuming a purely momentum transport mechanism. The $t \propto U_w^{-0.8}$ relation obtained experimentally by Hillmann and Hilzinger [15] is excellently matched. The good agreement may be fortuitous, considering the many simplifi-

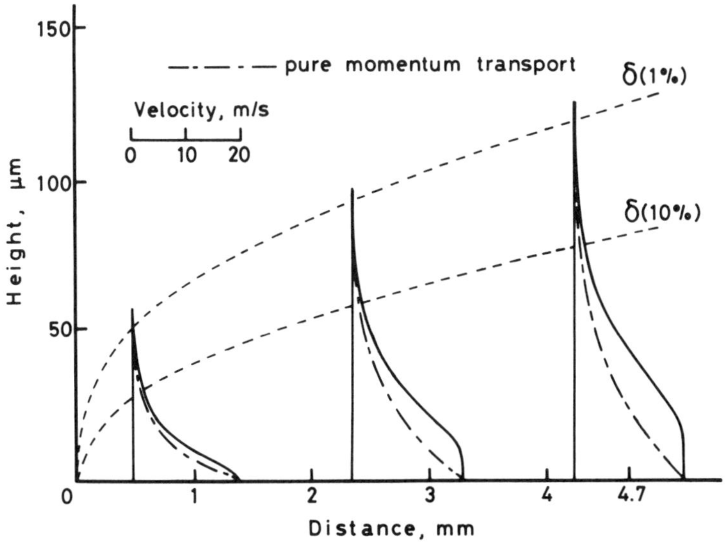

Figure 4 Distribution of fluid velocity within the flow boundary layer for the case of good thermal contact between the liquid alloy and the roller surface.

cations made for calculation; nonetheless, this result supports the importance of momentum transport in ribbon formation of amorphous materials.

In the case of crystalline material, the melt viscosity becomes almost infinitely large when crystallization takes place. Owing to the crystal growth shape and the direction of dendrite growth, the direction of solidification may be deduced in many

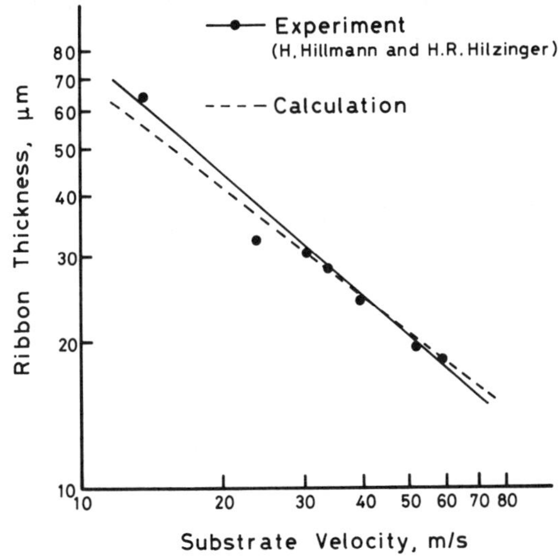

Figure 5 Relationship between the ribbon thickness and substrate velocity for glass-forming alloys. Ribbon thickness is roughly proportional to the −0.8 power of substrate velocity.

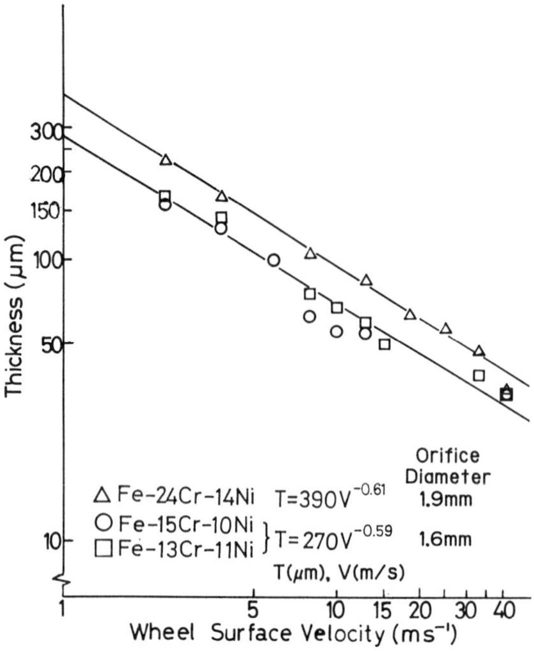

Figure 6 Relationship between the ribbon thickness and substrate velocity for crystalline materials. Ribbon thickness is roughly proportional to the -0.6 power of the substrate velocity.

cases. Such observations revealed that the crystallization mode varies, depending on the alloy species and composition. In most cases ribbons show the development of dendrites growing from the substrate contact to the free surface [16]. This indicates that the solidification proceeds in the same sense. An experimental study [17] of the roller surface velocity dependence on ribbon thickness resulted in the

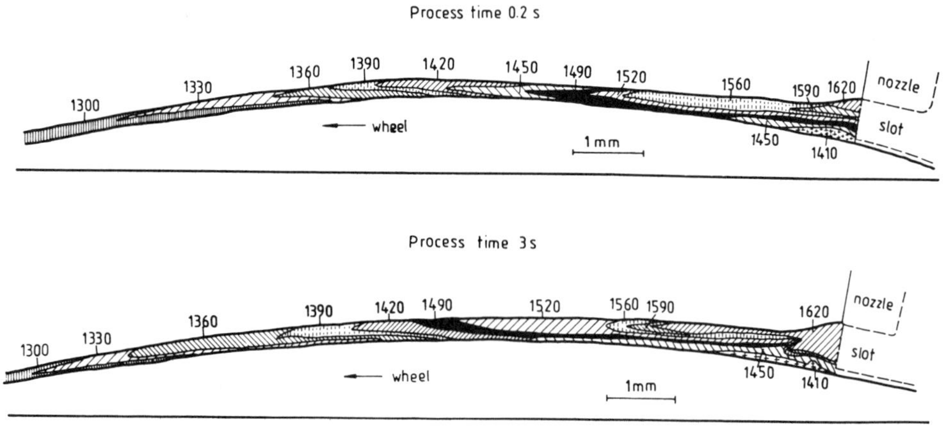

Figure 7 Infrared photograph of ribbon temperature distribution for a high-speed steel ribbon during casting. (Courtesy of H. Fiedler)

relation $t \propto U_w^{-0.6}$, as shown in Fig. 6. This exponent is slightly smaller than that for the case of amorphous alloys. Actual temperature measurement of crystalline materials [18], as shown in Fig. 7, clearly showed that solidification (crystallization) takes place far downstream from the melt puddle for relatively thick ribbon.

Thus, it may be concluded that the momentum transport mechanism is most likely operative both for amorphous and crystalline ribbon formation by the process of single-roller quenching. The momentum transport mechanism, which requires the formation of the ribbon shape while the alloy is in liquid state, may have the advantage of imparting good ribbon surface quality. A thermal transport-dominated process would result in ribbon being formed by the progression of solidification front. In this case, it is usually difficult to attain smooth, flat solidification front due mainly to the dendritic crystal growth.

III. TWIN-ROLLER METHOD

In comparison with the single-roller method, the twin-roller method seems more straight-forward, in principle, to cast thin ribbons from molten metals. Twin-roller casting is an invention of Bessemer, dating back to the late 19th century [19], for the purpose of continuously casting steel slab on a large scale. However, twin-roller casting has not been used as a standard method for the purpose of casting either large-scale steel slabs or rapid solidification of thin ribbons. The major advantage of the twin-roller method for ribbon making lies in the simultaneous heat extraction from both surfaces of the ribbon. As a consequence, the ribbon produced has identical surface conditions on both surfaces, whereas ribbon formed by single-roller methods typically have a shiny free frozen surface and less shiny roller-side surface. While a higher cooling rate might be expected for the twin-roller method in comparison with that of the single-roller method due to the two-sided cooling, many experiments proved this not to be the case. The formation of a puddle (melt pool) is necessary to make good continuous ribbon [20]. Solidification must take place within the narrow range of contact of the two rollers, otherwise the melt would break into globules and a ribbon would not be formed. Thus, the "contact length" of molten alloy against the substrate roller surface is much shorter for twin roller casting than that for single-roller methods, in which the ribbon usually departs from the substrate roller surface far downstream from the melt puddle. Thus, the basic difference between the single- and twin-roller methods is apparent. In the former case, ribbon shape is determined by the propagation of a liquid boundary layer, while in the latter case it is estimated by the nip between the two rollers. The twin-roller method is thus not a process suited for the casting of several tens to 100-micron-thick ribbon, but can be used to cast sheet having a thickness of several millimeters or more. Successful casting of relatively thick strips has been reported by Shibuya et al. [21].

IV. WIRE CASTING

As in the case of chill-block casting processes, the history of direct casting of wires with round cross section also dates back to the end of 19th century [22]. An excellent review on various techniques was given by Ohnaka [23]. Successful commercialization has only been achieved with INROLISP. Before the invention of this process, many attempts were made to cast liquid metals into a flowing coolant. In order to avoid breakdown of the shape of the liquid alloy stream which flows out of the orifice, it is crucially important to ensure laminar flow of the confluent cooling liquid. The essence of INROLISP is the utilization of a "laminar flow" of coolant liquid formed inside a rotating drum. The stream of molten metal should be neither stretched nor compressed by the coolant liquid during casting. Such condition can exist if the condition $|\vec{V}_D| = |\vec{V}_J| \cos \theta$ is satisfied, where $\vec{V}_D$ and $\vec{V}_J$ are, respectively, the velocities of the drum and molten metal jet, and θ is the angle between $\vec{V}_D$ and $\vec{V}_J$. The roundness of wire cross section is better for a small wire diameter. The roundness factor, η, which is the ratio of the minimum to maximum diameter of the wire cross section, obeys an empirical equation: $(1 - \eta) \propto (V_D \sin \theta)^2$. Wires having a cross-section of up to several hundred microns can be made by this technique. The cooling rate of INROLISP is similar or slightly less than that of chill-block melt casting methods. The cooling rate usually is in the range of 10^5 K/s.

INROLISP is particularly effective for the formation of fine amorphous wires because continuous reduction of melt viscosity and no evolution of latent heat during cooling result in good constancy of the wire diameter over extended lengths and allows good surface smoothness. Amorphous wire may be cold-drawn after casting by INROLISP to increase the homogeneity of cross-sectional roundness.

Crystalline wire may also be made by INROLISP. In the case of Fe–6.5 Si alloy, the grain diameter of as-cast wire is much larger than that of a ribbon sample cast by the single roller method [24]. In as-cast Fe–6.5 Si wire of 90-μm diameter, many parts of the wire were found to have a "bamboo" structure which means that single-crystal growth events took place for some lengths of the wire. Using heat flow analysis, Ohnaka concluded that such coarse-grained wire resulted from a solidification front which is convex in shape. INROLISP is not applicable for metals and alloys which do not form an oxide film when liquid is ejected from the orifice. This oxide film apparently reduces the tendency for the melt stream to break down into globules. A small addition of an oxide forming element such as cesium is effecitve in stabilizing the shape of the melt stream.

V. COMPACTION OF AMORPHOUS ALLOYS

In the previous section, processes by which liquid metals and alloys can be quickly formed into ribbon and wire were discussed. The topic to be discussed in this section is founded on the same basic concept as that of the previous section. Amorphous alloys may be regarded as highly viscous liquids since the viscosity can be continually reduced by raising temperature. In the case of rapid quenching from the melts, the molten alloy typically has high fluidity and high surface tension, so that special techniques are necessary to keep the shape of the alloy in either ribbon or wire form until the material becomes sufficiently solid by the lowering temperature. Special techniques must also be used for the forming of amorphous alloys, albeit for different reasons. The fluidity (workabilities) of an amorphous alloy must be increased by raising the temperature so that forming by ordinary mechanical working techniques such as rolling, extruding, or forging can become applicable [25]. While higher temperature lowers amorphous alloy viscosity, it also shortens relaxation time for crystallization. Hence, techniques for rapid heating and quick processing while at elevated temperature must be developed for the compaction of amorphous alloys. Processing should be carried out at temperatures near T_g, the glass transition temperature. The nature of amorphous alloy as a glassy material and the kinetics of crystallization near T_g must hence first be described.

A. Glass Transition Temperature and Viscosity of Amorphous Alloys

Thermodynamically, amorphous alloys may be regarded as being in a glassy state. Glass is defined as an undercooled liquid, the structure of which is frozen at T_g, the glass transition temperature. This glass transition temperature has been defined as that at which the viscosity becomes 10^{13} poise, the value which is generally regarded to divide the fluid and solid states of materials.

Thermal relaxation measured by thermal analysis is one of several evidences which show that amorphous alloys are indeed in a glassy state. One of the characteristics of a glassy state is structural relaxation, which occurs when a sample is held at temperatures close to but below T_g. Experimentally, the total heat which gradually evolves from an amorphous material due to relaxation can be observed as heat absorption when the relaxed sample is reheated above T_g [26]. Such heat absorption has been confirmed for many amorphous alloys and the kinetics of relaxation by heat treatment well explains the reversible "relaxation" and "melting" of amorphous alloys by repeatedly holding below T_g and then heating above T_g several times in a cyclical manner. An example of such a thermal measurement is given in Fig. 8.

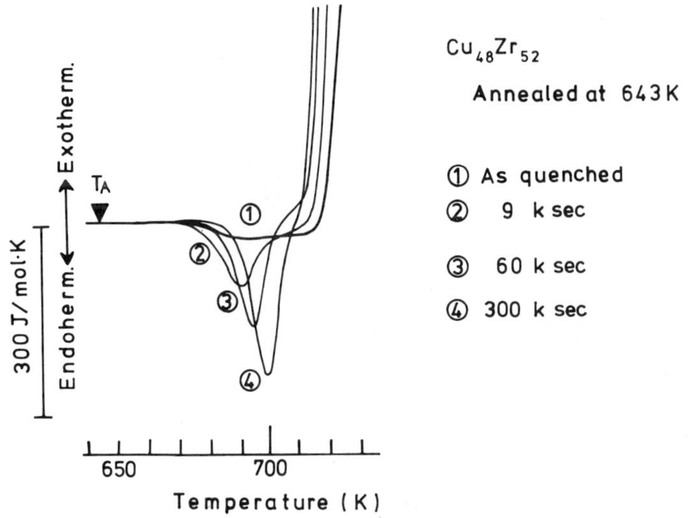

Figure 8 Experimental evidence of heat absorption at T_g upon heating a "relaxed" $Cu_{48}Zr_{52}$ glass samples (held at 643 K). Curve 1: as-quenched; curve 2: holding time of 9 ks; curve 3: holding time of 60 ks; curve 4: holding time of 300 ks. The longer the holding time at T_A, the larger is the amount of heat absorbed at T_g, indicating greater structural relaxation at T_A.

Evidence of "melting" of amorphous alloys at T_g suggests that they should behave as fluids above T_g. Superplastic behavior of many amorphous alloys is commonly observed when mechanically deformed at a temperature near T_g. Figure 9 shows a photo of such highly ductile deformation of an amorphous ribbon sample.

An example of the viscosity measurement of an amorphous alloy is given in Fig. 10. The temperature dependence of viscosity, as given by Eq. (12), is much stronger than that of a common thermally activated process. However, for a narrow temperature interval near T_g, a nearly exponential relation is followed, as shown in Fig. 10 [26].

A temperature-viscosity relation, such as the one shown in Fig. 11, readily gives the temperature range at which the compaction of amorphous alloys should be conducted in order to obtain good density with the use of some common compacting method. It should be noted that the melting temperature of the crystalline phases which are formed from an amorphous alloy usually are far higher than T_g. A sample in the crystalline state is therefore too hard to be compacted by usual compacting method at temperatures near the T_g of an amorphous alloy.

B. Consolidation Process

Once the temperature range at which the compaction process should be performed is determined, the process time allowed must next be estimated from the relaxation time for crystallization of the particular amorphous alloy to be compacted. The relaxation time for crystallization vs holding time, the TTT (time-temperature-transformation) curve, can experimentally be determined from the heating rate

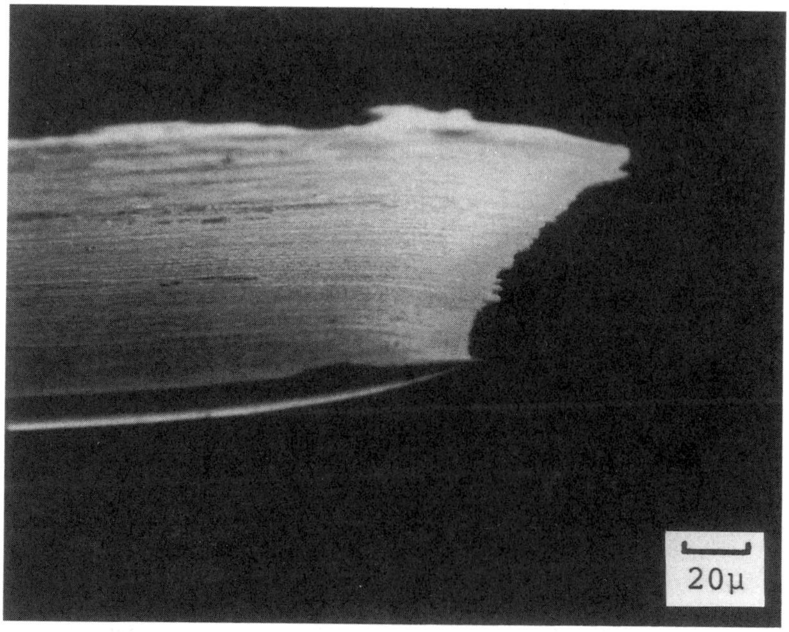

Figure 9 Evidence of the viscous flow at the tip of a creep-broken sample of $Fe_{77}P_{15}C_8$ amorphous sample tested near T_g.

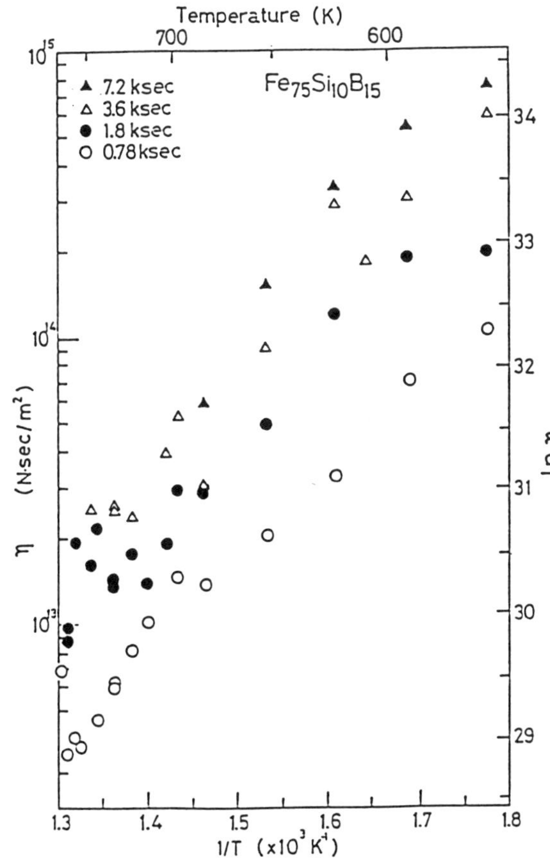

Figure 10 Viscosity values of $Fe_{75}Si_{10}B_{15}$ amorphous samples measured by the creep tests plotted against $1/T$: ▲, holding time of 7.2 ks; △, holding time of 3.6 ks; ●, holding time of 1.8 ks; ○, holding time of 0.78 ks. The viscosity increases with increasing holding time at the test temperature.

dependence of crystallization temperature [26]. Figure 11 shows an example of such a TTT diagram. For this sample, the allowed process time for compaction at about 750 K is in the order of 100 s. An example of the procedure of rapid heating to the processing temperature and quick compaction of amorphous alloy powder, ribbon or wire into bulk shape is shown in Fig. 12.

For samples with sufficiently long relaxation times for crystallization when held at temperature near T_g, quick compaction processing is not necessary. Ordinary warm pressing or extrusion may then readily be applied at near T_g. The good formability, by the utilization of high fluidity characteristic of amorphous alloys, can in such a case be fully utilized. Such process may be regarded, in principle, to be the same as the glass working of oxide glasses. For samples which crystallize at a temperature far below T_g, even a quick processing method is not applicable since such samples do not exhibit the viscosity low enough for forming by ordinary compacting methods. A dynamic compaction process should be utilized for the compaction for such amorphous alloys [27]. Dynamic compaction processes utilize

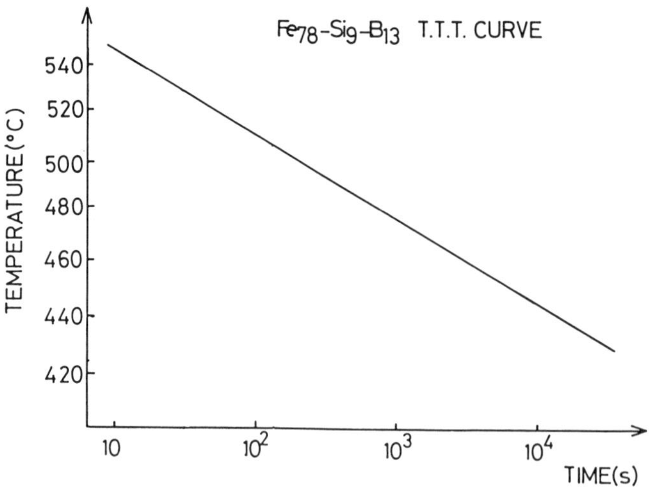

Figure 11 TTT curve for the crystallization of an $Fe_{78}Si_9B_{13}$ amorphous sample obtained from the heating rate dependent of T_x.

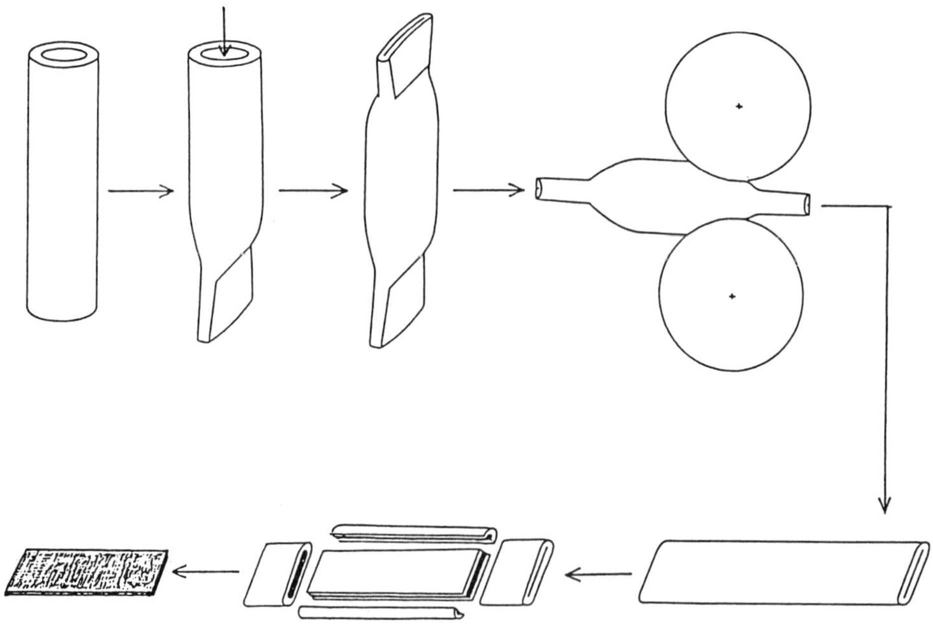

Figure 12 Schematic explanation of the forced powder rolling process for the consolidation of amorphous materials utilizing the high ductility of the metastable liquid state. The sheathed sample is rapidly heated to the desired temperature in a salt bath just before rolling and is subsequently quenched into water.

the extremely high pressure, shear, and adiabatic heating generated by the propagation of a shock wave produced by the explosion of explosives placed on the sample or by the impact of the collision of a projectile which was driven by the explosion. The shock wave pressure may reach several GPa, so that the powders can readily be compacted to full density.

VI. CONCLUSION

Rapid solidification processes to form ribbon and wire have reviewed both for the production of amorphous and crystalline materials. The essence of these processes lies on the forming or definition of ribbon or wire shape while the alloy is in the liquid state, where the fluidity is great enough, and to keep that shape until the temperature is sufficiently lowered to solidify a material in either the amorphous or crystalline state. The continuous increase of viscosity with lowering of temperature and no latent heat evolution, which are characteristic of amorphous alloys, make these materials particularly suited for the preservation of good shape definition and retention of the smooth surface condition of liquid to the solid state.

In the case of ribbon formation by the single-roller method, the formation of ribbon takes placed by momentum boundary layer propagation on the rotating wheel surface. This boundary layer is extracted from the melt puddle, which is formed at the position of molten alloy supply on the substrate roller, to solidify, keeping the film which separates from the roller far downstream.

In the case of wire formation, a method of utilizing laminar flow of liquid coolant formed inside the wall of a rotating drum ensures the maintenance of the shape of a melt stream ejected from an orifice until the sample is solidified into wire.

The compaction of amorphous alloy is performed, in principle, the same way as is the formation of ribbon or wire, stated above. The essence here, again lies in the shaping of the material while in the "liquid" state. "Liquid" state in this case means the highly ductile state characteristic of amorphous alloys near T_g. For amorphous alloys which exhibit long relaxation times for crystallization, an ordinary warm pressing is enough for compaction. For samples with shorter relaxation times for crystallization, a quick heating and quick processing method must be utilized where in a quantitative estimation of process temperature and process time becomes necessary.

REFERENCES

1. H. Jones, *Rapid Solidification of Metals and Alloys* (The Institution of Metallurgists, London, 1982).
2. R. Pond, Jr., and R. Maddin, *Trans. AIME*, *245*, 2475 (1969).
3. T. Masumoto and R. Maddin, *Acta Metall.*, *19*, 725 (1971).
4. R. E. Maringer and C. E. Molbey, in *Rapidly Quenched Metals III*, edited by B. Cantor (The Metals Society, London, 1978), p. 46.
5. G. F. Taylor, *Phys. Rev.*, *23*, 655 (1924).
6. I. Ohnaka and T. Fukusako, *J. Japan Inst. Metall.*, *42*, 1095 (1982).
7. F. H. Daniels, U.S. Patent No. 359, 348 (1887).
8. S. Kavesh, in *Metallic Glasses*, edited by J. J. Gilman (ASM, OH, 1978), p. 36.

9. H. H. Liebermann and R. L. Bye, in *Rapid Solidified Crystalline Alloys*, edited by S. K. Das, B. H. Kear, and C. M. Adam (TMS-AIME, 1985), p. 61.
10. H. Schlichting, *Boundary Layer Theory*, 6th ed. (McGraw-Hill, New York, 1968), p. 265.
11. K. Takeshita and P. H. Shingu, *Trans. Japan Inst. Metall.*, *24*, 529 (1983).
12. K. Takeshita and P. H. Shingu, *Trans. Japan Inst. Metall.*, *27*, 141 (1986).
13. H. Vogel, *Z. Phys.*, *22*, 645 (1921).
14. G. S. Fulcher, *J. Amer. Ceram. Soc.*, *6*, 339 (1925).
15. H. Hillmann and H. R. Hilzinger, in Ref. 4, p. 35
16. H. H. Liebermann, *J. Cryst. Growth*, *70*, 497 (1984).
17. M. Umemoto, Toyohashi University of Technology, private communication, 1988.
18. G. Stephani, H. Mühlbach, H. Fiedler, and G. Richter, *Mater. Sci. Eng.*, *98*, 29 (1988).
19. H. Bessemer, *Stahl Eisen*, *11*, 921 (1891).
20. K. Miyazawa and J. Szekely, *Metall. Trans.*, *12A*, 1047 (1981).
21. K. Shibuya, F. Kogiku, M. Yukumoto, S. Miyake, M. Ogawa, and T. Kan, *Mater. Sci. Eng.*, *98*, 25 (1988).
22. E. Small, U.S. Patent No. 262, 625 (1882).
23. I. Ohnaka, *Int. J. Rapid Solidification*, *1*, 219 (1984–5).
24. I. Ohnaka, in *Rapidly Solidified Materials*, edited by P. W. Lee and R. S. Carbonara (ASM, 1985), p. 91.
25. H. H. Liebermann, *Mater. Sci. Eng.*, *46*, 241 (1980).
26. P. H. Shingu, *Mater. Sci. Eng.*, *97*, 107 (1988).
27. T. Negishi, T. Ogura, T. Masumoto, T. Goto, K. Fukuoka, and H. Ishii, *J. Mater. Sci.*, *20*, 399 (1985).

5

Rapid Solidification Processing of Composites

E. R. Murray* and D. G. Ast
Cornell University, Ithaca, New York

I. INTRODUCTION

Rapid solidification technology (RST) holds much promise for the production of composite materials. The time for which the different phases of the composite are in contact at high temperature is limited to milliseconds. Combinations of materials can, therefore, be used, which would react or degrade when held at these temperatures for the length of time required to form composites by conventional techniques. Rapid quenching to room temperature also allows the production of matrix alloys with nonequilibrium concentrations of elements. Subsequent segregation may improve matrix properties [1] or improve the integrity of the bond at the matrix/reinforcement interface [2]. Additionally, RST is a commercially viable method for producing amorphous alloys in large volumes.

In this discussion of rapidly solidified composite materials, both the so-called in situ composites, and those composites which consist of a rapidly solidified matrix or reinforcement, which is then formed into a composite material using traditional processes, are excluded from consideration. For example, rapidly solidified Al-Li-Mg-Cu-Zr alloys may be aged to produce cored, or "composite," precipitates [3]. While these precipitates, which consist of Al_3Li around a Zr-rich core, may be composite precipitates, the material is not what is usually considered a composite material. Similarly, rapidly solidified Ti-Al ribbon may be reinforced with carbon fibers [4] by stacking alternating layers of the Ti-Al ribbons and the carbon fibers, and subsequently pressing the stack in an evacuated furnace to cause the metal to flow superplastically around the fibers. Although the matrix has been rapidly solidified, the composite has not. Thus, our examination is restricted to those materials with introduced dissimilar phases which are present at the time of rapid solidification.

**Present affiliation*: Haverford College, Haverford, Pennsylvania.

Defined this way, materials are possible with all the advantages listed above. The work that has been done with these materials falls into two major categories. First, there are the laminated rapidly solidified composites, which are comprised of bonded dissimilar rapidly solidified metallic ribbons. Second, there are the reinforced rapidly solidified composites, in which some reinforcing medium is embedded in the rapidly solidified metallic ribbon. (Reinforcement is simply a convenient term—it may be that the purpose of the included material is not to reinforce, but, for example, to alter the wear characteristics.) There are two classes of reinforcing media—discrete or particulate; and continuous, or filamentary.

II. LAMINATED RAPIDLY SOLIDIFIED COMPOSITES

One type of rapidly solidified (RS) composite is the laminates. These composites consist of several layers of rapidly solidified metal which bond together in the casting process. Thus, the advantages of rapidly solidified alloys are combined with the advantages of a bimetallic (or even multilayered) form. Casting the composite as a whole has advantages over casting each alloy separately and subsequently processing them into a composite. For example, composites containing amorphous alloys that crystallize at temperatures below those used in traditional composite forming processes could be produced.

The first laminated rapidly solidified composite [5] was produced with a double-roller rapid solidification apparatus in which the second layer of metal was ejected upon the first (Fig. 1). It was thought that the thermal contact between the ribbon and a single roller would be insufficient to allow the requisite thermal conduction necessary for producing an amorphous alloy ribbon. The distance from each crucible of molten alloy to the nip of the rollers was selected so that the first alloy ribbon

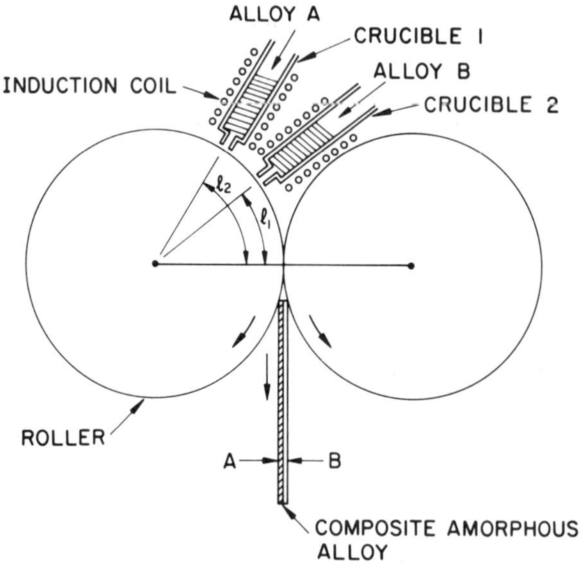

Figure 1 Double-roller apparatus for production of laminated RS composite ribbon [5]. (Courtesy S. Shimanuki)

was solid before the second alloy melt was cast onto it, and both ribbons were solid before they reached the nip of the rollers. Experiments were performed in which a single ribbon was cast to find the distance (and thus, for a given roller speed, the time) required for each alloy to be solidified. The alloys were still very hot when the nip of the rollers was reached, so good bonding occurred between the ribbons. It was necessary to take care, however, not to place the crucibles so close to the roller nip that the ribbons were very near their melting point when it was reached. Failure to take this precaution would result in a boundary between the alloy layers which was not sharp.

Several combinations of laminated ribbons were produced with this apparatus [6]: a narrow $Pd_{80}Si_{20}$ ribbon was bonded upon a wider $Ni_{59}Fe_{19}Si_8B_{14}$ ribbon, a $Co_{75}Si_{10}B_{15}$ ribbon was bonded upon an $Ni_{75}Si_{10}B_{15}$ ribbon, and a $Pd_{83.5}Si_{16.5}$ ribbon was bonded upon an $Fe_{75}Si_{10}B_{15}$ ribbon. Both layers were amorphous in these cases, as determined by x-ray diffractometry. Differential thermal analysis (DTA) showed that the crystallization temperatures were the same as those of the individual alloys, which suggests that no significant amount of a third alloy had formed. X-ray microanalysis was used to determine that the boundary between the alloy ribbons was sharp. In bending tests, these laminates could be bent 180° without separation. Another laminate, Permalloy $Fe_{55}Ni_{45}$ upon $Pd_{80}Si_{20}$ also had a sharply defined boundary between the constituent layers and could be bent 180°. This laminate, however, had a crystalline layer (the Permalloy) bonded to an amorphous layer. While an attempt was made to bond $Fe_{55}Ni_{45}$ upon $Co_{75}Si_{10}B_{15}$, the layers of this laminate had a sharply defined boundary but separated upon bending to 180°.

It is, however, possible for composites to be rapidly solidified on a single chill roller [7]. Both amorphous and amorphous/crystalline composites were produced using a single chill roller, much the same as the double-roller method examined above, except that the alloys were allowed to bond without the aid of the mechanical pressing of the second chill roller (Fig. 2). This was accomplished by selecting the roller surface speed and the distance between crucibles such that the first alloy was

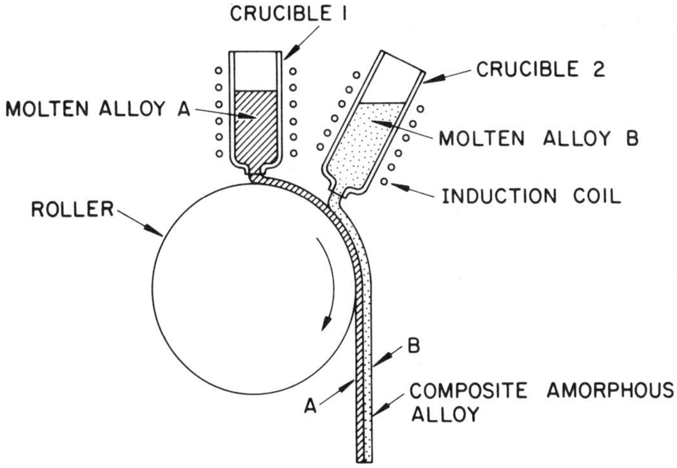

Figure 2 Single roller apparatus for production of laminated RS composite ribbon [7]. (Courtesy S. Shimanuki)

still a liquid when the second alloy was ejected onto it. Thus, the first alloy must be a highly viscous liquid at the time of second alloy deposition so that mixing between the two layers will be minimized. The technique was demonstrated in an entirely amorphous system by casting a narrow $Pd_{80}Si_{20}$ ribbon upon a wider $(Ni_{0.75}Fe_{0.25})_{78}Si_8B_{14}$ ribbon. A wide $(Ni_{0.75}Fe_{0.25})_{78}Si_8B_{14}$ ribbon was cast onto a narrow phosphor-bronze alloy ribbon to demonstrate this technique with an amorphous/crystalline system. As before, DTA, x-ray diffraction, x-ray profiling, and backscattered electron imaging were used to demonstrate that the alloys were not mixed at the interface (i.e., the boundary between the constituent ribbons was sharp) despite the fact that the boundary was mechanically strong. It was found that the parameter most important to producing a uniform composite ribbon was the temperatures of the melts, and thus the relationship between the roller speed and the distance between crucibles. Other parameters, such as the standoff distance of the crucible nozzles from the chill roller were also important, but less so. If, for example, the nozzles were too close to the roller (<0.1 mm), the flow of the alloy ejected upstream disturbed that of the alloy ejected downstream, while if they were too far (>0.5 mm), the surface of the composite was rough.

A modification of this technique [6] in which the molten alloys were ejected onto a Be-Cu belt that ran over two rollers (Fig. 3) was used to prepare a laminate of an $Fe_{75}Si_{10}B_{15}$ ribbon upon a $Pd_{83.5}Si_{16.5}$ ribbon. This modification is especially useful if one does not want to introduce a curvature into the composite due to the shape of the roller. Both layers of metal remained amorphous in this experiment, and the boundary between them was sharp. Some separation occurred in a 180° bending test, but none was observed when the laminate was wound around a 5-mm rod.

An interesting application of the single-chill-surface technique [8] has been used to solve the problem of brittleness in melt-spun semiconducting material. Melt-spinning semiconductors has been considered a low-cost route to photovoltaics, but in practice the semiconducting ribbon produced has been too brittle and of too small a grain size for practical applications. However, the problem of brittleness has been circumvented by casting a rapidly solidified laminate consisting of a metal bottom layer and a semiconducting top layer. The technique was demonstrated by casting InSb on top of amorphous $Cu_{50}Zr_{50}$, on the inside of a chill drum (Fig. 4).

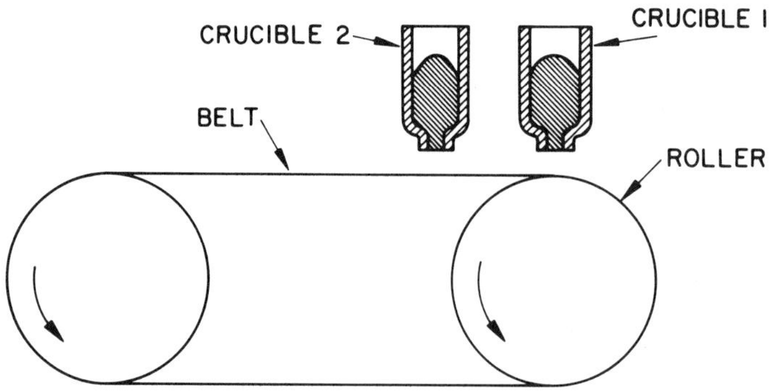

Figure 3 Flat belt modification of single-roller apparatus for production of laminated RS composite ribbon [6]. (Courtesy S. Shimanuki)

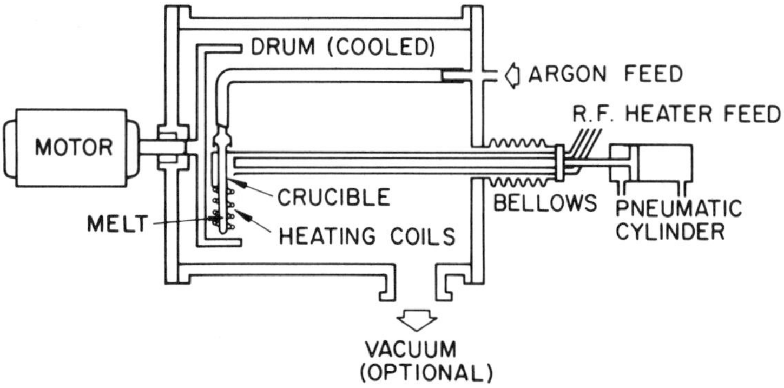

Figure 4 Drum interior apparatus for production of laminated RS composite ribbon [8]. (Courtesy R. V. Raman)

While the boundary was sharp, there were some entrapped gasses at the interface, as the process was not carried out in vacuum.

Another study [9] has been done involving both laminated composites and "incremental casting" of layers of the same alloy, which, although it does not produce a composite, points out some issues important to the production of laminated composites. The goal in these experiments was to produce an amorphous ribbon much thicker than previously possible, by casting up to five ribbons upon one another. When all the ribbons were of the same material, a single crucible with several nozzles, only about 3 mm apart, was used. When casting a crystalline Ni-based superalloy, epitaxy was found between the grains in each layer cast. Casting a two-layer amorphous $Ni_{40}Fe_{40}B_{20}$ ribbon was done so successfully that no boundary could be observed between the individual layers. When casting a four-layer ribbon from a single nozzle, letting the ribbon wrap itself around the chill roller for additional layers to be added, poor bonding was found between the third and fourth layers, and some crystallization was found at the interfaces. These layers were each quite thick ($\approx$50 μm) and thus contained too much heat to be conducted away in the short time necessary to prevent crystallization. Magnetic coercivity measurements revealed that some degree of crystallization had occurred, even in ribbons that had no observable crystallites. This is not a problem unique to multiple layers of the same alloy, but must also be considered when attempting to cast a multilayered composite. Finally, an experiment was performed which demonstrated that the top of a ribbon can be deformed by the next ribbon cast onto it. A narrow stream of $Fe_{81.5}B_{14.5}Si_4$ was cast onto a much wider ribbon, and the result was not a thicker piece where the narrow stream was directed, but, instead, a groove due to its deformation of the first ribbon (Fig. 5).

An interesting variation on this technique [10] can be used to create a composite in which the individual ribbons are not bonded face to face, but edge to edge. In this scheme, the crucibles do not cast each subsequent ribbon on top of the previous one, but, rather, to the side of the previous one (Fig. 6). Bonding is at the edge contact. The boundary zone between the ribbons is not necessarily sharp, and, from hardness data, may be as much as 1 mm wide. This technique may be used

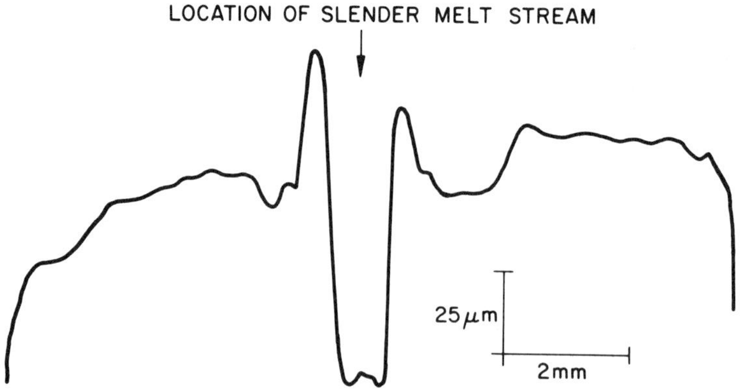

Figure 5 Transverse surface profile of $Fe_{81.5}B_{14.5}Si_4$ laminated RS composite ribbon made by ejecting a narrow melt stream upon a ribbon formed from a wider melt stream [9]. (Courtesy H. H. Liebermann)

to make wide ribbons of the same material, or composites, limited only by the width of the chill surface.

Spray forming [11] offers another route to laminated rapidly solidified composites. It is a method similar to that used for the production of rapidly solidified alloy powders, except that the trajectory of the metal droplets is interrupted by a substrate to form a cohesive deposit as they solidify, instead of cooling to a solid before they are collected (Fig. 7). It would be very easy to spray deposit additional layers onto a strip formed by a spray-rolling technique. As each particle undergoes rapid solidification upon striking the layer before it (or the substrate, if it is the first layer), there should be no alloying at the interface, and a sharp boundary should be the result.

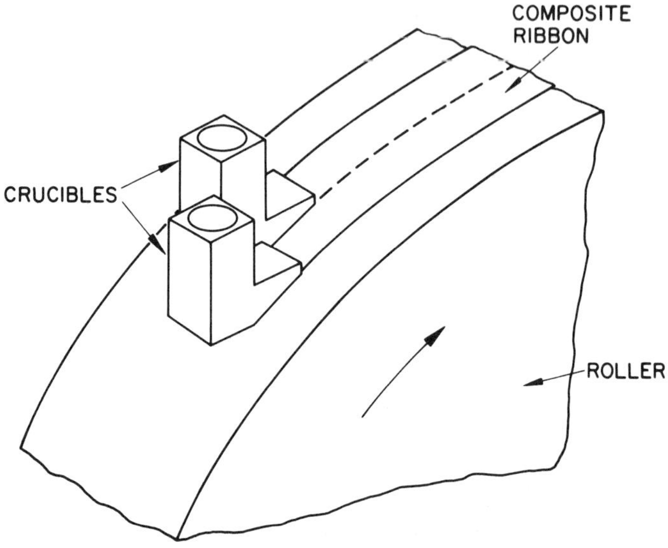

Figure 6 Apparatus for producing edge-bonded RS composite ribbon [10].

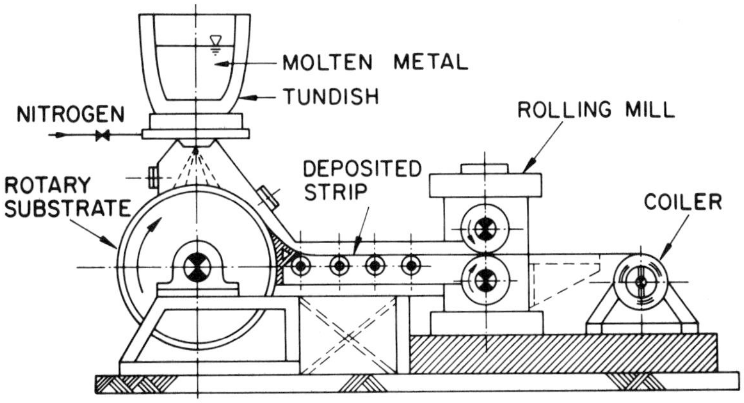

Figure 7 Apparatus for spray-forming a metallic strip [11]. (Courtesy A. R. E. Singer)

III. DISCRETELY REINFORCED RAPIDLY SOLIDIFIED COMPOSITES

There are two major methods for the preparation of particle-reinforced composite materials. One may place the particles in the melt while it is held in the crucible, where they will be distributed by either the stirring action of the induction heater, or by a stirring apparatus, before ejection onto the chill roller. Alternatively, one may inject the particles into the melt puddle while it is still liquid on the chill roller. In this case the particles are distributed by the flow of the liquid matrix material in the puddle and by the continuous addition of particles as the ribbon is produced. A third method, in which the particles are injected into the melt in the crucible just before it is ejected onto the chill roller, may be viewed as a modification of either of the two methods above, depending on whether one is concerned with where the particles are added or when they are added.

The simplest and, therefore, the first method [12] used to prepare particle-reinforced rapidly solidified composite materials is simply to add reinforcing particles to the melt before it is ejected onto the chill roller (Fig. 8). Although the particles may be dispersed through the melt by means of an agitator, the particles may be kept dispersed by the stirring action of the induction coils used to heat the melt, if the particles' density is not too different from that of the melt.

In practice, using MoB_2 particles and an $Fe_{40}Ni_{40}B_{20}$ matrix, it was found that many particles in the ribbon protruded from the top surface, while none were found at the surface that had been in contact with the chill roller, although the particles were well dispersed in the melt. This phenomenon was later determined [13] to be a result of turbulence in the melt puddle. Such a particle distribution was considered an advantage, as it made it possible for the strip to be used as an abrasive tool, the hard particles being firmly embedded in the matrix despite their protrusion. It is interesting that the particles, in this case, were formed in the melt by adding Mo and B in the proper stoichiometric ratio at 1773 K, the temperature at which MoB_2 forms but remains dissolved, and reducing the temperature to 1383 K, causing the MoB_2 to precipitate out in the form of particles. The particle size could be controlled by adjusting the cooling rate. Rapid quenching from 1383 K resulted in a thin

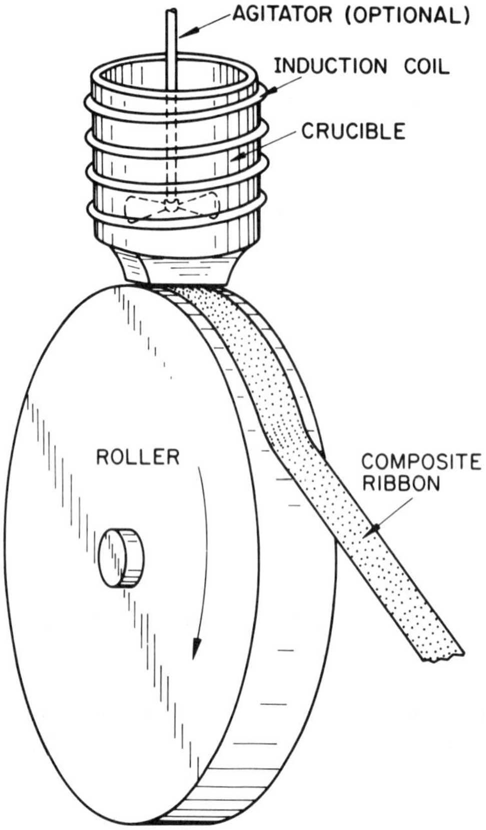

Figure 8 Apparatus for producing a discretely reinforced RS composite ribbon by adding particles to the melt [12].

$Fe_{40}Ni_{40}B_{20}$ ribbon with MoB_2 particles evenly distributed and firmly embedded in one surface. Some requirements should be noted for this sort of scheme to work with other alloy systems. Wetting of the particles by the melt is not necessary, but facilitates processing. The particles ought not react with the melt at the processing temperature. Obviously, the particles must not be of a lower melting temperature than the melt, which may be a severe restriction.

Much work has been done with WC as the reinforcing particles, as they are hard and have a very high melting point (3143 K). The effect of adding 4–5 μm WC particles to molten $Ni_{78}Si_{10}B_{12}$ on the tensile yield stress of the resulting ribbon has been investigated [14]. For ribbons containing about 7% WC particles by volume, it was found that these particles were distributed uniformly throughout the matrix. Some crystallization of the matrix was induced by the particles, as a crystalline metastable phase was observed at the interface between the matrix and some of the particles.

The presence of WC particles in ribbons was observed to improve the mechanical properties over those of an unreinforced $Ni_{78}Si_{10}B_{12}$ ribbon. The reinforced ribbon could be bent 180° without fracturing and the yield stress was 1.4 times that of the unreinforced metal (280 kg/mm^2 compared to 203 kg/mm^2). This is near the theoretical upper bound of a predicted increase, based on an assumption of approximate proportionality between the Young's modulus and the yield stress of an amorphous material, and the linear rule of mixtures which assumes equal strain in both the matrix and the reinforcement phases. A lower bound is set by a model which assumes equal stress in both the matrix and reinforcement. Thus, a small volume fraction of particles can make a large difference between the yield stress of the composite and that of the unreinforced matrix material.

The upper bound for the expected Young's modulus of a composite is found by assuming that the reinforcement and the matrix are parallel elements subject to the same strain [15]. The result of this assumption is that

$$E_c = E_m(1 - V_r) + E_r V_r \tag{1}$$

where E_c, E_m, and E_r are the Young's moduli of the composite, the matrix material, and the reinforcing material, respectively, and V_r is the volume fraction of the reinforcement. This is often called the linear rule of mixtures. A lower bound is calculated by assuming that the reinforcement and the matrix are serial elements, and thus are subject to the same stress. This assumption leads to the relationship

$$E_c = \frac{1}{(1 - V_r)/E_m + V_r/E_r} \tag{2}$$

which is appropriate when the reinforcement is poorly bonded to the matrix material, while the assumption of equal strain is more appropriate for a well-bonded composite.

The yield stress of an amorphous alloy has been observed [16] to be roughly proportional to the Young's modulus, which may lead one to expect that the yield stress of a reinforced amorphous alloy may be predicted from the Young's modulus derived via the linear rule of mixtures. Although both the Young's modulus and the yield stress of these composites appear to follow the linear rule of mixtures, theory [17] indicates that, in the case of the yield stress, this is simply fortuitous. This theory also indicates that the linear rule of mixtures is obeyed by Young's modulus only when it is determined at low stresses, due to the fact that the length and diameter of a particle are comparable. An elementary analysis shows that efficient stress transfer from the matrix to an element of reinforcement requires this element to have a length-to-diameter ratio that exceeds some critical value, as some finite length is required for the full transfer of stress if the maximum shear stress of the matrix is not to be exceeded. The critical aspect ratio is equal to the fracture strength of the reinforcement element divided by twice the maximum shear stress of the matrix. Typically, the value of this ratio is about 10.

The particles reinforcing these rapidly solidified materials have an aspect ratio of about unity, so the full stress on the matrix cannot be transferred to the reinforcement through the interface. It may be expected that the linear rule of mixtures will be obeyed by the Young's modulus only up to the low stress that may be transferred without exceeding the maximum shear stress of the matrix. An improvement of the yield stress brought about by the presence of particles should not be expected at all. The observed improvement must, then, be brought about by some mechanism other than the transfer of stress from matrix to reinforcement.

This mechanism may be that of work hardening by geometrically necessary dislocations [18] introduced to accommodate the differences in deformation behavior between the matrix and the reinforcing material. The additional stress required to generate these dislocations increases the yield stress of the composite. There is, however, no reason to expect this increased yield stress to obey the linear rule of mixtures. Although this theory was developed for crystalline matrices, it can be legitimately applied to amorphous matrices as well [19], as the deformation takes place locally in a shear band that is narrow compared to the particle diameter.

This empirical effect was verified in studies of the effect of volume fraction of reinforcing 4–5 μm WC particles on the mechanical properties of $Ni_{78}Si_{10}B_{12}$ [15,20]. It was shown that both the yield stress and the Young's modulus increase essentially according to the upper bound set by the linear rule of mixtures, up to 18.2% (by volume) of particles, which was the limit of the study. At this point, the Young's modulus (as determined by an ultrasonic method) was 2.2 times that of the unreinforced alloy (Fig. 9). An implication of this behavior of the Young's modulus is that there is excellent bonding between the particles and the matrix, and thus complete transference of stress and strain between them without debonding at the low stress levels produced by the ultrasonic method used in the study. This is not surprising, as WC is wetting by liquid Ni, the primary component of the melt.

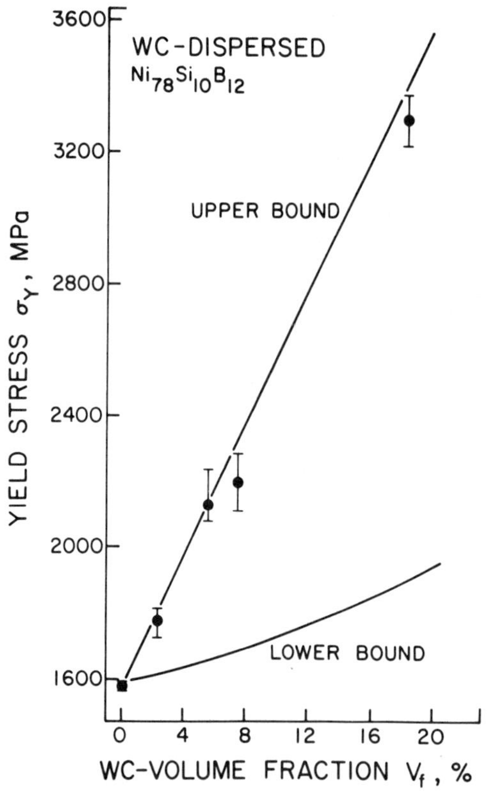

Figure 9 Yield stress for an amorphous $Ni_{78}Si_{10}B_{12}$ composite as a function of volume fraction 4–5 μm WC particles [15]. (Courtesy H. Kimura)

Examination of the slip bands in specimens plastically deformed in bending tests revealed that, up to a particle volume fraction of about 7%, the bands are straight and bypass the particles. At higher volume fractions the slip bands appear more complex—they occur in sets near the particles, and are of many different orientations. These observations are explained by a bypass mechanism, in which plastic deformation occurs solely in the matrix.

The effect of much smaller (0.6 μm) WC particles on Ni-based amorphous alloys has also been investigated [21]. Ribbons were prepared by, again, adding WC particles to the molten $Ni_{75}Si_8B_{17}$ and $Ni_{78}Si_{10}B_{12}$ in volume fractions up to 15%. These particles were uniformly distributed and had the same effect on the Young's modulus and the yield stress as the larger particles had, namely, that the mechanical properties were improved according to the linear rule of mixtures. The small size of the particles improved the composite's ductility, however, increasing the volume fraction of WC at which the ribbons may be bent back 180° without decohesion from 6% to 10%.

Additionally, these smaller particles were used to reinforce both amorphous $Ni_{75}Si_8B_{17}$ spheres and $(Ni_{0.78}P_{0.12}B_{0.10})_{98}Al_2$ wires. The spheres were produced by ejecting the melt into a rotating water bath in which the parameters were controlled so that the flow was laminar, which produces spheres with a median diameter larger than that of the nozzle hole (Fig. 10). In this case an 80-μm orifice was used to produce spheres having a median diameter of 174 μm and containing 4% WC by volume. For comparison, unreinforced spheres had a median diameter of 98 μm. The fact that the median particle sizes are larger than the nozzle hole is evidence that the flow was, indeed, laminar, as turbulent flow results in particles smaller than the hole diameter. Additionally, it should be noted that the particle size distributions were normal in both cases, although the distribution was wider in the case of the reinforced material. This difference may be explained as an increased

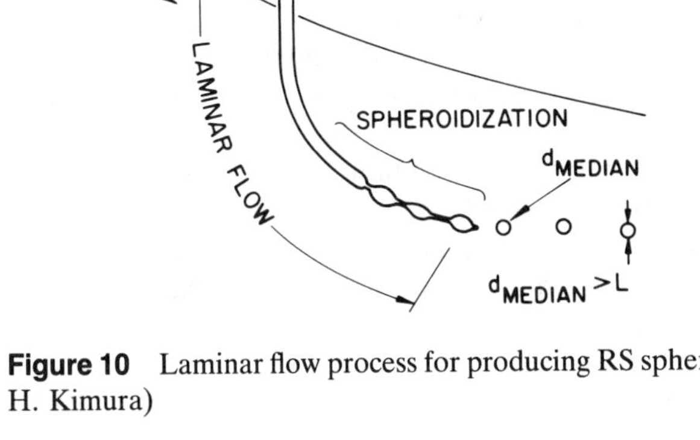

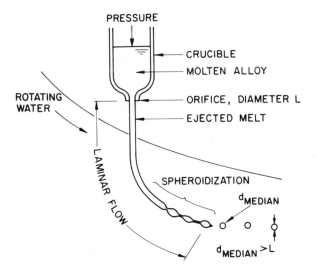

Figure 10 Laminar flow process for producing RS spheres in rotating water [21]. (Courtesy H. Kimura)

kinematic viscosity of the melt caused by the presence of particles, and thus is not necessarily a result of a difference in the process of solidification.

The reinforced wires were cast by conventional in-rotating-water spinning, at up to 4% WC by volume. As was the case with the reinforced ribbons, the Young's modulus was improved and the yield stress was increased in accordance with the linear rule of mixtures.

Other properties besides the mechanical can be affected by the inclusion of particles in an amorphous matrix. The effect on the magnetic properties of the magnetic tape head material $Co_{70.5}Fe_{4.5}Si_{10}B_{15}$ were examined [22] by the addition of up to 2% 2–3 μm WC particles by volume. Note that WC is well wetted by liquid Co. Unreinforced alloys were made under exactly the same conditions for comparison, on a single chill roller. No reaction occurred between the particles and the melt, and no other phases were found at the particle/matrix interface. Both the apparent saturation magnetic induction and the Curie temperature showed little change from that of the unreinforced material, certainly no more than would be predicted by considering that the matrix volume fraction has been reduced from 100% to 98%. Additionally, comparing the reinforced alloy to an alloy in which WC has been dissolved (which causes a significant reduction in the Curie temperature) demonstrated that no WC had dissolved in the matrix. The effective permeability was also investigated both as a function of annealing temperature at 10 mOe and 1 kHz and as a function of frequency at 10 mOe after annealing for 5 min at 723 K. Both the reinforced and the unreinforced materials showed the same effective permeability after having been annealed at various temperatures—the effect was that it was reduced except for annealing at temperatures just above the Curie temperature. The reinforced material, however, showed a 20–40% improvement in effective permeability at frequencies greater than 10 kHz compared to the unreinforced samples, each of which had been annealed for 5 min at 723 K (which is just above the Curie temperature) (Fig. 11). This suggests that the WC particles may reduce domain size, and thus also reduce eddy current losses during high-frequency magnetization.

Oxide particles of Cr_2O_3 and ZrO_2 are not well wetted by $Co_{70.5}Fe_{4.5}Si_{10}B_{15}$ alloy, but composite ribbons were produced [23] anyway by adding the particles, not to the melt, but to the melt stream of a conventionally cooled molten ingot as it was poured. This ingot was then remelted in a crucible for the melt spinning of a rapidly solidified composite on a single chill roller. The particles dissolved to some extent in the alloy during the conventional cooling. Thus, the particles included in the ingot were very fine. In this experiment, the Cr_2O_3 particles were reduced in diameter from 1 μm to an average of 61.8 nm, while the ZrO_2 particles went from 1 μm to an average of 56.2 nm due to the outer portions being dissolved. There was, however, no detectable size difference between the particles in the crystalline ingot and those in the melt-spun amorphous ribbon, indicating that no further dissolution took place during the remelting. The volume fraction of particles in these experiments was very low: only 0.76% for Cr_2O_3 particles or 0.52% for ZrO_3 particles. This technique succeeded in producing an amorphous ribbon which included very fine nonwetted particles, which would have floated and agglomerated if simply added to the melt before melt spinning.

Not only does the inclusion of particles affect the properties of the amorphous matrix of the composite, but it may also affect the crystallization process of the

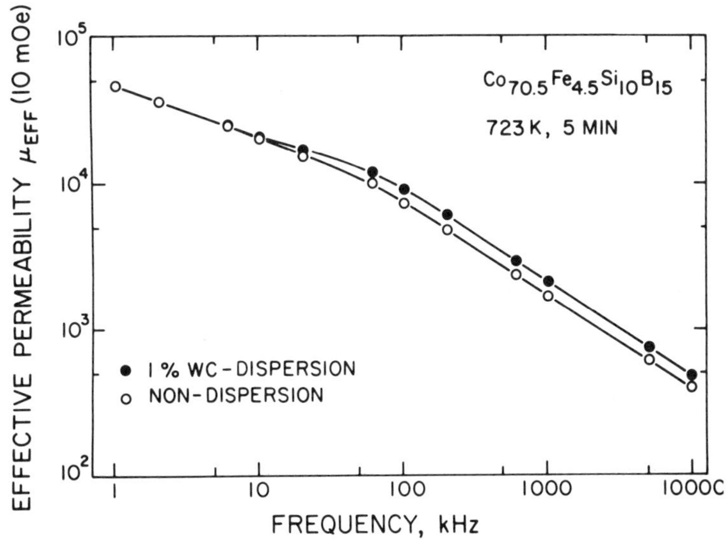

Figure 11 Effective permeability as a function of frequency for a $Co_{70.5}Fe_{4.5}Si_{10}B_{15}$ ribbon reinforced with 1% WC particles by volume, compared with an unreinforced $Co_{70.5}Fe_{4.5}Si_{10}B_{15}$ ribbon, both annealed for 5 min at 723 K [22]. (Courtesy H. Kimura)

matrix [24], which could have important technological ramifications. The effect of particles on the crystallization process has been investigated by adding 1-μm-sized WC to melt-spun $Pd_{77.5}Cu_6Si_{16.5}$. The WC was kept dispersed by induction stirring. The resulting composites ranged from 2% to 12% in volume fraction of WC. Differential scanning calorimetry, transmission electron microscopy, and x-ray diffractometry studies led to the conclusions that the activation energy for crystallization was reduced with increasing volume fraction of WC, that crystallization proceeds more rapidly in reinforced composites than in the unreinforced counterpart, and that the presence of WC particles may alter the crystallization mechanism from one of nucleation and growth to one of growth of embryos containing a WC particle.

Graphite particles have been embedded in a melt-spun LM-13 Al-Si alloy [25] by adding particles to the melt. This composite was investigated as it may have desirable wear characteristics. The volume fraction of graphite was not reported, but it was observed that the inclusion of the graphite did not affect the microstructure. Comparison with Al-Si material produced in the same manner, but without the graphite particles, showed that there was no difference in either the spacing of the dendritic α-Al arms or the size and shape of precipitated Si particles.

The other method of introducing particles into a rapidly solidified ribbon, injecting them into the melt puddle after the metal is ejected from the crucible is not quite so simple as adding them to the melt in the crucible and, therefore, has not been practiced for as long. This method [26,27] consists of adding the particles to a stream of gas flowing between the nozzle and the chill roller, thus injecting them into the melt just before solidification (Fig. 12). The flowing gas has the added effect of stabilizing the quenching of the metal, as the turbulence in the

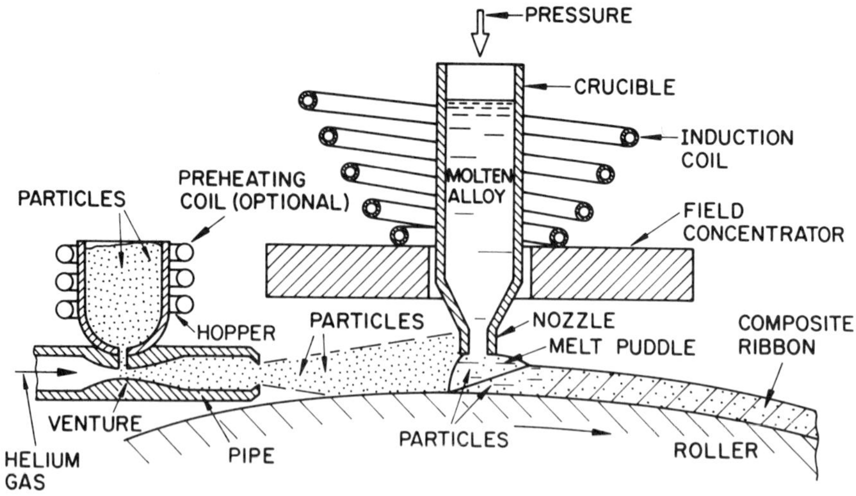

Figure 12 Apparatus for producing a discretely reinforced RS composite ribbon by blowing particles into the melt puddle [27].

boundary layer at the roller's surface is reduced. This technique was demonstrated using He gas to embed WC, TiC, and Mo particles in amorphous $Ni_{80}Si_{14}B_6$, WC and TiC particles in amorphous $Ni_{75}Si_8B_{17}$, and WC in polycrystalline Sn ribbons. It was discovered that the volume fraction of embedded particles increased linearly with the pressure of the gas stream, once a certain critical pressure had been reached. The results of these demonstrations were ribbons with up to 7% particles by volume homogeneously distributed throughout. Attempts to increase the volume fraction of particles beyond this point by further increasing the gas stream pressure resulted in a lower volume fraction and a poor ribbon surface quality, due to gas pockets formed at the ribbon/chill roller interface during melt spinning. It should be noted that homogeneity can be improved by preheating the particles to prevent them from forming clumps in the ribbon.

Particles introduced into a rapidly solidified ribbon in this fashion had the same effect on its mechanical properties as particles introduced by mixing them into the melt [19], namely, that the Young's modulus and the yield stress increased in accordance with the linear rule of mixtures. It was found, for example, that a volume fraction of 6.7% WC in $Ni_{75}Si_8B_{17}$ increased both its Young's modulus and its yield stress by a factor of about 1.4.

Experiments with larger particles, however, showed that the wettability of the particle by the liquid matrix material determines the effect that the inclusion of particles has on the mechanical properties of the ribbon [28]. Various kinds of particles (17 μm and 70 μm SiC, plain and coated with 2 μm of Ni, and 100 μm B_4C) were injected into the melt puddle of an $Ni_{75}Si_8B_{17}$ alloy. Tensile tests performed to determine the yield stress showed that the particles that were well wetted by the matrix (the Ni-coated SiC) could increase the yield stress by 20–30%. Particles that were not well wetted (plain SiC and B_4C), however, actually reduced the measured yield stress by about 25%. This is due to a difference in the deformation mechanisms between the ribbons with wetted particles and those with unwetted particles. When the particles are wetted, one expects the propagating slip bands to initiate cracking along themselves and across the particles. Unwetted particles, however, have pores at their particle/matrix interface which can concentrate stress when a slip band arrives at the particle. This stress is relieved by the formation of new slip bands. Thus, the pores grow rapidly and degrade the mechanical properties of the composite.

Because the particles are not in contact with the melt, except for the very short time before it solidifies, it is even possible to use particles having a melting point below the temperature of the melt [29]. For example, this technique has been used to add Ag particles to $Ni_{75}Si_8B_{17}$. In this case, some Ag was dissolved in the ribbon before solidification took place, but Ag particles could still be found in the ribbon, though, no doubt, reduced in size.

Even more dramatic differences in materials may be accommodated using this technique. Polycrystalline aluminum ribbon containing a uniform dispersion of polystyrene spheres has been produced [30] in an attempt to increase the mechanical damping. In this work, however, it was impossible to determine whether the shrunken spheres observed were still polystyrene or had been charred into carbon, and the volume fraction of particles was so low that no measurable effect on the damping could be expected.

Spray forming [11], which has been suggested for the production of RS laminates, may also be used for the production of discretely reinforced composites. This approach has been investigated by embedding SiC particles and graphite flakes in an Al–5 Si matrix. The composite was produced by adding the particles to the gas stream that carried them in molten form to the chill surface. The graphite flakes were observed to be in good interfacial contact with the SiC particles.

IV. CONTINUOUSLY REINFORCED RAPIDLY SOLIDIFIED COMPOSITES

Reinforcing a rapidly solidified material with continuous wires or fibers raises the prospect of significantly improved mechanical properties, at least in the direction of the reinforcement. As the fibers, if continuous, must be very long and neatly ordered, one can not simply add the reinforcing material to the melt or blow it into the melt puddle, as with the particle reinforcement of rapidly solidified materials. A method must be found to introduce the reinforcing material into the melt puddle at the same rate that solid ribbon is being produced to prevent either breakage or tangling, depending on whether the ribbon is moving faster or slower, respectively, than the reinforcement.

A method was devised [31] for producing fiber-reinforced metal-matrix composites, in which rapid solidification was considered a solution to the problem of undesired chemical reaction between the fiber and the liquid matrix material. This would result from the greatly reduced time for which the fibers are in contact with the matrix material in its liquid state.

In this scheme, fibers are first temporarily attached to a single chill roller in a rapid solidification apparatus (Fig. 13). When the roller was up to casting speed, the metal would be ejected onto the fibers, and the temporary attachment of the fibers to the roller broken, the fiber feed being maintained by the pulling of the solidified ribbon. Various modifications of this scheme were also proposed, such

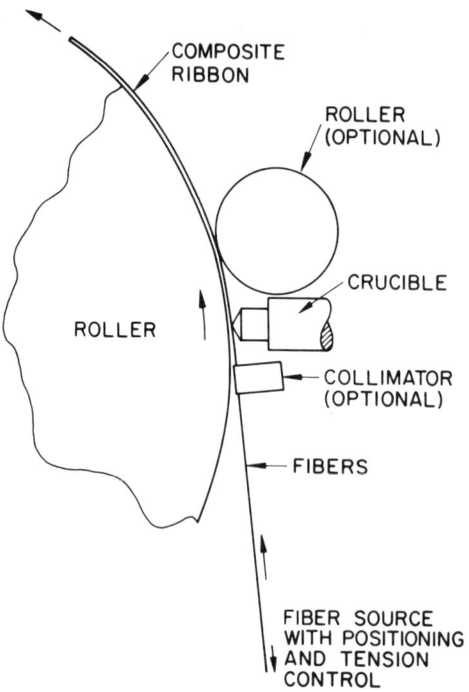

Figure 13 Apparatus for producing continuously reinforced RS composite ribbon by attaching reinforcing elements such as fibers or wires to chill roller [31].

as adding a second roller for more rapid solidification and surface finishing, or adding a collimator to insure the proper placement of the fibers before they enter the melt puddle. If the metal flow and the fiber feed could be maintained, this could be a continuous process, with no theoretical limit to the length of composite ribbon produced.

No examples of suitable fiber/matrix systems were suggested by the developers of this method, so it is unclear what, if any, systems were actually attempted. Some system was probably investigated, as it was claimed that the minimum volume fraction of fibers was reduced below the 60–70% required by the conventional method of pulling fibers through molten metal. It has been demonstrated [29], however, that it is important with this sort of apparatus, for the fibers and the matrix material be selected so that the fibers are wetted by the molten metal. If this is not the case, the metal may ride on top of the fibers and only a few chance fibers will be incorporated into the matrix.

One system in which the liquid matrix material wets the reinforcement is that of $Ni_{75}Si_8B_{17}$ ribbons reinforced with W wires. It has been shown [29] that excellent bonding can be achieved between the wire and the matrix, probably due to the good wetting of the wire by the matrix. In this case a length of 10-μm-diameter wire was incorporated into a matrix using a single-roller technique similar to that discussed above, except that the wire was not attached to the chill roller, but was simply draped over it while the other end was lightly fastened to a support. The motion of the first metal to solidify around the wire, imparted by the chill roller, tore the wire free and brought it forward as more ribbon was produced. The excellent quality of the fiber/matrix bond was demonstrated with bending tests, in which debonding did not occur until considerable necking had occurred. This technique can, obviously, only be used with such reinforcing media as can withstand the stress produced by being suddenly accelerated from stationary to the speed of the chill roller's surface; W wire is well suited to this. It should be noted that this technique could not incorporate C fibers into $Ni_{75}Si_8B_{17}$, probably due to the poor wetting of the fibers by the melt.

Further investigation of the $Ni_{75}Si_8B_{17}$/W system [32] showed that multiple reinforcing wires could be introduced with a method that, additionally, allowed for continuous production (Fig. 14). In this scheme, the wires were pulled from a supply wheel, along the surface of a single chill roller, and wound up on a driven takeup wheel. The speed of the wire was matched to the surface speed of the chill roller. A slight resistance to rotation of the supply wheel kept the wires taut. It also keeps them tangle-free if each wire is wound onto a separate section of the supply wheel. This method was used to produce $Ni_{75}Si_8B_{17}$ ribbon with either two 26-μm-diameter W wires, one 60-μm wire, or one 10-μm wire. The single 10-μm wire, being unable to withstand the acceleration of the drive wheel by itself was incorporated into a ribbon by winding a 60-μm wire onto the supply wheel separately and attaching it to the drive wheel, passing to the side of the melt puddle, while the 10-μm wire was parallel to it, but passed through the melt puddle. The 60-μm wire, then, accelerated the supply wheel, but only the 10-μm wire was embedded in the ribbon. Examination of the ribbon with the embedded wires revealed that the ribbons sometimes split or had holes and brittle zones along the wire along the wires if the roller speed was too high and thus the ribbon too thin. Slowing the roller's surface speed down to 20 m/s eliminated these problems, but the wire(s) still protruded from the free surface (the side not in contact with the chill roller) at places. Some areas of the surface in contact with the chill roller near

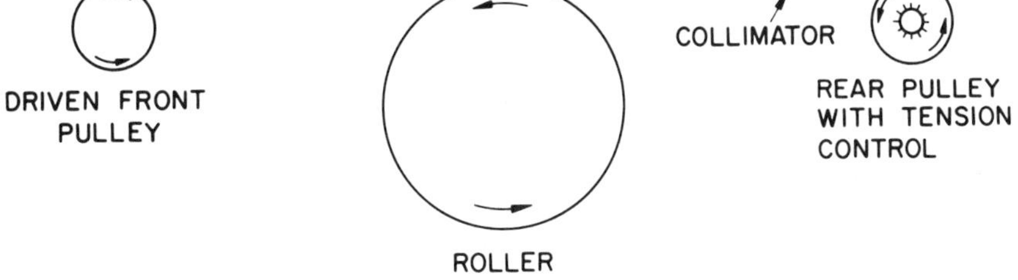

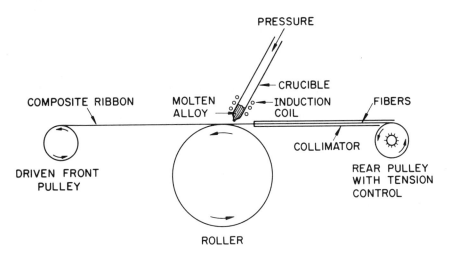

Figure 14 Apparatus for producing continuously reinforced RS composite ribbon by pulling reinforcing elements such as fibers or wires across chill roller surface [32].

the wire appeared as shiny as the free surface, indicating a reduced cooling rate there, and suggesting that crystallization and embrittlement may occur. X-ray analysis showed that the boundary between the wire and the matrix was sharp, indicating that no reaction or diffusion occurred between the wire and the melt in the very short time before solidification.

V. CONCLUSION

Rapid solidification is the route to many different novel composite materials. RS laminates may be produced from amorphous metallic alloys that would crystallize if hot-pressed. Reinforced RS composites may be prepared with amorphous metallic matrices and with reinforcing media that would react with the matrix material or be degraded by the temperature if prepared by conventional methods. These reinforced RS composites can have significantly improved mechanical or other properties compared to the unreinforced matrix material. The technology for the production of discretely reinforced RS composites is reasonably well developed, but that for the production of continuously reinforced composites will require much more research before it is of practical interest. The primary limitation on the selection of reinforcing and matrix materials, both from the standpoint of production and that of properties, is the wettability of the reinforcement by the matrix. Solving this class of problems will enable more combinations of reinforcements and matrices to be used, yielding composites with improved properties tailored to their intended use.

REFERENCES

1. M. B. Arbhavi and K. Chattopadhyay, *Z. Metallkde.*, *78*, 259 (1987).
2. E. R. Murray, Ph.D. thesis, Cornell University (1992).
3. P. J. Gregson and H. M. Flower, *J. Mater. Sci. Lett.*, *3*, 829 (1984).
4. F. H. Froes and D. Eylon, U.S. Patent No. 4,746,374 (1988).
5. K. Inomata, S. Shimanuki, and M. Hasegawa, *Japan J. Appl. Phys.*, *19*, L625 (1980).
6. S. Shimanuki and K. Inomata, U.S. Patent No. 4,428,416 (1984).
7. S. Shimanuki, H. Yoshino, and K. Inomata, in *Proc. Fourth Int. Conf. on Rapidly Quenched Metals*, Vol. I, edited by T. Masumoto and K. Suzuki (Japan Institute of Metals, (1982), p. 15.
8. R. V. Raman and A. F. Witt, in *Rapidly Solidified Amorphous and Crystalline Alloys*, edited by B. H. Kear, B. C. Giessen, and M. Cohen (Elsevier, New York, 1982), p. 141.
9. H. H. Liebermann, *Int. J. Rapid Solidification*, *1*, 103 (1984–85).
10. B. L. War, U.S. Patent No. 4,409,296 (1983).
11. A. R. E. Singer, *Mater. Design*, *4*, 892 (1983).
12. M. C. Narasimhan, U.S. Patent No. 4,330,027 (1982).
13. P. G. Zieliński and D. G. Ast, *Scripta Metall.*, *17*, 291 (1983).
14. H. Kimura, B. Cunningham, and D. G. Ast, in Ref. 7, Vol. II, p. 1385.
15. H. Kimura, T. Masumoto, and D. G. Ast, *Acta Metall.*, *35*, 1757 (1987).
16. L. A. Davis, C.-P. Chou, L. E. Tanner, and R. Ray, *Scripta Metall.*, *10*, 937 (1976).
17. A. G. Guy and J. J. Hren, *Elements of Physical Metallurgy*, (Addison-Wesley, Reading, MA, 1974), p. 502.
18. M. F. Ashby, in *Strengthening Methods in Crystals*, edited by A. Kelley and R. B. Nicholson, (Elsevier, Amsterdam, 1971), p. 137.

19. P. G. Zieliński and D. G. Ast, *Acta Metall.*, *32*, 397 (1984).
20. H. Kimura and T. Masumoto, *J. Non-Cryst. Solids*, *61–62*, 835 (1984).
21. H. Kimura, T. Masumoto, W.-N. Myung, and S. Furukawa, *Rapidly Solidified Materials*, edited by P. W. Lee and R. S. Carbonara (ASM, 1986), p. 303.
22. H. Kimura, T. Masumoto, A. Makino, and T. Sasaki, *J. Non-Cryst. Solids*, *61–62*, 1335 (1984).
23. H. Kimura, T. Masumoto, and M. Hasegawa, *J. Mater. Sci. Lett.*, *3*, 1003 (1984).
24. W.-N. Myung, S. Yang, H. Kimura, and T. Masumoto, *Mater. Sci. Eng.*, *97*, 259 (1988).
25. S. Das, A. H. Yegneswaran, and P. K. Rohatgi, in *Int. Conf. on Aluminum* (Indian Institute of Metals, 1987), p. 309.
26. P. G. Zieliński and D. G. Ast, *J. Mater. Sci. Lett.*, *2*, 495 (1983).
27. D. G. Ast, U.S. Patent No. 4,523,625 (1985).
28. J. Latuszkiewicz, P. G. Zieliński, and A. Załuska, *Mater. Sci. Eng.*, *97*, 181 (1988).
29. P. G. Zieliński and D. G. Ast, in *Rapidly Solidified Metastable Materials*, edited by B. H. Kear and B. C. Giessen (Elsevier, New York, 1984), p. 189.
30. E. R. Murray and D. G. Ast, Cornell University, unpublished data, 1988.
31. J. F. Williford and J. P. Pilger, U.S. Patent No. 3,776,297 (1973).
32. G. Nussbaum and D.G. Ast, *J. Mater. Sci.*, *22*, 23 (1987).

6

Production Techniques of Alloy Wires by Rapid Solidification

M. Hagiwara
Unitika Research Center, Unitika Ltd., Uji, Japan

A. Inoue
Institute for Materials Research, Tohoku University, Sendai, Japan

I. INTRODUCTION

For the last two decades, rapid solidification of metals and alloys has attracted great attention because of the following advantages: (1) the development of new functional properties caused by structural modification, such as the formation of nonequilibrium phases; (2) the near-net shape formation of metallic materials in which plastic deformation is very difficult, accompanied by reductions of energy, time and labor; and (3) an improvement of mechanical properties caused by decreased segregation and the refinement of grain size. Although rapid solidification can be achieved by various kinds of cooling techniques using solid, liquid, and gaseous cooling media, the shape of rapidly solidified materials is usually limited to that with at least one small cross-sectional dimension, such as thin sheet, wire, and powder because of the necessity of achieving high cooling rates during processing. Of the rapidly solidified materials in the forms described, continuous alloy wires with circular cross section have attracted increasing interest as a new type of engineering material, owing to their unique functional properties and simplified fabricating process. This chapter deals with the techniques and dominating factors for producing metallic wires with circular cross section directly from the melt, and with the alloy systems and morphologies of the resultant wires.

II. COMMON FACTORS FOR PRODUCTION OF METALLIC WIRES DIRECTLY FROM THE MELT

A fundamental process for producing metallic wires consists of pressure expulsion of a stream of melt through an orifice into a cooling fluid, followed by rapid solidification of the ejected stream before it breaks into droplets, and then the winding of the continuous wire formed. The principle of this process is the same as that of the melt-spinning process for the production of nylon and ester fibers. However, physical properties of metallic melt are significantly different from those of polymetric materials, e.g., for metallic melt, the viscosity is smaller by 10^4–10^6 times and the specific gravity and surface tension are larger by several and 10–200 times, respectively. This significant difference results in a high tendency for geometric destabilization of a metallic melt stream. Rayleigh [1] and Weber [2] estimated the distance over which a melt stream remains stable, using surface tension–wave theory, and pointed out that the distance at which the melt stream changes into liquid droplets is several centimeters for metallic melts and 10–100 m for molten polymers. One of the necessary conditions for the production of wires is to solidify the melt stream within a distance where its surface remains stable. However, it has been pointed out [3] that the stability of an ejected stream of molten lead, which has high specific heat and large surface tension, cannot be always explained on the basis of the surface tension–wave theory. The following two additional factors are also important for the production of continuous lead wire: (1) an optimum ratio of the ejected melt-stream velocity to the confluent cooling fluid velocity, and (2) a melt ejection condition having a small loss coefficient near the nozzle orifice in order to avoid the rupture of the melt stream caused by turbulence in cooling fluid. On the other hand, Liebermann [4] has reported for an $Fe_{40}Ni_{40}B_{20}$ alloy, that the stability of the melt stream can be described by analyses similar to those of Rayleigh [1] and Weber [2].

As described above, since a stream of metallic melt ejected through an orifice is apt to change into droplets before solidification, the development of a technique leading to the increased cooling rate and suppressed tendency to allow melt stream breakup into droplets is essential for the production of a continuous metallic wire. Finally, common factors for the direct production of a metallic wire with circular cross section from melt are summarized as follows:

1. Solidification of the metallic melt stream at a high cooling rate
2. The use of a cooling fluid with a high degree of "softness" (that is, with low viscosity and low surface tension)
3. Highly stable, turbulence-free cooling fluid flow at a high velocity

III. TECHNIQUES OF PRODUCING METALLIC WIRES DIRECTLY FROM THE MELT

A great deal of effort has been devoted to produce continuous metallic wire with circular cross section directly from melt. Success has been reported using the following four melt-spinning methods:

1. Glass-coated melt-spinning method (Taylor method) [5–7]
2. Method of injecting a melt stream into a confluent cooling fluid (Kavesh method) [8–10]
3. In-rotating-water spinning method [11,12]
4. Conveyer belt method [13,14]

All these techniques enable the production of wires with diameters of 1 to 300 μm at an estimated cooling rate of about 10^4–10^6 K/s. Accordingly, the resultant wires are expected to have quenching-induced structural modifications such as refinement of grain size, homogenization of constituent atom distribution which leads to the reduction of segregation, and formation of supersaturated solid solution and nonequilibrium crystalline, quasicrystalline and amorphous phases. Such structural modifications allow us to anticipate metallic wires exhibiting favorable mechanical, chemical, magnetic, and electronic properties. A detailed explanation of the four techniques to produce continuous metallic wire directly from melt will be given as part of the historical order of development in subsequent sections. The discussion will focus on amorphous alloy wires which have already found practical applications.

IV. GLASS-COATED MELT-SPINNING METHOD (TAYLOR METHOD)

The basic concept of this method was proposed by Taylor [5] in 1924. Subsequently, Ulitovskiy [6] and Wagner [7] tried to produce metallic wires on the basis of the Taylor concept and developed the technique which is presently known as glass-coated melt spinning. Figure 1 shows a schematic illustration of the glass-coated melt-spinning method, in which melt contained in a glass tube is drawn rapidly to a very fine wire, together with the coating glass which was softened by heating, by using a drawing machine. After the drawing operation, metallic wires are obtained by chemical dissolution of the coating glass in hydrofluoric acid. This method can produce a very fine wire with diameters of 2 to 20 μm and can be applied to alloy systems with low wire-forming capacity. It is necessary, however, that the melting

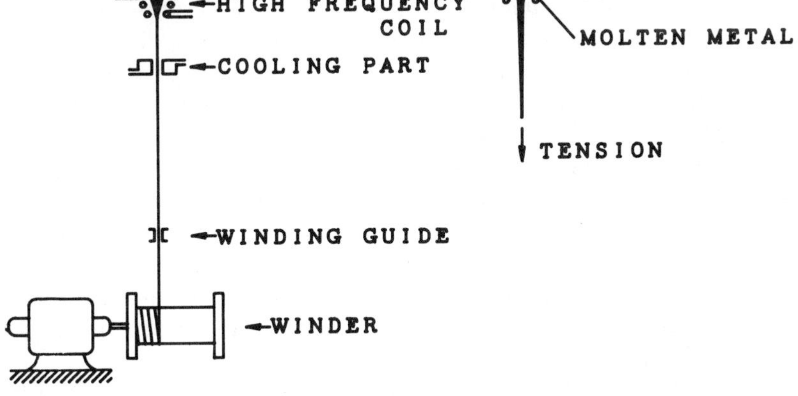

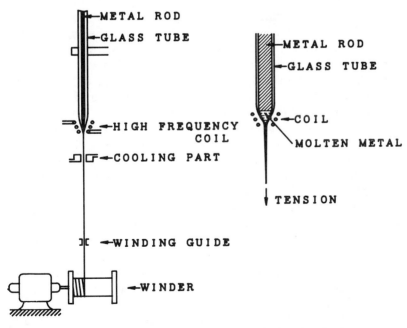

Figure 1 Schematic illustrations of the glass-coated melt-spinning apparatus for the production of a metallic wire from the melt.

temperature of metallic materials processed is nearly equal to the softening temperature of the coating glass; the melt should also wet the coating glass well. Goto [15] has analyzed the glass-coated melt-spinning process for metals and alloys by using equations for volume, energy, and stress balance, which had already been applied to the melt spinning of polymers. As a result, it has been pointed out that the solidification of metallic wires is complete within a distance of 0.5 cm from the point of drawing and that the cooling rate lies in the range of 10^5 to 10^6 K/s, depending on drawing velocity. Such a high cooling rate has been confirmed from the result [15] that amorphous phase in Fe-P-C-B and Fe-Si-B systems is produced for wires with diameters of several micrometers at a drawing speed above about 4 m/s. In addition to the Fe-based amorphous wires, wires of pure metals (Cu, Au, Ag, Fe, etc.) [16], Fe-based crystalline alloys (stainless steels [17], In-856 [17], maraging steel [18], etc.) and Ni-based superalloys (Hastelloy X [17]) have been produced by this method. It has further been reported that these metallic wires have diameters in the range of 1 to 10 μm and exhibit very high mechanical strengths through the formation of nonequilibrium phases and fine-grained structures. In addition, the Fe-Si wires have been reported to exhibit unique magnetic properties, probably because of the fine-wire geometry and an oriented crystalline structure [19].

V. INJECTION METHOD OF MELT INTO A CONFLUENT FLUID (KAVESH METHOD)

Chen [8] demonstrated for the first time the formation of a Pd-Cu-Si amorphous rod ≃2 mm diameter by simply injecting a stream of melt into water. This process of forming an amorphous wire has since been pursued further by Kavesh [9,10]. In this technique, melt is injected in a confluent cooling fluid with viscosity nearly equal to that of the molten metal and solidified before the changing into droplets. A schematic illustration of the equipment used by Kavesh is shown in Fig. 2. The ejected melt stream is rapidly solidified in a fluid (water or aqueous solution) which is flowing into a standpipe. The important points of producing a continuous wire by this technique are as follows: (1) the matching of the ejected melt stream velocity with that of the flowing fluid in the standpipe, (2) the suppression of turbulence in the fluid flowing in the standpipe, (3) an accurate control of melt temperature (optimum melt stream temperature) is about 50 to 100 K higher than the alloy melting temperature, (4) an accurate adjustment of the distance between the orifice and the fluid surface (0.25–0.50 cm). It is particularly important to control the velocity of flowing cooling fluid in the standpipe and to suppress the generation of turbulence in the standpipe. Fluid flow control and turbulence suppression have been made by adjusting the diameters and lengths of the sparger ring and standpipe, as well as their mutual positions [9,10]. In addition, in order to produce metallic wires with diameters of 20 to 600 μm, an appropriate cooling fluid must be used; e.g., water at temperatures of 273 to 293 K for metallic melts with T_m below 980 K and a refrigerated brine quenchant for the melts with T_m in the range of 980 to 1500 K [9,10].

The metallic wires produced by this method have diameters from 20 to 600 μm. Alloy systems investigated include gray cast iron in Fe-C-Si system [9], pure zinc and silver metals [9], amorphous Fe-Ni-Cr-P-B [9], and Fe-Ni-P-C-Al [20] alloys. As an example, the condition for producing an amorphous $Fe_{38}Ni_{38}P_{14}C_6Al_3$ wire

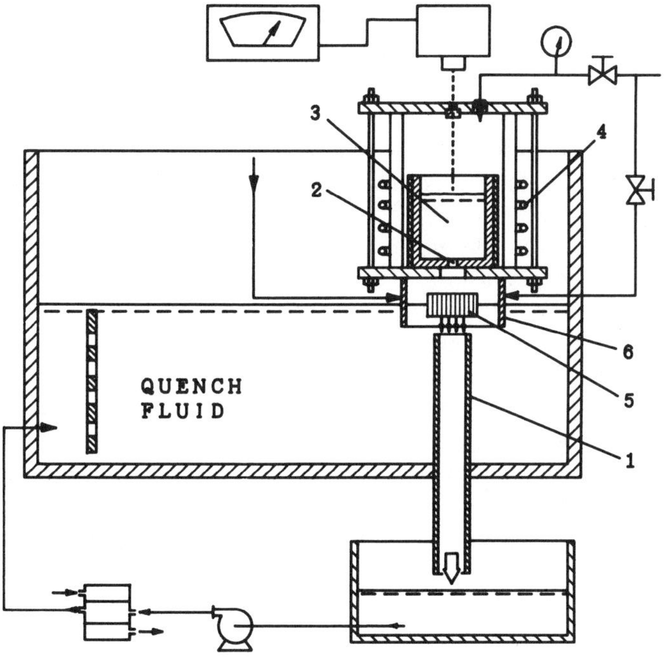

1. STANDPIPE
2. NOZZLE
3. MOLTEN METAL
4. INDUCTION HEATER
5. SPARGER
6. VORTEX BAFFLE

Figure 2 Schematic illustration of an apparatus to produce a metallic wire by injecting a melt stream into a flowing fluid.

has been described as follows [20]: The alloy was melted at 1323 K under helium atmosphere and extruded through an orifice of 0.2 mm diameter at a velocity of 200 cm/s. The molten jet was quenched in refrigerated 21.6% $MgCl_2$ brine at 243 K. Brine velocity in the standpipe was 195 cm/s. Continuous amorphous wires of 0.15-mm-diameter were thereby produced. Although the production of the amorphous wire by this method is made by the relatively simple equipment shown in Fig. 2, the method has some disadvantages: (1) it is very difficult to control the stability of flowing cooling fluid; (2) the cooling rate of melt stream is rather low because the velocity of the ejected stream is nearly the same as that of the flowing fluid; and (3) a relatively slow production rate of approximately 260 cm/s is realized. Production rate is inherently difficult to increase because of the limited free-fall velocity of the flowing fluid. The elimination of these disadvantages has been unsuccessful and hence this method has not been used at present for the production of continuous metallic wires with circular cross section.

VI. IN-ROTATING-WATER MELT-SPINNING METHOD

A. Process and Equipment

The concept of the in-rotating-water melt-spinning method was proposed by Ohnaka et al. [11,12] in 1978. They examined first the stability of a molten lead stream in air and water [3] and subsequently presented the in-rotating-water spinning method in which a molten lead stream is ejected through an orifice into a water layer which was formed by the centrifugal force of a rotating drum [21]. In 1980, the present authors noticed [12] the possibility that this method might be applied to the production of amorphous alloy wires. This possibility was examined in detail by constructing equipment which enabled us to solidify molten alloys with higher melting temperatures at high cooling rates.

Figure 3 shows a schematic illustration of the in-rotating-water spinning equipment [12] constructed by the present authors. The equipment consists of a heating element for the alloy, an ejection means for the molten alloy, and a cooling means of the melt stream ejected into the rotating water. The inner diameter of the drum is approximately 500 mm. The alloy melted in a quartz nozzle is ejected through an orifice with diameter 0.08 to 0.3 mm by an argon gas pressure of about 0.4 MPa into a water layer with a depth of about 25 mm formed by centrifugal force in the rotating drum which has a tangential velocity of 10 m/s. This process enables us to produce continuously an amorphous wire which is accumulated in the rotating water layer. At present, Unitika Ltd. has commercially produced amorphous metallic wires by this technique and tried to extend their fields of application through the clarification of fundamental properties of the wires.

The above-described method is a batch type. Accordingly, after a given amount of metallic wire is accumulated at the inner side of the drum, it becomes necessary to stop the drum and to wind the wire. This had prevented the enhancement of the producibility of amorphous wires. In order to resolve this problem, Unitika Ltd. [22] has constructed the melt-spinning equipment which enables us to produce continuously amorphous alloy wires and to wind simultaneously them. A schematic illustration of the equipment is shown in Fig. 4. The mechanism for the continuous winding of wires is due to the use of a magnetic roller which rotates at the same

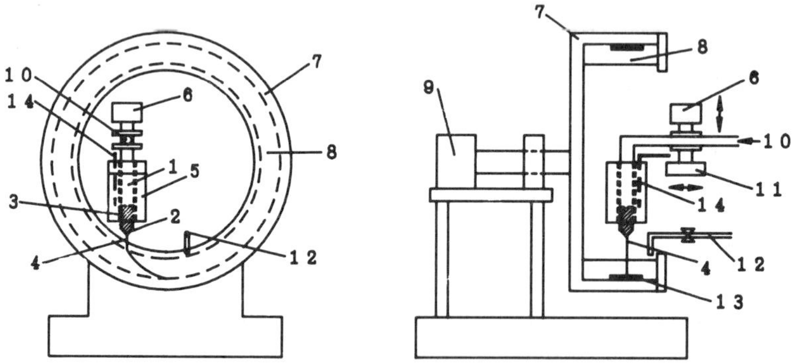

Figure 3 Schematic illustrations of the in-rotating-water spinning apparatus for the production of a metallic wire.

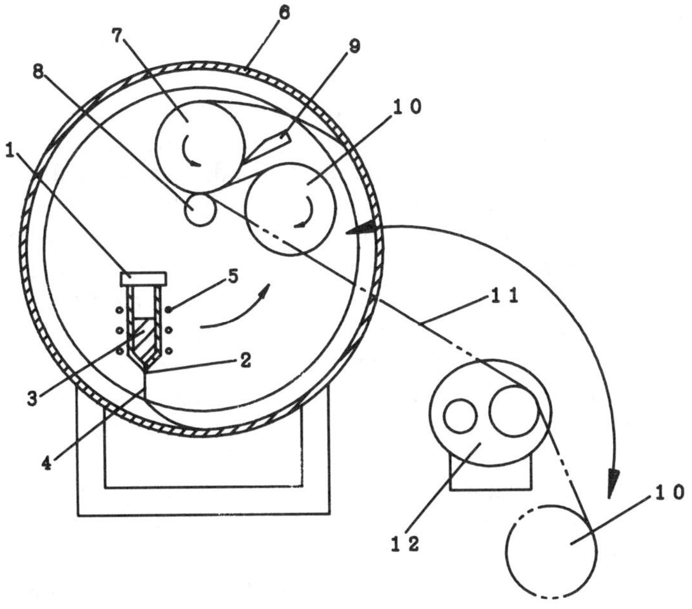

1. MELTING CRUCIBLE
2. SPINNING NOZZLE
3. MOLTEN ALLOY
4. EJECTED ALLOY
5. HIGH FREQUENCY COIL
6. ROTATING DRUM
7. FIRST MAGNET ROLLER
8. NIP ROLLER
9. SCRAPER
10. SECOND MAGNET ROLLER
11. WIRE SPECIMEN
12. WINDER

Figure 4 Schematic illustration of the in-rotating-water spinning apparatus combined with a winder.

velocity as the circumferential velocity of the drum. The magnetic roller, which is arranged at a higher position in the drum as shown in Fig. 4, catches by magnetic force the amorphous alloy wire produced at a lower position in the drum and carries the wire up to the winder which is situated at a fixed position. Although the winding mechanism of this equipment is relatively simple, its use is limited to magnetic materials and the winding is necessarily stopped by rupture of the wire. Accordingly, it is concluded that this winding technique is useful for handling continuous ferromagnetic wire in which no rupture of the wires takes place.

B. Necessary Condition for the Production of Amorphous Alloy Wire

In order to produce amorphous alloy wire having smooth surface and uniform diameter by the in-rotating-water melt-spinning method, it is necessary that the following three factors are satisfied simultaneously [23]: (1) high supercooling capacity of the ejected stream without the precipitation of crystalline phase in the temperature range between melting and glass transition, (2) high stability and rapid solidification of the ejected stream without breakup into droplets as it comes into contact with and penetrates the rotating cooling water layer, and (3) smooth water-free surface in the rotating drum. The first factor is strongly dependent on the glass-forming ability of alloys, the diameter of the ejected melt stream, and the cooling capacity of rotating fluid. Water cooled to approximately 278 K has been reported [24] to be the best cooling medium for reasons of economy and ease of treatment, etc., following experimental data using many kinds of cooling fluids. Furthermore, the diameter of the ejected melt stream is nearly the same as that of the orifice (0.08–0.3 mm) and hence the diameter of the amorphous wires produced by this technique is limited to a range of about 80 to 300 μm [24]. The lower limit of orifice diameter results from difficulties of injecting a molten alloy stream with a small diameter into the rotating water layer. Another factor requiring the use of a certain nozzle size is the very high ejection pressures which would be required to form a coherent melt stream at the risk of fracturing the quartz nozzle. On the other hand, the limitation of the maximum orifice diameter is due to the difficulties of the production of wires without crystalline phase having smooth surface and uniform diameter. The minimum diameter is about two to four times larger than the thickness of amorphous ribbon with uniform shape. The large difference of sample morphology implies that the cooling rate of the wires produced by this technique is considerably lower than that of the ribbon produced by a single-roller melt-spinning method, though the in-rotating-water spinning technique has an ability of achieving a high cooling velocity comparable to that for the melt-spinning method in the samples with a nearly same thickness or diameter. Accordingly, it can be concluded that the formation of amorphous alloy wires by this technique is limited to the alloys with high-glass-forming ability. It is therefore important to know the glass-forming ability of alloys for which the formation of amorphous alloy wires is desired.

The present authors have reported systematic data of glass-forming ability for Fe-Si-B [25], Co-Si-B [25], Ni-Si-B [26], and Fe-P-C [27] alloys with engineering importance. A quantitative evaluation of glass-forming ability defined by a critical cooling rate is rather difficult. Instead, a new method [26] of evaluating the glass-forming ability of alloys was developed: casting a ribbon with continuously decreasing thicknesses in one process operation through rapid deceleration to a complete stop of the rapidly rotating wheel (within 1–2 s) and then examining the as-quenched structure at the freely solidified side of the ribbon as a function of ribbon thickness by optical microscopy and x-ray diffractometry. The choice of the free side intends to avoid the influence of a preferential (inhomogeneous) surface crystallization on the glass-forming ability of alloys because the preferential surface crystallization is apt to occur on the substrate-contact side of ribbon. As examples, Figs. 5 and 6 show the compositional dependence of the critical thickness for the formation of an amorphous phase in Fe-Si-B [25] and Co-Si-B [25] ternary alloys. Although the glass-forming ability is strongly dependent on the alloy composition, alloys in rather wide compositional ranges around $Fe_{75}Si_{10}B_{15}$ and $Co_{72.5}Si_{12.5}B_{15}$ have a critical ribbon thickness above 100 μm, indicating the possibility of forming amorphous wires with diameters above about 100 μm in Fe-Si-B, Co-Si-B, and Ni-Si-B alloys by using the in-rotating-water spinning method. The significant compositional dependence of the glass-forming ability could not be successfully explained by the ratio of crystallization to liquidus temperature probably because of the difference of the mechanism of crystallization and glass transition.

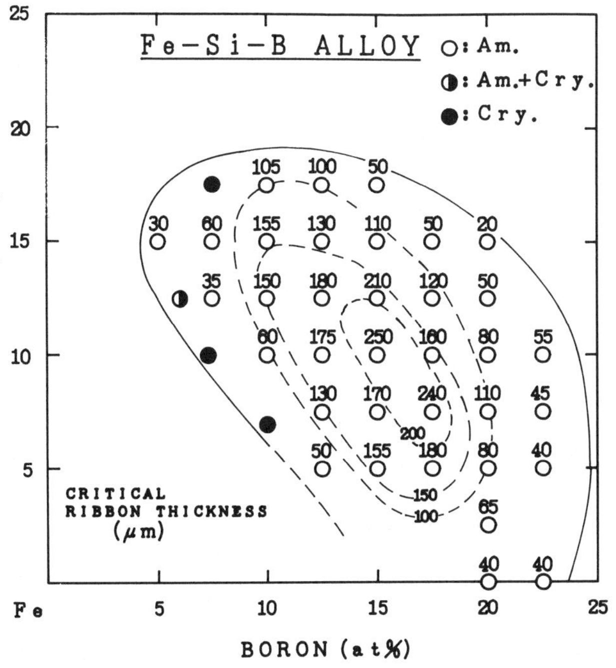

Figure 5 Compositional range and the critical ribbon thickness for the formation of an amorphous single phase in Fe-Si-B ternary alloys.

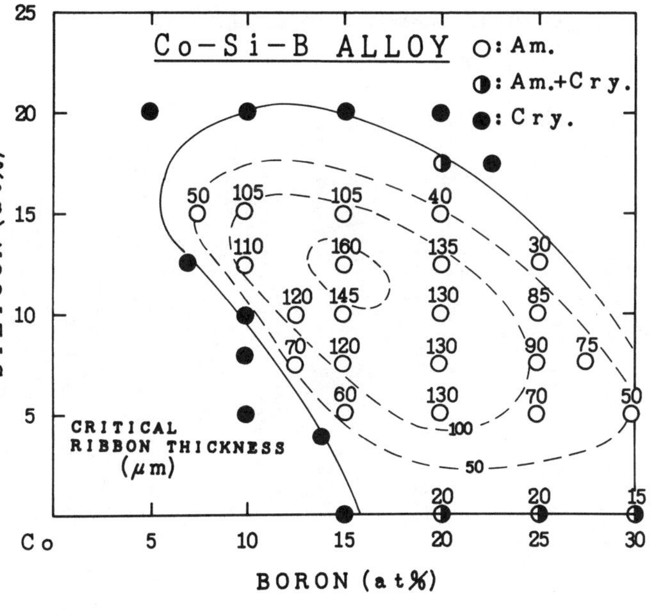

Figure 6 Compositional range and the critical ribbon thickness for the formation of an amorphous single phase in Co-Si-B ternary alloys.

In addition to the glass-forming ability of alloys, it has been reported [28] that the high resistance of the ejected melt stream against breakup into droplets is also necessary for the formation of an amorphous wire having circular cross section. This resistance has also been found to be strongly dependent on alloy composition [28]. As exemplified in Fig. 7, the alloy containing silicon and phosphorus has a higher wire-forming ability than alloys containing boron and carbon. Furthermore, the wire-forming ability appears to be much greater for Fe- and Co-based alloys than for Ni-based alloys for a given metalloid composition. This has been evidenced from the result [28] that no amorphous wire is formed in Ni-Si-B ternary alloys in spite of the experimentally determined high glass-forming ability. The addition of 1–3 at% Al has been found [29] to be very effective for the enhancement of the wire-forming capacity, leading to the formation of a continuous amorphous wire, as shown in Fig. 8. It has been confirmed by XPS analysis that the remarkable improvement for the wire formation by the addition of aluminum is due to the formation of Al_2O_3 thin film on the wire surface [29]. The stabilization of molten alloy streams by the formation of such stabilizing films has been reported for the melt spinning of steel wire [30]. While the resistance against the breakup of an ejected melt stream into droplets seems to be strongly dependent on the viscosity and surface tension of melt as well as on the ease of oxide surface film formation, but no quantitative interpretation on their mutual relation has been given.

The importance of the three factors: high glass-forming ability, high resistance of the molten stream to break into droplets, and high stability of the water surface layer achieved by the utilization of centrifugal force has previously been described. Although these factors may be satisfied, spinning conditions must also be optimized as well in order to form a continuous amorphous wire with circular cross section and smooth surface. The important factors in the spinning operation are [23] (1) ratio of ejected melt stream velocity (V_j) to that of the rotating water (V_w), (2) ejection temperature of molten alloy, (3) shape and size of the orifice, (4) distance

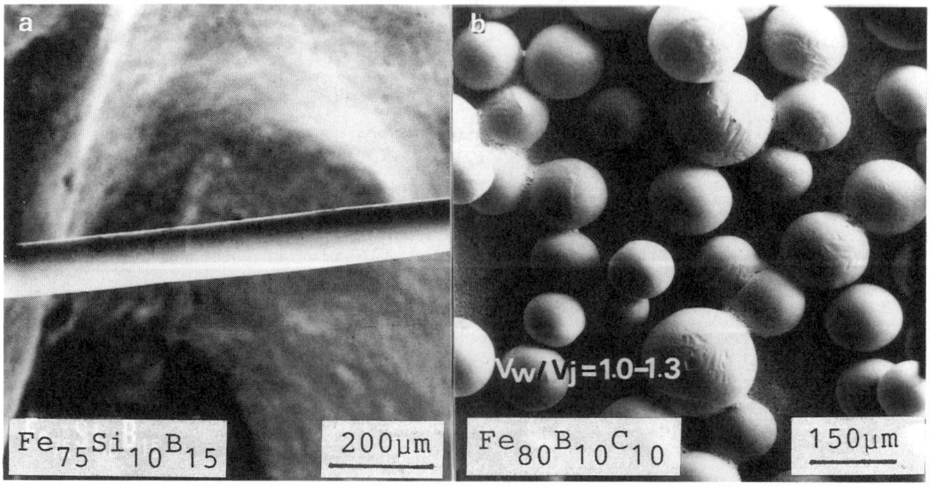

Figure 7 Scanning electron micrographs showing the form of (a) $Fe_{75}Si_{10}B_{15}$ and (b) $Fe_{80}B_{10}C_{10}$ alloys produced by the in-rotating-water spinning method.

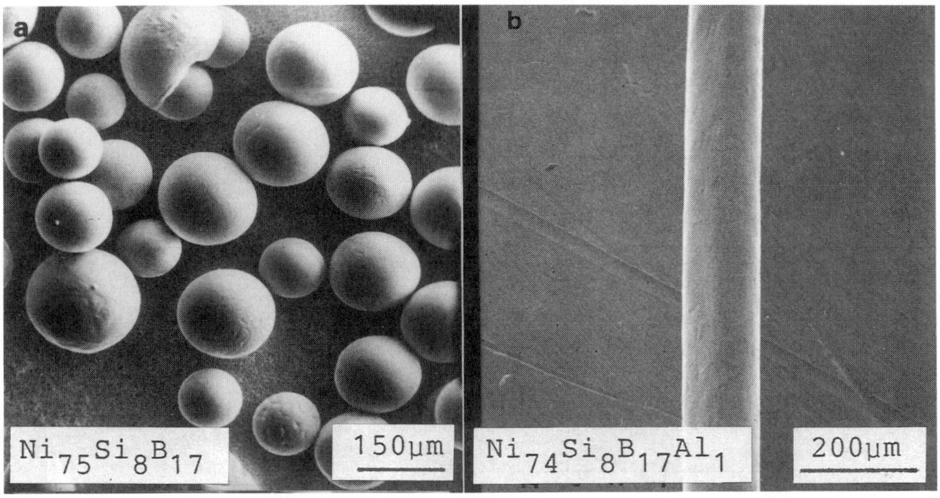

Figure 8 Scanning electron micrographs showing the form of (a) $Ni_{75}Si_8B_{17}$ and (b) $Ni_{74}Si_8B_{17}Al_1$ alloys produced by the in-rotating-water spinning method.

and angle between orifice and water surface, (5) stability of the water surface, and (6) amount and temperature of water. As an example, Fig. 9 shows the change of the wire morphology with the change of V_j/V_w for $Fe_{75}Si_{10}B_{15}$ alloy when using an orifice diameter of 0.30 mm [24]. Continuous wire with good geometric definition was obtained when V_j/V_w was approximately 0.80. This result indicates that the spinning condition is optimum when the ejected melt stream is solidified in a slightly pulled condition by the rotating water. The wire morphology and surface become rogose for V_j/V_w above about 1.0; wire is formed in short pieces for V_j/V_w below about 0.80. The optimum range of the V_j/V_w ratio is slightly different, depending on alloy component and orifice size, etc.

As described above, the number of the factors which must be optimally controlled is much greater for wire formation than for the ribbon formation. Also, the composition range over which an amorphous wire can be formed is much narrower in wire casting, probably because of the considerably lower cooling capacity in the in-rotating-water spinning method relative to that of a ribbon casting method. The main alloy systems in which the formation of an amorphous wire by the in-rotating-water spinning method has been reported to date are limited to Fe-Si-B [24], Fe-P-C [27], Co-Si-B [31], Ni-Si-B-Al [29], Ni-P-B-Al [29], Pd-Cu-Si [12], Pd-Ni-P [32], Pt-Ni-P [33], Cu-Zr [34], and Cu-Zr-Nb [34]. As an example, a photograph showing the form of $Co_{68}Fe_{4.5}Si_{12.5}B_{15}$ amorphous wire is shown in Fig. 10. The wire shown has a smooth surface and circular cross section with a diameter of about 120 μm. Minimum and maximum diameters of the amorphous wire produced by the in-rotating-water quenching method are about 80 to 300 μm, respectively, and their wire length reaches several kilometers. The approximate production rate of amorphous wires with a nearly constant diameter is of the order 13 m/s. Furthermore, these amorphous wires can easily be cold-drawn to a 15-μm-diameter without intermediate annealing [35]. Fe- and Co-based amorphous wire 0.5–3 km long in as-quenched state has been sold [35] by Unitika Ltd. as ferro-

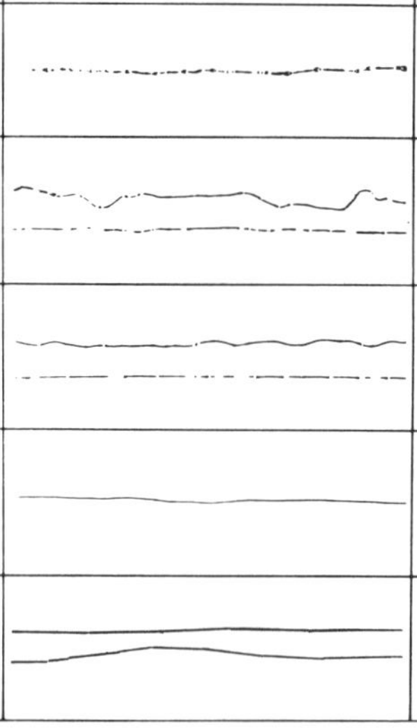

V_j/V_w	Optical micrographs	Structure
≃1.18		Am.+ Cry.
≃1.00		Am.+ Cry.
≃0.87		Am.
≃0.81		Am.
≃0.77		Am.

Figure 9 Variation in the macroscopic appearance of as-quenched $Fe_{75}Si_{10}B_{15}$ amorphous wire with the ratio of melt jet velocity to water layer velocity. Orifice diameter of the nozzle used was about 0.30 mm, and temperature of the molten alloy just before ejection was about 1450 K.

Figure 10 Amorphous $Co_{68}Fe_{4.5}Si_{12.5}B_{15}$ wire produced by the in-rotating-water spinning method.

magnetic materials with huge Barkhausen jump [36] and large Matteucci effect [37], and as high-strength and high-toughness materials combined with high fatigue strength and good corrosion resistance [38]. In addition to these applications, Fe- and Co-based amorphous wire has been used in many functional applications such as security sensors, rotation sensors, current sensors, rotary encoders, data tablets, cartridge displacement sensors, mechanocardiogram sensors and field sensors, reinforcement material in fishing rods and rubber sheet, fishing fiber, and cutting wire, etc.

In addition to amorphous alloy wire, crystalline phase wire has also been produced by the same spinning technique for alloy systems including Fe-Ni-Cr-Al-C [39], Fe-Mn-Al-C [40], Fe-Ni-Si-C [41], Mn-Al-C [42], Ni-Al-(Cr, Mn, Fe, Co or Cu) [43], and Fe-Si [44]. As exemplified by the $Fe_{79}Ni_{15}Si_4C_2$ wire in Fig. 11, these crystalline wires also have a circular cross section with a diameter of about 120 μm; the microstructure consists of very fine equiaxed crystalline grains. The diameter of such wire can be reduced to about 40 μm by cold-drawing, resulting in a significant increase of tensile fracture strength (from about 1000 MPa to about 4000 MPa). The remarkable increase of tensile fracture strength has been interpreted to result mainly from the fibrous formation of transformation-induced martensite from the metastable austenite. Similarly, Ni_3Al-type compound wires having high strength and good ductility have been produced in Ni-Al-X (X = Cr, Mn, Fe, Co, or Cu) systems by the in-rotating-water spinning method. There is no significant difference in the spinning conditions between crystalline and amorphous wire. However, it is more difficult to produce continuous crystalline wires because of high reactivity of molten alloy in the nozzle. Furthermore, the smoothness of the wire surface is much worse for crystalline wire than for the amorphous wire because of the existence of grain boundaries and the longer times required for completion of solidification.

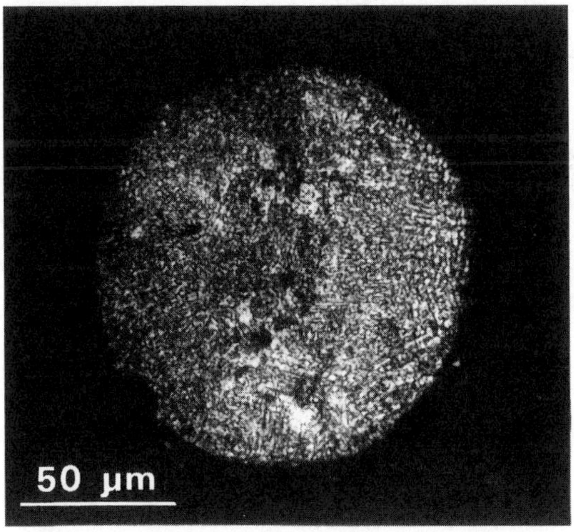

Figure 11 Optical micrograph showing microstructure of a longitudinal section of a $Fe_{79}Ni_{15}Si_4C_2$ wire produced by the in-rotating-water spinning method.

VII. CONVEYER BELT METHOD

This method was developed [13,45] by combining the advantages of both the in-rotating-water spinning method and the injection of a melt stream into confluent water (Kavesh method). Figure 12 shows a schematic illustration of the conveyer belt-type equipment. In this method, the melt stream is ejected and solidified rapidly in the water layer which is rapidly carried on a conveyer belt with grooves. The melt can be ejected at either the straight or the curved region of the belt, but wire with more uniform morphology is obtained only in the latter case. That is, in the former case, the ejected melt stream does not always lie in the running water and some regions of wire which happen to fly out from the water layer become brittle because of the precipitation of crystalline phase. In contrast, in the latter case, the melt stream always lies in the water layer due to centrifugal force. An amorphous wire with uniform morphology has continuously been produced when the centrifugal force is 30 to 50 g. It is also important in the melt spinning that the water surface has high stability without turbulence and that the water layer not contain bubbles. These conditions are strongly dependent on the shape and size of the nozzle used to supply cooling water, the ratio of ejected cooling water velocity to the velocity of the conveyer belt, and the length and vibration level of the conveyer

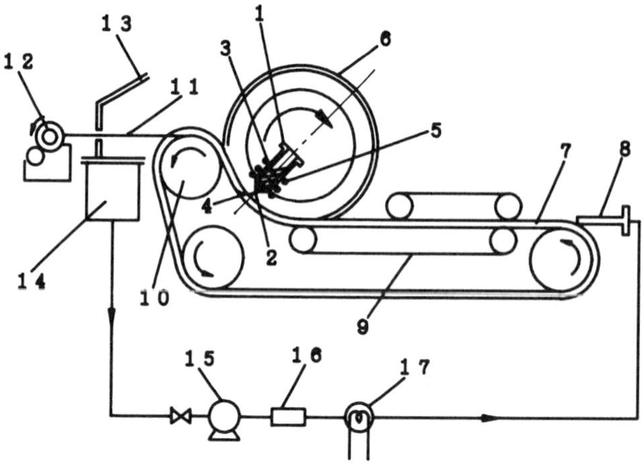

1. MELTING CRUCIBLE
2. SPINNING NOZZLE
3. MOLTEN ALLOY
4. EJECTED ALLOY
5. HIGH FREQUENCY COIL
6. GUIDE DRUM
7. GROOVED CONVEYER BELT
8. SUPPLY NOZZLE OF COOLING WATER
9. GUIDE BELLT
10. DRIVE PULLEY
11. WIRE SPECIMEN
12. WINDER
13. LIQUID BAFFLE
14. RECEPTACLE TANK
15. TRANSFER PUMP
16. FLOWMETER
17. CONDENSER

Figure 12 Schematic illustration of the conveyer belt-type spinning apparatus for the production of metallic wires.

belt. Only when these factors are optimally adjusted, can one produce continuous amorphous wire with morphology as uniform as that of the wires produced by the in-rotating-water spinning method. The wire diameter produced by the conveyer belt method is 80 to 300 μm, being the same as that for wire produced by the in-rotating-water spinning method. After solidification of an ejected melt stream, the cooling water is removed from the grooves of the belt and then wound by winder. The conveyer belt method has the following advantages over that of the in-rotating-water spinning method: (1) mass production of amorphous wires is possible because winding can be continued even after individual wires were broken off; (2) high throughput is achieved because of multiple ejection of molten alloy; (3) a higher cooling rate can be obtained because the cooling water is continuously exchanged. This method has the disadvantage that the equipment is on a large scale and that it is expensive.

VIII. COMPARISON OF WIRE-FORMING TENDENCY BETWEEN THE CONVEYER BELT METHOD AND THE IN-ROTATING-WATER SPINNING METHOD

Table 1 summarizes the continuity, toughness as defined by the number of breaks when wound around a 25-mm-diameter pin, cold-drawability, scatter of diameter, tensile fracture strength, and elongation for $Fe_{31}Co_{40}Cr_7Si_{10}B_{12}$ amorphous wire produced by the conveyer belt and the in-rotating-water spinning methods. There is no appreciable difference in the continuity, tensile strength and elongation, but the wire produced by the former method tends to have higher toughness and slightly larger scattering of diameter as compared with that produced by the latter method. The differences probably occur because the conveyer belt method has a higher cooling rate, caused by the continuous contact of the melt stream with fresh water, and a lower degree of the stability of the water surface. In addition to these fundamental properties, it was previously pointed out that the conveyer belt method has higher throughput and facilitated winding.

Table 1 Fundamental Characteristics for Amorphous $Fe_{31}Co_{40}Cr_7Si_{10}B_{12}$ Wire Produced by the Conveyer Belt and the In-Rotating-Water Spinning Methods

	Conveyer belt method	In-rotating-water melt-spinning method
1. Continuity		
No rupture	54%	78%
Below two ruptures	95%	96%
2. Toughness	0.9/100 m	1.6/100 m
3. Cold-drawability	1.3/1000 m	2.1/1000 m
4. Scattering of diameter	2.3%	1.3%
5. Fracture strength	3180 MPa	3150 MPa
6. Fracture elongation	2.2%	2.2%

IX. CONCLUDING REMARKS

In this chapter were described the four methods of producing wire of circular cross section directly from the melt, production conditions, and the continuity, shape, morphology, and toughness of the resultant Fe- and Co-based amorphous wires. In particular, the in-rotating water spinning method and the conveyer belt method are very useful for the production of continuous amorphous alloy wire. The Fe- and Co-based amorphous wires produced by these methods have been used practically in many application fields by utilizing the favorable characteristics of mechanical, magnetic, and chemical properties as well as the fine geometry of wire. Although the techniques to produce continuous amorphous alloy wire in Fe- and Co-based systems have almost completely been developed in the two methods compared, there is great difficulty in applying the processes to reactive metals and alloys because water is used as a cooling medium. Further progress seems to be attributed to the development of the production technique as well as to alloy composition of wire materials with new functional properties.

REFERENCES

1. L. Rayleigh, *Proc. Roy. Soc.*, *29*, 71 (1879); *34*, 130 (1882).
2. C. Weber, *Z. Angew. Math. Mech.*, *11*, 136 (1931).
3. I. Ohnaka and T. Fukusako, *J. Japan Inst. Metals*, *42*, 415 (1978).
4. H. H. Liebermann, *J. Appl. Phys.*, *50*, 6773 (1979).
5. G. F. Taylor, *Phys. Rev.*, *23*, 655 (1924).
6. A. V. Ulitovskiy, *Prekory Tech. Eksper.*, *3*, 115 (1957).
7. H. Wagner, *Wire J.*, *6*, 871 (1961).
8. H. S. Chen, *Acta Metall.*, *22*, 1505 (1974).
9. S. Kavesh, *AIChE, Symp. Ser.*, *74*, 1 (1978).
10. S. Kavesh, in *Metallic Glasses* (ASM, Metals Park, OH, 1977), p. 36.
11. I. Ohnaka, T. Fukusako, and T. Matui, *J. Japan Inst. Metals*, *45*, 751 (1981).
12. T. Masumoto, I. Ohnaka, A. Inoue, and M. Hagiwara, *Scripta Metall.*, *15*, 293 (1981).
13. A. Tanimura and M. Hagiwara, *Function Mater.*, *6*, 32 (1986).
14. T. Hamashima, H. Tubata, and M. Hagiwara, U.S. Patent No. 4,607,683 (1986).
15. T. Goto, *Metals*, *54*, 31 (1984).
16. T. Goto, *Bull. Japan Inst. Metals*, *15*, 633 (1976).
17. T. Goto, *Bull. Japan Inst. Metals*, *20*, 176 (1981).
18. J. Nixdorf, *Proc. R. Soc. London*, *A-319*, 17 (1970).
19. K. Narita, *Bull. Japan Inst. Metals*, *18*, 8 (1978).
20. S. Kavesh, U.S. Patent No. *3*, 960,200 (1976).
21. I. Ohnaka, Japanese Patent No. *64*, 948 (1980).
22. H. Tsubata, S. Tamamura, and A. Tanimura, U.S. Patent No. *4*, 617,983 (1986).
23. A. Inoue and T. Masumoto, *Eng. Mater.*, *30*, 47 (1982).
24. M. Hagiwara, A. Inoue, and T. Masumoto, *Metall. Trans.*, *13A*, 373 (1982).
25. M. Hagiwara, A. Inoue, and T. Masumoto, *Sci. Rep. Res. Inst. Tohoku University*, *A-29*, 351 (1981).
26. M. Hagiwara, A. Inoue, and T. Masumoto, *Metall. Trans.*, *12A*, 1027 (1981).
27. A. Inoue, M. Hagiwara, and T. Masumoto, *J. Mater. Sci.*, *17*, 580 (1982).
28. T. Masumoto, A. Inoue, M. Hagiwara, I. Ohnaka, and T. Fukusako, in *Proc. Fourth Int. Conf. on Rapidly Quenched Metals*, edited by T. Masumoto and K. Suzuki (Japan Institute of Metals, Sendai, 1981), p. 24.

29. A. Inoue, S. Furukawa, M. Hagiwara, and T. Masumoto, *Metall. Trans.*, *18A*, 621 (1987).
30. R. E. Cunningham, L. F. Rakestraw, and S. A. Dunn, in Ref. 9, p. 20.
31. M. Hagiwara, A. Inoue, and T. Masumoto, *Mater. Sci. Eng.*, *54*, 197 (1982).
32. Y. Masumoto, A. Inoue, N. Yano, A. Kawashima, K. Hashimoto, and T. Masumoto, *J. Mater. Sci.*, *20*, 97 (1985).
33. A. Inoue, H. S. Chen, J. T. Krause, T. Masumoto, and M. Hagiwara, *J. Mater. Sci.*, *18*, 2743 (1983).
34. A. Inoue, T. Masumoto, and N. Yano, *J. Mater. Sci.*, *19*, 3786 (1984).
35. Unitika Catalog, Amorphous Metals Fibers (1986).
36. H. S. Chen, R. C. Sherwood, S. Jin, G. C. Chi, A. Inoue, T. Masumoto, and M. Hagiwara, *J. Appl. Phys.*, *55*, 1796 (1984).
37. K. Mohri, *J. Mag. Soc. Japan*, *8*, 233 (1984).
38. M. Hagiwara, A. Inoue, and T. Masumoto, in *Rapidly Quenched Metals*, Vol. II, edited by S. Steeb and H. Warlimont (Elsevier, Amsterdam, 1985), p. 1779.
39. A. Inoue, H. Tomioka, and T. Masumoto, *J. Mater. Sci.*, *20*, 2603 (1985).
40. A. Inoue, H. Tomioka, M. Hagiwara, and T. Masumoto, *Trans. Japan Inst. Metals*, *23*, 341 (1982).
41. A. Inoue, N. Yano, H. Tomioka, and T. Masumoto, *Trans. Iron Steel Inst. Japan*, *26*, 993 (1986).
42. A. Inoue, H. Tomioka, and T. Masumoto, *Met. Trans.*, *14A*, 2319 (1983).
43. A. Inoue, H. Tomioka, and T. Masumoto, *J. Mater. Sci. Lett.*, *1*, 377 (1982).
44. I. Ohnaka, T. Ichiryu, I. Yamauchi, H. Tateoka, and T. Fukusako, in *Rapidly Solidified Materials*, edited by P. W. Lee and R. S. Carnonara (ASM, Metals Park, OH 1985), p. 91.
45. T. Hamashima, H. Tsubata, and M. Hagiwara, U.S. Patent No. 4,607,683 (1986).

7

Synthesis and Consolidation of Amorphous Metallic Powders

R. B. Schwarz
Center for Materials Science, Los Alamos National Laboratory, Los Alamos, New Mexico

I. INTRODUCTION

Metallic atomic bonding and the lack of long-range order combine to give amorphous metallic alloys unique and useful properties not found when the same alloy is in a crystalline state [1]. Advantageous properties include low magnetic losses; high corrosion resistance; low degradation of mechanical, electrical, and superconducting properties in irradiation environments; high strength; and improved catalytic properties. Considerable interest in amorphous alloys has arisen from the realization that many of these alloys can be prepared from relatively inexpensive elemental materials and that single-phase amorphous alloys can exist over broad composition ranges. This opens the possibility of developing new materials with physicochemical properties tailored to meet specific engineering requirements.

Since 1960, amorphous metallic alloys have been prepared by the rapid solidification (RS) techniques, whereby molten alloy is cooled to a temperature below the crystallization temperature, T_x, of the amorphous phase at rates of 10^4 to 10^{11} K s^{-1} [2]. Because of the high quenching rates required for amorphization, foils and ribbons prepared by RS techniques typically have a maximum thickness of 30 to 50 μm. Since 1983, amorphous metallic alloys have been also produced by controlled chemical interdiffusion reactions at the interface between two pure metals [3]. The amorphous layer thickness that can be formed by this solid-state reaction within reasonable times is only 0.2 μm or less, but the superposition of alternating thin foils of pure metals in principle allows for the synthesis of bulk amorphous alloys. In a related discovery, amorphous alloy powders have been produced by mechanical alloying (a form of high-energy ball milling) of a mixture of elemental powders [4,5].

Among the attractive properties of amorphous metallic alloys, none has a greater promise for use in applications than the low magnetic losses. Amorphous $Fe_{80}B_{20}$, for example, has 60-Hz hysteresis losses that are approximately 30% those of the best-oriented crystalline Fe–3 Si alloys now used in distribution transformers. None of the present methods of synthesis, however, is able to produce the bulk amorphous alloys needed for many applications. For example, the thickness of the foils now prepared by RS techniques (30–50 μm) is much less than the optimum thickness required for magnetic cores that operate at 60 Hz (the optimum thickness is determined by a balance between eddy currents, which increase with foil thickness, and packing losses, which decrease with foil thickness). Because of these and other applications, techniques are being developed to consolidate the amorphous powder, flake, and thin ribbon now available.

The goals to be achieved by the consolidation of amorphous powder or flake depend on the particular application of the product. For example, for magnetic applications, the consolidation must result in a product which has three characteristics: (1) the amorphous structure of the starting powder; (2) low porosity; and (3) free of second-phase inclusions such as oxides. Porosity and second-phase inclusions can act as magnetic pinning centers, increasing the magnetic hysteretic losses of the product. This restriction also means that the starting amorphous powder or flakes must be free of surface oxides. Because amorphous metallic alloys typically have very high strengths at temperatures up to 100 K below the glass transition temperature, T_g, achieving these goals introduces stringent requirements on the temperature-time-pressure path to be followed during consolidation. These restrictions have led researchers to explore the use of nonconventional consolidation techniques such as dynamic compaction [6,7].

The present chapter reviews the methods of synthesizing amorphous metallic alloy flake and powder, and the methods for consolidating these material forms. Regarding synthesis, we discuss in detail the newer methods based on solid-state reactions. The techniques based on the RS of melts are covered by other contributors to this volume and are only treated briefly here.

II. METHODS OF SYNTHESIS OF AMORPHOUS ALLOYS

A. Synthesis by Rapid Solidification of Melts

Amorphous alloy phases are thermodynamically metastable with respect to one or more crystalline phases. Thus, a logical procedure for preparing amorphous alloys is to start with the alloy in an "energized state" and to choose a reaction path that leads the system into the amorphous metastable state, while kinetically restricting it from reaching crystalline states of lower free energy. For synthesis methods based on the RS of melts, the energized state is the molten alloy state. Heat is removed from the melt at a rate sufficiently high to prevent crystallization until the temperature of the undercooled melt becomes lower than T_g. Below T_g, the undercooled melt becomes "frozen" into the glassy (amorphous) state.

Well-known ionic and covalent glasses are easy to form by the RS methods. In these alloys, the directional nature of the interatomic bonds limits the rate at which the atoms or molecules can rearrange to maintain thermodynamic equilibrium during cooling; thus, crystallization can be avoided even at cooling rates as low as 1 K s^{-1}. In contrast to covalent alloys, interatomic bonds in metallic alloys are largely nondirectional. Therefore, metallic melts exhibit far less resistance to crystallization when undercooled and do not form glasses unless cooled at much higher rates. Amorphous metallic alloys were produced in the 1950s by the condensation of metal vapors onto cryogenically cooled substrata [8]. About the same time, Brenner et al. reported the formation of amorphous Ni-P alloys by electrodeposition [9]. Research on amorphous metallic alloys increased rapidly following the experiments of Duwez and co-workers [2], who in 1960 developed a variety of methods for the RS of melts at cooling rates of 10^6 to 10^8 K s^{-1}. These rates are obtained, for example, by rapidly spreading molten alloy, in a thin overlay, onto the surface of a massive heat sink with high thermal conductivity. Variants of these methods are being used to manufacture thin magnetic amorphous ribbon for the cores of electrical transformers and other electromechanical devices [10]. Techniques based on the RS of melts are described elsewhere in this volume and are not discussed in detail here. We will return to this topic, however, in Section 7.2.3 to discuss the composition range over which amorphous alloys can be prepared by RS techniques.

B. Synthesis by Isothermal Solid-State Reactions

The synthesis of amorphous alloy by a solid-state amorphization reaction (SSAR) is based on a chemically driven interdiffusion reaction. Here the starting energized state has excess chemical energy and, as in the RS of melts, the (metastable) amorphous state is reached by controlling the reaction kinetics to prevent the formation of unwanted crystalline phases, this time in an isothermal environment. The first examples of such crystal-to-amorphous transformations were probably the hydriding experiments of Oesterreicher et al. [11], who in 1976 noted that amorphous hydrides are produced when crystalline $LaNi_2$, $LaNi_3$, and La_2Ni_7 are reacted with hydrogen at about 100 atmospheres and ambient temperature. Soon thereafter, van Diepen and Buschow [12] and Malik and Wallace [13] reported the formation of amorphous $CeFe_2H_x$ and $GdNi_2H_{4.35}$ by the adsorption of hydrogen into crystalline $CeFe_2$ and $GdNi_2$, respectively. In 1983, Yeh et al. [14] found that Zr_3Rh would also amorphize when hydrogenated, and a systematic study was undertaken to understand this reaction. These studies showed that chemical energies can drive a crystal-to-amorphous transformation in the solid state. The hydrogen atom, being relatively small, can easily diffuse in certain crystalline intermetallics, at temperatures lower than that necessary for the diffusion of the metal atoms. Thus, the reaction can occur at temperatures below the crystallization temperature of the amorphous hydride. Hydrogen diffusion also takes place at higher temperatures, but the product is then a crystalline hydride [14].

Amorphization by solid-state reactions is not limited to hydrogen diffusing into crystalline intermetallics. In 1972, Bolotov and Kozhin [15] observed the formation of an amorphous alloy by the reaction of vacuum-deposited thin films of amorphous tellurium and crystalline silver. This SSAR was later confirmed in the same system by Hauser [16]. These experiments, and similar experiments by Herd et al. [17], showed that metals can diffuse at low temperatures into amorphous semiconductors such as selenium, tellurium, and silicon without causing the amorphous semiconductor to crystallize. Schwarz and Johnson [3] reported in 1983 the first example of two pure, crystalline metals reacting to form a single-phase amorphous alloy. In this experiment, vacuum-deposited thin films of pure gold and lanthanum were thermally reacted at 70°C for a few hours to form a single-phase amorphous Au-La alloy.

The fundamentals of the SSAR can be easily explained with the help of Fig. 1. Part (a) of this figure shows the equilibrium phase diagram for a hypothetical binary alloy, A-B. Phases α and β are crystalline terminal solutions and phase γ is a crystalline intermetallic. Part (b) of this figure shows the Gibbs free energy of phases α, β, and γ, and of the liquid phase λ, evaluated at the reaction temperature T_r. At this temperature, phase λ is thermodynamically metastable with respect to the crystalline phases α, β, and γ, or mixtures of these phases. We assume that curve λ, the free energy of the undercooled liquid, gives a good representation for the free energy of the amorphous alloy phase. The binary system chosen here is

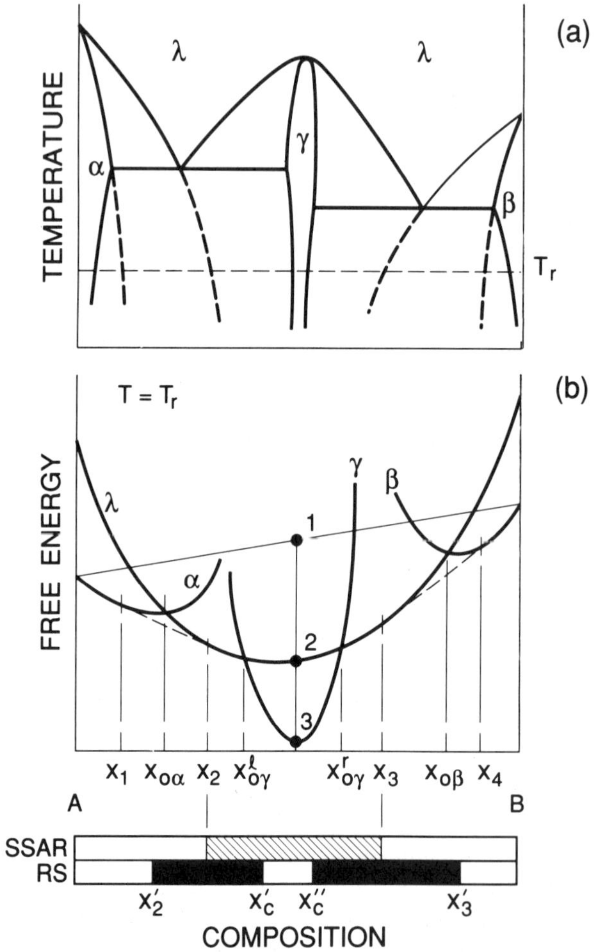

Figure 1 (a) Schematic phase diagram for a binary system with a negative heat of mixing in the liquid state. (b) Free-energy diagram for the phases in (a) at the temperature T_r. The dashed and black bars at the bottom of the figure give the homogeneity range(s) of the amorphous alloy phase that can be formed by the SSAR and RS methods.

particular in that the liquid (amorphous) phase has a negative heat of mixing, as indicated by the strongly concave curvature of curve λ. The negative heat of mixing of phase λ is a requirement for the SSAR method [3].

In the SSAR method the initial high-energy state is a mixture of crystalline metals A and B. For an equimolar mixture of A and B, the free energy of the initial state is at point 1 in Fig. 1b, half way along the straight line joining the free energies of crystals A and B. If A and B are allowed to interdiffuse freely, they will tend to form the stable crystalline intermetallic γ which has the lowest free energy (point 3). The essence of the SSAR method is preventing reaction 1 → 3 from occurring, and instead kinetically favoring reaction 1 → 2. This favoring is achieved by a proper choice of metals A and B, the reaction temperature T_r, and the reaction time, t_r.

Schwarz and Johnson [3] proposed two conditions for the formation of an amorphous binary alloy by a SSAR: (1) the two elements must have a large negative heat of mixing in the liquid (amorphous) state; and (2) the two elements must have largely different diffusion rates into each other and in the amorphous alloy phase. The negative heat of mixing provides the thermodynamic driving force for the reaction and experience has shown that all binary systems where the SSAR has been observed [18,19] obey this condition. The large difference in diffusivities is necessary for favoring the kinetics of reaction path 1 → 2 in Fig. 1b over that of reaction path 1 → 3. Binary systems that obey these two conditions include alloys of an early transition metal (ETM), typically elements from groups IVA and VA, and a late transition metal (LTM) or a noble metal. For these pairs of metals, the tracer diffusivity of the LTM in a crystal of the ETM is usually much larger than the self-diffusivity of either element [20]. This asymmetry in diffusivities has been correlated with the difference in atomic sizes, as shown in the inset to Fig. 2 for the diffusion of LTMs in crystalline zirconium [21]. Researchers [20] believe that the fast diffusion of LTMs in ETMs occurs because some of the LTMs dissolve interstitially and that their diffusion is dominated by some type of interstitial migration.

The dependence of diffusivities on atomic size found in crystalline materials appears also in the amorphous state. Figure 2 shows the measurements of Hahn and co-workers [22,23], where the tracer diffusivities of transition metals in amorphous $Ni_{50}Zr_{50}$ is a strong but smooth function of their atomic size. In particular, the diffusivity of nickel in amorphous Ni-Zr alloys exceeds that of titanium by more than three orders of magnitude. Hahn and co-workers suggest that an interstitial diffusion mechanism similar to that suggested for crystalline ETM-LTM alloys must operate for the diffusion of nickel in amorphous Ni-Zr alloy. There is no data for zirconium diffusion in amorphous $Ni_{50}Zr_{50}$. Taking the metallic radius of zirconium to be 0.16 nm, and assuming that its diffusivity in amorphous $Ni_{50}Zr_{50}$ conforms to that of the other transition elements in Fig. 2, we can estimate that its diffusivity will be about four orders of magnitude lower than that of nickel. This difference in the diffusivities of the two elements in the binary amorphous alloy is crucial to the operation of the SSAR mechanism, as we explain next.

The SSAR requires clean interfaces between metals A and B. One way to obtain such interfaces is by depositing metals A and B as thin films in a high vacuum [3]. Figures 3 and 4 help explain why at the reaction temperature, T_r, reaction path 1 → 2 is preferred over reaction path 1 → 3. Figure 3 shows the electrical resistance,

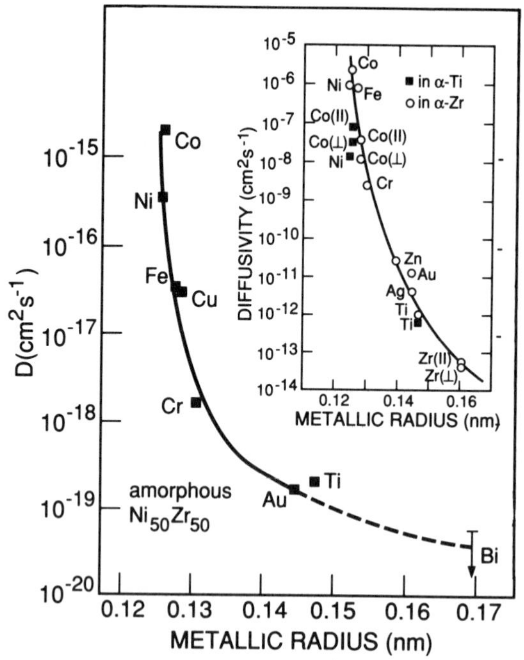

Figure 2 Dependence of tracer diffusion on atomic radius in amorphous Ni-Zr at T = 573 K (after Ref. 21) and in crystalline α-Zr and α-Ti at 1100 K (after Ref. 22).

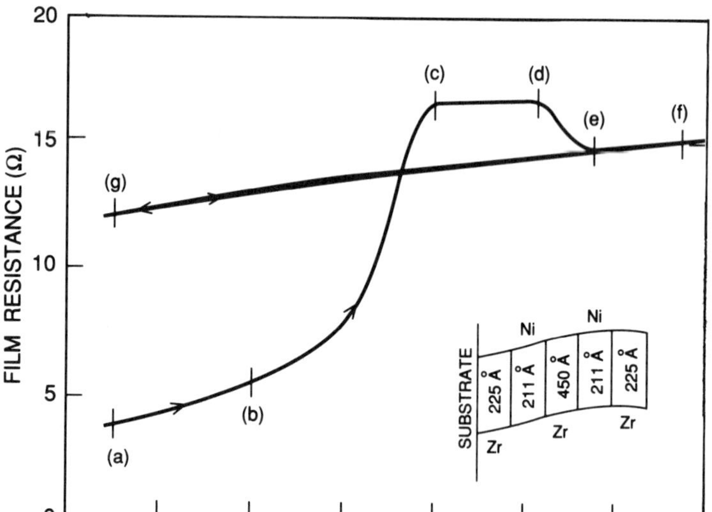

Figure 3 Resistance (R) of a multilayer system of nickel and zirconium thin films during the continuous heating and cooling at 10 K min^{-1}. The thermal cycle was repeated twice. The letters (a) to (e) correspond to those in Fig. 4.

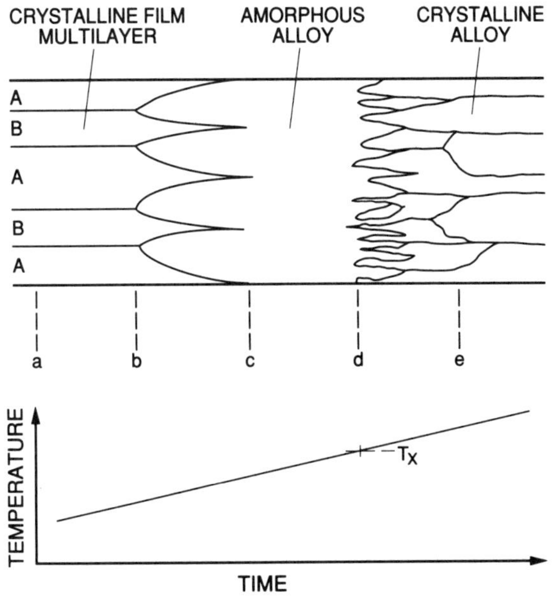

Figure 4 Schematic of the interdiffusion reactions taking place when the thin film multilayer in the inset of Fig. 3 is annealed at a constant heating rate. The letters a to e correspond to those in Fig. 3.

R, of a stack of alternating thin films of nickel and zirconium as a function of temperature. This data was obtained [24] during continuous heating from 300 to 1000 K, followed by cooling to 300 K, at the rate of 10 K min^{-1}. Figure 4 schematically shows the structural changes occurring in the films during this thermal cycle, as evidenced by x-ray diffraction. The initial thickness of the films, indicated in the inset of Fig. 3, are those necessary to form an equimolar amorphous alloy film. Letters have been added to both figures to indicate intermediate steps in the reaction. Between a and b the temperature is too low for interdiffusion and *R* increases almost linearly with increasing temperature, as expected for pure crystalline metals. At b, amorphous alloy begins to form at each of the Ni-Zr interfaces. Because the resistivity of the amorphous alloy is higher than that of the pure nickel and zirconium used to form it, the resistance of the multilayer film increases. The SSAR ends at c, when all the Ni and Zr have been consumed. Resistance is constant from c to d and agrees with the usual observation that the resistivity of amorphous alloys is largely temperature-independent. At d, a crystalline alloy phase nucleates in the amorphous alloy previously formed and the resistance starts to drop as long-range order develops. From d to e, the resistance decreases as the transformation proceeds and the crystallites grow in size. At e, the alloy has reached the thermodynamically stable crystalline state that is impervious to further temperature variations. For further cooling and heating cycles, e-f-g, the resistivity of the crystalline alloy has the expected positive linear temperature dependence.

It is clear from Fig. 3 that the temperature at (b), T_b, at which the amorphization commences must be significantly lower than the temperature at (d), T_d, at which crystallization of the amorphous alloy starts. In our current understanding of the

SSAR, we identify T_b as the temperature where nickel, the smaller of the two atoms, becomes mobile in the amorphous alloy formed at each of the Ni-Zr interfaces. This mobility is sufficient to enable nickel atoms to diffuse from the nickel film to the zirconium film—across the amorphous alloy already formed—to form additional amorphous phase. At T_d, zirconium also becomes mobile in the amorphous phase and this enables the alloy to crystallize. Thus, the temperature difference, $T_d - T_b$, is the "temperature window" for the SSAR. Certainly, this temperature window depends on heating rate. In isothermal experiments, T_r is chosen somewhere near the lower-end of this window. At the completion of the SSAR, the amorphous layer has a modulated composition which becomes homogeneous through further annealing.

Choosing crystalline films of different thickness enables us to prepare amorphous alloys over a broad composition range. A nice feature of the SSAR is that the products obtained can be predicted from free-energy diagrams [3]. Returning to our schematic diagram in Fig. 1b, the common-tangent rule applied to phases α, β, and λ predicts five metastable reaction products: a crystalline solid solution, α, for $0 < x < x_1$; a two-phase mixture of $\alpha(x_1)$ and $\lambda(x_2)$; a single-phase amorphous alloy, λ, for $x_2 < x < x_3$; a two-phase mixture of $\lambda(x_3)$ and $\beta(x_4)$; and the crystalline solid solution, β, for $x_4 < x < 1$.

The synthesis of amorphous Ni-Zr thin films by a SSAR between alternating layers of pure nickel and zirconium, as just described, may seem a convoluted procedure because the same amorphous film can be prepared by simply codepositing the alloy from two separate evaporation sources. The importance of the SSAR, however, is that a three-dimensional analog of the thin-film multilayer diffusion experiments should offer a means of synthesizing bulk amorphous alloys. Experiments with this goal have been performed with elemental foils and powders. Schultz [25] prepared 1-mm-diameter wires of amorphous $Ni_{68}Zr_{32}$ by a SSAR in rolled foils of elemental nickel and zirconium. The foils had initial thickness of tens of micrometers, which is too thick for the SSAR. Schultz then rolled and swaged these foils until the individual layers attained a thickness of about 0.1 μm. These composites were annealed in vacuum at temperatures between 500 K and 525 K. In related experiments, Atzmon et al. [26] reacted stacks of alternating foils of Ni and Zr, Cu and Zr, Ni and Er, and Cu and Er which had been rolled to decrease the thickness of the individual foils and to improve the metal-to-metal contact by breaking the surface oxides. Following repeated rollings and annealings, they obtained foils 20- to 200-μm thick which were mostly amorphous, with traces of the starting crystalline materials. These authors also prepared a Ni-Zr alloy [27] by cold-rolling a stainless steel tube containing a mixture of crystalline powders of nickel and zirconium. Following thermal annealings the product was about 80% amorphous. These experiments illustrate some of the difficulties encountered in scaling up the laboratory (thin film) SSARs to a single-reaction, bulk production of amorphous alloy.

C. Synthesis of Amorphous Powders by Mechanical Alloying

A large variety of amorphous alloy powders can be prepared by mechanical alloying (MA). MA is a high-energy ball-milling process originally developed to produce oxide-dispersion-strengthened nickel-based alloys [28,29]. In the synthesis of amorphous alloy powders by MA, the starting material is a mixture of pure elemental powders [4,5,30]. The MA process repeatedly deforms, fractures, and cold-welds the particles trapped between the colliding balls in the mill. This causes a refinement of the microstructure and an intimate mixing of the constituents. After only a short attrition time, the particles have a layered structure consisting of alternating layers of the starting crystalline elements, as shown in Fig. 5. This structure resembles that of the thin films used in the SSAR experiments (see inset to Fig. 3). With increasing attrition time, the thickness of the individual layers decreases as an amorphous layer forms at each metal-metal interface. For a sufficiently long attrition time, the process leads to the formation of a homogeneous amorphous powder. The efficiency of the attrition process and the rate of amorphous alloy formation depends on the attrition apparatus (attrition energy, number of balls, etc.) and milling temperature. The x-ray diffraction patterns in Fig. 6 show the evolution of the structure of an equimolar mixture of aluminum and hafnium powders ball-milled in a laboratory-size Spex model 8000 mixer [31]. For this combination of powder compositions and ball mill unit, and using hexane as a dispersant, 5 hr of MA were sufficient for the formation of the amorphous phase. To date, the majority of the amorphous alloy powders that have been prepared by this technique are binary alloys of the early transition metals, titanium and zirconium, and 3*d* transition metals: Mn, Fe, Co, Ni, Cu [18,19,30,32]. All of these metal-metal systems fulfill the two requirements for the SSAR we discussed previously. Other systems in which amorphous alloy powder has been prepared by MA are Ti-Pd, Nb-Ni, Hf-Al, Fe-Al, Cu-Nb-Sn, Sn-Ni, and Ti-Pd-Cu [19,33].

Although the synthesis of amorphous metal-metal alloy powders by MA is quite straightforward, the synthesis of amorphous metal-metalloid alloy powders, such as $Fe_{80}B_{20}$, has been found more difficult. The author knows of no example for the synthesis of amorphous $Fe_{80}B_{20}$ by a SSAR in thin films. Schultz and Co-workers

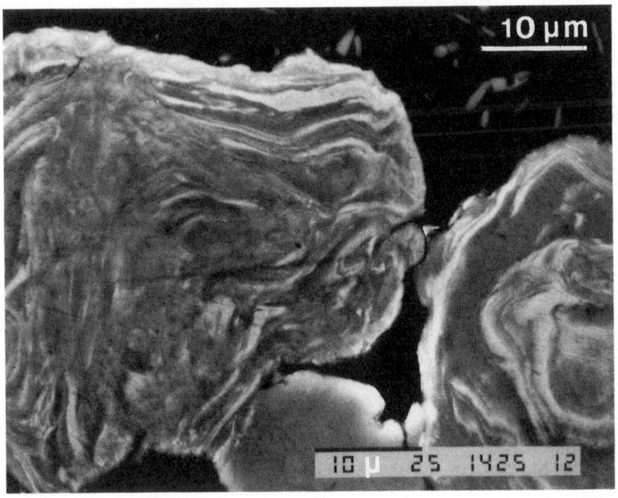

Figure 5 Scanning electron micrograph of sectioned powder particles after mechanically alloying a mixture of nickel and titanium powders for 2 hr.

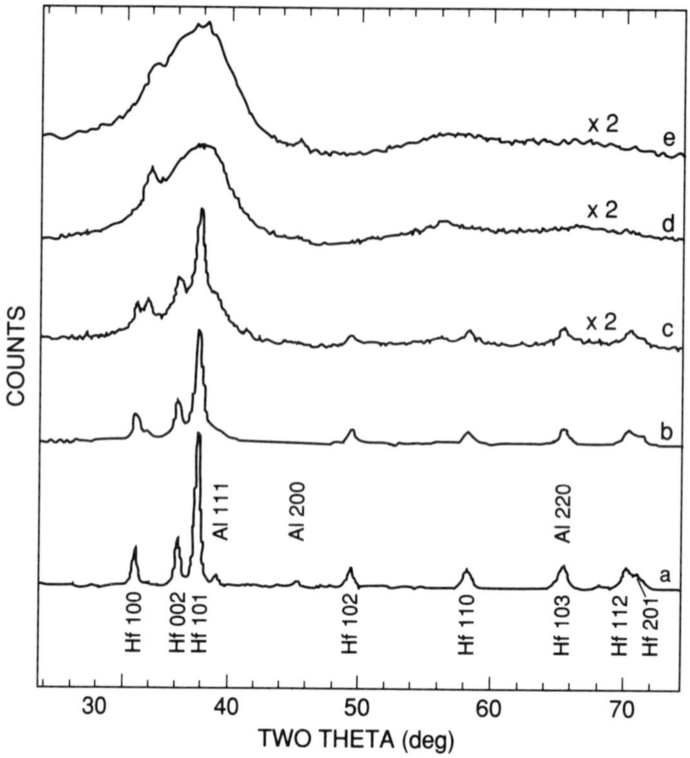

Figure 6 X-ray diffraction intensity as a function of scattering angle for a mixture of aluminum and hafnium powders after mechanically alloying for (a) 1 hr, (b) 2 hr, (c) 3 hr, (d) 4 hr, and (e) 5 hr (after Ref. 31).

[34] studied the formation of amorphous $(Fe_{0.75}Zr_{0.25})_{100-x}B_x$ by MA starting from elemental powders. The milling first produces a layered microstructure of Fe and Zr with small B particles embedded at the layer interfaces or within the layers. Further milling leads to an amorphous Fe-Zr phase with finely dispersed B particles. During additional diffusion treatment at temperatures below the T_x of the Fe-Zr phase, boron diffuses into the alloy to form the ternary amorphous phase. However, the authors could not form amorphous $Fe_{80}B_{20}$ by MA. The MA process was quite different in the absence of Zr. As for the ternary alloy, the initial milling of elemental Fe and B produced fine boron particles dispersed in crystalline Fe. However, contrary to the ternary alloy, the B particles did not dissolve upon annealing the alloy unless the annealing temperature was significantly higher than T_x of amorphous $Fe_{80}B_{20}$. The difficulty in forming amorphous $Fe_{80}B_{20}$ by MA, or by a SSAR, seems to be a consequence of the similar values of diffusivity of Fe and B in Fe-based amorphous alloys, a condition contrary to the second requirement for a SSAR [3]. Greer [35] noticed that the metal and metalloid diffusivities in Fe-based glasses are remarkably close and suggested that, contrary to previous belief, metalloid atoms in metal-metalloid glasses do not diffuse interstitially, but that their movement is linked to that of the metal atoms. Structural analyses of amorphous alloys such as $Fe_{80}B_{20}$ show no metalloid-metalloid nearest neighbors. Struc-

tural models of these alloys [36] suggests that each metalloid atom is surrounded by a trigonal prism of metal atoms, as in crystalline compounds of similar composition. It has been suggested [35] that defects in the network (analogous to broken bonds in silicates) are necessary for diffusion and that at these defects, diffusional jumps of both metal and metalloid atoms are possible. These observations seem to explain the difficulty in forming amorphous $Fe_{80}B_{20}$ by a SSAR or by MA. The presence of 25 at% Zr (which also has a large negative heat of mixing with Fe) seems to prevent the mobility of the metal atoms (Fe and Zr) at the temperature necessary for the diffusion of B in amorphous Fe-Zr.

In trying to understand the amorphization mechanism during MA, it is also natural to think that the severe plastic deformation of the powder particles trapped between milling balls may locally raise the temperature of the particles sufficiently to form thin layers of alloy melt. Following the collisions, these melt pools would solidify rapidly by heat conduction into the less deformed, and thus cooler, interior regions of the particles. Thus, the amorphous alloy would form by a cumulative RS process. However, two observations contradict this assumption: (1) although there is no direct measurement of the temperature increase in the particles during ball collisions, calculations indicate that the increase is at most a few hundred kelvins [37], and (2) the composition range of the single-phase amorphous alloys obtained by MA does not agree with the range obtained by RS at cooling rates of 10^6 to 10^8 K s^{-1} [38,39]. Researchers now believe that amorphization during MA is not a purely mechanical process and that a SSAR similar to that observed in thin films occurs during MA [30]. The conditions for a SSAR to occur are certainly present since the plastic deformation, cold-welding, and fracturing of the powder particles during MA generates a large density of clean metal-metal interfaces. Furthermore, the chemical interdiffusion necessary for the SSAR is enhanced by the momentary increase in the temperature of the particles trapped in ball collisions and by the point and other structural defects generated by plastic deformation. In agreement with this model, the alloy products obtained by MA are close to those predicted by free-energy diagrams evaluated at the average ball-milling temperature [30].

The synthesis of amorphous alloy powders can also be accomplished by a combination of MA and SSAR. A few hours of MA is usually sufficient to form particles having a layered structure of the starting elemental powders. If this structure is sufficiently fine, then annealing the powder leads to the completion of the reaction for the formation of the amorphous phase [31]. This observation further supports the view that the amorphization during MA occurs by SSARs at the clean interfaces created by MA processing.

Most of the experiments on MA have been done using laboratory-scale attritors or planetary ball mills capable of producing approximately 10 g of alloy powder per batch. A few researchers have used rotating-arm attritors capable of producing a few hundred grams of powder [40]. There is no doubt, however, that the milling conditions can be duplicated in rotating-drum ball mills having diameters of one or more meters which are capable of producing a ton or so of amorphous powder per batch [41]. Thus, provided methods are developed to consolidate these powders, MA may become an important tool for the industrial production of bulk amorphous alloys.

Several other solid-state methods for synthesizing amorphous metallic alloys have been developed over the last 10 years. These are based on: (a) reacting certain crystalline intermetallics with hydrogen to form amorphous hydrides, followed by the removal of the hydrogen [13,14]; (b) disordering crystalline lattices by irradiation with energetic particles [42,43]; (c) disordering crystalline lattices by mechanical deformation [44,37]; and (d) forming metastable crystalline phases at high-pressures, followed by a depressurization at low temperatures [45,46]. These methods have been summarized in various reviews [18,47–49] and are not discussed here. In practice, method (c) is quite similar to MA, the difference is that the ball milling is done on intermetallic powders [50,51] rather than on mixtures of elemental powders. The processing time for method (c) and MA are also similar [52].

D. Comparison of the RS and Solid-State Methods of Synthesis

The composition range for single-phase amorphous alloy formation, known as the glass-forming range (GFR), depends on the synthesis method. Although the GFRs of metal-metal amorphous alloys prepared by the RS of melts and by SSARs may be different, for those compositions at which the GFRs for the two methods overlap experience has shown that the alloys prepared by the two methods are structurally equivalent. The comparison is based on properties such as thermal stability (crystallization temperature, enthalpy of crystallization, and apparent activation energy for crystallization) [5,52,53], atomic structure (as described by atomic-par radial distribution functions) [30,54], superconducting transition temperature [32,55–57], Mössbauer spectroscopy [58], and hydrogen storage characteristics [59]. In a few instances, differences in properties were found, but these could be traced back to differences in composition, notably to the oxygen content in the alloy [57]. The general conclusion from these studies is that the "amorphous" state in metallic alloys is a unique and well-defined metastable state that may be reached by a variety of methods, including isothermal solid-state methods such as SSARs and MA. In making a structural comparison between amorphous alloys prepared by different techniques, the alloys must be in the same state of internal relaxation.

The factors determining the GFRs for both the RS and the SSAR synthesis methods are now reasonably well understood. For the RS of melts method, the GFR is mainly determined by the competition of kinetic factors (melt cooling rate versus rate of crystallization), whereas for the SSAR method the GFR is determined by metastable thermodynamic equilibrium. We now discuss how the GFRs can be predicted for each synthesis method.

The prediction of the GFR for the RS of melts has been the subject of many investigations over the last 20 years [1]. It has long been recognized that the retention of the amorphous structure on undercooling the liquid involves a combination of thermodynamic and kinetic factors. Crystallization may be polymorphous or it may involve solute redistribution, either by partitioning between melt and crystal or by local reordering at the melt-crystal front. If solute partitioning is necessary for crystallization, then the melt-crystal interface velocity is diffusion-limited and for low values of undercooling this velocity is on the order of meters per second [60,61]. Because impurity transport in the melt (and glass) is thermally activated, crystallization involving solute partitioning can be suppressed relatively easily by increasing the melt cooling rate.

Polymorphous (partitionless) crystallization of an alloy $A_{1-x}B_x$ becomes thermodynamically possible when the liquid is undercooled below the temperature $T_0(x)$ at which the free energies of the melt and a crystal of identical composition are equal. Figure 1b shows that at T_r there is a thermodynamic driving force for partitionless crystallization into three different phases: the α-phase for $0 < x < x_{0\alpha}$, the γ-phase for $x^l_{0\gamma} < x < x^r_{0\gamma}$, and the β-phase for $x_{0\beta} < x < 1$. Because the melt-crystal interface velocity during polymorphous crystallization can reach values close to the speed of sound in the liquid [62], that is, approximately three orders of magnitude higher than that for partitioning crystallization, the cooling rate needed to suppress partitionless crystallization is much higher than that needed to suppress partitioning crystallization. Thus, if the cooling rate is sufficiently high to avoid polymorphous crystallization, then the undercooled melt will most likely be trapped in the glassy (amorphous) state before it has a chance to crystallize.

Recently, Nash and Schwarz [63] used the classical theory of homogeneous nucleation and a numerical modeling of phase equilibrium data to calculate the temperature $T'(x, z, dT/dt)$ at which a volume fraction , z, of crystalline material of composition $A_{1-x}B_x$ was formed by polymorphous crystallization during continuous cooling of the melt at a given cooling rate, dT/dt. They defined the undercooled melt as being amorphous if $z < 10^{-6}$ at $T = T_g$. This z value is well below the volume fraction of crystalline material which can be detected with most experimental techniques (for example, x-ray diffraction). With this criterion they calculated $T'(x, z = 10^{-6}, dT/dt)$. Finally, they deduced the GFR from the compositions at which the $T'(x, z = 10^{-6}, dT/dt)$ and $T_g(x)$ curves intersect. The essence of these calculations is that for those alloy compositions for which $T' < T_g$, as the temperature of the molten alloy is lowered the alloy will become trapped in the glassy state before it has a chance to crystallize polymorphously.

Figure 7 shows the $T'(x)$ curves for the terminal solutions (solid curves) and the three equilibrium intermetallic compounds of Ni-Ti (dashed curves) evaluated for $z = 10^{-6}$ and a cooling rate of $dT/dt = 10^6$ K s^{-1}. Because the thermodynamic modeling treats the three intermetallics as line compounds, only one point (triangle) has been calculated for each intermetallic T' curve. These $T'(x)$ curves were given, arbitrarily, a width $\Delta x = 0.06$ at $T = T_x$. The GFR predicted by Fig. 7 is in good agreement with the measured GFR from RS experiments. In particular, the model predicts the difficulty in synthesizing amorphous Ni-Ti alloys by rapidly quenching melts with compositions close to those of the congruent-melting intermetallics, NiTi and Ni_3Ti, and the two terminal solutions. This occurs because the undercooled melt has little resistance to crystallize polymorphously at those compositions. As a consequence, the GFR for Ni-Ti, and for most ETM-LTM alloys prepared by RS methods, consists of one or more narrow composition ranges centered near the compositions of deep eutectics in the equilibrium phase diagram (that is, *away* from the compositions of high-melting-temperature intermetallics and terminal solutions).

With increasing cooling rate the $T'(x)$ curves shift downwards; the various regions of the GFR increase in width and eventually overlap. For cooling rates of 10^{12} K s^{-1} or higher, the GFR for ETM-LTM alloys prepared by RS methods is

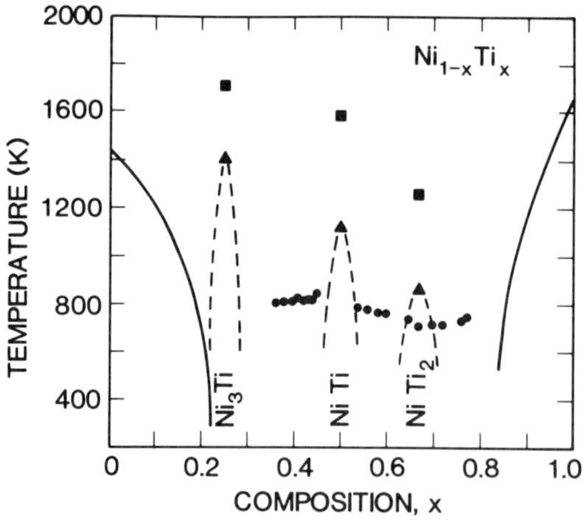

Figure 7 Temperature T' for which a volume fraction $z = 10^{-6}$ of crystalline material is formed by partitionless solidification of Ni-Ti melt as a function of composition, x, at the cooling rate of 10^6 K s^{-1}. The solid lines are the $T'(x)$ curves for the terminal solid solutions. The dashed $T'(x)$ curves for the intermetallic compounds are schematically drawn through the calculated single T' value (solid triangle). The solid circles are the crystallization temperatures of amorphous Ni-Ti alloys prepared by RS techniques (after Ref. 63).

usually continuous and covers a wide range of compositions. Lin and Spaepen [64] have shown, for example, that the glass $Ni_{60}Nb_{40}$ (eutectic composition) can be prepared using cooling rates of 10^6 K s^{-1}, and making the glass $Ni_{0.75}Nb_{0.25}$ (composition of the congruent-melting compound Ni_3Nb) requires cooling rates of approximately 10^{13} K s^{-1}. Such high cooling rates are obtained during the solidification of molten alloy overlays formed in situ on metal surfaces by short high-energy laser pulses. The GFR for the RS methods may be even narrower than dictated by thermodynamics and kinetics because of the difficulty in finding crucibles which will not react with melts of congruent-melting intermetallics prior to their rapid solidification.

As with the RS method, amorphization by a SSAR is restricted to certain alloy compositions. The GFR for the SSAR method is determined by metastable equilibrium thermodynamics, as discussed in Section IIB. The bars at the bottom of Fig. 1b schematically show the GFR for the RS and SSAR synthesis methods when applied to a binary system of two transition metals having an equilibrium phase diagram similar to that in Fig. 1a. The GFR for the SSAR method, and for MA in particular, is the range (x_2, x_3). This range is continuous and is located near the center of the composition regime. In contrast, the GFR for the RS method is fragmented, with two relatively narrow ranges (x_2', x_c') and (x_c'', x_3'), located near the compositions of the two eutectics in the equilibrium phase diagram. These differences are clearly illustrated in Fig. 8, which compares the GFRs of various ETM-LTM amorphous alloys prepared by both RS and MA techniques [39]. We have added to Fig. 8 recent data of Rubin and Schwarz [65] giving the GFR of Ni-Zr alloys prepared by codeposition in ultrahigh vacuum. This method of synthesis is akin to a RS technique at a cooling rate on the order of 10^{12} K s^{-1}. Crystallization can be avoided at all compositions for codeposited films, except for $x < 0.1$ and

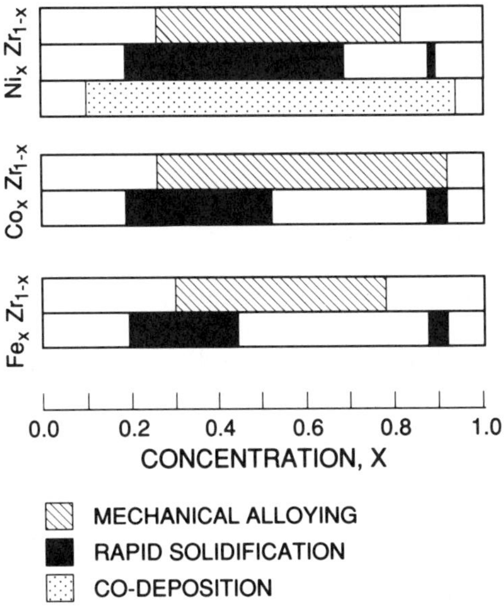

Figure 8 Homogeneity range(s) of the amorphous phase of M-Zr (M = Ni, Co, Fe) alloys prepared by mechanical alloying, rapid solidification of melts, and codeposition of the pure metals in vacuum (after Refs. 39 and 65).

$x > 0.95$, where the formation of the terminal solid solutions cannot be avoided even at cooling rates of the order of 10^{12} K s^{-1}.

III. CONSOLIDATION OF AMORPHOUS POWDERS

Powders can be compacted and consolidated by a variety of well-known techniques. These include cold-pressing followed by high-temperature sintering, cold and hot extrusion, hot isostatic pressing (HIP), and injection molding. The use of one or another method is usually dictated by the morphology of the powders, their intrinsic characteristics (yield strength, chemical diffusivity), and the need to prevent or induce structural changes in the compacted product brought about by grain growth or phase transformations. Amorphous alloys, being thermodynamically metastable with respect to the crystalline state, cannot be heated to temperatures close to T_g for long times without running the risk of inducing crystallization. In addition, amorphous alloys have high yield strengths. These two characteristics of amorphous alloys severely limit the number of consolidation techniques that can be used. The densification and consolidation of amorphous alloy powders requires special temperature-pressure-time paths which avoid the formation of crystalline phases. In the absence of heterogeneous nucleants, amorphous alloys crystallize homogeneously when heated above T_g, following the spontaneous formation of crystalline embryos. The classic homogeneous nucleation theory [66] predicts that the rate of homogeneous nucleation per unit volume, I_v, follows the expression:

$$I_v = N_v f_0 \exp\left(-\frac{K\gamma^3 V_m^2}{(\Delta G)^2 kT}\right) \tag{1}$$

where

N_v = number of atoms per unit volume of amorphous alloy
f_0 = frequency at which atoms attach to the critical nucleus
K = shape factor ($=16\pi/3$ for a spherical nucleus)
γ = amorphous-crystal interfacial energy
V_m = molar volume of the amorphous alloy
ΔG = free-energy difference between amorphous and crystal phases

The factor f_0 depends on the type of nucleation process. For partitionless crystallization, $f_0 = (kT/\hbar)\exp(-\Delta G_{\text{vis}}/kT)$, where $\hbar = h/2\pi$ and ΔG_{vis} is the activation energy for viscous flow [67]. For crystallization involving solute partitioning, f_0 must also consider the activation energy for diffusion. The volume fraction of crystal, ζ, formed at time t after heating instantaneously to temperature T is given by

$$\zeta = \int_0^t I_v(1 - \zeta)\frac{4\pi}{3}[r^* + u(t - \tau)]^3\, d\tau \qquad (2)$$

where $r^* = 2\gamma/\Delta G$ is the radius of the critical nucleus at this temperature and u is the growth velocity. Whereas the factor $1 - \zeta$ in Eq. (2) considers explicitly the decrease in the untransformed volume in which further nucleation may take place, it does not allow for the impingement between growing nuclei; therefore it is only valid for small fractions transformed.

When the temperature of an amorphous alloy is suddenly increased close to T_g, there may be a delay, t_0, in the formation of the crystalline embryos at which nucleation can occur. This delay is not described by classical nucleation theory. During time t_0 the nucleation is said to be transient. For small values of undercooling, t_0 is negligible and steady-state nucleation is established almost instantly. At temperatures close to T_g, however, t_0 can be as large as 10^2 to 10^3 s and give rise to a measurable *incubation time* [68]. Below T_g, t_0 is expected to become so long that steady-state nucleation is unlikely to be achieved. Crystallization can also take place heterogeneously, assisted by crystalline nuclei which may have been quenched-in in glasses prepared by the RS of melts or by crystalline surfaces (oxide inclusions or containment walls). These heteronucleants become much more important at temperatures significantly below T_g, where homogeneous nucleation is delayed by the incubation time. These various nucleation modes have been reviewed elsewhere [69,70]. Taking all these factors into consideration, the c-shaped time-temperature-transformation (TTT) curve in Fig. 9a schematically shows the time t, required to form an arbitrarily selected small volume fraction of crystalline material, ζ, after heating instantaneously to temperature T. Time, t, increases as T increases and approaches T_m (because of the decrease in ΔG) and also as T decreases and approaches T_g (because of the increase in the viscosity). The TTT curve in Fig. 9a is for an isothermal transformation at zero pressure. The consolidation of amorphous powders, however, is a dynamic process in which temperature and pressure vary as a function of time. Empirical transformation curves which explicitly consider the temperature-pressure-time path are not available, but are likely to have the main features of the TTT curve. Although consolidation processes are neither isothermal nor isobaric, we will use this TTT curve to discuss the consolidation of amorphous alloy powders.

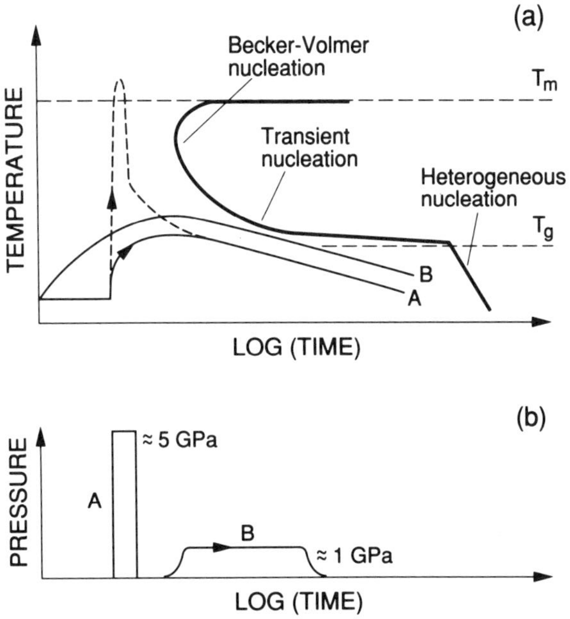

Figure 9 (a) Schematic temperature versus time and (b) pressure versus time paths followed during the consolidation of powders with metastable structures by dynamic consolidation (path A) and quasi-static consolidation (path B). The darker trace curve in (a) gives the time needed to form a given small fraction of crystalline material under isothermal conditions.

Two extreme temperature-pressure-time paths, shown schematically in Figs. 9a, b may be followed to consolidate amorphous powders: the dynamic path A and the quasistatic path B. In the dynamic process A, a shock wave is sent through the powder, with full densification occurring during the shock rise time. The work of deformation heats the powder heterogeneously and thus curve A in Fig. 9a is double-valued: the more deformed regions may reach the melting temperature and the less deformed regions attain much lower temperatures. Following the passage of the shock wave, the cooler regions serve as heat sinks for the melted regions. If this energy balance is properly chosen, then the hotter regions cool sufficiently fast to solidify back into the amorphous phase. In the quasistatic process B, the powder is heated homogeneously along a temperature-time path that reaches the highest possible temperature without crossing the TTT curve. As soon as the temperature reaches this limit, the pressure is rapidly increased to the highest value possible. The application of the pressure brings the hot powder in thermal contact with colder and massive elements (such as punches or rolling cylinders) which rapidly cool the product. Rolling mills and crank presses are appropriate for this method of consolidation.

Effective powder consolidation requires densification of the powder and bonding of the particles. Amorphous metallic alloys can be plastically deformed over a wide range of stresses, temperatures, and strain rates. Spaepen [71] summarized the deformation modes of amorphous alloys in a *deformation map*, following the

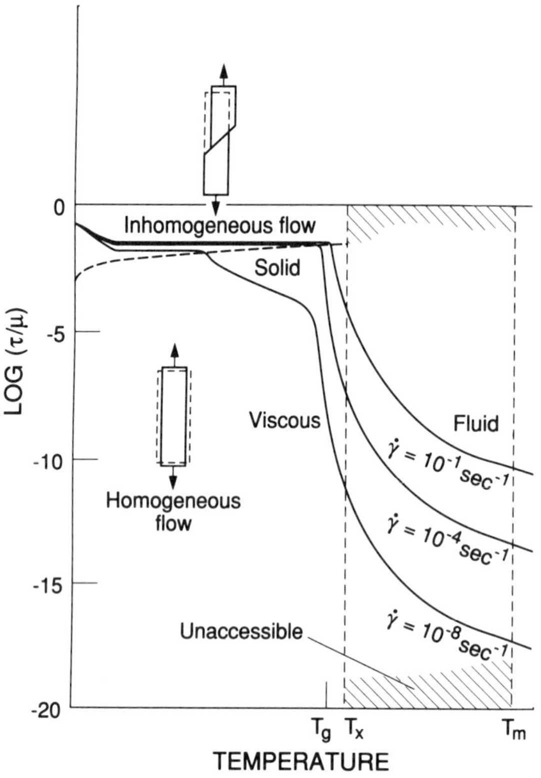

Figure 10 Schematic deformation mechanism map for an amorphous alloy (after Ref. 71).

method introduced earlier by Ashby and Frost for crystalline materials [72,73]. The deformation map shown in Fig. 10 delineates deformation modes of amorphous metallic alloys. Each deformation mode is represented by a steady-state constitutive law of deformation:

$$\dot{\gamma} = f(\tau, T, \text{structure}) \tag{3}$$

where $\dot{\gamma}$ is the shear strain rate, τ is the shear stress, and "structure" represents all the relevant structural parameters of the material, which are assumed to remain constant during a flow test at given values of τ and T. The deformation map was constructed [71] using information derived from a large number of deformation tests which were represented as contours of constant $\dot{\gamma}$ plotted on a τ-versus-T graph.

Plastic deformation in amorphous metallic alloys can be homogeneous or heterogeneous. In homogeneous deformation, each volume element of the material contributes approximately equally to the strain. This deformation mode is prevalent over most of the deformation map. Heterogeneous deformation takes place for large applied stresses (stresses exceeding $\mu/100$, where μ = shear modulus), and for temperatures below T_g. In this mode the deformation is localized into discrete shear bands, leaving most of the material plastically undeformed. Here $\dot{\gamma}$ is the average plastic strain rate, which is several orders of magnitude lower than the deformation rate within each band.

A. Shock-Wave Consolidation of Powders with Metastable Structures

In the shock-wave consolidation technique, the densification and bonding of the particles is achieved by a shock wave passing through the powder. The shock wave is generated by a projectile (a thin piece of sacrificial material) hitting the powder at velocities on the order of km s^{-1}. Two main methods, explosives and guns, are used to accelerate the projectile [74]. In the explosive-compaction method, the projectile, in the form of a metal plate placed a short distance from the powder, is accelerated by the controlled burning of explosives. In the gun method, compressed gas or a chemical propellant is used to accelerate a cylindrical projectile along the length of a gun barrel. The projectile hits the powder placed at the other end of the gun. The explosive method is preferred for the synthesis of large samples at a reduced cost. The gun method, on the other hand, enables a better control and measurement of the shock parameters and is preferred for studying the physics of shock-wave consolidation.

The processes that occur during the shock consolidation of powders are illustrated in Fig. 11. As the shock front penetrates the powder the material is compressed to several gigapascals, which largely exceeds the maximum shear stress that can be developed within the material. Thus, the shocked material behaves essentially as an inviscid, compressible fluid and reaches full density by the end of the shock rise time. In solids, the width of the shock front is of atomic dimensions.

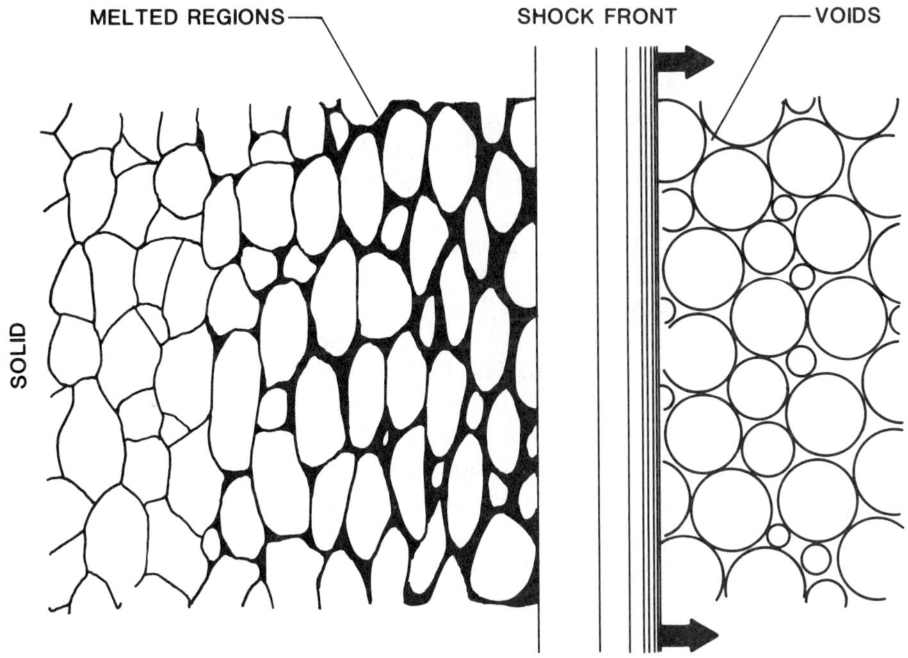

Figure 11 Schematic of thermal processes during the shock-wave consolidation of powder. The shock energy is that necessary for the formation of a small volume fraction of interparticle melt.

In porous media, and for shock pressures of several gigapascals, the width equals a distance of the order of the porosity [75]. Thus, a powder aggregate with particle diameters of 30–100 μm traversed by a shock wave moving at a velocity of 1.5 km s^{-1} is densified in approximately 40–80 ns. Metallographic observations of shock-compacted specimens [76] and numerical simulations [77] have shown that the plastic deformation during consolidation is heterogeneous. During the shock rise time the plastic strain for the material surrounding the original cavities is of order one and thus the plastic strain rate for this material can be as high as 10^7 s^{-1}. The bulk of the particles also deform, but to a lesser degree. For optimum shock parameters, the heavily deformed material melts, whereas the less deformed material (away from the cavities) reaches a much lower temperature, as indicated schematically in Figs. 9a and 11.

The formation of interparticle melt under highly dynamic conditions has the additional beneficial effect of breaking up and dispersing any oxide that may be coating the powder particles. Oxide layers are usually visible in shock-compacted specimens that do not form melt, especially between particles that were in close contact over large areas before compaction. Thus, the morphology of the particles is an important factor in shock-wave consolidation. Particles such as flakes are, in general, easier to compact but more difficult to bond [78].

Following the passage of the shock front, the localized melt pools solidify by heat conduction into the less deformed (cooler) regions of the compact (Fig. 11). This quenching lasts approximately $t_s = d^2/4D$, where d is the particle diameter and D is the thermal diffusivity of the material. Thus, for $d \approx 60$ μm and $D \approx 10^{-5}$ m^2 s^{-1}, $t_s \approx 90$ μs. Therefore, for internal temperature differences of a few hundred to 1000 kelvins, the melt pools cool at rates of the order of 10^7 K s^{-1}. For many alloys this quenching rate is sufficient to prevent the nucleation and growth of crystalline phases and thus the melt solidifies into a glass. After time t_s, the compact reaches a homogeneous temperature, T_h (point where the two branches of curve A in Fig. 9a join), and from then on the cooling rate is much slower because the heat must be dissipated externally. Certainly, T_h must be lower than T_x.

Direct evidence of local melting along particle boundaries and at sites of pore coalescence have been provided on the basis of metallographic analysis of the compacts [79–83]. The experiments of Thadhani and co-workers [82,83] illustrate interparticle melting quite convincingly. Figure 12 shows a scanning electron micrograph of flakes of amorphous MARKOMET 1064 alloy ($Ni_{55.8}Mo_{25.7}Cr_{9.7}B_{8.8}$) prepared by ball-milling commercial, rapidly solidified ribbons. These flakes were first crystallized by a high-temperature anneal, and were then shock-compacted at a shock energy of 414 J g^{-1}. Figure 13 shows an optical micrograph of the etched compact. The dark regions correspond to crystalline material and the light regions are amorphous. The amorphous alloy was formed when melt, formed between the particles during shock consolidation, was able to solidify at a fast rate by heat conduction into the interior of the adjacent particles.

Schwarz et al. [75] studied the kinetics of the shock consolidation of powders by measuring the thermal electromotive force that develops when a shock wave traverses the interface between copper and constantan powders of similar particle size. The experiments were done in a gun assembly, with the copper-constantan powder interface placed parallel to the shock-wave front, as shown in Fig. 14.

Figure 12 SEM micrograph of flakes obtained by grinding rapidly solidified ribbons of amorphous $Ni_{55.8}Mo_{25.7}Cr_{9.7}B_{8.8}$ (MARKOMET 1064). The powder was annealed above its crystallization temperature (after Refs. 82 and 83).

Figure 15a shows the voltage-time trace for one of the tests, recorded photographically from an oscilloscope screen and Fig. 15b identifies the amplitude and characteristic times measured. Ancillary equipment measured the projectile velocity prior to impact. This data, and thermodynamic and Hugoniot data on the projectile and powder [84], were used to calculate the shock pressure, $P = 3.5$ GPa, and

Figure 13 Optical micrograph of an etched compact prepared by shock-wave consolidating the crystalline powder in Fig. 12. The white regions are amorphous alloy which formed when the interparticle melt was rapidly solidified by heat conduction into adjacent cooler particles (after Refs. 82 and 83).

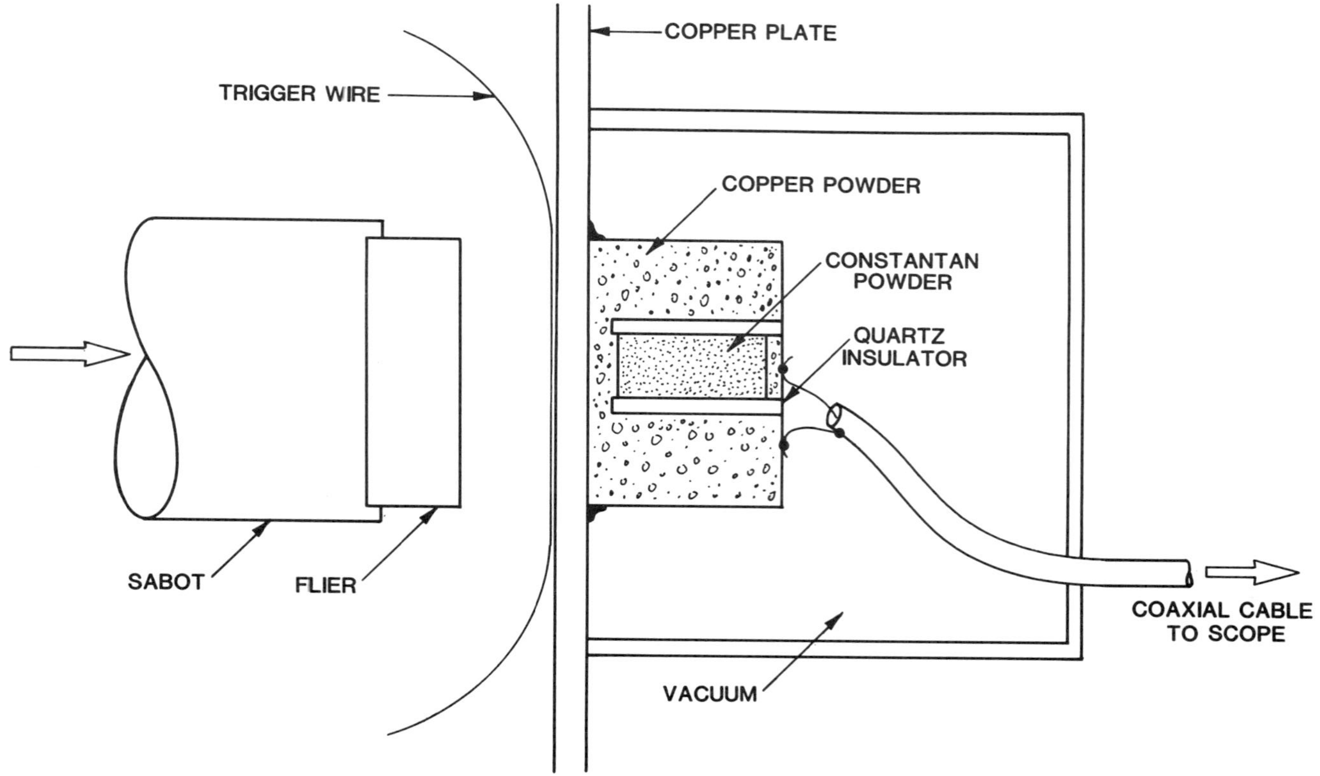

Figure 14 Schematic cross-sectional view of the projectile and shock assembly used to measure powder temperatures during shock consolidation (after Ref. 75).

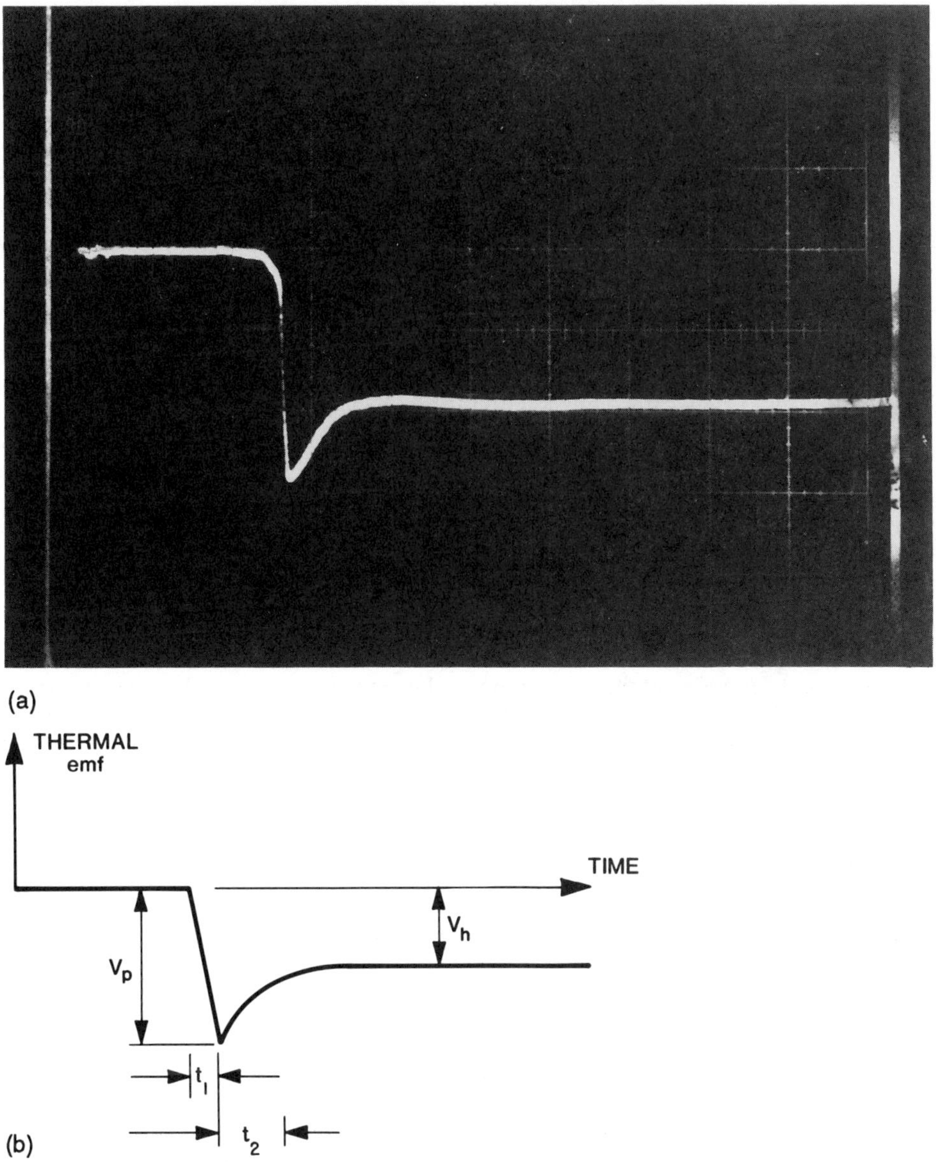

Figure 15 (a) Photograph of an oscilloscope screen recording the emf signal generated at the interface of copper and constantan powder being traversed by a shock wave; (b) Schematic diagram of the trace in (a), defining the voltage signals V_p and V_h, and the times t_1 and t_2 (after Ref. 75).

shock velocity, $V_s = 1.23$ km s^{-1} in the copper powder, and the similar values of P and V_s in the constantan powder.

The oscilloscope was triggered by the shorting of the thin enamel-coated wire placed in front of the buffer plate (Fig. 14). During the time the shock wave is traversing the buffer plate and the copper powder in front of the copper-constantan interface, the emf signal is zero. Because the copper powder is placed at ground potential, a negative pulse appears when the shock front arrives at the copper-constantan interface and raises its temperature. The rise time, t_1, is 81 ± 13 ns. From the known shock velocity and the average particle diameter, 70 μm, it follows that t_1 equals the transit time of the shock wave across one or two particles diameters. The densification of the powder occurs during t_1. During this short time the shock wave dissipates into the powder an energy density $E = E = 0.5P\,\Delta V = 135\ \mathrm{J\,g^{-1}}$, where $\Delta V = 0.77 \times 10^{-4}\ \mathrm{m^3\,kg^{-1}}$ is the difference between the specific volumes of powder and bulk copper. For a derivation of the energy dissipation equation see, for example, McQueen et al. [85]. This shock energy is dissipated heterogeneously, preferentially in heating the material that deforms plastically to fill the interparticle voids. The emf signal reaches the maximum value, V_p, near the completion of the densification. Because not all the copper-constantan powder interfaces reach the same temperature, V_p cannot be easily interpreted.

After t_1 the emf signal decreases during time t_2 to a value V_h. During t_2, of the order of 1 μs, the hotter interparticle regions cool by heat conduction into the cooler (less deformed) particle interiors. The temperature of the whole compact equilibrates at the homogeneous temperature T_h and from then on the cooling proceeds more slowly because it must rely on heat dissipation to the environment. The temperature T_h is obtained directly from the signal V_h using an emf-temperature calibration measured for a thermocouple made from sintered rods of the same copper and constantan powders. T_h was also calculated from the dissipated energy density, $E = 0.5P\,\Delta V$, and the known thermal properties of copper and constantan. Figure 16 shows the measured (four shock consolidation tests) and calculated T_h values as a function of E [75]. The agreement between the measured and calculated T_h gives support to the model for the shock-wave densification of powders. It also shows that T_h can be accurately predicted. Accurate T_h values are needed to

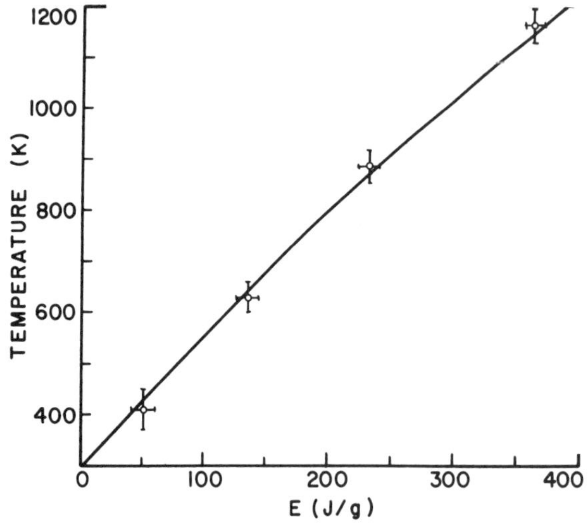

Figure 16 The dots are measured homogeneous temperatures at the interfaces of copper and constantan powders being traversed by shock waves of different shock energy. The curve gives the calculated homogeneous temperature (after Ref. 75).

estimate the maximum shock energy that can be applied to the amorphous powder without running the risk of crystallizing part of the compact.

The purpose of the experiments just described was to study the kinetics of the shock consolidation phenomenon. These were not recovery experiments, and once the shock wave reached the back of the porous sample (Fig. 13), it reflected as a rarefaction wave, causing spalling and the disintegration of the sample. To avoid spalling, the powder must be surrounded by *momentum traps* designed to carry away most of the surviving shock-wave energy. Because perfect momentum trapping is very difficult, a few tensile waves usually arrive at the compacted sample by the end of the consolidation. These waves put further restrictions on the shock parameters, which we discuss next.

Two parameters need to be controlled for the shock consolidation of powders: the shock pressure and the shock duration. The shock pressure P is clearly an important parameter because it determines T_h, as discussed above. Indeed, for shocks of increasing amplitude, a powder of given composition, size distribution, and porosity may be compacted to near zero porosity, compacted with interparticle melting, or totally melted [80]. To obtain strong compacts, the consolidation requires the formation of melt, but not in excess. The shock duration must exceed the time required to solidify the melt pools formed at the shock front because, otherwise, the compact will disintegrate during the release of the pressure and the unavoidable arrive of tensile waves. Because the volume fraction of melt formed at the shock front is proportional to P, the minimum shock duration must also increase with P. Experimentally, P is determined by the projectile velocity and thus can be controlled by varying the amount of explosives or propellant used. The shock duration is proportional to the thickness of the projectile.

Several researchers have investigated the conditions for the shock-wave consolidation of powders with metastable structures [80,86,87]. Figure 17 shows the calculation of Schwarz et al. [87] for the optimum regime of shock pressure and shock duration needed to consolidate rapidly solidified iron-based powders. The theory predicts that strong compacts are obtained when combinations of shock energy (proportional to shock pressure) and shock duration (both variables normalized by thermodynamic properties of the powders) are within the shaded area. This area is bounded by curves deduced from three criteria: (1) the shock energy must be sufficiently high to provide the energy dissipation required to form a critical amount of interparticle melt (curve 1); (2) the shock duration must exceed the time required for the solidification of the melt pools (curve 2); and (3) the shock energy must not exceed a limit beyond which the homogeneous temperature T_h is too high and thus causes unwanted metallurgical changes (curve 3). The data in Fig. 17 correspond to shock consolidation tests performed on an iron-based rapidly solidified (crystalline) AISI 9310 powder: the circles denote experiments at a constant shock duration and varying shock energy [88] and the squares are experiments at constant shock energy and varying shock duration [89]. Filled symbols denote samples whose ultimate tensile strength exceeded half the tensile strength of wrought AISI 9310 alloy. For both series of measurements the strength of the alloy increases as the consolidation variables move into the optimum regime predicted by the model.

Figure 18 shows the ultimate tensile strength (UTS) of the compacts obtained in constant-shock-duration experiments as a function of shock energy [88]. For

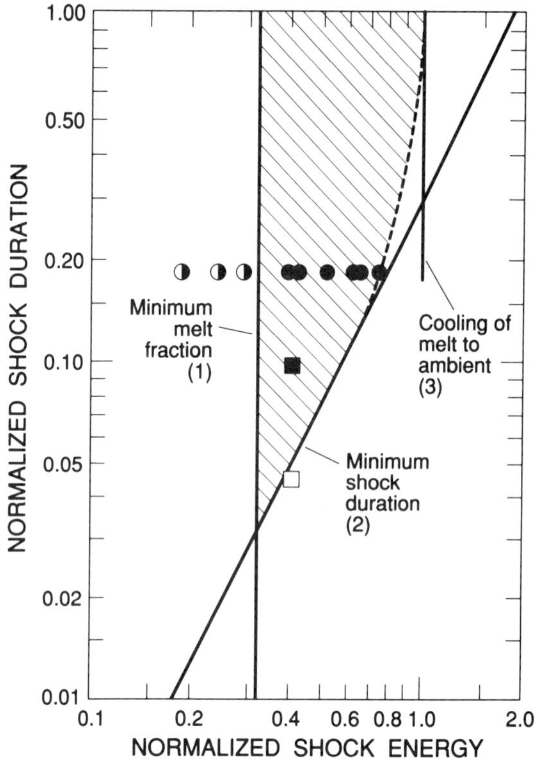

Figure 17 Map of dynamic consolidation for an iron-based powder. The theory predicts that compacts with good mechanical properties are obtained when operating within the shaded area. The points correspond to shock consolidation tests of rapidly solidified AISI 9310 powder (after Refs. 87, 89, and 88).

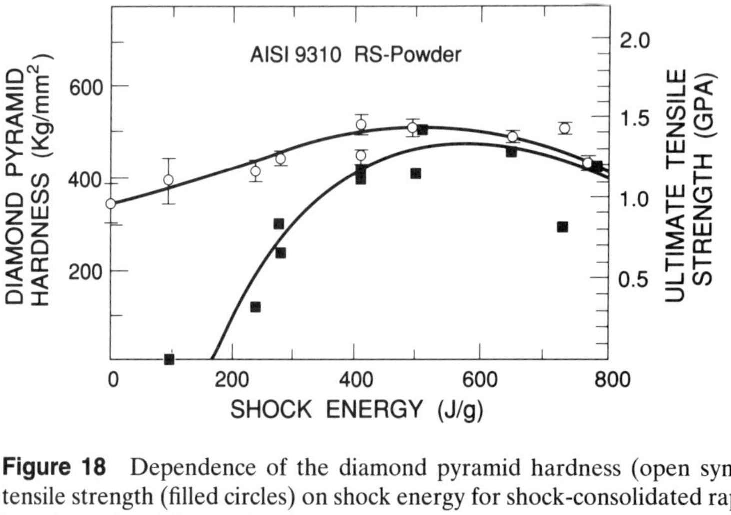

Figure 18 Dependence of the diamond pyramid hardness (open symbols) and ultimate tensile strength (filled circles) on shock energy for shock-consolidated rapidly solidified iron-based powder (after Ref. 88).

shock energies below 170 J g^{-1} (corresponding to $P < 4.4$ GPa) the compacts are fully dense but have no mechanical strength, indicating poor interparticle bonding. For $E > 170$ J g^{-1} ($P > 4.4$ GPa) the UTS increases rapidly, reaching a maximum near $E = 500$ J g^{-1} ($P = 13$ GPa). The maximum UTS, 1.3 GPa, exceeds that of wrought AISI 9310 alloy. This higher strength results from the fine grain size of the rapidly solidified powders, which was preserved during the shock consolidation process. For even higher P-values, however, the UTS decreases because some of the melt fails to solidify within the 2-μs duration of the shock state. Some melt pockets break apart during the unloading of the shock wave, resulting in cracked specimens. The open symbols in Fig. 18 show the diamond pyramidal hardness (DPH). The DPH of the uncompacted particles is 350 kg mm^{-2}. There is a slight increase in DPH with shock energy because of the plastic deformation of the particles during consolidation. Notice that although simple to perform, the DPH measurements provide a poor gauge of the consolidation process. Tensile tests are required to qualify the consolidation.

The shock-wave consolidation model described in this section was originally developed with the goal of obtaining specimens of optimum mechanical strength from rapidly soldified powders of metastable structures. The evaluation of the model in Fig. 17 was done for a metastable iron-based crystalline powder. The thermodynamic and kinetic criteria used for the crystalline powders may need to be modified when dealing with amorphous powders to ensure the absence of crystalline phase formation. In particular, the criterion for the maximum shock energy may need to be made more stringent, which would displace curve 3 in Fig. 17 to the left, to lower shock energies. Some applications of amorphous metallic alloys may require fully dense samples, but not high mechanical strength, in which case the minimum melt fraction criterion could be relaxed, moving curve 1 in Fig. 17 also to the left.

The various experiments on the shock-wave consolidation of amorphous powders and flakes prepared by the RS of melts and by mechanical alloying are summarized in Table 1. In most of these tests the shock energy and shock duration were chosen primarily by trial and error. Two works address the question of melt formation explicitly. Negishi et al. [81] report the formation of interparticle melt and discuss the narrow shock-pressure window within which the interparticle melt solidifies into the amorphous phase. Prümmer [98,99] also found that the formation of interparticle melt and its subsequent consolidation into an amorphous phase requires a careful estimate of the shock pressure. Because most of the shock energy is dissipated within the powder material that melts, an upper limit for the fraction F of molten material can be estimated from the following relation [87]:

$$F = \frac{P(V_o - V)/2}{s + \int_0^{T_o} c_p \, dT} \tag{4}$$

where V_o and V are the molar volumes of the powder and bulk amorphous product, respectively, s is the molar heat of fusion, and c_p is the molar specific heat. For amorphous $Ni_{65}Zr_{35}$ and $Fe_{60}Zr_{40}$ powders prepared by MA, Prümmer et al. [99] used Eq. (4) to estimate that the value $P = 4.4$ GPa would give $F = 0.26$. Such an F-value should be sufficient to obtain overall interparticle bonding without reaching too-high a T_h temperature. However, in trying to verify the model they

Table 1 Consolidation of Amorphous Alloy Powders by Shock Waves

Alloy	Powder, morphology[a]	Consolidation technique[b]	Product dimensions[a]	Product characterization[c]	Ref.
$Pd_{77.5}Cu_6Si_{16.5}$	Rods[d]	Explosives	Rods $L = 10.5$, $D = 0.8$		6
METGLAS 2826	RS flakes $L/D = 10$ to 30	Gas gun	Disks $L = 1$, $D = 2.5$	SEM, TEM FT = 1–4	7
METGLAS $Fe_{81}B_{13.5}Si_{3.5}C_2$	Flakes $L/t = 20$	Explosives	Disks $L = 1$, $D = 2.5$	B-H $d = 0.91$	90
MARKOMET 1064 $Ni_{55.8}Mo_{25.7}Cr_{9.7}B_{8.8}$	RS flakes	Propellant gun $E = 400$ $(P = 8.9)$	Disks $D = 0.5$, $t = 0.1$	$d = 8.9$ g cm^{-2}	91
$Ni_{77.5}Si_{7.5}B_{15}$ $Ni_{70}Si_{10}B_{20}$	RS flakes	Gas gun		$0.8 < d < 0.98$ $10 <$ BS < 1000	78
MARKOMET 1064 $Ni_{53.5}Mo_{38}Cr_8B_{1.5}$	RS flakes	Propellant gun	Disks $D = 1.5$, $t = 0.5$	Friction coef.	92 93
$Ni_{75}Si_8B_{17}$	RS flakes	Propellant gun	Disks	$d = 0.97$	81
$Pd_{78}Cu_6Si_{16}$	RS flakes $L/t = 30$ to 100	Propellant gun	Disks	$H_v = 450$–900	81
$Pd_{78}Cu_6Si_{16}$	Spherical powder $D = 0.01$–0.03	Propellant gun	Disks	$H_v = 350$–500	81
$Co_{70.3}Fe_{4.7}Si_{15}B_{10}$	Rs spherical (48–100 mesh)	Propellant gun $P = 5.5$	Disks $D = 2.5$, $t = 0.4$	T-Curie $H_v = 600$–700 $d = 0.96$ CS = 1400	94
$Fe_{40}Ni_{40}P_{14}B_6$	RS ribbons	Explosives	Thin foil		95
$Ni_{89}P_{11}$	Flakes	Explosives	Rod $L = 2.5$, $D = 0.7$		98
$Ni_{65}Zr_{35}$ and $Fe_{60}Zr_{40}$	MA powder	Explosives $P = 4.4$	Rod		99
$Cu_{75}Sn_{25}$	MA powder	Explosives, cylindrical	Rod $L = 4.0$	$d = 0.98$ $H_v = 300$–650 CS = 650, TS = 610	96

[a] L = Length (cm); D = diameter (cm); t = thickness (cm)
[b] E = Shock energy (J kg^{-1}); P = shock pressure (GPa)
[c] d = density (relative to bulk)
S(*T*)EM = Scanning (transmission) electron microscopy
B-H = magnetic hysteresis curves
H_v = Vickers pyramidal hardness (Kg mm^{-2})
BS = bending strength (MA)
CS = compressive strength (MA)
TS = tensile strength (MPa)
FT = fracture toughness (MN m$^{-3/2}$)
[d] Powder-like material consisting of small rods produced by melt extraction.

found that the molten volume fraction in these alloys could not be measured because the recovered specimens were fully amorphous and thus lacked metallographic contrast. Thus, they consolidated crystalline $Ni_{89}P_{11}$, a good glass former, using a P-value corresponding to $F = 0.26$. In this case metallographic etching clearly revealed that a volume fraction 0.25 of the material melted. Both experiments [81,99] show that the range of shock energies needed to consolidate amorphous powders into bulk amorphous products is rather narrow. Indeed, Prümmer found that with $P = 3.7$ GPa the compacts were porous and for $P = 5.1$ GPa the compact had crystalline inclusions. These presumably formed when the largest melt pockets formed during shock compaction cooled too slowly, enabling them to crystallize.

B. Quasistatic Consolidation of Powders with Metastable Structures

Taking advantage of the fast decrease in the viscosity η of amorphous alloys as T approaches T_g, several authors have used quasistatic pressing techniques to consolidate amorphous powders. These experiments are summarized in Table 2. If η

Table 2 Consolidation of Amorphous Alloy Powders by Quasistatic Methods

Alloy	Powder, morphology	Consolidation technique[a]	Product dimensions[b]	Product characterization[c]	Ref.
$Ni_{77.5}Si_{7.5}B_{15}$	RS flakes	Hot pressing $T = 873$–1073 K		$0.78 < d < 0.98$ BS = 100–300	7
$Ni_{77.5}Si_{7.5}B_{15}$ $Cu_{60}Zr_{40}$	RS flakes	Hot pressing $T = 720$ K	Cylinder $L = 0.95$, $D = 0.64$	$d = 7.5$ g cm^{-2} $H_v = 6000$ (?) TS = 800, TEM	101
$Fe_{40}Ni_{40}B_{20}$ and $Fe_{81.5}B_{14.5}Si_4$	RS flakes	Hot pressing $T = 580$ K	Toroid, disk	$d = 0.69$	100
$Fe_{81.5}B_{14.5}Si_4$	RS flakes	Extrusion $T = 640$ K	Rod	$d = 0.93$	100
$Fe_{78}Si_9B_{13}$ and $Fe_{71.3}Cr_{10}Mo_9P_8C_{1.7}$	RS flakes	Hot rolling	Plates: $L = 4$, $W = 2$, $t = 0.2$	$d = 6.8$ g cm^{-2}, $H_v = 950$; $d = 6.9$ g cm^{-2}, $H_v = 450$	102
$Al_{50}Hf_{50}$	MA powder	Hot pressing $T = 923$ K	Rod	$d = 8.34$ g cm^{-2} $H_v = 1025$	31
$Ti_{50}Cu_{50}$ and $Hf_{35}Ni_{65}$	MA powder	Pressing $P = 0.1$; $T = 300$ K	Disk	$d = 0.95$ (?)	97
$Hf_{35}Ni_{65}$	MA powder	Hipping $P = 0.1$; $T = 603$ K		Open porosity	97

[a] P = pressure (GPa); T = temperature (K)
[b] L = Length (cm); D = diameter (cm); t = thickness (cm); W = width (cm)
[c] d = density (relative to that of the bulk); S(T)EM = scanning (transmision) electron microscopy; B-H = Magnetic hysteresis curves; H_v = Vickers pyramidal hardness (kg mm^{-2}); BS = bending strength (MA); TS = Tensile strength (MPa); (?) Numerical results open to question; possible misprint(s).

can be decreased sufficiently while avoiding crystallization, then strains of the order 1, which are needed to consolidate the powder, may be achieved in short times. Liebermann [100] used uniaxial die pressing at pressures of 0.7 GPa and T/T_g = 0.81 to consolidate rapidly quenched ribbons of various metal-metalloid amorphous alloys into disks and toroids. Although the densities of some of the compacts were as high as 96% that of the bulk, there was little or no bonding between the amorphous foils. Significantly more interparticle bonding was observed in extrusion tests performed at similar values of pressure and temperature. Liebermann [100] conclude that large amounts of shear must occur between the foils for interdiffusion and bonding to occur. The relative displacements presumably break the oxide layers and other contaminant films.

Miller and Murphy [101] used warm uniaxial pressing to consolidate amorphous $Cu_{60}Zr_{40}$ and $Fe_{81.5}B_{14.5}Si_4$ powder prepared by gas atomization. Contrary to the flakes produced by RS techniques, the gas-atomized particles have aspect ratios close to one. These powders were pressed at 1.7 GPa for 150 s, followed by water-quenching. The authors were thus able to approach T_g closely. The density and hardness of the compacts increased monotonically with increasing consolidation temperature. Although most of the original boundaries of the particles were apparent in the compacts, the disappearance of some boundaries indicated extensive plastic flow. $Cu_{60}Zr_{40}$ compacted at 430°C had a compressive fracture stress of 800 MPa and the fracture was mainly between particles. The importance of localized strain in warm consolidation was clearly demonstrated by Morris [78] who compared the mechanical properties of amorphous Ni-Si-B alloys prepared by warm pressing, offset forging, and dynamic compaction.

Because warm consolidation involves the relatively slow heating of dies, the pressing temperature must be kept significantly below T_g. Decreasing the heating and cooling times allows us to increase the consolidation temperature and thus take advantage of the exponential decrease in η with increasing temperature. Shingu [102] solved this problem as shown in Fig. 19. The amorphous powder is encapsulated in a steel tube which is cold-pressed to increase the powder density. Following evacuation, the tube is heated by dipping it in molten salt, which provides

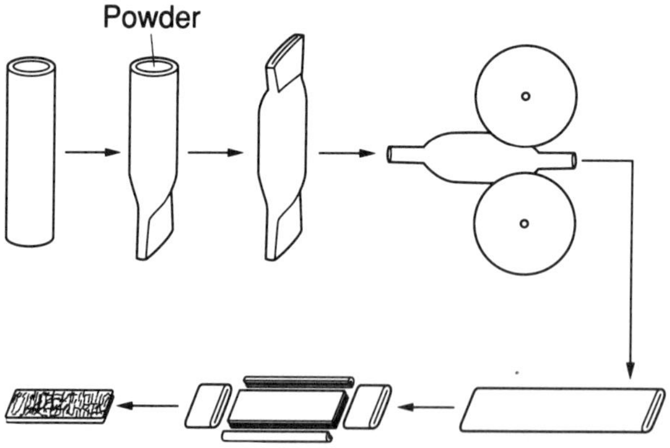

Figure 19 Schematic of the quasi-static consolidation of amorphous powders by rolling (after Ref. 102).

a fast uniform heating. Densification is achieved by rolling which ensures a large amount of plastic deformation. The tube is then water-quenched. The method was applied to amorphous $Fe_{78}Si_9B_{13}$ and $Fe_{71.3}Cr_{10}Mo_9P_8C_{1.7}$, producing centimeter-sized samples which were almost fully dense (see also Chapter 4 of this volume).

A similar technique was used by Schwarz et al. [31] to consolidate amorphous $Al_{50}Hf_{50}$ powders prepared by MA. As in the Shingu technique, the powder was encapsulated in a stainless steel tube but the heating was achieved by Joule heating while evacuating the tube. The compaction wass done in a crank press which delivers higher pressures and deformation rates than a rolling mill. Optical and scanning electron microscopy showed that the compacted samples were fully dense. Their density was 8.34 ± 0.01 g cm^{-3}, which is 7% lower than the density of the crystalline alloy (8.97 g cm^{-3}). The edges of the indents formed by a diamond pyramid indenter were crack-free, indicating that the amorphous alloy is ductile in compression. The DPH measurements were 1025 ± 25 and 950 ± 25 kg mm^{-2} on surfaces perpendicular and parallel to the compaction direction, respectively. Fracture was through the particles.

The temperature-time and pressure-time paths for the quasistatic consolidation experiments in Refs. 102 and 31 are shown in Fig. 9. The temperature-time path approaches closely the transformation curve to achieve the maximum possible decrease in viscosity. The general description of $\eta(T)$ for glass-forming alloys requires three temperature regimes [103]: $T > T_m$, $T_g < T < T_m$, and $T < T_g$. The viscosity of glass-forming alloys at T_m is low, typically between 10^{-1} and 10^1 Pa s. For $T > T_m$, η decreases according to an Arrhenius equation with an activation energy of approximately 40 kJ/mole. In the regime $T_g < T < T_m$, the viscosity of the undercooled liquid is still described by an Arrhenius-type equation, but with an activation energy that increases with decreasing temperature. This trend has classically been described by the Fulcher-Vogel equation [104,105]

$$\eta = A \exp\left[\frac{B}{T - T_o}\right] \tag{5}$$

where A and B are constants and T_o is an ideal glass transition temperature. For $T < T_g - 20$ K, the viscosity depends on deformation history. Taub and Spaepen [106] measured η for $Pd_{82}Si_{18}$ from creep experiments at constant load in the regime $0.66 < T/T_g < 0.84$. At the test onset, η was approximately 10^{15} Pa s. However, the viscosity was observed to increase almost linearly with time at all temperatures, with no evidence of asymptotic behavior, even after a month of isothermal annealing. This increase in η is caused by a structural relaxation of the amorphous alloy assisted by the strain and is accompanied by the annihilation of free volume.

Cold-pressed amorphous-powder compacts have densities approximately 50% that of the bulk; thus, filling the interparticle voids requires local plastic strains γ^c of the order of one. Because the consolidation process in a rolling or a crank press lasts approximately 0.1 s, the powder particles must deform plastically at rates of the order of 10^1 s^{-1}. A simple estimate, given below, shows that these values of strain and strain rate cannot be achieved by homogeneous deformation. Heterogeneous (localized) deformation is apparently necessary for the quasistatic consolidation of amorphous powders and this suggests a lower limit for the applied stress necessary to achieve an effective consolidation.

If we consider simply that at temperature $T^* \approx T_g - 20$ K an amorphous particle of viscosity $\eta(T^*)$ is compressed uniaxially by the stress σ, while being free to flow laterally during time t^*, we must have

$$\frac{\sigma/3}{\eta(T^*)} t^* > \gamma^c \approx 1 \tag{6}$$

With $\eta(T^*) \approx 10^{15}$ Pa s, and $t^* = 10^{-1}$ s, this condition becomes $\sigma > 10^{16}$ Pa. This stress exceeds by seven orders of magnitude the stress that can be developed in rolling mills or crank presses. The deformation mechanism map of Fig. 10, however, tells us that for $T \approx T_g - 20$ K, as the applied stress increases and approaches $\mu/100$ (μ = shear modulus), the deformation becomes heterogeneous, being localized within narrow shear bands. Within each deformation band, the deformation rate largely exceeds that which can be achieved by diffusional processes alone, enabling the densification of the powder during times on the order of 0.1 s. Thus, to achieve consolidation using rolling mills or crank presses, the maximum applied stress must exceed the critical stress at which the deformation becomes heterogeneous. This stress limit ($\mu/100$, typically between 2 and 4 GPa) has a weak temperature and strain rate dependence (Fig. 10).

If the applied stress is below the homogeneous → heterogeneous deformation limit, the densification must rely only on diffusion. Equation 6, with σ = 1 GPa and $\eta \approx 10^{15}$ Pa s, gives $t^* \approx 10^6$ s. At T_g, $\eta \approx 10^{13}$ Pa s, and t^* is reduced to 10^4 s. Thus, it seems unlikely that amorphous powders can be fully densified in the homogeneous deformation regime.

The consolidation time t^* and temperature T^* are further constrained by the requirement of avoiding crystallization. As seen in Fig. 9a, the temperature-time path approaches the C-shaped TTT curve near regions where crystallization is strongly affected by transient and heterogeneous nucleation. Because these nucleation rates are difficult to predict, the transformation curve must be measured a priori for each amorphous powder being consolidated. Such empirical approach has been followed by Shingu [102] with good results.

IV. CONCLUSION

The discovery of new solid-state methods for the synthesis of amorphous metallic alloys has resulted in an extraordinary growth of this field over the last eight years. These advances include

1. a better understanding of the thermodynamic and kinetic conditions that must be satisfied to access metastable alloy states by solid-state reactions
2. predictive capabilities for the glassforming ranges of binary amorphous alloys, both for synthesis by the RS of melts and by solid-state reactions
3. the realization that amorphous metallic alloys synthesized by RS and solid-state techniques are structurally equivalent (when in the same state of internal relaxation)
4. the view that crystal-to-amorphous transformations in the solid state are generalizations of a melting phenomenon [107–109]

Several recent international conferences and workshops have been devoted almost exclusively to new synthesis methods [110–113]. Among these new methods,

MA has shown the fastest rate of growth mainly because it requires a relatively low capital investment.

Interest in MA has also increased because it can be used to prepare powders with micro- or nanocrystalline grain sizes [114–120]. Mechanically alloyed powders, either amorphous or microcrystalline, are being used as precursors for the synthesis of intermetallic alloys, notably low-density high-modulus aluminum-based intermetallics [116,121]. As demonstrated by the original work of Benjamin [28,29], the MA process allows us to add a fine and uniform dispersion of precipitates (carbides, oxides, etc.) to the alloy powder. An even finer dispersion of carbide precipitates, approximately 7 nm in size, can be added to alloys powders which contain transition metals such as Ti, Zr, or Hf by mechanically alloying these powders in the presence of an organic dispersant. Organic dispersants such as stearic acid $[CH_3(CH_2)_{16}CO_2H]$ or hexane (C_6H_{14}) are commonly used to prevent the agglomeration of ductile powders such as aluminum on the walls of the milling vial. During the MA of Al_3X (X = Ti, Zr, or Hf) in the presence of hexane, part of the dispersant decomposes, and carbon and hydrogen are incorporated into the powder as solutes [116]. The hydrogen can be easily removed by a vacuum anneal. The carbon, however, reacts with the ETMs forming ultrafine carbides. These precipitates have the beneficial effect of adding strength to the alloy and limiting grain growth during its consolidation.

MA enables us to prepare almost any alloy powder in a short time. There are, however, potential complications arising from contaminations. In binary alloys between transition metals, the inclusion of oxides, even at levels of few atomic percent, favors the formation of the amorphous alloy phase. Erosion of the vial and balls may also contribute with impurities which can change the properties of the alloy. Thus, reports on alloy powders prepared by MA should always include a chemical analysis of the product, especially when claiming the formation of a yet unreported metastable alloy phase.

Although a variety of powders with metastable structures can now be produced by MA on a routine basis, techniques to consolidate these powders while preventing unwanted structural changes during high-temperature anneals need much further development. Only after such developments are done will we be able to tailor these alloys to specific applications and thus capitalize on the long-known potential uses of amorphous and microcrystalline alloys [122].

ACKNOWLEDGMENTS

This work was supported by the U.S. Department of Energy, Office of Basic Energy Sciences. The author thanks J. Rubin and S. Srinivasan for a critical reading of the manuscript.

REFERENCES

1. *Amorphous Metallic Alloys*, edited by L. E. Luborsky (Butterworths, London, 1983).
2. P. Duwez, R. H. Willens, and W. Klement, Jr., *J. Appl. Phys.*, *31*, 1136 (1960).
3. R. B. Schwarz and W. L. Johnson, *Phys. Rev. Lett.*, *51*, 415 (1983).

4. R. L. White, Ph. D. dissertation, Stanford University (1979).
5. C. C. Koch, O. B. Cavin, C. G. McKamey, and J. O. Scarbrough, *Appl. Phys. Lett.*, *43*, 1017 (1983).
6. C. F. Cline and R. W. Hooper, *Scipta Metall.*, *11*, 1137 (1977).
7. D. G. Morris, *Metal Sci.*, *14*, 215 (1980).
8. W. Buckel and R. Hilsch, *Z. Phys.*, *138*, 109 (1954).
9. A. Brenner, D. E. Couch, and E. N. Williams, *J. Res. Nat. Bur. Stand.*, *44*, 109 (1950).
10. D. Raskin and C. H. Smith, in *Amorphous Metallic Alloys*, edited by F. E. Luborsky (Butterworths, London, 1983), p. 381.
11. H. Oesterreicher, J. Clinton, and H. Bittner, *Mat. Res. Bull.*, *11*, 1241 (1976).
12. A. M. van Diepen and K. H. J. Buschow, *Solid State Comm.*, *22*, 113 (1977).
13. S. K. Malik and W. E. Wallace, *Solid State Comm.*, *24*, 283 (1977).
14. X. L. Yeh, K. Samwer, and W. L. Johnson, *Appl. Phys. Lett.*, *42*, 242 (1983).
15. I. Ye. Bolotov and A. V. Kozhin, *Fiz. Metal. Metalloved.*, *35*, 383 (1973).
16. J. J. Hauser, *J. Phys. Colloq.*, *42*, C4-943 (1981).
17. S. R. Herd, K. N. Tu, and K. Y. Ahn, *Appl. Phys. Lett.*, *42*, 597 (1983).
18. W. L. Johnson, *Prog. Mater. Sci.*, *30*, 81 (1986).
19. R. B. Schwarz and P. Nash, *J. Metals*, *41*, 27 (1989).
20. A. D. Le Claire, in *Properties of Atomic Defects in Metals*, edited by N. L. Peterson and R. W. Siegel (North-Holland, Amsterdam, 1978), pp. 70–96. Also published in *J. Nucl. Mater.*, *69*, *70*, 70–96 (1978).
21. H. Ninomiya, M. Koiwa, Y. Minonishi, and S. Ono, *Trans. Japan Inst. Met.*, *24*, 665 (1983).
22. H. Hahn and R. S. Averback, *Phys. Rev. B*, *37*, 6533 (1988).
23. H. Hahn, R. S. Averback, and V. M. Shyu, in *Solid State Amorphizing Transformation*, edited by R. S. Schwarz and W. L. Johnson (Elsevier, Lausanne, 1988), p. 345. Also published in *J. Less-Common Metals*, *140*, 345 (1988).
24. J. B. Rubin and R. B. Schwarz, *Appl. Phys. Lett.*, *55*, 36 (1989).
25. L. Schultz, in *Amorphous Metals and Non-Equilibrium Processing*, edited by M. von Allmen (Les Editions de Physique, Paris, 1984), p. 135.
26. M. Atzmon, K. M. Unruh, and W. L. Johnson, *J. Appl. Phys.*, *58*, 3865 (1985).
27. M. Atzmon, J. D. Verhoevem, E. D. Gibson, and W. L. Johnson, *Appl. Phys. Lett.*, *45*, 1052 (1984).
28. J. S. Benjamin and T. E. Volin, *Metall. Trans.*, *5*, 1929 (1974).
29. J. S. Benjamin, *Sci. Amer.*, *234*, 40 (1976).
30. R. B. Schwarz, R. R. Petrich, and C. K. Saw, *J. Non-Cryst. Solids*, *76*, 281 (1985).
31. R. B. Schwarz, J. W. Hannigan, H. Sheinberg, and T. Tiainen, in *Modern Developments in Powder Metallurgy*, Vols. 18–21 (Metal Powder Industries Federation, Princeton, NJ), p. 415.
32. J. Eckert, L. Schultz, and K. Urban, *J. Less-Common Metals*, *145*, 283 (1988).
33. A. W. Weeber and H. Bakker, *Physica B*, *153*, 93 (1988).
34. L. Schultz, E. Hellstern, and G. Zorn, *Z. Phys. Chem.*, *157*, 203 (1988).
35. A. L. Greer, *J. Non-Cryst. Solids*, *61*, *62*, 737 (1984).
36. P. H. Gaskell, *J. Non-Cryst. Solids*, *32*, 207 (1979).
37. R. B. Schwarz and C. C. Koch, *Appl. Phys. Lett.*, *49*, 146 (1985).
38. R. B. Schwarz, in *Non-Equilibrium Solid Phases of Metals and Alloys,* Suppl. *Trans. Japan Inst. Metals*, *29*, 49 (1988).
39. E. Hellstern, L. Schultz, and J. Eckert, *J. Less-Common Metals*, *140*, 93 (1988).
40. H. Kimura, M. Kimura, and F. Takada, *J. Less-Common Metals*, *140*, 113 (1988).
41. J. J. Fischer and J. H. Weber, *Adv. Mater. Process.*, *138*, 43 (1990).

42. J. R. Brimhall, H. E. Kissinger, and L. A. Charlot, in *Metastable Materials Formation by Ion Implantation*, edited by S. T. Picraux and W. J. Choyke (Elsevier, New York, 1982), p. 235.
43. H. Mori, H. Fujita, M. Tendo, and M. Fujita, *Scipta Metall.*, *18*, 783 (1984).
44. A. Ye. Yermakov, Ye. Ye. Yurchinov, and V. A. Barinov, *Phys. Met. Metall.*, *52*, 50 (1981).
45. E. G. Ponyatovski, I. T. Belash, and O. I. Barkalov, *J. Non-Cryst. Solids*, *117–118*, 679 (1980).
46. E. G. Ponyatovski and O. I. Barkalov, *Mater. Sci. Eng.*, *A132*, 726 (1990).
47. R. B. Schwarz, J. B. Rubin, and T. Tiainen, in *Science of Advanced Materials*, edited by H. Wiedersich and M. Meshii (ASM, Metals Park, OH, 1990), p. 1.
48. P. R. Okamoto and M. Meshii, in *Science of Advanced Materials*, edited by H. Wiedersich and M. Meshii (ASM, Metals Park, OH, 1990), p. 33.
49. W. J. Meng, P. R. Okamoto, and L. E. Rehn, in *Science of Advanced Materials*, edited by H. Wiedersich and M. Meshii (ASM, Metals Park, OH, 1990), p. 99.
50. P. Y. Lee and C. C. Koch, *J. Mater. Sci.*, *23*, 2837 (1988).
51. A. W. Weeber, H. Bakker, and F. R. de Boer, *Europhys. Lett.*, *2*, 445 (1986).
52. R. B. Schwarz and R. R. Petrich, *J. Less-Common Metals*, *140*, 171 (1988).
53. R. B. Schwarz, *Mater. Sci. Engng.*, *97*, 71 (1988).
54. W. Biegel, H. U. Krebs, C. Michaelsen, H. C. Freyhard, and E. Hellstern, *Mater. Sci. Eng.*, *97*, 59 (1988).
55. L. Schultz, E. Hellstern, and A. Thomä, *Europhys. Lett.*, *3*, 921 (1987).
56. L. Schultz, *Mater. Sci. Engng.*, *97*, 15 (1988).
57. R. Brüning, Z. Altounian, J. O. Strom-Olsen, and L. Schultz, *Mater. Sci. Engng.*, *97*, 317 (1988).
58. C. Michaelsen and E. Hellstern, *J. Appl. Phys.*, *62*, 117 (1987).
59. J. H. Harris, W. A. Curtin, and L. Schultz, *J. Mater. Res.*, *3*, 872 (1988).
60. D. Turnbull, *Met. Trans.*, A *12A*, 695 (1981).
61. D. Turnbull, in *Laser Solid Interactions and Laser Processing*, edited by S. D. Ferris, H. J. Leamy, and J. M. Poate (American Institute of Physics, New York, 1979), p. 73.
62. S. R. Coriel and D. Turnbull, *Acta Metall.*, *30*, 2135 (1982).
63. P. Nash and R. B. Schwarz, *Acta Metall.*, *36*, 3047 (1988).
64. C.-J. Lin and F. Spaepen, *Acta Metall.*, *54*, 1367 (1986).
65. J. B. Rubin and R. B. Schwarz, *Mat. Res. Soc. Symp. Proc.*, *230*, 21 (1992).
66. D. A. Porter and K. E. Esterling, *Phase Transformations in Metals and Alloys* (Van Nostrand Reinhold, New York, 1983).
67. D. Turnbull, *J. Appl. Phys.*, *21*, 1022 (1950).
68. M. G. Scott, G. Gregan, and X. D. Dong, in *Proc. 4th Int. Conf. Rapidly Quenched Metals*, edited by T. Masumoto, and K. Suzuki (Japan Institute of Metals, Sendai, Japan, 1982).
69. U. Köster and U. Herold, in *Glassy Metals I*, edited by H.-J. Güntherodt and H. Beck (Springer-Verlag, Berlin, 1981), p. 225.
70. M. G. Scott, in *Amorphous Metallic Alloys*, edited by F. E. Luborsky (Butterworths, London, 1983), ch. 10.
71. F. Spaepen, *Acta Metall.*, *25*, 407 (1977).
72. M. F. Ashby, *Acta Metall.*, *20*, 887 (1972).
73. M. F. Ashby and H. J. Frost, in *Constitutive Equations in Plasticity*, edited by A. S. Argon (MIT), Cambridge, MA, 1975), p. 117.
74. A. J. Cable, in *High-Velocity Impact Phenomena*, edited by Ray Kinslow (Academic Press, New York, 1970), ch. 1.
75. R. B. Schwarz, P. Kasiraj, and T. Vreeland, Jr., in *Metallurgical Applications of*

Shock-Wave and High-Strain-Rate Phenomena, edited by L. E. Murr, K. P. Staudhammer, and M. A. Meyers (Marcel Dekker, New York, 1986), p. 313.
76. D. Raybould, *J. Mater. Sci.*, *16*, 589 (1981).
77. R. A. Berry and R. L. Williamson, in *Metallurgical Applications of Shock-Wave and High-Strain-Rate Phenomena*, edited by L. E. Murr, K. P. Staudhammer, and M. A. Meyers (Marcel Dekker, New York, 1986), p. 167.
78. D. G. Morris, in *Rapidly Quenched Metals*, edited by S. Steeb and H. Warlimont (Elsevier, Amsterdam, 1985), p. 1751.
79. D. G. Morris, *Metals Sci.*, *15*, 116 (1981).
80. D. Raybould, in *Shock Waves and High-Strain Rate Phenomena in Metals*, edited by M. A. Meyers and L. E. Murr (Plenum, New York, 1981), p. 845.
81. T. Negishi, T. Ogura, T. Masumoto, T. Goto, K. Fukuoka, Y. Syono, and H. Ishii, *J. Mater. Sci.*, *20*, 399 (1985).
82. N. N. Thadhani, A. H. Mutz, P. Kasiraj, and T. Vreeland, Jr., in *Metallurgical Applications of Shock-Wave and High-Strain-Rate Phenomena*, edited by L. E. Murr, K. P. Staudhammer, and M. A. Meyers (Marcel Dekker, New York, 1986), p. 247.
83. N. N. Thadhani and T. Vreeland, Jr., *Acta Metall.*, *34*, 2323 (1986).
84. *LASL Shock Hugoniot Data*, edited by S. P. Marsh (Univ. California Press, Berkeley, 1980).
85. R. G. McQueen, S. P. Marsh, J. W. Taylor, J. N. Fritz, and W. N. Carter, in *High-Velocity Impact Phenomena*, edited by R. Kinslow (Academic Press, New York, 1970), chap. 7.
86. W. H. Gourdin, *J. Appl. Phys.*, *55*, 172 (1984).
87. R. B. Schwarz, P. Kasiraj, T. Vreeland, Jr., and T. J. Ahrens, *Acta Metall.*, *32*, 1243 (1984).
88. P. Kasiraj, T. Vreeland, R. B. Schwarz, and T. J. Ahrens, *Acta Metall.*, *32*, 1235 (1984).
89. R. B. Schwarz, P. Kasiraj, T. Vreeland, and T. J. Ahrens, in *Shock Waves in Condensed Matter*, edited by J. R. Asay, R. A. Graham, and G. K. Straub (Elsevier, New York, 1984), p. 439.
90. R. D. Caligiuri, P. S. DeCarli, and R. E. Tokheim, Report EL-2449 (SRI International, Menlo Park, CA, 1982).
91. P. Kasiraj, D. Kostka, T. Vreeland, Jr., and T. J. Ahrens, *J. Non-Cryst. Solids*, *61*, *62*, 967 (1984).
92. T. Vreeland, Jr., P. Kasiraj, A. H. Mutz, and N. N. Thadhani, in *Metallurgical Applications of Shock-Wave and High-Strain-Rate Phenomena*, edited by L. E. Murr, K. P. Staudhammer, and M. A. Meyers (Marcel Dekker, New York, 1986), p. 231.
93. T. Vreeland, N. N. Thadhani, and A. H. Mutz, *J. Mater. Res.*, *1*, 661 (1986).
94. Y. Toda, T. Ogura, T. Masumoto, K. Fukuoka, and Y. Syono, in *Rapidly Quenched Metals*, edited by S. Steeb and H. Warlimont (Elsevier, Amsterdam, 1985), p. 1755.
95. A. H. Clauer, R. V. Raman, R. S. Carbonara, and R. E. Maringer, in *Rapid Solidification Processing, Principles and Technologies, II* (Claitor's, Baton Rouge, LA, 1980), p. 399.
96. N. N. Gorshkov, E. Yu. Ivanov, A. V. Plastinin, V. V. Sil'vestrov, T. M. Sobolenko, and T. S. Teslenko, *Engl. Trans. from Fizika Goreniya i Vzryva*, *25*, 125 (Plenum, New York, 1989).
97. W. Krauss, C. Politis, and P. Weimar, *Metal Powder Rep.*, *43*, No. 4 (1988).
98. R. Prümmer, *Mater. Sci. Engng.*, *98*, 461 (1988).
99. R. Prümmer, J. Eckert, and L. Schultz, in *Proc. Int. Seminar High Energy Working of Rapidly Solidified Materials*, Novirsibirsk, USSR, 1988.
100. H. H. Liebermann, *Mater. Sci. Engng.*, *46*, 241 (1980).
101. S. A. Miller and R. J. Murphy, in *Rapid Solidification Processing, Principles and*

Technologies, II, edited by R. Merhabian, B. H. Kear, and M. Cohen (Claitor's, Baton Rouge, LA, 1980), p. 385.
102. P. H. Shingu, *Mater. Sci. Engng.*, *97*, 137 (1988).
103. F. Spaepen and A. I. Taub, in *Amorphous Metallic Alloys*, edited by F. E. Luborsky (Butterworths, London, 1983), ch. 13.
104. H. Vogel, *Z. Phys.*, *22*, 645 (1921).
105. G. Fulcher, *J. Amer. Ceramic Soc.*, *6*, 339 (1925).
106. A. I. Taub and F. Spaepen, *Acta Metall.*, *28*, 1781 (1981).
107. P. R. Okamoto, L. E. Rehn, J. Pearson, R. Bhadra, and M. Grimsditch, *J. Less-Common Metals*, *140*, 231 (1988).
108. H. J. Fecht, P. J. Desré, and W. L. Johnson, *Phil. Mag. B*, *59*, 577 (1989).
109. D. Wolf and S. Yip, *Mater. Res. Soc. Symp. Proc.*, *230*, 3 (1992).
110. *Proc. Conf. Solid State Amorphizing Trans.*, edited by R. B. Schwarz and W. L. Johnson (Elsevier, Lausanne, 1988). Also published in *J. Less-Common Metal*, *140*.
111. Workshop on Metastable Metallic Phases: Principles and Applications, Istituto Elettrotecnico Nazionale, G. Ferraris, Torino, Italy, March 8–10, 1989. Proceedings published in *Phil. Mag.*, *61*.
112. Symposium on the Preparation and Properties of Metastable Alloys, Europ. Mater. Res. Soc., Strasbourg, May 31–June 2, 1988.
113. *Mechanical Alloying*, Proc. International Symp. on Mechanical Alloying held in Kyoto, Japan, May 7–10, 1991; published in Materials Science Forum Vols. 88–90 (1992).
114. H. J. Fecht, E. Hellstern, Z. Fu, and W. L. Johnson, in *Advances in Powder Metallurgy* (Metals Powder Industries Federation, Princeton, NJ, 1989), p. 111.
115. J. S. C. Jang and C. C. Koch, *J. Mater. Res.*, *5*, 325 (1990).
116. R. B. Schwarz, P. B. Desch, and S. Srinivasan, *Scripta Metall. et Materialia*, *25*, 2513 (1991).
117. A. Calka, A. P. Radlinski, and R. Shanks, *Mater. Sci. Engng.*, *A133*, 555 (1991).
118. M. Oehring and R. Bormann, *Mater. Sci. Engng.*, *A134*, 1330 (1991).
119. M. L. Trudeau and R. Schulz, *Mater. Sci. Engng.*, *A134*, 1361 (1991).
120. P. H. Shingu, B. Huang, S. R. Nishitani, and S. Nasu, *Trans. Japan Inst. Metals*, *29*, 3 (1988).
121. R. B. Schwarz, S. Srinivasan, and P. B. Desch, Mater. Sci. Forum Vols. 88–90 (1992) pp. 595–602.
122. *Amorphous and Metastable Microcrystalline Rapidly Solidified Alloys*: *Status and Potential*, Report of Technical Committee, National Materials Advisory Board (National Academy of Science, Washington, DC, 1980). Publication NMAB-358.

8

Rapidly Solidified Surface Layers by Laser Melting

E. Schubert
ATZ-EVUS, Vilseck, Bavaria, Germany

H. W. Bergmann
University of Erlangen Nürnberg, Erlangen, Germany

I. INTRODUCTION

Nowadays we realize more and more that the availability of natural resources is limited and that, therefore, all kinds of materials have to be used more efficiently. Regarding materials science, this implies that highly technological materials with optimized property-to-weight ratio have to be developed. Due to complex service requirements, modern materials and components typically have to possess a combination of varied properties. For example, not only strength but also ductility, corrosion resistance, etc., are required. Sometimes it is even necessary for the surface of a component to have properties other than those of the bulk. Nowadays, materials with properties graded into the surface are quite popular. In order to modify the surface properties and to arrive at the surface/bulk property combination desired, there is also a possibility of generating nonequilibrium phases in a surface layer.

In general, surface treatments can be subdivided into thermal, mechanical, and chemical categories as well as mixed categories such as thermochemical and thermomechanical. The present chapter reports on thermal and thermochemical surface remelting treatments requiring steep thermal gradients, therefore allowing rapid solidification.

Surface remelting requires high power densities in order to generate a sufficiently steep thermal gradient which allows surface melting. This is possible by either a focused electron or a laser beam directed at a target component. During recent years, tremendous technological developments in the area of high power lasers have led to the situation wherein the majority of applications involve laser beam rather than electron beam technology. One of the main reasons for this occurrence is the greater flexibility of laser treatments, which are not restricted to vacuum technology, and hence are more cost-effective as they allow higher throughput. Rapidly solidified surface layers obtained by laser remelting are already currently applied in some areas of industry. In other areas, they are expected to be implemented into production operations in the near future. This is one of the reasons why an entire chapter in a book on rapidly solidified metals is dedicated to this only subject.

II. HISTORICAL BACKGROUND

In the mid-seventies, two significant developments were introduced in materials science. These resulted in some spectacular results heretofore unknown. One of them was the generation of materials in states under extreme situations [1], especially in the sense of an extreme state far from equilibrium. The various quenching techniques resulting in such states and their historical development are reported in Chapter 1 of this book. The second development was material processing by photons. In 1960, Mayman [2] had experimentally demonstrated the use of lasers, the first CO_2 laser was built in 1964 [3], and it only took a few years more for the first high-power CO_2 lasers to become available for industrial purposes in the early seventies. During the following years, the possibility of contact-free manufacturing by means of light resulted in a high degree of automation in automotive and other mass product manufacturing. In the beginning, cutting [4,5] and later on also welding applications [6,7] were the main achievements; more recently, surface treatment [8–11] has been of particular industrial interest. The modification of surface layers by laser and other high-energy beams takes place in the overlapping region of the two developments mentioned above, and it is there where aspects regarding materials and requirements of manufacturing meet. It was C. W. Draper who collected and chronologically registered the major contributions to the current state of knowledge [12–14].

III. FUNDAMENTAL ASPECTS

The following section summarizes some aspects important for the understanding of rapidly solidified surface layers.

A common advantage of both spray methods (atomization, splat quenching, spray deposition) and casting techniques (melt spinning, planar flow casting, melt drag) is that the melt is stabilized at constant temperature prior to quenching. The melt is then quenched using a suitable coolant, that is, either a gaseous (atomization), a liquid (gas-liquid atomization) or a solid (substrate) medium. This means that high cooling rates can only be obtained if the resolidifying volume is arranged in such a way that at least in one spatial dimension of the sample is small (powder, wire, foil). In all of these cases, the temperature gradient is positive, which means that heat is extracted through the surface to the surrounding coolant. Surface treatments with high energy beams are methods which are characterized by an initial rapid heating step. This heating rate can be varied by adjusting laser intensity. The resulting quenching rate is a consequence of the generated temperature gradient in and the thermal properties of the material. In a first approximation, heating rates achievable are proportional to laser intensity, which may be explained by the following calculations.

While there are a number of currently available simulation models based on various approximations we shall just consider a one-dimensional transient model for a flat semi-infinite body [15]. If we assume that t is exposure time, p power flux density, and if k and K represent thermal conductivity and thermal diffusivity, respectively, the temperature, T in a depth, z, is then given by

$$T(z, t) = \frac{2p}{K} \sqrt{kt}\, i\, \mathrm{erfc}\left(\frac{z}{2/\sqrt{kt}}\right) \tag{1}$$

The surface temperature is then

$$T(0, t) = \frac{2p}{K} \sqrt{kt} \tag{2}$$

and the heating rate at the surface can hence be expressed as

$$\dot{T}(0, t) = \frac{2p}{K} \sqrt{\frac{k}{t}} \tag{3}$$

As mentioned above, the heating rate at the surface is in a first approximation directly proportional to power flux density and to the reciprocal square root of interaction time. For the temperature gradient at the surface ($z = 0$) a value of

$$\left.\frac{\partial T}{\partial z}\right|_{z=0} \sim \frac{2p}{K} \tag{4}$$

will be obtained.

The above-mentioned approximations neglect the occurrence of transformations and assume heating only in the solid state. However, there is already a number of simulations for surface treatments published by various authors that take into account complex geometries, temperature-dependent material constants and the presence of internal heat sources (see [16–25]). A detailed description for three-dimensional models is given in [26–28]. In the present context, however, only a rough approximation of the maximum quenching rates will be of interest. This can be derived from the following approximation:

$$\dot{T}_Q \sim \frac{T(0, t) - T(0, 0)}{2t} \tag{5}$$

Quenching rates of 10^{10} K/s or 10^6 K/s may be obtained for a temperature rise of 10^3 K caused by a pulse of either 20 ns (excimer laser) or 6 ms (CO_2 laser), respectively. Again it will be noted that the relevant parameters are interaction time and power flux density. All these calculations, however, neglect convection in the liquid state. Therefore, other authors suggest a homogeneous temperature inside the melt (fast convection). On the other hand, however, the approximation used does not change the principal considerations about the quenching rate and its dependence on power flux density and interaction time. More recently, melt-bath dynamics [29] have been calculated taking into account the intensity distribution over the cross section of the laser beam and the temperature dependence of the surface tension, resulting in Marangoni convection.

The complex dynamic processes inside the melt and the resulting inhomogeneous distribution of alloying elements can lead to inhomogeneous microstructures. For quite some time, this has been one of the obstacles to the application of laser melting and alloying in production technology on a large scale.

So far, it may be appreciated why it is necessary to couple high power flux densities into the workpiece for the generation of high quenching rates. This procedure sounds easy, but it is difficult to perform as it requires high energies, suitable focusing units, and a good absorption of the laser light by the exposed component, as described below.

A. High Power

High-power lasers are required for surface treatments. In Table 1, three types of industrially used lasers are listed; the development of increased power output over a decade is evident. In addition, this table contains the maximum heating rates for pure iron that can be obtained using these lasers. Table 1 also indicates that CO_2 lasers emitting in the infrared at 10.6-μm wavelength with output powers of 5 kW were already available 10 years ago. Presently maximum output powers are up to 10 times as high. The most common solid state laser, for example, the Nd:YAG lasers, emit at the near-infrared at 1.06-μm wavelength. These are on the market up to a power range of 2 kW, while 10 years ago only 0.2 kW was available. As to excimer lasers emitting at a variety of wavelengths in the UV (Table 2) units of approximately 1 kW are now available. These can be regarded as the latest "offspring" in the family of industrial lasers, unknown only a decade ago. Interaction times are also of great importance. Regarding high power CO_2 lasers, it is possible to choose between CW and pulsed mode types down to 0.1 ms, while short pulse lasers such as TEA-CO_2, excimer, some Nd:YAG, and copper vapor lasers (CVL) exhibit pulse times of several nanoseconds. These interaction times are still of interest for surface modifications while pico- and femtopulsed lasers are mainly used for communications applications.

Table 1 Mean Power of Different Lasers over a Decade

Type	Max. CW or mean power (kW)	Pulse duration heating time	Year	Max. power flux density (W/cm^2)	Quenching rate for pure iron
CW CO_2	5	1 ms	1980	10^6	1000 K/1 ms
Pulsed CO_2	50	1 ms	1990	10^7	1000 K/0.1 ms
CW Nd:YAG	0.2	1 ms	1980	10^7	1000 K/0.1 ms
Pulsed Nd:YAG	2	1 ms	1990	10^8	1000 K/10 μs
Excimer	0.08	40 ns	1980	10^9	40 K/40 ns
Excimer	0.8	40 ns	1990	10^{10}	400 K/40 ns

Quench rates are calculated for the shown pulse durations or, in case of CW lasers, for the given heating times. Maximum power flux density not only depends on maximum power but also on absorption, focusability, and plasma formation during irradiation.

Table 2 Technical Data for Commercially Available Lasers

Type	Max. CW or mean power (kW)	Pulse duration	Repetition rate (kHz)	Energy/pulse (J)	Mode
CO_2	1.0	>0.1 ms	5	0.25	TEM_{00}
CO_2	5.0	>1.0 ms	5	0.5	TEM_{0*}
CO_2	25.0	CW	—	—	Multi
Nd:YAG	0.05	0.2–20 ms	0.02–0.2	25	TEM_{00}
Nd:YAG	0.5	0.2–20 ms	0.2	50	TEM_{00}
Nd:YAG	2.5	CW	0.1	—	Multi
Excimer	0.2	40 ns	0.1	4	Multi
Excimer	1.0	10–30 ns	2	2	Multi

B. Focusability

The focusability of a laser beam is much greater than that of incoherent white light. Theoretically, a laser beam can be focused down to a minimum diameter of limited diffraction, d, which is determined by wavelength, λ, the focal length, f, and the initial diameter, d_0, of the beam:

$$d = \frac{4\lambda f}{\pi d_0} \tag{6}$$

For practical reasons, however, extreme efforts are required to achieve this value. Therefore, 10 times the wavelength is a typical technical value for the focusability of a beam. It will be obvious, however, that focusability increases as wavelength decreases and that, with regard to equivalent pulse energy and pulse duration, much higher power flux densities can be achieved by means of excimer lasers.

C. Absorption

Metallic surfaces are not transparent for the laser radiation of the wavelength in question. Therefore, the incoming beam is partially absorbed and partially reflected. Figure 1 shows the reflectance for several metals as function of the wavelength. As is shown, only a small amount of the intensity is absorbed in the majority of cases. The measurement, however, is carried out at room temperature and for light of low intensity. The absorptivity however, increases with increasing temperature, surface oxidation, and laser intensity (Fig. 2). Oxidation by preheating is, of course, not a suitable method to improve absorptivity for the generation of rapidly quenched surface layers. To a certain extent, this is also the case if coatings such as graphite are required for sufficient absorptivity. Rather, the application of a sufficiently high critical beam intensity to the sample, which is characteristic for a particular material, should be preferred. Moreover, if intensity plasma formation occurs, then almost 100% of the radiation is coupled into the material. Physical principles involved in these phenomena are discussed elsewhere [30,31]. The above statements allow the conclusion that the highest quenching rates for a given material

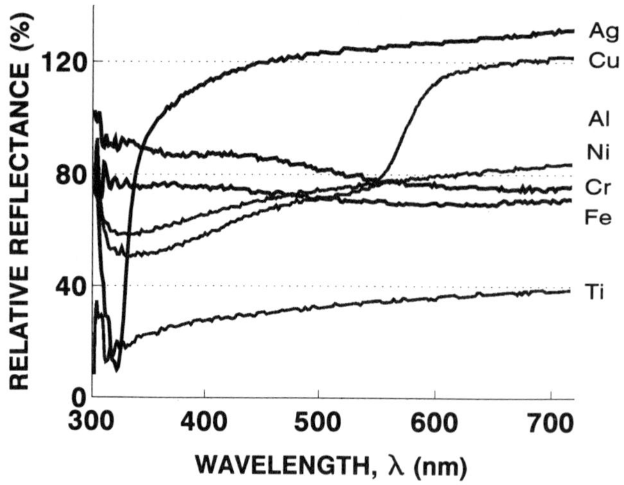

Figure 1 Reflectance of some metals as function of wavelength.

can be obtained if the smallest wavelength, the maximum intensity, and the maximum power flux density are employed. Taking into account available output power values and the damage thresholds of optical elements, a collection of maximum quenching rates (Table 1) can be gathered. The highest value found so far is in the range of 10^{11} K/s. In principle, it is possible to achieve higher values, but their technical utility is certainly doubtful.

IV. METALLURGICAL ASPECTS

Another important consideration is the reaction of the melt to the high quenching rate. In principle, there are several possibilities: either the melt freezes homoge-

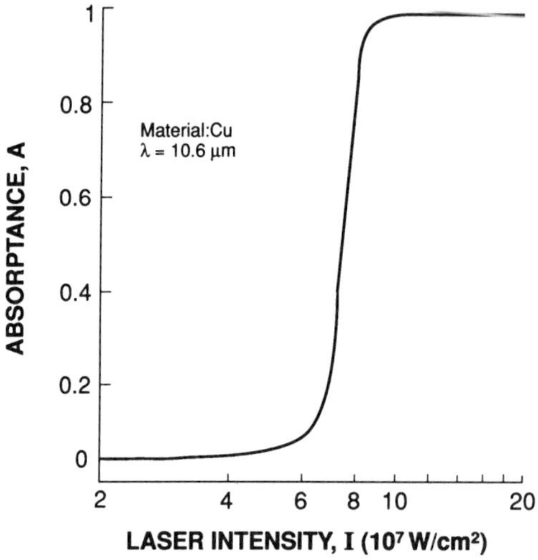

Figure 2 Absorptance of metals as function of the laser intensity.

neously or demixing occurs. If it freezes without decomposition, it becomes a supersaturated solid solution of stable, metastable, or glassy structure. In cases wherein quenching is fast enough to prevent decomposition, the melting temperature of the alloy may be represented by the T_0 curve, at which the Gibbs free energy of the liquid and solid state is equal. In a system with solid solution below T_0, the driving force for crystallization is ΔG. The Ag-Cu alloy system may be taken as a model system, as discussed in Chapter 2 of this book. If T_0 is extremely low or if no continuous T_0 curve exists, glass formation can be found (as in the Fe-B system).

To describe glass formation, the time-temperature regime in which crystallization occurs after quenching from the liquid (indirect method) has to be calculated. As to processes wherein the melt is quenched from a sufficiently high temperature above the liquidus, it may be described by the steps of nucleation and growth. By means of the values for nucleation frequency and growth velocity, experimental results can be described sufficiently well by the theory according to Johnston, Avrami, and Mehl. It was David who modified the equations for crystallization kinetics derived by Uhlmann and Turnbull in a slightly modified way for oxide systems. In essence, David's considerations aim at the time at which, for a given temperature, the crystallized volume fraction is less than 10^{-6}. If this is done for all temperatures, time-temperature crystallization curves for the alloys can be created. This model, however, has to be critically reconsidered if laser glazing is to be described. A fundamental empirical fact is that the resolidifying melt shows epitaxial or topotaxial crystallization on the substrate. Therefore, nucleation is not required in many cases, and the indirect definition of glass formation is not very helpful. A direct definition could be a separation of self-regulating and autocatalytic processes. If we consider cooling in a volume element at a temperature, T, below the melting temperature and the heat released by freezing is less than the heat conducted away into the next volume element, the temperature in the first element is reduced and the mobility of the constituent atomic species is also reduced. As the thermal diffusivity and the thermal conductivity increase with decreasing temperature, this results in a monotonically decreasing crystallization rate. In the next time increment, less material is solidified and hence temperature decreases further. Finally, this will lead to a certain volume fraction which will undergo glass formation.

If crystallization is avoided, the freezing velocity, $R = dx/dt$, with which the solid/liquid interface moves is determined by the temperature gradient, dT/dx, and the quenching rate, $\varepsilon = dT/dt$. If crystallization does occur, the crystallization velocity is determined by

$$u = u_0 \exp\left(-\frac{\Delta G_W}{RT}\right) \tag{7}$$

where ΔG_W is the activation enthalpy for crystal growth. If crystallization cannot be avoided, the morphology of the microstructure which forms is also determined by the three parameters ε, R, and G. Hereafter, it is possible to generate so-called solidification microstructure maps, indicating the regions in which planar, cellular, or dendritic crystallization occurs. The major effect is that, apart from the interface stability, there is an absolute stability, indicating that decomposition in the solid/

liquid interface can no longer occur above a certain solidification velocity. This condition is shown in Fig. 3a for a Fe-C-Si alloy at highest solidification rates while there is a supersaturated solid solution that still shows some carbon segregation, the formation of carbide, however, is avoided. At lower solidification velocities (Fig. 3b), a dendritic morphology is found. Variations in the quenching rate with constant G/R ratio result in the same morphology but with a different size of the relevant microstructural dimensions (for example, a different size of secondary dendrite arm spacing or lamellar spacing).

Speaking of rapidly solidified surfaces, a certain contradiction should not be left unmentioned. On the one hand, a dendritically solidifying system will exhibit a smaller dendrite arm spacing, and hence show increased hardness. On the other hand, large grains are found quite often. At shallow melting depths, epitaxial crystallization is responsible for an almost constant grain size in both substrate and

a

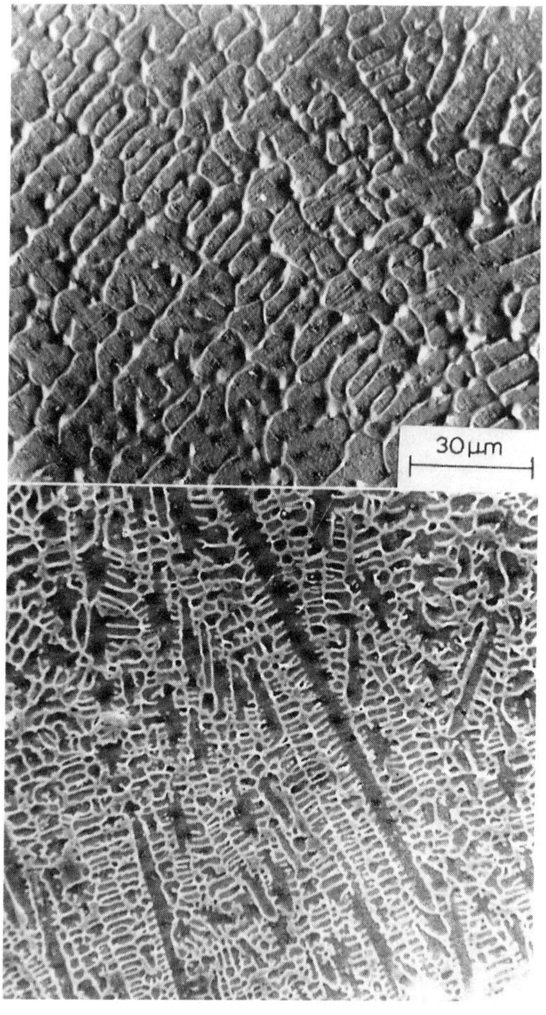

b

Figure 3 Microstructure of a laser remelted Fe-C-Si alloy.

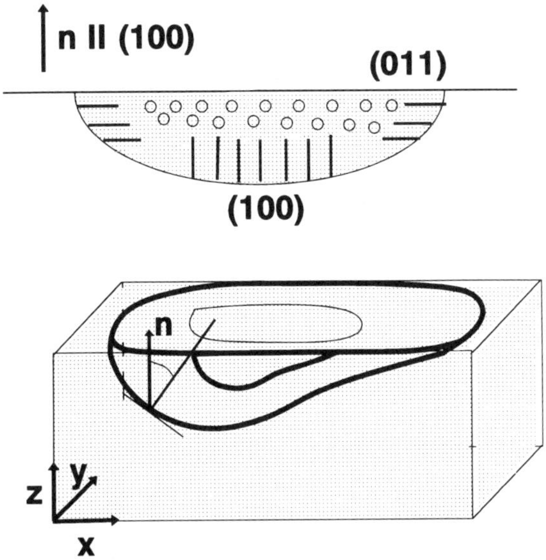

Figure 4 Formation of textures during laser remelting.

resolidified layer if no glass formation occurs. At greater melt depths, however, mainly suitably oriented nuclei will preferentally grow. "Suitable" nuclei are those with a direction of fastest solidification favorably oriented. In case of cubic crystals, this is the (100) direction, parallel to the temperature gradient. On the one hand, this effect results in large crystals, on the other hand it results in the formation of a crystallization texture. For single-spot melting, a symmetric situation prevails; if, however, the complete surface is remelted, that is, in a trackwise manner, asymmetric situations may occur. For cubic systems, the resultant texture is characterized by a preferred orientation at which the unit cell is tilted in comparison to the sheet normal, depending on the actual prevailing temperature gradient (asymmetric deviation of the liquid/solid interface during crystallization), as shown in Fig. 4.

During the laser treatment process, a volume element will undergo a different thermal cycle, depending on its distance from the surface. The resulting varying thermal expansion and contraction together with the temperature-dependent flow stress leads to both the presence of plastic deformation and the generation of internal thermal stresses, or to crack formation, if they exceed the strength of the material.

V. TECHNOLOGICAL ASPECTS

Surface treatment by lasers can be performed in a quite flexible manner regarding the size of the treated area of a workpiece. Depending on wavelength, different beam guiding and beam shaping systems can be used. For CO_2 and excimer lasers, mirror systems are required, while for Nd:YAG lasers and those radiating in the visible range, fiber optics may be used. Beam shaping is at least as important for the process as beam delivery with regard to quality of the layer and technological capability. For CW treatment, lateral intensity shaping is of primary importance, for pulsed lasers however, the temporal development of the pulse and the pulse-to-pulse stability are important. Quite often, a quick rise in intensity is desired, which favors a high absorption. Subsequently, however, a long tail of the pulse with reduced intensity may be required in order to obtain sufficient depth and to avoid the effect of plasma shielding. Lateral shaping is necessary in order to give the beam the desired cross-sectional shape and, furthermore, to supply the required power flux density. For surface treatments, rectangular cross sections are more suitable than the round ones (symmetric Gauss and TEM_{001}) used for cutting and welding applications. In the past, a constant intensity across the beam was desired; nowadays, it is possible to produce a constant temperature over a certain illuminated area by using deformable and adaptive mirrors.

Surface treatments may be carried out either by illuminating the component as a whole for a certain time period, the limitation of which is due to available power and is used only for small areas, or by a relative movement between beam and substrate with or without overlapping, respectively (Fig. 5), if treated patterns of points and lines as well as larger areas are desired. For short pulse lasers with a sufficiently large cross section, beam shaping via masking is a further alternative.

The technical use of surface treatments requires a high degree of reproducibility. Therefore, on-line quality control and, wherever possible, on-line process control are used. Two examples are shown for CO_2 and excimer lasers, respectively, in Figs. 6 and 7. Inhomogeneities in a working surface and fluctuations in the output power of about 5% are sufficient to result in a chaotic or at least inadequate surface quality after remelting. It is, therefore, necessary to monitor the in-coupling of the beam with a high temporal and spatial resolution. Real time recording could then be used for controlling the process; examples include the pyrometric measurement of surface temperature (suitable for CW processes) or differential reflectometry, plasma spectroscopy or other optical techniques that allow also to describe the result of pulsed laser/solid interactions.

During recent years, the number of laser surface treatments developed has considerably increased, which includes both fundamental investigations and industrial applications. Table 3 indicates a number of possible processes and the

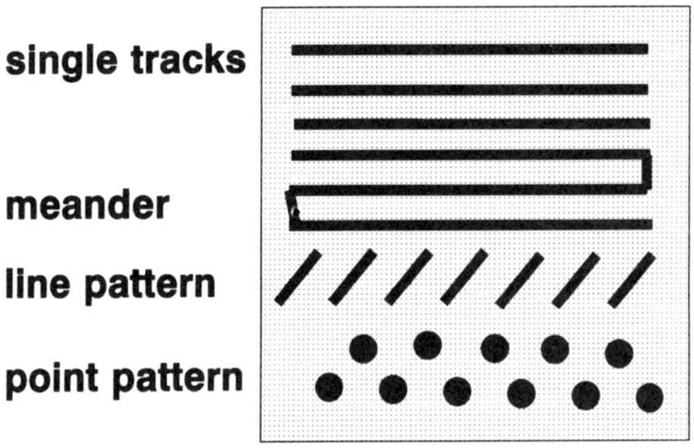

Figure 5 Various types of laser processing patterns.

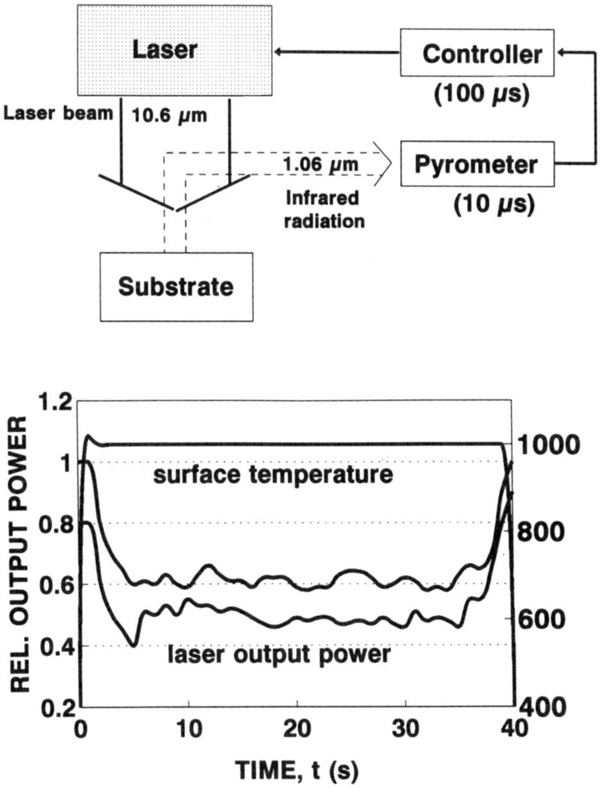

Figure 6 On-line temperature-controlled CO_2 laser hardening.

lasers to be used for their realization, particularly mentioning those processes that are already commercially used. In this table, the most important processes and obtainable modifications are summarized for gaseous, liquid, and solid treatments. For rapid quenching phenomena, of course, only the liquid treatments are relevant. A first distinction can be made between those processes where the surface melting is simply applied and others wherein additional alloying elements are involved, the latter might be added to the melt either by preplacement or via feeding while processing.

Simple remelting may be performed for various reasons. One of the reasons could be the destruction of impurities; nonmetallic inclusions, for instance, exhibit a high absorptivity and low thermal conductivity. They may, therefore, easily be evaporated. On the other hand, the rapid convection inside the melt might destroy such inclusions. Furthermore, simple remelting may result in a modified surface roughness. Such a roughness is quite characteristic for the applied wavelength and laser parameters. Process controllability for a wide range of roughness values by the settings of the laser is an attractive area of application, since tribological and corrosion properties may be improved as a result of such treatments. The majority of remelting processes, however, is carried out in order to generate a different

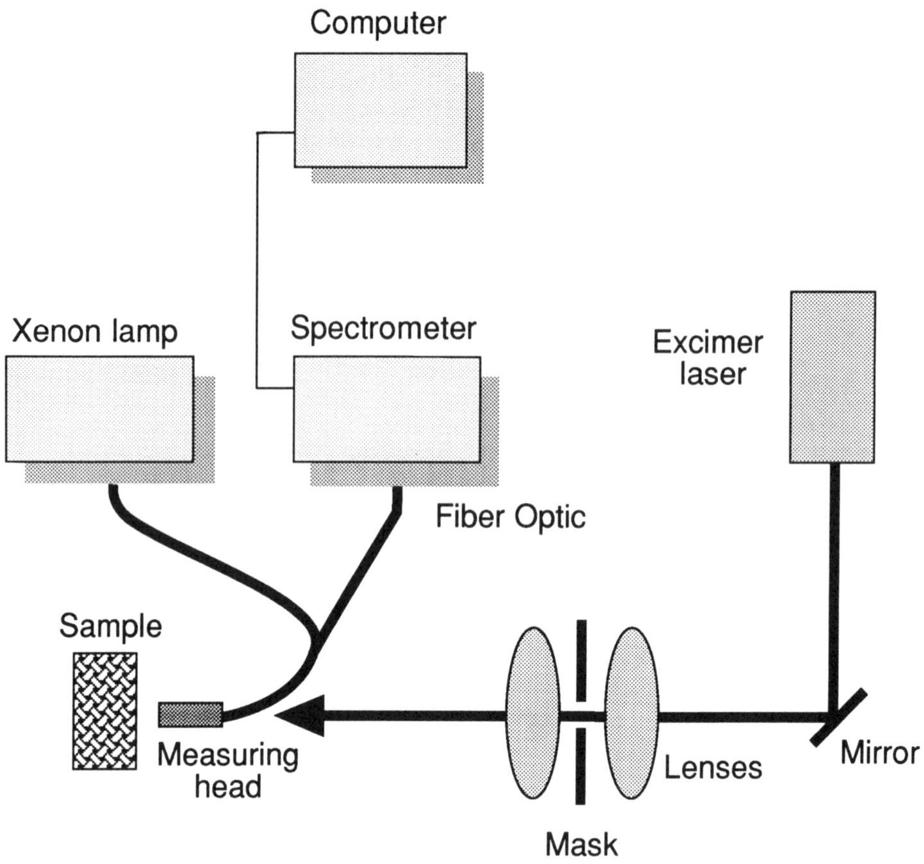

Figure 7 On-line optically controlled excimer laser surface treatment.

microstructure in the surface, which is metastable in most cases. This topic will be dealt with in the following sections.

VI. TYPES OF SURFACE MELTED LAYERS

The processes of laser surface melting and alloying may be subdivided into several groups, depending on different criteria. Such criteria could be process-related or others, differentiated according to the quenching rate or the solidification morphology (Fig. 8).

Table 3 Types of Laser Surface Treatments

Process type		Possible with CO_2	Possible with YAG	Possible with Exc	Commonly used laser	Modified depth
Thermal/ solid phase	Hardening	X	X	X	CO_2	0.2–3 mm
	Soft annealing	X	X	X	CO_2, Nd:YAG	0,1–2 mm
	Annealing	X	X	X	CO_2, Nd:YAG	0.1–3 mm
	Shock hardening		X	X	Exc, Nd:YAG	0.1–10 μm
Thermal/ liquid phase	Remelting	X	X	X	All	0.1 μm–2 mm
	Glazing	X	X	X	Excimer	0.1 μm
	Smoothing		X	X	Excimer	0.1–1 μm
Thermochemical/ liquid phase	Alloying	X	X	X	All	0.1 μm–2 mm
	Cladding	X			CO_2	1–3 mm
	Spraying	X			CO_2	1–3 mm
	Gas alloying	X	X	X	CO_2	0.01–1 mm
Thermochemical/ gas phase	Laser-CVD					
	Pyrolithic	X	X		CO_2, Nd:YAG	0.1–20 μm
	Photolithic			X	Excimer	0.1–20 μm
	Laser-PVD	X	X		Nd:YAG	0.1–1 μm
	Ablation and deposition			X	Excimer	0.1–1 μm
	Optical etching			X	Excimer	0.1–1 μm
	Optical cleaning			X	Excimer	0.1–1 μm

A. Processes

From the technical point of view, a required composition may already be present in the surface, but not the desired microstructure. In such cases, simple remelting is sufficient. Al-Si and cast iron represent this kind of materials. In other cases, it might be necessary to add suitable alloying elements to form a surface layer structure which is most suitable for the required application. The addition of boron, carbon, or nitrogen to titanium alloys is typical of such treatments. A third type of processing is the simultaneous feeding of alloying elements during laser treatment. Depending on the application envisaged, it might be necessary to either melt the additions in the laser beam and clad them onto the substrate with as little dilution as possible to generate a homogeneous composition in the surface layer or to leave one component unmolten and disperse it in the melt pool. Chromium plating on steels, surface coatings with stellites, or the dispersion of diamond

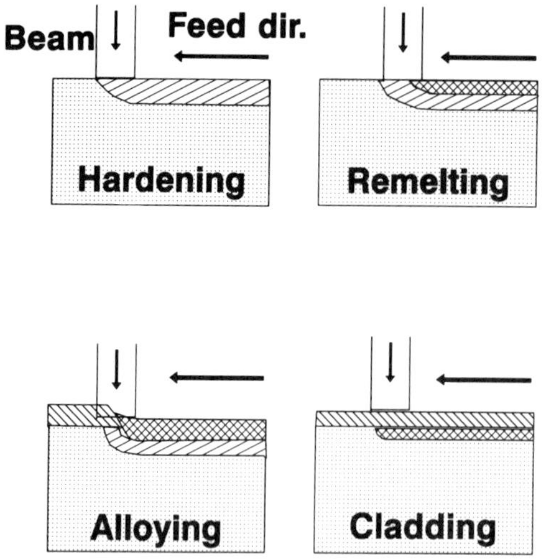

Figure 8 Schematic drawing of various laser surface treatment techniques.

particles respectively, are examples for the above-mentioned processes, respectively.

B. Quenching Rates

For a fixed composition a large variety of different microstructures and morphologies in surface remelted layers is strongly dependent on layer thickness and the temperature across the liquid/solid interface. Both are determined (Fig. 9) by the processing parameters: power flux density and interaction time. In this context, it

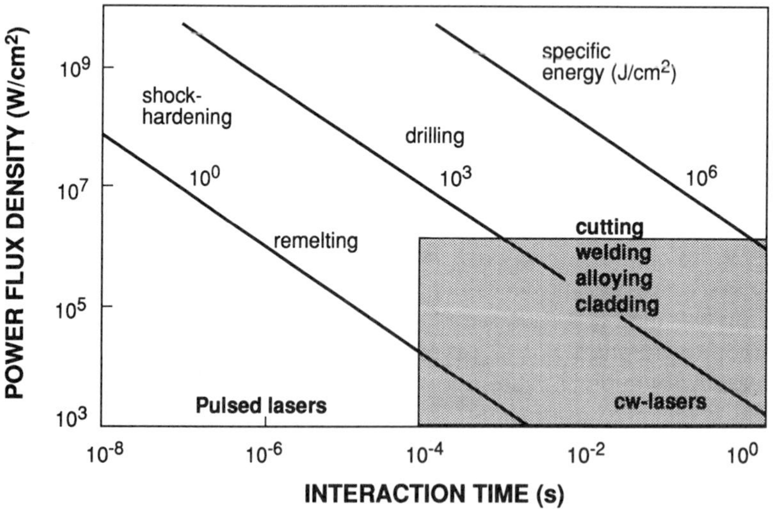

Figure 9 Interaction time–power flux density diagram for laser processing of materials.

is always the absorbed power flux density to which we refer as materials exhibit different absorption behavior in the UV, visible, and infrared wavelength ranges. Remembering that, in a first approximation, surface temperature is proportional to power flux density and that layer thickness increases parabolically with the exposure time, it will certainly be appreciated that their influence is of utmost importance, taking into account that both interaction time and power flux density may easily be varied over six or nine orders of magnitude. In Fig. 9, surface remelting and the generation of rapidly quenched layers cover the areas of welding and glazing. To demonstrate how far the temperature gradient across the solid/liquid interface may be altered, the following example may be of some assistance. With power flux densities of 10^3 W/cm^2 and interaction times of some seconds, laser transformation hardening can be obtained to the depth of about 3 mm, eventually causing a thin melted layer of a few micrometer thickness atop of the sample. If, on the other hand, a metal surface is exposed to an excimer laser, surface melting of about 1 μm can be achieved, where the heat-affected zone is not deeper than 2 or 3 μm. This allows a variation in temperature gradient of at least three orders of magnitude, which may be extended even more by either preheating the sample (lower values) or by applying even shorter interaction times and higher power flux densities (Fig. 9).

Not only the temperature gradient can be changed drastically; also the resolidification velocity may be varied over three to four orders of magnitude. Presently, fast welding processes use welding velocities in the range of 100 m/min; on the other hand, welding with defocused beams can be brought down to a speed range of several centimeters per minute. Long interaction times and large layer thicknesses that are obtainable by CW CO_2 lasers result from the high average power on the one hand and from the continuous mode of the sources on the other. Running such lasers in a pulsed mode, down to some fraction of a millisecond, can be regarded as pseudocontinuous as the time constants for heating up and cooling down are in the same range. Things are different for so-called short-pulse lasers which are driven in the pulsed mode of a certain frequency but with pulse duration times in the nanosecond region. TEA-CO_2 lasers, some of the Nd:YAG lasers, and all metal vapor and excimer lasers belong to this group. These lasers make quenching rates of some thousand kelvins in a few nanoseconds obtainable, which are fast compared to the motion of individual atoms by diffusion. Some solid state lasers allow the bridging of the two time regimes, as they allow pulse durations in the microsecond range.

C. Types of Solidification

Finally, it is possible to categorize the processes by resulting solidification microstructures. First of all, freezing can result in either crystallization or in glass formation. Crystalline phases formed may be quasi-stable or metastable. If crystallization occurs, the resulting microstructure is influenced by decomposition in the liquid or solid state. The diffusion of carbon in the melt forming supersaturated γ-crystals and Fe_3C is an example of the first case, and the spinoidal decomposition of supersaturated eutectic copper-silver alloy is an example of the other. Finally, the solidification microstructure may be characterized by the stability of the liquid/solid interface during crystallization, e.g., planar, cellular, or dendritic.

VII. EXAMPLES OF SURFACE MELTING AND ALLOYING

As mentioned above, iron-graphite and Al-Si are alloying systems in which simple surface remelting could cause a considerable change in microstructure and in the resultant properties. Because both have successfully been applied in the automotive industry for many years, they will be discussed in more detail.

A. Fe-C Alloys

The iron-carbon system is represented by two types of phase diagrams. On the one hand, there is the equilibrium phase diagram depicting iron and graphite, and then there is the metastable system $Fe\text{-}Fe_3C$ on the other hand. Graphite is formed if, due to sufficiently slow cooling, undercooling is avoided. Solidification in the stable system is favored by graphite-stabilizing elements such as silicon. Usually, technical Fe-C alloys have a near-eutectic composition and solidify during casting in the metastable system. When laser surface remelting such a sample, the resulting rapid quenching by the cold substrate will cause a certain undercooling, and subsequently a "white iron" solidification (stable phases) will occur. The typical microstructure of such layers consists of primarily solidified γ-Fe dendrites and an interdendritic iron carbide eutectic. The considerable amount of γ-Fe supersaturated with carbon is one of the reasons for the appearance of primary iron dendrites, even though the alloy has a nominally eutectic composition. From a practical point of view, however, it should be considered that a certain amount of graphite loss can occur by both floatation of the low density graphite and by its oxidation. Therefore, a eutectic microstructure can only be achieved by laser remelting if a hypereutectic equilibrium alloy composition is used, because the carbon content can change during laser remelting. At present, however, this latter procedure can only be carried out in laboratory experiments, as the decreased castability of hypereutectic cast iron causes manufacturing difficulties on an industrial scale.

Cast irons can be advantageously used as model materials for surface remelting processes because of the variety of morphologies that can be obtained by using different laser parameters. If, however, sufficient preheating is used and if certain quality criteria (composition, applied inoculants, mold and coatings) are met, it is possible to generate crack- and porosity-free layers several millimeters deep.

Depending on the values of R, G, and ε, different microstructures can be produced on solidification. This is demonstrated in Fig. 10 for cast iron. Increasing the quenching rate, ε, leads to a finer structure of the same morphology. If the solidification rate, R, is increased, the morphology of the solidification changes. This means that different microstructures in a given material can be produced by altering the solidification conditions. Working with a constant G/R ratio will give similar solidification conditions for various quenching rates. Higher quenching rates lead to an increase in the undercooling, resulting in a higher nucleation frequency and therefore in a finer structure of the same morphology (for example, secondary

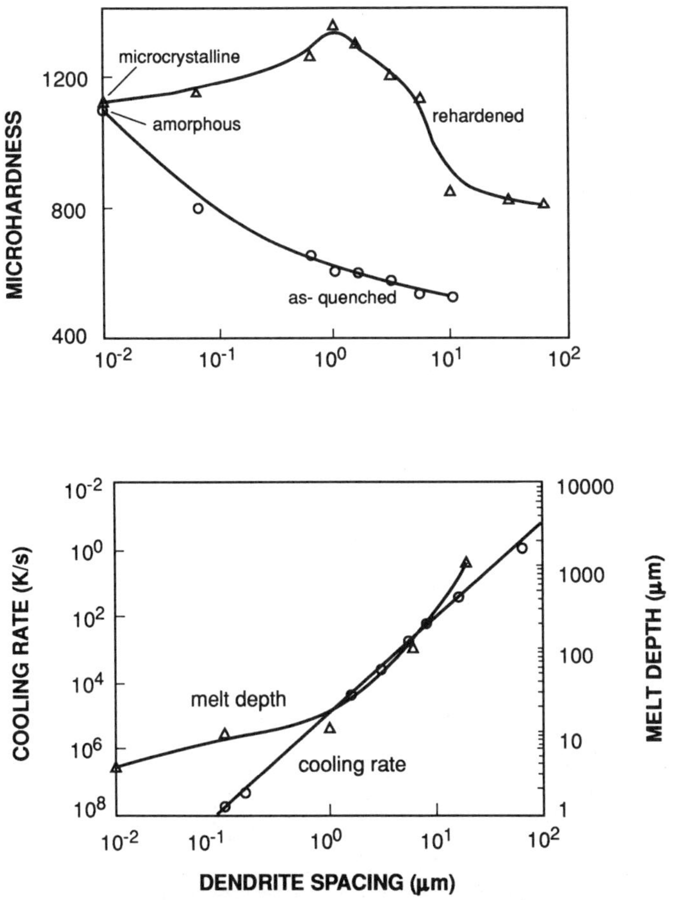

Figure 10 Relationship among microhardness, cooling rate, and dendrite arm spacing for laser-remelted Fe-C alloys.

dendrite arm spacing (Fig. 11). Keeping quench rate constant, a constant grain size results, but the morphology of solidification changes. Here, the nucleation frequency is constant, but with increasing solidification rate, the time for decomposition is reduced. Such behavior may be summarized in so-called solidification microstructure maps (see Chapter 2 of this book). Eutectic cyrstallization is observed at low solidification velocities and high-temperature gradients. An increase in the solidification velocity changes the microstructure first to a cellular and then to a dendritic structure. At high solidification velocity, there is a certain critical value, above which the material cannot be decomposed fast enough to form an interdendritic carbide phase. In this case only one phase is present after the solidification. This can be rationalized by considering that in the iron solid solution a certain amount of decomposition does occur, resulting in an inhomogeneous carbon distribution. This segregation, however, is not sufficient to lead to the formation of a second phase.

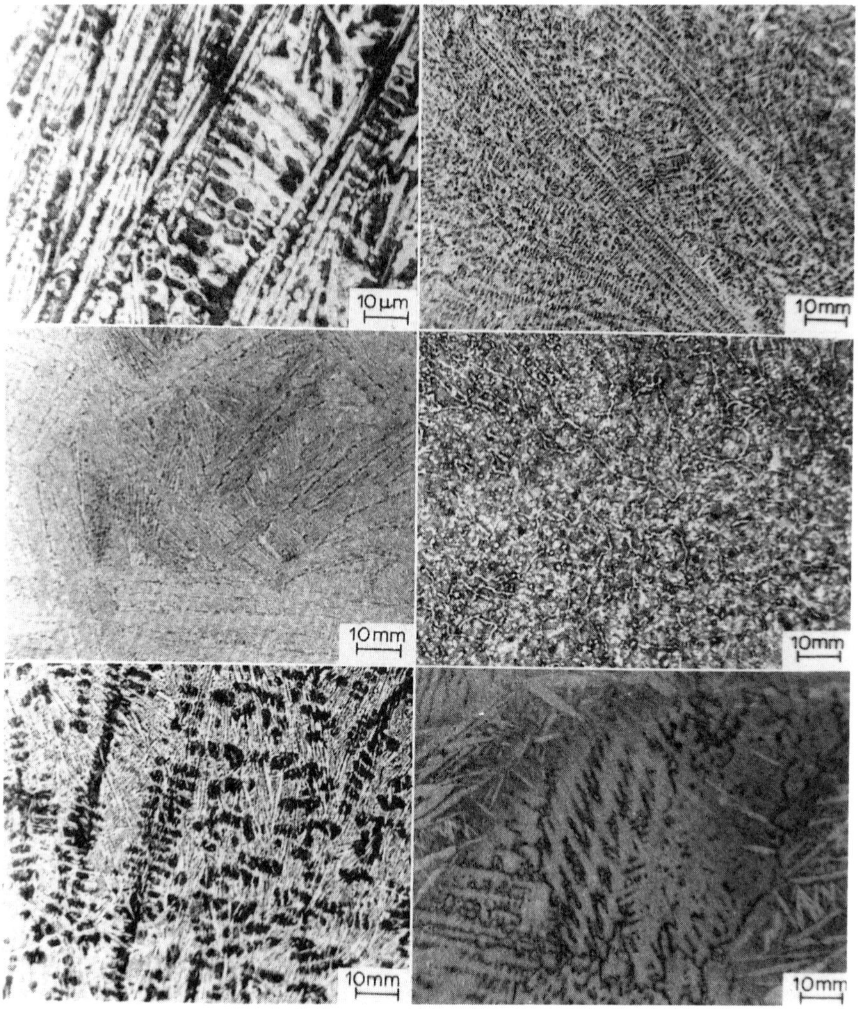

Figure 11 Microstructures of Fe-C alloys laser-remelted at different solidification rates.

For investigations of properties and for practical applications, melt depths of 0.7 to 1.4 mm are required with typical remelting velocities in the rate of 2–6 cm/s; a dendritic structure with interdendritic eutectic is representative if a Gaussian or a line source is used. Other types (square and rectangular cross sections) result in the production of different morphologies. These, however, are due to the large volume of the melt bath; they are technically irrelevant.

A specific feature of tensile testing of specimens which were laser surface melted is that elongations of more than 1% have to be avoided. Otherwise, case fractures and an equidistant crack pattern will occur. Yield stress and UTS are slightly reduced (5–10%) as compared to samples that were not laser-surface-melted; more consideration has to be given to the reduction in elongation to fracture (20–25%). The hardness within the layer is between 600 and 800 HV10, depending on the volume fraction of iron and carbide; furthermore, it depends on the type of transformation, i.e., whether γ-Fe undergoes relatively slow cooling, e.g., pearlite, bainite, or transforms rapidly to martensite. Compression strength can considerably be improved for samples which were laser surface melted. The same will apply if the fatigue behavior under compression is taken into account (push-push regime). If, however, the push-pull or pull-pull region is considered (Fig. 12), the results will be different. A tremendous advantage of such white iron layers is the improvement of wear behavior. Laser remelted samples show better behavior under abrasive, adhesive, and fatigue wear conditions. This fact could not only be demonstrated on a laboratory scale, but it was also proved by engine tests of cast iron parts (rocker arms, camshafts, cylinder walls), which have been laser surface remelted. The finely crystallized Fe-Fe_3C eutectic allows for a certain microductility. Scanning electron microscopy of rocker arm surfaces showed that cementite could be deformed to a certain extent. In the surface, laterally deformed cementite platelets build up during service and cover the majority of the component; this seems to be the major reason for the improved performance. As the hardness of coaked coal particles or of spoiled-off fragments is not higher than that of those carbide platelets, the abrasive wear is in the lower plateau region. The improvement of the sliding wear is often explained by the fact that the softer iron solid solution matrix phase is "washed away" in the surface, generating pockets into which the oil is pressed when the two components are rolled against each other. Tribooxidation may also be involved.

Besides the Fe-C-Si system (cast irons), another of the most widely investigated systems is Fe-Cr-C. Due to the rapid solidification, iron solid solution dendrites form or, depending on the solidification velocity, interdendritic eutectic or an area-

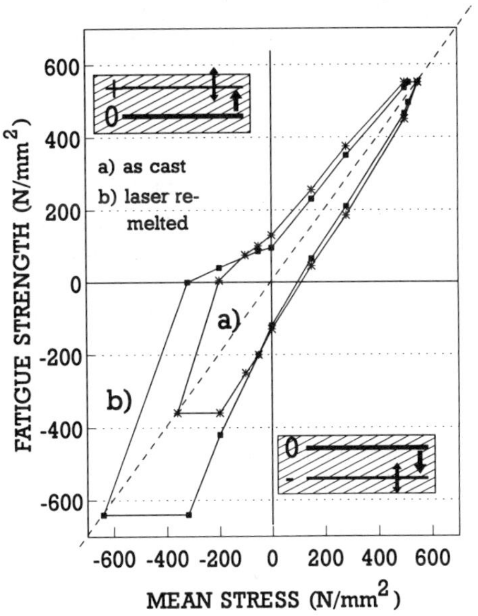

Figure 12 Smith diagram of a pearlitic cast iron before and after laser remelting.

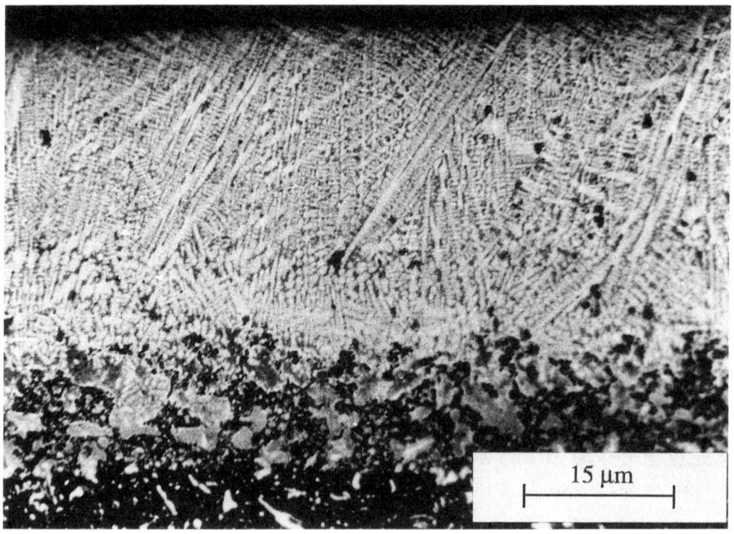

Figure 13 Microstructure of a laser-remelted Fe-Cr-C alloy.

covering chromium-carbide-phase Cr_7C_3 forms (Fig. 13). For this system, different secondary dendrite arm spacings over a wide range and therefore different quenching rates could be realized.

As in the as-quenched state, austenite hardness rises with increasing quenching rate (smaller dendrite arm spacing), whereas a hardness maximum occurs on subsequent hardening annealing treatment. Regarding abrasive or adhesive wear, no improvement could be observed after remelting. With respect to fatigue wear, and especially with pushing and knocking wear, significant improvement are obtained due to the microductility of the fine but hard microstructure. This is true for both model wear tests and real wear of components.

B. Fe-B

Another alloy system that has often been investigated by surface remelting experiments is Fe-B. By this we mean to include the large variety of ternary and quaternary modifications of this basic system as well as similar yet other metal-metalloid glass forming systems. It is well known that the iron-boron system shows a compositional range of glass formability in the near-eutectic region. Different from the iron-carbon phase diagram, in which extended solid solution of carbon in α-Fe is possible, the Fe-FeB system shows only neglegible solid solubility in both terminal phases.

The required quench rates ($\sim 10^6$ K/s) can be achieved by laser remelting. This is the reason why further investigations were performed to find out whether the attractive properties found by melt-spinning of such materials could also be generated in surface layers of near-net-shape components. Two major directions were chosen for these investigations. One of them was the exploration of suitable time windows for such a treatment, while the other aimed to develop techniques by which a suitable composition could be generated at the surface. While the first aim has now been achieved using a variety of different lasers, the second is still unsatisfactory.

C. Al-Si

Aluminum forms a eutectic with silicon that solidifies relatively coarsely compared to other eutectic alloys. In the microstructures of eutectic Al-Si alloys, primary Al and Si crystals are observed in addition to the eutectic microstructure. The latter are responsible for the good wear properties in the field of engines. Some years ago it was discovered that a fine dendritic mictrostructure with interdendritic eutectic forms when remelting a eutectic alloy not only in binary Al-Si alloys, but also in technically used alloys with more components; remelting may double the hardness of these alloys. For a binary Al-Si alloy in the as-quenched state values of about 100–120 HV was the result; for a commercial alloy it was 130–150 HV. Subsequent heat treatment may lower these values. This is important because especially overlapping melt tracks may cause aging and recrystallization. Ibe and co-workers [32] could show that interesting improvements in mechanical strength and dynamic properties can be achieved for surface remelted, finely crystallized samples. For engine applications, especially for cast cylinder liners, a remelting can be applied advantageously where the strength has to be increased, for example, in the area of the valve seats. To reduce wear at the cylinder wall, large Si crystals are preferred because of the better compressive strength due to the increased load-bearing capability of the hard Si phase. Of course, while a huge amount of primary Si crystals may be obtained in hypereutectic alloys, their use would result in reduced castability and mechanical processability. Because both of these factors are associated with immense costs, many research groups are searching for methods to alloy the cylinder surfaces to form an eutectic composition in the highly loaded areas. It was possible to achieve primary Si crystals in a rapidly quenched surface layer by either melting in high Si contents (up to 30%) or by melting in mean Si contents (~20%) with the aid of inoculants. Hardness and wear behavior could be improved significantly by this technique. The results achieved were, however, not as good as assumed so that further increases in hardness were sought by alloying other additional elements. From fundamental investigations it was known that the solid solution of binary alloys with *d*-band metals may be increased by rapid quenching. On the other hand, Al forms stable intermetallics (sometimes having high melting points) with these metals. Table 4 lists the obtainable hardness values for some alloy systems. The values obtained are rather spectacular but technically not applicable. Realistic is the use of Al-Si-Ni and Al-Si-Mn alloys with a hardness of about 350 HV in the as-quenched state. After a heat treatment just below the melting point (with regard to overlapping tracks) values of 200 to 250 HV result, exceeding the hardness of conventionally Al alloys by a factor of 4.

D. Ti Alloys

Ti alloys are somewhat special because of their strategic significance due to their applications, resulting in a high and extremely varying price on the world market. Essential properties of Ti and Ti alloys are high yield strength, good strength-to-weight ratio and excellent chemical properties. What is missing is an effective

Table 4 Alloying of Aluminum Alloys with Different Metals

Substrate	Alloying elements	Melt depth (mm)	Hardness (as lasered)	Hardness (after heat treatment)
Al-Si	—	0.6–0.7	80–120	80–90
Al-Si	Fe	0.6–0.7	80–120	80–120
Al-Si	Si	0.6–0.7	120–300	100–120
Al-Si	Mn	0.6–0.7	300–400	400–750
Al-Si	Fe + Si	0.6–0.7	200–300	200–300
Al-Si	Mn + Si	0.6–0.7	300–400	400–750
Al-Cr-Zr	—	0.3–0.5	55–65	37–42
Al-Cr-Zr	Mn	0.4–0.6	180–370	180–400
Al-Cr-Zr	Si	0.4–0.6	120–300	80–90
Al-Cr-Zr	Cr	0.4	240–260	250–360
Al-Cr-Zr	Co	0.4–0.7	150–350	150–350
Al-Cr-Zr	V	0.2	300–700	450–750
Al-Cr-Zr	Mo	0.2	370–750	800–900

hardening mechanism, especially for surface hardening because the tribological properties of these alloys are poor and the materials are prone to seizure.

Fortunately Ti forms hard, high melting point compounds with carbon and nitrogen. These compounds are used as cladding materials for steels and cementites due to their excellent wear resistivity. According to the formation process these layers are thin and require adequate support by the substrate material. However, such a support cannot be achieved using Ti alloys because of the lack of a hardening mechanism. For this reason attempts have been made to produce thick layers consisting of hard materials by laser-gas alloying. This process is based on a gradual remelting of the Ti material in a desired area and simultaneously blowing a lamellar gas stream of inert gas containing the desired quantity of reaction gas (for example, nitrogen or methane) into the melt bath.

In the melt bath, spheroidal to platelike cementite precipitates form which turn into a dendrite skeleton embedded in a metallic binder matrix. The depth of the melt bath, the process gas concentration and the interaction time are the parameters that determine the proportion of the hard phase produced. Figure 14 shows the principle of the process together with a typical microstructure. When using nitrogen, a higher hardness may be achieved than when using methane; on the other hand, nitrogen-remelted samples tend to form microcracks. This can be avoided by an appropriate heating cycle during the process. A subsequent relaxation heat treatment allows the achievement of tolerable strengths even for clad devices. The combination Ti-TiC leads to compressive stresses after the remelting because of the more favorable differential thermal expansion situation compared to that of the Ti-TiN system. Furthermore, combinations of subsequently or simultaneously carried out carburizing and nitriding processes also known from the literature are the melting in of cementites such as titanium boride, boron carbide, or boron nitride that dissolve in the melt bath and form hard, high-melting-point phases with

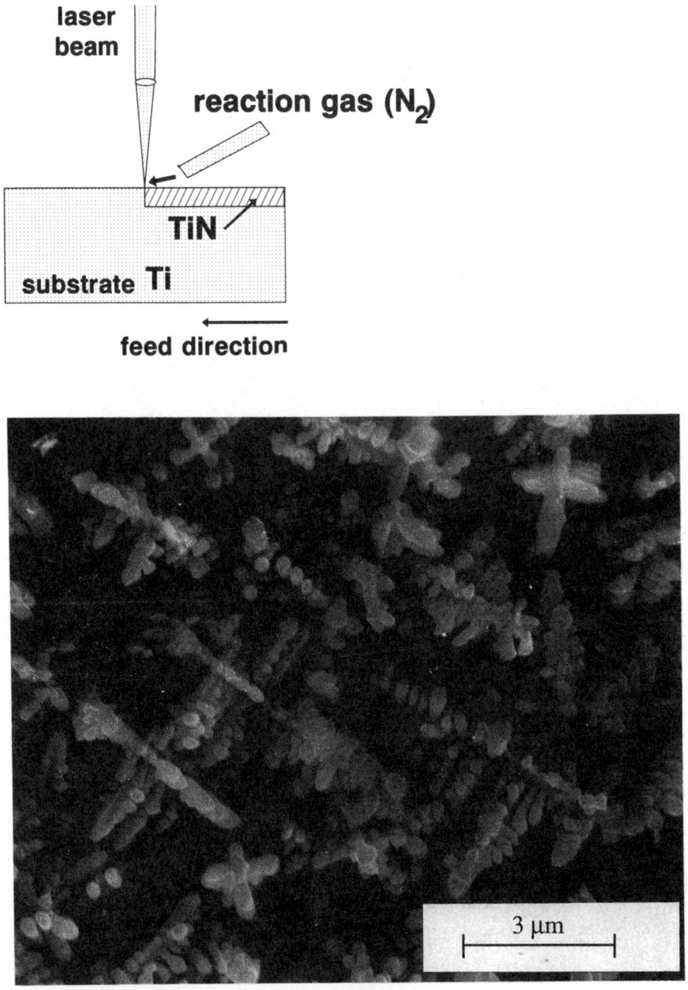

Figure 14 Schematic drawing of laser-gas nitriding and typical microstructure of a nitrided Ti sample with TiN dendrites.

titanium (Fig. 15). Worth mentioning is the possibility of melting Ti or Ti alloys (for example, a binary eutectic Ti-Fe alloy) into the surface of other base materials, simultaniously transforming the Ti metal into a hard phase by supporting a process gas. A similar process is possible if the substrate itself contains the relevant reaction partner, for example, graphite in cast iron. This enables the achievement of an effective hardness increase due to a two-phase hardening.

E. Short-Pulsed Lasers

Within the last four years so-called short-pulsed lasers have achieved importance for surface treatments. Due to their set-up these lasers allow pulses with high energy but of a short duration and with a repetition rate low enough to guarantee that

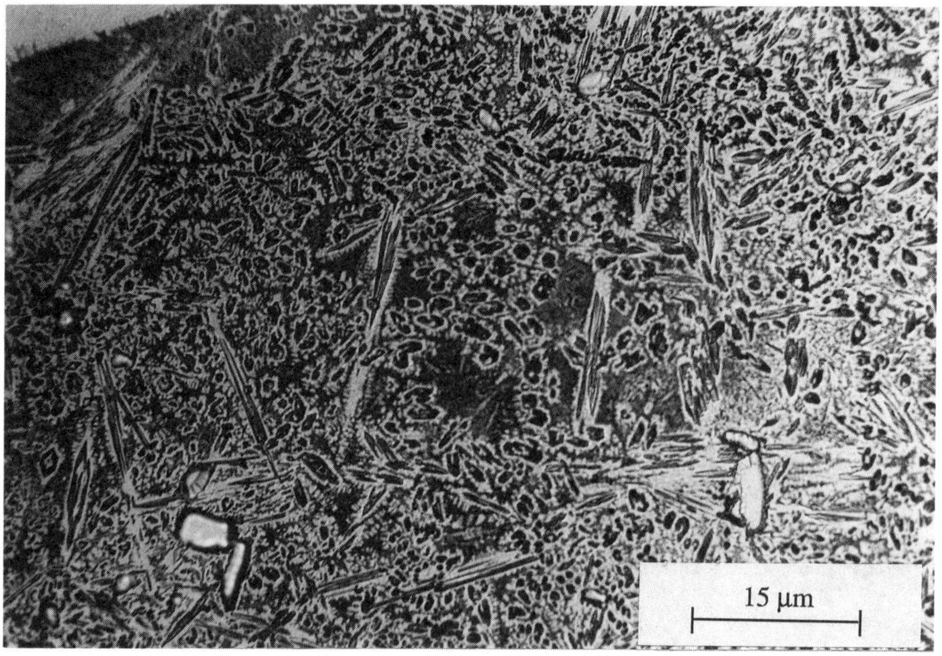

Figure 15 Typical microstructure produced by melting in of cermets into Ti substrates.

the sample cools down to the initial temperature between pulses. Included in this type of lasers belong the excimer lasers that cover the wavelength from 351 to 157 nm, the copper vapor lasers emitting in the green-yellow range and the TEA-CO_2 lasers. We shall restrict ourselves only to the excimer lasers because they are most suitable for surface treatments due to their large, rectangular beam cross section and their good absorptivity behavior in metals. Some important technical data on excimer lasers are summarized in Table 5. Applying an appropriate wavelength of the excimer lasers may result in direct nonthermal breaking of chemical bonds for some nonmetallic materials. The influence of different wavelengths will not be discussed, because only the amount of absorbed light is changed. Therefore, the influence of wavelength may be compensated with different focus points during the treatment of metals, resulting in different absorption as well. For this reason the following examples are restricted to the readily obtainable wavelength of 308 nm (laser active medium: XeCl).

Irradiation of the sample is carried out using a slide projector type of setup. Due to the short pulse duration, heating is possible only to depths up of several micrometers, resulting in micrometer-thin layers on relatively thick substrates. Assuming a pulse duration of 40 ns, a variation in the power flux density results in melt depths of 0.01 to 3 μm. Therefore, it is possible to control processing by optical methods, e.g., with differential reflectometry.

Earlier in this chapter we mentioned a processing diagram: interaction time–power flux density. Figure 16 shows a subdivision of this diagram for CW, pulsed, and short-pulsed lasers. The shallow penetration of excimer lasers makes it nec-

Table 5 Technical Data of Commercially Available Excimer Lasers for Materials Processing

Name/ Comp.	γ (nm)	E_{max} (mJ)	t_p (ns)	Rep (Hz)	P_m (W)	S_Q (cm^2)	PP (%)	Pre- ion.	t_L
LPF 205/L	157	60	10	50	3	1	8	UV-F	10^9
2000/Q.	157	30	20	100	14	—	3	UV-S	—
2000/Q.	193	700	20	250	40	—	5	UV-S	$>10^9$
LPX 205i/L	193	275	14	200	45	1.3	5	UV-F	$>10^9$
LPX 305i/L	193	800	20	50	25	4	5	UV-F	$>10^9$
Exc.600/Lu	193	225	12	300	55	175	6	UV-F	$>10^7$
2000/Q	222	200	20	100	8	—	5	UV-S	—
LPX 305i/L	248	1200	20	50	25	4	5	UV-F	$>10^9$
LPX 325i/L	248	800	25	250	160	3	3	UV-F	$>10^9$
GX500/P	248	0.007	40	2000	0.01	0.04	5	Cor.	10^{10}
2000/S	248	700	20	250	40	—	5	UV-S	$>10^9$
LPX 305i/L	308	800	30	50	25	3	3	UV-F	$>10^9$
LPX 605i/L	308	500	250	250	25	3	3	UV-F	$>10^9$
LPX 325i/L	308	600	30	250	120	3	3	UV-F	$>10^9$
L 3000/L	308	500	30	300	150	3	6	UV-F	2×10^8
EMG401/L	308	200	35	50	100	16	4	X-ray	5×10^8
GX500/P	308	0.02	50–250	2500	0.05	0.006	1	Cor.	10^{10}
XP2020/S	308	2000	50	20	40	20	<1	X-ray	10^8
XP4010/S	308	4000	50	10	40	25	<1	X-ray	10^8
5100/X	308	500	25	300	150	5.6	5	Cor.	10^9
LPX 305i/L	351	650	25	50	15	3	3	UV-F	10^9
LPX 325i/L	351	400	25	250	70	3	3	UV-F	10^9
2000/Q	351	400	20	500	50	—	3	UV-S	10^9

Laser producers: L = Lambda Physik; Q = Questek/Spectra Physics GMbH; S = Siemens KWU; X = XMR, Lu = Lumonics; P = Potomac Photonics Inc./Laser 2000 GmbH.
Abbreviations: λ = wavelength, E_{max} = max. pulse energy, t_p = pulse duration, Rep = max. repetition rate, P_m = max. mean power, S_Q = beam area, PP = pulse-to-pulse stability, Pre. = type of preionization, UV-F = UV-arc preionization, UV-S = UV-spark preionization, Cor. = UV-corona preionization, x-ray = x-ray preionization, t_L = min. lifetime (in terms of laser pulses).

essary to consider the fact that, technically, surfaces are treated. Surface contaminants in form of solid or liquid absorbates, adherent and pressed-in grinding and polishing materials, oxide layers, and deformation layers at the surface arising from the production can lead to artifacts and misinterpretations. Such extraneous layers can be removed using an excimer laser with low energy density, where low means a small fraction of the material-dependent metal plasma formation threshold. A whole series of phenomena can be observed by remelting various metals with different crystal structures. Face-centered-cubic (fcc) metals, e.g., copper (Fig. 17) show strong tendancy for mechanical twinning in the resolidified layer. This may be explained by the formation of internal stresses that cannot be compensated by the usual dislocation glide processes because of the extremely high quenching rate during the rapid cooling. Similar phenomena are found in the irradiation of hexagonal metals, in which there are much fewer slip systems. If the metals undergo a transformation in the solid state during the cooling, diffusion-controlled transformations will be suppressed by rapid cooling. Stress-induced transformations such

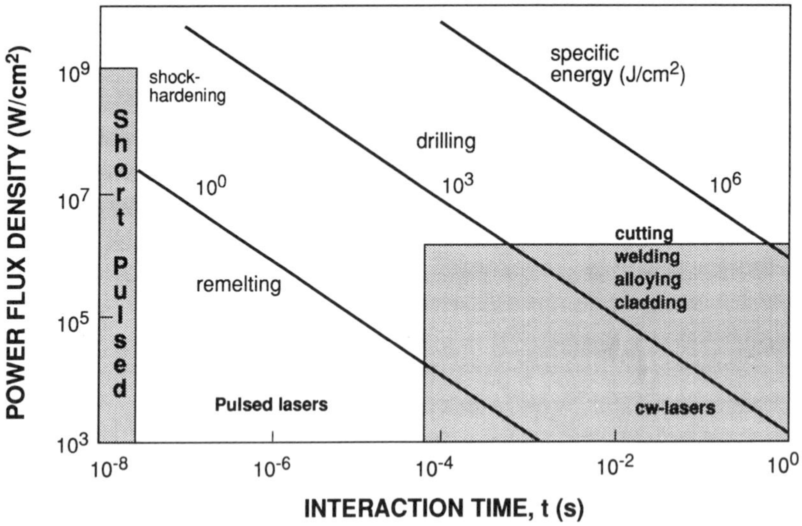

Figure 16 Processing windows for CW, pulse and short-pulse lasers in an interaction time–power flux density diagram.

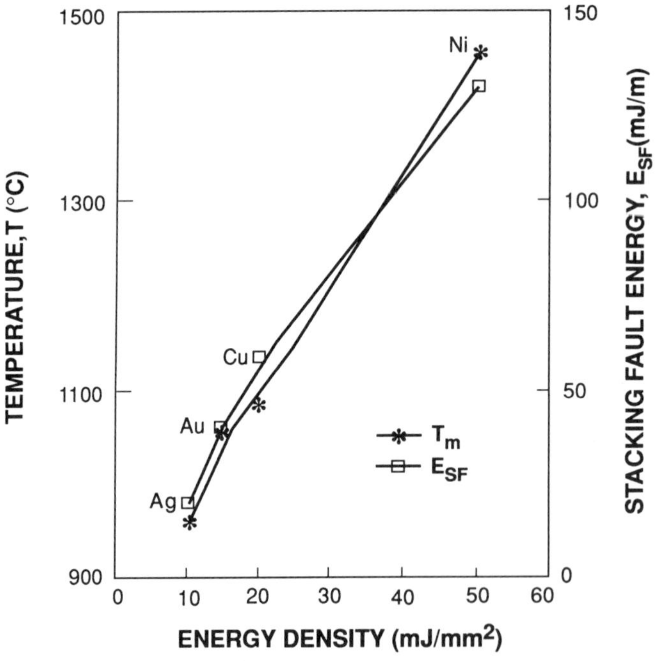

Figure 17 Start of twinning for fcc metals under excimer laser irradiation.

as martensite formation, however, will occur due to the thermal stresses generated. Irradiation with an excimer laser enables both weakening by recrystallization and strengthening by the deformation caused by the thermal stresses, depending on the power flux density used. Depositing metals on the surface, for example, by sputtering, a homogeneous alloy can be produced by excimer laser irradiation inspite of the short pulses. Segregation can be avoided due to the high self-quenching rate. The solidification structure can either be finely crystalline or amorphous. In fact, a glassy solidification structure detectable by Mössbauer spectroscopy is possible by remelting amorphous or crystallized amorphous alloy ribbons to depths up to 0.1 μm.

F. Time Windows

As CO_2 lasers of sufficiently high power have been available for quite a long time already [33], they were also used for glazing experiments. In the beginning, lasers with dc-excited transverse gas transport suffered from the lack of pulse mode and beam quality. In order to achieve sufficiently short interaction times at the required power flux densities, a fast-moving substrate had to be used so that the melt is partially ejected from the pool. Instabilities resulting from nonuniform intensity distributions were further problems that had to be addressed. It was, however, possible to demonstrate that glazing is achievable. RF-excited fast axial flow lasers offered the possibility of pulsed mode operation. These pulses, however, should be overlapped if entire surface areas have to be completely covered. The Gaussian intensity profile of these lasers, however, which had been developed for cutting applications is most unsuitable for a glazing process. It was only recently that the use of homogenized beam profiles with a cylindrical head intensity profile and a good pulse-to-pulse stability have been introduced. Due to the development of high-power Nd:YAG lasers it was possible for the first time to generate continuously-glazed surface layers resulting from overlapping pulses. Pulse durations in the microsecond range or shorter, which are possible by q-switch operation, allow repeated exposures without running the risk of crystallization in the heat affected zone. The most satisfying results of laser glazing, however, have only been realized during the past two years by means of short-pulse lasers (excimer, CVL, TEA).

VIII. POSSIBLE APPLICATIONS

A. CO_2 Laser Treatment Applications

Remelting of a material surface does not result in a modification of the bulk properties but rather in changes confined to a shell-like casing. Hence, possible applications will have to start in such areas in which wear, corrosion, and fatigue of the surface limit the lifetime of a component. On the other hand, this kind of processing is very expensive, as the surface of each and every component will have to be remelted sequentially one after another. Therefore, due to the expense of photons and sometimes inefficient systems, the economic use of surface remelting will only be justified in very few instances, as when other more cost-effective coating methods (CVD, PVD, etc.) are not successful. A suitable range of application, however, will always be found where—as in case of mechanical power transmission components—engine output is critical or a reduction of fuel consumption has to be achieved. In this respect, the camface of a cast-iron camshaft valve seat (remelting) and valve collar (remelting/coating), rocker arm, drag lever (remelting/coating), cylinder wall (remelting/alloying), and piston ring (remelting/coating) are components of interest. The following figures will demonstrate some examples in the area of automotive components.

Figure 18a shows the remelting process using laser radiation which is line-focused by means of a zone mirror, so that the whole cam face can be remelted at a time. The resultant cam lobes are shown in Fig. 18b.

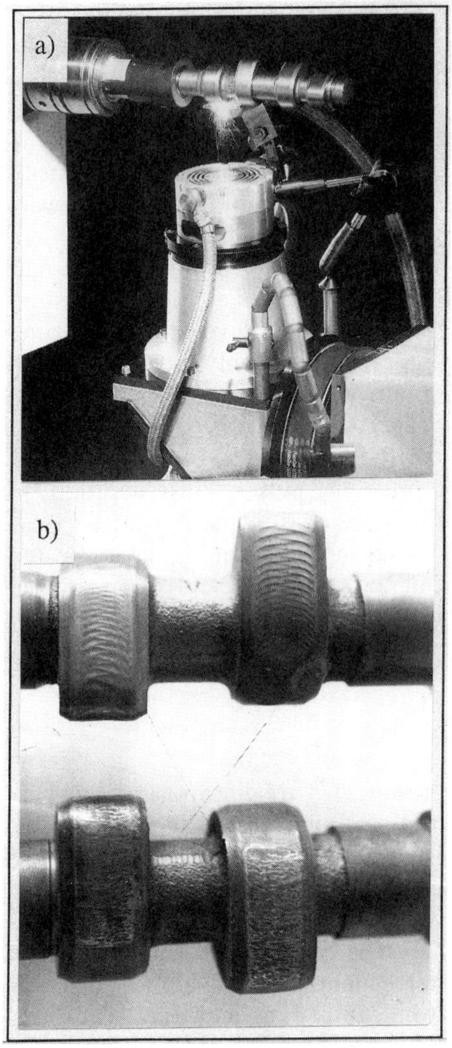

Figure 18 (a) Remelting of cam shafts using a 12-kW Co_2 laser unit with a six-axis handling system. (b) Surface of laser-remelted cams.

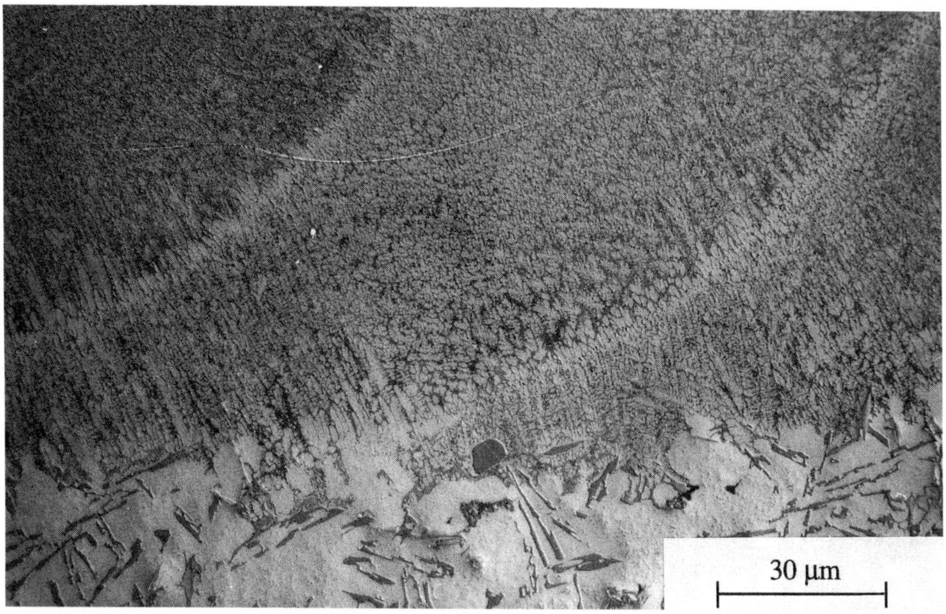

Figure 19 Alloying of Si in Al-Si cylinder sleeves.

Implants are a second group of components that are of particular importance for surface remelting. For example, according to Fig. 19, Si was alloyed in an eutectic Al-Si cylinder sleeve. For such components, the complex requirements imposed on the tribosystem as well as on corrosion resistance and biocompatibility are of such an outstanding importance that the inherent costs incurred due to processing are regarded as acceptable. This is a field wherein alloying by remelting of titanium implants, for instance, could find widely used application in future.

Besides the automotive examples, other possible ranges of applications will also be found as extremely expensive materials are able to be replaced by less expensive materials through the use of surface remelting. Typical examples are contact materials of different kinds ranging from plug and relay contacts to lead frames and metallic ceramic chip carriers. It was Draper, for example, who presented functioning samples of corrosion-resistant contacts made of nonprecious metals. Because processing and precious metal costs are subject to variations, manufacturers have not yet favored such an application on a larger scale.

B. Excimer Laser Treatment Applications

1. *Contact Materials*

It can be shown that excimer laser treatment results in improved properties of contact materials. One reason for this is the cleaning and smoothing of surfaces by evaporation of contaminations and melting of the surface. Another reason is the removal of nuclei for the growth of corrosion products. Performance improvements can be evaluated by evaluating mechanical properties, such as tensile strength, of bonded contacts. The best tensile testing results are obtained by using

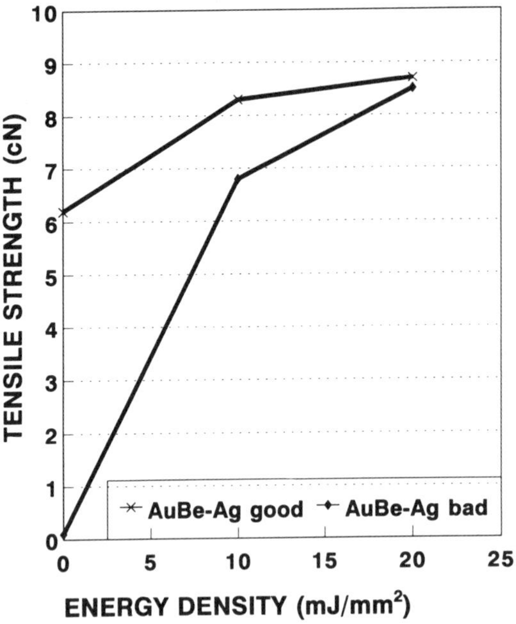

Figure 20 Influence of an excimer laser treatment on the bonding strength of lead frames.

power flux densities just below that resulting in plasma formation during excimer laser surface treatment, as can be seen in Fig. 20.

Besides mechanical properties, the corrosion behavior of excimer laser-treated contacts is also improved. Figure 21 shows examples for three precious metal contact ribbons for which the contact resistivity after an anneal in a corrosive atmosphere is one order of magnitude improved by laser treatment. Applying laser surface alloying to contact materials enables the production of alloys with improved mechanical properties and corrosion resistance. By laser irradiation of PVD-sputtered Cr layers on Cu substrates, thin homogeneous layers (up to 1 μm) of a solid solution of Cu and Cr can be obtained. The structure of the alloy is of fcc type with an increased lattice parameter due to the solution of Cr in Cu. The hardness is therefore improved by a factor of 2 because of the solid solution hardening (Fig. 22). The electrical conductivity of the layers is sufficiently high and the corrosion resistance of the alloys exceeds the values of unalloyed Cu substrates. Comparable investigations with pulsed Nd:YAG-lasers are known from literature. Applying this technique previous metal-free contacts have been successfully used in telephone dialing kits [34].

2. *Automotive Parts*

Possible applications for excimer lasers in the automotive industry may be in marking, writing, or even removal processes so long as they are restricted to small volumes. In the area of surface treatment, mainly a change of the tribological properties and in the corrosion behavior are of interest for application. In the

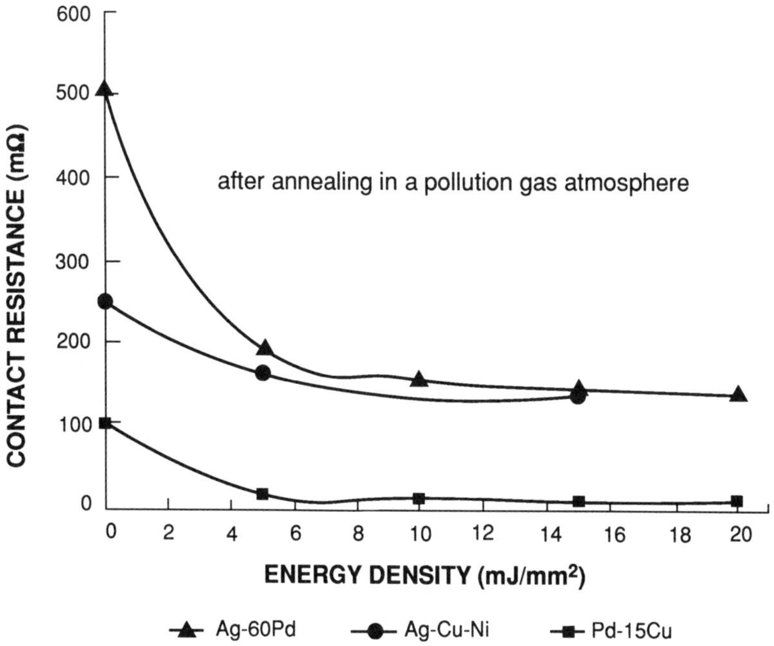

Figure 21 Increased corrosion resistance of excimer-laser-irradiated contact materials.

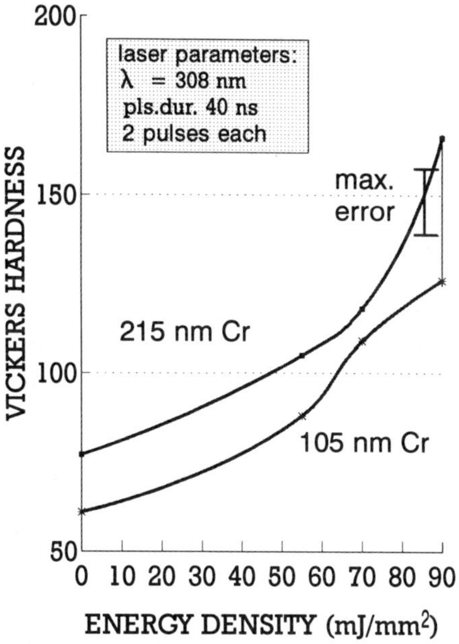

Figure 22 Microhardness of Cu-Cr alloy produced by excimer laser surface alloying.

production of internal combustion engines, engineers always yearn for further improvement of some critical parts. Especially the modification of tribological properties by excimer laser treatment of power-transmitting elements in the areas of camshafts and cam followers would be advantageous. The wear properties of the break-in phase of the engine could be improved by diminishing the roughness of these elements. Figures 23 to 24 show examples for different laser-treated devices. Figure 23 shows the results for a WIG-remelted camshaft that was irradiated with an excimer laser after final grinding. Parts (a)–(c), show macrographs of the camshaft and the overlapping pulses. In the upper part of (d) the grinding marks can easily be detected. On the other hand, the lower part of (d) shows that remelting of the surface leads to a rounding of the machine striae and in some cases a total elimination of the grinding furrows, as can be seen in (e). Etching the surface enables detection of the dendritic morphology of the WIG-remelted camshaft, parts (f) and (g). The etching is less effective in the treated area. The reason for this is most likely the generation of a homogeneous layer consisting either of an extremely supersaturated austenite or of an understoichiometric iron carbide. Both are known in the literature [35,36]. Only after some time of etching is it possible to remove the homogeneous layer generated by excimer laser remelting; a region with finely dispersed carbides is then detected underneath.

In this application, an excimer laser treatment fulfills the followng different tasks:

Residues of coolants, lubricants, and abrasives are removed from the grinding marks

Sharply delineated machining marks are rounded, leading to reduced surface roughness

A thin surface layer with higher hardness and improved corrosion resistance is produced

The two main advantages imparted by the use of excimer laser treatments, reduced surface roughness and diminished corrosion, are of particular interest for devices where a smoother surface prevents the adhesion of combustion residues. In addition, a sufficiently high corrosion resistance is required to avoid corrosion by hot combustion gases. Figure 24 shows the results of the irradiation of a valve stem. The transition region between the mechanically machined parts of a valve shaft and the laser-treated area are shown in part (b). The corrosion behavior of the untreated and the irradiated zone can be judged from parts (c) through (g). In Fig. 24c, a deep-etched nontreated area with a dark matrix and the light Cr carbides is seen. Grinding marks are still visible in spite of massive etching. Figure 8.24d (2 pulses, 40 mJ/mm^2) shows the formation of nonstoichiometric Cr carbides that retard the etching process. This effect can be seen more distinctively in Fig. 24e. While the formation of pitting corrosion is visible in the lower part of the picture (laser-treated), corrosion behavior is still much better than in the nontreated area (upper part of the same picture), in which the surface was much more attacked by the etchant. Figure 24f shows a macrograph of the valve after a motor test (200 hr at high temperature). The irradiated areas were much more corrosion resistant than the untreated ones. The cross section shown in Fig. 24g proves that the attack of the hot combustion gases was drastically reduced in the treated area.

Irradiating cast-iron devices that consist mainly of a phase mixture of graphite and a corresponding iron matrix microstructure (ferritic, pearlitic, bainitic, or glob-

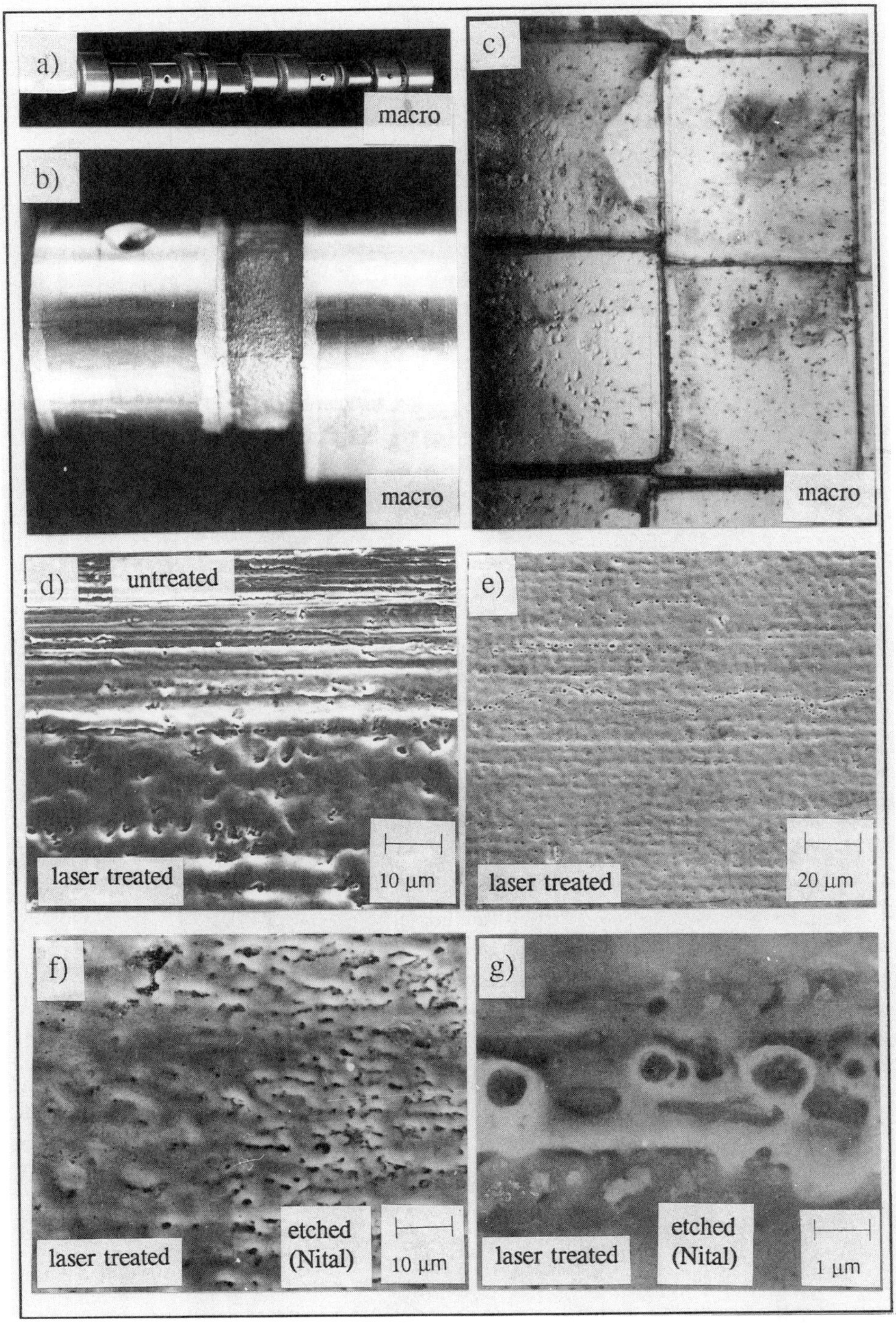

Figure 23 Influence of an excimer laser treatment on a ledeburitic cast-iron cam shaft.

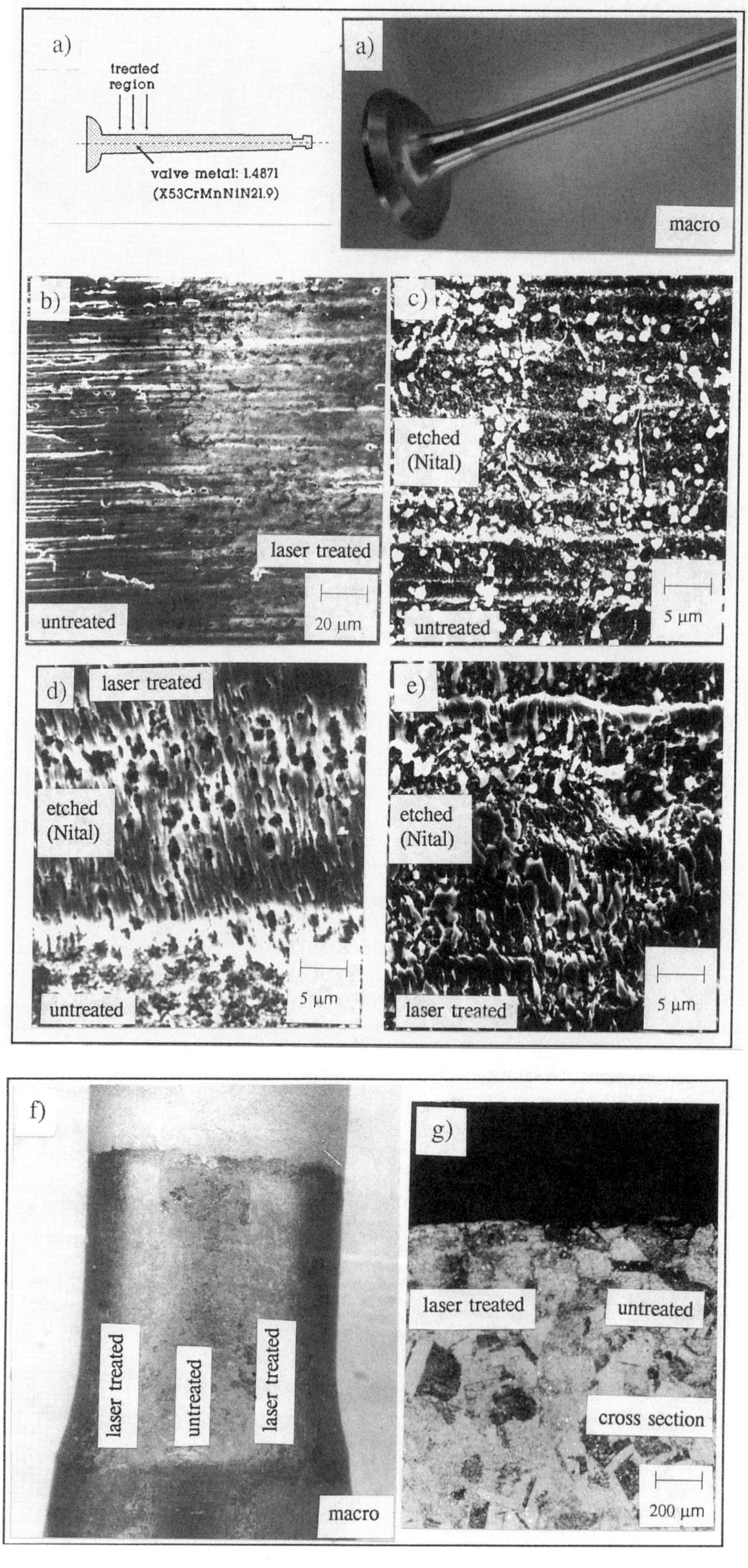

Figure 24 Metallographic investigation of an excimer-laser-treated valve.

ular graphite); a continuous ledeburitic surface will not form. Only near the graphite-iron interface may the lateral diffusion be sufficient to form a eutectic microstructure (diffusion time ~40 ns). However, the deformation layer produced by mechanical machining can be removed and the graphite can be laid bare. Mechanical machining often leads to a strong deformation of the ductile iron, resulting in its covering the graphite. Such strong deformation can limit the ductility of the iron and lead to cracks, which finally result in the spall-off of flakes during service. Subsequently, these flakes can corrode and cause the so-called fretting corrosion. On the other hand, smeared-over graphite regions are undesirable for two reasons: diminished self-lubricating properties of the part and the unavailability as lubrication bore reliefs and later as oil reservoirs. This is especially advantageous in critical engine running situations because a sufficiently high lubrication can be guaranteed.

IX. CONCLUSIONS

Laser surface melting is a technique that allows the heating of surfaces in a very short time scale. The higher the power flux density, the steeper the temperature gradients generated in the surface layer. Due to this steep temperature gradient, the temperature over the whole substrate is quickly leveled out after irradiation, which results in a rapid self-quenching of the surface layer. Rapid resolidification will occur if the surface is actually melted, rather than simply heated. A metastable or a glassy solidification event will occur, depending on quenching rate and chemical composition. In contrast to other rapid solidification techniques, in which heat is extracted through the surface to a surrounding coolant, the temperature gradient in laser remelting occurs between the melt and solid substrate. Another difference is the influence of the mostly crystalline substrate, as in most cases the solidification occurs epitaxially to the substrate so that no nucleation will be necessary. The ratio between crystallization velocity and quench rate is decisive in determining the microstructure.

Technologically, different layer thicknesses have to be distinguished. For power-transmitting components, layers in the range of 0.1 to 1 mm are technologically applied. An example is the surface remelting of cast iron with CO_2 lasers. For the remelting and laser surface alloying of electrical contact devices, rapidly solidified layers in the range of 1 to 10 μm are used. The Nd:YAG laser is the most suitable laser for this application. Excimer lasers have recently been used for special application purposes. The remelted layer thickness is typically between 0.1 and 2 μm.

An important advantage of beam techniques as compared to other rapid solidification techniques is the wide range of quenching rates that can be obtained, which is especially of interest for fundamental investigations in rapid solidification. Another advantage is the flexibility of the beam techniques allowing the treatment of three-dimensional objects at locally defined areas.

REFERENCES

1. P. Duwez, R. H. Willens, and W. Klement, *J. Appl. Phys.*, *31*, 1136 (1960).
2. Th. Mayman, *Nature*, *187*, 933 (1960).
3. C. K. N. Patel, *Phys. Rev. Lett*, *12*, 617 (1964).

4. Y. Arata and I. Miyamoto, *Jap. Weld. Rep. I*, *8*, 15 (1978).
5. F. O. Olsen, *DVS-Beri.*, *63*, 197 (1980).
6. T. VanderWert, in *The Industrial Laser Annual Handbook*, edited by C. Belforte, M. Levitt (Pennwell, Tulsa, OK, 1986), p. 56.
7. D. Banas, ibid; p. 69.
8. P. R. Strutt, B. G. Lewis, and B. H. Kear, in *Rapidly Solidified Amorphous and Crystalline Alloys*, edited by B. H. Kear, B. C. Giessen, and M. Cohen (North-Holland, Amsterdam, 1982), p. 945.
9. H. W. Bergmann and B. L. Mordike, in Ref. 8, p. 463.
10. C. W. Draper, J. M. Vandenberg, C. M. Preece, and C. R. Clayton, in Ref. 8, p. 529.
11. B. L. Mordike and H. W. Bergmann, in *Rapidly Solidified Metastable Materials*, edited by B. H. Kear and B. C. Giessen (North-Holland, Amsterdam, 1984), p. 45.
12. C. W. Draper, *J. Metals*, *6*, 24 (1982).
13. C. W. Draper and J. M. Poate, *Int. Metals Rev.*, *30*, 85 (1985).
14. C. W. Draper, J. M. Vandenberg, C. M. Preece, and C. R. Clayton, in Ref. 8, p. 945.
15. H. S. Carslaw and J. C. Jaeger, *Conduction of Heat in Solids*, (Clarendon Press, Oxford, 1990).
16. D. Farias, S. Denis, and A. Simon, in *Proc. European Scientific Laser Workshop on Math. Simulation*, edited by H. W. Bergmann (Sprechsaal, Coburg, 1989), p. 31.
17. M. Cantello, D. Cruciani, and G. Molino, in Ref. 16, p. 45.
18. M. Schellhorn, in Ref. 16, p. 161.
19. A. Zambon, L. Giordano, B. Badan, and E. Ramous, in Ref. 16, p. 136.
20. A. Barreiros, M. G. Carvalho, M. Freitas, and J. C. F. Pereira, in Ref. 16, p. 61.
21. M. F. Ashby and K. E. Easterling, *Acta Metall.*, *32*, 125 (1984).
22. W. B. Li, M. F. Ashby, and K. E. Easterling, *Acta Metall.*, *34*, 753 (1986).
23. J. Mazumder, in *Appl. Sciences NATO ASI Series E 11* (1986), p. 18.
24. P. Gay, in Ref. 23, p. 201.
25. H. U. Fritsch and H. W. Bergmann, in Ref. 16, p. 73.
26. H. W. Bergmann and E. Geissler, in Ref. 16, p. 73.
27. E. Geissler and H. W. Bergmann, *Opto Elektron. Mag.*, *4*, 396 (1988).
28. H. W. Bergmann and E. Geissler, *Maschinenmarkt*, *7*, 40 (1988).
29. J. Mazumder and A. Kar, in Ref. 16, p. 9.
30. H. W. Bergmann and S. Z. Lee, *Proc. MIOP '88* (Network GmbH, 1988), p. 16.
31. E. Schubert, H. W. Bergmann, and S. Rosiwal, *Opto Elektron. Mag.*, *5*, 17 (1989).
32. B. Grzemba, G. Ibe, and W. Hiller, *Forsch. BMFT*, *T79–20*, 6 (1979).
33. H. W. Bieler, A. Gasser, G. Herziger, E. W. Kreutz, J. Seelhorst, W. Sokolowski, and K. Wissenbach, *Proc. ECLAT '90* (DVS-Verlag, Düsseldorf, 1988), p. 46.
34. C. W. Draper, *Opto Elektron. Mag.*, *4*, 630 (1988).
35. G. Barton, M. Koschlig, and H. W. Bergmann, *Z. Werkstofftechnik*, *14*, 257 (1983).
36. H. W. Bergmann, G. Barton, B. L. Mordike, and H. U. Fritsch, in Ref. 11, p. 29.

9

Atomic Structure of Rapidly Solidified Alloys

T. Egami
University of Pennsylvania, Philadelphia, Pennsylvania

I. INTRODUCTION

The atomic structure of rapidly solidified alloys tends to be quite distinct from the structure of alloys in equilibrium: they are often in a metastable noncrystalline state, such as the amorphous and quasicrystalline state. Characterization and description of these noncrystalline structures is an interesting and challenging problem which requires specialized equipments and theoretical knowledge beyond those required for crystalline solids. Owing largely to the recent advances in the experimental methods, such as the advent of synchrotron radiation and pulsed neutron sources, the structure of noncrystalline solids is now determined with an accuracy high enough to allow discussion of various structure-properties relationships. At the same time, some progress has been made in theoretical understanding of these very complex systems. In this chapter we first review the experimental methods of structural analysis available today, then proceed to the description of the structure of quasicrystalline and amorphous alloys, and conclude with a discussion of the effects of atomic short-range order on the physical properties of amorphous alloys.

II. EXPERIMENTAL METHODS OF STRUCTURAL ANALYSIS

The product of rapid solidification can be a fine-grain polycrystalline solid, a noncrystalline solid including a quasicrystal and an amorphous alloy, or a mixture of these. The first step most commonly employed to characterize the product is to test the sample in an x-ray powder diffractometer. An alternative is to use an electron microscope, but sample preparation is not trivial in this case. If the sample is crystalline, the use of x-ray diffractometry is often sufficient in identifying known crystalline phases by indexing the observed Bragg peaks. However, when the sample is noncrystalline, a standard powder x-ray diffractometer equipped with a sealed x-ray tube having a copper target can only tell whether or not the sample is indeed noncrystalline, and is not capable of specifying any structural detail of the sample. Much more involved analysis is necessary for quantitative characterization of the atomic structure of noncrystalline materials. Owing to recent advances in diffraction techniques, such as the advent of synchrotron radiation and pulsed neutron sources, the structure of amorphous materials can now be studied with a much improved accuracy. In this section we explain why it is more difficult and involved to study the structure of noncrystalline solids than that of crystalline solids, and describe some of the advanced techniques of structural investigation available today.

A. Structural Characterization by Powder X-Ray Diffraction

A standard x-ray powder diffractometer consists of an x-ray tube, usually with a copper target, a θ-2θ goniometer, a curved analyzer crystal, a scintillation detector, and electronics associated with the detector. If the diffraction pattern consists of several sharp Bragg diffraction peaks, the sample is usually polycrystalline, and it is frequently possible to identify the phases present in the sample by indexing the peaks and calculating the lattice constants. Many modern diffractometers are computer-controlled and are equipped with software to search through known diffraction patterns to facilitate phase identification. However, the identification of phases may not be always so straightforward because of complications introduced by rapid solidification. For instance, rapid solidification tends to yield mestastable phases which do not appear in the standard (equilibrium) phase diagrams. Even when the diffraction peaks are sharp, the sample may not be crystalline, but quasicrystalline, with icosahedral (3-*d*) or decagonal (2-*d*) symmetries. If the sample is microcystalline or amorphous the diffraction peaks are broad and the characterization is even more tenuous.

The x-ray diffraction intensity from a solid may be presented as a function of the scattering vector, $\vec{Q}$,

$$\vec{Q} = \vec{k}_s - \vec{k}_i$$

$$Q = |\vec{Q}| = \frac{4\pi}{\lambda} \sin\theta = \frac{4\pi}{hc} E \sin\theta \tag{1}$$

$$\text{or} \quad Q = 1.014E \sin\theta \qquad \text{if } Q \text{ is in Å}^{-1} \text{ and } E \text{ in keV} \tag{2}$$

where $\vec{k}_i$ and $\vec{k}_s$ are the wave vectors of the incident and scattered beams respectively, θ is the diffraction angle, λ is the x-ray wavelength, h is Planck's constant, c is the velocity of light, and E is the energy of the x rays. For crystalline materials, $\vec{Q} = \vec{K}$, where $\vec{K}$ is the reciprocal lattice vector, constitutes the Bragg law. Once the diffraction peaks are indexed, the lattice constants can be obtained readily by the Bragg law. Identification and indexing of the quasicrystals will be discussed in a later section. For amorphous materials, on the other hand, the procedure is more involved. The diffraction pattern of amorphous materials consists of several broad maxima, but their positions do not directly relate to the interatomic distances, although in many cases, $5\pi/2Q_p$, where Q_p is the position of the main peak in the

Q-space, approximately correspond to the average interatomic distance. Therefore, structure determination means more than just determining the peak position: the peak intensity and shape, thus the whole Q-dependence of the scattering intensity has to be carefully determined in order to evaluate the structure of amorphous materials.

The total intensity of the scattered x-ray beam is given by

$$I_{\text{total}}(Q) = I_{\text{el}}(Q) + I_{\text{inel}}(Q) + I_{\text{mult}}(Q) + I_{\text{air}}(Q) \tag{3}$$

where

$I_{\text{el}}(Q)$ = intensity of elastically scattered x rays
$I_{\text{inel}}(Q)$ = intensity of Compton scattering
$I_{\text{mult}}(Q)$ = multiple scattering intensity
$I_{\text{air}}(Q)$ = air scattering intensity, including background due to scattering from a sample holder.

The structural information is contained in the first term in Eq (3), so that the other contributions have to be subtracted from the total intensity in order to evaluate this term. The air scattering intensity can be determined experimentally by performing a scattering measurement without a sample. A part of the Compton scattering intensity can be eliminated when a postsample analyzer crystal is used. Otherwise, the Compton intensity has to be subtracted during data processing, usually based upon theoretical values. The multiple scattering intensity is calculated for the particular sample geometry. This requires considerable computer time because of multiple integrations involved. The elastic scattering intensity is then corrected for polarization and absorption, and normalized with respect to the incident beam intensity to the electron units [1,2]. The most reliable way to carry out the last step of normalizing the data is to use the data obtained for high values of Q, usually 20 to 30 Å^{-1}, where the interatomic interference is negligible so that the scattering intensity per atom is approximately equal to the scattering intensity of an isolated atom. After these steps, the atom-atom interference function, or the total structure factor, $S(Q)$, is obtained by

$$S(Q) = \frac{I(Q)}{\langle f(Q)\rangle^2} + 1 - \frac{\langle f(Q)^2\rangle}{\langle f(Q)\rangle^2} \tag{4}$$

where $f(Q)$ is the atomic scattering factor and $\langle\cdots\rangle$ denotes the composition average. In the step of normalization mentioned above, $S(Q)$ is normalized to unity in the limit of $Q \to \infty$. The most useful information is obtained by Fourier-transforming $S(Q)$ to obtain the atomic pair distribution function (PDF), $\rho(r)$, by

$$\rho(r) = \rho_0 + \frac{1}{2\pi^2 r}\int_{Q_{\min}}^{Q_{\max}} Q[S(Q) - 1]\sin(Qr)\, dQ \tag{5}$$

where ρ_0 is the average atomic number density. The quantity, $4\pi r^2\rho(r)$, is called the radial distribution function (RDF), and the area under the first peak in the RDF gives the atomic coordination number. Theoretically the integration limits in Eq. (5) should be $Q_{\min} = 0$ and $Q_{\max} = \infty$; however, because of experimental limitations these limits are usually finite values. If $Q_{\max}$ is not sufficiently high, it introduces an error called the termination error. Examples of $S(Q)$, in the form

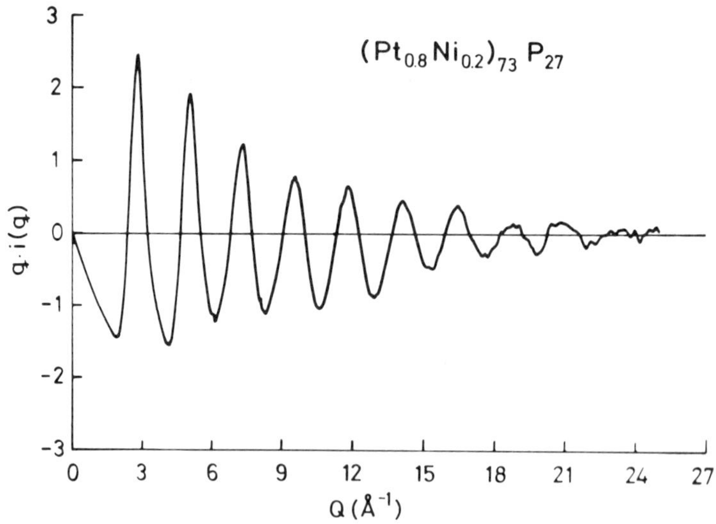

Figure 1 X-ray structure factor, $S(Q)$, of amorphous $(Pt_{0.8}Ni_{0.2})_{73}P_{27}$ presented in the form of $Q[S(Q) - 1]$ as it appears in the Fourier integral in Eq. (5) [3].

of $Q[S(Q) - 1]$ as it appears in Eq. (5), and $G(r) = 4\pi r[\rho(r) - \rho_0]$ are shown in Figs. 1 and 2 [3], as determined by the energy dispersive x-ray diffraction technique which is described in the next section. As shown here, $S(Q)$ oscillates up to 20–25 Å^{-1}, thus Q_{max} has to be at least 25 Å^{-1} for the Fourier integral to be accurate. The value of Q_{max} is determined by the wavelength of the incident x rays according

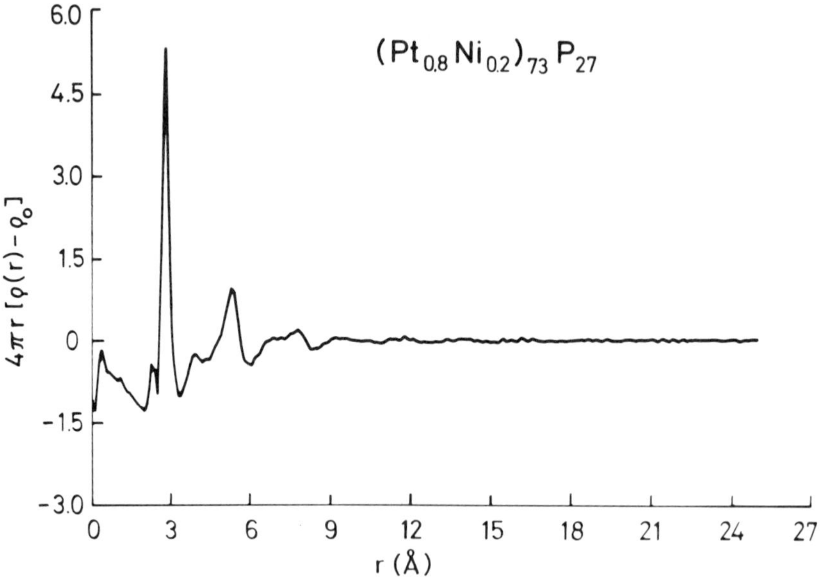

Figure 2 Atomic pair distribution function of amorphous $(Pt_{0.8}Ni_{0.2})_{73}P_{27}$ obtained from $S(Q)$ shown in Fig. 1, by Eq. (5), in the form of $G(r) = 4\pi r[\rho(r) - \rho_0]$ [3].

to Eq. (2). If we use Cu-K_α radiation (8.01 eV) which is most commonly used in standard diffractometers, the value of Q_{max} is only about 7.5 Å^{-1}, so that it is impossible to obtain $\rho(r)$ with any acuracy using Cu-K_α radiation. A sealed x-ray tube with a Mo target ($Q_{max} \sim 16.5$ Å^{-1}) is only marginally satisfactory. Thus, merely a semi-quantitative structural characterization of amorphous materials can be carried out with a standard power x-ray diffractometer, while more sophisticated instrumentation is required for more accurate structural analysis.

B. Energy-Dispersive X-ray Diffraction Technique

A method of determining $S(Q)$ up to higher values of Q than is possible with a standard diffractometer, using equipments which are relatively easily available, is the energy-dispersive x-ray diffraction (EDXD) method. In this method $S(Q)$ is determined by x-ray spectroscopy using white radiation (usually provided by a W or Au tube) and an energy-sensitive semiconductor detector (usually a pure Ge detector). Recent advances in the energy resolution in these detectors (typically 150 eV at 7 keV, 250 eV at 20 keV) made it possible to use them for structural analysis [4,5]. The principle is very simple: according to Eq. (2), Q is proportional to E for a constant value of θ; thus the spectrum of the diffracted white x rays, properly normalized with respect to the spectrum of the incoming beam, gives the structure factor. The actual process is a little more complicated because of the Compton shift and the difficulty of determining the incoming beam spectrum [6,7]. The principle and the data processing procedure of this method are similar to those of the neutron time-of-flight method [8].

The main advantage of this technique is that $S(Q)$ can be determined up to high values of Q. A standard sealed x-ray tube can generate white x rays up to 50 keV, so that with the EDXD method Q values up to 40 Å^{-1} are attainable. In practice, however, since $f(Q)^2$ decreases quickly with Q, it is difficult to determine $S(Q)$ by x rays much beyond 25 Å^{-1}. In fact the polynominal expression of $f(Q)$ found in the standard source [9] is valid only up to about 25 Å^{-1}. An important additional advantage of EDXD is that the total intensity of the white radiation is greater than that of the characteristic radiation by an order of magnitude [10]. Thus, structure can be determined with much greater statistical accuracy, and small changes due to structural relaxation [6] or structural anisotropy [11] can be determined accurately by this technique. In Fig. 3 we show the EDXD spectrum of amorphous $Fe_{40}Ni_{40}Mo_3Si_{12}B_5$ alloy obtained at $\theta = 15°$ and its anisotropy induced by mechanical creep at 300°C for 24 hr under a tensile stress of 800 MPa [11]. The anisotropy, $\Delta I(E)$, is the difference in the intensities obtained with Q parallel to and perpendicular to the stress axis. The analysis of this anisotropy led to an important understanding of the mechanism of anelastic deformation (polarization) in amorphous alloys in terms of the bond-orientational anisotropy as we discuss later.

C. Synchrotron Radiation

The advent of synchrotron radiation is producing a major impact in all fields of x-ray scattering and spectroscopic studies. Synchrotron radiation is emitted by electrons or positrons orbiting in the synchrotron ring, and has a wide continuous energy spectrum. Because of the relativistic effect the radiation is very much com-

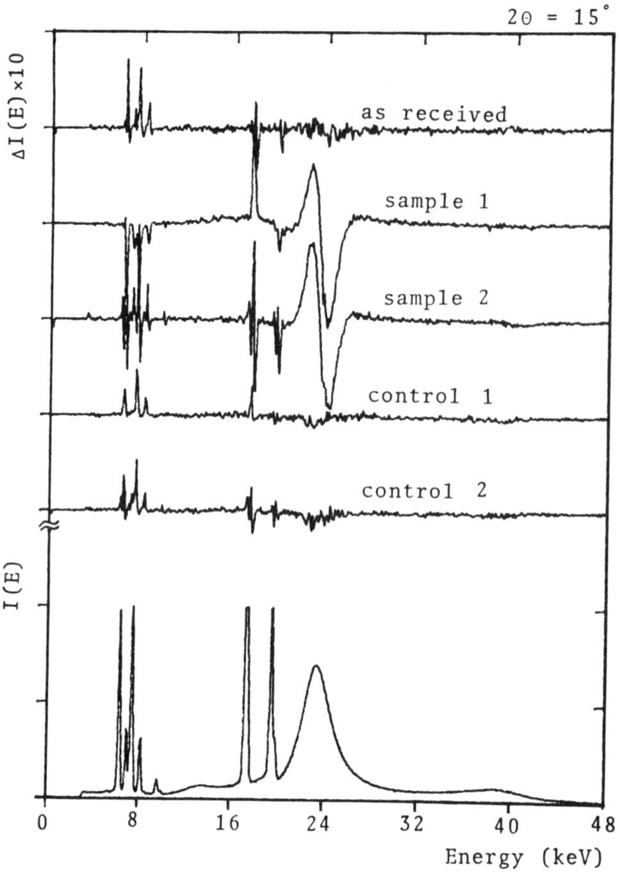

Figure 3 EDXD spectrum of amorphous $Fe_{40}Ni_{40}Mo_3Si_{12}B_5$ at $\theta = 15°$ (bottom), and the difference in the spectrum, $\Delta I(E)$, between Q being parallel to the ribbon direction (direction of the stress) and Q being perpendicular (above five). From the top, $\Delta I(E)$ for as-received sample, mechanically creep-deformed (two samples), and annealed without a stress (2 samples). The sharp lines below 12 keV are fluorescences from the sample, while the lines at 17.4 and 19.8 keV are due to the Mo K_α and K_β radiation from the Mo radiation source [11].

pressed to the direction of the orbiting particles, and has an extremely high intensity and excellent collimation. For the purpose of enhancing the energy of the radiation even further, insertion devices such as wigglers and undulators are often added. High-intensity monochromatic radiation of any energy within the spectrum of the beam can be obtained by using a crystalline monochromator. This continuous tunability as well as its high intensity are the main merit of synchrotron radiation.

At present, three synchrotron x-radiation sources are available in the United States: the National Synchrotron Light Source (NSLS) of Brookhaven National Laboratory, the Cornell High Energy Synchrotron Source (CHESS), and the Stanford Synchrotron Radiation Laboratory (SSRL); two more sources, the Advanced Light Source (ALS) of the University of California, Berkeley and the Advanced

Photon Source (APS) of Argonne National Laboratory, are under construction. In Europe x-ray sources are in operation in Duresburry, U.K. (SRS), and in Hamburg, Germany (HASYLAB, DESY). In Asia synchrotron rings are generating x rays in Novosibirsk, USSR, and Tsukuba, Japan (Photon Factory). A number of new sources are either under construction or being planned. Thus synchrotron radiation is no longer an exotic new method of experiment, but is becoming a centrally important tool for advanced structural research. In the following subsections we discuss four principal uses of synchrotron radiation in the study of rapidly solidified materials, namely the high-resolution scattering study, the scattering study using high-energy monochromatic radiation, extended x-ray absorption fine structure (EXAFS), and differential anomalous x-ray scattering (DAS).

1. High-Resolution Scattering Measurement

By using both a crystal monochromator and a postsample crystal analyzer, a very high Q resolution ($\Delta Q \sim 10^{-3}$ Å^{-1} or better) can be achieved. The use of synchrotron radiation is particularly advantageous in increasing the resolution because of its high collimation. With such a high resolution a detailed study of peak shape becomes feasible. The grain size and its distribution in microcrystalline solids can be determined by such a measurement. The Q-dependence of the peak width in icosahedral solids studied by this method [12,13] showed the nature of defects (phasons) in these solids very clearly, and contributed in part to showing that icosahedral solids are indeed quasicrystalline, as we discuss later.

2. Scattering Study Using High Energy Monochromatic Radiation

By placing a crystal monochromator before the sample one can obtain a strong monochromatic x-ray beam with energies much higher than generally available from sealed tubes, thus enabling $S(Q)$ to be determined up to high values of Q. The energy spectrum of synchrotron radiation depends upon the ring energy (2.5–5 GeV) and the magnetic field of the bending magnet, and the beam intensity rapidly decreases beyond a certain energy. At high-energy radiation sources such as CHESS and SSRL, or with wigglers, photons up to 50 keV or more are available, while with a bending magnet at the NSLS the practically useful energy range is below 30 keV, limiting the Q range to below 25 Å^{-1} or so. However, as we discussed earlier $f(Q)$ becomes too small and less accurate beyond 25 Å^{-1}, so that this range is in practice sufficient for the structural study.

3. EXAFS

Small oscillations in the absorption coefficient above the absorption edge are known as the extended absorption edge fine structure (EXAFS). These oscillations are due to the modulation of the transition probability of electrons from the core state to the extended states due to the presence of near neighbour atoms. In other words, the backscattering of the wave function by the neighboring atoms of the photoelectrons emmitted by the absorption of x-ray photons interfere with the transition probability [14]. Thus, by analyzing these oscillations, it is possible to determine the near neighbor distances and the coordinations, at least in principle. Measurement of EXAFS is very straightforward and fast, taking only about 10–20 min, an order of magnitude faster than a typical scattering measurement. Also the environment of a specific element at which absorption edge the measurement is made can be determined, while in standard diffraction measurement the total scattering from all the elements in the solid is collected. However, the EXAFS data analysis is complex, involving many adjustable parameters. Unless these parameters are calibrated using a known structure of similar composition the analysis is too often unreliable. Thus very critical judgements are required in making use of this technique.

4. *Differential Anomalous Scattering Measurement*

The scattering amplitude of an atom is known to become sharply reduced when the energy of the x-ray photon approaches the absorption edge from either side due to the electron-photon resonance. Making use of this anomalous dispersion phenomenon, it is possible to compositionally differentiate the scattering amplitude from a solid [15]. The normalized elastic scattering intensity from a multicomponent system is, after the corrections discussed earlier [2,16]

$$I(Q) = I_c(Q) + \langle f(Q)\rangle^2 - \langle f(Q)^2\rangle \tag{6}$$

$$I_c(Q) = \sum_{\alpha,\beta} c_\alpha c_\beta f_\alpha(Q) f_\beta(Q) S_{\alpha\beta}(Q) \tag{7}$$

where c_α is the composition of element α, $f_\alpha(Q)$ is the atomic scattering factor of element α, and $S_{\alpha\beta}(Q)$ is the compositionally resolved partial structure factor. By varying $f_\alpha(Q)$ through changing the x-ray photon energy near the α-absorption edge, we can determine the differential stucture factor [17]:

$$\begin{aligned} S_\alpha(Q) &= \frac{1}{2c_\alpha\langle f(Q)\rangle}\frac{\partial I_c(Q)}{\partial f_\alpha(Q)} \\ &= \sum_\beta \frac{c_\beta f_\beta(Q) S_{\alpha\beta}(Q)}{\langle f(Q)\rangle} \end{aligned} \tag{8}$$

Since $f(Q)$, or the real part of the anomalous dispersion, $f'(Q)$, varies appreciably only very near the absorption edge, it is quite important that the energy of the incident beam can be brought to the close vicinity of the absorption edge. Thus, the use of synchrotron radiation is the only way to carry out this experiment reliably. Furthermore, the absorption edge has to be situated at a convenient position in the energy scale for this measurement to be feasible. That means elements lighter than the 3*d* transition metals are not suitable for analysis by this technique. For elements heavier than the rare earth elements, lower edges such as L edges can also be used. The Fourier transform of $S_\alpha(Q)$, $\rho_\alpha(r)$, is called the differential distribution function (DDF), and describes the atomic distribution around the α atoms. The uranium and palladium DDFs of quasicrystalline $Pd_{58.8}U_{20.6}Si_{20.6}$ are shown in Fig. 4 [18]. By further differentiating the intensity by $f_\alpha(Q)$ one can in principle obtain $S_{\alpha\beta}(Q)$. However, experimental accuracy of determining the second derivative is usually not sufficiently high enough to carry out such a procedure reliably. This technique has been used only in a small number of applications mainly because of the high degree of skills needed to carry out the measurement successfully, but it is expected that the method will gain wide popularity as the use of synchrotron radiation becomes more commonly practiced.

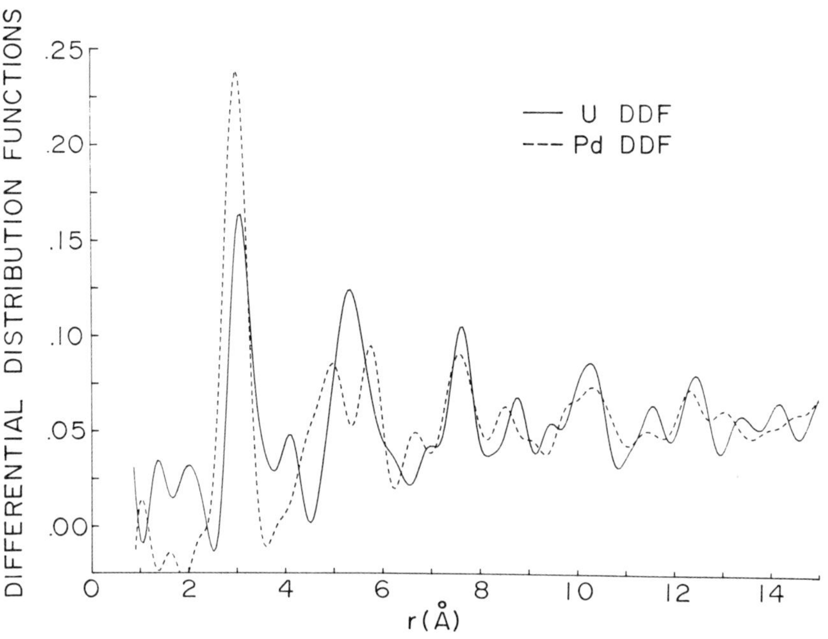

Figure 4 Uranium and palladium differential distribution functions of icosahedral $Pd_{58.8}U_{20.6}Si_{20.6}$ determined by a differential anomalous x-ray scattering experiment performed at the L_{III} absorption edge of uranium [18].

D. Neutron Scattering

Neutron scattering method has several distinct advantages in studying the structure of amorphous materials. One is that the nuclear scattering length of neutrons is almost independent of Q, so that it is easier to determine $S(Q)$ accurately up to high values of Q with neutrons than with x rays, particularly when one uses a spallation source (pulsed neutron source) which provides neutrons with higher energies than does a thermal source. Another advantage is that the neutron scattering factor, or the neutron scattering length b, does not scale as the atomic number as the x-ray scattering factor does, so that lighter elements such as oxygen can be more easily seen by neutrons than by x rays. Yet another advantage is that some of the elements (notably Li, Ti, Mn, and isotopes of a number of elements) have a negative scattering length, resulting in a strong compositional contrast. Thus, the compositionally resolved partial structures in Eq. (7) can be more easily determined either by isomorphous substitution, or better yet, by isotopic substitution [2,19]. Neutron scattering can also be used in detecting magnetic correlations which cannot be detected easily by other means [8]. Although the availability of neutron soucres is still limited, it is no more restricted than synchrotron sources. Through the submission of proposals, measurements can be performed with pulse neutrons at the Intense Pulsed Neutron Source (IPNS) of Argonne National Laboratory and the Los Alamos Neutron Scattering Center (LANSCE) of Los Alamos National Laboratory, and with thermal neutrons at the reactors of Brookhaven National Laboratory, Oakridge National Laboratory, and the National Institute of Standard

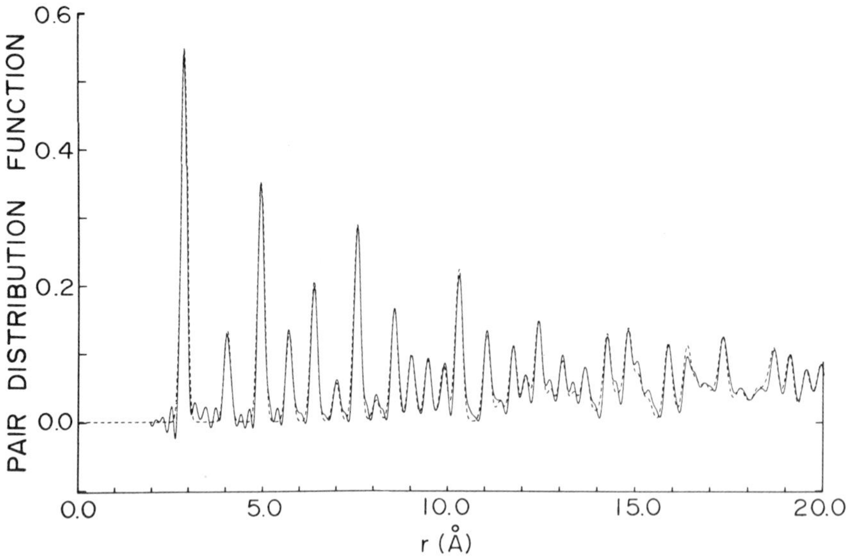

Figure 5 Atomic pair distribution function (PDF), $\rho(r)$, of fcc aluminum determined at $T = 10$ K by pilsed neutron scattering. The dashed line is the calculated PDF with a Gaussian broadening to account for zero-point lattice vibration [20].

and Technology. The most intense and developed thermal neutron source in the world is the reactor at the Institut-Laue-Langevin (ILL) in Grenoble, France.

As an example of the ability to determine $S(Q)$ up to high Q-values, Fig. 5 shows the PDF of fcc aluminium at $T = 10$ K obtained by the Fourier transformation of $S(Q)$ determined up to 28 Å^{-1} by pulsed neutron scattering performed at the IPNS [20]. By making use of this capability of determining the PDF accurately even for crystalline materials, pulsed neutron scattering was recently used successfully in resolving small atomic displacements in high-T_c oxide superconductors [21–23]. It was found that in many of the oxide superconductors some atoms (notably oxygen) are statically or quasi-statically displaced from a position of high crystallographic symmetry by a fraction of angstrom; this may be of importance in understanding the mechanism of superconductivity in this class of materials. If such displacements are correlated over a long range, they would induce a structural transition. Instead, they are correlated only over short ranges without causing a change in the long range symmetry of the oxides.

Figure 6 shows an example of neutron scattering detecting compositional short-range order in amorphous $(Mn_xNi_{1-x})_{75}P_{16}B_6Al_3$ [24]. In this case, since the scattering length of Mn is negative, the short-range ordering between Mn and Ni atoms results in a peak at around 1.5 Å^{-1}, which is very much like the superlattice peak in ordered crystalline alloys. In this alloy system Mn atoms carry magnetic moment which freeze into a spin-glass state at low temperatures. Using magnetic crystals, in this case a Heusler alloy, for both a monochromator and an analyzer, neutron scattering experiment can be carried out with the spin of the neutron polarized. By comparing the scattering intensities with and without the spin flopped, with the scattering vector parallel or perpendicular to the magnetization, magnetic scattering

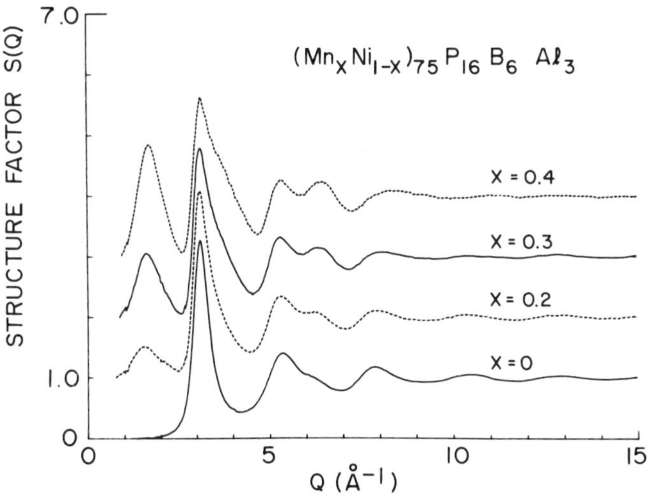

Figure 6 Neutron structure factor of amorphous $(Mn_xNi_{1-x})_{75}P_{16}B_6Al_3$ [24]. The intensity of the pre-peak at 1.5 Å^{-1} increases with x, indicating that this is due to the compositional short range order between Mn, which has a negative neutron scattering length, and Ni.

can be separated from the nuclear scattering [8,25]. The Fourier transform of the magnetic scattering intensity gives the spatial spin-spin correlation as shown in Fig. 7 [24]. It is seen that the spins at the nearest neighbors (at ~3.2 Å) are antiferromagnetically coupled, while those on the second neighbors (at ~4.1 Å) are ferromagnetically coupled, leading to the spin-glass behavior.

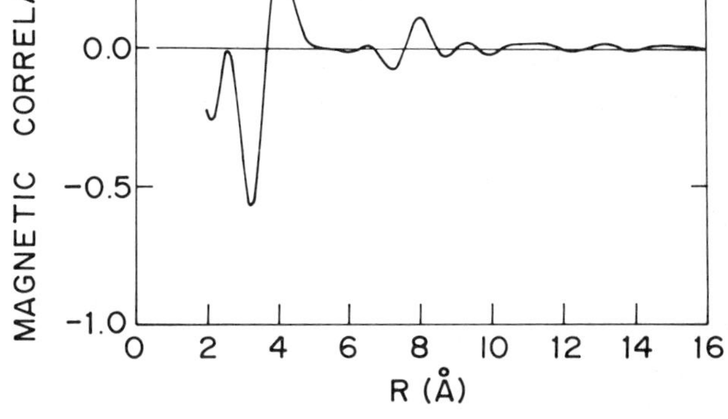

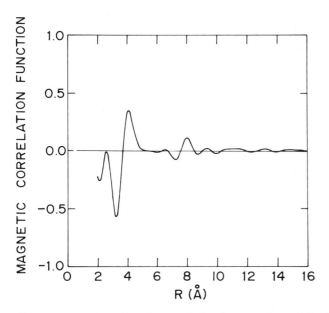

Figure 7 Spatial spin-spin correlation in amorphous $(Mn_{0.4}Ni_{0.6})_{75}P_{16}B_6Al_3$ determined by spin-polarized neutron scattering [24].

III. STRUCTURE OF QUASICRYSTALS

A. Icosahedral Solids

It is well known that the translational symmetry is compatible only with two-, three-, four-, and sixfold point symmetry. Therefore, when an electron diffraction pattern of rapidly quenched Al-Mn alloy exhibiting 10-fold symmetry was found by Shechtman et al. [26], much attention was focussed on the nature of the atomic arrangement in this alloy. The structure has the symmetry of icosahedron, or the point group $\bar{5}\bar{3}m$. The icosahedral local atomic arrangement is frequently found in complex crystals, such as elemental boron and the Frank-Kasper phases [27]. However, macroscopic icosahedral symmetry has never been observed other than in twinned polycrystals [28], while the icosahedral symmetry in this new alloy apparently is not due to multiple twins. Several proposals were immediately made on the structure of this alloy: Shechtman and Blech [29], and later Stephens and Goldmann [30] proposed that they are made of orientationally correlated but randomly packed icosahedral clusters (icosahedral glass), while Levine and Steinhardt [31] suggested that this alloy represents a new state of matter, characterized by quasiperiodicity rather than periodicity (quasicrystal) [32–35]. Quasiperiodicity is an established mathematical term describing a state with two periodicities which are mutually incommensurate. In other words the structure involves two length scales of which ratio is irrational. Other proposals include that of Pauling [36] that the icosahedral solids are cubic crystals with large unit cells, but as we discuss later, the idea of quasiperiodic solids has been most successful in explaining the structure of icosahedral solids.

Icosahedral phases were then found in a number of binary and ternary systems. They are compounds with relatively narrow ranges of solubility. While many of them are metastable so that they can be obtained only by rapid solidification and transform into a crystalline state upon annealing, some of them, such as $Al_{5.5}Li_{3.3}Cu$ and $Al_{65}Fe_{15}Cu_{20}$, are stable, and obtained by a slow annealing process [37,38]. Beautifully faceted single grains of several millimeters can be grown in these stable systems. The diffraction peak widths determined by high-resolution x-ray diffraction measurements are relatively wide, often of the order of 0.01 $Å^{-1}$ HFHM, indicating a high density of structural defects such as phasons as we discuss later [12,39,40].

B. Quasicrystalline Model

A quasiperiodic structure is generated by a set of two periodicities of which ratio is irrational. An example of a quasiperiodic structure in one dimension which in fact is observed in icosahedral and decagonal quasicrystals, is the Fibonacci series, discovered by a middle-age monk. This series is neither periodic nor aperiodic, but is generated by a set of two periodicities with the ratio of $\tau = (\sqrt{5} + 1)/2$, the golden mean. Such a structure in two dimensions is the Penrose tiling, shown in Fig. 8, found as a solution of a long-standing puzzle by Roger Penrose, a well-known cosmologist [32,41]. Its three-dimensional extention, or the three-dimensional Penrose tiling (3DPT) is the structure proposed for the explanation of the icosahedral solids. The structure consists of two units of rhombohedral shape which are arrayed in a quasiperiodic manner. The structure was coined "quasicrystal,"

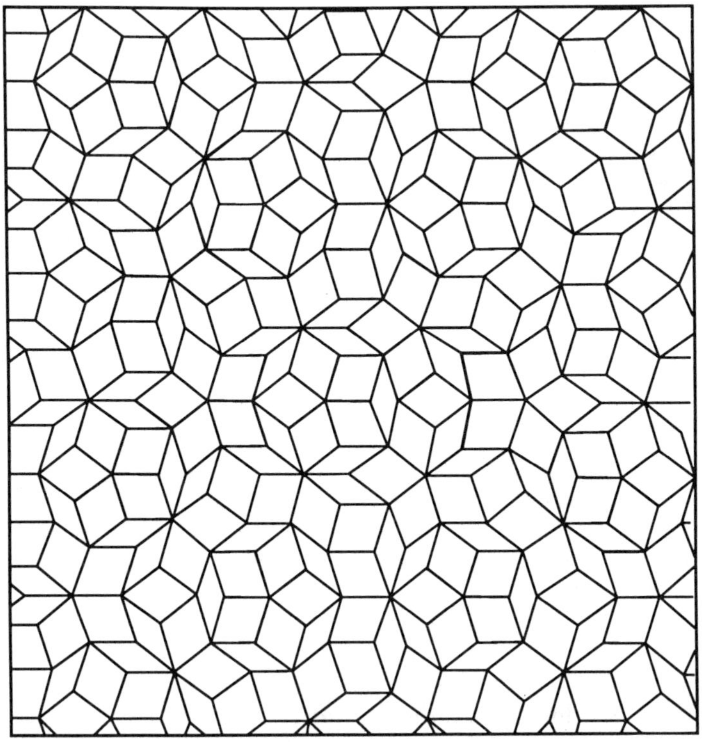

Figure 8 Two-dimensional Penrose tiling.

which is a little unfortunate, since this term was used earlier to describe a microcrystalline model [42].

An elegant way to produce a one-dimensional quasiperiodic structure is the projection from two dimensions as shown in Fig. 9. Similarly the 3DPT structure can be generated from a simple cubic structure in six dimensions (a hypercubic structure). A lattice point in six dimensions can be specified by the six unit vectors, $\vec{a}_\nu(\nu = 1, 2, \ldots, 6)$, by

$$\vec{R}_6 = \sum_\nu n_\nu \vec{a}_\nu \tag{9}$$

This can be decomposed into a sum of a vector in the real 3-d space, $\vec{R}_\parallel$, and a vector in the "perpendicular" space (since they are perpendicular to the real space), $\vec{R}_\perp$,

$$\vec{R}_6 = \vec{R}_\parallel + \vec{R}_\perp \tag{10}$$

By setting an appropriate window in the perpendicular space and projecting to three dimensions in such a way that the unit vectors in six dimensions correspond to the six vectors in three dimensions which form an icosahedron, the 3DPT structure can be readily produced [34,35,43,44]. Alternatively by "decorating" the lattice point by an atomic basis function and cutting the six dimensions by a 3-d hyperplane

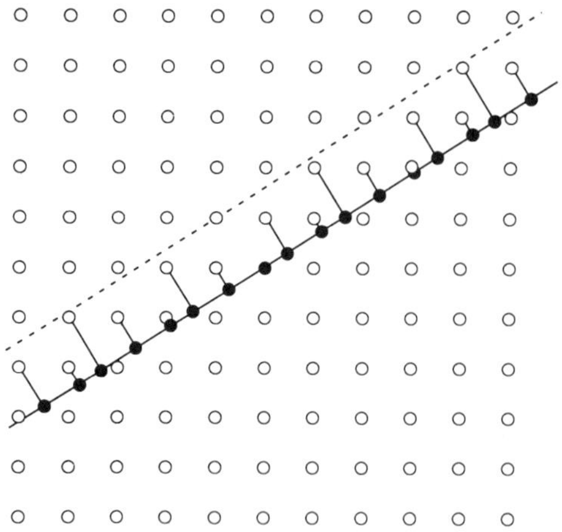

Figure 9 Construction of a quasiperiodic series by the projection method. A periodic 2-d lattice is projected to two axes (parallel and perpendicular) which are rotated by ϕ, such that tan ϕ is an irrational number, resulting in a quasiperiodic series. The Fibonacci series can be obtained by setting tan ϕ equal to τ or τ^{-1}. The window set for projection prevents the series from becoming continuous.

one obtains a quasicrystalline structure [45–47]. Because of the periodicity of the hypercubic structure in the six dimensions, the reciprocal lattice is formed in the 6-d Q space, $\vec{Q}_6$, which can also be decomposed or projected into the real 3-d space and the perpendicular 3-d space:

$$\vec{Q}_6 = \vec{Q}_{\parallel} + \vec{Q}_{\perp} \tag{11}$$

Thus the diffraction peaks are indexed by the reciprocal lattice points in the 6-d Q space, containing six indices such as (100000) [39,48]. The peak intensity is strongly dependent upon the magnitude of $\vec{Q}_{\perp}$, thus, even though every point in the real 3-d Q space can be indexed by the 6-d indexing, only a relatively small number of peaks can be experimentally observed. Recently it was found that many of the 6-d indices are missing from the diffraction pattern of quasicrystalline $Al_{65}(Fe,Ru)_{15}Cu_{20}$, resulting in an extinction rule applicable for the fcc structure (a bcc 6-d reciprocal lattice structure) [38,49]. This arises from the chemical ordering in six dimensions, forming the 6-d supperlattice with a fcc structure.

If the position of the window in the $\vec{R}_{\perp}$ space varies from place to place, it results in structural defects called phasons [12,35,50,51]. They are not produced by plastic deformation as are dislocations, but are produced by "mistakes" during the solidification process. If the phason strain is directional, perhaps due to the growth process, and linearly varying with position, shifts in the diffraction spots which are related linearly to $\vec{Q}_{\perp}$ should occur, as indeed experimentally observed [12,52]. Thus even though the orientational correlation is maintained throughout the grain, the positional correlation is limited by these phasons to below 1000 Å in many quasicrystalline solids, as mentioned earlier. When the phason density is very much increased the structure becomes the icosahedral glass. Since the phason strains appeared to be always present, it was suggested that the perfect quasicrystalline state is not the ground state and the icosahedral solids are orientationally correlated glasses after all [51]. However, recent discovery of quasicrystals almost without phason strains [40] strongly suggests that there is no limit in length for the quasicrystalline order to develop, just as for the regular crystalline order. The mechanism for the development in space for such quasicrystalline order, or the growth mechanism, however, is still being discussed, and is not clearly identified.

C. Atomic Structure of Quasicrystals

The 3DPT structure discussed above corresponds to the lattice structure for crystals rather than to the atomic structure, and the actual atomic positions have to be separately specified to describe the real structure of the solid. In the case of crystalline solids the identification of the lattice structure and the determination of the lattice constants can be made by simply measuring the Bragg peak positions in the *Q*-space, while the atomic positions can be specified only after a careful measurement of the peak intensities. Similarly, the icosahedral symmetry and the quasicrystalline lattice constant can be determined (within the ambiguity of τ^{3n} because of the self-similarlity of the structure) by the peak positions alone, but the actual atomic positions cannot be determined without evaluating the diffraction intensities. Furthermore, because of the absence of translational symmetry, the structure factor cannot be expressed as a product of the unit cell structure factor and the lattice structure factor as in the case of crystalline solids. Thus initially the structure was postulated from association with crystalline compounds with similar compositions. Elser and Henley [53] and Audier and Guyot [54] noticed that a rohmbohedral structure which is very similar to the appropriate quasicrystalline unit is found in crystalline α-(Al,Mn,Si). Similarly the local structure in the cubic Frank-Kasper (R) phase of $Al_{5.5}Li_{3.3}Cu$ can be used as a unit structure for the icosahedral (T2) phase of the same composition [55]. Indeed even though the structure factors of these two phases are rather different, reflecting differences in the long-range order (Figs. 10, 11 [56]), their PDFs are quite similar (Fig. 12 [56]), suggesting a strong

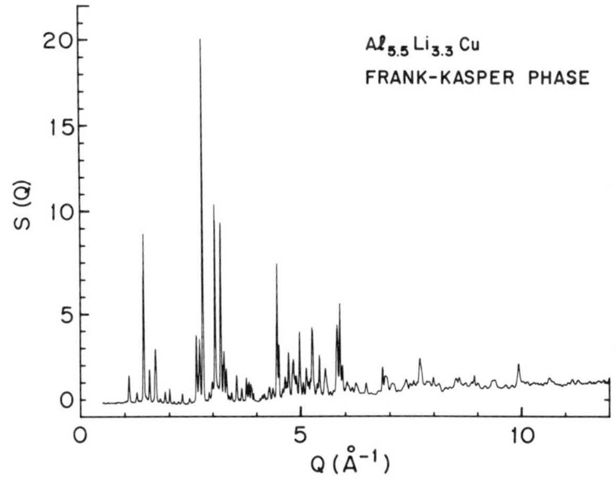

Figure 10 X-ray structure factor of crystalline (R-phase) $Al_{5.5}Li_{3.3}Cu$ [56].

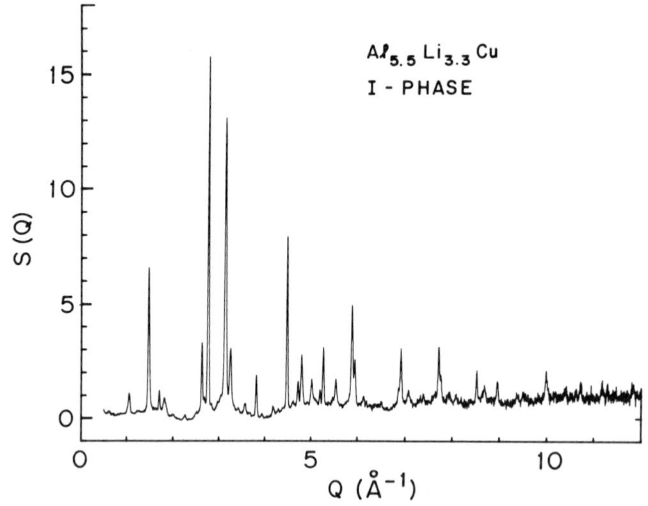

Figure 11 X-ray structure factor of icosahedral phase of $Al_{5.5}Li_{3.3}Cu$ [56].

similarity in the local atomic structure. The direct comparision of the experimental and calculated diffraction intensities also support this model [57].

One interesting approach recently used by Cahn et al. [58] is to use the translational symmetry in 6-d, and construct the 6-d Patterson function. For this purpose they used the cut method [45–47], and determined the basis function to decorate the 6-d lattice by using the structural similarity between certain crystalline com-

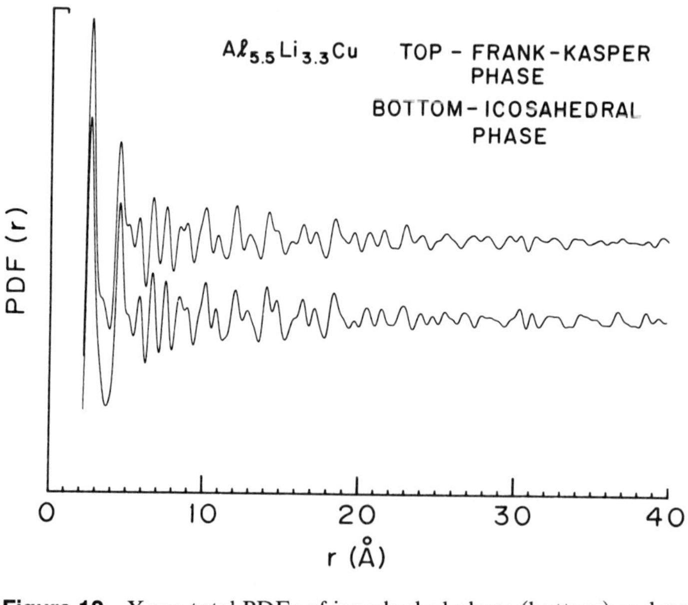

Figure 12 X-ray total PDFs of icosahedral phase (bottom) and crystalline (Frank-Kasper) phase (top) of $Al_{5.5}Li_{3.3}Cu$ [56].

pounds and the quasicrystals such as R-phase and T2-phase of Al-Li-Cu mentioned above. It is plausible that these two structures share the same 6-d structure, but are different in the cut. In the cut method, if the angle of the cut was rational, such that $\tan \theta = m/n$, then a periodic structure is obtained. Thus it is possible to generate a rational approximant of the quasicrystalline structure by a cut at an angle with m/n being the rational approximant of τ (Fibonacci numbers, 1/1, 2/1, 3/2, 5/3, 8/5, 13/8, 21/13, . . .). The Frank-Kasper phase is the 5/3 approximant of τ. The decorating function was determined so that a 5/3 rational cut will produce the crystalline structure, and then the cutting hyperplane was rotated to the icosahedral angle to produce the quasicrystal.

A related method, which is the application of the so-called "direct method" of crystallography to six dimensions, was proposed by Janot et al. [59,60]. In this method the phase of the crystallographic structure factor, which is either 0 or π for centrosymmetric system, is determined by an educated guess. These methods, however, require a number of assumptions such as the atomic decorating function being perpendicular to the real space and linear in the perpendicular space, which are not quite proven to be correct. Although these methods undoubtedly provide an excellent starting point the real structure needs to be determined by methods requiring fewer assumptions. More direct information is obtained by the PDF analysis of the diffraction data, as shown in Figs. 4, 12, and 13. These provide direct information regarding the local structure and suggest the atomic decoration of the unit cells. For instance in Fig. 13 the uranium DDF in icosahedral

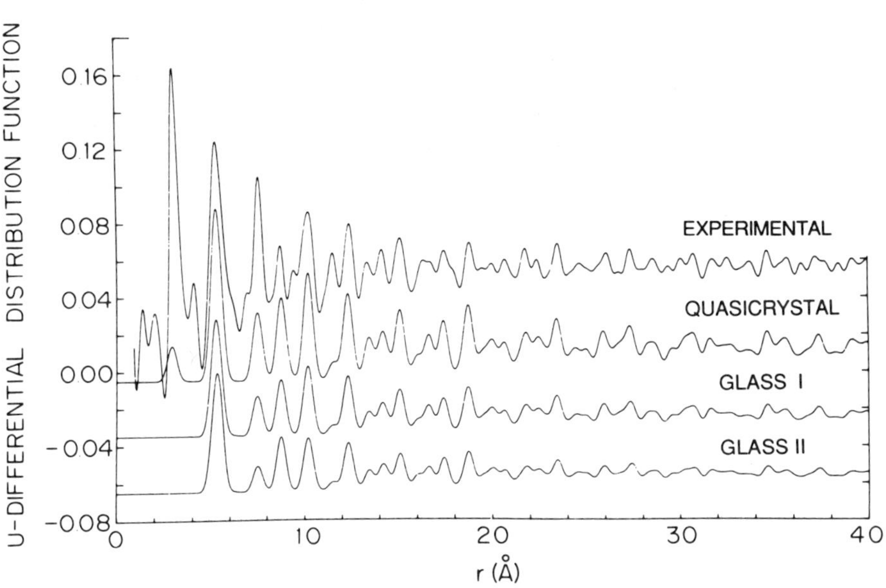

Figure 13 Uranium differential distribution function (DDF) determined by the differential anomalous x-ray scattering experiment (labeled experimental) compared with the simulated DDFs of the quasicrystalline (3-d Penrose lattice) model with uranium atoms decorating the vertices (quasicrystal) and of icosahedral glass models with uranium atoms at the vertices (glass I, glass II) [61].

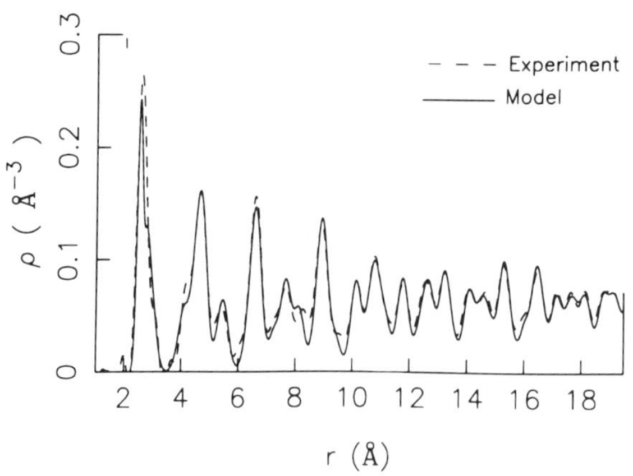

Figure 14 Pair distribution function of $Al_{65}Ru_{15}Cu_{20}$ determined by x-ray diffraction (dashed line) compared to the PDF of the model structure (solid line) [62].

$Pd_{58.8}U_{20.6}Si_{20.6}$ determined by the DAS method discussed earlier is shown up to 40 Å, and is compared with the uranium PDF of structural models decorated with uranium [61]. Good agreement suggest that in this solid uranium atoms are occupying the vertex positions of the quasicrystalline lattice, but the positions of palladium atoms are still underdetermined. Full determination of a 3-d structure requires a direct comparison through the construction of a large 3-d model. The atomic structure of $Al_{65}Ru_{15}Cu_{20}$ was recently determined by such an approach [62]. The structure of this solid is obtained by projecting the vertex positions of the 6-d lattice using a window which is larger than the one to produce the 3-DPT lattice by τ, and subdivided for each composition. This model explains quite well both the diffraction intensities in the Q-space as well as the PDF and DDF, as shown in Figs. 14 and 15. The Ru-DDF shows that the atomic distance between

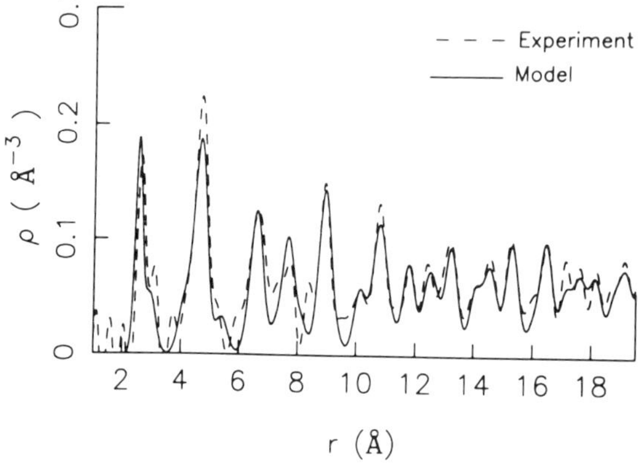

Figure 15 Ru-differential distribution function of $Al_{65}Ru_{15}Cu_{20}$ (dashed line) compared to the Ru-DDF of the model structure (solid line) [62].

Al and Ru is much shorter than the sum of the atomic radii, suggesting a strong covalent bonding between Al and Ru. Such a short aluminum-transition metal distance was found also in amorphous Al-Fe-Ce alloys [63,64]. This strong Al-Ru bonding and the special environment it requires is apparently stabilizing the quasicrystalline structure.

IV. STRUCTURE OF AMORPHOUS ALLOYS

As the grain size of a polycrystalline sample becomes smaller, the Bragg peaks of the diffraction intensity become broader because of the finite size effect. The diffraction peak width for a microcrystalline alloy with the grain size of 15 Å would be about 0.4 $Å^{-1}$, which is equal to a typical diffraction peak width of amorphous alloys. Therefore it is tempting to view the amorphous state simply being the microcrystalline state. This view, however, is *clearly incorrect.* In order to explain why it is incorrect, we have to discuss the structure of liquid metals to which the amorphous alloys are closely related. First of all, liquid metals are definite *not* microcrystalline, as suggested by many evidences. For instance, the entropy of melting, which is of the order of k_B per atom [65], is much larger than necessary to change a large grain into microcrystals by introducing grain boundaries at every 15 Å. Secondly, the phase change from liquid to crystalline state, and vice versa, is clearly first order and discontinuous. The liquid phase and crystalline phase are separated by a well-defined boundary the energy of which is higher than either of the two. Thus the liquid phase is a distinct condensed state with its own structure and organization. Now the diffraction intensities of amorphous alloys are rather similar to those of liquid alloys except for a slightly increased sharpness of the diffraction peaks, indicating the similarity of the atomic structure of these two states. The change from liquid to amorphous state is continuous and has all the characteristics of the glass transition [66] observed by thermal and mechanical measurements. These observations conclusively suggest that the amorphous solids are essentially frozen liquids, or glasses, just as ordinary inorganic or polymer glasses.

The atomic structure of crystalline solids can easily be described in terms of symmetry, lattice constants, and several atomic position parameters, and the procedure to determine these parameters by analyzing either the single-crystal diffraction data or in many cases the powder diffraction data is well established. Even in quasicrystals translational symmetry is preserved at least in the six dimensions. However, if the solid does not have the lattice periodicity, the procedure quickly breaks down since the standard procedure of crystallography is critically dependent upon the translational symmetry of the solid. In the previous section we discussed how diffraction experiments on noncrystalline materials yield information regarding the interatomic distances in terms of the PDF. The PDF, however, is a spherically averaged density function without orientational information, and cannot reproduce a 3-d structure directly. Therefore, structural models are always necessary to interpret the PDF. In this section we discuss atomistic models of amorphous structure and theories which attempt to explain the structure from a more general and fundamental point of view.

A. Models of Amorphous Alloys

1. Dense Random Packing Models

The first realistic structural model of a liquid was proposed by Bernal [67] and Scott [68] in 1960. The model simply consists of a large cluster of steel balls densely packed in a bag, and was called the dense random packing of hard sphere (DRP-HS) model. Steel balls packed in a cubic box would have "crystallized," or in other words, they would have taken the configuration of a close-packed periodic structure (fcc or hcp). But the nearly spherical boundary condition of a bag resulted in a nonperiodic stacking of spheres. Surprisingly, the physical density of the spheres was not very much lower than for a close-packed structure. This point is rather important, since the density change upon melting is rather small, typically a few percent for metals, and any realistic model should satisfy this condition. The density difference between the crystalline and amorphous states for the DRP-HS models is not so small (~16%), but by introducing the softness of the interatomic interaction (DRP-soft-sphere, or DRP-SS, model), for instance by using a pairwise interatomic potential, the difference can become much smaller [69,70]. This remarkable demonstration has clearly shown that the concept of DRP can be the basic idea for the description of the structure of both liquid and amorphous states. It was later on shown by Cargill [71] that the PDF of this model indeed agrees well with the PDF of amorphous alloys experimentally determined using x-ray diffraction. In Fig. 16 the PDF of amorphous $Fe_{40}Ni_{40}P_{14}B_6$ [72] is compared to the DRP-SS model with the Johnson potential for iron [70,73]. The agreement is quite satisfactory to small details. The DRP model was extended to binary and ternary systems both in HS and SS models [74].

The topology of the DRP model was investigated by various methods of analysis. The PDF of the DRP model, such as the one shown in Fig. 16, indicates that the first and second peaks are usually reasonably well separated by a valley. Thus it is possible to define the nearest neighbors by the first minimum in the PDF. The number of the nearest neighbors, or the coordination number, N_c, varies from site to site, with the average of 12 to 14. The average value of N_c reported for the DRP-HS model tends to be between 13 and 14 [74,75], but in the DRP-HS model the first minimum of the PDF is rather shallow, so that the value of N_c strongly depends upon the position of the valley. For DRP-SS models N_c is better defined, and in the model shown in Fig. 16 it is about $12.56 \approx 4\pi$ [76].

The topology of connectivity of atoms was studied by the Voronoi polyhedral analysis [75]. The Voronoi polyhedron is a space defined by planes bisecting a line connecting two atoms, and is equivalent of the Wigner-Seitz cell for a crystal. The faces of the Voronoi polyhedra were found to be predominantly pentagonal.

2. Local Cluster Models

The DRP model is purely a geometrical model, without a concern to the chemistry of the system. However, since all the stable amorphous metallic solids are alloys and not elements, it is reasonable to suspect that chemical interaction among the constituents may be playing some role. Accordingly, local cluster models which emphasize the local chemistry were advanced. In these models the local topology is dictated by chemistry, and random network of specific local clusters form the

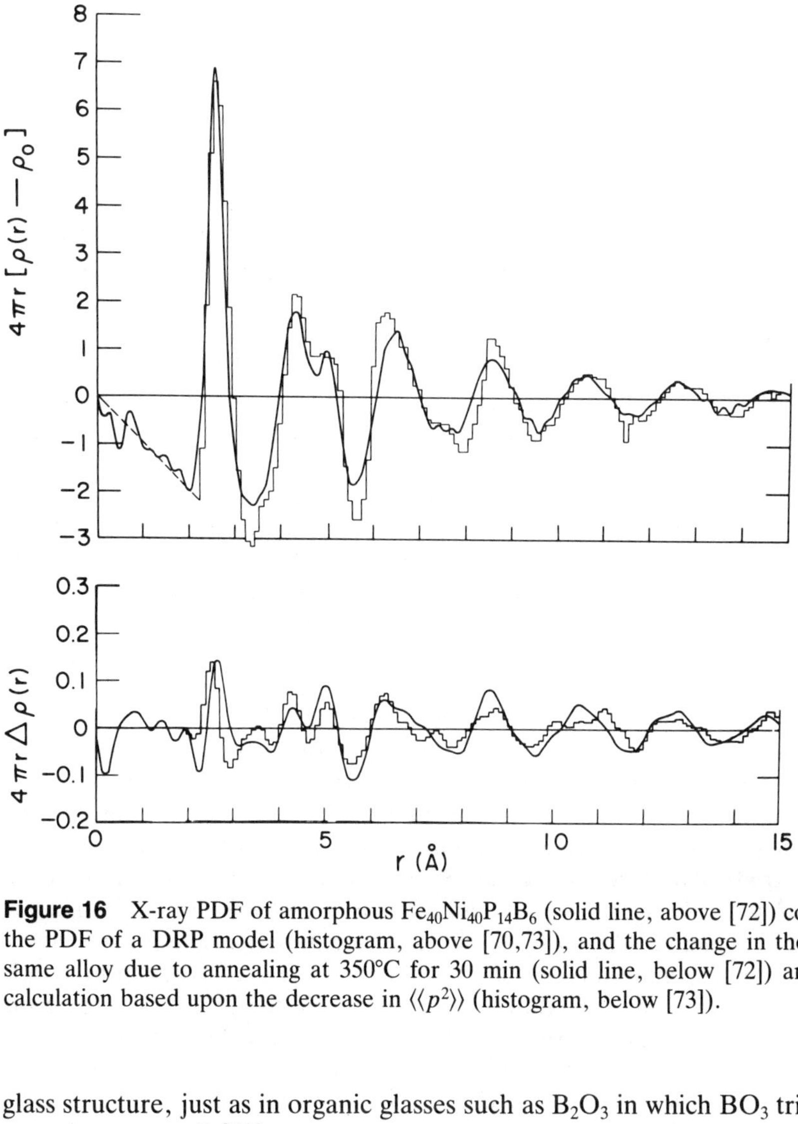

Figure 16 X-ray PDF of amorphous $Fe_{40}Ni_{40}P_{14}B_6$ (solid line, above [72]) compared with the PDF of a DRP model (histogram, above [70,73]), and the change in the PDF of the same alloy due to annealing at 350°C for 30 min (solid line, below [72]) and the model calculation based upon the decrease in $\langle\langle p^2 \rangle\rangle$ (histogram, below [73]).

glass structure, just as in organic glasses such as B_2O_3 in which BO_3 triangles form a random network [77].

Metallic glasses found up to mid-1970 were all transition metal–metalloid (TM-M) alloys such as Fe-B or Ni-P, and it appeared that the chemistry of this combination was important in glass formation. Gaskell noted that in TM-M compounds M atoms are usually located in the capped trigonal prisms, and proposed that the glasses are made of randomly connected TM-M prisms [78]. However, these clusters are not stabilized by directional chemical bonding, but is primarily the consequence of strong TM-M attraction and the relative size of TM and M atoms. Thus even within the framework of the DRP-SS model, it is possible to model the TM-M glasses in which the local environment of M atoms are dominated by trigonal prisms or Archimedian antiprisms, by using a strongly attractive but still isotropic TM-M potential and a repulsive M-M potential [74]. In addition, the discovery of many new metallic glasses comprised of only metallic elements such as Al-Ca [79] clearly ruled out special chemistry as the mechanism of glass formation.

B. Theories of Amorphous Structure

1. Icosahedral Cluster and Geometrical Frustration

It is instructive to consider why the microcrystalline model is not a good model of an amorphous structure. Perhaps the most important step is to challenge the idea that an atomic configuration found in the crystalline state is always preferred in any circumstance. Actually, the requirement of periodicity imposes a very severe constraint on the structure, and consequently the atoms often are forced to take on local configurations which are less stable in insolation. If the requirement of periodicity is removed, the atoms can assume different local configurations which are more stable, and consequently local configurations seldom found in crystalline states are seen in the liquid and amorphous state. An example is the icosahedral cluster. In 1952 Frank suggested that atoms might form icosahedral clusters in the liquid, since the lowest energy state of a cluster of 13 atoms interacting with the Lennard-Jones potential is an icosahedron, and not the close-packed fcc [80]. Indeed a large number of icosahedral clusters are found in the DRP structure [74,75]. The reason why an icosahedron is more stable is that it is composed of (20) tetrahedra alone, although they are slightly distorted, while the fcc structure is made of octahedra as well as tetrahedra. Thus, the 13 atoms in an icosahedron are connected by 42 bonds, while the fcc cluster has only 36 bonds.

Thus the *local* criterion of stability is to maximize the density of atomic bonds by forming triangles in two dimensions and tetrahedra in three dimensions. In two dimensions this local criterion is commensurate with the global criterion for the stability of a macroscopic system, since the entirety of space can be filled by triangles resulting in a triangular lattice. However, 3-d space cannot be filled by tetrahedra alone without severe distortions; thus, *the local and global criteria of structural stability are incompatible.* Adhering to the global criterion results in the close-packed structures which are made of both tetrahedra and octahedra. In the DRP structure the global stability criterion is discarded while the priority is placed on the local criterion. Thus the density of tetrahedra in the DRP structure is *higher* than in the close-packed structure, while the total energy of a monoatomic DRP structure is higher than that of the close-packed structure.

This incompatibility between the two criteria produces various levels of structural frustrations. For instance, in an icosahedron the distance between the neighboring apexes is longer than the distance between the center of the icosahedron and the apex by about 5%. Thus if one forms an icosahedron by 13 hard spheres, not all the spheres on the surface can touch each other. Thus the ground-state configuration is multiply degenerated, or geometrically frustrated since the atoms do not know which configuration they should take. This structural or geometric frustration is an important factor which contributes to the stability of the glassy state in 3 dimensions [81].

2. Curved Space Theories

One way to eliminate the frustration in the icosahedral cluster, as it may sound outlandish, is to introduce a curvature in the space [82,83]. If a 2-d surface is curved the circumference of a circle is smaller than $2\pi r$, where r is the distance measured on the curved surface from the center of the circle to the edge. In a similar way if the 3-d space is curved the surface area of a sphere is not equal to $4\pi r^2$. Therefore, by introducing the right curvature to the space one can make the 12 spheres on the surface of a 13-sphere icosahedron completely touching each other. Such a space is defined as the "surface" in four dimensions (S3), given by

$$x^2 + y^2 + z^2 + t^2 = R^2 \tag{12}$$

where t is the coordinate of the fourth dimension, and

$$\frac{R}{a} = \tau = \frac{1 + \sqrt{5}}{2} \tag{13}$$

where $a/2$ is the radius of the sphere and τ is the golden mean. This space is closed, since any coordinate cannot exceed R. It can be completely filled by 120 spheres locally forming perfect tetrahedra. This cluster is called a {3, 3, 5} polytope, and is an excellent starting point to develop a theory of the DRP structure. Since we live in a flat (Euclidean) space, this perfect polytope has to be brought back to our space somehow. In this process the polytope becomes distorted, and partially broken by the introduction of defects, primarily disclinations. A description of the amorphous structure in terms of the randomly distributed disclinations was discussed by Sachdev and Nelson [84].

3. *Atomic Level Stresses*

Foregoing discussions on the structural frustrations and the curved space highlighted the importance of distortions and strains in describing the structure of amorphous systems. Let us develop a very simple model to illustrate this situation. If we assume that the system can be described by a pairwise short range central force interatomic potential, $\phi(r)$, the total energy is given by

$$E_{\text{total}} = \sum_{i,j} \phi(r_{ij}) \tag{14}$$

where r_{ij} is the distance between the ith and jth atoms. This can be rewritten as

$$E_{\text{total}} = E_{\text{bond}} + E_{\text{dist}} \tag{15}$$

where E_{bond} is the bond energy,

$$E_{\text{bond}} = \left[\sum_{i} N_c(i)\right]\phi(a) \tag{16}$$

$N_c(i)$ is the local coordination number of the ith atom, a is the position of the minimum of $\phi(r)$, and E_{dist} is the distortion energy,

$$E_{\text{dist}} = \sum_{i,j} [\phi(r_{ij}) - \phi(a)] \tag{17}$$

E_{bond} decreases linearly with the increasing average coordination number, $\langle N_c \rangle$, and prefers denser packing. However, as $\langle N_c \rangle$ is increased the neighboring atoms start to push each other, and beyond a certain point E_{dist} begins to increase sharply,

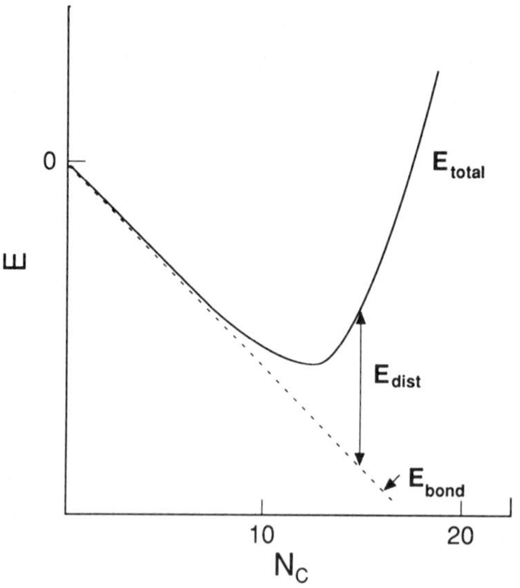

Figure 17 Schematic representation of energy-coordination number relationship.

resulting in a minimum as schematically shown in Fig. 17. According to the DRP-SS model with the short range Johnson potential, the minimum occurs at

$$\langle N_c \rangle = 12.57 \approx 4\pi \tag{18}$$

Arguments emphasizing topology focuses on E_{bond} alone. In order to describe the local structural distortion quantitatively the concept of atomic level stresses was introduced [85,86]. They were originally defined by Born and Huang [87] as

$$\sigma^{\alpha\beta}(i) = \frac{1}{v_i} \sum_j f_{ij}^{\alpha} r_{ij}^{\beta} \tag{19}$$

where $\sigma^{\alpha\beta}(i)$ is the $\alpha\beta$ component of the atomic level stress tensor of the ith atom, v_i is the local volume of the ith atom, f_{ij}^{α} is the α component of the two-body force between the ith and jth atoms, and r_{ij}^{β} is the β component of the separation between them. Even though the total force on the ith atom,

$$\vec{f}(i) = \sum_j \vec{f}_{ij} \tag{20}$$

is zero for an equilibrium structure, the local stress is not zero except in a perfect Bravais lattice. In the DRP-SS model the volume average of the local stress,

$$\langle \sigma^{\alpha\beta} \rangle = \sum_i v_i \sigma^{\alpha\beta}(i) \Big/ \sum_i v_i \tag{21}$$

is zero since it is equal to the external stress. However, the second moments are nonzero and is quite large. For a DRP-SS model of amorphous iron they are of the order of 100 MPa [86,88]. In terms of rotationally invariant quantities such as pressure, p,

$$p = \frac{1}{3}(\sigma_1 + \sigma_2 + \sigma_3) \tag{22}$$

where σ_n are the principal stresses, and the average (von Mises) shear stress, τ,

$$\tau = \left[\frac{1}{6}\{(\sigma_1 - \sigma_2)^2 + (\sigma_2 - \sigma_3)^2 + (\sigma_3 - \sigma_1)^2\}\right]^{1/2} \tag{23}$$

the rms values of the average local stain are

$$\varepsilon_v = \frac{\langle p^2 \rangle^{1/2}}{B} = 0.067 \tag{24}$$

$$\varepsilon_s = \frac{\langle \tau^2 \rangle^{1/2}}{2G} = 0.096 \tag{25}$$

where B and G are the bulk and shear moduli [88]. Thus the local environment of each atom is strongly distorted in the DRP structure. The atomic level stresses were used to elucidate the glass transition, mechanical deformation, and structural relaxation phenomena as we will discuss later.

V. ATOMIC SHORT-RANGE ORDER AND PROPERTIES OF AMORPHOUS ALLOYS

The discussion so far has been dealing with the structure itself, while the purpose of all these endeavors is to explain the properties of the solid based upon the structural information. To this end, however, the bare atomic positions or the PDF are not sufficient, and we need some more intermediate structural parameters to characterize the structure in a more direct and useful manner. We will introduce several short-range order parameters, starting with the compositional short-range order (CSRO) parameter, the bond orientational order parameter, and finally the atomic level stresses defined above, and discuss how these parameters help elucidate the physical properties of amorphous alloys.

A. Compositional Short-Range Order

In multicomponent alloys, both crystalline and amorphous, usually the components do not mix perfectly with each other because of chemical interactions among them, and their properties are affected by the degree of mixing. Some crystalline alloys, such as $Ni_{75}Fe_{25}$, develop compositional long-range order below a certain ordering temperature, and their magnetic, mechanical as well as electrical properties are affected by this ordering. Compositional long-range order can occur in quasicrystals as well, and recently found Al-Cu-Fe is considered to be ordered in the fcc lattice in the six dimensions. In liquid and amorphous alloys, on the other hand, only the CSRO can develop. In nonmetallic liquid and glasses the CSRO is usually quite strong, while in metallic amorphous alloys the degree of CSRO among the metallic elements can vary with temperature, resulting in very interesting relaxation phenomena.

1. Definition and Observation

In binary (A-B) crystalline alloys Warren and Cowley introduced a CSRO parameter [89,90]

$$\alpha^{\mathrm{CRSO}} = 1 - \frac{N_c^{\mathrm{AB}}}{c_{\mathrm{B}} N_c^{\mathrm{A}}} \tag{26}$$

where N_c^{AB} is the average number of B atoms around an A atom and N_c^{A} is the coordination number of an A atom. Thus if there is an association tendency between A and B due to chemical affinity, α^{CSRO} will be negative. This definition can be extended to the amorphous alloys by using the compositionally resolved coordination number

$$N_c^{\mathrm{AB}} = 4\pi c_{\mathrm{B}} \int_0^{r_{\mathrm{lm}}} \rho_{\mathrm{AB}}(r) r^2 \, dr \tag{27}$$

where $\rho_{\mathrm{AB}}(r)$ is the compositionally resolved partial PDF between A and B, which is the Fourier transform of the partial structure factor defined in Eq. (7), and r_{lm} is the position of the first minimum beyond the first peak in the PDF. In a simple crystal like Cu_3Au the total coordination numbers of Cu and Au are equal, however, in amorphous alloys N_c^{A} is usually different from N_c^{B} because of the size effect. Thus an improved definition was proposed by Cargill and Spaepen [91]. In liquids and glasses the CSRO parameter among metallic elements is small, and is in the range of $0{-}{-}0.2$ [92,93], but the CSRO between metalloid and metal elements can be much larger in magnitude [19,94,95].

For the determination of the CSRO parameters defined above the compositionally resolved PDF's have to be known rather accurately. This can be done most accurately by neutron scattering measurements with isotopic substitution, and less accurately with the differential anomalous x-ray scattering measurements as we discussed earlier.

2. Magnetic Properties and CSRO

One of the properties on which the CSRO has a significant effect is magnetic properties. In the molecular field approximation it can be readily shown that for small values of α^{CSRO}, the Curie temperature, T_c, depends linearly upon α^{CSRO}. Now if the value of α^{CSRO} is small, it depends inversely upon the annealing temperature, thus it is expected that T_c would depend upon the annealing temperature, T_a, as

$$T_c(T_a) = T_c(\infty)\left[1 + \frac{T_0}{T_a}\right] \tag{28}$$

where $T_c(\infty)$ is the Curie temperature of a totally random system without CSRO and T_0 is a constant. This relationship has been observed in Fe-Ni base alloys [96]. Based upon this relation, the magnetic fictive temperature of the as quenched alloy, or the temperature at which the configuration of the alloy froze during the rapid quenching, can be determined as shown in Fig. 18 [97]. It should be noted that in many alloys the equilibrium value of T_c can actually be attained without the onset of crystallization. This is somewhat surprising, since the change in T_c, thus the change in the CSRO, requires some atomic motion. The reason for this is that the

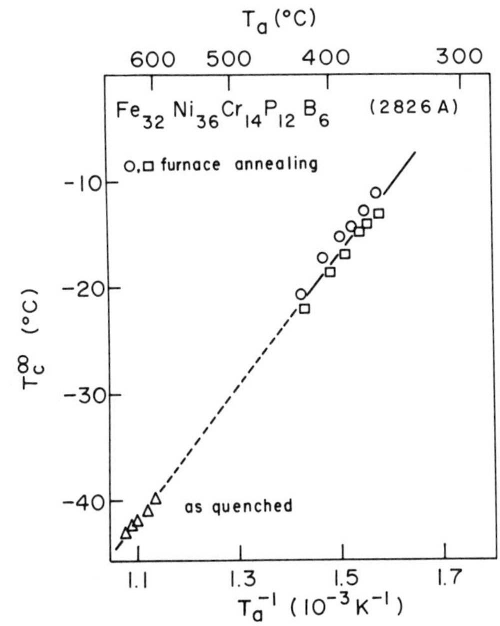

Figure 18 Curie temperature, T_c, and the inverse of the annealing temperature, T_a. The Curie temperature of the as-quenched materials (triangles) determines the fictive temperature, T_f [97].

CSRO can be changed by short range atomic rearrangement such as the bond-exchange mechanism, shown in Fig. 19 [11], while the crystallization requires nucleation and diffusion over longer ranges.

The Curie temperature discussed above is the equilibrium value which can be attained after annealing for a fairly long time. For shorter annealing times the value of T_c changes with the annealing time, t_a. The change in T_c is a thermally activated kinetic process which becomes faster with increasing temperature [98]. The activation energy of the process, however, is not a single value, as demonstrated nicely by the so-called crossover phenomenon [99]. The kinetics of the change, which approximately follows the ln t_a kinetics, can be described well as the superposition of the first-order kinetics with a spectrum of activation energies [100,101],

$$X(t_a, T_a) = \frac{T_c(t_a) - T_c(0)}{T_c(\infty) - T_c(0)}$$

$$= \int N(E_a)\Phi(t_a, T_a)\, dE_a \tag{29}$$

$$\Phi(t_a, T_a) = 1 - \exp\left(-\frac{t_a}{\tau(E_a, T_a)}\right) \tag{30}$$

$$\tau(E_a, T_a) = \tau_0 \exp\left(\frac{E_a}{kT_a}\right) \tag{31}$$

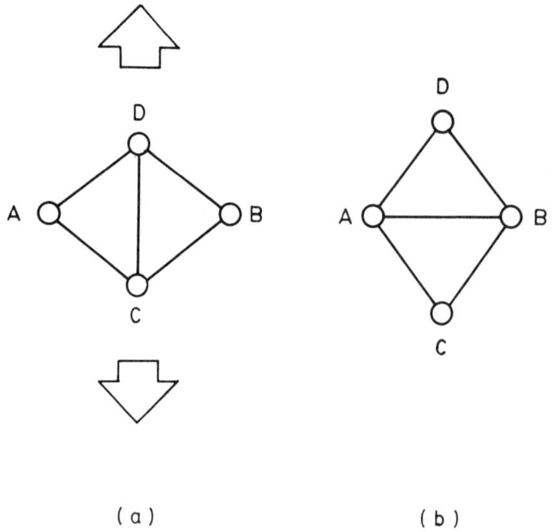

Figure 19 Bond exchange mechanism. (a) As a stress is applied, (b) the bond parallel to the stress is cut and a new bond is formed in a perpendicular direction [11].

Now further complication exists because of the change in $N(E_a)$ due to annealing, or the relaxation of the relaxation kinetics. This phenomenon will not be discussed here to save space, but the readers are referred to [102].

Another important consequence of the CSRO on magnetism is the induced magnetic anisotropy, K_c, due to the directional CSRO [103]. When the sample is annealed in a magnetic field below T_c, magnetic anisotropy is induced. This phenomenon is well known for crystalline alloys, and is believed to be due to the directional pairing [104], although the structural anisotropy necessary to result in such an anisotropy is quite small (of the order of 0.1%) and has not been directly observed. In amorphous alloys the directional CSRO can be defined in terms of the anistropic partial PDF, as described in [105]. Both crystalline field and pseudodipolar interaction can result in the magnetic anisotropy when such a directional CSRO is present [106]. The saturation value of K_u is dependent upon temperature, and is roughly proportional to $[m(T_a)]^2$, where $m(T_a)$ is the magnetization at T_a. The kinetics of the anisotropy induction, $K_u(t_a, T_a)$, shown in Fig. 20 [107], is very similar to those of the T_c kinetics, and again the saturation is obtained at much shorter times than the start of the crystallization. This is because crystallization is a first-order phase transformation requiring nucleation and growth, usually involving long-range diffusion, while the CSRO can be changed locally by displacements of only few atoms. Consequently the CSRO can reach local equilibrium, even though the glass as a whole may be unstable against crystallization.

B. Bond-Orientational Order

1. *Hexatic and Icosahedral Order*

As we mentioned earlier, in two dimensions the local criterion of close packing agrees with the global criterion, thus there is a strong tendency to restore local

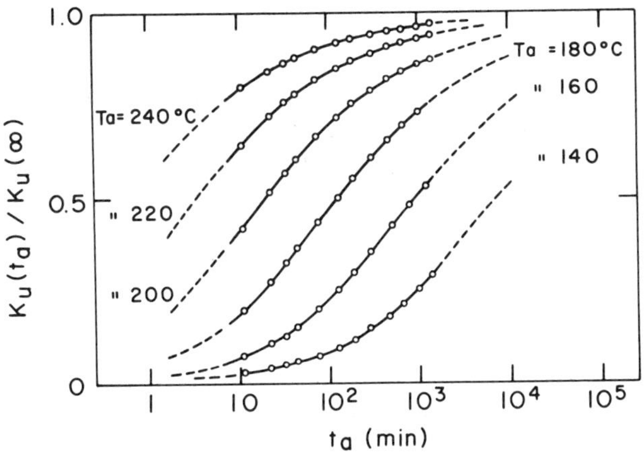

Figure 20 Kinetics of reorientation of the field-induced magnetic anisotropy, K_u, in amorphous $Fe_{40}Ni_{40}P_{14}B_6$ [107].

order. Consequently a 2-d liquid tends to become microcrystalline. Such a structure can be described well by a crystalline model with a very high density of dislocations. The stress field of a dislocation decreases as $1/r$, where r is the distance from the dislocation, so that the self-energy of an isolated dislocation diverges as $\ln(r)$. However, if two dislocations of opposite signs couple, they shield each other's stress field, so that the total energy of the pair is finite and is greatly reduced compared to that of a single dislocation. Thus at high temperatures each dislocation moves randomly, but at lower temperatures they may become paired while being still mobile, before they freeze into a solid state. Such a state of mobile paired dislocation, a hexatic phase, was shown to exist both theoretically [108] and experimentally [109]. If the dislocations of the same sign line up they form a (low angle) grain boundary. However, a coupled dislocation does not lead to tilting of the lattice, thus in the hexatic phase the orientational order is preserved while the translational order is lost since the dislocation pairs are mobile. The order parameter of such bond-orientational order (BOO) is

$$Q_6 = \langle e^{i6\theta} \rangle \tag{32}$$

where θ is the orientation of the atomic bond defined by a pair of atoms which are nearest neighbors to each other. The orientationally ordered state is characterized by Bragg-like diffraction peaks with a special power law peak shape.

Steinhardt, Nelson, and Ronchetti extended this idea to three dimensions, and showed that, using the molecular dynamics simulation, an orientationally ordered state can exist also in three dimensions, with the icosahedral order [110]. Whether the correlation they have seen is truly long range or not is debatable, and the work has not been reproduced. However, the observed tendency of icosahedral order beyond the size of the 13-atom icosahedron, which indicates the ordering tendency among the icosahedral clusters, is very intriguing. This observation certainly has triggered the development of the theory of quasicrystals.

2. *Bond-Orientational Anisotropy*

In the DRP structure, even when there is no long-range bond-orientational order, the orientation of the bonds can become anisotropic. For instance if amorphous polymers are stretched the polymer chains tend to line up in the direction of the stress, and the bond orientation becomes anisotropic. In the metallic DRP system, too, the bond-orientational anisotropy (BOA) can develop, although the consequence is just the opposite of the case of polymers, as we discuss later. BOA can be described by the anisotropic PDF,

$$\rho(\vec{r}) = \sum_{l,m} \rho_l^m(r) Y_l^m(\hat{r}) \tag{33}$$

Similarly the structure factor can be expanded using the spherical harmonics,

$$S(\vec{Q}) = \sum_{l,m} S_l^m(Q) Y_l^m(\hat{Q}) \tag{34}$$

They are connected through a spherical Bessel transformation,

$$\rho_l^m(r) = \frac{i^l}{2\pi^2} \int S_l^m(Q) \mathcal{J}_l(Qr) Q^2 \, dQ \tag{35}$$

If we neglect the terms with l higher than 2, ρ_0^0 and ρ_2^0 can be separated by performing the diffraction measurements with two orientations of $\vec{Q}$ with respect to the sample axis [11].

As is well known, there are three basic modes of mechanical deformation: elasticity, anelasticity, and plasticity. Elastic deformation is instantaneous and fully recoverable. Anelastic deformation is a delayed reaction, usually thermally activated, and is recoverable. Plastic deformation is not recoverable. By subjecting an amorphous alloy under stress and temperature, that is, by applying creep deformation, and cooling without removing the stress, it can be anelastically polarized. It was shown recently that the anelastically polarized state is characterized by BOA [11]. As shown in Fig. 3, the creep deformation results in the anisotropic PDF, which can be modeled well by the BOA with a higher density of bonds in the direction *perpendicular* to the stress, which is opposite to the case of amorphous polymers. This is because in metallic DRP structure the atomic bonds can be easily exchanged by the bond-exchange mechanism shown in Fig. 15, so that the bonds parallel to the stress are cut while the new bonds perpendicular to the stress are formed. This anelastic bond-exchange mechanism is the most basic common mechanism to bring about the changes in the SRO, including the CSRO. For this reason the kinetics of anelasticity, Curie temperature, and field-induced magnetic anisotropy are all very much the same [96,111].

C. Glass Transition and Structural Relaxation

The transition from a liquid to a glass, or the glass transition, is characterized by a rapid but continuous change in the viscosity, η. The glass transition is defined conventionally by the viscosity reaching 10^{13} poise, since the mechanical relaxation time of the system, $\tau = \eta/G$, becomes very long at such a high viscosity, and the system behaves like a solid. Whether this transition is a real transition or not, and if it is, whether it is a first-order transition or a second order transition, has been hotly debated. These arguments, however, concern an ideal fully relaxed glass which requires an infinite time to obtain, while in reality, because of finite cooling rates a liquid always freezes into a glass which is not fully relaxed. In this case the transition is not a real transition, but is merely a deviation from the equilibrium state (liquid) to an inequilibrium kinetically frozen state (glass). The liquid state is not a unique state, but is a function of temperature. Thus as temperature is changed the structure changes from an old equilibrium to a new. This change, however, is governed by viscosity, so that as viscosity becomes higher the change becomes more sluggish, and at one point (T_g) the structure no longer can follow the temperature change, and the structure becomes "kinetically arrested."

This process of kinetic arrest is dependent upon the cooling rate, since if the cooling rate is high the arrest occurs at a higher temperature while the viscosity is still low. The temperature at which this arrest occurred is called the "fictive temperature," T_f, and characterizes the glassy state [112]. A glass with a high T_f is less stable than the one with a lower T_f, and by annealing below T_g the system becomes more stable. This process of lowering T_f is the structural relaxation. The glass transition and structural relaxation are complex, highly collective phenomena which still are not fully explained theoretically, in spite of numerous attempts. In this section we discuss these phenomena from a particular point of view of atomic level stresses.

1. Thermal Fluctuation of Stresses and Glass Transition

The atomic level stresses were introduced in order to describe the deviations from a perfect geometry quantitatively, or in other words to account for the distortion energy, Eq. (17). The elastic self-energy of the atomic level stresses was calculated in the elastic approximation, as

$$E_{el} = v \sum_i \left[K_\alpha \frac{[p(i)]^2}{2B} + K_\phi \frac{[\tau(i)]^2}{2G} \right] \tag{36}$$

where v is the average atomic volume, K_α and K_θ are the renormalizing constants describing the effect of long-range stress field due to the atomic level stress at the ith atom, and is unity in the liquid state and around 2 in the glass [88]. The thermal fluctuations of the stresses can be calculated from this expression, and in the liquid state we obtain

$$\langle\langle p^2 \rangle\rangle = \frac{B}{2v} kT \tag{37}$$

where $\langle\langle \cdot\cdot \rangle\rangle$ implies thermal as well as volume average. This, however, extrapolates to zero at $T = 0$ K. Such a state, in which the atomic level pressure is zero everywhere, cannot be achieved in the glassy state. Therefore below a certain temperature $\langle\langle p^2 \rangle\rangle$ deviates from the equilibrium state given by Eq. (37), and saturates to a finite value [88,113]. This is the glass transition, and the glass transition temperature, T_g, is given by

$$T_g = \kappa v B \tag{38}$$

where

$$\kappa = \frac{2\langle p^2 \rangle_{\min}}{kB^2} \sim 60 \text{ K/eV} \qquad \text{(for metallic glasses)} \tag{39}$$

From this we obtain

$$T_g \sim 0.4\overline{T}_m \tag{40}$$

where $\overline{T}_m$ is the compositionally averaged melting temperature of the constituent element [88,114]. Equation (41) is valid for many glassy alloys as shown in Fig. 21, with a few exceptions to the rule in which it is suspected that B is not proportional to $\overline{T}_m$ [114].

2. *Structural Relaxation*

The structure of liquid can be characterized by the stress fluctuations as in Eq. (37). Then it is reasonable to assume that when a liquid is quenched the thermal fluctuations at the temperature at which the thermal arrest takes place, or the fictive temperature T_f, are more or less retained in the glassy state. In other words the value of T_f can be assessed from the magnitude of $\langle\langle p^2 \rangle\rangle$. Therefore the structural relaxation process can be understood in terms of the relaxation of the stress fluctuations. For instance in one estimate $\langle\langle p^2 \rangle\rangle$ changes as much as 30% during a typical relaxation [72]. Since the local pressure produces a phase shift in the local PDF, the PDF is dependent upon $\langle\langle p^2 \rangle\rangle$ by

$$\rho(r) = \rho_R(r) + \langle\langle p^2 \rangle\rangle \frac{\xi(r)^2}{2} \frac{\partial^2 \rho_R}{\partial r^2} \tag{41}$$

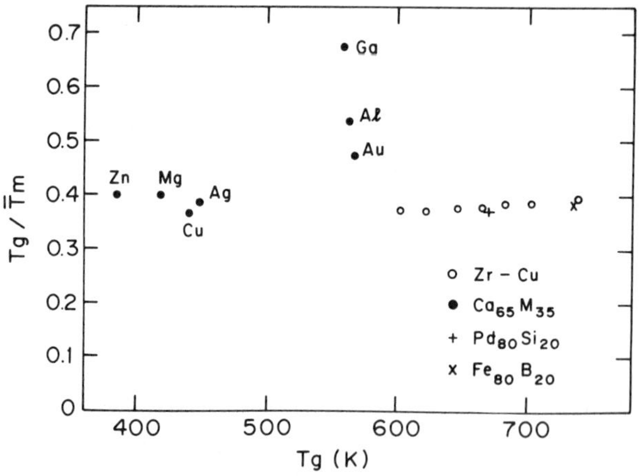

Figure 21 Glass transition temperature T_g and the ratio $T_g/\overline{T}_m$, where $\overline{T}_m$ is the compositional average of the melting temperature of constituent elements. Note that the ratio is always close to 0.4, except for a few cases of compound forming alloys for which the estimate of the bulk modulus by $\overline{T}_m$ is not likely to be accurate [114].

where $\rho_R(r)$ is the PDF from atoms with $p = 0$, and $\xi(r)$ is the phase shift factor which is weakly dependent upon r. This predicts that the change in the PDF due to structural relaxation is proportional to the second derivative of the PDF, and indeed the experimentally observed change is very closely related to the second derivative of the PDF as shown in Fig. 16 [72]. The magnitude of $\langle\langle p^2\rangle\rangle$ is related to the volume expansion through the anharmonicity [115] and therefore to the free volume, and hence viscosity [116] . The relaxation of these quantities is basically irreversible below T_g, and usually the sample crystallizes before reaching the equilibrium state.

The relaxation of other quantities such as magnetic properties and anelastic properties reach equilibrium much more quickly, and when the annealing condition is reversibly changed they follow also reversibly. The observation of the BOA in an anelastically polarized sample indicates that this type of relaxation phenomena are due to the reorientation of local shear stresses which apparently happens more easily than the change in the magnitude of the pressure fluctuation.

D. Composition Dependence and Glass Formability

While many structural variables, such as the PDF, depends very much upon composition and the atomic sizes of the constituent elements, the atomic level strains, or the stresses normalized by the elastic constants, were shown by a computer simulation to be quite independent of composition and size. Thus they describe a "universal" feature of the glassy state, upon which many basic arguments can be built. One such argument is the theory of glass forming composition. Various criteria have been proposed in attempting to predict the glass forming compositions. Among them the atomic size ratio appears to be the most important of all. By using the elasticity argument, the minimum concentration of B element needs to be added to A to form a glass upon quenching, $c_{\mathrm{B}}^{\mathrm{min}}$, was shown to be [117]

$$c_{\mathrm{B}}^{\mathrm{min}} \sim \frac{0.1}{|\Delta v|/v_A} \tag{42}$$

where

$$\Delta v = v_B - v_A \tag{43}$$

This result is in good agreement with experimental data as shown in Fig. 22. In this argument, a glass is formed not because a particular composition makes the energy of the glassy state lower, but adding a second element and forming a solid solution increases the atomic level stresses in the crystalline state, which leads to the collapse of the lattice structure.

VI. CONCLUDING REMARKS

Nonequilibrium solid phases produced by rapid solidification are often noncrystalline, and an attempt to characterize and understand the structure of these solids appear to lead to a foggy path. Owing to recent advances in experimental and theoretical methods, however, significant progress has been made in this field, and we are beginning to accomplish more quantitative description of the structure of

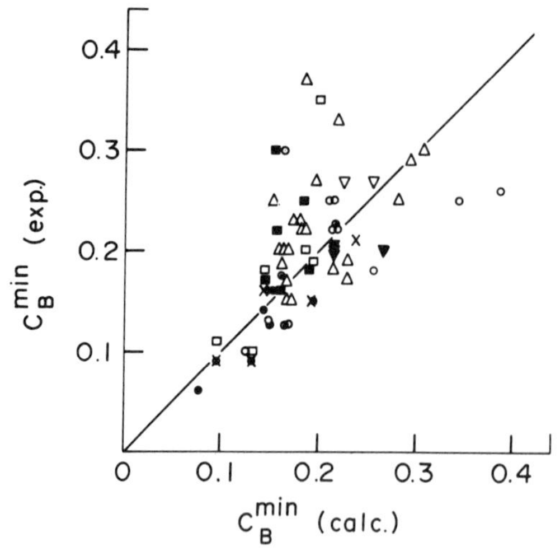

Figure 22 Minimum glass forming concentration, c_B^{min}, vs. λ which is inversely proportional to $\Delta v/v$ (Eq. (43)) for various binary alloy systems [117].

noncrystalline solids. The key for a successful experimental structure determination is to measure the scattering intensity accurately up to high values of scattering vector Q, using high-energy x-rays or neutrons. This usually requires the use of special facilities such as synchrotrons, rather than a short trip to the x-ray laboratory down the hall. Therefore an accurate characterization of glassy solids is a major research project by itself. But it is worth the effort to know such a structure, since unlike in the gaseous state, atoms in the glassy states are not really randomly distributed in space but are highly correlated and exhibit strong local order, which determines the properties of the solid. The recent discovery of a quasicrystalline solids has significantly changed our attitude toward the ordered state. It is now clear that the crystalline state is not always the state with the lowest energy, and once the restriction of periodicity, which, by the way, is a severe restriction, is removed a large variety of atomic configurations such as icosahedral clusters become attainable. Although many fundamental issues remain unsolved and await future research, a clear understanding of some of the important structural features of noncrystalline solids is not beyond reach.

REFERENCES

1. B. E. Warren, *X-Ray Diffraction* (Addison-Wesley, Reading, MA, 1969).
2. C. N. J. Wagner, *J. Non-Cryst. Solids*, *31*, 1 (1978).
3. S. Aur and T. Egami, *J. Phys. (Paris)*, *40*, C8-234 (1980).
4. B. C. Giessen and G. E. Gordon, *Science*, *159*, 973 (1968).
5. B. Buras, J. Chwaszczewska, S. Szarras, and Z. Szmid, Rep. 894/11/PS, Inst. Nucl. Res., Warsaw, 1968.
6. T. Egami, *J. Mater. Sci.*, *13*, 2589 (1978).
7. T. Egami, in *Glassy Metals I*, edited by H. Beck and H.-J. Güntherodt (Springer-Verlag, Berlin, 1981), p. 25.

8. D. L. Price and K. Sköld, in *Neutron Scattering*, Part A, edited by K. Sköld and D. L. Price (Academic Press, New York, 1986), p. 1.
9. *International Table for X-Ray Crystallography*, Vol IV, edited by J. A. Ibers and W. C. Hamilton (Kynoch Press, Birmingham, 1974).
10. T. Egami, R. S. Williams, and Y. Waseda, in *Rapidly Quenched Metals III*, Vol. 2, edited by B. Cantor (The Metals Society, London, 1978), p. 318.
11. Y. Suzuki, J. Haimovich, and T. Egami, *Phys. Rev.*, *B35*, 2162 (1987).
12. P. A. Heiney, P. A. Bancell, P. M. Horn, J. L. Jordon, S. LaPlaca, J. Angilello, and F. W. Gayle, *Science*, *238*, 660 (1987).
13. S. J. Poon, W. Dmowski, T. Egami, Y. Shen, and G. J. Shiflet, *Phil. Mag. Lett.*, *56*, 259 (1987).
14. J. Wang, in *Glassy Metals I*, edited by H. Beck and H.-J. Güntherodt (Springer-Verlag, Berlin, 1981), p. 45.
15. N. J. Shevchik, *Phil. Mag.*, *35*, 805 (1977).
16. Y. Waseda, *The Structure of Non-Crystalline Materials* (McGraw-Hill, New York, 1980).
17. P. H. Fuoss, P. Eisenberger, W. K. Warburton, and A. Bienenstock, *Phys. Rev. Lett.*, *46*, 1573 (1981).
18. D. D. Kofalt, S. Nanao, T. Egami, K. M. Wong, and S. J. Poon, *Phys. Rev. Lett.*, *57*, 114 (1986).
19. K. Suzuki, in *Neutron Scattering*, Part B, edited by K. Sköld and D. L. Price (Academic Press, New York, 1986), p. 243.
20. S. Nanao, W. Dmowski, T. Egami, J. W. Richardson, Jr., and J. D. Jorgensen, *Phys. Rev.*, *B35*, 435 (1987).
21. T. Egami, W. Dmowski, J. D. Jorgensen, D. G. Hinks, D. W. Capone II, C. U. Segre, and K. Zhang, *Rev. Solid State Sci.*, *1*, 247 (1987), and in *High Temperature Superconductivity*, edited by S. M. Bose and S. D. Tyagi (World Scientific, Singapore, 1987), p. 101.
22. W. Dmowski, B. H. Toby, T. Egami, M. A. Subramanian, J. Gopalakrishnan, and A. W. Sleight, *Phys. Rev. Lett.*, *61*, 2608 (1988).
23. B. H. Toby, T. Egami, J. D. Jorgensen, and M. A. Subramanian, *Phys. Rev. Lett.*, 64, 2414 (1990).
24. Y. Wu, W. Dmowski, T. Egami, and M. E. Chen, *J. Appl. Phys.*, *61*, 3219 (1987).
25. S. W. Lovesey, *Theory of Neutron Scattering from Condensed Matter*, Vol. 2 (Clarendon Press, Oxford, 1984).
26. D. Shechtman, I. Blech, D. Gratias, and J. W. Cahn, *Phys. Rev. Lett.*, *54*, 1951 (1984).
27. F. C. Frank and J. S. Kasper, *Acta Cryst.*, *11*, 184 (1958).
28. e.g., K. Heinemann, M. J. Yacaman, C. Y. Yang, and H. Poppa, *J. Cryst. Growth*, *47*, 177 (1979).
29. D. Shechtman and I. Blech, *Metall. Trans.*, *16A*, 1005 (1985).
30. P. W. Stephens and A. I. Goldman, *Phys. Rev. Lett.*, *56*, 1168 (1986).
31. D. Levine and P. J. Steinhardt, *Phys. Rev. Lett.*, *54*, 2144 (1984).
32. R. Penrose, *Math. Intell.*, *2*, 32 (1979).
33. N. G. de Bruijn, *Ned. Akad. Weten. Proc. Ser.*, *A43*, 39, 53 (1981).
34. P. Kramer and R. Neri, *Acta Cryst.*, *A40*, 580 (1984).
35. P. A. Kalugin, A. Yu. Kitaev, and L. S. Levitov, *JETP Lett.*, *41*, 145 (1985).
36. L. Pauling, *Nature*, *317*, 512 (1985).
37. P. Sainfort, B. Dubost, and A. Dubus, *C. R. Acad. Sci. Paris*, *10*, 689 (1985).
38. A. P. Tsai, A. Inoue, and T. Masumoto, *Japan J. Appl. Phys.*, *26*, L1505 (1987).
39. P. A. Bancel, P. A. Heiney, P. W. Stephens, A. I. Goldman, and P. M. Horn, *Phys. Rev. Lett.*, *54*, 2422 (1985).

40. C. A. Guryan, A. I. Goldman, P. W. Stephens, K. Hiraga, A. P. Tsai, A. Inoue, and T. Masumoto, *Phys. Rev. Lett.*, *62*, 2409 (1989).
41. M. Gardner, *Sci. Am.*, *236*, 110 (1977).
42. e.g., I. Vincze, D. S. Boudreaux, and M. Tegze, *Phys. Rev.*, *B19*, 4896 (1979).
43. A. Katz and M. Duneau, *J. Phys.* (*Paris*), *47*, 181 (1986).
44. V. Elser, *Acta Cryst.*, *A42*, 36 (1986).
45. T. Janssen, *Acta Cryst.*, *A42*, 261 (1986).
46. P. Bak, *Scripta Met.*, *20*, 1199 (1986).
47. P. Bak, *Phys. Rev. Lett.*, *56*, 861 (1986).
48. J. W. Cahn, D. Shechtman, and D. Gratias, *J. Mater. Res.*, *1*, 13 (1986).
49. S. Ebalard and F. Spaepen, *J. Mater. Res.*, *4*, 39 (1989).
50. T. C. Lubensky, J. E. S. Socolar, P. J. Steinhardt, P. A. Bancel, and P. A. Heiney, *Phys. Rev. Lett.*, *57*, 1440 (1986).
51. J. E. S. Socolar, T. C. Lubensky, and P. J. Steinhardt, *Phys. Rev.*, *B34*, 3345 (1986).
52. P. M. Horn, W. Malzfeldt, D. P. DiVincenzo, J. Toner, and R. Gambino, *Phys. Rev. Lett.*, *57*, 1444 (1986).
53. V. Elser and C. L. Henley, *Phys. Rev. Lett.*, *55*, 2883 (1985).
54. M. Audier and P. Guyot, *Phil. Mag.*, *B53*, L43 (1986).
55. C. L. Henley and V. Elser, *Phil. Mag.*, *B53*, L59 (1986).
56. W. Dmowski, T. Egami, Y. Shen, S. J. Poon, and G. J. Shiflet, *Phil. Mag. Lett.*, *56*, 63 (1987).
57. Y. Shen, S. J. Poon, W. Dmowski, T. Egami, and G. J. Shiflet, *Phys. Rev. Lett.*, *58*, 1440 (1987).
58. J. W. Cahn, D. Gratias, and B. Mozer, *J. Phys.* (*Paris*), *49*, 1225 (1988).
59. M. De Boissieu, J. Pannetier, C. Janot, and J. M. Dubois, *J. Non-Cryst. Solids*, *106*, 211 (1988).
60. C. Janot, J. Pannetier, J. M. Dubois, and M. de Boissieu, *Phys. Rev. Lett.*, *62*, 450 (1989).
61. D. D. Kofalt, I. A. Morrison, T. Egami, S. Priesche, S. J. Poon, and P. J. Steinhardt, *Phys. Rev.*, *B35*, 4489 (1987).
62. Ruizhong Hu, T. Egami, A.-P. Tsai, A. Inoue, and T. Masumoto, *Phys. Rev.*, *B46*, 6105 (1992).
63. H. Y. Hsieh, B. H. Toby, T. Egami, Y. He, S. J. Poon, and G. J. Shiflet, *J. Mater. Res.*, *5*, 2807 (1990).
64. H. Y. Hsieh, T. Egami, Y. He, S. J. Poon, and G. J. Shiflet, *J. Non-Cryst. Solids*, *135*, 248 (1991).
65. J. P. Hansen and I. R. McDonald, *Theory of Simple Liquids* (Academic Press, New York, 1976).
66. H. S. Chen and D. Turnbull, *J. Chem. Phys.*, *48*, 2560 (1968).
67. J. D. Bernal, *Nature*, *181*, 41 (1959); *188*, 410 (1960).
68. G. D. Scott, *Nature*, *188*, 408 (1960).
69. L. von Heimandahl, *J. Phys. F*, *5*, L141 (1975).
70. K. Maeda and S. Takeuchi, *J. Phys. F*, *8*, L283 (1978).
71. G. S. Cargill III, *J. Appl. Phys.*, *41*, 12 (1970).
72. T. Egami, *J. Mater. Sci.*, *13*, 2587 (1978).
73. D. Srolovitz, T. Egami, and V. Vitek, *Phys. Rev.*, *B24*, 6436 (1981).
74. D. S. Boudreaux, in *Glassy Metals*: *Magnetic*, *Chemical*, *and Structural Properties*, edited by R. Hasegawa (CRC Press, Boca Raton, 1983), p. 1.
75. J. L. Finney, *Proc. Roy. Soc.*, *A319*, 479 (1970).
76. T. Egami, in *Amorphous Materials*: *Modeling of Structure and Properties*, edited by V. Vitek (TMS-AIME, Warrendale, 1983), p. 127.
77. e.g., R. Zallen, *The Physics of Amorphous Solids* (Wiley, New York, 1983).

78. P. H. Gaskell, *J. Non-Cryst. Solids*, *32*, 207 (1979).
79. B. C. Giessen, J. Hong, L. Kabacoff, D. E. Polk, R. Raman, and R. St. Amand, in *Rapidly Quenched Metals III*, Vol. 1, edited by B. Cantor (The Metals Society, London, 1978), p. 249.
80. F. C. Frank, *Proc. Roy. Soc.*, *A215*, 43 (1952).
81. D. R. Nelson, *Phys. Rev.*, *B28*, 5515 (1983).
82. M. Kleman and J. F. Sadoc, *J. Phys. Lett.* (*Paris*), *40*, L569 (1979).
83. R. Mosseri and J. F. Sadoc, *J. Phys. Lett.* (*Paris*), *45*, L827 (1984).
84. S. Sachdev and D. R. Nelson, *Phys. Rev.*, *B32*, 4592 (1985).
85. T. Egami, K. Maeda, and V. Vitek, *Phil. Mag.*, *B41*, 883 (1980).
86. D. Srolovitz, V. Vitek, and T. Egami, *Phil. Mag.*, *B44*, 847 (1981).
87. M. Born and K. Huang, *Dynamical Theory of Crystal Lattices* (Clarendon Press, Oxford, 1975).
88. T. Egami and D. Srolovitz, *J. Phys. F*, *12*, 2141 (1982).
89. J. M. Cowley, *J. Appl. Phys.*, *21*, 24 (1950).
90. B. E. Warren, B. L. Averbach, and B. W. Roberts, *J. Appl. Phys.*, *22*, 1493 (1951).
91. G. S. Cargill III and F. Spaepen, *J. Non-Cryst. Solids*, *43*, 91 (1981).
92. P. Chieux and H. Ruppersberg, *J. Phys.* (*Paris*), *41*, C8-145 (1980).
93. M. Sakata, N. Cowlan, and H. A. Davies, *J. Phys. F*, *11*, L157 (1981).
94. P. Lamparter and S. Steeb, in *Rapidly Quenched Metals*, Vol. 1, edited by S. Steeb and H. Warlimont (North-Holland, Amsterdam, 1985), p. 459.
95. T. Fukunaga, N. Watanabe, and K. Suzuki, *J. Non-Cryst. Solids*, *61/62*, 349 (1984).
96. N. Morito and T. Egami, *IEEE Trans. Mag.*, *MAG-19*, 1901 (1983).
97. T. Jagielinski and T. Egami, *J. Appl. Phys.*, *57*, 3578 (1985).
98. H. H. Liebermann, C. D. Graham, Jr., and P. J. Flanders, *IEEE Trans. Mag.*, *MAG-13*, 1541 (1977).
99. A. L. Greer and J. A. Leake, *J. Non-Cryst. Solids*, *38/39*, 379 (1980).
100. M. R. J. Gibbs, J. E. Evetts, and J. A. Leake, *J. Mater. Sci.*, *18*, 278 (1983).
101. W. Primak, *Phys. Rev.*, *100*, 1677 (1955).
102. T. Jagielinski and T. Egami, *IEEE Trans. Mag.*, *MAG-21*, 2002 (1985).
103. B. S. Berry and W. C. Pritchet, *Phys. Rev. Lett.*, *34*, 1022 (1975).
104. S. Chikazumi and C. D. Graham, Jr., in *Magnetism and Metallurgy*, Vol. 2, edited by A. E. Berkowitz and E. E. Kneller (Academic Press, New York, 1969), p. 577.
105. T. Egami, in *Amorphous Metallic Alloys*, edited by F. E. Luborsky (Butterworths, London, 1983), p. 100.
106. G. S. Cargill III and T. Mizoguchi, *J. Appl. Phys.*, *49*, 1753 (1978).
107. W. Chambron and A. Chamberod, *Solid State Commun.*, *33*, 157 (1980).
108. D. R. Nelson and B. I. Halperin, *Phys. Rev.*, *B19*, 2457 (1979).
109. R. Pindak, D. E. Moncton, S. C. Davey, and J. W. Goodby, *Phys. Rev. Lett.*, *46*, 1135 (1981).
110. P. J. Steinhardt, D. R. Nelson, and M. Ronchetti, *Phys. Rev. Lett.*, *47*, 1297 (1981).
111. P. J. Flanders, N. Morito, and T. Egami, *IEEE Trans. Mag.*, *MAG-19*, 1907 (1983).
112. A. Q. Tool, *J. Am. Ceram. Soc.*, *29*, 240 (1946).
113. S.-P. Chen, T. Egami, and V. Vitek, *Phys. Rev.*, *B37*, 2440 (1988).
114. T. Egami, *Rep. Prog. Phys.*, *47*, 1601 (1984).
115. T. Egami, K. Maeda, D. Srolovitz, and V. Vitek, *J. Phys.* (*Paris*), *41*, C8-272 (1980).
116. T. Egami, V. Vitek, and D. Srolovitz, in *Rapidly Quenched Metals IV*, Vol. 1, edited by K. Suzuki and T. Masumoto (Japan Institute of Metals, Sendai, 1981), p. 517.
117. T. Egami and Y. Waseda, *J. Non-Cryst. Solids*, *64*, 113 (1984).

10

Structural Relaxation and Atomic Transport in Amorphous Alloys

A. L. Greer
University of Cambridge, Cambridge, England

I. INTRODUCTION

The possibility of a range of infinitesimally different structural states does not exist in single-phase crystalline materials, which at a given composition exhibit at most a few distinct structures. On the other hand, amorphous alloys, like amorphous materials in general, can have a variability in their structure. The smooth change from one structure to another in a single amorphous phase is generally termed *structural relaxation*. The relaxation must involve transport of the atomic species comprising the amorphous structure, and the atomic transport is in turn affected by the degree of structural relaxation. The closely interrelated topics of structural relaxation and atomic transport are covered in this chapter.

Structural relaxation is important because in principle any property of the amorphous phase may be affected. One of the most basic properties of a structure is its density. Structural relaxation of an amorphous alloy causes density changes of typically less than ~0.5%. Yet the property of viscosity, which is very closely related to density, can in the same relaxation change by five orders of magnitude. Properties of direct technological relevance, such as fracture strain and magnetostriction, are affected also, and control of structural relaxation may be important not only in optimizing the property for a particular application but also in maintaining the property through the life of the component. Knowledge of atomic transport processes is important in understanding structural stability both to relaxation and to crystallization (the reversion of amorphous alloys to the crystalline state being always thermodynamically preferred). Furthermore, some properties of amorphous alloys, for example their creep behavior and their ability to act as diffusion barriers, are directly related to atomic transport.

II. THE GLASSY STATE AND STRUCTURAL RELAXATION

The possibility of a range of amorphous structures can be appreciated by considering how the amorphous phase is formed. One method of production, and the one of most relevance in this book, is formation on cooling the liquid phase. If the liquid is cooled sufficiently rapidly to avoid crystallization, a *glass* is formed in which the liquidlike structure is preserved with a solidlike viscosity. It is well verified that the formation of amorphous alloys on rapid quenching of the liquid exhibits all the major characteristics of the conventional glass transition known in oxide glasses and polymers. Figure 1 is a schematic illustration of the densities of the liquid, crystalline and glassy phases in a hypothetical material in which compositional partitioning does not occur. Although in different form, the figure could be plotted using almost any material property (including fundamental thermodynamic quantities such as enthalpy) to illustrate the same points. It is clear from the figure that the temperature dependence of the liquid density is much greater than for the crystal. Thus as the liquid is undercooled below the equilibrium freezing point, its density approaches that of the crystal. Kauzmann [1] first pointed out the potential paradox in which the liquid at reasonable temperature could attain the density (and entropy) characteristic of the crystal. The paradox is avoided in practice by the intervention of the glass transition. The strong temperature dependence of the liquid density arises because its thermal expansion is due not only to the anharmonicity of atomic vibrations (as in the crystal) but also to structural changes. The changes in liquid structure as it is cooled require atomic mobility, yet that mobility decreases markedly as the liquid densifies. Eventually a temperature is reached on cooling at which the liquid can no longer change structure rapidly enough to stay in internal equilibrium; at this *glass transition temperature*, T_g, a glass is formed, the structure is frozen, and the thermal expansion is *isoconfigurational* (exhibiting only the anharmonic contribution) and very similar to that of the crystal. The point

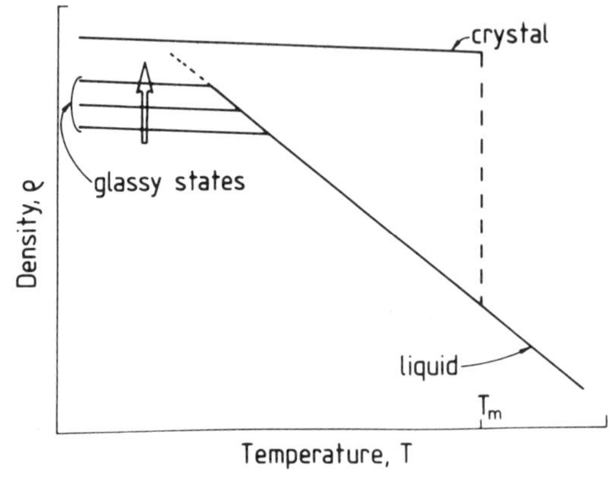

Figure 1 Schematic representation of the variation of density with temperature in a system undergoing polymorphic phase changes (i.e., without change in composition). Below the equilibrium melting point, T_m, the liquid is metastable and can lead to different glassy states, depending on the cooling rate. The arrow shows the effect of relaxation, in which a glassy state tends toward one characteristic of slower cooling.

at which the glass is formed must depend on the rate of cooling. If the liquid is cooled more slowly, the structural changes of the liquid can be followed to a lower temperature and a denser glass is formed. A range of glassy structures is illustrated in Fig. 1 by a series of *isoconfigurational* lines.

When structural relaxation occurs well within the glassy state (that is, far from the liquid line on Fig. 1), either at room temperature or on annealing, the structure evolves (open arrow) to one of higher density, which could be considered characteristic of formation at a slower cooling rate. The property changes at higher temperatures in the vicinity of the glass transition can be investigated in amorphous alloys, but with difficulty because of the propensity of the glass to crystallize. Figure 2a shows schematically the density variation expected around the glass transition. As the temperature and consequently atomic mobility increase, the density deviates from the isoconfigurational line, joining the liquid line after some densification has occurred (curve a). This relaxation before reaching the liquid line is characteristic of the situation, well exemplified by amorphous alloys, in which the heating rate is much less than the cooling rate originally used to form the glass. On subsequent cooling (at a rate presumed equal to that used for heating), a denser glass is formed (curve b), which on reheating and cooling cycles would show some hysteresis, following curves b and c in Fig. 2a. Reversible changes in glass properties are straightforwardly demonstrated by isothermal annealing treatments at two temperatures. On rapid heating from temperature T_1 to T_2 the glass property may approximately follow the isoconfigurational line, as shown for density in Fig. 2b.

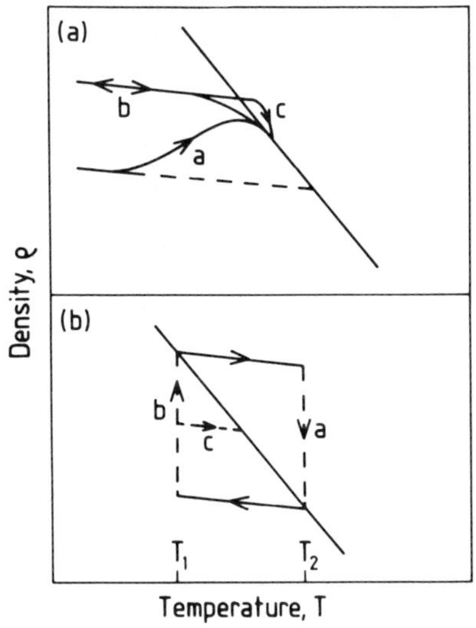

Figure 2 (a) Schematic illustration of the density changes expected around the glass transition on heating and cooling at a rate considerably slower than the cooling rate used to form the glass. Curve a, initial heating; curves b and c, subsequent cooling and heating cycles. (b) Schematic illustration of density changes on cycling between two internal equilibrium states of an undercooled liquid. Arrows a and b, isothermal relaxation after rapid, isoconfigurational heating and cooling; arrow c, rapid isoconfigurational heating to a state suitable for demonstrating the memory effect (Fig. 3).

On holding at T_2 the density evolves (arrow a) toward the value characteristic of equilibrium at T_2. After equilibrium has been attained the temperature can be changed rapidly to T_1. The approximately isoconfigurational behavior during cooling is followed by the isothermal evolution at T_1 (arrow b) of the density back to the value characteristic of T_1.

The concept that any glassy structure is characteristic of a given cooling rate suggests that a single parameter could be used to describe the range of structure. Thus, a glass structure and any of its properties would be a function of three variables: temperature and pressure (as for the crystal) and a parameter characterizing the cooling. The third parameter is normally taken to be the *fictive temperature*, T_f, defined by Tool [2] as the temperature at which the glass formed. Measurement of a glass property at any temperature, followed by extrapolation of the property along the appropriate isoconfigurational line to its point of intersection with the liquid line on a diagram such as Fig. 1 will give T_f. The effectiveness of T_f in specifying a glassy structure can be tested by comparing glasses with the same value of a property but with different thermal histories. For example, the isothermal evolution at temperature T_1 indicated by arrow b in Fig. 2b could interrupted. If the material is heated to the fictive temperature corresponding to the density value at the point of interruption (arrow c) then on holding at T_f the density should remain unchanged. This is found not to be the case. The density (or in modified form, any other property) changes as shown in Fig. 3, first reverting to an apparently less relaxed state before returning to the equilibrium value. This *memory effect* shows that the thermal history of the glass is important and that glass samples with the same property value can have different structures and consequently different behavior on annealing. The memory effect, first found in oxide and polymeric glasses [3], has been demonstrated for a number of properties of amorphous alloys

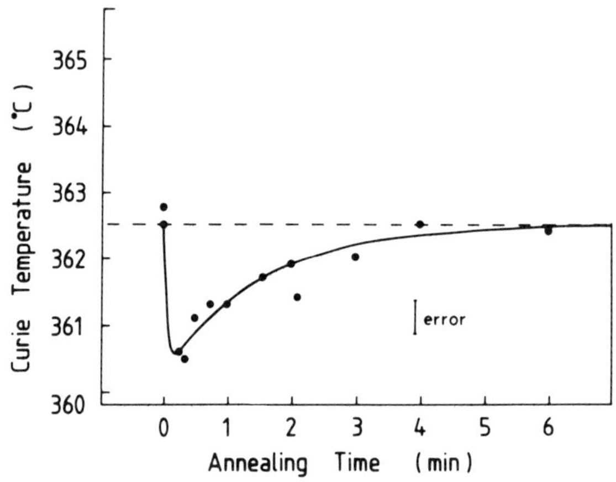

Figure 3 The change of Curie temperature, T_C, on annealing the metallic glass $Fe_{80}B_{20}$ at 400° C after a pre-anneal at 300°C. The preanneal of the as-quenched glass was for a duration sufficient to raise its initial T_C to the value that would be attained in equilibrium at 400°C. That T_C does not remain constant at that value demonstrates a memory effect. (From Ref. 42.)

also. The fictive temperature is not an adequate description of the glassy state away from equilibrium.

The failure of the concept of the fictive temperature means that the different possible glassy states cannot be fully represented by points on a diagram such as Fig. 1. Behavior not represented by lines in Figs. 1 and 2 can be observed. For example, some properties exhibit reversible changes, apparently similar to those represented in Fig. 2b, between states which must lie far from the equilibrium liquid line. The types of relaxation behavior found in amorphous alloys are summarized in Section IIIA.

The variability of amorphous structure for a given material has been considered in relation to glass formation. Amorphous alloys, however, can be made by techniques other than formation of a glass from the liquid. Other techniques include thin-film deposition (evaporation, sputtering, electrodeposition, electroless deposition), defect accumulation (electron irradiation, ion irradiation, mechanical working), and solid state reaction. In each case a range of amorphous structures is possible depending on the production conditions, and structural relaxation showing the same phenomena as for true glasses can be observed.

III. PROPERTY CHANGES ON STRUCTURAL RELAXATION

A. Types of Structural Relaxation Induced by Annealing

Property changes on annealing amorphous alloys can arise for a number of reasons. The alloy may crystallize, partially or completely, or may undergo separation into two amorphous phases. Through viscous flow there may be relief of internal stresses. Annealing may also cause superficial changes such as oxidation which may affect property measurements. Before interpreting a property change as arising from structural relaxation it is important to be sure that these other effects are not contributing. If the change can be induced by annealing a homogeneous, stress-free, single amorphous phase, then it can be attributed to relaxation.

The structural relaxation behavior found in amorphous alloys can be classified as follows:

1. Monotonic relaxation
2. Reversible relaxation near T_g
3. Reversible relaxation below T_g
4. Relaxation of relaxation kinetics
5. Memory effects
6. Effects of production conditions

1. Monotonic Relaxation

This is observed for almost any property of an amorphous alloy when the as-produced structure evolves toward equilibrium on annealing (open arrow in Fig. 1). The monotonic change has mostly been described as irreversible. While this is true for heat treatments confined to temperatures below T_g, reversibility should be possible (though difficult in practice) by heating above T_g. Table 1 indicates the properties for which monotonic changes have been recorded and the directions of change. For some properties such as density (a decrease) and Young's modulus (an increase), the direction of change on annealing is always the same. For others, for example, ferromagnetic Curie temperature, both types of variation have been observed. The property changes typically occur far from equilibrium and are larger and occur faster at higher annealing temperatures. As is expected for relaxation, the changes are initially rapid and then slow down. For many properties the variation is roughly linear with the logarithm of the anneal time, as shown, for example, for Young's modulus in Fig. 4. This log-time behavior (the relationship with relaxation mechanisms is discussed in Section VIIB) is not obeyed at the start of relaxation, or if the structure approaches equilibrium and consequent stabilization of the property value. It should be noted, however, that the rate of a property change may decrease so much on annealing that structures still far from equilibrium may appear to have stabilized.

2. *Reversible Relaxation Near* T_g

As illustrated in Fig. 2b, the states attained on annealing near T_g for reasonable times may be in true internal equilibrium, and it is possible to move reversibly between them. Because of rapid crystallization near T_g, however, the reversibility can be demonstrated only by careful experiments on alloys exceptionally resistant to crystallization. Figure 5 shows the example of viscosity attaining equilibrium, starting from conditions of higher and lower viscosity (i.e., lower and higher values of fictive temperature).

3. *Reversible Relaxation Below* T_g

Egami [4] first reported this result for Curie temperature, T_C, and it has since been found for a number of properties at temperatures down to $\sim 0.75T_g$. Figure 6, from Egami's work, shows the result of alternate anneals at 250°C and 300°C on the T_C of $Ni_{53}Fe_{27}P_{14}B_6$ glass. Clearly it is possible to move reversibly between two T_C

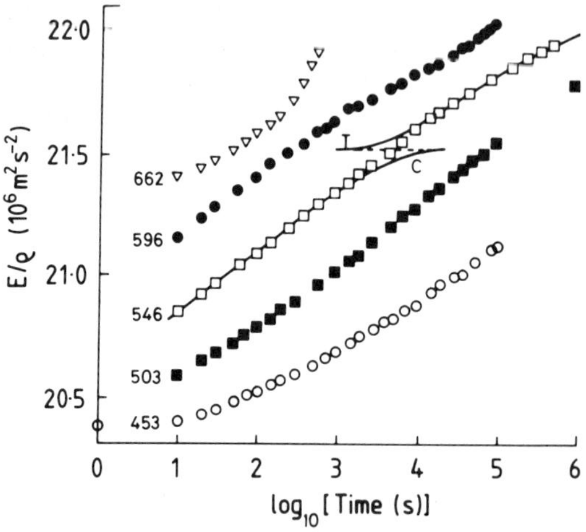

Figure 4 The isothermal evolution of E/ρ (E Young's modulus, ρ density) for as-quenched $Fe_{40}Ni_{40}B_{20}$ metallic glass annealed at the temperatures shown (K). For the anneal at 596 K calculated contributions of compositional short-range ordering (C) and topological short-range ordering (T) are shown. (Adapted from Ref. 35.)

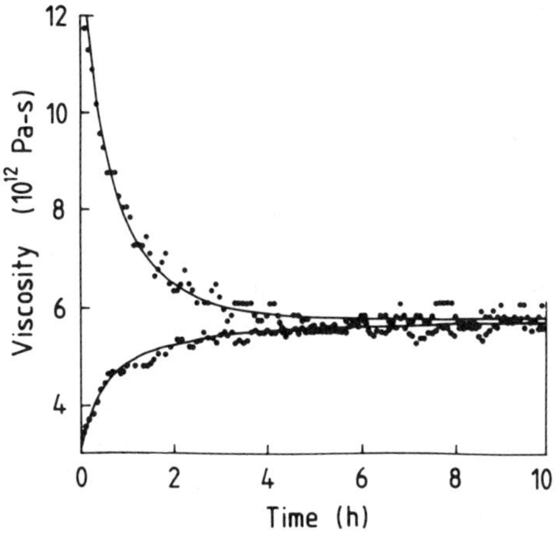

Figure 5 The viscosity of glassy $Pd_{40}Ni_{40}P_{19}Si_1$ as a function of annealing time at 300°C for an as-quenched sample (solid circles) and a sample previously annealed to equilibrium at 280°C. The lines show fits to the data assuming unimolecular kinetics for the viscosity decrease and bimolecular kinetics for the viscosity increase. (Adapted from Ref. 87.)

values, and because crystallization was not a problem for these anneals well below T_g, repeated reversals were possible. The states attained in the anneals cannot be internal equilibrium (Fig. 2b cannot apply), and Chen [5] has shown that the pseudoequilibrium values of T_C attained on annealing below T_g are not necessarily repeatable but can drift upward due to continuing relaxation of the "monotonic" type. Being far from equilibrium, the reversible relaxation does seem distinct from that occurring near T_g. Nonetheless, the reversibility suggests the existence of constrained pseudoequilibrium states for some process, which Egami suggested was

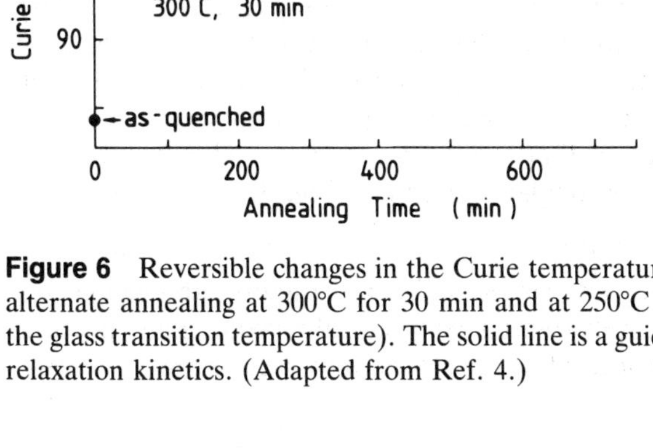

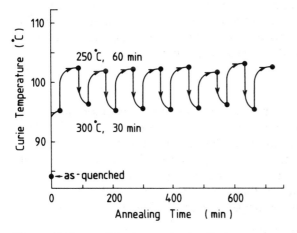

Figure 6 Reversible changes in the Curie temperature of glassy $Fe_{27}Ni_{53}P_{14}B_6$ induced by alternate annealing at 300°C for 30 min and at 250°C for 60 min (in each case well below the glass transition temperature). The solid line is a guide for the eye, not showing measured relaxation kinetics. (Adapted from Ref. 4.)

compositional short-range ordering. He distinguished two types of relaxation behavior: compositional short-range ordering (CSRO) and topological short-range ordering (TSRO). TSRO involves a rearrangement of atomic sites in the amorphous structure and is most naturally associated with relaxation near T_g, requiring considerable atomic mobility and affecting such properties as density. CSRO involves rearrangement of chemically distinct species on fixed sites, and it is assumed to be possible at lower temperatures. Further consideration of relaxation mechanisms is deferred until Section VII but it is useful to note here that there may be a distinction between relaxation processes at high and low temperature, and to adopt TSRO and CSRO as convenient labels.

The reversible relaxation near T_g is considered to involve TSRO. The monotonic relaxation may involve both TSRO and CSRO, with relative contributions depending on the property considered, but is likely to be dominated by TSRO. Without a demonstration of reversibility below T_g, it is difficult to separate the contributions of CSRO and TSRO to the monotonic relaxation, especially as both types of ordering usually lead to the same direction of property change. This is true for example of Young's modulus [6] (Fig. 4), but for electrical resistivity the two contributions can be of opposite sign [7], and in such a case the "irreversible" relaxation might not be monotonic, but exhibit an extremum.

4. *Relaxation of Relaxation Kinetics*

If the various relaxation processes in an amorphous alloy are not independent of each other, not only may the pseudoequilibrium values in an experiment such as that illustrated in Fig. 6 vary as a result of another relaxation process, but the kinetics of the relaxation between the values may vary. It is expected that the kinetics of reversible CSRO would be slower the greater the degree of TSRO, and there is evidence for this in measurements of a number of properties, particularly Young's modulus [8]. The effect is disputed by other authors [9]. The atomic rearrangements involved in developing anisotropy in an amorphous structure may be similar to those involved in CSRO. The kinetics of the development of anisotropy (e.g., anelastic strain on application of a tensile stress [10], internal friction [11], or magnetic anisotropy on application of a magnetic field [12]) may be slower in materials with a higher degree of TSRO. This is further evidence for the effect of relaxation on the kinetics of relaxation. Anisotropy in amorphous alloys is considered more in Section IIIB.

5. *Memory Effects*

Figure 3 shows the first demonstration of a memory effect in an amorphous alloy, and such effects have since been found for a number of properties besides Curie temperature. When as a result of differing thermal histories two samples of an amorphous material with the same value of a property and subject to the same anneal show different relaxation behavior for the property, a memory effect has been demonstrated. It is not required that the initial property value be the same as the final value, as shown in the particular example in Fig. 3.

6. *Effects of Production Conditions*

Different production conditions may give rise to different structures and properties. The origins of the differences are closely related to structural relaxation and memory effects. For melt-quenched amorphous alloys the most important production con-

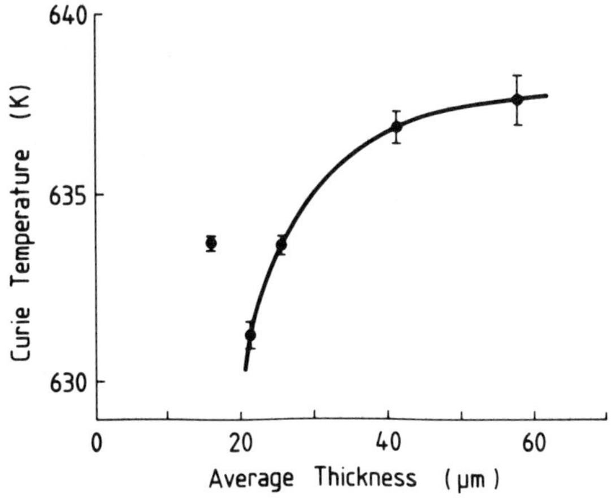

Figure 7 The Curie temperatures, T_C, of as-quenched glassy $Fe_{81.5}B_{14.5}Si_4$ produced in different ribbon thickness. Thinner ribbons are cooled faster, and generally have lower T_C, indicating a less relaxed structure. (From Ref. 127.)

dition is the cooling rate which can be varied in a controlled way by varying the wheel speed and consequently the thickness of the melt-spun ribbon. An example of the effect of this is shown in Fig. 7 for T_C, which shows, as expected from Fig. 1, that faster quenched (that is, thinner) ribbons have properties characteristic of less relaxed states. This range in property tends to shrink on subsequent annealing as the various as-produced states converge on the same final equilibrium state. For melt-spun samples the differences in as-produced states as a result of different wheel speeds are not as great as might be supposed. This is because the ribbons are annealed to some extent during the relatively slow air cooling which typically follows the rapid quench on the wheel surface. The ribbons may still be quite hot when they leave the wheel, particularly when they are thick.

B. Other Types of Induced Structural Relaxation

1. Isotropic Property Changes

So far consideration has been given only to structural relaxation induced by thermal treatments, but there are other possibilities. The effects on amorphous alloys of irradiation by electrons or light ions [13], or by neutrons have been studied. Most work has been on the effects of neutron irradiation [14,15]. This can cause density decreases of ~0.1%, and can reverse the density increase of similar magnitude in samples annealed at low temperature [14,15]. The embrittlement accompanying density on annealing (Section IIIC8) is also reversed by the irradiation [14]. Mechanical deformation of amorphous alloys can cause density decreases [16] similar to those caused by irradiation and also can reverse embrittlement [17].

2. Anisotropy in Amorphous Alloys

Although amorphous structures are normally considered to be isotropic, they can be anisotropic. As-produced amorphous alloys do show some anisotropic properties which tend to become isotropic on annealing. On the other hand, anisotropy can be induced in isotropic alloys by annealing in a magnetic field or a tensile stress. Either process may be useful in optimizing properties for applications. In either case, the changes in the alloy are another form of structural relaxation. Control of anisotropy is especially important for soft magnetic properties, affecting permeability, after-effect and losses [18]. The magnetic anisotropy in as-produced amorphous alloys may arise through magnetostrictive effects from quenched-in stresses, as well as from actual structural anisotropy. Anisotropy induced by annealing in a magnetic field is reversible, and in altering its orientation memory effects are observed [12,19]. It can be induced also by annealing under a tensile stress and is related to the frozen-in anelastic strain [20].

There is direct evidence for frozen-in anelastic strain in amorphous alloys produced by melt spinning. Sinning et al. [21] found on annealing a melt-spun ribbon that the length contraction in the casting direction was up to three times that in the transverse direction in the plane of the ribbon. It has been suggested that the anelastic strain causing this anisotropy could arise from the tensile stress in the casting direction during production [22]. Anisotropy in melt-spun ribbons is also found for tensile strength, which is greater in the casting direction. This anisotropy can be reduced by annealing without stress, or enhanced by annealing with tension in the casting direction [23]. Also it was found that annealing in a magnetic field parallel to the casting direction enhances the anisotropy, while a transverse field reduces it.

Clearly there is a close link between the magnetic and mechanical anisotropies in amorphous alloys. Anisotropy arises from some type of SRO, and the kinetics of changes in anisotropy are likely to be similar to those of property changes induced by low-temperature, purely thermal anneals [24] and attributed to CSRO. In changes of anisotropy, pair ordering of metal atoms may be important [25] (as for crystals [26]), but a mechanism based on carbon diffusion has been reported [27].

C. Properties Affected by Structural Relaxation

The types of relaxation observed for the properties affected in amorphous alloys are summarized in Table 1. Only property changes due to thermal annealing are included, and changes induced by other means are not considered. In this section some of the property changes are described. The selected properties are important either for illustrating the types of relaxation, or for elucidating relaxation mechanisms.

1. Density and Length Changes

As illustrated by the discussion in Section II, density is an important parameter in considering amorphous structures and their relaxation. Unfortunately it is difficult to measure (e.g., by the Archimedes method) with a precision better than 0.05% [28], and typical density increases on monotonic relaxation from the as-produced state are at most ~0.5% and often rather less. Density is therefore not a suitable property for the measurement of relaxation kinetics. Although density changes would be expected to relate mainly to TSRO and not to CSRO, reversible changes in density have been found on annealing below T_g [29]. For comparison, it should be noted that typical density increases on crystallizing an amorphous alloy are 1 to 3% [28].

Table 1

Property affected by relaxation	Type of relaxation i	ii	iii	iv	v	vi	Selected references
Density	I		√				15, 29
Length changes	D		√				30
Thermal expansion coefficient	D						31
Enthalpy	D		√	√		√	32, 33
Young's modulus	I		√	√	√	√	117, 118
Hardness/yield stress	I					√	37, 38
Viscosity/relaxation time for stress relief	I	√			√	√	87
Diffusivity	D					√	64
Curie temperature	I, D	√	√	√	√	√	
Electrical resistivity	I, D		√	√			46, 47
Temperature coefficient of resistivity	D, I					√	46, 47, 119
Internal friction	D						
Strain at fracture	D		√	√		√	49, 33
Saturation magnetization	D					√	120
Magnetic after-effect	D		√		√	√	121
Induced magnetic anisotropy			√	√	√		121
Structure factor (x-ray and neutron)	√					√	57
EXAFS	√						59
Positron lifetime	D						6
Superconducting transition temperature	D		√				
Mössbauer spectra	√	√				√	60, 122
Corrosion resistance	I						123
Density of states at Fermi level	I, D					√	124
Debye temperature	I					√	124
Magnetostriction coefficient	I					√	125, 126

The types of relaxation are (i) monotonic relaxation; (ii) reversible relaxation near T_g; (iii) reversible relaxation below T_g; (iv) relaxation of relaxation kinetics; (v) memory effects; and (vi) effects of production conditions. I, D indicate an increase or decrease of the property value in monotonic relaxation.

Because of the difficulty with direct density measurements, there has been much work on the more precise measurement of length changes. These have enabled detailed kinetic analysis [30] which confirms the major influence of TSRO on density and the existence of a small reversible component attributable to CSRO of about 0.4 ppm K^{-1}. However, because of the anisotropy discussed in Section IIIB, the derivation of density changes from length changes is not straightforward, though the anisotropy does decrease markedly on annealing. The effects of creep under the stress used to keep the sample straight, and of the relaxation of anelastic strain, must be considered. For similar reasons, and because relaxation may occur during the measurement, apparent changes in thermal expansion coefficient must be treated with caution [30,31].

2. *Enthalpy and Specific Heat*

Enthalpy is a fundamental thermodynamic parameter, and changes in it are readily calculated from integration of effective specific heat measured by differential scanning calorimetry. The data are sufficiently precise to permit detailed kinetic analysis [32]. The total enthalpy reduction on monotonic relaxation is about 1 kJ mol^{-1} [33], which can be compared with decreases of 3 to 8 kJ mol^{-1} on crystallization [34]. Reversible changes of enthalpy of about 50 J mol^{-1} have been found for relaxation below T_g [32]. Because heat is evolved or absorbed during relaxation, the specific heat measured at any instant in the calorimetric experiments is only an effective value and is not characteristic of the structural state of the amorphous alloy at that instant.

3. *Young's Modulus*

The Young's modulus of amorphous alloys is sensitive to the degree of structural relaxation, increasing by up to ~5% on annealing. Reversible changes of about 1% are observable on annealing below T_g. The modulus can be determined with sufficient precision, for example from measurements of sound velocity using the pulse echo technique (as used in Fig. 4 [35]), to permit kinetic analysis. The rise in Young's modulus due to relaxation may be compared with the rise of approximately 40% found on crystallization [36].

4. *Hardness/Yield Stress*

Because of the lack of slip systems, the yield stress of amorphous alloys tends to be rather high, for example about 3.5 GPa for iron-based alloys. The yield stress can be measured by mechanical testing [37], but accurate experiments on melt-spun ribbon are difficult. More commonly microhardness measurements are used [38], though even with these the precision is not sufficient for kinetic analysis. Hardness increases of up to 14% result from relaxation.

5. *Viscosity/Relaxation Time for Stress Relief*

Atomic transport in amorphous alloys, whether manifested as diffusion or viscous flow, is strongly affected by structural relaxation and is discussed in detail in Sections IV and V. The decrease in atomic mobility associated with relaxation has been particularly well characterized for viscous flow. The relief of stresses (internal or externally applied) in samples occurs through viscous creep and has been much studied in connection with the optimization of magnetic properties [39]. Magnetic properties, because of the strong magnetostrictive effects typical of amorphous alloys, can be adversely affected by quenched-in internal stresses.

6. *Curie Temperature*

Amorphous alloys containing a sufficiently high proportion of magnetic species are ferromagnetic, with a well-defined Curie transition. The Curie temperature is sensitive to the degree of structural relaxation and is readily measured either magnetically [40] or calorimetrically [41]. Changes in T_C have been much studied, particularly as the property first revealed the reversible relaxation below T_g (Section IIIA and Fig. 6) [4]. Measurements of T_C, however performed, require heating through the Curie transition. This heating to determine T_C could itself cause relaxation, and to minimize this most studies, e.g. [40], have been on alloys with rather low T_C. However, by rapid heating it is possible to extract useful data from alloys with high T_C which show some relaxation during the T_C determination [42]. Typically the T_C of as-produced amorphous alloys rises on annealing, by up to ~15% [40], though decreases have also been reported [43]. The reversible changes in T_C on annealing below T_g are about 2%, and reversibility has also been reported near T_g [44].

7. *Electrical Resistivity*

Amorphous alloys have high electrical resistivities, typically 100 to 300 μΩ-cm, associated with temperature coefficients of resistivity which are small in magnitude, and either positive or negative. Resistance changes can be measured with great precision. In calculating the corresponding changes in resistivity, density changes must be taken into account, otherwise errors of up to 20% can arise. The high precision obtainable with resistometric methods has enabled reliable relaxation time spectra to be extracted from the data [45]. Resistivity changes due to structural relaxation on annealing have been reviewed by Kelton and Spaepen [46] and by Balanzat et al. [47]. Monotonic relaxation can give increases or decreases in resistivity of magnitude up to ~1.5%. Alloys showing increases (decreases), show a negative (positive) temperature coefficient of resistivity which becomes more negative (positive) on annealing by ~5%. All of this behavior can be understood qualitatively on the basis of the Ziman model which assumes nearly free electrons [46]. The model predicts that, approximately, the resistivity is proportional to the magnitude of the structure factor (interference function) at the diffraction vector which is twice the Fermi vector. Systematic additions of vanadium to $Pd_{82}Si_{18}$ amorphous alloy change the Fermi level and alter the monotonic resistivity change from negative to positive [46]. The application of the Ziman theory is valid provided the resistivity changes are mainly due to TSRO. Small (~0.2%) reversible changes in resistivity attributable to CSRO have been identified [7,48]. These also can be of either sign. In $Fe_{40}Ni_{38}Mo_4B_{18}$ amorphous alloy the resistivity changes on increased ordering are of opposite sign for TSRO (decrease) and CSRO (increase). It appears also that Matthiessen's rule does not apply to the resistivity contributions from CSRO and TSRO.

8. *Strain at Fracture*

Unannealed melt-spun ribbons of amorphous alloys can often be bent double without fracture. One of the most undesirable consequences of annealing amorphous alloys is the embrittlement which in most cases occurs, reducing the strain to fracture to ~1%, or less. The embrittlement accompanies other property changes, and exhibits features of structural relaxation such as reversibility on annealing below T_g [29,49]. However, the embrittlement can show a dependence on small additions of solute which is not typical of other property changes due to relaxation. For example, small (0.2 at%) additions of segregating elements (Te, Se, and Sb) promote embrittlement [50]. On the other hand, embrittlement is hindered by additions (50 at ppm) of cerium which may act to remove oxygen and sulfur in solution by forming precipitates [51].

9. *Structural Measurements*

In elucidating the nature of the structural changes brought about by relaxation and the mechanisms of those changes, direct structural studies should be of prime importance. Unfortunately the structural changes are rather subtle, and difficult to quantify. Nonetheless it has been possible to detect the monotonic changes found on annealing as-produced samples. In an early study of $Fe_{40}Ni_{40}P_{14}B_6$ amorphous alloy, Egami [52] used energy-dispersive x-ray diffraction. On annealing, the radial distribution function (RDF) of the alloy sharpened, with the peaks becoming higher and the troughs deeper. The peak positions, however, remained essentially unaltered. The first and second peaks increased in height by ~2%, and subsequent peaks by up to 10%. Particularly significant was the increased splitting of the second peak, with the shoulder on the larger distance side increasing by up to 10%, a change which is opposite to that which would be caused by any incipient crystallization. Waseda and Egami [53] confirmed these results on other metal-metalloid alloys. For amorphous alloys without metalloids (e.g., Ti-Ni-Cu) Jergel and Mrafko [54] found similar effects, but of larger magnitude, with the shoulder on the second peak of the RDF increasing by up to ~40%. These results, together with the densification which can also be detected by x-ray diffraction [55], confirm that the relaxation can be considered to be TSRO. The basic structural units of the amorphous structure are little affected. The nearest neighbor distance is not affected, but the increase in density is associated with a small increase in average coordination number. The increase in the coordination number is confirmed by considering changes in ferromagnetic Curie temperature, T_C. In alloys showing an increase in T_C on annealing, compression causes a decrease [56]. This shows that the densification on compression, approximately a uniform shortening of bond lengths, is quite different from that due to annealing.

Diffraction studies have not proved sufficiently sensitive to detect reversible changes in structure, and there is little direct evidence for changes in chemical order (CSRO). Jergel and Mrafko [54] concluded from analysis of a pre-peak in their diffraction patterns that there was an increased tendency for unlike nearest neighbors as a result of annealing. In neutron diffraction studies of $Fe_{40}Ni_{40}B_{20}$ [57] the split second peak in the RDF was resolved into contributions from metal-boron spacings and metal-metal spacings. Changes in the peak shape suggested that the degree of CSRO (as quantified by the Cargill-Spaepen parameter [58]) was less in faster-quenched samples.

There is some evidence that EXAFS studies may be more sensitive to structural changes than diffraction studies [59], but so far the only clear evidence for reversible structural changes has come from Mössbauer scattering [60]. Although the nature of the structural changes is not clearly revealed, the changes which occur reversibly appear to be different from the monotonic changes attributable to TSRO.

IV. ATOMIC DIFFUSION

A. Methods of Measurement and Relaxation Effects

Diffusion measurements in amorphous alloys are restricted by the need to remain at temperatures low enough to avoid crystallization. This generally means that measurements must be made below, and possibly well below, the glass transition temperature. The diffusivities measured are consequently rather small, typically 10^{-17} to 10^{-24} $m^2 s^{-1}$ [61]. Composition profiling techniques (reviewed in [61]) must be very sensitive. The diffusing component is added to the specimen surface by coating (typically by electrodeposition or sputtering) or by ion implantation. To measure the composition profile generated by a diffusion anneal the specimen is sectioned by ion milling and analyzed using Auger electron spectroscopy, secondary ion mass spectroscopy, or radiotracer analysis. The profiling may also be nondestructive by using Rutherford backscattering analysis, or nuclear reaction analysis. The profiling techniques used on amorphous alloys have a depth resolution of approximately 3 nm and a minimum measurable diffusivity of approximately 10^{-24} $m^2\ s^{-1}$. Yet greater sensitivity, down to diffusivities of approximately $10^{-27}\ m^2\ s^{-1}$, can be achieved by measuring interdiffusion (homogenization) in compositionally modulated thin films produced by deposition [62]. In the modulated thin films the diffusion distances can be very short (a single layer thickness can be as small as ~0.4 nm) and the decay of the composition modulation can be monitored using x-ray diffraction.

The anneals necessary to make a diffusion measurement using the composition profiling techniques can be sufficiently short to avoid crystallization, but structural relaxation is inevitable. Typically the sample will undergo 200 jumps per atom during the diffusion anneal, leaving it in a condition far from its original state [63]. The detection of the effects of structural relaxation on diffusivity is made still more difficult by the relatively slow cooling to which a melt-spun ribbon is subjected after it has left the wheel. Even in their as-produced state, melt-spun samples may be substantially relaxed (Section IIIA6). Nonetheless, the effects of further relaxation due to annealing have been clearly shown in the most precise profiling measurements made using sputter sectioning and Auger [64] or radiotracer [65] analysis. A rapid initial decrease in diffusivity of about an order of magnitude was found followed by a stabilization. Similar effects have been found using the much more sensitive modulated thin film technique, but since this can yield a value of diffusivity after only about 0.2 jump per atom, the initial values of measured diffusivity are characteristic of less-relaxed structures [62]. A decrease in diffusivity of about three orders of magnitude due to relaxation has been found, and such changes have in addition been correlated with changes in density also determined using x-ray diffraction from the same modulated thin film [66].

Diffusivities may also be estimated by analyzing reaction kinetics in amorphous alloys. For example, the diffusion-controlled growth of crystals in a variety of metal-metalloid glasses has been measured. The extraction of diffusion coefficients may be difficult, and, in general, the results should be treated with caution [67]. Nevertheless, in some cases, for example in an Fe-Ni-B glass, there is excellent agreement with profiling measurements on boron diffusion [68]. In this case also there is good agreement with the boron diffusivity estimated from the kinetics of boriding and deboriding.

Associated with the difficulty of making diffusion measurements in amorphous alloys, most experiments have determined chemical interdiffusivities in significant concentration gradients. Except for hydrogen, tracer diffusivities are not available, though there has been one measurement of isotopic boron interdiffusion in an alloy of approximately uniform chemical composition [69].

B. Mechanisms of Diffusion

The diffusion mechanisms for crystalline metals can be revealed by the form of the correlations between the activation energy for diffusion, Q, and the diffusion prefactor, D_0. Sharma et al. [70] have applied this method to data on amorphous metals and their plot is shown in Fig. 8. This shows that hydrogen behaves quite differently from other diffusing species, and implies that most diffusion (other than

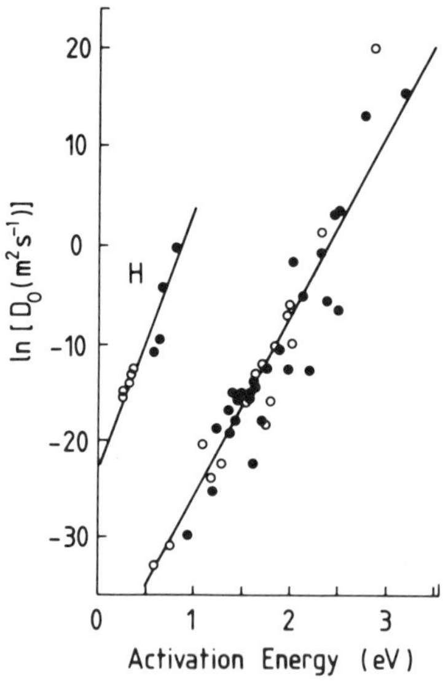

Figure 8 Correlations of the activation energy for atomic diffusion with the corresponding preexponential factor for most of the published data on a variety of species in binary alloys. The open circles are for diffusion in metal-metalloid glasses, the closed circles for metal-metal glasses. When hydrogen is the diffusing species, a separate correlation is found. (Adapted from Ref. 70.)

of hydrogen) in amorphous alloys may have a similar mechanism. Here we are concerned only with the diffusion of species which could be components of amorphous alloys, and specifically exclude consideration of hydrogen diffusion. It should be noted, however, that hydrogen diffusion in amorphous alloys has been much studied: it has been used as a structural probe, and its dependence on concentration has been measured [71]. Also, it has been predicted that with the range of atomic sites in an amorphous alloy, the temperature dependence of the hydrogen diffusivity should be non-Arrhenius [71].

Perhaps the most striking feature of other diffusivities in amorphous alloys is the almost universal Arrhenius temperature dependence [67]. This seems surprising in view of the expected range of atomic environments and therefore of local activation energies. However, the analysis of the data in Fig. 8 suggests that diffusion for species other than hydrogen is very highly correlated [70]. With the highly cooperative nature of the diffusion, each jump may average over several atomic environments to give a narrow range of activation barriers and the observed Arrhenius behavior. The temperature range in which it is possible to make measurements is rather narrow and values of activation energy, and more particularly prefactor, are subject to large errors [67]. The measured activation energies over all systems vary from 145 to 350 kJ mol^{-1} (similar to values for substitutional diffusion in crystalline alloys) and the prefactors from 10^{-12} to 10^{12} m^2 s^{-1}. The latter range is almost certainly not physically reasonable.

Many diffusion measurements have now been made in amorphous alloys, and a large fraction of the data is plotted in Fig. 9. This shows that typical diffusivities

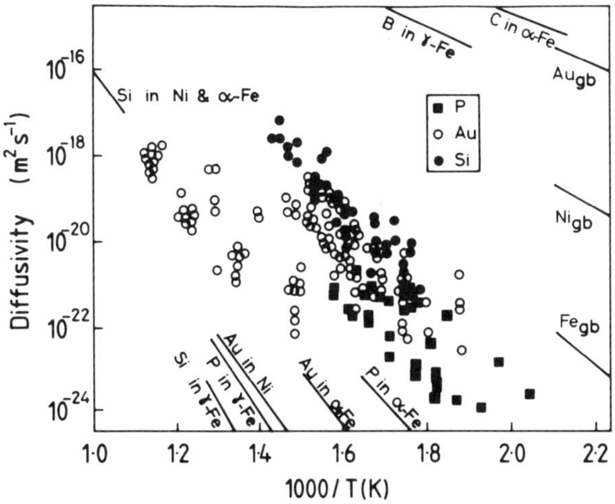

Figure 9 Diffusion coefficients of P, Au, and Si in glassy alloys. For comparison, grain boundary (gb) diffusivities are shown (for Au, Ni, and Fe), together with representative interstitial (C in α-Fe) and substitutional (Au in α-Fe) diffusivities in crystals. (Adapted from Ref. 67.)

in amorphous alloys are smaller than interstitial and grain boundary diffusivities in crystals, but larger than substitutional diffusivities in crystals. Amorphous alloys have consequently been of interest as diffusion barriers in semiconductor devices, because fast grain boundary diffusion is more readily avoided using amorphous thin films than single crystals [72]. Direct comparison of diffusivities in corresponding amorphous and crystalline materials is made difficult by the coresponding crystalline state normally being at least two phase. The materials obtained by crystallizing amorphous alloys can have diffusivities greater or less than the original amorphous alloy [61]. A direct comparison should be possible in some cases where amorphous alloys can be formed at the compositions of intermetallic or metal-metalloid compounds, but diffusivities in such compounds have not been measured.

Clearly there is a wide range of diffusivity in amorphous alloys at a given temperature. In some experiments there is evidence that solute diffusivities may be lowered by chemical interaction (in effect compound formation) with impurities (e.g., oxygen) in the alloy [73], or with an alloy component (e.g., silicon) [74]. In the absence of strong chemical effects, however, there is a well-established correlation of faster diffusion with smaller size of the diffusing species. This is seen in both the main types of amorphous alloy matrix, early-late transition-metal glasses (e.g., Ni-Zr) [75–77] and transition or noble metal-metalloid glasses (e.g., Fe-B) [74], but is clearer in the former type [70], presumably because chemical effects are weaker. Figure 10 shows diffusivities of a variety of solutes in Ni-Zr glasses at 573 K. For a given solute the diffusivity is not significantly a function of alloy composition in the range 40–60 at% Ni, and there is a single correlation of diffusivity with atomic sizes for these glasses. The diffusivity of one solute in different amorphous alloys of a given type shows a single correlation with temperature, which is particularly good if the temperature is normalized with respect to the glass transition temperature of the host alloy. For example, at the same reduced temperature

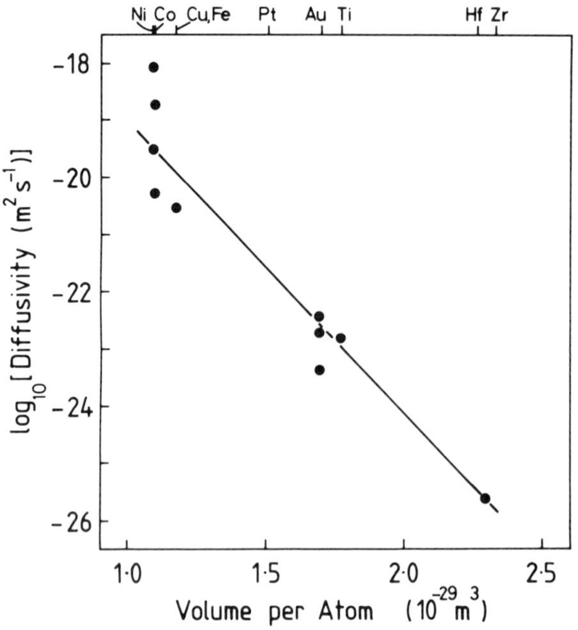

Figure 10 Tracer diffusivities at 573 K of various species in Ni-Zr glasses with >50 at% Ni. The line is a guide for the eye only. (Adapted from Ref. 128.)

(T/T_g) gold diffuses at similar rates in a range of metal-metalloid amorphous alloys, and at similar, but about 100 times greater, rates in a variety of metal-metal alloys [67].

A size dependence of diffusivities is found in crystalline solid solutions, in which small solutes may be interstitial and diffuse 6 to 15 orders of magnitude faster than substitutional solutes. Such large differences are not found in metal-metalloid amorphous alloys for solutes which might on the basis of their size be expected to be interstitial and substitutional, in particular for metalloid and metal solutes. The comparison with crystalline solid solutions may be flawed in any case because such solutions do not ordinarily exist at the compositions of amorphous alloys. More similar to the amorphous alloys are metal-metalloid compounds. Unfortunately, there are almost no data on diffusivities in such compounds, but one value, for carbon diffusion in Fe_3C, shows that metalloid diffusion can be very slow. In the compound, the bonding is quite different from that in a solid solution of carbon in iron, and the diffusivity is not interstitial-like, but rather similar to values typical for metalloids in amorphous alloys [63]. Even if clear interstitial diffusion of metalloids does not exist, the smaller solutes, boron and silicon, do diffuse about one hundred times faster than the metallic species in metal-metalloid alloys [67].

In metal-metal amorphous alloys, particularly at roughly equiatomic composition, Fig. 10 shows that there is a large variation, over about six orders of magnitude, in diffusivity with size. Although this variation is reminiscent of the interstitial/substitutional distinction in crystalline solid solutions, the diffusivity values do not seem to fall into two distinct classes. The behavior can be crudely understood in the free-volume theory of diffusive transport in liquids and amorphous alloys: a larger atom needs a larger, and therefore less likely, local fluctuation in free volume for transport to occur [63]. The variation shown in Fig. 10 is very significant in the formation of roughly equiatomic metal-metal amorphous alloys by the process of solid-state amorphization in which the two elemental metals react in the solid state. The two components of the alloy forming in the reaction lie at the extremes of the behavior shown in the figure. For example, amorphous Ni-Zr can be formed in the solid state [78], and it can be seen that whereas nickel is a fast diffuser in the amorphous alloy, zirconium is slow. This asymmetry in diffusion behavior has been verified by marker experiments in the amorphous alloy [79]. The fast diffusion of nickel allows the amorphous alloy layer formed between the crystalline elements to continue to grow, while the rival reaction of transformation to the crystalline intermetallic compound may be hindered by the slow diffusion of zirconium.

There is a clear need for more diffusion data, particularly on the dependence of diffusion coefficients on the composition of alloys. There has been little work on this, mostly suggesting that in the middle of glass-forming ranges, the composition dependence is not strong. However, as the extremes of the glass-forming composition range are approached, the diffusivity of a given solute can change markedly, usually upward, perhaps reflecting a structural change in the alloy correlated with lower stability. This behavior is exemplified by the diffusion of copper in Ni-Zr alloys [75].

V. VISCOUS FLOW

In analyzing glass formation the liquid viscosity is an important parameter. Conventionally, for very slow cooling, the glass transition is assumed to occur at a viscosity of approximately 10^{12} Pa-s. The critical cooling rate for formation of metallic glasses is much higher than for conventional glasses, and consequently the liquid viscosity at the point where internal equilibrium is lost is closer to 10^{10} Pa-s [80]. Good glass-forming ability of a liquid is found when there is a rapid rise in viscosity on cooling. The viscosity rise is accompanied by the development of ordering or association in the liquid, reflected in an excess specific heat. In general, metallic liquids of whatever composition (elemental, eutectic, or corresponding to a compound) have a viscosity which at the equilibrium melting point fits the Andrade [81] formula. Alloys with relatively good glass-forming ability, however, have melting-point viscosities well in excess of those given by the formula [80].

Viscosity is of interest not only in the liquid for analyzing glass formation, but also in the glass itself. It is a very sensitive indicator of the degree of structural relaxation and provides valuable further insight into atomic transport mechanisms. Because of the difficulty of avoiding crystallization, most viscosity measurements in amorphous alloys have been at temperatures well below the glass transition temperature. The measurements are then best made by monitoring creep under a tensile stress. The viscosity, η, is related to the plastic strain rate, $\dot{\varepsilon}$, and the applied stress, σ, by

$$\eta = \frac{\sigma}{3\dot{\varepsilon}} \tag{1}$$

The extraction of the plastic strain from the sample elongation is not straightforward as there are also contributions from the instantaneous elastic strain on application of the load, from a time-dependent but reversible (i.e., anelastic) strain [82], and from changes in density [83] and in Young's modulus [84] due to structural relaxation. The applied stress may have a significant contribution from the surface tension of the sample (usually in thin ribbon form) [85]. Notwithstanding these problems, viscosities can be obtained with great accuracy, not only in equilibrium but also in metastable unrelaxed states because in a typical measurement only one atom in 10^4 need jump [63]. The main results on viscous flow in amorphous alloys are summarized below.

For moderate stresses, the viscosity estimated from Eq. (1) is independent of the applied stress, i.e., the flow is Newtonian. By analogy with the flow of crystals, this has been modeled by assuming that the flow occurs by local shearing at defects. The shearing is taken to be thermally activated and biased by the application of a stress. Newtonian behavior is predicted, at low stresses according to

$$\eta = \frac{kT}{(\gamma_0 v_0)^2 k_0 n_f} \tag{2}$$

where k is Boltzmann's constant, T the temperature, γ_0 the local shear strain resulting from the shearing at a defect, v_0 the defect volume, k_0 the intrinsic jump frequency at a defect, and n_f the number of defects per unit volume [86].

Some amorphous alloys, notably $Pd_{40}Ni_{40}P_{20}$ and its variants, are remarkably resistant to crystallization. In such alloys it has been possible to measure η at temperatures close to T_g, corresponding to structural states which are in internal equilibrium. These states all have higher viscosity than the as-quenched glass. Strictly, once the equilibrium has been established, the alloy is no longer a glass, but a liquid (albeit with very high viscosity). At a given temperature, the viscosity stabilizes to a value which is independent of prior thermal history. If a sample is first annealed up to, or close to, equilibrium and then annealed at a higher temperature, the viscosity at the higher temperature decreases toward the new equilibrium value [87]. Such a reversibility, illustrated in Fig. 5, is important in confirming that internal equilibrium has been attained. It is found that the temperature dependence of the equilibrium viscosity has, to a good approximation within the limited temperature range of measurement, the Fulcher-Vogel form [86]

$$\eta = A \exp\left\{\frac{B}{T - T_0}\right\} \tag{3}$$

An expression of the form of Eq. (3) is predicted by both the free-volume and configurational entropy models of liquid viscosity. The free-volume model [88–90] has, almost without exception, been the model of choice in analyzing viscosity measurements in amorphous alloys. In the model, the concentration of sites at which atomic rearrangement can occur, i.e., the concentration of flow defects, is given by

$$n_f = \frac{1}{\Omega} \exp\left\{-\frac{\gamma' v^*}{v_f}\right\} \tag{4}$$

where Ω is the atomic volume, γ' a geometrical overlap factor (approximately 1/2), v^* the critical fluctuation volume for rearrangement, and v_f the average free volume per atom. The temperature dependence of the equilibrium viscosity is then dominated by the temperature dependence of the concentration of flow defects, n_f. It is then natural to associate changes in viscosity on structural relaxation with changes in the nonequilibrium population of flow defects. The feature of the free-volume model that makes it most attractive for analyzing structural relaxation is that it provides a link between density and viscosity. Strong quantitative justification for the model is provided by its correct prediction that density changes on relaxation are small while the associated viscosity changes are large.

On annealing as-quenched amorphous alloys, and alloys which although partially relaxed are still far from equilibrium, it is well established that the viscosity rises linearly with time [91]. This behavior is consistent [86] with a population of defects, n_f, being far in excess of the equilibrium population, $n_{f,\mathrm{eq}}$, for the temperature concerned, and decreasing with time according to a bimolecular law:

$$\frac{d(n_f - n_{f,\mathrm{eq}})}{dt} = -\kappa(n_f - n_{f,\mathrm{eq}})^2 \tag{5}$$

Just this behavior is found for the relaxation of as-quenched oxide glasses, and the bimolecular kinetics in that case arises from the recombination of pairs of broken bonds [86]. The relaxation behavior in the amorphous alloys is independent of the applied stress [86]. The linear rise in viscosity arises from the negligible value of $n_{f,\mathrm{eq}}$ in Eq. (5) in comparison with n_f. As equilibrium is approached, however, $n_{f,\mathrm{eq}}$, is no longer negligible and the rise in viscosity slows down as shown in Fig. 5. The form of the viscosity rise is still well described by Eq. (5). Decreases in viscosity in oxide glasses are governed by unimolecular kinetics, arising from the breaking of bonds. In amorphous alloys the kinetic law governing viscosity decreases (for example as in Fig. 5) is not so well established, but unimolecular behavior is consistent with the data [86,87].

The strong dependence of viscosity on degree of relaxation makes difficult the determination of an activation energy. If measurements are made at a series of temperatures it is likely that at higher temperatures the sample or samples will be more relaxed. The comparison of viscosity values is not valid, and the determined activation energy is erroneously low. However, since the viscosity rises linearly (by up to five orders of magnitude) on isothermal annealing, its relative rate of change, and therefore that of n_f, decreases noticeably with time. After some annealing a series of viscosity measurements can be conducted in which n_f barely alters, i.e., the measurements are isoconfigurational [91]. If the isoconfigurational measurements are at different temperatures a true activation energy of viscous flow can be determined. The isoconfigurational nature of the experiments can be checked by returning to the temperature of the initial measurement of the series and verifying that the viscosity has not changed after the intervening anneals. Such isoconfigurational measurements have now been made on several metal-metalloid amorphous alloys (for diffusivity [63] as well as viscosity). In states relatively far from equilibrium they show an Arrhenius temperature dependence, with activation energies in the range 190–250 kJ mol^{-1}. The activation energies are independent of the degree of relaxation. This corresponds to the behavior indicated by the parallel series of isoconfigurational lines in Fig. 1.

A simple interpretation of isoconfigurational flow behavior in the free-volume model is that along an isoconfigurational line the population of defects is constant and that the activation energy is that for the defect jump frequency k_0. However, the measured activation energies are too high to be fully consistent with the free-volume model. The temperature dependence of the equilibrium viscosity would be significantly affected by the variation of k_0, and not dominated by the variation in n_f as implied by Eq. (3). The equation must be modified by the inclusion of an Arrhenius term, as has been done for other amorphous materials [92]:

$$\eta = \eta_0 \exp\left\{\frac{Q}{kT}\right\}\exp\left\{\frac{B}{T - T_0}\right\} \tag{6}$$

Unfortunately the data available are so limited that Eqs. (3) and (6) provide equally good fits. There remain problems with discrepancies in the value of η_0.

An alternative view is that the temperature dependence of k_0 is very weak and that the isoconfigurational activation energy includes a major contribution from reversible structural changes, due for example to the redistribution of free volume arising from thermal expansion [93]. Additional effects of this type may have to be invoked to explain the isoconfigurational behavior found close to equilibrium. In contrast to the previously cited results, the isoconfigurational temperature dependence is markedly non-Arrhenius and dependent on thermal history. Also close to equilibrium it has been possible to demonstrate memory behavior arising from such effects. Figure 11 is an example, showing that a sample annealed at 280°C to attain the viscosity value characteristic of equilibrium at 300°C will on relaxing at 300°C show a transient decrease in viscosity before regaining the equilibrium value. The simple free-volume model does not admit of such behavior as it associates a

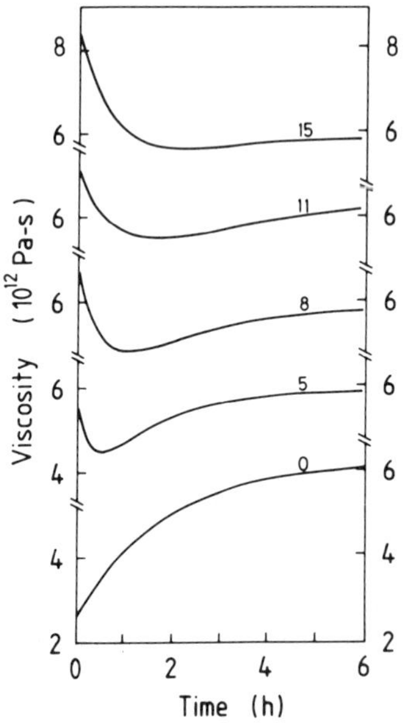

Figure 11 The viscosity of glassy $Pd_{40}Ni_{40}P_{19}Si_1$ as a function of annealing time at 300°C for samples previously annealed at 280°C for the times shown (hr). (Adapted from Ref. 87.)

given viscosity value with one value of average free volume and always assumes an equilibrium distribution of free volume.

Notwithstanding the potential complications arising from free-volume redistribution near equilibrium, the free-volume model modified by the incorporation of an Arrhenius term in Eq. (6) has been used with great success to model quantitatively a large body of data on viscosity relaxation safely below T_g. For example [94] for the alloy $Fe_{40}Ni_{40}B_{20}$ with $T_g \approx 720$ K, viscosity measurements in the range 520–600 K show Arrhenius isoconfigurational behavior and no memory effects. The average free volume in the sample appears to be an adequate parameter for characterizing the glassy structure. In this temperature range all the viscosity relaxation behavior could be classified under topological short-range ordering (TSRO). On the other hand, for the same alloy in the temperature range 450–520 K, it appears that the free volume is not adequate to specify the structural state. This is attributed to differing degrees of chemical short-range order (CSRO) (Section VIIB). In the higher temperature range relaxation effects from chemical short-range order do not arise because the ordering is sufficiently fast to remain in equilibrium. This is still more certainly the case near T_g, and thus chemical short-range ordering cannot be invoked as an explanation of memory and crossover effects such as in Fig. 11 [87].

VI. RELATIONSHIPS BETWEEN ATOMIC TRANSPORT PROCESSES

The relationship between the rates of atomic diffusion and viscous flow depends on the atomic transport mechanisms. In the free-volume model, for example, it is assumed that local volume fluctuations permit atomic rearrangements which contribute both to diffusion and shear deformation (i.e., viscous flow). In that case the atomic diffusivity, D, and the viscosity, η, should be linked through the Stokes-Einstein relation

$$\eta D = \frac{kT}{6\pi r} \tag{7}$$

where r is a characteristic particle radius [63]. Although the soundness of this relation at an atomic level is not clear, it does work remarkably well ($\pm \sim 25\%$) for several liquid metals when an ionic radius is substituted for r [80].

In crystalline alloys creep can occur by atomic diffusion through the grains (Nabarro-Herring creep). In contrast to the ideal liquid behavior, however, not every atomic jump contributing to diffusion contributes also to deformation of the sample. Only creation and annihilation of vacancies at sources and sinks, to be found predominantly in the grain boundaries, contribute to deformation. In this regime the relationship between D and η is

$$\eta D = \frac{kTd^2}{4\Omega} \tag{8}$$

where Ω is the atomic volume and d the grain size. For a given viscosity, the atomic diffusivity is much higher than would be expected from the Stokes-Einstein equation. This is also the case for another type of low-temperature creep, Coble creep, in which the atomic transport is along grain boundaries and only jumps to and from the boundaries contribute to deformation.

It is clearly of interest to know the relationship between D and η in amorphous alloys which have liquidlike structure but solidlike resistance to deformation. The relationship is not easy to establish, however. In alloys there is not a single diffusion coefficient for comparison, and as we have seen in Section IIIB the different species can diffuse at rates which differ by orders of magnitude. Such large disparities seem confined to metal-metal glasses, whereas comparisons of D and η have so far only been attempted in metal-metalloid glasses for which such problems may not be so severe. A second problem is that both D and η can be very sensitive to the degree of relaxation. For a valid comparison they should be measured in samples which are equally relaxed.

A comparison has been attempted for $Pd_{82}Si_{18}$ and $Fe_{40}Ni_{40}B_{20}$ alloys. In the former case [63], viscous flow in $Pd_{82}Si_{18}$ was compared with Pd/Fe interdiffusion which was measured in a multilayered sample. It appeared that on annealing $(1/D)$ rose linearly with time, as did η. The rates of increase of η and of $1/D$ (which are then independent of the degree of relaxation) were used in place of the absolute values to make the comparison. It was found that D was up to $\sim$600 times faster than predicted from the viscosity using the Stokes-Einstein relation and substituting an ionic radius for the characteristic length. On the other hand, the isoconfigurational activation energies for diffusion and for viscous flow were the same within experimental error. For the $Fe_{40}Ni_{40}B_{20}$ alloy [95] direct comparison of η and D values was possible because similar annealing treatments were used in each type of measurement. The diffusivity of iron was up to 4×10^5 times faster than predicted from the viscosity, but again the activation energies for diffusion and for viscous flow were very close. It seems clear that the Stokes-Einstein relation does not hold in these amorphous alloys, at least below the glass transition temperature. Even the metal diffusivities which have been measured, and are expected to be lower than metalloid diffusivities, are much higher than would be expected from the viscosity. This, together with the observed correspondence of activation energies suggests a mechanism similar to that of Nabarro-Herring creep. If Eq. (8) is applied it gives "grain sizes" in $Pd_{82}Si_{18}$ of 3–5 mm and in $Fe_{40}Ni_{40}B_{20}$ of 30–100 nm. In amorphous alloys these distances cannot be regarded as grain sizes, and the use of Eq. (8) is not intended to support a microcrystalline model for the glassy structure. Rather, the distances may represent the lengths of diffusion paths for a defect from source to sink. The distance is a parameter characterizing the relative populations of sites at which diffusive rearrangement can occur and sites (much less plentiful) at which the rearrangement can contribute to shear deformation. If the populations of the two types of site do not stay in a fixed proportion during structural relaxation, the relaxation kinetics of η and D will be different. This is suggested by data such as in Fig. 12, showing that the quantity ηD is not constant [95]. In the early stages of the anneal the diffusivity decreases more rapidly than the viscous flow. Later the diffusivity appears to stabilize while the viscous flow continues to slow down—behavior which would be expected in Nabarro-Herring creep of crystalline materials and which would arise from grain growth during the experiment. In amorphous alloys it can be interpreted as the shear defects relaxing out faster than the diffusion defects, presumably because they are greater departures from the equilibrium structure.

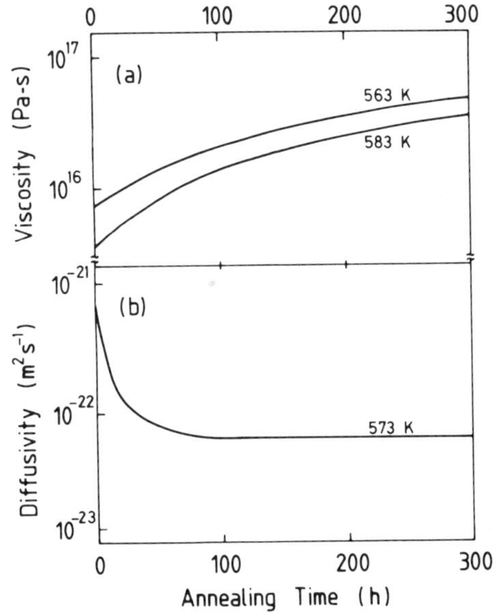

Figure 12 Relaxation of viscosity, η, and Fe self-diffusivity, D, on annealing $Fe_{40}Ni_{40}B_{20}$ glass at the temperatures shown, starting in the as-quenched condition. Clearly the product ηD is not constant. (Adapted from Ref. 95.)

VII. MODELING OF RELAXATION AND ATOMIC TRANSPORT

A. General Kinetics of Relaxation

Structural relaxation has long been studied in conventional, oxide glasses, and mathematical descriptions of the resulting property changes have been developed. Description is simplest in the idealized case of the isothermal property change after a sample at equilibrium at one temperature is subject to an instantaneous temperature change, ΔT. Here we take the property to be the fictive temperature, T_f (defined in terms of a particular property as described in Section II). Its variation with time t at the new temperature is

$$T_f(T) = T_f(0) + \Delta T[1 - \phi(t)] \tag{9}$$

where $\phi(t)$ is the relaxation function. If the material property is completely specified by the fictive temperature (in addition to the standard thermodynamic variables), and if there is only one relaxation function, then it must have the form

$$\phi(t) = \exp\left(-\frac{t}{\tau}\right) \tag{10}$$

i.e., there is a single relaxation time τ. The existence of memory effects (Section IIIA) shows that the fictive temperature is not an adequate description, and a spectrum of relaxation times is required, as treated by Narayanaswamy [96]:

$$\phi(t) = \sum_i g_i \exp\left(-\frac{t}{\tau_i}\right) \tag{11}$$

Tool, who first formulated the description in terms of fictive temperature [2], also suggested that the rate constant governing the kinetics of relaxation should be inversely proportional to the viscosity of the amorphous material. Even without a spectrum of relaxation times, this link is sufficient to give hysteretic behavior around T_g (as shown by curves b and c in Fig. 2a). Justification for the link between relaxation kinetics and viscosity has been discussed by Rekhson [97]. Both the property changes on relaxation and the viscoelastic response to an applied stress can be described by the same mathematics. For a simple relaxation, the single relaxation time is proportional to the viscosity, linked through thermodynamic parameters. Provided the temperature dependence of the thermodynamic parameters is weak compared to that of the viscosity, the temperature dependence of the relaxation time is the same as that of the viscosity. Even if the property change shows a spectrum of relaxation times, the times will all have the same temperature dependence, and the behavior can be described using a single well-defined activation energy. This is indeed found to be the case for structural relaxation of oxide glasses near T_g.

But well below T_g, far from equilibrium, it is found that a spectrum of activation energies is necessary for the description of a property change, and that the spectrum is different from property to property. The spread in activation energies reflects the variability in atomic environments in an amorphous material. The relaxation processes at low temperature (that is below T_g), are expected to involve only local rearrangements and to be sensitive to the variability in a way that the more cooperative, longer-range rearrangements near T_g are not.

B. Activation Energy Spectrum Model

For metallic glasses the activation energy spectrum (AES) model of Gibbs et al. [98–100] (based on that of Primak [101]) has been widely used. In this model it is assumed that in the amorphous structure a variety of atomic rearrangements is possible on annealing; these rearrangements or "processes" affect the properties of the material and have relaxation times governed by a constant prefactor (approximately the Debye frequency, ν) and characteristic activation energies, Q. The processes have a number density, $n(Q)$, which on annealing evolves toward an equilibrium distribution $n_e(Q)_T$ characteristic of the annealing temperature, T. In an anneal at temperature T for time t, to a good approximation, only processes with an activation energy less than $kT \ln(\nu t)$ can occur. Figure 13a shows (bold line) the distribution of processes obtained when a glass in equilibrium at T_1 is annealed for some time (not to equilibrium) at a lower temperature T_2. In the figure $n_e(Q)_{T_1} - n_e(Q)_{T_2}$ is assumed to be independent of Q. If it is further assumed that a property of the material is linearly related to $n(Q)$, then the form of the annealing function implies that the property will change linearly with the logarithm of the annealing time, as is commonly observed (Fig. 4).

The model can describe both irreversible and reversible property changes, and the memory effect (Fig. 3) can readily be predicted. Figure 13b shows the effect of annealing at an intermediate temperature, T_3, following the anneal corresponding to the distribution in Fig. 13a. As the glass structure relaxes toward equilibrium at T_3, the first processes to occur involve an increase in $n(Q)$, while later processes (at higher Q) involve a decrease; in this way the extremum in a material property

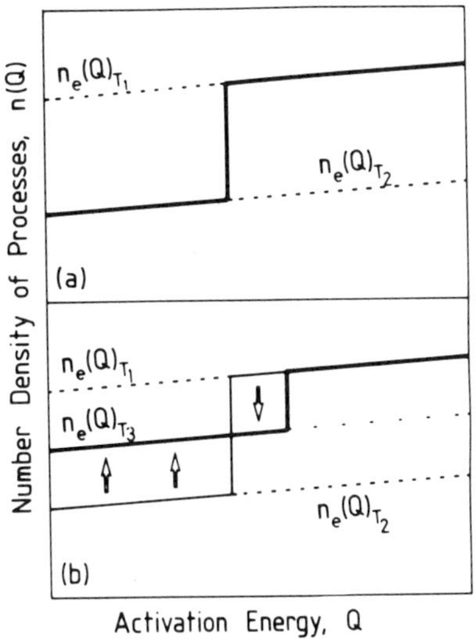

Figure 13 Schematic diagram showing the application of the activation energy spectrum model to property changes on annealing. The quantities $n_e(Q)_{T_{1,2,3}}$ are the equilibrium populations of processes at temperatures T_1, T_2, and T_3. Temperature T_2 is less than T_1, while T_3 is intermediate. (a) The bold line indicates the distribution of processes after a glass in equilibrium at T_1 is annealed for some time at T_2. (b) The bold line indicates the distribution of processes after a further anneal at T_3. The arrows indicate that in the early stages of the anneal at T_3 the number density $n(Q)$ rises, while later $n(Q)$ falls.

(Fig. 3) can arise. The AES model has also been applied to the kinetics of structural polarization [19].

Quantitative application of the AES model has been mainly by van den Beukel and co-workers. They use the model in its simplest form, assuming that relaxation processes exist only in a well-defined range of Q, with a population inside that range independent of Q. They show that the relaxation of, for example, Young's modulus cannot be explained by the AES model alone [102]. They conclude that the AES model can be used to describe relaxation well below T_g (CSRO, see Section IIIC9), while relaxation near T_g (TSRO) is best described by the free-volume model. The free-volume model applied to viscosity was discussed in Section 10.5 (Eqs. (2)–(4)). In their modeling of TSRO, van den Beukel and co-workers assume that the free volume evolves on annealing in a manner that would give a linear rise in viscosity. The relaxation is governed by a single activation energy, typically 250 kJ mol^{-1}. In contrast, CSRO is modeled with a spectrum of activation energies, typically 150 to 250 kJ mol^{-1} extending up to that for TSRO [35]. In most anneals the effects of CSRO and TSRO overlap; for example, in Fig. 4, the two calculated contributions are indicated for one of the curves. Van den Beukel and co-workers have shown that quantitative fits are possible to relaxation data for Young's modulus [35], electrical resistivity [103], enthalpy [92], and length changes [30].

In the AES model as usually applied the processes are assumed to be independent; the glass is modeled as an assembly of isolated atomic two-level systems. If the processes are independent, it is not possible to explain why processes in one part of the activation energy spectrum can affect the relaxation kinetics in a different part of the spectrum (Section IIIA4). The independence of relaxation processes is difficult to justify in terms of the structure of the glass, and is in any case not essential to the AES model, as discussed by Evetts [104]. He shows that in a strongly coupled multilevel system the effects of annealing can be described by a reversible relaxation with a broad spectrum of activation energies and an irreversible relaxation with a single activation energy (thus being similar to the description based on CSRO and TSRO). Furthermore, it should be noted that the distribution of relaxation processes (i.e., their population as a function of activation energy) could show peaks, corresponding to separable relaxation types. Some evidence for this is provided by the work of Cost and Stanley [45], who found that the relaxation time spectrum for electrical resistivity changes on annealing $Fe_{40}Ni_{40}P_{14}B_6$ showed four peaks.

It may be concluded that the two main approaches to modeling structural relaxation, based on discrete mechanisms or a continuous spectrum, are not incompatible, but are merely emphasizing different aspects of the true situation.

C. Compositional Short-Range Ordering

As described earlier (Section IIIA3) the low temperature (well below T_g) relaxation processes, giving reversible property changes, were attributed by Egami [4] to compositional short-range ordering (CSRO). Direct structural evidence for such ordering is minimal (Section IIIC9), though indirect methods, such as Mössbauer spectroscopy [60], show that CSRO at low temperature is different from the type of ordering (topological, TSRO) observed near T_g. Interestingly, as the temperature is raised, the CSRO-induced changes become similar to those produced by TSRO (reminiscent of the observation that the activation energy spectrum for CSRO extends up to the value for TSRO). It has been pointed out that the kinetics of CSRO are consistent with a diffusional process with on average about one jump per atom as might be expected for chemical ordering [63]. In the absence of direct structural information, attempts have been made to obtain evidence for CSRO by studying the effects of composition changes.

It is well known that metal-metalloid glasses, typically with 20 at% metalloid, show strong chemical order, manifested by an absence of metalloid-metalloid nearest neighbors [105]. It is not expected that the degree of this metal-metalloid ordering would be strongly temperature-dependent. In support of this idea, Inoue et al. [106] have shown that reversible relaxation (monitored using enthalpy changes) is a much smaller effect in such a glass when there is just one metal than when there are two metals. In the latter case, chemical ordering between the metals on their sites would be possible. The results of Inoue et al. have been questioned by Brüning et al. [107] who find that the composition dependence of crystallization kinetics may play a role in the results. Nevertheless, Balanzat et al. [47], using electrical resistivity measurements, have found little or no reversible relaxation in monometal-metalloid glasses, but a substantial effect when there is more than one metal. In both electrical resistivity measurements in $(Co_xFe_{1-x})_{75}Si_{10}B_{15}$ glasses [108] and Young's modulus measurements in $(Co_xFe_{1-x})_{78}Si_9B_{13}$ glasses [109], there

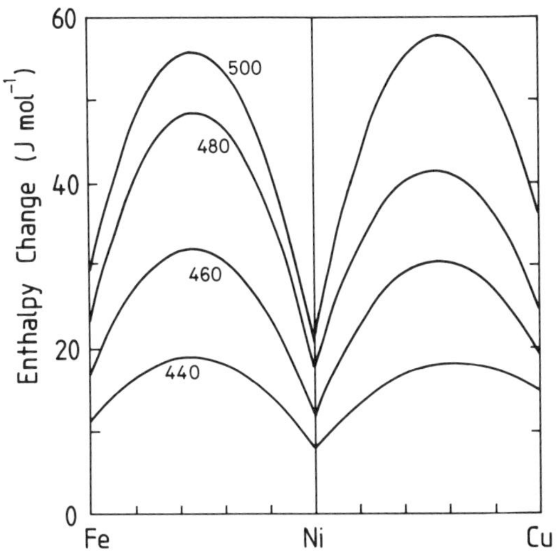

Figure 14 The magnitude of reversible enthalpy changes on annealing $(Fe_xNi_{1-x})Zr_2$ and $(Cu_xNi_{1-x})Zr_2$ metallic glasses at the temperatures (K) shown. (Adapted from Ref. 60.)

is a clear maximum in the reversible property changes at $x = 0.75$, suggesting ordering to form a Co_3Fe unit.

Binary metal-metal glasses, typically combinations of an early and a late transition metal, show less chemical order than in the metal-metalloid case [58,110]. However, the degree of ordering may have a temperature dependence suitable to give reversible effects. Such effects have been found in the electrical resistivity of binary glasses, but are especially marked in ternary glasses when there are two late transition metals between which CSRO can occur (Fig. 14) [60]. The data in Fig. 14 probably constitute the clearest evidence that reversible relaxation below T_g is due to, or at least greatly augmented by, chemical order effects.

D. Structural Defects and Relaxation

It is expected that there would be some difficulty in defining structural defects in a metallic glass because of the lack of a lattice as a reference state. Nevertheless, the concept of defects may be useful in describing the structure of a glassy material and structural relaxation. Also, defects may help to explain some material properties.

For atomic diffusion, the mediating defect most commonly considered for crystals is the atomic vacancy. The prefactors and activation energies for atomic diffusion in metallic glasses, insofar as they can be determined accurately, are consistent with vacancy mediation. Early molecular dynamics simulations on a one-component Lennard-Jones glass showed that a vacancy would not be stable [111]. However, ths conclusion may not remain valid when there is more than one component, and there is strong chemical ordering (as in metal-metalloid glasses); the metallic glass structure may then have some similarity to the networks in oxide glasses. Some evidence for a network structure is provided by the kinetics of viscosity decrease and increase which can be interpreted as reflecting bond breaking and bond forming (Section V).

In general, the viscous flow of metallic glasses is well modeled by the free-volume theory, and the idea of distributed free volume may be more useful than that of discrete vacancies. There is evidence from positron annihilation studies for excess free volume which can be removed by annealing [6]. When the local free volume exceeds a critical value, atomic rearrangement is possible and this may be considered to be a defect. However, the breakdown of the Stokes-Einstein relationship (Section VI) shows that it is necessary to propose different defects for flow and for diffusion. Recently, van den Beukel [112] has shown that the population of diffusion defects is proportional to the square root of the population of flow defects $(n_f)^{1/2}$. The rates of CSRO and of anelastic relaxation are also proportional to $(n_f)^{1/2}$. These relationships are as yet unexplained. There have been only a few tentative attempts as yet to relate flow and diffusion defects to specific structural features [95,113].

A microscopic description of the glassy structure and its relaxation has been attempted by Egami and co-workers [114–116]. Local stresses, both hydrostatic and shear, at the atomic level are proposed and are applied to the description of topological short-range order. An increase in order, such as would occur on annealing an as-quenched state, is described by the mutual annihilation of positive and negative deviations from the average density. The shear stress fluctuations have been related to, for example, internal friction and its relaxation [11]. More details on the description of glassy structures using atomic level stresses may be found in the chapter by Egami in this volume. Although powerful as a structural description, much remains to be done to relate the local stresses to the kinetics of structural change.

VIII. CONCLUSION

Structural relaxation, atomic diffusion, and viscous flow in metallic glasses have all been studied extensively. The main forms of behavior have been characterized. While ab initio calculation of properties is not possible, the variations in behavior are predictable within a given family of alloys. For most purposes, the quantitative modeling of relaxation, diffusion and viscous flow is adequate to permit commercial exploitation with confidence. It is important to bear in mind, though, that there are additional processes, notably surface crystallization, which might affect properties unexpectedly and limit useful lifetime.

In developing a fundamental understanding of the mechanisms involved, much remains to be done. It has been shown that the concept of defects is useful in modeling atomic transport and structural relaxation. On the other hand, the relationships between different types of defects are not understood, and there is not a clear relationship of defects to structure. Molecular dynamics studies of realistic glass-forming systems may prove useful here.

There are some forms of behavior general to metallic glasses. It would be reasonable, however, to expect that with the wide range of metallic glasses now available there may also be distinct categories of behavior. Atomic mechanisms of structural relaxation or atomic transport may be particular to some compositions and temperature ranges.

It is striking that many of the structural relaxation phenomena observed in metallic glasses have their counterparts in other glasses—oxide and polymeric. The history of research on the other glasses suggests that the task of relating structural relaxation to precise structural models is difficult and far from completed.

REFERENCES

1. W. Kauzmann, *Chem. Rev.*, *43*, 219 (1948).
2. A. Q. Tool, *J. Am. Ceram. Soc.*, *29*, 276 (1946).
3. L. Boesch et al., *J. Am. Ceram. Soc.*, *53*, 148 (1970).
4. T. Egami, *Mater. Res. Bull.*, *13*, 557 (1978).
5. H. S. Chen, *J. Appl. Phys.*, *52*, 1868 (1981).
6. J. de Vries et al., *Scripta Metall.*, *22*, 637 (1988).
7. A. Keupers et al., *J. Non-Cryst. Solids*, *72*, 267 (1985).
8. G. W. Koebrugge and A. van den Beukel, *Scripta Metall.*, *22*, 589 (1988).
9. T. Egami and T. Jagielinski, in *Rapdily Quenched Metals*, edited by S. Steeb and H. Warlimont (North-Holland, Amsterdam, 1985), p. 667.
10. G. J. Leusink and A. van den Beukel, *Acta Metall.*, *36*, 3019 (1988).
11. N. Morito and T. Egami, *Acta Metall.*, *32*, 603 (1984).
12. W. Chambron and A. Chamberod, *Solid State Commun.*, *35*, 61 (1980).
13. M. Nait Salem and A. Audouard, *Scripta Metall.*, *16*, 125 (1982).
14. R. Gerling et al., *Acta Metall.*, *36*, 575 (1988).
15. B. Toloui et al., *Scripta Metall.*, *19*, 947 (1985).
16. Deguo Deng and Banghong Lu, *Scripta Metall.*, *17*, 515 (1983).
17. H. A. Davies and C. J. Small, *J. Less-Common Metals*, *140*, 185 (1988).
18. T. Egami, in *Amorphous Metallic Alloys*, edited by F. E. Luborsky (Butterworths, London, 1983), p. 100.
19. P. D. Hodson and J. E. Evetts, *J. Magn. Magn, Mater.*, *59*, 81 (1986).
20. H. R. Hilzinger, in *Rapidly Quenched Metals*, Vol. 2, edited by T. Masumoto and K. Suzuki (Japan Institute of Metals, Sendai, 1982), p. 791.
21. H.-R. Sinning et al., *Int. J. Rapid Solidification*, *1*, 175 (1985).
22. A. L. Greer, in *Rapid Solidification: Materials, Processing and Applications*, edited by B. Cantor (North-Holland, Amsterdam), in press.
23. S. Arakawa, in Ref. 9, p. 1389.
24. T. Egami, in *Amorphous Metals and Semiconductors*, edited by P. Haasen and R. I. Jaffee (Pergamon, Oxford, 1986), p. 222.
25. L. Néel, *C. R. Hebd. Seanc. Acad. Sci. Paris*, *237*, 1613 1953).
26. H. Fujimori, in Ref. 18, p. 300.
27. W. Chambron et al., *J. Non-Cryst. Solids*, *61*, *62*, 895 (1984).
28. G. Konczos et al., *Key Eng. Mater.*, *13–15*, 19 (1987).
29. R. Gerling et al., *Scripta Metall.*, *22*, 1291 (1988).
30. E. Huizer and A. van den Beukel, *Acta Metall.*, *35*, 2843 (1987).
31. T. Komatsu et al., *J. Non-Cryst. Solids*, *72*, 279 (1985).
32. E. Woldt, *J. Mater. Sci.*, *23*, 4383 (1988).
33. A. Inoue et al., *Scripta Metall.*, *17*, 1205 (1983).
34. L. Battezzati and E. Garrone, *Z. Metallk.*, *75*, 305 (1984).
35. A. van den Beukel et al., *Acta Metall.*, *32*, 1895 (1984).
36. H. Kimura and T. Masumoto, in Ref. 18, p. 187.
37. L. Bresson et al., *Scripta Metall.*, *16*, 499 (1982).
38. D. Deng and A. S. Argon, *Acta Metall.*, *34*, 2011 (1986).
39. F. E. Luborsky and J. L. Walter, *Mater. Sci. Eng.*, *35*, 255 (1978).

40. H. H. Liebermann et al., *IEEE Trans. Magn.*, *MAG-13*, 1541 (1977).
41. A. L. Greer, *Thermochim. Acta.*, *42*, 193 (1980).
42. A. L. Greer and J. A. Leake, *J. Non-Cryst. Solids*, *33*, 291 (1979).
43. Y.-N. Chen and T. Egami, *J. Appl. Phys.*, *50*, 7615 (1979).
44. A. L. Greer and J. A. Leake, in *Rapidly Quenched Metals III*, Vol. 1, edited by B. Cantor (The Metals Society, London, 1978), p. 299.
45. J. R. Cost and J. T. Stanley, *J. Non-Cryst. Solids*, *61*, *62*, 799 (1984).
46. K. F. Kelton and F. Spaepen, *Phys. Rev. B*, *30*, 5516 (1984).
47. E. Balanzat et al., *Acta Metall.*, *33*, 785–796 (1985).
48. J. R. Cost et al., *Mater. Sci. Eng.*, *97*, 523 (1988).
49. A. L. Mulder et al., *Scripta Metall.*, *17*, 1399 (1983).
50. H. H. Liebermann and F. E. Luborsky, *Acta Metall.*, *29*, 1413 (1981).
51. D. M. Kroeger et al., *Acta Metall.*, *35*, 989 (1987).
52. T. Egami, *J. Mater. Sci.*, *13*, 2587 (1978).
53. Y. Waseda and T. Egami, *J. Mater. Sci.*, *14*, 1249 (1979).
54. M. Jergel and P. Mrafko, *J. Non-Cryst. Solids*, *85*, 149 (1986).
55. E. Chason et al., *Phys. Rev. B*, *32*, 3399 (1985).
56. J. Kamarad et al., *J. Magn. Magn. Mat.*, *15–18*, 1409 (1980).
57. R. Caciuffo et al., *Physica B*, *156*, *157*, 220 (1989).
58. G. S. Cargill and F. Spaepen, *J. Non-Cryst. Solids*, *43*, 91 (1981).
59. D. V. Baxter, *J. Non-Cryst. Solids*, *79*, 41 (1986).
60. J. O. Ström-Olsen et al., *J. Less-Common Metals*, *145*, 327 (1988).
61. B. Cantor and R. W. Cahn, in Ref. 18, p. 487.
62. A. L. Greer, *Ann. Rev. Mater. Sci.*, *17*, 219 (1987).
63. A. L. Greer, *J. Non-Cryst. Solids*, *61*, *62*, 737 (1984).
64. D. Akhtar and R. D. K. Misra, *Scripta Metall.*, *20*, 627 (1986).
65. J. Horvath and H. Mehrer, *Cryst. Latt. Def. and Amorph. Mater.*, *13*, 1, 15 (1986).
66. E. H. Chason and T. Mizoguchi, *Mater. Res. Soc. Proc.*, *80*, 61 (1987).
67. B. Cantor, in Ref. 9, p. 595.
68. U. Köster, in *Amorphous Metals*, edited by H. Matyja and P. G. Zielinski (World Scientific, Singapore, 1986), p. 113.
69. R. W. Cahn, J. E. Evetts, J. Patterson, R. E. Somekh, and C. Kenway-Jackson, *J. Mater. Sci.*, *15*, 702 (1980).
70. S. K. Sharma, S. Banerjee, Kuldeep, and A. K. Jain, *J. Mater. Res.*, *4*, 603 (1989).
71. R. Kirchheim and U. Stolz, *Acta Metall.*, *35*, 281 (1987).
72. M.-A. Nicolet, I. Suni, and M. Finetti, *Solid State Tech.*, 129–133 (1983).
73. J. Bøttiger, K. Dyrbye, K. Pampus, and B. Torp, *Int. J. Rapid Solidification*, *2*, 191 (1986).
74. J. Bøttiger, K. Dyrbye, K. Pampus, B. Torp, and P. H. Wiene, *Phys. Rev. B*, *37*, 9951 (1988).
75. H. Hahn, R. S. Averback, and H.-M. Shyu, *J. Less-Common Metals*, *140*, 345 (1988).
76. D. Akhtar, B. Cantor, and R. W. Cahn, *Scripta Metall.*, *16*, 417 (1982).
77. H. Hahn, R. S. Averback, and S. J. Rothman, *Phys. Rev. B*, *33*, 8825 (1986).
78. B. M. Clemens, W. L. Johnson, and R. B. Schwarz, *J. Non-Cryst. Solids*, *61*, *62*, 817 (1984).
79. Y.-T. Cheng, W. L. Johnson, and M.-A. Nicolet, *Appl. Phys. Lett.*, *47*, 800 (1985).
80. L. Battezzati and A. L. Greer, *Acta Metall.*, *37*, 1791 (1989).
81. E. N. da C. Andrade, *Phil. Mag.*, *17*, 497 (1934).
82. A. I. Taub and F. Spaepen, *J. Mater. Sci.*, *16*, 3087 (1981).
83. E. Huizer et al., *Acta Metall.*, *34*, 493 (1986).
84. A. L. Mulder et al., *Scripta Metall.*, *18*, 515 (1984).
85. C. A. Volkert and F. Spaepen, *Mater. Sci. Eng.*, *97*, 449 (1988).

86. S. S. Tsao and F. Spaepen, *Acta Metall.*, *33*, 881 (1985).
87. C. A. Volkert and F. Spaepen, *Acta Metall.*, *37*, 1355 (1989).
88. M. H. Cohen and D. Turnbull, *J. Chem. Phys.*, *31*, 1164 (1959).
89. D. Turnbull and M. H. Cohen, *J. Chem. Phys.*, *34*, 120 (1961).
90. D. Turnbull and M. H. Cohen, *J. Chem. Phys.*, *52*, 3038 (1970).
91. A. I. Taub and F. Spaepen, *Acta Metall.*, *28*, 1781 (1980).
92. A. van den Beukel and S. Radelaar, *Acta Metall.*, *31*, 419 (1983).
93. S. S. Tsao and F. Spaepen, *Acta Metall.*, *33*, 891 (1985).
94. A. van den Beukel et al., *Acta Metall*, *34*, 483 (1986).
95. Y. Limoge and G. Brebec, *Acta Metall.*, *36*, 665 (1988).
96. O. S. Narayanaswamy, *J. Am. Ceram. Soc.*, *61*, 146 (1978).
97. S. M. Rekhson, *J. Non-Cryst. Solids*, *73*, 151 (1985).
98. M. R. J. Gibbs et al., *J. Mater. Sci.*, *18*, 278 (1983).
99. M. R. J. Gibbs et al., *J. Non-Cryst. Solids*, *61*, *62*, 925 (1984).
100. J. A. Leake et al., *J. NOn-Cryst. Solids*, *61*, *62*, 787 (1984).
101. W. Primak, *Phys. Rev.*, *100*, 1677 (1955).
102. A. van den Beukel and E. Huizer, *Scripta Metall.*, *19*, 1327 (1985).
103. A. van den Beukel, *Scripta Metall*, *20*, 783 (1986).
104. J. E. Evetts, in Ref. 9, p. 607.
105. P. H. Gaskell, in *Glass—Current Issues*, edited by A. F. Wright and J. Dupuy (Martinus Nijhoff, Dordrecht, 1985), p. 54.
106. A. Inoue et al., *Sci. Rep. Res. Inst. Tohoku Univ. Ser. A*, *32*, 116 (1985).
107. R. Brüning et al. *J. Appl. Phys.* , *62*, 3633 (1987).
108. T. Komatsu et al., *Acta Metall.*, *34*, 1899 (1986).
109. J. Filipecki and A. van den Beukel, *Scripta Metall.*, *21*, 1111 (1987).
110. F. Spaepen and G. S. Cargill, in Ref. 9, p. 581.
111. C. H. Bennett et al., *Phil. Mag. A*, *40*, 485 (1979).
112. A. van den Beukel and J. Sietsma, *Mater. Sci. Eng.*, *A134*, 935 (1991).
113. A. van den Beukel, *Scripta Metall.*, *22*, 877 (1988).
114. T. Egami, in Ref. 18, p. 100.
115. T. Egami and V. Vitek, in *Proc. Symp. on Amorphous Materials*: *Modeling of Structure and Properties*, edited by V. Vitek (TMS-AIME, Warrendale, 1983), p. 499.
116. T. Egami and V. Vitek, *J. Non-Cryst. Solids*, *61*, *62*, 499 (1984).
117. G. W. Koebrugge and A. van den Beukel, *Scripta Metall.*, *22*, 589 (1988).
118. H. S. Chen et al., *Scripta Metall.*, *17*, 1413 (1983).
119. P. Allia et al., *J. Less-Common Metals*, *145*, 375 (1988).
120. Y. Nishi et al., *Scripta Metall.*, *20*, 1099 (1986).
121. H.-Q. Guo et al., *Scripta Metall.*, *20*, 185 (1986).
122. T. E. Tsai et al., *J. Non-Cryst. Solids*, *81*, 147 (1986).
123. Y. Masumoto et al., *J. Non-Cryst. Solids*, *86*, 121 (1986).
124. F. Zougmoré et al., *J. Less-Common Metals*, *145*, 367 (1988).
125. Ch. Cunat et al., *Mater. Sci. Eng.*, *97*, 497 (1988).
126. T. Tarnóczi et al., *Mater. Sci. Eng.*, *97*, 509 (1988).
127. A. L. Greer, *J. Mater Sci.*, *17*, 1117 (1982).
128. A. L. Greer, *Mater. Sci. Eng.*, *A134*, 1268 (1991).

11

Phase Transformations in Rapidly Solidified Alloys

Uwe Köster
University of Dortmund, Dortmund, Germany

Uwe Schünemann
Flachglas AG, Gelsenkirchen, Germany

I. INTRODUCTION

Rapid solidification is known to result in the formation of metastable material characterized by (1) a supersaturation in defects (i.e., refinement of the grain size in micro- and nanocrystalline material, high density of vacancies, or a lack of ordering), (2) a metastable extension of solid solubility above the equilibrium limit, (3) the formation of nonequilibrium crystalline, quasicrystalline or glassy phases. At temperatures high enough for thermally activated processes to be operative, the defects can anneal out by relaxation processes, recovery or recrystallization and the metastable phases can transform by phase transformations into the stable equilibrium. The aim of this chapter is to give an overview of the principal mechanisms of thermally activated phase transformations in rapidly solidified materials and to give a few examples of how this knowledge can even be used to produce improved materials. All the following discussion on phase transformations in rapidly solidified alloys is based on a general theory of transformations in metals and alloys as given for example by Christian [1].

Rapid solidification processes usually results in the formation of fine powders, flakes, or thin ribbons which usually cannot be used in the as-produced form. Any subsequent consolidation technique has to avoid crystallization, decomposition, or significant coarsening of the microstructure which would result from annealing at high temperatures and long exposure times. On the other hand, appropriate annealing conditions of some metallic glasses, for example, might result in partially or fully crystallized metallic glasses with unique engineering properties.

II. MICROSTRUCTURAL CHARACTERIZATION OF RAPIDLY SOLIDIFIED ALLOYS

During rapid solidification by melt spinning, for example, cooling occurs mainly from one side and a gradient in the effective cooling rate arises, thus resulting in an inhomogeneous microstructure over the thickness of the ribbon (see Fig. 1). Such an anisotropy of the microstructure can be observed for grain size, shape, and texture, and for the density of point defects as well as the concentration of the alloy [2–5]. In cross sections of rapidly solidified Al-Mn alloys, a strong gradient in the diameter of the quasicrystals with about 0.1 μm near the contact side and several micrometers near the free surface has been observed; such a difference in the diameter of the quasicrystals influences the stability during decomposition, as will be shown later [6]. In other alloys not only the crystal diameter depends on the distance to the contact side, but also the kind of phases formed; in $Fe_{66}Ni_{10}B_{24}$, for example, an amorphous structure has been found at or near the contact side, followed by a zone with few relatively large crystals; the grain size near the free surface is about 0.1 μm. In a $Au_{45.5}Cu_{45.5}La_9$ glass the intensity of phase separation has been found to increase with distance from the contact side [7].

Defects in crystalline alloys can act as nucleation sites or can accelerate growth; any difference in the defect density will influence the kinetics of precipitation or decomposition of metastable phases. Therefore, due to their anisotropic microstructure, rapidly solidified ribbons will exhibit a complicated transformation behavior because of the occurrence of phase transitions having different kinetics in conjunction with different types of phase transformations which are possible.

III. THERMODYNAMICS OF PHASE TRANSFORMATIONS

As in other metastable materials at sufficiently high temperatures, phase transformations in rapidly solidified materials have been observed to proceed thermally

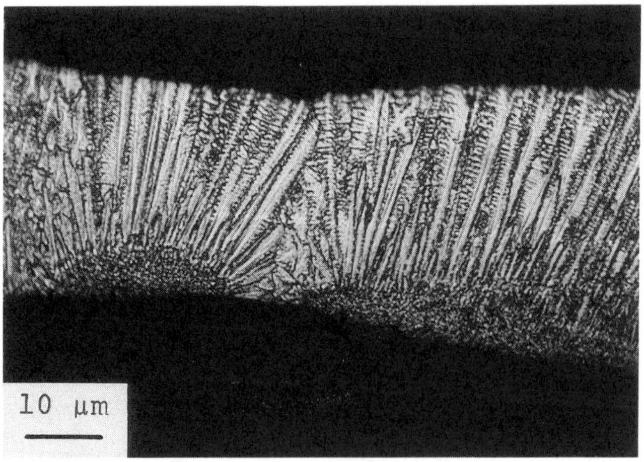

Figure 1 Anisotropic microstructure in a longitudinal section of a melt-spun $Ni_3(Al, Nb)$ ribbon melt-spun at 15 m/s.

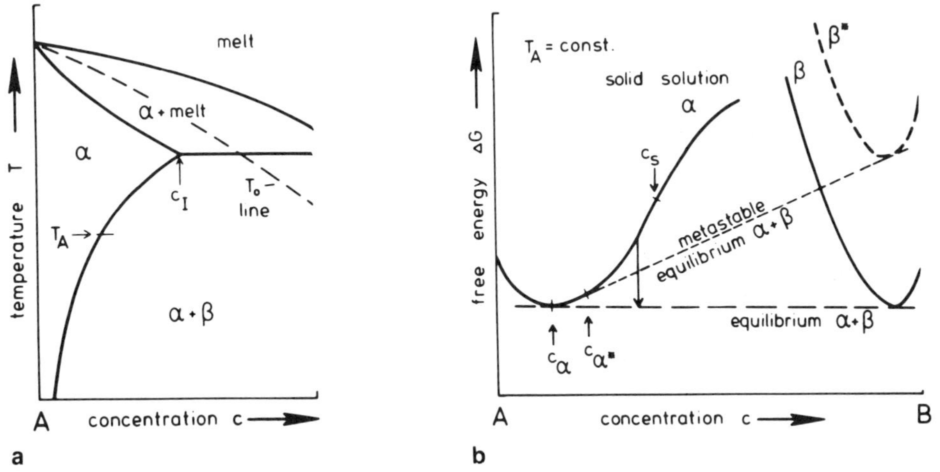

Figure 2 Thermodynamics for precipitation from a supersaturated solid solution: (a) typical phase diagram; (b) schematical diagram of the free energy of the different phases involved.

activated by nucleation and growth processes. The driving force is always the difference between the free energy of the metastable and stable equilibrium states. In order to gain an overall picture of all thermally activated transformation reactions, a schematic free-energy hierarchy of stable and metastable phases of the system has been found to be very useful. Figures 2–4 show a number of hypothetical free-energy diagrams for different types of metastable materials as produced by rapid solidification.

Figure 2 exhibits the case for precipitation from extremely supersaturated solid solutions. From the phase diagram one can conclude that the highest solubility which can be reached by quenching from the solid solution is given by c_I. The free-energy diagram in Fig. 2b indicates that alloys with a concentration less than c_α are stable as a solid solution, whereas alloys with a concentration higher than c_s are unstable and decompose by spinodal decomposition immediately after preparation. Alloys with a composition within the concentration range between c_α and c_s, however, will decompose during annealing by nucleation and growth of the phase β until the stable equilibrium $\alpha + \beta$ is reached.

The extent of supersaturation can usually be increased significantly by rapid solidification, but only as much as the melt can be undercooled to temperatures below the T_0-line before solidification. The higher solute content, i.e., supersaturation, in melt-spun alloys can lead to a propensity to precipitate a metastable phase β^* or to undergo spinodal decomposition; these reactions may not be possible in a particular system for alloys quenched from solid solution. For example, if the maximum solubility in the solid solution field, c_I, is less than or about equal to c_α^* in Fig. 2b, there is neither a driving force for the formation of the metastable phase β^* nor the possibility for spinodal decomposition. If c_I is in the range between c_α^* and c_s, the metastable phase can be formed by nucleation and growth, but again the alloy has no ability to undergo spinodal decomposition. Therefore, we should be able to observe precipitation of so-far-unknown metastable phases or spinodal decomposition in some supersaturated solid solutions solidified from the melt.

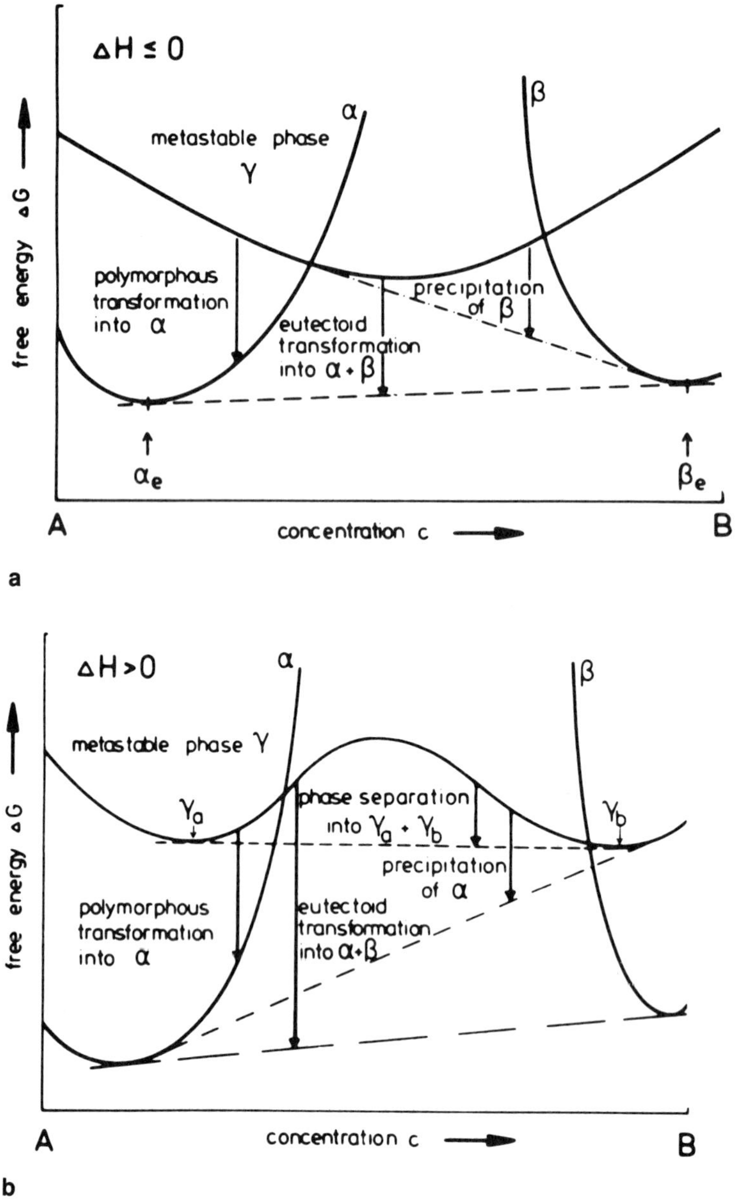

Figure 3 Free-energy-versus-composition diagrams indicating possible phase transformations assuming (a) the metastable phase γ as regular solution with negative heat of mixing; (b) the metastable phase γ as regular solution with positive heat of mixing.

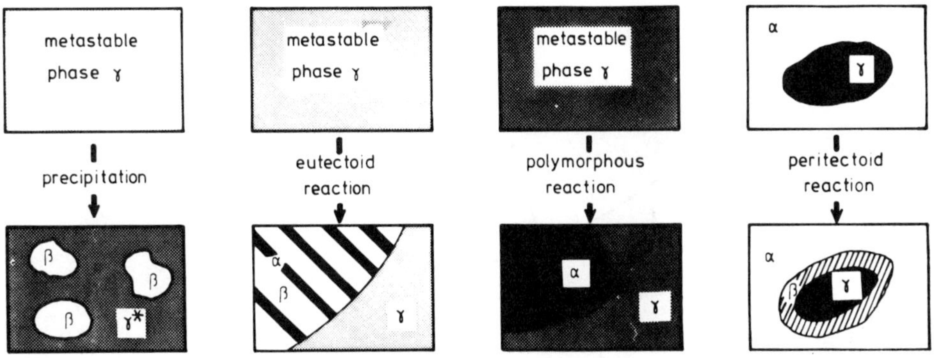

Figure 4 Schematic diagram of typical phase transformations in rapidly solidified alloys.

The situation for phase transformations of metastable phases is shown in Fig. 3. Four types of transformations are possible due to a gain in free energy: precipitation of one of the stable phases, eutectoid reaction, polymorphous reaction as well as a peritectoid reaction.

If the heat of mixing is positive (Fig. 3b), there exists a concentration range between $c(\gamma_a)$ and $c(\gamma_b)$ where in concurrence to the reactions just mentioned, separation into two metastable phases γ_a and γ_b can occur by nucleation and growth or in a narrower concentration range by spinodal decomposition. From a thermodynamic point of view there is no difference between decomposition of an amorphous, metastable crystalline or quasicrystalline phase. Let us consider a metastable phase which forms over the entire concentration range an ideal or regular solution with negative heat of mixing and two stable crystalline phases, as indicated in the free-energy diagram. Depending on the concentration of the alloy, phase transformations can occur by the following reactions.

A. Polymorphous Transformation

A polymorphous transformation (Figs. 3 and 5) involves partition-less growth of one phase with different structure but the same composition as the matrix. This reaction can occur only in concentration ranges near those of the pure elements or stable compounds and needs only single jumps of atoms across the crystallization front. Such a transformation may lead to a supersaturated solid solution, which itself can undergo further precipitation until the stable equilibrium is reached. A special type of a polymorphous transformation with very rapid movement of the interface is called massive transformation and has been observed to occur in a number of Cu or Fe alloys [8]. Kinetically, a massive transformation, which is thermally activated, is intermediate between an equilibrium reaction and a possible martensitic transformation.

B. Eutectoid Transformation

During a eutectoid transformation (Figs. 3 and 4) two crystalline phases grow cooperatively by a discontinuous reaction. There is no difference in the overall concentration across the reaction front; diffusion takes place parallel to the reaction

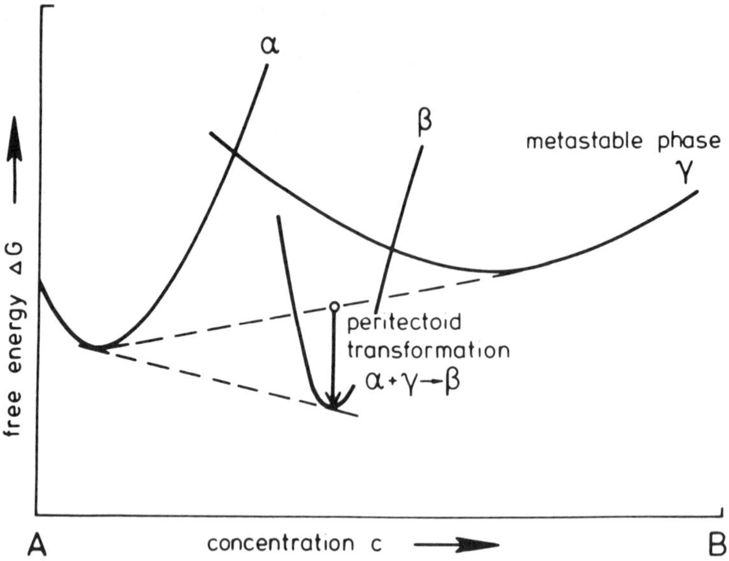

Figure 5 Free-energy-versus-composition diagram for the various phases as for a peritectoid reaction.

front (in, ahead, or behind the reaction front) and the two components have to separate into two phases, thus reducing the growth rate of this reaction compared with that during polymorphous transformation. The formation of pearlite by a eutectoid transformation of γ-iron is the best known example of this type of transformation.

C. Precipitation or Primary Crystallization

In the case of a glassy material (Figs. 3 and 4) particles of one crystalline or quasicrystalline phase, β, with a composition different from the matrix are produced. During this reaction, a concentration gradient is built up ahead of the particle interface and the matrix becomes enriched in the component A until the reaction reaches the metastable equilibrium. Subsequently, the A-enriched matrix transforms by one of the other mechanisms discussed here.

D. Peritectoid Reaction

A microstructure consisting of two phases, the metastable phase γ and the stable phase α, for example, can be produced by rapid solidification. This microstructure can transform by a peritectoid reaction into the stable phase β (Figs. 4 and 5). The metastable phase γ reacts with the embedding α-matrix and thus forms the new phase β. This reaction soon results in a microstructure consisting of γ-crystals surrounded by β-phase, all of which is embedded in the remainder of the α-matrix. Further conversion of the γ-phase needs diffusion through the β-layer, thus leading to a decrease in the reaction rate with increasing thickness of the β-layer. So far, such a peritectoid reaction has been observed in rapidly solidified alloys only during the decomposition of quasicrystals in Al-Mn-alloys with about 10 at% Mn (see Section IVC).

Appropriate free-energy-versus-composition diagrams can be constructed from thermodynamic data available on an alloy system, including phase diagrams. Only very few free-energy data have been measured for metallic glasses [9,10]; usually the amorphous alloy is considered as an extension of the liquid phase to temperatures in the undercooled region. If not enough thermodynamic data are available, regular solution behavior might be assumed and the heat of mixing of the liquid phase could be estimated using method of Miedema [11]. The driving force, i.e., the difference in free energy is such that in most cases transformation can occur by two or even more different reactions. Which reaction does occur depends not only on its thermodynamic driving force, but on its kinetics.

IV. THERMALLY ACTIVATED NUCLEATION AND GROWTH PROCESSES

Whereas thermodynamics selects those phase transformations in rapidly solidified material which are possible, i.e., which lead to a reduction in free energy, kinetics will determine which of the possible reactions will occur and which phases will form the resulting microstructure. In addition, investigations of phase transformations in microstructures formed by rapid solidification will help to understand their formation.

A. Growth

Let us assume that a stable nucleus is available and starts growing just at the beginning of the isothermal annealing. It is known that, depending on the mode of transformation, growth will proceed in a linear or a parabolic manner.

Linear Growth

Linear growth has been found to occur in transformation reactions that do not involve any compositional change (polymorphous reactions) or in a cooperative transformation into two crystalline phases involving long-range diffusion in such a way that the mean composition of the crystallized region is equal to that of the matrix (eutectoid reactions).

Once a stable nucleus has formed, it will continue to grow by a polymorphous reaction during isothermal annealing at a rate, u_1, which is given by

$$u_1 = u_0 \exp\left[-\frac{Q_s}{RT}\right]\left(1 - \exp\left[-\frac{\Delta G}{RT}\right]\right) \tag{1a}$$

where Q_s is the activation energy for growth. For a polymorphous reaction the preexponential factor, u_0, should be of the order of 5×10^3 m/s. In a eutectoid reaction the preexponential factor exhibits the interlamellar spacing, which itself depends on the temperature, and might be much larger.

At low temperatures, with $\Delta G \ll RT$, Eq. (1a) changes to a typical Arrhenius-type equation:

$$u_1 = u_0 \exp\left[-\frac{Q_s}{RT}\right] \tag{1b}$$

Growth as given by these equations will continue until the crystals touch each other; any diffusion-controlled Ostwald ripening cannot occur.

When the phase transformation proceeds by a eutectoid reaction, long-range diffusion is necessary for the redistribution of the components into the two phases and the growth rate u_l is expected to decrease with increasing interlamellar spacing, S, which itself depends on temperature. The growth rate during lamellar eutectoid growth can be described by an equation very similar to Eq. (1):

for volume diffusion $$u_e \approx 4\,\frac{D_V}{S} \tag{2a}$$

for interfacial diffusion $$u_e \approx 4D_I\,\frac{\Delta_f}{S^2} \tag{2b}$$

where Δ_f is the thickness of the reaction front. Since S is a function of temperature, an Arrhenius-plot which might fit the observed growth data does not result in an activation energy for diffusion; rather, it leads only to an effective activation energy somewhat related to that activation energy.

Parabolic Growth

Parabolic growth generally occurs in volume diffusion-controlled reactions, that is, during precipitation, primary crystallization, or peritectoid transformation. The temperature dependence of the diffusion coefficient, D, is given by the following Arrhenius-type equation:

$$D = D_0 \exp\left[-\frac{Q_D}{RT}\right] \tag{3}$$

If we assume the diffusion rate D is concentration independent, the radius, r, of a spherical particle during precipitation or the thickness of the intermediate layer during peritectoid transformation will be proportional to time, t, as shown below:

$$r = \alpha\sqrt{D\,t} \tag{4}$$

and the growth rate u_p will be given by

$$u_p = \frac{1}{2}\,\alpha D r^{-1} \tag{5}$$

where α is a dimensionless parameter evaluated from the composition at the particle interface and the composition of the sample. This time-dependent growth rate will be reduced drastically at larger particle diameter as soon as the diffusion fields overlap and the regime of Ostwald ripening is achieved. The final equation for the coarsening rate when the steady distribution has been achieved is given by the Lifshiftz-Wagner equation:

$$\bar{r}^3 - r_0^3 \approx \frac{D\sigma c}{RT}\,t \tag{6}$$

Figure 6 shows the expected distribution of particle size for such a diffusion-controlled coarsening.

Neither linear nor parabolic growth can usually be observed in situ during phase transformations because surfaces may change the reaction mode or the kinetics [12]. Growth rates have to be calculated from the diameter of the largest crystal

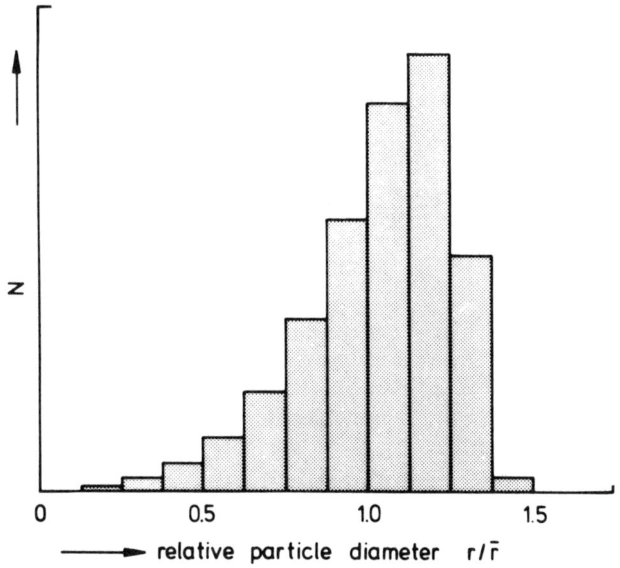

Figure 6 Distribution of particle size for diffusion-controlled coarsening (Ostwald ripening).

or particle found after each heat treatment. As will be shown later, simulation of the crystal diameter distribution can also be used for estimating growth rates.

B. Nucleation

Precipitation, peritectoid, and polymorphic transformations require nucleation of only one phase whereas a eutectoid reaction has been observed to be a process with two nucleation steps. In classic nucleation theory the steady-state homogeneous nucleation rate, I_{st}, is given by

$$I_{st} = I_0 \exp\left(\frac{-L\,\Delta G_c}{RT}\right)\exp\left(\frac{-Q_N}{RT}\right) \tag{7}$$

where I_0 is a constant factor, L the Loschmidt number, Q_N the activation energy for the transfer of atoms across the surface of the nucleus, and ΔG_c the free energy required to form a nucleus of critical size. For heterogeneous nucleation at a limited number, Z, of nucleation sites, the activation energy ΔG_c is reduced due to the gain in surface energy, thus leading to a significant higher nucleation rate I^*_{st}. During precipitation, defects are known to accelerate nucleation [13], for example grain boundaries reduce the amount of surface energy necessary for nucleation, and dislocations can reduce the stress around a nucleus, thus reducing the nucleation barrier. Vacancies usually accelerate diffusion and thereby mainly the growth of particles.

At the very beginning of annealing a finite period is expected, during which the steady-state distribution of clusters assumed in classical nucleation theory is established. This involves a transient or time-dependent nucleation rate $I(t)$ given by the following equation [14]:

$$I(t) = I_{st}\left[1 + 2\Sigma(-1)^n \exp\left(-\frac{n^2t}{\tau}\right)\right] \tag{8}$$

where τ is the time lag, which is expected to increase significantly with decreasing temperature. The duration of this time-lag can be estimated to be given to an order of magnitude by

$$\tau = \frac{s^2}{\pi^2 D} \tag{8a}$$

where s is the diameter of a critical nucleus. Kelton et al. [15] have shown that this equation is a very good approximation of the nonstationary nucleation problem in the crystallization of metallic glasses.

Nucleation in all phase transformations can be studied in some detail by statistical analysis of particle or crystal size distribution in partially transformed specimens, that is, from the fit of calculated and measured particle or crystal diameter distributions [16,17]. Let us assume polymorphous crystallization of spherical crystals and a constant growth rate u, and let us apply the calculation below only for the initial states of transformation, when noninterfering crystals are formed. The operating nucleation mode that applies determines how the number of crystals, ΔN_i, nucleated during the time interval, Δt, can be calculated (Fig. 7). In cases where the homogeneous mode applies, such a calculation may be carried out with the equation

$$\Delta N_i = I(t)(1 - X_{i-1})\,\Delta t, \qquad i = 1, \ldots, \frac{t}{\Delta t} \tag{9}$$

with

$$X_i = \frac{4\pi}{3} u^3 \Sigma\, \Delta N_j [\Delta t(i + 1 - j)]^3 \tag{10}$$

In many cases where the heterogeneous mode applies, such a calculation may be carried out with the equation

$$\Delta N_i = I^*(t)(1 - X_{i-1})[1 - Z^{-1}\Sigma\, \Delta N_j]\,\Delta t \tag{11}$$

for i where $\Sigma\, \Delta N_j \leq Z$; $\Delta N_j = 0$ for all other i.

If there is a limited number of quenched-in postcritical nuclei, Z, they will start growing as soon as the annealing temperature is reached. This will lead to a very narrow diameter distribution, as shown in Fig. 7a.

Computer simulation of crystal diameter distributions for precipitation as well as for primary crystallization is complicated due to the radius dependence of the growth rate (parabolic growth). In addition, Ostwald ripening is expected to lead after some coarsening to a diameter distribution very similar to those typical for transient type heterogeneous nucleation.

It could happen that two nucleation reactions overlap and proceed simultaneously. For example, a combination of heterogeneous transient and homogeneous transient nucleation would lead to a histogram as shown in Fig. 8.

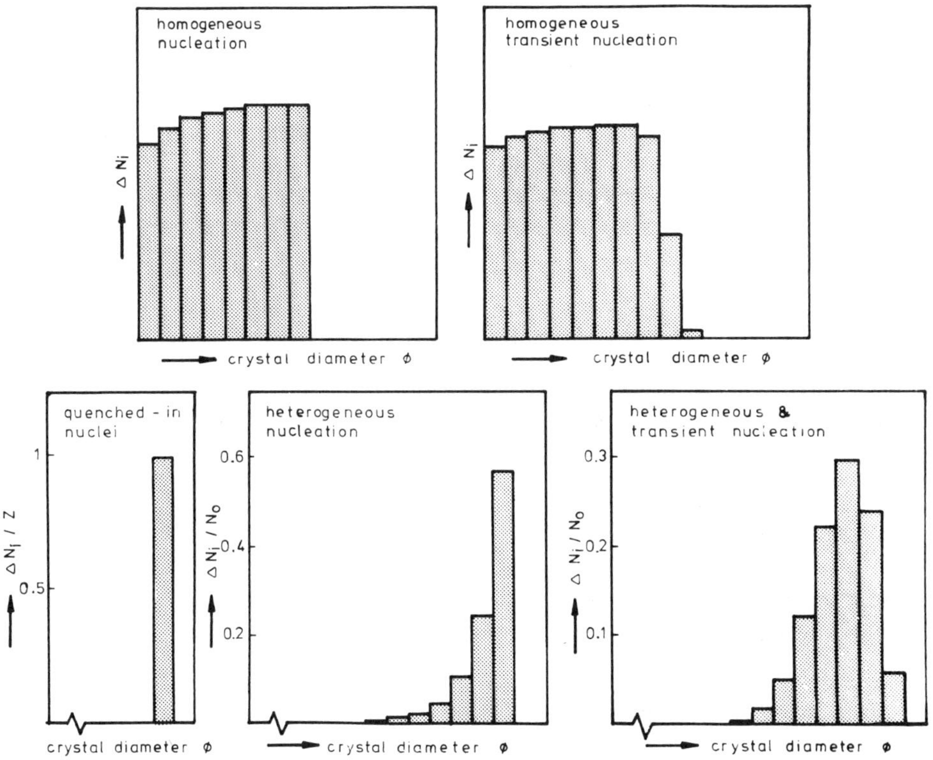

Figure 7 Schematic histograms for the crystal diameter distribution depending on the nucleation mode.

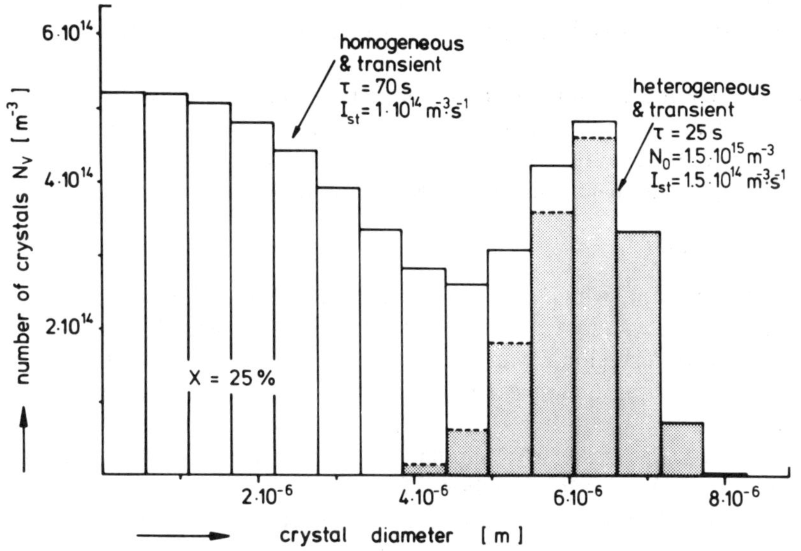

Figure 8 Schematic histogram for a crystal diameter distribution resulting from a combination of nucleation modes.

C. Kinetics

A detailed understanding of growth and nucleation mechanisms during precipitation, phase transformation in metastable crystalline, quasicrystalline or amorphous alloys renders possible to control or impede phase transformations or to design very special crystalline microstructure.

For a polymorphous as well as for a eutectoid phase transformation with isotropic constant growth rate u the volume fraction transformed, X, during isothermal annealing is given by [1]

$$X = 1 - \exp\left[- u^3 \int I(\tau)(t - \tau)^3 \, d\tau \right] \tag{12}$$

This equation may be integrated only by making specific assumptions about the variation of I with time. For homogeneous nucleation under steady-state conditions we find

$$X = 1 - \exp\left(-\frac{\pi}{3} I_{st} u^3 t^4 \right) \tag{13}$$

Suppose that, instead of homogeneous nucleation, there are Z postcritical quenched-in nuclei. Then, the volume fraction X is given by

$$X = 1 - \exp\left(-\frac{4\pi}{3} Z u^3 t^3 \right) \tag{14}$$

With increasing temperature, however, the validity of these various assumptions may falter. This indicates the risk in analyzing activation energies for phase transformation estimated from changes of physical properties at different temperatures, without any microscopical control of the nucleation and growth reactions which may have occurred.

Similar kinetic equations can be given for isochronal heating of rapidly solidified metallic alloys.

The analysis by computer simulation is not limited only on crystal diameter distributions as shown in Figs. 7 and 8. Usually phase transformations, e.g., crystallization of metallic glasses, are studied by measuring the amount of crystallinity, e.g., by DSC. The popular Johnson-Avrami-Mehl-equation as given by

$$X(t) = 1 - \exp[-(kt)^n] \tag{15}$$

where k is the rate constant, which can be represented by an Arrhenius equation, and n the Avrami-Mehl-exponent, is often used in analysis. Instead, we can use our knowledge about observed micromechanisms to simulate the amount of crystallinity present during isothermal or isochronal annealing to characterize the crystallization kinetics.

It would be of particular interest to simulate the crystallization kinetics in the case of a combination of nucleation reactions. For example by using the Johnson-Avrami-Mehl equation Budurov et al. [18] were able to demonstrate that crystallization in some Cu-Zr glasses can be described much better by assuming a combination of two nucleation modes. We can do that on a much better basis by starting with equations based on observed micromechanisms. For example, for growth of

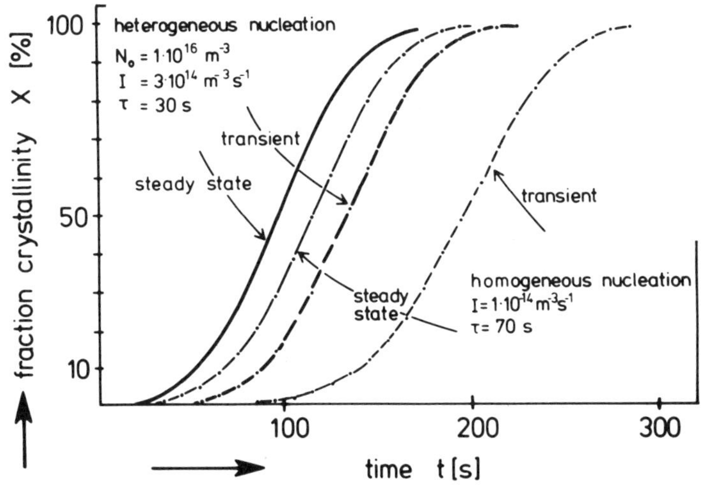

Figure 9 Influence of the nucleation mode on crystallization kinetics.

Z quenched-in postcritical nuclei combined with homogeneous nucleation one would find

$$X(t) = 1 - \exp\left[-\left(\frac{\pi}{3} I_{st}u^3t^4 + \frac{4\pi}{3} Zu^3t^3\right)\right] \tag{16}$$

Figures 8 and 9 exhibit a number of crystal diameter distributions which result from combination of two nucleation modes as well as plots for fraction crystallinity versus annealing time simulated by the method described so far.

V. EXAMPLES OF PHASE TRANSFORMATIONS IN RAPIDLY SOLIDIFIED METALLIC ALLOYS

A. Precipitation in Rapidly Solidified Supersaturated Solid Solutions

Of special interest for this chapter is to concentrate on effects resulting from the rapid solidification process instead of discussing the large number of alloys investigated so far; otherwise, one has to refer to literature on precipitation reactions from solid-state-quenched solid solutions, e.g., by Kelly [19].

As mentioned in Section II, particular *metastable phases* or zones cannot precipitate from solid solutions quenched from the solid in a number of alloy systems; the formation of metastable precipitates do need a higher supersaturation, which can be achieved by rapid solidification from the melt.

In Al-Fe alloys, for example, the maximum equilibrium solid solubility of Fe in Al is about 0.026 at% at 615°C. During precipitation, the metastable Al_6Fe and the stable Al_3Fe phases are usually formed. From rapidly solidified solid solutions with higher iron content, the formation of coherent iron-rich particles about 3 nm in diameter as well as a metastable bcc phase with lattice parameters similar to those of AlFe have been reported [20,21]; during further annealing these phases transform stepwise towards the equilibrium Al matrix + Al_3Fe.

The maximum solubility of Ge in aluminum is 2.8 at%. In Al-Ge alloys, solid solubility has been reported to be increased up to 7 at% [22]. So far, however, it is not known whether precipitation from such an extremely supersaturated solid solution will result in the formation of one of the metastable Al-Ge phases which are formed directly by rapid solidification of alloys with higher germanium content.

Most of the work on *spinodal decomposition* of supersaturated solid solutions has been limited to cases where solute concentration is within the solubility limit of the equilibrium phase diagram. Therefore, spinodal decomposition has been only found in particular alloy systems (e.g., Fe-Mo [23]). If significantly enhanced solute supersaturation is realized by rapid solidification from the melt, many more alloys should fall within the spinodal region of the miscibility gap (see Fig. 2). Such a behavior has been reported by Kozakai et al. [24] to occur in Fe-W alloys. Solid quenched α-(Fe, W) alloys are limited to 14.3 at% W and do not exhibit any spinodal decomposition. In rapidly solidified alloys with 17 at% W, however, a modulated structure has been observed after annealing at 550°C, typical for spinodal decomposition. From the wavelength, the coherent spinodal temperature could be evaluated to be about 790°C. Furthermore, the calculated coherent spinodal line confirms that solid quenched solid solution are too dilute for spinodal decomposition.

The *small grain size* usually observed in rapidly solidified alloys has a strong influence on precipitation. It is well known that grain boundaries exhibit a very strong influence on nucleation of incoherent particles [13]. The grain boundary provides part of the interface necessary for nucleation, thus reducing the nucleation barrier. Figure 10a shows a very typical microstructure resulting from such an influence of grain boundaries as preferred nucleation sites in an Al–4 at% Hf alloy; the grain boundaries are not only preferred nucleation sites, but also act as vacancy sinks, thus reducing the ability for nucleation of coherent particles. These influences of a grain boundary lead to the formation of a soft precipitation-free zone along the grain boundary which results in very bad mechanical properties of the alloy even though there may be very fine particles precipitated in the bulk of the grains.

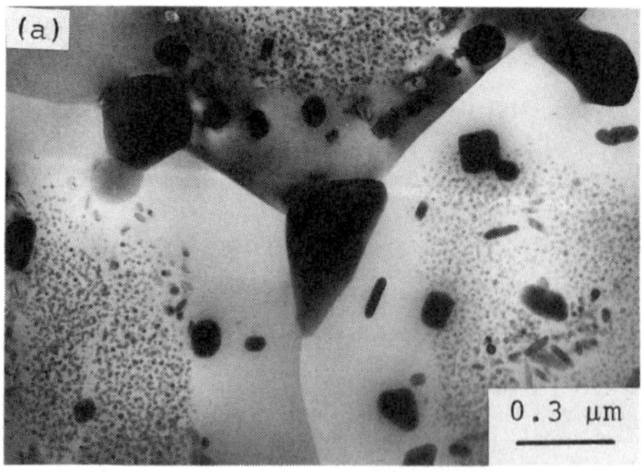

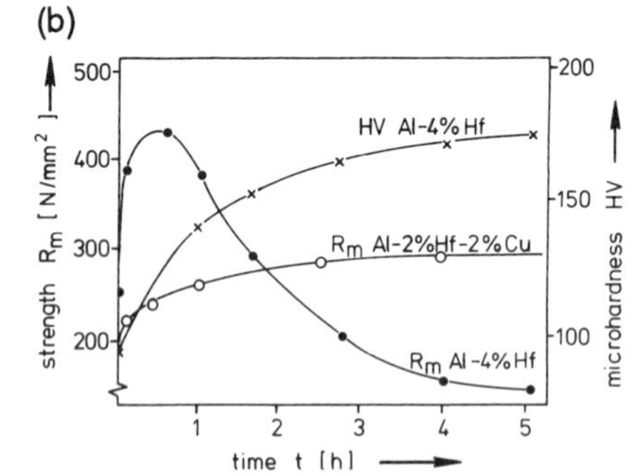

Figure 10 Precipitation in rapidly solidified Al–4% Hf alloys: (a) microstructure with a particle-free zone along the grain boundaries after annealing: 168 h at 450°C; (b) influence of annealing on mechanical properties.

As shown in Fig. 10b, the strength of the alloy deteriorates very early on due to the soft layers along the grain boundaries which occupy a quite large volume fraction in such rapidly solidified alloys with small grain diameters. The extension of these soft layers can be reduced by further alloying, e.g., by copper additions. Microhardness, a property which usually averages over larger areas, will not follow such a trend immediately.

Quenched-in vacancies usually exhibit a strong influence on nucleation and growth of precipitates. As an example, Fig. 11a shows a rapidly solidified Al–4 at% Hf alloy which has been annealed for 1 hr at 450°C; there is a lack of the coherent Al_3Hf particles close to the grain boundaries or in channels along dislocation lines. Whereas a similar behavior is known in a number of solid-state-quenched supersaturated solid solutions, e.g., Al (Ge) [25], the fanlike microstructure shown in Fig. 11b, which either might be formed by discontinuous precipitation of Al_3Hf or just reflects the vacancy distribution as produced during the rapid solidification process, seems to be very special for rapidly solidified material.

Quenched-in clusters are assumed to be of utmost importance for the nucleation process during precipitation in supersaturated solid solutions; rapid solidification from the melt is expected to change number and size of such clusters drastically.

Precipitation in Al–1.3 at% Ge alloy after solid-state quenching has been found to depend very strongly on annealing temperature [25] as well as on the temperature of homogenization [26]. The formation of small triangular plates with a parallel orientation relationship between matrix and particles (Fig. 12a) has been observed only at relatively low annealing temperatures of 150°C or even lower. This microstructural modification can only occur if the alloy was quenched from sufficiently high temperatures. It was concluded, that vacancies or vacancy-germanium clusters are responsible.

Quenching from the melt can result in a quite different morphology [27]. Instead of the small triangular plates, much larger particles (Fig. 12b) are formed even at this low annealing temperature; such large particles are typical for precipitation in solid-state-quenched alloys at higher temperatures, i.e., when most of the clusters are too small to become nuclei at this temperature or when prior to precipitation vacancies anneal out too fast. Figure 12c indicates that the formation of these

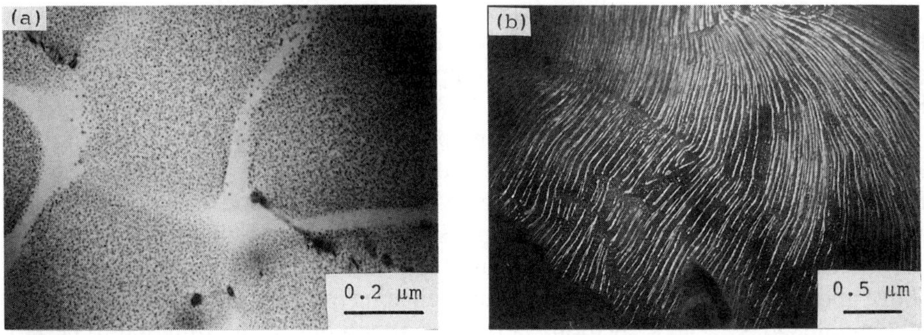

Figure 11 Precipitation in melt-spun Al–4 at% Hf alloys: (a) continuous precipitation of coherent Al_3Hf particles: 1 hr at 450°C; (b) discontinuous precipitation of Al_3Hf: 24 hr at 450°C.

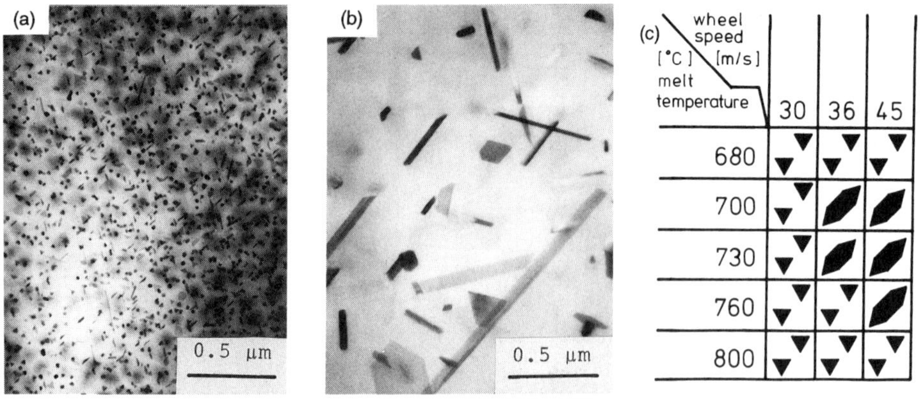

Figure 12 Precipitation in rapidly solidified Al–1.3 at% Ge alloy: 30 hr at 150°C [27]: (a) solid-state quenched from 440°C; (b) rapidly solidified from 730°C (wheel speed 45 m/s); (c) influence of melt temperature and wheel speed: ▼ high density of small particles as in (a); ◆ large particles as in (b).

large particles in rapidly solidified alloys happens only over a narrow range of melt temperatures and quenching rates. Otherwise, the quench rate is probably not fast enough to eliminate cluster formation or the melt itself contains clusters.

Similar effects have been reported from annealing of melt-spun Cu(Fe) solid solutions which exhibit a higher stability than solid-state-quenched alloys with even lower iron content [28]. In addition, there is some evidence of a gradient in the particle size which reflects not only the mentioned gradient in the concentration across the thickness, but probably also differences in the vacancy or iron-cluster density due to the decrease in the quench rate.

B. Phase Transformation of Metastable Crystalline Phases

Rapid solidification can be used to produce metastable crystalline phases which will transform into the stable phases by a polymorphous reaction or eutectoid decomposition during annealing. A precipitation reaction similar to primary crystallization has not been observed so far in rapidly solidified materials; such a reaction, however, is known in the decomposition of the high temperature phase FeO [29].

Figure 13 shows the eutectoid decomposition of (Fe, Ni)$_3$B; the temperature dependence of the growth rate indicates that this boride is a metastable phase. Khan et al. [30] suggested that the phase Fe_3B exists in thermodynamic equilibrium at temperatures above 1150°C. In comparison, the decomposition of Co_3B [31] is shown; Co_3B is a high-temperature phase [32] and can be produced by annealing above about 860°C as well as by moderate quenching from the melt or crystallization of a metallic glass. The activation energy for growth has been estimated to be 221 kJ/mole; the activation energy for grain boundary diffusion in boron-segregated α-Fe is known to be 220 kJ/mole. In both cases nucleation of the eutectoid reaction

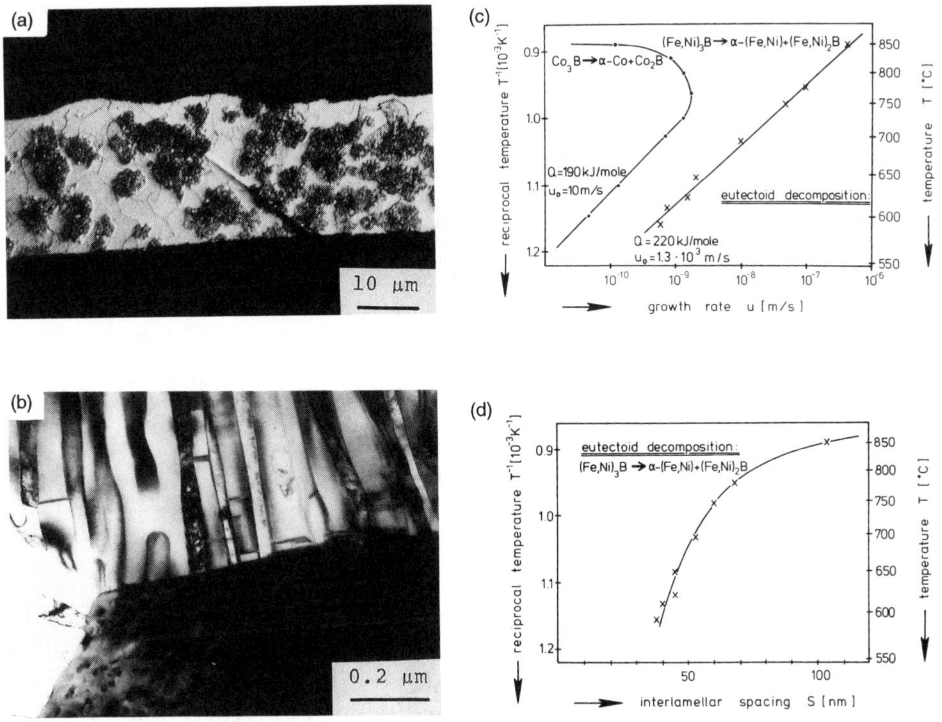

Figure 13 Eutectoid decomposition of (Fe, Ni)$_3$B in $Fe_{66}Ni_{10}B_{24}$: (a) 3 hr at 390°C + 10 min at 700°C; (b) 22 hr at 380°C + 10 min at 770°C; (c) temperature dependence of the eutectoid growth rate and (d) of the interlamellar spacing.

has been observed to occur at the original grain boundaries of the borides, as can be seen in Fig. 13. However, no systematic studies have been conducted so far and not all grain boundaries act as nucleation sites.

The Al-Ge system is well known for its large number of metastable phases formed by rapid solidification; these phases transform into the stable equilibrium ones on further annealing [33–38]. Figure 14 shows the eutectoid decomposition of metastable AlGe which can be formed by rapid solidification from the melt as well as by crystallization from vapor-deposited amorphous films. It is of interest to realize that one of the product phases, i.e., the Ge crystals, do not show any contact with the parent phase. Any accidental contact results in slowing down the decomposition. Such a reaction is possible because the diffusion necessary for the decomposition occurs in one of the product phases, i.e., the aluminum, and not in or ahead of the interface. The authors believe that this happens due to the difference in the bonding of the germanium, which is metallic in the aluminum, but covalent in the metastable phase, thus resulting in easy diffusion.

The polymorphous transformation from γ_4 into γ_2 [35] is controlled by diffusion of Ge through a thin aluminum layer, as one can see in Fig. 15; the reaction can proceed only by this mechanism since diffusional reactions at the direct interface between γ_4 and γ_2 are much too slow.

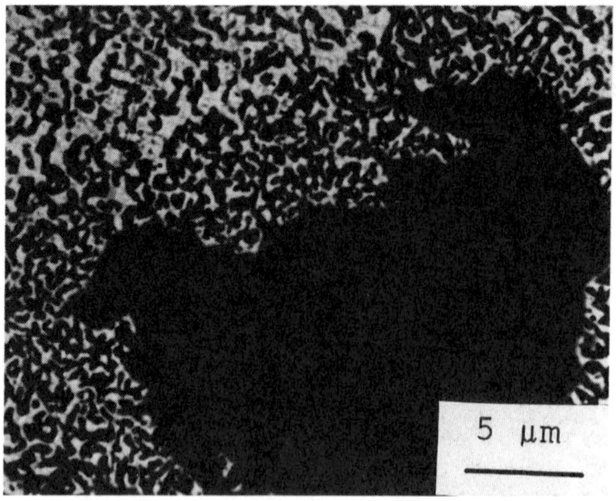

Figure 14 Eutectoid decomposition of metastable γ_2 (AlGe) into Al + Ge.

Interlamellar spacing, S, is usually known to decrease with the temperature. In contrast, the spacing has been observed to increase with decreasing annealing temperature during the decomposition of the metastable phases in the Al-Ge system [36]. A similar temperature dependence has been found to occur during eutectoid crystallization of glassy $Fe_{80}B_{20}$. In Fig. 16 a model is shown which explains such a behavior to occur due to an increased difficulty of branching of the germanium crystals, as well as increased difficulty to nucleate the α-Fe at the boride surface with decreasing temperature.

The polymorphic transformation of the tetragonal boride Fe_3B into the orthorhombic mentioned in an early paper [39] has been found to occur only in thin-film material.

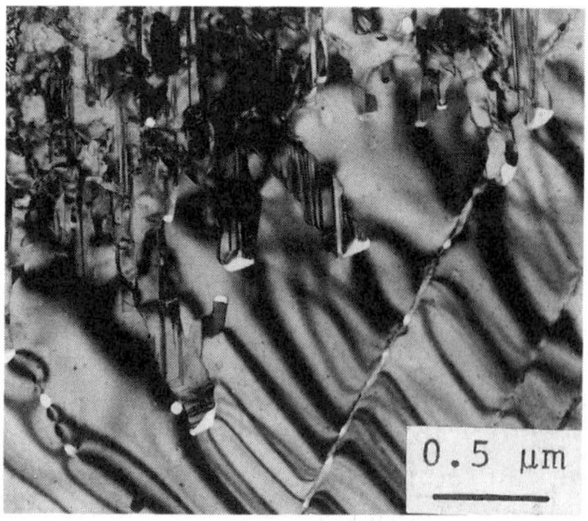

Figure 15 Polymorphous transformation of γ_4 into γ_2 in rapidly solidified Al-Ge.

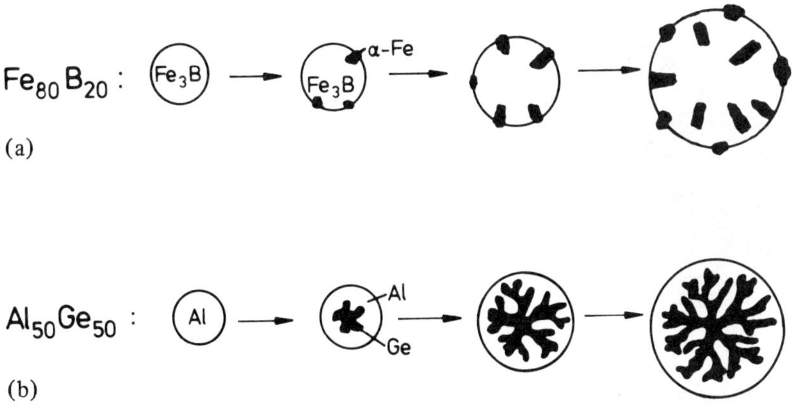

Figure 16 Eutectoid transformation (schematically) in (a) glassy $Fe_{80}B_{20}$; (b) crystalline or amorphous AlGe.

C. Decomposition of Quasicrystals

Quasicrystalline phases have been produced by rapid solidification in a number of aluminum alloys containing late transition metals [40,41]. The transformation of these metastable phases into the stable compounds has been studied in some detail in Al-Mn [42–45] as well as Al-Fe-Cu [46] alloys:

The transformation of the quasicrystalline phase has been observed to proceed either by a polymorphous or a eutectoid reaction or by precipitation, i.e., by the same reactions as observed during transformation of metastable crystalline phases or crystallization of metallic glasses. In addition, however, a peritectoid transformation can take place which will be described in rapidly solidified Al-Mn alloys [42–45]. During this peritectoid reaction the icosohedral phase (I-phase) reacts with the embedding Al matrix thus transforming into Al_6Mn. Nucleation has been observed to occur heterogeneously at the *I*-phase/Al matrix interface; growth seems to be controlled by long-range diffusion, probably along the *I*-phase/Al_6Mn interface or through the growing Al_6Mn layer. The reaction is shown schematically in Fig. 17a. Figure 17b shows a partially transformed icosahedral phase surrounded by a layer of Al_6Mn. The activation energy has been estimated to be about 208 kJ/mole which equals about the activation energy for Mn diffusion in aluminum (211 kJ/mole [47]).

Another mode of transformation has been observed in the case of very small quasicrystals embedded in the supersaturated aluminum matrix. The reaction starts with nucleation of Al_6Mn at the interfaces between quasicrystals and matrix or by precipitation of Al_6Mn in the supersaturated matrix. Decomposition of the quasicrystals proceeds by their dissolution and Mn diffusion toward the Al_6Mn particles. There is a driving force for this reaction because the equilibrium Mn content near the quasicrystal, c_I, is greater than the concentration near the Al_6Mn particle, c_β, as shown in Fig. 17d. Whether the transformation of quasicrystals proceeds by one or the other mechanism probably depends on their size. In both cases, reactions kinetics seems to be controlled by diffusion of the late transition metal in the aluminum.

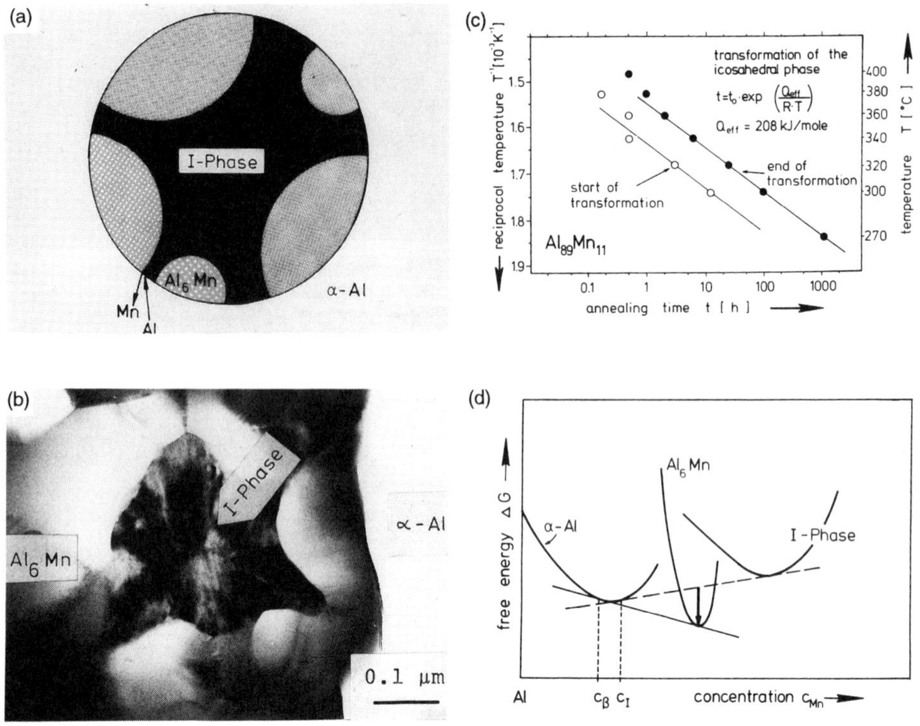

Figure 17 Transformation of the icosahedral phase in melt-spun $Al_{89}Mn_{11}$ (u = 30 m/s): (a) model for the transformation; (b) microstructure of partially transformed I-phase: 1 hr at 365°C; (c) influence of temperature on the start and termination of the reaction; (d) schematic free-energy-vs.-concentration diagram for the peritectoid transformation: I-phase + α-Al into Al_6Mn.

In some quasicrystalline alloys, e.g., $Al_{60}Cu_{30}Fe_{10}$, transformation into the stable crystalline compounds has been observed to proceed by a continuous reaction similar to the spinodal decomposition, i.e., by a transformation without any nucleation step and without a reaction front separating the transformed from the untransformed material [48].

D. Crystallization of Metallic Glasses

The crystallization of metallic glasses has been reviewed by a large number of authors over the years [49–53]. Crystallization of metallic glasses has been observed to proceed by nucleation and growth reactions in the bulk or at the surfaces (see Section V). The growth may be primary, eutectic, or polymorphic. Below the glass transition temperature, T_g, growth rates are controlled by diffusivity; nucleation occurs heterogeneously at several quenched-in nucleation sites (metal-metalloid-glasses); nucleation rates are transient for all crystallization reactions. Above the glass transition temperature, there is evidence for a transient homogeneous nucleation process; growth as well as nucleation rates have been found to be controlled by viscous flow in the undercooled melt. Close to the glass transition one can observe an overlap of both reactions.

Typical microstructures after primary crystallization are shown in Fig. 18. Because primary crystals are embedded in an isotropic, glassy matrix, it may be of interest to study the development of their morphology in more detail [53]. In Fe-based metallic glasses, for example, the morphology can be described by spheres only for very small crystal diameters. During further growth the interface becomes more and more irregular, facetted or even dendritic; controlling parameters may be perturbations in the diffusion fields, the anisotropy of surface energy, preferred growth directions, as well as the anisotropy of Young's modulus of the bcc iron crystals. A systematic investigation of the different influences on the morphology of primary crystals is still missing.

While carbon and silicon additions have been found to increase the diffusivity of the metalloid and to accelerate thereby the growth rate for primary crystallization, these additions have also been found to decrease the number of nucleation sites [54]. An increased silicon content may even change growth kinetics from parabolic to linear behavior [55,56] (see also Fig. 18d). On the other hand, Cu, or even better Au additions increase the number of nucleation sites by orders of

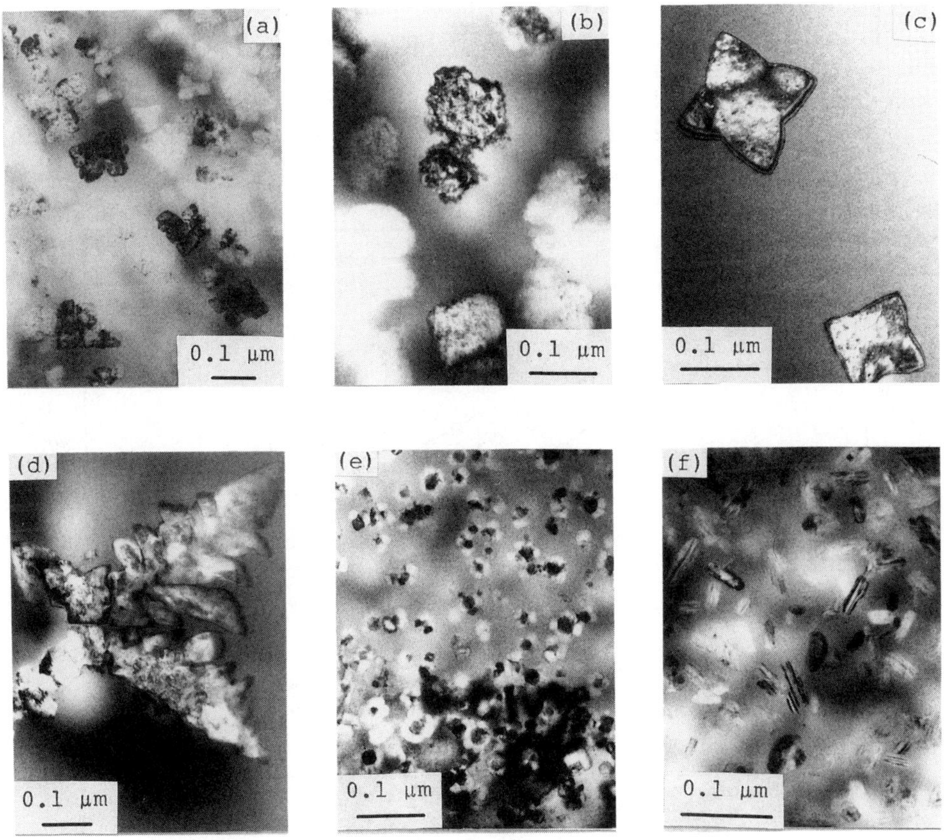

Figure 18 Morphology of primary α-Fe crystals [52]: (a) $Fe_{86}B_{14}$: 30 min at 380°C; (b) $Fe_{78}Zr_2B_{20}$: 5 hr at 430°C; (c) $Fe_{62}Ni_{22}B_{16}$: 3 hr at 350°C; (d) Metglas 2605SC: 1.5 hr 420°C; (e) $Fe_{78}Au_2B_{20}$: 5 hr at 360°C; (f) $Co_{81}B_{19}$: 90 min at 300°C.

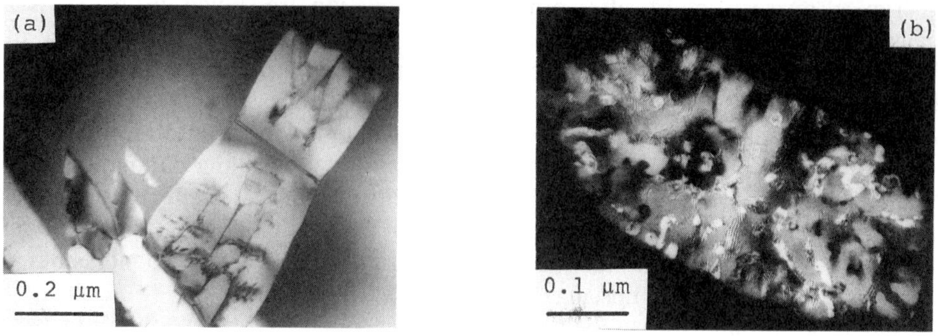

Figure 19 Morphology of partially crystallized metallic glasses after polymorphous or eutectoid crystallization.

magnitude (compare Fig. 18a and d); Cr, Co, Ni, or Pd additions have only minor influences on the kinetics of primary crystallization [54].

Typical microstructures after eutectoid as well as polymorphous crystallization of metallic glasses are shown in Fig. 19; the temperature dependence of the growth rate is shown in Fig. 20. In a number of metal-metalloid glasses, e.g., $Fe_{80}B_{20}$, an increase of interlamellar spacing has been observed with decreasing annealing temperature. In these glasses, eutectic crystallization starts usually with the formation of an iron-boride (Fig. 16a and Fig. 19b). After some probably diffusion-

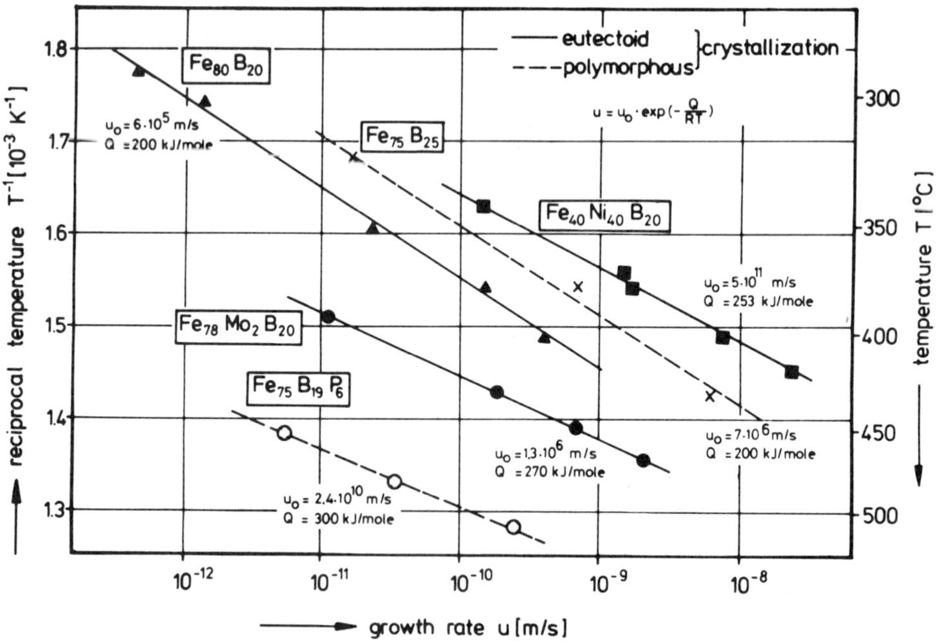

Figure 20 Temperature dependence of crystal growth rates during eutectoid and polymorphous crystallization of metallic glasses.

controlled growth of the boride phase small iron crystals are nucleated at the boride/glass interface. Since the nucleation rates for this second reaction are expected to decrease with the temperature, this reaction can be assumed to be responsible for the significant decrease in interlamellar spacing as the annealing temperature is increased.

For eutectoid as well as polymorphous crystallization below the glass transition temperature, growth can be described by an Arrhenius-type equation; prefactors and activation energies evaluated for the growth rate are quite often too large by orders of magnitude. For a eutectoid reaction this might happen due to the temperature dependence of interlamellar spacing, S, but in the case of polymorphous crystallization there must exist other reasons. Extremely large preexponentials for the growth rate are always correlated with very high activation energies and might therefore indicate growth by simultaneous transfer of a group of atoms from the glassy to the crystalline phase rather than growth by the transfer of individual atoms. Other explanations assume that impurities are present and that the temperature dependence of the growth rate does not only give the activation enthalpy for boundary movement.

Because large prefactors have only been observed in association with anisotropic growth, growth probably occurs by the lateral motion of steps. If the fraction of these active sites is assumed to be x_a, the growth rate in Eq. (1) has simply to be multiplied by x_a. Because x_a increases with annealing temperature, as indicated by a decrease of the length-to-thickness ratio of the prolate crystals, this will lead to an effectively higher activation energy and a larger prefactor. Figure 21 shows typical histograms for crystal diameter distributions.

At temperatures much above the glass transition temperature, the measured nucleation rates and growth velocities are orders of magnitude higher than extrapolated data from below T_g. The activation energies of 800–1100 kJ/mole, estimated by assuming an Arrhenius-type equation, are far from understood.

However, there exists a reasonable possibility of describing the experimental data. Assuming that above T_g, the temperature range of the undercooled melt, atomic transport is controlled by viscous flow rather than by diffusion, the diffusivity in Eqs. (1) and (7) has to be replaced by a linear combination of viscosity and diffusivity. The temperature dependence of viscosity is given by the Vogel-Fulcher-Tammann-equation:

$$\eta = A \exp\left(\frac{B}{T - T_0}\right) \tag{17}$$

with T_0 of the order of T_g; A, B are constants.

The temperature dependence of the nucleation rate and the growth rate is then given by

$$I_{St} = \left[I_{0Diff} \exp\left(-\frac{Q_N}{RT}\right) + I_{0n} \exp\left(\frac{-B}{T - T_0}\right)\right] \exp\left(-L_N \frac{\Delta G_c}{RT}\right) \tag{18}$$

$$u = \left[u_{0Diff} \exp\left(-\frac{Q_W}{RT}\right) + u_{0n} \exp\left(\frac{-B}{T - T_0}\right)\right]\left[1 - \exp\left(-\frac{\Delta G}{RT}\right)\right] \tag{19}$$

Figure 22a shows the measured data and data calculated from equations 18 and 19 assuming homogeneous nucleation above T_g and heterogeneous nucleation below

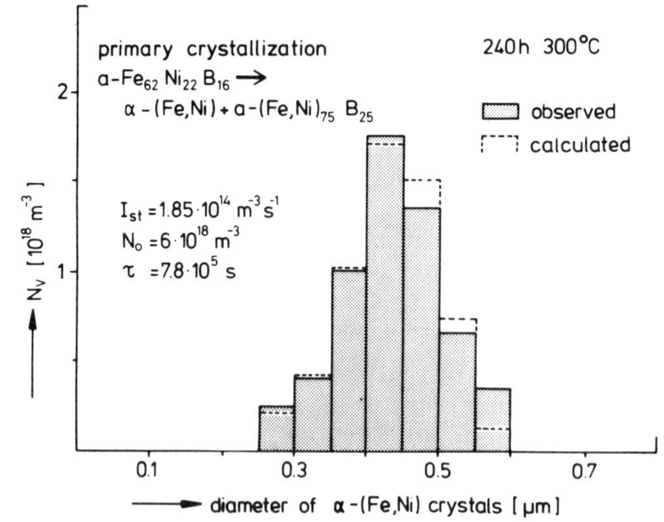

a

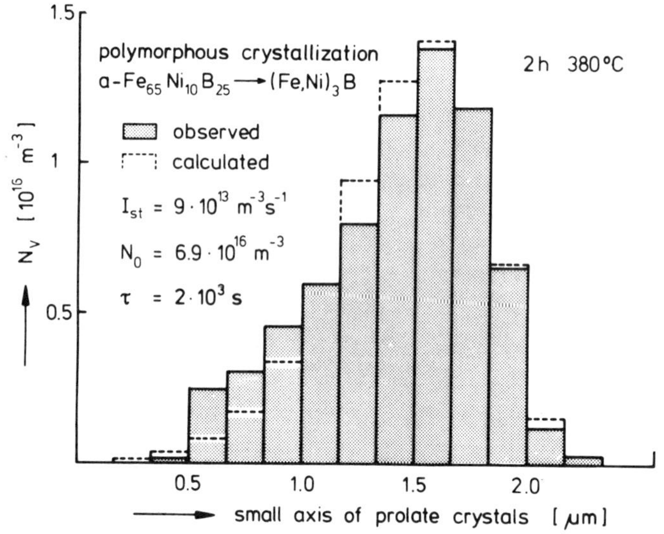

b

Figure 21 Typical histograms of crystal diameter distributions in partially crystallized metallic glasses: (a) primary crystallization in $Fe_{62}Ni_{22}B_{16}$ [53]; (b) polymorphous crystallization in $Fe_{65}Ni_{10}B_{25}$ [17].

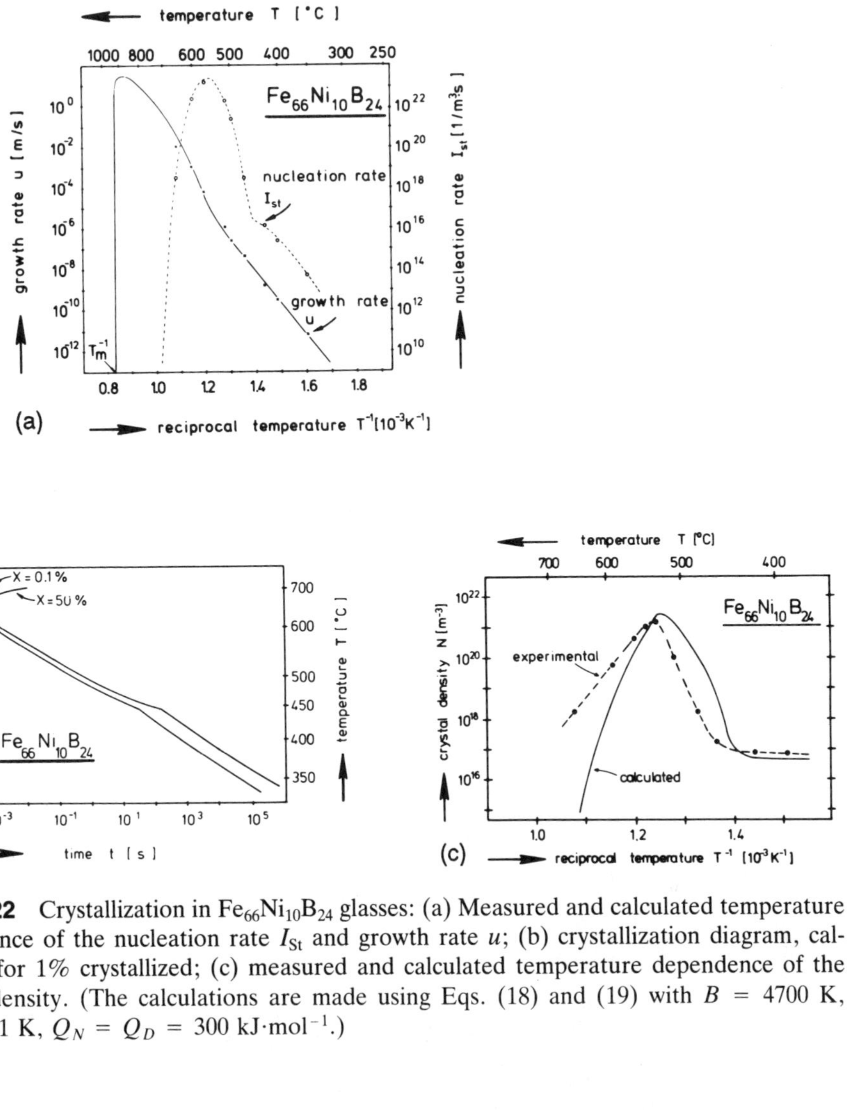

Figure 22 Crystallization in $Fe_{66}Ni_{10}B_{24}$ glasses: (a) Measured and calculated temperature dependence of the nucleation rate I_{St} and growth rate u; (b) crystallization diagram, calculated for 1% crystallized; (c) measured and calculated temperature dependence of the crystal density. (The calculations are made using Eqs. (18) and (19) with $B = 4700$ K, $T_0 = 631$ K, $Q_N = Q_D = 300$ kJ·mol^{-1}.)

T_g. Using these data one can calculate a crystallization diagram, e.g., for 1% crystallized volume, as shown in Fig. 22b as well as the temperature dependence of the number of crystals in fully crystallized $Fe_{66}Ni_{10}B_{24}$ as given in Fig. 22c. It is of large interest to realize that the smallest grain size can be achieved only at temperatures far below that of the nose in the crystallization diagram. In such materials it may be easier to produce a fine-grained material by crystallization of the glass as a precursor than directly by rapid solidification.

VI. SURFACE-INDUCED TRANSFORMATIONS

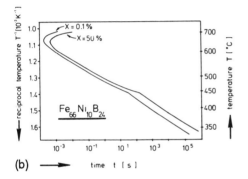

Rapidly solidified materials are usually produced as very thin ribbons or flakes, thus leading to a quite large surface-to-volume ratio. There are a number of reasons to expect phase transformations to occur preferentially at a surface. During the process of rapid solidification itself as well as during annealing treatments of such material surfaces are expected to accelerate nucleation as the new phase replaces a portion of the original surface, thus reducing the total surface energy required for nucleation. Stresses building up during the phase transformation due to the change of volume can be more easily relieved at a surface, for example, by reduction in the thickness of the crystallizing phase, as shown in Fig. 23, during the crystallization of an amorphous selenium film [57]. Surface crystallization may also be accelerated by faster atomic transport on a free surface, but probably this effect will be ruled out in most cases by oxidation. Segregation and selective oxidation of one of the components can change locally the surface composition and thus the driving force for a phase transformation.

For precipitation from solid-state-quenched supersaturated solid solutions, it is known that the sequence of precipitation reactions, their kinetics, as well as the morphology and the orientation relationship with the matrix can be changed at the surface. For example, precipitation in Al-Cu alloys [12,13,58]: GP-zone formation even at room temperature, Θ''-precipitation at higher temperatures followed by Θ'-precipitation on (100) planes of the aluminum matrix, and Θ with more complex orientation relationships. Figure 24 shows the typical microstructure after annealing of bulk and thinned Al–4% Cu: the orientation relationship $(001)_{Al}$ parallel to $(001)_{\Theta'}$ and the platelike shape are known to be due to a minimum of the strain energy and the very low energy of the semicoherent interface. No such strain energy is generated during precipitation at a free surface. On the other hand, the semicoherent $(001)_{Al}/(001)_{\Theta'}$ interface has a lower energy than an interface with random orientation after surface nucleation. However, replacing parts of the surface, there will be again a reduction in total surface energy. Due to these reductions in total free surface and strain energy, and the probable occurrence of more rapid surface diffusion, nucleation, and growth at the surface should be favored, which is in agreement with the observations.

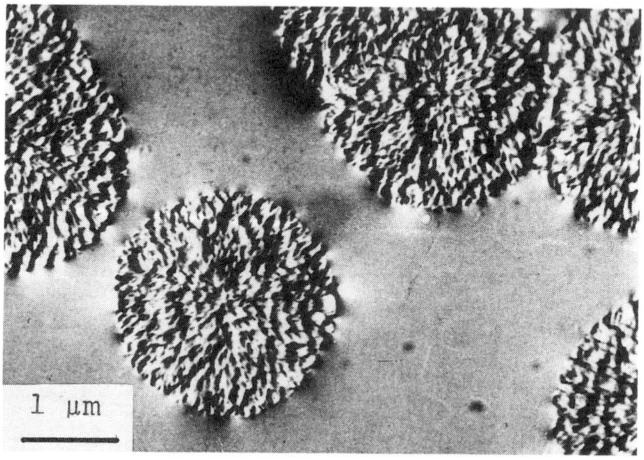

Figure 23 Crystallization of an amorphous selenium film [57] (thickness: 800 nm/65 hr at 80°C).

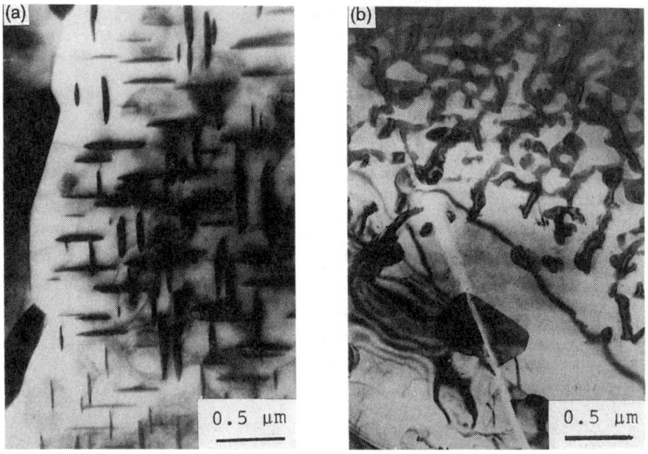

Figure 24 Θ′-precipitation in Al–4% Cu [12]: (a) 15 min 320°C annealed in the bulk; (b) 20 min 280°C annealed in situ in the microscope.

Grain boundaries as well as surfaces are sinks for vacancies, which are necessary for the formation of Θ″-precipitates. Precipitation-free zones with typical widths of about 0.5 μm are observed along grain boundaries, explaining the impossibility of Θ″ precipitation close to a surface.

Kang et al. [58] studied in detail the growth kinetics of Θ particles grown in thin foils. Precipitation normally occurs as surface precipitation having quite different morphology than in the bulk. By direct observations of interfacial structures at high temperatures, the authors found that ledges move very quickly during growth, while interfacial dislocations are more or less fixed. As in bulk material the growth kinetics of Θ-phase can be explained by models based on these interfacial structures and the related morphologies. However, at the surface, the effective diffusivity is about one order of magnitude larger than in the bulk because the surface provides a short-circuit diffusion path.

Because precipitation reactions are usually controlled by volume diffusion, the influence of surface-induced nucleation on the overall transformation is limited to relative thin layers of a few micrometers or even less.

Whereas grain boundaries in crystalline materials exhibit potency as nucleation sites similar to surfaces, surface crystallization in metallic glasses has been found to have an extremely large influence on crystallization kinetics. Among all reasons for preferred nucleation mentioned for crystallization reactions, local changes of surface composition by selective oxidation or segregation are assumed to have the strongest influence, as reviewed only recently [59]. Understanding surface crystallization, which can transform easily the whole ribbon prior to any nucleation in the bulk (Fig. 25), is of utmost importance for a number of applications of metallic glasses, e.g., corrosion resistance or catalysis; surface crystals may act as pinning centers and increase magnetic eddy current losses in high frequency applications [60]. Magnetic anisotropy after annealing is assumed to originate from compressive stresses due to the formation of crystalline surface layers with a higher density [61].

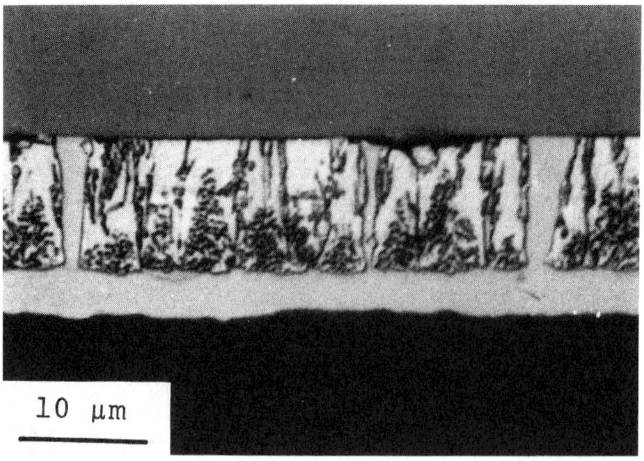

Figure 25 Surface-induced crystallization in $Ni_{66}B_{34}$ glasses (1 hr at 380°C/cross section).

Designing microstructures by controlled crystallization may require the use or suppression of surface-induced crystallization.

Surface crystallization of Zr-based glasses is controlled by the high affinity of the zirconium for oxygen. Glassy Cu-Zr ribbons are well known to change color to a copperlike red during storage at room temperature. Besides large ZrO_2 plates, an extremely fine-grained microstructure consisting of ZrO_2 and copper particles has been found by transmission electron microscopy at the surface of eight-year-old $Cu_{30}Zr_{70}$ ribbons (see Fig. 26); this glass usually crystallizes into one or more intermetallic compounds in the temperature range above 250°C.

In amorphous $Zr_{70}Au_{30}$ alloys an even more pronounced oxidation tendency has been reported [62]. These glassy alloys oxidize during room temperature aging in air and even pulverize into a very fine duplex structure of monoclinic ZrO_2 matrix

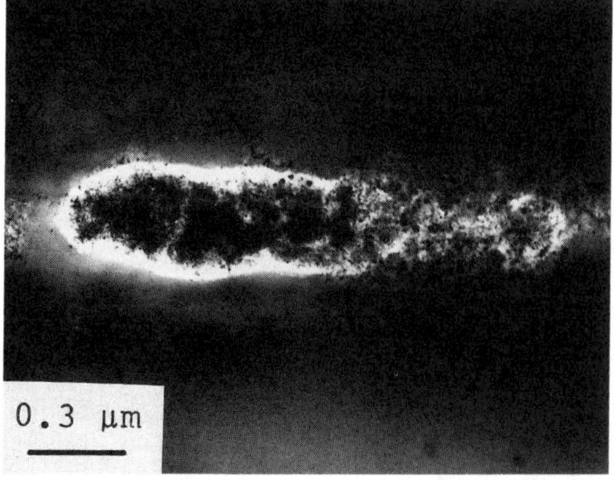

Figure 26 Partially oxidized contact side of a glassy $Zr_{70}Cu_{30}$ ribbon (eight years at room temperature/air).

and gold particles with a size as small as about 2 nm. This oxidation tendency may be interpreted as resulting mainly from a combination effect of a weak attractive interaction between zirconium and a late transition element, the high catalytic ability of gold as well as the large negative free energy for the formation of ZrO_2. Due to the local reduction in zirconium content in $Fe_{90}Zr_{10}$ metallic glasses, the ribbon surface has been observed to crystallize into only α-Fe and at temperatures about 300°C lower than that at which primary crystallization in the bulk [63] occurs.

In Fe-B glasses with boron contents of about 20 at%, selective oxidation of the boron at the surface (and thus enrichment of the surface iron content) is assumed to be responsible for the excessive primary crystallization of α-Fe at the surface instead of the expected eutectic reaction found to occur in the bulk. Such a behaviour has been studied in some detail by Mössbauer spectroscopy [64]. In $Pd_{80}Si_{20}$ glasses, surface oxidation and silicon outdiffusion lead to a silicon-depleted zone and thus primary crystallization of palladium [65]. Phosphorus is known to segregate at the surface of Fe-Ni-based metallic glasses [66]. In $Fe_{86}P_{14}$ glasses, sublimation of phosphorus has been observed during annealing in a vacuum, but enrichment in phosphorus and thus crystallization of Fe_2P during annealing in an argon atmosphere [64]. On the other hand, in Metglas 2826 ($Fe_{40}Ni_{40}P_{14}B_6$) surface crystallization has not been observed at all.

$Fe_{75}B_{25}$ metallic glasses are known to crystallize by a polymorphous reaction into tetragonal Fe_3B. Replacing iron by nickel leads to a decrease in the crystallization temperature and finally to crystallization of orthorhombic $(Fe, Ni)_3B$ instead of the tetragonal boride. In $Fe_{65}Ni_{10}B_{25}$ glasses preferred iron oxidation may lead locally to a nickel-enriched surface, thus promoting the nucleation of orthorhombic $(Fe, Ni)_3B$, typical for nickel-rich glasses instead of the tetragonal boride which is still observed in the bulk.

The difference in the potency of enhancing surface crystallization between the free surface and the substrate contact side in many metal-metalloid glasses might be explained by assuming the presence of thin, slightly protective oxide films (probably B_2O_3 in the boron-containing glasses) which probably slow down further selective oxidation [59, 67]. At the free surface, heat extraction during the solidification process is slower and a continuous B_2O_3 film can be formed, perhaps even without any boron-depleted zone underneath; the thickness of this layer decreases with increasing cooling rate. At the substrate contact side, and more especially along the lines of best contact with the casting wheel, only an inhomogeneous film consisting of B_2O_3 and an iron oxide with porous structure can be formed. Therefore, along this line of intimate contact easy selective oxidation can occur and lead to an enrichment in nickel, thus accelerating crystallization of orthorhombic $(Fe, Ni)_3B$ along these lines, as shown in Fig. 27. Another reason might be a physical damage of the surface during lifting of the ribbon off the substrate wheel during casting. Faster quenching rates made this effect even more pronounced. The existence of such native oxide layers of different thickness and composition was proposed by Kisfaludi et al. [68] in a study on catalytic activity of iron-based metallic glasses. In Metglas 2605S-2 ($Fe_{78}B_{13}Si_9$) after annealing at 420°C for 2 hr bands of dendritic primary α-Fe crystals have been observed by Liebermann et al. [69] near the substrate-contact surface of the ribbon. It is not clear whether these bands which are parallel to the ribbon casting direction and only 1 μm apart are due to the same reason.

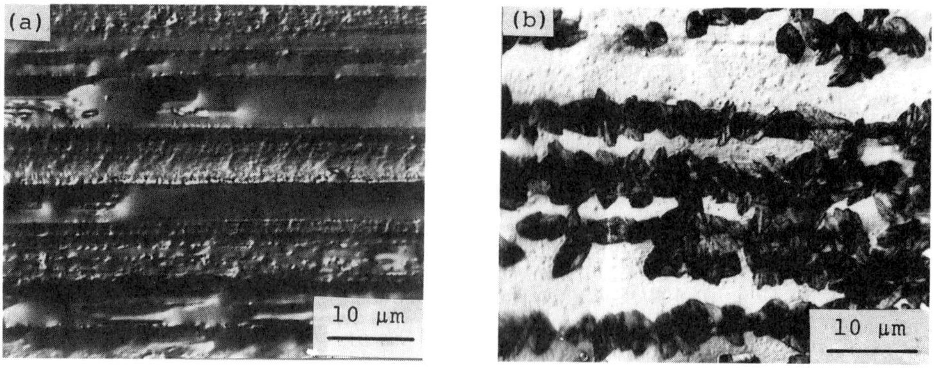

Figure 27 Influence of surface topography of the contact side on nucleation in $Fe_{39}Ni_{39}B_{22}$: (a) as cast; (b) partially crystallized along the groove lines (20 min at 380°C).

If surface heterogeneous nucleation is important, it is of interest to study crystallization of ribbons with their original surface removed. In a $Ni_{66}B_{34}$ glass such a treatment eliminates totally the quenched-in Ni_2B nuclei and thus the characteristic crystallization of Ni_2B with its strong (001) texture. Fig. 28 shows the surface of a ribbon which was annealed after remelting parts of the original surface with laser pulses (Nd:YAG-laser: 200 ns, 90 mJ/pulse) [70]. Whereas most of the surface is fully crystallized after the annealing treatment, no crystallization of Ni_2B or other phases took place in the remelted area. Remelting of a thin surface layer followed by the extreme fast self-quenching has been found to eliminate all quenched-in nuclei and therefore to increase thermal stability significantly.

The influence of surface treatments on crystallization can be summarized for glassy $Fe_{39}Ni_{39}B_{22}$ ribbons [71]: ion beam milling has been found to suppress any

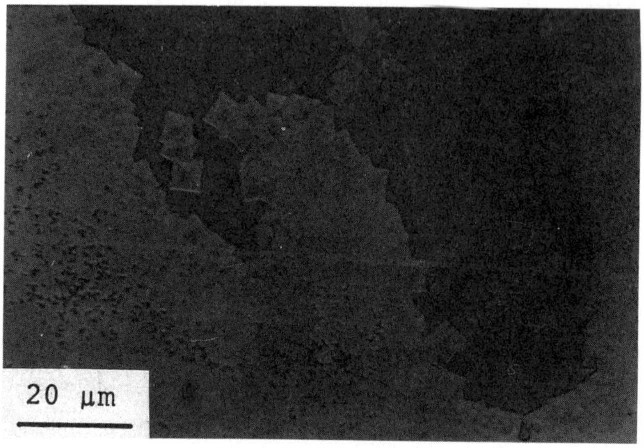

Figure 28 Surface crystallization in a $Ni_{66}B_{34}$ glass [70]: 16.5 hr at 325°C after remelting a surface area by Nd:YAG-laser irradiation.

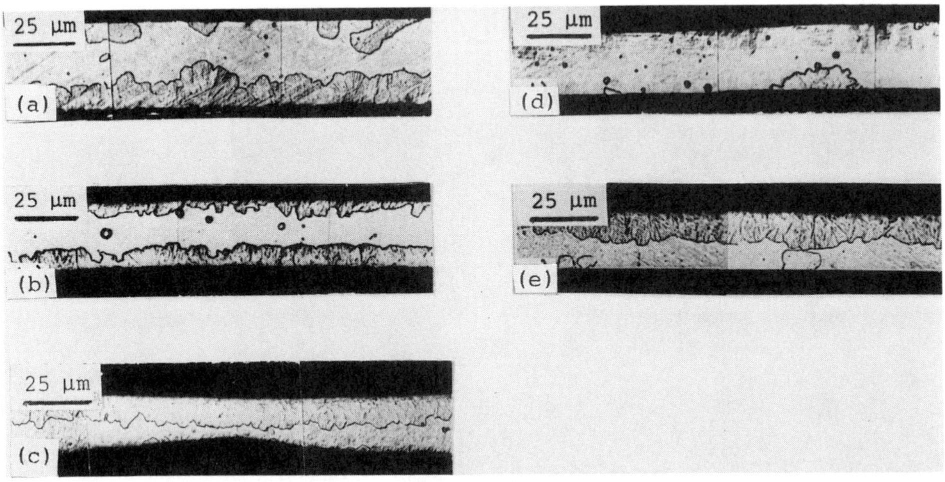

Figure 29 Influence of surface treatments on surface crystallization of melt-spun $Fe_{39}Ni_{39}B_{22}$ [71] (70 min at 390°C): (a) as-cast ribbon; (b) annealed after electropolishing; (c) annealed after electropolishing and iron deposition onto the bottom (substrate) side) ion beam milling; (d) annealed after ion beam milling; (e) annealed after ion beam milling and nickel deposition onto the top (free) surface of the ribbon.

surface-induced crystallization as it removes the nickel-enriched layer after selective oxidation. There is nearly no surface-induced crystallization remaining as shown in a cross section of the ribbon in Fig. 29d. A similar effect has been found after deposition of a thin iron layer onto an electropolished surface which would otherwise exhibit severe surface induced crystallization; however, the diffusion of the iron into the nickel-enriched surface layer destroys this ability. A nickel film deposited onto a ribbon surface cleaned by ion beam milling induces again severe crystallization in accordance with the assumption made.

VII. DESIGN OF PARTICULAR NOVEL AND USEFUL MICROSTRUCTURES BY PHASE TRANSFORMATION IN RAPIDLY SOLIDIFIED ALLOYS

A. Finemet [72]

Fe-B as well as Fe-B-Si glasses with an iron content less than 20 at% crystallize by a primary reaction. Copper additions are known to increase the number of α-iron crystals by orders of magnitude, whereas additions such as Zr reduce the ability to form the boride [54]. Only recently, a Japanese group combined these effects in producing a unique material with extremely fine iron crystals embedded in an amorphous matrix, leading to a soft magnetic material with low magnetostriction [72]. Such a microstructure with primary crystals 5 nm in diameter is shown in Fig. 30.

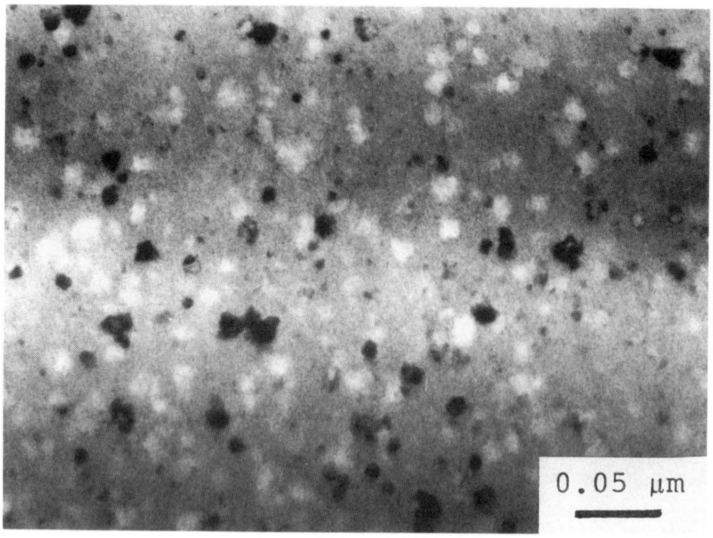

Figure 30 Microstructure of Finemet (partially crystallized $Fe_{73.4}Cu_1Nb_{3.1}Si_{13.4}B_{9.1}$: 1 hr at 470°C).

B. Al-Ge

Rapid solidification of Al-Ge alloys results, as shown earlier in this chapter, in the formation of a brittle ribbon containing a number of metastable crystalline phases; an amorphous microstructure has been obtained only after ultrarapid solidification, e.g., by splat cooling or by vapor deposition. The metastable crystalline as well as the amorphous phases transform into the stable equilibrium phases by a eutectoid reaction into an extremely fine microstructure which is associated with ductile behavior as, for example, in the Al–30 at% Ge alloy shown in Fig. 31. Such a microstructure, with hard germanium particles embedded in a soft aluminum ma-

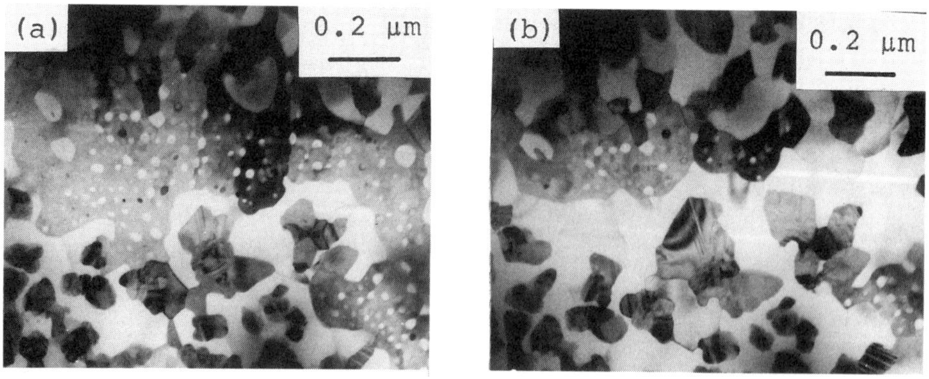

Figure 31 Decomposition of metastable γ_1 into Al + Ge [34] (both micrographs show the same area after further annealing at 180°C).

trix, exhibit reasonable ductility and has been proposed only recently to be used as a brazing foil [73]. The eutectic temperature is only 430°C.

REFERENCES

1. J. W. Christian, in *The Theory of Transformations in Metals and Alloys*, 2nd ed. (Pergamon, Oxford, 1975)
2. U. Köster, U. Herold, and D. Krause, in *Rapidly Solidified Amorphous and Crystalline Alloys*, edited by B. H. Kear, B. C. Giessen, and M. Cohen (Elsevier, New York, 1982), p. 179.
3. R. K. Garrett, Jr., and T. H. Sanders, Jr., in *Chemistry and Physics of Rapidly Solidified Materials*, edited by B. J. Berkowitz and R. O. Scattergood (AIME, New York, 1983), p. 11.
4. S. C. Huang and A. M. Ritter, in Ref. 3, p. 25.
5. J. A. van der Hoeven, P. van Mourik, and E. J. Mittemeijer, *J. Mater. Sci. Lett.*, *2*, 158 (1983).
6. B. Schuhmacher and U. Köster, in *Metallic and Semiconducting Glasses, Key Engineering Materials*, Vol. 13–15, edited by A. K. Bhatnager, (Trans. Tech. Publ., 1987), p. 257.
7. M. Atzmon and W. L. Johnson, CALTECH-Report, CALT-822-138, Pasadena, 1982.
8. C. S. Barrett and T. B. Massalski, *Structure of Metals*, 3rd ed. (McGraw-Hill, New Yori, 1966).
9. R. Bormann, F. Gärtner, and P. Haasen, *Z. Phys. Chem*, *157*, 29 (1988).
10. K. Zöltzer and R. Bormann, *J. Less-Common. Metal*, *140*, 335 (1988).
11. A. R. Miedema, *Philips Tech. Rev.*, *36*, 217 (1976).
12. U. Köster, *Kristall Techn.*, *14*, 1369 (1978).
13. E. Hornbogen, in *Nucleation*, edited by A. C. Zettlemoyer (Marcel Dekker, New York, 1969), p. 309.
14. I. Gutzow and S. Toschev, in *Advances in Nucleation and Crystallization of Glasses*, edited by L. L. Hench (American Ceramic Society, Columbus, 1971), p. 10.
15. K. F. Kelton, A. L. Greer, and C. V. Thompson, *J. Chem. Phys.*, *79*, 6261 (1983).
16. U. Köster and H. Blanke, *Scripta Metall.*, *17*, 495 (1983).
17. H. Blanke and U. Köster, in *Rapidly Quenched Metals*, Vol. I, edited by S. Steeb and H. Warlimont (North-Holland, 1985), p. 227.
18. S. Budurov, P. Zielinski, K. Marchev, in *Proc. 1st Int. Conf. Rapidly Quenched Metallic Alloys* (Akademie der Wissenschaften, Dresden, 1988), p. 377.
19. A. Kelly and R. B. Nicholson, *Prog. Mater. Sci.*, *10*, 149 (1963).
20. M. H. Jacobs, A. G. Dogget, and M. J. Stowell, *Fizika 2*, Suppl. 2, 18.1 (1970).
21. E. Blank, in Ref. 20, p. 24.1.
22. M. Laridjani, K. D. Krishnanand, and R. W. Cahn, *J. Mater. Sci.*, *11*, 1643 (1976).
23. T. Miyazaki, S. Takagishi, H. Mori, and T. Kozakai, *Acta Metall.*, *28*, 1143 (1980).
24. T. Kozakai, J. Takabatake, and T. Miyazaki, in *Proc. 4th Int. Conf. Rapidly Quenched Metals*, Vol. II, edited by T. Masumoto and K. Suzuki (The Japan Institute of Metals, Sendai, 1982), p. 1573.
25. U. Köster, *Mater. Sci. Eng.*, *5*, 174 (1969/70).
26. M. Beller, Ph.D. Dissertation, MPI Stuttgart (1971).

27. M. Schmücker and U. Köster, *Mater. Sci. Eng.*, *A133*, 486 (1991).
28. U. Köster and C. Caesar, in *Rapidly Solidified Metastable Materials*, edited by B. H. Kear and B. C. Giessen (Elsevier, New York, 1984), p. 419.
29. B. Ilschner and E. Mlitzke, *Acta Metall.*, *13*, 855 (1965).
30. Y. Khan, E. Kneller, and M. Sostarich, *Z. Metallkde.*, *73*, 624 (1982).
31. U. Köster, U. Herold, and H. Weissenberg, in *Conf. on Metallic Glasses: Science and Technology*, Vol. 2, edited by C. Hargitai, I. Bakonyi, and T. Kemény (Central Research Institute for Physics, Budapest, 1981), p. 253.
32. S. Omori and Y. Hashimoto, *Trans. Jpn. Inst. Metals*, *17*, 571 (1976).
33. C. Suryanarayana and T. R. Anantharaman, *J. Mater. Sci.*, *5*, 992 (1970).
34. U. Köster, in Ref. 20, p. 12.1.
35. U. Köster, *Acta Metall.*, *20*, 1361 (1972).
36. U. Köster, *Z. Metallkde.*, *63*, 472 (1972).
37. A. J. McAlister and J. L. Murray, *Bull. Alloy Phase Diagrams*, *5*, 341 (1984).
38. M. J. Kaufmann and H. L. Fraser, *Acta Metall.*, *33*, 191 (1985).
39. U. Köster and U. Herold, *Scripta Metall.*, *12*, 75 (1978).
40. D. Schechtman, I. Blech, D. Gratias, and J. W. Cahn, *Phys. Rev. Lett.*, *53*, 1951 (1984).
41. D. Schechtman, in *Quasicrystals*, *Materials Science Forum*, Vols. 22–24, edited by K. H. Kuo (Trans. Tech. Publ., 1987), p. 1.
42. U. Köster and B. Schuhmacher, *Mater. Sci. Eng.*, *99*, 417 (1988).
43. M. Harmelin, in *Quasicrystalline Materials*, edited by Ch. Janot and J. M. Dubois (World Scientific, Singapore, 1988), p. 19.
44. K. Yu-Zhang, M. Harmelin, A. Quivy, Y. Calvayrac, J. Bigot, and R. Portier, *Mater. Sci. Eng.*, *99*, 385 (1988).
45. K. Chattopadhyay and N. K. Mukhopadhyay, in Ref. 41, p. 639.
46. W. Liu and U. Köster, *Mater. Sci. Eng.*, *A133*, 388 (1991).
47. G. M. Hood and R. J. Schultz, *Phil. Mag.*, *23*, 1479 (1971).
48. W. Liu and H. Köster, *Phase Transformation*, (1992), in press.
49. U. Köster and U. Herold, in *Glassy Metals I*, *Topics in Applied Physics*, Vol. 46, edited by H.-J. Güntherodt and H. Beck (Springer-Verlag, Berlin, 1981), p. 225.
50. M. Scott, in *Amorphous Metallic Alloys*, edited by F. E. Luborsky (Butterworths, London, 1983), p. 114.
51. U. Köster, *Z. Metallkde.*, *74*, 691 (1984).
52. U. Köster, *Ann. New York Acad. Sci.*, *484*, 39 (1986).
53. U. Köster, in *Proc. Phase Transformations '87*, edited by G. W. Lorimer (The Institute of Metals, Cambridge, 1987), p. 597.
54. U. Köster, R. Abel, and H. Blanke, *Glastechn. Ber.*, *K56*, 584 (1983).
55. J. C. Swartz, R. Kossowsky, J. J. Haugh, and F. F. Krause, *J. Appl. Phys.*, *52*, 3324 (1981).
56. A. Zaluska and H. Matyja, in Ref. 24, Vol. I, p. 683.
57. U. Köster, *Adv. Colloid Interface Sci.*, *10*, 129 (1979).
58. S. K. Kang and C. Laird, *Acta Metall.*, *22*, 1481 (1974); *Acta Metall.*, *23*, 35 (1975).
59. U. Köster, *Mater. Sci. Eng.*, *97*, 233 (1988).
60. A. Datta, N. J. DeCristofaro, and L. A. Davis, in Ref. 24, p. 1007.
61. H. N. Ok and A. H. Morrish, *Phys. Rev.*, *B23*, 1835 (1981).
62. H. Kimura, A. Inoue, T. Masumoto, and S. Itabashi, *Sci. Rep. Res. Inst., Tohoku Univ.*, *A34*, 183 (1986).
63. M. Fujinami and Y. Ujihira, *J. Appl. Phys.*, *59*, 2387 (1986).
64. H.-G. Wagner, M. Ackermann, R. Goa, and U. Gonser, in Ref. 17, p. 247.
65. M. Maeda and M. Taniwaki, in Ref. 24, Vol. I, p. 643.

66. D. R. Baer, L. R. Pederson, and M. T. Thomas, *Mater. Sci. Eng.*, *48*, 283 (1981).
67. A. Garcia Escorial and A. L. Greer, *J. Mater. Sci.*, *22*, 4388 (1987).
68. G. Kisfaludi, Z. Schay, L. Guczi, G. Konczos, L. Lovas, and P. Korvacs, *Appl. Surf. Sci.*, *28*, 111 (1987).
69. H. H. Liebermann, J. Marti, R. J. Martis, and C. P. Wong, *Met. Trans.*, *20A*, 63 (1989).
70. G. Steinbrink, Ph.D. Dissertation, Universität Dortmund, 1992.
71. U. Herold, Ph.D. Dissertation, Ruhr-Universität Bochum, 1982.
72. Y. Yoshizawa, K. Yamauchi, and S. Oguma, European Patent EP 0,271,657 A2 (Oct. 1987).
73. L. Jllgen, H. Mühlbach, W. Löser, H.-G. Lindenkreuz, E. Alius, D. Rühlücke, M. Müller, *Mater. Sci. Eng.*, *A133*, 738 (1991).

12

Microstructure-Property Relations in Rapidly Solidified Crystalline Alloys

S. K. Das
AlliedSignal Corporate Research and Technology, Morristown, New Jersey

F. H. Froes
Institute for Materials and Advanced Processes University of Idaho, Moscow, Idaho

I. INTRODUCTION

The recent progress in rapid solidification processing (RS) of crystalline alloys has allowed a degree of microstructural control which has never before been achieved by conventional ingot metallurgy methods. The microstructural refinements include extended solid solubility, increased chemical homogeneity, refined grain size, and formation of metastable phases. These attributes of RS have been utilized in many alloy systems in designing alloys which exhibit improved mechanical properties. Particular examples of alloy systems which have received wide attention are high-strength corrosion-resistant magnesium alloys, reduced-density aluminum-lithium alloys, high-temperature aluminum alloys, and titanium alloys. Other alloy systems, such as copper-, nickel-, iron-, niobium-, and tungsten-based alloys, which show potential benefits of RS are also of interest.

This chapter will review advances that have occurred during the past few years in the understanding of microstructure-property relationship in rapidly quenched crystalline alloy systems. The information presented will be drawn from many sources including recent review articles on various crystalline alloy systems [1–3]. The alloy systems reviewed are discussed with emphasis on the variation of mechanical property sets with rapidly solidified microstructure. In general, commercial application of a new alloy is possible only if an important material property is significantly improved without concurrently sacrificing important secondary properties. For example, for structural alloys the primary mechanical properties such as strength/ductility, density, and stiffness should be improved while at least maintaining the secondary properties such as fatigue, creep, fracture toughness, corrosion, and oxidation. There are numerous examples of cases where attempts to improve one property results in degradation of other properties. For example, improvement in strength is generally associated with a decrease in toughness.

Although the design of new alloys to take advantage of RS will vary depending on the particular property requirements, there is one basic requirement for the development of high-performance structural alloys by RS, i.e., retention of beneficial modifications of the as-cast microstructures during subsequent powder metallurgy (P/M) processing. In general, the as-cast microstructure is difficult to preserve if one uses conventional P/M consolidation techniques. Thus, in order to take advantage of RS for the development of high-performance alloys, thoughtful alloy design is necessary such that the desirable microstructure present in the rapidly solidified alloys is not lost during high-temperature consolidation. This aspect of the alloy microstructure will be emphasized during the discussion of various alloy systems in this chapter.

II. ALLOY SYSTEMS

Studies of rapidly solidified crystalline alloys range from those on beryllium at one end of the periodic table to those on tungsten and uranium, near the other. However, we will limit our discussion to some selected systems where real benefits of RS on mechanical properties have already been demonstrated. Emphasis is placed on systems such as magnesium-, aluminum-, and titanium-based alloys which have reached commercial significance. Other alloy systems such as iron, nickel, niobium, and tungsten, which have been widely investigated but have not reached commercial significance, are also briefly discussed.

A. Magnesium Alloys

Interest in rapid solidification of magnesium alloys stems from the fact that conventional ingot metallurgy alloys exhibit poor strength, ductility, and corrosion resistance. Such properties can be improved by microstructural refinement via rapid solidification processing.

Table 1 summarizes some of the relevant magnesium alloy systems studied by RS. It is not meant to be an exhaustive survey but rather highlights the variety of processes used and the properties investigated. A bibliography on RS magnesium alloys for 1950 to 1988 has been prepared by Jones [22]. A comprehensive review on observations of constitutional changes produced by rapid solidification of magnesium alloys has been recently given by Hehmann and Jones [23]. These include solid solubility extension [10,11] and formation of amorphous [4,5] and quasicrystalline [23] phases. As can be seen in Table 1, the majority of the early studies on RS magnesium alloys have been on either commerical alloys or simple binary alloys; for example, Calka et al. [4] studied amorphous alloys of the composition $Mg_{70}Zn_{30}$ made by melt spinning. Microcrystalline $Mg_{100-x}Zn_x$ alloys with $x = 26–32$ at% have been produced by crystallization of amorphous splats prepared by a gun technique [5]. Masur et al. [6] studied microstructure of microcrystalline magnesium alloy ribbons containing 1.7 to 2.3 at% Zn made by melt spinning. The homogeneous solid solution range of such ribbon was found to be limited to a chill zone (the ribbon surface next to the quenching substrate) of 10 to 20 μm wide, beyond which a two-phase region was observed [6]. Electron transport properties of Mg-Zn metallic glasses have been recently studied by Mizutani and Matsuda [8]. In all of the aforementioned studies, no attempt was made to determine the mechanical

Table 1 Magnesium Alloys Investigated Using Rapid Solidification Process

Alloy system	RS technique used	Properties studied	Reference
$Mg_{70}Zn_{30}$	Splats, gun technique	Metallic glass formation	4
$Mg_{bal}Zn_{26-32}$	Splats, gun technique	Crystallization of metallic glass	5
$Mg_{bal}Zn_{1.7-2.3}$	Melt spinning/ extrusion	Mech. prop.	6, 7
Mg-Zn	Melt spinning	Electron transport properties	8
Mg-Ca	Melt spinning	Solid solubility extension	9
Mg-Al	Splats	Solid solubility extension	10
Mg-Mn	Splats	Solid solubility extension thermal stability	11
Mg-Zr	Splats	Solid solubility extension	10, 11
AZ91	Melt spinning	Mech. prop.	12, 13
ZK60	Melt spinning, double extrusion	Mech. prop.	7
	Atomization, rotating electrode process	Mech. prop.	14
	Atomization, rotating disk	Mech. prop.	15, 16
EZ33	Melt spinning/ extrusion	Mech. prop.	7
Mg–9 Li–2 X (X = Si, Ce)	Twin-roll quenching, melt spinning	Microstructure, mech. and corrosion prop.	9, 17
$Mg_{bal}Al_{8-10}Si_{1-4}$	Melt spinning/ extrusion	Microstructure, mech. and corrosion prop.	18, 19
$Mg_{bal}Al_{5-8}Zn_{1-2}Si_{0-2}$	Melt spinning/ extrusion	Microstructure, mech. and corrosion prop.	18, 21
$Mg_{bal}Zn_{1-2}Al_{5-8}$ $RE_{0.5-2}$ (RE:Nd,Pr,Ce,Y)	Melt spinning/ extrusion	Microstructure, mech. and corrosion prop.	9, 20, 21

properties of either the amorphous or microcrystalline alloys. Much of the earlier work on RS magnesium alloys involved conventional commercial alloys. Only recently has RSP been exploited to design novel magnesium alloys with fine stable-phase dispersoids which act as effective grain-refining agents and provide dispersion strengthening for improved mechanical properties [9,18,20,21].

1. Commerical Magnesium Alloys

One of the earliest works on RS magnesium alloys was by Busk and Leontis [24], who investigated three commercial alloys (AZ32, M1, and ZK60A) made from atomized powder. For the same extrusion conditions, tensile and compressive yield strengths were up to 60% higher for powder extrusions (299 MPa) compared with ingot extrusions (184 MPa).

Isserow et al. studied commercial ZK60A magnesium alloy powder made by a rotating electrode process [14]. The mechanical properties of the room temperature extrusions were significantly better than those obtained by Busk and Leontis [24]. However, care must be exercised in comparing their mechanical properties in the longitudinal direction from room temperature extrusions since they observed significant delamination on the fracture surfaces; i.e., the properties may be highly inferior in the transverse direction. For extrusions made at 394 K (250°F) the improvement in the mechanical properties over those of conventionally prepared alloy was only marginal. This is due to the fact that, in general, the as-quenched microcrystalline structure is difficult to preserve if one uses conventional consolidation techniques such as hot pressing, hot isostatic pressing (HIP) or hot extrusion to make bulk shapes.

Recent work by Nussbaum et al. on the extrusion of rapidly solidified AZ91HP alloy powder made by melt spinning shows a 70% improvement in yield strength (391 MPa) over conventionally processed AZ91HP (226 MPa) using the same extrusion conditions [13]. The improvement in yield strength is mainly due to the refinement of grain size.

In order to take advantage of RS, prudent alloy design is necessary such that thermally stable fine dispersoids can be formed that resist coalescence and growth during high-temperature consolidation. Since the grain size of as-quenched microcrystalline alloys is usually extremely fine, the stable dispersoids inherently pin the grain boundaries and prevent their coarsening during high-temperature consolidation. The dispersion-strengthened alloy systems that have shown significant mechanical property improvements are Mg-Al-Zn-Si and Mg-Al-Zn-rare earth alloys.

2. *Mg-Zn-Al-Si Alloys*

One of the thermally stable intermetallic compounds in magnesium base alloys in Mg_2Si. It has a high melting point (1375 K), low density (1.98×10^3 kg/m^3), and limited solid solubility in magnesium (0.004 wt%), and is a good candidate for exploitation by rapid solidification. Interest in Mg-Zn-Al-Si alloys stems from the beneficial effect of zinc in providing solid solution and precipitation strengthening.

Table 2 summarizes the tensile properties together with Rockwell B (R_B) hardness measured at room temperature for a number of Mg-Zn-Al-Si alloys. Both the yield strength and ultimate tensile strength are higher than the commerical high-strength wrought alloy ZK60A and high-purity corrosion-resistant casting alloy (AZ91D). For example, $Mg_{91}Zn_2Al_5Si_2$ has a yield strength of 427 MPa, ultimate tensile strength of 455 MPa, elongation of 5.7%. Table 2 also includes the mechanical properties of an alloy having a composition of $Mg_{89.5}Zn_{0.3}Al_{8.2}Si_{0.2}Mn_{1.5}$ which is similar to the commercial alloy AZ91D except that it contains silicon. This RS alloy shows a yield strength of 393 MPa, ultimate tensile strength of 448 MPa, elongation of 9.4%.

The benefit of silicon content on the properties of RS Mg-Zn-Al alloys (for the same amount of Zn and Al, the strength increases as the Si content increases) correlates well with the as-extruded microstructures. The limited solubility of silicon in magnesium and the high melting point of Mg_2Si help retain the fine precipitates of Mg_2Si, and a fine grain size of the matrix during hot compaction and extrusion, as illustrated in Fig. 1. The alloy $Mg_{91}Zn_1Al_8$ containing no silicon, shows the largest grain size (Fig. 1a) while $Mg_{90.5}Zn_{1.0}Al_8Si_{0.5}$ has a finer grain size (Fig. 1b) and $Mg_{89.5}Zn_{1.0}Al_8Si_{1.5}$ has an even finer grain size (Fig. 1c).

Table 2 Properties of Rapidly Solidified Mg-Zn-Al-Si Alloy Extrusion

Composition nominal (at%)	Density (g/cm^3)	Hardness (R_B)	TYS (MPa)	UTS (MPa)	El. (%)
$Mg_{91}Zn_1Al_8$	1.85	55	272	372	9.5
$Mg_{90.5}Zn_1Al_8Si_{0.5}$	1.85	65	337	411	5.4
$Mg_{90}Zn_1Al_8Si_1$	1.86	71	365	418	5.3
$Mg_{89.5}Zn_1Al_8Si_{1.5}$	1.84	73	390	429	2.8
$Mg_{89}Zn_1Al_8Si_2$	1.88	78	448	468	1.7
$Mg_{91}Zn_2Al_5Si_2$	1.80	65	427	455	5.7
$Mg_{89.5}Zn_{0.3}Al_{8.2}Si_{1.5}Mn_{0.5}$	1.84	68	393	448	9.4
$Mg_{91}Zn_1Al_8$	1.85	55	272	372	9.5
$Mg_{90.5}Zn_1Al_8Si_{0.5}$	1.85	65	337	411	5.4
$Mg_{90}Zn_1Al_8Si_1$	1.86	71	365	418	5.3
$Mg_{89.5}Zn_1Al_8Si_{1.5}$	1.84	73	390	429	2.8
$Mg_{89}Zn_1Al_8Si_2$	1.88	78	448	468	1.7
$Mg_{91}Zn_2Al_5Si_2$	1.80	65	427	455	5.7
$Mg_{89.5}Zn_{0.3}Al_{8.2}Si_{1.5}Mn_{0.5}$	1.84	68	393	448	9.4
	Commercial Alloys				
$Mg_{97.7}Zn_{2.1}Zr_{0.2}$ (ZK60A)	1.83	50	303	365	11
$Mg_{91.7}Al_{8.0}Zn_{0.2}Mn_{0.1}$ (AZ91D)	1.81	50	131	276	5

From Ref. 18.

3. *Mg-Zn-Al-RE (Rare Earth) Alloys*

Rare earth elements and yttrium have electronegativities which are negligibly different from that of magnesium. Additions of rare earth elements to magnesium offer the possibility of improving the corrosion resistance by reducing the galvanic local cell effect. Table 3 summarizes the tensile properties and Rockwell B (R_B) hardness measured at room temperature for a number of Mg-Zn-Al-RE alloys. Although not shown, the compressive strengths of the alloys are greater than the tensile strengths. Both the yield strengths and ultimate tensile strengths are higher than those in the commercial high-strength wrought alloy ZK60A and high-purity corrosion-resistant casting alloy (AZ91D). For example, $Mg_{91}Zn_2Al_5Y_2$(EA65-RS) has a yield strength of 456 MPa, ultimate tensile strength of 513 MPa, elongation of 5%; $Mg_{92}Zn_2Al_5Nd_1$ (EA55B-RS) has a yield strength of 434 MPa, ultimate tensile yield strength of 476 MPa, ultimate tensile strength of 516 MPa, elongation of 5.0%. EA55B-RS shows the best combination of strength, ductility, and good castability. The strengthening effect of the various rare earth elements on RS Mg-Zn-Al alloys is comparable. The superior tensile properties of rapidly solidified Mg-Zn-Al-RE alloy relative to those of commercial alloys are illustrated in Fig. 2.

Figure 3 shows the typical microstructure of a rare earth containing rapidly solidified alloy EA55B-RS. The major dispersoid in this alloy is Al_2Nd. The formation of Al_2Y (melting point temperature 1758 K) or Al_2Nd (melting point temperature of 1733 K) dispersoids instead of Mg-RE dispersoids, which have lower melting point temperatures, in RS Mg-Zn-Al-RE alloys is quite interesting. These thermally stable and fine dispersoids help to pin the grain boundaries during hot extrusion. The high yield strength of Mg-Zn-Al-RE alloys can be attributed to fine

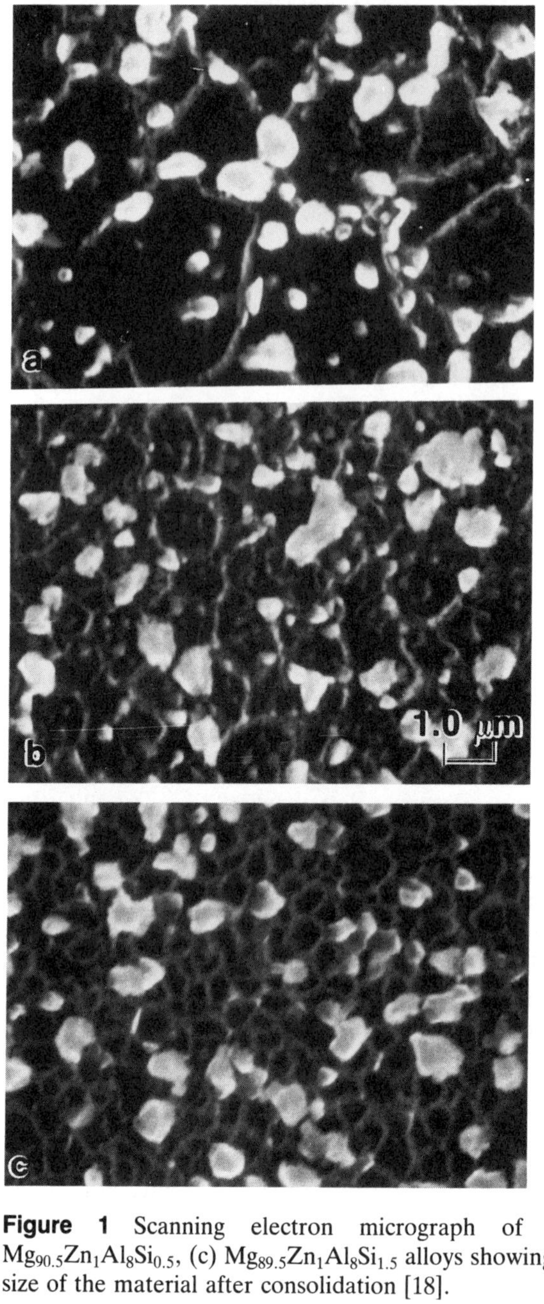

Figure 1 Scanning electron micrograph of as-extruded (a) $Mg_{91}Zn_1Al_8$, (b) $Mg_{90.5}Zn_1Al_8Si_{0.5}$, (c) $Mg_{89.5}Zn_1Al_8Si_{1.5}$ alloys showing the effect of silicon content on grain size of the material after consolidation [18].

grain size, dispersion strengthening by Al_2Nd (Al_2Y) and MgZn and solid solution strengthening.

4. Corrosion Properties of RS Magnesium Alloys

The poor corrosion resistance of magnesium alloys has often been cited as the reason for not using magnesium for structural application. Although magnesium

Table 3 Properties of Rapidly Solidified Mg-Zn-Al-RE Alloy Extrusions

Composition nominal (at%)	Density (g/cm^3)	Hardness (R_B)	TYS (MPa)	UTS (MPa)	El. (%)
$Mg_{92.5}Zn_2Al_5Ce_{0.5}$	1.89	66	359	425	17.5
$Mg_{92}Zn_2Al_5Ce_1$	1.93	77	425	487	10.1
$Mg_{92.5}Zn_2Al_5Pr_{0.5}$	1.89	65	352	427	15.9
$Mg_{92}Zn_2Al_5Pr_1$	1.94	81	447	491	3.5
$Mg_{91.5}Zn_2Al_5Si_{0.5}Pr_1$	1.94	82	476	516	5.0
$Mg_{92}Zn_2Al_5Y_1$	1.90	80	448	496	4.3
$Mg_{91}Zn_2Al_5Y_2$	1.93	81	456	513	5.0
$Mg_{92}Zn_2Al_5Nd_1$	1.94	80	434	475	13.8
$Mg_{91.5}Zn_2Al_5Mn_{0.5}Nd_1$	1.96	80	441	476	14.0
$Mg_{89.5}Zn_1Al_8Si_1Nd_{0.5}$	1.88	82	465	490	1.6

From Ref. 25.

has reasonable corrosion properties under regular atmospheric conditions, it is susceptible to attack by chloride-containing environments. It is well documented [26] that heavy-metal impurities such as Fe, Ni, Co, and Cu have a profound accelerating effect on the saltwater corrosion rate. Recently attempts have been made to improve the corrosion resistance of magnesium alloys by reducing the impurity levels; thus, high-purity alloys such as AZ91HP have been introduced in the marketplace. However, these alloys still fall short of the corrosion resistance required in the more aggressive corrosive environments, e.g., aerospace applications. Rapidly solidified alloys, in general, exhibit an improvement in corrosion

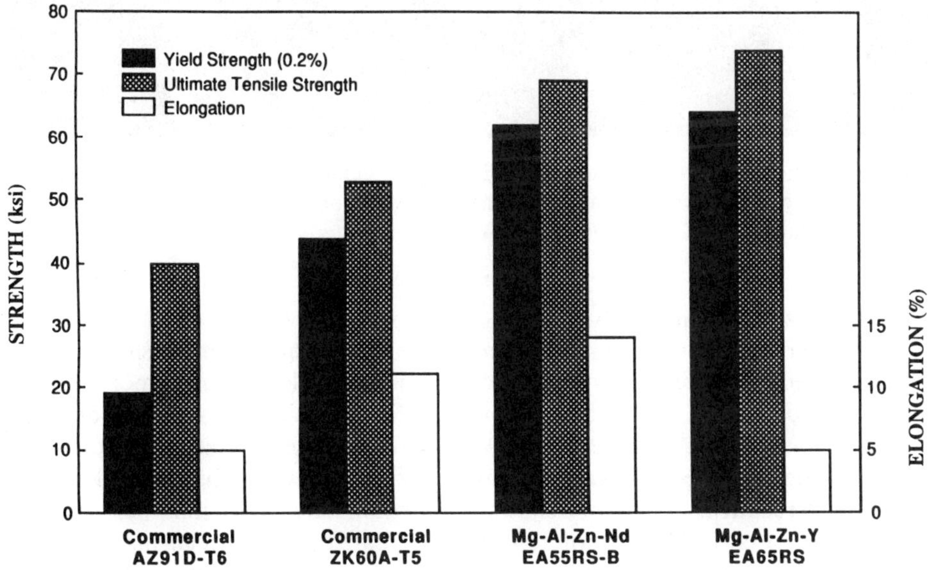

Figure 2 A comparison of mechanical properties of as-extruded rapidly solidified Mg-Zn-Al-RE alloys with two commercial alloys.

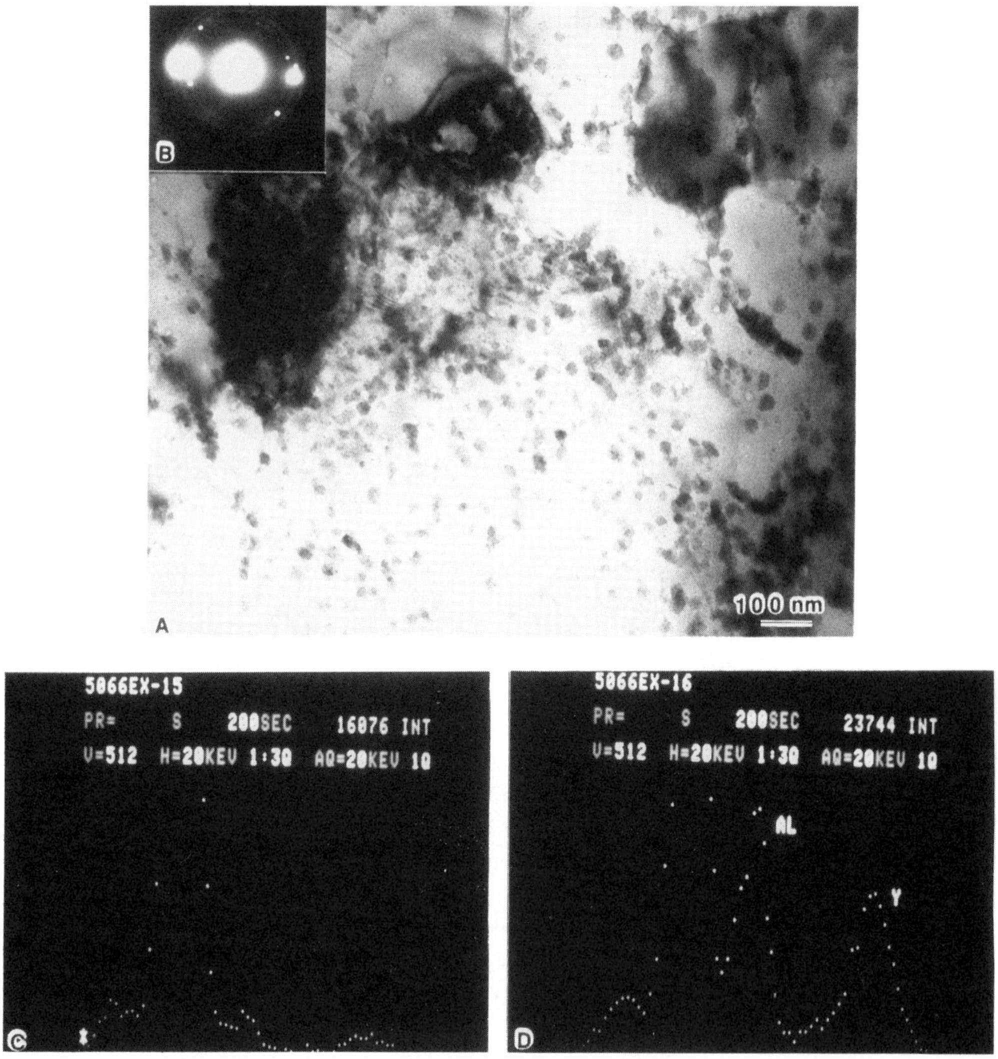

Figure 3 Microstructure of as-extruded EA65-RS including (A) bright field micrograph showing general microstructure and (B) microdiffraction of Al_2Y phase and EDS analysis of (C) Mg matrix and (D) precipitate.

resistance over conventionally produced alloys, primarily due to finer microstructure and increased homogeneity. Recent corrosion rate measurements on the rapidly solidified magnesium alloys estimated from weight loss measurements in 3% NaCl solution at 25°C show an improvement in corrosion resistance over commercial alloys [18,20]. In general, the Mg-Zn-Al-Si alloys exhibit lower corrosion rate than the Mg-Al-Si alloys. The alloy $Mg_{91}Zn_2Al_5Si_2$ which has a good combination of strength and ductility shows a corrosion rate of 94 mils per year, as compared to a corrosion rate of 104 mils per year for commercial high-strength alloy ZK60A tested under similar conditions [18]. Addition of small amounts of manganese has been found to further improve the corrosion resistance. For ex-

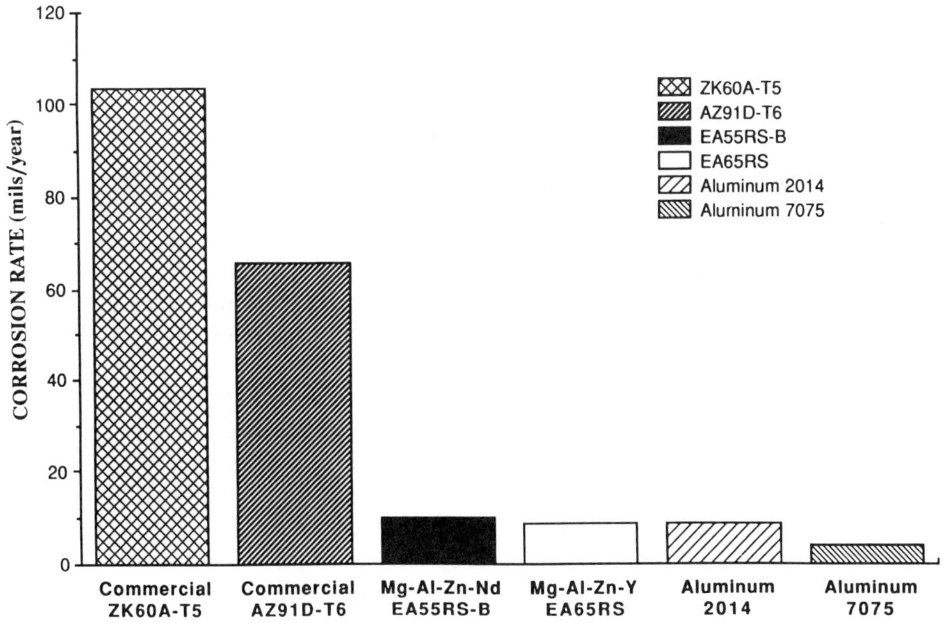

Figure 4 Corrosion rate of as-extruded rapidly solidified alloys in 3% NaCl solution at 298 K, as compared to some commercial alloys.

ample, the alloy $Mg_{89.5}Zn_{0.3}Al_{8.2}Si_{1.5}Mn_{0.5}$ has a corrosion rate of only 15 mils per year. The corrosion resistance of RS Mg-Al-Zn-RE alloys is even better. Figure 4 compares the corrosion rate of RS magnesium alloys with a number of commercial magnesium and aluminum alloys tested under the same condition. The significantly lower corrosion rate of the RS magnesium alloy can be attributed, in part, to the finer dispersion of intermetallics in rapidly solidified alloy as compared to the ingot cast material. The improvement in corrosion resistance of Mn-containing alloy is due to the combination of refined microstructure obtained through rapid solidification processing and formation of Al_3(Fe, Mn) compound which minimizes the deleterious effect of local action cell as a result of reduction potential differences between Fe and Mg in solid solution. On the other hand, the improvement in corrosion resistance of rare-earth-containing alloys is probably due to the formation of protective film on the surface of sample as the result of a reaction of the saline solution with the rare earth element of the small reduction potential difference with Mg.

B. Aluminum Alloys

The RS aluminum-based alloys have addressed a broad range of systems, including a variety of commercial alloys (e.g., aluminum alloy 2024) [27], with or without minor additional elements. The properties and processing characteristics of high strength powder metallurgy (P/M) aluminum alloys made by inert gas atomization techniques have been extensively reported (e.g., Ref. 28). In dealing with these P/M alloys, it is possible to achieve improved levels of chemical homogeneity and grain refinement, which can lead to high strength. However, the high-strength P/M alloys 7090 and 7091 developed in the 1970s and the second generation alloys CW67 and 7064 have achieved only limited commercial status. In this section we will discuss only the alloy systems, both dispersion-strengthened and precipitation hardened, for which significant advantages of RS has been demonstrated, heading to commercial significance.

1. *Dispersion-Strengthened Aluminum Alloys*

Beginning with the work of Jones [29] and Tonejc and Bonfacic [30], it was demonstrated that rapid cooling of liquid to solid could result in the extension of the solid solubility of iron in aluminum of orders of magnitude (from 0.05 to about 10 wt%). It was further shown by Thursfield and Stowell [31] that controlled nucleation and growth of metastable second phases (for example, Al_6Fe) [32] during extrusion of splat-cooled Al-Fe alloys leads to high stiffness and high strength, because of the obstacles provided to dislocation motion by dispersoids, and to the retention of high levels of strength to high temperatures (about 320°C (608°F)), because of the relative resistance of the Al-Fe dispersoids to Ostwald ripening.

Work on Al-TM alloys has also included studies of Al-Mn, Al-Cr, and Al-Zr alloys. A study on the former alloys led to the identification of quasicrystalline isosahedral phases [33], opening up a new chapter in the understanding of crystal structure. Studies of Al-Cr alloys have shown that chromium is retained in solution relatively easily to produce an "as quenched" alloy which is moderately soft [34]. In principle, subsequent processing can occur at modest stresses and temperatures, after which time the alloy can be strengthened by heat treatment at higher temperatures to form dispersoids. As a practical matter, however, it appears difficult to avoid aging during consolidation and forming, so that relatively high stresses and temperatures are still required (as they are for other Al-Tm alloys). On the basis of diffusivity data alone, Al-Zr alloys appear to offer the greatest thermal stabilities of binary Al-Tm alloys. Microstructural coarsening rates are, in fact, very low, when zirconium is present in the form of cubic ($L1_2$) Al_3Zr. However, on working in the vicinity of 400°C transformation of the cubic phase to a very rapidly coarsening tetragonal phase occurs [35].

Since all the binary Al-TM alloys are moderately soft and/or they possess only moderate thermal stabilities, continuing development of optimized high-temperature alloys has focused on ternary and quaternary alloys, with the most prominent of these being Al-Fe alloys, containing additions of cerium [36], molybdenum [37] or vanadium and silicon [38], which promote the formation and retention of favorable metastable phases.

2. *Al-Cr-Zr Alloys*

Alloys based on Al-Cr-Zr and Al-Cr-Zr-Mn have been extensively studied by Alcan International [34,39]. This alloy system, after appropriate rapid solidification, can be produced in bulk quantities with both elements in supersaturated solid solution. Both chromium and zirconium precipitate independently during subsequent thermomechanical consolidation; however, solute atom interactions have been reported which retard the aging kinetics normally observed for the respective binary alloys. The addition of manganese to the alloy Al–4.8 Cr–1.4 Zr–1.4 Mn has produced attractive property combinations [39] when consolidated by hydrostatic extrusion.

Yield strengths of 484 MPa (70 ksi) have been attained at toughness levels of 35 MPa$\sqrt{m}$ (33 ksi$\sqrt{in}$) from laboratory-scale billets. Subsequent annealing at 300°C for 100 hr, and at 350°C for 1 hr, produced a modest increase (4%) in tensile strength and a significant increase in tensile elongation from 8% to 12–15%, although the short-transverse fracture toughness remained relatively unaffected at 14 MPa$\sqrt{m}$ (12 ksi$\sqrt{in}$) it remains to be demonstrated that hydrostatic extrusion of 230–450 mm (9–18 in) diameter billets will produce similar properties. Consolidation by conventional extrusion [40] produced very inferior properties. For example, fracture toughness, K_{IC}, of only 18 MPa$\sqrt{m}$ (15.4 ksi$\sqrt{in}$) indicates precipitates overaging during thermomechanical processing. Problems resulting from the generally longer times at elevated temperature associated with adiabatic heating during consolidation by conventional extrusion and forging are endemic for most rapidly solidified alloys.

3. *Al-Fe-X (X = Ce, MoV) Alloys*

Subtle differences in the strength and toughness of Al-Fe alloys with different ternary and quaternary alloying additions are somewhat obscured by the overriding influence of consolidation-related variabilities. The effect of stress on coarsening kinetics during thermomechanical processing is understood in general terms, but specific effects on each of the dominant alloys in the Al-Fe-Ce, Al-Fe-Mo, and Al-Fe-V and derivative systems remain unexplored. For many structural applications two considerations remain important: the strength at temperature is obviously important, but the residual strength and toughness at room temperature after elevated temperature exposure is also critical for many applications involving protracted thermal cycling. This has been assessed as a percentage of the room temperature strength retained after either 100 or 1000-hr exposure to temperature [41]. Whereas ingot aluminum alloy 2219 loses more than half its room-temperature strength after exposure to 315°C (600°F) for 100 hr, the RS Al-Fe-Mo, Al-Fe-Ce, and Al-Fe-V alloys retain 100% of their strength after such exposure. Little has been published regarding the influence of thermomechanical processing variables on the available property data sets for Al–8 Fe–2 Mo or Al–8 Fe–4 Ce alloys. The predominent strengthening phases in the alloy Al–8 Fe–2 Mo is Θ' Al_3Fe [37]. In the Al–8 Fe–4 Ce alloy the metastable intermetallic phases have been identified as Al_6Fe, $Al_{10}Fe_2Ce$, $Al_{20}Fe_5Ce$ [42]. These intermetallics are generally 100–300 nm in size and provide the dislocation obstacles necessary for strengthening. However, upon annealing for 24 hr at 700 K (~800°F) these metastable phases decompose to equilibrium phases, indicating their limited stability at the high temperatures [42]. Both alloys are produced as powder by gas atomization techniques which restricts iron contents to about 8 wt%. At higher concentrations the coarser powder size fractions, 80–100 μm in diameter, frequently contain small primary Al_3Fe particles, and a preponderance of the dendritic solidification structures typical of 10^4 K/s cooling rates. Couper and Singer [43] estimated that powder particles smaller than 20 μm are necessary in the Al–8 Fe–4 Ce system for attainment of featureless solidification structures typical of the Jones' Zone A [29] structures, which is now understood to result from massive solidification at substantial droplet undercoolings, and involving liquid cooling rates of 10^6 K/s. The presence of relatively coarse Al_3Fe particles (in relationship to the 10–50 nm obstacles required for both strength and toughness after consolidation) is a major problem for all atomization process involving alloys with high transition element concentration, although the rotary atomization device is capable of producing featureless solidification structure in Al–8 Fe–2 Mo alloys at 50-μm particle diameters [37]. In contrast, the generally higher liquid cooling rates and undercoolings and subsequent solidification rates available from planar flow casting [44] eliminate this particular problem [45].

4. Al-Fe-V-Si Alloys

Alloys in the Al-Fe-V-Si system have been extensively investigated through melt spinning (both jet casting and planar flow casting) with an effective cooling rate of approximately 10^6K/s. Whilst a quantitative analysis of the effect of cooling rate on the microstructures has not been performed on this alloy system, it is known that sufficiently high cooling rate is necessary to produce an alloy which will successfully suppress the formation of the Al_3Fe phase (and the metastable Al_6Fe phase). The microstructures of the as-rapidly solidified ribbon thus produced by the melt-spinning processes can be characterized as either microcellular or discrete silicide dispersoids. These structures are shown in Figs. 5a and b, respectively where the matrix is shown by STEM/EDX analysis) to be supersaturated with Fe, V, and Si and the intercellular regions to be of quaternary silicide phase. The discrete silicide structure is indicative of sufficient recalescence occuring during the cooling of the ribbon to decompose the intercellular regions into discrete silicide particles. In certain alloys and under unusual casting conditions the icosahedral symmetry "quasicrystal" ("O"-phase) can also be found (Fig. 6a). If sufficient cooling rate is not maintained the microstructure can be relatively coarse (Fig. 6b). The occurrence of this O-phase is indicative of the recent speculations that the icosahedral and silicide phases have similar structures whereby the icosahedral phase can be

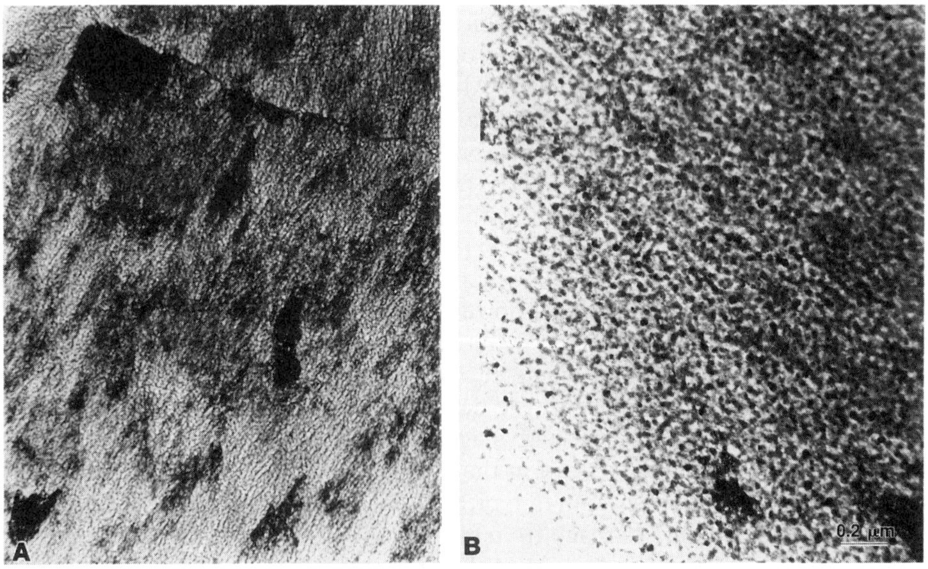

Figure 5 Microstructures of planar flow cast Al–8.7 Fe–1.0 V–1.7 Si alloy showing (A) microcellular and (B) cellular structure with discrete silicide dispersoids.

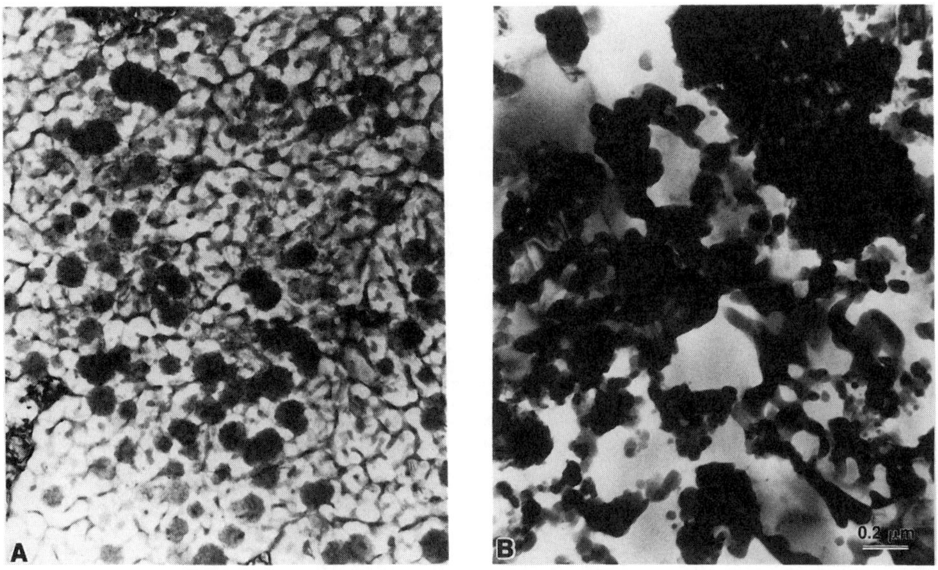

Figure 6 Microstructures of Al-Fe-V-Si alloys showing (A) quasicrystalline phase (B) coarse cellular structure under improper casting conditions.

described as a disordered silicide phase [46]. The major advantage of this alloy system over others is that when the ribbon has been cast at its optimum cooling rate (and thus contains no O-phase) the as-cast microstructure contains no metastable precipitates other than the silicides. This is an advantage during consolidation because the dispersoid size can be controlled through a coarsening process rather than a transformation sequence. During this consolidation (via vacuum hot pressing and subsequent deformation processes) the as-cast microstructure coarsens (intercellular regions) and additional silicide dispersoids precipitate from the supersaturated matrix. Thus the as-consolidated microstructure consists of near spherical dispersoids of composition approximately $Al_{13}(Fe, V)_3Si$ decorating both grain and subgrain boundaries as well as within these grain interiors (Fig. 7). The size of the dispersoids obviously varies with the processing/consolidation time and temperature but typically ranges in the 20- to 100-nm-diameter size. There is a large volume fraction of silicide particles uniformly distributed in the matrix, but there is a tendency to form agglomerates (A in Fig. 7a). This is because in the rapidly solidified material the silicides are primarily located at the cell boundaries. Figure 7b shows the selected area diffraction pattern from the region in Fig. 7a. The particles are randomly oriented and do not exhibit any consistent crystallographic orientation relationship to the matrix. The grain size is in the range from 0.5 to 2 μm. High-resolution electron microscopy (Fig. 8) and diffraction show that the silicide particles have the bcc structure with lattice parameter a = 1.25 nm [47].

The general features of this alloy system illustrating how variations in the volume fraction of the cubic $Al_{13}(Fe, V)_3Si$ phase present in the alloy affects mechanical properties have been recently described [48]. A broad group of such alloys can be processed to plane-strain fracture toughness levels of 20–25 $MPa\sqrt{m}$ at yield

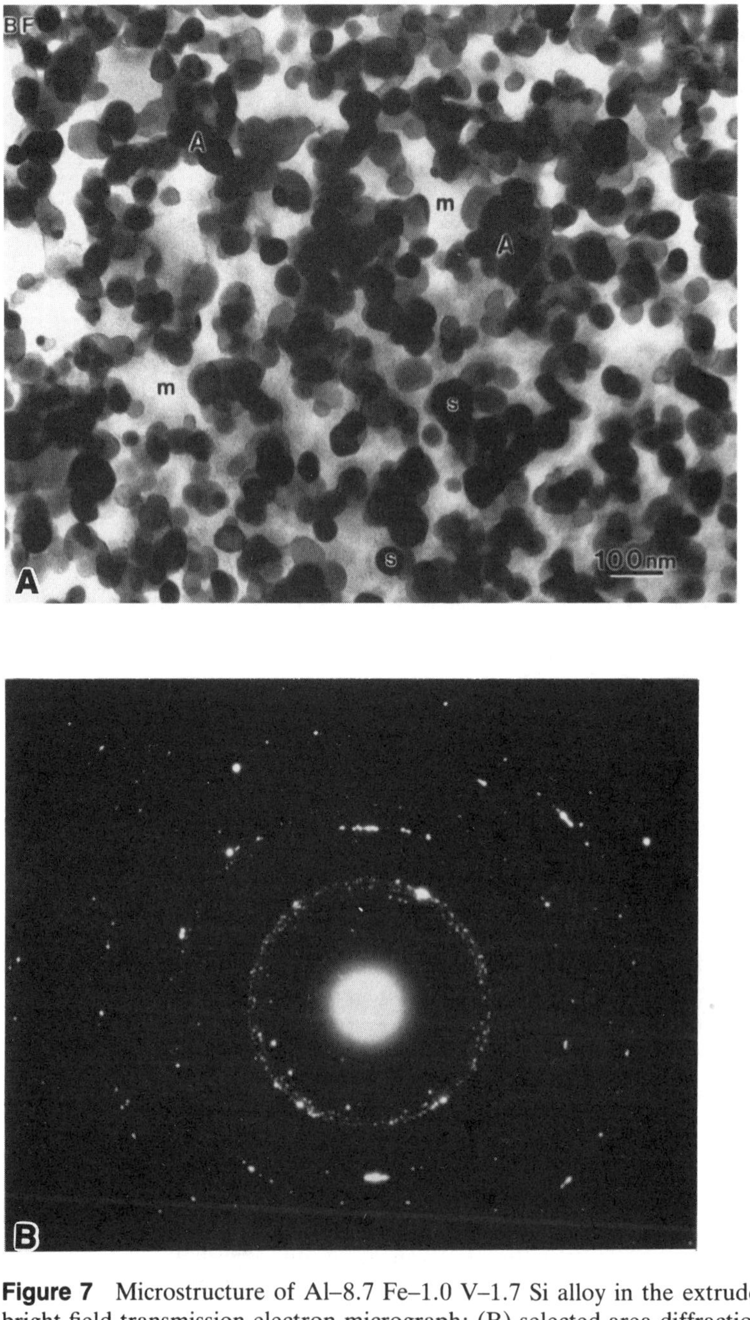

Figure 7 Microstructure of Al–8.7 Fe–1.0 V–1.7 Si alloy in the extruded condition: (A) bright field transmission electron micrograph; (B) selected area diffraction pattern.

strengths between 420 and 450 MPa (60–65 ksi) by both extrusion and forging. The toughness-optimized alloy Al–8.7 Fe–1.0 V–1.7 Si (FVS0812, commercially designated as 8009 contains about 27 vol% of the cubic silicide phase, corresponding to an easily processable alloy with current production equipment. It has been forged into various aerospace and automotive products; sheet, plate, and profiled extru-

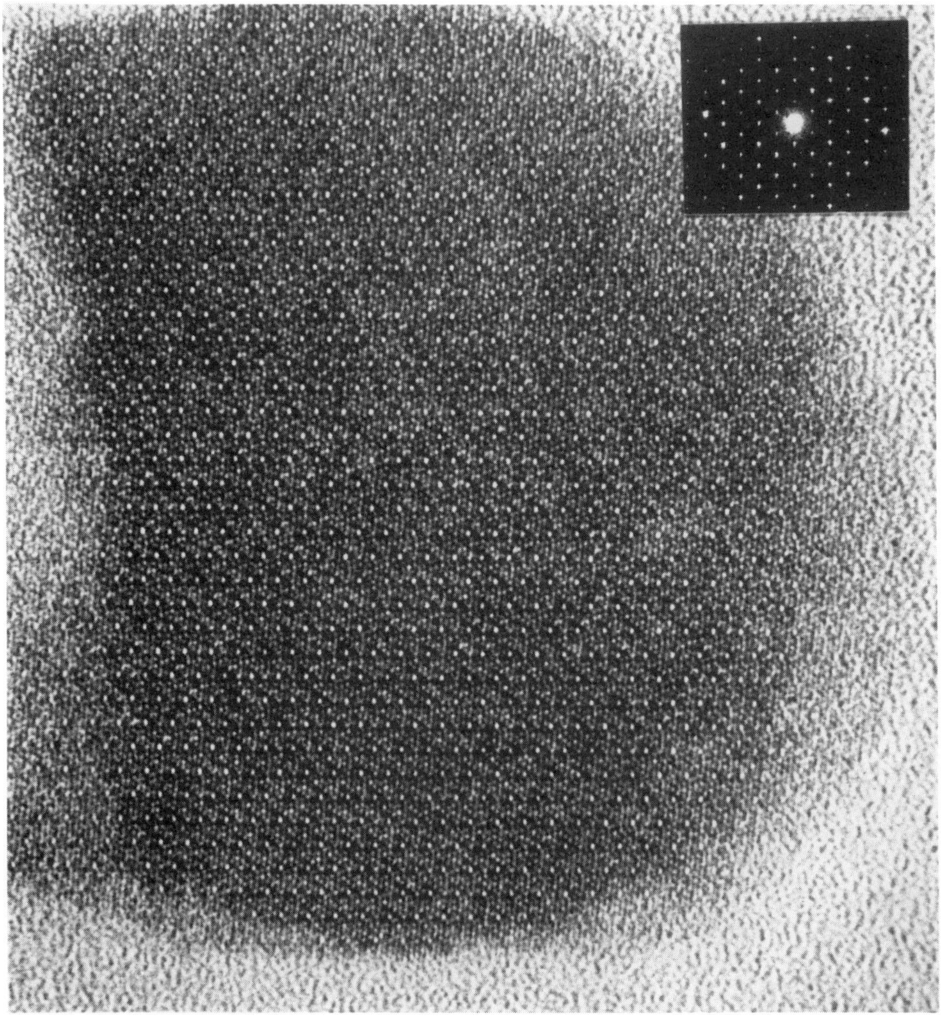

Figure 8 High-resolution transmission electron micrograph of an as-extruded Al–8.7 Fe–1.0 V–1.7 Si alloy.

sions have also been made. Another alloy containing higher iron content, Al–12.4 Fe–1.2 V–2.3 Si (FVS1212) containing 37 vol% silicide, was designed for high stiffness applications (room-temperature dynamics modulus ~ 14 Msi (95.5 GPa)). The plane-strain fracture toughness of this alloy with 550 MPa (80 ksi) yield strength is somewhat low (15 MPa$\sqrt{m}$). It has been fabricated into product forms such as sheet and extrusions. It is primarily of interest for sheet applications. Alloy FVS0611 (commercially designated as 8022), the lowest dispersoid-containing (16 vol%) alloy, has very high formability and has been extruded into wire.

The tensile properties of the three alloys are shown in Fig. 9. It may be noted the difference in strength of these three alloys is not as large at elevated temperatures as it is at room temperature. All the alloys show ductility minima at ~300°F (145°C) due to dynamic strain aging. This behavior has been recently studied in

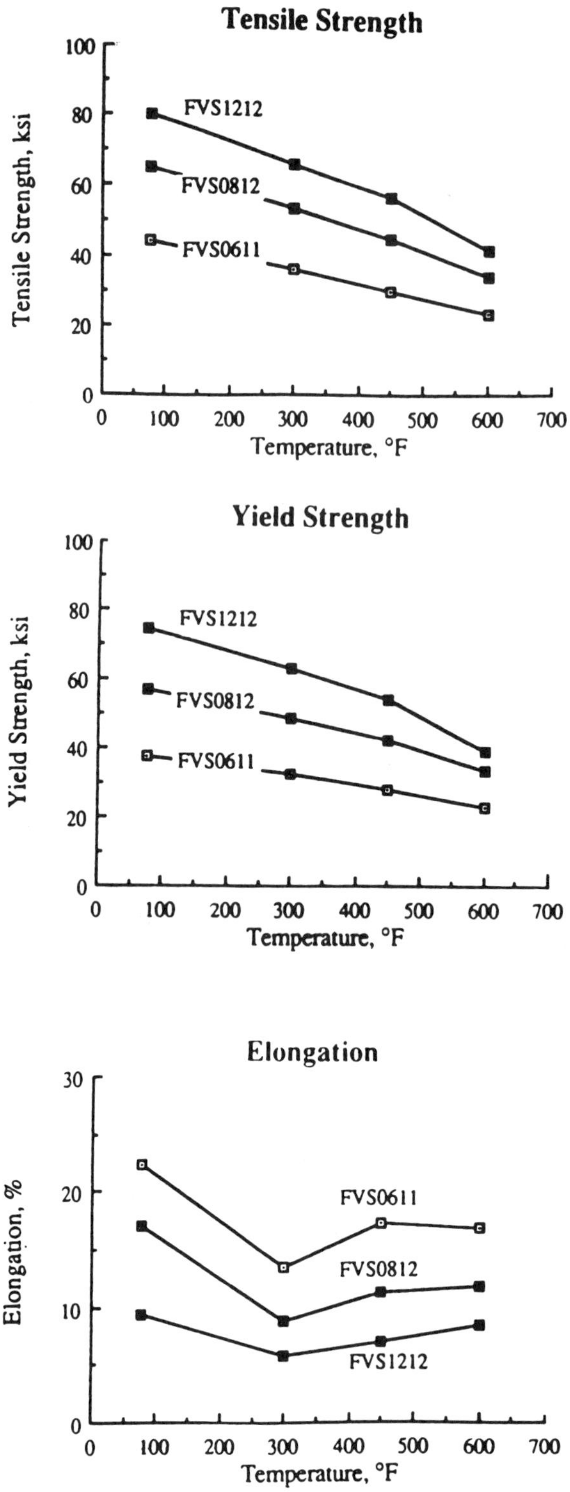

Figure 9 Tensile properties of Al-Fe-V-Si alloys as a function of temperature.

some detail [46] because of its importance in forming. The aforementioned temperature regime, where the ductility goes through a minimum, must be avoided during component fabrication. A large body of mechanical property data had been generated on these alloys by various potential users of these materials, particularly the aerospace community.

A desirable feature of this alloy system is that the structural silicide coarsening rate is low, in comparison to both the Al-Cr-Zr-based alloys and the other Al–8 Fe–X alloys produced by atomization processes [3]. Table 4 summarizes the strengthening phases and their coarsening rates. This thermal stability allows silicide-containing Al-Fe-V-Si alloys to be rolled into sheet after prolonged exposure at times and temperatures, which result in unacceptable coarsening in Al–8 Fe–X alloys, without severe degradation of properties [49]. Thin sheet (0.065 in.) of monolithic Al-Fe-V-Si alloy such as alloy FVS0812 (8009), consistently demonstrates plane-stress fracture toughness, *Kc*, between 100 and 150 ksi$\sqrt{\text{in}}$. As a consequence of the thermal stability of the silicide phase, the Al-Fe-V-Si alloys also exhibit no degradation of strength after elevated temperature exposure to 1000 hr, up to 800°F [50].

It has been observed that the primary intermetallics such as Al_3Fe, which form as needlelike particles, are detrimental to obtaining high fracture toughness. Plane-strain fracture toughness of the order of 24 go 30 MPa$\sqrt{\text{m}}$ may be achieved in Al-Fe-V-Si alloys; toughness observed for Al-Fe alloys with acicular dispersoids are less than 20 MPa$\sqrt{\text{m}}$ at high strength levels. This high toughness is due to uniform fine spherical (~50 nm diameter) silicide dispersoids of $Al_{13}(Fe, V)_3Si$. If the alloy chemistry is not properly balanced, that is, primary phases such as Al_3Fe or Al_7V are present in the consolidated microstructure, the toughness can decrease. Figure 10 represents fracture toughness data versus yield strength for a wide variety of alloys within the alloy system with regard to dispersoid volume fraction, dispersoid sizes, and particle size distribution [46]. The only clear difference between the two curves is that the microstructure of all the alloys in the lower curve are from a three-phase region (aluminum + silicide + Al_3Fe or Al_7V), where the third phase is typically either blocky or acicular in morphology. The upper curve represents a similiar set of alloys except that they are all within the two-phase region (aluminum + silicide). As may be seen, considerable improvement in fracture toughness has been achieved by the removal of extraneous phases through proper alloy chemistry modifications.

Table 4 Metastable Phases in RS Aluminum–Transition Metal Alloys

Alloy system	Process	Strengthening phases	Intermetallic coarsening rates at 425°C ($m^3\ h^{-1}$)
Al-Fe-Ce	Inert gas atomization	Al_3Fe, Al_6Fe, Al_8Fe_4Ce, $Al_{10}Fe_2Ce$	4.2×10^{-23}
Al-Fe-Mo-V	Centrifugal helium atomization	Al_3Fe, Al_6Fe. $Al_{12}Fe(Mo, V)$	1.7×10^{-23}
Al-Cr-Zr	Helium atomization	Al_3Zr, $Al_{13}Cr_2$	—
Al-Fe-V-Si	Melt spun ribbon	$Al_{13}(Fe, V)_3Si$	2.9×10^{-26}

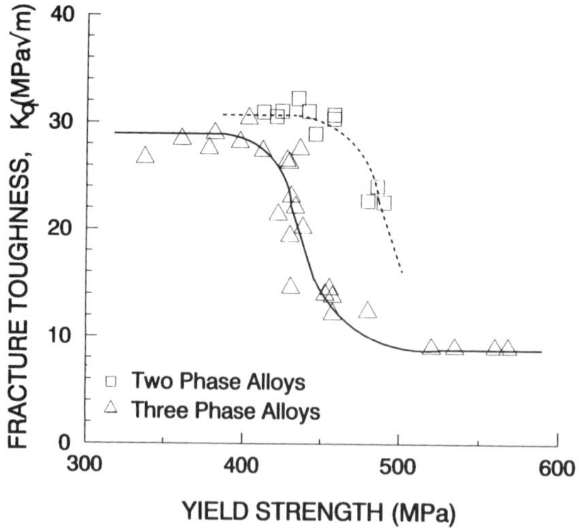

Figure 10 Effect of third phase on the fracture toughness of Al-Fe-V-Si alloys.

The phase relationships for these metastable phases are largely unknown. Recently considerable effort has been devoted to determine these phase relationships in Al-Fe-V-Si alloys, so that optimum alloy chemistries can be designed such that deleterious phases can be avoided in these complex alloy systems. Figure 11 is an example of isothermal section at 450°C for quarternary Al-Fe-V-Si alloy system [46]. This experimental isothermal section agrees quite well with the theoretically calculated phase diagrams [51]. Only through the determination of such metastable phase diagrams can one design alloys with a balanced set of properties.

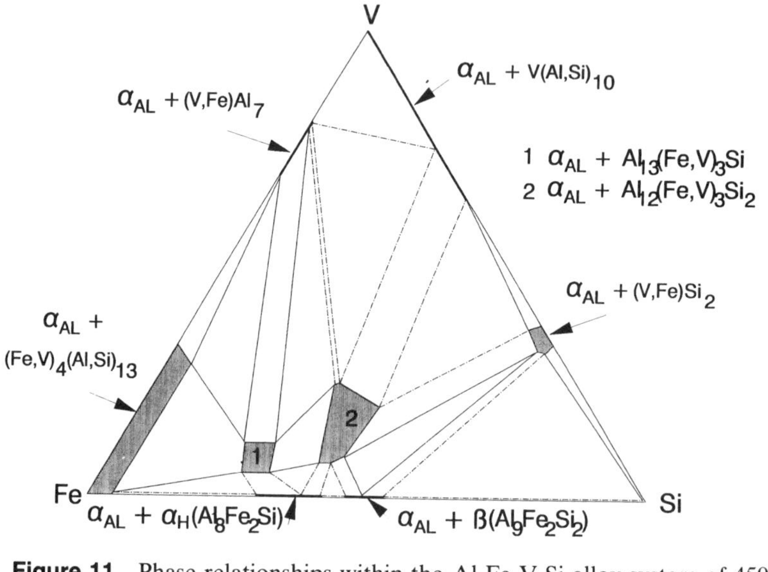

Figure 11 Phase relationships within the Al-Fe-V-Si alloy system of 450°C.

5. *Precipitation Strengthened Aluminum-Lithium Alloys*

The interest in rapidly solidified Al-Li alloys comes from the several beneficial effects offered by lithium addition in aluminum alloys (see for example Refs. 44, 52, 53). The addition of 1 wt% Li (about 3.5 at% Li) decreases the density of aluminum by about 3% and increases its elastic modulus by about 6%, hence giving a substantial increase in specific modulus [54]. Unlike the transition metal elements used for high-temperature aluminum alloys, lithium has a relatively high solid solubility in aluminum (about 4 wt% at the eutectic temperature), which means that, in principle, sufficient quantities can be alloyed using conventional ingot cating technology. However, while there has been much progress in the development of ingot metallurgy Al-Li alloys in recent years, severe segregation problems are encountered which limit the maximum lithium content to about 2.5 wt%, giving a weight-saving potential of less than 10%; if the lithium content is increased beyond this level, reductions in ductility and fracture toughness usually result. It has been shown that lithium contents of up to 5 wt% can be added to RS aluminum without forming deleterious coarse second-phase particles [55].

Additions of zirconium (about 0.5 wt%), have been shown to significantly improve the mechanical properties of rapid solidified Al-Li alloys [56]. There are several interesting features associated with this observation.

1. Zirconium reacts with aluminum to form metastable (Li_2) Al_3Zr, which is isostructural with the major Al-Li strengthening phase, Al_3Li.
2. Metastable Al_3Zr is quite resistant to dislocation shear.
3. During aging treatments, Al_3Li precipitates around coherent Al_3(Li, Zr) precipitates (Fig. 12).

Since the Al_3Li "precipitates" are present in association with shear-resistant Al_3Zr, they do not allow pronounced planar slip; rather the Al_3(Li, Zr) composite precipitates induce Orowan bypassing of dislocations, resulting in an improvement in ductility [56,57]. Figure 13 is an example of typical properties of RS Al-Li-Cu-Mg-Zr alloy.

These RS low-density Al-Li-Zr-based alloys possess additional advantages over conventional aluminum alloys. Ingot Al-Li alloys exhibit considerable anisotropic properties whereas the RS alloys exhibit very little anisotropy. The RS Al-Li alloys are particularly suited for forging applications, since no prior cold work is necessary to develop good mechanical properties [58]. Because of their fine grain size RS/PM alloys usually show poor fatigue crack growth resistance. However, recent studies show that RS Al-Li alloys have good superior fatigue crack growth resistances comparable to ingot metallurgy Al-Li alloys [59].

6. *Corrosion Properties*

In general, structural aluminum alloys show severe saline pitting corrosion, requiring substantial protection either by an electrolytic anodic film deposition (anodization) or by chemical protection followed by polymeric coating. All RS aluminum alloys have shown significantly improved saline corrosion resistance with reduced weight loss during ASTM B-117 salt fog testing. Exceptional performance, for which several Al-Fe-V-Si and Al-Li-Cu-Mg-Zr alloys are typical, is shown in Fig. 14 indicating that the best of these alloys is virtually free from pitting corrosion attack after some 70 days exposure. These RS aluminum alloys are especially

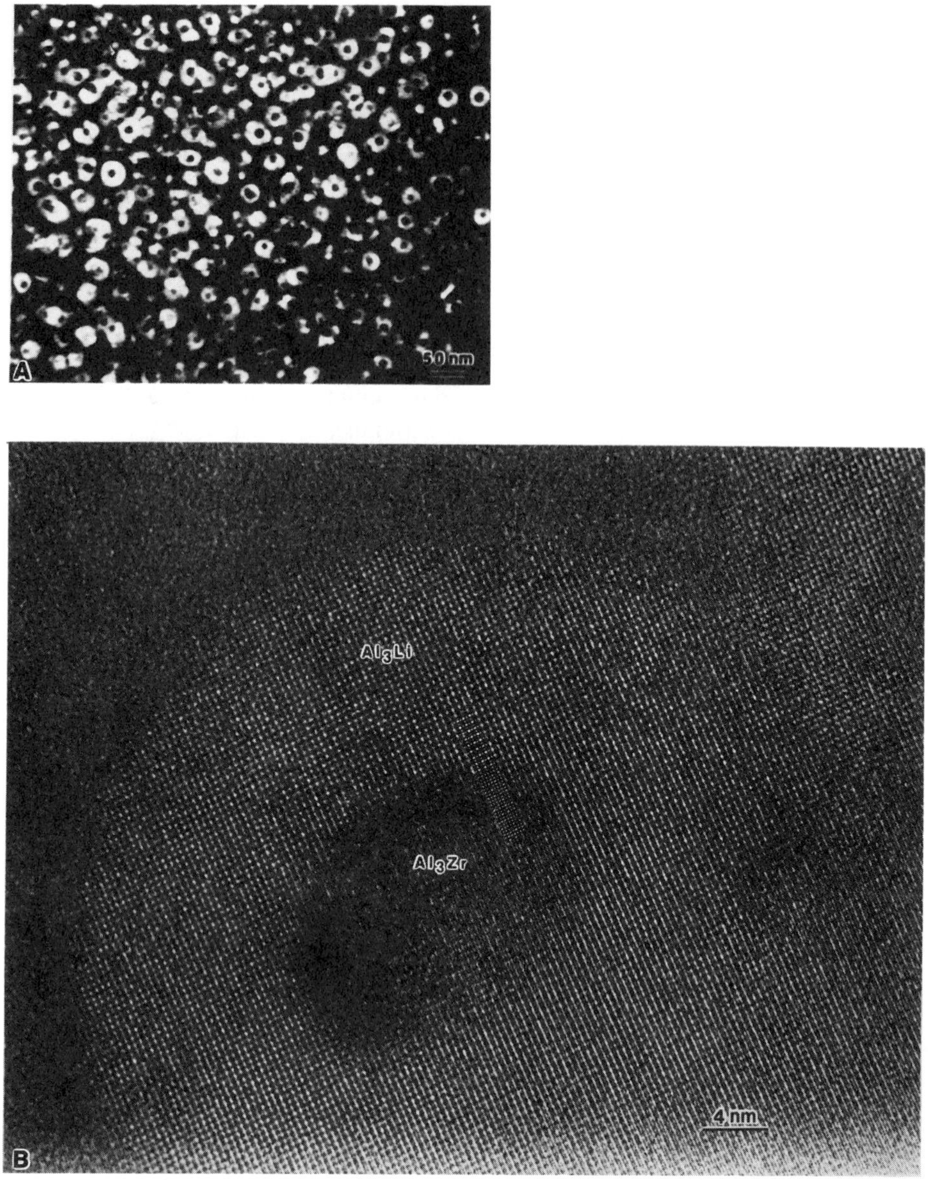

Figure 12 Microstructure of Al–2.7 Li–1 Cu–0.5 Mg–0.5 Zr alloy after aging showing composite $Al_3(Li, Zr)$ precipitates: (A) dark field micrograph; (B) high-resolution transmission electron micrograph.

suitable for a wide range of applications which involve exposure to saline environments for long periods of time before use.

C. Titanium Alloys

Titanium-alloying behavior depends in large part of the ability of elements to stabilize either the low-temperature alpha (HCP) or high-temperature beta phase

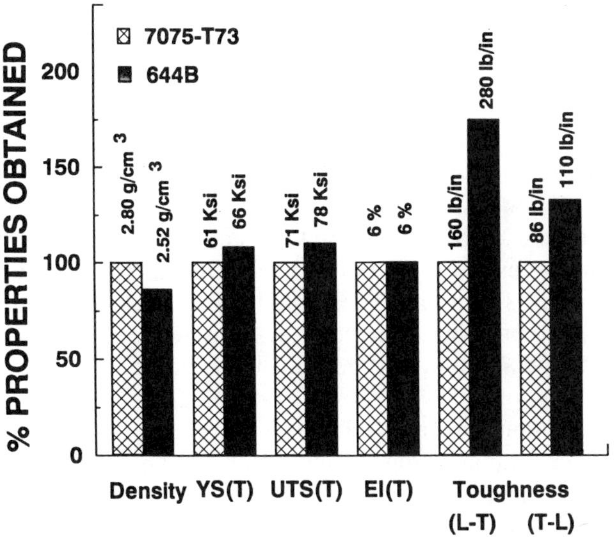

Figure 13 A comparison of the mechanical properties of rapidly solidified Al–2.5 Li–1 Cu–0.5 Mg–0.6 Zr alloy (644B) with ingot metallurgy (7075-T73) alloys.

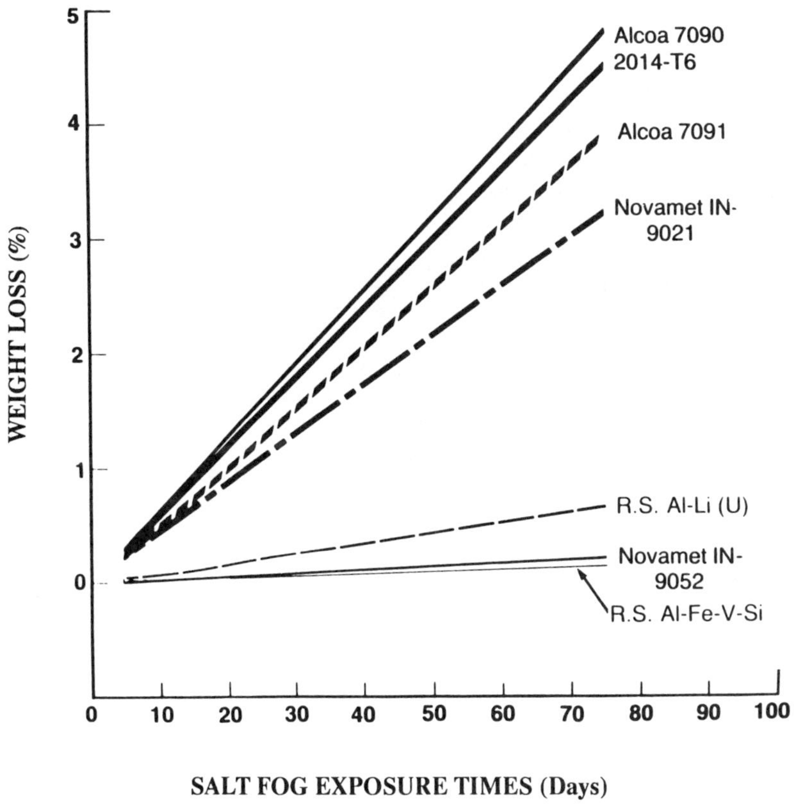

Figure 14 Corrosion rates (ASTM B-117 test) of rapidly soldified Al–8.5 Fe–1 V–1.7 Si (8009) and Al–2.5 Li–1 Cu–0.5 Mg–0.62 Zr alloys, compared with aluminum alloy 2014-T6 and other powder metallurgy alloy; aluminum alloys 7090 and 7091 are produced by gas atomization.

(BCC) which relates to the number of bonding electrons [60,61]. Titanium alloy phase diagrams are conveniently categorized into the two major subdivisions of alpha and beta stabilizers [41,62–64,66]. The alpha stabilizers are divided into those having complete stability, a simple peritectic reaction (e.g., Ti-O and Ti-N), and those which have limited alpha stability, with a peritectoid reaction into beta plus a compound (for example, Ti-B, Ti-C, and Ti-Al). The beta stabilizers are divided into two categories: beta isomorphous and beta eutectoid. In the former system an extreme beta solubility range exists with only a restricted alpha solubility range (e.g., Ti-Mo, Ti-Ta, and Ti-V). For the beta eutectoid systems (e.g., Ti-Cr and Ti-Cu) the beta phase has a limited solubility range and decomposes into alpha and a compound.

Investigation of the RS of titanium-base alloys has trailed that of aluminum-base materials for a number of reasons including the extreme reactivity of molten titanium [41]. However, in recent years the potential for use of RS to enhance the behavior of titanium systems has been recognized and extensive studies have been conducted [41,66–73]. However, the vast majority of this work has been in the research base rather than applications oriented, as in the case recently with aluminum alloys, and the following review will therefore attempt to catch this flavor of the current RS titanium scenario.

A summary of the solubility extensions determined in titanium alloy systems is shown in Table 5. Here the maximum equilibrium solubility is the solubility in the high-temperature beta phase. The extended solubility level indicated is the value judged [41,69] to be the level to which the matrix in question did not exhibit second-phase particles involving the solute in the as-RS condition. As a transition metal titanium is characterized by its ability to provide greater than 1 at% solid-state solubility for more than 50 elements [74]. Thus the potential for extended alloying

Table 5 Solubility Extensions in Titanium Alloy

Element	Matrix	Maximum equilibrium solubility[a] (at%)	Extended solubility[b] (at%)	Technique
B	Binary	0.5	6.0	Splat
C	Binary	3.1	10.0	Splat
Si	−10at%Zr	5.0	6.0	Splat
Ge	−5Al-1.5Sn	8.2	>5.0	EBSQ[c]
Nb	Binary	~0.3	>1.0	EBSQ
Y	−8Al	~0.5	>2.0	Laser
Gd	Binary	~0.3	>0.5	EBSQ
La	−5Al, −5Sn	~1.5	≥1.5	Splat
Ce	Binary	~1.5	>1.0	Splat
Er	−5Al	~0.3	≥1.5	Splat
Dy	Binary	~0.3	>0.6	EBSQ

[a]In beta titanium.
[b]As-rapidly-solidified product.
[c]Electron beam splat quench.
From Ref. 41.

by RS is not as great as for the aluminum system where the corresponding number of elements is only eight [75].

However, the classes of alloy for which RS does offer the potential for improved behavior are

1. Conventional
2. Rare earth containing (including Y) which normally exhibit low solubility
3. Metalloids (B, C, Si) which exhibit low solubility
4. Eutectoid formers, which although they exhibit quite high solubility are very segregation-prone.
5. Beta alloys
6. Intermetallic titanium aluminides
7. Various other systems including low-density alloys

Metal containment complicates the production of RS titanium alloys [41], although considerable attention has been given recently to circumventing this problem [41,69,71,76]. Even the most chemically stable ceramic compounds are dissolved in molten titanium [41] and to date no scheme to minimize the reaction has become commercially viable, although significant progress is being made. Thus the melt processing for this system is different to that which has been used in other metal systems [76]. However, melt spinning, melt extraction, and atomization processes have been developed at the laboratory and pilot-plant scale [41,69,71,72,76].

1. Conventional Titanium Alloys

The most comprehensive study done to date on RS conventional titanium alloys has been on the Ti–6 Al–4 V alloy [41,66,69,74] using cooling rates varying from 10^4 to 10^7K/s. From this work [73] two significant effects were noted; the beta grain size decreased as the cooling rate was increased (Fig. 15), and for material made

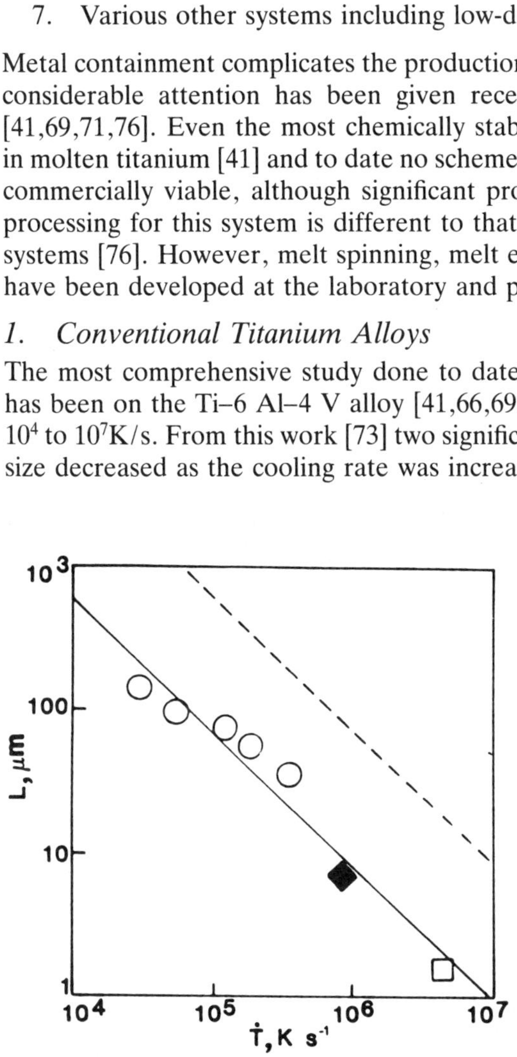

Figure 15 Variation of beta grain size in Ti-6A1-4V with cooling rate [73]. Dashed line is predicted for aluminum. ○, Radius for REP particulate; □, half-thickness for HA splat-cooled from both sides; ■, EBSQ flake.

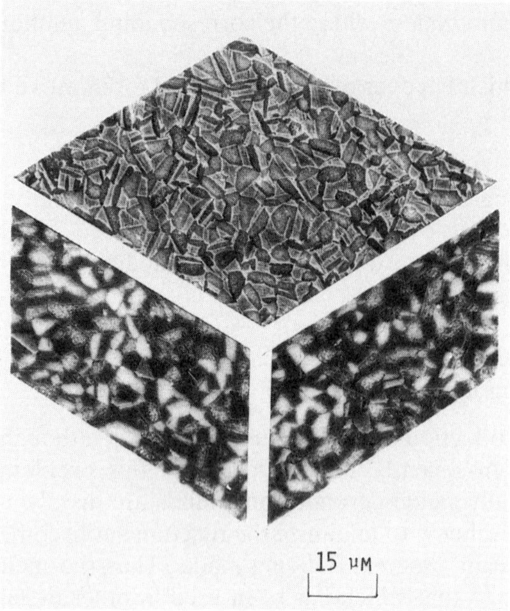

Figure 16 Equiaxed alpha morphology in Ti–6 Al–4 V produced after annealing material with a small beta grain size which resulted from rapid solidification.

at the higher cooling rate studied, equiaxed alpha was produced on subsequent high-temperature annealing (Fig. 16).

Other work on conventional alloys has shown the advantage to be gained from RS. In an alloy similar to Ti–6 Al–4 V, but containing Mo instead of V, both strength and ductility were increased as the cooling rate was increased from 10^3–10^4 to 10^6–10^7 K/s (Table 6) [77].

A study of the Beta III alloy (Ti–11.5 Mo–6 Zr–4.5 Sn) indicated that the beta grain size decreased as the cooling rate increased (~40 μm in PREP, 1–4 μm, in EBSQ [78]). These grain sizes were significantly smaller than in corresponding Ti–6 Al–4 V product (~135 μm and ~7 μm, respectively), indicating the effect of increasing solute content in decreasing grain size for a given cooling rate [75]. A further observation in Beta III was that alpha precipitated interdendritically in PREP material and at grain boundaries in the EBSQ alloy upon aging.

Table 6 Effect of Cooling Rate on Tensile Properties of a Ti-Mo-Al Alloy

Production method	UTS (MPa)	RA (%)
Casting	895	12
Cast + hot worked	1070	22
Atomized powder 10^3–10^4 K/s cooling rate + HIP	1070	22
Flake 10^6–10^7 K/s cooling rate + HIP	1135	30

2. *Rare-Earth-Containing Titanium Alloys*

The RS of titanium alloys containing a rare earth element produced an ultrafine dispersion of stable rare earth oxides with particle sizes much finer than that produced in earlier work on ingot metallurgy (IM) material [79–81], even when mechanical alloying was used [82]. Both Er_2O_3 [83–87] and Y_2O_3 [83,88] were found to produce a fine dispersoid in rapidly solidified titanium alloys. Virtually no dispersoid was present in the RS condition [87–89] provided solid-state cooling was-sufficiently fast. Aging at various temperatures produced extremely fine dispersoids, with aging at 500°C (930°F) producing the finest dispersoid [87]. Fabrication of components requires exposure to consolidation cycles at temperatures up to 950°C (1740°F). The stability of the dispersoid is a critical issue to temperatures considerably higher than the service temperature.

A comparison of the dispersoid distribution in a series of ternary titanium– and rare earth–oxygen alloys indicated that alloys containing La, Nd, Dy, and Er had the greatest resistance to coarsening at 800°C (1470°F) [83]. In these alloys, dispersoids had an average diameter range of 650–980 Å, with interparticle spacings from 0.3 to 0.8 μm. This translated to an Orowan strengthening increase of up to 85 MPa (12 ksi).

The stability of the Er_2O_3 dispersoid has been shown to be very good in alpha Ti-Al alloys as long as the alloy remains in the alpha titanium phase field. The alloy Ti–6 Al–2 Sn–4 Zr–2 Er (wt%), rapidly solidified by melt extraction, was annealed for 1 hr at 950°C (1740°F) [90,91]. A dispersoid 200–400 Å in diameter, with an interparticle spacing of ~0.3–0.4 μm, was maintained in the grain interiors of this alloy. The rare earth sulfides and oxysulfides form a fine dispersion in alpha titanium alloys and exhibit good resistance to a coarsening even after exposure to temperatures as high as 1000°C (1830°F) [91–94].

The rare earth dispersoid in titanium alloys appears to affect the behavior of the material in a number of ways. First, the dispersoid produces hardening due to the Orowan effect [95]. There are also indirect effects due to a grain size reduction [83,91–93,96] and scavenging of interstitial elements thereby reducing interstitial strengthening. A fine dispersion of erbia particles is shown in Fig. 17 [97]. The net result in ternary Ti-RE-O alloys was a strengthening on the order of 40MPa (6 ksi) at 700°C (1290°F) [96] and a significant creep rate reduction. An example of the creep behavior of a dispersion-strengthened alloy is shown in Fig. 18 [98,99].

3. *Metalloid-Containing Titanium Alloys*

The metalloid elements C, B, Si, and Ge have the potential for good chemical stability exceeded only by the rare earths. However, the stability of these elements is only moderate in titanium alloys, and the useful temperature range of titanium alloys containing them is below that of rare earth oxide dispersion-strengthened alloys [100], that is, below 600°C (1110°F). Fine, needlelike TiB [90] precipitates develop at temperatures above 500°C (930°F) (Fig. 19) [72,100–102] and coarsen dramatically at 800°C (1470°F) [90,101,102].

Rapidly solidified titanium alloys containing carbon have been shown to have good strength and ductility as-quenched. However, they lose both strength and ductility after aging at 700°C (1290°F) due to precipitation and rapid coarsening of titanium carbide [102]. Annealing carbon-bearing alloys at 900°C (1650°F) produced 0.05-μm-diameter carbides.

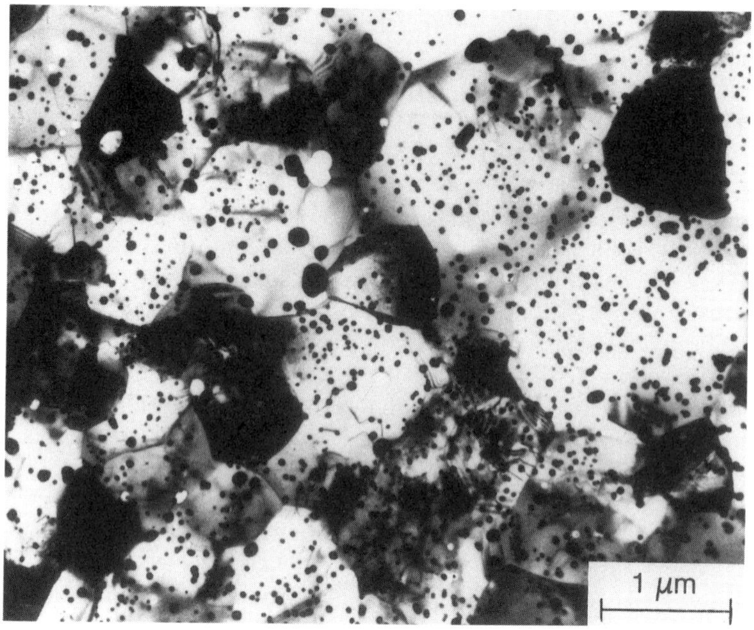

Figure 17 Fine dispersion of erbia particles in a Ti–6 Al–1 Sn–4 Zr–2 Mo–0.1 Si–2 Er alloy after rapid solidification (PREP ≅ 10^4 K/s cooling rate) and extrusion.

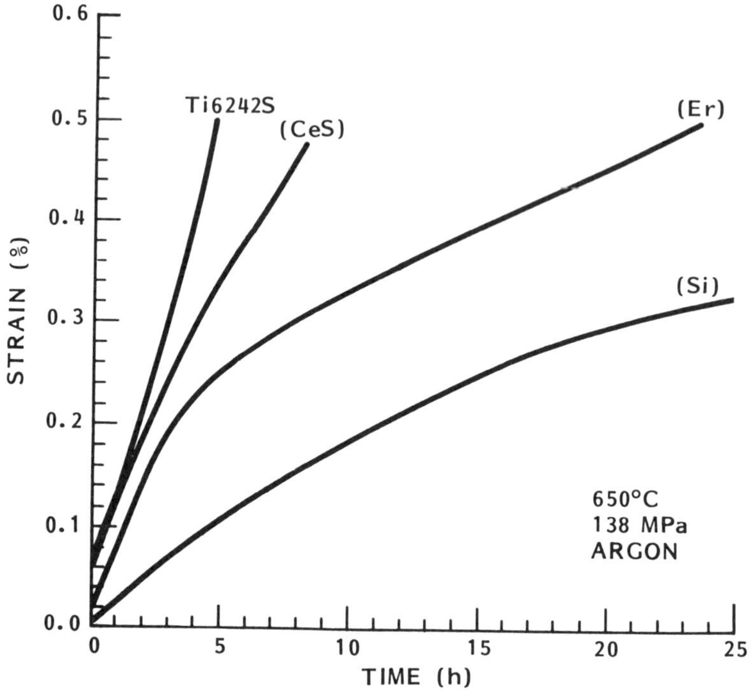

Figure 18 Creep behavior of a dispersion-strengthened alloy [98,99].

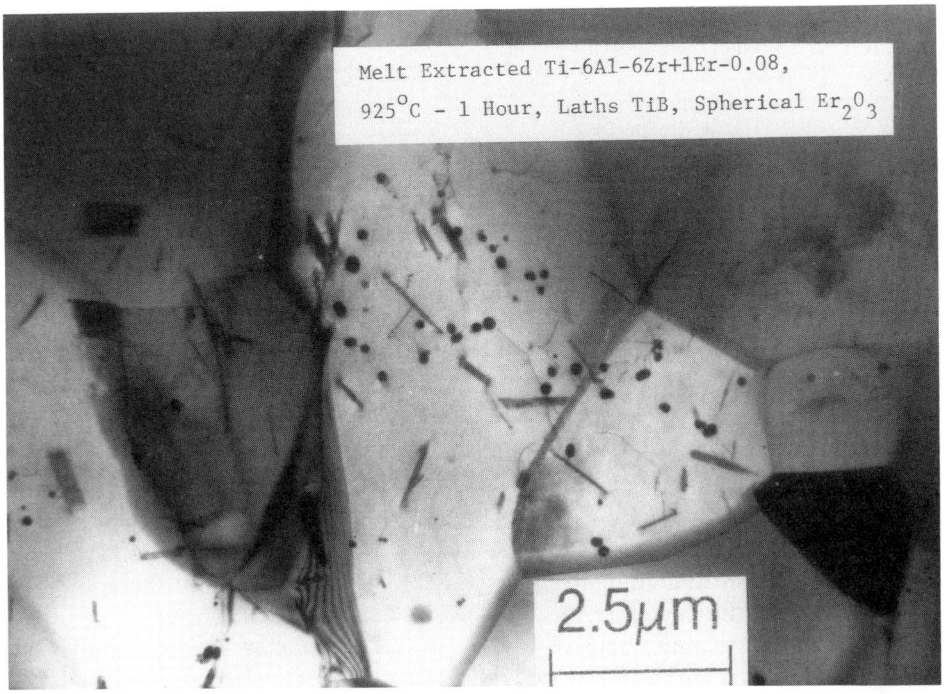

Figure 19 Fine needlelike TiB precipitates formed at temperatures above 500°C (930°F).

The inherent instability of the metalloid compounds in titanium alloys limits their upper temperature use to about 500–600°C (930–1110°F). However, RS is an effective means of introducing relatively high volume fractions of fine titanium-metalloid compounds. The potential for alloys based upon metalloid additions is yet to be fully identified.

4. *Eutectoid Former Titanium Alloys*

The eutectoid formers (e.g., Ni, Cu, Co, Fe, Cr, Si, and W) are characterized by a phase diagram at the titanium-rich end which involves both the eutectoid reaction ($\beta \rightarrow \alpha$ + compound), and often at higher temperatures and solute content, a eutectic decomposition ($L \rightarrow \beta$ + compound) [41]. This class of alloy has been studied in detail using ingot metallurgy (IM) [103]. It is normally possible to obtain relatively high levels of solute in the beta phase and lower though still significant levels in solution in the alpha phase.

A problem which arises, however, is that the eutectoid formers normally show a high partitioning coefficient between the liquid and the solid phase and a large temperature difference between the liquidus and solidus, meaning that they are very segregation-prone. Thus, though there have been a number of attempts to develop titanium alloys containing eutectoid formers using conventional IM, only limited success has been achieved due to ingot segregation and also poor workability [104].

In one comprehensive study a few percent of Be, Cu, Co, Fe, Ni, and Si were added to stable beta-based compositions, but with a significant adverse effect on

Table 7 Tensile Data from Experimental PM/RS High Strength Ti–1 Al–8 V–5 Fe Alloy

Alloy	Tensile properties YS (MPa)	UTS (MPa)	El. (%)
Ti-185	1390	1480	8
Ti-6Al-4V[a]	895	965	14

[a]Conventional.
From Ref. 113.

hot workability and rapid aging kinetics in the case of the Ni- and Si-containing alloys [105]. The commercial alloy Ti–10 V–2 Fe–3 Al which ultimately developed from this study is now seeing use in high-strength applications. The addition of Si is used to enhance high-temperature behavior of commercial alloys, and 2.5 Cu is used in the British moderate strength alloy IMI-230. Apart from these isolated instances, use of the eutectoid formers has not been realized.

By far the most comprehensive study of the RS characteristics of titanium eutectoid former alloys has been carried out by Krishnamurthy and co-workers [106–110]. The eutectoid former system for which RS showed the most significant advantage was the Cr-containing system with the addition of 4 wt% Al. Here a 15 wt% Cr alloy with the Al addition was evaluated in the RS (chill-block melt-spun) and conventionally prepared conditions. The RS material showed a much finer microstructure than the beta grain size conventional material (30 μm compared with 300 μm) and more rapid hardening, which was maintained for a considerably longer time. This enhanced behavior was related to the finer grain size and much more uniform precipitation, the latter effect perhaps due to a higher vacancy concentration. Preliminary results on a Ti–1 Al–8 V–5 Fe [111] PREP powder product indicate that strengths in excess of 1380 MPa (200 ksi) can be obtained with ductility levels close to 10% elongation [112,113], as shown in Table 7. These alloys would have applications in components such as landing gear.

5. *Beta Titanium Alloys*

Work on the beta alloys has included studies of the conventional alloys such as Beta III, Ti–15 V–3 Al–3 Sn–3 Cr and Ti–6 Al–15 V–2 Er [114], microstructural studies of Ti–25 V–4 Ce–0.6 S [115], and a heavily stabilized Ti–24 V–10 Cr alloy with and without Er additions [116,117]. The Ti–25 V–4 Ce–0.6 S alloys showed a very fine dispersion, leading to a predicted strength in excess of 1725 MPa (250 ksi). The Ti–6 Al–15 V–2 Er alloy showed no dispersoid coarsening to temperatures as high as 760°C (1400°F) (beta transus estimated at 830°C (1525°F)). Second-phase particles occurred in relatively large grains (~15 μm) in the Ti–24 V–10 Cr alloy, suggesting that this system may have potential for high-strength applications.

6. *Titanium Aluminides*

The titanium aluminides are potentially useful alloys for high-temperature application because of their ordered structure. However, the planarity of slip results in low ductility at room temperature. Rapid solidification offers the potential for improved ductility in these intermetallics by disordering [75], grain refinement, and development of fine dispersoid particles. When these fine dispersoids are oxides an additional gain in ductility can result from deoxidation of the matrix.

Early work on RS of this alloy class utilized TiAl-W alloys atomized by the PREP process. This material exhibited good creep resistance due to a fine precipitate of tungsten-rich beta phase in the alpha-2 plus gamma lath structure of the TiAl composition [118].

Rapid solidification of alloys with a base composition of Ti_3Al with the addition of 0.4 at% Er has been shown to produce a fine dispersoid of Er_2O_3 (Fig. 20 [93,94]). This dispersoid does not coarsen in grain interiors to temperatures as high as 1000°C (1830°F) [119]. Under HIP conditions, Er_2O_3 dispersoid coarsening in a Ti_3Al-Nb alloy was comparable to that observed during an anneal at the same temperature. However, extrusion caused rapid coarsening which was contrasted with cerium sulfide or cerium oxysulfide-containing alloys in which virtually no coarsening occurred [93,94]. The presence of a dispersoid in Ti_3Al-Nb alloys also produced a refinement in grain size. The ductility and fracture toughness of dis-

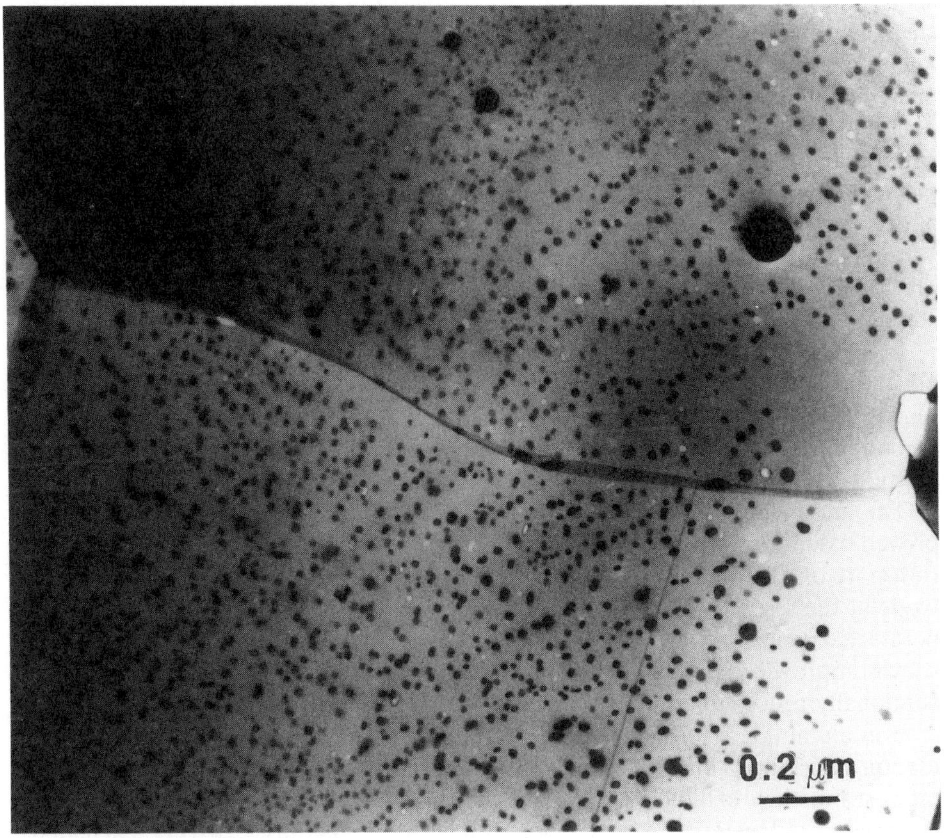

Figure 20 Fine dispersion of erbia particles formed in a Ti_3Al–11 Nb–0.5 Er (at%) after HIP'ing at 950°C (1740°F).

persoid-containing alloys was dependent on the test used. Hardness indentation ductility was improved by grain refinement but fracture toughness was not [93]. Much further work is required in this generic system to optimize the microstructure and hence mechanical properties.

7. Other Systems

Other alloying elements that could be added to titanium to give attractive characteristics are the low-density elements Li and Mg (Al and Si have a similar but smaller effect). The challenge here is to successfully add these elements which have a boiling point below the melting point of titanium.

D. Iron-Based Alloys

Rapidly quenched iron-based alloys fall into three broad categories: tool steels, structural alloys (e.g., ferritic, tempered martensitic and stainless steels, iron aluminides) and magnetic alloys. RS crystalline magnetic alloys are discussed elsewhere in this book. Powder metallurgy (P/M) high-speed tool steels (for example, T-15 and M2) that take advantage of modest solidification rates obtained in conventional gas atomization have been a commercial reality for over a decade. Recently centrifugal helium atomization has been used to refine grain size in high-speed steels [120]. The resistance to grain growth in RS M-1 and M-50 high-speed steels is primarily due to grain boundary pinning by very stable dispersoid phase. Such steels, if conventionally processed, undergo excessive grain growth at the austenitizing temperatures when the carbides are fully dissolved. For example, in M-2 tool steel the grain size reaches 100 μm after austenitizing for 1 hr at 1180°C whereas in rapidly quenched material it is only 25 μm. Some of the principal phases participating in grain boundary pinning are MnS, SiO_2, Al_2O_3, aluminum silicate and complex oxysulfides normally considered as undesirable inclusions. In conventionally processed steel some of these dispersoids are either not observed or are too coarse to effectively pin the grain boundaries. This fine grain size of RS steel results in superior toughness.

In some ultra-high-strength structural steels (0.25 C–4 Mo and 9 Ni–4 Co steels), advantage has been taken of the retention of finer grain size at higher austenitizing temperature, to improve fracture toughness and obtain lower ductile-brittle transition temperature. The grain-refining dispersions discussed above involve volume fractions of the order of only 0.1% [121].

The fine inclusion dispersions and grain-coarsening resistance also provide improved oxidation resistance in RS stainless steel through enhanced grain boundary transport of Cr to the protective oxide, together with mechanical keying of the oxide at the dispersoid inclusions and closely spaced grain boundaries. High temperature oxidation studies of Fe-Cr-Ni-Mn-Si alloys in pure O_2 and SO_2 gases have shown improved oxidation resistance of fine-grained RS alloys compared to conventionally processed alloys of the same composition (Fig. 21).

Iron aluminides (for example, Fe_3Al and FeAl), because of their excellent oxidation resistance, lower density, and high-temperature strength, have been of interest but they exhibit very low room-temperature ductility. These alloys have been processed by RS to improve room-temperature ductility [120,122]. Particularly in Fe_3Al fine dispersions of TiB_2 have been achieved which give a fine grain size and improve room-temperature ductility Fig. 21 [123].

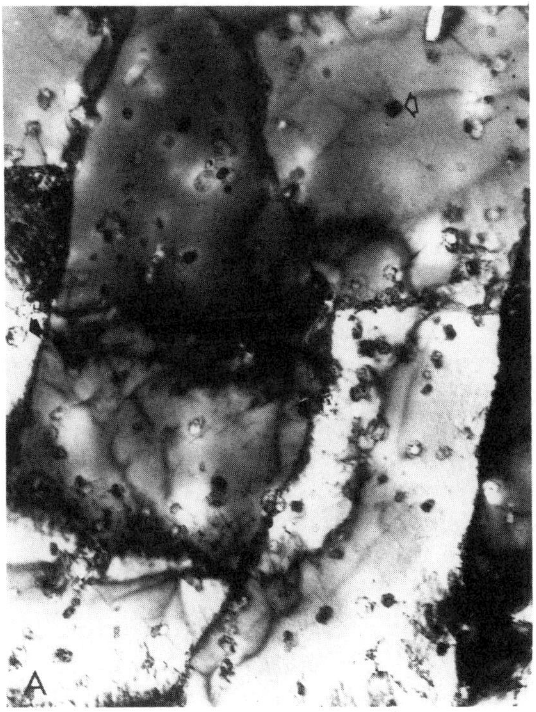

Figure 21 Fine dispersion of TiB_2 in as-extruded RS Fe_3Al; the RS powder was made by centrifugal gas atomization [123].

E. Nickel-Based Alloys

The application of rapid quenching to nickel-base high-temperature superalloys has been in progress for a number of years [124]. For this alloy family, the primary advantage offered by RS processing is the preparation of homogeneous alloys with a wider range of solute contents and strengthening phases having a significantly finer size and distribution than obtainable by conventional ingot metallurgy practice. In most cases powder atomization processes that have cooling rates of 10^3–10^4 K/s appear adequate to achieve homogeneity. Powder metallurgy superalloy disks so produced have been in use in gas turbine engines for the past two decades [125]. Much of the early effort on RS crystalline alloys started with nickel-based superalloys; the aim was to develop advanced air foil material that will have a higher temperature capability than the state-of-the-art directionally solidified Mar M200 alloy containing hafnium [126]. Several Ni-Al-Mo-Ta-W alloys were developed [127] using centrifugal helium-atomized powder, followed by hot consolidation and directional recrystallization. These alloys had a high propensity to grow along a ⟨100⟩ direction during recrystallization, an orientation which may be difficult to grow from the melt. Since superalloy single crystals exhibit anisotropic creep behavior, i.e., higher creep strength for ⟨111⟩ orientation versus ⟨100⟩ orientation, it appears that the improved creep properties are largely due to a recrystallized ⟨111⟩ orientation, rather than any effect of enhanced cooling rate [128,129]. The anisotropic creep behavior of superalloy single crystals, exhibiting higher creep strength for ⟨111⟩ orientation versus ⟨100⟩ orientation, is now well documented [130]. More recently, attention has focused on RS processing of intermetallics such as NiAl and Ni_3Al, primarily to improve their room-temperature ductility. The work in this area has been recently reviewed by Koch [131]. For Ni-Al-based alloys refined grain size appears to enhance ductility [132]. It has been discovered that doping Ni_3Al with boron either via RS or by microalloying also improves room-temperature ductility [133,134].

In addition to the above alloys, which require modest cooling rates achieved in atomization processes, nickel-based alloys containing fine stable dispersions have been prepared from amorphous alloy precursors [135,136]. In this novel approach, brittle intermetallic compounds such as Ni_3Mo and $Ni_3(Al, Ti)$ have been alloyed with sufficient metalloid (e.g., boron) so as to form an amorphous phase up on rapid quenching. Subsequent devitrification results in a fine dispersion of borides. Microcrystalline Ni-Mo-X-B and Ni-W-X-B alloys have been so synthesized to produce bulk shapes strengthened by dispersion of refractory borides, which are stable at temperatures as high as 1100°C. The extremely fine (submicron) grain size of these alloys does limit their high-temperature use when creep is the dominant deformation mode.

F. Copper-Based Alloys

The research on copper-based alloys has focused on several areas: dispersion-strengthened high-thermal-electrical-conductivity–high-strength copper alloys, high-strength corrosion-resistance structural alloys with improved damping capacity, and alloys which exhibit shape memory effects [137,138]. The dispersion-strengthened alloys are based on ZrO_2 dispersoids in Cu-Zr and Cu-Cr-Zr alloys. Zirconium has an equilibrium solubility of only 0.15 wt% and oxygen has high solubility in molten copper. Upon rapid quenching supersaturated Zr combines with oxygen to form ZrO_2. Thus, these alloys maintain high conductivity while providing dispersion strengthening [139]. They are candidate materials for replacement of Cu-Be alloys and also their elevated temperature strength makes them attractive for rocket thrust chambers. The ingot metallurgy damping alloys such as Cu-Mn-Al, although suitable for some structural marine components, are subject to stress corrosion cracking and have low fracture toughness due to extensive segregation. RS of these alloys primarily reduces segregation effects [137]. Cu-Al and Cu-Al-Si alloys, which display shape memory effects, have been studied to explore the effect of fine grain size obtained by rapid quenching. The martensitic transformation temperature associated with shape memory behavior is lowered in comparison to cast and wrought products. This depression is in part due to the refinement of grain size [140].

The effect of a RS melt-spun ribbon product on the subsequent microstructural stability of hot extruded (700°C to 900°C, 10:1 extrusion ratio) Cu–5 at% Cr alloy has been studied by Morris and Morris [141]. Since the solubility of Cr in Cu is about 1 at%, the excess Cr is found as discrete particulate within the Cu matrix. The results for the RS ribbon material indicated that the best room-temperature tensile properties were obtained using the 700°C extrusion temperature, producing material with a grain size of 1.0–1.5 μm and a hardness level of 161 VHN. Higher extrusion temperatures yielded significantly coarser Cr particle sizes and grain sizes.

By comparison, the hot extruded (700°C) mechanically alloyed powder yielded 0.1-μm grains, with a hardness level of 250 VHN. Subsequent annealing of the mechanically alloyed material for 1 hr at 900°C produced only a limited amount of decrease from the as-extruded properties, but 1 hr at 1000°C led to signficantly coarser grains and reduced room-temperature properties [141].

Patel and Diamond [142] studied the effect of annealing temperature on the stability of Cu–5 Cr and Cu–5 Cr–1 Zr alloys which were produced from RS/PM ribbon which was subsequently HIP'ed at 550°C, 240 MPa. The RS/PM Cu–5 Cr–1 Zr alloy exhibits higher as-extruded hardness than the RS/PM Cu–5 Cr alloy, but both show a strong decrease in hardness at annealing temperatures of 550°C to 600°C. One problem with the use of Cr as a dispersoid second phase in Cu is that there is solubility of Cr in Cu, hence diffusion-controlled Ostwald ripening ought to occur as a result of extended exposure at elevated temperatures.

G. Niobium-Based Alloys

Interest in rapidly solidified niobium alloys has been revived in this decade due to the resurgence of space nuclear power programs and new aerospace vehicles such as National Aerospace Plane (NASP). During the past few years, there has been progress in the use of the plasma rotating electrode process (PREP) followed by HIP consolidation to obtain mechanical properties in commercial C103 (Nb–10 Hf–1 Ti) and WC-3009 (Nb–30 Hf–9 W) alloys which are equivalent to their wrought counterparts [143–145]. The latter alloy, although it has a fairly high density for Nb alloys, is promising in so far as it has oxidation kinetics which are slow enough to withstand possible local loss of a protective oxidation coating [143]. A study by Himmelblau et al. [144] examined component fabrication using powders from PREP processes; the properties of the PREP + HIP consolidated material also matched the wrought C103 alloy. In a recent paper, Wojcik [145] has demonstrated that PREP versions of WC-3009 alloy powder possess both room-temperature and high-temperature (to 1425°C) mechanical properties which are equal to the wrought counterpart of this alloy.

The development of new RS Nb alloys for turbine applications will undoubtedly require either alloys with inherent oxidation resistance, or alloy/coating systems which will not fail catastrophically if portions of the coating are repeatedly lost during service. In order to have inherent oxidation resistance, Nb must be alloyed with sufficient amounts of protective scale-forming elements such as Al or Si. The addition of other bcc metals with low oxygen solubility (such as Cr, Mo, or W) should help to decrease the solubility for oxygen in the niobium alloy matrix. Recently Perkins and co-workers [146] have investigated a series of niobium aluminides which form protective oxide scales at temperatures of 1400°C. Based on 1-hr exposure times at 1400°C, the best aluminide investigated was a Nb–24 Ti–3 Cr–4 V–46 Al (at%) alloy. However, one consequence of the large Al additions to Nb is that the melting point is substantially reduced. For example, the above aluminide composition has a melting point less than 1650°C. Other investigators [147–148] have found $NbAl_3$ to be the most oxidation resistant of all the binary Nb aluminides. Unfortunately, $NbAl_3$ also has a relatively low melting point (~1600°C). Nb silicides are also interesting experimental materials since they have the potential for oxidation resistance by means of forming protective SiO_2 scale.

However, the Nb silicides are extremely brittle, and probably cannot be used for structural applications.

Thus, it appears that the concept which has the most potential for success in niobium alloy development is the use of alloy/coating systems, which have the capability of withstanding local loss of coating without catastrophic oxidation. It is also probable that the melt-spinning RS techniques will eventually be used to produce powders which can be consolidated by means of HIP. Such attempts are just beginning [149].

H. Tungsten-Based Alloys

Rapid solidification of variations of tungsten-nickel-iron-boron alloys has been explored [150,151] for possible improvement over conventional liquid-phase-sintered (LPS) tungsten-nickel-iron alloys. Tungsten alloys, because of their high density and strength properties, are good candidates for applications such as projectiles, counterweights, and radiation shielding where high density is a primary property requirement. LaSalle et al. [151] studied melt-spun ribbons containing 40 to 75 wt.% W and bulk material, which was produced from the ribbon after pulverization and consolidation via hot extrusion. It was found that the as-cast ribbon was either amorphous or microcrystalline and showed increasing crystallization temperature with increasing tungsten content. Upon extrusion the amorphous structure transformed to a fine multiphase structure of W and W_X (Ni,Fe)B_y particles in a continuous nickel-based solid solution matrix. The mechanical properties of bulk-extruded alloys showed that the alloys containing less than 75 wt.% W exhibit high impact energy values, the impact energy decreasing and tensile strength increasing with increasing levels of tungsten. For example, the unnotched impact strength for a 65 wt% W alloy (after heat treatment) was 214 J. The 75 wt% W alloy, by contrast, had relatively high tensile strengths with UTS values at 1390 MPa. Impact energies for the 75 wt% W, however, were rather low (19 J). This may be due to the higher volume fraction of brittle boride phases in this alloy, a result of the higher levels of boron required to depress the liquidus temperature in this alloy.

III. CONCLUSIONS

Improved materials will play a key role in the development of future aerospace systems. Development of such materials with enhanced physical and mechanical characteristics requires processing far from equilibrium. One such technique is that of rapid solidification processing. The present paper has discussed recent developments in the microstructure and properties of eight types of rapidly solidified alloys: those based on aluminum, magnesium, titanium, iron, nickel, copper, niobium, and tungsten. For commercial application of any new engineering material, some important material property must be significantly improved without sacrifice in the levels of secondary properties. It is relatively easy to generate increased levels of strength through formation of dispersoids, but it is difficult to produce a fine uniform dispersion which will yield, simultaneously, a high strength, a high toughness, and corrosion resistance. For chemistries based on Al-Fe, Al-Li, and Mg-Al-Zn-RE, the high solidification rates provided by substrate quenching have provided rapidly solidified alloys with fine-scale uniform microstructures and outstanding sets of balanced properties. Similar attempts are under way to produce other high performance alloys; these efforts show promise, but much remains to be done.

ACKNOWLEDGMENTS

The authors would like to acknowledge their co-workers for helpful discussions and for supplying information used in this review.

REFERENCES

1. *Rapid Solidified Crystalline Alloys*, edited by S. K. Das, B. H. Kear, and C. M. Adam (TMS-AIME, Warrendale, PA, 1985).
2. *Rapid Solidification Processing: Principles and Technolgies IV*, edited by R. Mehrabian and P. A. Parrish (Claitor's, Baton Rouge, LA, 1988).
3. S. K. Das and L. A. Davis, *Mater. Sci. Eng.*, *98*, 1 (1988).
4. A. Calka, M. Madhava, D. E. Polk, B. C. Giessen, H. Matyja, and J. Vander Sande, *Scripta Metall.*, *11*, 65 (1977).
5. P. G. Boswell,*Mater. Sci. Eng.*, *34*, 1 (1978).
6. J. J. Masur, J. T. Burke, T. Z. Kattamis, and M. C. Flemings, in *Rapidly Solidified Amorphous and Crystalline Alloys*, edited by B. H. Kear, B. C. Giessen, and M. Cohen (Elsevier, New York, 1982), p. 185.
7. M. C. Flemings and A. Mortensen, Report No. AMMRC TR 84-37 (Army Materials and Mechanics Research Center, Watertown, MA 1984).
8. M. Mizutani and T. Matsuda, in *Rapidly Quenched Metals*, Vol. 1, edited by S. Steeb and H. Warlimont (North-Holland, Amsterdam, 1985), pp. 10, 35.
9. A. Joshi, R. E. Lewis, and H. Jones, *Int. J. Rapid Solidification*, *4*, 251 (1989).
10. N. I. Varich and B. N. Litvin, *Phys. Metals Metall.*, *16*, 29 (1963).
11. H. L. Luo, C. C. Chao, and P. Duwez, *Trans. TMS-AIME*, *230*, 1488 (1964).
12. H. Gjestland, G. Nussbaum, and G. Regazzoni, in *Light-Weight Alloys for Aerospace Applications*, edited by E. H. Chia and N. J. Kim (TMS-AIME, Warrendale, PA, 1989), p. 139.
13. G. Nussbaum, H. Gjestland, and G. Regazzoni, in *Proc. 45th Annual World Magnesium Conf.* (International Magnesium Association, 1988), p. 19.
14. S. Isserow and F. J. Rizzitano, *Int. J. Powder Metall. Powder Tech.*, *10*, 217 (1974).
15. G. S. Foerster and H. A. Johnson, *Product Eng.*, *29*, 80 (1958).
16. G. S. Foerster, *Metals Eng. Q.*, *12*, 22 (1972).
17. P. J. Meschter and J. E. O'Neal, *Metall. Trans. A.*, *15A*, 237 (1984).
18. S. K. Das and C. F. Chang, in Ref. 1, p. 137.
19. P. J. Meschter, *Metall. Trans. A.*, *18A*, 347 (1987).
20. C. F. Chang, S. K. Das, and D. Raybould, in *Rapidly Solidified Materials*, edited by P. W. Lee and R. S. Carbonara (ASM, Metals Park, OH, 1986), p. 129.
21. C. F. Chang, S. K. Das, D. Raybould, R. L. Bye, and E. Limoncelli, in *Advanced Powder Metallurgy 1* (Metal Powder Industries Federation, Princeton, NJ, 1989), p. 331.
22. H. Jones, *Int. J. Rapid Solidification*, *4*, 297 (1989).
23. F. Hehmann and H. Jones, in *Rapidly Solidified Alloys and Their Mechanical and Magnetic Properties*, edited by B. C. Giessen, D. Polk, and A. Taub (Materials Research Society, Pittsburgh, PA, 1986), p. 259.
24. R. S. Busk and T. I. Leontis, *Trans. AIME*, *188*, 297 (1950).

25. C. F. Chang, S. K. Das, D. Raybould, and A. Brown, *Metal Powder Rep.*, *41*, 302 (1986).
26. J. D. Hanawalt, C. E. Nelson, and J. A. Peloubet, *Trans. AIME*, *147*, 273 (1942).
27. K. Sankaran and N. J. Grant, *Mater. Sci. Eng.*, *44*, 213 (1980).
28. G. J. Hildeman, L. C. Labarre, A. Hafeez, and L. M. Angers, in *High Strength Powder Metallurgy Aluminum Alloys II*, edited by G. J. Hildeman and M. J. Koczak (The Metallurgical Society, Warrendale, PA, 1986), p. 25.
29. H. Jones, *Mater. Sci. Eng.*, *5*, 1 (1969–70).
30. A. Tonejc and A. Bonefacic, *J. Appl. Phys.*, *40*, 419 (1970).
31. G. Thursfield and M. J. Stowell, *J. Mater. Sci.*, *9*, 1644 (1974).
32. M. H. Jacobs, A. G. Doggett, and M. J. Stowell, *Mater. Sci. Eng.*, *9*, 1631 (1974).
33. D. S. Schechtman, I. Blech, D. Gratias, and J. W. Cahn, *Phys. Rev. Lett*, *53*, 3984 (1951).
34. I. R. Hughes, G. J. Marshall, and W. S. Miller, in Ref. 8, p. 1743.
35. M. Zedalis, Allied-Signal Inc., Morristown, NJ, unpublished data, 1987.
36. G. H. Hildeman, U.S. Patent 4,379,719 (1983).
37. C. M. Adam, in Ref. 6, p. 411.
38. D. J. Skinner, R. L. Bye, D. Raybould, and A. M. Brown, *Scripta Metall.*, *20*, 867 (1986).
39. I. G. Palmer, M. P. Thomas, and G. J. Marshall, in *Dispersion Strengthened Aluminum Alloys*, edited by Y. W. Kim and W. M. Griffith, (The Minerals, Metals and Materials Society, Warrendale, PA, 1988), p. 217.
40. D. Y. Lee and D. E. Zupon, in Ref. 39, p. 265.
41. F. H. Froes and R. G. Rowe, in Ref. 23, p. 309.
42. R. Ayer, L. M. Angers, R. R. Mueller, J. C. Scalon, and C. F. Klein, *Metall. Trans. A.*, *19A*, 1645 (1988).
43. M. J. Couper and R. F. Singer, in *Processing of Structural Metals by Rapid Solidification*, edited by F. H. Froes and S. J. Savage (ASM, Metals Park, OH, 1987), p. 273.
44. *Aluminum-Lithium Alloys II, Proc. 2nd Int. Conf. Aluminum-Lithium Alloys*, edited by T. H. Sanders, Jr., and E. A. Starke, Jr. (AIME, New York, 1984).
45. R. L. Bye, N. J. Kim, D. J. Skinner, D. Raybould, and A. M. Brown, in Ref. 42, p. 283.
46. D. J. Sinner, in Ref. 39, p. 181.
47. V. Radmilovic and G. Thomas, University of California, Berkeley, CA, to be published.
48. A. M. Brown, D. J. Skinner, D. Raybould, S. K. Das, R. L. Bye, and C. M. Adam, in *Aluminum Alloys and Their Physical and Mechanical Properties*, edited by E. A. Starke and T. H. Sanders (EMAS, West Midlands, UK, 1986), p. 1029.
49. M. Zedalis, D. Raybould, D. J. Skinner, and S. K. Das, in Ref. 43, p. 347.
50. P. S. Gilman and S. K. Das, in *Proc. Int. Conf. PM Aerospace Materials, A Metal Powder Report Conf.*, Luzern, Nov. 1987 (MPIF, APMI, Princeton, NJ, 1988), p. 27.1.
51. L. Chandrasekaran, N. Saunders, and A. P. Miodownik, University of Surrey, unpublished data (1989).
52. *Aluminum-Lithium Alloys I, Proc. 1st Int. Conf. on Aluminum-Lithium Alloys*, edited by T. H. Sanders, Jr., and E. A. Starke, Jr. (AIME, New York, 1981).
53. *Aluminum-Lithium Alloys*, edited by C. Baker, P. J. Gregson, S. J. Harris, and C. J. Peel (Institute of Metals, London, 1986).
54. C. M. Adam and R. E. Lewis, Ref. 1, p. 157.
55. P. J. Meschter, R. J. Lederick, and J. E. O'Neal, in Ref. 50, p. 85.

56. N. J. Kim, D. J. Skinner, K. Okazaki, and C. M. Adam, in *Aluminum-Lithium Alloys III, Proc. 3rd Inst. Conf. Aluminum-Lithium Alloys*, edited by C. Baker, P. J. Gregson, S. J. Harris, and C. J. Peel (Institute of Metals, London, 1986), p. 78.
57. N. J. Kim and S. K. Das, *Scripta Metall.*, *20*, 1107 (1986).
58. N. J. Kim, D. Raybould, R. L. Bye, and S. K. Das, in *Proc. 5th Int. Conf. Aluminum-Lithium Alloys*, edited by T. H. Sanders, Jr., and E. A. Starke, Jr. (Materials and Components Engg. Publication, Birmingham, UK, 1989), p. 123.
59. K. T. Venkateswara and R. O. Ritchie, *Metall. Trans.*, *22A*, 191 (1991).
60. H. Margolin and J. P. Neilson, *Modern Materials Advances in Development and Application*, Vol. 2, edited by H. H. Hauser (Academic Press, New York, 1960), p. 225.
61. S. S. Joseph and F. H. Froes, *Light Metals Age*, *46*, (1988).
62. I. I. Kornilov, in *The Science, Technology and Application of Titanium*, edited by R. I. Jaffee and N. E. Promisel (Pergamon, Oxford, 1970), p. 407.
63. I. I. Kornilov, in *Titanium and Titanium Alloys*, Vol. 2, edited by J. C. Williams and A. F. Belov (Plenum, New York, 1982), p. 1281.
64. E. K. Molchanova, *Phase Diagrams of Titanium Alloys*, (Israel Program for Scientific Translations, Jersusalem, 1965).
65. D. E. Polk and B. C. Giessen, *Metallic Glasses* (ASM, Metals Park, OH, 1978), p. 1.
66. F. H. Froes and D. Eylon, *Titanium, Science and Technology*, Vol. 1, edited by G. Lutjering, U. Zwicker, and W. Bunk (DGM, Oberursel, West Germany, 1985), p. 267.
67. S. H. Whang, Polytechnic University, New York, private communication, 1985.
68. *Titanium, Rapid Solidification Technology*, edited by F. H. Froes and D. Eylon (TMS-AIME, Warrendale, PA, 1986).
69. F. H. Froes and R. G. Rowe, in Ref. 68, p. 1.
70. H. B. Bomberger and F. H. Froes, in Ref. 68, p. 21.
71. R. G. Rowe and F. H. Froes, *Rapid Solidification of Crystalline Alloys*, edited by J. V. Wood (Institute of Metals, London), in print.
72. S. M. L. Sastry, T. C. Peng, P. J. Meschter, and J. E. O'Neil, *J. Metals*, *35*, 21 (1983).
73. T. E. Broderick, A. G. Jackson, H. Jones, and F. H. Froes, *Metall. Trans.*, *16A*, 1951 (1985).
74. F. H. Froes and J. R. Pickens, *J. Metals*, *36*, 14 (1984).
75. H. Jones, *Rapid Solidification of Metals and Alloys*, Monograph No. 8 (Institution of Metallurgists, London, 1982).
76. S. J. Savage and F. H. Froes, *J. Metals*, *36*, 20 (1984).
77. A. F. Belov and I. S. Polkin, *Modern Trends in Titanium Production and Processing* (Germany Metallurgical Society Workshop, University of Nuremberg, Erlangen, 1982).
78. T. F. Broderick, F. H. Froes, and A. G. Jackson, in *Rapidly Solidified Metastable Materials*, edited by B. H. Kear and B. C. Giessen (Elsevier, New York, 1984), p. 345.
79. T. C. Peng, S. M. L. Sastry, and J. E. O'Neil, *Lasers in Metallurgy*, edited by K. Mukherjee and J. Mazumder (TMS-AIME, Warrendale, PA, 1981), p. 279.
80. J. E. O'Neil, T. C. Peng, and S. M. L. Sastry, *Proc. 39th Annual EMSA Meeting*, edited by G. W. Bailey (Claitor's, Baton Rouge, LA, 1981), p. 66.
81. B. W. Muddle, D. G. Konitzer, and H. L. Fraser, *Strength of Metals and Alloys*, Vol. 1, edited by R. C. Gifkins (Pergamon, Oxford, 1983), p. 313.
82. I. G. Wright and B. A. Wilcox, Battelle Columbus Laboratories Report No. AD-781-133, 1974.

83. S. M. L. Sastry, P. J. Meschter, and J. E. O'Neil, *Metall. Trans.*, *15A*, 1451 (1984).
84. D. B. Snow, *Laser Processing of Materials*, edited by K. Mukherjee and J. Mazumder (TMS-AIME, Warrendale, PA, 1984), p. 83.
85. D. G. Konitzer, B. C. Muddle, R. Kirchheim, and H. L. Fraser, in Ref. 8, p. 953.
86. K. G. Konitzer, B. C. Muddle, and H. L. Fraser, *Scripta Metall.*, *17*, 963 (1983).
87. D. G. Konitzer, B. C. Muddle, H. L. Fraser, and R. Kirchheim, in Ref. 66, p. 405.
88. D. G. Konitzer, B. C. Muddle, and H. L. Fraser, *Metall. Trans.*, *14A*, 1979 (1983).
89. D. G. Konitzer and H. L. Fraser, in *Modern Development in Powder Metallurgy*, edited by E. N. Aqua and C. I. Whitman (MPIF, Princeton, NJ, 1984), p. 607.
90. R. G. Rowe, T. F. Broderick, E. F. Koch, and F. H. Froes, in Ref. 20, p. 107.
91. M. F. X. Gigliotti, R. G. Rowe, G. E. Wasielewski, G. K. Scarr, and J. C. Williams, in Ref. 23, p. 343.
92. R. G. Rowe and E. F. Koch, in Ref. 20, p. 115.
93. R. G. Rowe, J. A. Sutliff, and E. F. Koch, in Ref. 23, p. 359.
94. J. A. Sutliff and R. G. Rowe, in Ref. 23, p. 371.
95. L. M. Brown and R. K. Ham, *Strengthening Methods in Crystals*, edited by A. Kelly and R. B. Nicholson (Elsevier, New York, 1971), p. 9.
96. S. M. L. Sastry, T. C. Peng, and L. P. Beckerman, *Metall. Trans.*, *15A*, 1465 (1984).
97. R. G. Vogt, D. Eylon, F. H. Froes, unpublished work, 1983–1986.
98. F. H. Froes, in *Materials Edge*, No. 5, May/June 1988, p. 19.
99. F. H. Froes, in *Space Age Metals Technology*, Vol. 2, edited by F. H. Froes and R. A. Cull (SAMPE, Covina, CA, 1988), p. 1.
100. S. H. Whang, *J. Metals*, *36*, 34 (1984).
101. S. H. Whang, Polytechnic University, New York, private communcation, 1985.
102. S. M. L. Sastry, T. C. Peng, and J. E. O'Neil, in Ref. 66, p. 397.
103. G. W. Franti, J. C. Williams, and H. I. Aaronson, *Metall. Trans.*, *9A*, 1641 (1978).
104. H. B. Bomberger and F. H. Froes, AFWAL Report No. TR-84-4164, 1985.
105. D. B. Hunter, AFWAL Report TR-405/2-15, Part II, 1966.
106. W. A. Baeslack III, S. Krishnamurthy, and F. H. Froes, *Strength of Metals and Alloys*, Vol. 2, edited by H. J. McQueen, J.-P. Bailon, J. I. Dickson, J. J. Jonas, and M. G. Akben (Pergamon, Oxford, 1985), p. 1633.
107. S. Krishnamurthy, A. G. Jackson, D. Eylon, R. R. Boyer, and F. H. Froes, in Ref. 8, p. 945.
108. S. Krishnamurthy, I. Weiss, D. Eylon, and F. H. Froes, in Ref. 106, p. 1627.
109. W. A. Baeslack III, L. Weeter, S. Krishnamurthy, P. R. Smith, and F. H. Froes, *Rapidly Solidified Metastable Materials*, edited by B. H. Kear and B. C. Giessen (Elsevier, New York, 1984), pp. 375–379.
110. S. Krishnamurthy, R. G. Vogt, D. Eylon, and F. H. Froes, in Ref. 78, p. 361.
111. J. Gross, Mallory-Sharon Titanium Corporation Internal Report No. 1000R164, 1956.
112. R. G. Vogt, P. R. Smith, D. Eylon, and F. H. Froes, unpublished work, 1983–1986.
113. F. H. Froes and R. G. Rowe, in *Proc. Sixth World Conf. Titanium*, edited by P. Lacombe et al. (Les Ulis Cedex, Les Editions de Physique, France, 1989), p. 979.
114. T. F. Broderick, F. H. Froes, and J. A. Snide, Air Force Materials Lab, WPAFB, Dayton, OH, unpublished work, 1985–1986.
115. T. F. Broderick and F. H. Froes, Air Force Materials Lab, WPAFB, Dayton, OH, unpublished work, 1985.
116. F. H. Froes and P. R. Smith, AFML, WPAFB, Dayton, unpublished work, 1985.
117. M. Gutierrez, I. A. Martorell, and F. H. Froes, unpublished work, 1986.
118. P. L. Martin, M. G. Mendiratta, and H. A. Lipsitt, *Metall. Trans.*, *14A*, 2170 (1983).
119. D. G. Konitzer, and H. L. Fraser, University of Illinois-Urbana, private communication, 1985.
120. G. B. Olson and R. G. Bourdeau, in Ref. 1, p. 185.

121. G. J. Yurek, D. Eisen, and A. J. Garrett-Reed, *Metall. Trans.*, *13A*, 473 (1982).
122. G. B. Olson, in Ref. 2, p. 82.
123. E. R. Slaughter and S. K. Das, *Rapid Solidification Processing Principles and Technologies II*, edited by R. Mehrabian, B. H. Kear, and M. Cohen (Claitor's, Baton Rouge, LA, 1980), p. 354.
124. R. Patterson II, A. Cox, and E. Van Reuth, in *Rapid Solidification Technology Source Book*, edited by R. L. Ashbrook (ASM, Metals Park, OH, 1983), p. 414.
125. M. M. Allen, R. L. Athey, and J. B. Moore, *Metals Eng. Q.*, *10*, 20 (1970).
126. R. J. Patterson II, A. R. Cox, and E. C. Van Reuth, *J. Metals.*, *32*, 34 (1980).
127. A. Giamei and J. Larson, in *Mechanical Behavior of Rapidly Solidified Materials*, edited by S. M. L. Sastry and B. A. McDonald (TMS-AIME, Warrendale, 1986), p. 103.
128. H. Chin and A. Adair, in *Superalloys 1984*, edited by M. Gell, C. Kortovich, R. Bricknell, W. Kent, and J. Radavich (ASM, Metals Park, OH, 1984), p. 335.
129. A. F. Giamei, in Ref. 1, p. 203.
130. R. McKay, R. L. Dreshfield, and R. D. Maier, in *Superalloys 1980*, edited by J. K. Tien, S. T. Wlodek, H. Morrow III, M. Gell, and G. E. Maurer (ASM, Metals Park, OH, 1980), p. 385.
131. C. C. Koch, *Int. Mater. Rev.*, *33*, 201 (1988).
132. E. M. Schulson and D. R. Barker, *Script. Metall.*, *17*, 519 (1983).
133. A. I. Taub and M. R. Jackson, in Ref. 23, p. 389.
134. S. C. Huang, K. M. Chang, and A. I. Taub, in Ref. 20, p. 255.
135. S. K. Das, L. A. Davis, J. R. Y. Wang, and D. Kapoor, *Rapid Solidification Processing Principles and Technologies III*, edited by R. Mehrabian (National Bureau of Standards, Washington, DC, 1982), p. 559.
136. S. K. Das, K. Okazaki, and C. M. Adams, in *High Temperature Alloys: Theory and Design*, edited by J. O. Stiegler (TMS-AIME, Warrendale, PA, 1984), p. 451.
137. I. E. Anderson and B. B. Rath, in Ref. 1, p. 219.
138. I. E. Anderson, K. M. Rusin, and B. B. Rath, in Ref. 2, p. 114.
139. V. K. Sarin and N. J. Grant, *Metall. Trans.*, *3*, 875 (1971).
140. S. Fujiwara and S. Miwa, *Mater. Sci. Eng.*, *98*, 509 (1988).
141. D. G. Morris and M. A. Morris, *Mater. Sci. Eng.*, *A104*, 201 (1988).
142. A. N. Patel and S. Diamond, *Mater. Sci. Eng.*, *98*, 329 (1988).
143. J. Wadsworth, C. A. Roberts, and E. H. Rennhack, *J. Mater. Sci.*, *17*, 2539 (1982).
144. C. D. Himmelblau, M. Kibrick, J. Runkle, A. Joshi, J. Wadworth, and J. Moncur, *Progress in Powder Metallurgy*, Vol. 39, edited by H. S. Nayar, S. M. Kaufman, and K. E. Meiners (MPIF, Princeton, NJ, 1984), p. 525.
145. C. C. Wojcik, *Modern Developments in Powder Metallurgy*, Vol. 21, edited by P. W. Gummeson and D. A. Gustafson (MPIF, Princeton, NJ, 1988), p. 187.
146. R. A. Perkins, K. T. Chiang, and G. H. Meier, *Scripta Metall.*, *22*, 419 (1988).
147. R. C. Svedberg, *Properties of High Temperature Alloys*, edited by Z. A. Foroulis and F. S. Pettit (Electrochemical Society, New York, 1976), p. 331.
148. M. G. Hebsur and I. E. Locci, NASA Tech. Memo 100264, NASA Lewis Research Center, Cleveland, OH, 1988.
149. D. M. Bowden, *Adv. Mater. Manufact. Process*, *3*, 79 (1988).
150. D. Vujic, S. H. Whang, and S. Cytron, in *Science and Technology of Rapidly Quenched Alloys*, edited by M. Tenhover, W. L. Johnson, and L. E. Tanner (Materials Research Society, Pittsburgh, PA, 1987), p. 307.
151. J. LaSalle, D. Raybould, E. V. Limoncelli, S. K. Das, and S. Cytron, *Mater. Sci. Eng.*, *98*, 165 (1988).

13

Recent Studies of the Mechanical Properties of Amorphous Alloys

J. C. M. Li
University of Rochester, Rochester, New York

I. EARLY REVIEWS ON MECHANICAL PROPERTIES

Ever since Klement et al. [1] discovered the possibility of rapidly quenching a molten alloy to form an amorphous solid, many researchers followed to study this interesting state of matter. While glassy state has been known for a long time, the glassy metallic state as obtained by rapid solidification has a history beginning only from that discovery in 1960.

Intensive study of the mechanical properties of amorphous alloys started in the early 1970s when a group of people at Allied Chemical succeeded making long ribbons of such material. Pampillo [2] wrote the first review article in which he showed among other things the nature of inhomogeneous deformation, intersecting shear bands and shearing steps, large surface steps at the emergence of shear bands, reverse shear of shear bands and the retraction of surface steps, preferential etching of shear bands after polishing, the vein pattern developed on the fracture surface, stress corrosion cracking, stress-strain curves and serrated flow, stress-temperature relations and the activation enthalpy, and a dislocation model of the amorphous structure.

The next comprehensive review on mechanical properties was written by Li [3] in which I summarized the studies on more than 200 compositions and tabulated the density, glass temperature, crystallization temperature, Young's modulus, shear modulus, bulk modulus, Poisson ratio, hardness, yield stress, and the fracture stress of all these materials available at the time. In addition, the following properties are described: (1) elastic properties including the effects of stress, composition, magnetization, cold-rolling, irradiation, and temperature; (2) anelastic properties including internal friction and strain recovery after deformation; (3) yielding behavior including serration and work hardening and the effects of annealing, composition, strain rate, temperature, and hydrostatic pressure; (4) shear localization and shear banding; (5) stress relaxation, rolling, and creep; (6) effects of the deformation on stored energy, Young's modulus, density, and magnetization; (7) fracture behavior including toughness, annealing embrittlement, *H* embrittlement, and the effects of strain rate, temperature, and hydrostatic pressure; (8) fatigue behavior; and (9) micromechanisms of flow and fracture.

Kimura and Masumoto [4] summarized the mechanical properties from the viewpoint of mechanics models. They covered the statics of plastic deformation, the dynamics of inhomogeneous plastic flow, and the mechanics of fracture. They showed evidence of elastic–perfectly plastic deformation, Dugdale model for the plastic zone, von Mises yield criterion with pressure dependence, slip band angle following the direction of zero extension rate, a power law for serrated flow (amplitude of serration versus aging time), etc.

Spaepen and Taub [5] reviewed mostly their own work on flow and fracture. They began by differentiating homogeneous and inhomogeneous flow in a deformation map. Then they emphasized the effect of structural relaxation on viscous flow, defined an isoconfigurational flow (reproducible upon increasing or decreasing temperature), and described several flow laws. Free volume model of flow was used to describe the viscosity-temperature relations, to relate viscosity with diffusivity, and to explain structural relaxation, shear localization, and annealing embrittlement.

Chaudhari et al. [6] reviewed the atomic transport and computer simulation of point and line defects in amorphous metals. They differentiated stable equilibrium (above the melting temperature), metastable equilibrium (between melting and glass temperatures) and nonequilibrium (below glass temperature) states. In the metastable and nonequilibrium regions, measurements can be made after sufficient relaxation so that properties do not change much in the laboratory time scale. These are called isoconfigurational measurements. Computer simulation of line defects (screw and edge dislocations) using Lennard-Jones potentials is reviewed in much detail. The detection of the existence of line defects is through their stress fields.

Künzi [7] reviewed elastic and anelastic behavior of amorphous alloys as well as some experimental techniques. He tabulated the elastic moduli and Poisson ratio of about 20 compositions. Magnetoelastic behavior was discussed including magnetostriction, and the ΔE (change of Young's modulus upon magnetization) effect. The temperature dependence of ΔE and the effect of annealing were reviewed also. For anelastic behavior, he covered internal friction, thermoelastic relaxation, magnetoelastic relaxation, and relaxation due to mobile impurities such as *H* and *D*.

In 1983, I supplemented [8] my earlier review [3] by including subsequent studies and those reported in two more (RQ3 and RQ4) international conferences.

Glezer et al. [9] reviewed some aspect of mechanical properties of amorphous alloys including some test methods such as tension, loop bending, tearing, and microhardness. They also briefly summarized the strength, Young's modulus, hardness, plastic flow, and fracture. The effects of quenching rate, applied magnetic field, cold-rolling, irradiation, hydrogen charging, and annealing are briefly mentioned also. In the atomistic mechanism of inhomogeneous plastic deformation, they mentioned that the radial distribution function of amorphous state computed from x-ray data cannot be described in terms of a dislocation model even though the dislocation density may be as high as 10^{14}–$10^{15}/cm^2$. While they did not cite the source for this statement it was in direct contradiction to the findings of Ninomiya and Koizumi [10,11] who introduced 10^{14}–$10^{15}/cm^2$ dislocations into a fcc lattice and relaxed with a Morse potential. The calculated radial distribution agreed with x-ray data and the bond angle distribution agreed with that of a dense random-packing model of the amorphous structure.

Rao and Radhakrishnan [12] made a survey of mechanical properties of metallic glasses. They tabulated the hardness, ultimate strength, and Young's modulus of about 40 compositions. They also listed the fracture toughness of 32 compositions. Fatigue of $Pd_{80}Si_{20}$ was reviewed in some detail. An elastic–crack-opening displacement (COD) approach was found applicable to the crack extension per cycle for this material and for $Ni_{39}Fe_{38}P_{14}B_6Al_3$. Exponents for power law creep were listed for 14 compositions. For $Fe_{40}Ni_{40}P_{14}B_6$, they found that the steady state creep rate is inversely proportional to the rupture time raised to a 0.8 power, based on the data collected by Gibeling and Nix [13]. They also briefly reviewed stress relaxation, inhomogeneous flow, and computer simulation of micromechanisms. About the existence of dislocations, they said that well-defined line defects are unlikely in metallic glasses because of the lack of long range order. However, the type of dislocations envisioned is of Volterra of Somigliana [14] type and not the Burgers dislocations and hence long-range order is not needed.

Egami and Vitek [15] reviewed their calculations on the atomic level stresses. They commented on the current free-volume concept as not being precise enough to allow any quantitative calculations. For example, the temperature dependence of free volume is only obtained by curve fitting to experimental data. In reality free volume cannot be the only parameter characterizing the amorphous structure. By using the atomic level stresses, atoms having a low pressure correspond to free volume. But now there are atoms having a high pressure or shear stresses. Upon annealing the distribution of these stresses may change, and those which disappear can be considered as defects. They also reviewed the concepts of dislocation and disclination.

Takeuchi [16] summarized the characteristics of homogeneous and inhomogeneous deformation and commented on the micromechanisms on the atomic level including the results of his own computer simulation.

Davis [17] started his review by emphasizing the commercial importance of mechanical properties in the processing of ribbons and in magnetic applications. Then he showed some examples of the composition dependence of elastic moduli followed by a discussion of deformation modes in plastic response and work hardening. Micromechanisms involving dislocations and free volumes were critically reviewed in some detail. The hardness, yield stress, and Young's modulus were tabulated for nine compositions. Then fracture mode, fracture toughness, annealing embrittlement, and fatigue behavior were described. Finally magnetoelastic effects including magnetostriction, effect of magnetization on Young's modulus, and magnetic annealing were discussed.

Argon [18] reviewed the inelastic deformation mechanisms in glassy alloys. For the structure, free volume, atomic-level stresses, and dislocations are mentioned. For the atomic modes of shear, results of computer simulation, soap bubble simulation, and activation energy spectrum modeling are summarized. About the pressure dependence of plastic resistance, only the effect of pressure on shear modulus is noted. The effect of pressure on the structure and the activation energy is not considered. For example, if shear localization at low temperatures is to be understood by the shear-induced flow dilatations, such processes should be affected by pressure. Experimental probes of mechanical response included internal friction, viscous flow, and the effect of aging on both hardness and strain to fracture. The measurements of Megusar et al [19] on flow dilatation are of doubtful validity. The results of Deng and Lu [21] are more believable. The flow dilation is too small to be the cause for shear localization.

Koch [22] reviewed the testing techniques for mechanical characterization of amorphous alloys. He covered the pulse-echo technique, vibrating reed, impulse induced resonance, piezoelastic ultrasonic composite oscillator, microhardness, creep, stress relaxation, fracture toughness, and fatigue testing.

Davis and Ramanan [23] first summarized the strength, hardness, stiffness, deformation, ductility, and toughness of glassy alloys. Then they mentioned briefly the mechanical aspects of ribbon production, magnetic core annealing, and the production of stacked cores. Magnetomechanical effects were discussed regarding mechanical vibrations induced by varying magnetic fields, ΔE effects, theft detection targets, and a metallic glass digitizer.

Other reviews which touched upon mechanical properties only briefly include Chaudhari and Turnbull [24], Duwez [25], Warlimont [26] and Cahn [27].

II. ELASTIC PROPERTIES

A. Effect of Magnetization

Jagielinski et al. [28] measured Young's modulus as a function of magnetic field of $Fe_{70}Si_{20}B_{10}$, $Fe_xSi_{90-x}B_{10}$ for $x = 68$–80 and $Fe_{75}Si_yB_{25-y}$ for $y = 0$–15. The dynamic Young's moduli E_H at constant magnetic field and E_B at constant magnetic induction and the magnetomechanical coupling coefficient k were calculated from

$$E_H = (2lf_r)^2\rho, \quad E_B = (2lf_a)^2\rho, \quad \text{and} \quad k^2 = 1 - \left(\frac{f_r}{f_a}\right)^2 \tag{1}$$

where f_r is the resonant and f_a the antiresonant frequency for the free longitudinal vibrations of the sample, l is the length, and ρ the density of the material. For $Fe_{70}Si_{20}B_{10}$ as a function of H (kA/m), $10^{-7}E_H/P$ changed from 2.17 m^2/s^2 at $H = (0)$ through a minimum of 2.16 (0.1) and increased to 2.23 (0.5), 2.3 (1), and 2.33 (1.5); $10^{-7}E_B/P$ increased from 2.19 m^2/s^2 at $H = 0$ to 2.35 at $H = 0.5$–1.5; k changed from 0.05 at $H = (0)$ though a maximum of 0.26 (0.3) and decreased to 0.22 (0.5), 0.14 (1), and 0.08 (1.5). For $Fe_xSi_{90-x}B_{10}$, $10^{-7}E_s/\rho$ (at magnetic saturation) decreased from 2.45 m^2/s^2 at $x = (68)$ to 2.15 (80). For $Fe_{75}Si_yB_{25-y}$, $10^{-7}E_s/\rho$ decreased from 2.47 m^2/s^2 at $y = (0)$ 2.25 (15); and $\Delta E/E_s = (E_s - E_0)/E_s$ changed from 0.07 at $y = (0)$ through a minimum of 0 (2) and increased to 0.2 (15).

Ishio et al. [29] reported Young's modulus measurements of $Fe_{100-x}B_x$ ($x = 12$–25), $Fe_{100-x}P_x$ ($x = 14$–21), $Co_{100-x}B_x$ ($x = 17$–33), $(Fe_{1-x}Ni_x)_{77}B_{13}Si_{10}$ ($x = 0$–1) and $(Fe_{1-x}Co_x)_{77}B_{13}Si_{10}$ ($x = 0$–1) amorphous alloys. First the effect of applied magnetic field (0–5 kOe) was measured. Usually a 10% increase was observed between 0 field and the saturation field. For Fe-B alloys, E_s (at saturation) changed from 138 GPa for x = (12) to 185 (25). For Fe-P alloys, it changed from 130 (14) through a minimum at 118 (16) to 130 (21). As a function of temperature, there was about a 10% decrease for a 400°C increase. A relation between Young's modulus and molar volume seemed to exist.

Fukamichi et al. [30] measured Young's modulus of $Fe_{100-x}Zr_x$ (x = 7–12) as a function of temperature (100–600 K) for both 0 and saturation magnetization. The modulus increased (less than 10%) from 100 K to a maximum at 250–350 K and then decreased (less than 10%) until crystallization (about 550 K) before it increased again. For the x = 7 alloy, T_c was about 180 K and yet the moduli at 0 and saturation magnetization did not come together until 270 K. This was attributed to magnetic clusters present above T_c. For the x = 9 alloy, annealing at 420 K for 2 hr increased the modulus by about 1%.

Hausch and Török [31] measured thermal expansion of $Fe_{82}B_{18-x}Si_x$ (x = 0, 3, and 6) after stress relief for 1 hr at 620 K and found $\alpha = 4.5 \times 10^{-6}$ at 200 K decreased to 2×10^{-6} at the Curie temperature (590 K for x = 0, 610 K for x = 3 and 6) and suddenly increased to 11×10^{-6} at 670 K. T_{x1} = 685 K, T_{x2} = 780 K for both x = 3 and 6). Young's modulus of $Fe_{82}B_{18}$, stress relieved for 1 hr at 620 K, was 125 GPa without any magnetic field and 160 with 100 Oe at room temperature. At 400 K these numbers were 120 and 150, respectively. Largest internal friction was 0.033 at 420 K. Under 10^3 Oe magnetic field, the internal friction became negligible.

B. Effect of Temperature and Annealing

Bothe and Neuhäuser [32] measured Young's modulus, E, of METGLAS alloy 2826A ($Ni_{36}Fe_{32}Cr_{14}P_{12}B_6$) by the vibrating reed technique. The small change of resonance frequency f was $2(\Delta f)f = \Delta l/l + \Delta E/E$ and since usually $\Delta l/l < 0.1(\Delta E/E)$, the fractional change of resonance frequency was almost proportional to the fractional change of Young's modulus. The resonance frequency was 119.5 Hz for the as-received material, decreased to 118.5 when heated to 420 K at 1°C/min. Isothermal annealing at 420 K for 24 hr increased it to 119.5 again. Heating to 470 K decreased it to 118.8. Isothermal annealing at 470 K for 24 hr increased it to 120. Heating to 520 K decreased it to 119.2 and isothermal annealing at 520 K for 24 hr increased it to 120.3. Heating to 570 K decreased it to 119.2 and isothermal annealing at 570 K for 24 hr increased it to 121. Cooling back to 300 K increased it further to 125 which was a 4.6% increase from the original value. During isothermal annealing, f increased but df/dt decreased with no indication of saturation. The results were analyzed based on a spectrum of activation energies.

Van den Beukel et al. [33] measured isothermal changes of sound velocity in $Fe_{40}Ni_{40}B_{20}$ (VITROVAC alloy 0040) at 453, 503, 546, 596, and 662 K for as-quenched specimens. Specimens preannealed at 503, 453 and 541 K for various times (10^4–10^6 s) were also tested. The sound velocity (all measured at room temperature) went through a minimum for preannealed specimens. The results were analyzed using a model involving a spectrum of activation energies.

Mulder, et al. [34] reported the change of Young's modulus in terms of v^2, the square of the sound velocity (pulse-echo), during isothermal annealing of $Fe_{40}Ni_{40}B_{20}$ (VITROVAC alloy 0040) at 45, 503, 546, 597, and 662 K from the as-quenched state. An increase of 3% in 105 s was shown and there was no sign of saturation. The rapid increase in v^2 at 662 K was attributed to the onset of crystallization. A specimen was preannealed at 503 K for 1.1 Ms and then annealed at 578 K; v^2 first decreased about 1% until 1 ks and then increased about 2% at 1 Ms. For the same preannealed condition, the maximum decrease in v^2 increased and the time to reach the minimum v^2 decreased with the temperature for subsequent annealing. For the same preannealing and subsequent annealing temperatures, both the maximum decrease of v^2 and the time to reach the minimum v^2 increased with the time of preannealing.

By using the model proposed by van den Beukel and S. Radelaar [35], an activation energy of 180 kJ/mole was obtained for the disappearance of free volume.

Dietz and Heinen [36] measured the change in shear modulus of $Ni_{100-x}P_x$ for $x = 13$, 18, and 21 upon continuous heating at 1°C/min. The sample was in the form of a tube (7.6 mm id, 80 μm thick, 2 cm long) excited for torsional vibration. The shear modulus decreased with increasing temperature until crystallization; then it increased sharply. For $x = 13$, the shear modulus was 42 GPa at 10°C, 40.3 at 118, 40 at 250, and 40.8 at 290°C. For $x = 18$, it was 40.8 at 10°C, 39.2 at 118, 38.5 at 300, and 40.2 at 320°C. For $x = 21$, it was 40.7 at 10°C, 39 at 100, 37.2 at 250, and 39.8 at 270°C. Isothermally the shear modulus increased with time.

Bothe [37] studied the change of Young's modulus and internal friction of as-quenched $Cu_{67}Ti_{33}$ by a step annealing program. Each isothermal step was 24 hr or longer and between steps the heating rate was 3°C/min. The modulus increased during each isothermal step and decreased between steps because of the increasing temperature. On the other hand, the internal friction decreased during each step and increased between steps.

Bothe and Neuhäuser [38] reported Young's modulus and internal friction of METGLAS alloy 2826A ($Ni_{36}Fe_{32}Cr_{14}P_{12}B_6$) after 30% cold-rolling and a program of step annealing (24 hr at one temperature and then the temperature was increased at 3°C/min to the next step). The cold-rolled sample was stored at −25°C before use. Room temperature was the first step during which the modulus increased 5.4%. The next step was 60°C (5.4%), 111° (2.8%), and 160° (1.4%). Then the modulus decreased during the following steps: 210°C (−3.0%), 250° (−6.8%), and 295° (−9.0%). After that the sample began to crystallize and the modulus increased again. The modulus was measured at temperature by the resonant frequency of a vibrating reed. For the same step-annealing program, internal friction decreased with every step. The increase between steps followed an Arrhenius law. The simultaneous decrease of modulus and internal friction is inconsistent with the concept of free volume as the only structural parameter.

Taylor et al. [39] reported Young's modulus of $Cu_{78}Sn_9P_7Ni_6$ alloy as a function of temperature between room temperature and 190°C in a Rheometrics dynamics spectrometer (Piscataway, N.J.). At about 5°C/min the modulus fluctuated between 65 and 69 GPa from 25 to 120°C, increased to 79 GPa at 150°C and then suddenly dropped at about 170°C.

C. Atomistic Calculations

Knuyt et al. [40] calculated elastic moduli of a pure amorphous metal using the Johnson potential for Fe. The bulk modulus B is 0.765 eV/Å^3 (0.80 for bcc Fe and

0.69 for fcc Fe) and the shear modulus G is 0.340 eV/Å^3 (0.48 for bcc and 0.41 for fcc). Somewhat lower values (B = 0.60 and G = 0.24 eV/Å^3) were obtained previously by Lancon et al. [20] using the same potential. The relation between amorphous and crystalline elastic properties was similar to that obtained by Weaire et al. [41].

III. DAMPING AND INTERNAL FRICTION

A. Effect of Temperature and Annealing

Hettwer and Haessner [42] reported internal friction measurements of $Ni_{36}Fe_{32}Cr_{14}P_{12}B_6$ (METGLAS alloy 2826A) ribbons (55 μm × 1.8 mm × 40 mm) with a torsion pendulum. At 0.1 Hz a maximum which varied between 70 and 100°C was observed. Annealing lowered the maximum and shifted it to higher temperatures.

Bothe and Neuhäuser [43] measured internal friction as well as Young's modulus of METGLAS alloy 2826A ($Ni_{36}Fe_{32}Cr_{14}P_{12}B_6$) in successive isothermal steps at 3°C/min. The internal friction decreased and Young's modulus increased in each isothermal step of about 24 hr. During heating the internal friction increased and Young's modulus decreased. Some internal friction ($10^4\ Q^{-1}$) values were as follows: heating to 600 K, 20; isothermal at 600 K, 20–13; heating to 625, 30; isothermal at 625, 30–22; heating to 650, 53; isothermal at 650, 53–25; heating to 675, 48; isothermal at 675, 48–23; heating to 700, 44; isothermal at 700, 44–28. The results were analyzed by using a spectrum of activation energies.

Posgay et al. [44] reported on the effect of quenching rate on internal friction and hardness of $Co_{65.6}B_{13.8}Si_{9.1}Fe_{4.7}Ni_{3.4}Nb_{2.2}Cr_{1.1}$ amorphous alloy prepared by planar flow casting with different (7, 13, 23, 39 ms) ribbon-substrate sticking times. Hardness decreased (about 10%) and internal friction increased (about 50% at 100°C and no change at 300–400°C) with the sticking time.

Sinnema et al. [45] measured internal friction of $Pd_{77.5}Si_{16.5}Cu_6$ (1-mm wire) at 4 and 15 Hz over a temperature range of 385 to 476 K. For the same internal friction, the change of frequency versus temperature gave an activation energy of 2.75 eV/atom. However, a plot of logarithm of internal friction with $1/T$ gave 0.7 eV/atom for both frequencies. The discrepancy must arise from having many relaxation processes operative.

Chen and Morito [46] measured internal friction of $Pt_{58.4}Ni_{14.6}P_{27}$, $Pd_{48}Ni_{32}P_{20}$, and $Pd_{77.5}Si_{16.5}Cu_6$ amorphous alloy rods (~0.5 mm dia.). The T_g for these were 487, 575, and 632 K respectively at 5°C/m. Q^{-1} increased from 10^{-4} at room temperature to 0.1 near T_g and to greater than 1 over T_g. A loss peak was found at 20°C above T_g. Later Chen and Morito [47] analyzed the data of $Pd_{77.5}Si_{16.5}Cu_6$ by a spectrum of activation energies.

Deng and Argon [48] reported the aging behavior of $Cu_{59}Zr_{41}$ and $Fe_{80}B_{20}$. For $Cu_{59}Zr_{41}$, the as-quenched configuration was a two-phase mixture. The Zr-rich phase in the form of about 10-nm-diameter particles were dispersed randomly in the Cu-rich phase. Aging at 710 K ($0.92T_x$) caused clustering of such particles into a space network of cell walls surrounding the Cu-rich pockets of 50–100 nm diameter. After 1 hr aging the cell wall thickness was about 8 nm. Crystallization started in the Zr-rich cell wall after 10 hr and completed after about 32 hr of aging. Internal friction (tan δ) of as-quenched material at 300 K and 0.14–0.38 Hz was about 3×10^{-3} which increased to 5×10^{-3} at 500 K and to 15×10^{-3} at 700 K at a heating rate of 5°C/min. The internal friction dropped precipitously after crystallization (771 K). Aging at 710 K for 2 and 8 hr or at 738 K for 0.5, 2, and 8 hr reduced the internal friction (lowest 2×10^{-3} at 300–600 K). Aging at 710 K increased the hardness from 470 kg/mm^2 (0 hr) to 520 (8), 500 (15), 530 (30), and 550 (40 hr) for the shiny side of ribbon. The dull side was about 20 kg/mm^2 harder. Aging at 710 K reduced the strain to fracture from 50% (2 hr) to 1.5% (5 hr) and 1.1% (40 hr). Upon the onset of crystallization, the hardness first decreased and then increased and the strain to fracture decreased further. For $Fe_{80}B_{20}$ the as-quenched state was nearly a single phase. Aging at 658 K ($0.92T_x$) showed phase separation at 1.5 hr in the form of glassy particles of about 3 nm diameter and at about 3 hr, crystal nuclei were formed but seemed unrelated to the phase separation. Internal friction (tan δ) of as-quenched material at 0.16 to 0.62 Hz was about 3×10^{-3} at 300–500 K, increased to a peak of 17×10^{-3} (0.62 Hz) at 605 K, dropping down somewhat before it increased again to 30×10^{-3} at 730 K (T_x = 714 K). Aging at 633 K ($0.89T_x$) for 0.5 hr completely removed the damping peak without crystallization.

Bonetti [49] reported internal friction measurements on $Ni_{84}P_{16}$, $Fe_{67}Co_{18}B_{14}Si_1$, $Ni_{36}Fe_{32}Cr_{14}P_{12}B_6$, and $Fe_{40}Ni_{38}B_{18}Mo_4$ amorphous alloys. Internal friction at 300 Hz increases with increasing temperature (300–600 K) but decreases upon annealing. The opposite is true for the dynamic Young's modulus. Plastic elongation of 1–2% increased internal friction and annealing decreased it.

He and Li [50] found four internal friction peaks in $Pd_{77.5}Si_{16.5}Cu_6$ at 628 K, 667 K, 709 K, and 744 K for a heating rate of 5°C/min. The activation energies were 6.6, 1.8, 2.9, and 2.9 eV/atom obtained from the peak shift with heating rate. A change of frequency from 0.173 to 0.752 Hz shifted the first peak to a higher temperature by about 8°C. The effect of frequency on peak temperature gave 6.2 eV/atom as the activation energy for the first peak. The other three peak temperatures were not affected by frequency. Furthermore, the first peak could be repeated on the same specimen provided that the temperature did not exceed the peak temperature by 10°C. The motion of atomic clusters was proposed to be the process for this peak.

Sinning and Haessner [51] reported internal friction of $Ni_{40}Pd_{40}P_{20}$ at 0.07–0.1 Hz and 0.03–0.85°C/m heating rate between 250 and 340°C. They suggested that the temperature at which the slope dQ^{-1}/dT changes could be considered as T_g. Q^{-1} was between 0.03 and 0.1 below T_g and between 0.1 and 2 above T_g. The T_g determined this way seemed to agree with DSC results using the end of C_p increase. The results showed T_g of 562 K at 0.02°C/m and 601 K at 100°C/m for a straight-line relationship.

Sinning and Haessner [52] plotted Q^{-1} versus T for $Ni_{60}P_{20}Pd_{20}$ (0.08 and 450 Hz) and $Ni_{78}B_{14}Si_8$ (0.095 Hz and 400 Hz) at 0.3°C/m. The first alloy (T_g = 300°C) showed a peak (Q^{-1} = 1.1 for 0.08 Hz and 0.04 for 450 Hz) at 335°C. The second alloy (no T_g) showed a small peak at 420°C. Repeated scanning (first alloy, 0.08 Hz) after Q^{-1} reached maximum and the specimen cooled down to room temperature moved the maximum to higher temperatures. After the third scan the specimen showed crystallization in TEM. Thus the Q^{-1} maximum may be a result of viscoelastic damping which increases with temperature and damping due to crystallization which decreases with temperature.

Yue and He [53] measured internal friction of $Pd_{77.5}Si_{16.5}Ni_6$ by an inverted torsion pendulum in 10^{-2} Torr. The specimen dimensions were 0.04 × 1.5 × 30 mm^3. Three internal friction peaks were observed: 645 K (Q^{-1} = 0.155), 690 K (0.105) and 781 K (0.107) at 10°C/m. DSC at the same heating rate showed T_g at 646 K, one large exothermic peak at 687 K and a small one at 783 K. By using a heating stage. TEM observation showed crystallization to begin at about 660 K. The appearance of Pd_4Si, Pd_3Si, Pd_2Si and Ni_3Si phases during heating was shown.

Bonetti et al. [54] measured internal friction of $Fe_{67}Co_{18}B_{14}Si_1$ alloy from room temperature and up using a Bordoni-type apparatus operating between 100 and 1000 Hz for a vibrating reed in a saturating magnetic field. The anelastic strain amplitude was less than 10^{-6}. The as-cast sample showed damping about 10^{-3} (Q^{-1}) at room temperature increasing to 5 × 10^{-3} at 510 K. These values were reduced to almost half when the damping measurements were made again on the same sample. Such values after stress relaxation following loading to 1030 MPa at a strain rate of 10^{-3}/s became 9 × 10^{-4} and 2.2 × 10^{-3} respectively. Even lower values of damping were obtained when the strain rate in loading before stress relaxation was 10^{-6} instead of 10^{-3}/s.

Sinning [55] reported internal friction for six Ni- and Pd-based metallic glasses using a low frequency of 0.1 Hz and at a constant heating rate of 0.3°C/min through the glass transition. The amorphous alloy, its glass transition temperature in °C, the temperature in °C for maximum internal friction and the peak Q^{-1} at that temperature are as follows: $Ni_{60}Pd_{20}P_{20}$, 303, 324, 1.2; $Ni_{40}Pd_{40}P_{20}$, 301, 326, 1.5; $Ni_{40}Pd_{40}P_{18.5}Si_{1.5}$, 306, 329, 0.6; $Pd_{77.5}Si_{16.5}Cu_6$, 355, 369, 0.5; $Pd_{77.5}Si_{16.5}Ag_6$, 350, 375, 1; and $Pd_{80}Si_{12}Ge_8$, 350, 371, 0.22. A Collette pendulum was used and the importance of instrument correction was pointed out. The internal friction peak was suspected to arise from a visoelastic mechanism which increases Q^{-1} with temperature and a mechanism due to crystallization which decreases Q^{-1} with temperature. The peak Q^{-1} temperature was close to onset of crystallization.

B. Effect of Cold Working

Zolotukhin et al. [83] found an internal friction peak at about 270 K for $Cu_{50}Ti_{50}$ cold-rolled 2–3% and measured at 260–300 Hz with a strain amplitude of 2 × 10^{-5} in the form of a freely vibrating cantilever (8 mm × 1–2 mm × 30–40 μm). The peak height was the largest (Q^{-1} = 0.01) for 3% deformation, decreased with increasing deformation and disappeared after 16% deformation. Electron irradiation (2 MeV for $10^{19}/cm^2$) of the 3% deformed sample also removed the effect of rolling. The peak shifted to 220 K if the amplitude of vibration was (6–10) × 10^{-5}. An increase of frequency from 366 to 2477 Hz increased the peak height from 0.004 to 0.017 (6% rolled) but did not change the position (280 K). Annealing at 383 to 473 K reduced the peak height. For $Ni_{78}B_{14}Si_8$ the peak was also at about 270 K and the peak height was the largest also after about 2–3% rolling. Aging at room temperature of the 2% deformed samples first increased the peak height about a factor of 3 in 50 days followed by a steady decrease to the original height (Q^{-1} = 0.008) in about 285 days. The effect was explained by postulating the existence of dislocations in deformed specimens.

C. Effect of Magnetic Field

Hausch and Török [31] showed in $Fe_{82}B_{18}$ an internal friction peak of 0.033 at 420 K which became negligible when a magnetic field of 10^3 Oe was applied. The behaviors of $Fe_{82}B_{15}Si_3$ and $Fe_{82}B_{12}Si_6$ were similar.

D. H Peak

Künzi et al. [56] reported some internal friction measurements using the vibrating reed technique (40 mm × 1 mm × 2 mm). For $Co_{70.4}Si_{15}B_{10}Fe_{4.6}$, Young's modulus decreased from 195 GPa at −200°C to 160 GPa at 500°C (T_g); Q^{-1} increased from 5×10^{-6} at −200°C (1.5×10^{-4} at 0°C, 2×10^{-4} at 200°C, 4×10^{-4} at 300°C, 4×10^{-3} at 400°C) to 1.5×10^{-2} at 450°C. A thermoelastic peak for $Mg_{70}Zn_{30}$ at 104 Hz was shown. A H peak in $Pd_{80}Si_{20}$ (1–2% H at saturation) at 175 K (156 Hz) was found. The peak shifted to 185 K for 979 Hz and to 197 K for 2.71 kHz. A plot of peak temperature ($^1/_T$) versus frequency gave an activation energy of 0.31 eV/atom. Other activation energies were 0.55 eV/atom for H in $Nb_{75}Ge_{25}$ and 0.31 eV/atom for H in $Pd_{77.5}Si_{16.5}Cu_6$. Hydrogen desorption measurements from $Pd_{80}Si_{20}$ between room temperature and 100°C gave an activation energy of 0.25 eV/atom. Results of EXAFS measurement indicated that H prefers to sit between two metal atoms than between a metal atom and a metalloid atom. Internal friction studies of H in $Zr_{76}Ni_{24}$ which dissolves a large amount of H, showed that, with increasing H concentration, the internal friction peeks become broader and shifted to lower temperatures.

IV. YIELDING, FLOW, AND SHEAR BANDS

A. Yielding Criteria

Donovan [57] did compression of $Ni_{40}Pd_{40}P_{20}$ cylinders (2 mm dia) and examined the angle between the shear band and the compression axis (40° at yielding, increased to 55° after 75% length reduction). She concluded that yielding in $Ni_{40}Pd_{40}P_{20}$ obeys a Mohr-Coulomb or a three-parameter (Li and Wu [245]) Tresca criterion. The normal stress coefficient, 0.11 ± 0.05, was comparable with the value for amorphous polystyrene. More recently she studied the yield behavior of $Ni_{40}Pd_{40}P_{20}$ (rods of 1–2 mm dia or sheets of 0.2 mm thick) by testing in uniaxial compression, plane-strain compression, and pure shear [58]. The material was found to obey the Mohr-Coulomb yield criterion (not the pressure-modified Tresca criterion) with a normal stress coefficient of 0.113 ± 0.005. The pure shear test using the antisymmetric four-point bending method gave 0.795 ± 0.025 GPa as the shear yield strength. The lack of a pressure dependence of yield stress is inconsistent with the notion that plastic flow requires some "free volume."

B. Dispersion Hardening

Zielinski and Ast [59] injected 2–3 μm TiC or WC particles with a high velocity He gas jet into the melt puddle of $Ni_{75}B_{17}Si_8$ during melt spinning into ribbon. The yield stress and Young's modulus were determined in bending . For TiC, Young's modulus (GPa) and yield stress (GPa) as a function of volume fraction (%) of particles were 56, 1.60, 0; 59, 1.72, 1.2; 63, 1.85, 2.3; 75, 2.19, 5 and 78, 2.32, 6.4. For WC, they are 67, 1.77, 1.7; 74, 1.92, 3.4; 82, 2.08, 4.5; 85, 2.13, 5.0 and 97, 2.31, 6.7. The yield stress was linear with $1/\sqrt{x}$ where x was either the mean free path or the average distance between particles. However, the correlation with the mean free path had a more reasonable intercept at the zero volume fraction. SEM observations showed parallel surface steps in the particle-free region and wavy ones around particles. TiC particles were cut by shear bands whereas the WC particles were not.

Kimura et al. [60] studied the effect of WC (4–5 μm size, added before melt spinning on a single roller) dispersions in the $Ni_{78}B_{12}Si_{10}$ amorphous alloy. Young's modulus increased from 90 GPa (0 vol% WC) to 138 (8) and 195 (18). Yield stress in bending increased from 1.6 GPa (0 v/o WC) to 1.8 (2), 2.1 (6), 2.2 (8) and 3.3 (18). The ratio of Young's modulus to yield stress was about 60, independent of the WC content. Essentially the same results were presented in 1984 by Kimura and Masumoto [246].

C. Effects of Deformation and Stress Annealing

Deng and Lu [21] measured the density change of $Pd_{77}Si_{16.5}Cu_{6.5}$ wires (0.3–0.5 mm dia) after drawing. The specific gravity was 10.541 when the diameter was 0.340 mm before drawing. These numbers after drawing were 10.545, 0320; 10.537, 0.305; 10.530, 0.285; 10.524, 0.257; 10.526, 0.240; 10.526, 0.215; and 10.525, 0.186. The present results are much more reasonable than some previous ones reported by Megusar, et al. [19].

Vianco and Li [61] annealed (120–245°C) METGLAS alloy 2826B ($Ni_{49}Fe_{29}P_{14}B_6Si_2$) for 530, 1060 and 1590 min under a tensile stress of 20–400 MPa. The increase of microhardness (from 760 to 810 DPN without stress) was much less under stress. Further annealing of stress-annealed samples, this time without stress, could increase the microhardness for some samples. An additional exothermic peak in the DSC thermogram of stress-annealed samples indicated the presence of a more disordered structure which can correlate with the lower hardness values.

Radlinski and Calka [62] used DSC to determine T_g (385°C) and T_x (397°C) of $Zr_{76}Fe_{24}$ amorphous alloy. Mechanical grinding (600 grade abrasive paper) lowered T_x to 180°C (surface crystallization) but the bulk crystallization was still at 397°C. X-ray diffraction showed a surface layer of α-Zr and Zr_2Fe with an amorphous structure below after the sample was heated to 380°C. After heating to 400°C, the remaining amorphous phase crystallized into Zr_3Fe. Samples etched with aqua regia from an original thickness of 30 μm to about 10 μm showed the onset of surface crystallization at 130°C with broad exothermic peak at 311°C and a small peak at 392°C.

Audouard et al. [63] reported irradiation of $Fe_{85}B_{15}$ amorphous alloy (21 μm thick) with 2.8 GeV Xe ions at 80 K. An in situ electrical resistance measurement was used to monitor the effect of irradiation. One sample had a tensile stress of 600 MPa applied in the direction of the dc measuring current. The other sample had no stress. $\Delta R/R_0$ was 0.025 for the unstressed sample at a fluence of 10^{13} ions/cm^2, as compared with 0.11 for the stressed sample. At a fluence of 1.7×10^{13} ions/cm^2, $\Delta R/R_0$ was 0.06 (unstressed) and 0.30 (stressed). If n_g =

$[d(\Delta R/R_0)/d\phi]Na$, where N is the atomic density and a the interatomic spacing, then $n_g = 8$ for the unstressed and much higher for the stressed sample.

Nasu et al. [64] measured EXAFS (extended x-ray absorption fine structure) spectra of $Pd_{83}Si_{17}$ during tensile deformation. The Pd-Si distance increased from 0.165 nm at (0) strain, 0.165 (0.3), 0.169 (0.6) to 0.170 (0.8) and then decreased upon unloading from 0.167 (0.6), 0.169 (0.3) to 0.169 (0). On the other hand, the Pd-Pd distance decreased from 2.45 (0), 2.44 (0.3), 2.43 (0.6), 2.42 (0.8), 2.41 (0.6), 2.42 (0.3) to 2.41 (0).

Nishi et al. [65] mechanically peened $Pd_{77.5}Si_{16.5}Cu_6$ by 0.4 ± 0.1 mm SuJ 2 steel (HRc 64) balls of about 0.7 mg each supplied at 2.66/mm/s. An 8-mm-dia nozzle with an air velocity of 200 m/s was 2 cm away from the specimen at a peening angle of 30°. As a function of peening time, the Vickers microhardness changed from 480 at (0) s, 495 (10^2), 455 (10^3) to 425 (10^4) for a slow-quenched sample. For a fast-quenched sample, the hardness changed from 467 (0), 460 (10^2) 440 (10^3) to 425 (10^4).

D. Shear Bands

Kimura and Masumoto [4] observed the distribution of shear bands under various loading conditions. A cylinder of $Pd_{78}Si_{16}Cu_6$ under torsion showed elastic-perfectly plastic deformation with a pure shear stress of 91 ± 9 kg/mm² or 892 MPa. Two sets of straight intersecting shear bands along planes of maximum shear stress were observed as predicted by the simple torsion theory. The density of these shear bands increased with the torsional strain. For a notched plate specimen under tension, the slip bands were curved and the plastic zone size as revealed by the slip lines obeyed the Dugdale relationship for a rigid perfectly plastic material:

$$R = a\left\{\frac{W}{2a}\sin^{-1}\left(\sin\frac{\pi\alpha}{2W}\sec\frac{\sigma}{2\sigma_y}\right) - 1\right\} \tag{2}$$

where R is plastic zone size, a the crack length, W the width of the specimen, σ the net stress (load divided by the cross-sectional area without the notch) and σ_y the tensile yield stress.

Neuhäuser and Stössel [66] reported stress relaxation measurements on $Fe_{40}Ni_{40}B_{20}$ (VITROVAC alloy 0040) (3 mm × 10 mm × 40 μm gage section) in the Instron. Temperature was controlled by using electric current and a noncontact pyrometer. The nonelastic strain rate was plotted against stress for stress levels between 600 and 1100 MPa at 608 K. The stress exponent varied from 50 to 100 for less than 900 MPa to 2–4 for more than 1000 MPa. The activation strain volume varied from (1–2) × 10^3 A^3 for low stresses to 20–40 A^3 for high stresses. They also did bending experiments on $Ni_{36}Fe_{32}Cr_{14}P_{12}B_6$ (METGLAS alloy 2826A, 2 mm × 55 μm cross section) between vertical plates moving at a constant speed. The specimen was heated by an electric current. In dark-field illumination the appearance of shear bands was recorded by a high speed (Hycam K20 SAE, Red Lake Labs) camera at 7000 frames/s. The bending was stopped as soon as the first shear band appeared. The residual angle, α, remaining after unloading was recorded and α/π was considered as the fraction of plastic strain and that fraction of the applied strain rate was taken as the plastic strain rate. Such plastic strain rate for the appearance of the first shear band was plotted versus T^{-1} and an activation enthalpy of 0.32 ± 0.05 eV/atom was obtained for a range of strain rates 10^{-6}–10^{-4}/s and in a temperature range of 370–667 K.

Cao and Li [67] studied the reverse shear of shear bands in METGLAS alloy MBF-20 ($Ni_{69.2}B_{13.7}S_{7.9}Cr_{6.6}Fe_{2.6}$). Shear bands produced by bending could be sheared in reverse by reverse bending before or after isothermal annealing or continuous heating before the onset of embrittlement. After the onset of embrittlement, some shear bands still could be reversed outside the crack area. Fully embrittled samples would fracture before any shear bands could be reversed. During reverse bending, all the strain was produced by reverse shearing of old shear bands without the creation of any new ones.

Manokhin et al. [68] stretched $Fe_{40}Ni_{40}P_{14}B_6$ alloy inside the JSM-SI electron microscope at 20–400°C. Homogeneous deformation took place between 250 and 400°C and shear bands appeared between 20 and 250°C. The shear-band velocity measured from video tapes was 0.04 to 0.1 cm/s. The onset of microcracks followed local microplastic deformation. As a rule, the primary crack always appeared in the region of the broadest and most developed shear band or at the intersection of several bands. Then the crack grew along the thickest shear band.

Vianco and Li [69] etched the shear bands (after polishing to remove the steps) in METGLAS alloy 2826 ($Fe_{40}Ni_{40}P_{14}B_6$) of Allied-Signal by immersion for 5 s in a solution of 50 ml HCl, 10 g $CuSO_4$ and 5 ml H_2O. Microscopic observation of etched shear bands showed micro cracking resembling something caused by stress corrosion. The etching characteristic of the shear bands can be eliminated by annealing at some temperature (257–273°C) for a certain time period. The temperature-time data obeyed an Arrhenius relation from which an activation energy of 250 ± 40 kJ/mole was obtained. This activation energy is similar to that of stress relaxation during annealing [247].

Alpas and Embury [70] made a laminated composite by first electroplating melt-spun (55 μm thick) $Ni_{78}Si_{10}B_{12}$ ribbon, a layer of copper about 70 to 150 μm on both sides. Alternate layers of such plated ribbons and annealed copper sheets of various thickness were diffusion-bonded at 560 K under a pressure of 250 MPa. The ultimate tensile strength of the composite is linear with the volume fraction of the amorphous alloy between 2 and 10 vol%. More shear bands were observed in the amorphous ribbon of the composite than in the monolithic ribbon when tested alone. As a result the fracture strain is increased by a factor of 25. When the amorphous ribbon fractured in the composite, the stress-strain curve showed serration. Further increases in load caused fragmentation of amorphous ribbons into 1-mm lengths. Two sets of shear bands were developed before cracking along any one of them. One was parallel to the width direction and the other to the thickness direction. The second set was rarely seen in monolithic ribbons under tension.

Lakshmanan and Li [71] reported on the structure of magnetic domains surrounding shear bands formed by bending of $Fe_{81}B_{13.5}Si_{3.5}C_2$ (METGLAS alloy 2605SC). The domain patterns developed on the two sides of a shear step were different, indicating different residual stresses which were represented by edge dislocations emitted from the free surface and distributed along the shear bands. The effect of an external magnetic field and that of an applied stress on the structure of magnetic domains showed the existence of internal stresses of opposite signs, implying inhomogeneous deformation within a single shear band as in the case of crystalline materials.

Kabacoff et al. [72] reported on the effect of cold drawing of $Co_{45}Fe_{23}B_{13}Cr_{10}Si_9$ wires on their tensile strength. For the 130-μm-diameter wire, the tensile strength increased from 434 ksi at 0 to 514 ksi at 45% reduction of area. The failure mode also changed from a tensile type to shear type. The authors believed that the shear bands introduced during drawing greatly influenced the direction of shear bands which could form during tensile failure. It seemed that the new shear bands have difficulty propagating through the old shear bands, resulting in strain hardening. As a result, it was easier to shear along the old shear bands produced in cold drawing than to create new ones.

Donnadieu et al. [73] deformed silica glass between 1000 and 1400°C (T_g = 1200°C). Thin slices, cut from samples deformed at 1000°C for 10–20%, exhibited birefringent zones in the TEM sometimes in the form of regularly arranged crosses. Later Donnadieu [248,249] studied these zones or glide fronts and found that they were 1–2 μm apart and 0.1 μm thick for samples deformed 20% by 500 bar stress at 1000°C. Deformation at 1100°C resulted in more glide planes but fewer glide fronts in each. Thus the density of glide fronts was $10^{11}/m^2$ in both cases. However, for samples deformed at 1100°C, the glide fronts were curved or bulged between pinning points spaced about 800 nm to 2 μm apart. Since the spacing between pinning points were similar to that between glide planes, the pinning points were possibly created by glide front intersections. The behavior of glide fronts deformed at 1200°C was similar to that of 1100°C except that the fronts were wavy (wavelength of 0.2 μm) between the pinning points.

Harbert and Wolfenden [74] bent Ni-3.2B-4.5Si (Allied-Signal) 5 mm × 55 mm × 76 μm strips in liquid nitrogen and the electric resistance measured by the four-point method. The diameter of the loop was 20 mm to start with. The resistance increased linearly with decreasing diameter (0.012% at 10 mm). With another slope it increased again (0.029% at 5 mm) but erratically due to the appearance of shear bands. For a specimen preannealed at 200°C, the increase in resistance was faster (0.025% at 12 mm, 0.035% at 7 mm and 0.058% at 4 mm, three nearly straight segments; the last one was erratic due to shear bands). A special stage mounted inside a Jeol JSM-35CF II scanning electron microscope was used to bend the specimen and observe the shear bands directly. See some more of their work [135] in Section VIII.

V. STRESS RELAXATION AND CREEP

A. Stress Relaxation

Van't Spijker [75] reported relaxation experiments on METGLAS alloys 2826, 2605, and 2605A by maintaining an initial radius of curvature during annealing and then measuring the final radius of curvature after annealing. The following equation was found applicable between 100 and 300°C:

$$\sigma = \sigma_0 \exp[-(at)^b] \tag{3}$$

with

$$-\ln a = \frac{p}{T} + q \tag{4}$$

The value of b was 0.185, 0.150 and 0.168 for 2826, 2605, and 2826A, respectively. The temperature dependence of a resulted in activation energies of 56.4, 47.1, and 48.5 kcal/mole for 2826, 2605, and 2826A respectively.

B. Creep and Viscous Flow

Anderson and Lord [76] determined the glass transition temperature (T_g) of $Fe_{40}Ni_{40}P_{14}B_6$ in two ways: DSC and creep experiments on 20 μm × 2 mm × 1 cm samples under an initial tensile stress of 10–50 MPa and at a temperature which was increasing at 1–25°C/min in which an upward increase of length was taken as T_g. Results of the two methods were in mutual agreement. A Kissinger plot and an Ozawa plot were used to obtain an activation energy of 164 ± 1.5 kcal/mole. Both plots gave the same slope.

Tsao and Spaepen [77] did "isoconfigurational" (reproducible upon temperature cycling) measurements of the viscosity of $Pd_{77.5}Si_{16.5}Cu_6$. The sample was first annealed at 592 K for 457 hr. Then a series of measurements were done by small temperature increases and decreases and a set of 20 viscosities was obtained between 500 and 588 K. The specimen was then allowed to relax at 588 K for 40 hr before a second set of 10 measurements was made between 535 and 613 K. Yet another set was performed on a wire specimen between 518 and 595 K. Five more sets were obtained for a ribbon specimen between 524 and 615 K. The log η versus $1/T$ plots were generally nonlinear. While the activation energy for $T > 580$ K was about the same (3.6 ± 0.3 eV/atom), that for $T < 560$ K decreased with decreasing temperature. The authors concluded that the "isoconfigurational" viscosity cannot be a function of only a single structural parameter such as the average "free volume" of the specimen.

Van den Beukel et al. [78] reported on the change of viscosity during structural relaxation of $Fe_{40}Ni_{40}B_{20}$. The as-quenched material was first annealed at 523 K during which the viscosity increased faster in the beginning and then slowed down to a linear relation with time. Subsequent annealing at 543, 563, and 583 K all showed linear increase of viscosity with time. A plot of $\dot{\eta}/T$ versus $1/T$ gave an activation energy of 54 kJ/mole. Another specimen was first annealed at 470 K for 10^6 s and then its viscosity was determined by heating to a temperature (455–502 K, eight steps) for a few 10^4 s stay at each temperature. Then the same specimen was annealed at 502 K for a long time before a second set of viscosity measurements (502–533 K, five steps) was made. This went on until 601 K after six sets of measurements. A plot of η/T versus $1/T$ gave the following activation energies: 235 kJ/mole (455–502 K), 243 (502–533), 247 (533–550), 304 (550–571), 302 (571–591), and 315 (591–601).

Ma and He [79] studied elongation of $Pd_{77.5}Si_{16.5}Ni_6$ ribbons (2 mm × 75 μm) under continuous heating (12°C/min) and constant load (1, 2, and 4 MPa). The temperature at which the flow rate was maximum increased slightly with applied stress at a heating rate of 12°C/min and it increased with the heating rate at an applied stress of 2 MPa. A Kissinger plot of the latter data yielded 453 ± 11 kJ/mole which was compared with 427 kJ/mole for crystallization. However, using the temperature of the maximum flow rate to be the crystallization temperature is questionable. Nevertheless, heating at 688 K which was between this temperature (683 K) and the temperature at which the elongation was maximum (698 K) (usually regarded as the onset of crystallization) did cause crystallization.

Volkert and Spaepen [80] did creep measurements on $Ni_{40}Pd_{40}P_{19}Si_1$. The creep rate increased (or the viscosity decreased) at 573 K for a sample preannealed at 563 K but the creep rate decreased at 573 K for the as-quenched sample or that preannealed at 593 K. The final viscosity was the same independent of the thermal history. The strain rate was linear with stress at 583 K and was zero at 1.4 MPa due to surface tension. The change of viscosity with time was fit to a flow defect model [81,82]. For decreasing viscosities a slightly better fit was consistently obtained using first order kinetics. For increasing viscosities below 565 K, from either the as-quenched state or an annealed state (above 565 K), the second order kinetics gave a better fit. For increasing viscosities above 565 K, neither kinetics gave particularly good fit.

Tsao and Spaepen [82] measured viscosity changes near equilibrium for $Pd_{77.5}Si_{16.5}Cu_6$. The change of viscosity is nonlinear with time and can increase or decrease to approach a stationary value, depending on whether the specimen was preannealed at a higher or lower temperature. Such stationary value was about an order of magnitude higher than previously measured values [84]. A plot of such viscosity with $1/T$ showed a curve which was well described by the Fulcher-Vogel equation.

Van den Beukel [85] compared the diffusion data [86] of ^{59}Fe in $Fe_{40}Ni_{40}B_{20}$ and the viscosity after annealing at 593 K for various times. He found that the product of viscosity η and the diffusivity D was not a constant but increased by a factor of 3.7 between 0.5 and 7.5 hr annealing time. D decreased by a factor of 3.7 and η increased by a factor of 13.7. However, $D\eta^{1/2}$ seemed constant within $\pm 1\%$. Van den Beukel attributed this to the different defects responsible for self-diffusion and flow.

Zheng [87] studied the flow behavior of amorphous $Zr_{70}Ni_{30}$ under continuous heating from 0.8°C/min to 30°C/min. The stress range was 1–20 MPa. By assuming that the role of stress was to cause flow only and other changes such as structural relaxation, glass transition, and crystallization were not affected by stress, the length changes of the specimen (5 cm × 0.6 mm × 20 μm) over and above that due to a small stress of 0.16 MPa were taken as being flow behavior. For 10°C/min heating rate, the flow stopped at about 400°C and the total flow elongation was proportional to the applied stress. The maximum flow rate was at 380°C for all stress levels and was proportional to the stress. At 5 MPa, the temperature at which the flow stopped increased from 370°C at 0.8°C/min to 450°C at 30°C/min. Some results are shown in Table 1.

Wang et al. [88] showed the effect of heating rate ϕ on the viscous flow start temperature, steady viscous flow temperature, and the crystallization temperature

Table 1 Continuous Heating Experiments [87] of $Zr_{70}Ni_{30}$ at 5-MPa Stress

Heating rate	T_g	T_x	$T_{\max\,\dot{\varepsilon}}$	$\dot{\varepsilon}_{\max}$	$\Delta L/L_{0\,\max}$
0.8°C/min	340°C	340°C	360°C	$0.31(10^{-5}/s)$	$4.5(10^{-3})$
3	361	376	375	2.26	7.52
10	372	390	385	6.50	10.17
30	385	408	395	22.80	12.00

and attempted to extrapolate to $\phi = 0$. However, the variation of these temperatures with ϕ was very large near $\phi = 0$ so that this extrapolation was difficult.

Bhatti and Cantor [89] measured elongation of $Fe_{78}B_{13}Si_9$ amorphous ribbons, 30 μm × 1.5mm × 2 mm, in an argon atmosphere under constant stress (13 and 26 MPa) and continuous heating (15–75°C/min) conditions. Flow started at about 450°C and reached a maximum rate at about 530°C. These temperatures were not affected by prior annealing (375°C, 8–11 hr) although the maximum rate decreased with prior annealing of the sample. For a given starting material, the maximum flow rate increased with increasing heating rate. When the elongation stopped and the length started to decrease, that temperature was assigned to the onset of crystallization. This temperature was about 560°C and was found to decrease with prior annealing.

Taub [90] obtained essentially the same result on the same material a few years earlier. He pointed out that, at 100°C/min heating rate, the viscosity prior to the onset of crystallization was only 10^{10} Pa-s, well below an accepted definition of glass transition at which the viscosity would reach about 10^{12} Pa-s. However, DSC tests at rates up to 80°C/min failed to show a glass transition. Other alloy systems should be used to clarify this issue.

Volkert and Spaepen [80] made viscosity measurements on $Ni_{40}Pd_{40}P_{19}Si_1$ under a constant uniaxial stress (less than 20 MPa). Viscosity was defined as $\eta = \sigma/3\dot{\varepsilon}$ where $\dot{\varepsilon}$ was the strain rate (about 10^{-6}/s). Let the test temperature be T and the preannealing temperature be T_a. Then if $T > T_a$, the viscosity increased with time and if $T < T_a$ it decreased with time. However, both approached the same viscosity value at T. Now if the time for preannealing at $T_a < T$ was not sufficiently long, the viscosity at T first decreased with time, reached a minimum and then increased to approach the value at T. For example, a sample preannealed at 260°C for 25 hr and then tested at 300°C, the viscosity started at 5.5 (10^{12} N-s/m^2), decreased to a minimum of 4.3 at 0.5 hr, increased to 5.5 at 2.3 hr and to 6.0 at 4 hr. Such behavior was difficult to understand by using a single parameter such as the free volume to describe the amorphous structure.

C. Negative Creep and Negative Stress Relaxation

Negative creep [91] is defined as the shrinkage of a specimen under a tensile load. As much as 5×10^{-4} negative strain was observed in BNi_2 (METGLAS alloy MBF-20 $Ni_{68.8}B_{14.1}Si_{7.9}Cr_{6.6}Fe_{2.6}$). In this case the as-quenched specimen was first precrept at 853 MPa and 178°C for 300 min and then annealed at 268°C for 500 min. Final creep testing was done at 123°C and 570 MPa where the negative strain was observed. There was no length change without the applied stress. This is different from length contraction without stress to be discussed later in this chapter. The maximum negative strain increased with increasing applied stress until 570 MPa and then decreased to zero at about 850 MPa. Such phenomenon is a sure indication of the instability of metallic glass structure.

Bonetti et al. [54] observed negative stress relaxation in $Fe_{67}Co_{18}B_{14}Si_1$ amorphous alloy supplied by Ist. Elettrotecnico Nazion. 'G. Ferraris' Torino, Italy. When the as-cast sample was loaded to 1030 MPa at room temperature and then the straining stopped, the stress decreased first in the initial 100 s and then increased in the next 10^4 s to reach almost the initial applied stress. The time needed to reverse the stress rate (to start the negative stress relaxation) decreased with increasing strain rate during the first loading. Increased strain rate also increased damping after relaxation.

The instability was interpreted [93] as a mechanochemical spinodal decomposition of the amorphous structure under tension. The system can be considered to have three independent variables, the temperature T, the strain ε, and the concentration of a minor component, c. The system is unstable in the sense that if F is the free energy per unit volume of the system, some of the following inequalities may be valid:

$$\left(\frac{\partial^2 F}{\partial c^2}\right)_{T,\varepsilon} < 0 \tag{5}$$

$$\left(\frac{\partial^2 F}{\partial \varepsilon^2}\right)_{T,c} < 0 \tag{6}$$

$$\left(\frac{\partial^2 F}{\partial c^2}\right)_{T,\varepsilon}\left(\frac{\partial^2 F}{\partial \varepsilon^2}\right)_{T,c} < \left(\frac{\partial^2 F}{\partial c \partial \varepsilon}\right)_T \tag{7}$$

If the first inequality is valid, the system is a chemical spinodal. The second is a mechanical spinodal. But most likely it is a chemically induced mechanical spinodal or mechanochemical spinodal, the third inequality, which is operative.

Consider for example a combined effect of dc and $d\varepsilon$ such as

$$d\varepsilon = \alpha\, d\zeta, \qquad dc = \beta\, d\zeta \tag{8}$$

where ζ is now a new variable. If ζ is dimensionless, α will have units of strain and β units of concentration. Along the ζ coordinate, $d\varepsilon/dc = \alpha/\beta$. Then

$$\left(\frac{\partial^2 F}{\partial \zeta^2}\right)_T = \alpha^2\left(\frac{\partial^2 F}{\partial c^2}\right)_{T,e} + \beta^2\left(\frac{\partial^2 F}{\partial \varepsilon^2}\right)_{T,c} + 2\alpha\beta\left(\frac{\partial^2 F}{\partial c \partial \varepsilon}\right)_T \tag{9}$$

which shows that $\partial^2 F/\partial \zeta^2 = 0$ has two roots of α/β. Any ratio of α/β between the two roots will make $\partial^2 F/\partial \zeta^2$ negative. Then the system may decompose along the ζ coordinate. For example if α/β is

$$\alpha\left(\frac{\partial^2 F}{\partial c^2}\right)_{T,e} + \beta\left(\frac{\partial^2 F}{\partial c \partial \varepsilon}\right)_T = 0 \tag{10}$$

then

$$\left(\frac{\partial^2 F}{\partial \zeta^2}\right)_T + \beta^2\left\{\left(\frac{\partial^2 F}{\partial \varepsilon^2}\right)_{T,c} - \frac{(\partial^2 F/\partial c \partial \varepsilon)_T^2}{(\partial^2 F/\partial c^2)_{T,\varepsilon}}\right\} < 0 \tag{11}$$

on account of Eq. (7). The system may decompose along the ζ coordinate in which

$$\left(\frac{\partial s}{\partial e}\right)_T = \left(\frac{\partial s}{\partial e}\right)_{c,T} + \left(\frac{\partial s}{\partial c}\right)_{T,e}\frac{a}{b} \tag{12}$$

$$= \left(\frac{\partial \sigma}{\partial \varepsilon^2}\right)_{c,T} - \left(\frac{\partial^2 F}{\partial \varepsilon \partial c}\right)_T^2 \Big/ \left(\frac{\partial^2 F}{\partial c^2}\right)_{T,e} < 0 \tag{13}$$

and hence there will be negative creep.

VI. ANNEALING EMBRITTLEMENT

Liebermann [94] measured the surface tension of the melt of $Fe_{81.5}B_{14.5}Si_4$ (1.55 N/m at 1200°C, 1.3 at 1600°C), $Ni_{81.5}B_{14.5}Si_4$ (1.4 N/m at 1100°C, 1.25 at 1600), $Fe_{40}Ni_{40}B_{20}$ (1.3 at 1100 and 1.5 at 1600) and $Fe_{81.5}B_{14.5}Si_{3.9}Te_{0.1}$ (0.8 at 1200 and 0.5 at 1600°C). He also studied the effect of element additions on the melt surface tension such as La, Sb, Se, and Te in $Fe_{80.5}B_{15}Si_4C_{0.5}$, Te in $Fe_{81.5}B_{14.5}Si_4$, La in $Fe_{39}Ni_{39}P_{14}B_6Si_2$, La, Sb, Se, and Te in $Fe_{40}Ni_{40}B_{20}$, and Sb, Se, and Te in $Ni_{81.5}B_{14.5}Si_4$. Most importantly he found a correlation between the melt surface tension and the amorphous ribbon embrittlement temperature for 1-hr anneals (roughly a straight line between 0.5 N/m at 250 K and 1.2 N/m at 600 K).

Zielinski and Ast [95] attempted to relate annealing embrittlement with the heat content of $Fe_{40}Ni_{40}B_{12}Si_8$ amorphous alloy. Isochronal anneals (15 min every 25°C) showed a sudden increase of Curie temperature at 550 K and a drop of fracture strain at 600 K. Specific heat measurements did not reveal anything unusual.

Gerling et al. [96] studied the effect of specimen thickness on the annealing embrittlement of $Fe_{40}Ni_{40}B_{20}$ (VITROVAC alloy 0040). For isochronal annealing of 43 hr, the embrittling temperature for the 30-μm sample was 285°C, 40 μm (245), and 50 μm (165). Accompanying the embrittlement transition was a 0.1% density increase. The embrittled samples restored their ductility by 2×10^{17} thermal n/cm^2 irradiated below 70°C. However, all samples became brittle after 10^{18}–10^{19} thermal n/cm^2 irradiation with the thinner specimen requiring more neutron flux. The density of the material decreased linearly with the logarithm of the neutron flux (1% for 3×10^{19} n/cm^2).

Arakawa et al. [97] measured the tensile strength of as-cast $Ni_{72}B_{20}Si_8$ and $Co_{75}Si_{15}B_{10}$ (50-mm-wide ribbons) in both longitudinal and transverse directions. In GPa, they were 2.68 and 2.54 for the Ni alloy and 2.52 and 2.34 for the Co alloy, respectively. The longitudinal direction had a higher strength than the transverse direction. For the Ni alloy after stress relief for 1 hr at 400°C, the strengths were 2.61 and 2.51, respectively. Stress relief for 1 hr at 330°C for the Co alloy did not change the strength much (2.51 and 2.34). The ratio of these strengths (longitudinal/transverse) was similar after stress annealing and annealing in a magnetic field.

Yamasaki et al. [98] found some interesting composition dependence of annealing embrittlement (see eutectic toughening).

Cao and Li [67] measured the activation energy for the onset of embrittlement of METGLAS alloy MBF-20 ($Ni_{69.2}B_{13.7}Si_{7.9}Cr_{6.6}Fe_{2.6}$) obtained from the times required for the appearance of a crack upon reverse bending during isothermal annealing at several temperatures between 658 and 733 K to be 290 kJ/mole. Instead of isothermal annealing, continuous heating at a constant heating rate was used until the temperature for the onset of embrittlement T_E, was reached; a Kissinger plot between the heating rate and T_E showed an activation energy of 300 kJ/mole. X-ray diffraction patterns for samples at the onset of embrittlement after either isothermal annealing or continuous heating showed no sign of crystallization, being similar to that of the as-received material. However, Kissinger plots using several heating rates to reach 69 ± 2 and 43 ± 2 mm x-ray first peak heights showed the same activation energy of 280 kJ/mole for both peak heights. Another Arrhenius plot using the time needed during isothermal annealing to reach the x-ray first peak height of 41 ± 2 mm gave an activation energy of 285 kJ/mole. Because of the similarities in all these activation energies, annealing embrittlement of MBF-20 could be caused by incipient crystallization.

Wu and Spaepen [99] reported on the ductile-brittle transition temperature T_{DB} of $Fe_{79.3}B_{16.4}Si_{4.0}C_{0.3}$ (97° ± 20° after 2 hr anneal at 350°C, −50°C ± 30° after 2 hr at 250°C, T_x was about 380°C). T_{DB} increased linearly with log(time) for isothermal anneals at 350°C, 321°C, 281°C, and 251°C. Small-angle x-ray scattering showed no evidence of a phase separation during the embrittlement process. By using partially crystallized specimens as calibration, an upper limit of 0.9 vol% of a possible second phase like Fe_3B was estimated in the most embrittled samples (Wu and Spaepen [100].

Misra and Akhtar [101] reported on the annealing of $Ni_{60}Nb_{40-x}Al_x$ ($x = 0$, 5, and 10) made from homogenized ingots which were melt-spun onto a rotating (3000 rpm) polished copper wheel (22.5 cm dia) using a quartz nozzle (1 mm dia) under 10 psi Ar. The ribbon were 2–3 mm wide and 30–40 μm thick. T_x from DSC at 20°C/min was 918, 922, 915, and 902 K for $x = 0$, 5, 7.5, and 10 respectively. Kissinger plots of the peak crystallization temperature gave 4.7 eV/atom ($x = 10$) and 5.1 eV/atom ($x = 5$) as activation energies. Annealing for 2 hr at up to 820 K did not change the yield strain of about 0.02 much. For $x = 5$, the fracture strain was 1.0 (300 K for 2 hr), 1.0 (725 K), 0.1 (775 K), 0.05 (825 K), and 0.007 (875 K). For $x = 10$, the fracture strain was 1.0 (300 K), 0.12 (675 K), 0.045 (725 K), 0.01 (775 K), 0.0028 (825 K), and 0.002 (875 K). Annealing at 873 K for up to 40 min did not change the yield strain (0.02), but the fracture strain decreased starting at about 5 min for $x = 10$ and 10 min for $x = 5$. Time to reach 0.02 fracture strain at various temperatures gave activation energies 1.75 eV/atom for $x = 5$ and 1.69 eV/atom for $x = 10$. Hardness began to increase at 620 K for $x = 10$ and at 650 K for $x = 5$ after annealing for 2 hr at these temperatures. Hardness began to increase after 10 min for $x = 10$ and after 40 min for $x = 5$ for annealing at 873 K. Stress relief began at 500 K for $x = 5$ and 550 K for $x = 10$ after annealing for 2 hr at these temperatures. The variation of the first slope with $1/T$ gave 0.53 eV/atom as the activation energy for stress relaxation. The later slope gave 0.66 eV/atom. Both slope versus T^{-1} relations are the same for $x = 5$ and $x = 10$.

Yavari et al. [102] measured fracture strain (by bending) at several temperatures between 300 and 600 K for $Fe_{78}B_{13}Si_9$ (METGLAS alloy 2605S-2) after Joule heating annealing (10 s at 800 K) and compared with that of the same material after conventional annealing (2 hr at 665 K). The fracture strain was somewhat larger (about a factor of 2) in the case of Joule heating.

Yamasaki et al. [103] measured the DBTT (ductile-brittle transition temperature) of $Fe_{90-x}B_xSi_{10}$ ($x = 10$–15) as a function of annealing time (0–500 min) at 623 K. The DBTT for as-quenched specimens was less than 70 K. It increased rapidly with the time of annealing and leveled off at 295 K for $x = 13$, 420 K for $x = 14$ and 470 K for $x = 15$. For $x = 10$, the DBTT was 165 K after 200 min and 195 K after 500 min and there was no indication of leveling off. Similarly these temperatures were 183 and 216 for $x = 11$ and 200 and 227 for $x = 12$. For the $x = 13$ sample, the DBTT seemed to level off at 290 K for anneals at 593, 603, 613, and 623 K but kept on increasing even after 500 min if annealed at 633, 643, and 653 K. (DBTT was 395 K for 633 K, 410 K for 643 K, and 425 K for 653 K). This two-stage embrittling process with the transition at about 620 K was supported by the effect of annealing on Curie temperature (appreciable increase beginning after annealing at 620 K) and small-angle x-ray scattering (the increase of total scattering intensity with temperature had a break at 620 K and another at 850 K which was due to crystallization). A Guinier plot followed by the Fankuchen analysis suggested that the second stage embrittlement (starting at 620 K) might be caused by the decomposition of the amorphous phase (spinodal?) resulting in the development of microscopic regions of 4–10 (average 7) nm throughout the specimen (confirmed by electropolishing).

Liebermann et al. [104] examined the effect of annealing $Fe_{78}B_{13}Si_9$ amorphous ribbon in a 10-Oe dc magnetic field on the strain to fracture, fraction of stress relief, Curie temperature and dc coercive field. The authors then attempted to relate these properties to the microstructural changes observed by TEM. Two temperatures are of interest. The strain to fracture began to decrease after annealing at 600 K for 2 hr. For this annealing the stress relief was almost complete and the Curie temperature began to increase. This was also when surface crystallization appeared to start. Annealing at 700 K for 2 hr caused the dc coercive field to increase, after reaching a minimum and the heat of crystallization to begin to decrease. This occurred when bulk crystallization began and surface crystallization appeared complete. Extrapolation of the crystallite size data, as shown in Fig. 1, suggested that annealing embrittlement might be caused by very small surface crystals or crystal nuclei as suspected also by Cao and Li [67]. By using spectroscopy and positron life time measurements Janot et al. [105] also thought that annealing embrittlement could be due to the onset of precrystallization of microcrystalline layers developed on the surfaces of the ribbon. Guo and den Boer [106] studied

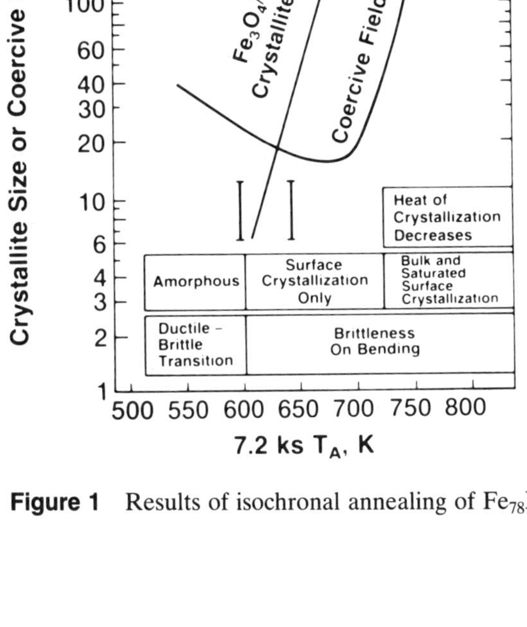

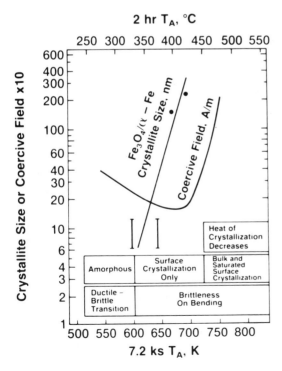

Figure 1 Results of isochronal annealing of $Fe_{78}B_{13}Si_9$ (from Ref. 104).

crystallization of $Fe_{40}Ni_{32}P_{12}B_8Cr_8$ by EXAFS measurements and the results implied that crystallization begins at the surface before it occurs in the bulk. Piotrowski and Finocchiaro [92] measured fracture strain after 30 min isochronal annealing every 50°C, the embrittling temperature was 275°C for $Fe_{74.7}B_{12}Mo_{8.3}Si_5$ and 375°C for $Fe_{76.8}B_{12}V_{6.2}Si_5$.

VII. ALLOY STRENGTHENING AND EMBRITTLEMENT

A. Hydrogen Embrittlement

Namboodhiri et al. [107] studied hydrogen embrittlement of $Fe_{40}Ni_{40}B_{20}$ (VITROVAC alloy 0040), $Fe_{39}Ni_{39}B_{12}Si_6Mo_4$ (VITROVAC alloy 4040), and $Ni_{40}Co_{20}B_{16}Cr_{12}Fe_6Mo_6$ (Allied-Signal) alloys. Cathodic charging in 0.1 *N* H_2SO_4 reduced the fracture stress to 20–40% of that in air. A large number of microvoids of various sizes were observed in hydrogen embrittled specimens.

Lee and Lee [108] charged $Pd_{80}Si_{20}$ with hydrogen and then desorbed the samples under a constant heating rate. The peak desorption (71×10^{17} H/g-min) temperature was 300 K for hydrogen charged under 1 atm at 325 K. It was 320 K (peaked at 55×10^{17} H/g-min) for 0.25 atm charging and 325 K (peaked at 35×10^{17} H/g-min) for 0.1 atm charging. The three curves (evolution rate versus temperature) merged at the high temperature side. It showed that hydrogen fills the low-energy sites (high activation energy to desorb) first. However, the situation was not that clear when the hydrogen was charged at higher temperatures (373 and 473 K). This reviewer suggests that high-fugacity hydrogen probably can create or alter the energy sites at high temperatures to favor its absorption.

B. Boron Strengthening

Hillenbrand et al. [109] studied the effect of B on hardness and tensile strength of $Co_{100-x}B_x$ alloys with $x = 17$–40. In the low B range ($x = 17$–27), the hardness data (900–1200 HV) agreed with those of Chen [110] but higher than those of Masumoto [111] (70–800). In the high B range ($x = 30$–40) the hardness (1250–1350) was much higher than those of Inoue et al. [112] (95–1000 for $x = 35$–40). Tensile strength increased from 3.1 GPa ($x = 17$) to 3.5 ($x = 25$), decreased to 3.3 ($x = 30$), and suddenly increased to 250 MPa ($x = 34$) and 100 MPa ($x = 35$).

Graf et al. [113] measured the tensile strength of $Ni_zSi_8B_{92-z}$ as a function of z (66–79). While the UTS seemed to increase from 1.2 GPa at $z = (68)$ through 1.5 (71) to 1.6 (74, 76, and 79), the microhardness decreased linearly from 1130 at $z = (66)$ to 870 (79). The T_x also decreased linearly from 810 K at $z = (6)$ to 750 K (79). These T_x values were for 20°C/min heating rate. Increase of heating rate from 1 to 50°C/min increased T_x about 40°C.

Yavari [114] studied annealing embrittlement of several Fe-Ni-B alloys. For $(Fe_{0.5}Ni_{0.5})_{100-x}B_x$, T_{DB} were 400 K (free side in tension) and 440 K (wheel side in tension) for $x = 23$ and were 315 K (free side) and 335 K (wheel side) for $x = 20$. The $x = 17$ alloy was ductile at 290 K. After 2 hr annealing at 573 K, the T_{DB} (wheel side in tension) were 450 K ($x = 23$), 338 (20), and 230 (17). For $Fe_{100-x}B_x$

annealed 2 hr at 553 K, T_{DB} were 640 K ($x = 25$), 390 (23), and 220 (17). T_{DB} for the following alloys annealed 2 hr at 653 K were 474 K for $Fe_{79.5}$ $(SiB)_{20.5}$, 454 K for $Fe_{77}Ni_4$ $(SiB)_{19}$, 450 K for $Fe_{74}Ni_7$ $(SiB)_{19}$, and 435 K for Fe_{82} $(SiB)_{18}$. Thus the metalloid content played an important role in annealing embrittlement.

C. Phosphorous Embrittlement

Rudkowski et al. [115] studied the effect of P substitutions for B in $Fe_{78}B_{13}Si_9$ alloy. Both crystallization temperatures increased with P substitution. T_{x1} increased from 825 K for 0 to 830 K for 1.5% P. T_{x2} increased from 847 K for 0 to 859 K for 1.5% P. The enthalpy for the first crystallization event increased from 6 to 8 cal/g but that for the second event decreased from 18 to 11 cal/g in the same range. The tensile strength (phosphorous atom %) are as follows: 2.6 GPa (0.1 at%), 2.3 (0.2), 2.4 (0.3), 1.9 (0.5), 1.3 (0.8), 1.0 (1.0), 1.1 (1.2), and 1.2 (1.6). After 2 hr annealing at 633 K, the tensile strengths (P%) are as follows: 2.7 GPa (0.1 at%), 1.4 (0.8), 1.1 (1.0), and 1.4 (1.6). Small amount (0.1–0.2 $^a/_o$) of P substitution also improved soft magnetic properties.

Bresson et al. [116] added 1.8 at% P or 0.04 at% S to $Fe_{78}B_{13}Si_9$ amorphous alloy and did bending test with either the wheel side or the free side in tension. During bending, the strain at the apex was $\varepsilon = 2\alpha d/(D - d)$ where $\alpha = 0.59907$, d was specimen thickness, and D was platen separation. If $\dot{\varepsilon}$ is 10^{-4}/s and constant, $\varepsilon = \dot{\varepsilon} t$ so the D, t relation is $D = d + 2\alpha d/\dot{\varepsilon} t$, which had a 30-nm resolution. Yielding behavior was similar between doped and undoped alloys. Annealing (20 hr at 523 K) embrittlement was improved by P doping. B doping showed no annealing embrittlement only when the free side was in tension. Thermal analysis of doped and undoped alloys were similar in both the as-quenched condition and after annealing. Hence there was no correlation of embrittlement with the thermal behavior.

D. Embrittling Effects of Sb, Se, and Te

Liebermann and Luborsky [117] reported on the embrittling effect of minor additions of Sb, Se, and Te to $Fe_{81.5}B_{14.5}Si_4$, $Fe_{40}Ni_{40}B_{20}$, and $Ni_{81.5}B_{14.5}Si_4$ amorphous alloys. In each case either Si or B was replaced. The as-cast thickness below which the material was ductile and the temperature above which 1 hr annealing of ~25 μm as-cast sample became brittle were as follows: first alloy: 0 substitution, 58 μm, 600 K; 0.1 Sb, 40 μm, 510 K; 0.1 Se, <20 μm, <RT; and 0.1 Te, <20 μm, <RT. Second alloy: 0 substitution, 42 μm, 600 K; 0.1 Sb, 40 μm, 550 K; 0.1 Se, 28 μm, 400 K; 0.1 Te, <25 μm, <RT. Third alloy: 0 substitution, 40 μm, 500 K; 0.1 Sb, 40 μm, 500 K; 0.1 Se, 40 μm, 340 K; 0.1 Te, 40 μm, 30 K. Alloys with embrittling additions also showed faster stress relaxation. However, DSC and magnetic measurements were not affected.

E. Cesium Toughening

Kroeger et al. [118] fractured $Fe_{80}B_{16}C_2Si_2$ samples by bending in ultrahigh vacuum of the Auger spectrometer, after annealing for 25 hr at 250°C or 47 hr at 300°C. The second sample showed more oxygen on the fractured surface. Addition of 100 at. ppm Ce almost eliminated the annealing embrittlement. However, addition of 500 at. ppm Ce only slowed down the annealing embrittlement by about 5 hr.

Kroeger et al. [119] reported the improvement of annealing embrittlement by microadditions of Ce. For $Fe_{80}B_{16}C_2Si_2$, the bend fracture strain at room temperature after 200 hr annealing at 300°C was 0.03 for (0) ppm Ce, 0.09 (25). 0.15 (100), 0.4 (75), and 1 (50). For 1 hr annealing, the ductile/brittle transition temperature (bend test at room temperature) was 350°C for (0 or 25) ppm Ce, 375 (75 or 100), and 400 (50). Note the maximum improvement was for 50 ppm Ce. Another set of experiments on the same alloy but annealed at 325°C for 500 hr showed the bend fracture strain at room temperature to be 0.04 for (0) ppm Ce, 0.04 (25), 0.07 (50), 1 (75), 0.07 (100), and 0.04 (260). Here the maximum effect was at 75 ppm. The material with this optimal doping, unlike the under- or overdoped materials, had a domain structure without the maze domains, indicating the absence of quenched-in stresses. The authors thought that addition of Ce in excess of that required to combine with the available O and S in the material would enhance embrittlement.

Kroeger et al. [120] reported the effect of 50–100 ppm Ce additions on the annealing embrittlement of $Fe_{80}B_{20}$, $Fe_{80}B_{16}C_2Si_2$, and $Fe_{78}B_{13}Si_9$. As a function of x in $Fe_xB_{13}Si_{87-x}$, the embrittlement temperature of 1 hr anneals started at 315°C for $x = (0)$ increased to 385 (81) and then decreased to 275 (83). Addition of 75 ppm Ce increased the embrittlement temperature by about 10° for $x = 74–80$, 40° for $x = 81–82$, and 80–100° for $x = 82–83$. Oxygen (arc melting in Ar + 2% O_2 before melt spinning) did not have any effect for $x = 74–80$ but the embrittlement temperature decreased about 70–80° for $x = 80–82$.

Lamparter et al. [121] did small-angle x-ray scattering (SAXS) of $Fe_{80}B_{16}C_2Si_2$ with and without 75 ppm Ce and after annealing at 300 and 325°C for up to one month (longest was 227 days). The structure factor $S(Q)$ increased with annealing and so did embrittlement and there seemed to be a correlation between the mean squared fluctuation of electron density (the integral of $Q^2S(Q)\,dQ$) and the bending fracture strain. The doping for Ce was to retard the increase of $S(Q)$. The SAXS results could be understood by (1) 7–30% volume separation of amorphous phases of sizes larger than 10 nm, (2) a few ppm of surface crystallization or amorphous phase separation (>10 nm), and (3) 0.004 volume fraction of α-Fe in the bulk (>10 nm). The latter two possibilities could be affected by Ce addition.

F. Eutectic Toughening

Yamasaki et al. [98] studied annealing embrittlement of some Fe-B-Si alloys. After isothermal annealing at 623 K, the time needed to cause fracture strain of 1 was 55 s for $Fe_{75}B_{15}Si_{10}$ ($T_x = 813$ K), 147.6 ks for $Fe_{78}B_{13}Si_9$ ($T_x = 783$ K), and 6 Ms for $Fe_{79.5}B_{12}Si_{8.5}$ ($T_x = 758$ K) and $Fe_{81}B_{11}Si_8$. For Fe-Bi-Si (Cr or Ti) the embrittlement resistance increased in the following order: $Fe_{65}B_{15}Si_{10}Cr_{10}$ ($T_x = 827$ K, 5 s for $\varepsilon_f = 1$), $Fe_{68}B_{13}Si_9Cr_{10}$ (300 s), $Fe_{72.5}B_{10}Si_{7.5}Cr_{10}$ (1500 s), and $Fe_{78.5}B_{12}Si_{8.5}Ti_1$ (2×10^5 s). The alloys of the highest embrittlement resistance were of near eutectic compositions as evidenced from microstructural observations on slowly solidified specimens since they consisted of very fine mixtures of two or more phases. The hypoeutectic structures were composed of the eutectic mixtures and other precipitate forms such as dendrites.

VIII. FRACTURE, FATIGUE, AND WEAR

A. Fracture

Henderson et al. [122] measured the fracture toughness of METGLAS alloy 2826MB ($Fe_{40}Ni_{38}B_{18}Mo_4$) specimens (25 mm × 12.5 mm × 50 μm) precracked midway along the long side by scissors. Sharpening of the crack was by fatigue between 80 and 240 MPa. Fracture toughness was measured by tension to have values of 49.6 MPa $\sqrt{m}$ ± 20% considerably higher than those obtained by Davis [123,124] for plane-strain failure. The present failure region was slanted 45° which is characteristic of plane-stress failure.

Wetzig et al. [125] reported on some in situ tensile experiments in the SEM to relate the fracture toughness with the crack tip localized deformation. Isochronal annealing (1 hr at each temperature every 25°C) of $Fe_{40}Ni_{40}P_{14}B_6$ and $Co_{70}Si_{15}B_{10}Fe_5$ reduced the fracture strength from 1.4 GPA at 260 K to 1 GPa at 520 K while the hardness increased about 10% for both materials. For $Fe_{40}Ni_{40}P_{14}B_6$, the fracture toughness increased from 38 MPa $\sqrt{m}$ at 260 K to 48 at 350 K and then decreased to 32 at 550 K. During the increase, the shear-band density increased from 300/mm to 750/mm after which it decreased to 300/mm at 500 K. No shear bands and no river pattern appeared on the fracture surface at higher temperatures. Instead, pores (less than 1 μm) emerged and the crack grew by the coalescence of these pores. Isochronal annealing (1 hr in oil) at T and then testing at room temperature gave a fracture toughness of 50 MPa $\sqrt{m}$ for T = 300 K, 70 for 375 K, and 55 for 425 K. The shear-band density at the crack tip also showed a maximum (120/mm for 325 K, 400 for 375 K, and 230 for 425 K) at 375 K. For $Co_{70}Si_{15}B_{10}Fe_5$ isochronal annealing (1 hr in oil) and room temperature testing showed a maximum toughness (120MPa $\sqrt{m}$) for the 450 K anneal (80 for 300 K and 40 for 725 K). The shear band density also showed a maximum (900/mm) at this temperature (500 at 300 K and 300 at 575 K).

Bengus et al. [126] measured the tensile strengths of $Ni_{78}B_{14}Si_8$ (VITROVAC alloy 0080) and $Fe_{40}Ni_{40}B_{20}$ (VITROVAC alloy 0040) at low temperatures. For the first alloy, the tensile strengths were 1 GPa at (4.2) K, 1.2 (77), 1.9 (300), 2.12 (400), and 2.06 (450). For the second alloy, they were 2.32 (4.2), 2.72 (77), and 2.44 (300).

Henning et al. [127] studied the effect of crack length on fracture toughness and the R curves of $Fe_{40}Ni_{40}B_{20}$ (2.3 mm wide, 40 μm thick) and $Co_{70.43}Cr_{21}W_{4.5}B_{2.4}Si_{1.6}C_{0.07}$ (25 mm wide and 50 μm thick). A center-cracked tensile specimen was used for each toughness determination. For the first alloy K_c was 35 MPa $\sqrt{m}$ for a half-crack length of (0.14)mm, increased to 45 (0.35 and decreased to 35 (0.9). Specimens stiffened by Teflon platelets had larger K_c by up to about 10 MPa $\sqrt{m}$ for a half-crack lengths longer than 0.3 mm. For the second alloy, K_c scattered between 20 and 77 MPa $\sqrt{m}$, independent of the crack length (a/b ratio of crack length to specimen width, changed from 0.1 to 0.6). However, the high values were associated with the vein pattern on the fracture surface and low values with the chevron pattern. Intermediate values show mixed patterns. An attempt to plot the R curves was done only for the first alloy.

Kimura and Masumoto [128] defined a plastic stress concentration factor $Q = \sigma_f/\sigma_y$ between the microfracture stress σ_f and the yield stress σ_y in tension. For $Fe_{40}Ni_{40}P_{14}B_6$, σ_y was calculated based on a flow law and σ_f was estimated by extrapolating to the most brittle situation in which $Q = 1$. Then by assuming constant σ_f (about 2.8 GPa), Q can be calculated and found to saturate at about 1.30.

Söder et al. [129] measured the fracture strength of $Fe_{86}B_{14}$ as a function of temperature in Ar. In GPa, it was 3.5 at 25°C, 3.4 at 170°C, 3.2 at 260°C, 2.8 at 310°C, 2.3 at 350°C (T_{x1}) 2.6 at 390°C, 2.5 at 410°C (T_{x2}), 3.0 at 450°C, and 2.7 at 500°C. For $Fe_{82}B_{18}$ in Ar, it was 2.1 at 25°C, 1.8 at 180°C, 1.5 at 310°C, 0.9 at 400°C(T_x), 1.1 at 420°C, and 0.7 at 450°C. Values were somewhat higher if tested in air.

Calvo et al. [130] measured fracture toughness of $Ni_{80.8}Cr_{15.2}B_4$ glassy ribbons of 30 μm thick. As a function of the ratio of crack length to specimen width, the toughness was 30–45 MPa $\sqrt{m}$ for 0.1 ratio, 60 for 0.2. 63 for 0.3, 61 for 0.4, and 40 for 0.7. Similarly for $Co_{70.43}Cr_{21}W_{4.5}B_{2.4}Si_{1.6}C_{0.07}$ ribbons of 47–58 μm thick and 12–13 mm wide, the toughness was 15 MPa $\sqrt{m}$ for 0.05 ratio, 90 (max) for 0.4 ratio, and 80 for 0.7 ratio. These were for the fracture surfaces with veins. For the fracture surfaces with chevrons, the toughness was much lower (between 15 and 30). At 77 k, the toughness was about 15 independent of the crack length/specimen width ratio. For $Fe_{40}Ni_{40}B_{20}$ ribbons of 40 μm thick, the toughness (for 0.2–0.35 ratio) was between 42 and 52 independent of temperature (50 to 300 K) while the fracture strength increased from 1 GPa at 50 K to 2.4 GPa at 290 K.

Hunger and Mordike [131] measured the microhardness of $(Fe_{0.75}M_{0.25})_{83}B_{17}$ glasses and found a relation to the melting temperature of M borides (M = Co, Ni, Fe, Mn, Cr, Mo, and W). At the low end for M = Co and Ni, HV (0.05) ~ 900 and T_m = 1150°C. AT the high end for M = W, Hv (0.05) ~ 1900 and T_m = 2700°C. Fracture toughness in units of $N/mm^{3/2}$, 650 for M = Co, 600 for M = Fe, 420 for M = Mn, 250 for M = Cr, 100 for M = Mo and 50 for M = W. Isochronal (60 min) annealing of $Fe_{40}Ni_{40}B_{20}$ increased the hardness only slightly (Hv (0.05) from 1050 to 1100) but the fracture toughness decreased from 1000 $N/mm^{3/2}$ at room temperature to 100 after annealing at 650 K. Isothermal annealing at 500 K caused a decrease of fracture toughness from 100 to 350 after 100 min (almost linear with log t). The decrease was faster at 600 K with 100 being the lowest value.

Kimura and Masumoto [132] reviewed their studies in the mechanics of fracture and the effect of annealing on the ductile brittle transition. The alloys were $Fe_{40}Ni_{40}P_{14}B_6$, $Fe_{80}P_{20}$, $Fe_{80}B_{20}$, and $Fe_{78}B_{12}Si_{10}$. The fracture criterion was a critical tensile stress σ_f in bending. This fracture stress was related to the yield stress σ_y by $\sigma_f = Q\sigma_y$ with Q being the plastic-stress concentration factor. Q varied from 1 (brittle) to about 1.3 (ductile). Decrease of fracture strain ε_f in bending as a function of annealing temperature showed a two-stage behavior. The first sharp drop of ε_f was at 400 K and the second at 550 K. For stage I, the decrease of ε_f paralleled the enthalpy of $Fe_{40}Ni_{40}P_{14}B_6$. An activation energy of 0.94 eV/atom was the same as that for structural relaxation. For stage II, the activation energy was 2.47 eV/atom.

Ocelik et al. [133] reported on the fracture toughness of $Fe_{40}Ni_{40}B_{20}$ (43MPa $\sqrt{m}$ for 20 μm thickness, 50 for 28 μm, 60 for 36 μm, and 57 for 37 μm), $Ni_{40}Fe_{30}B_{20}Cr_{10}$ (15 MPA $\sqrt{m}$ for 18 and 20 μm), $Ni_{80}B_{10}Si_{10}$ (39 MPa $\sqrt{m}$ for 23 μm), and $Ni_{80}B_{15}Si_5$ (42 MPa $\sqrt{m}$ for 27-μ thickness). The effect of thickness was explained by the triaxial state of stress. A circular hole, 0.1–0.5 mm diameter, was put in the middle of the sample by electric discharge machining. A longitudinal ultrasonic wave of 22.3 kHz was used to produce sharp cracks if the length of the specimen was one-half wavelength of the stress wave. A final length of the crack ($2a$ = 0.7 to 2.3 mm) was obtained after 1–3 million cycles of 2–3 μm amplitude. A linear relation between the fracture stress and the square root of crack length was obtained from which the fracture toughness was calculated. The failure of all specimens was caused by shear and fracture took place by a combination of tearing and opening modes.

Ocelik et al. [134] also reported on the fracture toughness of $Fe_{40}Ni_{40}B_{20}$ (10 mm × 36 μm cross section) by tensile test (5 × 10^{-5}/s strain rate) of specimens with a centrally located sharp crack (longer than 0.1 but shorter than 0.25 of the width of the specimen). The toughness was between MPa $\sqrt{m}$ at 4.2 K to 60 MPa $\sqrt{m}$ at 300 K. The transition temperature depended on the time kept at room temperature as follows: 70 K for 5 months, 120 K for 14 months and 230 K for 20 months. The toughness of $Ni_{80}B_{10}Si_{10}$ and $Ni_{80}B_{15}Si_5$ was about 40 MPa $\sqrt{m}$ independent of temperature between 4.2 to 300 K.

Harbert and Wolfenden [135] bent Ni-3.2B-4.5Si amorphous alloy (5 mm × 76 μm × 25 mm) and observed in SEM. Shear bands were observed at a critical bending diameter of 4.2 mm. They started in the middle of the tensile surface and propagated toward the edges. Sometimes additional bending was required to further propagate the shear bands. With additional bending the shear-band density increased as well as the step height until fracture. The fracture surface consisted of the slipped region, the vein pattern region, the rapid fracture region and the catastrophic rupture region.

Pak et al. [136] ground with a beater mill $Fe_{78}B_{13}Si_9$ (METGLAS alloy 2605-S2) 28 μm ribbons to form a powder with a size distribution peaked at 0.25–0.5 mm. X-ray diffraction of the powder showed α-Fe and Fe_3B bct phases. DSC of the fine powder revealed greatly reduced exothermic peaks at 547 (α-Fe) and 565°C (Fe_3B). Powders finer than 53 μm were round probably produced by extensive plastic deformation. These powders were 75% crystallized.

B. Fatigue

Chaki and Li [137] studied the overload effect of fatigue crack propagation in $Co_{72.15}B_{15}Fe_{5.85}Si_5Mo_2$ (METGLAS alloy 2705M) and $Co_{76.05}B_{12}Si_6Mn_4Fe_{1.95}$ (METGLAS alloy 2705MN). In the power law region the fatigue crack propagation per cycle was found to be

$$\frac{da}{dN} = 1.2 \times 10^{-5}(\Delta K)^{3.3} \ \mu\text{m/cycle} \quad \text{for METGLAS alloy 2705M} \tag{14}$$

and

$$\frac{da}{dN} = 1.4 \times 10^{-7}(\Delta K)^{5.7} \ \mu\text{m/cycle} \quad \text{for METGLAS alloy 2705MN} \tag{15}$$

respectively with ΔK in MPa $\sqrt{m}$. For METGLAS alloy 2705M, when ΔK was 8.77 so that $d\alpha/dN$ was about 0.02 μm/cycle, 5 cycles of overload of 1.8 overload factor was applied and then ΔK returned to 8.77. The da/dN suddenly reduced to about 0.001 μm/cycle and it took about 10,000 cycles for the da/dN to return to normal. Such overload effect suggests strong interaction between the plastic zone and the propagating fatigue crack.

Jost et al. [138] studied fatigue of $Fe_{41}Ni_{41}B_{18}$ in the amorphous and semicrystalline states. Tensile test of $Fe_{42}Ni_{42}B_{16}$ as a function of annealing time at 350°C started at 2.5 GPa, decreased to 1.6 after 1 min due to annealing embrittlement in the amorphous state, increased to 2.2 after 10 min due to partial crystallization and decreased again to 1.4 after 100 min and 0.3 after 1000 min due to complete crystallization. For $Fe_{41}Ni_{41}B_{18}$, fatigue testing showed similar effects: 1 min annealing at 400°C decreased the fatigue life by a factor of 2–3 but 10 min annealing increased it by a factor of 10. The primary effect of partial crystallization was to increase the number of slip steps and to decrease the shear displacement of each step. The consequence was to reduce crack nucleation and increase the strength and fatigue life.

Le et al. [139] reported on the fatigue life of $Fe_{70}B_{15}Si_{10}Cr_5$ and $Fe_{67}B_{15}Si_{10}Cr_8$ wires in bending. Deleterious environmental effects occurred in the following order of severity: 65% humidity, H_2O, 3.5 wt% NaCl, and 1 *N* H_2SO_4. For the low Cr alloy in humid air, the strain versus cycle to failure relation was as follows: 0.016 (10^3 cycles), 0.008 (10^4), 0.004 (10^5), and 0.003 (10^6). The same alloy in NaCl solution gave 0.008 (10^3), 0.003 (10^4), and 0.002 (10^5). In 1 *N* H_2SO_4, 0.007 (10^3) and 0.002 (5×10^3). The high Cr alloy in humid air showed the following strain versus cycle to failure relation 0.015 (3×10^3), 0.013 (10^4), 0.011 (10^5), and 0.010 (3×10^6). In water 0.015 (10^3), 0.012 (10^4), 0.009 (10^5) and 0.008 (2×10^5). In 3.5 wt% NaCl, 0.008 (10^3), 0.0056 (3×10^3), and 0.0044 (10^4 and 10^5). In 1 *N* H_2SO_4, 0.006 (2×10^3), 0.004 (3×10^3), 0.003 (5×10^3), and 0.002 (10^4).

Kabacoff et al. [140] reported on the tensile strength as a function of the area reduction by drawing of $Co_{45}Fe_{23}B_{13}Cr_{10}Si_9$ (120 μm and 130 μm original dia) wires manufactured by Unitika, Inc. and sold by Allied-Signal. It increased from 430–495 ksi at 0% to 514 ksi at 40–45%. Bending fatigue in 3.5 wt% NaCl showed a strain and life relation as follows: 0.0072–0.009 (3×10^3 cycles), 0.0056–0.0063 (10^4), 0.0045–0.005 (3×10^4), and 0.0035 (3×10^5).

Hagiwara et al. [141] studied the fatigue strength of Fe-based amorphous wires. The tensile strength (GPa), elongation to fracture (%) and Young's modulus (GPa) were 3.4, 2.3, and 151 for $Fe_{75}B_{15}Si_{10}$ wire, 2.87, 2.1, and 142 for $Fe_{77.5}P_{12.5}C_{10}$ wire, 3.02, 2, and 181 for piano wire and 0.72, 30, and 132 for SUS 304 wire, respectively. For 10^6 cycles to failure at 3.2 Hz in 65% RH air (293 K), the bending strains for these wires were 0.0032, 0.0030, 0.0039, 0.0042 respectively. The strains for amorphous ribbons of the same compositions were 0.0019 and 0.0017 respectively. The fracture surface of $Fe_{75}B_{15}Si_{10}$ after fatigue failure (bending strain = 0.0040, N_f = 5.4×10^5 cycles) showed surface nucleation of a crack, which propagated toward the middle of the thickness accompanied by striations (~0.7-μm separation). The final failure surface was very smooth. Addition of Cr to $Fe_{75-x}Cr_xB_{15}Si_{10}$ (x = 0–30) resulted in an increase of fatigue limit from 0.0035 (strain) at x = 0 to 0.007 at x = 5. In water, the fatigue limit increased in the following order, $Fe_{75}B_{15}Si_{10}$, piano wire, SUS 304 wire (0.004 strain) and $Fe_{67}B_{15}Si_{10}Cr_8$ (0.0075).

Alpas et al. [142] did fatigue testing of $Ni_{78}B_{12}Si_{10}$ in the form of single edge-notched tensile specimens (4 cm long, 2 cm wide, 57 μm thick) with a starting notch of 3 mm deep (by scissors). Crack length was measured optically with a traveling microscope. For R = 0.1 (stress ratio) the threshold ΔK is 0.05 MPa

$\sqrt{m}$ where da/dN was 10^{-7}–10^{-8} mm/cycle. The unstable ΔK was 50 MPa $\sqrt{m}$ or $0.75K_c$ where da/dN was (0.5–1.5) $\times$ 10^{-3} mm/cycle. Annealing at 440°C for 7 min caused partial crystallization (Ni crystals of 150 nn size spaced 3 μm apart) and increased the threshold ΔK to 2 MPa $\sqrt{m}$ (da/dN = (0.3–3) $\times$ 10^{-7} mm/cycle) but reduced the unstable ΔK to 10 MPa $\sqrt{m}$ (da/dN = 10^{-4}–10^{-5} mm/cycle). Some microstructural features of the fracture surfaces were described. Crack growth in the amorphous alloy, and in the partially crystallized alloy for low and intermediate ΔK's seemed to follow shear bands the number of which increased with ΔK. Crack growth in the partially crystallized alloy for high ΔK's seemed to follow the weak particle-matrix interfaces.

C. Wear

Wong and Li [143] studied dry sliding wear of METGLAS alloy 2605 SC ($Fe_{81}B_{13.5}Si_{3.5}C_2$) against WC coated rolls of 3, 8, and 15 μm average roughness. The wear mechanism changed with increasing roughness of the rolls from ploughing to microcutting to crack nucleation and growth. Similar to crystalline metals, the mass loss increased linearly with time after an incubation period and the wear increased proportionally with the applied load. When sliding at 25 mm/s on 15 μm roughness rolls, the heat of crystallization (from DSC) of the debris implied that they were almost completely crystallized. However at 75 mm/s the heat of crystallization on the debris showed an increase. The size of the debris was also larger (80 versus 30 μm) in the latter case. These findings suggested that the hot spot was about 30 μm in size. Together with the result of computer simulation which estimated the actual contact area was only about 10^{-5} of the apparent contact area, there existed perhaps only one hot spot per mm^2 at any time.

Sudarsan et al. [144] also studied dry sliding wear of METGLAS alloy 2605SC ($Fe_{81}B_{13.5}Si_{3.5}C_2$) by using a pin (6 mm disk of 30 μm thick metallic glass glued to a test pin)-on-disc (hardened AISI 52100 steel disk) technique. The speed ranged from 1–4.5 m/s and the load ranged 2.5–50 N. At a sliding speed of 2.9 m/s and under a load of 22.5 N, the friction coefficient started at 0.28, dropped to 0.25 at 200 m increased to 0.32 at 1 km and remained at 0.32 for 1–2 km. At 3.2 m/s, the steady-state friction coefficient was 0.7 at (3) N load, dropped to 0.28 (13) and 0.2 (32–47). For lower speeds the friction coefficient was higher. TEM of debris showed fine-grained crystalline structure in agreement with Wong and Li [143]. The ploughing groove and elongated debris indicated an abrasive type of wear.

Dolezal and Hausch [145] studied adhesive wear of four (Fe, Ni, Co)-SiB alloys, VITROVAC alloy 0080, 6025, 6030, and 7505. Experiments were carried out using a pin (3-mm sphere of 100 Cr6 tool steel)-on-plate (with amorphous ribbon glued on it) wear tester. The Co-based alloys had the lowest friction coefficients (0.3–0.5) as compared with (0.5–1.0) for the Fe- and Ni-based alloys. The Ni-based alloy being the most ductile exhibited the largest wear (12.2) while all others showed little wear (0.6–1.7).

Morris [146] studied wear of $Ni_{77.5}B_{15}Si_{7.5}$ and $Ni_{70}B_{20}Si_{10}$ alloys by pressing the flat specimen onto a rotating hardened steel wheel (dry sliding wear) or a rotating disk of 1000-grade silicon carbide paper (abrasive wear). For sliding wear, the order of decreasing wear resistance was second alloy (950 HV), WC-8Co (2000), first alloy (890), US 58 (820), S700 (1100), and VS 35 (360). For abrasive wear the amorphous alloys were much poorer (a factor of 3 or 4 more wear) than the steels of the same hardness. Annealing increased wear first and then decreased it to approach the wear resistance of steels. The wear resistance of amorphous alloys does not necessarily increase with hardness.

Kishore et al. [147] measured the friction coefficient, μ, of $Fe_{81}B_{13.5}Si_{3.5}C_2$, $Fe_{67}Co_{18}B_{14}Si_1$, and $Ni_{78}B_{14}Si_8$ on 52100 steel disk (hardness RC 63). In general μ (0.3 to 0.35) increased with the sliding distance and reached a steady value of 0.35 after 500 m. The steady state had a "transfer" layer on the disk due to the build up of debris. Removal of the transfer layer reduced μ to the initial value. As a function of normal load, μ was 1.2–1.3 at 2N, 0.8 at 5, 0.5 at 10, 0.4 at 20, and 0.3 at 30–45N (Fe_{67} alloy). Wear tracks on the surface of Fe alloys had shear bands and vein patterns in them. TEM of the wear surface showed glassy structure but the debris contained extremely fine grains. TEM of the wear surface of the Ni alloys showed primary crystallization (715 K) and the debris were fcc textured fine grains.

IX. ANNEALING AND RELAXATION

A brief review of earlier work on atomic transport and structural relaxation was written by Greer [148] in 1984. One conclusion is that the Stokes-Einstein relation is not valid for metal-metalloid glasses. Measured diffusivities are significantly greater than expected from measured viscosities. There are diffusive jumps which do not contribute to flow. By using the Nabarro-Herring creep equation, the effective grain size, in the range of nanometers, seems to increase with decreasing temperature.

A. Length Contraction

Mulder et al. [149] reported accurate length change measurements of $F_{40}Ni_{40}B_{20}$ (VITROVAC alloy 0040) of 2.85 × 0.041 mm^2 cross section and about 200 mm long (length of heating zone) under a stress of 0.4 MPa and a heating rate of 1.25°C/min. Temperature was increased to 623 K and immediately lowered at the same rate to 273 K. Then the heating and cooling cycle was repeated a second time. The difference between the two curves was taken as the length contraction during continuous heating. Since the overall length change of the first run was −0.124% and that of the second run was only −0.004%, the second run had virtually no length contraction. From the difference of the two curves, the contraction started at about 400 K. However, the contraction did not seem to have completed at 623 K before the specimen was cooled down. The authors also reported isothermal length changes at 487 K, 523 K, 546 K, and 578 K. Their results were different from those reported by Kursomovic et al. [250,251]. Creep curves at 546 K and 88 MPa were also shown, from which a $d\eta/dt = 4 \times 10^{10}$ Pa was calculated. However, a contraction correction of about −7% and a Young's modulus correction of about −10% reduced this value to 3.3×10^{10} Pa.

Van den Beukel et al. [33] reported isothermal length changes of $Fe_{40}Ni_{40}B_{20}$ (VITROVAC alloy 0040) at 546 and 578 K up to 10^6 s. The final fractional length shrinkage was about 4.5×10^{-4} before the measurement was stopped. The length-log(t) plot was still straight at 10^6 seconds. Length changes at a constant heating rate of 1.25°C/min were recorded also between room temperature and 623 K. Specimens preheated 2.2×10^5 s at 487 K and 525 K were used also to measure length changes at constant heating rates. The results were analyzed by using a spectrum of activation energies.

Leonardsson [150] measured the length shrinkage of Ni-Nb amorphous alloys during isothermal annealing under a 3 mN (0.27 MPa) force. Curves were given for $Nb_{59}Ni_{41}$ at 647 K, 633 K, and 626 K. Most curves showed negative strain rate up to 10^5 s. Only the high load and the high temperature experiments resulted in a change to positive strain rate at 3×10^4 s and 7×10^4 s, respectively. By assuming the negative strain as the reduction of free volume, the Taub and Spaepen equation [81] could fit the curves only with a wrong time scale. The author concluded that the free-volume theory of creep was clearly not valid in this case.

Huizer and van den Beukel [151] claimed that the length changes in $Fe_{40}Ni_{40}B_{20}$ VITROVAC alloy 0040 could be separated into an irreversible part and a small reversible part. A specimen was first preannealed at 533 K for 325,000 s and then the annealing temperature was cycled between 565 K and 533 K. Upon heating from 533 to 565 K, the specimen length first increased suddenly due to thermal expansion, then increased slowly for a few thousand seconds before it irreversibly decreased. Similarly, upon cooling from 565 to 533 K, the specimen length first decreased suddenly due to thermal expansion and then decreased slowly before it settled down to an irreversible rate of decrease. However, due to some experimental uncertainties, this reversible part of change over and above that due to thermal expansion seemed to need further investigation.

B. Density and Hardness

Zedler and Lehmann [152] measured the density of $Fe_{40}Ni_{40}P_{14}B_6$ and $Co_{70}Si_{15}B_{10}Fe_5$ alloys at room temperature after 1 hr isochronal annealing at each temperature. For the first alloy the as-quenched density was 7.60 ± 0.02 g/cc. The density first decreased about 0.005% then increased to about 0.2% before crystallization. After crystallization the increase was about 1%. For the second alloy, the as-quenched density was 7.76 ± 0.01 g/cc. The annealing behavior was similar to that of the first alloy.

Wagner et al. [153] measured the density and the strain to fracture of $Fe_{40}Ni_{40}B_{20}$ (VITROVAC alloy 0040, 3 mm wide, 40 μm thick) after neutron irradiation. If the density of as-quenched material were 1.000, annealing for 43 hr at 280°C would increase it to 1.001. Irradiation of the annealed material to 10^{17} n/cm^2 reduced it back to 1.000. Further irradiation of 2×10^{19} n/cm^2 reduced it further to 0.992. Postirradiation annealing could recover only about 50% of the density loss (Gerling and Wagner [154]). Let the strain to fracture, ε_f, of the as-quenched material be 1.0. Irradiation up to 2.5×10^{18} n/cm^2 did not change it. However, 6×10^{18} n/cm^2 reduced it to 0.4, 10^{19} to 0.02, and 10^{20} to 0.01. Specimens preannealed 43 hr at 280°C had a strain to fracture of 0.015; 10^{17} n/cm^2 increased it only to 0.025. However, 2.5×10^{17} n/cm^2 restored the ductility completely ($\varepsilon_f = 1$). The ductility remained at $\varepsilon_f = 1$ for 10^{18} and 2.5×10^{18} n/cm^2 and then dropped to 0.012 after 4×10^{19} n/cm^2 irradiation.

Zedler and Lehmann [155] determined the density and hardness of $Fe_{81.9}B_{12.6}Si_{3.4}C_{1.9}Mn_{0.2}$ and $Fe_{79.3}B_{12.6}Si_{3.8}C_{2.5}Al_{1.1}Cu_{0.7}$. The densities were (23°C) 7.387 ± 0.027 and 7.232 ± 0.022 g/cc and the hardnesses were 980 ± 85 and 1080 ± 120 HV respectively. Annealing of as-quenched material in vacuum (1.3×10^{-3} Pa) for 1 hr at various temperatures changed both density and hardness. For the first alloy, the density fluctuated between −0.2% to +0.1% up to 350°C, at which the crystallization started causing a sharp increase in density with a total density increase of 2.6% upon full crystallization. The hardness fluctuated ±5% with a maximum increase of 20% at 375°C and then decreased for higher temperature annealing to even below the as-quenched value (the hardness after 575°C annealing was 6% below). For the second alloy, the crystallization temperature was 450°C. The fluctuation in both density (±0.3%) and hardness (±10%) were larger.

Hunger and Mordike [131] measured the microhardness of $(Fe_{1-x}Co_x)_{83}B_{17}$ for x = 0 to 0.9. The HV (0.05) for x = 0, 0.1, 0.15, 0.25, 0.5, 0.7, 0.75, 0.85, 0.9, 0.95, and 0.98 were 1050, 1000, 920, 840, 830, 860, 1020, 1050, 700, 920, and 820 respectively. Isothermal annealing at 600 K for x = 0.93, 0.95, and 0.99 alloys showed a decrease of hardness from 920 (x = 0.95) to 850 after 500 min.

Kohmoto [156] reported on the microhardness of $(Co_{0.5}Fe_{0.5})_{75}$ $(B_{1-z}Si_z)_{25}$. It was 920 kg/mm^2 for z = (0.2), 820 (0.4), and 790 (0.6) in the as-quenched state. For comparison, the microhardness was 1100 HV for $Co_{74}B_{20}Fe_6$ (Chou et al. [253]), 960 for $Co_{78}B_{12}Si_{10}$ (Donald et al. [254]), 910 for $Co_{73}Si_{15}B_{12}$ (Masumoto [111]) and 810 for $Co_{75}B_{15}Si_{10}$ (Doi et al. [255]). As a function of annealing temperature, z = 0.2 alloy gave 880 for 200°C, 860 (250°C), 830 (300°C), 850 (350°C), 970 (400°C), and 700 (500°C). The z = 0.4 alloy gave 770 for 200°C, 830 (250°C, 870 (300°C), 830 (350°), and 900 (400°C). The z = 0.6 alloy gave 780 for (100°C), 810 (200°C), 860 (250°C), 870 (300°C), 860 (350°C), and 920 (400°C).

Elshafie et al. [157] found both density and microhardness increased with time of annealing, temperature of annealing, and Sb content in $Se_{100-x}Sb_x$ (x = 5, 10, 15 at%) systems. The melt was heated to 850°C for 8 hr and quenched in ice water. The temperature of annealing were 80, 90, 100, 110, and 120°C (between T_g and T_x). A typical density increase was 15% and a typical microhardness increase was a factor of 2 to 3. The data were analyzed in the light of Kolmogorov-Avrami equation:

$$\frac{d_c - d_t}{d_c - d_0} = 1 - \exp(-kt^n) \tag{16}$$

where d is density or hardness, d_0 is initial value, d_c is final value, and d_t is the value at t, k, and n are adjustable constants which may vary with temperature and composition. The activation energies obtained from ln k versus $1/T$ plots were 31.8, 39.8, and 49.7 kcal/mole for x = 5, 10, and 15, respectively, from the density data. The values from the microhardness data were 43.8, 49.7 and 54.7 kcal/mole, respectively.

Kim et al. [158] measured microhardness of $Fe_{67.5}Al_{22.5}Zr_{10}$ at room temperature after 20 min isochronal annealing from 100 to 900°C for every 100°C. Maximum hardness of 1150 HV (free side of ribbon) or 1350 HV (wheel side of ribbon) were found after annealing at 700°C due to crystallization.

Fenge et al. [159] made 75 different compositions of Fe-(0–22 at% P)–(0–10 at% Si) and mapped out the region of glass formation ($14.5 \le P \le 18.5$, $1.5 \le Si \le 6.5$ at%). For $Fe_{78.2}P_{17}Si_{4.8}$, the crystallization kinetics were studied also by microhardness. For 400°C annealing, the hardness started at 6.6 GPa increased to 8 GPa in 5 min (first peak) and 7.6 GPa in 20 min (second peak). At 300°C these numbers were 6.8, 7.9 (10 min), and 7.3 (120 min). Three more sets of data were at 350, 250, and 200°C. Another five sets were for $Fe_{80.7}P_{14.5}Si_{4.8}$. The range of hardness was 5–8.7 GPa.

Friedrichs and Neuhäuser [160] studied structural relaxation of $Fe_{40}Ni_{40}B_{20}$ (VITROVAC alloy 0040), $Co_{66}Fe_4(MoSiB)_{30}$ (VITROVAC alloy 6025) optimized for near-zero magnetostriction, and $Ni_{78}B_{14}Si_8$ (VITROVAC alloy 0080) in a newly designed stress-free dilatometer. The data were analyzed by a spectrum of activation energies.

C. Enthalpy and Specific Heat

Kupicar et al. [161] measured specific heat and thermal diffusivity of $Pd_{82}Si_{18}$ and $Fe_{85}B_{15}$ glassy systems by using a heat pulse method. For $Fe_{85}B_{15}$, at 0.5°C/min, both specific heat and thermal diffusivity increased about 30% between 350 and 525 K but there was a sudden drop of about 7% at 430 K. For $Pd_{82}Si_{18}$ the sample was annealed at 540 K for 12 hr, cooled down to 280 K in 3 hr and then the temperature was maintained at 280 K for 2 hr, during which both specific heat and thermal diffusivity practically did not change. Then the temperature was raised to 300 K in 30 min maintained 300 K to do the measurements. Both specific heat and thermal diffusivity increased first and then decreased gradually to a value lower than those at 280 K.

Suzuki et al. [162] suggested that the kinetics for enthalpy relaxation during isothermal annealing could be analyzed based on the assumption that the rate was proportional to the free-energy difference from the equilibirium state. The data were for $Te_{85}Ge_{15}$, and the agreement was good for isothermal anneals of as-quenched specimens at 363, 377, 386, and 392 K.

Ding and He [163] studied enthalpy relaxation in $Pd_{77.5}Si_{16.5}Cu_6$ by annealing the as-quenched sample at 563 K for various times. The annealed specimens were heated in the DSC using five different heating rates 40, 50, 60, 80, and 100°C/min and the temperature of an enthalpy relaxation peak near T_g was noted. Kissinger plots gave apparent activation energies of 470 kJ/mole at short times (2 hr), increased to 574 kJ/mole (8 hr), and then decreased to 422 kJ/mole at long times (64 hr).

D. Curie Temperature and Magnetic Properties

Gibbs and Evetts [164] measured the coercive field of $Fe_{40}Ni_{40}B_{20}$ (VITROVAC alloy 0040) during isothermal annealing at various temperatures (97–375°C) after cold-rolling with and without preannealing. The coercive field decreased linearly with the logarithm of time. The data were analyzed by a distribution of activation energies.

Leake et al. [165] reported on the Curie temperature T_c of $Fe_{80}B_{20}$ (METGLAS alloy 2605) after a two-stage annealing treatment, first at 593 K for a time t_1 (14–

180 min) and then at 623 K for a time t_2. For a long t_1 such as 180 min, T_c decreased with t_2 initially (6°C in 10 min) and then much slower with a sign of leveling off to a constant value. However, for a short t_1 (14–45 min), T_c decreased slightly with t_2 (1–2°C in 1–2 min), passing through a minimum and then increased to approach a constant value which was the same for all t_1. DSC heating rate was 100°C/min for all annealing temperatures and times. Similar behavior was reported by Boesch et al. [166] for the change of refractive index of B_2O_3 after a two-stage annealing at 498.5 K and 532.7 K. These behaviors were explained by a model involving an activation energy spectrum.

Gibbs et al. [167] preannealed $Fe_{40}Ni_{40}B_{20}$ (VAC alloy 0040) at 438 K for 2 hr and then annealed at 475 K isothermally. The coercive field H_c (A/m) stayed at 1550 for 3 min during the isothermal anneal and then decreased almost linearly with log(t) to 1150 after 80 min. Preannealing at 460 K for 2 hr and then annealing at 550 K isothermally, H_c changed from 1100 at 1 min to 350 at 150 min, again almost linearly with log(t). The activation energy spectrum model was used to understand the behavior.

Brüning et al. [168] studied the λ transition at the Curie temperature of $(Fe_{1-x}Ni_x)_{80}B_{10}Si_{10}$ as a function of x. DSC scans of as-quenched samples at 40°C/min showed a ΔC_p of 5.6–7.8 J/mole-K independent of x(0–1). Preannealing by heating the as-quenched sample at 40°C/min to a temperature 100°C above the λ transition and then cooling at 320°C/min to room temperature could reduce the ΔC_p to zero. However, the ΔC_p for the λ transition, 1–3 J/mole K, increased with x and also with the temperature T_a (450–600 K) for the second preannealing (1 hr) except when the sample was crystallized during annealing. A 3% crystallization (mostly on the surface) reduced the ΔC_p for λ transition by more than 30%. Removal of the crystallized surface layer restored the ΔC_p somewhat. A good correlation was found between the ΔC_p for λ transition and T_a/T_x. Kissinger plots of the λ transition temperature with heating rate (10–160°C/min) gave an activation energy in the range of 1.35–1.57 eV/atom.

Balasubramanian et al. [169] annealed $Fe_{65}B_{27}Cr_8$ metallic glass (40 μm thick, 2 mm wide) 1 hr at various temperatures between 100 to 350°C. Stress relaxation (annealed in a quartz tube of 8 mm id and the final radius of curvature measured), fracture strain (by compressing a U-shaped length) and the uniaxial anisotropy energy $K_u = H_k M_s/2$ (where M_s is saturation magnetization and $H_k = (H_h - H_e)/2$ is the difference in resonance fields along the hard and easy axes respectively measured by the ferromagnetic resonance technique at 9.5 GHz using an x-band Varian E-12 ESR spectrometer) were determined. K_u varied linearly with annealing temperature from 2.71×10^{-5} J/m^3 for the as-quenched material to 1.64×10^{-5} J/m^3 after 350°C annealing. In the same range fracture strain changed from 6% as-quenched to 3.7% after 100°C annealing and 1.86% after 350°C annealing. Stress relaxed 33% after 100°C annealing and 67% after 350°C annealing. No quantitative relation among the three quantities was advanced. A comparison between the K_u relaxation and anelasticity was made by Egami and Jagielinski [170].

Bourrous and Kronmüller [171] measured the initial susceptibility of $Co_{58}B_{16}Si_{11}Ni_{10}Fe_5$ ribbons (20 μm × 1.8 mm cross section) during isothermal annealing at three different temperatures. The data were analyzed by using a two-level, two-configuration model, and a spectrum of activation energies.

E. Electrical Resistivity

Komatsu et al. [172] reported that resistivity changes after preannealing at a higher temperature could fit into the following kinetic law:

$$\frac{P - P_\infty}{P_0 - P_\infty} = \exp\left[-\left(\frac{t}{\tau_m}\right)^n\right] \tag{17}$$

where P(P is at any time t, P_0 is at $t = 0$, P_∞ is at $t = \infty$) is any property such as the resistivity τ_m is relaxation time and n is a constant. Amorphous alloys of $(Co_{1-x}Fe_x)_{75}Si_{10}B_{15}$ ($x = 0.067, 0.1, 0.25$, and 0.5) and $(Co_{0.75}Fe_{0.25})_{100-x}(Si_{0.4}B_{0.6})_x$ ($x = 20$, 25, and 30) were examined. The value n was about 0.4 and τ_m obeyed an Arrhenius equation with an activation energy of 220 kJ/mole.

Antonione et al. [173] measured electric resistivity of $Fe_{81}Si_xB_{19-x}$ ($x = 3$, 7.5, and 10). At 2°C/min the resistivity increased linearly with temperature (about 2% per 100°C) until crystallization, at which point it decreased abruptly. Resistivity decreased about 0.5% when annealed isothermally at 309°C. The results were analyzed by using an activation energy spectrum.

Hygate and Gibbs [174] measured isothermal resistance changes at 253°C for $Pd_{82}Si_{18}$, $Pd_{81}V_1Si_{18}$ and $Pd_{80}V_2Si_{18}$ amorphous alloys. After 10^4 s, the fractional resistance changes were -8.70×10^{-3}, -2.06×10^{-3}, and 1.67×10^{-3} respectively. Pressurized the as-received materials to 2 kbar caused the following resistance changes: -2.6×10^{-4}, 0.0, and 0.0 respectively.

F. Positron Lifetime

De Vries et al. [175] measured position lifetime in $Fe_{40}Ni_{40}B_{20}$ (VITROVAC alloy 0040) after various isothermal anneals between 475 and 600 K. The range of lifetime was 144.5–147.5 (±0.6) ps. The results seemed to show an earlier onset of crystallization than indicated by other techniques such as DSC, elastic moduli or electric resistivity. They agree with those of Tanigawa et al. [256], who attributed them to surface crystallization.

G. Hydrogen Solubility

Lee and Lee [176] studied the effect of annealing of $Pd_{80}Si_{20}$ on hydrogen solubility. The solubility at 333 K under 1 atm H_2 decreased about 1.2% after 30 min annealing at 425 K, 2.8% if annealed at 473 K, 5.5% at 531 K, 8.1% at 573 K, and 8.4% at 597 K. As a function of time, the decrease of solubility after isothermal annealing at 473 K was 2% at 30 m, 4% at 100 m, 6% at 10^3 m, and 7% at 10^4 m. These numbers for isothermal annealing at 573 K were 3% at 5 m, 5% at 10, and 9% at 60–300 m. After the onset of crystallization at about 500 m, the solubility after 10^3 m was reduced to 86% of the as-quenched value. Hydrogen evolution during heating at 3°C/min of as-quenched specimens after 333 K charging had a peak at 305 K. This peak was shifted to lower temperatures for annealed specimens. Hence annealing removes deep energy sites for hydrogen resulting in a lower activation energy for hydrogen removal.

H. Models and Mechanisms

Van den Beukel and Radelaar [35] proposed a model based on the free-volume theory of liquids and the solid-state diffusion mechanism. Relaxation toward equilibrium was assumed to be first order with a single relaxation time. Later when the model was compared with experimental results (DSC) by Majewsak-Glabus et al. [177] it was found necessary to assume a spectrum of activation energies for the relaxation process (a Gaussian distribution). A spectrum of activation energies (AES) was used also by Gibbs et al. [178] to describe the change of coercive field with time upon annealing $Fe_{40}Ni_{40}B_{20}$ at 607 K after cold rolling and preannealing at other temperatures (data of Gibbs [179]). AES was used by Gibbs et al. [178] to the data of Greer and Spaepen [180] on the "crossover" effect. Van den Beukel et al. [33] used AES to explain the length and modulus changes of $Fe_{40}Ni_{40}B_{20}$ during annealing. The data were analyzed more extensively in 1986 by Van den Beukel [181], who used a box distribution of activation energies. Gibbs [182] analyzed the data of Mulder et al. [34] of sound velocity using the AES model. Using the same model Gibbs and Sinning [183] analyzed the dilatometric data of Van den Beukel et al. [33] and those of others on $Fe_{40}Ni_{40}B_{20}$ as well as the length changes of $Ni_{64}Zr_{36}$ and $Cu_{66}Ti_{34}$ metal-metal glasses. Leusink and Van den Beukel [184] did elastic strain relaxation of $Fe_{40}Ni_{40}B_{20}$ and analyzed their data using the AES model provided that the relaxation time is taken inversely proportional to the flow defect concentration (whatever that is).

Perez [185,186] proposed as the basic mechanism for deformation the nucleation of shear microdomains not unlike a dislocation loop of the Volterra or Somigliana type. The nucleation event was considered heterogeneous rather than homogeneous and could be thermally activated. The free energy of activation has a maximum at the critical radius of the nucleus. After nucleation, the microdomain can grow in size. Let τ_1 be the mean time for nucleation and τ_2 be the mean time for growth; equations were derived relating the creep rate to these times. The expansion of the microdomain was considered as a diffusion process. Anelastic behavior comes about because the activated sites for nucleation may decrease upon removal of the applied stress.

Egami [187] attempted to define a local fictive temperature and suggest the idea of a distribution of fictive temperatures. It appears to be another attempt to force macroscopic concepts to have microscopic implications such as the famous free volume which, according to Egami, now may have different sizes and shapes.

Stanley and Cost [188] analyzed the so-called crossover effect by using a log-normal distribution of relaxation times and compared with four discrete relaxation times. The difference was less than 0.5% except at long times where it could reach a maximum of 5%.

X. NEW ALLOYS AND PROCESSES

A. Amorphous Al Alloys

In the search for Al-rich icosahedral quasicrystals, a new class of Al-rich metallic glasses was discovered (He et al. [189], He and Poon [190, 191], and Inoue et al. [192]). The alloys were made by arc-melting of elements in Ar and melt spinning

Table 2 Some Mechanical Properties of Amorphous Al-Based Alloys

	Young's modulus (GPa)	Tensile strength $\varepsilon = 1.3 \times 10^{-4}$/s (MPa)	T_x (°C)
$Al_{90}Ce_5Fe_5$	66	940	250
$Al_{87}Fe_{8.7}Gd_{4.3}$	53	835	310
190°C, 5 min anneal	43	450	
$Al_{87}Fe_{6.7}Gd_{6.3}$	58	810	
190°C, 5 min anneal	46	450	
$Al_{87}Fe_{6.7}Ce_{6.3}$	59	670	300
$Al_{87}Fe_{7.7}Gd_{5.3}$	51	850	
$Al_{87}Ni_{8.7}Ce_{4.3}$	46	500	300
$Al_{87}Ni_{8.7}Y_{4.3}$	50	880	260
$Al_{87}Co_{8.7}Ce_{4.3}$	63	790	300
$AL_{87}Co_{8.7}Y_{4.3}$	55	740	260
$Al_{87}Rh_{8.7}Ce_{4.3}$	63	730	290

in He atm using a 20-cm-diameter Cu wheel with a circumferential velocity of 40 m/s. Some properties are listed in Table 2.

Inoue et al. [192] reported Al-Y-Ni amorphous alloys having tensile strengths above 980 MPa. For $Al_{95-x}Y_xNi_5$ and $Al_{90-x}Y_xNi_{10}$ alloys, T_x increased from 400 K for $x = 2$ to 650 K for $x = 15$. For $Al_{95-x}Y_5Ni_x$, T_x increased from 490 K for $x = 10$ to 620 K for $x = 17$. For $Al_{90-x}Y_{10}Ni_x$, T_x increased from 520 K for $x = 3$ to 710 K at $x = 20$. Some mechanical properties are shown in Table 2.

B. Some Processing Techniques

Morris [193] did compaction of amorphous $Ni_{77.5}B_{15}Si_{7.5}$ and $Ni_{70}B_{20}Si_{10}$ ribbons and powders. Good strengths of the final product were obtained by forging the powders. The powder was prepared by using a slotted melt-spin roller or by grinding the embrittled ribbon. Forging preforms were made by hot pressing the powder to 1.5 GPa at 500°C for 15 min. Forging reduced the height/diameter ratio from 0.5 (preform) to 0.1. The best results were obtained by loading the powder into a cylinder with two pistons made of high-speed tool steel of Nimonic 105 alloy, pressing at 0.6 GPa and 700°C until the cylinder was bulged, and quenching in water.

Sato [194] constructed a water atomization unit to make amorphous powders. Electrolytic iron, ferroboron, ferrophosphorous, metallic silicon, and vacuum-melted high-carbon iron were used as raw materials to melt mixtures of $Fe_{80}P_{13}C_7$ and $Fe_{75}B_{15}Si_{10}$ compositions in an induction furnace under Ar in a MgO crucible at 1673 and 1723 K respectively. The molten metal stream flowed downward freely through a 3-mm orifice to meet a 12-MPa water jet at a flow rate of 4.67 kg/s. The particles were of irregular shape and had a mean size of 60 μm in a range of 50–150 μm. The transverse rupture strength was 9.9 MPa for $Fe_{80}P_{13}C_7$ and 9.3 for $Fe_{75}B_{15}Si_{10}$.

Hays and Naugle [195] reported the structure and properties of $Cu_{60}Zr_{40}$ arc-melted and injected at high velocities into a circular channel (0.5–1 mm dia) by a gas/vacuum pressure gradient. DSC analysis yielded 10.52 cal/g as the heat of crystallization with T_x = 513°C, showing that the material was amorphous. Microhardness measurements showed 375 kg/mm^2 at the center of the sample and 425 at the edges.

Otooni [196] consolidated $W_{67}Ni_{23}C_6B_4$ particulates at 732 K and under 3.5 MPa uniaxial pressure. The product had high yield strength and ductility (no data). The strengthening was attributed to plastic deformation at boundaries between the particulates.

Wang and Yan [197] deposited Ti on bulk Ni and annealed in vacuum at 300°C for 24 hr. This process resulted in an amorphous layer of Ni-Ti.

Schneeweiss [198] found in a rapidly quenched $Fe_{83}Si_{17}$ alloy produced by a double-roller technique the structure was partially amorphous and partially crystalline. Powders made by mechanical crushing of the ribbons in an agate mortar showed 6% amorphous in the 54-μm powders and 17% amorphous in the 29-μm powders (obtained by Mössbauer spectra and model fitting). Amorphization due to plastic deformation was thought to originate from regions with a high density of defects.

C. Amorphization by Energetic Particles

Ahmed and Potter [199] implanted 180-keV Al^+ to fluences 4 × 10^{18} ions/cm^2 onto Ni substrates. Depth profile showed a maximum of 75 at% Al at 160 nm. The penetration depth (below 5 at%) was 520 nm. The solubility of Al in Ni was extended to 30 at% Al from 10 at% under equilibrium conditions. An amorphous phase was observed at implanted compositions exceeding 60 at% Al, consistent with $NiAl_3$ rendered amorphous by the radiation damage accompanying implantation.

Ossi [200] reviewed previous work (100 references) on the amorphization of crystalline metals or alloys after bombardment with energetic ions. He collected about 7 alloy systems which became amorphous through ion implantation and about 12 alloy systems through ion mixing. The segregation charge transfer model for the amorphization process was discussed with examples. The formation of intermediate quasi-crystalline phases was discussed also.

Seidel et al. [201] implanted 200-keV Mn^+ into Al thin films and single crystals. The concentration of Mn needed for complete amorphization was about 12 at% at liquid N_2 temperature and 19 at% at room temperature. Denanot et al. [257] implanted 160-keV Ni^+ into Al thin foils. For fluences between 8 × 10^{16} and 2 × 10^{17} Ni^+/cm^2 under less than 1 mA/cm^2 ion current, an amorphous phase formed. Under high current (20 mA/cm^2), a fcc $Al_{85}Ni_{15}$ precipitate appeared in a supersaturated $Al_{95}Ni_5$ solid solution.

Thome et al. [202] implanted 70-keV B at 15 K onto a Ni [100] surface and obtained a 70% amorphous $Ni_{90}B_{10}$ surface layer. Annealing of this layer led to the formation of a multiphased alloy in which Ni_3B would become amorphous by B postirradiation.

Delage et al. [203] implanted 250-keV Ni^+ ions into crystalline Ni_xTi_{100-x} (x = 30, 43, 50, 56, and 68, crystallized from amorphous ribbons). Amorphization could be complete after 10^{15} ions/cm^2 at 77 K for all x. The temperature of implantation could be higher (up to 500 K) if x was between 43 and 56.

Aaen Anderson et al. [204] evaporated sequentially Cu and Nb multilayers (each layer less than 10 nm) of a total of 80 nm thick and ion mixed by 500-keV Xe^+ ions with doses of 2×10^{16} ions/cm^2. Amorphous structures were obtained between 20 and 60 at% Cu and below 300 K.

Bottiger et al. [205] prepared alternating layers of Ni-Ti, Ni-Nb, Fe-Zr, and Ni-Zr paris (each layer about 10 nm thick) of about 80 nm thick and ion mixed with 500-keV Ar^+ and Xe^+ ions with fluences 7×10^{16} and 2×10^{16} ions/cm^2, respectively. The glass-forming range was found to be Ni-Ti, 35–70% Ni, 300–600 K; Ni-Nb, 40–80% Ni, 300–800 K; Fe-Zr, 15–47% Zr at room temperature and 10–80% Zr at 30 K; Co-Zr, 17–65% Zr at *RT* and 13–87% Zr at 30 K; Ni-Zr, 10–80% Zr at *RT* and 7–80% Zr at 77 K.

Lin et al. [206] studied in situ the effects of 1 MeV electron irradiation on crystalline Ge/Al and Si/Al bilayer specimens at 10 K by high-voltage electron microscopy. A crystalline-to-amorphous transition took place at the Ge/Al interface only if the Ge layer faced the electron beam, while the amorphization occurred in Si/Al regardless of the beam direction. Diffuse rings first appeared at a fluence of 3×10^{23}/cm^2 (about 30, 19, or 18 displacements per atom in Ge, Si, and Al respectively) and the amorphous volume fraction increased with fluence.

D. Amorphization by Mechanical Alloying

Koch et al. [207] made amorphous $Ni_{60}Nb_{40}$ by mechanical alloying of elemental powders in a Spex Mixer/Mill Model 8000. ASI 52100 steel balls (9.5 mm dia) were used in a cylindrical hardened tool steel vial (76 mm × 57 mm dia) with a ball to powder ratio of 30 g to 10 g. About 14 hr were needed for the milling process.

Schwarz et al. [208] mechanically alloyed the elements at 240 K to form Ni_xTi_{100-x} for $28 < x < 72$. Outside this range, it was a mixture of amorphous phase and the terminal solution of the major element. More recently for $x = 50$, Cocco et al. [258] showed amorphization in 30 hr starting from pure elements. If the pure elements were milled separately first for 25 hr each (Warren-Averbach analysis showed saturation of crystallite size, 18 nm for Ni, 40 nm for Ti, and of strain content, 6×10^{-4} for both, at about 5 hr and then mixed and subjected to mechanical alloying, amorphization still required about 30 hr.

Schwarz and Koch [209] produced amorphous powders of $Ti_{68}Ni_{32}$ and $Nb_{55}Ni_{45}$ by mechanical alloying of pure elements or crystalline intermetallics $NiTi_2$ and Ni_9Nb_{11} respectively. For both alloys, the peak temperature increase was about 38°C over the average temperature of −40°C. The time needed was about 14 hr for all situations.

Politis and Johnson [210] made amorphous $Ti_{100-x}Cu_x$ (x = 10 to 87) by mechanical alloying of the elemental powders in a WC (+6% Co) vial. WC balls of 8 and 10 mm dia were used with the ball-to-powder weight ratio varied from 3 to 1 to 10 to 1. The ball mill was sealed in argon with an O-ring made of Viton.

Thompson and Politis [211] mixed elemental powders of Ti and Pd and loaded with WC balls in a WC vial (ball/powder = 5:1) under Ar at room temperature. Amorphous alloys $Pd_{100-x}Ti_x$ with x = 42–85 were obtained in about 17 hr. Outside this range such as $x = 35$, a crystalline component persisted after 23 hr. Temperature rise during mechanical alloying was estimated to be a few tens of degrees and no fast-diffusing species were known.

Kim and Koch [212] made amorphous $Nb_{75}Sn_{25}$ by mechanical alloying of the elemental powders. X-ray diffraction showed the formation of intermetallic Nb_3Sn before it was transformed into the amorphous phase. More recently Tiainen and Schwarz [213] also mechanically alloyed pure elements into $Ni_{100-x}Sn_x$ and found that complete amorphization was possible only for $x = 25$ in the range of 20–40 studied. X-ray diffraction showed the formation of intermetallic Ni_3Sn_2 first and then the formation of amorphous $Ni_{75}Sn_{25}$ at the expense of Ni_3Sn_2 (for $x < 25$) or Ni ($x > 25$) or both ($x = 25$). Amorphous $Ni_{75}Sn_{25}$ powder hot-pressed at 523 K and 0.86 GPa had a density 98% of crystalline Ni_3Sn and a DPH of 750 kg/mm^2. Amorphous $Ni_{75}Sn_{25}$ crystallized polymorphously into Ni_3Sn at 608 K.

Lee et al. [214] mechanically alloyed mixtures of intermetallic compounds in a Spex Mixer/Mill Model 8000 using a hardened tool steel vial and 440C martensitic stainless steel balls. The vial was sealed with an elastomer O-ring in Ar. The ball to powder weight ratio was 10 to 1. The results are shown in Table 3. There was no apparent relation between the ability of amorphization and the estimated heat of mixing of the elements or the free-energy difference between the amorphous phase (higher) and the intermetallic compounds.

Enzo et al. [215] investigated amorphous $Ni_{50}Ti_{50}$ produced by mechanical alloying using x-ray diffraction and extended x-ray absorption fine structure and found evidence of progressive increase of short-range ordering with the time of milling.

Ennas et al. [216] mechanically alloyed pure iron (five 9's) and Zr (three 9's), 5–40 μm powders in the proportion of $FeZr_2$ and FeZr in a conventional planetary ball mill (ball-to-powder weight ratio: 10 to 1) under pure Ar. It took about 23 hr of milling to make the $FeZr_2$ alloy amorphous (17 hr for Fe_2Zr). They [217] also mixed elements in the compositions $Pd_{75}Si_{25}$ and $Pd_{80}Si_{20}$ and mechanically alloyed them by ball milling. X-ray diffraction showed that the disappearance of Si peaks was accompanied by both the appearance of Pd_3Si peaks and the reduction of Pd peaks. Then the broadening of both Pd_3Si peaks and Pd peaks merged them into a continuous halo.

Table 3 Amorphization by Mechanical Alloying of Mixtures of Intermetallic Compounds

Average composition	Mixture of intermetallics	Amorphization
$Cr_{63}Ti_{37}$	$Cr_{60}Ti_{40}$(C15) + $Cr_{65}Ti_{35}$(C36)	Partial
$Mn_{62}Ti_{38}$	Mn_2Ti + $Mn_{52}Ti_{48}$	Yes
$Cu_{43}Ti_{57}$	$CuTi_2$ + CuTi	Yes
$Fe_{60}Ti_{40}$	FeTi + Fe_2Ti	Yes
$Co_{60}Ti_{40}$	CoTi + Co_2Ti	Yes
$Ni_{60}Ti_{40}$	NiTi + Ni_3Ti	Yes
$Cu_{60}Zr_{40}$	$CuZr_2$ + $Cu_{51}Zr_{14}$	Yes
$Ni_{40}Zr_{60}$	$NiZr_2$ + $Ni_{11}Zr_9$	Yes
$Ni_{50}Zr_{50}$	$NiZr_2$ + $Ni_{11}Zr_9$	Yes
$Mn_{54}Si_{46}$	MnSi + Mn_5Si_3	No

Calka and Radlinski [218] prepared $Mg_{70}Zn_{30}$ by melting in a Ta crucible enclosed in a fused silica capsule under pure He. The alloy was crushed into small pieces and placed inside a planar-type ball mill cell at 200 rpm. It took about 58 hr to make the alloy amorphous.

Suzuki [219] made amorphous NiZr by mechanical alloying of elemental powders or crystalline NiZr. The first process proceeded directly without any intermediate steps such as compound formation. The second process seemed faster than the first. EXAFS showed more Ni-Ni and Zr-Zr pairs in mechanically alloyed materials than in rapidly quenched alloys from the melt. He also made amorphous Co_5Y by mechanical alloying crystalline Co_5Y. The magnetic properties at 4.2 K was identical to sputter-deposited amorphous Co_5Y. He also made a glass by mechanical alloying of 3 $BiFeO_3$ and $ZnFe_2O_4$ crystals.

E. Amorphization by Hydrogenation

Yeh et al. [220] made amorphous $H_{5.5}Zr_3Rh$ by hydrogenation (1–50 atm H_2) of the crystalline intermetallic compound Zr_3Rh at 200°C. Partial dehydrogenation was possible by heating the amorphous hydride at 180°C in 10^{-6} Torr and resulted in the formation of fcc ZrH_2 (approx.).

Aoki et al. [221] hydrogenated the laves phase of the type RNi_2 with R = Y, La, Ce, Pr, Sm, Gd, Tb, Dy, Ho, and Er and formed amorphous compound at 323 K and under 5 MPa (50 atm) hydrogen for about one day. Two examples H_3Ni_2Ce and H_3Ni_2Gd were described in more detail. However, hydrogenation of $NdNi_2$ did not form an amorphous structure but yielded a mixture of NdH_2 and $NdNi_5$ instead. The same mixture would result from crystallization of an amorphous structure formed by hydrogenation of amorphous Ni_2Nd.

Aoki et al. [222] started with a single laves phase of $CeFe_2$ prepared by arc-melting the elements and annealing at 873 K for five days. Pulverized crystalline samples (<100 mesh) were then reacted with hydrogen below 5 MPa pressure and at a temperature in the range 293–673 K for one day. Complete amorphization took place only below 500 K. The magnetization curve of a fully hydrogenated crystalline sample was identical with that of similarly hydrogenated amorphous film (dc sputtering) confirming the amorphization process.

Pareja et al. [223] made positron life time measurements on $Co_{58}B_{16}Si_{11}Ni_{10}Fe_5$, $Fe_{78}B_{13}Si_9$ (METGLAS alloy 2605S-2), and $Fe_{81}B_{13.5}Si_{3.5}C_2$ (METGLAS alloy 2605SC). For the first alloy, the positron lifetime spectrum has either a single component or is resolvable into two exponential terms depending on the production parameters of the ribbons. METGLAS alloy 2605S-2 showed a two-component spectrum and METGLAS alloy 2605SC exhibited a single-component spectrum. After hydrogen charging, the intensity of the long-lived component decreased and the life time of the short-lived component increased in the two-component spectrum. The single-component spectrum was not affected by hydrogen charging.

Aoki et al. [224] hydrogenated the laves phases of RFe_2 (R = Y, Sm, Gd, Tb, Dy, Ho, or Er) between 450 and 550 K and formed the amorphous alloys containing hydrogen. Heating above 550 K, the amorphous alloys decomposed into RH_2 and α-Fe.

Chung and Lee [225] arc melted the pure elements to form the intermetallic compounds LaNi, $LaNi_2$, and La_2Ni_7. They were crushed below 200 mesh and reacted with <60 atm H_2 at 200–573 K. Both $LaNi_2$ and $LaNi_{1.5}$ were transformed into amorphous state after hydrogenation at room temperature. La_2Ni_7 were transformed after 10 cycles of absorption and desorption of H_2. Hydrogenation of LaNi changed into a new orthorhombic structure.

F. Amorphization by Chemical Reaction

Newcomb and Tu [226] deposited a layer (about 0.2 mm thick) of Ni onto a thermally oxidized silicon substrate followed by a similar layer of Zr. The bilayer film was heated at 300°C for 1 hr in flowing He. A layer of amorphous $Ni_{60}Zr_{40}$ (about 25 nm thick) was formed between the two layers. Voids were seen at the Ni side of the Ni-NiZr interface showing that Ni was the dominant diffusing species. Other systems investigated include Rh/Si (Herd et al. [227]) La/Au (Schwarz and Johnson [228]), Zr/Co (Schroder et al. [229]) Ni/H_f (Van Rossum et al. [230]), and Ni/Zr (Clemens et al. [231]).

XI. MICROSTRUCTURAL STUDIES

A. Atomic Level Stresses

Vitek et al. [232] used molecular dynamics to examine atomic level stresses at high temperatures. They found that the atomic level pressure $\langle p^2 \rangle$ increased by a factor of 6 from 0 to 3000 K in a constant volume simulation of Fe glass (modified Johnson potential). Defining $\langle p_{\rm int}^2 \rangle = \langle p^2 \rangle - \langle p \rangle^2$, they found that $\langle p_{\rm int}^2 \rangle$ increased by a factor of 4 over the same temperature range. Similarly, the atomic level shear stress $\langle \tau^2 \rangle$ increased by a factor of 5. Cooling from the liquid state showed the change of pair distribution functions toward the glassy state. However, the glass transition was not very obvious.

B. Point Defects

Laakkonen and Nieminen [233] reported on a molecular dynamics simulation of point defects in amorphous structures (single-component L-J potential solid). The conclusion was that both the vacancy (one atom removed and then based on the cavity distribution) and the interstitial (one atom added and then based on the distribution of pressure centers) were unstable.

Similar conclusions were obtained by Chaki and Li [234] who used the same Lennard-Jones amorphous solid. A cube containing 685 atoms was first equilibrated thermally and then one of the atoms in the middle of a cube face was given additional kinetic energy so as to inject it into the cube. The Voronoi volume distribution was followed afterwards. Both vacancies and interstitials were produced by the energetic atom. Interstitials disappeared rapidly within a few picoseconds; vacancies took a longer time to disappear.

Experimentally Chaki and Li [235] irradiated METGLAS alloy MBF-20 ($Ni_{69.2}Si_{13.7}B_{7.9}Cr_{6.6}Fe_{2.6}$) with 70-MeV Ni^{+6} ions at a dose about $10^{16}/cm^2$. Surface swelling was shown but no cavities were found. Annealing for 3 hr at each 50°C increment between 500 and 800 K did not remove the swelling. The swelling was attributed to the fast disappearance of interstitials at the irradiated spot. The gradual disappearance of vacancies may recover the density to the value before irradiation but may not regain the original shape. Hence swelling remained without cavitation.

C. Dislocations

In 1985, I discussed [236] the question whether microscopic line defects such as dislocations can exist in amorphous metals. Gilman [237] suggested in 1968 that such defects may exist in vitreous silica. The line defect was envisioned to have slip vectors varying in both direction and magnitude along the line although they may have a well-defined average. The concept was modified by myself [14] after it was realized that the Gilman dislocation cannot move without leaving something behind so that the slipped area is really a Somigliana dislocation. The motion of dislocation or the expansion of slipped area is to take place by nucleation of small Volterra loops along the dislocation line or inside the slipped area. The Burgers vector can be defined by using embedded markers of lattice points and can have a distribution in both magnitude and direction. The stress field can be used also as identification. Experiments on shear bands, shear step distribution, reverse shear phenomena, chemical etching properties, and magnetic domains surrounding shear bands are consistent with models using dislocation distributions. Computer simulation by Chaudhari and his co-workers [238–241] showed the stability and mobility of such dislocations.

D. Amorphous Phase Separation

Piller and Haasen [242] reported that $F_{40}Ni_{40}B_{20}$ decomposed into two amorphous phases, $(FeNi)_3B$ and FeNiB upon annealing below 370°C by atom probe field ion microscopy. The kinetics of decomposition were faster than those of crystallization apparently because of the lack of a nucleation barrier. The diameter of decomposed regions was about 4 nm and the separation between them was about 20 nm. The growth of the decomposed regions (to 30–40 nm) seemed to cause embrittlement probably due to the B-rich phase. Crystallization showed two exothermic peaks. Kissinger plots gave activation energies of 353 and 370 kJ/mole. The ductile-brittle transition temperature was 200 ± 25°C for all annealed specimens which were brittle at room temperature such as those annealed 30 min at 370°C, 60 m at 350°C, or 120 min at 300°C. The ductile-brittle transition temperature was −200°C for as-quenched and −50°C for other annealed specimens (5 hr at 250°C).

Grüne et al. [243] discovered spatial variation of chemical composition of as-quenched $Ti_{50}Be_{40}Zr_{10}$ by FIM (field ion microscopy) atom probe to a depth of about 50 nm (250 atom layers). Both ±10 at% Ti and ±10 at% Be with a wavelength of about 28 nm (or 140 layers) were detected with a Ti peak corresponding to a Be valley, and vice versa. In these regions the Zr (10 at%) content was the same). However, there was another region corresponding to $Be_{50}Ti_{35}Zr_{15}$. Ruptured FIM specimens showed Be enrichment near the fracture surface. Annealing of $Ti_{50}Be_{40}Zr_{10}$ at 350°C caused the Ti concentration to increase in the Ti-rich region with a corresponding decrease of Be concentration. As a result, the volume fraction of Be-rich regions increased. Another region corresponding to (Zr, Ti)Be_2 was formed also. The annealing embrittlement of this alloy showed two stages, the first had a density increase and the second had this composition variation.

Yavari et al. [244] did DSC of $Pd_{46}Ni_{36}P_{18}$ and showed an exothermic peak at 610 K before T_X (640 K). X-ray diffraction revealed no trace of crystallinity for samples annealed 1 min at 612 and 618 K and some surface crystallization (amorphous after surface removal) for samples annealed 1 min at 630 K. DSC thermograms for samples annealed 1 min at 612 K showed no exotherminic peak at 610

K but two T_g's at 580 and 615 K. However, TEM showed nothing for this sample. For samples annealed 1 min at 630 K, TEM and STEM did show very fine scale granulation (1 nm). EDX peaks suggested that the darker and brighter zones were richer in P and Pd respectively. All these findings confirm the possibility of phase separation (both amorphous) within an amorphous structure.

XII. SUMMARY AND CONCLUSIONS

While many papers have been published in recent years attempting to understand the mechanical properties of amorphous alloys, the excitement is not as great as before. A few highlights are mentioned below. They are necessarily biased by my own interests and understandings.

1. Zolotukhin et al. [83] found some interesting damping behavior of cold-rolled $Cu_{50}Ti_{50}$ and $Ni_{78}B_{14}Si_8$ ribbons. An internal friction peak characteristic of cold work was shown. The peak height first increased with the amount of cold work and then decreased. Increasing the strain amplitude of vibration shifted the peak to lower temperatures. Increasing the frequency of vibration increased the peak height but did not change the peak position (temperature). Aging at room temperature first increased the peak height followed by a steady decrease. These measurements seem to have the potential of characterizing the defects created by deformation.

2. Zielinski and Ast [59] and Kimura et al. [60] dispersed WC particles (4–5 μm size) into an amorphous metal ($Ni_{78}B_{12}Si_{10}$) and found a large effect on Young's modulus (doubled for 18% volume fraction of particles) and yield stress (1/60 of Young's modulus independent of WC volume fraction). This phenomena of dispersion hardening is very interesting and may shed some light on the mechanism of plastic deformation. It was interpreted from both viscous flow and dislocation viewpoints [259]. More work should be done such as the temperature and strain rate dependences of stress, the use of particles of different strengths, sizes, and shapes, and the use of different amorphous matrices. The effect is also large enough to be technologically attractive.

3. The study of magnetic domain structures [71] around shear bands have the potential of revealing the defect structure inside the shear bands provided that a high-powered microscope such as the atomic force microscope can be used to reveal the atomic size domains which may help locate a line defect such as a dislocation.

4. Donnadieu et al. [73,248,249] deformed silica glass between 1000 and 1400°C (T_g = 1200°C) and observed birefringent zones in the TEM. These zones exhibit line-defect features such as bulging between pinning points. In situ observations are needed to completely examine their behaviors.

5. Donovan [57,58] did uniaxial compression, plain-strain compression and pure shear of $Ni_{40}Pd_{40}P_{20}$ and concluded that the material obeys the Mohr-Coulomb yield criterion with a normal stress coefficient of 0.113 ± 0.005. The lack of a pressure dependence of yield stress is inconsistent with the notion that plastic flow requires some free volume.

6. Bothe and Neuhäuser found simultaneous decrease of Young's modulus and internal friction of Metglas 2826A during some step-annealing program. This is inconsistent with the free volume as the only structural parameter. The same conclusion was reached by Vokert and Spaepen [80] who, after annealing

$Ni_{40}Pd_{40}P_{19}Si_1$ at $T_a < T$, measured its viscosity at T and found the viscosity decreased first, through a minimum and then increased to approach the value at T.

7. Several people measured properties during continuous heating of a specimen. Among them Anderson and Lord [76], Taub [90], Ma and He [79], Zheng [87], and Bhatti and Canton [89] made length measurements under a constant load. Wang et al. [88] noted the viscous flow start temperature. Cao and Li [67] watched the onset of embrittlement. Yavari [102] determined the fracture strain. Antonione et al. [173] measured the electric resistivity. Huang and Li [260] observed the Curie temperature at high heating rates and also studied stress relief and the change of magnetic properties after rapid Joule heating [261]. Rapid heating (~200°C/s) annealing is a technique for improving magnetic properties without annealing embrittlement. More information on the change of properties and on the behavior of structural relaxation under very fast heating conditions is still needed.

8. Negative creep [91] is the shrinkage of a specimen under a tensile load. It is not due to volume contraction because there is no shrinkage without the tensile load. A similar effect is the negative stress relaxation [54]. It was interpreted [93] by a mechanochemical spinodal decomposition. The internal stresses resulting from such decomposition could be the cause for annealing embrittlement. Both alloy weakening and strengthening may have their origin in affecting such decomposition. Experimentally, phase separation (both amorphous) within a homogeneous phase has been observed [242–244]. Yamasaki et al. [103] used a Guinier plot followed by the Fankuchen analysis and suggested that the embrittlement might be caused by spinodal decomposition resulting in the development of microscopic regions of 4–10 (average 7) nm throughout the specimen (confirmed by electropolishing). This series of observations and interpretations may well pave the way for our understanding of aging brittleness.

9. On the other hand, several investigators suggested surface crystallization as a possible cause of annealing embrittlement. Liebermann et al. [104] extrapolated their crystallite size data to the onset of embrittlement and found the size to be about 10 nm which would escape detection. Cao and Li [67] based on the activation energy and Janot et al. [105] based on the positron lifetime also indicated a correlation between precrystallization and embrittlement. Guo and Boer [106] used EXAFS and implied that crystallization began at the surface. Wu and Spaepen [99,100] suspected that embrittlement may be caused by only 1% crystallization.

10. The overload effect [137] observed in fatigue crack propagation suggests a strong interaction between the plastic zone and the crack tip, implying the existence of long-range stress field of the plastic zone. It seems in direct contradiction to the expectations of the free volume theory which would suggest that it is easier for the crack to extend into a larger plastic zone with more free volumes created by overloading.

11. Mechanical alloying is a new technique [207] for making amorphous alloys. The collected references [207–219] are only a part of the published literature in this area. It deserves a lot more studies on the micromechanisms as well as the potentials and limitations of such mixing process.

12. Finally I want to make a plea for extra effort in the study of micromechanisms of deformation of amorphous metals. Like people in oxide glasses and polymers we have fallen into the "free-volume" trap and it is going to take some time to get out of it. We need to know the atomic configurations and motions

involved in the deformation processes. The existence of dislocations is still a distinct possibility, but we need new techniques of detection and observation, new ways of studying shear bands and new ideas for understanding the nucleation and propagation of localized deformation.

ACKNOWLEDGMENT

The work is suported by US DOE, Basic Energy Sciences, through DE-FG02-85-ER45201. The skillful typing of Fremonia Wallace is appreciated.

REFERENCES

1. W. Klement IV, R. H. Willens, and P. Duwez, *Nature* (*London*), *187*, 869 (1960).
2. C. A. Pampillo, *J. Mater. Sci.*, *10*, 1194 (1975).
3. J. C. M. Li, in *Treatise on Materials Science and Technology*, Vol. 20, *Ultra-Rapid Quenching of Liquid Alloys*, edited by H. Herman (Academic Press, New York, 1982), p. 325.
4. H. Kimura and T. Masumoto, in *Amorphous Metallic Alloys*, edited by F. E. Luborsky (Butterworths, London, 1983), p. 187.
5. F. Spaepen and A. I. Taub, in Ref. 4, p. 231.
6. P. Chaudhari, F. Spaepen, and P. J. Steinhardt, in *Glassy Metals II*, *Atomic Structure and Dynamics*, *Electronic Structure*, *Magnetic Properties*, edited by H. Beck and H. J. Grüntherodt (Springer-Verlag, New York, 1983), p. 127.
7. H.-U. Künzi, in Ref. 6, p. 169.
8. J. C. M. Li, in *Chemistry and Physics of Rapidly Solidified Materials*, Ed. by B. J. Berkowitz and R. O. Scattergood (Metall. Soc., 1983), p. 173.
9. A. M. Glezer, B. V. Molotilov, and O. L. Utevskaya, *Phys. Metals*, *5*(1), 46 (1984).
10. H. Koizumi and T. Ninomiya, *J. Phys. Soc. Japan*, *49*, 1022 (1980).
11. T. Ninomiya and H. Koizumi, *Proc. Fourth Int. Conf. on Rapidly Quenched Metals*, edited by T. Masumoto and K. Suzuki (Japan Institute of Metals, 1982), p. 259.
12. P. R. Rao and V. M. Radhakrishnan, in *Metallic Glasses*: *Production*, *Properties and Applications*, edited by T. R. Anantharaman (Trans. Tech. Publ., 1984), p. 225.
13. J. C. Gibeling and W. D. Nix, *Scripta Metall.*, *12*, 919 (1978).
14. J. C. M. Li, in *Metallic Glasses* (ASM, 1978), p. 224.
15. T. Egami and V. Vitek, *J. Non-Cryst. Solids*, *61*, *62*, 449 (1984).
16. S. Takeuchi, *Solid State Physics* (*Japan.*), *20*, 123 (1985).
17. Lance A. Davis, in *Glass—Current Issues*, edited by A. F. Wright and J. Dupuy (Martinus-Nijhoff, Dordrecht, 1985), p. 94.
18. A. S. Argon, *Rapidly Quenched Metals*, *Vol. II*, edited by S. Steeb and H. Warlimont (North-Holland Elsevier, Amsterdam, 1985), p. 1325.
19. J. Megusar, A. S. Argon, and N. J. Grant, *Mater. Sci. Eng.*, *38*, 63 (1979); see also *Rapidly Solidified Amorphous and Crystalline Alloys*, edited by B. H. Kear et al. (North-Holland, Amsterdam, 1982), p. 283, and Ref. 2, Vol. II, p. 1411.
20. F. Lancon, L. Billard, and A. Chamberod, in Ref. 18, Vol. II, p. 1337.
21. D. Deng and B. Lu, *Scripta Metall.*, *17*, 515 (1983).
22. C. C. Koch, in *Mechanical Behavior of Rapidly Solidified Materials* (Ed. by S. M. L. Sastry and Bruce A. MacDonald (The Metallurgical Society, 1986), p. 41.
23. L. A. Davis and V. R. V. Ramanan, *Key Eng. Mater.*, *13–15*, 733 (1987).
24. P. Chaudhari and D. Turnbull, *Science*, *199*, 11 (1978).
25. Pol Duwez, *Proc. Indian Acad. Sci.*, *C2*, Pt 1, 117 (1979).
26. H. Warlimont, *Phys. Technol.*, *11*, 28 (1980).

27. R. W. Cahn, *Contemp. Phys.*, *21*, 43 (1980).
28. T. Jagielinski, T. Walecki, W. Dmowski, and H. Matyja, in *Metallic Glasses: Science and Technology*, Vol. II, edited by C. Hargitai, I. Bakonyi, and T. Kemeny (Central Res. Inst. Phys., Budapest, 1980), p. 49.
29. S. Ishio, Y. Sato, T. Ikeda, and M. Takahashi, *J. Non-Cryst. Solids*, *61*, *62*, 995 (1984).
30. K. Fukamichi, M. Kikuchi, and T. Masumoto, *J. Non-Cryst. Solids*, *61*, *62*, 961 (1984).
31. G. Hausch and E. Török, in Ref. 18, p. 1341.
32. K. Bothe and H. Neuhäuser, in Ref. 18, Vol. I, p. 735.
33. A. Van den Beukel, S. Van der Zwaag, and A. L. Mulder, *Acta Metall.*, *32*, 1895 (1984).
34. A. L. Mulder, S. Van der Zwaag, E. Huizer, and A. Van den Beukel, *Scripta Metall.*, *18*, 515 (1984).
35. A. Van den Beukel and S. Radelaar, *Acta Metall.*, *31*, 419 (1983).
36. G. Dietz and W. Heinen, in Ref. 18, Vol. I, p. 727.
37. K. Bothe, in Ref. 18, Vol. I, p. 731.
38. K. Bothe and H. Neuhäuser, *Scripta Metall.*, *16*, 1053 (1982).
39. T. J. Taylor, Y. P. Khanna, R. Kumar, and H. H. Liebermann, *J. Mater. Sci. Letters*, *8*, 1165 (1989).
40. G. Knuyt, L. De Shepper, and L. M. Stals, *Mater. Sci. Eng.*, *98*, 527 (1988).
41. D. Weaire, M. F. Ashby, J. Logan, and M. J. Weins, *Acta Metall.*, *19*, 779 (1971).
42. K. J. Hettwer and F. Haessner, in Ref. 28, p. 37.
43. K. Bothe and H. Neuhäuser, *J. Non-Cryst. Solids*, *56*, 279 (1983).
44. G. Posgay, F. J. Kedves, B. Albert, S. Kiss, and I. Z. Harangozo, in Ref. 18.
45. S. Sinnema, J. Groote Schaarsberg, E. Pennings, P. M. Bronsveld, and J. Th. M. De Hosson, in Ref. 18, Vol. I, p. 719.
46. H. S. Chen and N. Morito, in Ref. 18, p. 715.
47. H. S. Chen and N. Morito, *J. Non-Cryst. Solids*, *72*, 287 (1985).
48. D. Deng and A. S. Argon, in Ref. 18, Vol. I, p. 771.
49. E. Bonetti, *Phil. Mag.*, *B56*, 185 (1987).
50. Yizhen He and Xiao-Guang Li, *Phys Stat. Sol.* (*a*), *99*, 115 (1987).
51. H. R. Sinning and F. Haessner, *J. Non-Cryst. Solids*, *93*, 53 (1987).
52. H. R. Sinning, and F. Haessner, *Z. Phys. Chem. Neue Folge*, *156*, S115 (1988).
53. L. Yue and Y. He, *J. Non-Cryst. Solids*, *105*, 33 (1988).
54. E. Bonetti, E. G. Campari, L. Ferrari, and G. Russo, *J. Phys. F: Metal. Phys.*, *18*, 1351 (1988).
55. H. R. Sinning, *J. Non-Cryst. Solids*, *110*, 195 (1989).
56. H. U. Künzi, E. Armbruster, and K. Agyeman, in Ref. 28, Vol. I. p. 107.
57. P. E. Donovan, *Mater. Sci. Eng.*, *98*, 487 (1988).
58. P. E. Donovan, *Acta Metall.*, *37*, 445 (1989).
59. P. G. Zielinski and D. G. Ast, *Acta Metall.*, *32*, 397 (1984).
60. H. Kimura, T. Masumoto, and D. G. Ast, *Acta Metall.*, *35*, 1757 (1987).
61. P. T. Vianco and J. C. M. Li, *J. Mater. Res.*, *2*(4), 461 (1987).
62. A. P. Radlinski and A. Calka, *J. Appl. Phys.*, *64*, 423 (1988).
63. A. Audouard, E. Balanzat, J. C. Joussett, G. Fuchs, D. Lesueur, and L. Thomé, *J. Phys. F. Metal Phys.*, *18*, L101 (1988).
64. T. Nasu, K. Nagaoka, M. Sakurai, T. Fukunaga, F. Itoh, and K. Suzuki, *Mater. Sci. Eng.*, *98*, 553 (1988).
65. Y. Nishi, H. Harano, H. Ishizuki, M. Kawakami, and E. Yajima, *Mater. Sci. Eng.*, *98*, 505 (1988).
66. H. Neuhäuser and R. P. Stössel, in Ref. 18, p. 1349.
67. Xingguo Cao and J. C. M. Li, *Acta Metall.*, *33*, 449 (1985).

68. A. I. Manokhin, V. P. Alyokhin, and M. Kh. Shorshorov, in Ref. 18, p. 1443.
69. P. T. Vianco and J. C. M. Li, *J. Mater. Sci.*, *22*, 3129 (1987).
70. A. T. Alpas and J. D. Embury, *Scripta Metall.*, *22*, 265 (1988).
71. V. Lakshmanan and J. C. M. Li, *Mater. Sci. Eng.*, *98*, 483 (1988).
72. L. T. Kabacoff, A. Le, and N. Y. Martinez, in *Rapidly Solidified Materials: Properties and Processing* edited by P. W. Lee and J. H. Moll (ASM, 1988), pp. 139–144.
73. P. Donnadieu, O. Jaoul, and M. Klèman, *Phil. Mag.*, *A52*, 5 (1985).
74. S. J. Harbert and A. Wolfenden, *J. Mater. Sci.*, *24*, 2923 (1989).
75. F. Van't Spijker, in Ref. 28, p. 349.
76. P. M. Anderson III, and A. E. Lord, *Mater. Sci. Eng.*, *43*, 267 (1980).
77. S. S. Tsao and F. Spaepen, *Acta Metall.*, *33*, 891 (1985).
78. A. Van den Beukel, E. Huizer, A. L. Mulder, and S. Van der Zwaag, *Acta Metall.*, *34*, 483 (1986).
79. Xueming Ma and Yizhen He, *Phys. Stat. Sol. (a)*, *103*, K13 (1987).
80. C. A. Volkert and F. Spaepen, *Mater. Sci. Eng.*, *97* 449 (1988).
81. A. I. Taub and F. Spaepen, *Acta Metall.*, *28*, 17818 (1980).
82. S. S. Tsao and F. Spaepen, *Acta Metall.*, *33*, 881 (1985).
83. I. V. Zolotukhin, V. I. Belyavskii, V. A. Khonik, and T. N. Ryabtseva, *Phys. Stat. Sol. (a)*, *116*, 255 (1989).
84. H. S. Chen and M. Goldstein, *J. Appl. Phys.*, *43*, 174 (1971).
85. A. Van den Beukel, *Scripta Metall.*, *22*, 877 (1988).
86. J. Harrath and H. Meherer, *Crystal Lattice Defects Amor. Mater.*, *13*, 1 (1986).
87. Fu-Qian Zheng, *Mater. Sci. Eng.*, *97*, 487 (1988).
88. Jingtang Wang, Bingzhe Ding, Dexing Pang, Shuling Li, and Gusung Li, *Mater. Sci. Eng.*, *97*, 483 (1988).
89. A. R. Bhatti and B. Cantor, *Mater. Sci. Eng.*, *97*, 479 (1988).
90. A. I. Taub, in *Rapidly Quenched Metals*, in Ref. 18, p. 1365.
91. P. T. Vianco and J. C. M. Li, *Mater. Sci. Eng.*, *95*, 175 (1987).
92. C. Piotrowski and R. S. Finocchiaro, *Mater. Sci. Eng.*, *A119,* 239 (1989).
93. J. C. M. Li, *Mater. Sci. Eng.*, *98*, 465 (1988).
94. H. H. Liebermann, *J. Mater. Sci.*, *19*, 1391 (1984).
95. P. G. Zielinski and D. G. Ast, *J. Non-Cryst. Solids*, *61–62*, 1021 (1984).
96. R. Gerling, F. P. Schimansky, and R. Wagner, in Ref. 18, p. 1377.
97. S. Arakawa, H. Harada, G. Hausch, H. R. Hilzinger, F. J. Menges, and H. Warlimont, in Ref. 18, p. 1389.
98. T. Yamasaki, M. Takahashi, and Y. Ogino, in Ref. 18, p. 1381.
99. T. W. Wu and F. Spaepen, in *Mechanical Behavior of Rapidly Solidified Materials*, edited by S. M. L. Sastry and B. A. McDonald (TMS, Warrendale, PA, 1986), p. 293.
100. T. W. Wu and F. Spaepen, *Acta Metall.*, *33*, 2185 (1985).
101. R. D. K. Misra and D. Akhtar, *Mater. Sci. Eng.*, *92*, 207 (1987).
102. A. R. Yavari, R. Barrue, M. Harmelin, and J. C. Perron, *J. Magn. Magn. Mater.*, *69*, 43 (1987).
103. T. Yamasaki, Y. Ogino, T. Honda, and Y. Anemiya, *Scripta Metall.*, *23*, 1963 (1989).
104. H. H. Liebermann, J. Marti, R. J. Martis, and C. P. Wong, *Metall. Trans.*, *20A*, 63 (1989).
105. Chr. Janot, B. George, D. Teirlinck, G. Marchal, C. Tete, and P. Delcroix, *Phil. Mag.*, *A47*, 301 (1983).
106. T. Guo and M. L. den Boer, *J. Non-Cryst. Solids*, *110*, 111 (1989).
107. T. K. G. Namboodhiri, T. A. Ramesh, G. Singh, and S. Seghal, *Mater. Sci. Eng.*, *61*, 23 (1983).
108. S. M. Lee and J. Y. Lee, *J. Appl. Phys.*, *63*, 4758 (1988).

109. H. G. Hillenbrand, U. Köster, and G. Tersteegen, in Ref. 28, Vol. I, p. 383.
110. H. S. Chen, *J. Appl. Phys.*, *49*, 462 (1978).
111. T. Masumoto, *Sci. Rept. Res. Inst. Tohoku U. Sec. A.*, *26*, 246 (1977).
112. A. Inoue, A. Kitamura, and T. Masumoto, *Trans. JIM*, *20*, 404 (1979).
113. K. H. Graf, W. Lohman, and A. Ribbens, in Ref. 18, p. 1411.
114. A. R. Yavari, *Mater. Sci. Eng.*, *98*, 491 (1988).
115. P. Rudkowski, J. O. Strom-Olsen, R. Schulz, and R. Roberge, in *Science and Technology of Rapidly Quenched Alloys*, edited by M. Tenhover, W. L. Johnson, and L. E. Tanner (Elsevier, New York, 1987), p. 171.
116. L. Bresson, M. Harmelin, and J. Bigot, *Mater. Sci. Eng.*, *98*, 495 (1988).
117. H. H. Liebermann and F. E. Luborsky, *Acta Metall.*, *29*, 1413 (1981).
118. D. M. Kroeger, C. C. Koch, C. G. McKamey, J. O. Scarbrough, and R. A. Padgett, in Ref. 18, p. 1369.
119. D. M. Kroeger, G. S. Canright, C. G. McKamey, D. S. Easton, and J. O. Scarbrough, *Acta Metall.*, *35*, 989 (1987).
120. D. M. Kroeger, G. G. McKamey, and J. O. Scarbrough, in Ref. 115, p. 131.
121. P. Lamparter, D. M. Kroeger, and S. Spooner, *Scripta Metall.*, *21*, 715 (1987).
122. S. Henderson, J. V. Wood, and G. W. Weidmann, *J. Mater. Sci. Letters*, *2*, 375 (1983).
123. L. A. Davis, *J. Mater. Sci.*, *10*, 1557 (1975).
124. L. A. Davis, *Metall. Trans.*, *10A*, 235 (1979).
125. K. Wetzig, W. Pompe, H. Fiedler, and V. P. Aljochin, *Cryst. Res. Technol.*, *18*, 1181 (1983).
126. V. Z. Bengus, E. D. Tabachnikova, and V. I. Startsev, *Phys. Stat. Sol.*, *81*(*a*), K11 (1984).
127. W. Henning, M. Calvo, and F. Osterstock, *J. Mater. Sci.*, *20*, 1889 (1985).
128. H. Kimura and T. Masumoto, in Ref. 18, p. 1373.
129. B. Söder, R. Ziemann, B. F. Wortmann, and S. Methfessel, in Ref. 18, p. 1393.
130. M. Calvo, W. Henning, and F. Osterstock, in Ref. 18, p. 1385.
131. G. Hunger and B. L. Mordike, in Ref. 18, p. 1371.
132. H. Kimura and T. Masumoto, in Ref. 22, p. 71.
133. V. Ocelik, P. Diko, K. Csach, V. Hajko, V. Z. Bengue, E. D. Tabachnikova, E. B. Korolkova, and P. Duhaj, *J. Mater. Sci.*, *22*, 3732 (1987).
134. V. Ocelik, P. Diko, V. Hajko, Jr., J. Miskuf, and P. Duhaj, *J. Mater. Sci.*, *22*, 2305 (1987).
135. S. J. Harbert and A. Wolfenden, in Ref. 72, p. 121.
136. Han-Ryong Pak, Jinn Chu, R. J. Deangelis, and Kenji Okazaki, *Mater. Sci. Eng.*, *A118*, 147 (1989).
137. T. K. Chaki and J. C. M. Li, *Scripta Metall.*, *18*, 703 (1984).
138. N. Jost, H. G. Hillenbrand, and E. Hornbogen, in Ref. 18, p. 1417.
139. A. Le, L. T. Kabacoff, and N. Y. Martinez, in Ref. 115, p. 137.
140. L. T. Kabacoff, A. Le, and Y. Martinez, in Ref. 72, p. 139.
141. M. Hagiwara, A. Inoue, and T. Masumoto, in Ref. 18, p. 1779.
142. A. T. Alpas, L. Edwards, and C. N. Reid, *Mater. Sci. Eng.*, *98*, 501 (1988).
143. C. J. Wong and J. C. M. Li, *Wear*, *98*, 45 (1984).
144. U. Sudarsan, K. Chattopadhyay, and Kishore, in Ref. 18, p. 1439.
145. N. Dolezal, and G. Hausch, in Ref. 18, p. 1767.
146. D. G. Morris, in Ref. 18, p. 1775.
147. Kishore, U. Sudarsan, N. Chandran, and K. Chattopadhyay, *Acta Metall.*, *35*, 1463 (1987).
148. A. L. Greer, *J. Non-Cryst. Solids*, *61*, *62*, 737 (1984).
149. A. L. Mulder, S. Van der Zwaag, and A. Van den Beukel, *J. Non-Cryst. Solids*, *61*, *62*, 979 (1984).

150. L. Leonardsson, in Ref. 18, p. 1353.
151. E. Huizer and A. Van den Beukel, *Acta Metall.*, *35*, 2843 (1987).
152. E. Zedler and G. Lehmann, in Ref. 18, Vol. I, p. 743.
153. R. Wagner, R. Gerling, and F. P. Schimansky, *J. Non-Cryst. Solids*, *61*, *62*, 1015 (1984).
154. R. Gerling and R. Wagner, *Scripta Metall.*, *16*, 963 (1982).
155. E. Zedler and G. Lehmann, *Phys. Stat. Sol. (a)*, *81*, 445 (1984).
156. O. Kohmoto, *Phys. Stat. Sol. (a)*, *104*, K101 (1987).
157. A. Elshafie, M. M. El-Zaidia, A. Abdel-Aal, and M. Abou-Ghazala, *J. Appl. Phys.*, *64(1)*, 103 (1988).
158. M. C. Kim, M. J. McNallan, C. C. Cheng, and W. E. King, *J. Mater. Sci. Letters*, *8*, 793 (1989).
159. B. Z. Feng, Y. Q. Cao, Y. Zheng, S. L. Li, X. Liu, G. L. Jin, and Q. L. Tan, *J. Non-Cryst. Solids*, *110*, 81 (1989).
160. H. Friedrichs and H. Neuhäuser, *J. Phys: Condens. Matter*, *1*, 8305 (1989).
161. L. Kubicar, E. Illekova, and P. Mrafko, in Ref. 18, Vol. I, p. 655.
162. R. O. Suzuki, R. Kita, and P. H. Shingu, in Ref. 18, Vol. I, p. 651.
163. X. Z. Ding and Y. Z. He, *Phys. Stat. Sol. (a)*, *110*, K67 (1988).
164. M. R. J. Gibbs and J. E. Evetts, in Ref. 11, p. 474.
165. J. A. Leake, M. R. J. Gibbs, S. Vryenhoef, and J. E. Evetts, *J. Non-Cryst. Solids*, *61*, *62*, 787 (1984).
166. L. Boesch, A. Napolitans, and P. B. Macedo, *J. Am. Ceram. Soc.*, *53*, 148 (1969).
167. M. R. J. Gibbs, D. W. Stephens, and J. E. Evetts, *J. Non-Cryst. Solids*, *61*, *62*, 925 (1984).
168. R. Brüning, Z. Altounian, and J. O. Ström-Olsen, *J. Appl. Phys.*, *62(9)*, 3633 (1987).
169. G. Balasubramanian, A. N. Tiwari, and C. M. Srivastava, *Phys. Stat. Sol. (a)*, *111*, K229 (1989).
170. T. Egami and T. Jagielinski, in Ref. 18, Vol. I, p. 667.
171. M. Bourrous and H. Kronmüller, *Phys. Stat. Sol. (a)*, *113*, 383 (1989).
172. T. Komatsu, K. Iwasaki, S. Sato, and K. Matusita, *J. Appl. Phys.*, *64(10)*, 4853 (1988).
173. C. Antonione, M. Baricco, and G. Riontino, *J. Mater. Sci.*, *23*, 2225 (1988).
174. G. Hygate and M. R. J. Gibbs. *J. Phys. Condens. Matter*, *1*, 1021 (1989).
175. J. de Vries, G. W. Koebrugge, and A. Van den Beukel, *Scripta Metall.*, *22*, 637 (1988).
176. J. Y. Lee and S. M. Lee, in Ref. 72, p. 127.
177. I. Majewska-Glabus, B. J. Thijsse, and S. Radelaar, *J. Non-Cryst. Solids*, *61*, *62*, 553 (1984).
178. M. R. J. Gibbs, J. E. Evetts, and J. A. Leake, *J. Mater. Sci.*, *18*, 278 (1983).
179. M. R. J. Gibbs, in Ref. 28, p. 37.
180. A. L. Greer and F. Spaepen, in *Structure and Mobility in Molecular and Atomic Glasses*, edited by J. M. O'Reilly and M. Goldstein; *Ann. NY Acad. Sci.*, *371*, 218 (1981).
181. A. Van den Beukel, *J. Non-Cryst. Solids*, *83*, 134 (1986).
182. M. R. J. Gibbs, in Ref. 18, Vol. I, p. 643.
183. M. R. J. Gibbs and H. R. Sinning, *J. Mater. Sci.*, *20*, 2517 (1985).
184. G. J. Lusink and A. Van den Beukel, *Acta Metall.* , *36*, 3019 (1988).
185. J. Perez, *Acta Metall.*, *32*, 2163 (1984).
186. J. Perez, in Ref. 17, p. 125.
187. T. Egami, in Ref. 18, Vol. I, p. 611.
188. J. T. Stanley and J. R. Cost, in Ref. 18, p. 647.
189. Y. He, J. S. Poon, and G. J. Shiflet, *Science*, *241*, 1640 (1988).
190. G. J. Shiflet, Y. He, and S. J. Poon, *J. Appl. Phys.*, *64(12)*, 6863 (1988).

191. G. J. Shiflet, Y. He, and S. J. Poon, *Scripta Metall.*, *22*, 1661–4 (1988).
192. A. Inoue, K. Ohtera, A. P. Tsai, and T. Masumoto, *Japan. J. Appl. Phys.*, *27*, L479 (1988).
193. D. G. Morris, in Ref. 18, p. 1751.
194. T. Sato, *Scripta Metall.*, *20*, 1801 (1986).
195. C. C. Hays and D. G. Naugle, in Ref. 115, p. 77.
196. M. A. Otooni, in Ref. 115, p. 121.
197. Wenkui Wang and Zhihua Yan, *J. Mater. Sci. Lett.*, *8*, 817 (1989).
198. O. Schneeweiss, *J. Phys: Condens. Matter*, *1*, 4749 (1989).
199. M. Ahmed and D. I. Potter, *Acta Metall.*, *33*, 2221 (1985).
200. P. M. Ossi, *Mater. Sci. Eng.*, *A 115*, 107 (1989).
201. A. Seidel, S. Massing, B. Strehlau, and G. Linker, *Mater. Sci. Eng. A 115*, 139 (1989).
202. I. Thomè, A. Benyagoub, F. Pons, E. Ligeon, J. Fontenille, and R. Danieloun, *Mater. Sci. Eng. A*, *115*, 127 (1989).
203. J. Delage, O. Popoola, J. P. Villain, and P. Moine, *Mater. Sci. Eng. A*, *115*, 133 (1989).
204. L. U. Aaen Anderson, J. Bottiger, and K. Dyrbye, *Mater. Sci. Eng. A*, *115*, 123 (1989).
205. J. Bottiger, K. Dyrbye, K. Pampus, and R. Poulsen, *Phil. Mag.*, *59*, 569 (1989).
206. X. W. Lin, J. Koike, D. N. Seidman, and P. R. Okamoto, *Phil. Mag. Lett.*, *60*, 233 (1989).
207. C. C. Koch, O. B. Cavin, C. G. McKamey, and J. O. Scarbrough, *Appl. Phys. Lett.*, *43*, 1017 (1983).
208. R. B. Schwarz, R. R. Petrich, and C. K. Saw, *J. Non-Cryst. Solids*, *76*, 281 (1985).
209. R. B. Schwarz, C. C. Koch, *Appl. Phys. Lett.*, *49*, 146 (1986).
210. C. Politis and W. L. Johnson, *J. Appl. Phys.*, *60*, 1147 (1986).
211. J. R. Thompson and C. Politis, *Europhys. Lett.*, *3*, 199 (1987).
212. M. S. Kim and C. C. Koch, *J. Appl. Phys.*, *62*, 3450 (1987).
213. T. J. Tiainen and R. B. Schwarz, *J. Less-Common Metals*, *140*, 99 (1988).
214. P. Y. Lee, J. Jang, and C. C. Koch, *J. Less-Common Metals*, *140*, 73–83 (1988).
215. S. Enzo, L. Schiffini, L. Battezzat, and G. Cocco, *J. Less-Common Metals*, *140*, 129 (1988).
216. G. Ennas, M. Magini, F. Padella, P. Susini, G. Boffitto, and G. Licheri, *J. Mater. Sci.*, *24*, 3053 (1989).
217. G. Ennas, M. Magini, F. Padella, F. Pompa, and M. Vittori, *J. Non-Cryst. Solids*, *110*, 69 (1989).
218. A. Calka and A. P. Radlinski, *Mater. Sci. Eng. A*, *118*, 131 (1989).
219. Suzuki, K., *J. Non-Cryst. Solids*, *112*, 23 (1989).
220. X. L. Yeh, K. Samwer, and W. J. Johnson, *Appl. Phys. Lett.*, *42*, 242 (1983).
221. K. Aoki, T. Yamamoto, and T. Masumoto, *Scripta Metall.*, *21*, 27 (1987).
222. K. Aoki, T. Yamamoto, Y. Satoh, K. Fukamichi, and T. Masumoto, *Acta Metall.*, *35*, 2465 (1987).
223. R. Pareja, J. M. Riveiro, and A. Jerez, *J. Mater. Sci.*, *22*, 4523 (1987).
224. K. Aoki, A. Yanagitani, X-G. Li, and T. Masumoto, *Mater. Sci. Eng.*, *97*, 35 (1988).
225. U-In Chung and Jai-Young Lee, *J. Non-Cryst. Solids*, *110*, 203 (1989).
226. S. B. Newcomb and K. N. Tu, *Appl. Phys. Lett.*, *48*, 1436 (1986).
227. S. R. Herd, K. N. Tu, and K. Y. Ahn, *Appl. Phys. Lett.*, *42*, 597 (1983).
228. R. B. Schwarz and W. L. Johnson, *Phys. Rev. Lett.*, *51*, 415 (1983).
229. H. Schroder, K. Samwer, and U. Koster, *Phys. Rev. Lett.*, *54*, 197 (1985).
230. M. Van Rossum, M. A. Nicolet, and W. L. Johnson, *Phys. Rev. B*, *29*, 5498 (1984).
231. B. M. Clemens, W. L. Johnson, and R. B. Schwarz, *J. Non-Cryst. Solids*, *61–62*, 817 (1984).

232. V. Vitek, S. P. Chen, and T. Egami, *J. Non-Cryst. Solids*, *61*, *62*, 583 (1984).
233. J. Laakkonen and R. M. Nieminen, *J. Phys. C: Solid State Phys.*, *21*, 3663 (1988).
234. T. K. Chaki and J. C. M. Li, *Phil. Mag. B*, *51*, 557 (1985).
235. T. K. Chaki and J. C. M. Li, *J. Nucl. Mater.*, *140*, 180 (1986).
236. J. C. M. Li, *Metall. Trans.*, *16A*, 2227 (1985).
237. J. J. Gilman, in *Dislocation Dynamics*, edited by A. R. Rosenfield et al. (McGraw-Hill, New York, 1968), p. 3.
238. P. Chaudhari, A. Levi, and P. J. Steinhardt, *Phys. Rev. Lett.*, *43*, 1517 (1979).
239. P. Chaudhari, and P. J. Steinhardt, *Phil. Mag.*, *A46*, 25 (1982).
240. P. J. Steinhardt and P. Chaudhari, *Phil. Mag.*, *A44*, 1375 (1981).
241. L. T. Shi and P. Chaudhari, *Phys. Rev. Lett.*, *51*, 1581 (1983).
242. J. Piller and P. Haasen, *Acta Metall.*, *30*, 1 (1982).
243. R. Grüne, M. Öhring, R. Wagner, and P. Haasen, in Ref. 18, Vol. 1, p. 761.
244. A. R. Yavari, S. Hamar-Thibault, and H. R. Sinning, *Scripta Metall.*, *22*, 1231 (1988).
245. J. C. M. Li and J. B. C. Wu, *J. Mater. Sci.*, *11*, 445 (1976).
246. H. Kimura and T. Masumoto, *J. Non-Cryst. Solids,*, *61*, *62*, 835 (1984).
247. D. G. Ast and D. J. Krenitsky, *J. Mater. Sci.*, *14*, 287 (1979).
248. P. Donnadieu, *J. Non-Cryst. Solids*, *105*, 280 (1988).
249. P. Donnadieu, *J. Non-Cryst. Solids*, *111*, 7 (1989).
250. A. Kursumovic, R. W. Cahn, and M. G. Scott, *Scripta Metall.*, *14*, 1245 (1980).
251. A. Kursumovic, E. Girt, E. Babic, B. Leontic, and N. Njuhovic, *J. Non-Cryst. Solids*, *44*, 57 (1981).
252. J. C. M. Li, in Ref. 19, p. 267.
253. C. P. Chou, L. A. Davis, and M. C. Narasimhan, *Scripta Metall.*, *11*, 417 (1977).
254. I. W. Donald, S. H. Whang, H. A. Davies, and B. C. Giessen, in Ref. 11, Vol. II, p. 1377.
255. M. Doi, K. Sugiyama, T. Tono, and T. Imura, in Ref. 11, Vol. II, p. 1349.
256. S. Tanigawa, K. Shima, and T. Masumoto, in Ref. 11, p. 501.
257. M. F. Denanot, O. Popoola, and P. Moine, *Mater. Sci. Eng.*, *A115*, 145 (1989).
258. G. Cocco, S. Enzo, L. Schiffini and L. Battezzati, *Mater. Sci. Eng.*, *97*, 43 (1988).
259. J. C. M. Li, in Ref. 11, Vol. II, p. 1335.
260. D.-R. Huang and J. C. M. Li, *Effect of Heating Rate on the Curie Temperature of an Amorphous Alloy* (TMS: Fall Meeting Abstracts, 1989), p. 57.
261. D.-R. Huang and J. C. M. Li, *Scripta Metall. Mater.*, *24*, 1137 (1990).

14

Electronic Transport Properties

T. Richmond and H.-J. Güntherodt
University of Basel, Basel, Switzerland

I. INTRODUCTION

Over the last two decades considerable effort has been directed toward obtaining an understanding of the nature of electronic transport in disordered systems. As amorphous, liquid, or glassy alloys lack long-range periodicity, the concept of the Brillouin zone is no longer applicable. The understanding of electronic transport in such disordered systems therefore demands a totally new approach compared to that used for crystalline materials.

The main distinguishing characteristic of the resistivity in metallic glasses is that it is several times larger than that of the corresponding alloy systems in the crystalline state, and the change in resistivity is usually small and often negative. There exist a number of anomalies which have initiated a great deal of research, including the occurrence of a resistivity minimum at low temperatures for alloys with a positive temperature coefficient of resistivity (TCR), and also a positive Hall coefficient in some amorphous alloys. Although much has been achieved in understanding the possible mechanisms that may play a role in electronic conduction in disordered systems, as yet clear explanations to the above anomalies are lacking. The aim of this chapter is to survey relevant experimental reports concerning all rapidly quenched and amorphous alloys in light of appropriate theories and above all, to provide a general guide to the subject of electronic transport in metallic glasses.

II. THEORY

A. Historical Background

The Ziman theory was the first to provide a quantitative explanation for the electrical transport properties of liquid metals [1,2]. In this model, the conduction electrons in a liquid metal are assumed to form a degenerate free electron gas with a spherical Fermi surface. Ziman argued that, although the ions are positioned in an irregular manner in a liquid, a strong short-range correlation exists which gives rise to coherent interference between electron waves (assumed as plane waves) diffracted by adjacent ions, leading to the existence of a mean free path. Representing the arrangement of ions in the liquid state in terms of a pair correlation function, $g(r)$, the structure factor, $a(K)$, is then a Fourier transform of $g(r)$. Within the concept of a pseudopotential, together with the assumption that a mean free path exists, a relationship for electrical resistivity may be obtained using the Boltzmann conductivity equation. This relationship is greatly simplified by assuming that the electron waves are only weakly scattered, and that it is sufficient to use first order perturbation; that is, the Born approximation is applicable.

The Ziman formula is

$$\rho = \frac{3\pi\Omega}{4e^2\hbar v_F^2 k_F^2} \int_0^{2k_F} a(K)|U(K)|^2 K^3 \, dK \tag{1}$$

where Ω is the atomic volume, v_F is the Fermi velocity, k_F is the Fermi radius, K is the scattering vector, and $U(K)$ is the pseudopotential.

According to this relation, $a(K)$ determines the temperature dependence of resistivity, predicting a positive TCR in the case of monovalent simple metals, where $2k_F$ lies to the left of K_P (wave number corresponding to the first peak in the structure factor), and a negative TCR in the case of $2k_F$ coinciding with K_P (Fig. 1). This is one of the most important predictions of the Ziman theory, which offered for the first time an explanation for the observed negative TCR of liquids such as Zn and Cd. Excellent agreement was also found between the resistivity values calculated by Eq. (1) and those found experimentally for the liquid alkali metals Na and K, using the available pseudeopotentials. Later, in the Faber-Ziman theory [3] this was extended to liquid alloys of simple metals by taking into account the partial structure factors.

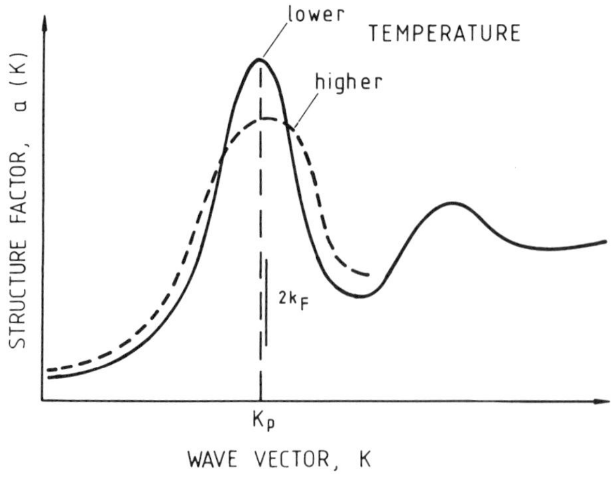

Figure 1 Structure factor versus scattering vector for lower and higher temperatures.

It was surprising to observe that liquid alloys containing transition metals also showed negative temperature coefficients of electrical resistivity. As the scattering by transition metals is considerably stronger than in the case of simple liquid metals, the resonance scattering due to the *d*-states should be taken into consideration and the Born approximation of the Ziman theory is no longer appropriate. In explaining the electrical resistivity of liquid transition metals, Evans et al. [4] derived a formula in the spirit of the Ziman theory without the use of either Born approximation or the Boltzmann equation. Using a muffin-tin model of transition metals, they assumed the ions to be nonoverlapping spheres and the potential in the interstitial regions as constant. This model greatly simplifies the scattering problem by replacing the pseudopotential in Eq. (1) with a single-site on-shell *t*-matrix, where the scattering can be calculated exactly. The *t*-matrix was expressed in terms of phase shifts as

$$t(K, E_F) = -\frac{2\pi\hbar^3}{m^*\Omega_0(2mE_F)^{1/2}} \sum_l (2l + 1)\sin \eta_l e^{i\eta_l} P_l(\cos \Theta) \tag{2}$$

where m^* is the effective electron mass, Ω_0 is the average atomic volume, η_l is the lth phase shift, and P_l is the lth Legendre polynomial. The results calculated using this method for liquid noble metals are in good agreement with experimental results.

A different approach was taken by Mott [5], who emphasized the importance of *s-d* scattering. According to this model, in the case of transition metal liquids with mean free paths short enough to be on the order of interatomic spacing, high resistivities occur mainly due to scattering of electrons into vacant *d*-states. In this model the resistivity is given in terms of the *d*-density of states (DOS) rather than the structure factor, thereby explaining the rather small change in resistivity upon melting. The negative TCR is explained by the broadening of the *d*-band at the Fermi level as the temperature is increased. This method also predicts a positive thermopower for a nearly empty *d*-band. However, calculations based on this method seem to lead to overestimations of the resistivity values, although an expression by Brown et al. using Mott's method within a muffin-tin model does give nearly correct thermopower values [6].

With the possibility of freezing of molten alloys and the advent of "metallic glasses," interest in this field started to grow. The electrical resistivity of these materials showed magnitudes similar to those of liquid metals, and the high-temperature liquid-state resistivity data could be extrapolated to the glassy state, where again a negative TCR was often observed. It was shown that the extended Ziman theory could be applied to amorphous alloys [7,8]. It was noted, however, that for temperatures below the Debye temperature, the assumption that the spectrum of density fluctuations in the liquid is limited to energies less than those of thermal excitations of the conduction electrons could no longer apply. By using a dynamic structure factor $S(K, w)$ [19], and including the effect of thermal vibrations on the elastic structure factor by taking into consideration the Debye-Waller damping factor [10], Cote and Meisel [7] demonstrated that such a model was able to account for resistivity behavior of simple metallic glasses. This model is known as the generalized Ziman theory. The behaviors explained include the variation of resistivity with concentration, the $+T^2$ dependence of the resistivity at low temperatures, and also the resistivity maximum observed in some alloys, corresponding to the crossover from a positive and quadratic temperature dependence to a negative and linearly dependent one (in the case of $2k_F \sim K_P$) with the increasing temperature.

However, not all experimental results could be explained by the above theory. A number of anomalies remained unresolved, particularly in the case of high-resistivity metallic glasses. Some amorphous alloys exhibit negative TCR, where the back scattering condition was not satisfied, and observed TCRs tend to be larger than the calculated values. Also, the maximum predicted by the Ziman theory is not observed in most glassy alloys. Such anomalies were shown to have rather universal features referred to as saturation effects or Mooij laws [11]. The desire to explain such anomalies stimulated further experimental and theoretical work in this field, which continues to this day.

Subsequently it was suggested that a more constrained consideration of the electron-phonon interactions in the high-resistivity disordered systems would help to describe these anomalies [12]. Phonons with wavelengths longer than the electron mean free path are ineffective scatterers, and this produces a low-frequency cutoff to the integrals over the phonon spectrum which occur in the expressions for the resistivity static structure factor. This led to a modified expression for the inelastic part of the static resistivity structure factor, depending explicitly on the electron mean free path. This modification was shown to improve the agreement of theory with experimental data in the case of low resistivity metallic glasses such as Mg-Zn [13]. However, it was found that a quantitative agreement could not be reached for the case of high-resistivity amorphous alloys.

A great deal of research has been carried out since on the possible mechanisms involved in conduction in high-resistivity metallic glasses. A comprehensive review of all pertinent theories is beyond the scope of this chapter. What follows is a summary of those theories considered most relevant, except for that of two-level systems, which is discussed in the following chapter.

B. *d*-Band Conduction

Many authors have found *d* electrons to contribute considerably to the conductivity in transition metal-containing glassy metals. The conductivity in a transition metal may be thought of as consisting of *s*-band and *d*-band conduction. In the Ziman model the *s* electrons contribute the most. However, for an amorphous transition metal, the mean free path of the *s* electrons is very short and the number of such states at the Fermi level is reduced because of hybridization, and the *s*- and *d*-band conductions are at least comparable in magnitude. In a paper by ten Bosch and Bennemann [14], it was shown that the contribution of the *d* electrons was of the same order of magnitude as that of the *s* electrons in the case of a nearly half-filled *d*-band of liquid transition metals, after accounting for the *s*-*d* hybridization effects and treating the *d* electrons within the tight binding approximation. Weir and Morgan [15] computer-simulated electron diffusion and conduction in a two- and three-dimensional amorphous metal. With a random phase model (RPM), they also arrive at the importance of the *d*-conduction in amorphous metals. Bose et al. [16], on the other hand, used the linear combination of atomic orbitals (LCAO) method to study the electronic structure of liquid and amorphous Fe. They found that the *d* electrons are responsible for up to 84% of conductivity. Calculations using a similar method were carried out for liquid La by Ballentine et al. [17]. Their results also point to the dominant role of *d* electrons in the conductivity of liquid transition metals.

C. Phonon-Assisted Tunneling

Belitze and Götze [18], proposed that in strongly disordered systems an additional contribution to conductivity may arise from random tunneling processes due to spatially fluctuating effective mass. Belitz and Schirmacher [19] used this theory to derive an explanation for Mooij laws [11,20] in high-resistivity conductors by taking account of electron-phonon interactions in the mode-coupling formalism of Götze [21]. The result is a general expression for the conductivity as

$$\sigma = e^2\left(\frac{n}{m(M_0 + M_T)} + L_0 + L_T\right) \tag{3}$$

where M_0 and M_T are current relaxation rates due to static disorder and due to phonons, respectively, and L_0 and L_T are the tunneling rates. The first term is in fact the normal Drude expression for conductivity; i.e., $ne^2\tau/m$. As expected,

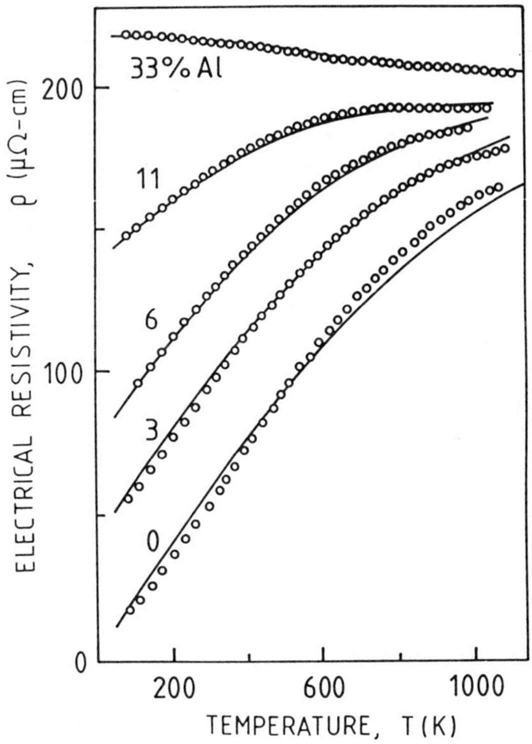

Figure 2 Theoretical fits to resistivity data of Ti-Al using a phonon-assisted tunneling model [22].

scattering rate increases as temperature increases, but phonon-assisted tunneling processes also increase proportionally with conductivity, therefore allowing for the possibility of occurence of a negative TCR. The tunneling terms, however, become important only when the sum of the current relaxation rates are sufficiently large; this is the case when the mean free path is comparable to the Fermi wavelength. This model was found to describe the saturation effects in Ti-Al alloys [22], as can be seen in Fig. 2.

D. Quantum Interference and Correlation Effects

Quantum interference effects (QIEs) have been shown to be of importance as corrections to the Boltzmann conductivity in high-resistivity amorphous alloys [23]. Although there exist a number of excellent reviews on the subject of incipient localization and electron correlation effects in disordered systems, for example, that of Lee and Ramakrishnan [24], it is worthwhile to summarize the development of this theory and its relation to liquid, amorphous, or glassy metals.

The concept of localization was first introduced by Anderson [25], who pointed out that if the disorder in a system is very strong the electron wave function may become localized; i.e., the probability of the electron's return to the same site from which it started to diffuse is equal to one over a long period of time. The weak localization regime acts as the precursor for complete localization. While the probability of the return in this region is less than unity, it is not negligible. Thousless [26] was the first to try to formulate localization using a scaling model. Abrahams et al. [27], based on this scaling approach, were able to show that the conductance (not the conductivity) of a sample is a dimensional quantity. Abrahams et al. [27] pointed out that it is possible to calculate quantum corrections for the Boltzmann conductivity in the weak localized region. This correction is obtained by evaluating the Kubo-Greenwood formula using Feynman diagrams. An intuitive picture put forward by Larkin and Khmel'niltskii [28] and by Bergmann [29], avoids the use of Feynman diagrams. The propagation of electrons in a disordered system is considered to be diffusive and similar to a random walk of length l_e, the elastic mean free path. The probability, P, of an electron diffusing between two points can then be expressed quantum mechanically in terms of the probability amplitudes, A_i, for each possible path between two points as

$$P = \sum |A_i|^2 + \sum_{i \neq j} A_i^* A_j \tag{4}$$

The first term represents the sum of probabilities for the particle to pass by in any way. The second term is the interference term, which usually averages to zero as the intersecting paths usually differ in phase due to their varying path lengths; the above relation is then equivalent to the classical Boltzmann equation. However, in a weakly localized system there is a finite probability of the electron returning to its starting point. In this case, the two probability amplitudes associated with electrons traversing the loop in opposite directions are coherent and therefore lead to a nonvanishing contribution to the interference term; i.e., they contribute to the conductivity equation. This is the case when the inelastic mean free path, l_i, is much longer than the elastic one, l_e. This effect then gives a positive contribution to the resistivity due to enhanced backscattering. As temperature increases, coherence is destroyed and this contribution decreases with increasing temperature, so that the temperature coefficient of resistivity due to localization is negative.

A general equation for conductivity with localization corrections can be written as

$$\sigma(T) = \sigma_B(T) - (\sigma_L(0) - \sigma_L^T(T)) \tag{5}$$

where $\sigma_B(\mathrm{T})$ = Boltzmann conductivity

$\sigma_L(0)$ = conductivity due to localization at $T = 0$

$\sigma_L^T(T)$ = temperature-dependent part of conductivity due to localization

The theories of Larkin and Khmel'nitskii [28] and of Bergmann [29,30], and also the theories of Kaveh and Mott [31–33] arrive at the same expression for this correction [34]:

$$\sigma_L^T(T) \sim \frac{e^2}{\pi^2\hbar}\left(\frac{1}{l_e} - \frac{1}{L_i(T)}\right) \tag{6}$$

where L_i is the inelastic diffusion length with

$$L_i^2 = \frac{1}{2}\, l_i l_e \tag{7}$$

On the other hand, the Boltzmann conductivity may be written as

$$\sigma_B(T) = \sigma_B(0) + \sigma_B^T(T) \tag{8}$$

and the corresponding relation for the mean free path given as

$$\frac{1}{L} = \frac{1}{l_e} + \frac{1}{l_i(T)} \tag{9}$$

The temperature-dependent part in Eq. (8) is due to the weak temperature dependence of $l_i(T)$. In the case of $l_i(T) \ll l_e$, $\sigma_B(T) = \sigma_B$. However, in the case of highly disordered metals, $l_i(T) \gg l_e$, and so the temperature dependent part becomes significant. The latter condition together with Eq. (9) gives

$$L \simeq l_e\left(1 - \frac{l_e}{l_i}\right) \tag{10}$$

and therefore the Boltzmann conductivity becomes

$$\sigma_B(T) \simeq \frac{e^2}{3\pi^2\hbar}\,(k_F l_e)^2\left(\frac{1}{l_e} - \frac{1}{l_i(T)}\right) \tag{11}$$

Then the total conductivity is

$$\sigma(T) \simeq \sigma_B(0)\left[1 - \frac{3}{(k_F l_e)^2}\right] + \frac{e^2}{\pi^2\hbar}\left[\frac{1}{L_i(T)} - \frac{(k_F l_e)^2}{3}\,\frac{1}{l_i(T)}\right] \tag{12}$$

Considering now only the temperature-dependent part of the conductivity, that is,

$$\sigma^T(T) \simeq \frac{e^2}{\pi^2\hbar}\left[\frac{1}{L_i(T)} - \frac{(k_F l_e)^2}{3}\frac{1}{l_i(T)}\right] \tag{13}$$

it is clear that the first term always gives a nonmetallic positive contribution to $d\sigma/dT$, while the contribution of the second term is metallic and negative. As L_i is proportional to $\sqrt{l_i}$, and since the inelastic mean free path decreases with temperature, the first term is only significant at low temperatures. With increasing temperature, the second term increases appreciably, which means that a resistivity minimum at intermediate temperatures may occur, as has indeed been observed for a number of metallic glasses [35,36].

The critical resistivity, ρ_c, at which this change from negative to positive temperature coefficient occurs can be calculated from Eq. (12). According to Tsuei [37],

$$\rho_c^{3D} = \frac{3\pi^2\hbar}{e^2 k_F^2 l_e}\left(1 - \frac{3}{(k_F l_e)^2} + \frac{6.75}{(k_F l_e)^4}\right)^{-1} \tag{14}$$

Tsuei points out that this formula demonstrates that this critical resistivity is not universal, but depends on the specific material through k_F and the degree of disorder (l_e). Therefore, the competition between the quantum mechanical effects of incipient localization and the Boltzmann electron transport gives rise to the nonuniversality of the Mooji correlation.

The nonmetallic term in Eq. (13) predicted to dominate up to crystallization temperatures [34] in alloys with electron mean free paths on the order of an interatomic spacing. Therefore, by considering only the first term in such alloys and taking into account the electron-phonon scattering, Howson [34] predicted the following temperature dependence of conductivity in 3D:

$$\sigma_L \propto T, \qquad T < \Theta_D$$
$$\sigma_L \propto \sqrt{T}, \qquad T > \Theta_D$$

This behavior has already been observed in many systems as will later be discussed in relation to nonmagnetic strong-scattering metallic glasses, although the critical temperature is found to be in fact a third of the Debye temperature.

The above theory does not take into consideration many-body effects. As pointed out by Al'tshuler and Aronov [38], such effects are of importance in disordered systems as they are enhanced due to slow diffusion of electrons.* The temperature dependence of conductivity due to electron correlation effects in three dimensions is predicted [38,33] as

$$\sigma_{\mathrm{INT}}^T = f(x)\,\frac{e^2}{\pi^2\hbar}\,\frac{1}{L_{\mathrm{INT}}(T)} \tag{15}$$

*A more comprehensive account of electron correlation effects is given in Chapter 15.

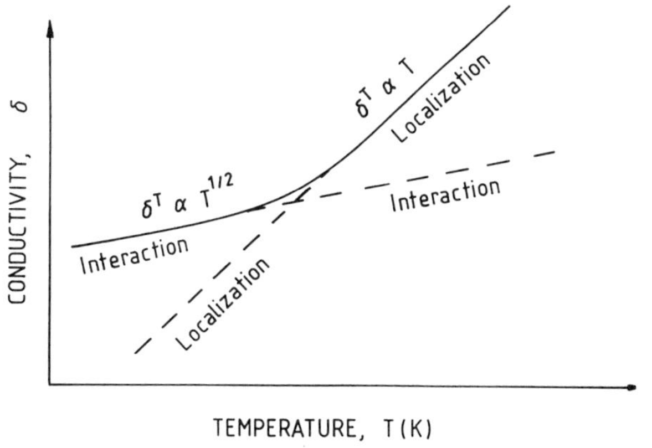

Figure 3 Conductivity as a function of temperature [33].

with

$$L_{\mathrm{INT}} = \sqrt{\frac{\hbar D}{k_B T}} \tag{16}$$

where D is the diffusion coefficient and $f(x)$ is given by Rosenbaum et al. [39] as

$$f(x) = \begin{cases} 1 & \text{for } k_F L \gg 1 \\ <0 & \text{for } k_F L > 1 \\ 1 & \text{for } k_F L \simeq 1 \end{cases}$$

From Eq. (15) it can be seen that σ_{INT}^T varies as $\sqrt{T}$. Incipient localization, on the other hand, gives rise to a T-dependence of conductivity at low temperatures. Therefore, electron correlation effects are expected to dominate at very low temperatures. This change from T to $\sqrt{T}$ behavior with decreasing of temperature has been observed experimentally in glassy metals such as Cu-Zr [40]. Figure 3 (adopted from the work of Kaveh and Mott [33]) depicts the expected conductivity behavior arising from a combination of incipient localization and electron correlation effects.

III. EXPERIMENTS

In this section we will summarize the main experimental results in the investigation of the transport properties of glassy metals. In carrying out experiments or in the interpretation of results, one should ensure the exclusion of effects derived from small fractions of magnetic impurities or small amounts of crystallinity. The effects of features such as these on the transport properties have been reviewed by Gallagher et al. [41]. Some alloys also need to be annealed before the true magnitude of resistivity, ρ, is measured [42]. The effects of thermal cycling and annealing on the Hall resistivity of magnetic amorphous alloys has been investigated by Malmhäll et al. [43].

A. Simple Amorphous Alloys

In this section we will consider the electronic transport properties of those amorphous metals where *d* electrons play no part in conduction. Although these types of alloys are invaluable in understanding the transport theory of amorphous metals, they do not as yet have potential applications. This means that the research in producing these alloys through rapid quenching has suffered and most of the studies of electronic transport involves those of thin films evaporated onto a cold substrate. Alloys so far prepared in the form of ribbons by melt-spinning techniques include Mg-Zn [44], Ag-Cu-X with X = Mg, Si, Sn, and Sb [45], Mg-Zn-Ga [46], and Mg-Cu-Zn [47].

The alloys prepared by evaporation belong to the family of noble and polyvalent metals [48,49]. A systematic study of these type of alloys has been carried out by Häussler and his co-workers, shedding new light on the Hume-Rothery-phase in amorphous metals [49–51].

The generalized Ziman theory has proved particularly successful in its application to the experimental results on simple amorphous alloys; i.e., a negative temperature coefficient of resistivity is observed for those alloys for which K_P roughly coincides with $2k_F$. Another feature that again points to the validity of the assumptions of Ziman theory for simple amorphous metals is that values of $2k_F$ derived using Hall effect measurements are often consistent with corresponding values calculated from the free electron formula, as shown by Mizutani and Yoshida [52] for Ag-Cu-Mg and Ag-Cu-Al alloy systems. An exception to this are those alloys with a structure factor close to $2k_F$ or for alloys with $K_P = 2k_F$ [51,49]. Mizutani discusses this topic in detail [53].

1. *Resistivity*

Simple glassy metals usually show relatively low resistivities, sometimes as low as 30 μΩ-cm, as in the case of magnesium-poor Ag-Cu-Mg alloys [45]. On the other hand, in replacing Mg with Si or Ge, the resistivities are found to increase to up to 150 μΩ-cm. Ca-Al glasses also show unexpectedly high resistivities. These alloys studied by different groups show resistivities as high as 400 μΩ-cm [54,55]. This has been attributed to the importance of the *d*-states of the Ca at the Fermi energy [55].

The temperature dependence of simple amorphous metals is seen to vary with the magnitude of resistivity. In alloys with low resistivities, the dominant mechanism is the inelastic electron-phonon interaction, and therefore the resistivity increases with increasing temperature. On the other hand, highly resistive simple amorphous metals show a negative temperature coefficient of resistivity. It should be said that although there seems to be a correlation between the magnitude of resistivity and the value of TCR, the value of the critical resistivity varies from 50 μΩ-cm in the case of amorphous Ag-Cu-Mg alloys [45], to 100 μΩ-cm in the case of Ca-Mg-Al alloys [56].

Regarding the temperature dependence of the high resistivity simple glasses, above the Debye temperature, Θ_D, where the assumption of elastic scattering is valid, the simple amorphous metals are usually found to behave according to the extended Ziman theory and show a linear dependence of resistivity with respect to temperature. However at low temperatures, the generalized Ziman theory is more appropriate. According to this theory the resistivity may be expressed in terms of the Debye-Waller factor as

$$\rho = \exp(-2W(T))(\rho_0 + \Delta\rho) \tag{17}$$

where ρ_0 is the residual resistivity at 0 K, and $\Delta\rho$ is the contribution due to inelastic electron-phonon interaction. This equation has been used by many authors to explain the temperature dependence at temperatures below Θ_D. Inelastic electron-phonon scattering causes an increase in resistivity with increasing temperature with a T^2 dependence, and the Debye-Waller factor reduces the resistivity, varying as $-T^2$. The temperature dependence at low temperatures is thus explained as the interplay of these two competing terms. Mizutani and Yoshino [45] found their experimental data on Ag-Cu-Mg metallic glasses (Fig. 4) can be satisfactorily explained by the above expression. The alloys with positive TCR correspond to cases in which electron-phonon interactions dominate. Similar results are also found for

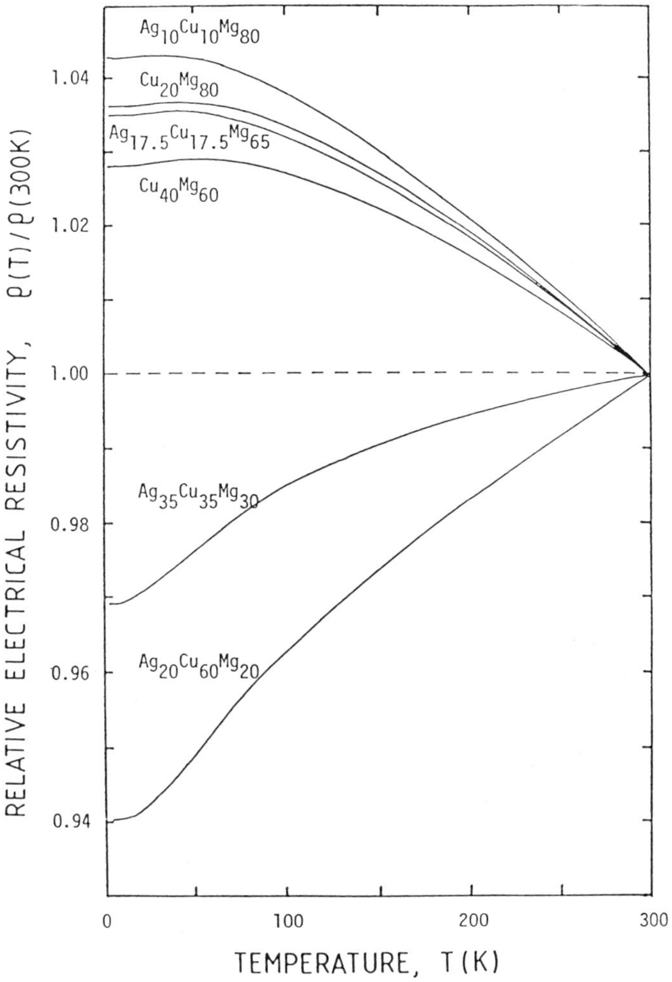

Figure 4 Temperature-dependence of resistivity in Ag-Cu-Mg metallic glasses [45].

Ca-Mg-Al [47]. In both cases a deviation from the T^2 dependence of resistivity to higher values is reported below ~10 K. In order to test whether this effect is due to magnetic impurities, Mizutani and Yoshino [45] deliberately increased the estimated concentration of the Fe impurity from 3 to 17 ppm. However, the upturn in resistivity remained identical. They concluded that this effect can only be a direct result of the disordered nature of the alloys.

Recently Mizutani et al. [57] have studied the temperature-dependence of the electrical resistivity of Ag-Cu-Ge amorphous alloys over a wide composition range. They found that quantum corrections become important in explaining the low-temperature behavior of samples with resistivities higher than 200 μΩ-cm.

2. *Hall Effect*

The Hall effect of simple liquid metals and alloys have been reviewed by Busch and Güntherodt [58] and by Künzi and Güntherodt [59] and that of the simple amorphous alloys by McGuire et al. [60] and by Mizutani [53]. The Hall coefficients of the simple liquid alloys are found to be quite close to their free-electron values R_0, obtained from the classical relation

$$R_0 = -\frac{1}{n|e|} \tag{18}$$

The Hall effects of simple glassy metals such as magnesium-rich Mg-Zn and Cu-Mg alloys were studied by Matsuda and Mizutani [61] where agreement with the free electron values is again obtained. The Hall coefficients are found to be negative and essentially temperature-independent.

In a systematic study of amorphous polyvalent element/noble metal alloys, Häussler and Baumann [49] and Häussler [51] have investigated the extent of the applicability of the above relation to amorphous (Au, Ag, Cu)-Sn systems. As can be seen from Fig. 5, large deviations from the free electron values to smaller magnitudes of the Hall coefficient occur for alloys with concentration ranges of $\overline{Z} = 1.8e/a$, for which K_P, the wave number of the maximum peak in the structure factor, coincides with $2k_F$. The Hall coefficients of alloys with the greatest deviations also exhibit a temperature dependence. Häussler [62] proposes that this temperature dependence might be related to changes of the minimum in the density of states at E_F with temperature. To explain the anomalies in the transport properties of liquid mercury, Mott [63] was the first to point to the concept of the structure-induced minimum in the density of states (DOS). A measure of the structure influence is the quantity g, the ratio of the DOS of the liquid system at the Fermi level to its free-electron value. Nagel and Tauc [64] extended this to amorphous metals and suggested that when $K_p = 2k_F$, a decrease in the DOS at the Fermi level is expected. This was proposed as a stability criterion for amorphous systems. Theoretical calculations in support of this prediction were later presented by Beck and Oberle [65] and by Nicholson and Schwartz [66]. Experimental minima in the DOS at the Fermi energy have been observed by Häussler [67] in noble metal/polyvalent element alloys. Häussler also observes a correlation between $(1 - g)^2$, the square of the depth of the minimum, and the Hall coefficient as shown in Fig. 6 for Au-Sn, Ag-Sn and Cu-Sn amorphous alloys [62]. It should be mentioned that recent photoemission measurements on liquid polyvalent metals such as Sn and Ge also point to deviations from free or nearly free-electron-like behavior [68].

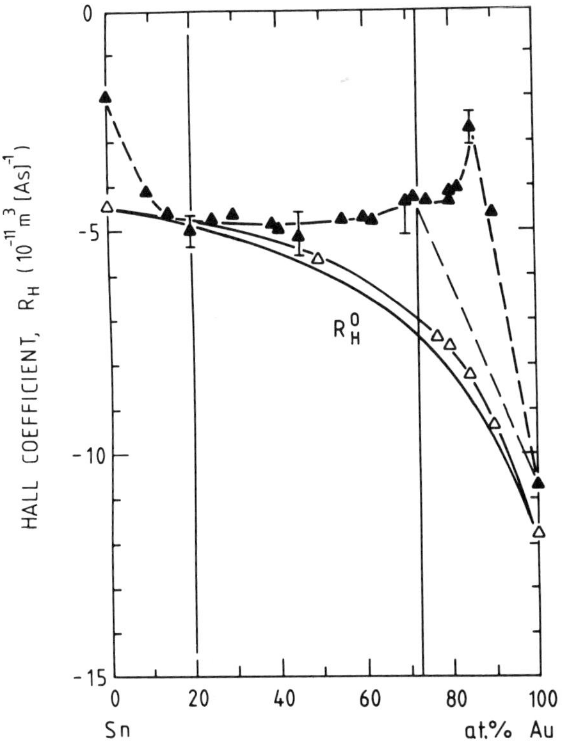

Figure 5 Hall coefficient of the as-quenched films (amorphous or partly crystalline) (solid prints) and molten alloy samples (open points) compared with the free electron values, R_H^0, versus composition [49].

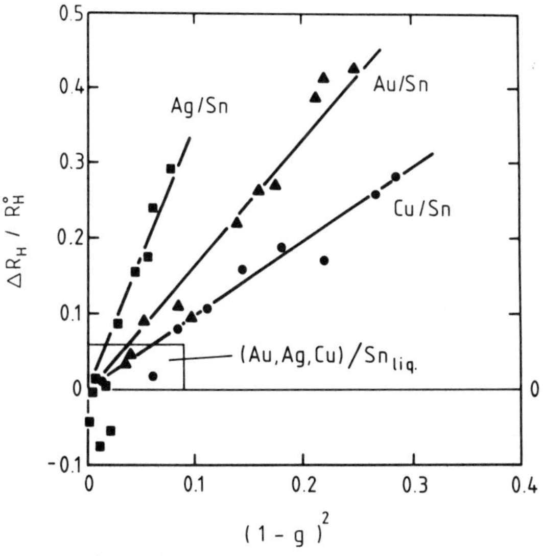

Figure 6 Relative deviation of the Hall coefficient from its free-electron value versus the square of the depth of the minimum in the density of states [62].

B. Late Transition Metal (LT)–Metalloid Metallic Glasses

Glassy alloys based on Fe, Ni, and Co have so far turned out to be the most industrially useful class of the metallic glasses. This has naturally promoted a great deal of research on this type of alloy. These glassy alloys are usually soft ferromagnets but their ordering temperature decreases as the amount of glass-forming metalloid elements is increased.

1. Resistivity

Late transition metal-based glassy alloys usually exhibit a positive but small TCR and the magnitude of their resistivity is usually between 100 and 150 μΩ-cm. Their temperature dependence with respect to resistivity usually exhibits a minimum at low temperatures. Beside this minimum, interesting changes appear by addition of Cr in this type of alloys, as studied by many authors such as Gudmundsson et al. [69], Olivier et al. [70], Jones et al. [71], and Rao [72]. The resistivity minimum is found to move to higher temperatures with increasing amounts of Cr, as in the case of $(Fe_{1-x}Cr_x)_{75}P_{16}B_6Al_3$ [69]; in the case of $Fe_{80-x}Cr_xB_{12}Si_8$ [71] and $(Fe\text{-}Cr)_{75}(P_{16}B_6Al_3)_{25}$ [72], a second minimum appears. Figure 7 demonstrates the change in resistivity behavior of Fe-B-Si metallic glasses with the addition of Cr. The work of Sadat-Akhavi et al. [73] on Fe-Ni-Cr-Si-B glassy metals seems to show that the observed anomalies caused by addition of Cr are indeed of magnetic origin. They found that, for large Fe concentrations, magnetism is ferromagnetic in character with local moments on Cr atoms coupling antiferromagnetically with Fe moments. For small or zero concentrations of Fe, the Cr local moments were found to couple with each other via conduction electrons and perhaps partially due to *d-d* interactions. They conclude that this coupling, which is on average antiferromagnetic in character, leads to susceptibility and resistivity anomalies at low temperatures. Using high-field-magnetization measurements, they showed the existence of magnetic clusters, some of which order below about 17 K. They consider the observed maximum for $Ni_{73}Cr_5Si_{10}B_{12}$ to be associated with a spin-freeze-out of the Kondo effect due to local moments on Cr atoms.

Initial investigations suggested that the increase in resistivity at lowest temperatures as logarithmic, which led to explanations based on the Kondo effect [74]. However, as the Kondo effect is associated with the presence of dilute amounts of magnetic impurities in otherwise nonmagnetic systems, it was surprising that this effect was also observable in concentrated ferromagnetic metallic glasses, for example, [75]. Grest and Nagel [76] explained this through computing the distribution of effective fields by introducing the superexchange caused by the metalloid atoms in such alloys in addition to Ruderman-Kittel-Kasuya-Yoshida (RKKY) and direct exchange interactions. The Monte Carlo calculations of the distribution function, $P(H)$, showed a long tail; i.e. $P(H)$ was found to be small but nonzero at small values of H. The existence of spins which are present at zero field were proposed to explain the existence of the Kondo effect and also the fact that there is no change in the effect by increasing the magnetic field in some cases.

Scattering from two-level systems (TLS) [77] as the other possible structural effect has been ruled out by some authors, as no change in temperature coefficient of resistivity is found upon annealing [70], whereas it is known that the number of TLS is drastically reduced by this treatment.

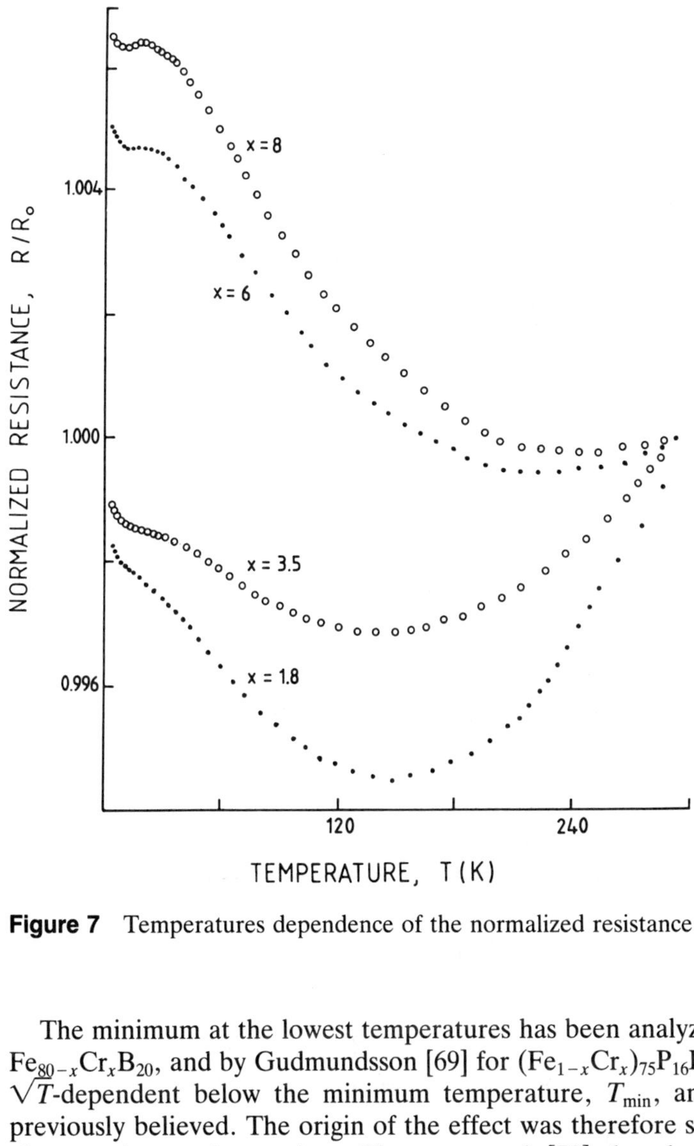

Figure 7 Temperatures dependence of the normalized resistance of $Fe_{80-x}Cr_xB_{12}Si_8$ [71].

The minimum at the lowest temperatures has been analyzed by Olivier [70] for $Fe_{80-x}Cr_xB_{20}$, and by Gudmundsson [69] for $(Fe_{1-x}Cr_x)_{75}P_{16}B_6Al_3$ and found to be $\sqrt{T}$-dependent below the minimum temperature, T_{min}, and not logarithmic as previously believed. The origin of the effect was therefore suggested to be due to electron-electron interactions. Thummes et al. [78] also observed such a behavior for the case of $Ni_{80-x}Fe_xB_{12}Si_8$. Figure 10 of Chapter 15 also shows the same phenomenon in the case of $Fe_{40}Ni_{40}P_{14}B_6$ and $Ni_{75}P_{16}B_6Al_3$ [79]. The second minimum in Fe-Cr-B alloys, on the other hand, has been found to be sensitive to applied magnetic field [70]. Therefore, a second mechanism, such as a Kondo effect or spin fluctuations, seems to be responsible for anomalies besides the characteristic $\sqrt{T}$ dependence due to electron correlation observed at very low temperatures.

2. *Hall Effect*

The Hall effect of this type of alloys is usually explained in terms of the anomalous Hall effect [80]. In addition to the ordinary Hall effect, R_0, which is a consequence of the Lorentz force, there can exist another contribution to the total Hall effect, called the spontaneous Hall effect, R_s. The latter arises when the spin-polarized carriers are scattered asymmetrically due to spin-orbit interaction. Whereas R_0 is proportional to the field $B = \mu_0(H + M)$, the anomalous part, R_s, is proportional to the magnetization. Then

$$\rho_H = R_0B + R_sM \tag{19}$$

A separation of the two contributions is possible at fields above those required to achieve magnetic saturation. In the ferromagnets, the slope of a ρ_H-B curve for a saturated region can be assumed to represent R_0 if the field dependence of the second term is neglected [81]. However, in the paramagnetic range, $M = \chi H$ and the total effect becomes proportional to B.

$$\rho_H = \left(R_0 + R_s\frac{\chi}{1 + \chi}\right) B \cong (R_0 + R_x\chi)B \tag{20}$$

A measurement of the total Hall effect, R_H, and susceptibility, χ, with respect to temperature then is sufficient to distinguish between the two components. A plot of R_H versus χ gives R_s from the slope and R_0 from the intercept of the data line extrapolated to $\chi = 0$. There have been a number of Hall effect measurements on magnetic glassy metals that have been interpreted using this explanation [82,83]. The positive Hall coefficients seen in Fe-rich alloys are thought to be due to the spontaneous part of the Hall effect that should disappear by application of a sufficiently high field. Figure 8 shows how the increase in Fe content in glassy $(Fe_xNi_{100-x})_{77}B_{13}Si_{10}$ alloys changes the sign of the Hall coefficient from negative to positive.

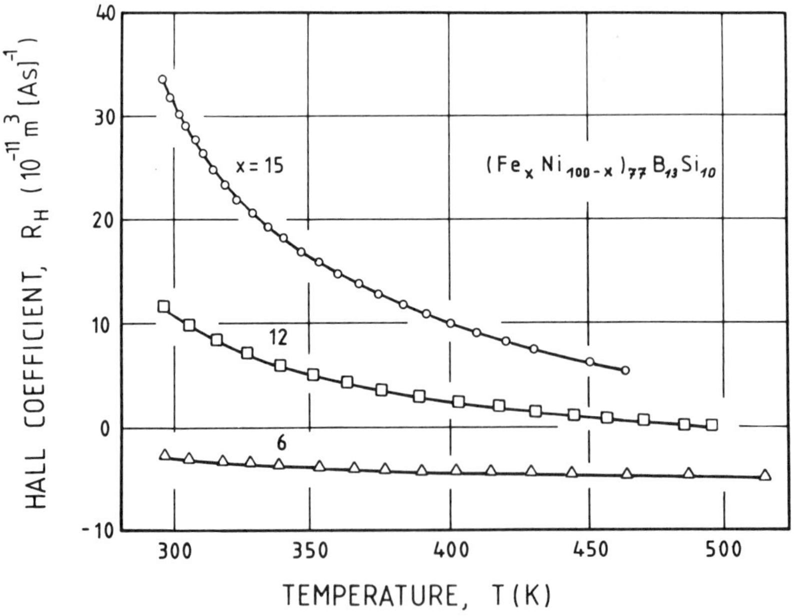

Figure 8 Hall coefficient versus temperature for $(Fe_xNi_{100-x})_{77}B_{13}Si_{10}$ metallic glasses [87].

C. Rare Earth (RE)–Transition Metal (TM) Amorphous Alloys

1. Resistivity

The electrical resistivity measurements for a series of rare earth elements in the liquid state has been reported by Güntherodt et al. [84]. In analysis of the temperature-dependence of resistivity, they found deviations from the usual linear behavior of the phonon resistivity at high temperatures. This was explained within the Mott *s–d* scattering model, which predicts such deviations when the electronic density of states has a pronounced structure in the neighborhood of the Fermi energy so that contributions to resistivity which arise from the expansion of the Fermi distribution function become important. The TCR values of rare earth/tin molten alloys were in qualitative agreement with those of the extended Ziman model, which showed a negative TCR for concentrations corresponding to two electrons per atom and $2k_F$ of the order of K_P.

The magnitude of resistivity, ρ, and also TCR values for a number of glassy $RE_{65}Co_{35}$ alloys in comparison to the values for pure RE liquids are shown in Fig. 9 [85,86,87]. As the figure indicates, the magnitude of resistivity of metallic glasses increases in value from $La_{65}Co_{35}$ to $Ho_{65}Co_{35}$, and the increase of this resistivity

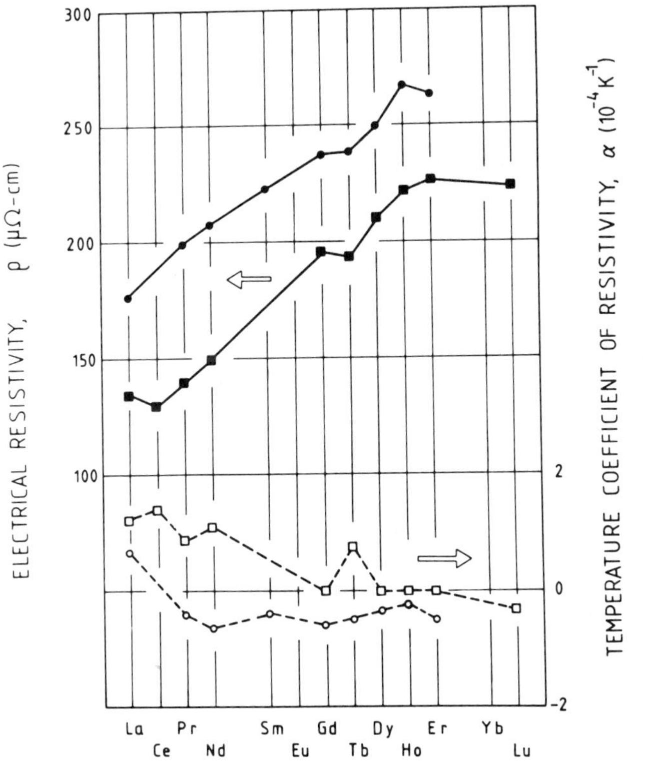

Figure 9 Resistivity (solid lines) and its temperature coefficient (dashed lines) values for pure molten rare earth (squares) and $RE_{65}Co_{35}$ metallic glass samples [85–87].

nearly parallels that of the corresponding pure liquids, indicating similarities in electronic transport in the two systems.

Other such alloys so far reported include amorphous Ni_3RE alloys obtained by sputtering [88,89] and TM_3Re_4 obtained by melt spinning [90]. These were shown to exhibit rather unusual $\rho(T)$ behaviors which should be studied in relation to the magnetic properties of these alloys, as metallic glasses containing rare earth metals exhibit a wide variety of magnetic properties ranging from ferromagnetism to diamagnestism; e.g., alloys rich in Gd show ferromagnetic behavior whereas non-magnetic alloys with dilute amounts of Gd exhibit spin glass or mictomagnetic properties (see also Chapter 16).

The magnetic model for these alloys is based on random magnetic anisotropy (RMA) which arises from the coupling of the non-*S*-state rare earth ions to their immediate environment. The magnetic properties of rare-earth-containing metallic glasses is also dependent on the nature of strong electrostatic fields ("crystal" fields) acting on the RE atoms, affecting the *f*-electron spins in much the same way as those in corresponding crystalline materials despite the lack of symmetry. The interrelated magnetic and electronic transport properties of a series of amorphous Ni_3RE alloys have been studied by Asomoza et al. using the RMA model [88,89]. Almost all samples show a minimum in resistivity at lowest temperatures. In the case of Ni-Gd, a maximum appears at temperatures above T_{min}. Nonmagnetic Ni-Y is also shown to exhibit a minimum in the resistivity. The Dy, Ho, and Er moments were found to order asperomagnetically at low temperatures (Chapter 16). To explain the resistivity anomalies, the importance of the additional contribution to resistivity from coherent exchange scattering by neighboring ions has been invoked [89]. The contribution from independent scattering by the magnetic ions has the same value above and below the magnetic ordering temperature. On the other hand, the contribution resulting from coherent exchange scattering by neighboring ions (although negligible above the magnetic ordering temperature) is nonzero where the spins are correlated. This can then give rise to a positive (when constructive) or a negative (when destructive) contribution to resistivity, in contrast to the behavior in crystalline alloys in which a drop in resistivity due to freezing of the spin-flip processes is exhibited. This argument therefore explains the anomalies in resistivity appearing at the magnetic ordering temperature. However, anomalies also occur at temperatures that do not correspond to the magnetic ordering temperature, such as the minima in resistivity for nonmagnetic $Ni_{3.5}Y$ and ferromagnetic Ni_3Gd. The origin of these has been proposed to be of structural origin [89].

Measurements on Ce-rich Ce-Au, Ce-Cu, and Ce-Al amorphous alloys showed low temperature resistivity and thermopower anomalies such as a pronounced maximum in the thermopower together with a steep rise in resistivity at the same temperature [91,92]. Felsch and Schröder [91] regard these anomalies to be due to magnetic effects. Other RE-rich amorphous alloys studied included $Yb_{80}X_{20}$ with X = Cu, Ag, and Au [93].

Cochrane et al. [94] have studied Ni-rich Ni-Y, also Fe_2Y and Fe_5Y amorphous alloys. They again found a minimum at ~10 K in Ni-Y alloys, which show a positive TCR at higher temperatures. The onset of magnetic order was seen in the $\delta\rho/\delta T$ curve as a sharp step around the magnetic ordering temperature. Fe-Y on the other hand showed a negative TCR at all temperatures and no effect was observed at the ordering temperature.

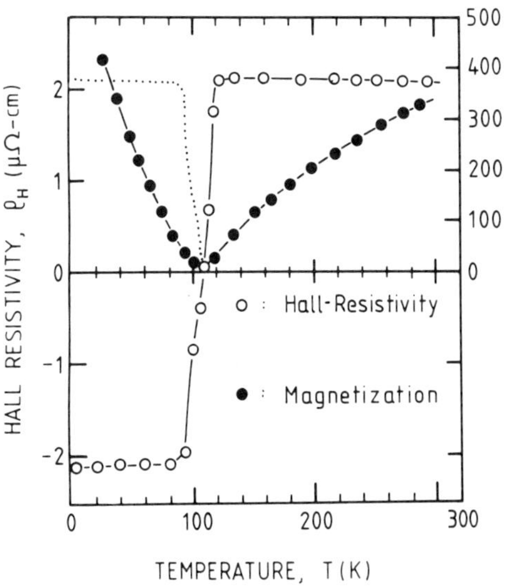

Figure 10 Temperature dependence of magnetization and Hall resistivity of a $Gd_{17}Co_{83}$ sputtered film [95].

2. Hall Effect

The Hall effect of rare-earth-containing amorphous alloys has been discussed by McGuire et al. [60]. In particular, they considered the case of Gd-Co in which the two ions seem to exhibit magnetic interaction. Shirakawa et al. [95] and Ogawa et al. [96] both observed a sharp reversal of the sign of the Hall voltage for $Gd_{17}Co_{83}$ and $Gd_{19}Co_{81}$, respectively. The results of Shirakawa et al. are shown in Fig. 10. This phenomenon is attributed to the reversal of Gd and Co magnetizations with respect to the field direction. McGuire et al. [60] therefore proposed that the rare earth and the transition metal elements both contribute to the Hall conductivity.

Asomoza et al. [89] obtained positive Hall coefficients for Ni_3Gd, Ni_3Dy, and Ni_3Ho. This is attributed to the extraordinary Hall effects exhibited by all these alloys.

D. Nonmagnetic Early–Late Transition Metal Amorphous Alloys

The body of research on nonmagnetic metallic glasses containing transition metals consists of experiments performed on those containing Ti, Zr, and Hf. These elements, in combination with late transition metals, may be prepared as metallic glasses over a wide concentration range so that systematic studies become possible. A detailed review of the transport properties of these materials has already been presented by Howson and Gallagher [97] only recently. Here we would like to consider the example of refractory amorphous alloys.

Refractory metallic glasses are of considerable interest due to their ceramic-like mechanical properties. Figure 11 shows the very high crystallization temperatures exhibited by this type of metallic glasses. Another interesting behavior of such alloys is their rather high superconducting temperatures. For certain alloys, the superconducting transition temperature is much higher than that of the crystalline

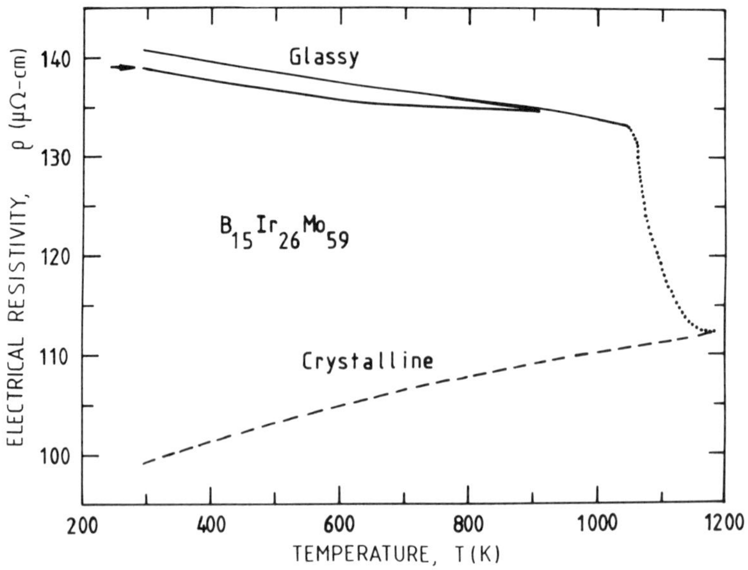

Figure 11 Electrical resistivity of glassy $B_{15}Ir_{26}Mo_{59}$ alloy versus temperature [101].

phase, as in the cases of Mo-Ru and W-Re alloys. This phenomenon has been discussed in detail by Collver and Hammond on vapor-quenched amorphous alloys [98]. They derived a general behavior for the superconducting transition temperature of amorphous alloys with respect to *e*/*a* ratio (Fig. 2, Chapter 15). Besides the measurements by Collver and Hammond [98], Johnson and Williams [99] have also studied the superconducting behavior of refractory metals such as glassy Mo-Ru-metalloid foils obtained by the piston-and-anvil method. Koch et al. [100] have made detailed studies on different aspects of the electronic transport properties of glassy (Nb-Ta)-Ir and (Nb-Ta)-Rh alloys at low temperatures. $(Nb\text{-}Ta)_{55}Ir_{45}$ and $(Nb\text{-}Ta)_{55}Rh_{45}$ metallic glasses have also been studied by Richmond et al. [101], who also report Hall coefficient measurements.

A new series of metallic glasses obtained over a wide concentration range is Ni-Ta [102]. The magnitude of resistivity, TCR values, the Hall coefficient and the magnetic susceptibility values at room temperature for this series of amorphous alloys are shown in Fig. 12. The inset shows that the Hall coefficient may be extrapolated to that of the molten Ni (-11.5×10^{-11} $m^3[As]^{-1}$) [58]. Apparently the maximum of resistivity, the minimum in TCR and also in the magnetic susceptibility, and the change of sign of R_H occur in the same concentration range. The reason is not yet clear and has to be further investigated.

Figure 13 shows the variation of electrical conductivity of three alloys over the temperature range between 4.2 and 300 K. According to discussions based on quantum interference effects, the temperature dependence of conductivity is predicted as linear below T' and proportional to $\sqrt{T}$ for $T > T'$, where T' is proposed to be a third of the Debye temperature [103]. Figures 13 and 14 seem to support this general picture, as a square-root dependence between 100 and 300 K is evident and a linear T-dependent between 55 and 120 K is also detectable [101]. Figure 14 also indicates the existence of a $\sqrt{T}$ behavior of conductivity at lowest tem-

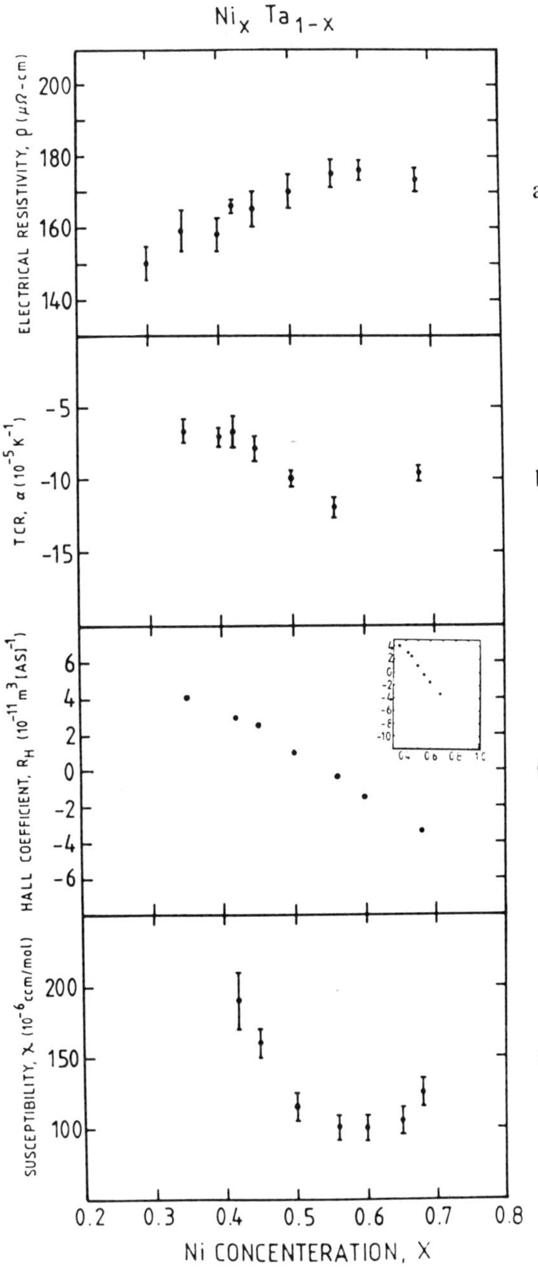

Figure 12 (a) Electrical resistivity, (b) temperature coefficient of resistivity, (c) Hall coefficient, and (d) magnetic susceptibility of Ni_xTa_{1-x} metallic glasses versus the Ni concentration, x [101,121]. The inset shows that the Hall coefficient values may be extrapolated to the value for molten Ni (-11.5×10^{-11} $m^3[As]^{-1}$) [58].

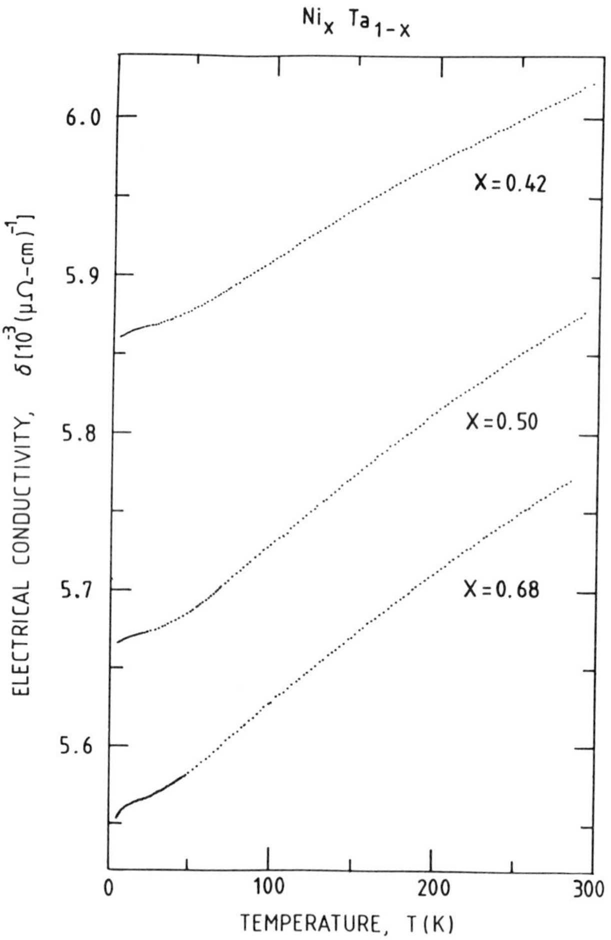

Figure 13 Electrical conductivity of Ni_xTa_{1-x} alloys versus temperature. The numbers denote the Ni concentration, x [102].

peratures, characteristic of electron correlation effects. However, while a square-root fit with a high correlation coefficient was obtained for all samples at temperatures below 20 K, a $\ln(T)$ fit was found to be more appropriate in the case of the sample with Ni concentration of 0.68. Kondo-type effects may be a cause for this deviation. The above behavior, as described in terms of weak localization and electron correlation effects, are reported by a number of authors for Cu-Ti and Ti-Be-Zr [34], Cu-Zr [103], and Ni, Co, Fe-Zr [104] metallic glasses.

1. *Hall Effect*

As yet there is no generally accepted explanation for the occurrence of positive Hall coefficients in glassy metals containing transition metals. One approach to the problem has been to calculate corrections to the nearly free electron result for R_H. Mott [63] proposed the expression

$$R_H = \frac{R_0}{N(E_F)/N_{FE}(E_F)} \tag{21}$$

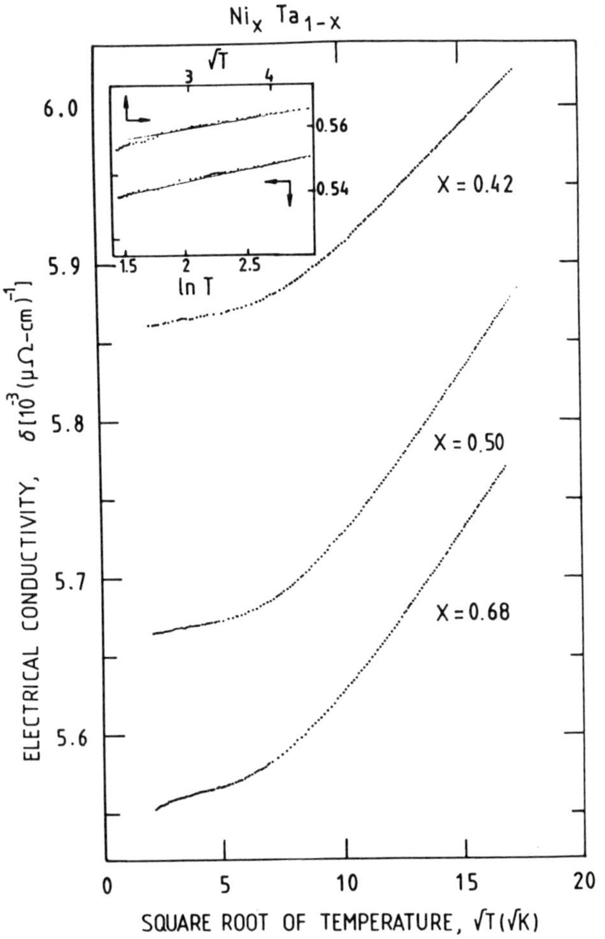

Figure 14 Electrical conductivity of Ni_xTa_{1-x} alloys versus square-root of temperature. The numbers denote the Ni concentration, x [102].

While this correction accounts for the deviations from free electron values, it cannot explain the positive sign.

Harris [105] drew a correlation between the Hall coefficient and the thermopower using the Faber-Ziman theory, and provided a qualitative explanation for the occurrence of the positive Hall coefficient. However Harris' calculations are based on the assumption that the scattering is weak. He suggested that a quantitative solution is possible if the strong scattering due to transition metals is taken into account.

Another approach involves recalculation of the electronic structure. Datta [106] generalized the concept of effective mass so that it can still be valid in cases where k is no longer a good quantum number. Datta obtained an expression for the Hall mobility involving matrix elements of the velocity operator. However this was not directly applied to the Hall coefficient.

Weir et al. [107], on the other hand, carried out a self-consistent perturbation calculation on the effects of hybridization on the electronic structure. Their calculations led to a S-shaped dispersion curve for cases where the s-band was modified at the Fermi level due to mixing with the d-band. The part of the $E(k)$ corresponding to negative group velocities was then suggested as a direct explanation for the positive Hall effects in transition metal-containing glassy alloys. From this theory, it is concluded that the dispersion relation, $E(k)$, is still valid for strong scattering systems and that it only slightly depends on the details of the structure. In systems where E_F lies near the top of the d-band in a region where $\delta E/\delta k$ is positive, a negative Hall coefficient is expected, whereas the E_F in the central regions causes a positive Hall effect. The calculations of Weir et al. [107] have since been challenged by Bose et al. [16], who disagree on the existence of a negatively sloped section in $E(k)$.

It seems to be the common consensus that the positive Hall effect is related to the electronic structure of the transition metal-containing metallic glasses and that it is not a structural effect. Electronic structure studies show that for the Zr-rich amorphous alloys, which show a positive Hall coefficient, the Fermi level lies in the Zr d-band region. The Fermi level shifts into the TL d-band region for alloys low in Zr, which exhibit negative Hall coefficients [109,110]. As regards Ni-Ta alloys, to our knowledge the only UPS study available has been performed on $Ni_{0.63}Ta_{0.32}$ by Oelhafen et al. [111], where they report a distinct shift of Ni d-band to higher binding energies relative to that of pure Ni. Therefore, as an approximation, it may be said that as Ni-Nb alloys show behavior similar to that of Zr alloys and since there is little change in the band structure by replacement of an early transition metal by another of the same group [111], it might be expected that the same argument can apply to Ni-Ta alloys.

The concentration at which the Fermi level shifts from the early transition metal d-band to the late transition metal d-band has been proposed by Jacobs [112], and later adopted by Ivkov et al. [109], to be given by

$$n_{TL}x + n_{TE}(1 - x) = 10x \tag{22}$$

where n stands for the number of d-electrons of a particular atom. Taking $n_{\mathrm{Ta}} = 3$ and $n_{\mathrm{Ni}} = 9$, the crossover value is 0.75, which is 0.21 higher than the crossover seen experimentally for the Hall coefficient as can be seen from Fig. 12. In studying glassy CoZr, NiZr and CuZr alloys, Ivkov et al. [109] also found that the R_H crossover concentration was 0.2 lower than that predicted by this formula. This constant difference was explained by the fact that the Zr subband remains almost unchanged around the Fermi level when one late 3d element is replaced with another.

According to Eq. (15), the Hall coefficient in simple metals is temperature-independent. Although most nonmagnetic metallic glasses do indeed show a temperature-independent Hall coefficient, there has recently been a number of reports wherein a small temperature dependence is detected in some systems. Specific alloys include $Pd_{30}Zr_{70}$, $Ni_{64}Zr_{36}$, $Fe_{24}Zr_{76}$, and $Cu_{50}Ti_{50}$, as studied by Sculte et al. [113], who reported a change in magnitude of the Hall coefficient ranging between 2% for Cu-Ti to 10% in the case of Pd-Zr over a temperature range of 300 K. It should be mentioned that as R_H is proportional to molar volume, it is sensitive to thermal expansion of the sample. A small increase in R_H upon heating is therefore expected. As estimated by Löbl [114], thermal expansion over a temperature range of 1.6–300 K corresponds to a relative change in R_H of about 3 ×

10^{-3} of the volume. The results of Schulte et al. [113] however show a temperature dependence of opposite sign; i.e., they report a decrease in the Hall coefficient with increasing temperature. Gallagher et al. [115] interpret this behavior as further evidence for the presence of electron correlation effects in disordered metals. The change in Hall coefficient is proposed to correlate with the change in conductivity as

$$\frac{\Delta R_H}{R_H} = \frac{2\,\Delta\sigma}{\sigma} \propto \sqrt{\frac{kT}{2\hbar D}} \tag{23}$$

Schulte et al. [116] find that their data indeed follow a $\sqrt{T}$ dependence.

In an analysis of Hall coefficient and magnetic susceptibility data on Fe-Zr metallic glasses, Trudeau et al. [117] proposed that the side jump effect is responsible for the temperature dependence of R_H in some paramagnetic systems. This argument had been previously only considered for ferromagnetic metals. However, Trudeau et al. [117] suggest that in a high-resistivity paramagnetic material with a large spin-orbit scattering, the side jump effect can become significant. A linear correlation was indeed observed between the Hall coefficient and magnetic susceptibility of Fe-Zr metallic glasses [117].

IV. CONCLUSION

The present survey of the main experimental results on different types of metallic glasses and amorphous alloys points to one common feature in the resistivity behavior of these materials: the $\sqrt{T}$-dependence with a negative TCR at lowest temperatures. Cochrane and Strom-Olsen [110] have demonstrated that this is a general behavior related to the high resistivity of the system and not to the amorphous state since the effect is observed in systems such as amorphous Nb-Si [118], crystalline Si-P [119], and granular Al [120] as well as in metallic glasses such as $Y_{77.5}Al_{22.5}$ and $Fe_{80}B_{20}$. The other feature pertaining to metallic glasses is the qualitative agreement of the experimental results with the predictions of the theory of weak localization in the case of the high-resistivity alloy chemistries. Although a number of additional effects such as spin-orbit interactions or superconducting fluctuations modify the overall picture, Fig. 3 seems to depict the general behavior observed in nonmagnetic high-resistivity metallic glasses. Similarities exist between the quantum interference effects and the phonon-assisted tunneling model, which also demonstrates very good agreement with the experimental data, as shown in Fig. 2. A detailed examination of these similarities is as yet lacking.

As regards the question of the positive Hall coefficient in the metallic glasses, it seems clear that this is an electronic structure-related phenomenon rather than a structural one. However, much remains to be done to confirm the interesting new explanations put forward for the cause of the positive sign and its temperature dependence.

ACKNOWLEDGMENT

We are indebted to D. Belitz, P. Häussler, D. F. Jones, M. Kaveh, U. Mizutani, N. F. Mott, W. Schirmacher, and T. Shirakawa for allowing us to present their figures in this article. We also thank our colleagues M. Liard, A. Tschumi, and Th. Zingg for their valuable contributions. The research in this field has been supported by the Swiss National Science Foundation.

REFERENCES

1. J. M. Ziman, *Philos. Mag.*, *6*, 1013 (1961).
2. J. M. Ziman, *Adv. Phys.*, *13*, 89 (1964).
3. T. E. Faber and J. M. Ziman, *Philos. Mag.*, *11*, 153 (1964).
4. R. Evans, D. A. Greenwood, and P. Lloyd, *Phys. Lett. A*, *35*, 57 (1971).
5. N. F. Mott, *Philos. Mag.*, *26*, 1249 (1972).
6. D. Brown, S. Fairbrain, and G. J. Morgan, *Phys. Stat. Sol.*, *93*, 617 (1979).
7. P. J. Cote and L. V. Meisel, *Phys. Rev. Lett.*, *39*, 102 (1977).
8. K. Froböse and J. Jäckle, *J. Phys. F: Metal Phys.*, *7*, 2331 (1977).
9. G. Baym, *Phys. Rev. A*, *135*, 1691 (1964).
10. D. Markowitz, *Phys. Rev. B*, *15*, 3617 (1977).
11. J. H. Mooij, *Phys. Stat. Sol. (a)*, *17*, 521 (1973).
12. P. J. Cote and L. V. Meisel, *Phys. Rev. Lett.*, *40*, 1586 (1978).
13. L. V. Meisel and P. J. Cote, *Phys. Rev. B*, *27*, 4617 (1983).
14. A. ten Bosch and K. H. Bennemann, *J. Phys. F: Metal Phys.*, *5*, 1333 (1975).
15. G. F. Weir and G. J. Morgan, *J. Phys. F: Metal Phys.*, *11*, 1833 (1981).
16. S. K. Bose, L. E. Ballentine, and J. E. Hammerberg, *J. Phys. F: Metal Phys.*, *13*, 2089 (1983).
17. L. E. Ballentine, S. K. Bose, and J. E. Hammerberg, *J. Non-Cryst. Solids*, *61–62*, 1195 (1984).
18. D. Belitz and W. Götze, *J. Phys. C: Solid State Phys.*, *15*, 981 (1982).
19. D. Belitz and W. Schirmacher, *J. Phys. C: Solid State Phys.*, *16*, 913 (1983).
20. P. B. Allen and Chakraborty, *Phys. Rev. B*, *23*, 4815 (1981).
21. W. Götze, *Philos. Mag.*, *43*, 219 (1981).
22. D. Belitz and W. Schirmacher, in Ref. 17, p. 1073.
23. Y. Imry, *Phys. Rev. Lett.*, *44*, 499 (1980).
24. P. Lee and T. V. Ramakrishnan, *Rev. Mod. Phys.*, *57*, 287 (1985).
25. P. W. Anderson, *Phys. Rev.*, *109*, 1492 (1958).
26. D. J. Thouless, *Phys. Rep.*, *13*, 93 (1974).
27. E. Abrahams, P. W. Anderson, D. C. Licciardello, and T. V. Ramakrishnan, *Phys. Rev. Lett.*, *42*, 673 (1979).
28. A. Larkin and D. E. Khmel'nitskii, *Sov. Phys.-Usp.*, *25*, 185 (1982).
29. G. Bergmann, *Phys. Rev. B*, *28*, 2914 (1983).
30. G. Bergmann, *Phys. Rev. B*, *28*, 515 (1983).
31. M. Kaveh and N. F. Mott, *J. Phys. C: Solid State Phys.*, *14*, L177 (1981).
32. M. Kaveh and N. F. Mott, *J. Phys. C: Solid State Phys.*, *15*, L697 (1982).
33. M. Kaveh and N. F. Mott, in Ref. 31, p. L707.
34. M. A. Howson, *J. Phys. F: Metal Phys.*, *14*, L25 (1984).
35. S. R. Nagel, J. Vassiliou, P. M. Horn, and B. C. Giessen, *Phys. Rev. B*, *17*, 462 (1978).
36. J. Kästner, H. J. Schink, and E. F. Wassermann, *Solid State Commun.*, *33*, 527 (1980).
37. C. C. Tsuei, *Phys. Rev. Lett.*, *57*, 1943 (1986).
38. B. L. Al'tshuler and A. G. Aronov, *Solid State Commun.*, *30*, 115 (1979).
39. T. F. Rosenbaum, K. Andres, G. A. Thomas, and P. A. Lee, *Phys. Rev. Lett.*, *46*, 568 (1981).
40. M. A. Howson and D. Greig, *Phys. Rev. B*, *30*, 4805 (1984).

41. B. L. Gallagher, D. Greig, M. A. Howson, and A. A. M. Croxon, *J. Phys. F: Metal Phys.*, *13*, 119 (1983).
42. H. Chen, A. A. M. Croxon, and D. Greig, *J. Non-Cryst. Solids*, *86*, 94 (1986).
43. R. Malmhäll, G. Backström, S. M. Bhagat, and K. V. Rao, *J. Non-Cryst. Solids*, *28*, 159 (1978).
44. J. Hafner, E. Gratz, and H.-J. Güntherodt, *J. Phys. Coll.*, *41*, C8 512 (1980).
45. U. Mizutani and K. Yoshino, *J. Phys. F: Metal Phys.*, *14*, 1179 (1984).
46. T. Matsuda, N. Shiotani, and U. Mizutani, *J. Phys. F: Metal Phys.*, *1974*, 1193 (1984).
47. U. Mizutani and T. Matsuda, *J. Non-Cryst. Solids*, *94*, 345 (1987).
48. D. Korn, H. Pfeifle, and G. Zibold, *Z. Phys.*, *270*, 195 (1974).
49. P. Häussler and F. Baumann, *Z. Phys. B*, *49*, 303 (1983).
50. P. Häussler, W. H.-G. Müller, and F. Baumann, *Z. Phys. B*, *35*, 67 (1979).
51. P. Häussler, *Z. Phys. B*, *53*, 15 (1983).
52. U. Mizutani and T. Yoshida, *J. Phys. F: Metal Phys.*, *12*, 2331 (1982).
53. U. Mizutani, *Progr. Mater. Sci.*, *28*, 97 (1983).
54. D. P. Love, F. C. Wang, D. G. Naugle, C. L. Tasi, B. C. Giessen, and T. O. Callaway, *Phys. Lett.*, *90A*, 303 (1982).
55. D. G. Naugle, R. Delgado, H. Armbrüster, C. L. Tsai, T. O. Callaway, D. Reynolds, and V. L. Moruzzi, *Phys. Rev. B*, *34*, 8279 (1986).
56. U. Mizutani, M. Sasaura, Y. Yamada, and T. Matsuda, *J. Phys. F: Metal Phys.*, *17*, 667 (1987).
57. U. Mizutani, K. Sato, I. Sakamoto, and K. Yonemitsu, *J. Phys. F: Metal Phys.*, *18*, 1995 (1988).
58. G. Busch and H.-J. Güntherodt, *Solid State Phys.*, *29*, 235 (1974).
59. H. U. Künzi and H.-J. Güntherodt, *The Hall Effect and Its Application* (Plenum, New York, 1980), p. 215.
60. T. R. McGuire, R. J. Gambino, and R. C. O-Handley, in Ref. 59, p. 137.
61. T. Matsuda and U. Mizutani, *J. Phys. F: Metal Phys.*, *12*, 1877 (1982).
62. P. Häussler, *Habilitationsschrift* (Karlsruhe Universität, 1989).
63. N. F. Mott, *Philos. Mag.*, *13*, 989 (1966).
64. S. R. Nagel and J. Tauc, *Phys. Rev.*, *B 35*, 380 (1975).
65. H. Beck and R. Oberle, in *Rapidly Quenched Metals III*, Vol. 1, edited by B. Cantor (The Metal Society, London, 1978), p. 416.
66. D. Nicholson and L. Schwartz, *Phys. Rev. Lett.*, *49*, 1050 (1982).
67. P. Häussler and F. Baumann,*Z. Phys. Chem. Neue Folge*, *157*, 471 (1988).
68. G. Indlekofer and P. Oelhafen, *Disordered Systems and New Materials* (World Scientific, Singapore, 1989), p. 707.
69. H. Gudmundsson, H. J. Hannesson, and H. U. Aström, *J. Appl. Phys.*, *57*, 3523 (1985).
70. M. Olivier, J. O. Strom-Olsen, and Z. Altounian, *Phys. Rev. B*, *35*, 333 (1987).
71. D. F. Jones, G. Stronik, Z. M. Stadnik, and R. A. Dunlap, *Mater. Sci. Eng.*, *99*, 207 (1988).
72. K. V. Rao, in *Amorphous Metallic Alloys*, edited by F. E. Luborsky (Butterworths, Boston, 1983), p. 401.
73. H. Sadat-Akhavi, G.Hadjipanayis, and D. J. Sellmyer, *Phys. Rev. B*, *24*, 5318 (1981).
74. R. Hasegawa and C. C. Tsuei, *Phys. Rev. B*, *2*, 1631 (1970).
75. O. Rapp, S. M. Bhagat, and Ch. Johannesson, *Solid State Commun.*, *21*, 83 (1977).
76. G. S. Grest and S. R. Nagel, *Phys. Rev. B*, *19*, 3571 (1979).
77. R. Cochrane, R. Harris, J. Ström-Olsen, and M. Zukermann, *Phys. Rev. Lett.*, *35*, 676 (1975).
78. G. Thummes, R. Ranganathan, and J. Kötzler, *Z. Phys. Chem. Neue Folge*, *157*, 699 (1988).

79. Ö. Rapp, S. M. Bhagat, and H. Gudmundsson, *Solid State Commun.*, *42*, 741 (1982).
80. L. Berger and G. Bergmann, in Ref. 59, p. 55.
81. G. Bergmann,*Phys. Lett.*, *60A*, 245 (1977).
82. Y. Marohnic, E. Babic, and D. Pavuna, *Phys. Lett.*, *63A*, 348 (1977).
83. K. V. Rao, R. Malmhäll, G. Bäckström, and S. M. Bhagat, *Solid State Commun.*, *19*, 193 (1976).
84. H. J. Güntherodt, E. Hauser, and H. U. Künzi, *Inst. Phys. Conf. Ser.*, *30*, 324 (1977).
85. B. Delley, H. Beck, H. U. Künzi, and H.-J. Güntherodt, *Phys. Rev. Letters*, *40*, 193 (1978).
86. M. Liard, Ph.D. thesis, Univ. Basel, Switzerland (1980).
87. A. Tschumi, Ph.D. thesis, Univ. Basel, Switzerland (1984).
88. R. Asomoza, A. Fert, I. A. Cambell, and R. Meyer, *J. Phys. F: Metal Phys.*, *7*, L327 (1977).
89. R. Asomoza, I. A. Campbell, A. Fert, A. Lienard, and J. P. Rebouillat, *J. Phys. F: Metal Phys.*, *9*, 349 (1979).
90. A. Apostolov, H. Sassik, L. Iliev, and M. Mikhov, in Ref. 78, p. 705.
91. W. Felsch and H. Schröder, *Solid State Commun.*, *45*, 1043 (1983).
92. H. Schröder and W. Felsch, *J. Non-Cryst. Solids*, *56*, 219 (1983).
93. W. Felsch, J. Tebbe, D. Weschenfelder, and G. Czjzek, in *Rapidly Quenched Metals*, edited by S. Steeb and H. Warlimont (Elsevier, New York, 1985).
94. R. W. Cochrane, J. Ström-Olsen, Gwyn Williams, A. Lienard, and J. P. Rebouillat, *J. Appl Phys.*, *49*, 1677 (1978).
95. T. Shirakawa, Y. Nakajima, K. Okamoto, S. Matsushita, and Y. Sakurai, *AIP Conf. Proc.*, *34*, 349 (1976).
96. A. Ogawa, T. Katayama, M. Hiramo, and T. Tsushima, *Appl. Phys. Suppl.*, *15*, 87 (1976).
97. M. A. Howson and B. L. Gallagher, *Phys. Rep.*, *170*, 265 (1988).
98. M. M. Collver and R. H. Hammond, *Phys. Rev. Lett.*, *30*, 92 (1973).
99. W. L. Johnson and A. R. Williams, *Phys. Rev. B*, *20*, 1640 (1979).
100. C. C. Koch, D. M. Kroeger, J. O. Scarbrough, and B. C. Giessen,*Phys. Rev. B*, *22*, 5213 (1980).
101. T. Richmond, Ph.D. thesis, Univ. Basel, Switzerland (1991).
102. T. Richmond, L. Rohr, P. Reimann, and H.-J. Güntherodt, *Mater. Sci. Eng.*, *A133*, 63 (1991).
103. K. Saub, E. Babic, and R. Ristic, *Solid State Commun.*, *53*, 269 (1985).
104. E. Babic and K. Saub, *Solid State Commun.*, *56*, 111 (1985).
105. R. Harris, *J. Phys. F: Metal Phys.*, *16*, 53 (1986).
106. S. Datta, *Phys. Rev. Lett.*, *44*, 828 (1980).
107. G. F. Wier, M. A. Howson, M. A. Gallagher, and G. J. Morgan, *Phil. Mag.*, *B 47*, 163 (1983).
108. L. E. Ballentine, in Ref. 59, p. 201.
109. J. Ivkov, E. Babic, and R. L. Jacobs, *J. Phys. F: Metal Phys.*, *14*, L53 (1984).
110. R. W. Cochrane and J. O. Ström-Olsen, *Phys. Rev. B*, *29*, 1088 (1984).
111. P. Oelhafen, 2Glassy Metals II (Springer, Berlin, 1983).
112. R. L. Jacobs, *Phys. Stat. Sol. (b)*, *107K*, 13 (1981).
113. A. Schulte, A. Roithmayer, G. Fritsch, and E. Lüscher, *Amorphous and Liquid Metals* (NATO ASI series, 1987).
114. P. Löbl, *Diploma Work* (München 1983).
115. B. L. Gallagher, A. B. Kaiser, and D. Greig, in Ref. 17, p. 1231.
116. A. Schulte, A. Eckert, G. Fritsch, and E. Lüscher, *J. Phys. F: Metal Phys.*, *14*, 1877 (1984).

117. M. Trudaeu, R. W. Cochrane, D. V. Baxter, J. O. Ström-Olsen, and W. B. Muir, *Phys. Rev. B*, 4499 (1988).
118. G. Hertel, D. J. Bishop, E. G. Spencer, J. M. Rowell, and R. C. Dynes, *Phys. Rev. Lett.*, *50*, 743 (1983).
119. T. F. Rosenbaum, K. Andrei, G. A. Thomas, and R. N. Bhatt, *Phys. Rev. Lett*, *45*, 1723 (1980).
120. R. C. Dynes and J. Garno, *Phys. Rev. Lett.*, *46*, 137 (1981).
121. G. Leemann, Ph.D. thesis, Univ. Basel, Switzerland (1991).

15

Low-Temperature Properties of Rapidly Solidified Alloys

H. v. Löhneysen
Institute of Physics, University of Karlsruhe, Karlsruhe, Germany

I. INTRODUCTION

In this chapter, the low-temperature properties of rapidly solidified alloys will be reviewed. As outlined in previous chapters, rapid solidification results in a variety of structures, for example, amorphous, microcrystalline, quasicrystalline. In this chapter, emphasis will be put on amorphous solids.* Many amorphous alloys exhibit superconductivity, and with the newly exploded interest in this field, it is only fair to begin with a brief elaboration of the superconducting properties of these materials, even if, nowadays, superconductivity is no longer restricted to very low temperatures. After a brief review of the basic features of BCS superconductors in Section IIA, Sections IIB and C will treat superconductivity in amorphous simple metals and in amorphous alloys containing transition elements. Critical fields and electron-localization effects will be the subject of Section IID. We will also discuss superconductivity of quasi- and microcrystalline alloys (Section IIE). In Section IIF we will briefly review the impact of rapid solidification techniques on high-T_c superconductors. In the spirit of this book, we will restrict ourselves to rapid solidification from the melt. Vapor deposition, sputtering, and laser ablation techniques which are, of course, also rapid solidification processes, are widely used to prepare high-T_c films. However, even a rough overview would be beyond the scope of this chapter, in view of the numerous papers and rapid advance in this subject.

Sections III and IV deal with low-temperature electronic and vibrational properties which are akin to the disordered state of matter. Section three treats electron localization in disordered solids. A fundamental theorem of solid-state physics, the

*A note on usage: here we use "amorphous" as a generic term applied to any solid lacking long-range order. Although "amorphous metal" and "metallic glass" are often used interchangeably, we use the latter term only for amorphous metals quenched from the melt.

Bloch theorem, states that the stationary electron states of a perfect crystal are extended. Breaking translational symmetry by introducing disorder ultimately leads to electron localization, as first recognized by Anderson [1]. A one-parameter scaling approach [2] in which a (dimensionless) conductance is the only relevant parameter, proved very fruitful. However, as the electron states become increasingly localized, that is, the spatial extent of the wave functions becomes smaller and smaller, electron–electron interactions become successively more important. These two aspects will be treated in Sections IIIA and B. In addition, Section IIIB contains the experimental observations in amorphous metals, including the problem of strong electron localization as one approaches the metal-insulator transition.

Section IV treats the atomic dynamics at low temperatures. It has long been known that amorphous dielectrics exhibit an enhanced specific heat and a much reduced (phonon) thermal conductivity at low temperatures, with respect to their crystalline counterparts. Zeller and Pohl [3] were the first to point out that these features are related, and suggested that the excess specific heat, varying roughly proportional to the absolute temperature T, is due to some extra low-energy excitations which scatter thermal phonons very effectively, thus leading to a reduced thermal conductivity varying as T^2. Again Anderson and co-workers [4], and independently Phillips [5], suggested a model which could account for these results. In this model, the low-energy excitations are attributed to tunneling of atoms or groups of atoms between two nearby positions (Section IVA). This model has subsequently been applied successfully in explaining not only thermal but also acoustic properties of insulating glasses at low temperatures. Amorphous alloys seemed to evade a simple investigation, due to the contributions by conduction electrons which mask that of tunneling states. However, several years later, much evidence for tunneling states in amorphous alloys had accumulated [6,7]; see Sections B and C. Since then, experiments have focused on the interaction of tunneling states with conduction electrons and on structural relaxation studies. These points will be treated in Sections IVC to E. Also, low-energy excitations have been observed in some quasicrystalline and microcrystalline alloys and were interpreted in terms of the tunneling model (Section IVF). Such experiments are useful to unravel the still unresolved question: which atoms tunnel? The thermal properties of amorphous alloys at "higher" temperatures will not be discussed. Here, the question of phonon localization arises, in analogy to the concept of electron localization.

As may have become clear from the above remarks, emphasis in this chapter will be on common aspects and concepts to treat the low-temperature properties of disordered solids. We therefore refrain from a detailed discussion of, often subtle, differences in these properties for materials prepared via different routes, and only give a few examples for such differences.

II. SUPERCONDUCTIVITY OF RAPIDLY SOLIDIFIED ALLOYS

In conventional crystalline or amorphous superconductors, superconductivity occurs via a weak phonon-mediated pairwise attraction of conduction electrons which are treated as a nearly free electron gas. For some classes of materials, notably "heavy-fermion" superconductors [8] and high-T_c superconductors [9] alternative models have been proposed. We restrict ourselves to a discussion in terms of the conventional pairing model which applies to the vast variety of amorphous alloys to be discussed below. Previous reviews of amorphous superconductors were given by Bergmann [10], Johnson [11], and Poon [12].

A. Basic Features of BCS-Type Superconductivity

In the original BCS theory of superconductivity [13], only three material-dependent parameters, which can in principle be determined from normal-state properties, are needed to characterize the superconducting state. For example, the superconducting transition temperature, T_c, is given by

$$T_c = 1.14 \frac{\hbar}{k_B} \langle \omega \rangle \exp\left(-\frac{1}{N(E_F)V} \right) \tag{1}$$

where $N(E_F)$ is the (bare) electronic density of states at the Fermi level, $\langle \omega \rangle$ is a suitably averaged phonon frequency (often taken as the Debye frequency, ω_D), and V an effective matrix element for the electron–electron interaction mediated by phonons. In the original BCS theory, it is assumed that $V \approx$ constant over an energy range $\hbar\langle \omega \rangle$ and, furthermore, that $T_c \ll \langle \omega \rangle / k_B \ll T_F$, where T_F is the Fermi temperature. Even in this *weak-coupling* approximation ($\lambda \approx N(E_F)V \ll 1$; λ is often called the electron-phonon coupling constant) the problem of a T_c analysis is hampered by the fact that the three parameters $N(E_F)$, $\langle \omega \rangle$, and V are not independent.

In the *strong-coupling* approximation, the electron-phonon coupling constant is defined as [14]

$$\lambda \equiv 2 \int \alpha^2(\omega) F(\omega) \frac{d\omega}{\omega} \tag{2}$$

In this expression, the spectral distribution $F(\omega)$ of phonon states and also the energy dependence of the inelastic electron-phonon matrix element $\alpha(\omega)$, averaged over initial and final electron states $\mathbf{k}_i$ and $\mathbf{k}_f$ on the Fermi surface, are taken into account. McMillan [14] showed that T_c can be approximated by

$$T_c = \frac{\Theta_D}{1.45} \exp\left(-\frac{1.04(1 + \lambda)}{\lambda - \mu^*(1 + 0.62\lambda)} \right) \tag{3}$$

Here μ^* is an effective repulsive Coulomb interaction between electrons, usually taken to be $\mu^* = 0.1$, and Θ_D is the Debye temperature. For $\lambda \ll 1$ and $\mu^* \to 0$, Eq. (3) reduces to the weak-coupling formula $T_c \sim \exp(-1/\lambda)$. McMillan further showed that λ can be factorized as

$$\lambda = \frac{N(E_F)\langle I^2 \rangle}{M\langle \omega^2 \rangle} \tag{4}$$

where M is the average ionic mass and $\langle I^2 \rangle$ is an average squared electron-phonon matrix element. Note that the average over phonon frequencies in Eq. (1) and (4) is different. Equations (3) and (4) form the basis of interpretation of superconductivity data for amorphous alloys.*

B. Simple Amorphous Metals and Alloys

The first systematic investigation of superconductivity in amorphous alloys dates back to 1954, when Buckel and Hilsch [15] published their seminal paper on superconductivity of vapor-quenched films condensed onto a substrate held close to 4 K ("quench-condensed") that were shown to be amorphous by in situ electron diffraction [16]. The field received further interest with the discovery of melt-quenched superconductors which are stable at room temperature [17]. Initially, most of the latter materials were alloys containing transition metals, while superconductivity of simple metals seemed to be restricted to materials which had to be quench-condensed. Of course, there is no fundamental difference between the two preparation methods (except the higher cooling rate and lower deposition rate and temperature in quench condensing). So it was no surprise when "bulk" superconductivity in melt-quenched $Mg_{70}Zn_{30}$ alloys was found at ~0.1 K [18]. Actually, superconductivity had been anticipated from superconducting fluctuations at $T >$ 1.2 K [11].

A discussion of the important parameters for superconductivity (see Eqs. (1)–(4)) has been given by Johnson [11] in terms of the free-electron model. Theoretical justification for this model comes from the observation that for electrons in a weak periodic potential, the only modification of their $E(\mathbf{k})$ dispersion relation occurs near the boundaries of the Brillouin zone. The effect on the electronic properties is particularly large when the Fermi surface is close to or crosses a zone boundary. The corresponding deviations from the free-electron model result in sharp features in $N(E)$, which however are "smeared out" in amorphous alloys. Indeed, many experiments have shown that the electronic properties of amorphous simple metals are free-electron-like [10]. It is only at the edges of the concentration range for stability of amorphous binary alloys that systematic deviations exist which can be ascribed to the equivalent of Hume-Rothery rules for amorphous phases [19].

In terms of the free-electron model, an increase of λ with average conduction-electron-per-atom ratio, z, is expected [11]. This trend is also seen in experiments [11] where λ has been determined from tunneling data from which $\alpha^2(\omega)F(\omega)$ can be obtained. Since tunneling data are scarce, estimates of λ where λ has been determined from Eq. (3) with the help of specific heat measurements (yielding Θ_D) are shown in Fig. 1. While for $\lambda \ll 1$ the agreement between both sets of data is quite good, it becomes increasingly poorer for λ values close to or exceeding 1. The reason for this discrepancy might lie in the various approximations used in deriving Eq. (3). It is interesting, however, to point out that the theoretical estimates for simple free-electron metals [11] (dashed line in Fig. 1) are much closer to the λ values determined from Eq. (3), indicating that the same type of approximations have been used, e.g., in averaging over the phonon frequencies.

*In amorphous metals, the description of electron and phonon states in terms of the crystal momentum $\hbar\mathbf{k}$ is problematic because $\mathbf{k}$ is not a good quantum number, due to the lack of periodicity. Nevertheless, it can be argued that the quantities discussed above retain significance also in amorphous metals when suitably redefined [11].

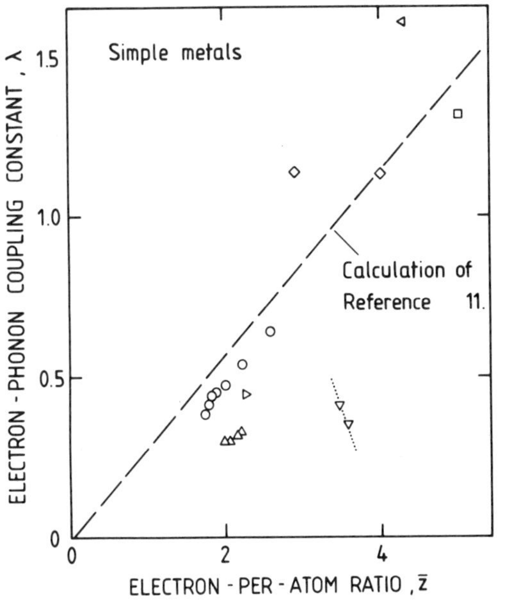

Figure 1 Electron-phonon coupling constant λ versus average electron-per-atom ratio $\bar{z}$ for simple amorphous superconductors: ○ Sn_xCu_{100-x} [20], △ $Mg_{70}Zn_{30-x}Ga_x$ [18,21], ◇ $Ga_{95}Ag_5$ [22], □ $Bi_{84}Sb_{16}$ [22], ◁ $Pb_{70}Bi_{30}$ [23], ▷ $Mg_{37.5}Zn_{37.5}Al_{25}$ [24], ▽ $Ge_{100-x}Au_x$ [25].

The increase of T_c (and hence of λ) has been observed in Sn_xCu_{100-x} [20] and $Mg_{70}Zn_{30-x}Ga_x$ [21] with increasing x; see Fig. 1. Independent of any model, this confirms the basic idea of an increase of T_c with increasing $\bar{z}$ in simple-metal alloys. This no longer holds when one approaches the metal-insulator transition in semiconductor-metal mixtures as $Ge_{100-x}Au_x$. Here, of course, T_c goes down when one approaches the $z = 4$ amorphous semiconductor [25]. The destructive influence of electron localization on superconductivity will be discussed further below.

Traditionally, simple amorphous metals have been thought to be strong-coupling superconductors [10,11]. Clearly, the lack of translational symmetry generally leads to an enhancement of the electron-phonon interaction. In addition, a pronounced softening of transverse phonons with respect to the crystalline state, which also tends to enhance λ, is seen [26]. However, as Fig. 1 reveals, there are materials which exhibit $\lambda \ll 1$ even in the amorphous state, indicative of weak-coupling behavior. A decisive test in this respect would be the determination of the gap parameter $2\Delta_0/k_BT_c$, where Δ_0 is the energy gap at $T = 0$. This quantity takes on the value 3.52 for BCS (that is, weak coupling) superconductors and reaches values up to ~5 for strong-coupling amorphous superconductors. In the lack of tunneling data for simple amorphous metals with $\lambda \ll 1$, the discontinuity ΔC in the specific heat at T_c can be taken to estimate the coupling strength. For $Mg_{70}Zn_{30}$, experimental data close to the BCS value $\Delta C/\gamma T_c = 1.43$ were found for liquid-quenched [18] and quench-condensed samples [27]. Here γ denotes the coefficient of the electronic contribution to the specific heat (Sommerfeld constant).

In the remainder of this section, we will briefly compare materials prepared differently and/or subjected to structural relaxation induced by annealing. It has been observed already very early that T_c changes very little when λ is already large [14]. However, when $\lambda \ll 1$, small changes in λ can cause large differences in T_c. Therefore, one would expect low-T_c materials to be much more susceptible to small structural changes which might be brought about by different preparation methods or thermal treatment. In general, T_c is higher for samples prepared with higher quenching rates. For instance, T_c for liquid-quenched $Mg_{70}Zn_{30}$ is 0.1 K [18], while for quench-condensed samples $T_c \approx 0.4$ K [27], with a difference in λ of only 10%. Upon annealing the quench-condensed $Mg_{70}Zn_{30}$ film up to room temperature where it still remains amorphous, T_c drops gradually to values below 0.1 K [27]. On the other hand, amorphous $Pb_{90}Cu_{10}$ alloys ($T_c \approx 6$ K) can be even annealed to yield a microcrystalline structure without a change in T_c [28].

C. Amorphous Transition Metals

An early systematic study of T_c of amorphous transition metals was published by Collver and Hammond [29]. They used the quench-condensation technique to produce alloys of neighboring elements, and plotted T_c versus the average group number (see Fig. 2). Loosely speaking, the two sharp maxima in T_c versus group number occurring for crystalline transition metal alloys [30] are washed out, and only a very broad maximum remains.

The *d* states of the transition metals are much more atomic-like than the *s-p* states in simple metals. The tight-binding approach is therefore usually a good starting point. Hence, the free-electron picture is not at all appropriate for amorphous transition metals, while it leads to considerable success in simple metals. The measured density of states at the Fermi level $N(E_F)$ as determined, e.g., from specific heat measurements and corrected for electron-phonon mass enhancement

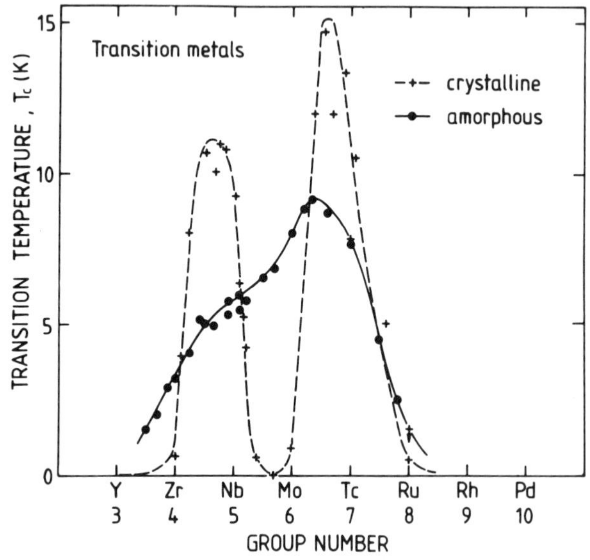

Figure 2 Superconducting transition temperature T_c versus average group number for 4*d* amorphous [29] and crystalline [30] alloys [11].

(see below) is usually three to five times larger than calculated from the free-electron model. Hence the average group number, or the average electron-per-atom ratio $\bar{z}$ should be taken as a parameter describing the *d*-band occupation.

In terms of a very general argument, the sharp features in the electronic density of states $N(E)$ arising from the narrow *d* bands are smeared out in amorphous transition metal alloys, in the spirit of the argument presented in Section B, and a broad featureless $N(E)$ remains. As will be discussed further below, $N(E_F)$ is decisive in determining T_c. Therefore, a featureless T_c versus $\bar{z}$ behavior results. The broad maximum of T_c then reflects the maximum of $N(E)$ occurring for amorphous metals in the middle of the *d* band.

The broadening of the features of the density of states can also account for the observation that for alloys with a high T_c in the crystalline state (due to a high $N(E_F)$), the corresponding amorphous phase tends to have a lower T_c, and vice versa. The most striking examples are Nb_3Ge ($T_c = 21.5$ K in the A15 phase and 3.9 K in the amorphous phase [31]) for the former case and Mo ($T_c = 0.92$ K in the bcc phase and 8.8 K as amorphous film [32]) for the latter case.

Many transition metal superconductors contain elements that are far apart in the periodic table, that is, "early" and "late" transition elements, such as Zr-Cu, Zr-Ni, or Zr-Fe. In these cases, the density of states shows distinct, albeit rounded features originating from the *d* bands of each of the constituents. This has been investigated in detail by photoelectron spectroscopy [33]. $T_c(\bar{z})$ for a given alloy system, for example, Zr_xCu_{100-x}, differs strongly from the Collver-Hammond curve [12].

The decisive influence of $N(E_F)$ on T_c can be inferred from systematic low-temperature specific heat studies. In the normal state, the specific heat can be expressed as

$$C = \gamma T + \beta T^3 \tag{5}$$

where

$$\gamma = \frac{1}{3}\pi^2 k_B^2 N_\gamma(E_F) \qquad \text{and} \qquad \beta = \frac{12\pi^4}{5}\frac{R}{\Theta_D^3} \tag{6}$$

Here $N_\gamma(E_F)$ is the electron-phonon–mass-enhanced density of states which is related to the bare density of states $N(E_F)$ by

$$N_\gamma(E_F) = N(E_F)(1 + \lambda) \tag{7}$$

In Eq. (5), nonelectronic linear contributions to C due to tunneling states (they usually amount to only a few percent of γ) and possible deviations from the Debye model are neglected (i.e., Θ_D is an "effective" Debye temperature), see Section IVB.

Measurements on amorphous Zr_xCu_{100-x} [34] and $(Mo_{1-x}Ru_x)_{80}P_{20}$ [35] over a wide concentration range indeed show the systematic dependence of T_c on γ, the latter study confirming previous arguments based on measurements of the Pauli susceptibility [11].

Varma and Dynes [36] argued that if the bandwidth B is constant for a given transition metal series, then $\lambda \sim N(E_F)$ within the lower (bonding) or upper (antibonding) halves of the *d* band, in good agreement with the experimental findings.

In the discussion of simple metals, we relied on the determination of λ in order to decide whether a material exhibits weak or strong electron-phonon coupling. In amorphous transition metals, a softening of transverse phonons also occurs [38]. However, because T_c is a few K and Θ_D is of the order of 200 to 300 K, $T_c \ll \Theta_D$ is a good approximation, and amorphous transition metal alloys exhibit weak to intermediate electron-phonon coupling. This is inferred from λ which is typical ≈ 0.5 for transition metal alloys with T_c of a few K. Tunneling data for several amorphous transition metals show that $2\,\Delta_0/k_BT_c$ is between 3.5 and 3.8, i.e., very close to the BCS value [32,39].

The role of spin fluctuations on T_c has been considered for Zr-based amorphous alloys with Cu, Ni, Co, and Fe [37]. They can be incorporated in the McMillan Eq. (3).

The influence of sample preparation (e.g., cooling rate) and structural relaxation on T_c for single-phase amorphous transition metal alloys is minor, provided that phase separation can be avoided. This is borne out by Fig. 3 for the system Zr_xCu_{100-x}, where data from many different groups have been collected [40]. In general T_c decreases irreversibly by a few tenths of a kelvin upon annealing. Also, T_c for evaporated and sputtered films is somewhat higher, indicating that these samples are in a less-relaxed state (because of the higher quenching rate). The origin of the irreversible changes of T_c upon annealing which are found in many amorphous superconductors and appear to be always negative, might be attributed to the release of the internal strains causing a hardening of the amorphous network and hence an increase of $\langle\omega^2\rangle$, that is, a (small) decrease of λ [41]. This implies that evaporated and sputtered films grow with a considerable amount of internal strain which has also been directly observed [42].

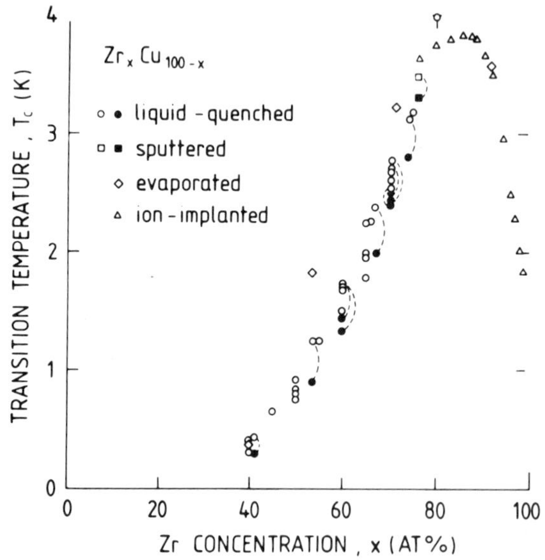

Figure 3 Superconducting transition temperature T_c versus concentration x for differently prepared Zr_xCu_{100-x} samples. Dashed lines and closed circles indicate changes in T_c occurring upon annealing [40].

An open question (in the light of the relatively small T_c differences) pertaining to sputtered Zr_xCu_{100-x} samples regards the high γ values (a factor 2–3 larger than melt-quenched samples [43]). Recent measurements on vapor-quenched Zr-Cu films have yielded γ's that tend to be *smaller* than those of melt-quenched films [44], suggesting that the atom-by-atom growth process does not cause the large γ's, rather they must be attributed to a structural difference caused by the bombardment of the growing film with energetic ions and atoms during the sputtering process.

Reversible changes of T_c have been observed in $Zr_{75}Rh_{25}$ upon cyclic annealing near the glass temperature [45]. These changes are much smaller than the irreversible changes mentioned above.

D. Upper Critical Field and Localization Effects in Amorphous Superconductors

Amorphous metals, due to their short electronic mean free path, l^e, and their resultant short Ginzburg-Landau coherence length, ξ, are type II superconductors, which are characterized by the condition $\kappa = \lambda_p/\xi > 1/\sqrt{2}$, where λ_p is the penetration depth.

In the Ginzburg-Landau-Abrikosov-Gorkov (GLAG) theory [46], ξ and λ_p are determined in the "dirty limit" ($l^e \ll \xi, \lambda_p$) by a few material-dependent parameters such as l^e, $N(E_F)$, the Fermi velocity, v_F, and the superconducting energy gap, Δ_0, at $T = 0$. ξ and λ_p both depend on temperature as $\sqrt{T_c/(T_c - T)}$. Hence, the ratio is to a good approximation temperature-independent and can be expressed as

$$\kappa = \frac{\lambda_p}{\xi} = 0.725\,\frac{\lambda_L}{l^e} \tag{8}$$

where $\lambda_L = 3/\mu_0 e^2 v_F^2 N(E_F)$ is the London penetration depth. Typical values are (for T approaching zero) $\lambda_L \approx 100$ nm, $\xi \approx 10$ nm and $l^e \approx 1$ nm for amorphous alloys, hence κ can reach values up to 100. Amorphous alloys are therefore extreme type II superconductors.

In the dirty limit, the upper critical field for temperatures approaching zero is given by $B_{c2}(0) = 1.2\sqrt{2}\,\kappa B_c(0)$ where B_c is the thermodynamic critical field. (The reader is reminded that B_{c2} is the largest field in which the superconductor remains in the mixed phase before becoming normal conducting.) Hence, the large κ leads to rather large $B_{c2}(0)$ values, typically of the order of 5 to 10 T for an amorphous superconductor with T_c of a few K. Of course, B_{c2} approaches zero as T approaches T_c. Typical examples of $B_{c2}(T)$ versus T are shown in Fig. 4 [12].

In the dirty limit, the $B_{c2}(T)$ curve can be calculated from normal-state parameters. For instance, the slope of $B_{c2}(T)$ at T_c, measured in T/K, is given by [47]

$$\left.\frac{dB_{c2}}{dT}\right|_{T=T_c} = 0.95\,\frac{e\hbar^2\rho_n}{m}\,N_\gamma(E_F) \tag{9}$$

Here ρ_n is the normal-state residual resistivity, e and m are electronic charge and mass, respectively. Here $N_\gamma(E_F)$ refers to a unit volume. Equation (9) does not include strong-coupling effects which yield a factor of order unity [48]. Equation (9) provides a useful basis for comparison with specific heat data (cf. Eq. (6)), and

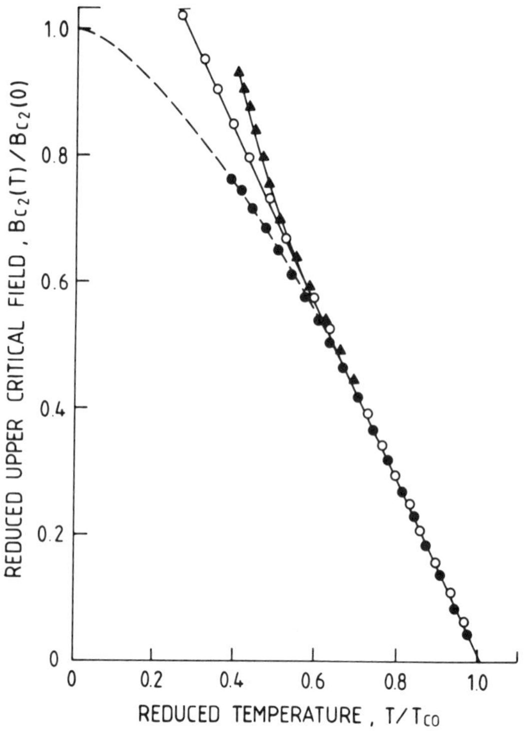

Figure 4 Reduced upper critical field $B_{c2}(T)/B_{c2}(0)$ versus reduced temperature T/T_c for amorphous alloys: ○ $Mo_{30}Re_{70}$, ● $Mo_{52}Ge_{48}$, ▲ $Pb_{25}Sb_{25}Au_{50}$. The value $B_{c2}(0)$ and the broken line represent the expected GLAG behavior [12].

thus for a check of sample homogeneity (see below). For $T \ll T_c$, B_{c2} is limited by the possibility to break the spin correlations of paired electrons (for low fields, the field acts mainly on the orbital correlation). This is known as the Clogston limit $g\mu_B B_{c2} \leq 2\Delta_0$. However, this limit can be exceeded when spin-orbit effects are taken into account.

While often $B_{c2}(T)$ for amorphous can be described in terms of the standard theory (cf. data for Mo-Ge in Fig. 4), there are examples where $B_{c2}(T)$ varies linearly over a very large T range which cannot be explained (Mo-Re), or even exhibits a positive curvature (Pb-Sb-Au). There are several possible reasons for this discrepancy. Sample inhomogeneities can severely affect the $B_{c2}(T)$ curve when distributions of T_c or B_{c2} are considered. In fact, the Pb-Sb-Au sample in Fig. 4 is phase-separated [12]. Weak-localization and electron–electron interaction effects, which will be treated in more detail in Section III have been advocated as possible reasons [49]. The basic physical idea is that upon incipient electron localization, the repulsive Coulomb interaction between electrons becomes progressively less screened, therefore the attractive phonon-mediated interaction is weakened [50]. Since these considerations are quite general, they should apply to any metal approaching electron localization, i.e., also to high-resistivity crystalline alloys. However, this interpretation, which essentially is equivalent to an increase in μ^*, has been questioned for the case of Nb_3Sn [51]. A recent calculation [52] shows that localization effects alter the dB_{c2}/dT at T_c because of field-induced delocalization, but do not alter the shape of the $B_{c2}(T)$ curve. This accounts for the fact that homogeneous amorphous superconductors apparently follow the standard dirty limit theory [52], but makes a direct comparison of dB_{c2}/dT with $N_\gamma(E_F)$ questionable.

Experimentally, it has been known for some time that T_c decreases when the resistivity increases strongly. A systematic theoretical analysis by Belitz [53] shows that the T_c degradation can be explained by localization effects together with electron–electron interactions induced by disorder. (For low-T_c materials T_c first increases upon introduction of disorder, as discussed in Section IIB.) An example is shown in Fig. 5, where T_c of thin homogeneous amorphous $Mo_{79}Ge_{21}$ films is plotted versus the normal-state sheet resistance $R_\square$ [54]. When $R_\square$ exceeds $\sim R_{\square c} \approx 2k\Omega$, T_c is practically zero. Here $R_\square$ can be regarded as a measure of electron localization. Such a behavior with $R_{\square c}$ of the same magnitude has also been seen in very thin films of crystalline metals, e.g., Pb [55] and Nb [56]. $R_{\square c}$ is close to the "quantum resistance for Cooper pairs" $R_0 = h/(2e)^2 = 6.45$ kΩ. This general $T_c(R_\square)$ dependence has been quantitatively interpreted in terms of electron–electron interactions [57,58].

Recently, several interesting results were reported for very thin evaporated Sn films which are presumably highly disordered [59]. In particular, it was found that for a certain range of sheet resistance $R_\square$ ($100\ k\Omega > R_\square > R_0$), a decrease of $R_\square$ around T_c of bulk Sn and a subsequent increase toward lower temperatures were found, resembling a "reentrant" superconductor. Superconductivity with zero resistance was seen only when $R_\square < R_0$. This observation was confirmed by several groups for different types of granular materials. Theoretically, an explanation in terms of a resistively shunted Josephson junction was invoked [60]. A physical realization of an array of identical Josephson junctions by means of microlithography has indeed reproduced the $R_\square$-T behavior found in Sn films [61].

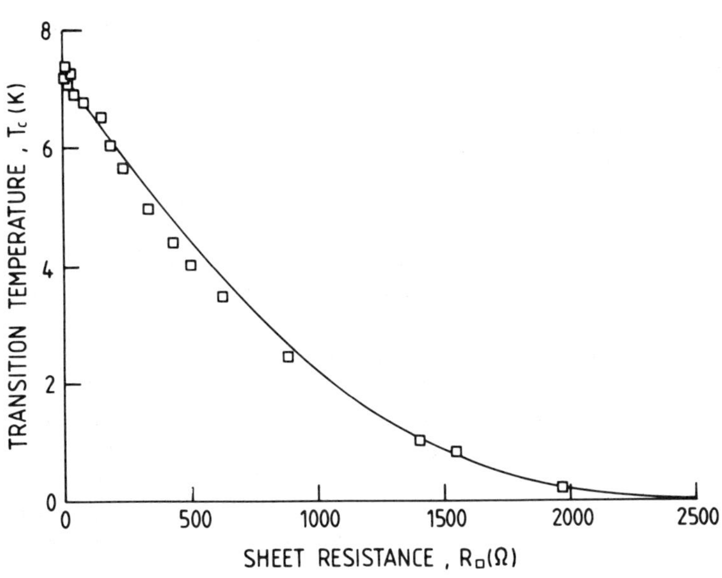

Figure 5 Superconducting transition temperature T_c of amorphous $Mo_{79}Ge_{21}$ films of different thickness versus sheet resistance $R_\square$ [54]. Solid line indicates calculation [57].

E. Superconductivity in Quasicrystalline and Metastable Crystalline Metals

Quasicrystals form a new state of matter which lacks long-range translational order but retains some orientational order. Although such structures, for example, the famous Penrose tiling, had been anticipated by mathematicians and theoretical physicists, the field received general interest only after the discovery by Shechtman et al. [62] that quasicrystalline materials exist in nature. They found that certain rapidly solidified Al-Mn alloys exhibit fivefold icosahedral symmetry. This was very exciting because it seemingly contradicted the textbook knowledge that no "fivefold symmetry" can exist in infinite crystals. This contradiction is resolved by noting the difference between topological and orientational order. A lot of discussion has gone on how to best describe the icosahedral and other quasicrystalline structures [63]. Also, many more materials have been found to exhibit quasicrystallinity.

The quasicrystalline state is usually obtained either by rapid quenching from the melt, by moderate annealing of an initially amorphous sample, or by solid-state reaction. Icosahedral alloys of the Al-Li-Cu system were even obtained by conventional casting [64], indicating that the icosahedral phase may be an equilibrium phase in a certain temperature, pressure, and composition range. Up to now, no grossly unusual properties have been found for this new state of matter. For instance, the vibrational properties of $Pd_{58.8}U_{20.6}Si_{20.6}$ which can be directly compared in the icosahedral and amorphous states, are much alike in both states as seen from inelastic neutron scattering [65] and specific heat [66]. Even for icosahedral samples with no magnetic contribution to the electrical resistivity, the latter can be as large or larger than that of amorphous metals. A calculation in the frame of the Ziman theory of resistivity has suggested that the scattering of electrons due to the quasicrystalline potential does not contribute to the resistivity [67]. The electronic properties of quasicrystals composed of simple metals can be described by Hume-Rothery considerations [68]. This brief summary might suffice as an introduction to these materials whose properties with respect to low-energy tunneling states will be discussed in Section IVF.

In view of the properties just described, the occurrence of superconductivity in certain icosahedral samples is not surprising. A comparative specific heat study for cubic, icosahedral and amorphous $Mg_3Zn_3Al_2$ showed T_c's of 0.32, 0.41 and 0.75 K, respectively [24]. Since the density of states is roughly the same for the three phases (corresponding to the free-electron model), the increase of T_c and hence of λ, see Eq. (3), can be attributed to an increasing lattice softening as directly observed from a decrease of Θ_D [24].

As a representative for metastable crystalline superconductors containing no transition metals, $Al_{100-x}Si_x$ alloys will be briefly discussed. These alloys (up to $x = 30$) can be obtained by rapid quenching or by pressure quenching. In the latter process, the samples are heated to 1000 K under a pressure of 4 GPa followed by a rapid quench, yielding a rather homogeneous Al-Si solid solution with Si occupying regular sites of the Al fcc lattice [69]. In these $Al_{100-x}Si_x$ alloys T_c goes up to ~6 K (compared to $T_c = 1.2$ K for pure Al). This might be attributed to the important shear-wave softening observed [70] in the alloys with respect to pure Al yielding an increase of $F(\omega)$ and hence of the electron-phonon coupling constant λ.

Metastable bcc $Zr_{100-x}Si_x$ alloys can be prepared by rapid quenching for $8 \leq x \leq 11$ [71]. For $12 \leq x \leq 24$ the alloys are amorphous. Starting from the amorphous phase, an apparently continuous increase of T_c with decreasing x through the (sluggish) amorphous bcc phase boundary was observed by resistivity measurements [71]. A detailed specific heat study, however, showed that this "high-T_c" phase is an amorphous minority phase, while for the majority bcc phase T_c decreases with decreasing x [72]. This puts a warning on using resistivity as the only probe for the check of superconductivity in metastable alloys, because a small superconducting percolating path is sufficient for zero resistance, while the specific heat is a true bulk property: the discontinuity of the specific heat at T_c gives an indication of how much of the sample actually becomes superconducting (an exact determination requires knowledge of the electron-phonon coupling strength; see Section IIB). In $Zr_{100-x}Si_x$, the increase of T_c to 2.5 K in the bcc phase and up to 3.2 K in the amorphous phase is mostly due to the increase in $N(E_F)$ which of course can be determined from the normal-state specific heat. No large phonon softening occurs, as inferred from the fact that Θ_D is almost concentration-independent [72].

F. Rapidly Solidified High-T_c Superconductors

Immediately after the discovery of high-T_c superconductivity by Bednorz and Müller [73], many routes to obtain high-T_c superconductors were followed. As mentioned in the introduction we will confine ourselves to the melt-quenching techniques.

$YBa_2Cu_3O_{7-\delta}$ has a T_c of $\approx$92 K [74]. The standard procedure consists of mixing and pressing powders of CuO, Y_2O_3, and $BaCO_3$ and giving specified heat treatments in air. In melt quenching [75], these powders are melted together at 1400°C. The melt is then poured quickly on an iron plate and pressed to a thickness of ~1–2 mm. A final heat treatment at 900°C results in samples with $T_c \approx 90$ K. However, the high value of the (almost temperature-independent) resistivity ρ_n above T_c (in some samples $d\rho/dT$ was found to be even slightly negative) suggests that the samples are not yet in an optimized state, although the ac Josephson effect was observed in these samples [75]. Melt-textured growth of samples leads to rather high critical current densities [76].

The twin roller technique was used to obtain ~50-μm-thick samples [77] which consist of metastable crystalline and amorphous regions [77]. After annealing in air, superconductivity with $T_c = 93$ K (midpoint of resistive transition) was observed. In contrast to the former method [75], densities close to the microscopic density of $YBa_2Cu_3O_{7-\delta}$ were found.

The $BiSrCaCu_2O_x$ superconductors ($T_c \approx 105$ K) have also been obtained via rapid quenching and subsequent annealing [78]. X-ray and differential scanning calorimetry (DSC) show that the roller-quenched samples are amorphous [78]. Upon annealing, only multiphase materials have been obtained so far with only very minor portions of the sample becoming superconducting above 100 K.

Experiments on the physical properties of high-T_c superconductors require well-characterized materials, preferably single crystals, in order to elucidate the mechanism responsible for superconductivity. Rapid solidification and subsequent annealing may nevertheless prove to be a reliable preparation process. Of course, only the future can show if this route is of importance for technical applications of high-T_c superconductors.

III. ELECTRON LOCALIZATION AND ELECTRON–ELECTRON INTERACTION EFFECTS IN DISORDERED METALS

In this section we will briefly discuss the low-temperature electronic transport properties of disordered metals. As will be shown, disorder can lead to a localization of electronic wave functions which has immediate consequences for the electrical resistivity. For a large degree of disorder, a transition to insulating behavior can occur. With increasing degree of localization, the electron–electron interaction becomes more important, which makes this problem rather difficult. The electron localization, in particular the "weak" localization to be discussed below, is a quantum mechanical effect which cannot be accounted for within the semiclassical Boltzmann transport theory. We will focus our attention on those effects which in amorphous and disordered metals are most eminent at low temperatures.

It should be mentioned that the electrical resistivity, ρ, at higher temperatures is often interpreted in terms of the Ziman theory [79] which is based on the Boltzmann transport theory and its extension, i.e., the Kubo formalism. The Ziman theory attributes the electrical resistivity to scattering by the disordered structure as represented by the static structure factor $S(Q)$, and is therefore often termed "diffraction model." It has been extended to account for strong scattering as in liquid and amorphous transition metal alloys [79]. Experimentally, the magnitude of ρ and its temperature coefficient, $\alpha = (1/\rho)d\rho/dT$, seem to be related [80]: highly disordered metals (whether amorphous or crystalline) with $\rho > 150$ μm tend to have $\alpha < 0$. However, a recent compilation of much data [81] indicates that this relation might not be as universal as initially thought. In view of the many different materials and scattering processes at hand, this might not be too surprising. However, regardless of the details, incipient electron localization in highly disordered metals will be one common feature which has to be taken into account.

A. Electron Localization of Noninteracting Particles

In a perfect crystal, the electronic wave functions (Bloch states) extend over the whole crystal. In usual transport theory, one regards wave packets of these states, which are scattered only by deviations from the perfect lattice such as imperfections and lattice vibrations. While the electron-phonon scattering is inelastic and gives rise to the temperature-dependent resistivity, the static imperfections lead to a finite d_c electrical conductivity even for temperatures approaching absolute zero, which in the case of a perfect metal would be infinite. In Boltzmann transport theory, the conductivity can be expressed through a relaxation time τ:

$$\sigma^B = \left(\frac{e^2 n}{m}\right)\tau \tag{10}$$

For weak disorder, that is, large τ, the wave functions are still extended in three dimensions, while for strong disorder they become localized. This was first shown by Anderson [1], who regarded the return probability of an electron on a lattice with random potential wells of depths V_i. If the return probability P to a fixed lattice site is zero for long "waiting" times ($t \to \infty$), the electron can be regarded as having diffused away, that is, the states are extended. If P is finite, the electron is regarded as localized. The electron will diffuse away more easily if the difference in energy of adjacent sites is small compared to the bandwidth, B. Therefore, if the width, W, of the distribution of V_i is smaller than B, diffusion is possible. If W becomes much larger than B, diffusion will be inhibited: the system undergoes (with increasing disorder, that is, with increasing W/B) a metal-insulator transition. Here we use the term "insulator" in the sense that $\sigma_0 = \sigma(T \to 0) = 0$ as temperature approaches zero, while for a metal σ_0 remains finite. This transition can also happen in the case of moderate disorder, where localized states exist at the edges of a band and extended states are in the center. The energy separating extended and localized states is called the "mobility edge." If, with changing electron concentration, the Fermi level is pushed through the mobility edge, a metal-insulator transition occurs.

The field received considerable impact when a one-parameter scaling theory was developed [2]. In this theory, the only relevant variable measuring the disorder is a generalized (dimensionless) conductance. This scaling analysis lead to the following results [2]: (i) all electron states are localized for arbitrarily weak disorder in one and two dimensions, two dimensions being the marginal case.* For weak disorder, the phrase weak or incipient localization is often used. (ii) In three dimensions, the metal-insulator transition is continuous, the conductivity goes smoothly to zero with increasing W/B; that is, there is no minimum metallic conductivity at which σ abruptly goes to zero.

Both consequences of scaling theory have been thoroughly tested experimentally, that is, (i) with work on thin films and wires, in particular in the weakly localized regime to be discussed below, and (ii) for different samples close to the metal-insulator transition. As an example, Fig. 6 shows σ_0 for amorphous Nb_xSi_{100-x} films (after [83]).

The quantum-mechanical origin of weak localization can be discussed in terms of a return probability. Consider the diffusion of an electron from A to B (Fig. 7a). There are many different trajectories i, and the probability W to reach B from A is given by the probability amplitudes A_i:

$$W_{A\to B} = \left|\sum_i A_i\right|^2 = \sum_i |A_i|^2 + \sum_{i\neq j} A_i A_j^* \tag{11}$$

The second term describes the interference between different amplitudes which arises from the wave nature of the electron, while the first term describes the usual Boltzmann transport. It is important to note that for a quantum mechanical particle, the "trajectories" are in reality "tubes" with, for conduction electrons, a width of approximately the Fermi wavelength $\lambda_F = 2\pi/k_F$. Since there are many trajectories, the amplitudes A_i have different phase relations which leads to an (average) cancellation of the interference term. Consider, however, a closed loop (Fig. 7b). The return probability is enhanced, because the electrons on the diffusion path 1 and 2 interfere coherently:

$$W_{A=B} = A_1^2 + A_2^2 + A_1 A_2^* + A_2 A_1^* \tag{12}$$

*For strong spin-orbit coupling, "antilocalization" is observed [82].

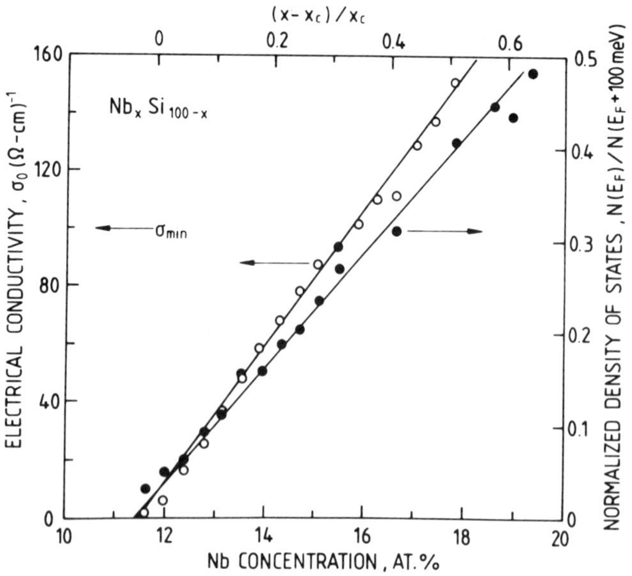

Figure 6 Electrical conductivity $\sigma_0 = \sigma\ (T \rightarrow 0)$ for amorphous Nb_xSi_{100-x} films versus concentration x (open symbols). Also shown is the normalized electronic density of states at E_F, $N(E_F)/N(E_F + 100\ \text{meV})$ obtained from tunneling (cf. Fig. 9). A minimum metallic conductivity, σ_{min}, as indicated, is not observed [83].

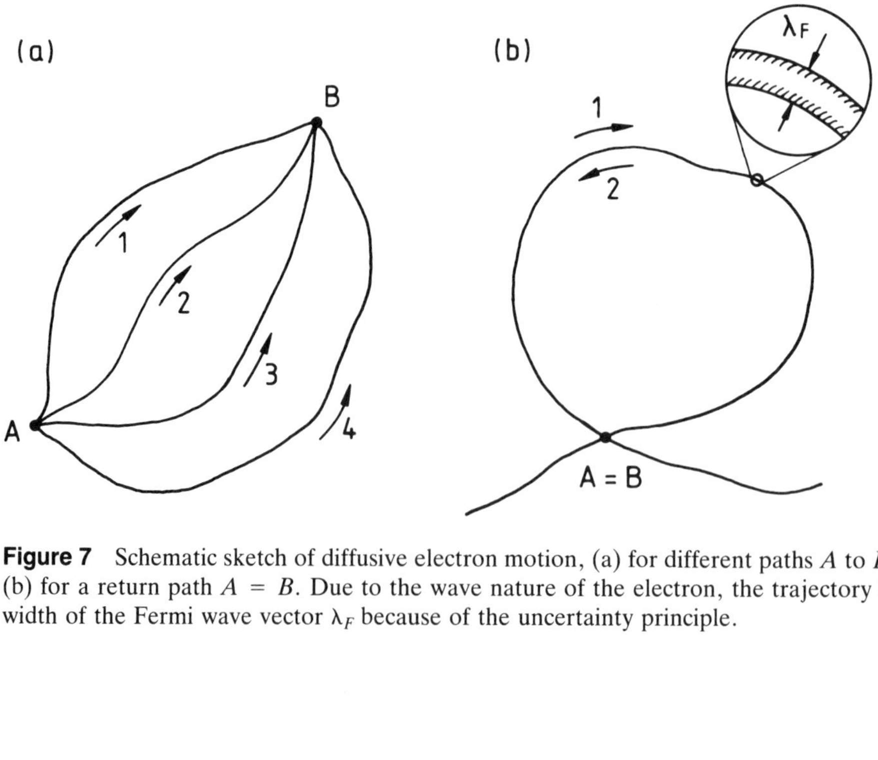

Figure 7 Schematic sketch of diffusive electron motion, (a) for different paths A to B and (b) for a return path $A = B$. Due to the wave nature of the electron, the trajectory has a width of the Fermi wave vector λ_F because of the uncertainty principle.

This enhanced return probability (in the simple case $A_1 = A_2$ it is twice the classical value) leads to a slower diffusion of quantum mechanical particles than of classical particles, that is, the electrical resistivity will be larger than in the classical case.

The requirement for the occurrence of this process is that the wave packets interfere coherently, that is, that the scattering is coherent. It is important to realize that elastic electron scattering from the static imperfections is in general coherent (an important exception is scattering from magnetic impurities), while for example, inelastic electron-phonon scattering is incoherent. Therefore, the basic requisite is that the coherent scattering time, τ, is much smaller than the time between incoherent scattering events τ_{in}, in order that a quantum mechanical correction to the resistivity be observed. Since inelastic electron-phonon scattering is frozen out at low temperature, these effects can be observed at low temperatures only.

These ideas can be put on a quantitative basis in terms of a perturbation theory in the small parameter $\Lambda = \hbar(E_F\tau)^{-1}$ where the elastic scattering rate τ^{-1} from crystal imperfections is a direct measure of the amount of disorder. Assuming that the inelastic scattering rate vanishes as $\tau_{\text{in}}^{-1} \sim T^p$ for $T \rightarrow 0$, one obtains a *temperature-dependent* correction $\delta\sigma$ to the conductivity σ^B, which depends on the dimension D of the sample. The exponent p is characteristic for the scattering process, for example, $p = 3$ for electron-phonon scattering in $D = 3$. The result for $\delta\sigma$ to lowest order in Λ is [84]

$$\begin{aligned}
\delta\sigma_{3D} &= \frac{e^2}{\hbar\pi^3}\frac{1}{a}\,T^{p/2} \\
\delta\sigma_{2D} &= \frac{e^2}{\hbar\pi^2}\frac{p}{2}\ln\frac{T}{T_0} \qquad (13)\\
\delta\sigma_{1D} &= -\frac{e^2}{\hbar\pi}\,aT^{-p/2}
\end{aligned}$$

In all cases, σ decreases with decreasing temperature because the relevant length scale $L \approx \sqrt{v_F^2\tau_{\text{in}}\tau/D}$ increases, hence the term "weak localization." In Eq. (13), a is the proportionality factor between L and $T^{-p/2}$. The logarithmic temperature dependence of $\delta\sigma$ in $2D$ has been observed in many experiments on thin films.* Fig. 8 gives an early example (after [85]). Experiments on $3D$ disordered metals are discussed in Section IIIB.

A magnetic field shifts the phases of the two return probability amplitudes A_1 and A_2 relative to each other through the vector potential, hence progressively destroys the coherent backscattering, resulting in a negative magnetoresistance. This simple argument, however, does not hold in the presence of spin-orbit scattering. For a detailed discussion of weak localization effects in thin films, in particular of the use of magnetoresistivity to obtain information about characteristic scattering times, the reader is referred to a review article by Bergmann [82].

*A film is regarded as $2D$ or $3D$ depending on whether its thickness is smaller or larger than a characteristic length scale, which in this case is L. In this sense, the film of Fig. 8 is $2D$ (3 nm), while that of Fig. 6 is $3D$ (100 nm).

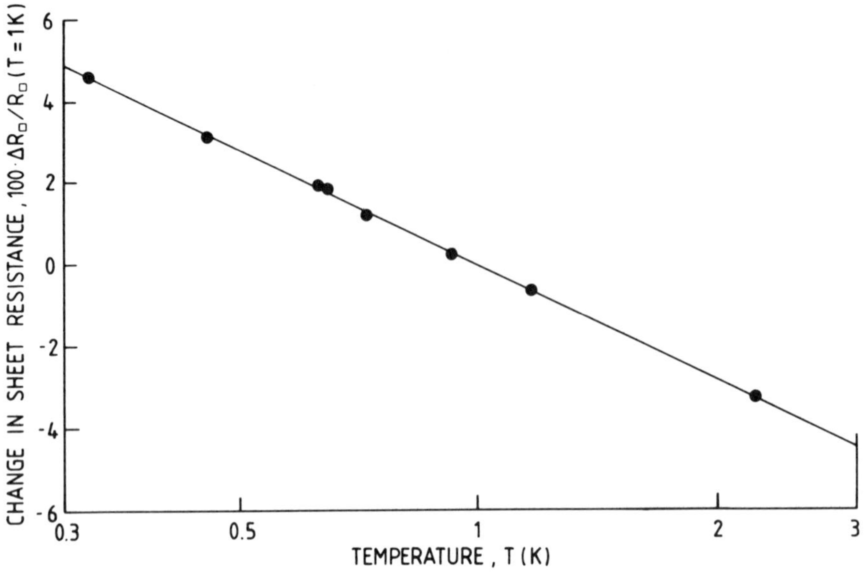

Figure 8 Change of electrical sheet resistance $\Delta R_{\square}/R_{\square}$ ($T = 1$ K) versus log of temperature T for a 3-nm-thick $Pd_{57}Au_{43}$ film [85].

B. Electron–Electron Interaction and Localization in Amorphous Metals

Before we discuss experiments on $3D$ disordered metals in detail, we have to treat electron–electron interactions. A simple argument for the importance of electron–electron interactions is as follows: In reasonably pure crystals, the electron–electron interaction is limited by the Pauli principle in conjunction with **k** conservation. In addition, local electron density fluctuations are screened by the other (mobile) electrons. Both effects work together in reducing the Coulomb repulsion of electrons to a small effective value. In disordered metals, **k** is no longer a good quantum number, hence one might expect that electron–electron interactions are more important. Also, as electron localization sets in, local fluctuations cannot be effectively screened. This enhancement of the electron–electron interaction is believed to be responsible for the T_c degradation in highly disordered superconductors discussed in Section IID.

Ultimately, the electron–electron interaction leads to an instability of the Fermi sea against electron-hole excitations which results in a gap at E_F. One way to see this is to regard the attractive interaction $e^2/4\pi\varepsilon_0 r$ between an electron-hole pair. In ordinary metals, this attractive interaction is screened by the mobile electrons. As the disorder becomes stronger, this screening becomes less effective and the system actually would lower its energy if the energy ΔE needed to form an electron-hole pair is smaller than $e^2/4\pi\varepsilon_0 r$. This instability of the Fermi sea against such excitations does not occur if a gap is formed at E_F such that a finite excitation energy for an electron-hole pair is required.

Such a Coulomb gap opens in disordered semiconductors (it results, e.g., in deviations from Mott's famous $T^{-1/4}$ law for the hopping conductivity, $\sigma \sim$

$\exp(-AT^{-1/4})$, toward a $T^{-1/2}$ dependence). A precursor of the Coulomb gap is predicted to occur [86] in disordered metals close to the metal-insulator transition, in the form of a dip in the density of states at E_F. This correction of $N(E)$ depends on the dimension of the sample, in $3D$ the single-particle density of states is [86]

$$N(E) = N(E_F)\left(1 + \sqrt{\frac{E - E_F}{\Delta}}\right) \tag{14}$$

Here Δ is an energy scale (correlation gap) that is very large far away from the metal-insulator transition and vanishes at the metal-insulator transition.

The corresponding corrections to the conductivity in the limit of weak electron-electron interactions have exactly the same temperature dependence in $2D$, e.g., for thin films as for the weak localization of independent electrons, of course, with different prefactors [84]. While this first came as a surprise to many workers, it is now clear that this is not by accident, but rather reflects the fact that electron–electron interactions become increasingly important with increasing disorder, i.e., localization, as discussed above. Indeed, two-parameter scaling theories [86] try to treat both effects (localization and interactions) on an equal footing. In these theories, the single-particle density of states $N(E_F)$ is predicted to vanish contin-

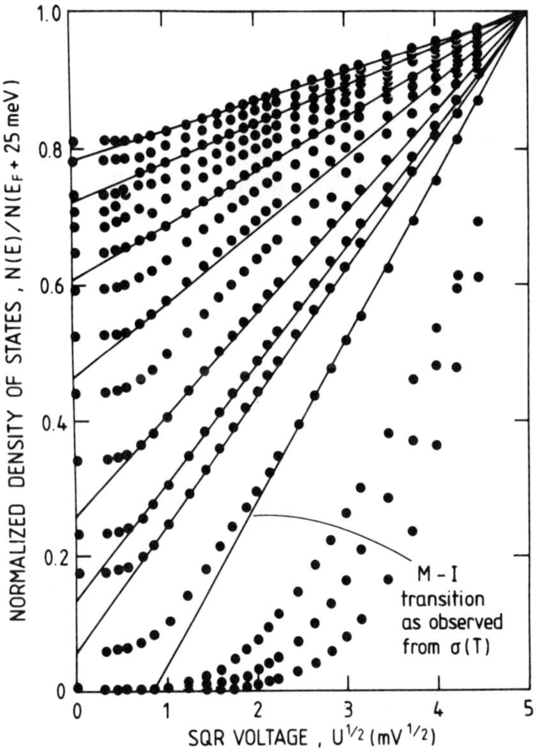

Figure 9 Electronic density of states $N(E)$ normalized to its value $N(E_F + 25\text{ meV})$ as obtained from tunneling measurements versus square root of tunneling bias voltage for amorphous Nb_xSi_{100-x} close to the metal-insulator (M-I) transition [83].

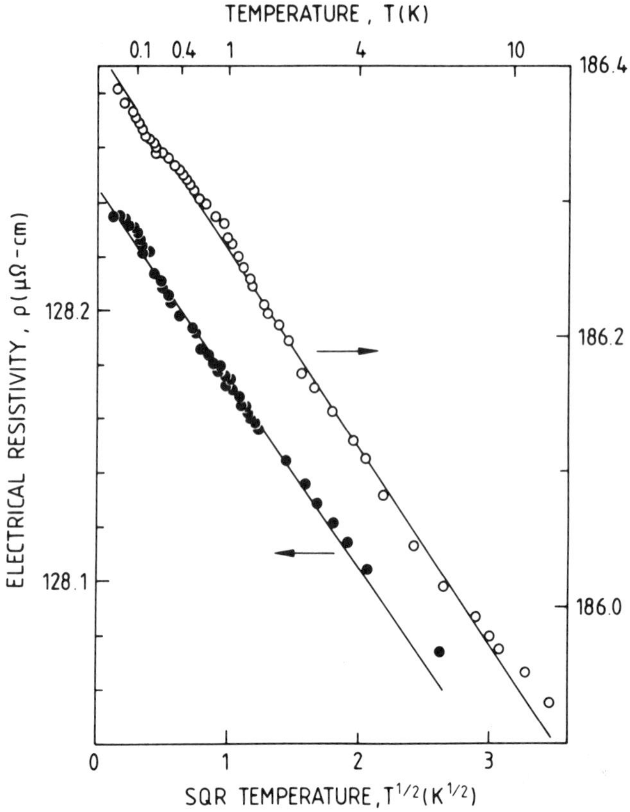

Figure 10 Electrical resistivity ρ versus $\sqrt{T}$ for amorphous alloys: $Fe_{40}Ni_{40}P_{14}B_6$ (open circles) and $Ni_{75}P_{16}B_6Al_3$ (closed circles) [92].

uously at the metal-insulator transition while for a metal-insulator transition of noninteracting particles $N(E_F)$ should not be affected.*

Two types of experiments have been carried out on amorphous metals close to and through the metal-insulator transition in order to elucidate the behavior of the density of states: electron tunneling and specific heat. The latter will be discussed at the end of this section. In the tunneling experiment, the single particle density of states $N(E)$ is explored. Figure 9 (after [83]) shows that $N(E)$ as measured for amorphous Nb_xSi_{100-x} in the vicinity of the metal-insulator transition (occurring for $x_c = 11.5$) exhibits the two features expected from the preceding discussion: (i) a decrease of $N(E_F)$ while approaching x_c; $N(E_F)$ varies linearly with $(x - x_c)/x_c$, as does the conductivity for similar samples (see Fig. 6), and (ii) a $\sqrt{V}$ dependence of $N(E)$ on the applied voltage V, which arises from the $\sqrt{(E - E_F)/\Delta}$ term in Eq. (14). The same behavior has also been observed for amorphous $Ge_{100-x}Au_x$ ($x_c \approx 12$) [89], $Ge_{100-x}Al_x$ ($x_c \approx 50$) [90], and also for granular A1 [91].

*For interacting particles, the single-particle density of states and the thermodynamic density of states $dn/d\mu$ have to be distinguished [87,88].

The incipient localization and electron–electron interaction effects can be observed in a large variety of amorphous metals, as first pointed out by Rapp et al. [92], through their resistivity rise as the temperature is lowered. In the course of several years, many explanations for this resistivity rise have been advocated, including scattering by two-level tunneling states [93]; see Section IVD. Figure 10 (after [92]) shows the resistivity ρ of amorphous metals over two decades in temperature, which can be expressed as

$$\rho(T) = \rho_0\left(1 - \sqrt{\frac{k_B T}{\Delta}}\right) \tag{15}$$

This idea was taken up by Cochrane and Strom-Olsen [94], who determined Δ for several amorphous metals from Eq. (15). These data fall on the same universal Δ versus ρ curve that had been determined earlier for materials close to the metal-insulator transition [83,89]; see Fig. 11. (Data for $Ge_{100-x}Al_x$ [90] appear to be somewhat smaller.) Experimentally, $\Delta \sim \rho^{-2}$. This important result shows that high-resistivity amorphous alloys exhibit incipient electron-localization and interaction effects.

Magnetoresistance measurements have been used to elucidate the relative roles of weak-localization and interaction effects in amorphous metals. While the former effect gives a negative magnetoresistance, as discussed in Section IIIB, the latter gives a positive contribution, both effects being additive. A complication arises

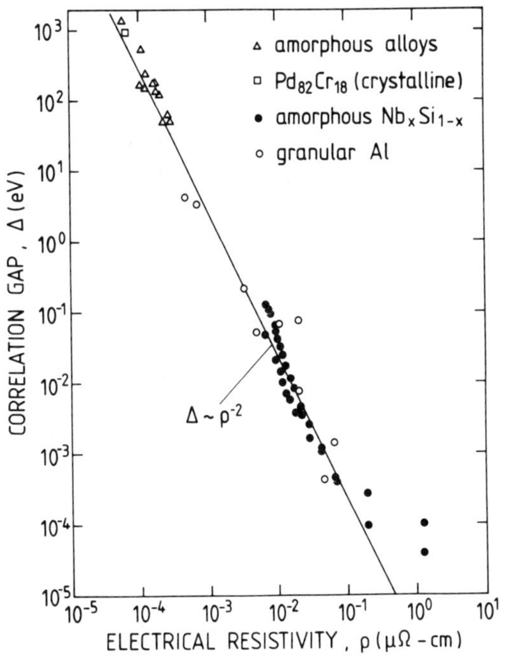

Figure 11 Correlation gap Δ versus resistivity ρ as obtained from the temperature-dependent resistivity of amorphous and disordered metals [94], and from tunneling for amorphous Nb_xSi_{100-x} [83] and for granular Al [91].

from the occurrence of superconductivity. Even well above T_c there is fluctuation conductivity and forward scattering of conduction electrons from these fluctuations [10–12]. Both effects are suppressed in a field, yielding a positive magnetoresistance. Still, several experiments have been published showing the feasibility of a detailed analysis [95,96].

Measurements of the specific heat in the vicinity of the metal-insulator transition yielding the thermodynamic density of states $dn/d\mu$ have been reported for only two amorphous alloy systems so far. While for $Mo_{100-x}Ge_x$ the coefficient γ of the linear specific heat (which is proportional to the density of states, for noninteracting as well as for interacting particles, provided that appropriate renormalizations are carried out) depends linearly on x and varies smoothly through the metal-insulator transition [97], for $Ge_{100-x}Au_x$ a strong decrease of γ was found when approaching the metal-insulator transition from the metallic side, that is, a factor 7 decrease of γ when going from $x = 18$ to $x = 14$ [25]. Further work appears necessary before definite conclusions concerning the thermodynamic density of states at the metal-insulator transition can be drawn. For completeness, we mention that the disorder-induced metal-insulator transition is investigated thoroughly in heavily doped crystalline semiconductors [98,99] where the disorder stems from the random distribution of donor atoms.

IV. LOW-ENERGY EXCITATIONS IN DISORDERED METALS

In the previous sections we have mainly dealt with the electronic properties of rapidly solidified amorphous and disordered metals, we now treat the atomic dynamics of these materials. As mentioned in the introduction, there are additional atomic degrees of freedom in amorphous and disordered solids at low temperatures which do not exist in perfect crystals. These give rise to excitations which can be observed in thermal and acoustic experiments.* One can rationalize the additional atomic degrees of freedom in an amorphous solid as follows.

The rapid quenching process usually necessary to produce an amorphous alloy ($\sim 10^6$ K/s) traps the sample in a particular local minimum of the potential-energy surface in configuration space. This minimum is separated by high-energy barriers from the crystalline ground state (which may be a phase mixture) and also from other minima in the disordered state which may have even lower energy; thus the amorphous alloy is said to be in a metastable state. Due to the random character of the amorphous state, there are also many minima in configuration space of roughly the same depth, hence there are many states which have the lowest energy for the glass (albeit higher than that of the true ground state of the crystal). This fact is often called the "high degeneracy of the ground state" of an amorphous solid.

If, due to the rapid quenching process, the glass is trapped in a state of higher energy, it can go to a minimum of lower energy via thermal activation. This process which can occur already during slow cooling from the melt, or during annealing at elevated temperatures, or during prolonged storage even at room temperature is called structural relaxation. Covalently bonded glasses which are obtained from

*An earlier review on this subject contains most references up to 1980 [7].

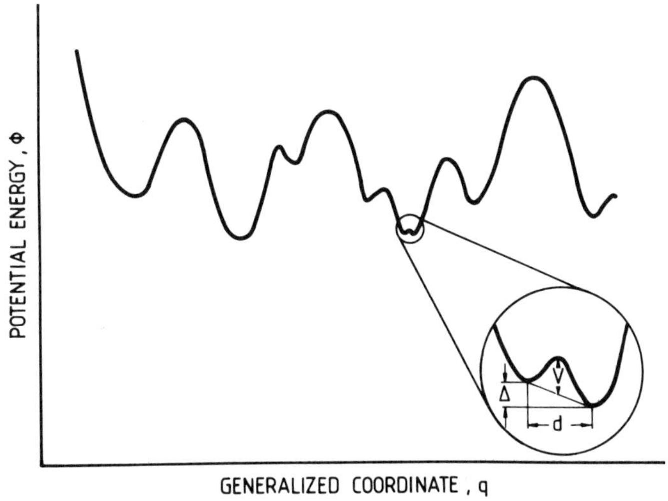

Figure 12 One-dimensional sketch of a potential-energy surface ϕ of a glass in configuration space along a generalized coordinate q. Enlarged view shows the pertinent parameters of a double-well potential.

the melt with rather low cooling rates are usually in a relaxed or "quasi-equilibrium" state. This also applies to metallic glasses such as $Pd_{77.5}Cu_6Si_{16.5}$, which can be prepared with cooling rates as low as $\sim 10^3$ K/s [100], and to amorphous metals prepared by solid-state reaction [101]. It is important to realize that these materials are nevertheless in a metastable state.

A. The Tunneling Model

At low temperatures the system is trapped in a local minimum, and the other minima of the potential surface ϕ in configuration space will be inaccessible because they are separated by high-energy barriers. However, for certain configurations there may be another local minimum nearby (in configuration space) which is accessible at low temperatures, involving the motion of a single atom or a few atoms only. This idea is sketched in a highly schematic way in Fig. 12. The existence of double-well potentials with a wide range of parameters V, d, and asymmetry Δ is the main ingredient of the model. At low temperatures thermally activated transitions between the two wells are frozen out and only atomic tunneling can occur. Considering only the lowest state in each well of the double-well potential, we have a two-level tunneling system (TLS) with an energy difference $E = \sqrt{\Delta_0^2 + \Delta^2}$ between the ground state and the excited state of the double-wall potential. These states are the symmetric and antisymmetric linear combinations of the single-well eigenstates.* Here $\Delta_0 = \hbar\omega_0 e^{-\lambda}$ is the tunneling splitting of the corresponding symmetric TLS with the same tunneling parameter $\lambda \approx \hbar^{-1} d\sqrt{2mV}$, and m is the mass of the tunneling particle; $\hbar\omega_0$ is of the order of the zero-point energy in each well.

*The energy difference E between the eigenstates of the double-well potential is obtained from diagonalization of the Hamiltonian, H_0, in the basis of the single-well states.

Transitions between the two states of the TLS can occur via interaction with thermal or acoustic phonons or with conduction electrons. We treat the interaction with phonons first. A deformation of the surroundings of a TLS by a (long-wavelength) phonon changes mainly the asymmetry Δ. Neglecting the change of Δ_0 with strain, the corresponding Hamiltonian is

$$H = H_0 + H_1 = \frac{1}{2}\begin{pmatrix} \Delta & \Delta_0 \\ \Delta_0 & -\Delta \end{pmatrix} - \begin{pmatrix} \gamma & 0 \\ 0 & -\gamma \end{pmatrix}\varepsilon \tag{16}$$

where ε is the strain tensor and $\gamma = (1/2)d\Delta/d\varepsilon$ is the deformation potential. The transformation diagonalizing H_0 leads to off-diagonal terms in H_1. With the notation $M = (\Delta_0/E)\gamma$ and $D = (2\Delta/E)\gamma$ and using the Pauli matrices, the Hamiltonian $H' = H_0' + H_1'$ can be written as

$$H' = ES_z - (2MS_x + DS_z)\varepsilon \tag{17}$$

For simplicity we have not included the dependence of γ on the polarization mode of the sound wave.

The off-diagonal term $2MS_x\varepsilon$ in Eq. (17) induces resonant transitions of the TLS from $(-1/2)E$ to $(+1/2)E$, and vice versa. The diagonal term $DS_z\varepsilon$ describes the change of the TLS energy splitting with the strain field.*

Equation (17) is equivalent to the Hamiltonian of a spin-1/2 particle in a magnetic field $\mathbf{B} = \mathbf{B}_0 + \mathbf{B}_1$:

$$H_B = -\hbar\gamma_G(\mathbf{B}\cdot\mathbf{S}) \tag{18}$$

E corresponds to the static magnetic field $\mathbf{B}_0 = B_z$, $E = -\hbar\gamma_G B_z$. The strain field, ε, which in the case of a sound wave oscillates at the TLS site, corresponds to the oscillating magnetic field, $(2M, \mathrm{O}, D)\varepsilon = \hbar\gamma_G\mathbf{B}_1$. Here γ_G is the gyromagnetic ratio. Equation (18) leads to the Bloch equations, which therefore can be used to describe the dynamics of a TLS in a strain field [102,103].

Transitions between the two states of a TLS due to one-phonon processes can be calculated from a rate equation [103] or within the Bloch formalism. The resulting TLS relaxation rate is† assuming a Debye model for the phonon spectrum [103]:

$$\tau_{1,\mathrm{ph}}^{-1} = \left(\frac{M_L^2}{v_L^5} + \frac{2M_T^2}{v_T^5}\right)\frac{E^3}{2\pi\rho_d\hbar^4}\coth\frac{E}{2k_BT} \tag{19}$$

Here v_L and v_T are the longitudinal and transverse sound velocities, M_L and M_T are the corresponding off-diagonal TLS–phonon coupling parameters, and ρ_d is the mass density.

In (normal-conducting) metals transitions between the two states of the TLS can also be mediated by conduction electrons. Physically, this interaction arises from the fact that the two states of the TLS correspond to different ionic positions. Hence the conduction electrons experience a different ionic potential which depends on the state occupied by the TLS. The TLS–conduction-electron coupling

*In the tunneling model, M and D are linked through the relation $E = \sqrt{\Delta_0^2 + \Delta^2}$ and hence are not independent of each other.

†In the Bloch formalism, τ_1 corresponds to the spin-lattice relaxation time.

leads to an additional term H_2 in the TLS Hamiltonian which when transformed into the basis of TLS states becomes [6]

$$H_2' = \frac{1}{N} \sum_{\mathbf{k},\mathbf{q}} (V_\perp(\mathbf{q})S_x + V_\parallel(\mathbf{q})S_z)c_\mathbf{k}^+ c_{\mathbf{k}+\mathbf{q}} \tag{20}$$

Here $c_\mathbf{k}^+$ and $c_{\mathbf{k}+\mathbf{q}}$ denote creation and annihilation operators of conduction-electron plane-wave states.* $V_\perp$ and $V_\parallel$ are the TLS–conduction-electron coupling parameters, corresponding to M and D for the TLS–phonon coupling. As in the case of phonons, it is assumed that the interaction with electrons affects mainly the asymmetry of the TLS. $V_\perp$ is proportional to Δ_0/E and induces transitions between the two TLS states and $V_\parallel$ describes the change in energy splitting. N is the total number of atoms in the sample. Equation (20) yields in lowest-order perturbation theory an expression for the TLS relaxation rate due to the interaction with conduction electrons [6]

$$\tau_{1,e}^{-1} = \frac{\pi}{4\hbar} (N(E_F)\Omega V_\perp)^2 E \coth \frac{E}{2k_B T} \tag{21}$$

where $V_\perp = \langle V_\perp(q) \rangle$ has been averaged over the Fermi surface and Ω is the atomic volume. Equation (21) corresponds to the Korringa relaxation of nuclear magnetic moments, again in analogy between TLS and spin-1/2 particle in a magnetic field.

The total relaxation rate of a TLS is given by

$$\tau_1^{-1} = \tau_{1,\text{ph}}^{-1} + \tau_{1,e}^{-1} \tag{22}$$

$\tau_{1,e}^{-1}$ and $\tau_{1,\text{ph}}^{-1}$ differ by their energy (or temperature) dependence, which arises from the E dependence of the density of states, $N(E_F) = \text{const}$ for electrons and $Z(E) \sim E^2$ for phonons. For low energies, i.e., at very low temperatures, $Z(E)$ is much smaller than $N(E_F)$, hence τ_1^{-1} is mainly determined by the conduction electrons and, therefore, much larger than in insulators. This fact accounts for most differences in TLS dynamics between amorphous insulators and (normal-conducting) metals.

In amorphous superconductors well below T_c, the TLS relaxation rate approaches that for phonons because of the opening of the gap in the quasiparticle density of states. The difference between superconducting and normal-conducting metals concerning TLS will be discussed further below.

Up to now we have only considered a single TLS. There is no microscopic theory from which to deduce the distribution of the TLS parameters Δ, Δ_0, γ, $V_\parallel$, $V_\perp$, and ω_0 for all TLS in a given amorphous material. The disordered structure suggests a wide distribution of Δ and Δ_0, while single (average) values for γ, $V_\parallel$, $V_\perp$, and ω_0 are assumed. Instead of a distribution of Δ_0 and Δ, a constant distribution of $\lambda \equiv \ln(\hbar\omega_0/\Delta_0)$ and of Δ, which are assumed to be independent of each other, is usually considered:

$$P(\Delta, \lambda) = \overline{P} = \text{const} \qquad (\lambda_{\min} < \lambda < \lambda_{\max}) \tag{23}$$

For a subset of TLS with given energy splitting E, λ is larger than $\lambda_{\min} = \ln(\hbar\omega_0/E)$ corresponding to the symmetric TLS ($\Delta = 0$). These TLS have the largest

*The assumption of plane waves is of course questionable in amorphous metals. However, the fact that the free-electron model works quite often in these materials [10], gives some justification.

tunneling splitting $\Delta_0 = E$ of this subset and couple most strongly to electrons and phonons. They have the shortest relaxation times. $P(\Delta, \lambda)$ can be transformed into a distribution [104]

$$P(E, r) = \frac{\overline{P}}{2r(1 - r)^{1/2}} \tag{24}$$

where $r \equiv \tau_1^{-1}/\tau_{1,\min}^{-1} = \Delta_0^2/E^2$; see Eq. (19). Equation (23) also contains an upper limit $\lambda_{\max}$ corresponding to TLS with very long relaxation times $\tau_{1,\max}$. The wide distribution of relaxation times within a subset of TLS of given energy E is inherent in the tunneling model and independent of the relaxation mechanism.

B. Thermal Properties of Amorphous Metals

The density of TLS excitations which can be observed within a time t is given by (assuming $t < \tau_{1,\max}$)

$$n(E, t) = \int_{r_{\min}}^{1} P(E, r)\, dr = \frac{\overline{P}}{2} \ln\left(\frac{4t}{\tau_{1,\min}}\right) \tag{25}$$

where $r_{\min} = t^{-1}/\tau_{1,\min}^{-1}$. Each TLS contributes a Schottky anomaly to the *specific heat*

$$\begin{aligned} C_{\text{TLS}}(T, t) &= \frac{k_B}{\rho_d} \int_0^{\infty} \left(\frac{E}{k_B T}\right)^2 \frac{\exp(-E/k_B T)}{(\exp(-E/k_B T) + 1)^2}\, n(E, t)\, dE \\ &= \frac{\pi^2}{12} k_B^2 \overline{P} \ln\left(\frac{4t}{\tau_{1,\min}}\right) T \end{aligned} \tag{26}$$

Equation (26) states that the specific heat is weakly time-dependent. Neglecting this for the moment, C_{TLS} is expected to vary approximately linearly with T:

$$C_{\text{TLS}} \approx \frac{\pi^2}{6} k_B^2 n_0 T \tag{27}$$

with a constant TLS density of states n_0.

In insulating glasses, the specific heat usually can be described as

$$C = aT^n + \beta T^3 \tag{28}$$

with $n \approx 1$ and $\beta = \beta_D + \beta_{\text{exc}}$, where $\beta_D = (2\pi^2/5)k_B^4/\hbar^2\rho_d v^3$ is the Debye contribution which can be determined from the average sound velocity $v = (v_L^{-3} + 2v_T^{-3})^{-1/3}$. The excess term β_{exc} over the Debye contribution, whose origin is not well understood, is roughly of the order of 0.1 to 1 β_D [105]. The first term obviously represents C_{TLS}.

In amorphous metals, the conduction electrons contribute a linear specific heat whose magnitude is typically a factor 10 to 100 larger than that of the TLS excitations. Therefore, one has to investigate amorphous superconductors in order to resolve a TLS-induced linear specific-heat term, because well below T_c the electronic quasiparticle excitations are frozen out.

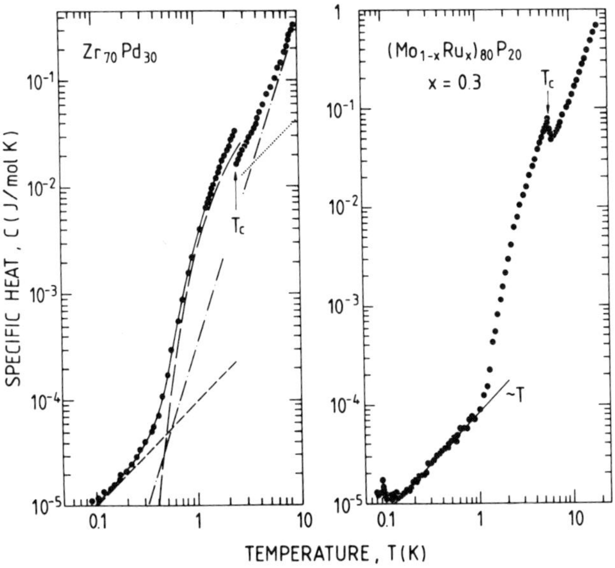

Figure 13 Specific heat C versus temperature T of amorphous superconductors: $Zr_{70}Pd_{30}$ [106], and $(Mo_{0.7}Ru_{0.3})_{80}P_{20}$ [35]. For $Zr_{70}Pd_{30}$, contributions by conduction electrons above T_c, quasiparticles below T_c, phonons and TLS are indicated by dotted, long-dashed, dash-dotted, and short-dashed lines, respectively.

Figure 13 shows two examples of C for amorphous liquid-quenched superconductors [106,35]. Well below T_c, a linear contribution due to TLS* is observed. Other materials include Zr_xCu_{100-x} [34], quench-condensed amorphous $Bi_{84}Sb_{16}$ and $Ga_{95}Ag_5$ films [22], and a variety of Zr-based sputtered amorphous superconductors where a sublinear variation of C_{TLS} was found with n as low as 0.4, which in some cases evolves toward a linear T dependence upon annealing [107].†

For virtually all amorphous metals C_{TLS} at 0.1 K is of the order of $\sim 10^{-5}$ J/mole-K [35] which is the same magnitude as found for vitreous silica and other insulating glasses [105]. Assuming $n_0 \approx$ const, this corresponds to $n_0 \approx 0.1$ eV^{-1} at^{-1}. Hence at 0.1 K, roughly 10^{18} TLS per mole are excited, or, a fraction of 1.5×10^{-6} of the atoms. Unfortunately, the TLS specific heat cannot be measured above 1 K because of the rapidly rising quasiparticle and phonon contributions.

The logarithmic time dependence of C expected from the tunneling model (Eq. (26)) has indeed been observed in some insulating glasses [108]. One has to carry out measurements over several decades in time, which has not been attempted for amorphous metals. Another way of observing the time dependence of C_{TLS} is by cooling a sample to a temperature T_f from a higher temperature T_i where thermal

*A critical discussion of other possible sources of a linear specific heat in amorphous metals for $T \ll T_c$ such as normal-conducting regions or impurities has been given elsewhere [34].

†In amorphous superconductors, $\tau_{min}^{-1} \sim T^3$ in the dominant phonon approximation (see Eq. (19)). Hence the weak time dependence should lead to a slightly stronger than linear T dependence of C ($\sim T \ln T$) instead of a sublinear variation which remains at present unexplained. We note that also the normal-state properties of sputtered metals, e.g., the electronic density of states, behave in an unexpected manner compared to liquid-quenched metals (see Section IIE).

equilibration was acquired. The relaxation of slow TLS to the new equilibrium population at T_f results in a heat flow $\dot{Q}$ out of the sample, the sample does not attain T_f by a simple exponential decay. Differentiation of C_{TLS}, Eq. (26), with respect to t and integration over T yields

$$\dot{Q} = \frac{\pi^2}{24\rho_d} k_B^2 m \overline{P}(T_i^2 - T_f^2)\frac{1}{t} \tag{29}$$

The trivial heat flow which results from cooling the sample (of mass m) from T_i to T_f is not included in Eq. (29). Such an experiment was performed first on vitreous silica by Zimmermann and Weber [109], and subsequently by Sahling et al. [110] on a number of metallic glasses. The expected T and t dependences were indeed observed at low T, giving substantial support to the tunneling model and yielding a direct determination of $\overline{P}$.

We now turn to the *thermal conductivity*. In (nonmagnetic) metals, heat is carried by conduction electrons and phonons:

$$\kappa = \kappa^e + \kappa^{ph} \tag{30}$$

The thermal transport is limited by various scattering processes. Often, the dominant phonon approximation is used and the thermal resistivity in the phonon channel is written as $(\kappa^{ph})^{-1} = \Sigma(\kappa_\nu^{ph})^{-1}$ where $(\kappa_\nu^{ph})^{-1}$ denotes the phonon thermal resistivity arising from the νth scattering process.

When the energy dependence of the scattering processes for electrons which lead to thermal and electrical resistance is the same (as holds, for example, for elastic processes) the Wiedemann-Franz law is valid for κ^e

$$\frac{\kappa^e \rho}{T} = \frac{\pi^2}{3}\left(\frac{k_B}{e}\right)^2 = 2.45 \times 10^{-8}\,\frac{W\Omega}{K^2} \tag{31}$$

This need not be the case for inelastic scattering or for *s-d* scattering in metallic glasses with transition metals. For simplicity, however, Eq. (31) is widely used for an analysis of the thermal conductivity in normal-conducting amorphous metals. In amorphous metals, due to their high electrical resistivity ρ, κ^e is generally small and may be only a minor fraction of the total thermal conductivity. Also, the temperature dependence of ρ is usually very small (see, for example, Fig. 10). In superconductors the electronic thermal conductivity vanishes quickly for $T < T_c$ because only quasiparticles carry energy. If elastic electron scattering dominates, κ_s^e/κ_n^e depends only on the reduced temperature $t = T/T_c$ and the gap parameter $2\,\tilde{\Delta}_0/k_B T_c$ (see [28], for example).* For $T \ll T_c$ only phonons contribute to thermal transport.

The conduction electrons are not only heat carriers but also act as scattering centers for phonons. In normal-conducting metals two cases must be distinguished: (1) If the product of the dominant phonon wave number q_{dom} and the electron mean free path l^e is much larger than unity (this is very often the case in crystals), then $\kappa_e^{ph} \sim T^2$. Here κ_e^{ph} denotes the phonon thermal conductivity as limited by scattering from electrons only. (2) When $q_{dom}l^e \ll 1$ as in most amorphous metals, $\kappa_e^{ph} \sim T$ [111]. In superconductors κ_e^{ph} rises steeply below T_c because the number

*In this chapter we write $\tilde{\Delta}_0$ for the superconducting energy gap (at $T = 0$) in order to distinguish it from the tunneling splitting Δ_0.

of quasiparticles, that is, scattering centers, decreases. For $T \ll T_c$ the phonon thermal conductivity is limited by other scattering processes. Superconductors well below T_c thus behave thermally like insulating materials: $\kappa \equiv \kappa_i^{ph}$ since $\kappa^e \to 0$ and $(\kappa_e^{ph})^{-1} \to 0$.

A quantitative analysis of κ is possible for superconductors by comparing κ in the superconducting state and in the normal state in a magnetic field exceeding B_{c2}. Such an experiment has been performed for $Zr_{70}Cu_{30}$ [112], the result is reproduced in Fig. 14a. In the superconducting state, κ increases just below T_c, indicating that κ^{ph} is indeed dominant. Well below T_c, $\kappa \equiv \kappa_i^{ph} = fT^m$ with $m = 1.7$. Figure 14b shows data [106] over an extended T region, where $m = 1.9$. As will be discussed shortly, $m = 2$ is expected for resonant phonon scattering by TLS.

In the normal state below T_c, κ is reduced in a considerable temperature range. This is of course due to the additional phonon scattering leading to $(\kappa_e^{ph})^{-1}$. The inset of Fig. 14a shows that $\kappa_e^{ph} = T/D$ in the normal state below 1 K, in agreement with the T dependence predicted by standard theories [111,112] for phonon-electron scattering in the limit $q_{dom}l^e \ll 1$. The larger scattering coefficient D (by a factor of ~3 compared to the free-electron model) is due to the large electronic density of states at E_F in $Zr_{70}Cu_{30}$ [112].

For normal-conducting amorphous metals where such an analysis is not possible, κ^{ph} is often *assumed* to vary below 1 K as

$$(\kappa^{ph})^{-1} = (\kappa_e^{ph})^{-1} + (\kappa_i^{ph})^{-1} = \frac{D}{T} + \frac{1}{fT^m} \tag{32}$$

For $m = 2$, a plot of T/κ^{ph} versus T^{-1} yields a straight line [100]. Alternatively, for simple metallic glasses with no transition-metal constituents where the free-electron model works one may calculate κ_e^{ph} from theory to obtain κ_i^{ph} over an extended temperature range [18].

The phonon thermal conductivity of insulating and metallic superconducting glasses below 1 K is limited by resonant scattering from TLS. The corresponding phonon mean free path which also determines the ultrasonic behavior (Section IVC) can be calculated from a rate equation or in the Bloch formalism, as outlined in Section IVA. The result is

$$(l_{res}^{ph})^{-1} = \pi \frac{\overline{P}\gamma^2}{\hbar \rho_d v^2} E \tanh \frac{E}{2k_B T} \tag{33}$$

which in the dominant phonon approximation becomes simply $l_{res}^{ph} \sim T^{-1}$. In this approximation $\kappa_i^{ph} = (1/3)C_D l^{ph} v$ with $v \approx$ const an average sound velocity and $C_D \sim T^3$ the Debye contribution to the specific heat. Hence $\kappa_i^{ph} \sim T^2$ in fair agreement with experiment. The same T dependence is obtained when the Debye model (involving integrals over phonon frequencies [113]) is employed:

$$\kappa_i^{ph} = \frac{\rho_d k_B^3}{6\pi\hbar^2}\left(\frac{v_L}{\overline{P}\gamma_L^2} + \frac{2v_T}{\overline{P}\gamma_T^2}\right)T^2 = fT^2 \tag{34}$$

The prefactor is of the order of $f \approx 10^{-4}$ to 10^{-3} W/cm-K^3 [105], that is, the thermal conductivity of amorphous materials at 1 K is typically a few tenths of a mW/cm-K, regardless whether metallic or insulating. This "universality" of phonon-TLS

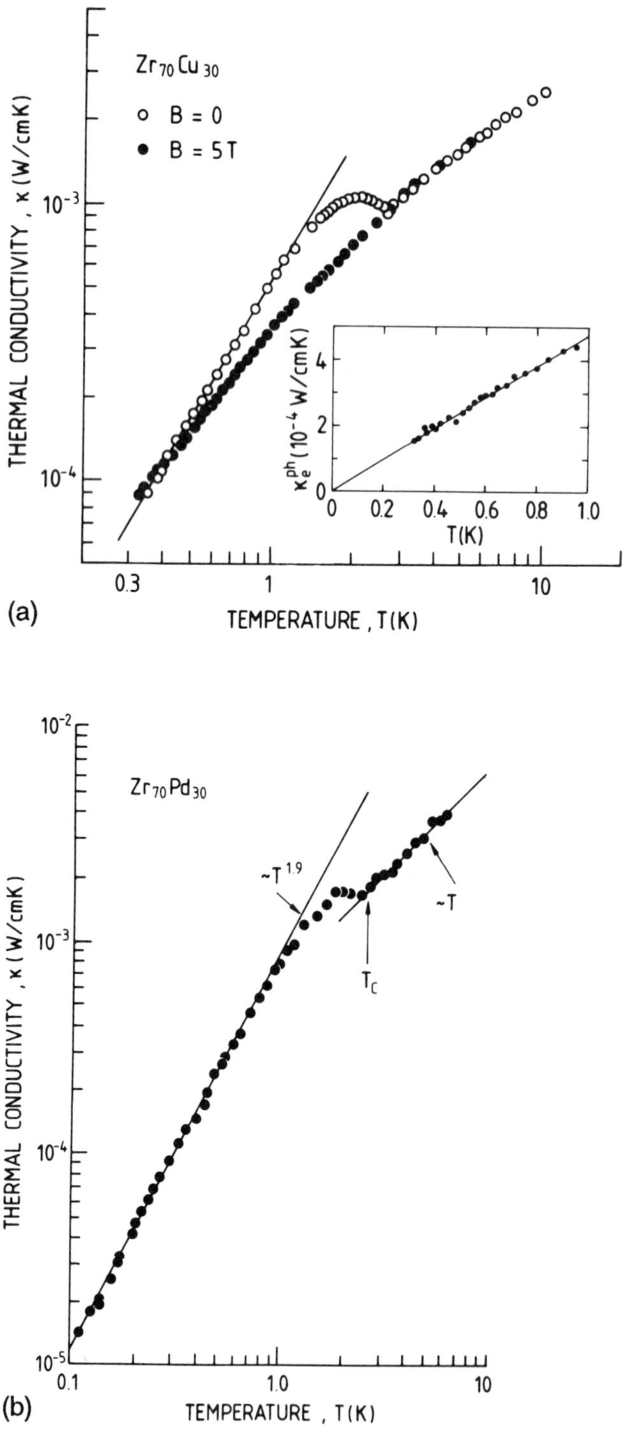

Figure 14 (a) Thermal conductivity κ of amorphous $Zr_{70}Cu_{30}$ as function of temperature for $B = 0$ and $B = 5$ T. Solid line represents $T^{1.7}$ dependence below 1 K. Inset shows phonon thermal conductivity, κ_e^{ph}, as limited by scattering by electrons only versus T [112]. (b) κ as function of T for amorphous $Zr_{70}Pd_{30}$ [106].

scattering corresponds to the results already encountered when discussing the specific heat, and shows that not only the TLS density of states but also the coupling to phonons is rather universal. The small but not negligible deviations of the experimentally determined exponents from 2 are unexplained at present.

At temperatures above 1K, κ^{ph} rises less steeply than T^2, which cannot be solely attributed to phonon-electron scattering. The "remaining" phonon conductivity κ_i^{ph} as limited by intrinsic scattering only, that is, scattering by conduction electrons subtracted, becomes almost independent of T at temperatures above a few K (see Fig. 15). The similarity in magnitude and temperature dependence to insulating glasses such as a vitreous silica is striking. When the contribution of electrons to κ, and also $(\kappa_e^{ph})^{-1}$ are small, as for $Pd_{77.5}Cu_6Si_{16.5}$, such a "plateau" in κ can be seen directly [114]. The origin of the plateau is not clear, other explanations besides TLS have been invoked [103,7].

We conclude this section with a brief discussion of *thermal expansion* measurements. Since TLS are extremely anharmonic excitations, they make an important contribution to the linear thermal expansion coefficient $\alpha = (1/3)(1/V)dV/dT$. α is related to the macroscopic Grüneisen parameter Γ through

$$\alpha = \frac{\Gamma}{3BC} \tag{35}$$

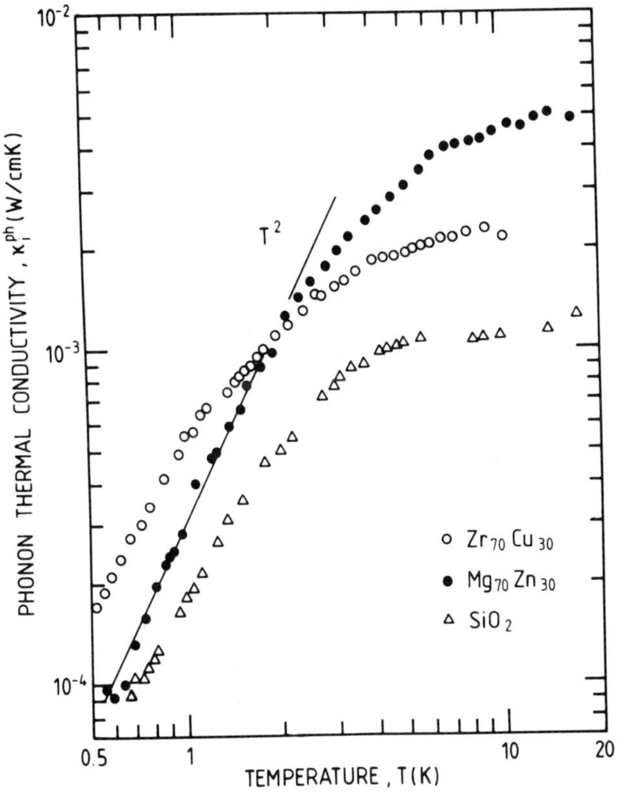

Figure 15 Phonon thermal conductivity, κ_i^{ph}, of metallic glasses as limited by "intrinsic" scattering only, i.e., conduction-electron contribution to phonon scattering subtracted, for $Zr_{70}Cu_{30}$ [112] and $Mg_{70}Zn_{30}$ [18]. κ for vitreous silica is shown for comparison [40].

where B is the bulk modulus and C the total specific heat. Γ is a weighted average over the Grüneisen parameters Γ_i of mode i with specific heat C_i:

$$\Gamma = \frac{\Sigma\, \Gamma_i C_i}{\Sigma\, C_i} \qquad \text{with} \qquad \Gamma_i = \frac{\partial \ln E_i}{\partial \ln V} \tag{36}$$

For amorphous materials, a negative $\Gamma \approx -2$ to -3 (only weakly temperature dependent) is expected as long as positive and negative contributions of the deformation potential γ cancel, due to the (small) dependence of the tunneling splitting Δ_0 on strain ε [115]. Partial noncancellation may lead to much larger positive or negative Γ's, since $\gamma \approx 1$ eV results in $\Gamma_i \approx -(1/\varepsilon)dE_i/E_i = 2\gamma/\Delta \approx 10^4$ for asymmetric ($\Delta \approx E_i$) TLS with $\Delta/k_B = 1$ K.

Thermal expansion measurements on $Pd_{77.5}Cu_6Si_{16.5}$ yielded largely different results. A negative Γ was reported [116] and attributed to TLS. A later measurement actually showed no effect by TLS (because Γ due to TLS is masked by the large contribution of conduction electrons) [117]. This difference might be due to different time scales used in the experiment. The largest contribution to Γ is expected from asymmetric TLS, as discussed above. These have rather long relaxation times, see Eq. (21), and possibly could have evaded detection in the experiment [117] with shorter time scale. A direct measurement of Γ by means of the elastocaloric effect yielded negative Γ's at low T for $Pd_{77.5}Cu_6Si_{16.5}$, $Zr_{70}Pd_{30}$, and $Zr_{60}Cu_{40}$ [115].

C. Acoustic Properties of Amorphous Metals

The *resonant absorption* of phonons* by TLS which limits the thermal conductivity has already been mentioned, see Eq. (33). It results from transitions between the two states with energy difference $E = \hbar\omega$ where ω is the frequency of the sound wave. This expression† for the phonon mean free path l_{res}^{ph} is valid only for low intensities of the sound wave. For high intensities, the population difference of the two levels of the TLS which in thermal equilibrium is given by $\tanh(\hbar\omega/2k_BT)$ will be altered markedly, fewer TLS will be in the ground state and the absorption will decrease ("saturation"), a phenomenon well known from magnetic resonance. This effect can again be quantitatively treated in the Bloch formalism [102,103]. Figure 16 gives an illustration of this effect for amorphous $Pd_{77.5}Cu_6Si_{16.5}$ [119]. For low intensities, α increases with decreasing T at low temperature, in agreement with Eq. (33), while at high intensities, this low-temperature upturn of α is suppressed. The saturation occurs at much lower temperatures and/or higher intensities than in insulating glasses, this is a direct consequence of the short TLS relaxation time τ_1 because of the conduction-electron relaxation. The intensity dependence itself is very important in that it proves the two-level character or, more generally, the strong anharmonicity of the low-energy excitations. A localized harmonic oscillator would not show saturation.

*Acoustic properties of amorphous solids, including metallic glasses have been reviewed recently by Hunklinger and Raychaudhuri [118].

†The inverse phonon mean free path $(l_{ph})^{-1}$ describing the damping of the energy of the sound wave and the (amplitude) absorption coefficient α are related by $(l_{ph})^{-1} = 0.23\alpha$ when $(l_{ph})^{-1}$ is measured in cm^{-1} and α in dB/cm.

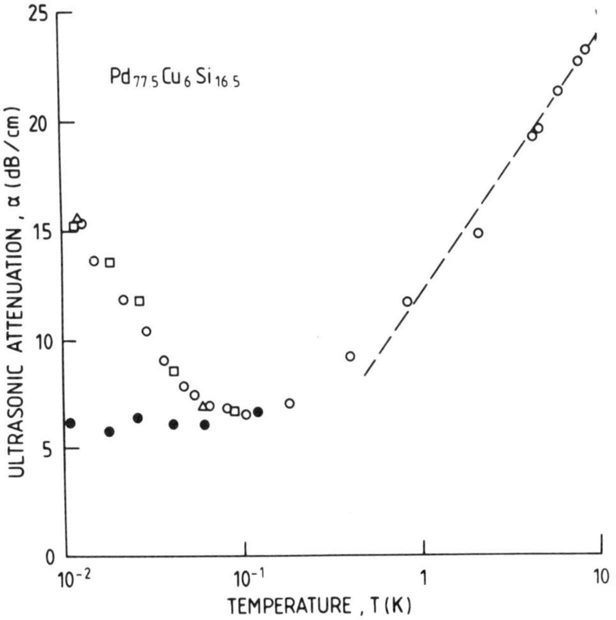

Figure 16 Ultrasonic attenuation, α, of amorphous $Pd_{77.5}Cu_6Si_{16.5}$ as a function of temperature, T, for transverse waves (0.96 GHz). Open symbols: sound intensity $I \approx 10$ mW/m^2. Closed symbols: $I = 10^5$ mW/m^2. Dashed line indicates $\alpha \sim \log T$, that is, leveling off for $\omega\tau_{1,min} < 1$ (data of Reference [119]).

In addition to the resonant interaction, there is a *relaxational absorption* which arises from the diagonal coupling $D = 2(\Delta/E)\gamma$. Its origin can be interpreted as follows. A strain field modulates the asymmetry of all TLS whose population is therefore no longer in thermal equilibrium. The ensemble of TLS returns to equilibrium by resonant emission or absorption of thermal phonons. Each TLS gives a contribution $\alpha_{rel} \sim D^2\omega^2\tau_1/(\omega^2\tau_1^2 + 1)$ to the relaxational absorption. The total relaxational absorption $(l_{rel}^{ph})^{-1}$ is obtained by integrating with respect to E and the relaxation rate τ_1^{-1} (or r). This integration can be performed analytically only in the limiting cases $\omega\tau_{1,min} \gg 1$ or $\ll 1$. We will discuss these two cases briefly.

At low temperatures and/or high frequencies ($\omega\tau_{1,min} \gg 1$), the energy dependence of τ_1^{-1} determines the T dependence of $(l_{rel}^{ph})^{-1}$. In amorphous metals, $(l_{rel}^{ph})^{-1} \sim T$, whereas in insulators $(l_{rel}^{ph})^{-1} \sim T^3$. At high temperatures and/or low frequencies ($\omega\tau_{1,min} \ll 1$) [118]

$$(l_{rel}^{ph})^{-1} = \frac{\pi\omega\bar{P}\gamma^2}{2\rho_d v^3} \tag{37}$$

independent of T! In principle, Eq. (37) offers the possibility to determine $\bar{P}$ in a temperature range where the specific heat is dominated by electronic and phonon contributions. However, often additional relaxation processes set in at higher temperatures, and the possibility of a quantitative analysis is limited.

The relaxational absorption is seen as the rise of α in Pd-Cu-Si above ~0.1 K in Fig. 16 [119]. Again, it sets in at much lower T than for insulating glasses, due to the small τ_1 in metallic glasses. Above a few K, α does not become temperature-independent as expected from Eq. (37), but continues to rise slowly ($\sim\ln T$). Even for low-frequency (vibrating reed) measurements ($f \approx 1$ kHz), where the condition $\omega\tau_{1,\min} \ll 1$ is fulfilled down to much lower T, a truly constant $\alpha_{\rm rel}$ has never been observed in normal-conducting metallic glasses, see, for example, Fig. 18 (after [120]). This might be due to a nonconstant TLS density of states, or possibly, to a temperature-dependent coupling between TLS and electrons (see below).

Both resonant and relaxational absorption yield (via the Kramers-Kronig relation) a change in the dispersion, that is, a change in the *velocity of sound*, v. Although small (typically $\Delta v/v \sim 10^{-4}$), it can be measured very accurately and yields reliable additional information about TLS.

For the *resonant* interaction phonons with TLS and under the assumption $\hbar\omega \ll k_BT$ a very simple logarithmic temperature dependence results [102,103]

$$\frac{\Delta v_{\rm res}}{v} = \frac{\overline{P}\gamma^2}{\rho_d v^2} \ln \frac{T}{T_0} \tag{38}$$

where T_0 is an arbitrary reference temperature. The *relaxational* interaction gives a $\Delta v_{\rm rel}/v$ which decreases with T. For $\omega\tau_{1,\min} \gg 1$, that is, low T, $\Delta v_{\rm rel}/v$ is negligible compared to $\Delta v_{\rm res}/v$. For $\omega\tau_{1,\min} \ll 1$ one obtains [118]

$$\frac{\Delta v_{\rm rel}}{v} = -\frac{q}{2}\frac{\Delta v_{\rm res}}{v} \tag{39}$$

where q is the exponent of the energy dependence of τ_1^{-1}, that is, $q = 1$ for TLS-relaxation via conduction electrons and $q = 3$ for one-phonon processes. Hence at low T we expect a logarithmic increase* of v followed by a maximum at the temperature where $\tau_{1,e} \approx \tau_{1,\rm ph}$. This behavior has been observed in a number of metallic glasses, although the early interpretation of the increase of v was in terms of the resonant contribution only [121]. Figure 17 gives an example of measurements at two frequencies differing by a factor 10^6 (after [119,122]). The high-frequency data deviate from Eq. (38) at low temperatures where $\hbar\omega \ll k_BT$ is not valid.

From Eqs. (37) to (39), the coupling constant $\overline{P}\gamma^2$ can be obtained. The values derived from the relaxational absorption are in reasonable agreement with those derived from the change of sound velocity [122]. Also, the thermal conductivity data can be used to extract an averaged (over phonon polarizations) $\overline{P}\gamma^2_{\rm av}$. Again reasonable agreement is obtained, taking into account that the relaxational contribution to $(\kappa_l^{\rm ph})^{-1}$ has not been considered, that is, $\overline{P}\gamma^2_{\rm av}$ is overestimated when taken directly from experimental κ data with the help of Eq. (34) [7].

As the relaxational attenuation involves an integration over all TLS with energies up to k_BT, saturation would at first not be expected, in contrast to the resonant absorption. However, such an effect has been observed in metallic glasses [123] and subsequently been interpreted along two different lines within the tunneling model [124,125].

In this section, we have so far discussed only the ultrasonic properties of normal-conducting metallic glasses. Superconductivity of course modifies the TLS relax-

*This increase should have the slope given by Eq. (38) at very low T where $\omega\tau_{1,\min} \gg 1$ and crossover to half the slope when $\omega\tau_{1,\min} \approx 1$. However, this crossover has never been observed in normal-conducting metallic glasses.

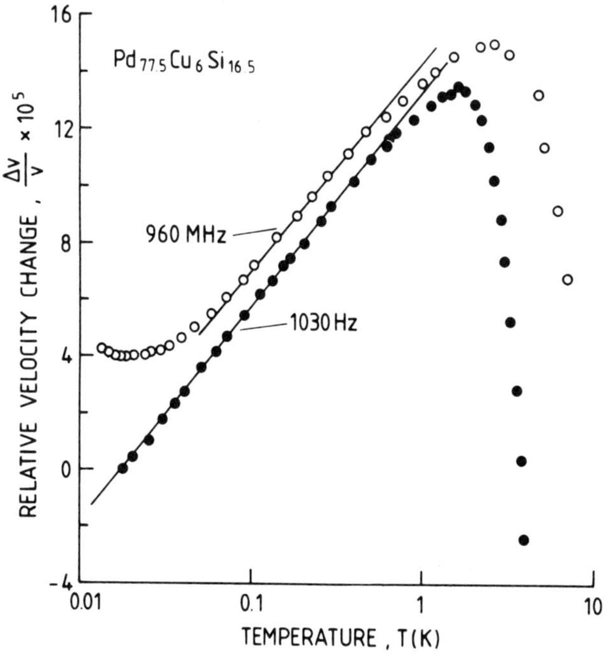

Figure 17 Variation of transverse sound velocity of amorphous $Pd_{77.5}Cu_6Si_{16.5}$ versus temperature for two frequencies: closed circles: $f = 0.96$ GHz [119], open circles: $f = 1030$ Hz [122].

ation via conduction electrons. Well below T_c, when no quasiparticles are excited, the TLS will couple only to phonons. Superconducting glasses therefore provide the possibility to study directly the influence of conduction electrons on the TLS dynamics by performing measurements in the superconducting state and in the normal state in a magnetic field exceeding B_{c2}.

The relaxation rate of a TLS energy $E \ll \tilde{\Delta}$ in the superconducting state via quasiparticles is given by [6]

$$\tau_{1,\mathrm{es}}^{-1} = \frac{\pi}{\hbar}\,(N(E_F)\Omega V_\perp)^2\,\frac{k_B T}{\exp(\tilde{\Delta}(T)/k_B T) + 1} \tag{40}$$

When $E > 2\tilde{\Delta}$, the possibility of an excited TLS to create two quasiparticles leads to an enhancement of $\tau_{1,\mathrm{es}}^{-1}$. Equation (40) shows that for $T \ll T_c$, $\tau_{1,\mathrm{es}}^{-1}$ becomes exponentially small due to the energy gap $\tilde{\Delta}$ of the superconductor. Hence the low-temperature ultrasonic behavior of amorphous metals should approach that of amorphous dielectrics. Qualitatively, this is indeed observed in high-frequency measurements [120,122]. A major unexpected result is the maximum of the relaxational absorption well below T_c in the superconducting state for low frequencies, where $\omega\tau_{1,\min} \ll 1$ is still valid although the relaxation occurs only via phonons. Here $\alpha_{\mathrm{rel},s}$ becomes much larger than $\alpha_{\mathrm{rel},n}$ (see Fig. 18, after [120]) indicating that $\bar{P}\gamma^2$ is larger in the superconducting than in the normal state, see Eq. (37). The TLS–conduction-electron coupling (in the normal-conducting state) may possibly cause a renormalization of the TLS tunneling splitting which perhaps can lead to

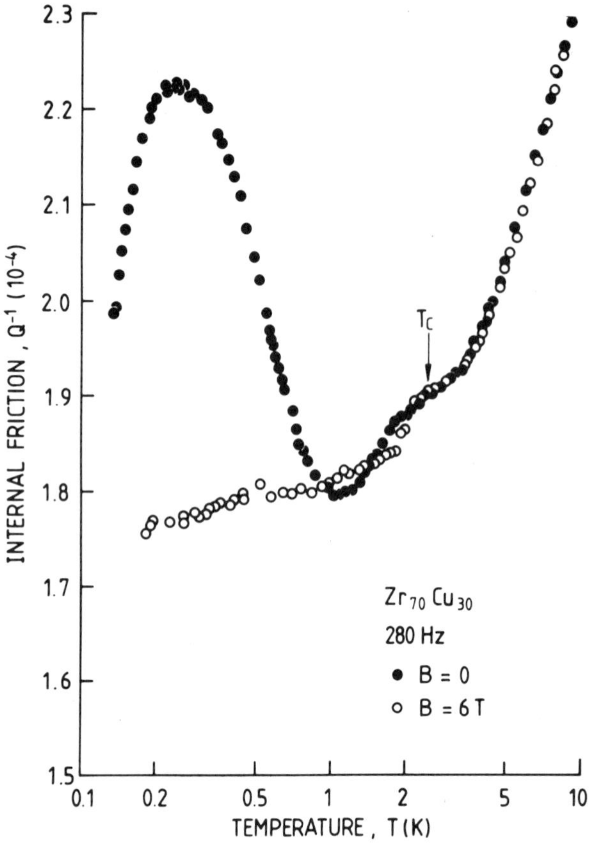

Figure 18 Internal friction Q^{-1} which is related to the sound absorption α by $Q^{-1} = \alpha v/\omega$ versus temperature for amorphous $Zr_{70}Cu_{30}$ for $B = 0$ and 6 T [120].

a change in the distribution function $P(E, r)$ [126,127]. In addition, $\alpha_{rel,n}$ varies slowly with temperature instead of being constant (see Eq. (37)). Further work appears necessary before definite conclusions can be drawn.

So far we have only considered the case that the duration of acoustic pulses τ_p is long compared to the relaxation time τ_1 and to the TLS-TLS relaxation time τ_2 originating from the indirect elastic interaction between TLS. For $\tau_p \ll \tau_1, \tau_2$ the Bloch equations yield TLS echoes (often called phonon echoes) as solutions [128]. These are the analog of spin echoes in nuclear magnetic resonance. For instance, a spontaneous echo after 2τ is generated following a first pulse at $t = 0$ and a second pulse at $t = \tau$. The requirement $\tau_p \ll \tau_1, \tau_2$ is rather stringent. In normal-conducting metals, e.g., Pd-Cu-Si, $\tau_1 \approx 5$ ns at 10 mK for $f = 1$ GHz [119]. Even in amorphous superconductors, due to the lower sound velocities, τ_1 is much shorter than in, say, vitreous silica. Therefore it took more than 10 years following the discovery of TLS echoes in the latter [129] until echoes in an amorphous superconductor, $Zr_{70}Pd_{30}$, were recently observed [130]. This directly confirmed the disappearance of Korringa relaxation in amorphous superconductors, cf. Eqs. (21) and (40). In addition, a direct determination of $\gamma = 1.1$ eV is possible, which is very close to the value in vitreous silica [129].

D. Electrical Resistivity Due to Scattering from Tunneling States

Having dealt in the previous section at some length with the interaction of TLS with conduction electrons, we can ask if in turn the TLS contribute measurably to the inelastic scattering of conduction electrons. This has been suggested early [131,93] to explain the negative temperature coefficient of the electrical resistivity at low temperatures occurring in many amorphous metals. However, as discussed in Section IIIB, the main origin of the low-temperature resistance rise is incipient localization. Nevertheless, the general problem of a TLS coupled to electrons remains intriguing for theoreticians [126,132].

On the experimental side, a convincing proof that TLS contribute only minor to electron scattering has been given by Herlach et al. [133], who showed that in $Pd_{77.5}Cu_6Si_{16.5}$, the thermal conductivity is changed by a factor of 2 upon annealing at 590 K, indicating a decrease of $\overline{P}\gamma_{av}^2$ by this factor, while the coefficient of the temperature-dependent resistivity ρ, which varies roughly as $-\sqrt{T}$, hardly changes. Ultrasonic measurements are reported [133] which suggest that γ is nearly unaffected by annealing, allowing the conclusion that the change in $\overline{P}$ does not affect $d\rho/dT$.

E. Structural-Relaxation Effects on Tunneling States

Structural relaxation studies have become an important tool in the investigation of amorphous alloys. The small structural changes towards a "more stable," yet still metastable, "relaxed" structure which are brought about by heat treatment at temperatures close to the glass transition temperature T_g, manifest themselves in the change of many physical properties of a given amorphous material, as discussed elsewhere in this book.

Usually the free volume* decreases irreversibly during a heat treatment below T_g. The volume change is typically ~0.5% [135]. Therefore one might expect that properties involving atomic motion, such as properties determined by TLS, are sensitive to structural relaxation. The effect of structural relaxation on superconducting properties of rapidly solidified alloys has been briefly discussed in Sections IIB and C.

Structural-relaxation effects on TLS-related properties [40] have been often studied through thermal conductivity measurements. In general, an irreversible increase in the low temperature κ is observed, which can be attributed to a decrease of $\overline{P}\gamma_{av}$. This has been inferred from measurements on $Pd_{77.5}Cu_6Si_{16.5}$ [136], and more directly for superconducting metallic glasses [137–139,35]. Sputtered metals, while different in their annealing behavior for other properties, show similar effects for κ [107].

Recently, the change of $\overline{P}\gamma_{av}^2$ in $Pd_{77.5}Cu_6Si_{16.5}$ upon annealing was attributed to the change in free volume [140] on the basis of a systematic investigation of κ for samples produced with different quenching rates [100]. However, this is not unambiguous since samples with similar $\overline{P}\gamma_{av}^2$ as determined from κ, do not have the same calculated free volume [140]. Similar changes in $\overline{P}\gamma^2$ on the same samples [100,140] were also determined by low-frequency absorption measurements [141].

*The free-volume model assumes that there is some "excess" volume in the undercooled liquids, which is distributed at random and may move freely above T_g, but is trapped at T_g [134].

In addition to the irreversible change in κ after a first annealing step, smaller reversible changes (by $\approx$10%) can be obtained by thermal cycling near T_g [142,140]. κ^{ph} decreases with increasing annealing temperature T_a, hence $\overline{P}\gamma^2_{av}$ increases with T_a. This is also interpreted in terms of the free-volume theory where the TLS are associated with atomic tunneling into voids, that is, agglomeration of free volume. Hence the total amount of free volume V_F in a glass determines the number of TLS [143]. As V_F of an undercooled liquid in thermal equilibrium increases with temperature, the total density of TLS (and thus also $\overline{P}$) should increase with annealing temperature T_a, as long as T_g remains unchanged. It should be mentioned that the prototype insulating glass, vitreous silica, shows an opposite effect of reversible structural relaxation on $\overline{P}\gamma^2_{av}$ [144]. The examples show that while the free model-volume may be a useful qualitative concept, a detailed understanding of the TLS is lacking.

The specific heat C_{TLS} of liquid-quenched superconducting metallic glasses well below T_c remains almost unaffected by structural relaxation [137,139,35], indicating that the TLS density of states n_0 remains unchanged. Since in the tunneling model n_0 and $\overline{P}$ are directly related, see Eqs. (26) and (27), the apparent discrepancy between C and κ can only be reconciled by invoking a change of the distribution $P(E, r)$ or of the deformation potential γ upon annealing. While initially the former possibility was favored [137], a recent detailed low-frequency acoustic study of annealing effects on $Zr_{70}Cu_{30}$ also pointed out the possibility of a change in γ [144]. Most importantly, however, these experiments support the different behavior of C and κ. As mentioned in Section IVK, sputtered amorphous Zr-based superconductors show large changes in C_{TLS} upon annealing [107], which indicates that these samples may have a different short-range order.

F. Low-Energy Excitations in Quasicrystalline and Crystalline Alloys

Tunneling states have been detected through their low-frequency acoustic properties (see Section IVC) in icosahedral ribbons of Al_5Li_3Cu and $Mg_3Zn_3Al_2$, and also in a single-grain quasicrystal $Al_{5.1}Li_3Cu$ [145]. A brief discussion of quasicrystalline materials was already given in Section IIE. In all these materials, the TLS-phonon coupling constant $\overline{P}\gamma^2$ as determined from the sound velocity and damping is roughly a factor of 5–10 smaller than in metallic glasses, and depends on the sample preparation. Interestingly, also the crystalline cubic Frank-Kaspar phases of the same composition show indications of TLS, yet with a somewhat smaller $\overline{P}\gamma^2$. The evaluation of specific heat measurements on $Mg_3Zn_3Al_2$ for $T \ll T_c$ gives an upper limit of $a = 6 \times 10^{-7}$ and 3×10^{-8} J/gK^2 for the icosahedral and cubic phases, respectively [24], in qualitative agreement with the acoustic data.

Thermal conductivity measurements performed on $Pd_{58.8}U_{20.6}Si_{20.6}$ between 0.1 K and 2 K show a very similar behavior for icosahedral and glassy phases with $\kappa^{ph} \sim T^{1.8}$ and $\kappa^{ph} \sim T^{1.7}$ respectively [146]. The specific heat in this case is dominated by large contributions originating from the uranium [66].

Low-temperature anomalies in the thermal [147] and acoustic behavior [148] were observed in $Zr_{100-x}Nb_x$ alloys rapidly quenched from the bcc β phase and were attributed to TLS arising from small-scale β-ω fluctuations. The ω phase results from the simultaneous displacement of two atoms in the β-phase unit cell by ~500 pm along the [111] direction. This could possibly serve as a microscopic model of atomic tunneling.

Evidence for tunneling states from ultrasonic measurements has been reported for Pd-Cu-Si samples that had been annealed to obtain the crystalline state [149]. Another interesting observation is the linear specific heat of similar size as in amorphous metals in pressure-quenched metastable superconducting Al-Si alloys [150].

Evidence for TLS has also been given for sintered high-T_c superconductors by thermal conductivity and acoustic measurements [151–155]. Whether these are due to atomic motion in grain boundaries, twin boundaries or at oxygen vacancies remains to be clarified.

All these examples show that TLS are not restricted to amorphous materials but can be found whenever there are local degrees of freedom which allow for some sort of atomic rearrangement. Of course, strong disorder is the most obvious way to provide these extra degrees of freedom. On the other hand, disorder alone is not sufficient for TLS to occur. A counterexample is amorphous Ge where the TLS density of states is very small compared to amorphous metals or vitreous silica [156].

Finally, impurity-related tunneling states have been observed in crystalline metals, for example, Nb. In particular, the tunneling of H atoms into various sites which are bound to interstitial O or N has been studied via thermal [157] and acoustic measurements [158], and also by inelastic neutron scattering [159]. These TLS exhibit a unique tunneling splitting Δ_0, but may experience strain fields, e.g., from other impurities, leading to a distribution of Δ. Hence the distribution function is different from that of TLS in amorphous metals. Nevertheless, these systems provide interesting insight into the TLS-electron coupling because they can be studied in the superconducting and normal states yielding information about a possible renormalization of Δ_0 due to the coupling to conduction electrons.

Turning back to amorphous alloys, it would be very interesting to relate some properties of TLS, for example, density of states $\overline{P}$, or the deformation potential γ, to other properties of the rapidly solidified metal. However, we are still far away from such a goal. For insulating glasses, empirical relations between $\overline{P}$ and the glass transition temperature T_g have been found as discussed by Hunklinger [141]. For amorphous metals, no such relation exists, not even for a single system where T_g can be varied with composition, such as $(Mo_{1-x}Ru_x)_{80}P_{20}$ [35] or Zr_xCu_{100-x} [34]. This is probably due to the dependence of short-range order on concentration. As the TLS presumably involve atomic motion, their properties in turn will depend critically on the short-range order. In addition, the particular state into which the alloy is frozen depends very much on the details of the quenching process.

In conclusion, the phenomenological tunneling model for the low-temperature anomalous properties of amorphous metals describes the experiments rather well, although extensions and modifications of the tunneling model may be necessary [160,161]. However, it may well turn out that the tunneling entities which as yet have not been identified, may differ in different materials. The vibrational properties at temperatures above a few kelvins, which we have not treated here, are much less understood, and it will take a lot of effort until some of the concepts currently being developed for disordered solids will be successfully applied to these phenomena.

V. ACKNOWLEDGMENTS

This chapter grew out of a collaboration over the years with many co-workers and colleagues whose contributions can be identified from the references cited. I am indebted to all of them. Thanks are also due to F. Baumann and P. Wölfle for a critical reading of the manuscript. Part of this work was done when the author was at the Rheinisch-Westfälische Technische Hochschule Aachen. Support of the Deutsche Forschungsgemeinschaft is gratefully acknowledged.

REFERENCES

1. P. W. Anderson, *Phys. Rev.*, *109*, 1492 (1958).
2. E. Abrahams, P. W. Anderson, D. C. Licciardello, and T. V. Ramakrishnan, *Phys. Rev. Lett.*, *42*, 673 (1979).
3. R. C. Zeller and R. O. Pohl, *Phys. Rev. B*, *4*, 2029 (1971).
4. P. W. Anderson, B. I. Halperin, C. M. Varma, *Phils. Mag.*, *25*, 1 (1972).
5. W. A. Phillips, *J. Low Temp. Phys.*, *7*, 351 (1972).
6. J. L. Black, in *Glassy Metals I*, edited by H.-J. Güntherodt and H. Beck (Springer, Berlin, 1981), p. 167.
7. H. v. Löhneysen, *Phys. Rep.*, *79*, 161 (1981).
8. F. Steglich, in *Theory of Heavy Fermions and Valence Fluctuations*, edited by T. Kasuya and T. Saso (Springer, Berlin, 1985), p. 23; K. Ueda and T. M. Rice, ibid., p. 267.
9. See, e.g., *Theories of High Temperature Superconductivity*, edited by J. W. Halley (Addison-Wesley, 1988).
10. G. Bergmann, *Phys. Rep.*, *27C*, 159 (1976).
11. W. L. Johnson, in Ref. 6, p. 191.
12. S. J. Poon, in *Amorphous Metallic Alloys*, edited by F. E. Luborsky (Butterworths, London, 1983), p. 432; a systematics of amorphous alloy T_c with only a minor second constituent ($<10\%$) has been given by S. J. Poon, *Physica*, *135B*, 259 (1985).
13. J. Bardeen, L. Cooper, and J. Schrieffer, *Phys. Rev.*, *108*, 1175 (1975).
14. W. L. McMillan, *Phys. Rev.*, *167*, 331 (1968).
15. W. Buckel and R. Hilsch, *Z. Phys.*, *138*, 109 (1954).
16. W. Buckel, *Z. Phys.*, *138*, 136 (1954).
17. W. L. Johnson, S. J. Poon, and P. Duwez, *Phys. Rev. B*, *11*, 150 (1975).
18. R. van den Berg, S. Grondey, J. Kästner, and H. v. Löhneysen, *Solid State Commun.*, *47*, 137 (1983).
19. P. Häussler and F. Baumman, *Z. Phys. B*, *49*, 303 (1983); P. Häussler, *Z. Phys. B*, *53*, 15 (1983).
20. J. Dutzi and W. Buckel, *Z. Phys. B*, *55*, 99 (1984).
21. R. van den Berg, H. v. Löhneysen, A. Schröder, U. Mizutani, and T. Matsuda, *J. Phys. F*, *16*, 69 (1986).
22. G. Kämpf, H. Selisky, and W. Buckel, *Physica*, *108B*, 1263 (1981).
23. S. Ewert and W. Sander, *Z. Phys.*, *247*, 21 (1971).
24. J. E. Graebner and H. S. Chen, *Phys. Rev. Lett.*, *58*, 1945 (1987).
25. T. Müller, M. Hofacker, and H. v. Löhneysen, *Jap. J. Appl. Phys.*, *26*, Suppl. 26–3, 735 (1987).
26. W. Dietsche, H. Kinder, J. Mattes, and H. Wühl, *Phys. Rev. Lett.*, *45*, 1332 (1980).
27. M. Hofacker, J. Kästner, H. v. Löhneysen, and W. Sander, *Z. Phys. Chem. NF*, *157*, 747 (1988).
28. H. v. Löhneysen and F. Steglich, *Z. Phys. B*, *29*, 89 (1978).
29. M. M. Collver, R. H. Hammond, *Phys. Rev. Lett.*, *30*, 92 (1973).

30. B. T. Matthias, in *Progress in Low Temperature Physics*, Vol. 2, edited by C. Gorter (Interscience, New York, 1957), p. 138.
31. C. C. Tsuei, S. von Molnar, and J. M. Coey, *Phys. Rev. Lett.*, *41*, 664 (1978).
32. D. B. Kimhi and T. H. Geballe, *Phys. Rev. Lett.*, *45*, 1039 (1980).
33. P. Oelhafen, in *Glassy Metals II*, edited by H. Beck and H.-J. Güntherodt (Springer, Berlin, 1983), p. 283.
34. K. Samwer and H. v. Löhneysen, *Phys. Rev. B*, *26*, 107 (1982).
35. C. Sürgers and H. v. Löhneysen, *Z. Phys. B*, *70*, 361 (1988).
36. C. M. Varma and R. C. Dynes, in *Superconductivity of d- and f-Band Metals*, edited by D. H. Douglass (Plenum, New York, 1976), p. 507.
37. Z. Altounian and J. O. Strom-Olsen, *Phys. Rev. B*, *27*, 4149 (1983).
38. B. Golding, B. G. Bagley, and F. S. L. Hsu, *Phys. Rev. Lett.*, *29*, 68 (1972).
39. S. Ewert, U. Roll, and W. Sander, in *Proc. Int. Conf. on Physics of Transition Metals*, Leeds 1980, *Inst. Phys. Conf. Ser.*, *55*, 639 (1981).
40. H. v. Löhneysen, in *Amorphous Metals and Non-Equilibrium Processing*, edited by M. von Allmen (Les Editions de Physique, Les Ulis, 1984), p. 153.
41. R. E. Elmquist and S. J. Poon, *Solid State Commun.*, *41*, 221 (1982).
42. M. Moske and K. Samwer, *Rev. Sci. Instrum.*, *59*, 2012 (1988).
43. F. Zougmore, J. C. Lasjaunias, O. Laborde, and O. Béthoux, *Key Engineer. Mat.*, *13–15*, 203 (1987).
44. U. Wallrabe, Diploma thesis, Universität Karlsruhe, 1988.
45. S. J. Poon, *Phys. Rev. B*, *27*, 55129 (1983).
46. For a theoretical review, see N. R. Werthamer, in *Superconductivity*, Vol. 1, edited by R. D. Parks (Marcel-Dekker, New York, 1969), p. 321; A. L. Fetter and P. C. Hohenberg, ibid., Vol. 2, p. 817.
47. N. R. Werthamer, E. Helfand, and P. C. Hohenberg, *Phys. Rev.*, *147*, 295 (1966).
48. G. Bergmann and D. Rainer, *J. Low Temp. Phys.*, *14*, 501 (1974).
49. L. Coffey, K. A. Muttalib, and K. Levin, *Phys. Rev. Lett.*, *52*, 783 (1984).
50. P. W. Anderson, K. A. Muttalib, and T. V. Ramakrishnan, *Phys. Rev. B*, *28*, 117 (1983).
51. J. Geerk, H. Rietschel, and U. Schneider, *Phys. Rev. B*, *30*, 459 (1984).
52. S. J. Poon, *Phys. Rev. B*, *31*, 7442 (1985).
53. D. Belitz, *Phys. Rev. B*, *36*, 47 (1987).
54. J. M. Graybeal and M. R. Beasley, *Phys. Rev. B*, *29*, 4167 (1984).
55. R. C. Dynes, A. E. White, J. M. Graybeal, and J. P. Garno, *Phys. Rev. Lett.*, *57*, 2195 (1986).
56. S. I. Park and T. H. Geballe, *Phys. Rev. Lett.*, *57*, 901 (1986).
57. A. M. Finkel'stein, in *Anderson Localization*, edited by T. Ando and H. Fukayama (Springer, Berlin, 1988), p. 230; A. M. Finkel'stein, *Sov. Phys. JETP Lett.*, *45*, 46 (1987).
58. U. Eckern and F. Pelzer, *J. Low Temp. Phys.*, *73*, 433 (1988).
59. B. G. Orr, H. M. Jaeger, A. M. Goldman, and C. G. Kuper, *Phys. Rev. Lett.*, *56*, 378 (1986); H. M. Jaeger, D. B. Haviland, A. M. Goldman, and B. G. Orr, *Phys. Rev. B*, *34*, 4920 (1986).
60. U. Eckern and A. Schmid, *Phys. Rev. B*, *39*, 6441 (1989).
61. L. J. Geerligs, H. S. J. van der Zant, and J. E. Mooij, *Phys. Rev. B*, *38*, 5154 (1988).
62. D. Shechtman, I. Blech, D. Gratias, and J. W. Cahn, *Phys. Rev. Lett.*, *53*, 1951 (1984).
63. See, e.g., *Proc. ILL/CODEST Workshop on Quasicrystalline Materials*, edited by C. Janot and J. M. Dubois (World Scientific, Singapore, 1988).
64. B. Dubost, J. M. Lang, M. Tanaka, P. Sainfort, and M. Audier, *Nature*, *324*, 48 (1986).

65. J.-B. Suck, H. Bretscher, H. Rudin, P. Grütter, and H.-J. Güntherodt, *Phys. Rev. Lett.*, *59*, 102 (1987).
66. J. Wosnitza, R. van den Berg, H. v. Löhneysen, and S. J. Poon, *Z. Phys. B*, *70*, 31 (1988).
67. J. B. Sokoloff, *Phys. Rev. Lett.*, *57*, 2223 (1987).
68. J. L. Wagner, B. D. Biggs, and S. J. Poon, *Phys. Rev. Lett.*, *65*, 203 (1990).
69. J. Chevrier, D. Pavuna, and F. Cyrot-Lackmann, *Phys. Rev. B*, *36*, 9115 (1987).
70. J. Chevrier, J.-B. Suck, J. J. Capponi, and M. Perroux, *Phys. Rev. Lett.*, *61*, 554 (1988).
71. A. Inoue, T. Masumoto, and H. S. Chen, *J. Phys. F*, *13*, 2603 (1983).
72. G. Flinspach, C. Sürgers, and H. v. Löhneysen, *J. Phys.: Cond. Matt.*, *2*, 4559 (1990).
73. J. G. Bednorz and K. A. Müller, *Z. Phys. B*, *64*, 189 (1986).
74. M. K. Wu, J. R. Ashburn, C. Torng, P. H. Hor, R. L. Meng, L. Gao, Z. J. Huang, Y. Q. Wang, and C. W. Chu, *Phys. Rev. Lett.*, *58*, 908 (1987).
75. T. Komatsu, K. Imai, K. Matusita, M. Takata, Y. Iwai, A. Kawakami, Y. Kaneko, and T. Yamashita, *Jpn. J. Appl. Phys.*, *26*, L1148 (1987); T. Komatsu, K. Imai, K. Matsusita, M. Ishii, M. Takata, and T. Yamashita, *Jpn. J. Appl. Phys.*, *26*, L1272 (1987).
76. S. Jin, T. H. Tiefel, R. C. Sherwood, R. B. van Dover, M. E. Davis, G. W. Kammlot, and R. A. Fastnacht, *Phys. Rev. B*, *37*, 7850 (1988).
77. J. McKittrick, M. E. McHenry, C. Heremans, P. Standley, T. R. S. Prasanna, G. Kalonji, R. C. O'Handley, and M. Foldacki, *Physica C*, *153–155*, 369 (1988).
78. V. Skumryev, R. Puzniak, N. Karpe, Z.-h. Han, M. Pont, H. Medelius, D.-X. Chen, and K. V. Rao, *Physica C*, *152*, 315 (1988); T. Komatsu, K. Imai, R. Sato, K. Matusita, and T. Yamashita, *Jpn. J. Appl. Phys.*, *27*, L533 (1988).
79. See, e.g., P. J. Cote and L. V. Meisel, in Ref. 6, p. 141.
80. J. H. Mooij, *Phys. Stat. Solidi. A*, *17*, 521 (1973).
81. C. C. Tsuei, *Phys. Rev. Lett.*, *57*, 1943 (1986).
82. G. Bergmann, *Phys. Rep.*, *107*, 1 (1984).
83. G. Hertel, D. J. Bishop, E. G. Spencer, J. M. Rowell, and R. C. Dynes, *Phys. Rev. Lett.*, *50*, 743 (1983).
84. For a review, see P. A. Lee and T. V. Ramakrishnan, *Rev. Mod. Phys.*, *57*, 287 (1985).
85. G. J. Dolan and D. D. Osheroff, *Phys. Rev. Lett.*, *43*, 721 (1979).
86. W. L. McMillan, *Phys. Rev. B*, *24*, 2739 (1981).
87. P. A. Lee, *Phys. Rev. B*, *26*, 5882 (1982).
88. A. M. Finkel'stein, *Sov. Phys. JETP*, *57*, 97 (1983).
89. W. L. McMillan and J. Mochel, *Phys. Rev. Lett.*, *46*, 556 (1981).
90. J. Lesueur, L. Dumoulin, and P. Nedellec, *Phys. Rev. Lett.*, *55*, 2355 (1985).
91. R. C. Dynes and J. P. Garno, *Phys. Rev. Lett.*, *46*, 137 (1981).
92. O. Rapp, S. M. Bhagat, and H. Gudmundsson, *Solid State Commun.*, *42*, 741 (1982).
93. For a review, see R. Harris and J. O. Strom-Olsen, in Ref. 33, p. 325.
94. R. W. Cochrane and J. O. Strom-Olsen, *Phys. Rev. B*, *29*, 1088 (1984).
95. M. A. Howson and D. Greig, *J. Phys. F*, *13*, L 155 (1983).
96. J. B. Bieri, A. Test, G. Creuzet, and J. C. Ousset, *Solid State Commun.*, *49*, 849 (1984); R. Richter, D. V. Baxter, and J. O. Strom-Olsen, *Mat. Sci. Eng.*, *99*, 183 (1988).
97. D. Mael, S. Yoshizumi, and T. H. Geballe, *Phys. Rev. B*, *34*, 467 (1986).
98. T. M. Rice, *Phil. Mag. B*, *52*, 419 (1985).
99. S. Sachdev, R. N. Bhatt, and M. A. Paalanen, *J. Appl. Phys.*, *63*, 4285 (1988).
100. D. M. Herlach, H. W. Gronert, and E. F. Wassermann, *Europhys. Lett.*, *1*, 23 (1986).
101. K. Samwer, *Phys. Rep.*, *161*, 1 (1988).

102. S. Hunklinger and W. Arnold, in *Physical Acoustics*, Vol. 12, edited by W. P. Mason and R. N. Thurston (Academic, New York, 1976), p. 155.
103. For a general review of the tunneling model, see *Amorphous Solids*: *Low Temperature Properties*, edited by W. A. Phillips (Springer, Berlin, 1981).
104. J. Jäckle, *Z. Phys.*, *257*, 212 (1972).
105. R. B. Stephens, *Phys. Rev. B.*, *13*, 852 (1976); R. O. Pohl, in Ref. 103, p. 27.
106. J. E. Graebner, B. Golding, R. J. Schulz, F. S. L. Hsu, and H. S. Chen, *Phys. Rev. Lett.*, *39*, 1480 (1977).
107. J. C. Lasjaunias and A. Ravex, *J. Phys. F*, *13*, L101 (1983); J. C. Lasjaunias, A. Ravex, O. Laborde, and O. Bethoux, *Physica*, *126* B, 126 (1984); J. C. Lasjaunias, F. Zougmore, and O. Bethoux, *Solid State Commun.*, *60*, 35 (1986).
108. M. T. Loponen, R. C. Dynes, V. Narayanamurti, and J. P. Garno, *Phys. Rev. Lett.*, *46*, 457 (1980); M. Meißner and K. Spitzmann, *Phys. Rev. Lett.*, *46*, 265 (1981).
109. J. Zimmermann and G. Weber, *Phys. Rev. Lett.*, *46*, 661 (1981).
110. S. Sahling, A. Sahling, and M. Kolác, *Solid State Commun.*, *65*, 1031 (1988), and references therein.
111. P. Lindenfeld and W. B. Pennebaker, *Phys. Rev.*, *127*, 1881 (1962).
112. H. v. Löhneysen, D. M. Herlach, E. F. Wassermann, and K. Samwer, *Solid State Commun.*, *39*, 591 (1981).
113. See, e.g., R. Berman, *Thermal Conductivity of Solids* (Clarendon, Oxford, 1976).
114. J. R. Matey and A. C. Anderson, *J. Non-Cryst. Solids*, *23*, 129 (1977).
115. H. Tietje, M. von Schickfus, and E. Gmelin, *Z. Phys. B*, *64*, 95 (1986).
116. W. Kaspers, R. Pott, D. M. Herlach, and H. v. Löhneysen, *Phys. Rev. Lett.*, *51*, 930 (1983).
117. D. A. Ackerman, A. C. Anderson, E. J. Cotts, J. N. Dobbs, W. M MacDonald, and F. J. Walker, *Phys. Rev. B*, *29*, 966 (1984).
118. S. Hunklinger and A. K. Raychaudhuri, in *Progress in Low Temperature Physics*, Vol. 9, edited by D. F. Brewer (North-Holland, Amsterdam, 1986), p. 265.
119. B. Golding, J. E. Graebner, A. B. Kane, and J. L. Black, *Phys. Rev. Lett.*, *41*, 1487 (1978).
120. P. Esquinazi, H. M. Ritter, H. Neckel, G. Weiss, and S. Hunklinger, *Z. Phys. B*, *64*, 81 (1986).
121. G. Belessa, P. Doussineau, and A. Levelut, *J. Phys. Lett.*, *38*, L-65 (1977).
122. A. K. Raychaudhuri and S. Hunklinger, *Z. Phys. B*, *57*, 113 (1984).
123. H. Araki, G. Park, A. Hikata, and C. Elbaum, *Phys. Rev. B*, *21*, 4470 (1980).
124. W. Arnold, P. Dussineau, and A. Levelut, *J. Phys. Lett.*, *42*, L-289 (1982).
125. Yu M. Galperin, V. L. Gurevich, and D. A. Parshin, *J. Phys. Lett.*, *45*, L-747 (1984).
126. K. Vladár and A. Zawadowski, *Phys. Rev. B*, *28*, 1564, 1582, 1596 (1983).
127. Yu. Kagan and N. V. Prokof'ev, *Solid State Commun.*, *65*, 1385 (1988).
128. B. Golding and J. E. Graebner, in Ref. 103, p. 107.
129. B. Golding and J. E. Graebner, *Phys. Rev. Lett.*, *37*, 852 (1976).
130. G. Weiss and B. Golding, *Phys. Rev. Lett.*, *60*, 2547 (1988).
131. R. Cochrane, R. Harris, J. Strom-Olsen, and M. Zuckermann, *Phys. Rev. Lett.*, *35*, 676 (1975).
132. A. Muramatso and F. Guinea, *Phys. Rev. Lett.*, *57*, 2337 (1986).
133. D. M. Herlach, A. B. Kaiser, and J. Kästner, *Europhys. Lett.*, *4*, 97 (1987).
134. M. H. Cohen and G. S. Grest, *Phys. Rev. B*, *20*, 1077 (1979); G. S. Grest and M. H. Cohen, *Phys. Rev. B*, *21*, 4113 (1980).
135. T. Egami, *Ann. NY Acad. Sci.*, *371*, 238 (1981).
136. J. R. Matey and A. C. Anderson, *Phys. Rev. B*, *17*, 5029 (1978).
137. S. Grondey, H. v. Löhneysen, H. S. Schink, and K. Samwer, *Z. Phys. B*, *51*, 287 (1983).

138. P. Esquinazi, M. E. de la Cruz, A. Ridner, and F. de la Cruz, *Solid State Commun.*, *44*, 941 (1982).
139. H. W. Gronert, D. M. Herlach, A. Schröder, R. van den Berg, and H. v. Löhneysen, *Z. Phys. B*, *63*, 173 (1986).
140. D. M. Herlach, H. W. Gronert, E. F. Wassermann, and W. Sander, *Z. Phys. B*, *65*, 353 (1987).
141. S. Hunklinger, *Phils. Mag. B*, *56*, 199 (1987).
142. E. J. Cotts, A. C. Anderson, and S. J. Poon, *Phys. Rev. B*, *28*, 6127 (1983).
143. M. H. Cohen and G. S. Grest, *Solid State Commun.*, *39*, 143 (1981).
144. P. Esquinazi and J. Luzuriaga, *Phys. Rev. B*, *37*, 7819 (1988).
145. N. O. Birge, B. Golding, W. H. Haemmerle, H. S. Chen, and J. M. Parsey, Jr., *Phys. Rev. B*, *36*, 7685 (1987).
146. J. J. Freeman, K. J. Dahlhauser, A. C. Anderson, and S. J. Poon, *Phys. Rev. B*, *35*, 2451 (1987).
147. L. F. Lou, *Solid State Commun.*, *19*, 335 (1976).
148. N. Thomas, W. Arnold, G. Weiss, and H. v. Löhneysen, *Solid State Commun.*, *33*, 523 (1980); G. Weiss, S. Hunklinger, and H. v. Löhneysen, *Phys. Lett.*, *85* A, 84 (1981).
149. G. Ciluzar, A. Hikata, and C. Elbaum, *Phys. Rev. Lett.*, *53*, 326 (1984).
150. J. Chevrier, J. C. Lasjaunias, F. Zougmore, and J. J. Capponi, *Europhys. Lett.*, *8*, 173 (1989).
151. B. Golding, N. O. Birge, W. H. Haemmerle, R. J. Cava, and E. Rietman, *Phys. Rev. B*, *36*, 5606 (1987).
152. M. Nunez Regueiro, D. Castello, M. A. Izbizky, D. Espaza, and C. Ovidio, *Phys. Rev. B*, *36*, 8813 (1987).
153. P. Esquinazi, C. Duran, C. Fainstein, and M. Nunez Regueiro, *Phys. Rev. B*, *37*, 545 (1988).
154. M. Nunez Regueiro, M. A. Izbizky, and P. Esquinazi, *Solid State Commun.*, *67*, 401 (1988).
155. A. Hikata, M. J. McKenna, C. Elbaum, R. Kershaw, and A. Wold, *Phys. Rev. B*, *40*, 5247 (1989).
156. H. v. Löhneysen and H. J. Schink, *Phys. Rev. Lett.*, *48*, 1121 (1981); R. van den Berg and H. v. Löhneysen, *Phys. Rev. Lett.*, *55*, 2463 (1985).
157. H. Wipf and K. Neumaier, *Phys. Rev. Lett.*, *52*, 1308 (1984).
158. J. L. Wang, G. Weiss, H. Wipf, and A. Magerl, in *Phonon Scattering in Condensed Matter*, edited by W. Eisenmenger, K. Laßmann and S. Döttinger (Springer, Berlin, 1984), p. 401.
159. H. Wipf, A. Magerl, S. M. Shapiro, S. K. Satija, and W. Thomlinson, *Phys. Rev. Lett.*, *46*, 247 (1981).
160. V. G. Karpov, M. I. Klinger, and F. N. Ignat'ev, *Sov. Phys. JETP*, *57*, 439 (1983); M. I. Klinger, *Phys. Rep.*, *165*, 275 (1988).
161. C. C. Yu and A. J. Leggett, *Comments Cond. Mat. Phys.*, *14*, 231 (1988).

16

Intrinsic Magnetic Properties of Amorphous Alloys

T. Mizoguchi
Gakushuin University, Tokyo, Japan

I. INTRODUCTION

The basic magnetic properties of amorphous alloys may be described in terms of magnetic moments, exchange interactions, and magnetic anisotropies, which determine the ground-state spin configuration of magnetic systems.

Structural disorder introduces a distribution both in magnetic moments and exchange couplings between interacting pairs of atoms. The local anisotropy for each atomic moment must also have not only a distribution in size but also in direction; that is, easy axis for each moment differs randomly from site to site depending on the local environment in amorphous materials.

The distributions of magnetic moments may be recognized by those of hyperfine fields in different types of amorphous magnetic materials [1]. The distribution is extremely narrow, with a relative width of only 1% for the rare earth Dy in a-$Dy_{21}Ni_{79}$ (a- represents "amorphous"), which has well-defined localized moment by unpaired electrons in the inner 4*f* shell. For a ferric compound insulator, $Fe(OH)_3$, it is moderately broad (about 10%) due to variation in Fe-O bond length. There is a much greater spread (about 50% in half-width) of the hyperfine field in an amorphous 3*d* alloy, for example, Co in a-$Co_{81}P_{19}$. This comes from the situation that the 3*d* metals are near the limit for the appearance of magnetism, and their moments are quite sensitive to the change in overlap integrals.

In a one-electron picture, the delocalized electrons in a band which is formed by overlap of 3*d* wave functions will be conductive if the band is partially filled; but in fact, electron-electron correlation interaction may be so strong as to open up a correlation gap [2]. Many transition metal compounds, which should be narrowband metals according to classical band theory, becomes Mott insulators because of electron correlation effects. As the extent of overlap increases, the Mott-Hubbard subbands broaden and the moment is reduced. Ultimately the subbands cross at the metal-insulator transition where the greatly reduced localized moment becomes itinerant.

The simplest possible Hamiltonian [3] for a metal that will show magnetic effects may consist of two terms: a term allowing electron transfer from one site to the next which gives the one-electron band structure of bandwidth W, and the other term of Coulomb correlation which is represented with U, the average intra-atomic Coulomb interaction of two electrons on the same site.

In a solid, the d orbitals of different symmetry will overlap to varying extents, giving bands of different widths. Magnetism in real $3d$ metals is greatly complicated by the degeneracy of the d bands and their overlap with the $4s$ band.

Although the observed dependence of average magnetic moment on averaged electron concentration of the $3d$ alloys is fairly well explained by a simple spin-polarized rigid-band model, there is little direct evidence in favor of the required large splitting (roughly 1 eV) between each spin band, and photoemission experiments do not usually support the rigid-band picture.

It is particularly instructive to refer an electronic state of a single impurity in a broad band metal since this is a simple case where the condition for the existence of a moment on the impurity can be calculated [2,4]. Mixing the impurity wave function with the conduction band results in a local density of states, $N(E)$, for each of the two spin states, separated by the energy, nU, where n is a number of unpaired impurity electrons, which may be expanded as a power series for small n as follows:

$$n = \int_0^{E_F} N\left(E + \frac{nU}{2}\right) dE - \int_0^{E_F} N\left(E - \frac{nU}{2}\right) dE$$
$$= nUN(E_F) + \frac{1}{24}(nU)^3 N''(E_F) + \cdots \tag{1}$$

Then we get

$$n = \left[\frac{24\{1 - UN(E_F)\}}{N''(E_F)U^3}\right]^{1/2} \tag{2}$$

The condition for formation on magnetic moment ($n > 0$) is

$$U > \frac{1}{N(E_F)} \tag{3}$$

since $N''(E_F)$ is negative. $N(E_F)$ is proportional to $1/W$, where W is the bandwidth of the local density of states, and this depends in turn on the overlap integral between the impurity electrons and the conduction band. The existence and value of impurity moment will depend sensitively on the nearest-neighbor environment through $N(E)$. If $N(E_F)$ varies smoothly with concentration of the impurity, x, then the magnetic moment on the impurity just above the critical concentration, x_c, will vary as $(x - x_c)^{1/2}$, as seen from Eq. (2). Qualitative results of the above discussion are not restricted to isolated impurities alone, but are generally applicable to amorphous alloys.

The appearance of a magnetically ordered state requires interactions to couple the magnetic moments. The dominant mechanism is the exchange interaction which is represented phenomenologically by the isotropic Heisenberg Hamiltonian for a pair of spins as

$$H = -2J_{ij}S_iS_j \tag{4}$$

where J_{ij} is the exchange constant between spins at sites i and j. There are various exchange mechanisms and the calculation for J_{ij} from the first principle is by no means simple, however, it can be accepted as a phenomenological parameter.

The dominant magnetic interaction in transition metals is direct exchange due to the overlap of the wavefunction of nearest-neighbor atoms, which is quite sensitive to the distance between $3d$ atoms. Empirically the exchange interaction changes sign from negative to positive between atomic spacing of γ-Fe and that of α-Fe. The distribution of interatomic separations in an amorphous solid necessarily leads to a distribution of exchange interactions, which may sometimes include interactions of either sign especially in Fe alloys.

The RKKY interaction is the major source of exchange coupling between $4f$ shells and is also important in dilute $3d$ alloys where direct exchange cannot operate. This is the interaction of a localized spin with the polarization of the conduction band induced by other localized spins. The Fermi cutoff of conduction electrons in k-space induces the oscillation of RKKY interaction in sign as a function of $x = 2k_Fr$.

In order to consider a magnetic order the coupling between spins at sites i and j must be summed over all pairs in a solid. It would be useful to introduce the probability, $P(J)$, of finding an exchange interaction of a given magnitude and sign. For a crystalline solid this consists of one or more delta functions thanks to the regular atomic lattice. In an amorphous solid the disorder will broaden the peaks, even to the extent that interactions of both signs may be included in the distribution.

The molecular field approximation is an useful method for describing magnetism even in amorphous materials. There the interactions of any atom with all its neighbors may be replaced by an effective field.

$$H_{\text{eff}} = \left(\frac{2}{g\mu_{\text{B}}}\right)J_{ij}\langle S_j\rangle \tag{5}$$

which acts on the magnetic moment $-g\mu_{\text{B}}S_i$ to produce an interaction

$$H_i = g\mu_B S_i H_{\text{eff}} \tag{6}$$

In amorphous magnetic materials, not only the magnitude of effective field but also its direction will have a distribution, except in ferromagnets, where $P(J)$ is exclusively or predominantly positive and all spins are parallel within a domain.

Although much weaker than the exchange interaction, the dipolar interaction introduces the magnetic anisotropy in amorphous ferrimagnetic alloys with anisotropic atomic pair distribution.

The electrostatic field created by the surroundings at an atomic site in a solid acts to remove some or all of the orbital degeneracy for $3d$ ions and to impose preferred directions on the moment in the ground state through spin-orbit coupling. For $4f$ electrons, spin-orbit coupling is much greater than the electrostatic interaction, so the total angular momentum, J, becomes a good quantum number and the magnetocrystalline anisotropy is created directly by the latter for non-S-state ions.

The interaction can be expressed by the single-ion Hamiltonian which is expanded by spherical harmonics. For crystalline materials the number of terms in the expansion is greatly reduced by symmetry with an appropriate choice of axes. In contrast, no simplification by symmetry will occur in an amorphous solid. According to the direction of the orbital moment, a complicated three-dimensional energy contour may be given with no particular symmetry except for an inversion. Still, a uniaxial easy direction may be defined for each atom in amorphous solids, for which the single ion Hamiltonian is represented as

$$H_i = -D_i S_{zi}^2 \tag{7}$$

where D_i has a distribution in magnitude and defines easy axes, z_i, which differ in direction at every site i.

Interesting new types of magnetic order may occur in amorphous magnetic materials depending on the intensity of the random local anisotropy energy relative to the exchange energy.

II. TYPES OF MAGNETIC ORDER

Unless magnetic moments are completely independent, there are interactions between them which result in some kind of magnetic order below a critical temperature at which the second-order transition, in general, takes place. The type of magnetic order can be predicted with $J(q)$, which is the Fourier transform of exchange interactions,

$$J(q) = \sum_j J_{ij} \exp[iq(r_i - r_j)] \tag{8}$$

Generally speaking, the ground-state spin configuration is given as a helical spin structure of propagation wave vector, Q, at which the $J(q)$ is maximum. When all J_{ij} are positive, $J(q)$ becomes maximum at $q = 0$, at which collinear ferromagnetism is realized without any incompatibility due to the structural disorder. There are no fundamental differences, in principle, between crystalline and amorphous ferromagnets [5].

In contrast to positive exchange interactions, negative (antiferromagnetic) interactions in a disordered structure can possibly induce quite complex situations. Frustration arises with the negative interactions for a structure wherein the neighbors of a given atom are themselves neighbors of each other. It is then impossible to find a configuration for the spins in which all the interactions are simultaneously satisfied. Whenever frustration is present it will be accompanied by degeneracy or near-degeneracy of the ground state. Many different spin configurations are close in energy, but separated by energy barriers; that is, there are many local minima in configurational space.

It is plausible that frustration in a disordered structure will result in random noncollinear magnetic ordering below a critical temperature. The essential feature of any magnetic order is that a component of magnetic moment of each atom becomes constant with respect to time, $\{S_i\} \neq 0$, where $\{\ \}$ donates the time average. The term "spin glass" has been given to a random noncollinear spin configuration with no short-range correlations, that is, $\langle\{S_i\}\{S_j\}\rangle = 0$, where $\langle\ \rangle$ denotes the average for all *ij* pairs. Thermal fluctuations may be suppressed as the spins freeze on cooling, but there is little evidence that this occurs at a well-defined conventional phase transition. The following characteristics are usually associated with the spin freezing temperature, T_f.

1. A sharp peak appears in the low-field (<1 Oe) susceptibility [6], which becomes rounded in high ac or dc fields.
2. The peak shift to higher temperature with increasing frequency in the ac susceptibility measurement.
3. Irreversible behavior including remanence and coercivity appear near T_f. Remanent magnetization obtained by cooling in the absence of an applied field is generally less than that after cooling through T_f in an applied field.
4. The remanence, which is on the order of a few percent of the collinear magnetization, decays linearly approximately with the logarithm of time.

The random anisotropy due to structural disorder is also one of the important factors for spin configuration. Amorphous solids with significant single-ion anisotropy must be also frustrated in a broader sense, because it is impossible to fully satisfy the competing effects of exchange interaction and single-ion anisotropy at each atomic site which favors different orientations of the spins; then some sort of random noncollinear spin configuration may appear. These are referred to as "asperomagnetic" or "speromagnetic," depending on whether they have a net magnetization. While there seems to be no clear distinction between the speromagnetism and spin glass state, the possibility of very short range (first or second neighbor) average spin correlation in the former may be admitted. There is some long range preferred orientation of the moments in an asperomagnet.

The local anisotropy may be expressed by Eq. (7) as a first-order approximation. If the local anisotropy energy, D, is much smaller than the exchange energy, J, ($D/J \ll 1$), as in 3*d*-based amorphous alloys, the magnetic moments hardly respond to local anisotropy because of strong exchange coupling. Hence, collinear ferro- or ferrimagnetism is realized.

On the other hand, for the case of $D/J \gg 1$, as in amorphous alloys containing 4*f* rare earth elements, magnetic moments are more strongly coupled to the local anisotropy than to the neighboring moments through the exchange interaction. It is important to note here that the lowest order of local anisotropy is, in general, not unidirectional but uniaxial so that either direction along the local easy axis gives the same minimum energy of single-ion local magnetic anisotropy. If exchange interactions are introduced it becomes more advantageous for an easy axis to have only slight deviation from one of directions of the single-ion easy axis to minimize the sum of the local anisotropy and exchange energies.

An interesting category of amorphous alloys consists of two magnetic subnetworks which are specified on chemical basis, e.g., 3*d* transition metals and 4*f* rare earth metals. The exchange interactions between 3*d* atoms are about one order of magnitude stronger than those between 3*d* and 4*f* atoms, while those between 4*f* atoms are much smaller. Thus the magnetic moments of 3*d* atoms align parallel, while those of 4*f* atoms are oriented by negative spin-spin interactions along one of the easy directions which are distributed in the hemisphere, the axis of which is parallel to the magnetization of the 3*d* atoms. This type of spin configuration is called "sperimagnetism."

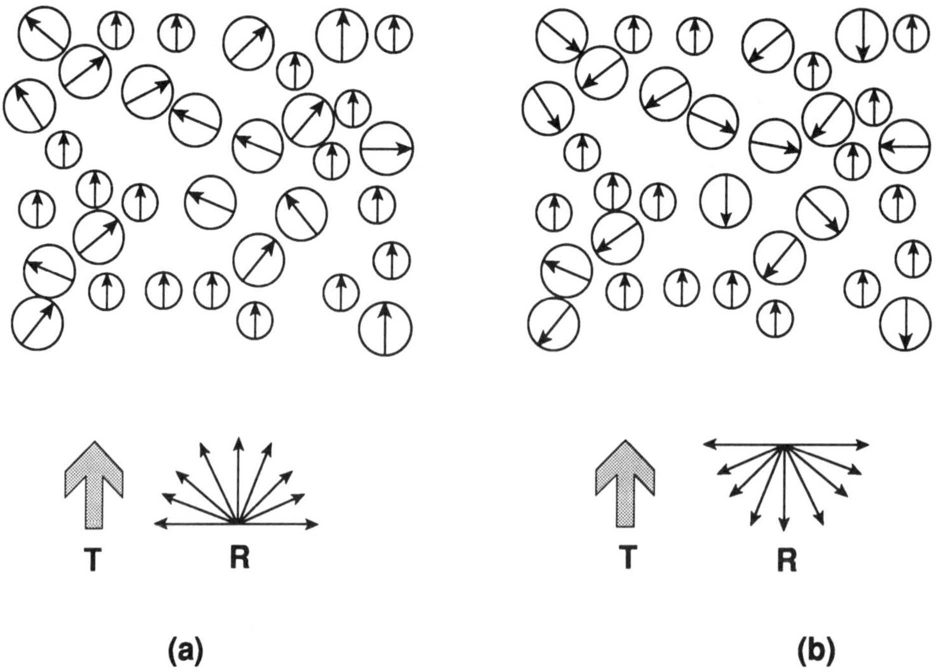

Figure 1 Two-subnetwork sperimagnetic structure in amorphous rare earth–transition metal alloys: (a) spferromagnetism, (b) spferrimagnetism.

In the case of rare earth elements such as Tb or Dy, the total angular momentum, J, is parallel to S. Then the heavy rare earth moments are oriented within a hemisphere of opposite sense with respect to the direction of $3d$ magnetic moments. For light rare earth elements such as Nd or Sm, in which J is antiparallel to S, the moments distribute within a hemisphere oriented along the direction of the $3d$ moments. In some literature [7], "speromagnetism" is defined for the latter spin configuration. In order to avoid confusion, new names "spferromagnetism" or "spferrimagnetism" would be proposed to classify the two subnetwork sperimagnet, depending on the net parallel or net antiparallel coupling between the subnetwork magnetizations, as shown in Fig. 1.

Weaker coupling of the rare earth subnetwork to the random local field occurs for smaller values of atomic orbital angular momentum. For these intermediate cases the dispersed moments fill a cone of narrow angle rather than a hemisphere. Such a spin configuration may be called "asperimagnetism."

III. MAGNETIC MOMENTS AND EXCHANGE INTERACTIONS

In this section the magnetic moments of atoms and the exchange interactions between them which are given from the magnetization and its temperature dependence will be described in different classes of amorphous alloys.

A. Metal-Metalloid System

1. Magnetic Moment

Fe and/or Co base amorphous alloys with about 15–30 at% of light metalloid elements of group IIIA, IVA, and VA (B, C, Si, Ge, and P) exhibit ferromagnetism (reviewed in [8–10]). In the metal-metalloid amorphous alloys small metalloid atoms (B, C) likely occupy relatively larger holes as interstitials in the dense random-packed structure of the metal atoms, lowering the mass density weakly, while others (P, Si, Ge) substitute metal atom sites and thereby lower the density appreciably [7]. The presence of metalloid atoms in these amorphous alloys causes the reduction of average magnetic moment of transition metal atoms compared with those in pure metals. The interstitial metalloid atoms suppress the magnetic moments of transition metal atoms and Curie temperature more than the substitutional ones of same valence.

Such reduction also occurs in crystalline intermetallic compounds (for example, $\mu_{Fe} = 1.91\ \mu_B$ in Fe_2B). The chemical bonding of metalloids with metal atoms plays a significant role in changing the band structure, as does the relative size of the components of the alloy.

It would be interesting whether the changes in magnetic moments in these amorphous alloys can be ascribed entirely to the immediate chemical effects or whether the structural disorder of amorphous alloys have some influences. This can be investigated experimentally in favorable cases where the composition of an amorphous sample is the same as that of a crystalline compound obtained from careful crystallization from the amorphous phase without phase separation. The effects of structural disorder turn out to be generally small, being 1–3% for Co-based alloys and 7–10% for Fe-based alloys [11]. It should be noted, however, that structural disorder lead to a variety of chemical environments which sometimes produce drastic effects on the moments. This is seen vividly, for example, in cases of Fe-Si and Co-Si alloys. Their crystalline phases are nonmagnetic because the chemical environments are unfavorable for the appearance of magnetic moment on transition metal atoms, while in amorphous phase a-$Fe_{0.5}Si_{0.5}$ and a-$Co_{0.5}Si_{0.5}$ become magnetic since multiple environments are realized with some of them favoring the formation of magnetic moments [12].

The magnetic moment of Fe or Co in the metal-metalloid amorphous alloys increases toward that of the pure crystalline value as metalloid content decreases [13,14]. The magnetization of a-Fe_xSn_{1-x} or a-Co_xSn_{1-x} alloys disappear around $x = 0.3$ or 0.5, respectively [15]. The magnetic moment of Fe in a-$Fe_{80}B_{20-x}G_x$, where G = Ge, Si, C, and P, is shown in Fig. 2 [16]. By extrapolating to $x = 20$, Fe moment in a-$Fe_{80}G_{20}$ is largest for G = Ge and in the order of Si, B, C and smallest for G = P.

The average magnetic moment and Curie temperature of amorphous 3*d* transition metal alloys containing metalloids was systematically studied first in $(T_{1-x}M_x)_{80}B_{10}P_{10}$, where T represents Fe or Co, and M represents V, Cr, Mn, Fe, Co, or Ni [17]. The average magnetic moments of transition metal atoms are plotted against the average number of outer valence electrons, N, in Fig. 3 along with the data for corresponding crystalline 3*d* alloys which extend over both bcc and fcc phases. In the Fe-Co, Fe-Ni, and Co-Ni based amorphous alloys ($N > 8$) the average

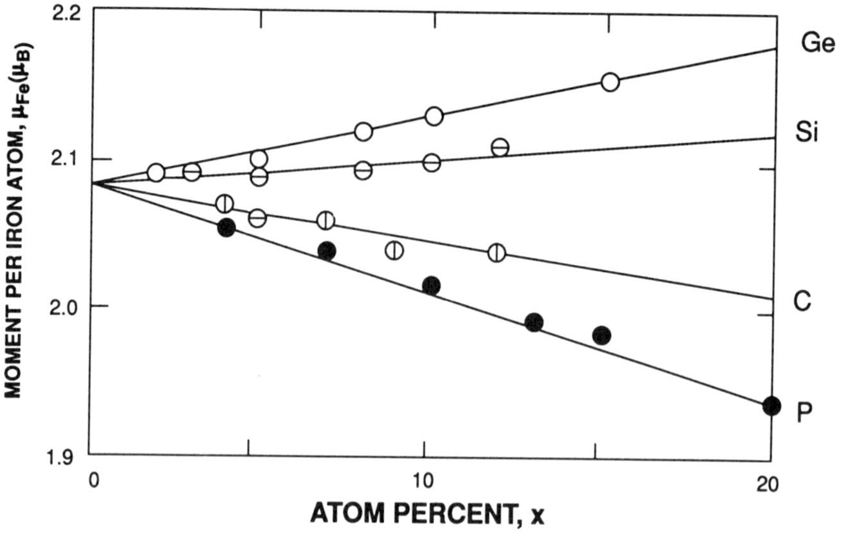

Figure 2 The average magnetic moment of Fe atoms in a-$Fe_{80}B_{20-x}G_x$ alloys, where G = Ge, Si, C, and P [16].

magnetic moments of transition metal atoms decrease linearly with increasing N, starting from $\mu/\mu_B = 2$ with the slope of -1 on a parallel line to the Slater-Pauling curve for the crystalline alloys without metalloid. This behavior can be interpreted if the magnetic moment of Fe, Co, and Ni in these amorphous alloys is roughly 2, 1, and 0 μ_B, respectively. Similar trends of the average magnetic moment of 3d atoms are observed in transition metal amorphous alloys with varieties of metalloids [18,24].

The addition of early transition metals lighter than Fe (V, Cr, or Mn) in Fe-based alloys causes a sharp drop of the average magnetic moment, which may be attributed not only to the reduction of Fe moment but also to antiferromagnetic coupling of the lighter transition metal atomic moments with those of Fe.

A detailed study on the behavior of 3d impurities in amorphous $Fe_{79}P_{13}B_8$ was interpreted in terms of Friedel's virtual bound state model [25]. The excess charge, Z, of impurity atoms is screened by either up- or down-spin electrons, creating the deviation of average magnetic moment from that of the matrix, $x(Z_\uparrow - Z_\downarrow)\mu_B$, where x is the impurity concentration, $Z_\uparrow$ and $Z_\downarrow$ represent the number of spin-up and spin-down screening electrons, respectively. From the observed change of averaged magnetic moments, together with the relation $Z = Z_\uparrow + Z_\downarrow$, the following numbers were obtained:

		V	Cr	Mn	Co	Ni
In amorphous	$Z_\uparrow$	−4.30	−4.20	−2.45	+0.15	+0.15
$Fe_{79}P_{13}B_8$	$Z_\downarrow$	1.30	2.20	1.45	0.85	1.85
In bcc Fe	$Z_\uparrow$	−3.15	−2.18	−1.55	1	1.65
	$Z_\downarrow$	0.15	0.18	0.55	0	0.40

In the table above the spin-dependent screening electron numbers for bcc Fe are also listed. It is pointed out that amorphous $Fe_{79}P_{13}B_8$ is a strong ferromagnet

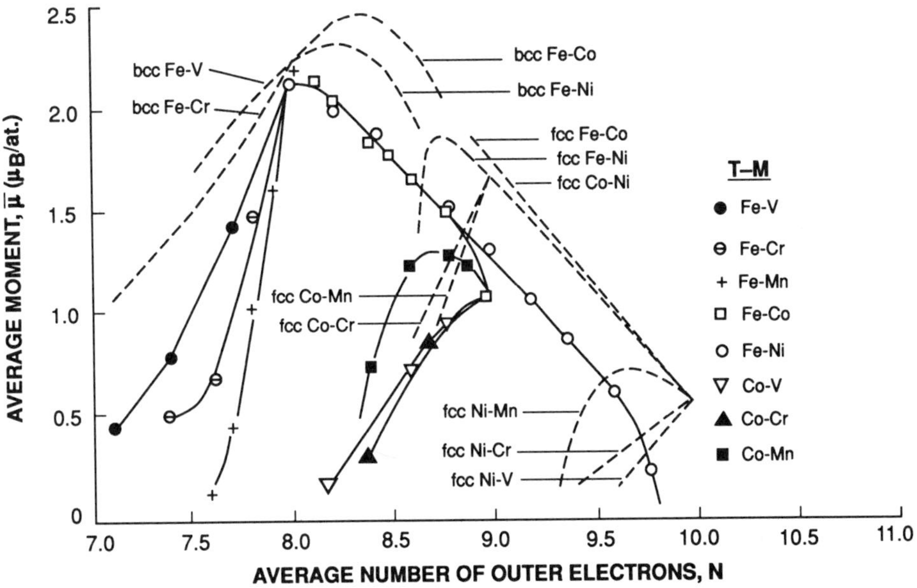

Figure 3 The average magnetic moment of transition metal atoms plotted versus the average number of outer valence electrons, N, in quasibinary a-$(T_{1-x}M_x)_{80}B_{10}P_{10}$ alloys, where T represents Fe or Co, and M represents V, Cr, Mn, Fe, Co, or Ni [17].

as spin-up electrons scarcely screen Co or Ni impurities in this alloy, while crystalline bcc Fe is a weak ferromagnet, where both spin bands have holes as $Z_\uparrow > 1$.

The behavior of the early transition metals in Co-based amorphous quasibinary alloys seems to resemble that in the crystalline Ni-based alloys. For Fe impurity, the majority spin $3d$ states remain below the Fermi level and the moment increases by 1 μ_B per Fe atom. The situation is similar for Mn impurity up to 6 at% Mn, but the rate of increase is now 2 μ_B per Mn [26,27]. Above the limit, the majority spin Mn $3d$ state begins to be pushed out above the Fermi level and electrons now overflow from spin-up to spin-down states, causing magnetic moment reduction with further increase in Mn content. For the case of Cr or V impurities, the magnetic moment decreases even at the lowest concentration since the majority spin $3d$ states intersect the Fermi level even at first. The effect of Nb in Co-B glasses seems to be more dominant and close to the VBS prediction of -6 (-5.5 or -7 μ_B per Nb in $Co_{94-x}Nb_xB_6$ or $Co_{78-x}Nb_xB_{22}$) than that of V (-4 μ_B per V in $Co_{80-x}V_xB_{20}$), suggesting that the Nb d states are more completely above the Fermi level than are those of V [28].

2. *Coordination Bonding Model*

The chemical effects on electronic structures due to the necessary presence of glass-forming species are often more dramatic than those due solely to the structural and/or chemical disorder. Covalent bonding occurs between orbitals of two atoms with similar electronegativities. In the covalent bond, charge is delocalized from each of the atomic sites and form bonding and antibonding hybrid orbitals between the atoms. Covalent bonds formed between partially occupied (near E_F) valence orbitals will often involve magnetic states. The *d* states become more delocalized as a result of covalent bonding. This delocalization results in loss of *d* character and hence weaker intra-atomic exchange, and in a suppression of the density of states at the Fermi level, both of which weaken moment formation.

To get insight into the differences among various metalloids in bonding with Fe, eigenvalues in M-centered tetrahedral clusters, Fe_4M with M = B, C, Si, N, are compared as shown in Fig. 4 [7,29]. The atomic eigenvalues for the various unbonded species are indicated in the right side of each column in the figure. The extent of covalent mixing of states by formation of an Fe_4M clusters will increase as the difference in energy between atomic states decreases. The degree of polarity in a bond will increase as the difference in electronegativity between the initial state increases. The metalloid atomic *p* states immediately above the Fe 3*d* state will form a predominantly *p-d* covalent orbital, with the greatest covalent mixing taking place between Fe and N. The bonding between the Fe 3*d* and the lower lying metalloid s states will be largely polar in character and again strongest for N, the *s* orbitals of which are most electronegative relative to Fe 3*d*. Polar interactions between Fe *sp-d* and metalloid *s*-like molecular orbitals are found to contribute strongly to chemical bonding. These interactions involve charge transfer to the lower-lying orbital with compensation by conduction electron transfer in the opposite direction to maintain overall charge neutrality.

The molecular orbital eigenvalues for an empty Fe_4 cluster are shown in the left side of the first column, where solid lines indicate occupied states and dashed line, empty states, and the cluster density of states DOS obtained by Gaussian broadening of cluster eigenvalue are drawn over the discrete states. Allowing the cluster states to spin split results in a moment for Fe_4 cluster of 2.5 or 3 μ_B/Fe, for touching or 20% overlapping atomic spheres, respectively.

The effect of bonding with various metalloids are shown in the next four columns. Bonding strength is determined mostly by the polar *s-d* interaction, which has,

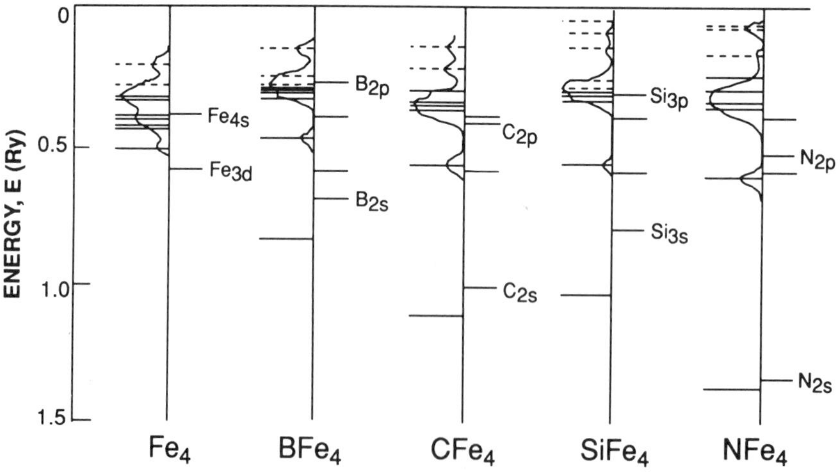

Figure 4 Molecular orbital states and their Gaussian broadened cluster DOS for tetrahedral Fe_4 and Fe_4M clusters, where M = B, C, Si, N (the left side of each column). The relevant Fe and M atomic levels are shown at the right side of each column [7].

however, a secondary impact on magnetic properties, which are affected more strongly by *p-d* hybridization. B and Si, the atomic *p* states of which lie well above the atomic Fe 3*d* states, show little hybridization and thus leave the DOS at the Fermi level high, and covalent mixing is not large. Charge distribution contours show that the *d* and *p* electrons for Fe_4B and Fe_4Si clusters tend to remain more localized within their atomic spheres than do the electrons involved in the *s-d* interaction. Consequently, the Fe_4B and Fe_4Si spin-unrestricted clusters have large moments, 2.6 and 2.4 μ_B/Fe atom, respectively.

For C and, especially, N the *p-d* covalent hybridization is large due to the small energy difference between the initial atomic states. This leaves DOS at Fermi level low and results in excess interstitial charge. The clusters containing C and N are calculated to have low moments, 0.2 and 0 μ_B/Fe, respectively. These calculations on Fe-metalloid clusters show the mechanisms behind the general observation that chemical bonding weakens magnetism.

The essence of the coordination bonding model is that for a given metalloid content in a 3*d*-base metal-metalloid alloy, $T_{1-x}M_x$, the metalloid atom, M, is more or less effective in suppressing the host magnetization, depending upon whether M more or less strongly bonds with transition metal atom, T [30–32]. The extent of *p-d* bonding is assumed to be proportional to the number of T atoms surrounding an M atom. Let us assume that each of the Z_{TM} nearest-neighbor T atoms around an M atom forms a bond with the central M atom and therefore loses, on average, one-fifth of its moment because one of its five 3*d* electron orbitals is tied up in a nonmagnetic covalent bond [29]. Then the average magnetic moment μ per T atom in the alloy is suppressed below that in the pure host n_B as follows:

$$\mu_T = n_B\left[1 - \frac{Z_{TM}N_M}{5N_T}\right] \tag{9}$$

where N_M/N_T, more commonly written as $x/(1 - x)$, is the ratio of the number of M atoms to T atoms. Experimental observation of average atomic moment of Co in metallic grasses as well as in crystalline solid solutions are shown in Fig. 5. The solid lines represent Eq. (9) for $Z_{CoB} = 6$ (top) based on the Co_3B cementite structure, and $Z_{CoP} = 9$ (middle), based on the Co_3P tetragonal structure, with Z_{CoP} varying continuously from 12 (appropriate to hcp solid solutions) at $N_M/N_{Co} = 0$ to $Z_{CoP} = 9$ (again as in Co_3P) at $N_M/N_{Co} = 0.3$. This implies that while a boron atom in the glassy alloy bonds with, and therefore suppresses the moment of six cobalt atoms, a phosphous atom in the glass bonds with nine cobalt atoms and therefore gives rise to a sharper moment reduction. For the crystalline Co-P alloys the coordination numbers used to fit the data (12 and 9) agree with the results of direct structural determination of Z_{TM}.

The coordination model illustrates a case in which metalloid valence seems to play no role in the rate of decrease in magnetic moment. For a common structure (hcp Co solid solution, $Z_{CoM} = 12$) three metalloid solutes Al, Si, and P, each with different valence, all produce the same moment suppression [31].

The model can be extended to Ni-based alloys [31] and, with further assumption, to metal-metal systems [33]. This simple model provides a clear and easily applied relation between magnetic properties and atomic structure in both crystalline and amorphous alloys.

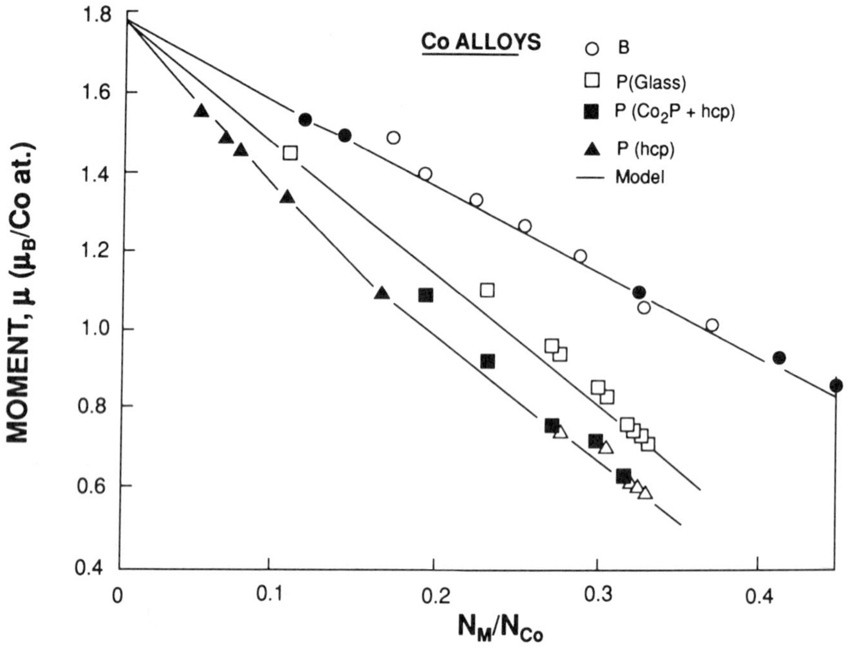

Figure 5 Average magnetic moment per transition atom as a function of N_M/N_T, the ratio of number of nonmagnetic metalloid atoms to that of magnetic transition metal atoms [7].

3. *Hyperfine Interactions*

Only microscopic methods can give information about the distribution of moments which is characteristic in an amorphous phase. Mössbauer absorption spectroscopy is a powerful tool for Fe-based alloys [34], while NMR [35] can be utilized for broader nuclear species. For transition metals a dominant part of hyperfine field is the Fermi contact term, which is proportional to the unpaired spin density at the nucleus. Although a proportionality between the hyperfine field and magnetic moment is by no means a general rule, approximate linear behaviors (125 kOe/μ_B, or 135 kOe/μ_B) are observed for number of crystalline Fe-Ge alloys [36] or FeB, Fe_2B, Fe_3C, and Fe_3P [37], respectively. Thus the probability distribution of magnetic moments may be indicated by that of the hyperfine field in amorphous alloys.

The hyperfine field distributions in amorphous $(Fe_xNi_{1-x})_{80}P_{14}B_6$ [38] at 4.2 K are shown in Fig. 6. The averaged value of hyperfine field, 280 ± 5 kOe in $Fe_{80}P_{14}B_6$ with the averaged moment 1.85 μ_B/Fe [39], leads the proportional constant 150 kOe/μ_B, with which one concludes that the magnetic moment distribution in this alloy has a range extending from 1.3μ_B to 2.4μ_B.

The shape of the distribution is affected little, but the peak shifts to lower fields with increasing Ni concentration. One of the origins for the decreasing hyperfine field with increasing Ni content may be attributed to the decreasing interatomic distance. Actually the density of a-$Fe_{40}Ni_{40}B_{20}$, 7.72 g/cm^3, is 5% larger than that of a-$Fe_{80}B_{20}$, 7.39 g/cm^3 [40], but the greater atomic mass of Ni accounts only 2% of the increase.

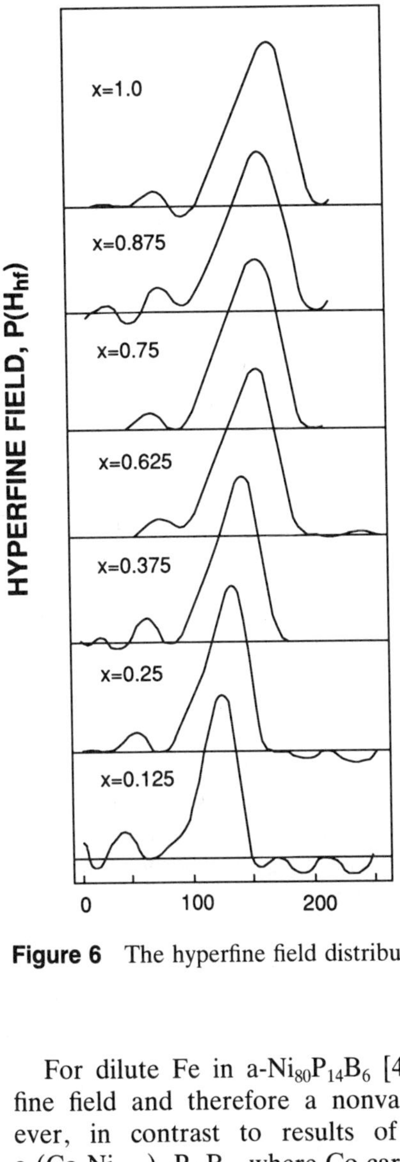

Figure 6 The hyperfine field distribution of Fe in a-$(Fe_{1-x}Ni_x)_{80}B_6P_{14}$ at 4.2 K [33].

For dilute Fe in a-$Ni_{80}P_{14}B_6$ [41] or a-$Ni_{79}P_{13}B_8$ [42] there is a finite hyperfine field and therefore a nonvanishing moment on Fe atoms. This is, however, in contrast to results of saturation magnetization measurements on a-$(Co_xNi_{1-x})_{78}P_{14}B_8$, where Co carries no moment for dilute concentration, develops a moment with increasing x to get its full moment of 1.15 μ_B/Co, and stays constant for $x > 0.15$ [43].

An extensive study of Mössbauer spectroscopy for a-$Fe_{80}B_{20-x}G_x$ (G = C, Si, Ge, P) [44] shows a similar profile of hyperfine distribution for all alloys despite considerable differences in chemical configurations. The change of average hyperfine field with x is qualitatively similar to the change of average Fe moment. The charge transfer or chemical valence alone will not explain the results, and the size of the metalloid atoms seems to play a role.

By sputter deposition, a wider range of a-Fe_xB_{1-x} $(0.35 < x < 0.7)$ thin films were prepared and studied by Mössbauer spectroscopy [45–47]. The absorption spectra for these sputtered films showed asymmetry, which is in contrast to those for liquid-quenched a-$Fe_{58}P_{12.5}C_{7.5}$ [48]. The latter showed a quadrupole doublet with greater line width than the natural line width in the paramagnetic state above T_c, and a considerably broadened symmetric six-line spectrum below T_c. This symmetric spectrum in the combined presence of magnetic hyperfine and electric quadrupole interaction can be explained by a random distribution of orientation of the electric field gradient axes, which broadens but does not shift the lines. The asymmetric Mössbauer spectra for the sputtered films may be due to correlation between magnetization and electric field gradient axes.

The distribution of hyperfine field in the sputtered films is centered at 222 kOe for $x = 0.71$ and shifts to lower fields as the Fe content is decreased. Near $x = 0.5$, nonmagnetic Fe appears and complete collapse of hyperfine field at 4.2 K is observed for $x = 0.4$. The critical concentration, x_c, may be different from the value at which the moment appears in a finite field. Some Fe atoms are observed to bear a moment down to $x = 0.2$ in the field needed for magnetic resonance studies [49].

Nuclear magnetic resonance is also a powerful tool which is used to obtain microscopic information, especially for Co-based amorphous alloys. The NMR spectrum of ^{59}Co in a-$Co_{70}B_{20}P_{10}$ at 4.2 K consists of a broad peak due to both the distribution of hyperfine field and electric field gradient [50]. NMR spectra for electrodeposited a-Co_xP_{1-x} $(0.74 < x < 0.94)$ [51,52] show considerable structure but the mean resonance frequency shifts to higher values with decreasing P concentration. The ratio of the average hyperfine field of Co to the magnetic moment is approximately 110 kOe/μ_B.

4. *Curie Temperature*

Experimentally, well-defined Curie temperatures and sharp phase transitions obeying scaling laws for critical behavior have been observed for a large number of homogeneous amorphous alloys. It is, however, not obvious whether there still will be a well-defined Curie point for a wider distribution of exchange interactions, which may induce local spacial fluctuations in the ordering temperature. It is found [53] that the sharp transition in the diluted bond model in three dimensions can occur only for $n > 2/3$, where n is a critical exponent for correlation length which grows proportionally with $|T - T_c|^{-n}$ as one approaches T_c from a higher temperature.

The molecular field expression for T_c is

$$T_c = \left[\frac{2S(S+1)}{3k_B}\right] \langle J(0) \rangle = \left[\frac{2S(S+1)}{3k_B}\right] \left\langle \sum_j J_{ij} \right\rangle_i \tag{10}$$

where $J(0)$ is the value at $q = 0$ of $J(q)$ in Eq. (8) and $\langle\ \rangle_i$ means the average over the centered atom i as $J(0)$ varies with i and has some distribution in amorphous alloys.

Curie temperatures of amorphous alloys are systematically lower than those of crystalline alloys containing no metalloid and even lower than T_cs of crystalline metal-metalloid compounds. The reduction of T_c of amorphous alloys seems to be due to not only the chemical environment but also structural disorder.

The variation of T_c with metalloid content is quite different for Co and for Fe-based amorphous alloys. The extrapolated value of T_c for $x = 0$ in a-$Co_{1-x}P_x$ is close to that for hcp Co (1,388 K), whereas in the case of a-Fe_xB_{1-x} the extrapolated

T_c is roughly one-third of that for bcc Fe (1043 K) as shown in Fig. 7 [13,14]. This may be due to the presence of some antimagnetic Fe-Fe pairs in the distribution of exchange interactions in the Fe-based amorphous alloys.

The Curie temperatures of a-$Fe_{80}E_{20-x}G_x$, where E = B or P, G = C, Si, P, Ge, is shown in Fig. 8 [16], which indicates clear correlation between the extrapolated T_c of a-$Fe_{80}G_{20}$ to the magnetic moment of Fe shown in Fig. 2. The T_c's are plotted versus the magnetic moment of Fe in Fig. 9 (open circles). The monotonic increase of T_c with increasing Fe moment is much stronger than expected from Eq. (10) (a broken line) taking $\langle J(0)\rangle$ as a constant. There is an interesting positive linear correlation between the exchange interaction and the moment of Fe in a-$Fe_{80}G_{20}$ solid circles).

From the observed T_c for a-$T_{80}B_{10}P_{10}$ shown in Fig. 10, we can estimate the averaged nearest-neighbor exchange constant, J, using an average spin value deduced from the saturation moments in Fig. 3. Taking $S = 0.5$ for a Co atom in a-$Co_{80}B_{10}P_{10}$ and 0.86 for hcp Co, the ratio of exchange interactions in these phases is calculated as the ratio of $T_c/S(S + 1)$ to be $\langle J_a(0)\rangle/J_c(0) = 1.1$; that is, the average nearest neighbor exchange interaction between Co atoms is a little greater in a-$Co_{80}B_{10}P_{10}$ than crystalline hcp Co, assuming the same number of interacting neighbors in each. The situation is quite different for Fe; that is, the ratio of $\langle J(0)\rangle$ is 0.7 and that of average nearest neighbor exchange may be 0.5 if we take z = 12 in a-$Fe_{80}B_{10}P_{10}$ compared to $z = 8$ in bcc Fe.

The exchange interaction is affected drastically by atomic arrangement; for example, there is great discontinuity between the Curie temperature of crystalline bcc and fcc Fe-Ni alloys. In an amorphous phase T_c varies continuously with composition of the constituent transition metals as shown in Fig. 8. Analysis by a simple

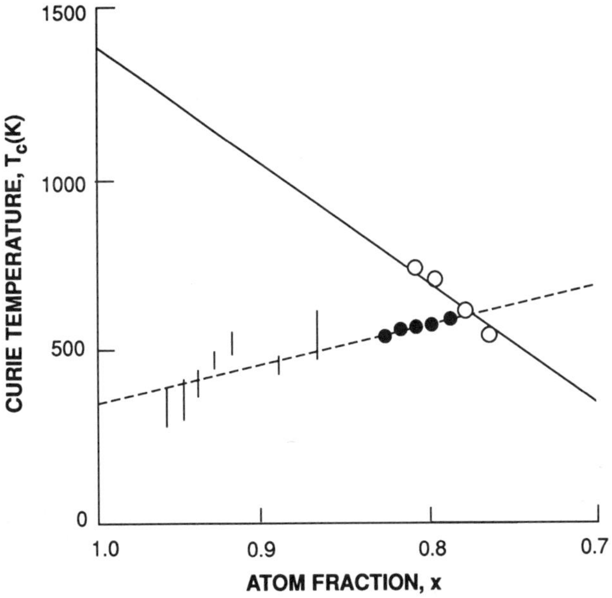

Figure 7 Curie temperatures of a-Co_xP_{1-x} [13] (open circles) and a-Fe_xB_{1-x} [14] (solid circles, bars denote data obtained from extrapolation of T_c of a-Fe-P-B alloys).

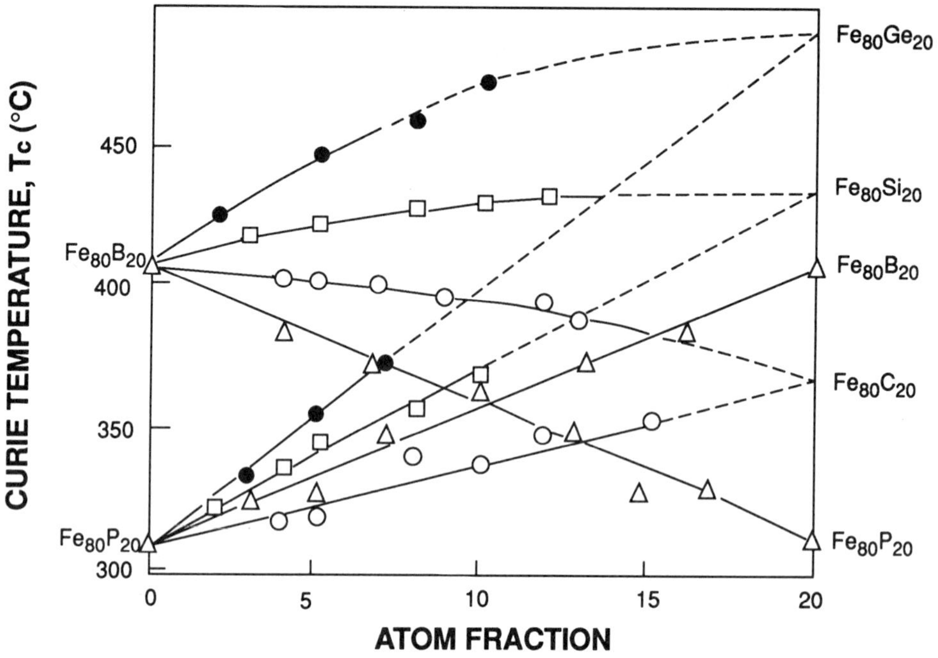

Figure 8 Curie temperatures of a-$Fe_{80}E_{20-x}G_x$, where E = B or P, G = B, C, Si, P, or Ge [16].

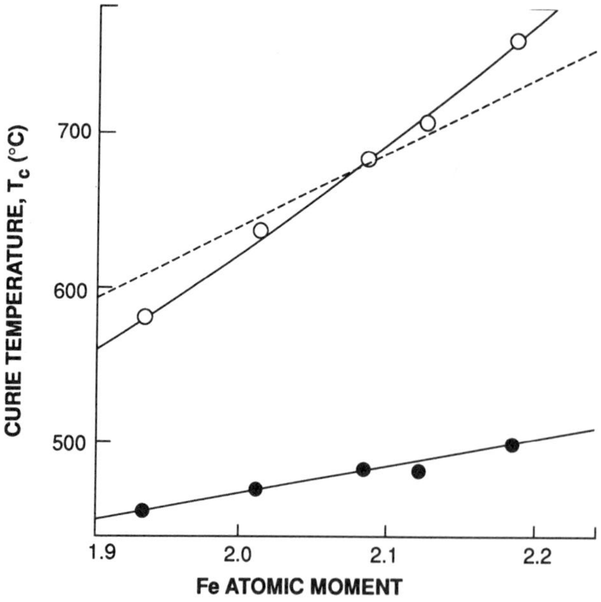

Figure 9 Curie temperature of a-$Fe_{80}G_{20}$, (G = B, C, Si, P, and Ge) plotted versus atomic magnetic moment of Fe (open circles) and $\langle J(0)\rangle/k_B$ (solid circles). A broken line represents calculated T_c with a constant $\langle J(0)\rangle/k_B$ = 480 K.

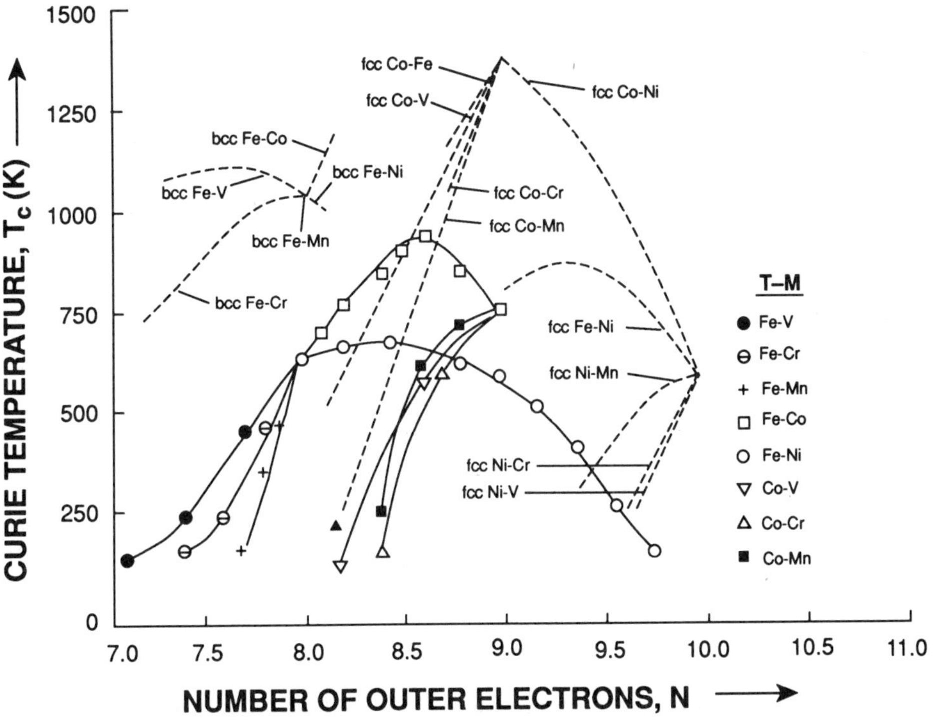

Figure 10 Curie temperature of a-$(T_xM_{1-x})_{80}B_{10}P_{10}$ alloys, where T = Fe or Co and M = V, Cr, Co, or Ni, plotted versus average outer electron numbers of transition metal atoms, N [17].

pair model gives the exchange constants as $651k_B$, $1055k_B$, and $-175k_B$ for Fe-Fe, Fe-Ni, and Ni-Ni pairs in a-$(Fe_xNi_{1-x})_{80}B_{20}$. It should be noted, however, that there must be distribution of exchange interactions for each kind of pairs.

Substantial reductions in T_c observed when Fe is replaced by Mn or Cr may be attributed to antiferromagnetic interaction with the host Fe moment (mainly in the case of Mn) and to the reduction of Fe moment (likely in the case of Cr).

5. *Temperature Dependence of Magnetization*

Some typical examples of the temperature dependence of magnetization of amorphous ferromagnetic alloys are shown in Fig. 11a along with that of two crystalline ferromagnets for comparison. While the reduced magnetization versus reduced temperature curves for amorphous alloys lie substantially below that for crystalline Fe, such an effect is also observed for crystalline Invar alloys.

The simplest theory describing the temperature dependence of magnetization is the molecular (mean) field theory. Although it inadequately describes the behavior at low temperature and near the Curie point, overall temperature dependence may be well reproduced. In this approximation the effective field acting on a single-spin S_i is represented as

$$\langle H_i \rangle = \frac{2}{g\mu_B} \sum_j J_{ij} \langle S_j \rangle + H \tag{11}$$

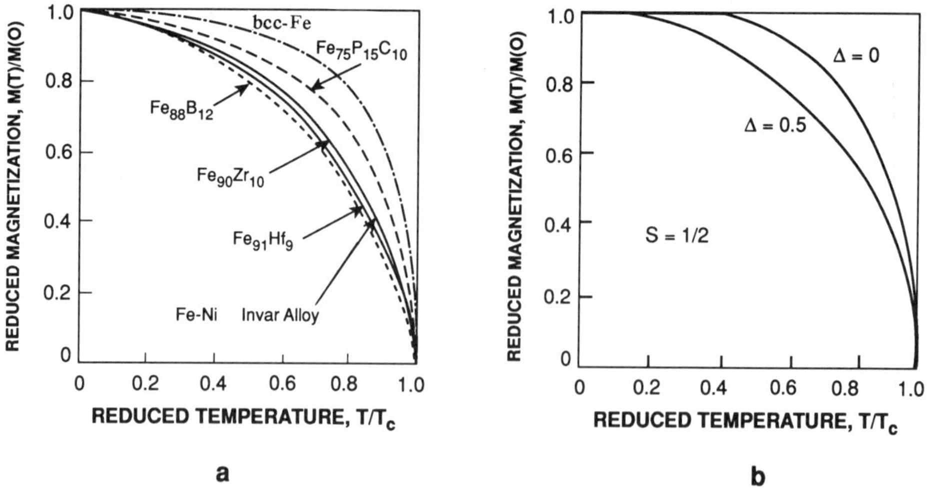

Figure 11 (a) Some typical examples of temperature dependence of reduced magnetization of amorphous ferromagnetic alloys along with crystalline bcc-Fe and Invar alloys [73]. (b) Calculated temperature dependence of reduced magnetization by Eq. (15) with $\Delta = \langle J_{ij}J_{jk}\rangle / J^2$ [54].

where $\langle\ \rangle$ means thermal average and H is an external field. Neglecting a correlation between spin S_i and the surrounding spins, the Heisenberg Hamiltonian is expressed

$$H = -g\mu_B \sum_i \langle H_i \rangle S_i \tag{12}$$

To apply the molecular field approximation to amorphous ferromagnets, the distribution of exchange interactions should be taken into account [54]. Representing the configurational average by { }, we assume

$$\{J_{ij}\langle S_j\rangle\} \simeq \{J_{ij}\}m = Jm + \{\Delta J_{ij}\}m \tag{13}$$

where $m = \langle S_j \rangle \simeq \{\langle S_j \rangle\}$ and $J_{ij} = J + \Delta J_{ij}$. Then the configurational average of Eq. (11) becomes

$$g\mu_B\{\langle H_i \rangle\} = 2zJm + g\mu_B H + 2m \sum_j \{\Delta J_{ij}\} \tag{14}$$

where z is the number of nearest-neighbor atoms. Putting $\mu_B = 1$ for simple case of $S = 1/2$, $g = 2$, we get

$$m = \tanh\left[\frac{\{zJm + H + m\,\Sigma_j \langle \Delta J_{ij}\rangle\}}{k_B T}\right] \tag{15}$$

which can be solved for m.

The temperature dependence of m obtained from Eq. (15) is shown in Fig. 11b. The distribution of exchange interactions reduces the magnetization, as observed experimentally.

A serious objective to the use of a molecular field approximation to amorphous alloys lies in its failure to treat the effect of fluctuation on the ordering temperature, T_c. Other effective field methods based on clusters of atoms (Bethe-Peierls-Weiss approximation [55,56]) have been applied to amorphous ferromagnets to investigate the effect of disorder [57], in which the presence of fluctuations in the exchange constant turn out to decrease T_c with respect to the transition temperature in the mean field theory.

6. *Spin Waves*

The reduction in low-temperature magnetization with increasing temperature is generally well interpreted by thermal excitation of long-wavelength spin waves which obey the dispersion relation

$$E(q) = Dq^2 \tag{16}$$

where D is the stiffness constant. The resulting effect of the spin waves on $M(T)$ is given by

$$M(T) = M(0)[1 - BT^{3/2}] \tag{17}$$

The coefficient B is related in the linear spin-wave theory with D as

$$B = \left[\frac{2.612 g\mu_B}{M(0)}\right]\left(\frac{k_B}{4\pi D}\right)^{3/2} \tag{18}$$

This temperature dependence is confirmed in amorphous ferromagnetic alloys, e.g., in a-Co_xP_{1-x} [58], a-$Fe_{80}B_{20}$ [59]. In amorphous alloys, $T^{3/2}$ variation of magnetization seems to persist over a much broader temperature range ($0 < T/T_c < 0.2 \sim 0.4$) than for crystalline ferromagnets, which show deviation at reduced temperatures. The value of the coefficient B in Eq. (17) is a few times larger in amorphous alloys than in crystalline ferromagnets having similar Curie temperatures.

The stiffness constant, D, can be obtained directly from the dispersion relation measured by inelastic neutron scattering. In Fig. 12 both D observed from neutron scattering and that calculated from B by Eq. (18) are plotted versus T_c for a-$(Fe_xNi_{1-x})_{75}P_{16}B_6Al_3$ alloys along with those of bcc Fe [60]. Agreement between them is good in this system.

In other cases, there is a substantial discrepancy between the D values obtained from inelastic neutron scattering and those calculated from B obtained from low-temperature magnetization data [61–63]. For example, in a-$(Fe_{93}Mo_7)_{80}B_{10}P_{10}$ well-defined long-wavelength spin waves were observed, but these account for only 70% of B. The mean spin-wave energy is well represented by the relation [61]

$$\langle \hbar\omega(q, T)\rangle = \langle(\hbar\omega(q, 0)\rangle\left[1 - 0.612\left(\frac{T}{T_c}\right)^{5/2}\right], \qquad (T_c = 450 \text{ K})$$

The ratio D/T_c is a measure of the range of exchange interaction. The presence of nonmagnetic metalloid atoms can also contribute to the reduction in D/T_c. The effect of a small concentration, y, of nonmagnetic impurity on D is derived to be proportional to $1 - 2y$ [64], whereas the effect on T_c is expected to vary as $1 - y$ in the mean field theory. Using the ratio $D/T_c = 0.27$ for bcc Fe, that for a-$Fe_{75}P_{16}B_6Al_3$ is expected to be 0.18 which is close to the observed ratio, 0.21 [60].

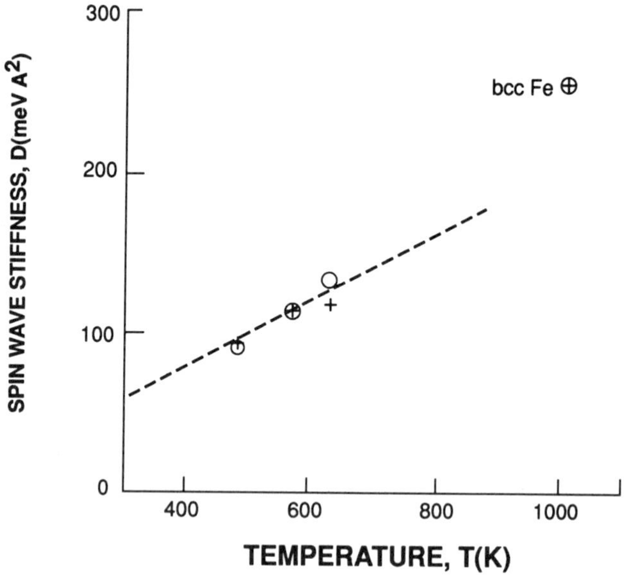

Figure 12 Spin-wave stiffness constant, D, measured by neutron scattering (open circles) and obtained from temperature dependence of magnetization (crosses) versus Curie temperatures in a-$(Fe_xNi_{1-x})_{75}P_{16}B_6Al_3$ ($x = 1.0, 0.65, 0.50$) and bcc-Fe [60].

7. *Critical Behavior*

It is interesting and important to examine whether a static scaling law is valid for amorphous ferromagnets in the critical region, because an affirmative answer would mean that amorphous magnetic alloys are magnetically isotropic and that systems with a distribution of exchange interactions can undergo a second-order phase transition. The static scaling hypothesis is based on the assumption that, in the neighborhood of the critical point, the thermodynamic potential is a generalized homogeneous function of two variables: the reduced temperature $t = (T - T_c)/T_c$ and field, which is conjugate to the order parameters, for example, the magnetization for ferromagnetic phase transition. For a ferromagnet, near T_c in the limit of $H \to 0$, the form of equation of state is

$$\frac{H}{|t|^{\beta+\gamma}} = f_{\pm}\left(\frac{M}{|t|^{\beta}}\right) \tag{19}$$

where the $\pm$ signs refer to the sign of t, β is the critical exponent describing the temperature dependence of spontaneous magnetization just below T_c,

$$M \propto (T_c - T)^{\beta} \qquad (H = 0) \tag{20}$$

and γ is the critical exponent for the temperature dependence of inverse susceptibility just above T_c:

$$\lim_{H\to 0}\left(\frac{H}{M}\right) = \chi^{-1} \propto (T - T_c)^{\gamma} \tag{21}$$

The scaling hypothesis leads to the following equality among the critical exponents:

$$\beta + \gamma - \beta\delta = 0 \tag{22}$$

where δ is the exponent in the expression for the critical isotherm:

$$M \propto H^{1/\delta} \qquad (T = T_c, H \rightarrow 0) \tag{23}$$

The first detailed study on the critical behavior of an amorphous ferromagnetic $Co_{70}B_{20}P_{10}$ alloy [65,66] showed the validity of the scaled magnetic equation of state, as shown in Fig. 13. Observed values of the critical exponents are $\beta = 0.38 \sim 0.43$, $\gamma = 1.30 \sim 1.35$, and $\delta = 4.4 \sim 4.5$ [65–71]. These are not far from the ones predicted from the three-dimensional Heisenberg model ($\beta = 0.365$, $\gamma = 1.39$, and $\delta = 4.80$).

The long-range nature of the critical fluctuation obliterates effects arising from atomic-scale inhomogeneity. The effective field approach concludes that the critical exponents are not affected by the presence of defects. However, when the correlation length of fluctuations become comparable or shorter than the extension

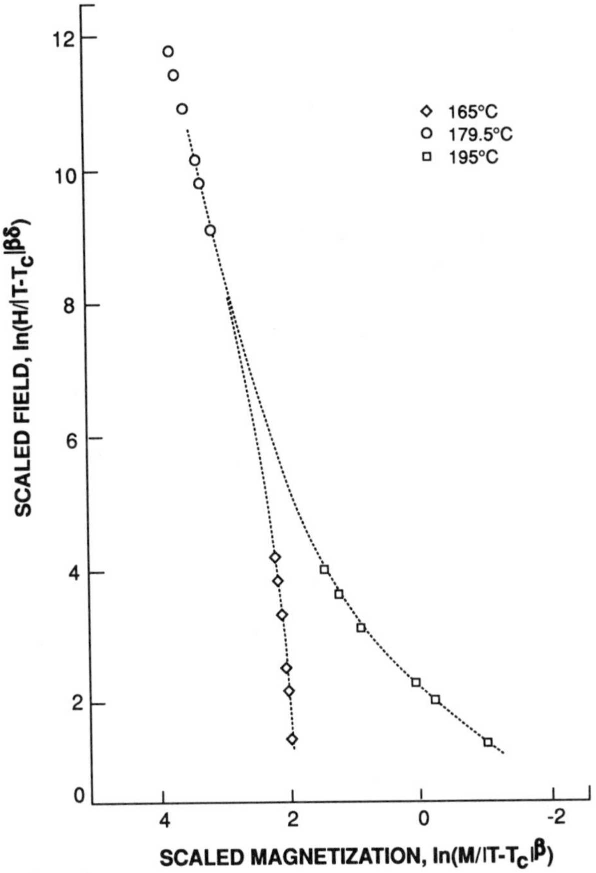

Figure 13 Scaled field versus scaled magnetization showing the critical behavior for a-$Co_{70}B_{20}P_{10}$ alloy [65, 66].

of defects (at temperatures departing from T_c) inhomogeneity will play a decisive role.

Long-range interactions in real systems, such as dipolar forces, will lead to disagreement between theory and experiment unless t is sufficiently larger than a critical value: $t_c = (g\mu_B SM(0)/k_B T_c)^{1/\beta(\delta-1)}$ [72]. For a-$Fe_{10}Ni_{70}B_{19}Si_1$ [68] t_c is found to be 2×10^{-3} and the critical region spans $|t| = 0.1$ ($T < T_c$) and 0.08 ($T > T_c$). In general, for transition metal-metalloid amorphous alloys, the extent of the critical region is found to be larger than that observed in crystalline materials, and therefore the values of the critical exponents in amorphous alloys are close to the ones found using the isotropic 3*d* Heisenberg model. Amorphous alloys can behave as an ideal ferromagnet in the critical region because atomic irregularity is smoothed out over the long-range critical fluctuation.

B. Early Transition Metal–Late Transition Metal Alloys

Ferromagnetism is realized in amorphous alloys of another category: T_xE_{1-x}, where T represents Fe, Co, or Ni, and E is an early transition metal element such as Zr, Nb, Hf, Ta, etc. The amorphous phase is available in a composition range less than 90 at% Co or Fe by melt quenching [73–75], but it is stabilized over a much wider range when prepared by sputter deposition [76]. The lower limits of the early transition elements in this type of amorphous phase are correlated to the size difference of constituent atoms.

The critical concentration of T atoms for the appearance of magnetism in a − T_xE_{1-x} is $x_c = 0.4$, 0.5, or 0.8 for T = Fe, Co, or Ni, respectively. This is understood on the basis of a certain critical number of transition metal atomic neighbors being necessary for the development of a magnetic moment.

The magnetic moment of Fe in a-Fe_xZr_{1-x} increases increasing with x for $x <$ 0.86, extrapolating to that of pure bcc Fe. However, above 86 at% it decreases toward zero with increasing x. The Curie temperature of amorphous Fe_xZr_{1-x} increases slightly as x increases to 86 at%, but above this limit it decreases toward zero as the magnetic moment does. The alloys are good ferromagnets for $0.4 < x < 0.8$, however, noncollinear structure seems to appear in the more concentrated Fe-based alloys where short Fe-Fe bonds with antiferromagnetic direct exchange are present [73,78].

In amorphous Co- or Ni-based alloys there are no negative direct exchange interactions. In Co_xZr_{1-x} amorphous alloy the magnetic moment of Co increases with $x > x_c = 0.6$ [77], extrapolating to approximately $1.7\mu_B$ at pure Co. Curie temperatures cannot be determined precisely as they are above a crystallization temperature in these alloys.

The average moment of a-$Fe_{90}Zr_{10}$ ($1.29\mu_B$) increases as one substitutes for Fe with other T metals, where T = V, Cr, Mn, Co, or Ni, independent of whether they have more or less *d* electrons than Fe, and reaches a maximum whose position and value depend on the substituted metal, then decreases as shown in Fig. 14 [73].

All the a-T_xZr_{1-x} alloys exhibit superconductivity in the range where the magnetic moment of T atoms disappears. Superconductivity and magnetism are adjacent in these series of amorphous alloys.

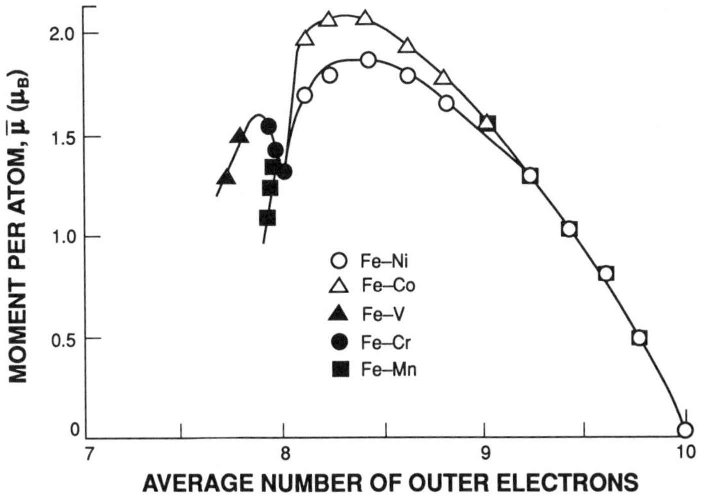

Figure 14 Average magnetic moment per 3*d* transition metal atom in a-$(Fe_{1-x}T_x)_{90}Zr_{10}$ (T = V, Cr, Mn, Ni, or Co) alloys versus average outer electron concentration of 3*d* transition metals [73].

C. Rare Earth–Transition Metal Alloys

An important class of magnetic amorphous alloys are the alloys of rare earth, R, with 3*d* transition metals, T = Fe, Co, or Ni. When the latter is the major component, exchange interactions become strong enough to raise the magnetic ordering temperature above room temperature, and a wide variety of soft and hard magnetic properties are available with suitable choice of the rare earth. This class of alloys may be divided into four categories according to the rare earth elements, the first is nonmagnetic R, that is, Y, La, or Lu; the second is R = Gd, which is *S* state with no orbital angular momentum; the third is light rare earth, and the fourth is heavy rare earth.

1. Nonmagnetic Rare Earth Alloys

The combination of the nonmagnetic rare earths with 3*d* transition metals may clearly exhibit the magnetic behavior of the 3*d* elements in these rare earth–transition metal alloys. A metallic radius of Y (1.80 Å) is closest to that of the trivalent magnetic rare earths (1.75–1.83 Å), while La (1.88 Å) is larger and Lu (1.73 Å) is smaller than all the intervening lanthanides.

Let us start with the Y-Fe system. There are four crystalline intermetallic compounds: Y_2Fe_{17}, Y_6Fe_{23}, YFe_3, and YFe_2, all of which are ferromagnetic [79]. The Fe moment falls from 2.2 μ_B in the pure bcc metal to 1.5 μ_B in YFe_2 [80]. Melt-spun as well as sputtered amorphous $Y_{1-x}Fe_x$ alloys can be prepared over a wide range of x as there is a deep eutectic in the phase diagram at $x = 0.35$.

While the magnetic properties of the Y-Fe system have been examined in some detail, they are unusually sensitive to the method and conditions of preparation. For example, melt-spun YFe_2 was reported to become ferromagnetic below T_c =

270 K [81], whereas evaporated [82] and sputtered [83,84] samples were spin-glass-like with T_f = 55–70 K, and the quite nonlinear Arrott plots of which might give T_c = 200 K [83]. It was clearly indicated by high-field-magnetization curves and Mossbauer spectroscopy [85] that the magnetic structure of Fe-rich a-$Y_{1-x}Fe_x$ alloys is not collinear. It is asperomagnetic when $x > x_c$, with coexisting magnetic and nonmagnetic Fe atoms, and the distribution of moment direction tends to narrow as x approaches unity.

The Fe magnetic moment in a-$Y_{1-x}Fe_x$ may be deduced from hyperfine field measured by ^{57}Fe Mössbauer spectroscopy, since a constant proportionality between hyperfine field and magnetic moment of 145 kOe/μ_B was established for crystalline Y-Fe alloys [86]. The following conclusions may be drawn from absorption spectra of a-$Y_{1-x}Fe_x$ at 1.6 K [87]: (i) an appreciable spontaneous Fe moment begins to appear near x_c = 0.4; (ii) at higher concentration, up to x = 0.8, magnetic and nonmagnetic Fe coexist; (iii) the average moment per magnetic Fe atom and the fraction of magnetic Fe both increases with increasing x.

The concentration dependence of the fraction of nonmagnetic Fe is analyzed by assuming a structure of the alloys as 12-coordinated random dense packing. This turns out to be same as the fraction of Fe having at least seven Y neighbors. It appears therefore that at least six Fe neighbors are needed for an Fe atom to carry a moment.

In the critical region for the appearance of magnetism, magnetization curves at low temperature show a large high-field slope, indicating induced Fe moment for $x > 0.2$. Internal exchange field can also induce moments on suitable neighboring atoms.

The direct exchange interaction between Fe atoms can be positive or negative depending on Fe-Fe distance. The first peak of radial distribution function of a-YFe_2 is centered at this critical distance (2.54 A) with a width of several tenths of an angstrom [84]. A broad distribution of Fe-Fe exchange interactions, which is predominantly positive but includes a significant proportion of antiferromagnetic interactions, is the origin of the asperomagnetism of a-$Y_{1-x}Fe_x$ alloys. It is also the cause of the depression of the magnetic ordering temperature compared with the Curie temperature of corresponding crystalline compounds, as shown in Fig. 15.

A negative pressure coefficient of the Curie temperature, dT_c/dP, is observed both for Fe-rich Y-Fe crystalline compounds [88] and amorphous alloys [89]. A drastic change of magnetic properties of amorphous Y-Fe alloys by hydrogenation may be also explained by means of dilated Fe-Fe distance. There is little change in the iron moment, but a Curie temperature of the amorphous hydride $Y_{12}Fe_{88}H_{36}$ rises up to more than 400 K and the material becomes soft magnetic with a room temperature coercivity of 1 Oe [90]. The amorphous La-Fe alloys are ferromagnetic as the interatomic distances are greater than in Y-Fe or Lu-Fe, both of which are not collinear ferromagnets.

Amorphous Co or Ni alloys with nonmagnetic rare earth show much simpler magnetic properties compared with Fe alloys. They are all good ferromagnets when the concentration of transition metals exceeds a critical value, x_c. The Curie temperatures of amorphous Y-Co and La-Co are much higher than those of their crystalline counterpart, at least for $x_c < x < 0.8$. For some Co-rich amorphous alloys, crystallization prevents the direct determination of T_c. The decrease in T_c with decreasing x is thought to be due to the reduction in coordination number,

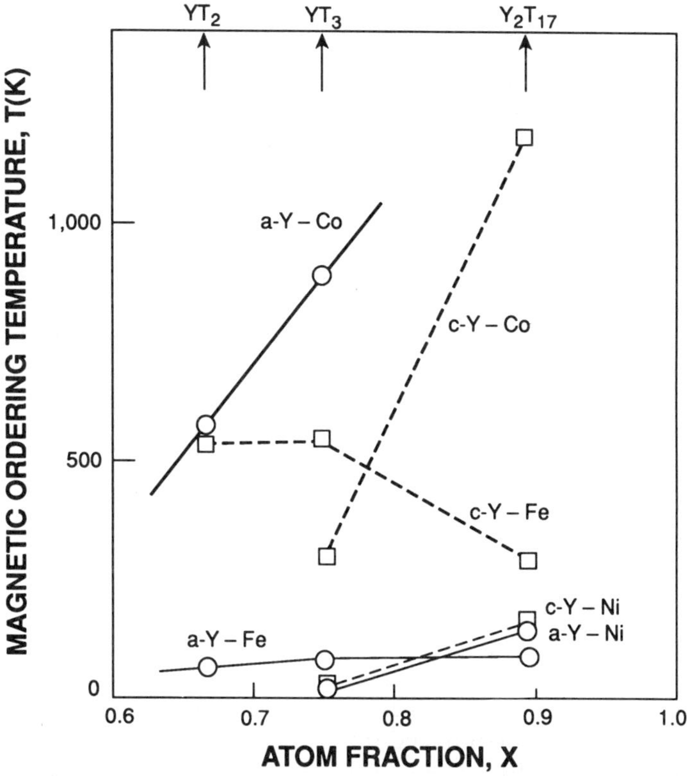

Figure 15 Magnetic ordering temperatures of amorphous (open circles) and crystalline (squares) Y-Fe, Y-Co, and Y-Ni alloys [83].

z, and Co moment. The Co-Co exchange interaction is manifest by a constant value of $J/k_B = 145$ K throughout the series [91], which is deduced from the molecular field theory formula $J = 3k_BT_c/2zS(S + 1)$, where $z = 12x$ assuming a random dense-packed structure, and S is determined empirically from the relation $\mu = g\mu_B S$ with $g = 2$.

The average Co moment in amorphous Y-Co alloys decreases more slowly with decreasing Co content than in their crystalline counterparts, that is, 1.7 μ_B in the pure metal to 0.5 μ_B in YCo_3 to zero in YCo_2 [92]. An amorphous alloy with $x = 0.55$ has a well-defined average moment of 0.55 μ_B which cannot be changed even by a large applied field [93], and it vanishes near $x_c = 0.45$ for both Y-Co [91,94,95] and La-Co [94]. The decrease in Co moment in the amorphous alloys was interpreted in terms of the Jaccarino-Walker model [96], which concludes that a Co atom must have at least eight Co neighbors in order to carry a moment.

The threshold for the appearance of magnetism is even higher in Y-Ni alloys. In crystalline intermetallic compounds, the Ni moment vanishes in YNi_5, but there is weak itinerant ferromagnetism for Y_2Ni_7 and YNi_3; in Y_2Ni_{17} a Ni atom on some sites has a well-defined moment whereas on others the moments are induced [93]. In amorphous $Y_{1-x}Ni_x$ alloys, the moment falls much more rapidly at first than in the Y-Fe and Y-Co series. Charge transfer of electrons from the rare earth into a rigid 3*d* band had been proposed as the simplest model with little direct support [97,98]. A better qualitative explanation is that hybridization of the 3*d* orbitals with the 5*d* and 6*s* rare earth (4*d* and 5*s* for Y) electrons broadens the local 3*d* density of states to the point where it can no longer support a moment. In a-$Y_{1-x}Ni_x$, the moment vanishes for $x < 0.78$, which suggests that a Ni atom needs about 10 Ni neighbors to carry a moment, although the concentration dependence does not fit the Jaccarino-Walker model.

Ni-rich amorphous Y-Ni alloys are excellent examples of weak itinerant ferromagnets [99], which shows T^2 temperature dependence of magnetization, as predicted by Stoner's theory of band ferromagnetism. The homogeneous itinerant ferromagnetic state breaks down near x_c into a random array of interacting clusters.

In summary, the critical concentrations x_c for the appearance of magnetism is 0.4, 0.5, and 0.8 for Fe, Co, and Ni, respectively, when alloyed with nonmagnetic rare earth elements. This will be somewhat lower in alloys with magnetic rare earth elements. Nearest-neighbor environments always produce some distribution in the magnitude of the atomic moments in the amorphous state. Close to the critical concentrations, atomic moments are strongly affected by applied field.

2. *Gadolinum-Transition Metal Alloys*: *Ferrimagnetism*

Both amorphous alloys and crystalline intermetallic compounds of Gd with 3*d* transition metals, Fe, Co, or Ni are ferrimagnetic due to a negative R-T exchange interaction if the transition metal carries a moment [100]. This is not due to RKKY coupling but to parallel spin polarization of the rare earth 5*d* electrons by the 4*f* shell and negative 5*d*-3*d* coupling [101]. The two subnetworks are defined chemically as the ensemble of Gd or transition metal atoms in amorphous alloys, in contrast to sublattices in crystalline phases where they are defined as crystallographic sites. Transition metal moments deduced from the saturation magnetization at 4.2 K follow the same trends as in a-Y-T alloys at least for $x \gg x_c$. Uncertainty in the precise value of Gd moment, which may be slightly different from $7\mu_B$ because of 5*d* contribution, prevents the reliable determination of transition metal moments near the critical concentration.

The exchange interactions J_{TT}, J_{TR}, and J_{RR} should be taken into account in order of decreasing importance. The Gd-Gd exchange interaction is relatively unimportant, and Gd-T interactions may be primarily responsible for the differences in magnetic properties between Y-T and Gd-T alloys. As shown in Fig. 16, which should be directly compared with Fig. 15 of Y-T, all the ordering temperatures are higher in the Gd alloys (except a-Gd-Co) than in their yttrium alloy counterparts. The increase is most marked for a-Gd-Fe alloys, in which coupling with Gd suffices to displace the effective Fe-Fe exchange distribution toward positive values, thereby eliminating the noncollinear spin structure of the Fe subnetwork. It also pushes x_c to somewhat lower values than for a-Y-Fe alloys [102].

The concentration dependence of the ordering temperature in the amorphous alloys is shown in Fig. 16 [89,103–107]. The downturn of T_c of a-Gd-Fe alloys at the Fe-rich end of the series may be due to the appearance of antiferromagnetic interactions between some pairs of Fe atoms, which reduces the average J_{FeFe}. The curves for Co and Ni alloys show clearly the changes of slope which correspond to the appearance of induced or permanent magnetic moment on the transition metals.

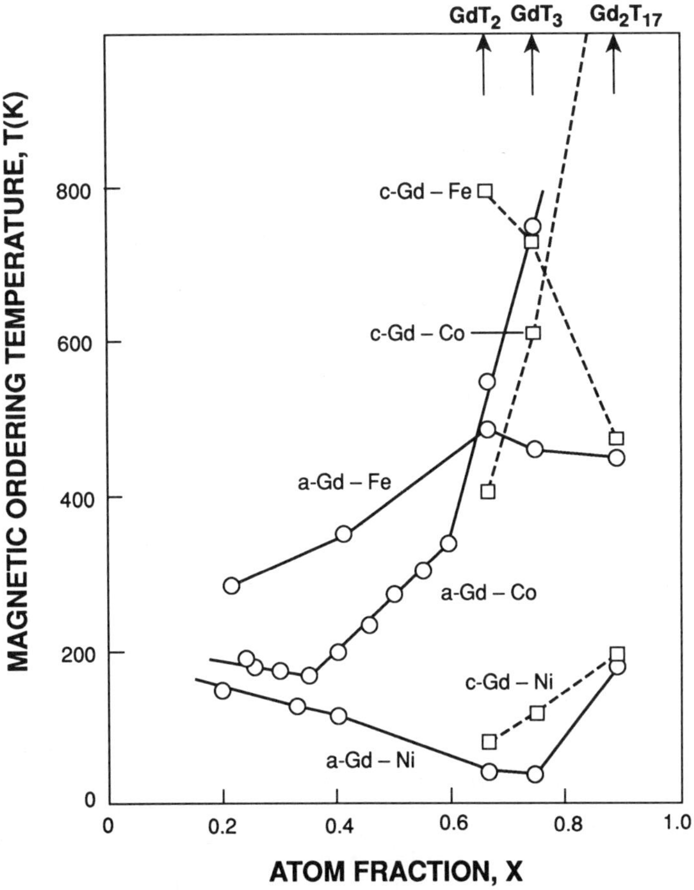

Figure 16 Magnetic ordering temperatures of amorphous (open circles) and crystalline (squares) Gd-Fe, Gd-Co, and Gd-Ni alloys [95].

The temperature dependence of net magnetization of ferrimagnetic a-Gd-T alloys is peculiar. The alloys in a certain range of concentration, where the Gd subnetwork magnetization at low temperature exceeds to that of the transition metal subnetwork, have a compensation temperature at which net magnetization disappears and above which it reappears. This is interpreted to be the result of the faster fall off of Gd subnetwork magnetization than that of the transition metals with increasing temperature due to the difference of J_{GdT} and J_{TT}. This is illustrated in Fig. 17. The direction of each subnetwork magnetization is reversed at the compensation temperature to maintain the direction of the net magnetization along applied external field. Then the hysteresis loop of Kerr effect or Hall effect above and below the compensation temperature is generally mutually inverted because those effects are dominated by either of the subnetworks rather than by the net magnetization.

The temperature dependence of the magnetization curve of a-$Gd_{1-x}Fe_x$ and a-$Gd_{1-x}Co_x$ have been analyzed using molecular field theory for ferrimagnet with g factors, 2.00, 2.15, and 2.22 for Gd, Fe, and Co, respectively. The concentration

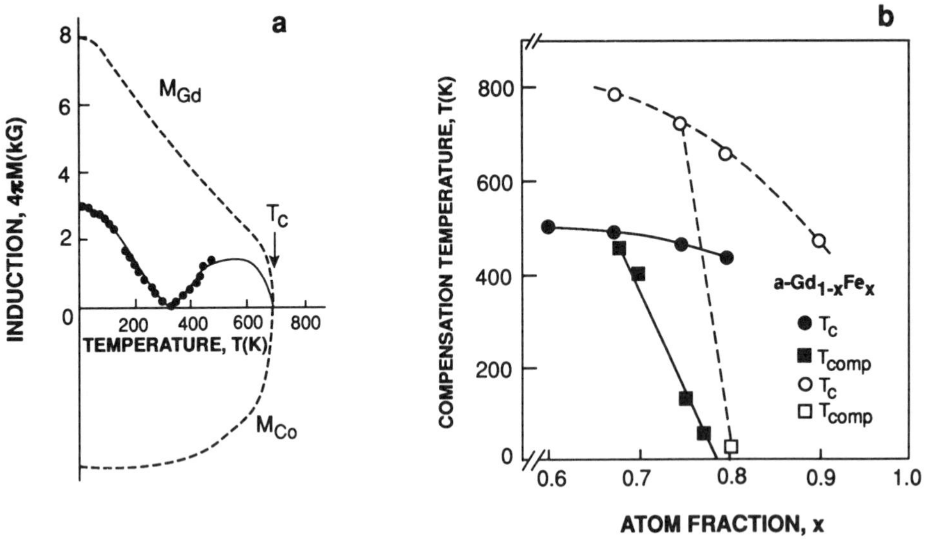

Figure 17 (a) Temperature dependence of net (solid line) and subnetwork magnetization (broken lines) of a-$Gd_{22}Co_{78}$ alloy [108]. (b) Compensation (squares) and Curie (circles) temperatures of amorphous and crystalline Fe_xGd_{1-x} alloys [120].

dependence of calculated exchange constants and magnetic moments are shown in Fig. 18.

The exchange stiffness parameter, A, is expressed by J_{ij} as follows:

$$A = \frac{2|J_{ij}|S_iS_jp_{ij}}{a_{ij}} \tag{24}$$

where p_{ij} is the relative probability of finding an ij pair, and a_{ij} is the interatomic distance. The value of A for a-$Gd_{0.2}Co_{0.8}$ is 6×10^7 erg/cm [110].

3. *Light Rare Earth and Transition Metal Alloys*

The spin orbit coupling constant of light rare earths, in which less than half of the 4*f* shell is filled by electrons, is positive. The ground-state total angular momentum is therefore $J = |L - S|$, which is antiparallel to the spin momentum because the orbital momentum exceeds the spin momentum. Then the negative spin-spin exchange interaction, J_{RT}, makes the coupling parallel between total atomic moment of rare earth and transition metal.

A list of rare earth elements in their most common 4*f* configurations is given in Table 1, where the magnetic moments of the free ion, $gJ\mu_B$, are cited. It should be noted, however, that the strong electrostatic fields modify the magnitude of the rare earth moments and influence their direction in the ordered magnetic structure. The atomic moments of rare earths, Pr, Nd, and Sm in a crystalline RAl_2 are 2.6, 2.5, and $0.2\mu_B$, respectively [100], in contrast to their free ion values of 3.20, 3.27, and 0.71.

Any exchange interaction, whether it is direct or indirect, between two 4*f* moments of type S_iS_j should be scaled by the de Gennes factor, $(g - 1)^2J(J + 1)$, listed in the last column of the table, as the spin component of the total moment is effective.

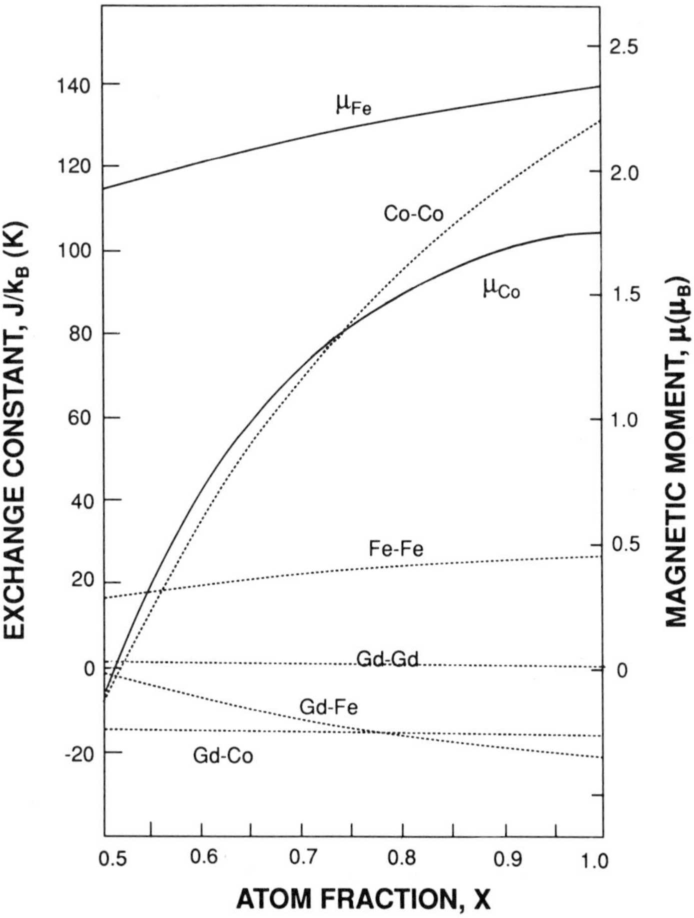

Figure 18 Magnetic moments of Fe and Co (solid lines), and exchange constants, J/k_B (dotted lines) in a-$Gd_{1-x}Fe_x$ and a-$Gd_{1-x}Co_x$ alloys calculated with mean field fitting parameters [106,108,109].

Melt-spun a-$Pr_{1-x}Fe_x$ alloys showed well-defined magnetic ordering temperatures [111], with a fall-off at higher Fe content, as shown in Fig. 19a. The magnetization could in no case be saturated in fields of up to 95 kOe. The spontaneous magnetization extrapolated to zero field, given in Fig. 19b is consistent with a collinear Fe subnetwork and random orientation of Pr moments with no resultant subnetwork magnetization.

The ordering temperature and magnetization curve for melt-spun Nd-Fe alloys are very similar as those for Pr-Fe alloys. A molecular field analysis of the temperature dependence of the magnetization shows that J_{RT} is about three times smaller in a-Nd-Fe or a-Nd-Co alloys than the Gd alloys [112]. Ordering temperatures of evaporated [112] and sputtered [113] Nd-Fe alloys are different from those of melt-spun materials, indicating a more disordered amorphous structure in the former.

Enormous coercivities as high as 2.8, 7.5, and 24 kOe in melt-spun ribbon samples of Fe with Nd, Pr, and Sm, respectively [114] are remarkable. These

Table 1 Rare Earth Elements in the Trivalent State

Element	$4f^n$	Ionic radius (Å)	Metallic radius (Å)	L	S	J	g	gJ (μ_B)	$(g-1)^2 J(J+1)$
La	0	1.14	1.83	0	0	0	—	—	—
Pr	2	1.06	1.83	5	1	4	4/5	3.20	0.80
Nd	3	1.04	1.82	6	3/2	9/2	8/11	3.27	1.84
Sm	4	1.00	1.80	5	5/2	5/2	2/7	0.71	4.46
Gd	7	0.94	1.80	0	7/2	7/2	2	7.00	15.75
Tb	8	0.93	1.78	3	3	6	3/2	9.00	10.50
Dy	9	0.92	1.77	5	5/2	15/2	4/3	10.00	7.08
Ho	10	0.91	1.77	6	2	8	5/4	10.00	4.50
Er	11	0.89	1.76	6	3/2	15/2	6/5	9.00	2.55
Tm	12	0.87	1.75	5	1	6	7/6	7.00	1.17
Yb	13	0.86	1.73	3	1/2	7/2	8/7	4.00	0.32
Lu	14	0.85	1.73	0	0	0	—	—	—

samples are not, however, entirely amorphous as the ribbons with maximum coercivity are too thick (more than 100 μm). These properties may be attributed to the formation of a fine microcrystalline [115] or glassy [116,117] phase having the optimum single domain size. The temperature dependence of the coercive field is expressed as [118,119]

$$H_c(T) = H_c(0)\exp\left(-\frac{T}{D}\right) \tag{25}$$

where $D = 125$ K and $H_c(0) = 79$ kOe for $Nd_{40}Fe_{60}$.

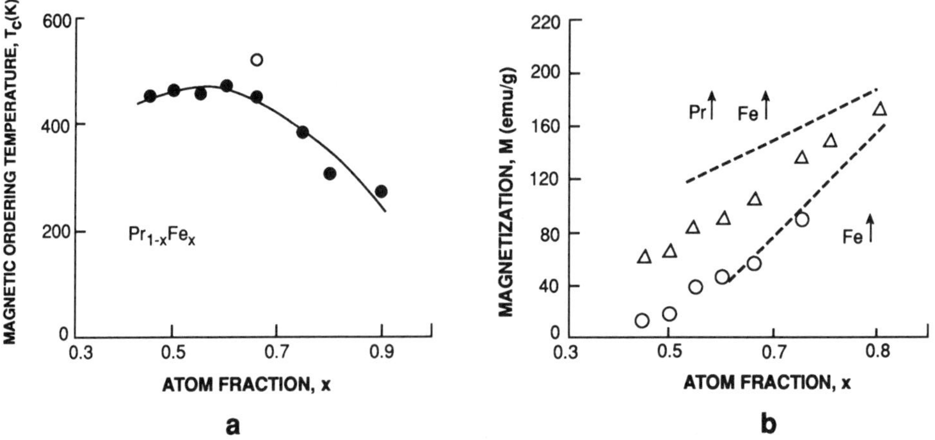

Figure 19 (a) Magnetic ordering temperatures of amorphous (solid circles) and crystalline (open circles) $Pr_{1-x}Fe_x$ alloys. (b) Spontaneous (open circles) and saturation (at 95 kOe) (triangles) magnetization of a-$Pr_{1-x}Fe_x$ alloys [111].

4. *Heavy Rare Earth Transition Metal Alloys*

$J = L + S$ is the ground state of a heavy rare earth atom since the spin orbit coupling constant is negative. Then negative J_{RT} ensures antiparallel coupling between rare earth and transition metal moments. A deep eutectic in the binary phase diagrams around $x = 0.3$ ensures good range of rare-earth-rich melt-spun amorphous alloys, but most attention has been focused on sputtered and evaporated films with x around 0.67 or 0.75, which have higher ordering temperature and can be compared directly with RT_2 or RT_3 crystalline intermetallic compounds.

The Curie temperatures of crystalline RFe_2 [100] and a-$R_{33}Fe_{67}$ [120,121] show a maximum at R = Gd in either phase and continue to decrease from Gd to Lu, as shown in Fig. 20, with decreasing de Gennes factor, $(g - 1)^2J(J + 1)$, listed in Table 1. Another factor which should be taken into account is the interatomic separations across the series. The T_c of a-$R_{33}Fe_{67}$ for R = La, Y, and Lu decreases, following the lanthanide contraction and suggesting an increasing proportion of negative Fe-Fe interactions at short interatomic distances, and also, perhaps, a decrease in the average Fe moment.

A clear decrease of Fe moment has been found in $R_{60}Fe_{40}$ across the lanthanide series [122]. The average ^{57}Fe hyperfine field decreases monotonically from 310 kOe in a-$LaFe_2$ to 200 kOe in a-$LuFe_2$ in amorphous alloys, while it shows a maximum in $GdFe_2$ in crystalline intermetallic compounds.

Random single-ion anisotropy tends to align the rare earth moment along random local easy directions, producing a sperimagnetic structure [123–125] which reduces the net moment per formula compared with the ferrimagnetic crystalline RFe_2, where R subnetwork magnetization exceeds Fe's, and also favors a noncollinear structure for the Fe subnetwork through J_{RT}. The temperature dependence of the spontaneous magnetization of crystalline and amorphous $TbFe_2$ is shown in

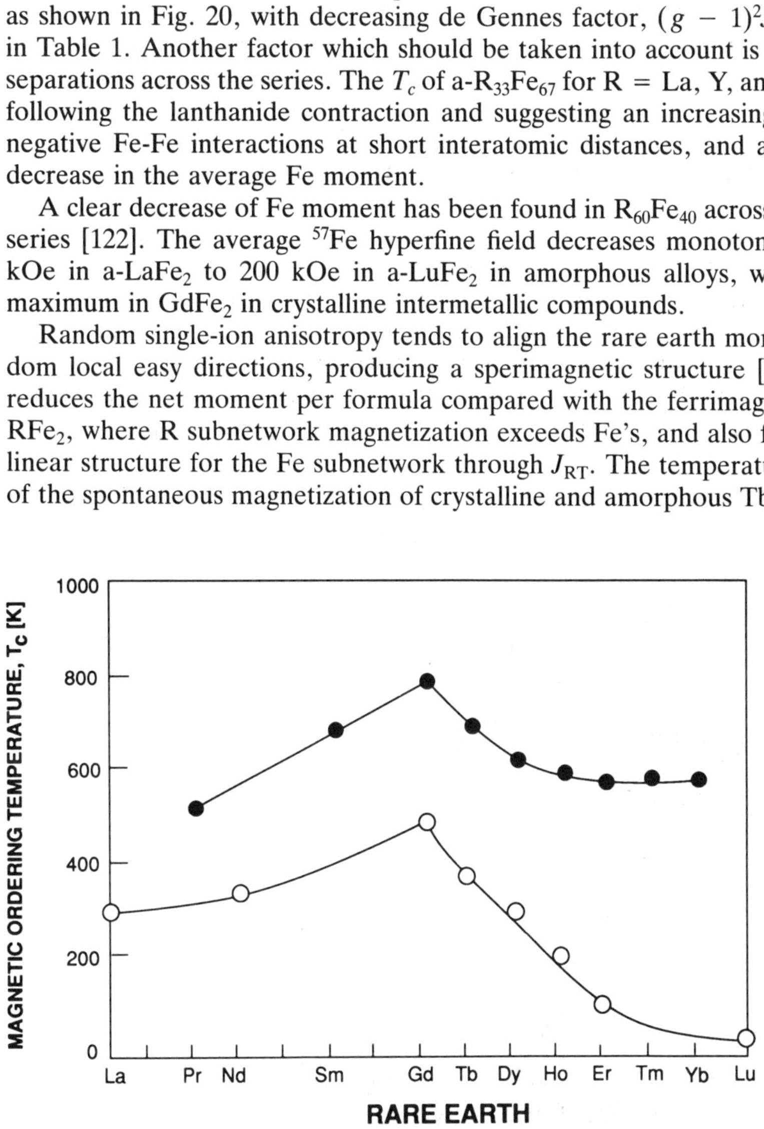

Figure 20 Magnetic ordering temperatures of amorphous (open circles) and crystalline (solid circles) RFe_2 alloys [100,120,121].

Fig. 21, where it is remarkable that both the magnetization and Curie temperature are reduced in the amorphous alloy [126]. Magnetization curves show the lack of saturation even at high fields with a large coercivity characteristic of a noncollinear spin structure, which accounts for the reduction of the net magnetization in terms of fanning of Tb moments due to single-ion anisotropy.

The Mössbauer spectra of ^{161}Dy in a-$Dy_{25}Fe_{75}$ are little different from those in crystalline $DyFe_3$. An analysis of the line widths of the absorption peaks which have a large quadrupole shift establishes that the local hyperfine field directions are strongly correlated with the local principal axis of the electric field gradient [127,128]. This verifies the base of random local anisotropy model, in which the direction of magnetic moments of Dy are correlated with the local anisotropy axis.

Amorphous R-Co alloys [91,125,129–132] are easier to understand than R-Fe alloys because ferromagnetic exchange interaction within the Co subnetwork is so strong that Co forms a collinear ferromagnetic subnetwork, as suggested from Mössbauer spectra of ^{57}Fe-doped alloys [128,131]. There is a clearly defined compensation point for Co-rich alloys with heavy rare earth and T_c exceeds the crystallization temperature when $x > 0.7$ [130].

Polarized neutron study on a-$Er_{33}Co_{67}$ [133,134] reveals a coherent contribution in the high angle scattering due to the erbium moments in a direction perpendicular to the net magnetization. While there is some tendency for Er anisotropy axes on neighboring sites to be parallel, and nearest-neighbor Er spin are coupled antiparallel when the anisotropy axes are perpendicular to z axis (the direction of Co moments) because of the relatively weak negative J_{RR} interaction, they are coupled parallel when their easy axes are close to z because of stronger negative J_{RT} in-

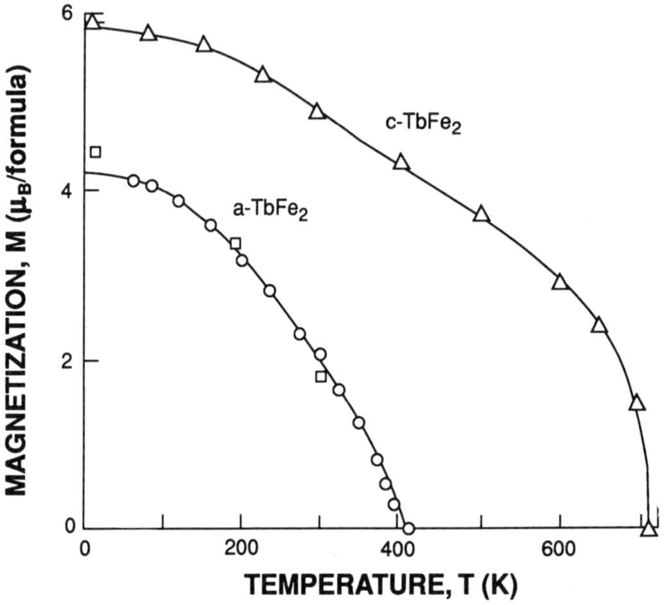

Figure 21 Spontaneous magnetization per formula unit for crystalline (triangles) and amorphous (open circles and squares) $TbFe_2$ [126].

teraction. Dynamic critical fluctuation have been observed in several amorphous R-Fe alloys [135–140]. It turned out that the correlation length remains finite at T_c in all alloys with non-S-state rare earths, which is in contrast to usual divergence of spin correlation length as T approaches T_c.

IV. ANISOTROPY

A. Single-Ion Anisotropy in Amorphous Alloys

Each atom or ion in a solid experiences an electrostatic field created by charges of all other atoms or ions as well as by the conduction electrons. This electrostatic potential stabilizes the particular spatial distribution of atomic electron density which is associated with a specific state of orbital angular momentum. By action of spin-orbit coupling, the magnetic moment of the atom is therefore subjected to a single-ion anisotropy.

The potential, $V(r)$, due to the charge distribution (R) outside a particular atom may be expanded as a series of spherical harmonics as follows,

$$V(r, \theta, \phi) = \int \frac{\rho(R)}{|R - r|} d^3R = \sum_{n=0}^{\infty} \sum_{M=-n}^{n} r^n \gamma_{nm} Y_n^m(\theta, \phi) \tag{26}$$

Interaction with the electrostatic field is normally stronger than the spin-orbit interaction for elements of the $3d$ series, so the potential, $V(r, \theta, \phi)$, acts to perturb the $(2L + 1)$-fold degenerate atomic orbitals. In the rare earth series, by contrast, the electrostatic field acting on the well-shielded $4f$ shell is greatly weakened, so the spin-orbit coupling must be treated first. In other words, to evaluate the matrix elements of the potential for appropriate states, $|M_L\rangle$ are taken as a set of basis functions for $3d$ transition metals, whereas $|M_J\rangle$ are used for the $4f$ rare earth elements. The sum in equation 26 is effective only for terms with $n \leq 4$ for d electrons or $n \leq 6$ for f electrons since the higher terms give zero matrix elements thanks to the orthogonality of the spherical harmonics.

Furthermore in crystals, the crystal field Hamiltonian is greatly simplified by site symmetry and a judicious choice of coordinate axes. No such simplification is expected in an amorphous solid as the only symmetry element present there in average is an inversion. A possible practical approach to this problem is to truncate the series expansion at the second-order term. If the second-order energy surface for each atom is visualized as an ellipsoid, then the coordinate axes required correspond to the three principal axes of the ellipsoid.

The method of operator equivalence is a convenient way of evaluating the matrix element of the electrostatic potential for basis corresponding to a particular value of L or J. A quantum mechanical operator is expressed as a power series of angular momentum operators [141]. We then get an equivalent form of the Hamiltonian for the second-order interaction as

$$H_2 = aJ_x^2 + bJ_Y^2 + cJ_z^2 \tag{27}$$

where $a + b + c = 0$.

If an asymmetry parameter is defined as $\eta = (a - b)/c$, and the axes are usually labeled so that $|c| > |b| > |a|$, then it follows that $0 < \eta < 1$. When $\eta = 0$, the single-ion Hamiltonian reduces to just one diagonal term and the energy levels are given by

$$E_2 = -DJ_z^2 \tag{28}$$

where $D = -(3/2)c$. This corresponds to states with axial symmetry about z, which is an easy axis for $D > 0$. The approximation is equivalent to replacing the complicated energy surface by a prolate ellipsoid of revolution with the same major axis. The local easy axis is expected to vary from site to site in an amorphous solid, leading to random single-ion magnetic anisotropy [142].

The distribution of second-order electrostatic field has been evaluated in models of amorphous materials, for example, large computer-generated clusters of hard spheres of two different sizes chosen to model of rare earth–transition metal alloys [138]. The local easy axes were found to be uniformly distributed in orientation with insignificant correlation among the directions for adjacent sites. The distribution of overall splitting caused by the perturbation of Eq. (26) is broadened but, for large J values, the ground state is shown to be effectively the doublet over most of the range of η [143]. The axial approximation with positive D is well justified for these ions.

The nuclear quadrupole interaction which can be measured by hyperfine techniques such as nuclear quadrupole resonance or Mössbauer spectroscopy gives important information about the second-order electrostatic field that is qualitatively different from those in the radial distribution function obtained by diffraction experiments. The contribution from charges around a given ion to the electric field gradient acting at its nucleus has a form identical to that of a second order electrostatic field acting on an atom with an antishielding factor $1 - \gamma$. The calculation of the field gradient due to a randomly occupied single shell of ions showed that zero second-order electrostatic fields have a vanishingly small probability and that field with $\eta \simeq 0$ have little chance of occurring at random [144].

From ^{155}Gd Mössbauer spectra on a-Gd-Ni alloys, the average principal electric field gradient is $\langle |V_{zz}| \rangle = 8.2 \times 10^{17}$ V/cm^2 [144]. Surprisingly, this value is close to that calculated in the point charge model with the first shell of neighbors, attributing $+3e$ to each Gd ion. The average values of $|c|$ may be estimated for any other ions; for instance, $\langle D \rangle = 6$ K is obtained with the constants appropriate for Dy^{3+} ($J = 15/2$). This is consistent with the value obtained by interpreting the magnetization curve of $DyNi_3$ using the random axial anisotropy model [145], and comparable values have been found for other heavy rare earth ions [142,146]. Overall splitting of the J manifold by the electrostatic field in rare earth alloys is of the order of hundreds of degrees, usually greater than that by exchange interactions or any laboratory external magnetic field.

B. The Random Axial Anisotropy Model

Although highly simplified, a model based on the idea of random axial anisotropy [147] must be important as a point of reference for understanding one of the characteristics of amorphous magnetism. The Hamiltonian contains terms of uniaxial anisotropy with randomly oriented easy direction, n_i, at each site in addition to the usual exchange interactions and the effect of an externally applied field, H, along a fixed z direction; that is,

$$H_{RAA} = -\sum_i D_i(n_i \cdot J_i)^2 - \sum_{i>j} j_{ij}J_iJ_j - \sum_i g\mu_B J_{iz}H \tag{29}$$

For simplicity we assume that D_i has the same value at each site and consider what happens at $T = 0$ in the absence of exchange interactions. The magnetization curve for an atom at site i can be deduced, in the classical case [148], by minimizing the energy with respect to θ_i,

$$E_i = -d\cos^2(\Theta_i - \theta_i) - mH\cos\theta_i \tag{30}$$

where d is the classical anistropy energy, m the magnetic moment on each atom, and Θ_i and θ_i are the polar angles of the local easy axis, n_i, and magnetic moment, m_i, at i site, respectively.

After averaging over all possible orientations of n_i, the reduced magnetization, $\mu = \langle J_z \rangle / J = \langle \cos\theta_i \rangle$, is obtained as a function of effective field, $h = (g\mu_B/J)H/D$ or $(m/d)H$, the ratio of magnetic to anisotropy energy, as shown in Fig. 22a, where $J = 5/2$ and $J = 8$ correspond the extreme values encountered in the rare earth series. $J = \infty$ is the classical limit where the equilibrium condition

$$\frac{dE_i}{d\theta_i} = -d\sin 2(\Theta_i - \theta_i) + mH\sin\theta_i = 0 \tag{31}$$

relates θ_i to Θ_i, and the reduced magnetization is given by

$$\mu = \frac{1}{2\pi m}\int_0^{\pi/2} m\cos\theta_i\, 2\pi\sin\Theta_i\, d\Theta_i \tag{32}$$

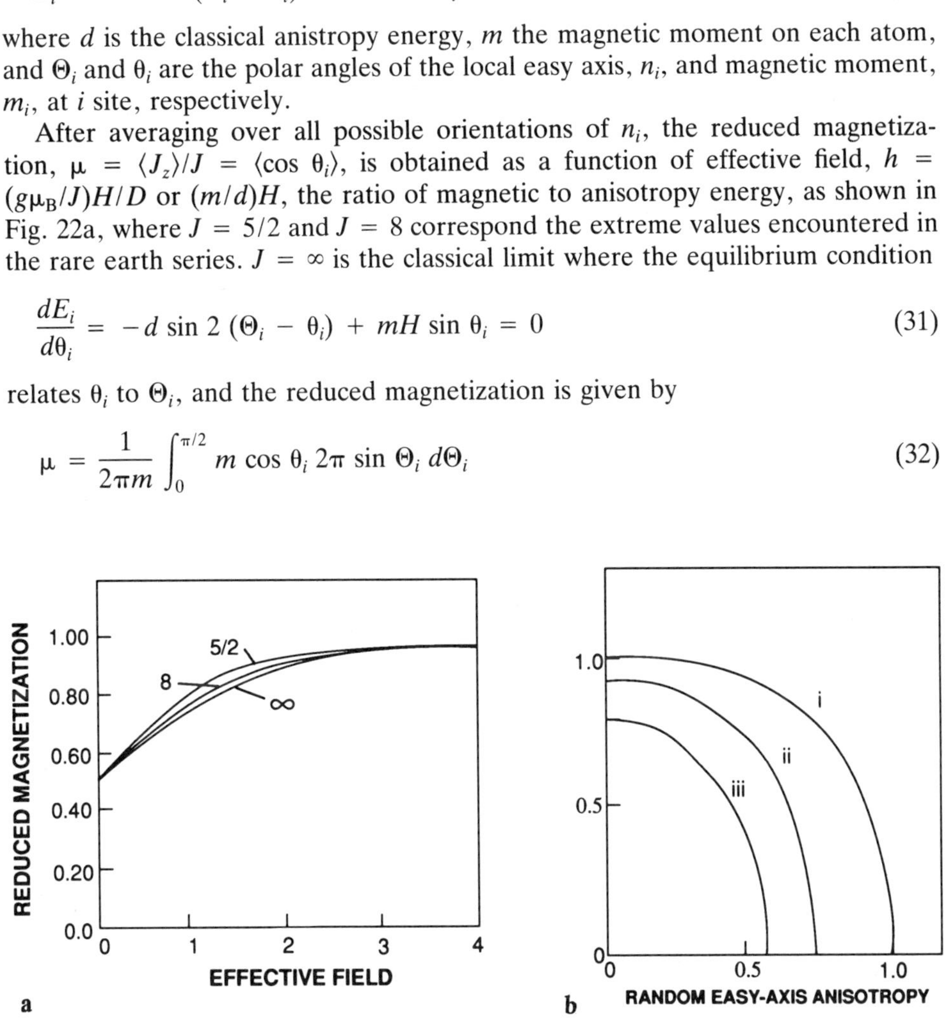

Figure 22 (a) Reduced magnetization for a magnetic system with random easy anisotropy as a function of the effective field, $h = (g\mu_B/JD)H = (m/d)H$, the ratio of magnetic to anisotropy energy [B4]. (b) Temperature dependence of the magnetization with random easy-axis anisotropy for (i) $D/jZ = 0$, (ii) 0.8, (iii) 1.6 [142].

For vanishingly small field ($H \simeq 0$) at $T = 0$, $\theta_i \simeq \Theta_i$ and the reduced magnetization $\mu \simeq 1/2$, because moments adopt an asperomagnetic configuration with their orientations distributed at random in a hemisphere around the positive z-axis. The increasing applied field narrows the hemisphere to a cone, giving an initial susceptibility $du/dh = 1/3$, and the magnetization approaches saturation as $\mu = 1 - 4/15h^2$ at very high field; for example, if $D = 6$ K for Dy^{3+}, $H = 1.2 \times 10^6$ Oe for $\mu = 0.95$. For reversed field, we get $\mu = -1/2$ for $H \simeq -0$, since the moments now distribute in a hemisphere around the negative z-axis. There are two minima of energy for each moment separated by an anisotropy barrier of order d.

Next, the exchange interaction must be taken into account. The simplest way is the molecular field approximation in which the exchange interactions are assumed to be equivalent to an internal field, H_m, proportional to $\langle J_z \rangle$ and directed along a unique z-direction everywhere, that is,

$$H_m = \frac{\langle Z_i \rangle j \langle J_z \rangle}{g\mu_B} = \lambda\mu \tag{33}$$

where the exchange interaction are all assumed to be j.

In this approximation we may only replace H in Eq. (30) by $H + H_m$. The magnetic structure in zero applied field will be asperomagnetic, with the magnetic moments distributed within a cone. The reduced remanent moment, μ_0, at $T = 0$ is obtained from the intersection of the curve in Fig. 22a with the straight line, $\mu = (d/m\lambda)h$ in the classical limit.

There is no hysteresis in the magnetization curve at $T = 0$ below a threshold value of anisotropy constant, $d = 0.6\lambda m$ [143]. At high values of d, coercivity becomes proportional to anisotropy, $H_c \simeq d/m$.

The temperature variation of the moment has been calculated in the molecular field approximation self-consistently [142]. At each temperature the thermodynamic average $\{J_{iz}\}$ is calculated in terms of Θ_i, and $\langle J_z \rangle$ is obtained by integrating over all sites. The procedure is then repeated until a self-consistent result for $\langle J_z \rangle$ is achieved. There is a well-defined phase transition in the molecular field approximation at a temperature lower than that which would be in the absence of anisotropy. Results for $J = 1$ are shown in Fig. 22b, where the magnetization $\mu(T)$ is always less than that without anisotropy and its shape is flatter.

The random axitial anisotropy model can be applied to a sperimagnet which is actually realized in rare earth–3d transition metal amorphous alloys. The slope of so-called Arrott (M^2 versus H/M) plots observed for a-$Tb_{72}Fe_{28}$ was found to be only one third that of a value estimated from molecular field approximation without random anisotropy [149]. The experimental results can be accounted for by introducing random axial anisotropy, $D = 9.6 \times 10^{-15}$ erg, which seems to be reasonable compared to $D = 1.8 \times 10^{-14}$ erg in the pure Tb crystal.

It is well known that under the influence of dipolar fields and magnetocrystalline anisotropy, crystalline ferromagnets generally break up into well-defined magnetic domains. Molecular field theory is applicable only within a domain where the magnetization is uniform with specific z axis. It is unlikely, even discounting the influence of dipolar fields, that the uniform asperomagnetic state throughout a whole sample with a common z axis is realized in the random axial anisotropy model in the absence of externally applied field. It is easy to visualize a magnetic configuration where the z axis is defined locally for group of spins but wanders over larger distance since each atom interacts with only a few shells of magnetic neighbors. A state of lower energy may be obtained if the local z axis is allowed to respond to the random local fluctuations in the correlation of the anisotropy axes. The problem is basically that of length scale; that is, the correlation length of local z axis, which depends on the range of the interaction or, equivalently, on the number of interacting neighbors.

First we will consider a case of the limit $D \gg J$, where moments are strongly pinned to anisotropy axes, and the local moment direction varies on a length scale comparable to the interatomic spacing. Define the z-axis as the direction of the exchange field acting on a central atom which has Z interacting neighbors. The exchange field at any site in this local asperomagnetic cluster is the resultant of z component proportional to $(1/2)Z$ and a perpendicular component in the xy plane proportional to $(\pi/4)Z^{1/2}$, where $1/2$ and $\pi/4$ are the average values of parallel and perpendicular components, respectively, to z axis of a unit vector randomly oriented within a hemisphere. Therefore, the exchange field at any other site is misoriented with respect to the z axis by an angle $q = \tan^{-1}(\pi/2\sqrt{Z})$ on average, for example, $q = 24°$ for $Z = 12$. These misorientations will accumulate in a random way on going out from a central atom in any direction, so that the sense of the original z direction will be lost after a distance of order $(\pi/2q)^2 a \simeq Za$, where a is the interatomic spacing and $q \simeq \tan q$ is assumed.

For the opposite limit, $D \ll J$, the spin directions are strongly correlated within a domain, which is assumed to contain $n \simeq (L/a)^3$ atoms where L is the linear size of a domain. The average anisotropy energy per atom, E_a, due to the fluctuations within the domain will be of order $-DJ^2n^{-1/2}$. When the magnetization rotates by $\pi/2$ over a distance of order L on passing from one domain to the next, the excess exchange energy per atom is $E_{ex} \simeq J^2 j(na/2L)^2$. Minimizing the total energy, $E_a + E_{ex}$, we get

$$L = \frac{1}{9}\pi^4\left(\frac{j}{D}\right)^2 a \tag{34}$$

Thus, the pure ferromagnetic state is destabilized by random anisotropies even if they are weak, although the correlation length, L, could be so long that the system may be indistinguishable from that of a normal ferromagnet.

The random axial anisotropy model of Eq. (28) has been examined directly by computer model calculations [150–153] which, however, are limited by the size of a model cluster less than the correlation length discussed above. There is an inconsistency between calculations which predict an intrinsic coercivity varying as D [148,154,155] and those which find it varies as j [150,152,156,157]. This is related to the way in which spins to relax toward their lowest energy state. The former is attained by allowing only small reorientation or thermal excitation, whereas the latter results when large reorientation or tunneling through energy barriers is permitted.

C. Overall Magnetic Anisotropy

Amorphous materials, in which there is no long-range atomic ordering, are likely to be considered as isotropic as a whole. Nature does, however, present us with some unexpected varieties. No one had expected so strong a perpendicular anisotropy in amorphous alloy films before the first report [158] that amorphous Gd-Co thin films prepared by sputter deposition had strong perpendicular magnetic anisotropies which exceeded the shape anisotropies of the thin films. This overall magnetic anisotropy has important implications for the practical application of amorphous materials, especially for thin films. Magneto-optical recording, for example, is possible only if there is sufficient perpendicular magnetic anisotropy.

Inversely, if there is no macroscopic overall magnetic anisotropy, it does not mean that each atomic moment sits at a completely isotropic site; that is, the free energy associated with each atom is dependent on the direction of its magnetic moment or spin. The important consequences of structural disorder for magnetic anisotropy are based on the concepts on random local anisotropy, as discussed in the previous section.

As the mechanism of the overall magnetic anisotropy we can cite the following:

1. Single-ion anisotropy which is effective for non-S-state ions in a local ligand field through the spin-orbit interaction. In order to get the macroscopic overall anisotropy there must be anisotropic distribution of the directions of the axis of the ligand field.
2. Classical magnetic dipolar interactions based on
   a. Anisotropic pair distribution of atomic moments
   b. Internal shape effect due to inhomogeneity or microstructures of thin films.
3. Anisotropic exchange interactions (including pseudodipolar interactions).
4. Inverse effect of magnetostriction with anisotropic distribution of internal stress.

The mechanisms cited above are not exclusive, but may coexist. It is difficult, however, to distinguish their contributions qualitatively. It should be noted that if the angular distribution of atomic pairs in an amorphous alloy system is completely random as it was thought to be in an ideal state, and the distribution of internal stress is also isotropic, there never occurs a macroscopic overall magnetic anisotropy in an amorphous material because the local anisotropies originated from any mechanism must be averaged to cancel out to zero.

In the following section we discuss the macroscopic overall magnetic anisotropy, especially a mechanism of magnetic anisotropy due to classical dipolar interactions so as to understand the relation between structure and magnetism. The magnetic anisotropy is quite sensitive to the structural anisotropy. Even when deviation from perfect isotropy in average atomic arrangement is so small that it cannot be detected by direct diffraction methods, there is a possibility to induce an appreciable magnetic anisotropy.

D. Magnetic Anisotropy due to Dipolar Interactions

The dipolar or pseudodipolar interaction between atomic magnetic moments m_i and m_j is expressed as

$$u_{ij} = f(r_{ij})\left[m_i m_j - \frac{3(m_i r_{ij})(m_j r_{ij})}{r_{ij}^3} \right] \tag{35}$$

where r_{ij} is the vector from atom i to atom j. For classical magnetic dipolar interaction, $f(r) = r^{-3}$, but it is expected to be shorter-ranged for pseudodipolar interaction.

The dipolar interaction is dependent not only on the angle between the magnetic moment, m_i and m_j, but also on angles between the moment and r_{ij}. This is different from the exchange interaction which depends only on the angle between the two moments, independent of the direction of r_{ij}. While spin configuration is determined by the overwhelming exchange interaction, the dipolar interaction still plays important role through magnetostatic energy, resulting in shape anisotropy. If magnetic materials are inhomogeneous the internal shape effect is derived from the dipolar interaction. Even in homogeneous amorphous alloys it produces magnetic anisotropy if there is anisotropy in the distribution of the atomic pair vector, r_{ij}. This mechanism can be enhanced for ferrimagnetic spin configuration which will be treated in the following section.

1. Dipolar Interaction in Ferrimagnetic Alloys

Consider an amorphous alloy film containing 3*d* transition atoms and heavy rare earth atoms (T and R), for example Co and Gd, which occur in ferrimagnetic subnetworks satisfying all exchange interactions, even for topologically disordered arrangement of atoms. The dipolar interaction energy for an atomic pair of parallel spins in a given same subnetwork is

$$u_{AA}(r_{ij}) = f_{AA}(r_{ij})m_A^2[1 - 3(\alpha x + \beta y + \gamma z)^2/r_{ij}^2] \tag{36}$$

where A = T or R, α, β, and γ are direction cosines of the magnetic moments, and x, y, and z are the components of r_{ij}. Let us take the z axis normal to the film plane. If atoms i and j belong to different subnetworks m_i and m_j are antiparallel to one another, giving the dipolar energy

$$u_{TR}(r_{ij}) = -f_{TR}(r_{ij})m_T m_R[1 - 3(\alpha x + \beta y + \gamma z)^2/r_{ij}^2] \tag{37}$$

The total dipolar energy for TT, RR, and TR pairs can be expressed in terms of probability functions $P_{TT}(r)$, $P_{RR}(r)$, and $P_{TR}(r)$, where $P_{TT}(r)\,dV$ is the probability of finding a T atom in volume element dV separated from an arbitrary central T atom by vector r, and so on. The average dipolar energy associated with TT pairs is given by integration over the vector r_i

$$U_{TT} = \int_0^{2\pi}\int_0^{\pi}\int_0^{\infty} u_{TT}(r)P_{TT}(r, \theta, \phi)r^2 \sin\theta\, dr\, d\theta\, d\phi \tag{38}$$

Using an analogous expression for RR and TR pairs the total dipolar energy per unit volume is

$$E(\alpha, \beta, \gamma) = \frac{1}{2}[N_T(U_{TT} + U_{TR}) + N_R(U_{RR} + U_{RT})] \tag{39}$$

where N_T and N_R are the number of T and R atoms per unit volume, respectively.

Probability functions may be expanded by using spherical harmonics as follows:

$$P_{AB} = \frac{N_B}{N} R_{AB}(r)[1 + p_{AB}(r)(3\cos^2\theta - 1)] \tag{40}$$

where AB = TT, RR, or TR, $(N_B/N)R_{AB}(r)$ is the isotropic partial radial distribution function of B atoms around A atoms, and p_{AB} is a measure of the anisotropy in the distribution of AB pairs. Inserting Eq. (40) into Eq. (39), the total dipolar energy is obtained as

$$E(\gamma) = E_0 + K_u(1 - \gamma^2)$$
$$= \frac{1 - 3\gamma^2}{5N}\left[M_T^2 G_{TT} - M_T M_R(G_{TR} + G_{RT}) + M_R^2 G_{RR}\right] \tag{41}$$

where

$$G_{AB} = 4\pi \int_0^\infty f_{AB}(r) R_{AB}(r) p_{AB} r^2\, dr \tag{42}$$

and K_u is the uniaxial anisotropy energy defined so that positive value corresponds to a perpendicular easy axis anisotropy.

In densely packed amorphous alloys the structural anisotropy parameters, $p_{AB}(r)$, are mutually coupled. It is plausible to expect that $P_{AA} + P_{AB}$ is isotropic. Then, assuming for simplicity that $p_{AB}(r) = 0$ for larger r than the nearest-neighbor distance, one obtains for the nearest-neighbor shell

$$p_{AB}Z_{AB} + p_{AA}Z_{AA} = 0 \tag{43}$$

where Z_{AB} and Z_{AA} are number of B and A nearest neighbors around an A atom. Introducing a single-pair distribution anisotropy constant, p, it follows that

$$p = -p_{TR} = -p_{RT} = \left(\frac{Z_{TT}}{Z_{TR}}\right)p_{TT} = \left(\frac{Z_{RR}}{Z_{RT}}\right)p_{RR} \tag{44}$$

Using the above relations, Eq. (42) turns out to be

$$G_{TR} = -Z\langle f_{TR}\rangle p, \qquad G_{TT} = Z\left(\frac{x_R}{x_T}\right)\langle f_{TT}\rangle p, \qquad G_{RR} = Z\left(\frac{x_T}{x_R}\right)\langle f_{RR}\rangle p \tag{45}$$

where $x_A = N_A/N$, $Z = Z_{RT}/x_T = Z_{TR}/x_R$, and

$$\langle f_{AB}\rangle = \int_0^\infty f_{AB}(r)R_{AB}(r)r^2\, dr \Big/ \int_0^\infty R_{AB}(r)r^2\, dr \tag{46}$$

The uniaxial magnetic anisotropy energy resulting from dipolar mechanism with slight structural anisotropy is now expressed as

$$K_u = \left(\frac{3}{5N}\right)Zx_Tx_Rp\left[\langle f_{TT}\rangle\left(\frac{M_T}{x_T}\right)^2 + \frac{2\langle f_{TR}\rangle M_TM_R}{x_Tx_R} + \langle f_{RR}\rangle\left(\frac{M_R}{x_R}\right)^2\right] \tag{47}$$

For classical dipolar interaction, $f_{ij}(r) = r_{ij}^{-3}$. If it is allowed to simplify as $\langle f_{AB}\rangle \simeq d^{-3}$, where d is average nearest-neighbor separation, Eq. (47) takes a simple form

$$K_u = \frac{3}{5}cZx_Tx_R\left(\frac{M_T}{x_T} + \frac{M_R}{x_R}\right)^2 p \tag{48}$$

where $c = N^{-1}d^{-3}$ is the packing fraction (about 0.7 for amorphous metallic alloys).

Since the net magnetization of this ferrimagnet is $M = |M_T - M_R|$, the condition for perpendicular magnetization in thin films, $Q = K_u/2\pi M^2 > 1$, can be satisfied with resonably small structural anisotropy parameter, p.

2. *Anisotropic Microstructure and Dipolar Interactions*

Classical magnetic dipolar interactions are also responsible for magnetic anisotropy due to anisotropic inhomogeneity or microstructure, for example, columnar structure, through a so-called internal shape effect. Since this inhomogeneity may extend more than several atoms it is convenient to treat the material as a magnetic continuum. There are several simple case for which the associated anisotropy energy can easily be calculated.

Consider a magnetic thin film the magnetization of which has columnar variation with x, y taking the z direction to be the normal to the film plane. The magnetization can be expanded in Fourier series as

$$M = M(x, y)\,u = M_0 + \sum_{k \neq 0} m_k \exp(ikr) \tag{49}$$

where u is a unit vector independent of x, y, z. The magnetic energy density in this inhomogeneous system turns out to be [159]

$$E = \pi(u_x^2 + u_y^2)\langle |M(x, y) - M_0|^2 \rangle \tag{50}$$

and an uniaxial anisotropy constant is given as

$$K_u = \pi \langle |M(x, y) - M_0|^2 \rangle \tag{51}$$

For the case of long columns aligned parallel to z and having magnetization M_1 in a matrix of magnetization, M_2, K_u becomes

$$K_u = \pi V(1 - V)|M_1 - M_2|^2 \tag{52}$$

where V is the volume fraction occupied by the columns. K_u is maximized for $V = 0.5$.

For a nonmagnetic matrix, the average magnetization is $M = VM_1$, and the condition for the perpendicular magnetization for the film

$$Q = \frac{K_u}{2\pi M^2} = \frac{1 - V}{2V} > 1 \tag{53}$$

is satisfied only for $V < 1/3$.

For the ferrimagnetic alloy T_xR_{1-x}, having subnetwork magnetizations M_T and M_R and atomic magnetic moments m_T and m_R, consider compositional inhomogeneities consisting of long columns of composition x_1 in a matrix x_2, and the volume fraction V. For small composition differences, $\Delta x = |x_1 - x_2|$,

$$|M_1 - M_2| = \left|\frac{dM}{dx}\right| \Delta x = N(m_T + m_R)\,\Delta x = \left|\frac{M_T}{x} + \frac{M_R}{1 - x}\right| \Delta x \tag{54}$$

The resultant anisotropy energy is given by

$$K_u = \pi V(1 - V)(\Delta x)^2 \left|\frac{M_T}{x_T} + \frac{M_R}{x_R}\right|^2 \tag{55}$$

This expression has similarity to Eq. (48) which describes the anisotropy due to dipolar interactions between atomic moments the pair distribution of which is slightly anisotropic.

3. Comparison with Experimental Observations

Macroscopic magnetic anisotropy in amorphous alloy films is sensitive to preparation method and conditions. Gd-Co-based sputtered films have perpendicular easy magnetic anisotropy only when prepared by biased negative with respect to the sputter plasma. Conditions for maximizing K_u depend upon sputter gas pressure, target voltage, and sputter geometry, as well as on the substrate bias. It would be instructive to examine how magnetic anisotropy observed in Gd-Co amorphous alloy films can be attributed to the classical magnetic dipolar interactions, although it does not mean to exclude other mechanisms of macroscopic anisotropy.

The magnetic anisotropy energy, $K_u(T)$, has been determined from ferromagnetic resonance field measurements in parallel and perpendicular orientations of film specimen, together with bulk magnetization measurements. It has also been extracted from magnetization curves in parallel and perpendicular fields [160]. In order to evaluate Eqs. (48) or (55), it is neccssary to know the subnetwork mag-

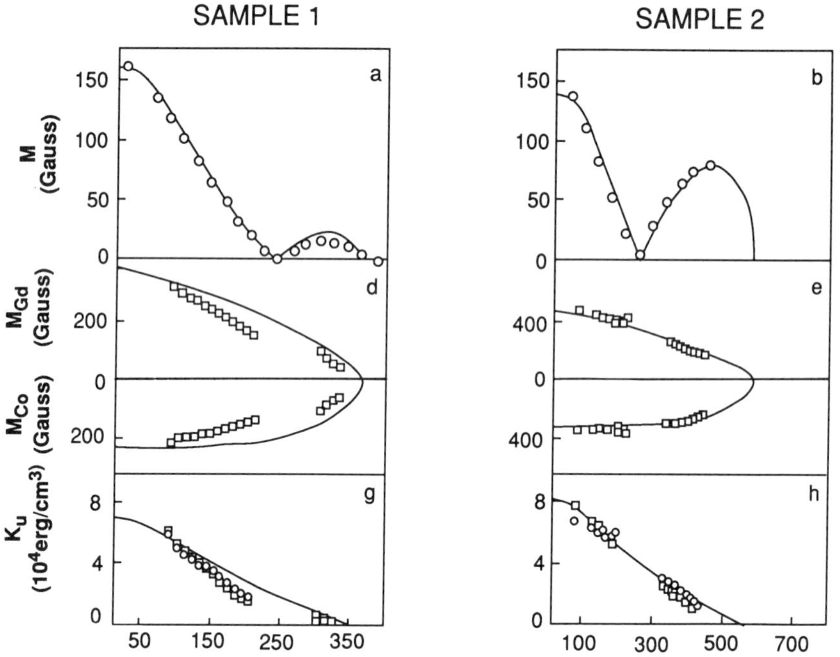

Figure 23 Observed and calculated magnetization and anisotropy energy as a function of temperature for sputtered films; sample 1: $Gd_{0.113}Co_{0.672}Mo_{0.159}Ar_{0.057}$, sample 2: $Gd_{0.145}Co_{0.720}Mo_{0.107}Ar_{0.029}$. Upper: magnetization, $M(T)$, observed (circles), and calculated with mean field approximation (MFA) (line). Middle: subnetwork magnetization, $M_{Co}(T)$ and $M_{Gd}(T)$, obtained from FMR (squares), and MFA (line). Lower: Anisotropy energy $K_u(T)$, observed (circles), and calculated using MFA subnetwork magnetization with p = 0.016 and 0.011 for sample 1 and 2, respectively (line), and calculated using FMR subnetwork magnetization with p = 0.02 and 0.010 for samples 1 and 2, respectively (squares) [159].

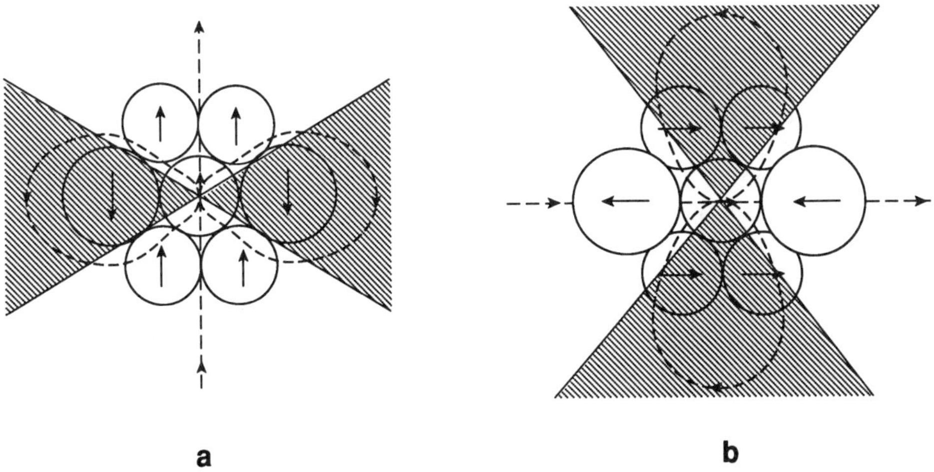

Figure 24 Illustration of spin alignment (a) favored and (b) opposed by magnetic dipolar interaction in a ferrimagnetic structure with anisotropic pair distribution; that is, there is a greater possibility of finding like atoms up and down, and different atoms side by side around a given atom in the films.

netization $M_T(T)$ and $M_R(T)$. These may be calculated by mean field theory, or deduced from observed g factor by ferromagnetic resonance experiments. By the Wangness formula [161] g factor of a ferrimagnetic material is related to the subnetwork magnetizations as

$$g = \frac{|M_T - M_R|}{|M_T/g_T - M_R/g_R|} \tag{56}$$

where g_T and g_R are g factors for the individual subnetworks which may be deduced from data for pure elements or remain as adjustable parameter to obtain physically reasonable values of M_T and M_R for $T = 0$. Experimental and calculated results for a-$Gd_{0.145}Co_{0.720}Mo_{0.107}Ar_{0.029}$ sputter-deposited films are shown in Fig. 23. The subnetwork magnetizations calculated by mean field theory and those obtained from g factor coincide reasonably with each other for this sample, though it was not so well for many other cases. The observed K_u can be interpreted by taking $p = 0.01$ for Eq. (48), or $\Delta x = 0.08$ by arbitrarily setting $V = 0.5$ in Eq. (55). The slight anisotropic pair distribution of order of a few percent, which is difficult to detect by direct diffraction experiments, seems to be plausible in thin films in which the growth condition is anisotropic. A positive p means there are more possibility of finding like atoms up and down, and different atoms side by side around a given atom in the films. The dipolar interactions between ferrimagnetically coupled atomic moments stabilize perpendicular easy-axis anisotropy for this pair distribution, as illustrated in Fig. 24.

V. CONCLUDING REMARKS

A great number of amorphous alloys contain transition elements and/or rare earth elements, which makes them, in most cases, magnetic. Amorphous magnetic materials have opened new attractive fields not only in application but also in intrinsic magnetism. It has been shown by studies in this field that local atomic environments play a dominant role to determine the basic magnetic properties such as magnetic moment or exchange interactions, to which the long-range periodicity of atomic arrangements characteristic of crystalline solids has little importance. There must be some distribution in magnetic moment and exchange interactions in amorphous alloys because of the varieties of local environments, but macroscopic ferromagnetism is affected little in most cases.

The situation regarding magnetic anisotropy is quite different in crystalline solids and in amorphous solids. Overall magnetic anisotropy in crystalline solids is directly comprised of a crystal field contribution at each atomic site; in amorphous solids, wherein local anisotropy can be considerably large, such effects cancel out because of random orientations of the local easy axis at each atom. Depending on the relative strength of exchange interaction and local anisotropy, new categories of magnetic ordering can appear in amorphous magnetic materials.

From a mesoscopic point of view, amorphous materials are rather more homogeneous than polycrystalline materials. They may offer ideal behaviors in some physical properties, for example, in critical behavior. They had been also believed to be isotropic before the discovery of considerable overall magnetic anisotropy which was sufficiently strong to support perpendicular magnetization in amorphous films. The importance of mechanisms to enhance the slight structural anisotropy into an appreciable overall magnetic anisotropy should be noted.

REFERENCES

Several books review the magnetism of amorphous materials:

B1. *Amorphous Magnetism I*, edited by H. O. Hooper and A. M. deGraaf (Plenum, New York, 1973).

B2. *Amorphous Magnetism II*, Edited by R. Hasegawa and R. A. Levy (Plenum, New York, 1977).

B3. *Metallic Glasses*: *Magnetic, Chemical and Structural Properties*, edited by R. Hasegawa (CRC, Boca Raton, FL, 1983).

B4. K. Moorjani and J. M. D. Coey, *Magnetic Glasses* (Elsevier, Amsterdam, 1984).

1. J. M. D. Coey, *J. Appl. Phys.*, *49*, 1646 (1978).
2. N. F. Mott, *Metal Insulator Transition* (Taylor and Francis, London, 1974).
3. J. Hubbard, *Proc. Roy. Soc.*, *A276*, 238 (1963); *A277*, 237 (1964); *A280*, 401 (1964).
4. P. W. Anderson, *Phys. Rev.*, *124*, 41 (1961).
5. A. I. Gubanov, *Fiz. Tver. Tela.*, *2*, 502 (1960).
6. R. Rammal and J. Souletie, in *Magnetism of Metals and Alloys*, edited by M. Cyrot (North-Holland, Amsterdam, 1982); P. J. Ford, *Contemp. Phys.*, *23*, 141 (1982); J. A. Mydosh and G. J. Nieuwenhuys, in *Ferromagnetic Materials*, edited by E. P. Wohlfarth (North-Holland, Amsterdam, 1980), p. 71.
7. R. C. O'Handley, *J. Appl. Phys.*, *62*, R15 (1987).
8. F. E. Luborsky, in *Ferromagnetic Materials*, Vol. 1, edited by E. P. Wohlfarth, (North-Holland, Amsterdam, 1980), p. 451.
9. C. D. Graham, Jr., and T. Egami, *Ann. Rev. Mater. Sci.*, *8*, 423 (1978).
10. J. Durand, in Ref. B3.
11. N. Kazama, M. Kameda, and T. Masumoto, *AIP Conf. Proc.*, *34*, 307 (1976).
12. G. Marchal, D. Teirlinck, P. Mangin, C. Janot, and J. Hubsch, *J. Phys.*, *41*, C8-662 (1980).

13. Durand and M. Yung, in Ref. B2, p. 275.
14. G. S. Cargill III, and R. W. Cochrane, *J. Phys.*, *35*, C4-269 (1974).
15. D. Teirlinck, These d'Etat Nancy (1981).
16. M. Mitera, M. Naka, T. Masumoto, and N. S. Kazama, *Phys. Stat. Sol.* (*a*), *49*, k163 (1978).
17. T. Mizoguchi, K. Yamauchi, and H. Miyajima, in Ref. 1.
18. H. Fujimori, H. Morita, Y. Obi, and Ohta, in Ref. B2, p. 393.
19. R. C. O'Handley, R. Hasegawa, R. Ray, and C. P. Chou, *Appl. Phys. Lett.*, *29*, 330 (1976).
20. R. C. O'Handley, R. Hasegawa, R. Ray, and C. P. Chou, *J. Appl. Phys.*, *48*, 2095 (1977).
21. N. Kazama, M. Kameda, and T. Masumoto, *AIP Conf. Proc.*, *34*, 307 (1976).
22. A. Amamou, *IEEE Trans. Mag.*, *MAG-12*, 325 (1976).
23. J. Durand, in Ref. B2, p. 305.
24. A. K. Shinha, *J. Appl. Phys.*, *42*, 338 (1971).
25. J. Durand, C. Thompson, and A. Amamou, in *Rapidly Quenched Metals III*, Vol. 2, edited by B. Cantor (Metals Society, London), p. 109.
26. R. C. O'Handley, *Sol. State Commun.*, *38*, 703 (1981).
27. Y. Obi, H. Morita, and K. Fukamichi, *IEEE Trans. Magn.*, *MAG-16*, 1132 (1981).
28. B. W Corb, R. C. O'Handley, and V. L. Moruzzi, *J. Magn. Magn. Mater.*, *31–34*, 1537 (1983).
29. A. Collins, Ph.D. thesis, MIT Dept. Mater. Sci. Eng. (1986).
30. B. W. Corb, R. C. O'Handley, and N. J. Grant, *J. Appl. Phys.*, *53*, 7728 (1982).
31. B. W. Corb, R. C. O'Handley, and N. J. Grant, *Phys. Rev.*, *B27*, 636 (1983).
32. B. W. Corb, *Phys. Rev.*, *B31*, 2521 (1985).
33. B. W. Corb and R. C. O'Handley, *Phys. Rev.*, *B31*, 7213 (1985).
34. C. L. Chien, in *Nuclear and Electron Resonance Spectroscopies Applied to Materials Science*, edited by E. N. Kaufmann and G. K. Shenoy (Elsevier North-Holland, Amsterdam 1981).
35. J. Durand, *J. Phys.*, *41*, C8-609 (1980).
36. O. Massenet, H. Daver, V. D. Nguyen, and J. P. Rebouillat, *J. Phys.*, *F9*, 1687 (1979).
37. J. Shinjo, F. Ito, H. Takaki, Y. Nakamura, and N. Shibazono, *J. Phys. Soc. Japan*, *19*, 1252 (1964).
38. C. L. Chien, D. Musser, F. E. Luborsky, and J. L. Walter, *J. Phys.*, *F8*, 2407 (1978).
39. J. J. Becker, F. E. Luborsky, and J. L. Walter, *IEEE Trans.*, *MAG-13*, 988 (1977).
40. R. C. O'Handley, R. Hasegawa, R. Ray, and P. C. Chou, *Appl. Phys. Lett.*, *29*, 330 (1976).
41. D. G. Onn, T. H. Antoniuk, T. A. Donnelly, W. D. Johnson, T. Egami, T. Prater, and J. Durand, *J. Appl. Phys.*, *49*, 1730 (1978).
42. J. Durand, in Ref. B2, p. 305.
43. A. Amamou, *IEEE Trans. Mag.*, *MAG-12*, 948 (1976).
44. H. G. Wagner, M. Ghafari, U. Gonser, and M. Naka, *J. Phys.*, *41*, C8-199 (1980).
45. N. A. Blum, K. Moorjani, F. G. Satkiewicz, and T. O. Poehler, *J. Appl. Phys.*, *52*, 1808 (1981).
46. N. A. Blum, K. Moorjani, F. G. Satkiewicz, and T. O. Poehler, *J. Appl. Phys.*, *53*, 2074 (1982).
47. C. L. Chien and K. M. Unruh, *Phys. Rev.*, *B24*, 1556 (1981); *B25*, 5790 (1982).
48. C. C. Tsuei, G. Longworth, and S. C. H. Lin, *Phys. Rev.*, *170*, 603 (1968).
49. D. J. Webb, S. M. Bhagat, K. Moorjani, T. O. Poehler, and F. G. Satkiewicz, *Solid State Commun.*, *43*, 239 (1982); *IEEE Trans. Mag.*, *MAG-19*, 1892 (1983).
50. T. Mizoguchi, *AIP Conf. Proc.*, *34*, 286 (1976).

51. K. Raj, J. I. Budnick, R. S. Alben, G. C. Chi, and G. S. Cargill II, in *Structure and Excitation of Amorphous Solids*, edited by G. Lucovsky and F. L. Galeener (AIP, New York, 1976), p. 390.
52. J. Durand and M. F. Lapierre, *J. Phys., F6*, 1185 (1976).
53. A. B. Harris, *J. Phys., C7*, 1671 (1974).
54. K. Handrich, *Phys. Stat. Solidi, 32*, K55 (1969).
55. H. A. Bethe, *Proc. Roy. Soc. London, Ser. A, 105*, 552 (1935).
56. P. R. Weiss, *Phys. Rev., 74*, 1493 (1948).
57. K. Moorjani and S. K. Ghatak, *J. Phys., C10*, 1027 (1977).
58. R. W. Cochrane and G. S. Cargill II, *Phys. Rev. Lett., 32*, 476 (1974).
59. C. L. Chien, *Phys. Rev., B18*, 1003 (1978).
60. R. J. Birgeneau, J. A. Tarvin, G. Shirane, E. M. Gyorgy, R. C. Sherwood, H. S. Chen, and C. L. Chien, *Phys. Rev., B128*, 2192 (1978).
61. J. D. Axe, G. Shirane, T. Mizoguchi, and K. Yamauchi, *Phys. Rev., B15*, 2763 (1977).
62. H. A. Mook, N. Wakabayashi, and D. Pan, *Phys. Rev. Lett., 34*, 1029 (1975).
63. J. W. Lynn, G. Shirane, R. J. Birgeneau, and H. S. Chen, *AIP Conf. Proc., 34*, 31 (1976).
64. F. Keffer, *Handbuch der Physik*, Vol. 18, Part 2, edited by S. Flugge (Springer-Verlag, Berlin, New York, 1966).
65. T. Mizoguchi, N. Ueda, K. Yamauchi, and H. Miyajima, *J. Phys. Soc. Japan, 34*, 1691 (1973).
66. T. Mizoguchi, K. Yamauchi, and H. Miyajima, *J. Phys., 5*, C4-287 (1974).
67. K. Yamada, Y. Ishikawa, Y. Endoh, and T. Masumoto, *Solid State Commun., 16*, 1335 (1975).
68. S. N. Kaul and M. Rosenberg, *Phil. Mag., B44*, 357 (1981).
69. S. N. Kaul, *Phys. Rev., B23*, 1205 (1981).
70. S. N. Kaul, *Phys. Rev., B22*, 278 (1980).
71. J. A. Tarvin, G. Shirane, R. J. Birgenau, and H. S. Chen, *Phys. Rev., B17*, 241 (1978).
72. L. P. Kadanoff, W. Gotze, D. Hamblen, R. Hecht, E. A. S. Lewis, V. V. Palciauskas, M. Rayl, J. Swift, D. Aspnes, and J. Kane, *Rev. Mod. Phys., 39*, 395 (1967).
73. T. Masumoto, S. Ohnuma, K. Shirakawa, M. Nose, and K. Kobayashi, *J. Phys., 41*, C8-686 (1980).
74. Y. Obi, L. C. Wang, R. Motsay, D. G. Onn, and M. Nose, *J. Appl. Phys., 53*, 2304 (1982).
75. K. Fukamichi, R. J. Gambino, and T. R. McGuire, *J. Appl. Phys., 53*, 2310 (1982).
76. Y. Shimada and H. Kojima, *J. Appl. Phys., 47*, 4156 (1976).
77. N. Heiman and N. Kazama, *Phys. Rev., B19*, 1623 (1979); *B17*, 2215 (1978).
78. J. M. D. Coey, D. Givord, A. Lienard, and J. P. Rebouillat, *J. Phys., F11*, 2707 (1981).
79. D. Giboord, F. Givord, and R. Lemaire, *J. Phys., 32*, C1-668 (1971).
80. D. GigNoux, D. Givord, and A. Lienard, *J. Appl. Phys. , 53*, 2321 (1982).
81. J. J. Croat and J. F. Herbst, *J. Appl. Phys., 53*, 2294 (1982).
82. N. Heiman and N. Kazama, *Phys. Rev., B19*, 1623 (1979).
83. J. M. D. Coey, D. Givord, A. Lienard, and J. P. Rebouillat, *J. Phys., F11*, 2707 (1981).
84. D. W. Forrester, N. C. Koon, J. H. Schelleng, and J. J. Rhyne, *J. Appl. Phys., 50*, 7336 (1979).
85. J. Chappert, J. M. D. Coey, A. Lienard, and J. P. Revouillat, *J. Phys., F11*, 2727 (1981).
86. P. C. M. Gubbens, J. H. F. Appeldorn, A. M. van der Kraan, and K. H. J. Bushow, *J. Phys., F4*, 921 (1974).

87. W. Kummerle and U. Gradmann, *Solid State Commun.*, *24*, 33 (1977).
88. K. H. Buschow and A. M. van Diepen, *Solid State Commun.*, *19*, 79 (1976).
89. J. W. M. Biesterbos, M. Brouha, and A. G. Dirks, *AIP Conf. Proc.*, *29*, 184 (1976).
90. J. M. D. Coey, D. H. Ryan, D. Gignoux, A. Lienard, and J. P. Rebouillat, *J. Appl. Phys.*, *53*, 7804 (1982).
91. N. Heiman and N/Kazama, *Phys. Rev.*, *B17*, 2215 (1978).
92. R. Lemaire, *Cobalt*, *33*, 201 (1966).
93. D. Gignoux, R. Lemaire, and P. Matho, *J. Magn. Magn. Mater.*, *21*, 119 (1980).
94. K. H. J. Buschow, M. Brouha, J. W. M. Biesterbos, and A. G. Dirks, *Physica*, *91B*, 261 (1977).
95. K. H. J. Buschow, *J. Appl. Phys.*, *53*, 7713 (1982).
96. V. Jaccarino and L. R. Walker, *Phys. Rev. Lett.*, *15*, 258 (1965).
97. M. Cyrot and M. Lavagna, *J. Phys.*, *40*, 763 (1979).
98. L. J. Tao, S. Kirkpatrick, R. J. Gambino, and J. Cuomo, *Solid State Commun.*, *13*, 1491 (1973).
99. A. Lienard and J. P. Rebouillat, *J. Appl Phys.*, *49*, 1680 (1978).
100. K. H. J. Buschow, *Rep. Prog. Phys.*, *40*, 1179 (1977).
101. I. A. Campbell, *J. Phys.*, *F2*, L47 (1972).
102. K. H. J. Buschow, *J. Less Comm. Met.*, *66*, 89 (1979).
103. R. Asomoza, I. A. Campbell, A. Fert, A. Lienard, and J. P. Rebouillat, *J. Phys.*, *F9*, 349 (1979).
104. K. Lee and N. Heiman, *AIP Conf. Proc.*, *24*, 108 (1975).
105. L. J. Tao, R. J. Gambino, S. Kirkpatrick, J. J. Cuomo, and H. Libensthal, *AIP Conf. Proc.*, *18*, 641 (1974).
106. R. C. Taylor and A. Gangulee, *J. Appl. Phys.*, *48*, 358 (1977).
107. T. R. McGuire and R. J. Gambino, *IEEE Trans. Magn.*, *MAG-14*, 838 (1978).
108. A. Gangulee and R. C. Taylor, *J. Appl. Phys.*, *49*, 1762 (1978).
109. A. Gangulee and R. J. Kobliska, *J. Appl. Phys.*, *49*, 4896 (1978).
110. R. Hasegawa, *J. Appl. Phys.*, *45*, 3109 (1974).
111. J. J. Croat, *J. Appl. Phys.*, *52*, 2509 (1982).
112. R. C. Taylor, T. R. McGuire, J. M. D. Coey, and A. Gangulee, *J. Appl. Phys.*, *49*, 2885 (1978).
113. H. A. Alperin, W. R. Gillmor, S. J. Pickart, and J. J. Rhyne, *J. Appl. Phys.*, *50*, 1958 (1979).
114. J. J. Croat, *IEEE Trans. Magn.*, *MAG-18*, 1442 (1982).
115. J. J. Croat, *J. Magn. Magn. Mater.*, *24*, 125 (1981).
116. S. G. Cornelison, D. J. Sellmyer, J. G. Zhao, and Z. D. Chen, *J. Appl. Phys.*, *53*, 2330 (1982).
117. G. C. Hadjipanayis, S. H. Wollins, R. C. Hazelton, K. R. Lawless, R. Prestikino, and D. J. Sellmyer, *J. Appl. Phys.*, *53*, 7780 (1982).
118. J. J. Croat, *IEEE Trans. Magn.*, *MAG-18*, 1442 (1982).
119. K. H. J. Buschow and A. M. van der Kraan, *J. Magn. Magn. Mater.*, *22*, 220 (1981).
120. N. Heiman, K. Lee, and R. I. Potter, *AIP Conf. Proc.*, *29*, 130 (1976).
121. J. J. Rhyne, *AIP Conf. Proc.*, *29*, 128 (1976).
122. K. H. J. Buscho and A. M. van der Kraan, *J. Magn. Magn. Mater.*, *22*, 220 (1981).
123. M. A. Alperin, J. R. Cullen, A. E. Clark, and E. Callen, *Physica*, *86–88B*, 767 (1977).
124. J. R. Cullen, G. Blessing, S. Rinaldi, and E. Callen, *J. Magn. Magn. Mater.*, *7*, 160 (1978).
125. G. Hadjipanayis, S. G. Cornelison, J. A. Gerber, and D. J. Sellmyer, *J. Magn. Magn. Mater.*, *21*, 101 (1980).
126. J. J. Rhyne, J. H. Schlleng, and N. C. Koon, *Phys. Rev.*, *B10*, 4672 (1974).

127. J. Chappert, L. Asch, M. Boge, G. M. Kalvius, and B. Boucher, *J. Magn. Magn. Mater.*, *28*, 124 (1982).
128. J. M. Coey, J. Chappert, J. P. Rebouillat, and T. S. Wang, *Phys. Rev. Lett.*, *36*, 1060 (1976).
129. K. Lee and N. Heiman, *AIP Conf. Proc.*, *24*, 108 (1975).
130. H. Jouve, J. P. Rebouillat, and R. Meyer, *AIP Conf. Proc.*, *29*, 97 (1976).
131. R. Arrese-Boggiano, J. Chappert, J. M. D. Coey, A. Lienard, and J. P. Rebouillat, *J. Phys.*, *37*, C6-771 (1976).
132. K. H. J. Buschow and A. M. van der Kraan, *Phys. Stat. Solidi*, *a53*, 665 (1979).
133. B. Boucher, A. Lienard, J. P. Rebouillat, and J. Schweizer, *J. Phys.*, *F9*, 1421 (1979).
134. B. Boucher, A. Lienard, J. P. Rebouillat, and J. Schweizer, *J. Phys.*, *F9*, 1433 (1976).
135. S. J. Pickart, J. J. Rhyne, and H. A. Alperin, *Phys. Rev. Lett.*, *33*, 424 (1974).
136. H. A. Alperin, W. R. Gillmor, S. J. Pickard, and J. J. Rhyne, *J. Appl. Phys.*, *50*, 1958 (1979).
137. S. J. Pickart, J. J. Rhyne, and H. A. Alperin, *AIP Conf. Proc.*, *24*, 177 (1975).
138. S. J. Pickart, in Ref. B2, p. 479.
139. S. J. Pickart, H. A. Alperin, and J. J. Rhyne, *Phys. Lett.*, *64a*, 337 (1977).
140. J. J. Rhyne, S. J. Pickart, and A. H. Alperin, *J. Appl. Phys.*, *49*, 1691 (1978).
141. A. R. Edmonds, *Angular Momentum in Quantum Mechanics* (Princeton, 1957).
142. R. Harris, M. Plischke, and M. J. Zuckermann, *Phys. Rev. Lett.*, *31*, 160 (1973).
143. R. W. Cochrane, R. Harris, and M. Plischke, *J. Non.-Cryst. Solids*, *15*, 239 (1974).
144. G. Czjzek, J. Fink, F. Gotz, M. Schmidt, J. M. D. Coey, J. P. Rebouillat, and A. Lienard, *Phys. Rev.*, *B23*, 2513 (1981).
145. J. P. Rebouillat, A. Lienard, J. M. D. Coey, R. Arrese-Boggiano, and J. Chappert, *Physica*, *86–88B*, 773 (1977).
146. R. Ferrer, R. Harris, D. Zobin, and M. J. Zuckermann, *Phys. Rev. Lett.*, *31*, 160 (1973).
147. R. Harris, M. Plischke, and M. J. Zuckermann, *Phys. Rev. Lett.*, *31*, 160 (1973).
148. E. R. Callen, Y. J. Liu, and J. R. Cullen, *Phys. Rev.*, *B16*, 263 (1977).
149. S. Hatta, T. Mizoguchi, and N. Watanabe, in *Rapidly Quenched Metals*, Vol. 1 (Elsevier, New York, 1984); T. Mizoguchi, *IEEE Trans. Mag.*, *TJMJ-1*, 139 (1985).
150. J. D. Patterson, G. R. Gruzalski, and D. J. Sellmyer, *Phys. Rev.*, *B18*, 1377 (1978).
151. M. C. Chi and T. Egami, *J. Appl. Phys.*, *50*, 1651 (1979).
152. C. Jayaprakash and S. Kirkpatrick, *Phys. Rev.*, *B21*, 4072 (1980).
153. M. C. Chi and R. Alben, *J. Appl. Phys.*, *48*, 2987 (1977).
154. R. Harris and S. Sung, *J. Phys.*, *F8*, L299 (1978).
155. R. Harris and D. Zobin, *J. Phys.*, *F7*, 337 (1977).
156. R. Harris, *J. Phys.*, *F10*, 2545 (1980).
157. R. Alben, J. J. Becker, and M. C. Chi, *J. Appl. Phys.*, *49*, 1653 (1978).
158. P. Chaudhari, J. J. Cuomo, and R. J. Gambino, *IBM J. Res. Dev.*, *17*, 66 (1973).
159. T. Mizoguchi and G. S. Cargill III, *J. Appl. Phys.*, *50*, 3570 (1979).
160. P. Chaudhari and D. C. Cronemeyer, *AIP Conf. Proc.*, *29*, 113 (1976).
161. R K. Wangness, *Amer. J. Phys.*, *24*, 60 (1956).

17

Engineering Magnetic Properties

A. Hernando and M. Vázquez
Instituto de Magnetismo Aplicado, Madrid, Spain

I. MAGNETIC ORDER IN AMORPHOUS AND NANOCRYSTALLINE MATERIALS

A. Introduction

It is known that 3*d* metal-based amorphous alloys obtained by rapid quenching of the melt are excellent soft magnetic materials. The usual approach to the atomic structure of ordered magnetic solids assumes lattice periodicity. Such a framework idea is not applicable for metallic glasses, which are defined as solids in which the orientation of local symmetry axes fluctuate with a typical correlation length $l \approx$ 10 Å. Therefore, it has been and still it is an important challenge within the scientific knowledge of magnetic materials to unravel the fundamental physics that underlies their excellent technical properties.

There are some fundamental questions related to the existence of well-defined magnetic order in a material having structural disorder. In fact, if we consider ferromagnetic interactions of the magnetic moments we immediately think of a ferromagnetic structure. Magnetic anisotropy effects have been neglected in this quick impression. It is important to note that, in a single crystal, magnetic anisotropy is perfectly compatible with ferromagnetic order. Magnetic moments tend to arrange their orientations parallel to each other via exchange interactions; this they do when lying along a magnetic easy axis which is in the same direction at every point in the single crystal. However, if the easy-axis orientation fluctuates from site to site, a conflict between ferromagnetic coupling and anisotropy arises. As long as we imagine lattice periodicity, a ferromagnetic structure is a consequence of ferromagnetic exchange interactions, the strength of the anisotropy being irrelevant. But we are assuming a major simplification: the direction of the easy axis is uniform throughout the sample. This is a simple example of the questions related to the influence of an amorphous structure on magnetic order.

Presently, we know that magnetic order stems from two contributions, exchange and local anisotropy. The exchange arises from the electron-electron correlations. The mechanism of the electrostatic interactions between electrons has no relation to structural order, and is sensitive only to overlapping of the electron wave functions. Moreover, magnetic anisotropy also originates by the interaction of the local electrical field with spin orientation, through the spin-orbit coupling. Therefore, magnetic anisotropy is also a local concept. Nevertheless, the structural configuration of magnetic solids exerts an important influence on the macroscopic manifestation of the local anisotropy. As outlined above, in a single crystal there is a macroscopic anisotropy with a macroscopic easy axis oriented in the same direction as all the local axes. However, when the local axes fluctuate in orientation as a consequence of the structural fluctuation (amorphous structure for example), calculation of the resultant macroscopic anisotropy becomes a difficult task.

The following sections will show how the local fluctuations of the easy axes' orientation can result in ferromagnetic order with vanishing macroscopic anisotropy as well as in "frustration" of ferromagnetism in some cases. In fact, the correlation length of the magnetic moments, for a constant correlation length of orientation fluctuations of local easy axes, depends on the strength of the local anisotropy constant, D, relative to that of the exchange interactions. Hence, "frustration" of order is expected for high local anisotropy strength; in this case, the anisotropy overcomes the exchange, and the spins are oriented along the easy axes and therefore their orientations fluctuate with the same correlation length as that of the structure. When anisotropy is relatively weak, the exchange interactions dominate and ferromagnetism is achieved with correlation length, L, much larger than the structural correlation length, l. In this case the macroscopic or large-scale anisotropy undergoes a drastic lowering relative to the local anisotropy strength by a factor of $(l/L)^{3/2}$.

Since large-scale anisotropy is the key physical property in magnetic engineering, the considerations outlined in the previous paragraph outline the most relevant information for understanding the excellent soft magnetic properties of iron- and cobalt-based amorphous alloys. For a $3d$-based amorphous alloy, D ($\sim 10^5$ erg/cm^3) is a relatively low value in comparison to the strength of interaction between magnetic moments. As a consequence, these alloys are ferromagnetic on a large scale, $l \approx 1$ mm; macroscopic structural anisotropy is $10^{-9}D$ in other words 10^{-9} times the anisotropy constant of a single crystal.

With regard to influence on magnetic properties, the fluctuation of local electrical field orientation is the most relevant characteristic of an amorphous structure. This statement has been effectively demonstrated from observations about the properties of nanocrystalline ferromagnets. In these types of materials the magnetic softness is a dramatic function of the grain size, l. Consider the well-known alloy $Fe_{73.5}Cu_1Nb_3Si_{13.5}B_9$ [1], which can be rapidly quenched as an amorphous alloy ribbon. A subsequent heat treatment above its crystallization temperature produces an ultrafine, randomly oriented grain structure of α-Fe(Si) with grain sizes of $l \approx$ 10–20 nm. This grain size is remarkably less than that of grains usually obtained by crystallizing of metallic glasses ($l \approx 0.1$–1 μm). Crystals having grain size on the scale of nanometers are known as nanocrystals. The appearance of such small grains has been associated with the presence of Cu and Nb in Fe-based amorphous alloys. Cu helps nucleation of α-Fe(Si) grains whereas Nb limits grain growth. As annealing temperature increases, so does grain size, thereby causing the nanocrystalline structure to evolve towards a polycrystalline one. The ability to control the grain size (over a range of six orders of magnitude) by annealing at different temperatures has been successfully used to give evidence of the close relation between magnetic properties and microstructure.

In the amorphous state the initial permeability, μ_i, is 8000. After isochronal annealing of 1 hr at temperature, T_a, of 525°C, μ_i increases up to 10^5. This value does not change with T_a, ranging from 525 to 560°C. Above this range of T_a, μ_i varies continuously and drastically decreases to 10^2 for T_a = 610°C. It is worth noting that a relatively narrow range of T_a (85°C) gives rise to such remarkable variation in the magnetic softness.

The full explanation of the observed magnetic behavior can be summarized as follows. In the amorphous state there is no structural macroscopic anisotropy, however, internal stresses give rise to the appearance of magnetic anisotropy through magnetoelastic coupling, which is proportional to the saturation magnetostriction constant.

After annealing at 520°C, the onset of crystallization has been overcome and some nanocrystals coexist with the amorphous matrix. The grain size of the nanocrystals is smaller than the exchange correlation length and therefore there is not any structural macroscopic anisotropy contribution associated with them. However, the magnetostriction constant of the α-Fe(Si) crystallites is about -6×10^{-6}, whereas in the amorphous structure its value is $+20 \times 10^{-6}$. Hence, it is expected that for a ~0.23 volume fraction of the α-Fe (Si) the average magnetostriction should be very small, thus explaining the annealing temperature window for magnetic softness. For higher annealing temperatures the grain size increases, ultimately reaching the exchange correlation length. Consequently the effect of crystalline anisotropy becomes increasingly noticeable and the softness drastically drops.

B. Exchange and Local Anisotropy

Exchange and anisotropy are the properties determining the magnetic structure of materials. Several experiments have given evidence that amorphous alloys exhibit a well-defined short-range order in spite of the lack of translational periodicity of the atomic arrangement for distances above 10 Å. The local structure can be characterized by a few local configurations with icosahedral, octahedral, and trigonal symmetry. These structural units have randomly distributed orientations. The local magnetic anisotropy would be larger in those units with lower symmetry. If, for simplicity, we assume the local anisotropy to be uniaxial, the characteristic of the amorphous structure should be the orientational fluctuation of the easy axis. In general, amorphous nanocrystals or polycrystals, with local uniaxial anisotropy, are characterized by fluctuations of the orientation of local axis. What is different among these types of structures is the correlation length, l, of such fluctuation which is typically the correlation length of the structure and ranges from 10 Å (amorphous) to 10 nm (nanocrystals) to 1 mm (polycrystals).

Fluctuations in the interatomic distances associated with the amorphous structure should also contribute to some degree of randomness in the magnetic interactions of the magnetic moments. Nevertheless, such randomness is expected not to affect the magnetic behavior qualitatively [2]. On the other hand, *random distribution of the orientation of the easy axis drastically affects the magnetic properties*. Let us outline the general argument which points out how the correlation length, l, exerts a relevant influence on both magnetic structure and magnetic engineering properties.

Consider a system of spins coupled by positive exchange interactions, located in a solid structure in which the local symmetry axes fluctuate in direction with correlation length, l. If $\alpha(x)$ is the unit vector describing the direction of the magnetization it, should minimize the function

$$\varepsilon = \int d^3x \left[A\left(\frac{\partial \alpha}{\partial x}\right)^2 - D(\alpha n(x)) \right] \tag{1}$$

where $n(x)$ is a unit vector describing the local easy-axis direction which is randomly distributed in a range of dimension l; D, the local anisotropy strength is assumed to be everywhere uniform; A is the exchange stiffness parameter.

Chudnovski and Serota [3] have systematically analyzed the magnetic structures, or $\alpha(x)$ solutions, in topologically disordered systems by obtaining solutions of the differential equation found by minimizing ε in Eq. (1). As an example of this procedure let us consider a two-component random field, $n(x) = (\cos \zeta(x); \sin \zeta(x))$, where $\zeta(x)$ is a random scalar field. The magnetization should also have two components: $\cos \phi(x)$ and $\sin \phi(x)$; minimization of ε through Eq. (1) yields

$$A\left(\frac{\partial^2 \Phi}{\partial x^2}\right) = D \sin 2(\phi - \zeta) \tag{2}$$

The solution of Eq. (2) is

$$\phi(x) = \frac{D}{A} \int d^3x' \, G(x - x') \sin 2(\phi(x') - \zeta(x')) \tag{3}$$

G being Green's function.

By considering the correlation length, l, of the axes' orientation:

$$\langle \sin 2\zeta(x') \sin 2\zeta(x'') \rangle = \frac{1}{2} \exp\left(-\frac{|x' - x''|}{l} \right) \tag{4}$$

Eq. (3) leads to a correlation function of the magnetic moments: $C_d(|x|) = \langle (M(x) - M(0))^2 \rangle$, where $M(x) = M_s \alpha(x)$ and M_s is the saturation magnetization. The correlation length of the magnetic structure, L, can be defined as $|x|$, for which $C_d(|x|) \ll 1$.

Alben et al. [4] have shown that "random walk" considerations can be very useful in accounting for the influence of the random anisotropy on the magnetic structure. The important question is: What is the range of orientational correlation of the spins? If we assume $L \gg l$, the number of randomly oriented easy axes in a volume L^3 should be $N = (L/l)^3$. Let us call z the direction of the magnetization for which the total anisotropy energy is a minimum. If we call γ_i the cosine of the angle made by the ith axis and the overall easy axis (z axis), this minimum energy takes the value

$$F(l) = -D \sum \left(\gamma_i^2 - \frac{1}{3} \right) \tag{5}$$

After applying a saturating field along a direction perpendicular to z, labeled the y axis, the anisotropy energy becomes

$$F(F) = -D \sum \left(\beta_i^2 - \frac{1}{3} \right) \tag{6}$$

The experimental anisotropy constant will be obtained from the magnetization work as $D^* = F(F) - F(I)$, which after considering the equivalence of all directions perpendicular to the z axis ($\langle \beta_i^2 \rangle = 1/2 - \langle \gamma_i^2 \rangle$ can be written as

$$D^* = \frac{D \left(3 \sum [\gamma_i^2 - 1] \right)}{2N} \tag{7}$$

For a random distribution, the mean value $\langle \gamma^2 \rangle$ is 1/3 and the actual value $\Sigma\, \gamma_i^2$ deviates from the average $N/3$ as $\sqrt{N}$, according to the general rule of random walk. Therefore, we can write

$$\sum \gamma_i^2 = \frac{N}{3} + \alpha\sqrt{N} \tag{8}$$

For $N = 1$, $\gamma^2 = 1$; hence $\alpha = 2/3$.

By substituting the second term of Eq. (8) into Eq. (7) we finally obtain

$$D^* = \frac{D\sqrt{N}}{N} = \frac{D}{\sqrt{N}} \tag{9}$$

The problem then is reduced to evaluating the two terms of the local energy, given by the integrand in Eq. (1), after substituting D^* for D. The exchange term varies as A/L^2, whereas, according to Eq. (9), the anisotropy term can be written as $D(l/L)^{3/2}$. The following is obtained by minimizing the energy with respect to L;

$$L = \frac{16A^2}{9D^2l^3} \tag{10}$$

C. 3*d* and 4*f* Metal-Based Amorphous Alloys

If we consider $A = 10^{-11}$ J-m^{-1} and $l = 10^{-9}$ m, which are typical values of ferromagnetic metallic glasses [2], L in Eq. (10) becomes $10^5/D^2$. For 3*d*-based alloys we can take the value of D corresponding to crystalline samples ($\sim 10^4$ J-m^{-3}) while for 4*f* alloys it may reach 10^7 J-m^{-3}. The relation described in Eq. (6) shows that L is 10^{-3} m for 3*d* metallic glasses, which therefore exhibit long-range ferromagnetism. However for 4*f*-rich glasses the ferro- or antiferromagnetic coupling cannot give rise to ferro- or antiferromagnetic structures, respectively, because L attains an order of 10^{-9} m, which is equal to the structural correlation length of an amorphous material. The high value of the anisotropy in comparison to the exchange, which is a consequence of the large atomic number and thereby the large spin-orbit coupling of 4*f* metals, has been shown to be the cause of the "frustration" of the magnetic order in structurally disordered 4*f* compounds.

Notice that the "random anisotropy" affects not only the magnetic structure but also the magnetic properties of ferromagnetic structures. According to Eq. (5) and after considering the value of L given in Eq. (6) the following value for macroscopic anisotropy is obtained:

$$D^* = \frac{D^4 l^6}{A^3} \tag{11}$$

Equation (11) points out that the macroscopic structural anisotropy is negligible in $3d$ amorphous alloys ($D^* = 10^{-9}D$); this is a consequence of the averaging of several local easy axes which produces the reduction in magnitude. Since the correlation length of magnetic moments in $4f$ amorphous alloys is of the same order as the correlation length of the structure, the macroscopic anisotropy roughly coincides with the local anisotropy.

An example of the ideas outlined above has been recently published by Tejada et al. [5]. In this report, the dependence of the magnetic structure on both temperature and field has been determined in $Fe_{74}RE_6B_{20}$ rapidly solidified amorphous alloy ribbons, RE being a rare earth element. The magnetic behavior of these compounds at room temperature is similar to that observed in the well-known magnetically soft amorphous alloy $Fe_{80}B_{20}$, that is, low coercitivity, high susceptibility, and easy saturation at low applied field. The thermal variation of magnetization is illustrated in Fig. 1. It is seen that the magnetization shows a decrease as the temperature drops during cooling under low-field processes. This anomalous thermal dependence of the magnetization is explained by the presence of rare earth atoms diluted in the ferromagnetic matrix of $Fe_{80}B_{20}$. At high temperatures, on the other hand, the magnetic moments of the rare earth atoms are ferromagnetically (Ce, Nd) or ferrimagnetically (Gd, Dy) coupled to the iron atoms. As temperature is lowered anisotropy rises and becomes stronger than the exchange interactions at the atomic sites in which rare earth atoms are located. Hence a partial frustration of ferromagnetism takes place. The experimental work reported in [5] illustrates

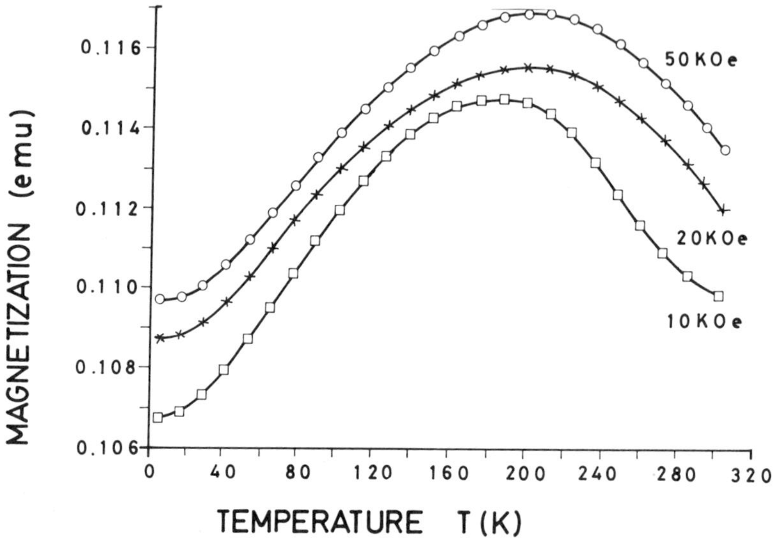

Figure 1 Temperature dependence of magnetization of amorphous $Fe_{74}Dy_6B_{20}$ alloy under various applied fields [5].

the influence of the ratio of exchange to anisotropy on the magnetic behavior of a structurally disordered system.

II. FERROMAGNETISM IN 3*d* METAL-BASED METALLIC GLASSES

A. Magnetic Properties Relevant for Engineering Applications

An important requirement for magnetic materials used in applications is that of high saturation magnetization. In general this property is limited in amorphous or nanocrystalline materials due mainly to the presence of nonmagnetic atoms. However, 3*d* metal-based alloys, rapidly quenched from the melt, have outstanding properties as soft magnetic materials [6], as well as field and stress-sensing elements [7]. Additionally, there are several examples of alloys with 4*f* elements mixed with 3*d* metals and metalloid atoms in suitable proportions which shows excellent hard magnetic properties when quenched at suitable cooling rates or after subsequent crystallization [8]. The relevant question we must try to answer is: What is the relation between the rapid solidification of these materials and their magnetic properties?

There are two important factors which help us to understand the influence of rapid solidification on magnetic properties. The first one has been thoroughly discussed above; that is the possibility of "tailoring" the correlation length of the orientational fluctuation of the easy axes. The second factor is closely related to the metastable character of solids obtained by rapid quenching from the melt. Metastability is associated with an anomalous high atomic mobility (high fictive temperature) which allows us to "tailor" macroscopic anisotropies by either structural relaxation or by annealing processes under a polarizing magnetic or stress field.

Random anisotropy affects the macroscopic anisotropy, as described in Eq. (11). Hence, 3*d* metal-based amorphous alloy or nanocrystalline materials with grain size smaller than 200 Å do not show any structural macroscopic anisotropy, as indicated in Fig. 2 [1]. As the strength of macroscopic anisotropy is the property governing the softness or hardness of a magnetic material it can be said that rapid solidification allows control of the most important property of engineering magnetism through the control of the structural correlation length. Note that the lack of long-range structural magnetic anisotropy does not imply the lack of long-range or macroscopic anisotropy of various origins. For instance, shape magnetic anisotropy, which is related to neither composition nor atomic structure, can also be significant in rapidly solidified alloys. Besides the usual anisotropies, rapid solidification also generates long-range stress fields which are coupled to the magnetization via magnetostriction, so giving rise to magnetoelastic anisotropies. Such residual stresses can be partially relaxed by annealing.

The existence of magnetostriction in metallic glasses is an interesting physical feature. Since magnetostriction can be defined as the derivative of the anisotropy with respect to the applied stress, it might initially be expected to average out as does the anisotropy. However, magnetostriction does not average out and, furthermore, it is the physical property which mainly determines the classification of

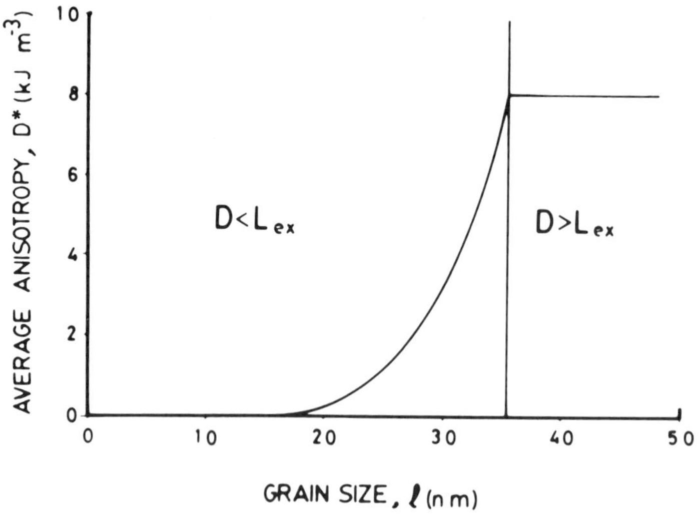

Figure 2 Theoretical estimate of the average anisotropy, D^*, for randomly oriented crystalline α-$Fe_{80}B_{20}$ as a function of grain size, l [1].

alloy compositions for various applications. The magnetic susceptibility of high magnetostrictive 3*d*-based metallic glasses ($\lambda = 10^{-5}$) is extremely sensitive to applied stress, with a figure of merit several orders of magnitude larger than that of the more sensitive conventional strain gages. The nonmagnetostrictive ferromagnetic metallic glasses ($\lambda \approx 10^{-7}$) are excellent soft magnetic materials, showing improvement in softness after suitable thermal treatment (stress relief).

The second important aspect of the rapid solidification related to the magnetic applications is the high atomic mobility in the metastable structures. While it is possible to achieve stress relaxation at relatively low temperatures, thus increasing the magnetic softness of the material, it is also possible to induce homogeneous macroscopic anisotropies by applying field or stress during annealing. Thermal treatment affects magnetic properties as a consequence of the structural changes produced within the amorphous state. Drastic changes in magnetic behavior can appear at higher annealing temperatures and from the onset of crystallization. It is also well established that many physical properties which can undergo evolution after thermal treatments can also be controlled by cooling rate during solidification [9]. We will deal with these concepts in more detail in Sections III and IV.

B. Curie Temperature, Saturation Magnetization, and Saturation Magnetostriction

Three main intrinsic magnetic parameters, namely Curie temperature, saturation magnetization, and saturation magnetostriction determine most of the engineering applications of metallic glasses. The higher the Curie temperature, the wider becomes the range of working temperature. A large saturation magnetization implies a large magnetic response to external agents such as magnetic and stress fields. Finally, while large magnetostriction is desired for magnetoelastic sensing applications, low magnetostrictive alloys exhibit almost no hysteretic losses. Furthermore, magnetic anisotropy also plays an important role in defining engineering properties. Nevertheless, as magnetocrystalline anisotropy has negligible values in metallic glasses, macroscopic anisotropies induced by applied stress and by thermal treatments become determinant.

Intrinsic magnetic properties depend on alloy composition. Most alloys range from 70 to 80 at% transition metal (TM) content (chiefly Fe, Co, and Ni, but also Cr, Cu, Mn, etc., in small amounts), while the rest consists of metalloids (M) as amorphizing elements (for example B, P, Si, Al, Ge, or C). Other interesting alloys also contain small additions of rare earth elements (Pr, Sm, Dy, Y, Lu, etc.). Basic magnetic properties as a function of composition have been reviewed in [10] and in Chapter 16 of this volume.

Magnetic moment per TM atom in many metallic glasses takes values similar to those of their crystalline counterpart. Nevertheless, it is found to be dependent on the quantity and type of TM and M atoms. In Co-based alloys, the magnetic moment increases toward that of crystalline Co as metalloid content decreases. Similarly, in Fe-rich alloys, it increases up to that of α-Fe as metalloid decreases. In the case of alloys containing Ni, magnetic moment decreases proportionally with the Ni content, suggesting that Ni atoms usually do not bear magnetic moment in metallic glasses. Interactions between TM and M atoms were evidenced by the variation of magnetic moment in metallic glasses with metalloid content [11]. Interpretation of this dependence first was in terms of charge transfer from metalloid elements to the transition metal *d* band. In particular, results performed with single TM-B and TM-P alloys led to consider the rigid-band model to explain the difference in magnetic moment between crystalline and amorphous alloys [12]. More recently, the reduction of the magnetic moment in metallic glasses has been interpreted by assuming bonding between metal and metalloid elements (*p*-*d* hybridation). The number of effective ferromagnetic *d* electrons and the magnetic moment arising from their balance are then reduced. In addition, the dependence of magnetic moment on metalloid content is explained by the extent of hybridation promoted by the particular kind of metalloid [13].

As regards the Curie temperature, in Co-base alloys Curie point decreases as the metalloid content increases. Moreover, it seems that exchange interactions between Co atoms becomes enhanced as the radius of neighboring metalloid atoms is smaller. Very different behavior is observed in some Fe-rich alloys, where Curie temperature decreases with decreasing metalloid content. This would indicate that the exchange interaction becomes weaker and even can change from ferromagnetic to antiferromagnetic as the distance between iron atoms becomes sufficiently small. This anomalous effect could be attributable to the difference between the SRO of the α-Fe and that of the Fe-rich metallic glasses.

Although magnetic anisotropy is macroscopically negligible, its stress derivative defines the magnetostriction coefficient, which depends on alloy composition. Saturation magnetostriction constant of Fe-base alloys exhibit large and positive values, while Co-rich alloys have negative magnetostriction. Vanishing magnetostriction is found for those alloys containing a ratio Fe to Co content of about 5%. The microscopic origin of magnetostriction is the same as that of magnetic anisotropy, and it can be understood by studying its temperature dependence. From such a study, it can be determined that single-ion anisotropy is the responsible mechanism for the magnetostriction of Fe-rich alloys. Deviations from this mechanism are found in case of Co-rich alloys in which a significant component corresponding to two-ion mechanism is presumed. In fact, nearly zero magnetostriction Co-rich alloys can exhibit a change of sign of the magnetostriction well below the Curie temperature. That has been interpreted by considering the opposite in sign contributions (a and b, respectively) arising from the two mentioned mechanisms. For these alloys, the temperature dependence of the magnetostriction can be related to that of the saturation magnetization, M_s, as

$$\lambda(T) = aM_s(T)^3 + bM_s(T)^2 \qquad (12)$$

From this functional relation, it is evident that a change of sign can appear below the Curie point [14,15].

C. General Aspects of the Magnetization Process

The lack of long-range periodicity, which characterizes an amorphous structure, gives rise to a vanishing magnetocrystalline anisotropy. Consequently, magnetoelastic and shape anisotropies are the main sources that determine the magnetization process in these kinds of materials [16]. Shape is not an intrinsic factor but a geometric one and can be conveniently chosen so that macroscopic demagnetizing field can be either neglected or have acceptable values. On the other hand, magnetoelastic anisotropy originates by intrinsic stresses coupled with the magnetostriction constant. Consequently, it depends on the strength and local arrangement of internal stresses and on the intensity of magnetoelastic coupling. While the magnetostriction value can be adequately controlled according to alloy composition, internal stresses are determined by the rapid quenching process and do not depend significantly on composition. These internal stresses are frozen-in during the quenching from the melt and their distribution is determined, among other things, by the local quench rate from point to point. In the case of ribbon-shaped metallic glasses with positive magnetostriction, the easy magnetization direction introduced by local magnetoelastic anisotropies changes almost continuously from point to point. Nevertheless, two main regions can be distinguished: one in which tensile stresses are predominant, so that, the easy axis can be considered to be in-plane with the ribbon and those regions where compressive stresses dominate and the easy magnetization direction is perpendicular to the ribbon plane. A simplified model considering a Gaussian distribution of internal stresses has been successful in the interpretation of the magnetic behavior of these samples [17]. Distribution of magnetoelastic anisotropy determines the as-quenched domain structure. Information about the distribution of easy axis can be inferred from to the domain patterns observed by magneto-optic Kerr effect and by the Bitter technique [18].

As an axial magnetic field is applied, the magnetization process takes place in two main steps. For low applied fields (typically up to tenths of an oersted), walls between domains where the main component of magnetization is in-plane move quickly, giving rise to a fast increase of the axial magnetization. In this range of applied fields, irreversible processes take place and susceptibility can reach values up to 10^5 while the coercive field can be as small as a few mOe. For applied fields up to some tens of Oe, magnetization within those regions where the easy axis is oblique to the ribbon axis rotates almost reversibly toward the field direction. For large applied fields, magnetization approaches to its saturation value.

While magnetic softness of these materials is a consequence of their slight spatial gradient of magnetic anisotropy, "structural defects" act as pinning centers for the wall displacements. Local gradients caused by compositional and topological fluctuations give rise to such pinning centers. Such centers primarily determine magnetic parameters of the hysteresis loop as coercive field, remanence, and the coefficients of the law of approach to saturation. Figure 3a shows the hysteresis loop of a particular alloy exhibiting typical as-cast sample magnetization behavior.

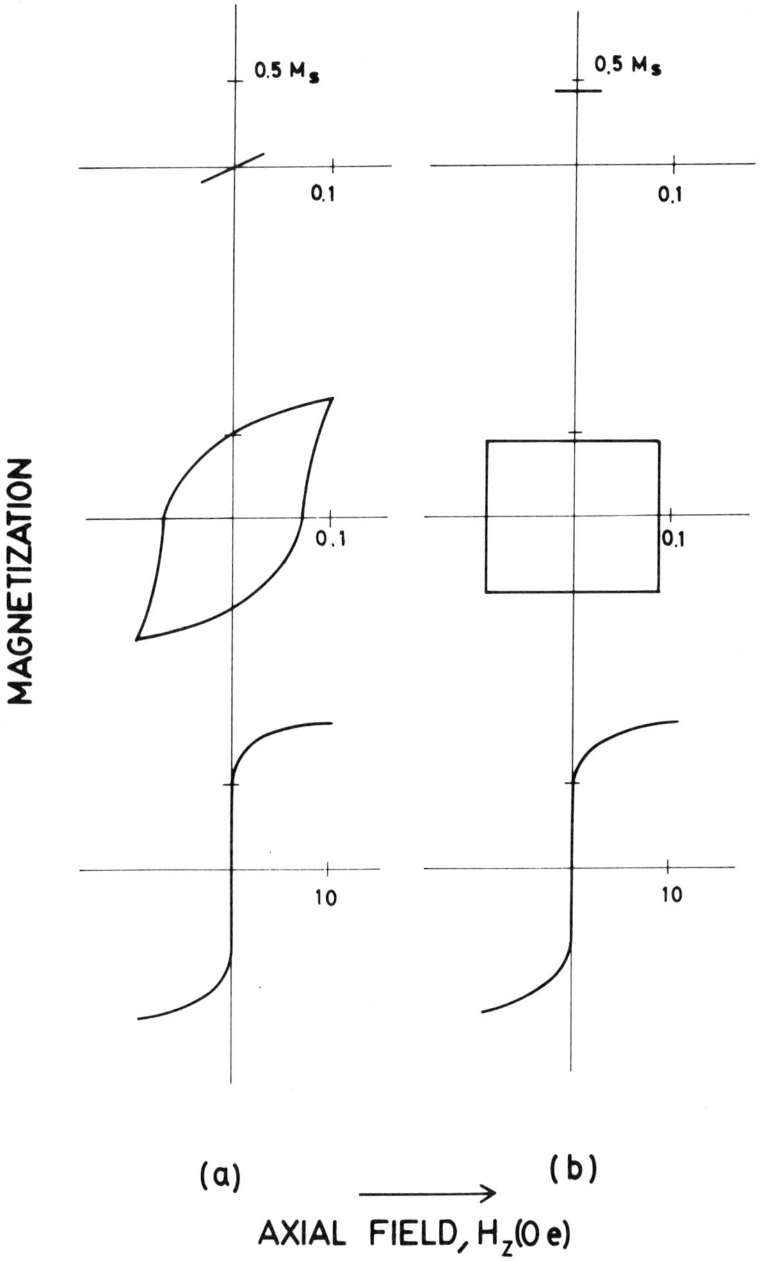

Figure 3 Axial hysteresis loops of $Fe_{40}Ni_{40}P_{14}B_6$ amorphous alloy ribbon (a) and $Fe_{77.5}Si_{7.5}B_{15}$ amorphous wire (b).

During recent years, the study of wire-shaped metallic glasses has been a topic of growing interest [19]. Wires exhibit quite outstanding magnetic behavior, particularly those of alloys with large magnetostriction, which makes them suitable for technological applications. The internal stress distribution produced during quenching is not yet well established. Nevertheless, two regions can be well defined from the easy magnetization point of view. At the internal inner core, magnetization is mainly axially oriented while at the outer shell, radial stresses originate a transverse easy axis [20,21]. Within the intermediate region as well as the wire surface, a closure domain structure should be formed to reduce local stray fields. Moreover, shear stresses are also present, as deduced from the domain observations and also from the existence of spontaneous Matteucci and inverse Wiedemann effects (see Section V).

As a consequence of the strong magnetic anisotropy at the inner core, magnetization within this region points along the axial direction. This gives rise to perfectly square loops. The switching field is determined by the field necessary to nucleate a reverse magnetization domain. Studies performed about this bistable behavior have led to the conclusion that the switching field is determined by the strength of the axial anisotropy (that is, equivalent internal stress) and by the demagnetizing field of the core (from the wire length dependence of the switching field). Figure 3b shows the typical hysteresis loop of a large magnetostriction wire.

III. ENGINEERING MAGNETIC PROPERTIES: STRUCTURAL RELAXATION AND SHORT-RANGE ORDER EFFECTS

A. Influence of the Structural Relaxation

As is known, as-quenched amorphous alloys have a metastable microstructure. Its relaxation towards more stable atomic configuration can be achieved by means of thermal treatments [16]. Thermal agitation during such treatments gives rise to slight displacements of atoms and a concurrent reduction of free volume. Such atomic rearrangements also reduce internal stresses to some extent. Thermal treatments in the presence of applied magnetic or stress fields can also induce macroscopic magnetic anisotropies, which will be discussed later. Extended annealing of the samples at elevated temperatures finally results in crystallization according to various mechanisms [22,23].

Let us focus here on the influence of relaxation of internal stresses on the macroscopic parameters of a hysteresis loop. A convenient technique to observe the influence of structural relaxation is to perform annealing treatments at temperature above the Curie point (when Curie point is below the crystallization temperature). In this way, superimposed magnetic effects are avoided. As structural relaxation proceeds, internal stresses are progressively reduced and a subsequent reduction of number and strength of pinning centers results as the material becomes magnetically softer.

Magnetization is roughly axially oriented after structural relaxation because of the shape anisotropy of ribbons. Domains with initial inplane axial anisotropy increase their volume at the expense of those in which magnetization was previously transversely oriented. In fact, this is equivalent to the induction of a macroscopic anisotropy along the axis of the ribbon. As a consequence of this, the magnetization process of fully relaxed samples consists predominantly of domain wall displacements, as described in the previous section. Once this process is complete, the final approach to saturation is completed by magnetization rotation in those local regions where residual stresses are still significant.

The degree of structural relaxation can be evaluated by studying the increase of remanence after different treatments. The remanence-to-saturation magnetization ratio can reach values close to unity for fully relaxed samples [24]. On the other hand, coercive field diminishes after treatments as a consequence of the reduced strength of pinning centers for domain wall movement. The final value of coercive field is determined by the pinning centers inherent to the amorphous structure itself, even in the relaxed state, and by local inhomogeneities at the ribbon surface [25,26]. Also, susceptibility is increased and its value can be considered as almost constant in the range of fields for which wall displacements take place. Finally, homogenization of the amorphous microstructure gives rise to a reduction of the coefficients of the law of approach to saturation [27]. These coefficients depend on the local inhomogeneities of the amorphous structure (i.e., density) and on the extent of structural "defects" (see Chapter 9 of this volume).

As regards the influence of structural relaxation in the case of amorphous wires exhibiting bistability, this phenomenon can disappear after full relaxation [28]. In fact, this is a consequence of the reduction of the axial anisotropy within the core since it arises from the magnetoelastic anisotropy coming from internal stresses. In this case, the propagation field becomes larger than the nucleation field so that reentrant flux reversal is lost.

Finally, let us comment on the effect of thermal treatments on a particular kind of Fe-rich metallic glasses containing small additions of elements such as Cu, Nb, Zr, etc. It has been recently reported that annealing treatments even above the crystallization temperature can give rise to a magnetic softening in such alloys [1]. In these alloys, small crystalline units with dimensions in the range of nanometers are segregated within an amorphous matrix. As the grain boundary is rich in atoms not soluble in the crystal phase, the increase of size of the nanocrystals is hindered. Magnetic softness of these nanocrystals arises from the competition between exchange interaction and strength of local anisotropy within each nanocrystal. In fact, a for grain size smaller than the exchange length, effective magnetocrystalline anisotropy is reduced via exchange interaction, as discussed in a previous section.

B. General Aspects of Short-Range Order and Structural Relaxation

As a consequence of the rapid quenching technology used for production, glasses are metastable not only with respect crystallization, but also with respect to structural relaxation within the amorphous phase. Atomic redistributions within the glass tend to gradually reach the "ideal" amorphous configuration. Many physical properties are sensitive to structural relaxation and change with time [29,30]. Although the rate of change is usually negligible at room temperature it increases at higher temperatures (below the crystallization temperature). Knowledge of the structural relaxation kinetics of glasses used for applications is important for two reasons: (1) to guarantee the stability of properties under specific working conditions and (2) to enhance the quality of the device by optimizing the value of the physical property upon which its performance is based. Properties which undergo changes after annealing include Curie temperature [31], permeability after effect or disaccommodation [32], macroscopic anistropy [33], and magnetostriction [34].

The dependence of a measured physical property on annealing time, t_a, and temperature, T_a, is a complex function of thermal history of the sample. The change of any property with t_a at a given T_a depends on a number of factors; also, different properties exhibit quite different sensitivity to structural relaxation.

The evolution of physical properties must be related to structural changes which take place within the amorphous configuration. Furthermore, these changes on an atomic scale could be viewed in terms of the presence of some kind of "structural defects" (Chapter 9 of this volume). Hence, study of the stuctural relaxation must be carried out to test concepts about atomic order and defects in amorphous materials.

Some kind of short-range order (SRO) seems to be present in the microstructure of metallic glasses. It is probably in the field of magnetic engineering properties in which the greatest indirect evidence about the existence of the SRO has been obtained. As is well known, uniaxial magnetic anisotropies are induced in metallic glasses when these alloys are annealed under the action of a magnetic field. For the case of transition metal-metalloid alloys with two types of metallic atoms, a mechanism of pair ordering has been invoked as the origin of the induced anisotropy. In amorphous alloys with only one type of metallic atom, the origin of the induced anisotropy has been explained as a consequence of the monoatomic directional order where a metalloid occupies interstitial positions [35]. However, there are some cases in which pair-ordering mechanisms have to be disregarded in order to account for the behavior observed. A study of the compositional dependence of stress induced anisotropy in (Co, Fe) and (Co, Fe, Ni) alloys has shown that a maximum anisotropy strength is achieved for the composition containing Co and Fe atoms in the proportion 3:1 [36]. Suran et al. [37] have studied the field induced anisotropy as a function of composition and deposition parameters of $Co_{1-x}Ti_x$ films obtained by sputtering. A compositional dependence which cannot be interpreted by a directional SRO mechanism was found. In these Co-rich compounds the induced anisotropies reach a maximum for annealing temperatures of about 600 K. It is worth noting that studies of structural relaxation monitored by differential scanning calorimetry, x-ray diffraction [38], magnetostriction [34], resistivity [39], and Young's modulus [14,40] have also shown a maximum of relaxation at temperatures close to 600 K.

O'Handley et al. have reported on the possibility of transforming reversibly from a fairly well defined local atomic order to another configuration in some metallic glasses [41]. Thermomagnetic hysteresis observed in magnetostriction and magnetic anisotropy in Co-rich glasses [42], suggested the existence of *first-order structural transformations*. More recently, Riveiro et al. [43] have detected reversible and nonreversible transformations at low temperatures. This reversible transformation exhibited characteristics of *martensitic transformations* typical of crystalline materials. All these results are evidence that the local structure can be characterized, at least partially, by well-defined local configurations. According to the picture given by Corb et al. [42], the amorphous structure of Co-rich glasses is assumed to be built up of clusters with icosahedral, octahedral, and trigonal symmetry. The low symmetry trigonal units would contribute largely to the magnetic anisotropy.

The transformations which occur at certain temperatures consist of a small distortion of a fraction of a type of clusters which takes the configuration corresponding to the other two types of clusters. By using these ideas, Suran et al. [37] have been able to explain the dependence of the induced anisotropy on the deposition parameters on $Co_{1-x}Ti_x$ films. Therefore, the topological symmetry of the stuctural units has been shown to be closely related to the induced anisotropy. However,

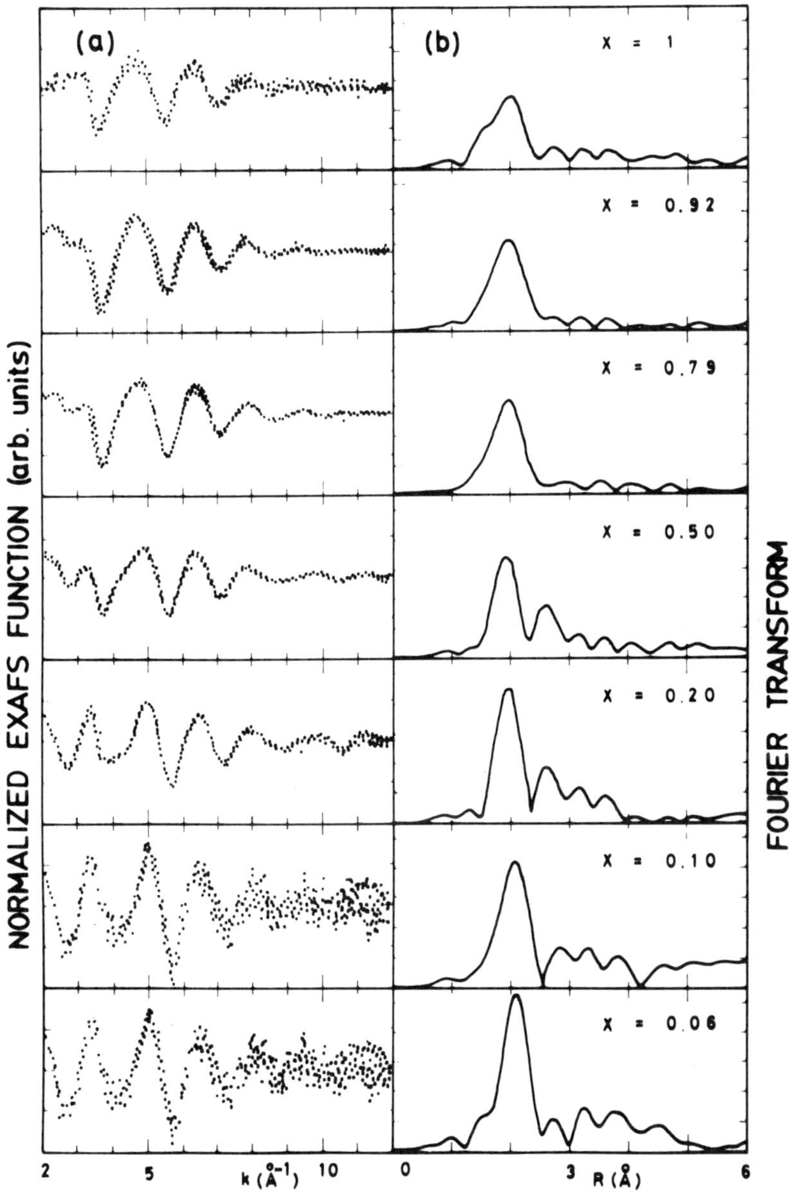

Figure 4 (a) Experimental normalized EXAFS functions, $\chi(k)$, of $(Fe_xCo_{1-x})_{75}Si_{15}B_{10}$ metallic glasses at the Fe *K*-edge. (b) Fourier transform of (a). The transformation range used was the same for the whole set of samples and extends from 2 to 12 $Å^{-1}$ in *k* space. Fe and Co environment are shown [44].

the compositional dependence observed for a large set of compositions indicates that chemical order, in addition to the topology, is the other component required to induce anisotropy. The way in which chemical and topological SRO are correlated remains to be elucidated through new experiments. As an example of such experiments, let us outline a recent observation carried out by using EXAFS spectroscopy on $(Co_{1-x}Fe_x)_{75}Si_{15}B_{10}$ alloys [44]. X-ray absorption spectra were recorded on Fe and Co K-edge using a Si (111) channel cut monochromator. The detection of the x-ray beam was achieved by two Ar-filled ionization chambers. Figures 4a and 4b show the spectra. It is clear that Fe and Co atomic enviroments are different in all samples studied, indicating different chemical short range order in the surroundings of both of these metallic atoms. Moreover, the Fe spectra show great changes as a function of composition. With decreasing Fe content a new oscillatory peak, at low k values, appears around $x = 0.5$. The radial distribution function around Fe shows that, for-Fe rich samples only, the first coordination shell gives an important contribution to the EXAFS spectrum. However for $x < 0.5$ a more ordered structure around Fe is observed, a four-shell contribution to the spectra being evident. In the Co environment, no special changes are observed with composition (Fig. 4). Fourier transformation of the EXAFS signals yields only a broad and nonsymmetric radial peak. This means that only the first-shell contribution is present, without any significant structure at high R values; this indicates a lack of longer-range order around Co in all compositions. Changes in metal-metalloid coordination occur as a function of composition. A preferred Fe-B and Co-Si coordination is suggested on the basis of the experiments. These results are in agreement with composition dependence studies on magnetostriction and induced anisotropy which already suggested changes of chemical SRO as a function of sample composition. Fluctuations of symmetry at magnetic atomic sites, such as those observed by EXAFS at Co and Fe sites, can explain such striking characteristics of Co-rich alloys as the stress dependence of magnetostriction, value of magnetostriction constant, and the capability of exhibiting a high induced anisotropy.

IV. TAILORING MAGNETIC PROPERTIES FOR APPLICATIONS

A. Field-Induced Magnetic Anisotropy

As described in Section II, as-quenched metallic glasses can be considered as macroscopically isotropic although macroscopic anisotropies arising from internal stresses change locally from region to region. Nevertheless, for some technological applications it is convenient for the sample to have a macroscopically well-defined domain structure. This leads to the consideration of inducing a macroscopic magnetic anisotropy with a predetermined easy magnetization direction [45]. Anisotropies can be induced by thermal treatment under convenient conditions. As it has been previously described, increasing the temperature of the sample gives rise to structural relaxation. However, when the annealing temperature is below the Curie temperature, interaction between magnetic moments and microstructural topology also favors the formation of atom pairs, the direction of which is related to the orientation of constituent magnetic moments. In fact, the easy magnetization direction inside each domain is thus reinforced by this locally induced anisotropy. However, such reinforcement can be somehow reduced by structural relaxation which leads to the reduction of internal stresses.

Macroscopic anisotropy can be then induced when a saturating magnetic field is applied during thermal treatment so that the domain structure consists of a single domain during that thermal treatment. Induced anisotropy after such a treatment is characterized by a preferred direction which is parallel to the field orientation during the annealing. The strength of the uniaxial induced anisotropy constant is a function of the time and temperature of treatment, and its kinetics follow the same trend as that of other quantities, as described in Chapter 10.

The origin of field-induced anisotropy has been ascribed to the generation of atom pairs aligned along a preferred direction determined by the magnetization orientation [46,47]. This follows from results reported by different authors performing various experiences [35]. As discussed above, the orientation of uniaxial structural units in alloys of some compositions have been invoked to account for the induced anisotropy [37]. Macroscopic anisotropies so-induced are reversible in the sense that subsequent thermal treatments above the Curie point destroy such induced anisotropies. Moreover, when further annealing in a magnetic field perpendicular to the previous one is performed, a new magnetic anisotropy is induced, the easy axis now being parallel to the that of the most recent field orientation. Strength of the field-induced anisotropy constant can reach values of a few hundreds of $J\text{-}m^{-3}$, depending on the time and temperature of annealing and on alloy composition. Also, this strength depends on the treatment history of the material. In fact, preannealing treatments lead to a reduction in the subsequent field-induced anisotropy strength as a consequence of the reduced atomic mobility arising from this previous treatment (reduction of the fictive temperature). Very homogeneous induced anisotropies of low strength are obtained when cooling down through the Curie temperature under an applied magnetic field [48]. Transverse induced anisotropies are useful for some applications, as will be shown in Section VI.

Regarding annealing techniques, conventional practice is to increase temperature by placing the sample inside an oven and applying a magnetic field by means of a solenoid, electromagnet, etc. Alternative techniques make use of an electric current to increase the sample temperature and eventually to produce magnetic field [49,50]. Flash annealing consists of a high density current pulse flowing through the sample so that, in spite of the instantaneously high annealing temperature, the short annealing time avoids crystallization. Current annealing technique uses the flow of lower-density electric currents to increase the sample temperature (usually below the Curie point) and to generate a transverse magnetic field, which in turn induces inhomogeneous transverse anisotropy. This field-induced inhomogeneity can be avoided by the action of a sufficiently strong superimposed homogeneous field or applied stress.

B. Stress-Induced Magnetic Anisotropy

Regarding magnetic anisotropies induced when thermal treatments are carried out in the presence of an applied stress, much more complex induction behavior has been reported. Mechanical properties of metallic glasses are quite outstanding at room temperature since, practically, they follow an elastic behavior up to the fracture. Nevertheless, as the experiments are performed at elevated temperature, viscoelastic deformations appear to be relevant. That leads to interesting magnetic consequences. Nielsen [51] has thoroughly studied the induction of anisotropies in Co-rich alloys when annealing at the presence of applied tensile stress [52,53]. Results can be summarized as follows. First of all, samples are usually preannealed above the Curie temperature in order to have a well-defined initial state. Subsequent annealing under applied stress leads to the induction of transverse magnetic anisotropy, the strength of which depends on annealing parameters (stress, time, and temperature of the annealing) as well as on alloy the composition. It is noteworthy that the easy magnetization direction induced is always transverse to the ribbon axis, irrespective of the sign of magnetostriction. Also, such stress-anneal-induced anisotropies can be induced by annealing even above the Curie temperature. These characteristics suggest that the source of the induced anisotropy is not related to magnetoelastic effects but rather to transformations of the microstructure probably connected to viscoplastic flow.

As further treatment is performed at the same annealing conditions but without applied stress, the resultant induced anisotropy changes the preferred direction toward that of the ribbon axis. This particular behavior has been interpreted by considering that the stress-induced anisotropy consists of two contributions, namely

$$K_{ind} = K_{an} + K_{pl} \tag{13}$$

The anelastic anisotropy, K_{an}, results in a transverse easy axis and is reversible while the plastic anisotropy, K_{pl}, results in an axial easy axis and is irreversible in the sense that it is mostly retained even if further treatments without applied stress are performed. When the first stress annealing is carried out, K_{an} predominates as its kinetics permits rapidly induced anisotropy, K_{pl} is induced more slowly as viscoplastic flow is being generated. After the second (stress-free) annealing, only plastic deformation and thereby plastic anisotropy remains. The strong influence of preannealing on the stress-induced anisotropy is notable. In general, preannealing favors the anelastic component to be outstanding.

The strength of stress-induced anisotropy can be much larger than in the case of field annealing, reaching values close to 10^3 J-m^{-3} for some Co-rich alloys. Fe-rich compounds exhibit much lower anisotropies, being comparable to the anisotropies induced magnetoelastically. Ni content seems to reduce strength of the induced anisotropies.

C. Stress-Field-Induced Magnetic Anisotropies

We will now summarize some recent results concerning the induction of magnetic anisotropies when thermal treatments are performed in the simultaneous presence of a saturating magnetic field and a tensile stress [54,55]. First of all, stress-field-induced anisotropies cannot be simply considered as the addition of induced anisotropies when sequential stress and field annealings are performed. In fact, induced anisotropy constants resulting form stress-field annealing can be significantly larger than that due to sequential single treatments. Maximum reported induced anisotropies reach values up to 1.5 kJ-m^{-3} in Co-rich alloys, while much lower induced anisotropies are reported for Fe-rich alloys. On the other hand, the preferred magnetization direction after these treatments strongly depends on alloy composition, being either parallel or transverse to the ribbon axis. Actual strength of the induced anisotropy depends also on the many treatment parameters as well as on preannealing conditions. Subsequent treatments carried out free of applied stress and field lead to remanence-induced anisotropies as in the case of single-stress annealing.

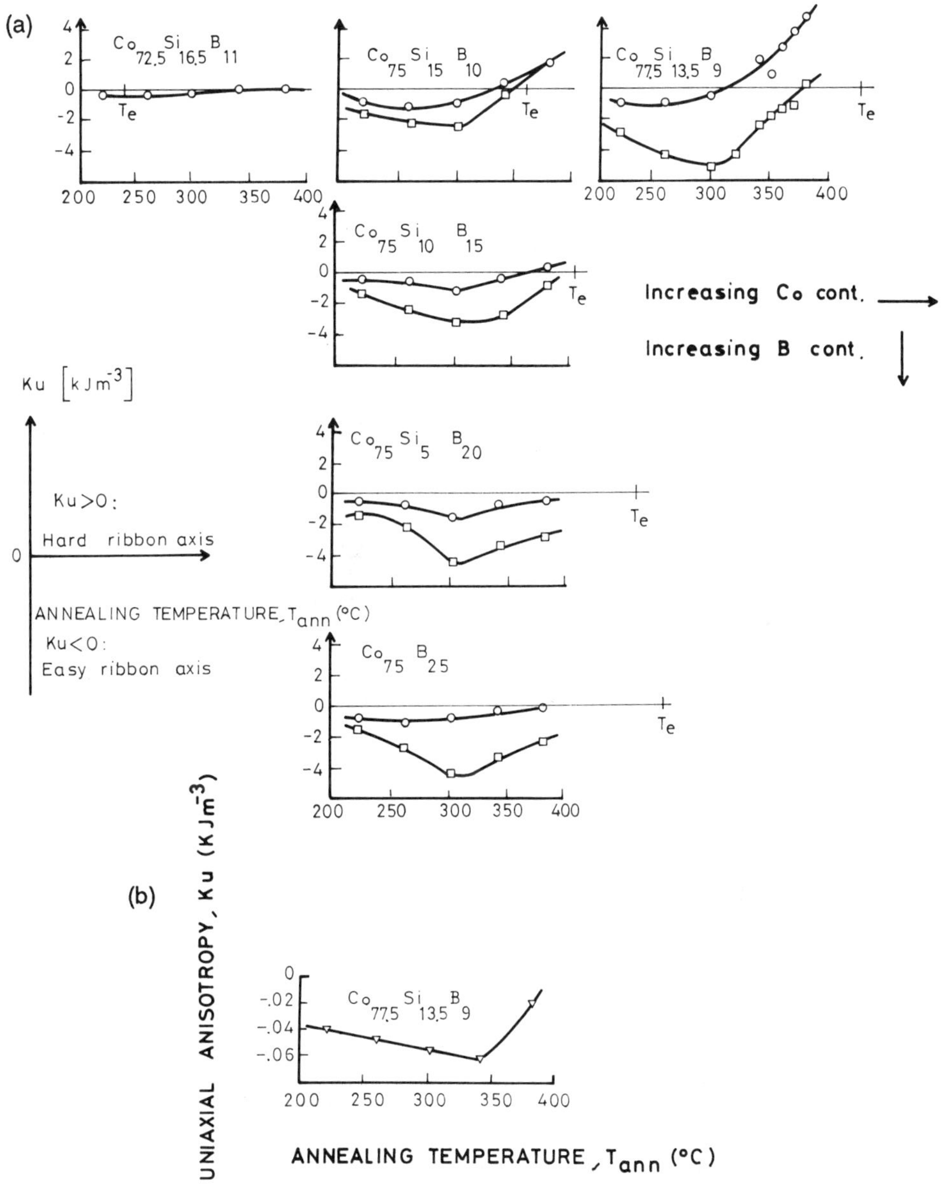

Figure 5 Induced magnetic anisotropy as a function of the annealing temperature for a series of Co-rich alloys. (a) Stress (○) and stress-field (□) annealing. (b) Field annealing for a particular composition. Annealing time was 30 min for each treatment; annealing stress was 615 MPa and axial annealing field was 95 Oe [54].

The complexity of these induced anisotropies mechanisms is evident. Nevertheless, some conclusions can be outlined. The effect of the applied stress during annealing is to modify short-range order, especially from the topological point of view. Such a modification can be important enough to give rise to viscoplastic deformations quenched in at room temperature after the treatments. Somehow, microstructural changes have effects equivalent to those of simply structural relaxation, described in Section III. In fact, these effects are superimposed on the formation of atom pairs or the directional order mechanism due to the magnetic ordering effect, which is prompted by the field and magnetoelastic contributions during annealing. Nevertheless, both effects are rather difficult to separate since they occur almost simultaneously; therefore, interactions between these effects must also be considered. For comparison, Fig. 5 shows three cases of induced magnetic anisotropies for the cases of field, stress, and stress-field annealings.

D. Dependence of Magnetostriction on Structural Relaxation

Magnetostriction constant depends first on the alloy composition. Figure 6 shows the dependence on composition of the magnetostriction constant at room temperature for Co-Fe and Co-Fe-Ni alloys and also for various metalloid contents. Magnetostriction constant also depends on temperature, although the experimental study of such a dependence has to be performed after preannealing at higher temperatures to avoid a structural relaxation contribution during the measurement. Figure 7 shows the temperature dependence of the magnetostriction for a Co-rich alloy which exhibits a compensation point. This anomalous behavior has been

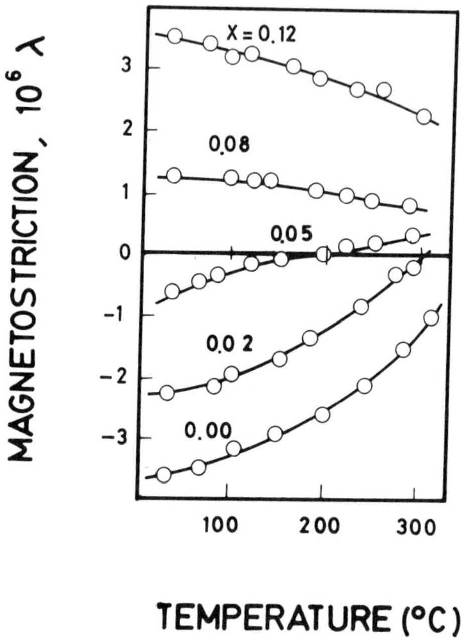

Figure 6 Temperature dependence of magnetostriction for the $(Fe_xCo_{1-x})_{75}Si_{15}B_{10}$ amorphous ribbon series [45].

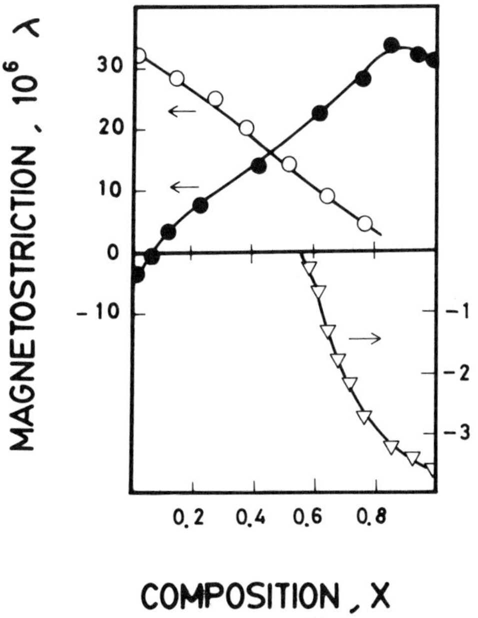

Figure 7 Composition dependence of magnetostriction for $(Fe_xCo_{1-x})_{80}B_{20}$ (●), $(Fe_xNi_{1-x})_{80}B_{20}$ (○), and $(Co_xNi_{1-x})_{75}Si_{15}B_{10}$ (▽) ribbons [45].

explained by assuming two contributions of opposite sign of magnetostriction and also different temperature dependences of these contributions. As commented previously, some Co-rich alloys have nearly zero magnetostriction at room temperature ($\sim10^{-7}$). Modification of the magnetostriction when the microstructure is somehow changed has been reported for these low magnetostriction alloys.

Change can be produced by the modification of sample fabrication procedure [9], for instance by changing the quenching rate. Such modifications can be used to show the dependence of magnetostriction on the local atomic arrangement only in so-called nearly zero magnetostriction alloys because magnetostriction changes are of the order of 10^{-7}. The dependence of magnetostriction on structural relaxation was conclusively evaluated in ribbon- and wire-shaped alloys after subjecting the samples to thermal treatments. It is noteworthy that structural relaxation moves the magnetostriction toward more positive values, so that a change of sign from negative to positive can also appear in some cases [34]. When annealing temperature reaches values of the order of 320–350°C, this trend is reversed and possibly related to transformations of the Co-rich short-range ordered units at such temperatures, as discussed previously in Section III. [41–43].

One question still remains open: whether the magnetostriction changes are due purely to structural relaxation or are somehow connected with directional ordering and giving rise, for instance, to field-induced anisotropies. Consensus on the parallelism between induced anisotropy and magnetostriction changes, or the lack thereof, does not exist. Although hard to distinguish due to overlapping effects, it seems that modification of the magnetostriction is proportional to induced anisotropy when stress and stress-field annealing are performed. A different trend is found in many cases for field annealing. Annealing with a magnetic field applied

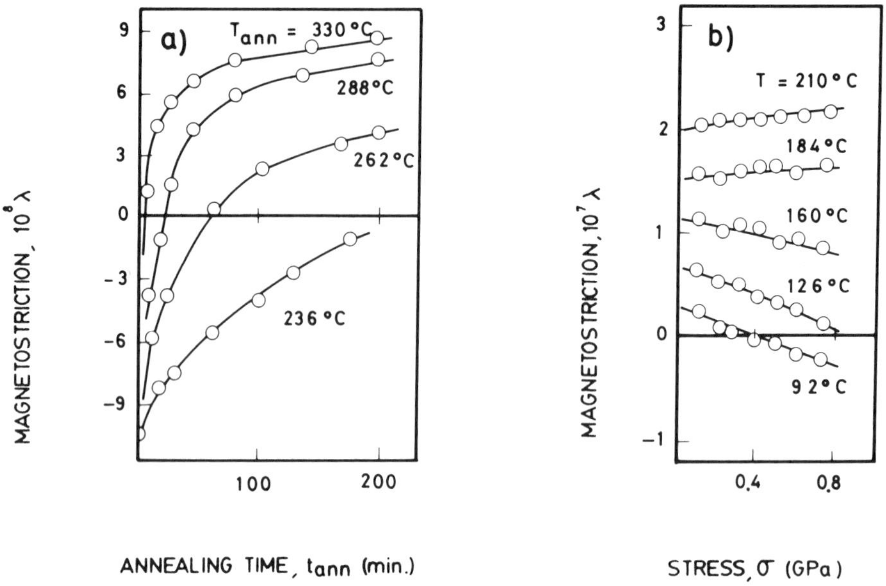

Figure 8 Dependence of magnetostriction on annealing parameters for the alloy $Co_{58}Fe_5Ni_{10}Si_{11}B_{16}$ (a) and on applied stress at different temperatures for the alloy $(Co_{0.94}Fe_{0.06})_{75}Si_{15}B_{10}$ (b) [34,57].

along different directions gives rise to some modifications of magnetostriction. Accordingly, the modification of the magnetostriction has to be connected with the structural relaxation or topological changes of short-range order. However, a different mechanism must be invoked for magnetostriction modifications in those Fe-rich alloys for which thermal treatments above the crystallization temperature result in a nanocrystalline structure. In this case, a drastic reduction of magnetostriction is observed as compared to the as-cast amorphous alloy. This reduction is due to the appearance of a nanocrystalline phase with negative magnetostriction constant which counterbalances the positive magnetostriction of the amorphous matrix [1]. It remains to be elucidated whether the coexistence of different amorphous phases within the amorphous structure could account for the general aspects of the magnetostriction behavior.

E. Applied Stress Dependence of Magnetostriction

Magnetostriction constant of low magnetostrictive Co-rich ribbon-and wire-shaped amorphous alloys depends on applied tensile stress. It follows the phenomenological law

$$\lambda(\sigma) = \lambda(0) - a\sigma \tag{14}$$

where $\lambda(0)$ is the extrapolated value at zero applied tensile stress. This linear behavior as well as the negative sign of the magnetostriction stress derivative ($a > 0$) is followed by all Co-rich compositions investigated. According to this law, magnetostriction can change sign with applied stress for alloys exhibiting a small positive magnetostriction. This effect has also been reported by the observation of the domain structure with the sample under applied stress. Moreover, while $\lambda(0)$ is altered by structural relaxation as commented previously, the a coefficient seems to be less dependent on these parameters [56–58].

Different mechanisms have been claimed to be the origin of such stress dependence of magnetostriction [59–61]. In particular, higher-order magnetoelastic effects as well as pair ordering or random anisotropy models can be considered to explain the observed behavior. It seems that although they are quite reasonable qualitatively, no data have ever been reported to be compared with experimental results. The origin for the negative stress derivative of the magnetostriction irrespective of the composition, λ value, and structural relaxation remain unclear. An alternative explanation, or just a superposition of the previous mechanisms, could be connected with the intrinsic fluctuations of the magnetostriction according to the amorphous microstructure, as could the addition of discrete different magnetostriction contributions from various amorphous phases [62]. Figure 8 shows the dependence of magnetostriction constant on structural relaxation and applied stress for some typical cases.

V. TORSION MAGNETOELASTIC EFFECTS

A. Influence of the Applied Torsion on the Magnetization Processes

Basic knowledge about the influence of applied torque to ribbon- or wire-shaped metallic glasses on the magnetic properties of these materials is very important because torsion is used in many technological applications [7]. Here, we will briefly review some aspects regarding this topic.

As is known, shear stresses introduced when a torque is applied on a rod are equivalent to the combination of tensile and compressive stresses of equal strength, mutually perpendicular, and making an angle of 45° with the rod axis. That produces easy magnetoelastic directions following a helical path (tensile or compressive lines according to the sign of the saturation magnetostriction constant). Hence, nondiagonal terms appear in the susceptibility tensor which can be written as

$$\begin{pmatrix} M_z \\ M_\phi \end{pmatrix} = \begin{pmatrix} \chi_{zz} & \chi_{z\phi} \\ \chi_{\phi z} & \chi_{\phi\phi} \end{pmatrix} \begin{pmatrix} H_z \\ H_\phi \end{pmatrix} \tag{15}$$

χ_{ij} is the susceptibility relating the ij component of magnetization to the magnetic field acting along the i direction. Z or ϕ mean axial or azimuthal direction, respectively.

This leads to a net axial magnetization, M_z, when applying an azimuthal field, H_ϕ (inverse Wiedemann effect, IWE) or alternatively, to a net azimuthal magnetization, M_ϕ, when applying a longitudinal field, H_z (Matteucci effect, ME) [63–66].

Let us consider the case of a ribbon-shaped sample with cross section $4ab$ (see Fig. 9a). If the ribbon is assumed to be infinitely wide, that is, $b/a \gg 1$, the complex stress distribution originated by torsion can be approximated considering only the following stress sensor terms:

$$\sigma_{yz}(x) = \mu\varphi x \tag{16a}$$

and

$$\sigma_{zz}(y) = \frac{1}{2} E\varphi^2\left(y^2 - \frac{b^2}{3}\right) \tag{16b}$$

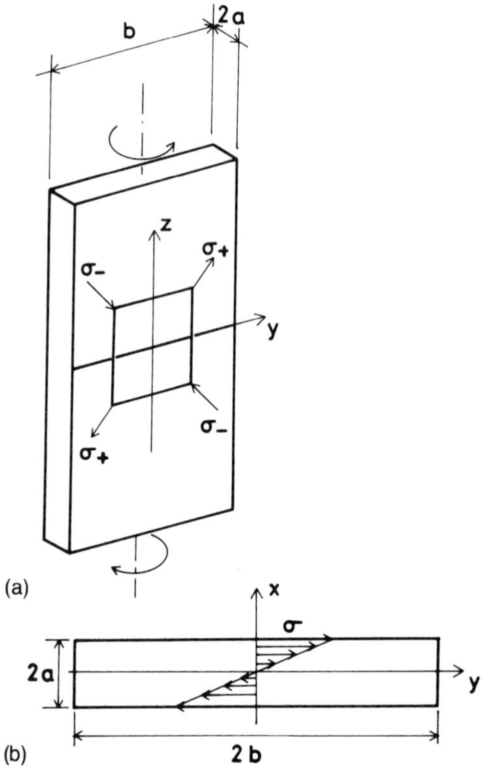

Figure 9 Schematic view of the stress distribution in a thin ribbon under torsion. (a) Tensile and compressive stresses acting on the surface element. (b) Projection of resulting stresses on the cross section of the ribbon.

where μ and E are the shear and Youngs modulus respectively, and φ is the torsion angle per unit length. The corresponding strain tensor terms are

$$e_{yz} = \varphi x \tag{17a}$$

and

$$e_{zz} = \frac{\varphi^2}{2}\left(y^2 - \frac{b^2}{3}\right) \tag{17b}$$

Note that shear stress, σ_{yz}, in Eq. (16a) increases linearly from the axis to the surface of the ribbon (Fig. 9b). On the other hand, tensile stress, σ_{zz}, introduced by torsion is a second-order term and can be neglected except for the case when either large torques are applied or in the presence of applied tensile stress.

Such stress distributions introduce elastic and magnetoelastic energy terms which general expressions for an isotropic medium are respectively

$$F_{\text{el}} = \left(\frac{\mu}{2}\right)(e_{xy}^2 + e_{yz}^2 + e_{zx}^2) + \left(\frac{E}{2}\right)e_{zz}^2 \tag{18a}$$

and

$$F_{\text{me}} = B(e_{xx}\alpha_1^2 + e_{yy}\alpha_2^2 + e_{zz}\alpha_3^2 + e_{xy}\alpha_1\alpha_2 + e_{xz}\alpha_1\alpha_3 + e_{yz}\alpha_2\alpha_3) \tag{18b}$$

where B is the magnetoelastic coefficient and α_i are the direction cosines defining the magnetization orientation.

In the first approximation, the second term on the right side of Eq. (18a) can be neglected while Eq. (18b) reduces to [67,68]

$$F_{\text{me}} = -\frac{3}{2}\lambda\mu\varphi x \sin 2\Theta \tag{19}$$

where Θ denotes the angle between magnetization and the axis of the ribbon and λ is the isotropic magnetostriction constant.

In the case where energy terms associated with torsion prevail, a kind of 90° wall centered at the $x = 0$ plane is formed. Its width is determined by the compromise between exchange energy and the strength of the applied torque, according to the equilibrium condition [63]

$$A\left(\frac{d^2\Theta}{dx^2}\right) + 3a^2\lambda\mu\varphi x \cos 2\Theta = 0 \tag{20}$$

where A is the exchange constant.

The magentization process of twisted samples is determined by the competition between applied torsion energies and Zeeman energy arising from the applied field. In general, anisotropy energy terms coming from internal stresses or induced anisotropies also have to be considered.

An azimuthal magnetic field is required when performing inverse Wiedemann effect experiments. Such a field can be obtained by making an electrical current flow along the sample. In the case of ribbon, the magnetic field originated by current, I, varies in a complicated way through the sample cross section. Nevertheless, it can be accurately described by considering only its component along the y direction according to

$$H_y(x) = \frac{Ix}{4ab} \tag{21}$$

Deviations from values given by Eq. (21) are only detectable in the neighborhood of the ribbon edges. This approximation is moreover favored by the low demagnetizing factors along the ribbon length (x direction). $H_y(x)$ is opposite in either half of the ribbon and, as the shear stress introduced by torsion, the strength of $H_y(x)$ increases linearly with the distance from the axis of the ribbon.

The Zeeman energy density when axial, H_z, and transverse, $H_y(x)$, magnetic fields are applied is then

$$F_H = -\mu_0 M_s(H_z \cos \Theta + H_y(x) \sin \Theta) \tag{22}$$

Experimentally, the influence of applied torsion on the magnetization curves is much more evident in the cases of cross effects (IWE and ME) than in the M_z-H_z curves. This is clearly a consequence of the lack of those effects in the absence of torsion.

In the case of IWE or ME, magnetization increases monotonically for a given applied field, with applied torsion levels sufficiently high to achieve helical saturation, for which $M = M_s \cos 45°$. While for low applied torque, the magnetization process takes place by domain wall displacement and magnetization rotation, the hysteresis loops become squared when the sample is helically saturated. In the latter case, bistable behavior is a consequence of the large nucleation field necessary to switch the magnetization into its opposite direction within the same helix. Figure 10 shows the IWE

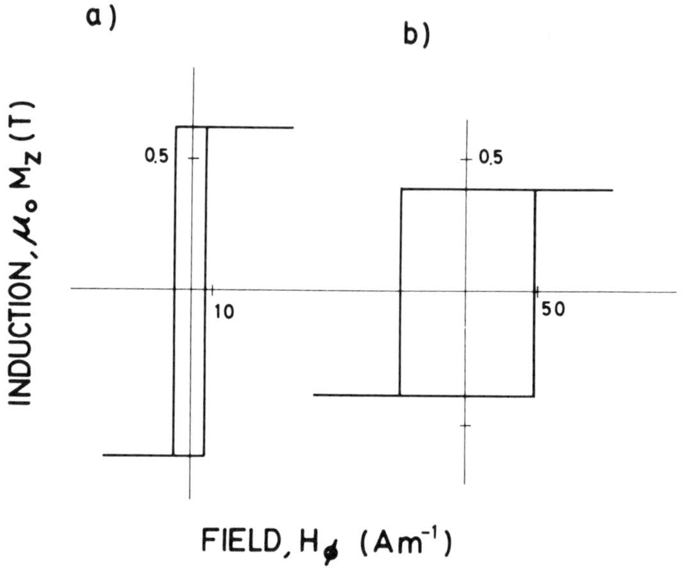

Figure 10 Inverse Wiedemann effect for the $Fe_{40}Ni_{40}P_{14}B_6$ ribbon under applied torsional strain ($\varphi = 12\pi$ rad m^{-1}) (a), and spontaneous IWE for $Fe_{77.5}Si_{7.5}B_{15}$ wire (b) [65].

loops of two samples with ribbon and wire samples for which bistability is observed [65,69,70].

The existence of spontaneous IWE or ME can be used to obtain information about the helical anisotropy of the sample. Such anisotropy can be macroscopically induced by a convenient annealing in the presence of an applied torque. In that case, the applied torsion which is necessary to cause the IWE or ME to disappear gives the strength of the induced helical anisotropy. On the other hand, shear stresses frozen-in during the fabrication process can also give rise to spontaneous IWE or ME without applying any torsion. From these spontaneous signals, information about distribution of internal stresses can be inferred. In fact, for example, spontaneous IWE in highly magnetostrictive amorphous wires has indicated the presence of such shear stresses in addition to the more familiar radial stress [69].

B. Discussion on a Method for Measuring Magnetostriction Using Torsion Effect

As in the axial case, the application of a helical magnetic field originates spontaneous deformation of the samples through magnetostrictive effects. In this case, magnetoelastic coupling leads to torsional strain (Wiedemann effect) which is proportional to the magnetostriction constant. Measurements of the angular deflection of the sample then allows the evaluation of the magnetostriction. This method for measuring magnetostriction has been developed because of its sensitivity to measure low λ_s values. Nevertheless, some experimental problems remain to be solved and also mathematical approaches have to be developed according to the actual experimental conditions [68].

Experimentally, this method implies the application of helical field and the simultaneous detection of the angular displacement. A helical component of magnetic field can be achieved by the simultaneous application of an axial field produced by a solenoid and a transverse field created by the flow of electric current through the sample. The sample is placed vertically along the common axis of the solenoid producing the axial field and of an oven heated with coaxial wire, which allows the study at different temperatures. As the sample must remain free to rotate, the flow of current causing the transverse field can be achieved by means of an electrolytic bath. While the sample is fixed by its upper end to a copper clamp, a thin-wall Al tube is clamped at its lower end. This tube is introduced into a conducting liquid solution through which the circuit is closed. A tiny mirror is cemented on the tube surface to allow the detection of angular displacement. Such detection is performed by measuring the displacement of a reflected laser beam on a calibrated screen. Figure 11 shows φ as a function of the current flowing through a ribbon at increased temperatures supplied by the oven.

The flow of current through the ribbon gives rise not only to the transverse field but it also increases the temperature of the sample. Consequently, such current must be kept below a certain intensity (usually 30–40 mA) so that the resulting increase of temperature can be dissipated. This limitation implies that the axial field is commonly much larger than the transverse field to obtain detected angular displacements.

For the isotropic case, when neither magnetoelastic anisotropy induced by internal stresses nor macroscopic induced anisotropies are considered, minimization

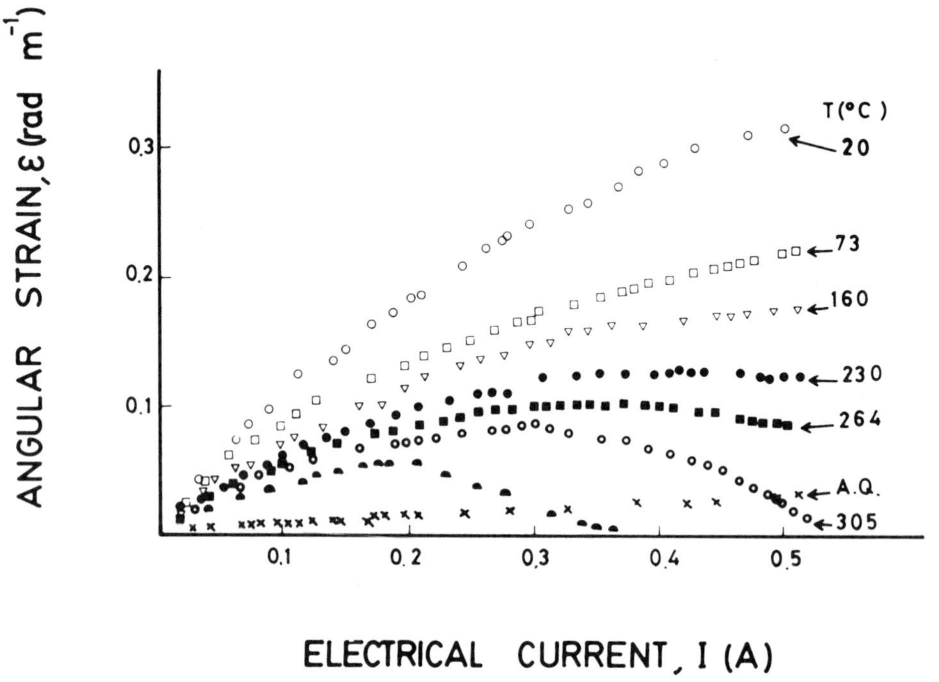

Figure 11 Angular displacement per unit length as a function of current flowing through as-cast and relaxed METGLAS alloy 2605SC amorphous alloy ribbon [68].

of Eqs. (18) and (22) leads to

$$\varphi = \frac{3\lambda H_y(x = a)}{aH_z} \tag{23}$$

where $H_z \gg H_y$ is taken into account.

Owing to experimental requirements, the ribbon is subjected to an applied tensile stress, σ_a, due to the tube hanging from the sample. While its weight can be modified, it cannot be reduced to zero. Introducing the tensile stress into the elastic and magnetoelastic energy terms and minimizing the involved energies leads to

$$\varphi = \frac{3\lambda H_y(x = a)}{aH_z(1 + \eta\sigma_a)(1 + \eta'\sigma_a)} \tag{24}$$

with $\eta = (2/3\sqrt{3})(b/a)^2$ and $\eta = \lambda/\mu_0 M_s H_z$.

A more realistic calculation would include the influence of internal stresses since these compete with magnetostrictive torsional stresses, resulting in reduced observed φ. This is clearly shown Figure 11, where comparison between as-cast and relaxed samples shows large difference in the detected φ values.

Introducing the parameters $\gamma = 3\lambda_s\sigma_a/\langle K\rangle a$ and $\gamma = \mu_0 M_s H_z/\langle K\rangle$, where $\langle K\rangle$ is the mean value of local anisotropy, a general expression for φ can be taken as

$$\varphi = \frac{3\lambda H_y(x = a)}{a(I + \eta\sigma_a)H_z} g(\gamma, \gamma') \tag{25}$$

g is a function accounting for the distribution of local anisotropies. According to Eqs. (23) and (24), this function verifies the limit conditions:

$$g(\gamma, \gamma') \approx 1 \qquad \text{for } \gamma \approx 0 \text{ and } \gamma' \gg 1$$

and

$$g(\gamma, \gamma') \approx \frac{1}{1 + \gamma/\gamma'} \qquad \text{for } \gamma \gg 1 \text{ and } \gamma' \gg 1$$

Saturation magnetostriction can be obtained as

$$\lambda_s = \frac{a}{3H_y(x = a)} \lim(\varphi H_z) \qquad \text{when } H_z \approx \infty \text{ and } \sigma_a \approx 0 \tag{26}$$

since the product γH_z tends to a saturation value for large H_z and vanishing σ_a. Although a Gaussian function seems to be adequate to account for the distribution of local anisotropies, that saturation value is independent of the chosen distribution of local anisotropies.

This method is particularly interesting when measuring samples with low magnetostriction. Its sensitivity can be observed in Fig. 12, in which the temperature dependence of the magnetostriction is shown for a nearly zero magnetostriction amorphous ribbon.

Owing to the sensitivity of the method, it has been used to measure the magnetostriction constant in the vicinity of the phase transition from ferro- to para-

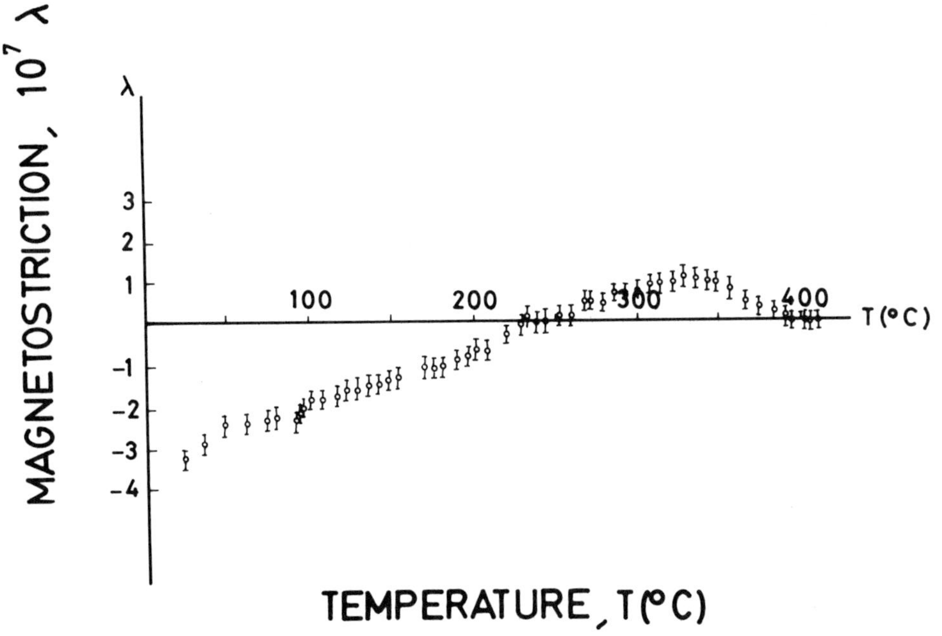

Figure 12 Temperature dependence of magnetostriction for the $(Fe_{0.05}Co_{0.95})_{75}Si_{15}B_{10}$ ribbon obtained through Wiedemann effect measurements [68].

magnetic states. These kinds of measurements were not previously carried out, probably because of experimental difficulties.

By analogy with other components, a critical exponent, κ, for the magnetostriction was introduced according to [71,72]

$$\lambda(T) \approx (T_c - T)^{\kappa} \tag{27}$$

This exponent was evaluated for a series of Fe-Ni- and Co-Ni-based amorphous alloys in the range of temperatures $0.05 < \varepsilon < 0.01$ ($\varepsilon = 1 - T/T_c$). The values obtained are within the range 1.0 to 1.1. If we compare these values with the critical exponent, β, of magnetization according to the Heisenberg model ($\beta = 0.36$), we obtain a factor 2.6–3.0 between κ and β. This indicates that the temperature dependences of λ and M_s follow a similar behavior at the phase transition range and at lower temperatures. A full theoretical interpretation remains to be done.

VI. ENGINEERING PROPERTIES OF HIGH-MAGNETOSTRICTION AMORPHOUS ALLOYS

A. Giant Magnetoelastic Effects

As discussed in Section 1, an interesting characteristic of ferromagnetic amorphous alloys is the lack of macroscopic structural anisotropy and the moderate values of the average magnetostriction constant. When the internal stresses are removed by means of thermal treatments macroscopic anisotropy vanishes and the sensitivity of the susceptibility to applied stress becomes enormous. Most of the applications of metallic glasses are based on inductance changes of a coil having the metallic glass as core element. A sensor or transducer can be defined as a device changing the magnitude of a given parameter or variable into another. Widely used sensors are based upon suitable processes and materials which give directly a voltage or an impedance change in response to a given stimulus. Ferromagnetic metallic glasses are used in these devices and changes in permeability produced by externally applied magnetic fields or mechanical stresses are the processes upon which these sensors are based. The large influence of magnetic fields and mechanical stresses in determining the magnetic permeability of metallic glasses is related to the amorphous structure. To show this, let us analyze the influence of the applied stress on the magnetic susceptibility.

Consider a ribbon of metallic glass in which a magnetic anisotropy, K, transverse to the sample axis has been previously induced. The dependence of the initial susceptibility, χ, on the applied tensile stress is described by the equation

$$\chi = \frac{\mu_0 M_s^2}{2K - 3\lambda_s\sigma} \tag{28}$$

where σ, λ_s, and $\mu_0 M_s$ are the applied stress, the saturation magnetostriction constant, and the saturation magnetization of the sample, respectively. The sensitivity of the susceptibility is then given by

$$\frac{d\chi}{d\sigma} = \frac{3\mu_0 M_s^2 \lambda_s}{(2K - 3\lambda_s\sigma)^2} \tag{29}$$

For $\sigma = 0$ the sensitivity becomes

$$\left(\frac{d\chi}{d\sigma}\right)_{\sigma=0} = \frac{3\chi_0^2\lambda_s}{\mu_0 M_s^2}, \qquad \chi_0 = \frac{\mu_0 M_s^2}{2K} \tag{30}$$

It is obvious that the higher the saturation magnetostriction value, the higher the stress sensitivity; however, the influence of K is decisive. Rare earth compounds, such as Terfenol, with giant λ ($\approx 10^{-3}$) values do not exhibit noticeable stress sensitivity as a consequence of the high values of the anisotropy, which appears as K^2 in the denominator of the second term, according to Eq. (29). Note that, in general, those substances with high magnetostriction constant also have high structural anisotropy since magnetostriction is the strain derivative of the anisotropy. Nevertheless, an amorphous structure has the characteristic of exhibiting zero macroscopic structural anisotropy but a nonzero macroscopic, or average, strain derivative of the anisotropy. In other words, a higher stress sensitivity is exhibited for iron-rich alloys with moderate λ ($\approx 10^{-5}$) values, in which very low and homogeneous anisotropies can be induced. In order to illustrate the significance of Eqs. (29) and (30), let us consider $3d$ and $4f$ crystalline magnetic substances and $3d$-based amorphous alloys. We define a relative stress sensitivity, S, as

$$S = \frac{1}{\chi_0}\left(\frac{d\chi}{d\sigma}\right)_{\sigma=0} = \frac{3\lambda_s}{2K} \tag{31}$$

For $3d$ crystalline materials, $\lambda_s \approx 10^{-6}$ and $K \approx 10^5$ J-m^{-3}, so $S \approx 10^{-11}$ Pa^{-1}. For $4f$ crystalline materials, $\lambda_s \approx 10^{-4}$ and $K \approx 10^7$ J-m^{-3}, so in crystalline materials $S \approx 10^{-11}$ Pa^{-1}. Since the ratio of magnetostriction to anisotropy is roughly constant, S does not change with composition in a noticeable way. Nevertheless, the residual as well as the induced anisotropy can be about 10 J-m^{-3} in amorphous materials in which the internal stresses have been released by thermal treatment.

If we then consider λ_s to range from 10^{-7} to 10^{-5}, it is found that S becomes three orders of magnitude larger in low magnetostriction amorphous materials than in crystalline samples, and five orders of magnitude higher in highly magnetostrictive alloys. This enormous stress sensitivity is the more remarkable technical aspect of the combination of the lack of structural anisotropy and the presence of a macroscopic magnetostriction. It can be said that the amorphous structure breaks up the linear relation between anisotropy and magnetostriction, which yields outstanding magnetoelastic properties.

Thus, it has been shown how metallic glasses with moderate values of magnetostriction constant attain higher stress sensitivity. These materials also exhibit a higher magnetoelastic coupling coefficient, which is an indication of the ability of a material to transfer elastic into magnetic energy, and vice versa. Because of the magnetostriction, the application of magnetic fields induces variations of both magnetization and elongation. The same result occurs when a tensile stress is applied. Young's modulus, E, and the magnetic susceptibility, χ, take on different values, depending on the values of the field or stress applied. One can define the following parameters:

$$E^{h,m} = \left(\frac{\partial\sigma}{\partial\varepsilon}\right)_{h,m} \tag{32}$$

$$\chi^{\sigma,\varepsilon} = \left(\frac{\partial m}{\partial h}\right)_{\sigma,\varepsilon} \tag{33}$$

Let us first illustrate the difference between the Young's modulus at constant field and at constant magnetization. Support a positive magnetostrictive amorphous ribbon with an easy axis transverse to the ribbon length. Under the effect of a tensile stress, at zero applied field, the ribbon elongates elastically and the magnetization rotates toward the ribbon length by magnetoelastic effect. The rotation of the magnetization results in an extra elongation along the ribbon axis due to the magnetostriction. Hence, the total elongation produced by the tensile stress is larger than that expected from the elastic contribution alone. The ribbon at zero field (constant field) exhibits a lower Young's modulus than when magnetically saturated. Notice that no magnetostrictive elongation changes take place when magnetization is fixed. A similar argument can be used to explain the difference between the susceptibility at constant stress and at constant strain. Suppose an axial field acting along the ribbon. The rotation of the magnetization toward the ribbon axis produces a magnetostrictive elongation when the applied stress is zero. However, if we try to eliminate such elongation (i.e., working at constant strain) we have to constrain the ribbon with a compressive stress. This stress χ reinforces χ the transverse anisotropy and therefore, under a given field strength the magnetization, would rotate an angle smaller than that obtained at zero retention stress. This example shows that the magnetic susceptibility at constant strain is lower than the susceptibility at constant stress.

It is easy to show that the magnetoelastic coupling coefficient, k, of a material which describes the energy transfer between the elastic and magnetic subsystems can be written as

$$k = \frac{E^m - E^h}{E^m} = \frac{\chi^\sigma - \chi^\varepsilon}{\chi^\sigma} \tag{34}$$

For crystalline materials, nickel was considered as a good material for magnetomechanical transducers ($k = 0.4$). Terfenol is a rare earth alloy with $k = 0.7$, which compares very favorably with piezoelectric transducer materials [7]. As-quenched metallic glasses can exhibit values of the magnetoelastic coupling coefficient around 0.6 and a carefully annealed sample of METGLAS alloy 2605SC (nominal composition $Fe_{81}B_{13.5}Si_{3.5}C_2$) has shown $k = 0.98$, the largest value ever found [73]. This makes metallic glasses ideal materials for transducers and sensors because all the elastic energy provided is transformed into magnetic energy (or vice versa).

B. Two Examples of Applications: Stress Sensor and an Experiment of Chaos

We have seen that the stress sensitivity of the magnetic susceptibility of metallic glasses is four or five orders of magnitude larger than that of magnetostrictive crystalline materials. This suggests the possibility of using metallic glasses as sensing elements of stress sensors. In order to compare the sensitivity of metallic glasses to the sensitivity of other standard sensing elements, let us consider a "gauge factor," which is defined in resistive strain gauges as

$$G = \frac{\Delta R/R}{\Delta \varepsilon} \tag{35}$$

where R is the unstressed resistance and ΔR the change in resistance when strained ($\Delta\varepsilon$). Gauge factors are around 2 for metallic and around 100 for semiconductor strain gauges. In metallic glasses an "impedance" figure of merit can be defined by using the material as the core of a coil. We have

$$F = \frac{\Delta Z/Z}{\Delta \varepsilon} \tag{36}$$

where Z and ΔZ are the impedance and the impedance change of the coil respectively. Values of F around 1500 have been found in a very simple device [74]. The highest possible sensitivity of a metallic glass device can be deduced from a susceptibility figure of merit. Since $\Delta Z/Z$ can be replaced by $\Delta\chi/\chi$, Eq. (36) can be rewritten from Eq. (31) as

$$F = ES \tag{37}$$

where S has been defined in Eq. (31). Using the equations above, we calculate a value $S = \sim 10^{-6}\ \mathrm{Pa}^{-1}$ for high magnetostriction metallic glasses; hence, by considering a reasonable E value in such materials ($E = 10^{11}$ Pa), we found an expected maximum F of 10^5. This gives three orders of magnitude better sensitivity than semiconductor strain gauges and much better temperature stability. Savage and Spano [73] have calculated and measured values of this figure of merit "intrinsic" to the material. Because of such characteristics, metallic glasses are by far the best-known materials for mechanical stress sensing. A general review on different devices as force, pressure, torque, and position sensors developed by using metallic glasses as sensing elements has been collected in Ref. 7. Figure 13 shows a schematic diagram of a torque sensor built with magnetostrictive amorphous wires.

As a second application of the giant magnetoelastic effects exhibited by metallic glasses, we describe an experiment of "chaos" based in the so-called ΔE effect. It has been shown that the Young's modulus, E, of a magnetostrictive material depends on magnetic field. Figure 14 shows the normalized modulus as a function of the field longitudinally applied to a ribbon of METGLAS alloy 2605SC, transversely annealed. This annealing under transverse field induces an homogeneous, low anisotropy with anisotropy field of about 1 Oe ($K \approx 100$ J-m^{-3}). It can be seen at Fig. 14 that the reversible changes of E with small fields (0–2 Oe) are large (factor of 10 or more) [75].

Let us outline the general aspects upon which a theory of the ΔE effect are based without going into details. Consider a ribbon with a transverse easy axis of anisotropy strength, K, magnetostriction constant, λ, saturation induction, $\mu_0 M_s$, and Young's modulus at constant magnetization, E_0. For a rotational model, the relations linking the applied stress, σ, and longitudinal field, H, with the strain, ε, and longitudinal magnetization, M, are

$$\varepsilon = \frac{\sigma}{E_0} - \frac{3\lambda}{2}\left(\frac{M}{M_s}\right)^2 \tag{38}$$

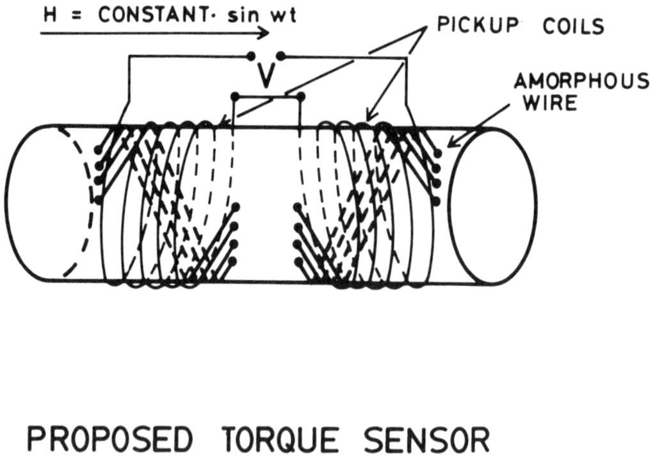

Figure 13 Schematic view of a torque sensor. Amorphous wires are disposed at 45° to the shaft axis [76].

$$M = \frac{\mu_0 M_s^2 H}{2K - 3\lambda\sigma} \tag{39}$$

Equation (39) was written as Eq. (28) to describe the rotation process of the magnetization. Equation (38) points out that, under applied stress, the strain contains an extra term coming from the intrinsic elongation of the material in the direction of the magnetization, which rotates as the stress rises. The magnetostrictive term (proportional to M^2) must be expressed as a function of σ, so allowing to find $E(H)$ as that coefficient dividing the stress in the general expression of ε.

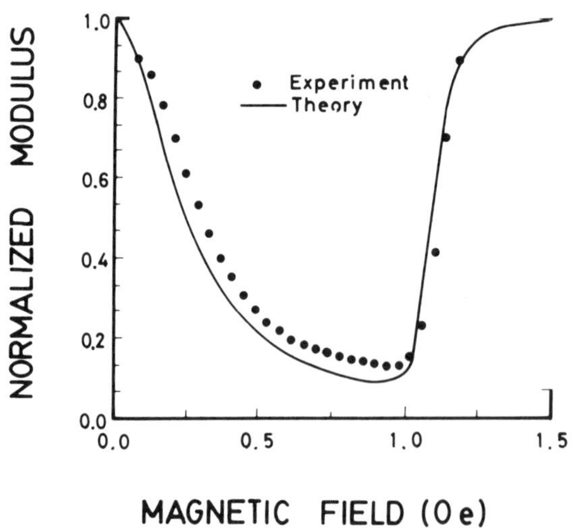

Figure 14 $E(H)/E_0$ versus H measured from experiments (dots) and from the theory (line) [74].

The enormous variations of the elastic constants with applied field over so a small a range of field suggested to Savage et al. [75] the possibility of performing the following experiment. As is well known, a vertically aligned ribbon, clamped at the bottom and free at the top, has a critical height, h_c, proportional to $E_0^{1/3}$, for which it will buckle under its own weight. Hence, if we use a transverse field annealed ribbon of METGLAS alloy 2605SC subjected to a variable magnetic field, is possible for a constant height to cross over h_c as the field varies due to the variation of E. In other words, the variation of field produces a variation of E; also, the critical height is altered above and below the actual height of the ribbon. This causes the ribbon gravitationally to buckle and unbuckle. The relevance of such an experiment is derived from the experimental observations about the ribbon motion. When the ribbon is driven with an alternating magnetic field, a variety of behaviors is observed; in particular, periodic motion, period-doubling cascades, quasiperiodic motion, transient chaotic motion, intermittent chaotic motion, and sustained chaotic motion. Savage et al. [75] have investigated in detail the intermittent temporal bursting (passage of the ribbon through a vertical position, buckling on the other side and returning). This type of behavior is known as crisis-induced intermittency. These experiments have allowed to compare their results with theoretical predictions about the critical exponents of the crisis-induced intermittency. They have also shown a direct way to control "chaos" with a magnetic field.

VII. CONCLUSION

Rapid quenching techniques have been shown to be quite suitable for "tailoring" magnetic properties. In particular, the cooling rate used during the production of a sample determines the correlation length of the structure and, therefore, the correlation length of the easy-axes orientation. For those structurally disordered alloys for which exchange interaction are sufficiently with respect the local anisotropy constant, a remarkable magnetic softness has been observed to overcome the misalignment of easy axes. This characteristic is due to the lack of structural magnetic anisotropy on a macroscopic scale. It is also notable that, in spite of the averaging out of magnetic anisotropy (second-rank tensor) in the amorphous structure, the magnetostriction which is the strain derivative of the anisotropy (fourth-rank tensor) does not average to zero. Consequently, magnetostriction is the only factor governing the technical magnetism of the amorphous ferromagnets. The outstanding magnetic softness, and other favorable characteristics such as frequency response of nearly zero magnetostriction alloys, make the amorphous materials ideal for magnetic cores of transformers and electronic devices. The only drawback of this kind of material is the relatively low saturation magnetization. Highly magnetostrictive alloys can be used as excellent sensing elements for detecting stresses. Because the magnetoelastic coupling coefficient nearly reaches values of unity, sensitivities greater than that exhibited by well-annealed magnetostrictive amorphous alloys is not expected for any other type of materials. Amorphous alloys have been successfully incorporated as magnetic field detectors, in encoding devices, and in the antitheft systems industry. The nanocrystalline magnetic materials, discovered more recently, also open a wide scope of possibilities in various technical applications.

REFERENCES

1. G. Herzer, in *Rapidly Quenched Materials* (Elsevier, Oxford, 1991), p. 1.
2. R. C. O'Handley, *J. Appl. Phys.*, *42*, R15 (1987).
3. E. M. Chudnovski and R. A. Serota, *J. Phys.*, *C16*, 4181 (1983).
4. R. Alben, J. J. Becker, and M. C. Chi, *J. Appl. Phys.*, *49*, 1653 (1978).
5. J. Tejada, B. Martínez, A. Labarta, R. Grossinger, H. Sassik, M. Vázquez, and A. Hernando, *Phys. Rev. B*, *42*, 898 (1990).
6. *Magnetic Properties of Amorphous Metals*, edited by A. Hernando et al. (North-Holland, Amsterdam, 1987).
7. A. Hernando, M. Vázquez, and J. M. Barandiarán, *J. Phys. E. Sci. Instrum.*, *21*, 1129 (1988).
8. *Concerted European Action on Magnets*, Commission of the Europe an Communities, edited by I. V. Mitchell et al. (Elsevier, New York, 1989).
9. V. Madurga, J. M. Barandiarán, M. Vázquez, O. V. Nielsen, and A. Hernando, *J. Appl. Phys.*, *61*, 3228 (1987).
10. A. R. Ferchmin and S. Kobe, in *Amorphous Magnetism and Metallic Magnetic Materials Digest*, edited by E. P. Wohlfarth (North-Holland, Amsterdam, 1983).
11. R. C. O'Handley, in *Amorphous Metallic Alloys*, edited by F. E. Luborsky (Butterworths, London, 1983), p. 257.
12. K. Yamauchi, and T. Mizoguchi, *J. Phys. Soc. Japan*, *39*, 541 (1975).
13. R. E. Watson and L. H. Benett, in *Theory of Alloy Phase Formation*, edited by L. H. Bennett (TMS-AIME, New York, 1980), p. 245.
14. R. C. O'Handley and M. D. Sullivan, *J. Appl. Phys.*, *52*, 1841 (1981).
15. A. Hernando, V. Madurga, C. Nuñez de Villavicencio, and M. Vázquez, *Appl Phys. Lett.*, *45*, 802 (1984).
16. H. Fujimori, in Ref. 11, p. 300.
17. M. Vázquez, W. Fernengel, and H. Kronmüller, *Phys. Stat. Solidi (a)*, *80*, 195 (1983).
18. H. Kronmüller, M. Fähnle, M. Domann, H. Grimm, R. Grimm, and B. Gröger, *J. Magn. Magn. Mater.*, *13*, 53 (1979).
19. F. B. Humphrey, K. Mohri, J. Yamasaki, H. Kawamura, R. Malmhall, and I. Ogasawara, in Ref. 6, p. 110.
20. I. Ohnaka, T. Fukusako, T. Ohmichi, T. Masumoto, A. Inoue, and M. Hagiwara, in *Rapidly Quenched Metals*, Vol. 1, edited by T. Masumoto and K. Suzuki (Japan Institute of Metals, Sendai, 1982), p. 31.
21. K. Mohri, *IEEE Trans. Magn.*, *MAG-20*, 942 (1985).
22. A. García Escorial and A. L. Greer, *J. Mater. Sci.*, *22*, 4388 (1987).
23. U. Köster and U. Herold, in *Glassy Metals I*, edited by H. J. Güntherodt and H. Beck (Springer, Berlin, 1981), p. 225.
24. F. E. Luborsky, in Ref. 11, p. 360.
25. H. Kronmüller, *J. Magn. Magn. Mater.*, *24*, 159 (1981).
26. F. E. Luborsky, H. H. Liebermann, J. J. Becker, and J. L. Walter, in *Rapidly Quenched Metals III*, Vol. 2, edited by B. Cantor (The Metals Society, London, 1979), p. 188.
27. H. Kronmüller, *Atomic Energy Rev.*, *Suppl. 1*, 255 (1981).
28. H. S. Chen, R. C. Sherwood, S. Jin, G. C. Chi, A. Inoue, T. Masumoto, and M. Hagiwara, *J. Appl. Phys.*, *56*, 1796 (1984).
29. M. R. J. Gibbs, J. E. Evetts, and J. A. Leake, *J. Mater. Sci.*, *18*, 278 (1983).
30. T. Egami, *Rep. Prog. Phys.*, *47*, 1601 (1984).
31. A. L. Greer and J. A. Leake, *J. Non-Cryst. Solids*, *33*, 291 (1979).
32. P. Allia and F. Vinai, *Phys. Rev.*, *B26*, 6141 (1982).
33. H. Fujimori, M. Morita, Y. Obi, and S. Ohta, in *Amorphous Magnetism II*, edited by R. A. Levy and R. Hasegawa (Plenum, New York, 1977), p. 393.

34. A. Hernando, M. Vázquez, V. Madurga, and H. Kronmüller *J. Magn. Magn. Mater.*, *37*, 161 (1983).
35. H. Fujimori, in Ref. 11, p. 300.
36. A. Hernando, O. V. Nielsen, V. Madurga, and J. M. Barandiarán, *Sol. Stat. Comm.*, *54*, 1059 (1985).
37. G. Suran, K. Ounadjela, and F. Machizand, *Phys. Rev. Lett.*, *57*, 24 (1986); and *36*, 73 (1986).
38. H. Bruning, Z. Altounian, and J. O. Strom-Olsen, *J. Appl. Phys.*, *62*, 3633 (1987).
39. T. Komatsu, S. Sato, and K. Matusita, *Acta Metall.*, *34*, 1891 (1986).
40. J. Filipecki and A. Van den Beukel, *Scripta Metall.*, *21*, 1111 (1987).
41. R. C. O'Handley, B. W. Corb, and M. J. Grant, *J. Appl. Phys.*, *55*, 1808 (1984).
42. B. W. Corb, R. C. O'Handley, J. Megusar, and N. J. Grant, *Phys. Rev. Lett.*, *51*, 1386 (1983).
43. J. M. Riveiro, V. Madurga, and A. Hernando, *Phys. Rev.*, *B39*, 1195 (1989).
44. M. L. Fdez-Gubieda, J. M. Barandiarán, F. Plazaola, A. Hernando, and S. Mobilio, *J. Phys. Con. Matt.* (in press).
45. A. Hernando, *Phys. Scripta*, *T24*, 11 (1988).
46. B. S. Berry and W. C. Pritchett, *Phys. Rev. Lett.*, *34*, 1022 (1975).
47. F. E. Luborsky and J. L. Walter, *IEEE Trans. Magn.*, *MAG-13*, 953 (1977).
48. C. Modzelewski, H. T. Savage, L. T. Kabacoff, and A. E. Clark, *IEEE Trans. Magn.*, *MAG-17*, 2837 (1981).
49. T. Jagielinski, *IEEE Trans. Magn.*, *MAG-19*, 1925 (1983).
50. M. Vázquez, J. Gonzalez, and A. Hernando, *J. Magn. Magn. Mater.*, *53*, 323 (1986).
51. O. V. Nielsen, *IEEE Trans. Magn.*, *MAG-21*, 2008 (1985).
52. F. E. Luborsky and J. J. Becker, *IEEE Trans. Magn.* , *MAG-15*, 1939 (1979).
53. H. R. Hilzinger, in Ref. 20, Vol. II, p. 791.
54. M. Vázquez, E. Ascasibar, A. Hernando, and O. V. Nielsen, *J. Magn. Magn. Mater.*, *66*, 37 (1987).
55. J. Gonzalez, M. Vázquez, J. M. Barandiarán, and A. Hernando, *J. Phys.*, *C8*, 1335 (1988).
56. G. Herzer, in *Proc. Conf. Soft. Magn. Mater. 7, Blackpool* (Waltson Center, Cardiff 8, 1986), p. 335.
57. J. M. Barandiarán, A. Hernando, V. Madurga, O. V. Nielsen, M. Vázquez, and M. Vázquez-López, *Phys. Rev.*, *B35*, 5066 (1987).
58. H. K. Lachowicz, *Ann. Fis.* (*B*), *86*, 129 (1990).
59. L. Kraus, *Acta Phys. Polon. A*, *76*, 147 (1989).
60. E. T. Lacheisserie, *J. Magn. Magn. Mater.*, *67*, 702 (1987).
61. M. Fähnle, J. Furtmüller, R. Pawellek, and G. Herzer, in *Proc. IV Int. Conf. Phys. Magn. Mater.* (Szczyrk-Bita, Poland, 1988).
62. A. Hernando, C. Gomez-Polo, E. Pulido, G. Rivero, M. Vázquez, A. Garcia-Escorial, and J. M. Barandiarán, *Phys. Rev.*, *B42*, 6471 (199).
63. A. Hernando, and J. M. Barandiarán, *Phys. Rev.*, *B22*, 2445 (1980).
64. J. D. Livingston and W. G. Morris, *IEEE Trans. Magn.*, *MAG-20*, 1397 (1984).
65. J. M. Barandiarán, A. Hernando, and E. Ascasibar, *J. Phys. D: Appl. Phys.*, *12*, 1943 (1979).
66. K. Mohri and S. Shirogi, *IEEE Trans. Magn.*, *MAG-19*, 215 (1982).
67. M. Liniers, V. Madurga, M. Vázquez, and A. Hernando, *Phys. Rev.*, *B31*, 4425 (1985).
68. C. Nuñez de Villavicencio, M. Vázquez, V. Madurga, and A. Hernando, *J. Magn. Magn. Mater.*, *59*, 333 (1986).
69. M. Vázquez, J. González, J. M. Blanco, J. M. Barandiarán, and A. Hernando, *J. Magn. Magn. Mater.*, *96*, 321 (1991).

70. K. Mohri, F. B. Humphrey, J. Yamasaki, and F. Kinoshita, *IEEE Trans. Magn.*, *MAG-21*, 2017 (1985).
71. M. Vázquez, C. Nuñez de Villavicencio, V. Madurga, J. M. Barandiarán, A. Hernando, and H. Kronmüller, in Ref. 6, p. 327.
72. M. Vázquez, A. Hernando, and H. Kronmüller, *Phys. Stat. Solidi* (*b*), *133*, 167 (1986).
73. H. T. Savage, M. L. Spano, *J. Appl. Phys.*, *53*, 8092 (1982).
74. E. E. Mitchell, R. de Mover, and J. Kranish, *IEEE Trans. Indust. Electron.*, *I.E. 33*, 166 (1986).
75. H. T. Savage, W. L. Ditto, P. A. Braza, M. L. Spano, S. N. Rauseo, and W. C. Spring, *J.Appl. Phys.*, *67*, 9 (1990).
76. H. T. Savage, A. E. Clark, M. Wun-Fogle, L. Kabakott, A. Hernando, and L. B. Beihoff, U.S. Patent (A.D. 71488, 1989).

18

Chemical Properties of Rapidly Solidified Alloys

K. Hashimoto
Institute for Materials Research, Tohoku University, Sendai, Japan

I. INTRODUCTION

Since it has been reported that amorphous Fe-Cr-P-C alloys have extraordinarily high corrosion resistance in acidic environments, including hydrochloric acids [1], many investigators have started to study the chemical properties of amorphous alloys and several reviews have appeared [2–15]. These reviews mostly deal with corrosion behavior. Proceedings of a recently held symposium on corrosion, electrochemistry and catalysis of metallic glasses have also been published [16].

From the chemical point of view the most important characteristic of amorphous alloys is their chemically homogeneous single-phase nature through extension of the solubility limit, whereas corresponding crystalline counterpart alloys consist of multiple phases. Consequently, new materials having a specific property can now be created by producing an alloy which contains the necessary amounts of elements necessary to achieve that property.

Even if a particular amorphous alloy cannot be formed, rapid solidification extends the solubility limit of microcrystalline alloys in addition to refining their grain size. These two factors modify the surface-related chemical properties of crystalline alloys.

II. CORROSION-RESISTANT ALLOYS

A. Extremely High Corrosion Resistance

In general, the corrosion resistance of alloys is based on the formation of a protective surface film (passive film) as a result of alloy surface oxidation by an environment. This passive film is able to separate the bulk of an alloy from an aggressive environment. The corrosion rates of amorphous iron-metalloid alloys decrease by additions of almost any kind of additional metallic elements [1,17–23]. The addition of chromium is particularly effective. For instance, amorphous $Fe_{72}Cr_8P_{13}C_7$ alloy spontaneously forms a passive film in 2 *N* HCl at ambient temperature [24]. Eben low-chromium alloys such as amorphous $Fe_{77}Cr_3P_{13}C_7$ alloys containing 2 at% of molybdenum, tungsten, or other elements become passive in 1 *N* HCl by increasing oxidizing power of the environment due to anodic polarization [25]. The combined addition of chromium and molybdenum to amorphous iron-metalloid alloys is further effective, and in some cases they passivate spontaneously in hot concentrated hydrochloric acids [26].

Boiling concentrated nitric acid attacks almost all alloys for which corrosion resistance is based on the presence of chromium; this exposure even removes the metallic luster of tantalum metal, for example. In such environments, however, amorphous Ni-Ta alloys are immune to corrosion, maintaining their metallic luster [27].

While phosphoric acids contain no particularly aggressive anions, concentrated phosphoric acids are quite aggressive because of their high boiling points. As shown in Fig. 1 [28], amorphous Ni-Ta alloys have a higher corrosion resistance than tantalum metal, and there are various amorphous alloys possessing corrosion resistance comparable to that of tantalum metal.

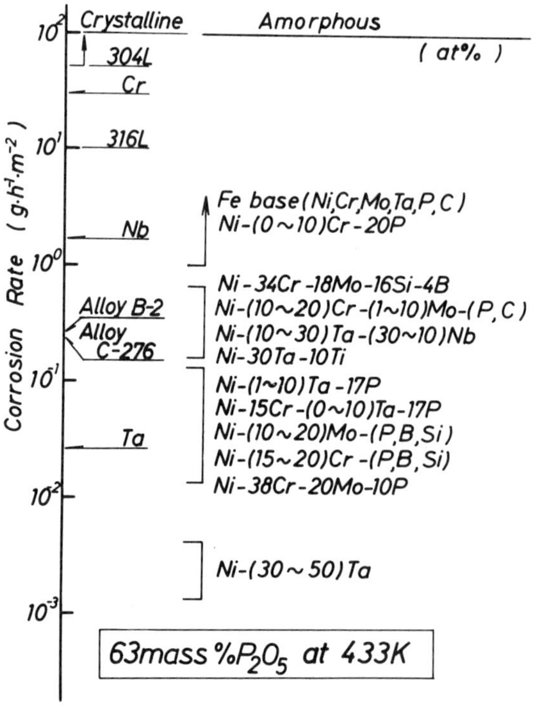

Figure 1 Corrosion rates of various amorphous and crystalline alloys and metals in a hot concentrated phosphoric acid [28].

B. Origins of High Corrosion Resistance

1. Rapid Formation of Passive Film Being Highly Concentrated with Beneficial Ions

According to an x-ray photoelectron spectroscopic study, 97% of cations in the film formed by spontaneous passivation of amorphous $Fe_{70}Cr_{10}P_{13}C_7$ alloy in 1 *N* HCl were chromic ions; the film consists mainly of hydrated chromium oxyhydroxide ($CrO_x(OH)_{3-2x}nH_2O$) [29]. According to subsequent studies, the passivation of amorphous and of crystalline alloys, in which corrosion resistance relies on the presence of chromium, occurs due to the formation of the passive films containing hydrated chromium oxyhydroxide as the main constituent [24,26,29–34]. As can be seen from the example that the pitting potential of austenitic stainless steels becomes higher with increasing chromic ion content of the surface film [35], a higher chromic ion content results in greater protective quality of the passive film.

Table 1 shows the fraction of chromic ions relative to the total number of cations in the passive films formed on various chromium-containing alloys in 1 *N* HCl. In comparison with crystalline stainless steels, amorphous alloys enable the extraordinary accumulation of chromic ions in the passive films. The composition of the alloy surface just under the chromic-ion-concentrated passive film is almost the same as that of the bulk alloy [32]. Accordingly, the passive hydrated chromium oxyhydroxide film is formed as a result of dissolution of alloy constituents unnecessary for the passive film formation. The thermodynamically metastable nature of the amorphous alloys is effective for rapid, active dissolution of alloy constituents into corrosive solutions, and hence favors the enrichment of chromic ions in the surface region with a subsequently rapid passivation [37]. The presence of large amounts of metalloid elements in the amorphous metal-metalloid alloys also accelerates chromic ion enrichment due to acceleration of active dissolution; phosphorus is particularly effective in enhancing the passive film formation [38].

Table 1 Concentration of Chromic Ion in Passive Films Formed on Amorphous Alloys and Ferritic Stainless Steel in 1 *N* HCl

Alloy	Cr^{3+}/total metallic ions	Method for passivation	Reference
Amorphous alloy			
$Fe_{70}Cr_{10}P_{13}C_7$	0.97	Spontaneous passivation	24
$Fe_{75}Cr_3Mo_2P_{13}C_7$	0.57	Anodic polarization	25
$Co_{70}Cr_{10}P_{20}$	0.95	Spontaneous passivation	31
$Ni_{70}Cr_{10}P_{20}$	0.87	Spontaneous passivation	36
Ferritic stainless steel			
Fe-30Cr-2Mo	0.75	Anodic polarization	33
Fe-19Cr-2Mo	0.58	Anodic polarization	34

On the other hand, alloys for which corrosion resistance relies on the presence of tantalum or niobium, such as Ni-Ta and Ni-Nb, form passive films in which almost 100% of the cations are Ta^{5+} or Nb^{5+} [28,39]. Phosphorus addition is still effective in passive film formation on amorphous Ni-Ta alloys when exposed to strong acids with a poor oxidizing power [39].

2. *Chemical Homogeneity of the Amorphous Alloys*

As will be described later, passive films formed on crystalline alloys having chemical heterogeneity are not uniform because the films formed on the chemically heterogeneous sites and on the phases with lower concentrations of passivating elements are weaker than those formed on the chemically homogeneous regions. The chemically homogeneous single-phase nature of amorphous alloys ensures the formation of uniform passive films without weak points, even in very aggressive environments. Accordingly, the formation of chemical hetereogeneity by crystallization heat treatment of amorphous alloys is known to lead to disappearance of the high corrosion resistance [37,40–44]. For instance, amorphous $Fe_{70}Cr_{10}P_{13}C_7$ alloys is spontaneously passive in 1 *N* HCl, but the formation of a microcrystalline phase in the amorphous matrix by isothermal annealing immediately leads to passivity breakdown [42]. The chemically homogeneous amorphous Fe-Mo-P-C alloys are passivated anodically in 1 *N* HCl by the formation of a passive hydrated iron oxyhydroxide film [43], which is the same as the passive film formed on crystalline iron in 2 *N* H_2SO_4 [32], while crystalline iron cannot be passivated in 1 *N* HCl as is shown in Figs. 2 and 3 [46].

Figure 2 Although a passive hydrated iron oxyhydroxide film is not unstable in acidic chloride solutions such as 1 *N* HCl, the stable passive film cannot be formed on chemically heterogeneous sites of crystalline iron and hence catastrophic corrosion occurs [45].

Figure 3 Amorphous Fe-Mo-metalloid alloys are chemically homogeneous and can be covered uniformly by a passive hydrated iron oxyhydroxide film even in 1 *N* HCl [45].

III. RAPIDLY SOLIDIFIED MICROCRYSTALLINE ALLOYS

Rapid solidification is also effective in enhancing the corrosion resistance of crystalline alloys. Microcrystalline stainless steels prepared by the consolidation of rapidly solidified parent material through hot extrusion have greater resistance to pitting corrosion in 0.5 *N* NaCl at 60°C than conventionally processed counterparts [47]. This is interpreted in terms of insufficient segregation or accumulation of impurities at pit growth sites because of a uniform, fine dispersion of impurities throughout the steels which results from rapid solidification processing.

Rapid solidification of conventional alloy surfaces has been carried out often by laser surface melting and subsequent self-quenching. Anthony and Cline [48] have attempted surface normalization of sensitized 304 stainless steel by laser treatment and found that intergranular corrosion of a sensitized steel due to chromium depletion is prevented as a result of the redistribution of chromium by the laser melt-quenching process. It is known [49] that pitting potential of weld metals of austenitic stainless steels are considerably raised due to prevention of $M_{23}C_6$ precipitation by laser melt quenching.

Figure 4 shows a comparison of corrosion rates due to pitting corrosion [50] of conventionally processed high-purity Fe-Cr alloys and of their rapidly solidified counterparts. Rapidly solidified alloys show very low corrosion rates in comparison with those of conventionally processed counterparts and of heat-treated rapidly

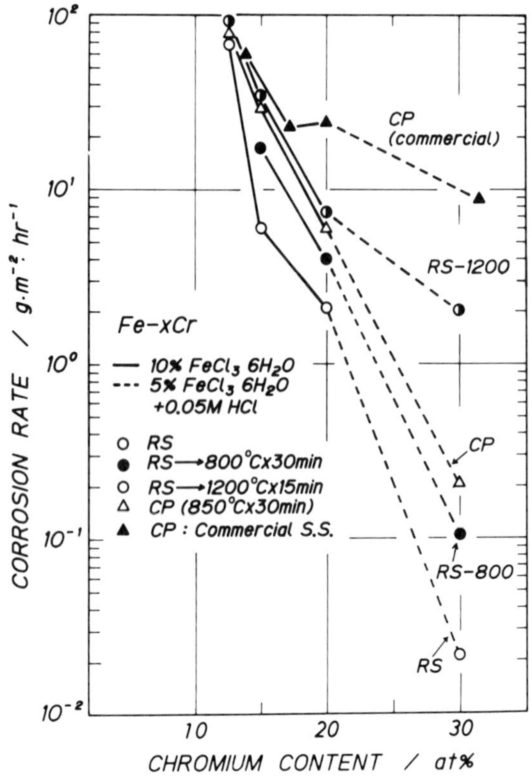

Figure 4 Average corrosion rates of rapidly solidified (RS) and conventionally processed (CP) Fe-Cr alloys measured in 10% $FeCl_36H_2O$ at 30°C except for Fe-30Cr alloy which was immersed in 5% $FeCl_36H_2O$ + 0.05 *M* HCl at 50°C. Included for comparison in the figure are average corrosion rates of commercial stainless steels and those measured after heat treatment of rapidly solidified alloys [50].

solidified alloys. In order to specify the detrimental surface sits for pitting corrosion, Fe-20Cr alloys were prepassivated at −0.2 V(SCE) in deaerated 0.5 *M* NaCl solution. Then, the potential was made more noble in order to observe the current oscillation during the induction time prior to pit propagation due to the nucleation and repassivation of micropits. The depth of micropits was estimated from the amount of charge for individual current oscillations on the basis of assuming the formation of the hemispherical micropits. Figure 5 shows the depth of micropits as a function of polarization time [50]. The arrows indicate pit propagation without repassivation. Before catastrophic damage by pit growth, a number of micropits several μm deep nucleated and repassivated. It is clear that the rapidly solidified specimen is more prone to micropitting, in spite of a higher resistance to pit propagation, than the conventionally processed counterpart. Because current density is decreased several orders of magnitude by prepassivation treatment, it can be said that the entire surface has been covered by the passive film, even at heterogeneous sites of the alloys. However, pitting corrosion often initiates at defective regions of the surface films on the alloys. These defective regions may

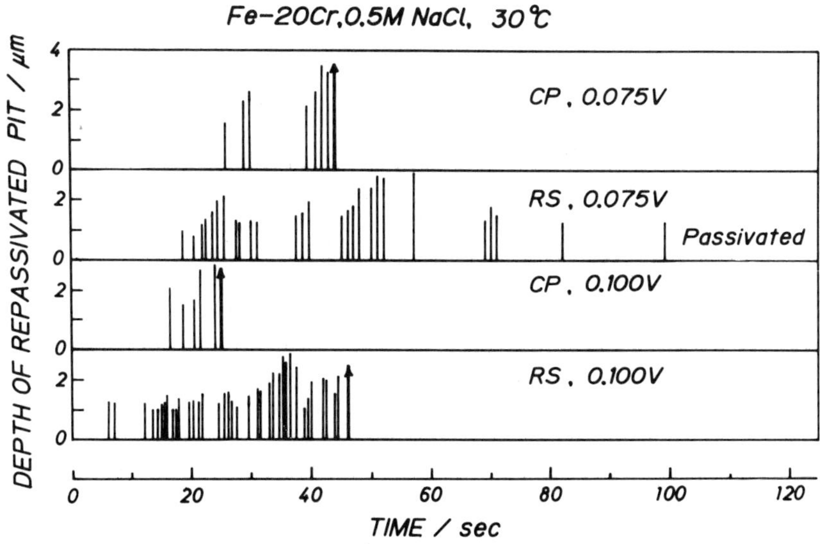

Figure 5 Depth of micropits formed on prepassivated Fe-20Cr alloys at different times of polarization at 0.075 and 0.1 V(SCE) in a deaerated 0.5 *M* NaCl solution of 30°C [50].

Figure 6 Even if passivation occurs, the stability of the passive film on chemically heterogeneous sites of crystalline alloy surfaces is not high [10].

be related to the compositional nonuniformity, such as various kinds of inclusions, or to compositional segregation. Rapid solidification result in the formation of microcrystalline alloys in which inclusions of small sizes are finely distributed, while conventional processing allows the formation of precipitates and segregates in which detrimental impurities are locally highly accumualted. Accordingly, a lower accumulation of detrimental impurities at the pit nucleation sites in the rapidly solidified alloys leads to easy repassivation of any micropits formed, although a high density of finely distributed impurities is responsible for frequent micropitting. Figures 6–8 interpret nucleation, repassivation and propagation of a pit [10].

Rapid solidification is also effective to improve resistance against high temperature oxidation. Oxidation of conventional wrought 18-8 stainless steel forms a less-protective iron oxide scale, whereas the rapidly solidified counterpart is covered by protective Cr_2O_3 and SiO_2 scales, promoted by intergranular diffusion in the fine-grained alloys [51]. Similarly, grain refining of laser melt-quenched surfaces of Type 430 and 304 stainless steels, Incoloy 800 and Hastelloy X at 800–1200°C significantly improved resistance against oxidation in air [52].

A. Effects of Stability and Amorphous Nature

Because of the thermodynamically metastable nature of amorphous alloys, the active dissolution rate is essentially faster than that of the thermodynamically more stable crystalline counterpart, provided that the crystalline alloy consists of a single

Figure 7 Passivity breakdown and repassivation occur at weak points of the passive film on chemically heterogeneous sites of the alloy surface [10].

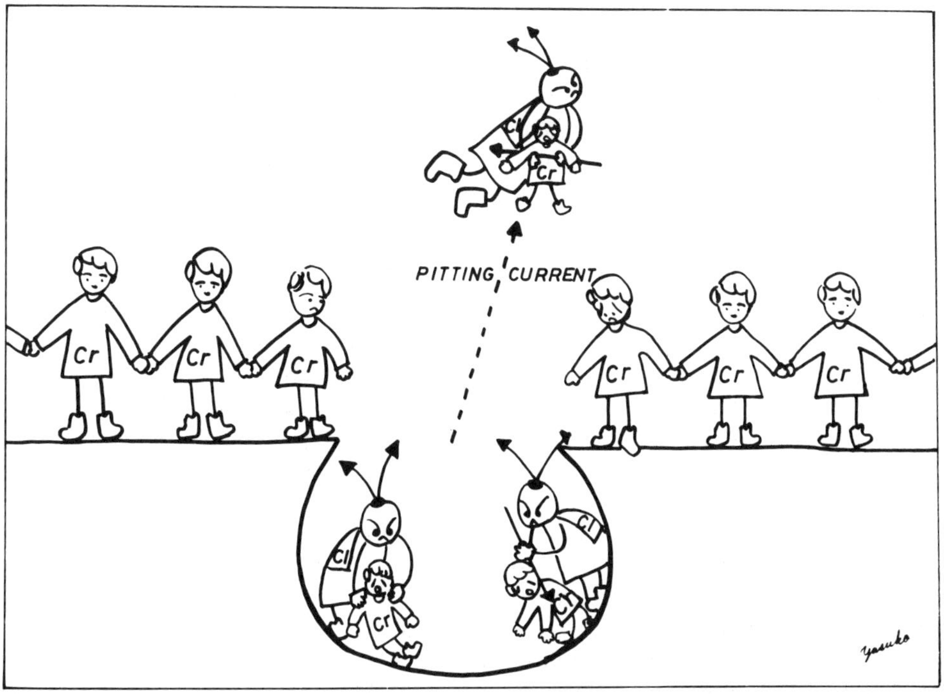

Figure 8 If a depression of the metal exceeds the critical depth such as about 10 μm for conventional crystalline stainless steels, repassivation no longer takes place, since the solution chemistry inside the depression became extremely aggressive due to lowering pH and increasing chloride concentration, and hence catastrophic corrosion occurs in the form of pitting corrosion [10].

phase [37]. Heusler and Huerta [53] have reported an interesting result: after heat treatment of amorphous $Co_{75}B_{25}$ alloy for 5 hr at different temperatures, they measured dissolution current density by potentiostatic polarization in the active region in an acidic sulfate solution where a stable passive film was not formed. The crystalline Co_3B phase having the same composition as the amorphous matrix was formed at 525 K or at higher temperatures, and the active dissolution current density decreased with increasing heat treatment temperature since the crystalline Co_3B phase is thermodynamically more stable than the amorphous phase. The dissolution current density was minimum at 625 K, at which the alloy transformed to the crystalline single Co_3B phase. At even higher temperatures, a three-phase alloy was formed due to decomposition of the Co_3B phase into α-Co and Co_2B phase, and hence the active dissolution current increased at these higher temperatures.

On the other hand, structural relaxation of amorphous alloys by heat treatment also affect corrosion behavior [54]. Not only by intentional relaxation treatment, but also by preparation of amorphous alloys under different conditions, the corrosion behavior of amorphous alloys having a given composition is affected since frozen-in structural disorders are different [55].

B. Preparation of Corrosion Resistant Coatings

Even if amorphous alloys exhibit extremely high corrosion resistance, practical utilization of melt-spun amorphous alloy ribbon is seriously restricted due to there not being available conventional methods for welding without crystallization and by their limited thickness, which is unavoidable from the requirement of rapid quenching of the molten alloy. One solution for this problem is the preparation of amorphous alloy surface layers on conventional bulk alloys.

1. Laser and Electron Beam Treatments

Instantaneous irradiation of high-energy density laser or electron beam, for example, results in instantaneous melting and self-quenching. Many investigators [56–61] have found vitrification after a single laser irradiation trace, provided that the surface composition is suitable for vitrification by rapid quenching from the liquid state. However, vitrification of a large surface area requires overlapping traverses by the laser beam, as shown in Fig. 9 [62]. Figure 10 [62] shows schematically the cooling sequence in laser processing. When a small volume (1) in the surface is instantaneously melted by the laser irradiation, the heat of the melt in this volume is rapidly absorbed by a large volume of the directly surrounding solid alloy. The absorption of this heat of the melt by the directly surrounding solid in the laser processing is much easier than that in conventional rotating-wheel melt-spinning methods because cooling in the latter must be accomplished over a very short time during the contact of the solidifying ribbon with the wheel surface. Therefore, once melted, the vitrification due to a single laser trace occurs for a relatively wide range of alloy compositions. However, when the laser beam is irradiated at an area neighboring the previously vitrified phase, a portion (2) of the previously vitrified phase is heated without melting by contact with the neighboring melt. Since the thermal history of the previously vitrified region 2 includes the heating curve during the laser irradiation to form the melt in neighboring pass, the cooling curve of region 2 often intersects the crystalline nose of the TTT curve with a consequent formation of crystalline phase in region 2. Thus, the vitrification of a wide surface

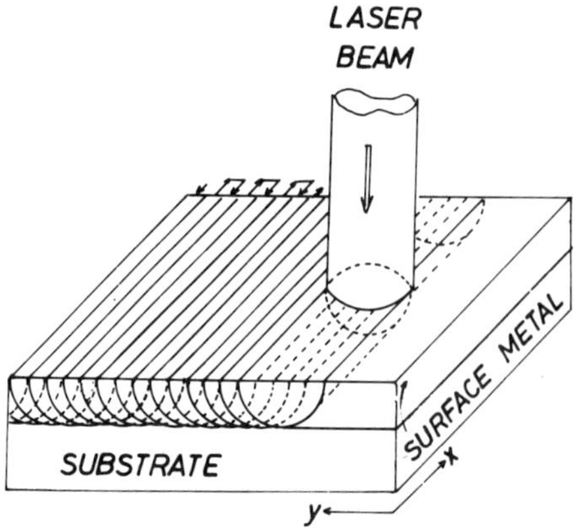

Figure 9 Schematic of laser processing for a crystalline nickel-based alloy on a mild steel [62].

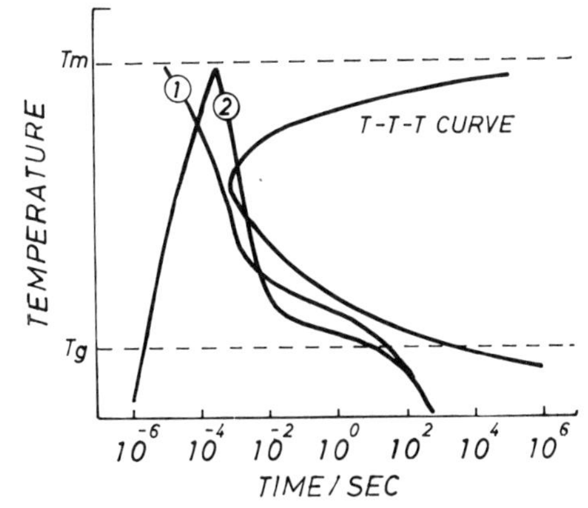

Figure 10 Schematic of cooling sequence for laser surface treatment [62].

area by laser processing requires an alloy having sufficiently long time for appearance of the crystalline nose in addition to a high crystallization temperature. In other words, crystallization of alloys vitrifiable by conventional melt-spinning technique often occurs in the heat-affected zone of previously vitrified phase in laser processing.

Figure 11 [62] shows a comparison of polarization curves measured before and after laser processing. The laser-processed surface alloy passivates spontaneously and shows rather lower anodic current density than the melt-spun amorphous counterpart; this is in contrast to a very high anodic current density of the crystalline surface alloy. Consequently, extremely corrosion-resistant amorphous surface layers have been prepared on mild steel by laser processing. The formation of a wide

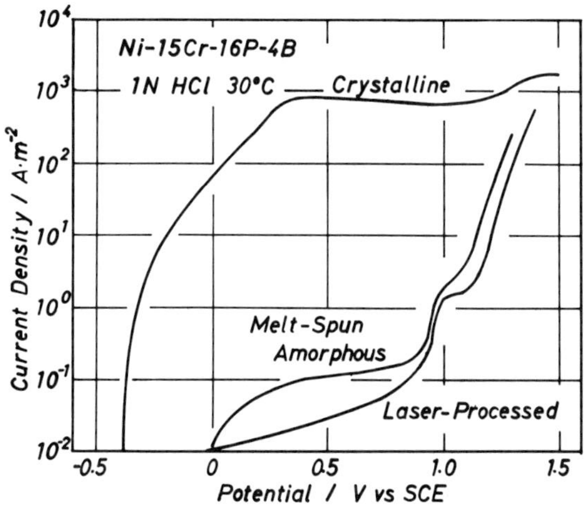

Figure 11 Anodic polarization curves of laser-processed and unprocessed crystalline $Ni_{65}Cr_{15}P_{16}B_6$ surface alloys on a mild steel and melt-spun amorphous $Ni_{65}Cr_{15}P_{16}B_6$ alloy measured in 1 *N* HCl at 30°C [62].

amorphous surface layer on a mild steel by laser processing was successful only for $Ni_{63-66}Cr_{14-17}P_{16}B_4$ alloys, in contrast to the fact that melt spinning can prepare amorphous Ni-Cr-P-B alloys in the chromium content range from 0 to 43 at%.

As discussed, laser processing is a difficult technique for preparing thermodynamically metastable amorphous alloys; but wide surface areas can be vitrified by laser processing when the alloy compositions having very high glass-forming abilities are chosen.

2. *Sputtering*

Sputtering is another potential method for the preparation of a variety of amorphous alloy surface layers on conventional crystalline alloys. Amorphous sputter deposits prepared for corrosion-resistant coatings have been reported to have high corrosion resistance [63,64]. It has, however, been found that these sputter deposits sometimes have defects responsible for corrosion of the substrate alloys [65] and compositional fluctuations causing increased anodic dissolution [66].

According to our investigations, the preparation of defect-free amorphous sputter coatings by which substrate alloys are not corroded during immersion for more than a month in aggressive environments such as boiling concentrated nitric acids, for example, requires significant technological improvements of sputtering technique. When sputtering is carried out by ordinary procedures using a conventional apparatus for the basic study of the physical properties of sputtered amorphous alloys, the resultant amorphous sputter deposits have never been defect-free, corrosion-resistant coatings. However, the challenge of forming defect-free amorphous sputter deposits can nearly be met by technological improvements.

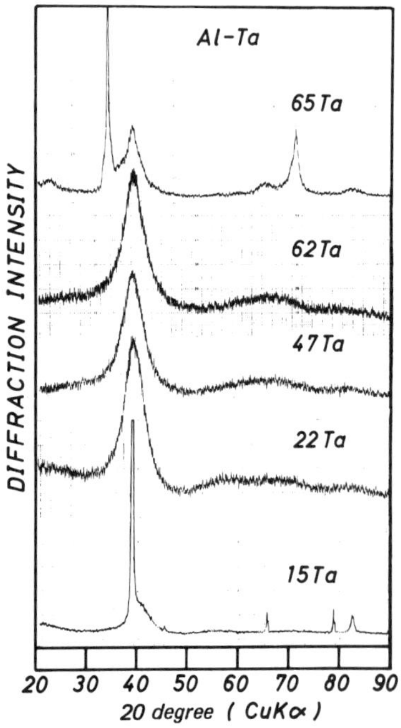

Figure 12 X-ray diffraction patterns of sputtered Al-Ta alloys [68].

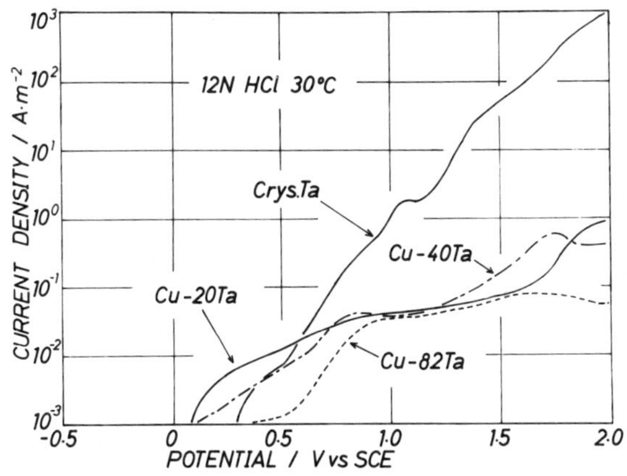

Figure 13 Anodic polarization curves of sputtered amorphous Cu-Ta alloys and crystalline tantalum metal measured in 12 *N* HCl at 30°C [67].

On the positive side, sputtering is known to form amorphous structures over the widest composition range of a given alloy family. Furthermore, sputtering does not require the use of alloy targets. For example, binary alloys cannot be prepared by conventional methods if the melting point of an alloy constituent exceeds the boiling point of the other alloy constituent. Sputtering enables the formation of this kind of alloy by using a target consisting of a metal plate onto which other alloy constituent is placed or embedded. It is further possible that these sputtered alloys are amorphous.

On the basis of the concepts above amorphous Cu-Ta [67], Cu-Nb [67], Al-Ta [68], Al-Nb [68], and Al-W [69] were prepared. As shown in Fig. 12 [68], these binary alloys become amorphous over considerably wide composition ranges. Figure 13 [67] shows potentiodynamic polarization curves of amorphous Cu-Ta alloys and crystalline tantalum metal in 12 *N* HCl. Because of the aggressive 12 *N* HCl anodic polarization of crystalline tantalum metal results in a considerable increase of current density. In contrast, these amorphous Cu-Ta alloys are spontaneously passive and show very low anodic current density over a wide potential range. Consequently, sputtering is quite a useful method to prepare corrosion-resistant coatings which cannot be prepared by any other techniques.

IV. ELECTRODES FOR ELECTROLYSIS

Electrode materials generally have a high catalytic activity and selectivity for a specific reaction, as well as the high corrosion resistance under the electrolytic conditions. Amorphous alloys having high concentrations of necessary elements in the form of the single-phase solid solution are quite suitable for tailoring new electrode materials by alloying. Furthermore, metallic anodes for electrolysis of aqueous solutions are anodically polarized at high potentials and hence are generally covered by a protective passive film on which electrocatalytic reactions take place. Consequently, electrocatalytically active elements must be contained as constituents which are able to form the protective passive film on the metallic anodes.

A. Anode for Electrolysis of Hot Concentrated Sodium Chloride Solutions

In the chlor-alkali industry the electrolysis of hot concentrated sodium chloride solutions produces sodium hydroxide and hydrogen on the cathode and chlorine on the anode. The anode materials are required to have a high electrocatalytic activity for chlorine evolution, and a low activity for parasitic oxygen evolution, along with a high resistance to corrosion in extremely aggressive chloride-containing environments of high oxidizing power. Although the anodic characteristics of platinum group metals in sodium chloride solutions have been extensively studied, these metals have not been used in the chlor-alkali industry because none of them possesses both the electrocatalytic activity and corrosion resistance for the chlorine evolution in hot concentrated sodium chloride solutions. Among these metals, palladium has excellent electrocatalytic properties for chlorine evolution but has a poor corrosion resistance. Attempts were made to improve the corrosion resistance of palladium without harming its superior electrode characteristics for the electrolysis of hot concentrated sodium chloride solutions on the basis of transformation to a corrosion-resistant amorphous structure containing corrosion-resistant alloying elements.

As shown in Table 2 [70,71] palladium metal dissolves actively without forming chlorine gas in hot sodium chloride solution, whereas amorphous palladium-based alloys exhibit corrosion resistance comparable with that of platinum group metals other than palladium. Table 3 gives a comparison of the current densities measured in NaCl and in Na_2SO_4 solutions. The currently used active RuO_2/Ti electrode is active in the hot concentrated sodium chloride solution but is also active in the

Table 2 Corrosion Rate of Amorphous Pd-P Alloys and Crystalline Platinum Group Metals Estimated from the Weight Loss during Galvanostatic Anodic Polarization at 1000 A-m^{-2} in 4 *M* NaCl Solution at 80°C

Electrode	Corrosion rate (g^{-2} h^{-1})
Amorphous alloys	
$Pd_{61}Rh_{20}P_{19}$	0.00
$Pd_{51}Rh_{30}P_{19}$	0.00
$Pd_{51}Pt_{30}P_{19}$	0.02
$Pd_{41}Pt_{40}P_{19}$	0.01
$Pd_{41}Ir_{40}P_{19}$	0.18
$Pd_{39}Ir_{40}Ti_2P_{19}$	0.11
$Pd_{46}Ir_{30}Rh_5P_{19}$	0.04
$Pd_{46}Ir_{30}Pt_5P_{19}$	0.01
Crystalline metal	
Pd	2027.08
Rh	0.02
Pt	0.03
Ir	0.02
Ru	0.18

Source: Refs. 70, 71.

Table 3 Current Densities for Amorphous Alloys, Crystalline Platinum Group Metals, and the Currently Used Electrodes at 1.15 V(SCE) Estimated from Galvanostatic Polarization Curves Measured in NaCl and Na_2SO_4 Solutions of pH 4 and 80°C

Electrode	Current density ($A\text{-}m^{-2}$): 4 *M* NaCl (chlorine and oxygen) evolution	Current density ($A\text{-}m^{-2}$): 1 *M* Na_2SO_4 (oxygen evolution)
Amorphous alloy		
$Pd_{51}Rh_{30}P_{19}$	220	7
$Pd_{41}Pt_{40}P_{19}$	400	6
$Pd_{41}Ir_{40}P_{19}$	2000	5
$Pd_{46}Ir_{30}Ti_5P_{19}$	3000	2
$Pd_{41}Ir_{30}Ti_{10}P_{19}$	1500	5
$Pd_{41}Ir_{30}Ru_{10}P_{19}$	2200	28
$Pd_{45}Ir_{30}Ti_5Rh_1P_{19}$	2600	15
$Pd_{45}Ir_{30}Ti_5Pt_1P_{19}$	1700	21
$Pd_{45}Ir_{30}Ti_5Ru_1P_{19}$	2000	15
Crystalline metal		
Ru	1300	1400
Rh	26	6
Pd	—	16
Ir	340	150
Pt	340	1
Graphite	41	20
RuO_2/Ti	1700	340

Source: Refs. 70, 71.

Na_2SO_4 solution. Therefore, chlorine formed on the RuO_2/Ti electrode is contaminated with oxygen; hence, liquification of chlorine is required to remove oxygen in order to avoid explosion of the mixture of hydrocarbon, chlorine and oxygen before this chlorine is used for chlorination of hydrocarbon. In contrast, there are various amorphous palladium-based alloys which have very high activities for chlorine evolution and low activities for oxygen evolution. Consequently, these amorphous alloys are capable of producing high-purity chlorine with a high efficiency.

B. Anode for Electrolysis of Seawater

In industrial plants using seawater as a coolant, sodium hypochlorite is introduced into the cooling system for the sake of protection against marine life, such as barnacles, which decreases the cooling efficiency and sometimes clogs the system. Sodium hypochlorite is produced by the reaction of chlorine and sodium hydroxide, both of which are produced by the electrolysis of seawater on the anode and cathode, respectively, as follows:

On the anode: $$2Cl^- = Cl_2 + 2e^- \quad (1)$$

On the cathode: $$2Na^+ + 2H_2O + 2e^- = 2NaOH + H_2 \quad (2)$$

In the solution: $Cl_2 + 2NaOH = NaClO + NaCl + H_2O$ (3)

The anodic reaction is the same as that in hot concentrated sodium chloride solutions used in the chlor-alkali industry. However, the concentration of chloride ions in seawater is about 33% and 1/8 of that of the solutions used in the chlor-alkali industry, and seawater is neutral in contrast to the acidic solutions used in the chlor-alkali industry. In this manner, the solutions used in the chlor-alkali industry are particularly prepared so as to favor chlorine evolution, while oxygen evolution is quite easy in electrolysis of seawater. Therefore, the anode materials for electrolysis of seawater must have a particularly high activity for chlorine evolution.

The electrocatalytic activity of amorphous palladium-based alloys mentioned above was found to be insufficient for chlorine evolution in seawater and should be increased about two orders of magnitude. Accordingly, a special treatment for surface activation was applied. In general, an increase in the activity of alloy catalysts is achieved by the selective dissolution of alloy constituents unnecessary for the catalytic activity into aqueous solutions for the sake of an increase in effective surface area and an enrichment of catalytically active elements at the alloy surface. However, useful reagents to dissolve unnecessary alloy constituents such as metalloids in the above-mentioned palladium-based alloys were not found. Accordingly, a new method based on surface alloying with a low-melting-point element easily diffusible into amorphous alloys, and subsequent leaching of this element into an aqueous solution, was used. The surface activation treatment of the palladium-based alloys consists of electrodeposition of zinc on the alloys, diffusion of zinc into the alloys by heat treatment at the temperatures lower than the crystallization temperatures of the alloys, and finally leaching of zinc from the alloys by immersion in a hot alkali solution [72]. This method can be applied only to amorphous alloys: The heat treatment of zinc-deposited crystalline metals leads preferentially to grain boundary diffusion of zinc, and hence subsequent leaching of zinc results in grain boundary degradation without enhancement of the electrocatalytic activity. In contrast, zinc diffusion uniformly takes place in amorphous alloys, which are disordered, and hence subsequent leaching of zinc results in a remarkable increase in the effective surface area of the amorphous alloys.

In Table 4 a comparison of the activities of amorphous alloys is shown for both chlorine and oxygen evolution, with those of currently used anodes in which a titanium substrate was covered by platinum, palladium oxide or platinum-iridium alloy. The activities of as-cast amorphous alloys for chlorine evolution are not high, but the surface-activated amorphous alloys exhibit very high activities [73,74]. Figure 14 shows a comparison of current efficiencies for chlorine production [74]. When galvanostatic polarization was carried out, the overpotential of the anode having a low activity for chlorine evolution became very high, and hence a large amount of oxygen was evolved with a subsequently low current efficiency for chlorine evolution. The current efficiency on the surface-activated amorphous $Pd_{56}Rh_{25}P_{10}Si_9$ alloy is 10–20% higher than that on the Pt-Ir/Ti anode, which is known to have the highest activity for seawater electrolysis among conventionally used anodes.

The amorphous alloys utilized are composed mainly of expensive platinum group elements, which are responsible for both high electrocatalytic activity and corrosion

Table 4 Current Densities for Amorphous Alloys and Currently Used Electrodes at 1.15 V(SCE) Estimated from Galvanostatic Polarization Curves Measured in NaCl and Na_2SO_4 solutions at 30°C

Electrode	Current density ($A\text{-}m^{-2}$) 0.5 *M* NaCl (chlorine and oxygen evolution)				0.25 *M* Na_2SO_4 (oxygen evolution)
		Surface-activated at			
Amorphous alloy	As-cast	200	250	300°C	Surface-activated at 300°C
$Pd_{56}Rh_{25}P_{10}Si_9$	18.0	28.0	200	800	35.0
$Pd_{46}Rh_{25}Ti_{10}P_{10}Si_9$	4.6	9.5	27.0	600	23.0
$Pd_{46}Rh_{25}Ti_5Ru_5P_{10}Si_9$	17.0	40.0	50.0	1300	22.0
$Pd_{46}Rh_{25}Ti_5Pt_5P_{10}Si_9$	9.0	27.0	45.0	260	22.0
Currently used electrode					
Pt/Ti		12.0			1.4
PdO/Ti		250			36.0
Pt-Ir/Ti		750			64.0

Source: Refs. 73, 74.

resistance. An attempt was made to prepare other electrochemically active alloys based on the good corrosion resistance of inexpensive amorphous alloys and the electrocatalytic activity of very low concentrations of platinum group elements in these alloys.

As described already, amorphous nickel-niobium and nickel-tantalum alloys have very high corrosion resistance even in hot concentrated acids [27,28]. In fact, amorphous Ni-Nb and Ni-Ta alloys are spontaneously passive and show a very high

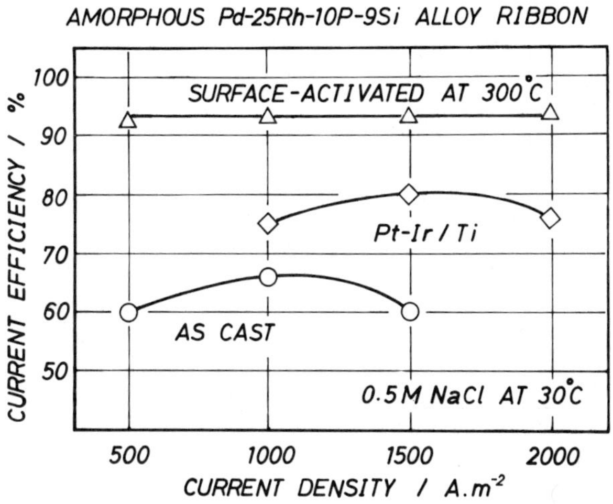

Figure 14 Current efficiencies for chlorine evolution in a 0.5 *M* NaCl solution at 30°C on surface-activated and as-prepared amorphous Pd-25Rh-10P-9Si alloys and on a currently used Pt-Ir/Ti anode [74].

corrosion resistance in 0.5 *M* NaCl solution, although their activity for chlorine evolution is not high [75]. When a few atom percent of platinum group elements are added, the activity of these alloys for chlorine evolution in 0.5 *M* NaCl solution is not greatly increased. Nevertheless, the above-mentioned surface activation treatment cannot be applied to these alloys, because the formation of a passive tantalum or niobium oxide film prevents diffusion of electrodeposited zinc into the amorphous alloys. On the other hand, since valve metals such as niobium and tantalum easily dissolve in hydrofluoric acids, the activation treatment of amorphous nickel-based valve metal alloys containing a few atom percent platinum group elements was performed by immersion in hydrofluoric acids. As is distinct from the presumption of selective dissolution of valve metals, immersion into hydrofluoric acid dissolves not only valve metals but also nickel. Thus, the activity of the amorphous alloys is increased almost four orders of magnitude by the surface activation treatment due to surface enrichment of platinum group elements from only a few atom percent to 80–90 at%, and due also to more than two orders of magnitude increase in effective surface area. An example is shown in Fig. 15 [75]. The current density at 1.2 V(SCE) is increased almost four orders of magnitude due to the surface activation treatment. The surface activation treatment is useful only for amorphous alloys since the uniformly distributed platinum group metals in the amorphous alloys act as the cathode for hydrogen evolution, which is effective for selective dissolution of valve metals and nickel in hydrofluoric acids. The crystalline heterogeneous counterparts cannot be activated by the same treatment.

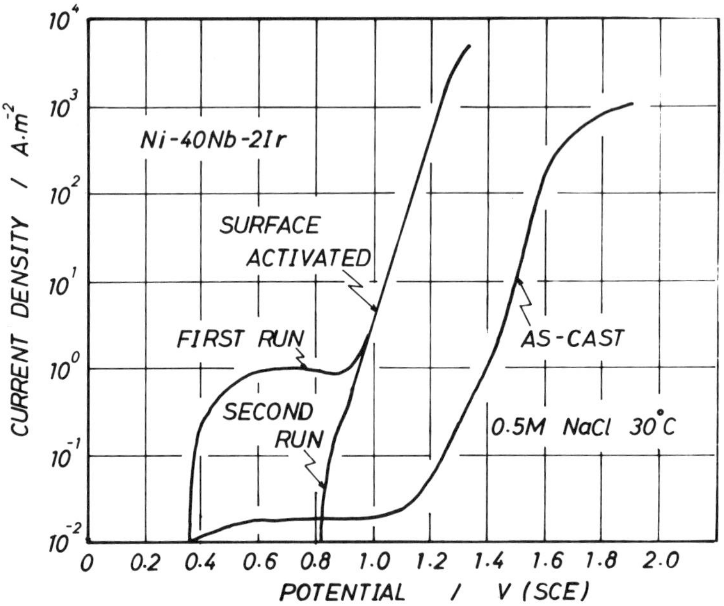

Figure 15 Change in anodic polarization curves by surface-activation treatment for the amorphous $Ni_{58}Nb_{40}Ir_2$ alloy measured in 0.5 *M* NaCl at 30°C. The first and second runs indicate the number of measurements of the polarization curves after surface activation treatment [75].

C. Anode for Electrowinning of Metals

An insoluble anode such as a Pb-Ag alloy is currently used for electrowinning of metals, mostly in strong sulfuric acids. Electrowinning consists of the electrodeposition of metal on the cathode and oxygen evolution at the anode. The lead alloy anode is known to have relatively high corrosion resistance but a considerably high overpotential for oxygen evolution. In addition, stress-induced bending of the lead anode occurs as a result of growth of PbO_2 film due to insufficient corrosion resistance, and hence reformation of the anode is carried out at regular intervals. Highly active and corrosion-resistant anode materials have been found in the surface-activated amorphous nickel-valve metal alloys containing a few at% platinum group elements (76).

D. Preparation of Electrodes

One serious restriction for practical utilization of melt-spun amorphous alloys for electrode materials is their high electric resistance, based on their limited thickness of several tens of μm, which is required to achieve quench rates sufficiently high to stabilize an amorphous structure during casting. For example, the specific resistivity of amorphous $Pd_{56}Rh_{25}P_{10}Si_9$ alloy is 101.2 μΩ-cm; in order to maintain the current density, due to voltage drop caused by the electric resistance of the anode, within 780–1000 A-m^{-2} the anode input terminals must be connected every 2 cm of the length of the anode of 5-μm thickness. Preparation of amorphous surface alloys having superior electrocatalytic activity on conventional crystalline bulk metals acting as an electric conductor is, therefore, performed by using CO_2 laser and electron beam processing techniques.

Preparation of active palladium-based surface alloys on conventional metals was carried out by the laser processing, starting from the crystalline ribbon-shaped alloys adhesively attached on the surface of bulk, metal by short time melting in a furnace [77]. The anodic characteristics of the electrodes were not different from melt-spun counterparts [78].

Amorphous Ni-Nb-P group metal surface alloys on a niobium substrate were prepared by laser processing specimens composed of a bulk niobium substrate covered by electrodeposited nickel and platinum group metals [79]. Because of high reflectivity to the infrared CO_2 laser beam (10.6-μm wavelength) more than 90% of the beam energy is reflected and hence laser processing of a wide surface area requires a very long time. In contrast, when an electron beam was used, amorphous surface alloys can be prepared with very low energy densities. For instance, when 2-kW CO_2 laser and 6-kW electron beam machines are used for the formation of 1-m^2 amorphous surface alloy from 15-μm-thick nickel-plated niobium specimen, the processing times are expected to be 22.5 hr and 22 min, respectively [80]. Consequently, electron beam processing is more convenient than laser beam processing for the preparation of amorphous surface alloys.

V. CATALYSTS

The thermodynamically metastable nature of amorphous alloys is expected to be quite effective in providing a high catalytic activity when the metallic alloy surfaces without surface films directly take part in catalytic reactions. However, the catalytic reactions in which the amorphous state of alloy catalysts is responsible have not been found so far. Instead, because of unique alloy compositions being available in the form of single-phase solid solutions, amorphous alloys have been found to be quite effective not only as catalysts but also as precursors of active catalysts.

A. Fuel Cell Reactions

Methanol-air fuel cells produce electricity during oxidation of methanol to carbon dioxide and water. For the operation of this fuel cell at ambient temperature, platinum is always required as a catalyst. However, because of oxidation, intermediates of methanol are strongly adsorbed to the platinum catalyst; this results in the high catalytic activity of platinum metal for electrochemical oxidation of methanol fading away after only a short time. The strength of adsorption depends upon the electronic state of the catalyst, and hence can be modified by alloying. However, alloying of platinum black powder catalysts is not easy. In contrast, the extension of the solid solubility in amorphous alloys to form a single-phase solid solution permits the creation of new alloys containing various necessary elements; an activation treatment is easily applied to such homogeneous single-phase alloys. Utilizing these superior characteristics of amorphous alloys, new catalysts for fuel cell electrodes have been prepared.

Some surface-activated amorphous palladium-based alloys have a high electrocatalytic activity for methanol oxidation in a concentrated alkaline solution [72]. It is, however, necessary to remove carbonate ions produced by methanol oxidation in alkaline solutions. Accordingly, an electrode capable of working in acidic so-

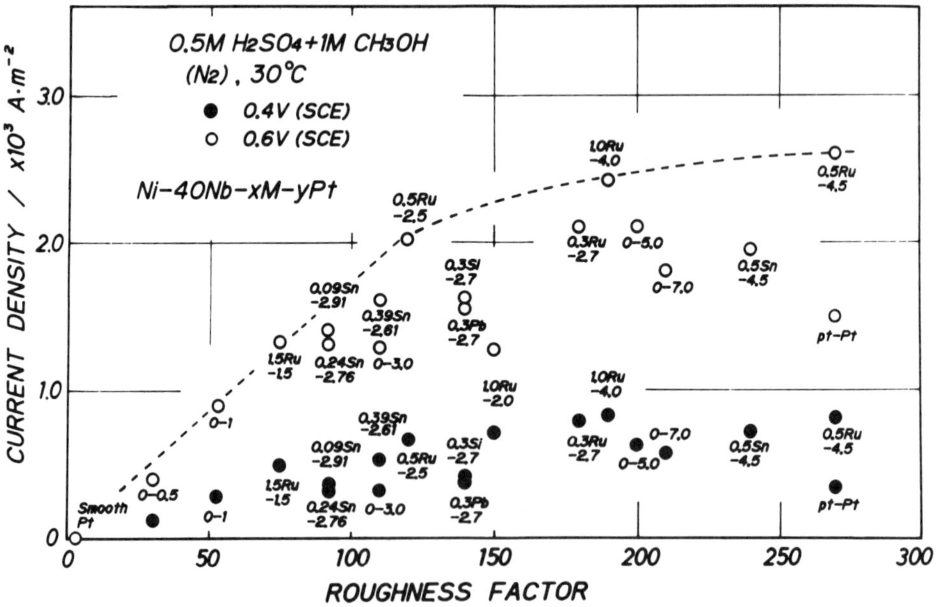

Figure 16 Change in catalytic activity of surface activated amorphous $Ni_{57}Nb_{40}Pt_3$ alloys for methanol oxidation in 0.5 *M* H_2SO_4–1 *M* CH_3OH at 0.4 V (SCE) (●) and 0.6 V (SCE) (○) and at 30°C as a function of roughness factor [81].

lutions has been desirable, but the surface-activated palladium-based alloys are not corrosion resistant in strong sulfuric acids. When amorphous nickel-based valve metal alloys containing a few atom percent platinum group elements are activated by immersion in hydrofluoric acids, several alloys show higher activities for methanol oxidation than the platinized platinum sheet [81]. Figure 16 shows the relationship between methanol oxidation current densities and the roughness factor of the specimens. The roughness factor corresponds to the surface area of platinum existing on the specimen surfaces [81]. Consequently, there is a number of amorphous alloys having higher activities for methanol oxidation than platinized platinum.

When immersion of amorphous nickel-based valve metal–platinum group metal alloys in hydrofluoric acids is continued for a few tens of minutes, they transform to powder catalysts [82]. The gas-permeable porous electrodes were prepared by coating the mixture of these powder catalysts with carbon black powder and PTFE on a porous carbon sheet, and then subsequently baking this material. The activities of the electrodes for methanol oxidation are significantly higher than the same type of electrode made of platinum black catalyst. Furthermore, the durability of the electrodes for methanol oxidation is excellent. Figure 17 shows an example of the difference between the durability of an electrode made of amorphous $Ni_{57}Nb_{40}Pt_3$ alloy and an electrode made of platinum black catalyst and platinized platinum sheet electrode [83]. The gas permeable porous electrode made of amorphous alloy catalyst is significantly superior to the platinum black electrode in terms of the durability not only for methanol oxidation but also for oxygen reduction [83]. According to XPS analysis, a slight charge transfer occurs from platinum to alloying element in the powder catalysts made from amorphous alloys [82]. This fact is assumed to be responsible for the long durability of the electrode made of powder catalysts.

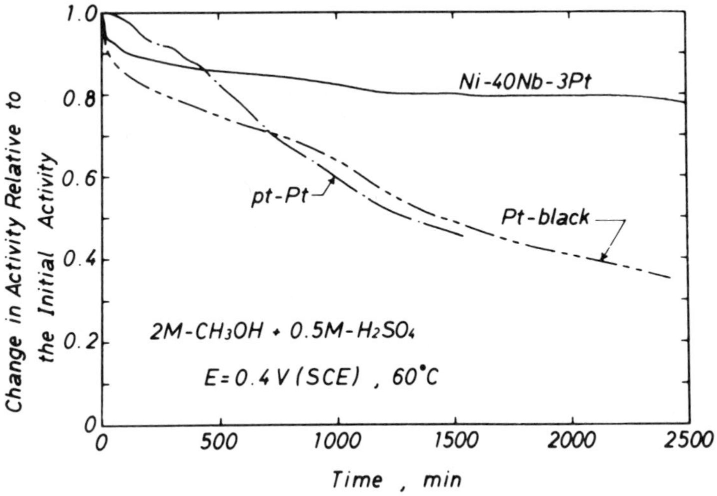

Figure 17 Change in activities of electrodes made of amorphous Ni-40Nb-3Pt alloy and platinum black and platinized platinum sheet for methanol oxidation with time [83].

B. Other Catalyst Reactions

The catalytic activity of amorphous alloys for gas-phase reactions has at first been examined in the hydrogenation of carbon monoxide [84]. A comparison of 15 different amorphous Ni-Fe-metalloid alloys and their crystalline counterparts reveals that amorphous alloys have higher activity, except for one composition. Because there is no difference in reaction mechanism between amorphous and crystalline alloys, the amorphous alloys seem to possess a higher density of active sites [84–86]. The catalytic activity of amorphous Pd-Zr [87] and Ni-Zr [88,89] is greater than that of alumina-supported ruthenium and nickel catalysts, which are known to have very high turnover frequency based on CO chemisorption. Their activity increases at steady state due to increase in the active sites during the catalytic reaction. This has been interpreted differently for each alloy family as follows: the activity increase of the amorphous Ni-Zr alloys seems due to significant increase in the effective surface area as a result of surface cracking, based on the formation of ZrO_2, and due to the formation of highly active NiZr compound. The catalytic reaction for the amorphous Pd-Zr catalysts leads to no increase in the effective surface area, but rather to the formation of catalytically active new crystalline phase consisting of palladium, zirconium and oxygen. Guczi and his co-workers [90–94] also studied the hydrogenation of carbon monoxide by using amorphous Fe-B and Fe-Ni-B alloy catalysts, and found that partially crystallized alloys have higher activity than crystalline counterparts due to stabilization of small iron particles and some oxides which prevent the formation of inactive graphite and carbide.

Other than hydrogenation of carbon monoxide, amorphous alloy catalysts have been used for hydrogenation of olefine [95,96], isomerization and deuteration of *cis*-cyclododecene [97] and (+) apopinene [98], hydrogenation of propenel [99], and methanol synthesis [100]. Amorphous alloys have been found to have higher catalytic activity and/or reaction selectivity in comparison with crystalline catalysts.

VI. CONCLUDING REMARKS

As briefly described in this chapter, rapidly solidified alloys possess a variety of very interesting chemical properties, and various novel materials having unique and attractive properties have been developed through rapid solidification techniques.

Corrosion resistant amorphous alloy coatings have already been available commercially, and electron beam processed amorphous alloy electrodes will become available soon.

Catalytic properties are still a novel field. There are a number of chemical and electrochemical reactions which require active catalysts. These reactions are waiting to be examined in regard to the applicability of new amorphous alloys as catalysts and catalyst precursors.

REFERENCES

1. M. Naka, K. Hashimoto, and T. Masumoto, *J. Japan Inst. Metals*, *38*, 835 (1974).
2. T. Masumoto and K. Hashimoto, *Ann. Rev. Mater. Sci.*, *8*, 215 (1978).
3. T. Masumoto, K. Hashimoto, and M. Naka, in *Rapidly Quenched Metals III*, Vol. 2 (The Metals Society, London, 1978), p. 435.

4. K. Hashimoto, *Suppl. Sci. Rep. Res. Inst. Tohoku Univ.*, *A*, 233 (1978).
5. K. Hashimoto, *Suppl. Sci. Rep. Res. Inst. Tohoku Univ.*, *A-28*, 201 (1980).
6. K. Hashimoto and T. Masumoto, in *Treatise on Materials Science and Technology 20*, edited by H. Herman (Academic Press, New York, 1981), p. 290.
7. K. Hashimoto and T. Masumoto, in *Glassy Metals*: Magnetic Chemical and Structural Properties, edited by R. Hasegawa (CRC Press, Boca Raton, FL, 1983), p. 235.
8. R. B. Diegle, N. R. Sorensen, T. Tsuru, and R. M. Latanision, in *Treatise on Materials Science and Technology 23*, edited by J. C. Scully (Academic Press, New York, 1983), p. 59.
9. K. Hashimoto, in *Amorphous Metallic Alloys*, edited by F. E. Luborsky (Butterworths, London, 1983), p. 471.
10. K. Hashimoto, in *Proc. 5th Int. Symp. on Passivity*, edited by M. Froment (Elsevier, Amsterdam, 1983), p. 235, 247.
11. R. M. Latanision, A. Saito, R. Sandenbergh, and S.-X. Zhang, in *Chemistry and Physics of Rapidly Solidified Materials*, edited by B. J. Berkowitz and R. O. Scattergood (TMS-AIME, Warrendale, PA, 1983), p. 153.
12. R. B. Diele, *J. Non-Cryst. Solids*, *61*, *62* (1984).
13. K. Hashimoto, in *Rapidly Quenched Metals*, Vol. II, edited by S. Steeb and H. Warlimont (Elsevier, Amsterdam, 1985), p. 1449.
14. M. D. Archer, C. C. Corke, and B. H. Harji, *Electrochim. Acta*, *32*, 13 (1987).
15. K. Asami, A. Kawashima, and K. Hashimoto, *Mater. Sci. Eng.*, *99*, 475 (1988).
16. *Proc. Symp. on Corrosion*, *Electrochemistry and Catalysis of Metallic Glasses*, edited by R. B. Diegle and K. Hashimoto (The Electrochemical Society, Pennington, 1988).
17. K. Hashimoto, M. Naka, and T. Masumoto, *Sci. Rep. Res. Inst. Tohoku Univ.*, *A-26*, 48 (1976).
18. M. Naka, K. Hashimoto, and T. Masumoto, *Sci. Rep. Res. Inst. Tohoku Univ.*, *A-26*, 283 (1977).
19. M. Naka, K. Hashimoto, and T. Masumoto, *J. Non-Cryst. Solids*, *29*, 61 (1978).
20. K. Hashimoto, K. Asami, M. Naka, and T. Masumoto, *Sci. Rep. Res. Inst. Tohoku Univ.*, *A-27*, 237 (1979).
21. M. Naka, K. Hashimoto, A. Inoue, and T. Masumoto, *J. Non-Cryst. Solids*, *31*, 347 (1979).
22. P. Cadet, M. Keddam, and H. Takenouti, in *Proc. 4th Int. Conf. Rapidly Quenched Metals*, Vol. II, edited by T. Masumoto and K. Suzuki (The Japan Institute of Metals, Sendai, 1982), p. 1477.
23. P. Kovacs, J. Farkas, L. Kiss, A. Lovas, and K. Tompa, in Ref. 22, p. 1471.
24. K. Asami, K. Hashimoto, T. Masumoto, and S. Shimodaira, *Corrosion Sci.*, *16*, 909 (1976).
25. K. Hashimoto, M. Naka, J. Noguchi, K. Asami, and T. Masumoto, in *Corrosion Monograph Series*, edited by R. P. Frankenthal and J. Kruger (The Electrochemical Society, Pennington, 1987), p. 156.
26. K. Hashimoto, K. Kobayashi, K. Asami, and T. Masumoto, in *Proc. 8th Int. Cong. Metallic Corrosion* (DECHEMA, Frankfurt, 1981), p. 70.
27. A. Kawashima, K. Shimamura, S. Chiba, T. Matsunaga, K. Asami, and K. Hashimoto, in *Proc. 4th Asian-Pacific Corrosion Control Conf.*, Vol. 2 (Tokyo, 1985), p. 1042.
28. A. Mitsuhashi, K. Asami, A. Kawashima, and K. Hashimoto, *Corrosion Sci.*, *27*, 957 (1987).
29. K. Hashimoto, T. Masumoto, and S. Shimodaira, in *Passivity and Its Breakdown on Iron and Iron Base Alloys*, edited by R. W. Staehle and H. Okada (National Association of Corrosion Engineers, Houston, TX, 1975), p. 34.
30. K. Hashimoto and K. Asami, in Ref. 25, p. 749.

31. K. Hashimoto, K. Asami, M. Naka, and T. Masumoto, *Corrosion Eng. (Boshoku Gijutsu)*, *28*, 271 (1979).
32. K. Asami, K. Hashimoto, and S. Shimodaira, *Corrosion Sci.*, *18*, 151 (1978).
33. K. Hashimoto, K. Asami, and K. Teramoto, *Corrosion Sci.*, *19*, 3 (1979).
34. K. Hashimoto and K. Asami, *Corrosion Sci.*, *19*, 3 (1979).
35. K. Asami and K. Hashimoto, *Corrosion Sci.*, *19*, 1007 (1979).
36. A. Kawashima, K. Asami, and K. Hashimoto, *Corrosion Sci.*, *24*, 807 (1984).
37. K. Hashimoto, K. Osada, T. Masumoto, and S. Shimodaira, *Corrosion Sci.*, *16*, 71 (1976).
38. K. Hashimoto, M. Naka, K. Asami, and T. Masumoto, *Corrosion Eng. (Boshoku Gijutsu)*, *27*, 279 (1987).
39. K. Shimamura, A. Kawashima, K. Asami, and K. Hashimoto, *Sci. Rep. Res. Inst. Tohoku Univ.*, *A-33*, 196 (1986).
40. R. B. Diegle and J. E. Slater, *Corrosion*, *32*, 155 (1976).
41. T. Kulik, J. Baszkiewicz, M. Kaminski, J. Latuszkiewicz, and H. Matija, *Corrosion Sci.*, *19*, 1001 (1979).
42. M. Naka, K. Hashimoto, and T. Masumoto, *Corrosion*, *36*, 679 (1980).
43. S. Kapsta and K. E. Heusler, *Z. Metall.*, *72*, 785 (1981).
44. R. B. Diegle, in Ref. 22, p. 1457.
45. K. Hashimoto, M. Naka, K. Asami, and T. Masumoto, *Corrosion Sci.*, *19*, 165 (1979).
46. K. Hashimoto, in Ref. 13, p. 1109.
47. T. Tsuru and R. M. Latanision, *J. Electrochem. Soc.*, *129*, 1402 (1982).
48. T. R. Anthony and H. E. Cline, *J. Appl. Phys.*, *49*, 1248 (1978).
49. Y. Nakao, K. Nishimoto, R. Ueno, and M. Kouso, in *JSCE '86* (Japan Soc. Corrosion Eng., Tokyo, 1986), p. 183.
50. A. Kawashima and K. Hashimoto, *Corrosion Sci.*, *26*, 467 (1986).
51. G. J. Yurek, D. Eisen, and A. Garratt-Reed, *Metall. Trans.*, *13A*, 473 (1982).
52. N. Wade, T. Koshihama, and Y. Hosoi, in *Proc. 4th Asian-Pacific Corrosion Control Conf.*, Vol. I (Tokyo, 1985), p. 61.
53. K. E. Heusler and D. Huerta, in *Proc. 9th Int. Cong. Metallic Corrosion*, Vol. I (National Research Council Canada, Ottawa, 1984), p. 222.
54. Y. Masumoto, A. Inoue, A. Kawashima, K. Hashimoto, A. P. Tsai, and T. Masumoto, *J. Non-Cryst. Solids*, *86*, 121 (1986).
55. H. Habazaki, S.-Q. Ding, A. Kawashima, K. Asami, K. Hashimoto, A. Inoue, and T. Masumoto, *Corros. Sci.*, *29*, (1989).
56. R. Becker, G. Sepold, and P. L. Ryder, *Scripta Metall.*, *14*, 1238 (1980).
57. S. Yatsuya and T. B. Massalski, in Ref. 22, p. 169.
58. K. Asami, T. Sato, K. Hashimoto, T. Shigematsu, and S. Kimura, in Ref. 22, p. 177.
59. H. W. Bergmann and B. L. Modike, in Ref. 22, p. 181.
60. R. Becker, G. Sepold, and P. L. Ryder, in Ref. 11, p. 235.
61. K. Asami, T. Sato, and K. Hashimoto, *J. Non-Cryst. Solids*, *68*, 261 (1984).
62. H. Yoshioka, K. Asami, A. Kawashima, and K. Hashimoto, *Corrosion Sci.*, *27*, 981 (1987).
63. R. B. Diegle and M. D. Merz, *J. Electrochem. Soc.*, *127*, 2030 (1980).
64. R. Wang, *J. Non-Cryst. Solids*, *61*, *62*, 613 (1984).
65. I. A. Anderson, E. A. Dobisz, J. H. Perepezko, R. E. Thomas, and J. Wiley, in Ref. 11, p. 169.
66. R. G. Walmesley, Y. S. Lee, F. Marshall, and O. A. Stevenson, *J. Non-Cryst. Solids*, *61*, *62*, 625 (1984).
67. K. Shimamura, K. Miura, A. Kawashima, K. Asami, and K. Hashimoto, in Ref. 16, p. 232.
68. H. Yoshioka, A. Kawashima, K. Asami, and K. Hashimoto, in Ref. 16, p. 242.

69. H. Yoshioka, A. Kawashima, K. Asami, and K. Hashimoto, in *Proc. MRS Int. Meeting on Advanced Materials*, Meterials Research Society, Pittsburgh, Vol. 3, p. 429 (1988).
70. M. Hara, K. Hashimoto, and T. Masumoto, *J. Appl. Electrochem*, *13*, 295 (1983).
71. M. Hara, K. Hashimoto, and T. Masumoto, *J. Non-Cryst. Solids*, *13*, 295 (1983).
72. A. Kawashima and K. Hashimoto, *Sci. Rep. Res. Inst. Tohoku Univ.*, *A-31*, 174 (1983).
73. N. Kumagai, A. Kawashima, K. Asami, and K. Hashimoto, in Ref. 13, p. 1795.
74. N. Kumagai, A. Kawashima, K. Asami, and K. Hashimoto, *J. Appl. Electrochem.*, *16*, 565 (1986).
75. N. Kumagai, Y. Samata, A. Kawashima, K. Asami, and K. Hashimoto, *J. Appl. Electrochem.*, *17*, 347 (1987).
76. K. Shimamura, K. Miura, A. Kawashima, K. Asami, and K. Hashimoto, in Ref. 16, p. 390.
77. N. Kumagai, K. Asami, and K. Hashimoto, *J. Non-Cryst. Solids*, *87*, 123 (1986).
78. N. Kumagai, K. Asami, A. Kawashima, and K. Hashimoto, *J. Electrochem. Soc.*, *133*, 1876 (1986); *87*, 123 (1986).
79. N. Kumagai, Y. Samata, A. Kawashima, K. Asami, and K. Hashimoto, *J. Non-Cryst. Solids*, *93*, 78 (1987).
80. N. Kumagai, Y. Samata, S. Jikihara, A. Kawashima, K. Asami, and K. Hashimoto, *Mater. Sci. Eng.*, *99*, 489 (1988).
81. A. Kawashima, T. Kanda, and K. Hashimoto, *Mater. Sci. Eng.*, *99*, 521 (1988).
82. A. Kawashima, T. Kanda, K. Asami, and K. Hashimoto, in Ref. 16, p. 401.
83. T. Kanda, A. Kawashima, K. Asami, and K. Hashimoto, in Ref. 16, p. 411.
84. A. Yokoyama, H. Komiyama, H. Inoue, T. Masumoto, and H. M. Kimura, *J. Catalysis*, *68*, 355 (1981).
85. A. Yokoyama, H. Komiyama, H. Inoue, T. Masumoto, and H. M. Kimura, *Scripta Metall.*, *15*, 365 (1981).
86. A. Yokoyama, H. Komiyama, H. Inoue, T. Masumoto, and H. M. Kimura, *Nippon Kagakukaishi*, 199 (1982).
87. A. Yokoyama, H. Komiyama, H. Inoue, T. Masumoto, and H. M. Kimura, *Chem. Lett.* 195 (1983).
88. Y. Shimogami, H. Komiyama, H. Inoue, T. Masumoto, and H. M. Kimura, *Chem. Lett.*, 66 (1985).
89. Y. Shimogami, H. Komiyama, H. Inoue, T. Masumoto, and H. M. Kimura, *J. Chem. Eng. Jpn.*, *21*, 293 (1988).
90. G. Kisfaludi, K. Lazar, Z. Schay, L. Guczi, Cs. Fetzer, G. Konczos, and A. Lovas, *Appl. Surf. Sci.*, *24*, 225 (1985).
91. L. Guczi, Z. Zsoldos, and Z. Scahy, *J. Vac. Sci. Technol. A*, *5*, 1070 (1987).
92. G. Kisfaludi, Z. Schay, L. Guczi, G. Konczos, A. Lovas, and P. Kovacs, *Appl. Surf. Sci.*, *28*, 111 (1987).
93. G. Kisfaludi, Z. Schay, L. Guczi, *Appl. Surf. Sci.*, *29*, 367 (1987).
94. Z. Zsoldos, Z. Schay, and L. Guczi, *Surf. Interfacial Anal.*, 12 (1988).
95. S. Yoshida, H. Yamashita, T. Funabiki, and T. Yonezawa, *J. Chem. Soc. Chem. Comm*, 964 (1982).
96. T. Imanaka, J. Tamaki, and S. Teranishi, *Nippon Kagakukaishi*, 1064 (1985).
97. W. E. Brower, Jr., M. S. Matyjaszczyk, T. L. Pettit, and G. V. Smith, *Nature*, *301*, 497 (1983).
98. G. V. Smith, O. Zahra, A. Molnar, M. M. Khan, B. Rihiter, and W. E. Brower, Jr., *J. Catal.*, *83*, 238 (1983).
99. M. Funakoshi, H. Komiyama, and H. Inoue, *Chem. Lett.*, 245 (1985).
100. M. Shibata, Y. Ohbayashi, N. Kawata, T. Masumoto, and K. Aoki, *J. Catal.*, *96*, 296 (1985).

19

Applications of Rapidly Solidified Soft Magnetic Alloys

Carl H. Smith
AlliedSignal Research and Technology, Morristown, New Jersey

I. INTRODUCTION

The development of soft magnetic applications of rapidly solidified alloys is a classical example of the development of an experimental technology (rapid solidification) preceding the concept of the eventual commercial applications by several years. The possibility of ferromagnetism in amorphous systems was predicted by Gubanov in 1960 [1], the same year that the first rapidly solidified amorphous alloys were reported by Pol Duwez's group at Caltech [2]. The first ferromagnetic rapidly solidified amorphous alloy, $Fe_{80}P_{13}C_7$, was reported by that same group in 1967 [3]. Efforts to develop rapidly solidified alloys for commercial applications began at Allied-Signal, Inc. (then Allied Chemical Corp.) in 1971 but concentrated on mechanical and strength applications. The soft magnetic properties of the first commercial metallic glass, METGLAS* alloy 2826, $Fe_{40}Ni_{40}P_{14}B_6$, which was introduced in 1973, were not recognized until measurements made by Graham's group at the University of Pennsylvania in 1974 [4]. The slow rate of development of soft magnetic applications of rapidly solidified alloys is understandable when considering the many precursory technologies that needed to be developed. A discussion of alloy and casting technology development for various rapidly solidified soft magnetic alloys (RSSMA) appears in other chapters of this book and in reviews and other books [5–9]. In this chapter I will discuss how these developments allowed various soft magnetic applications to be developed.

A. History of RSSMA Applications

The earliest attempts to measure magnetic properties of ferromagnetic metallic glasses and to find practical applications had to await not only their production,

*METGLAS is a registered trademark of AlliedSignal Inc. for amorphous metallic alloys.

but also their production in a more geometrically regular form than mere small splats. Perceived markets due to the mechanical properties of metallic glasses resulted in research and development at Allied Chemical, prompting the process development for casting continuous wires and narrow ribbons. The first commercially available ribbon, mentioned above, was chosen for its modest raw materials cost, ease of casting, and good tensile properties. METGLAS alloy 2826 was available as 1.7 mm × 50 μm ribbons. Its fortuitous combination of soft magnetic properties and attractive mechanical properties resulted in the development of METSHIELD,* the first commercial product sold utilizing METGLAS ribbon [10]. A pilot business was set up in 1976 to commercialize this flexible, woven magnetic shielding "cloth" made of narrow amorphous alloy ribbons.

Two developments which occurred in the mid-1970s allowed amorphous alloys to be considered for the large transformer steel market. First, iron-boron alloys with high saturation induction, low 60-Hz losses, and low raw material cost were demonstrated [11]. Second, the development of planar-flow casting technology made available ribbons several centimeters in width [12]. Needless to say, both the iron-based alloys and the casting technology have evolved through continuous development since then.

Early applications of metallic glasses were definitely influenced by the price of this new material. At first sold by the foot (later by the meter), amorphous alloys were used mainly in the production of research papers and graduate students. Even the first kilogram-quantity pricing in 1978, $300/kg, placed the cost of this material significantly above that of other soft magnetic materials. Only applications that required small quantities of magnetic material with large value added in manufacture and that required unique combinations of properties were economically practical at that stage of development. Magnetic tape heads using amorphous ribbon were one such early application [13].

At the same time, however, the promise of large quantities of material in the future at prices competitive with grain-oriented silicon iron resulted in several programs at transformer manufacturers to develop the technology of using these materials in distribution transformers. Several key strategic projects funded by the Electrical Power Research Institute (EPRI) helped spur the development of larger-scale production facilities, transformer core technology, and prototype transformer manufacture.

As production increased and the prices decreased due to economies of scale, metallic glasses became attractive in many more applications such as toroidal magnetic cores for high-frequency applications, targets in article surveillance systems, sensor and transducer applications, and high-energy pulsed-power applications. New alloys with improved properties were developed for specific applications such as low-magnetostriction cobalt-based alloys for tape heads and for very high frequency applications. In these applications RSSMA were able to displace nickel-iron and thin silicon-iron ribbons and cores as their prices and properties became competitive. The history of high-volume price trends for iron-based metallic glasses from Allied-Signal Inc. are shown in Fig. 1 [14]. It is interesting to note that early projections for prices have been closely matched by later actuality. Such predict-

*METSHIELD is a trademark of AlliedSignal Inc. for metallic fabric used in magnetic shielding.

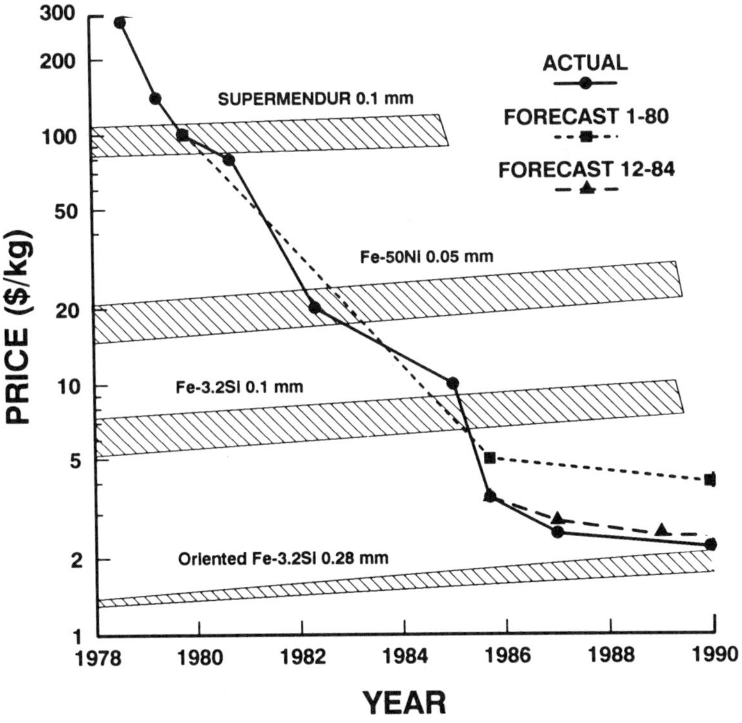

Figure 1 High-volume price trends of iron-based amorphous alloys from AlliedSignal Inc. together with historical forecasts. Prices are also shown for conventional soft magnetic materials [14,15].

ability is important when large investments are necessary both in development and in production facilities for large-scale applications such as distribution transformers. Also shown in Fig. 1 are typical prices for conventional polycrystalline soft magnetic materials [15].

The same rapid solidification technology originally developed for amorphous alloys has been more recently applied to crystalline materials for soft magnetic applications. Crystalline ribbons of Fe–6.5 Si with zero magnetostriction and isotropic in-plane properties have been produced by rapid solidification [16]. These materials appear to be of interest for high-frequency motors. Sendust (Fe-Si-Al alloy), another notoriously difficult alloy to produce in useful forms by ingot technology, has been rapidly solidified into ribbons for use in magnetic tape heads [17].

By the end of the decade of the 1980s, magnetic applications of rapidly solidified alloys are a commercial reality in the United States, Europe, and Asia. Large-scale production of rapidly solidified alloys have fulfilled the promise of lower prices. Ribbon as wide as 0.2 m is commercially available. A diversity of soft magnetic applications of rapidly solidified alloys are now on the market and will be discussed later.

B. Importance of Timing

The success of an application depends not only on the development of the required technology, but also on the readiness of the market. When research on rapidly solidified alloys began in the 1960s or even when commercial development began in the early 1970s, there did not exist a large demand for energy efficient 60-Hz transformers. Electricity was "penny cheap," and the demand was for smaller and more inexpensive transformers. It took the oil embargo of 1973 and oil prices increasing by an order of magnitude before one could consider a material which would result in an energy-saving transformer, albeit larger and more expensive. Fortunately, rapid solidification technology was already being developed based on other perceived markets.

In the late 1970s new requirements for high-energy pulsed-power systems for accelerators and lasers could not be easily met with conventional switching techniques. Magnetic pulse compression techniques using saturable-core reactors, first developed in 1951 by Melville [18], were resurrected and their capabilities elevated to new power levels using newly developed metallic glasses [19]. High induction alloys developed for 60- and 400-Hz transformer applications proved to be of great value in these pulsed power applications with magnetization rates many orders of magnitude higher than those of the original intended applications.

High-frequency magnetic cores of various soft magnetic materials have found increasing use in energy efficient, switched-mode power supplies (SMPS). These power supplies, which were virtually unknown until the mid-1970s, are now rapidly displacing less efficient linear power supplies. Millions of switched-mode power supplies are used today in electronic equipment from TVs to PCs. RSSMA have had an increasing role in this application area especially as frequencies and power levels have increased. Yet SMPSs is another market opportunity which was neither significantly developed nor predictable when metallic glasses were first developed and commercialized.

C. Overview

In this chapter I will first discuss the properties of RSSMA both amorphous and crystalline which are important in magnetic applications. In most applications it is not just one property, but a combination of properties, both magnetic and mechanical, which make rapidly solidified soft magnetic alloys attractive. Next, several applications of these materials, which offer historical, technical, or commercial significance, will be described. Finally, a few projections about the future will be risked.

II. APPLICATIONS-RELATED PROPERTIES

Soft magnetic materials, by definition, are easy to magnetize and demagnetize. A small change in applied field results in a large change in induction. The distinction between soft and hard magnetic metals was traditionally related to their mechanical properties and hence the names. Soft magnetic materials were annealed to minimize impediments to domain wall motion and were mechanically soft. Hard magnetic materials were precipitation-hardened to introduce pinning centers to impede domain wall motion [20]. RSSMA are unique among metals in that they are magnetically soft while mechanically hard. Figure 2 is a plot of hardness versus dc coercive field, a typical measure of magnetic "softness," for various classes of magnetic materials. The coercive field is the reverse magnetic field required to reduce the induction to zero after the material has been saturated. As a typical division of materials, soft magnetic materials have coercive fields less than 1000 A/m and hard magnetic materials have coercive fields greater than 10 kA/m [21]. Materials falling between these limits are called semihard.

An additional important consideration for soft magnetic materials in many applications is that they can be easily magnetized and demagnetized repetitively at a high frequency. While low-frequency properties are dominated by the mobility of domain walls due to imperfection and grain boundaries, high-frequency properties are dominated by the dynamics of domain wall motion. The high resistivity of RSSMA, which arises from both their composition and their lack of crystalline order, together with their thin ribbon form, result in less eddy current damping of domain wall motion than in conventional polycrystalline soft magnetic metallic alloys.

The homogeneous and isotropic character of amorphous alloys not only contributes to the ease of domain wall motion but also results in an absence of magnetocrystalline anisotropy energy—the energy required to change the direction of magnetization relative to crystalline axes. A small field-induced anisotropy energy can be created by applying a magnetic field during annealing, allowing a preferred or easy direction of magnetization to be oriented to suit a given application. Applying a longitudinal field in the circumferential direction in a core during annealing results in a square *B-H* loop with high remanence. A field transverse to the ribbon direction, on the other hand, produces a sheared *B-H* loop with low remanence

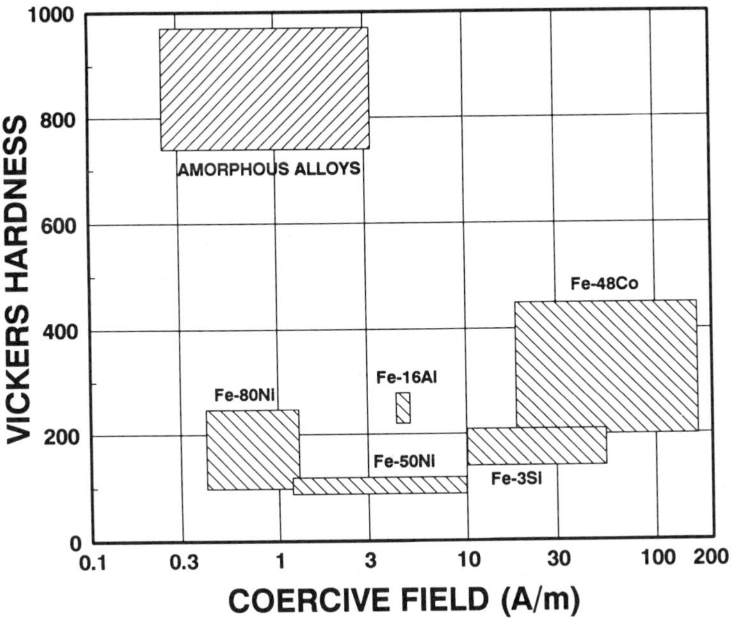

Figure 2 Mechanical hardness versus dc coercive field for several soft magnetic materials.

and relatively constant permeability. The energy required to create a domain walls in metallic glasses is correspondingly low due to their low anisotropy energies further contributing to their magnetic softness. See Chapter 17 for a more complete description of tailoring the magnetic properties to a given application.

The lack of crystallinity in amorphous magnetic alloys not only provides the intrinsic advantages mentioned above, it also allows the their compositions to be continuously varied over fairly wide ranges allowing many of their magnetic properties to be "tuned" to an extent not possible in crystalline alloys. Complete discussions of the magnetic and engineering properties of magnetic metallic glasses appear in Chapters 16 and 17, in a recent review article by O'Handley [22], and in books by Luborsky [7] and Hasegawa [23].

While amorphous alloys owe many of their properties to their amorphous nature, other RSSMA are useful due to their crystalline nature. Rapid solidification allows some materials which were formerly difficult to roll into thin sheets to be cast directly as thin ribbons. Fe–6.5 Si, a brittle material when conventionally prepared, can be rapidly quenched and annealed to form a thin ribbon with isotropic, in-plane magnetic properties [24]. This material has potential for application as magnetic cores in rotating machinery. Another recently discovered class of rapidly solidified soft magnetic material which depend upon their crystalline nature for their properties is nanocrystalline alloys [25]. In this chapter I will concentrate on properties of RSSMA relevant to applications.

A. Relationship Between Properties and Applications

When choosing a new material such as RSSMA for existing applications, one usually looks for a candidate material which most closely matches the properties of the conventional soft magnetic material being used, but one which will result in some advantage such as higher efficiency, less weight, smaller size, or lower cost. Very often the new material cannot be directly substituted into an existing design which was optimized for a previous material. A new design, and sometimes even new technology, must be considered.

The primary magnetic characteristic of interest can often be determined by the class of application. Devices which handle large quantities of power, such as transformers in power distribution systems, usually require materials with the highest saturation induction. High-frequency transformers, on the other hand, are usually limited by temperature rise during operation and require materials with the lowest losses. Shielding applications and many electromagnetic transducer applications require high-permeability materials. Applications with saturable-core reactors, such as magnetic amplifiers, require materials with very square *B-H* loops, while sensitive ground fault interrupters and pulse transformers for digital telecommunications require low remanence and high permeability. Therefore, several RSSMA have been developed for differing applications just as there are numerous conventional soft magnetic materials. The choice of the proper magnetic material is often not a trivial consideration for the magnetic design engineer.

B. Importance of Combinations of Properties

While soft magnetic materials for a given application are usually chosen primarily for a single property such as high permeability, high saturation induction, or low losses over some frequency range, additional characteristics such as cost, availability in usable configurations, and other magnetic characteristics determine whether a given material is practical for the application. On the other hand, some properties may limit a material's usefulness. For example, a low Curie temperature or the existence of a compensation temperature can limit its useful temperature range at the high and low ends, respectively. Stress sensitivity and mechanical properties of a material may affect its incorporation into a device or core or limit the methods of fabricating requisite laminations. It is usually the material with a combination of favorable properties which is best suited to a given application rather than the material with one "best" property.

There are several examples of applications which require two or more properties in combination. Magnetic tape heads utilize materials with high permeability at high frequencies as well as with wear resistance to the moving magnetic recording tape. In addition, low sensitivity to stress is essential. At the other extreme, sensor applications rely on combinations of properties such as Young's modulus, anisotropy energy, and magnetostriction to provide high sensitivity to changes in stresses imposed upon the sensing element. As mentioned above, RSSMA have unique combinations of properties, such as mechanical hardness while being magnetically soft. In addition, some of their magnetic properties can be tailored to a particular requirement by an appropriate heat treatment.

C. Important Properties and Comparisons

RSSMA can be divided into amorphous alloys and crystalline alloys, based on whether the structural form in which they are used is primarily amorphous or crystalline. This classification does not necessarily indicate the form in which they were cast. Some alloys such as the new nanocrystalline alloys are amorphous when they are cast but are crystallized during heat treatment. In addition, some amorphous alloys are annealed to produce a small volume fraction of crystallites to improve their high-frequency properties. Amorphous alloys are usually further divided into subfamilies based on the primary transition metal—iron, nickel, or cobalt. Rapidly solidified crystalline alloys are currently more limited in number than amorphous alloys, with Sendust (Fe-Al-Si) and related alloys, Fe–6.5 Si, and the new nanocrystalline alloys as the principal classes. The important properties of each of these groups of materials will be discussed below in comparing these families. Further properties will be discussed within sections on applications.

A discussion of properties of magnetic alloys needs to distinguish between intrinsic properties—those properties which depend upon alloy chemistry—and technical properties—those properties which can be significantly influenced by processing. Intrinsic properties such as saturation magnetization, resistivity, density, Curie temperature, and saturation magnetostriction are affected very little by annealing. Technical properties such as magnetic losses, permeability, coercive field, and remanence, on the other hand, can be changed by processing after the alloy has been rapidly solidified. In addition, technical properties such as losses and permeability are functions of the frequency and induction level at which they are measured. Comparisons at a single frequency and induction level are at best incomplete and at worst misleading.

1. Amorphous Alloys

Over the past two decades three main families of rapidly quenched ferromagnetic amorphous alloys have been developed and made commercially available: iron-based alloys, cobalt-based alloys, and iron-nickel-based alloys. All of these alloys belong to the class of amorphous alloys consisting of approximately 80 at% transition metals and 20 at% metalloids, especially B and the combination of Si and B. To be of commercial interest, the alloys must be reasonably ductile as-cast, have Curie temperatures and crystallization temperatures well above expected use temperatures, and have stable magnetic properties at use temperatures. These requirements somewhat limit the alloys of interest from the almost unlimited number of amorphous alloys which can be produced by rapid quenching. For example, Ni-based alloys without significant Fe or Co additions have Curie temperatures below room temperature [26].

Iron-based alloys have primarily been developed for applications demanding the highest saturation induction and the least expensive raw materials [14]. These alloys are of interest in the large market for transformers in electrical power distribution systems. Substitution of a few percent Co for Fe results in alloys with saturation induction as high as 1.8 T [27]. Proper annealing of selected Fe-based alloys refines the domain width by precipitating α-Fe crystallites and can produce materials with attractive high-frequency properties [28–30]. Fe-based alloys, however, have values of saturation magnetostriction, λ_s, as high as 35 ppm. Magnetostriction results in a coupling between the magnetic state of a material and its state of strain. Therefore, optimum properties are only achieved after annealing out the strains developed during rapid quenching and during winding cores. On the other hand, the same coupling between strain and magnetic state makes these materials useful in sensors of various types.

The quest for strain-insensitive alloys with near-zero magnetostriction has concentrated on Co-based alloys, as discussed in other chapters. The lowest losses and highest permeabilities at high frequencies are achieved in Co-based alloys [31]. Several alloys have been developed for differing applications. Excellent as-cast magnetic properties can be achieved with alloys with saturation inductions as high as 0.7 T. Other Co-based alloys with saturation induction as high as 1.2 T have been developed for tape heads for high-coercivity metal-particle tapes [32]. Rounding out the Co-based alloys are alloys with saturation induction around 0.55 T, extremely low losses, and square *B-H* loops. These alloys are useful in applications such as high-frequency magnetic amplifiers.

Fe-Ni-based alloys with roughly equal Fe and Ni combine high permeability and moderate saturation magnetostriction with corrosion resistance and high tensile strength. The earliest commercial amorphous alloy, METGLAS alloy 2826, is a member of this family. Much early research concentrated on this alloy due to its commercial availability; however, due to the relatively rapid aging of the magnetic properties of this phosphorus-containing alloy [33], it has been replaced by more stable alloys [34]. Limited applications for this family of alloys has resulted in fewer species of this family being available than of the Fe- and Co-based alloy families.

A variety of amorphous alloys from the families discussed above are listed in Table 1, as are intrinsic magnetic properties and other properties relevant to their typical applications. Magnetic properties such as losses and permeabilities over a variety of frequencies and induction levels will be given in graphical form in the sections on applications.

Table 1 Properties of a Variety of Commercially Available Ferromagnetic Amorphous Alloys

METGLAS alloy designation	Saturation induction B_s (T)	Curie temperature T_c (°C)	Crystallization temperature T_x (°C)	Electrical resistivity ρ ($\mu\Omega$m)	Saturation magnetostriction λ_s (ppm)	Density δ (g/cm^3)	Thermal expansion CTE (ppm/K)	Vicker's hardness H_v 50-g load
Iron-based amorphous alloys								
2605TCA	1.56	415	550	1.37	27	7.18	7.6	860
2605SC	1.61	370	480	1.35	30	7.32	5.9	880
2605CO	1.80	415	430	1.23	35	7.56	8.6	810
2605S-3A	1.41	358	535	1.38	20	7.29	6.7	860
2605SM	1.28	310	520	1.28	17	7.50	6.6	990
Cobalt-based amorphous alloys								
2705M	0.70	365	520	1.36	<1	7.80	12.1	900
2714A	0.55	205	550	1.42	<1	7.59	12.7	960
2705MN	1.2	407	560	1.00	<1	8.0	—	950
Iron-nickel-based amorphous alloys								
2826MB	0.88	395	410	1.38	12	7.90	11.7	740

Table 2 Properties of POWERCORE Strip, METGLAS Alloy 2605S-2, and Oriented Fe–3.2 Si Alloy

	METGLAS alloy 2605S-2 ribbon	POWERCORE strip 100 mm × 0.13 mm	Oriented silicon steel (0.23 mm)
Specific gravity	7.2	7.2	7.65
Lamination factor (%)	<90	90	95–97.5
Young's modulus (GPa)	180	150	122[a]
Thermal conductivity (W/m-K)	8	9	25
Electrical resistivity (μΩm)	1.3	—	0.48
Saturation induction (T)	1.58	1.58	2.00
Remanent induction (T)	1.40	1.20	1.30
Coercive field (A/m)	1.6	3.2	6.4
Curie temperature (°C)	415	415	740
Saturation magnetostriction (ppm)	27	27	27[b]

[a]Along down web direction.
[b]Along [100] direction.
Source: From Ref. 36

2. *POWERCORE* Strip*

The thickness of Fe-based amorphous alloys suitable for power-handling transformers is typically limited to approximately 25 μm or 1/1000 in. Thicker ribbon tends to be brittle due to slower cooling rates during casting. In comparison, conventional polycrystalline materials used in power-handling transformers are about of 10 times as thick. As an alternative to stacking many layers of 25-μm amorphous ribbons to make transformer cores, Allied-Signal, with funding from EPRI, developed techniques of laminating 5–10 layers of metallic glass ribbon into a single packet named POWERCORE strip. POWERCORE strip can be produced in continuous lengths from amorphous transformer ribbon by using pressure and temperature which results in local plastic flow around ribbon irregularities [35]. The strip retains its amorphous structure and can be wound around a 37-mm radius without fracture or delamination [36]. The lamination factor of the processed strip is about 0.90, which is greater than that of stacked or wound ribbon. Good magnetic properties are obtained without additional thermal processing. Properties of POWERCORE strip are shown in Table 2 and are compared to those of METGLAS alloy 2605S-2 (similar to current production transformer core alloy, METGLAS alloy 2605TCA) and oriented Fe–3.2 Si alloy. Core losses and exciting power at 60 Hz are shown in Fig. 3 as functions of compressive stress.

An additional potential advantage of POWERCORE strip in stacked-core transformers is due to the very low anisotropy energy of amorphous alloys. In-plane core losses measured parallel and perpendicular to the strip direction are essentially the same. Oriented Fe–3.2 Si alloy has significantly higher losses and much higher exciting power perpendicular to the direction of grain orientation. The directionality of the oriented Fe–3.2 Si alloy affects the path that the flux can take in the vicinity of joints and, therefore, increases the losses near the joints of stacked-core trans-

*POWERCORE is AlliedSignal's registered trademark for consolidated strip of amorphous metal.

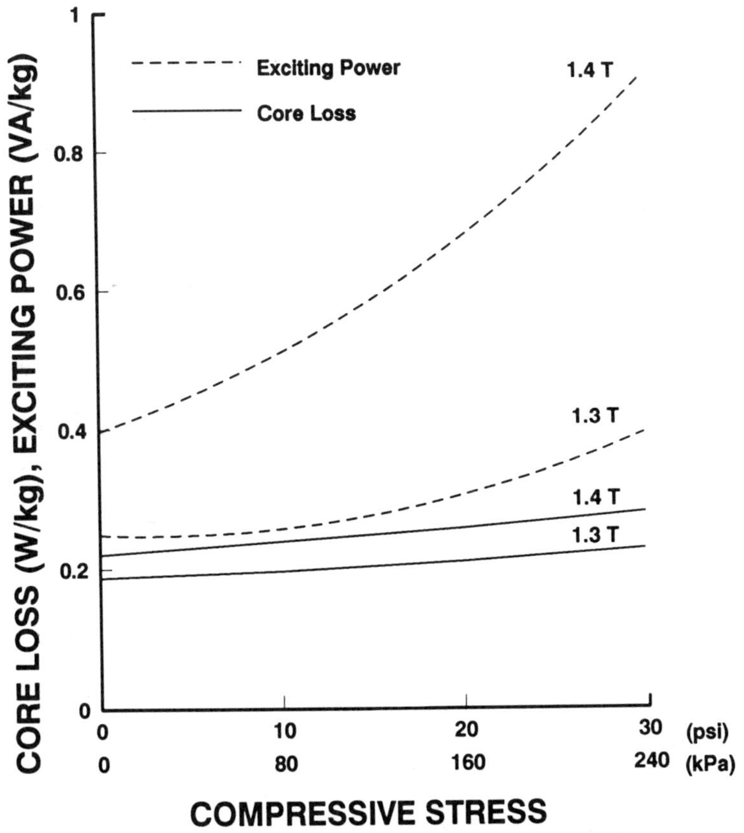

Figure 3 Core loss and exciting power at 60 Hz for POWERCORE strip under compression at 1.3- and 1.4-T induction levels [36].

formers. Similar effects have been observed in stacked cores made by amorphous alloy ribbon [37].

Two factors have thus far limited the use of POWERCORE strip in stacked-core transformers: stress sensitivity and hardness. The stress sensitivity of this material, as shown in Fig. 3, requires that cores be carefully supported and not subjected to extreme clamping stresses. The extreme hardness of metallic glasses when compared to conventional Fe–3.2 Si alloy requires innovative techniques to cut laminations. Due to the decrease in hardness as the glass transition temperature is approached [38], methods have been developed using localized heat during cutting [36,39]. POWERCORE strip has also been cut on a laboratory scale using lasers [40].

3. *Rapidly Quenched Crystalline Alloys*

The techniques developed in produce metallic glass ribbon by melt spinning have also been utilized to produce thin ribbon of some crystalline alloys which are otherwise difficult to manufacture to thin gage. Sendust (Fe–10 Si–5 Al) has long been known as an iron-based crystalline alloy with very high permeability [41]. The high permeability of this alloy is due to the possibility of simultaneously achieving near-zero magnetostriction and very low magnetocrystalline anisotropy energy in the ternary phase field. Rolling thin sheets of this material is impossible since it is very brittle due to atomic ordering. Rolling this material typically reduces it to a powder. The name Sendust, in fact, is due to this propensity. Consolidated parts made from the Sendust powder and binder have been used in choke cores. More recently, the high permeability and stress insensitivity together with the high wear resistance of this alloy have been utilized in tape heads. New high-coercivity metal-particle media recording tapes require write heads with higher saturation induction than is available in ferrites or Permalloys. Sendust has a desirable saturation induction of 1.2 T. The manufacturing process, however, involves grinding ingot material to a head width of approximately 100 μm. Rapid quenching of this alloy from the melt has offered a method of directly producing thin ribbons suitable for laminating into magnetic tape heads [17].

Another alloy which was historically recognized to have promising magnetic properties but was not producible in sheet form is Fe–6.5 Si. The increased level of Si compared to the conventional 2–4 wt% Si results in increased resistivity,

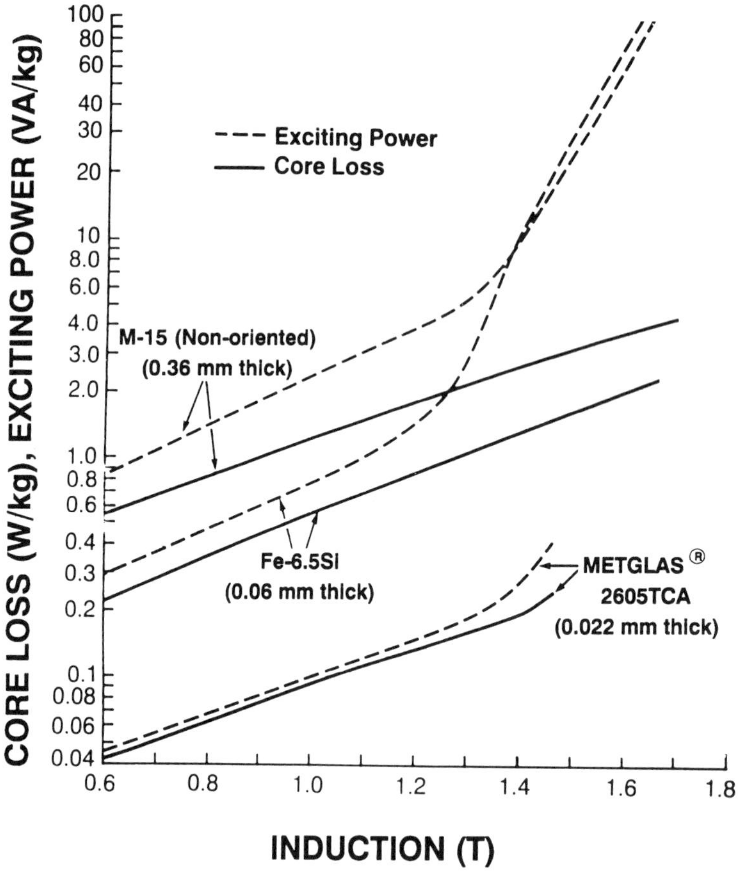

Figure 4 Core loss and exciting power at 60 Hz for Fe–6.5 Si, M-15 nonoriented Fe–3.2 Si alloy, and METGLAS alloy 2605TCA [14].

reducing eddy current losses. At 6.5 wt% Si the average magnetostriction is reduced to almost zero, corresponding to a minimum in losses [42]. At the same time saturation induction is reduced to 1.8 T compared to 2.0 T for conventional Fe–3.2 Si alloy. However, ordered DO_3 and B2 phases at Si levels above 5 wt% result in a brittle alloy which cannot be heat-treated and rolled to achieve sheet material necessary for most transformer applications. Rapid solidification of this alloy controls the ordering reaction and directly produces a 20- to 100-μm-thick ribbon. Annealing procedures have been developed which produce a (001)⟨*uv*0⟩ fiber texture with isotropic in-plane properties [43]. Annealing can grow the columnar grains from about 10 μm (as-cast) up to 1 mm in diameter. Two-stage annealing improves the magnetic properties by refining the size of the B2 domains and decreasing the coercive field, and therefore the hysteresis loss [24].

The 60-Hz core losses of this alloy are significantly lower than nonoriented M-15 electrical steel and are similar to those of oriented M-4 along its preferred direction. For example, the losses at 60 Hz and 1.4 T are approximately 1.3 W/kg. Losses and exciting power for these alloys at 60 Hz are given in Fig. 4. Losses and exciting power for METGLAS alloy 2605TCA, an amorphous alloy oriented by field annealing for transformer uses, are shown for comparison.

Due to its isotropic properties, Fe–6.5 Si alloy offers greatest promise as a low-loss core material in rotating electrical equipment. Figure 5 shows 60-Hz *B-H* loops taken at various directions to the ribbon direction. This alloy's high resistivity and thin ribbon form are especially attractive at frequencies above 60 Hz. At 2 kHz

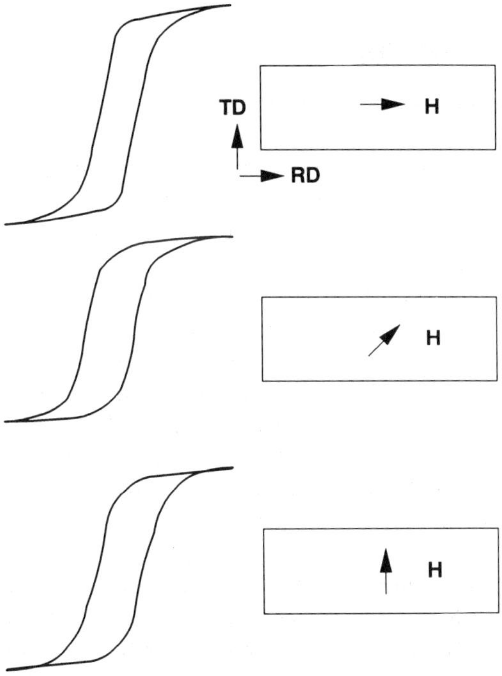

Figure 5 *B-H* loops for annealed Fe–6.5 Si ribbon taken at 60 Hz along, at 45° to, and perpendicular to the ribbon longitudinal direction [24].

and 1.0 T its losses are 10–20% those of conventional nonoriented M-15 Fe–3.2 Si alloy [44]. Recently, NKK Corporation has announced commercial-scale production of Fe–6.5 Si alloy made by rolling and by chemical vapor deposition [45]. Nonoriented sheets 400 mm wide and 0.1 to 0.5 mm thick are available in coil form.

A very recent addition to crystalline RSSMA is that of nanocrystalline alloys. One of the first alloys announced by Yoshizawa et al. was called FINEMET, $Fe_{73.5}Cu_1Nb_3B_9Si_{13.5}$ [25]. These alloys are rapidly quenched from the melt to form amorphous ribbon. Heat treatment above the crystallization temperature of about 510°C fully transforms their structure to bcc Fe solid solution with a homogeneous, extremely fine grain structure of the order of 10–20 nm [46]. These Fe-based alloys with values of B_s up to 1.2 T have coercive fields and high-frequency core losses comparable to those of Co-based amorphous alloys. Field annealing can produce either sheared or square *B-H* loops, depending upon the direction of the field [47]. Lowest losses at high frequencies are achieved by annealing without a field—lower than 40 W/kg at 100 kHz and 0.2 T. It appears that the microstructure results in an averaging of both saturation magnetostriction and anisotropy energy to very low values, resulting in their attractive high-frequency properties. Other nanocrystalline bcc alloys with even higher saturation induction have been recently announced by Suzuki et al. [48]. Rapidly quenched ribbons of $Fe_{91}Zr_7B_2$ were crystallized to yield materials with 15- to 20-nm grain size, saturation induction of 1.7 T, and permeability at 1 kHz of 14,000. Undoubtedly future developments will result in improved versions of these nanocrystalline materials as well as in high-frequency applications for them.

III. LINE-FREQUENCY TRANSFORMERS

Electrical power is usually generated in central locations and delivered over a power grid to the points of use at a low frequency—typically 50 or 60 Hz. The power system in the United States can serve as a typical example. Three-phase transmission line transformers near the generators boost the voltage for high-voltage transmission lines. The voltage of these transmission lines is as high as a megavolt to minimize losses. These large power transformers (PTs) are built in sizes up to 1500 MVA. Substation power transformers with sizes between 250 kVA and several thousand kVA reduce the transmission line voltage to the 2- to 12-kV level, at which it is widely distributed to the vicinity of its point of use. Near the point of use, the voltage is reduced to 240/120 V for household and business use by single-phase distribution transformers (DTs). These 10–100 kVA and larger pole- or pad-mounted transformers are ubiquitous throughout the American landscape. For industrial power, larger three-phase 208- to 580-V transformers are often the final step-down stage. In large metropolitan areas larger PTs may serve as the final step-down stage for the power consumer.

The many transformers which elevate the voltage for efficient high-voltage power distribution and reduce the voltage to its final level for use, in several-stage power distribution systems, constitute probably the largest use of soft magnetic material. Not only do these transformers offer a potentially large market for RSSMA, but also these new materials offer possibilities for significant improvements in the efficiency of these transformers. The extremely low 60-Hz losses of amorphous alloys can reduce the core losses of these transformers to 20–30% the values for current design transformers constructed from conventional grain-oriented Fe–3.2 Si alloy. The reduction in core losses is even larger if amorphous alloy cored transformers are compared to older, less efficient, in-service transformers. A review of progress in amorphous materials for electrical power distribution by Fish and Smith has been mentioned earlier [14].

A. Distribution Transformers

The final transformers in the electrical power distribution network, the distribution transformers, have the greatest installed capacity. Therefore they have the greatest total core material and operate at the smallest fraction of full load of all the transformers in the system. Their core losses constitute approximately 1% of the total power generated. By 1990 this loss is estimated to be 60 billion kW-hr/yr—worth over $4 billion at $0.07/kW-hr.

The installed base of DTs in the United States is 35–40 million units, and each year over one million new DTs are purchased by utility companies as replacements and for new service. Therefore, an excellent opportunity exists for installing a significant number of high-efficiency amorphous-cored transformers each year.

1. Economic Considerations

The current emphasis on energy efficiency by electrical utilities is a direct consequence of the oil embargo of 1973 and the subsequent increases in the costs of energy and energy production. Only in the latter half of the 1970s did total ownership costs rather than initial costs enter the DT purchase decisions of utilities. Total ownership costs take into account not only the purchase price, but also the cost of operating the device over its expected lifetime.

Loss evaluation, as practiced in over 80% of transformer purchases today, assigns an equivalent cost per watt to both load losses (conductor losses) and no-load losses (core losses). The total ownership cost (TOC) then becomes the purchase price plus the conductor losses times the load-loss cost factor plus the core losses times the no-load-loss cost factor. The calculation of the loss cost factors is a complicated econometric calculation which includes not only current cost of energy and money and expected transformer lifetime and loading, but also estimates of future costs, inflation, and changes in utilization. Utilities vary in their evaluations of these cost factors, but in 1981 load-loss cost factors ranged from $0 to $3 per watt and no-load-cost factors from $0.60 to $8 per watt [49]. The general trend has been for no-load losses to be valued at over twice the value of load losses, reflecting the average loading of DTs. This ratio ensues since a transformer experiences no-load losses continuously, while the load losses are proportional to the power delivered.

The average loss evaluation factors have been steadily increasing from approximately $0.70 and $1.50 per watt in 1978 to $1.20 and $3.80 per watt in 1984 [49]. Multiplying the more recent of these factors by typical losses for two 25-kVA transformers made with an amorphous core (15.4-W core loss, 328-W load loss) and a Fe–3.2 Si alloy core (57-W core loss, 314-W load loss) results in loss costs of $452 and $593, respectively [50]. In fact the TOC for a transformer can be twice the initial purchase price of the transformer.

Evaluation trends have not only been toward higher evaluation factors, but also additional components are beginning to be included in the equations. Some of the more progressive utilities include estimates of future load growth and costs of building future generating capacity when calculating loss evaluation factors. These additional factors can add as much as $2.00 per watt to the value of no-load losses in transformers.

Reduced losses in transformers are also reflected through the power generation system as saving of nonrenewable (and increasingly imported) fuel resources and reduction of emissions from power plants. These additional advantages to low-loss transformer core materials are beginning to be perceived by public utility regulatory agencies.

2. *Design Considerations*

Given the economic advantages of low core losses attainable with amorphous alloys, rapid adoption by transformer manufacturers might be expected even at somewhat higher initial prices than conventional materials. However, the thin and hard (Rockwell C-63) metallic glass ribbons are not easily substituted into the traditional transformer core manufacturing processes designed for Fe–3.2 Si alloy. Current DT transformer designs in the United States are typically "shell-form" transformers. Each layer of the core is cut while it is wound as a toroid prior to shaping and annealing. The cores are disassembled into subgroups and reassembled through previously wound coil assemblies. Alternative methods of transformer manufacture such as stacked laminations are impractical for amorphous alloys since the hardness and thinness of the ribbons make punching laminations difficult and expensive.

As mentioned above, field annealing of iron-based alloys minimizes losses and produces a square *B-H* loop. Best properties are achieved when the core is first wound and then annealed in its final configuration. A square *B-H* loop is necessary to minimize the volt-amp requirement of the core and also to minimize noise due to magnetostriction. Although these alloys have saturation magnetostriction of typically 30 ppm, little magnetostriction is evident in field-annealed cores. This phenomenon is due to the fact that magnetization proceeds by displacement of 180° domain walls within the vertical section of the *B-H* loop. A change in length in magnetostrictive materials occurs only when the direction of magnetization within a domain rotates. These rotations occur only above the "knee" of the *B-H* loop or the place at which H starts to increase rapidly with increasing B. Transducer measurements at 60 Hz of "engineering" (effective) magnetostriction of longitudinally field-annealed cores have shown essentially no magnetostriction up to 1.4 T along straight sections of the core and only 5 ppm at the corners [51]. Noise measurements on actual transformers have confirmed these results.

Amorphous alloys are somewhat more sensitive to overannealing than are conventional Fe–3.2 Si alloys. The annealing of large (50 kg or more) transformer cores requires adaption of annealing techniques developed for smaller test cores. Temperatures must be ramped up more slowly and core and skin temperatures monitored to avoid overannealing the outer and inner portions of the core [36]. While annealing amorphous Fe-based alloys relieves stresses, reduces losses, and creates a desired easy direction of magnetization, it also causes a certain degree of embrittlement. Whereas as-cast ribbon can be bent back on itself and creased, annealed ribbon may fracture at some nonzero bend radius. This embrittlement after annealing must be considered in transformer manufacturing methods. Alternate annealing methods have been explored. In-line annealing against a heated curved block while a core was being wound resulted in high-temperature, short-time anneals [52]. These cores had magnetic properties somewhat superior to cores produced with conventional no-field anneals. High-temperature anneals have also been developed by dipping cores in molten metal followed by a rapid quenching by immersion in a coolant fluid [53]. This technique proved to produce annealed material with improved ductility. Annealing has also been accomplished by passing a current through the amorphous ribbon to achieve high heating and cooling rates [54]. Such annealing studies serve to demonstrate the extensive research which has gone into adapting amorphous alloys to transformer manufacture.

Needless to say, several methods of utilizing amorphous alloys in DTs have been developed. One of the economic challenges to be solved is designing a transformer such that the core material can be annealed without also annealing the conductor coil windings and their insulation at the same time. Few insulation materials will withstand the 350–400°C anneal temperatures. On the other hand, it is not a simple task to reassemble an annealed amorphous alloy core through conductor coil windings. The earliest large (15 kVA) amorphous-cored transformer used the expedient of winding insulated wires through annealed toroidal cores. This prototype transformer was built at Allied-Signal, Inc. (then Allied Chemical) in 1979 for the MIT Solar Photovoltaic/Thermal Residence Experiment at the University of Texas at Arlington [55]. The high-efficiency transformer was installed as the interface between the photovoltaic power source and the house. Another method has utilized annealed cruciform cores with near-circular cross section so that the windings could be wound by driving a bobbin which had been installed on a straight section of the core. Figure 6 shows an illustration of transformer construction with a cruciform

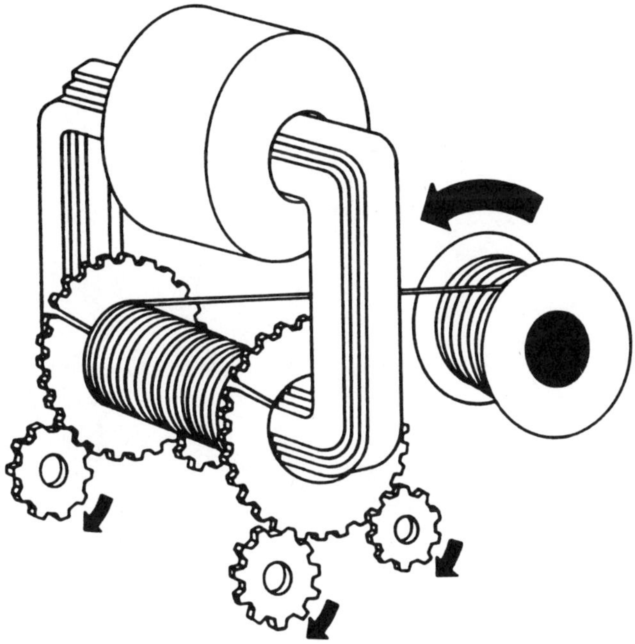

Figure 6 Transformer construction with cruciform cores.

core. Transformer manufacturers such as General Electric and Westinghouse have developed techniques for fabricating annealed cores with each layer cut and with sufficient ductility of the ribbon to open the core up and install prewound coils [56]. The cores are then reinterleaved and banded before being placed into transformer tanks. An example of this method is shown in Fig. 7. An ingenious method of winding a core of preannealed material into toroidal coil windings has been developed by Kuhlman Corporation [57]. A diagram of Kuhlman's TOROFORM design transformer construction is shown in Fig. 8.

Experience with amorphous-cored transformers in the field has been very good to date. Projects to determine the long-term performance of DTs with amorphous alloy cores were started as early as 1983. A joint project with General Electric Transformer Business Department, the Electrical Power Research Institute, and the Empire State Electric Energy Corporation (ESEERCO) was established to explore the feasibility of manufacturing amorphous-cored transformers and their stability in the field [58]. By now six years of stable performance have been documented on the original 25 preprototype 25-kVA pad-mounted transformers installed on utility systems. One thousand more 25-kVA, 15-kV pole-mounted prototype transformers were completed and shipped to utilities during 1985 in a follow-on project. These transformers were installed by 90 utilities across the United States. A comparison of the typical performance of these transformers and conventional transformers is given in Table 3. Test data taken after one and two years

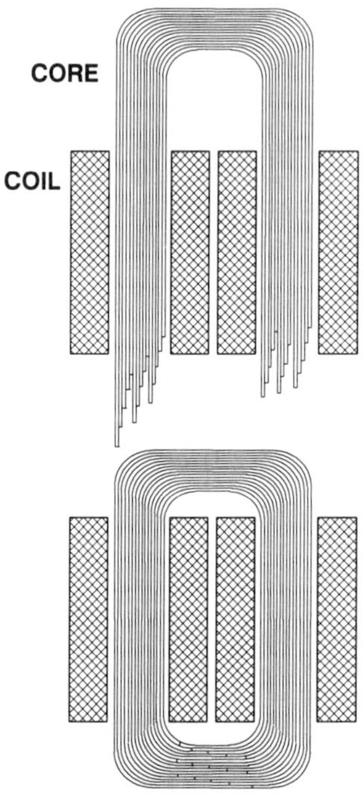

Figure 7 Coil assembly onto a lapped-joint amorphous-alloy-core transformer [56].

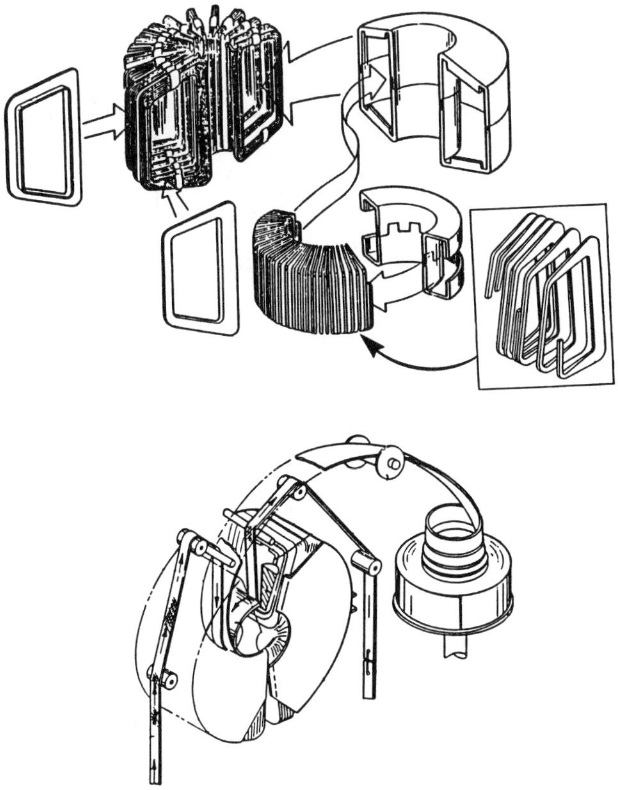

Figure 8 Construction of a Kuhlman TOROFORM transformer [57].

Table 3 Comparative Characteristics of Amorphous-Core and Fe–3.2 Si-Alloy-Core 25-kVA Transformers

Specification	Amorphous metal	Silicon steel
Core loss (W)	15.4	57
Load loss (W)	328	314
Exciting current (%)	0.14	0.36
Temperature rise (°C)	48	57
Audible noise (dB)	33	40
TIF[a] 100%/110% (IT/kVA)	2/10	5/25
Short-circuit test	40 times	40 times
In-rush magnetizing current:		
calc, 0.01 s/0.1 s	21/31 times	23/14 times
Weight (kg)	200	184

[a]Telephone interference factor at load.
Source: From Ref. 59.

in service indicated that the transformers were reliable and the core losses stable. No failures were reported attributable to the amorphous alloy cores [59].

By 1990 transformers were available in the United States from at least four companies: General Electric (GE), Asea Brown Bovari (ABB), formerly Westinghouse Transformer Division, Kuhlman Corporation, and Howard Industries. Single-phase transformers with ratings between 10 and 100 kVA and three-phase transformers with ratings between 75 and 750 kVA are available, with ratings up to 2500 kVA to be available soon [60]. These amorphous-core transformers typically offer 60–70% reduction in core loss compared to current-design silicon-iron-core transformers, comparable or lower conductor losses, and at most a 10% increase in weight. The first cost of amorphous-core transformers is typically 20 to 35% higher than conventional transformers [61]. By 1988 GE alone had sold 10,000 amorphous-cored transformers and expected sales to double each year for several years [62]. Faith in the future of amorphous alloys, which led Allied-Signal's Metglas Products to open a plant in 1989 with the infrastructure to support production of 60,000 tons of METGLAS alloy per year, appears to be well founded.

B. Power Transformers

The construction of power transformers (PT) has traditionally utilized cores of stacked laminations of silicon iron and has differed considerably from the wound-core construction of DTs. The height of these stacked cores can be over a meter. Direct substitution of thin amorphous alloy laminations does not appear to be a cost-effective method for manufacturing amorphous-cored PTs. However, as an EPRI/ESEERCO/Westinghouse demonstration project, a 500-kVA, 14,440/216Y/125 V transformer was constructed from laminations of 25-μm-thick amorphous alloy ribbon [63]. Operating losses of this unit were only one-fourth that of the 1-kW core losses typical of a conventional unit. This transformer is in service with the Niagara Mohawk Power Corp. in Syracuse, New York, in an underground network where low losses are especially important [64].

To address the challenges of stacked core transformers for PTs and large DTs especially in Europe, Allied-Signal, Inc., with funding from EPRI, developed POWERCORE strip. This material, the properties of which were discussed in an earlier section, is fabricated from 5 to 10 layers of amorphous transformer core alloy. The resultant strip, with a lamination factor of approximately 90%, can be cut at elevated temperatures and stacked into a transformer core as a direct replacement for silicon iron [36]. The advantage of this material over silicon iron comes not only from its lower losses, but also from its weaker in-plane anisotropy. Since amorphous alloys have low induced anisotropy energy, they are easy to magnetize in any direction. Stacked-core transformers inevitably have some sections, typically in corners, in which the direction of the magnetic flux is not in the easy direction of magnetization. This condition results in uneven flux sharing between laminations and, therefore, increased losses. Mitered cores with corners cut at angles are often used to minimize this effect. POWERCORE strip, with low anisotropy, provides better flux sharing in the vicinity of joints and allows the use of simpler rectangular plates in cores [65].

The losses of POWERCORE strip without postprocessing approach those for the constituent straight-strip amorphous alloy after annealing. The lack of the requirement for annealing facilities for POWERCORE strip results in the ease of initiating production of amorphous-cored transformers with this material. However, somewhat more care must be exercised with this material than with silicon iron to avoid increases in losses due to stress concentrations and edge shorting in these stacked cores.

IV. SPECIALTY MAGNETIC CORES

There exist a variety of specialized applications for magnetic materials in the form of small cores with one or more windings. The finished magnetic components are either transformers or inductors. The frequencies and voltage waveforms to which these cores are subjected vary over a considerable range as do the magnetic materials used. For want of a better term, these applications of RSSMA are grouped together in this chapter as specialty magnetic cores. Most of the applications in which RSSMA have been found useful utilize wound toroidal cores—the most easily fabricable core configuration using rapidly quenched ribbon. In some cases the cores are cut, either to control effective permeability or to facilitate installation of windings. RSSMA offer a variety of advantages in these application due to specific combinations of their properties, including low losses at high frequencies, high saturation induction, square *B-H* loops, high permeabilities, and high Curie temperatures. Several articles have been written on applications of amorphous alloys, including these specialty or electronic core applications [66–70].

A. Switched-Mode Power Supplies

High-efficiency switched-mode dc-dc power supplies (SMPS) have rapidly supplanted the less efficient traditional linear power supplies over the past decade [71]. These power supplies maintain a regulated output voltage by using feedback to change the duty cycle of a high-frequency switch between source and load. Additional components such as diodes, inductors, and capacitors insure a dc output with very little dissipation of energy by the circuit. SMPSs achieve highly accurate regulation over a wide range of variations in load resistance and input voltage. The quest for greater space and weight efficiency has resulted in ever-increasing switching frequencies in SMPS designs paced by the development of fast semiconductor switches. Higher frequencies also require magnetic materials for the various magnetic components, transformers, chokes, and magnetic amplifiers, which are capable of operating at these higher frequencies. In the past decade, switching frequencies in commercially available units have increased from 20 kHz to over 100 kHz, with laboratory SMPS designs utilizing amorphous alloy cores running at 1 MHz [72]. Recently, small 1-MHz, three-terminal board-mounted switching regulators with power densities up to 200 W/in^3 have become available (although not with amorphous cores) [73].

The uses of magnetic cores in components of a simple SMPS are illustrated in Fig. 9. Inductive components in SMPS include obvious applications such as power transformers and filter inductors or chokes on the outputs [74]. They also can include magnetic amplifiers to regulate the output, base-drive transformers for the switching transistors, current transformers for circuit protection, and even saturable-reactor snubbers on the output diodes [75]. Magnetic amplifiers are especially

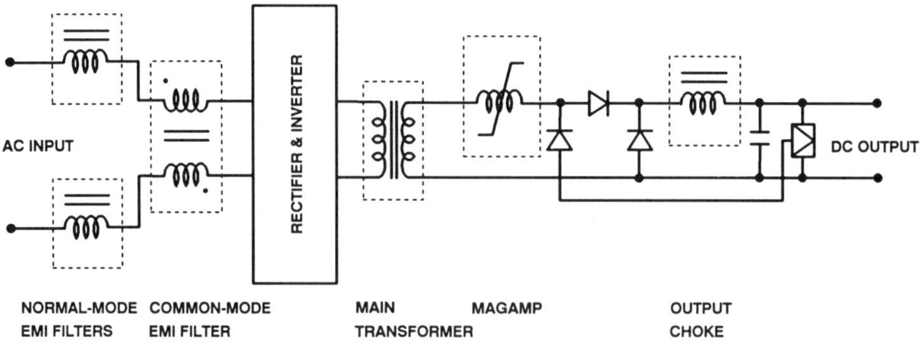

Figure 9 Uses of magnetic materials in a switched-mode power supply.

advantageous on multiple-output SMPSs. Increasing use is also being made on SMPS of EMI filters, discussed separately later. These uses of magnetic materials require differing properties in the magnetic cores.

Various magnetic material, including RSSMA, used for high-frequency applications are compared in Table 4. Representative losses and permeabilities are given as well as saturation induction and Curie temperature. The frequency dependence of losses of several of these materials are given in Fig. 10. The low-frequency region is dominated by the hysteresis losses which are proportional to frequency and the dc coercive field. At higher frequencies, the eddy current losses which are proportional to frequency squared start to dominate. Ferrites, which have much higher resistivities than metals, retain a lower slope to higher frequencies. For this reason, ferrites have lower losses than magnetic alloys at sufficiently high frequencies. However, the frequency of this crossover point varies with induction level, temperature, and the particular materials compared [76].

The quantity of magnetic material used in transformers and inductors is related to the flux-handling capabilities of the materials. This capability is determined by the product of the cross-sectional area of the core and the achievable change in flux density. For the typical core sizes used in switched-mode power supplies, the core area and, therefore, the core volume are inversely proportional to the usable flux swing. Compact transformers for SMPSs have been made from high-induction Fe-based alloys [77]. These transformers are especially suited to high-power (1 kW) switchers at relatively modest (20 kHz) frequencies. In many SMPS transformers, however, especially those used at high frequencies, less than full flux swing is used in order to limit core temperature rise. Therefore, cobalt-based amorphous alloys with near-zero magnetostriction, low anisotropy energy, and the lowest losses are typically used. Their comparatively low saturation inductions of 0.5 to 0.7 T are not a disadvantage when compared to conventional magnetic alloys, and are, in fact, greater than ferrites.

The same considerations which make amorphous alloys attractive for SMPS transformers—low losses and reduced core area—also apply to transformers used for induction heating. As an example, a 25-kHz, 150-kVA transformer was constructed using a Co-based amorphous alloy and compared to designs using Fe-based amorphous alloy and conventional magnetic alloys [78]. To keep core losses below 100 W/kg, maximum flux density had to be limited for all alloys at frequencies

Table 4 Properties of RSSMA and Conventional Magnetic Materials Used in High-Frequency Applications

Material	Elements	B_s (T)	T_c (°C)	λ_s (ppm)	ρ ($\mu\Omega$m)	μ_z (50 kHz, 0.1 T) (μ_0)	L (100 kHz, 0.2 T) (W/kg)	Ribbon thickness (μm)
Supermendur	FeCoV	2.33	950	80	0.35	1500	1200	100
Fe–6.5 Si	FeSi	1.8	700	<1	0.82	1000	1000	100
Deltamax	FeNi	1.6	480	25	0.45	4000	600	50
METGLAS 2605TCA	FeBSi	1.56	415	27	1.37	4000	700	25
Nanocrystalline	FeMoCuBSi	1.2	600	1	1.1	40000	50	25
Supermalloy	NiFeMo	0.82	400	1	0.55	15000	100	25
METGLAS 2714A	CoFeNiBSi	0.55	205	<1	1.42	30000	40	25
H7C4 ferrite	MnZnFeO	0.5	215	−2	10^7	2300	120	—

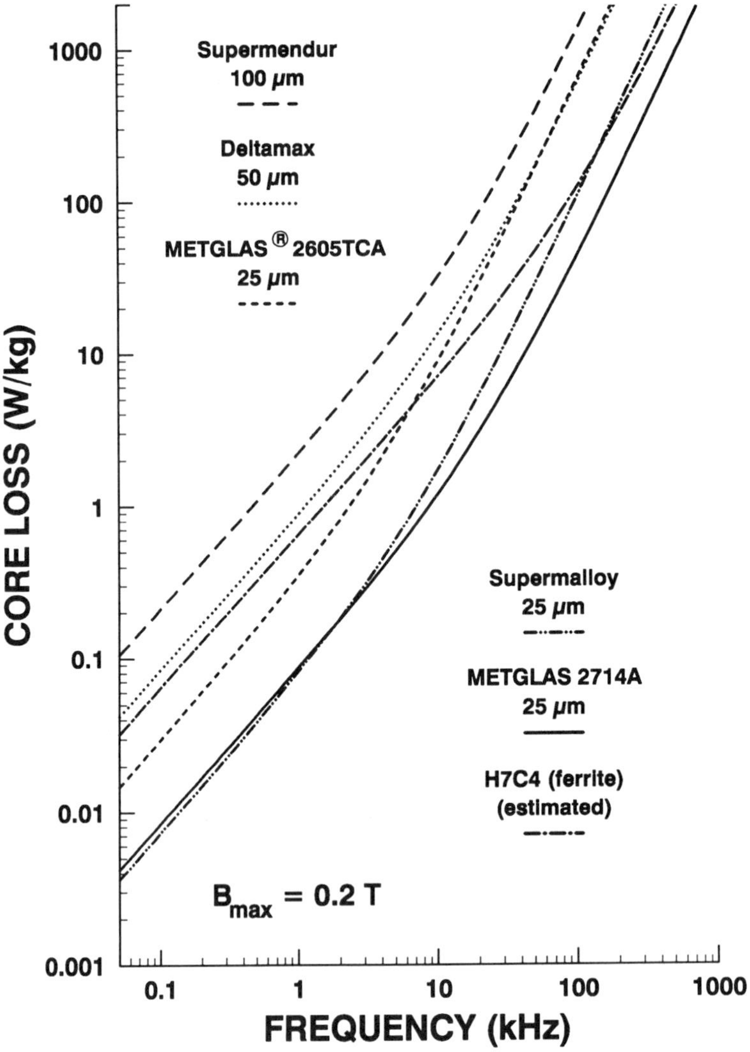

Figure 10 Frequency dependence of losses at 0.2 T for various core materials used in high-frequency applications [100].

above 15 kHz. Although initial projected costs were higher for transformers made from amorphous alloys, the total owning costs, including operating costs, were lower than costs for transformers with polycrystalline cores.

In inductors such as output chokes, on the other hand, the ac component to be filtered out is superimposed on a much larger dc current. These cores require relatively large saturation induction and a low effective permeability to avoid saturation due to the dc output current. Core losses are relatively less important than in transformers, because the core losses are caused by the small ac voltage which causes small flux excursions. Low permeability is achieved by cutting the core and introducing an air gap or by using a core with a distributed gap, such as with a powder core. Recent studies have indicated the superiority of cut cores of Fe-based amorphous alloys over conventional Molypermalloy powder cores and silicon-iron cut cores in high-frequency output chokes [79]. Figure 11 shows comparative losses of typical choke core materials. A related application of amorphous alloys in SMPSs is in cores for flyback transformers. The same magnetic considerations that govern choke design are also applicable to flyback transformers, the transformer used in some simple SMPS designs. A core with a sheared *B-H* loop allows the flyback transformer to operate with unipolar pulses, requiring simpler switching circuitry.

An additional method of shearing over the *B-H* loop for choke applications has been explored by several groups. As mentioned earlier, Fe-based amorphous alloys when annealed to precipitate out α-Fe crystallites have reduced high-frequency losses due to increased nucleation of domain walls. When this precipitation is increased beyond the amount required for minimum high-frequency losses, material with a highly sheared *B-H* loop suitable for chokes is produced [80,81]. Nanocrystalline alloys, transverse-field-annealed to produced sheared *B-H* loops, have also been suggested as core materials for chokes [25].

Magnetic amplifiers or magamps (called transductors in Europe) have regained popularity in the 1980s. Magnetic amplifiers were in common use 40 years ago for controlling 60- to 400-Hz power by the use of saturable core inductors. Extensive bibliographies exist of early publications [82]. The advent of semiconductor switches including SCRs and triacs largely obviated the need for magnetic amplifiers in low-frequency power control circuits. The simplicity, low part count, and inherently soft switching characteristics of self-saturating magnetic amplifiers have recently made them attractive at high frequencies in multiple-output SMPSs to regulate secondary outputs [83,84].

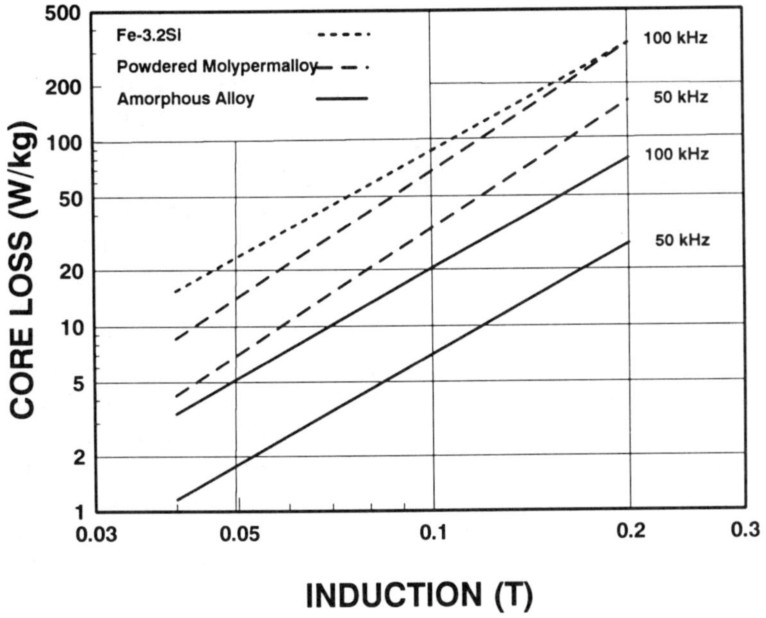

Figure 11 Core losses of commercially available high-frequency inductor materials [79].

B. EMI Filters

The increased use of high-frequency SMPSs, in everything from TVs to PCs, has necessitated regulations by the Federal Communications Commission (FCC) and other regulatory agencies governing the electromagnetic interference (EMI) generated by these devices. To limit high-frequency signals from being fed back into the ac power line, current-compensated chokes are installed in SMPSs on the incoming power lines. These chokes, also called common-mode filters, attenuate high-frequency interference in the 10-kHz to 30-MHz range while passing line-frequency power. Using paired windings on a common core prevents the input current from saturating the core material.

Performance over this broad range of frequencies requires a magnetic material with high permeability and high saturation induction in order to minimize the number of turns necessary for filtering at the low-frequency end. In addition, using a small number of turns minimizes winding capacitance and extends the high-frequency performance. Common mode filters made using Co-based amorphous alloys which have been annealed for sheared *B-H* loops with relative permeability of 100,000 at 10 kHz have demonstrated performance superior to that of filters made with ferrite or Permalloy cores. Insertion losses of up to 40 dB were achieved at 2 MHz. In addition, choke volume was reduced to one-half to one-third the original volume [79,85]. Recently developed nanocrystalline alloys, with saturation induction over twice that of the Co-based amorphous alloys, have also been used in common-mode filters [86]. Different nanocrystalline alloys and anneals were used to produce cores with permeabilities of 20,000 to 50,000 at 10 kHz and chokes with insertion losses of 35 dB at 4–6 MHz. The lower permeability core material, which had an extremely sheared over *B-H* loop, also had very good attenuation of high-voltage pulses such as noise produced by lighting.

C. Digital Telecommunications

The advent of digital telecommunication network service, designed to integrate voice and data communications, provides new opportunities for inductive components. Integrated Services Digital Network (ISDN) is an international set of standards for such services developed by the International Telegraph and Telephone Consultative Committee (CCITT). The basic ISDN interface would provide one data channel at 16 kilobits per second (kb/s) and two voice or data channels at 64 kb/s to every terminal in the system [87]. These channels would be carried to each terminal over two pairs of conductors—one pair for transmit and one pair for receive. Two pulse transformers, therefore, would be required at each terminal as well as two in the network termination. In addition, each pair of conductors in the terminal must be equipped with a current-compensated choke to limit EMI. These components would require a total of three magnetic cores per terminal or telephone (voice terminal) in the system. In addition, each network termination, which converts the information from the two-conductor format from the central exchange to the four-conductor format to the terminal, requires three cores plus one in the central exchange. The inductive components and transformers in a simple, basic-rate ISDN service between central office and terminal are shown in Fig. 12.

Among the ISDN standards that affect inductive components are impedance requirements for terminal equipment and voltage level requirements for pulses

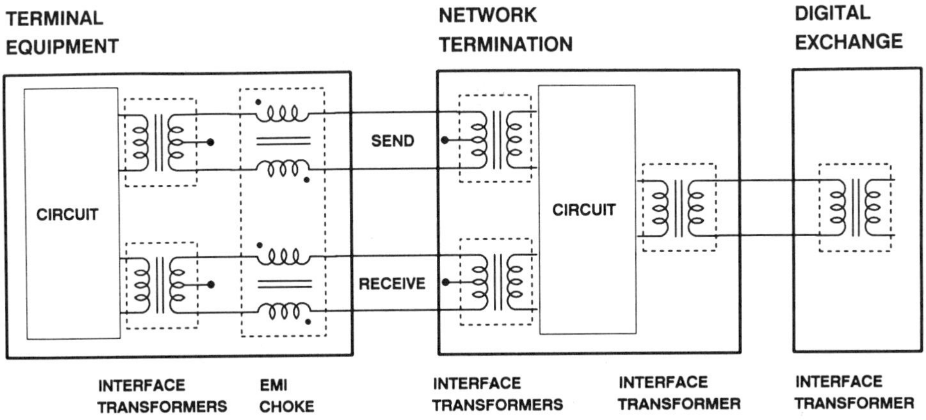

Figure 12 Inductive components and transformers in a basic-rate ISDN connection.

passed by pulse transformers [88]. The stringent pulse-voltage masks require minimization of self-capacitance and leakage inductance in the transformers to minimize ringing. Toroids of amorphous, Co-based alloys allow higher operating induction levels than ferrites and, hence, designs with fewer turns and lower self-capacitance. Toroidal-core transformers have inherently lower leakage inductance than do two-part cores. In addition, reduced transformer size with smaller footprint allows more communications circuits to be handled on the same board. These advantages have resulted in commercial availability of amorphous-core ISDN interface transformers [89].

D. Ground Fault Interrupters

Another application which requires a large quantity of small, toroidal cores is that of ground fault interrupters. These protection devices, often incorporated in electrical outlets in the United States, utilize a core to sense any difference in current between the two conductors supplying line voltage to a load. A difference between the two currents indicates a leakage path to ground, possibly through a person. The imbalance is amplified, either electronically or by transformer action, to activate a relay and deactivate the circuit. Cores of ferrites or polycrystalline Ni-Fe alloys are typically used, but Co-based amorphous alloys annealed for low remanence are gaining increased acceptance [90]. In fact, this application is reported by Yang [91] to be the largest use of amorphous alloys in China, consuming some 30 tons per year.

Co-based amorphous alloys are more expensive than ferrites or polycrystalline Ni-Fe alloys and, therefore, have not had much impact on this market in United States. Proper annealing of Fe-based amorphous alloys originally designed for high-frequency applications, such as METGLAS alloys 2605S-3A and 2605SM, however, has been shown by Fish [92] to produce high 60-Hz permeabilities suitable for this application. When these alloys are annealed with a longitudinal field at temperatures below those which cause precipitation of α-Fe crystallites, maximum impedance permeabilities of 300,000 to 400,000 are achieved. Cores annealed with transverse fields achieved permeabilities of over 100,000 with low values of dc remanence. The high permeabilities and low 60-Hz losses of these alloys are significantly better than those of higher saturation induction Fe-B-Si amorphous alloys designed for 60-Hz transformer applications.

E. Current Measurement Transformers

Metering current transformers (CTs) are used to measure the current in each phase of an ac transmission line. Toroidal magnetic cores are used in order to reduce the leakage flux between primary and secondary windings. The accuracy of current metering by a CT depends not only on the accuracy of the current ratio between primary and secondary but also on its accuracy in preserving the phase angle of

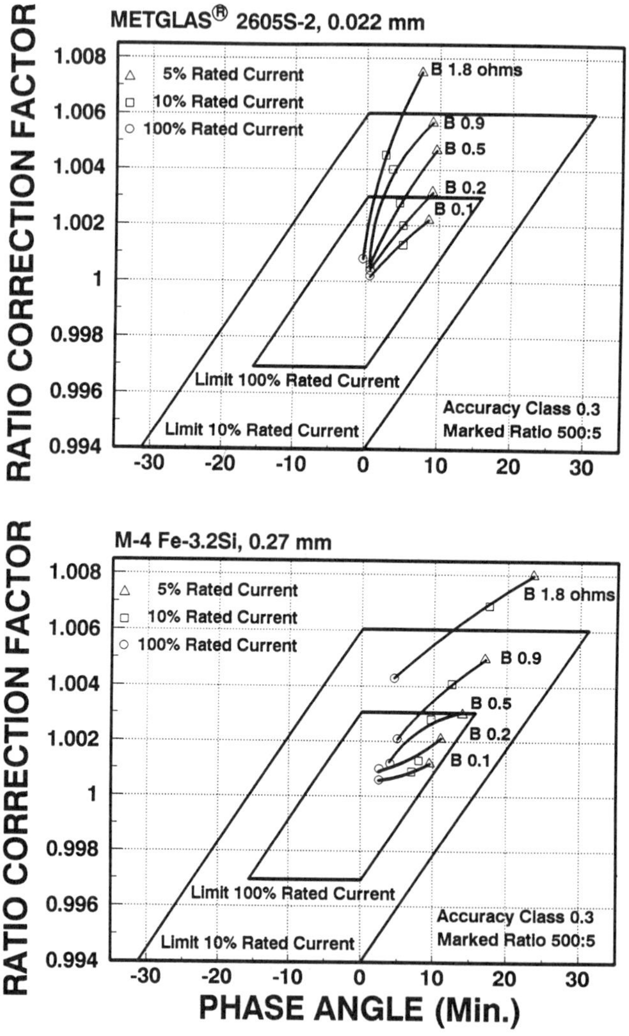

Figure 13 Ratio correction factor and phase angle as a function of percent rated current for various burden resistors for CTs using METGLAS 2605TCA and Fe–3.2 Si cores. Specification limits for 10% and 100% rated current are indicated [93].

the current relative to the voltage. The ratio correction factor depends mainly upon losses in the core material, while the phase error is minimized by using high-permeability material. Low-loss, inexpensive, Fe-based amorphous alloys are ideally suited for this application, the cores for which typically weigh a kilogram or more. Figure 13 shows ratio correction factor and phase angle as a function of percent rated current and burden resistor for CTs with METGLAS alloy 2605TCA and Si-Fe cores [93]. Notice the smaller phase angles in the CT with the amorphous core due to its higher permeability.

Other amorphous-core CTs have been designed with feedback amplifiers to ensure low losses by operating at virtually zero flux density [94]. A current transformer for measuring dc and ac currents has been constructed by using a switched-mode amplifier to symmetrically saturate an amorphous core [95]. The mean value of the magnetizing current is zero, and the mean value of the secondary current is just the mean value of the primary current times the turns ratio. An ac current can also be measured by this circuit if the switching frequency is large compared to the frequency of the ac current. In a very different current measurement application, radiation immunity is important. A core of α-Fe precipitated Fe-B-Si amorphous alloy exhibited a slower rate of decrease of permeability with neutron fluence than did a conventional Permalloy core, indicating promise for use in beam current monitors in particle accelerators [96].

F. Instrumentation Transformers

The same characteristics which make amorphous alloys attractive for CTs (low loss and high permeability) are also attractive for other instrumentation transformers. Instrumentation transformers are used, among other things, to match impedances between a sensor or transducer and its measuring device while preserving the fidelity of the signal. The most important considerations are preservation of phase angle (high permeability) and avoidance of attenuation of high-frequency components of the signal (low high-frequency losses). These are the same criteria as those for a current transformer core, albeit over a larger frequency range. Two examples of such transformers, both of which utilize Co-based amorphous alloys, are a matching transformer manufactured by Matsushita for use with a moving-coil phonograph pickup and a transformer in a dynamic microphone made by Sony.

Instrumentation transformers, more than power transformers, require core materials which approach ideal characteristics. Their accuracy is affected by their losses—due to nonideal transfer characteristics and to changes in temperature dependent properties from self-heating. An example of such an instrument transformer is a precision device for measuring the turns ratios of transformers. The transformer in this instrument uses a low-loss, square-loop core of METGLAS alloy 2605SC, a high-induction, Fe-based amorphous alloy.

V. TRANSDUCERS

Transducers utilize a variety of physical phenomena to convert energy from one form to another. Due to the ease of measuring voltage signals, many transducers used as sensors convert the condition to be sensed such as force, pressure, temperature, position, or acceleration into a voltage proportional to the condition sensed. Other transducers used as drivers cause a condition related to the voltage signal which they receive. Examples of such transducers are loudspeakers, ultrasonic vibrators, and even positioners.Transducers are usually judged by the fidelity and efficiency with with they transform the condition to signal or vice versa.

The high permeability and good frequency response of ferromagnetic amorphous alloys make them obvious candidates for replacing conventional soft magnetic materials in magnetic cores in various transducers. Lenz [97] recently reviewed magnetic sensors. In addition, amorphous alloys have other properties, and combinations of properties such as mechanical hardness, magnetostriction, thermal sensitivity, elasticity, and high tensile strength, which make them uniquely suited to some sensing application. Considerable development activity has recently occurred in sensors based on amorphous alloy ribbons or wires. Reviews on amorphous sensors have appeared by Mohri [98], as well as sections in more general review articles by Warlimont [99], Luborsky [69], Yang [91], and Fish [100]. The reader is referred to these articles for more complete details on this application.

A. Classification of Sensors

The properties of magnetic materials relevant to various sensors are best understood when sensors are classified by their operating principles. For the purpose of this chapter we will use the following classifications: electromagnetic effects, magnetomechanical effects, and all other effects.

Probably the largest class of sensors using magnetic materials is that which relies upon electromagnetic effects. In these sensors a change in magnetic induction in a magnetic core results in an induced voltage in windings around the core. The magnetic material is used, in effect, as a flux collector or flux guide. These sensors range from proximity sensors and magnetic phonograph cartridges to sensitive fluxgate magnetometers. The important characteristics of the magnetic material are high permeability and insensitivity to stresses incurred during manufacture. Co-based amorphous alloys with zero magnetostriction are excellently suited to these applications. Some of these sensors, such as magnetic tape heads, require abrasion resistance in addition to magnetic performance.

The second class of sensors, magnetomechanical sensors, include those which depend upon changes in magnetic induction caused by changes in the mechanical strain state of the material. The change in magnetic induction is again detected by the voltage induced in windings surrounding the magnetic material. The appropriate sensitivity figure is the relative change in permeability divided by the relative change in stress. The inverse effect can also be utilized. For example, a change in magnetic induction in the material caused by a change in the applied magnetic field results in a change in the strain or mechanical dimensions of the material. Although this change in dimensions is of the order of parts per million, it can be detected by optical devices such as interferometers. The magnetic properties required in materials for these applications are significant magnetostriction and low anisotropy energy. Fe-based amorphous alloys have saturation magnetostrictions in the 20- to 35-ppm range and induced anisotropy energies of 100 to 1000 J/m^3. To achieve maximum magnetomechanical effects, these Fe-based alloys are annealed in a magnetic field transverse to the ribbon to produce a sheared *B-H* loop. Magnetization in such a sample takes place principally by rotation rather than by domain wall motion. To ensure repeatable performance, the tensile strength of the material must be sufficient for it to remain within its region of elastic deformation over the measured range of strains. For higher-frequency applications, attenuation of the mechanical excitation of the magnetic material must be small. Magnetostrictive ultrasonic drivers, fiber-optic magnetometers, magnetostrictive delay lines, and sonar transducers are all examples of devices utilizing the magnetomechanical coupling between mechanical and magnetic properties.

The magnetostrictive properties of metallic glasses have opened up many new sensor applications, some of which are mentioned in Mohri's reviews [98]. A simplified model of magnetomechanical coupling in amorphous alloys has been given by Livingston [101], relating it to basic material properties. Magnetomechanical properties including magnetomechanical coupling coefficient k_{33}^2 of various magnetostrictive amorphous alloys have been measured by several groups [102–107]. Magnetomechanical coupling coefficients, the square of which equals the ratio of magnetic energy in to mechanical energy out, as high as 0.97 have been measured on METGLAS alloy 2605SC [108]. In contrast, conventional magnetostrictive magnetic materials have coupling coefficients of only 0.3 to 0.5 Related to these magnetomechanical properties is the giant ΔE effect in which the Young's modulus, and therefore the speed of sound in the material, is significantly reduced by an applied magnetic field [109].

Other magnetic, thermal, and mechanical properties of RSSMA can be utilized in sensors. Rapidly quenched Fe-Cr-(Al, Si) alloys have high resistivity and low-temperature coefficients of resistivity, making them useful for strain gages [110]. Amorphous alloy ribbons and wires can exhibit bistable flux reversal behavior due to either cast-in stresses or stresses caused by annealing and reforming the ribbon [111]. This bistable behavior, similar to the Wiegand effect [112] but occurring at a much lower applied field, causes sharp induced voltage pulses in coils surrounding the cores being magnetized. The high magnetomechanical coupling factor and low attenuation of sound waves exhibited by some amorphous alloys together with the variation in Young's modulus with applied magnetic field have suggested magnetically variable delay lines using magnetostrictive films on piezoelectric surface acoustic-wave substrates [113]. Since amorphous alloys can be prepared over a continuous range of compositions, the wide range of resultant Curie temperatures can be utilized in magnetically sensed thermostats.

So many sensor concepts using amorphous alloys have been developed and reported in the literature that only a few of them can be mentioned. I will discuss some of the more interesting sensors and those that are commercially available in entertainment electronics, measurement devices, and security systems.

B. Sensors in Entertainment

One of the earliest commercial applications of amorphous alloys was in audio magnetic tape heads available from TDK, Sony, and Matsushita as early as 1981. The high initial price per kilogram of this new material was not a significant barrier in an application using but a few grams per head. Hard, wear-resistant, Co-based alloys with zero magnetostriction and high permeability were natural choices for this application since fabrication of tape heads results in stresses on the magnetic material. Makino [114] has reviewed the use of amorphous alloys in audio, video, and digital recorder tape heads. Alloy development has continued to improve amorphous alloys for tape heads. This development has resulted in alloys of higher saturation induction for high-coercive-field tapes, improved wear resistance, and deposited thin-film heads. In addition, tape heads made from rapidly quenched ribbon Sendust (Fe-Al-Si) are being used.

Besides their use in matching transformers for dynamic microphones and phonograph cartridges mentioned earlier, amorphous alloys are used in phonograph cartridges themselves. Sony has manufactured moving-magnet phonograph cartridges using intricate laminations etched from Co-based amorphous ribbons. The laminations form a four-pole structure, with windings on each leg, surrounding a magnet attached to the needle. The high permeability and low phase angle result in improved fidelity and frequency response compared to similar cartridges made from polycrystalline Permalloy.

Even the ancient oriental strategy game of GO has yielded to modernization. In 1981 Matsushita commercialized a computerized version of this game in which the position of the "stones" is magnetically detected by amorphous alloy ribbons under the playing board. This flux-sensing determination of position was an early indication of the many sensing applications now being developed.

C. Measurement Sensors

Measurement sensors are designed to cover a wide range of properties. The most straightforward magnetic measurement sensors are those for measuring magnetic fields or anomalies in magnetic fields. These sensors are used in magnetic compasses, geological prospecting, and even submarine detection. Fluxgate magnetometers, which depend upon the nonlinearity of the *B-H* loop of their cores as they enter saturation, have been made from Co-based amorphous alloys [115,116]. Amorphous alloys allow higher excitation frequencies to be used than do conventional materials; therefore, higher-frequency magnetic field variations can be detected. Miniaturized fluxgate magnetometers for applications such as mechanocardiographs to sense cardiac motion have been constructed using Co-based amorphous wires [117]. Combining the tensile strength and magnetostrictive properties of Fe-based amorphous alloys with the measurement sensitivity of interferometers by bonding an amorphous ribbon to an optical fiber in a Mach-Zahn interferometer has resulted in extremely sensitive magnetometers [118,119].

Other sensors which depend upon an external field applied to a magnetic core include speed, proximity, and position sensors. In these sensors a magnet is either moved relative to the magnetic core or the presence of a ferromagnetic object such as a gear tooth concentrates the field from a permanent magnet attached to the core. A variation on this principle is used in a digitizing tablet made by Wacom Co. in Japan [120]. The presence of a magnet in a stylus is detected by the change in permeability in one of an array of amorphous wires under the tablet. The stylus needs no connecting wire to either the tablet or the computer. Speed sensors using bistable amorphous ribbons or wires mentioned earlier have uniform output voltages even at very low rotation rates. The height of the voltage pulse is determined by the switching speed of the bistable magnetic material rather than being proportional to the speed of the gear being sensed.

A large variety of sensors can be designed by using the magnetomechanical sensitivity of amorphous alloys together with their robust mechanical properties. By converting the physical property to be sensed into a force and applying it directly to an amorphous ribbon or toroid, the strain in the ribbon can be sensed as a change in permeability [121,122]. A variety of this type of sensor bonds the amorphous ribbon to a substrate that stresses the amorphous ribbon [123]. Depending on the condition to be sensed, this stress can be due to thermal contraction or due to flexing of the substrate. Another sensor which utilizes amorphous ribbon bonded to a substrate is a contractless torque sensor [124]. A torque applied to a cylindrical shaft causes a stress field in the shaft which results in positive and negative strains at the shaft surface in opposite helical directions. The magnitude and direction of the torque is sensed by pickup coils that respond to changes in the permeability of amorphous alloy ribbons bonded to the shaft below them.

Sensors depending on propagation of sound waves in amorphous ribbons or wires have been constructed. The speed of sound in a magnetostrictive amorphous ribbon varies with applied field and temperature [125]. The presence of a magnet, a force, or even frost on the ribbon can be detected by the reflection of a sound wave from the impedance mismatch caused by a region with a different sound velocity. Digitizing tablets have been constructed based on the reflection of sound waves from the anomaly caused by the presence of a permanent magnet in a cordless "mouse" [126]. An array of magnetostrictive ribbons under the tablet are periodically queried by launching magnetostrictive waves down each ribbon.

D. Article Surveillance Systems

A unique class of proximity sensors is that of electronic article surveillance (EAS) systems. In these systems inventory is monitored by attaching tags to the articles to be controlled, for instance library books. The tags are detected if they pass through an access gate beyond the checkout counter. In one type of system, tags consist of thin strips of a soft magnetic material. When excited into saturation by an ac magnetic field in the gate. the tags are detected by the presence of harmonic signals at multiples of the excitation frequency. These harmonics are produced by the nonlinearity in the response of a magnetic material when driven into saturation. The magnetic material needs to have low coercive field and a square *B-H* loop to produce copious harmonic signals in the presence of a small excitation field. Amorphous alloy ribbons with their attractive combination of soft magnetic and robust mechanical properties are supplanting the traditional Permalloy strips in these systems [127].

A more recent EAS system uses the magnetomechanical resonant frequency of a magnetostrictive amorphous alloy as a unique, detectable tag with extremely low probability of false alarms [128]. If a tag is excited by a pulsed-drive field, the frequency can be detected during the ringdown period between excitation pulses [129]. This detection scheme is possible since, due to low sound attenuation, the mechanical vibration persists after the excitation is removed. The alternating tensile and compressive strains caused by the mechanical resonance causes alternating changes in induction in a magnetically biased tag which, in turn, are easily detected by an antenna.

Even chaos has reached the EAS market. Bifurcations caused by chaos can result in a frequency division phenomenon in which submultiples of the drive frequency are observed. One EAS system using magnetostrictive amorphous ribbon targets utilizes this bifurcation to produce signals at one-half the drive frequency, which are then detected by an antenna in a gate [130].

VI. MOTORS

Since the manufacture of motors represents a large consumption of soft magnetic material, and since their operation represents a significant source of core losses, second only to 60-Hz transformers in the distribution system [131], it is not surprising that various efforts have been made to use amorphous alloys in 60-Hz motors, both in a 1/3-hp three-phase motor [132], and a 1/2-hp single-phase motor [133]. Both of these motors, with stators made from photoetched laminations of Fe-based amorphous alloy ribbon, achieved significant reductions in core losses compared to motors made using conventional electrical steels. The overall efficiencies of the motors, however, did not increase as significantly due to the dominance of the winding losses in the rotor and the stator over the stator core losses.

The results of these efforts have not come close to approaching the commercial successes of applying these materials to DTs. There are several significant differences between transformers and motors which contribute to this lack of success. Most 60-Hz motors are constructed from stamped laminations which have complicated shapes to accommodate the windings and form poles within the motor. Inherently thin, hard amorphous alloys are difficult to economically fabricate into laminations, either by punching or by photochemical etching. Thin laminations result in a large number of laminations for a given stack height, thereby increasing assembly costs. The scrap inherent in motor-lamination fabrication multiplies the price penalty of a premium-priced material. Finally, unless a motor is run at low load for a significant fraction of its operating time, reduced core losses do not command a price premium over reduced winding losses as they do in DTs.

The recent availability of Fe–6.5 Si alloy may cause RSSMA to make a more significant impact on motor materials usage than have amorphous alloys. This material is nonoriented, as are most steels used in motor laminations, and is available up to 0.5 mm thick and 400 mm wide. Its hardness is about halfway between that of Fe–3.2 Si and amorphous alloys, making it easier to punch. As mentioned earlier, the losses, especially at higher frequencies, make it especially attractive for high-frequency rotating equipment [44].

New, nonconventional motor designs may offer opportunities for amorphous alloys in the future. The advent of high-energy-product permanent magnet (PM) materials has allowed, and even encouraged, the design of PM motors with considerably larger gaps between rotor and stator [134]. In such designs, toothless-stator designs are possible which use wound toroids with windings as the stator. The fabrication problems with amorphous alloys are thus ameliorated.

Improvements in semiconductor devices have sped up development of brushless dc PM motors and efficient variable speed brushless dc PM motors. High-speed brushless motors offer opportunities for low-loss magnetic materials. Two prototype high-speed brushless dc motors were recently constructed by ARC of Virginia using laminations that were laser-cut from Fe-based amorphous alloy ribbons [40]. These disk-type motors, constructed from 500 laminations with a stacking factors of 84% and 87%, produced 3 oz-in of torque at 24,000 rpm. Peak efficiency of 70%, significantly higher than the efficiency of similar motors made with polycrystalline electrical steel, was achieved with a motor constructed from high-induction METGLAS alloy 2605CO.

VII. SHIELDING

As mentioned in the beginning of this chapter, the first commercially available product made from an amorphous alloy was METSHIELD, a woven magnetic shielding product made of narrow ribbons of a Ni-Fe-based amorphous alloy, METGLAS 2826 alloy [10]. The advantage of this new shielding material was that its properties were not degraded by mechanical work such as bending or vibration, and the material did not have to be annealed while in its final shape [135]. Shields made from this material were used in experiments launched on *Voyager I* and *Voyager II* in 1977. In addition to woven shielding fabric up to 1 m in width, braided cable shielding of the same amorphous alloy was fabricated and tested [136].

These early magnetic shielding materials using Ni-Fe-based amorphous alloys proved to be good low-frequency shielding materials with shielding ratios of 10 to 20 for a single layer of woven shielding with applied fields of a few hundred A/m. Their low field performance was limited by the low initial permeability of unannealed Fe-Ni-based amorphous alloys. For high initial permeability, one must use Co-based amorphous alloys with near-zero magnetostriction. These alloys have been used in another form of amorphous cable shielding in which narrow amorphous alloy ribbon was spiral-wound around a cable [137]. Near-zero magnetostriction Co-based alloys allow the amorphous ribbon to be wound around a small-diameter cable without decreasing its shielding factor.

Fabricated shields for specific devices have also been made from amorphous alloy ribbons. As miniaturization reduces the spacing between components, magnetic shielding of transformers and other inductive components becomes vital to reduce interference between components. Both Co- and Ni-based amorphous alloys have been tested for magnetic screening of inductive units [138]. Cylindrical shields of Co-based amorphous alloys are available for shielding small motors such as those used in computer disk drives [139].

The advent of wide, inexpensive, Fe-based alloys designed for distribution transformer applications has caused a renewed interest in amorphous alloys for shielding, especially in EMI-conscious Japan. Delta Research Co. [140] has developed the technology to copper-plate Fe-based amorphous alloys. This copper plating has several purposes: it serves as a high-conductivity layer for high-frequency electromagnetic shielding, provides corrosion protection, enhances solderability, and improves bonding to other materials. Shielding panels made from coated amorphous ribbons soldered edge to edge and combined with building materials are used to construct shielded rooms in hospitals for magnetic resonant imaging (MRI) devices. Magnetostatic shielding to protect external regions from the field of the superconducting magnet is provided by the amorphous metal ribbon. In addition, high-frequency electromagnetic shielding, to prevent external fields from interfering with the MRI device itself, is provided by this composite shielding material. Cu-plated, Fe-based amorphous alloy ribbon has also been used in a variety of other applications such as magnetically shielded floppy disk boxes and magnetically shielding venetian blinds.

VIII. PULSED-POWER APPLICATIONS

Requirements for high-reliability, high-power pulsed energy sources for accelerators and lasers arose in the late 1970s, at the same time that metallic glasses became available in significant quantities. Metallic glasses with their intrinsically high resistivities and thin, strong ribbon form had precisely the properties required for this application. In addition, metallic glasses were available with high saturation induction, allowing reasonably compact units to be designed. For these reasons, considerable research has been conducted over the past 10 years in the development of pulsed-power systems incorporating metallic glasses [19,141–145]. These systems have powered a variety of projects from particle accelerators to excimer lasers. All of these applications required very short pulses of very high power. Although much of this research has been done at National Laboratories, there is increasing interest in pulsed-power systems for industrial uses: in excimer lasers in medicine, in linear accelerators for materials processing, and in radiation sources for x-ray lithography.

Cores of soft magnetic materials are used in pulsed-power conditioning in several ways. The principle applications can be classed as pulse transformers [142], switch protection inductors [146], and magnetic pulse compression inductors, [141,143–145,147].

A. Explanation of Pulsed Power

A typical pulsed-power system obtains energy from a prime power source over a long acquisition time at a low peak power. This energy is then processed in intermediate or pulse-conditioning stages in which the energy is stored, shaped, and switched into the load. The energy is delivered to the load over a short time period and with a high peak power.

Magnetic material selection for pulsed-power systems is dominated by two conditions inherent to pulsed-power: high magnetization rates and high power densities. High magnetization rates are a consequence of output pulses, usually measured in microseconds or less. High power densities result from the high power levels required and the practical necessity of minimizing size. High magnetization rates in magnetic alloys are associated with high magnetic losses due to eddy currents. Losses not only reduce the efficiency of the system but they also increase the core temperature.

The quantity of magnetic material used in transformers and inductors in pulsed-power systems is related to the flux-handling capabilities of the materials. This capability is determined by the product of the cross-sectional area of the core and the achievable change in flux density. The volume of magnetic material used in a system varies inversely with the available change in flux density raised to as high as the second power [148]. Maximum change in flux density is achieved by magnetizing the core material from negative remanence, $-B_r$, to saturation, $+B_s$. Therefore, not only the saturation induction, but also the squareness of the loop or the remanent magnetization, are important in choosing magnetic materials. Since they lack magnetocrystalline anisotropy, metallic glasses can be annealed in a

Table 5 Properties of Square-Loop Magnetic Materials Used in Pulsed-Power Applications (all ribbon materials 25 μm thick)

Material	Elements	B_s (T)	B_r (T)	ΔB (T)	H_c (A/m)	ρ (μΩm)	T_c (°C)	λ_s (ppm)
Amorphous								
METGLAS 2605CO	FeCoBSi	1.8	1.7	3.5	3.2	1.23	415	35
METGLAS 2605SC	FeBSiC	1.6	1.5	3.1	2.4	1.35	370	30
METGLAS 2705M	CoFeNiMoBSi	0.7	0.7	1.4	0.8	1.36	365	<1
METGLAS 2714A	CoFeNiBSi	0.55	0.5	1.05	0.2	1.30	205	<1
Crystalline								
SUPERMENDUR	CoFeV	2.35	1.5	3.8	90	0.35	950	80
Fe–3 wt% Si	FeSi	1.97	1.4	3.4	50	0.50	730	3
50% Ni-Fe	NiFe	1.6	1.5	3.1	8	0.45	480	25
80% Ni-Fe	NiFeMo	0.8	0.7	1.5	2.2	0.55	460	1
Ni-Zn ferrite	NiZnFeO	0.33	0.25	0.58	80	10^{12}	>280	−10
Mn-Zn ferrite	MnZnFeO	0.51	0.12	0.63	12	10^{7}	>230	−2

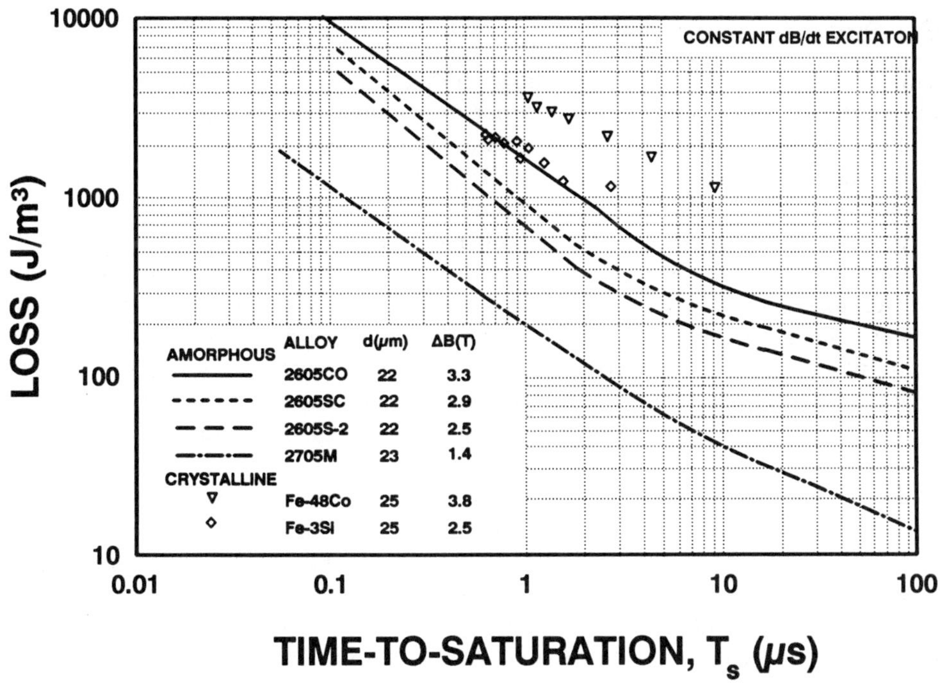

Figure 14 Core loss per pulse for various ferromagnetic materials as a function of time to saturation (constant dB/dt excitation) [148].

magnetic field to achieve very square B-H loops with their available flux swing (ΔB) approaching twice B_s.

Some of the important magnetic properties, along with resistivities (ρ), of amorphous alloys used in pulsed-power applications are given in Table 5 together with those of some conventional crystalline magnetic materials. The properties are all for nominally 25-μm-thick ribbon, except for those for the ferrites. The coercive field, H_c, is a measure of how difficult it is to reset the material to negative remanence before the next pulse. Although some polycrystalline materials have significantly higher saturation inductions than do amorphous alloys, much of this advantage is lost when ΔB is considered. It is quite difficult to achieve an extremely square loop with 25-μm-thick ribbons of either Supermendur or Fe–3.2 Si alloy.

Losses at high magnetization rates are a critical factor for magnetic materials used in pulsed-power systems. Losses determine efficiency, maximum repetition rate and, therefore, average output power. Losses have been measured on a variety of materials including amorphous alloys under pulse conditions. Usually constant-voltage excitation resulting in a constant magnetization rate, dB/dt, has been used [149–151]. Results for some typical materials are given in Fig. 14. The loss per pulse (J/m^3) is evaluated for magnetization from negative remanence to positive saturation. The average flux swing (teslas) for each material is given in the legend. In this figure, losses are plotted as a function of time to saturation rather than magnetization rate. Pulse applications are designed for specific pulse lengths; therefore, they require the same time to saturation despite differing saturation inductions for differing materials.

B. Pulse Transformers

Transformers are used to match voltage to that required by the load and to match pulsed-power source impedance in order to maximize energy transfer to the load and to minimize reflections from the load. Pulse transformers require high saturation induction core material to minimize their size and to minimize the required number of turns of conductor. The number of turns must be minimized in order to provide low inductance and allow fast rise times.

One use of pulse transformers is in linear induction accelerators. Linear induction accelerators are a class of particle accelerators which couple energy from pulsed-power sources to a beam of charged particles through magnetic induction cores surrounding the beam. The cores are effectively 1:1 transformers with the particle beam as the secondary of the transformer. Metallic glasses are finding application in linear inductors both in their pulsed-power supplies and in the beam inductor cores.

The high power available in a linear induction accelerator beam has been used in various applications. Heavy-ion linear induction accelerators are being developed for inertial confinement fusion [152]. Free-electron lasers (FELs) have been driven by linear induction accelerators with as much as 45% of the beam energy converted to microwave radiation [153]. The frequency of the FEL radiation is proportional to the energy (and therefore velocity) of the particles in the beam and to the period of the magnetic wiggler; therefore, the frequency can be tuned by changing the energy of the beam. This capability offers almost unlimited potential as a source of extremely high energy, tunable laser radiation. In more prosaic applications, the ability of linear induction accelerators to deposit high quantities of energy in small volumes, or to be defocused to irradiate larger volumes, has suggested their use in materials processing, in food irradiation, and even in waste sterilization [154]. Accelerators have a significant advantage over other sources of ionizing radiation used for processing. When not in use, accelerators do not continue to produce radiation as do radioisotopes used as radiation sources.

C. Switch Protection

Switch protection is accomplished by placing a saturable inductor in series with the switch. The high reactance of the unsaturated inductor limits the current through the switch until the core of the inductor saturates. This delay allows the switch to become fully conducting before it must pass a high current, therefore minimizing switch losses and increasing switch life. After saturation the inductor does not impede the current due to its low reactance. The length of the delay is determined by the product of the available ΔB, the cross-sectional area of the core, and the number of turns on the inductor divided by the voltage. A schematic diagram of a saturable inductor in series with a switch together with the voltage and current waveforms in the inductor are shown in Fig. 15. It is interesting to note that almost no energy is stored in the inductor because, at any time, either the current or the inductance is small.

The same principle of using saturable reactors to protect components is also used in SMPSs [155]. The overshoot on the leading edges of square-wave voltages produced in switching regulators results in noise and can also cause premature

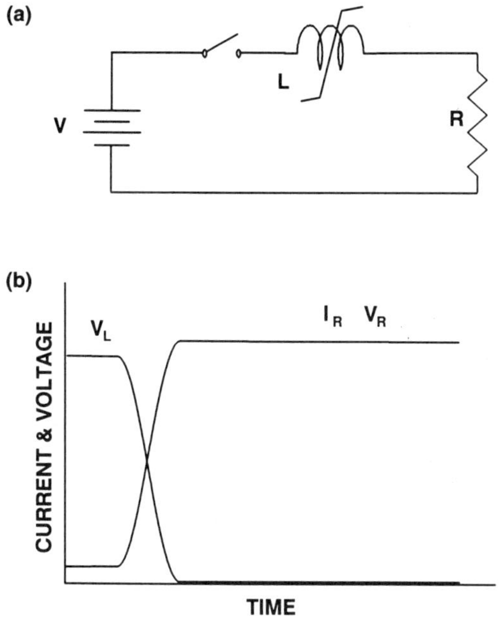

Figure 15 A saturable inductor used for switch protection: (a) circuit schematic, and (b) voltage across and current through inductor.

failures by exceeding component voltage ratings. An ultrasmall, properly sized saturable inductor or snubber can absorb this spike on the leading edge, hence the name given to them by Toshiba Corporation, "spike killer" [156]. The core is automatically reset to negative remanence before each use due to the bipolar nature of the waveform. These spike killers, which can operate at hundreds of kHz, are made with cores of low-loss, Co-based alloys which have been annealed to enhance the squareness of their *B-H* loops. The cores are similar in size and material to those used in high-frequency magamp regulators in SMPSs.

D. Magnetic Pulse Compression

The use of metallic glasses in pulsed-power systems which has resulted in the greatest increase in the state of the art in repetitive, high-energy pulses is due to the resurrection of the technique of magnetic pulse compression, originally developed for radar in the 1950s. Magnetic pulse compression utilizes saturable magnetic switches to compress the width of a pulse while retaining its voltage. Between each saturable reactor, capacitors are connected to ground to store the energy. Decreasing the energy transfer time between capacitors to successively shorter and shorter times results in an increase in the peak pulse power and, ultimately, a high peak-power pulse delivered to the load. A series magnetic pulse compression circuit is shown in Fig. 16 along with the voltage waveforms across successive capacitors. The inductors are sized with successively smaller values of saturated inductance such that a given capacitor is discharged in one-third to one-fifth the time in which it was charged. This sizing results in a gain in current, of 3 to 5 per stage. Once an inductor is saturated in the forward direction, it is still unsaturated in the reverse

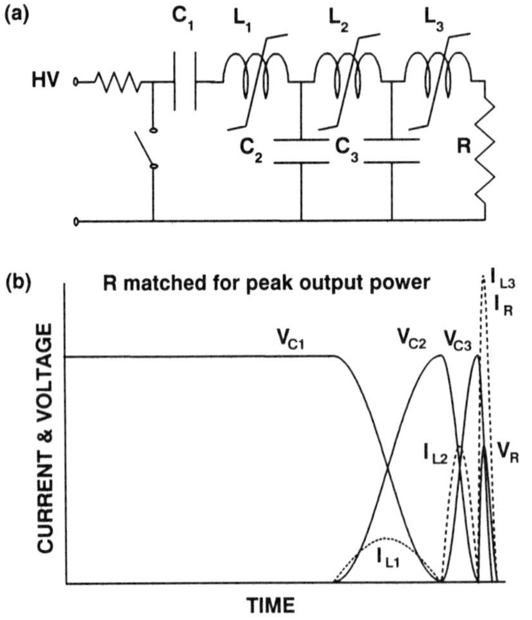

Figure 16 A series magnetic pulse compression circuit: (a) circuit schematic, and (b) voltage waveform across successive capacitors.

direction; therefore, the inductors act as diodes to prevent reverse current flow back into previous stages. A properly designed magnetic pulse compression system, sometimes called a magnetic modulator, can achieve high efficiency while delivering a pulse to the load with greatly increased current and peak power.

Research on magnetic pulse compression with metallic glasses at Lawrence Livermore National Laboratories in Livermore, California, has culminated in pulsed-power supplies called the MAG1-D drivers [157]. These magnetic modulators produce 125-kV, 20-kA, 50-ns pulses with 2.5 gigawatts (GW) peak power at repetition rates up to 5 kHz to drive linear induction electron accelerators such as the Experimental Test Accelerator II (ETA II). Each modulator has approximately 20 metallic glass cores in three magnetic pulse compression stages and one 1:10 step-up transformer. Linear induction accelerators operate at high beam current, and hundreds of successive acceleration gaps can allow the accelerators to reach extremely high final beam power.

IX. CONCLUSIONS

It should be obvious by now that the magnetic applications of RSSMA have developed significantly since amorphous alloy ribbons first became available almost two decades ago. There is no longer any question that these materials are commercially significant with growing uses. Some of their applications are evolutionary with RSSMA having improved properties or lower costs than conventional materials. Other applications are truly revolutionary and could not have been accomplished with conventional materials.

A. Current Status

Today, rapidly quenched amorphous alloys are being produced in significant quantities in several countries, including the United States, Germany, Japan, China, and the former Soviet Union. This scale of production is, in itself, an indication of the commercial significance of this relatively new class of magnetic material. The largest applications of amorphous metals to date have probably been in 60-Hz DTs, high-frequency electronic components, GFIs, and article surveillance tags. Each of these applications now consumes tens to thousands of tons of material per year, although not necessarily all in the same country.

The impact of amorphous alloys on the soft magnetic materials market is not limited to its displacement of conventional materials in the market. Competition from a new class of materials with improved properties has forced improvements in conventional materials such as thinner grades of oriented Fe–3.2 Si with laser scribing and improved texture to reduce their losses [158]. Rapidly quenched, nonoriented Fe–6.5 Si has inspired more conventional rolled nonoriented alloys with 4.5 and 6.5 wt% Si content [159]. By threatening complacency, new materials improve old materials.

In comparison to amorphous alloys, other classes of RSSMA, such as crystalline materials, have been available for much less time and, therefore, cannot yet be expected to have as large an impact. Hitachi, however, has already announced plans to open a plant for production of nanocrystalline alloys in 1991 with an initial capacity of one ton per month [160].

B. Future Projections

Any exposition on application of a new class of materials is required to make predictions as to the future—if for no other reason, to teach the author humility when read 10 years later. At the Third Joint Intermag—Magnetism and Magnetic Materials Conference on Montreal in 1982, I bravely predicted that pulse power would be the most significant market for amorphous alloys in the 1980s and would not be eclipsed by DTs until the 1990s. Despite SDI and some notably successful pulsed-power projects with amorphous cores, government funded pulsed-power projects have moved with glacial speed, especially those related to fusion energy. On the other hand, utilities are rapidly embracing the energy saving possible with amorphous-cored DTs.

Future developments in markets for RSSMA will, of course, be driven not only by property benefits afforded by the materials, but also by economic forces. Some of the demand-side forces, such as the price of oil, have not been reliably predicted in the past. The sudden quadrupling of the price of cobalt in the late 1970s due to political upheavals in Zaire [161] probably delayed significant adoption of Co-based, zero-magnetostriction alloys in many applications until cobalt prices dropped to one-half their peak value of $55/kg. Another force that will affect which markets RSSMA will enter and which countries will dominate these markets will be future support of research on magnetic materials. By plotting the number of publications per year on amorphous metals, Gonser [162], in 1984, predicted that based on future trends in research activity ". . . we have reached a crossroad where the amorphous metals are either going to become *the* material of the century or just turn out to be a dream." Seven years later we have proven that they are more than a dream, but the question of *where* the research activity will take place is still open. In 1985 the Committee on Magnetic Materials of the National Materials Advisory Board completed a study on opportunities for future research in magnetic materials [163]. Their conclusions included the warning that the U.S. industry based on magnetic technology is waning. To reverse this trend, they recommended increased research and development in selected magnetic technologies including amorphous magnetic materials. In a recent review of research opportunities for RSSMA, Fish [164] suggested improved production and manufacturing engineering as the most immediate opportunities in line-frequency applications. Better alloys, high-quality thin ribbons, and better surface coatings were identified as benefiting higher-frequency applications.

There is still much work to be done in basic research as well as in development of applications, alloys, and manufacturing technology before RSSMA reach their full potential. How soon they reach that potential, and in which countries will depend to a large part on the investment in that research and development.

ACKNOWLEDGMENTS

Any chapter such as this draws heavily on prior works, advice, and assistance of colleagues. I would like to particularly acknowledge the assistance of Dr. Gordon E. Fish, whose article "Soft Magnetic Materials" [100] was written concurrently; Dr. V. R. V. Ramanan and Mr. D. M. Nathasingh, with whom I have collaborated over the past several years on various projects and papers; and, of course, my patient editor, Dr. Howard H. Liebermann.

REFERENCES

1. A. I. Gubanov, *Sov. Phys. Solid St.*, *2*, 468 (1960).
2. W. Klement, R. H. Willens, and P. Duwez, *Nature*, *187*, 869 (1960).
3. P. Duwez and S. C. H. Lin, *J. Appl. Phys.*, *38*, 4097 (1967).
4. T. Egami, P. J. Flanders, and C. D. Graham, Jr., *AIP Conf. Proc.*, *24*, 697 (1975).
5. L. A. Davis, N. J. DeCristofaro, and C. H. Smith, in *Proc. Conf. Metallic Glasses: Science and Technology*, edited by C. Hargitai, I. Bakonyi, and T. Kemeny (Central Research Institute for Physics, Budapest, 1981), p. 1.
6. T. Masumoto, *Technocrat*, *15*(4), 27 (1982); *ibid.*, (5), 26 (1982).
7. *Amorphous Metallic Alloys*, edited by F. E. Luborsky (Butterworths, London, 1983).
8. H.-J. Güntherodt, in *Rapidly Quenched Metals*, Vol. II, edited by S. Steeb and H. Warlimont (North-Holland, Amsterdam, 1985), p. 1591.
9. H. R. Hilzinger, *J. Magn. Magn. Mater.*, *83*, 370 (1990).
10. L. I. Mendelsohn, E. A. Nesbett, and G. R. Bretts, *IEEE Trans. Magn.*, *MAG-12*, 924 (1976).
11. D. E. Polk, B. C. Giessen, and F. S. Gardner, *Mater. Sci. Eng.*, *23*, 309 (1976).
12. M. C. Narasimhan, U.S. Patent 4,142,571 (1979).
13. TDK Electronics Co., Tokyo, Japan, Preliminary product data sheet (1980).
14. G. E. Fish and C. H. Smith, in *Soft and Hard Magnetic Materials with Applications*, edited by J. A. Salsgiver, K. S. V. L. Narasimhan, P. K. Rastogi, H. R. Shepard, and C. M. Maucione (American Society for Metals, Metals Park, OH, 1986), p. 7.
15. F. E. Luborsky, in Ref. 7, p. 1.
16. N. Tsuya, K. L. Arai, K. Ohmori, H. Shimanaka, and T. Kan, *IEEE Trans. Magn.*, *MAG-16*, 728 (1980).
17. N. Tsuya, K. L. Arai, K. Ohmori, and T. Homma, *J. Appl. Phys.*, *53*, 2422 (1982).
18. W. S. Melville, *Proc. Inst. Elec. Eng. (London)* Pt. 3, *98*, (53), 1985 (1951).
19. D. L. Birx, E. J. Lauer, L. L. Reginato, D. Rogers, Jr., M. W. Smith, and T. Zimmerman, in *Proc. 3rd IEEE Int. Pulse Power Conf.*, edited by T. H. Martin and A. H. Guenther (IEEE, New York, 1981), p. 262.
20. R. M. Bozorth, *Ferromagnetism* (Van Nostrand Reinhold, New York, 1951).
21. *Soft Magnetic Materials*, edited by R. Boll (Heyden, London, 1978), p. 34.
22. R. C. O'Handley, *J. Appl. Phys.*, *62*, R15 (1987).
23. *Glassy Metals: Magnetic, Chemical and Structural Properties*, edited by R. Hasegawa (CRC Press, Boca Raton, FL, 1983).
24. C. F. Chang, R. L. Bye, V. Laxmanan, and S. K. Das, *IEEE Trans. Magn.*, *MAG-20*, 553 (1984).
25. Y. Yoshizawa, S. Oguma, and K. Yamauchi, *J. Appl. Phys.*, *64*, 6044 (1988).
26. S. Ohnuma, K. Watanabe, and T. Masumoto, *Phys. Stat. Sol.*, *A-4*, K-151 (1977).
27. A. Datta and C. H. Smith, in Ref. 8, p. 1315.
28. D. M. Nathasingh and C. H. Smith, *Proc. Powercon 7* (Power Concepts, Inc. Ventura, CA, 1980), p. B2-1.
29. R. Hasegawa, G. E. Fish, and V. R. V. Ramanan, *Rapidly Quenched Metals*, Vol. II, edited by T. Masumoto and K. Suzuki (Sendi, 1981), p. 929.
30. A. Datta, N. J. DeCristofaro, and L. A. Davis, in Ref. 29, p. 1007.
31. R. Hasegawa, *J. Appl. Phys.*, *53*, 7819 (1982).
32. V. R. V. Ramanan, *J. Appl. Phys.*, *53*, 7822 (1982).
33. F. E. Luborsky, *AIP Conf. Proc.*, *29*, 209 (1976).
34. R. Hasegawa, *J. Appl. Phys.*, *49*, 1712 (1978).
35. D. Raybould and A. C. Diebold, *J. Mater. Sci.*, *21*, 193 (1986).
36. D. M. Nathasingh and H. H. Liebermann, *IEEE Trans. Power Delivery*, *PWRD-2*(3), 843 (1987).
37. A. T. Moghadam and A. J. Moses, *IEEE Trans. Magn.*, *25*, 3964 (1989).

38. D. Raybould and C. F. Cline, in *Rapidly Solidified Crystalline Alloys*, edited by S. K. Das, B. H. Kear, and C. M. Adam (TMS-AIME, Warrendale, PA, 1985), p. 127.
39. A. I. Taub and P. G. Frischmann, in *Rapidly Quenched Metals 6*, edited by R. W. Cochrane and J. O. Ström-Olsen (Elsevier, New York, 1988), p. 403.
40. J. M. Glass, H. P. Groger, R. J. Churchill, and E. M. Norin, *J. Mater. Eng.*, *12*, 59 (1990).
41. H. Masumoto, *Sci. Rep. Tohoku Imp. Univ.*, *25*, 388 (1936).
42. Ref. 20, p. 81.
43. C.-F. Chang, R. L. Bye, S. K. Das, and C. H. Smith, in Ref. 14, p. 29.
44. G. E. Fish, C.-F. Chang, and R. Bye, *J. Appl. Phys.*, *64*, 5370 (1988).
45. M. Abe, Y. Takada, T. Murakami, Y. Tanaka, and Y. Mihara, *J. Mater. Eng.*, *11*, 109 (1989).
46. G. Herzer, *IEEE Trans. Magn.*, *25*, 3327 (1989).
47. Y. Yoshizawa and K. Yamauchi, *IEEE Trans. Magn.*, *25*, 3325 (1989).
48. K. Suzuki, N. Kataoka, A. Inoue, A. Makino, and T. Masumoto, *Mater. Trans., JIM 31*, 743 (1990).
49. D. J. Bailey and L. A. Lowdermilk, in Ref. 8, p. 1625.
50. L. A. Lowdermilk, in *Moving to Metglas*, Vol. 1, No. 1 (Metglas Products, Parsippany, NJ, 1988).
51. A. Datta, D. Nathasingh, R. J. Martis, P. J. Flanders, and C. D. Graham, Jr., *J. Appl. Phys.*, *55*, 1784 (1984).
52. A. I. Taub, *J. Appl. Phys.*, *55*, 1775 (1984).
53. J. Silgailis, D. Nathasingh, and C. Bruckner, U.S. Patent 4,668,309 (1987).
54. T. Jagielinski, *IEEE Trans. Magn.*, *MAG-19*, 1925 (1984).
55. *Inside R & D*, Vol. 8, No. 51 (December 19, 1979); D. Utroska, *Electrical Light and Power* (April 1980), p. 54.
56. D. E. Ballard and W. Klappert, U.S. Patent 4,790,064 (1988); M. D. Valencie and D.A. Schaffer, U.S. Patent 4,709,471 (1987).
57. Kuhlman Corporation, European Patent Application, Publication No. 0-083-567 (1987).
58. D. J. Bailey, L. A. Lowdermilk, and A. C. Lee, *J. Magn. Magn. Mater.*, *54–57*, 1618 (1986).
59. H. W. Ng, *IEEE Electrical Insulation Mag.*, Vol. 5, No. 3, 36 (1989).
60. *The Ultra Efficient Alternative*, Bull. AMT 101B (General Electric, Hickory, NC, 1989).
61. W. D. Nagel, in *Moving to Metglas*, Vol. 2, No. 1 (Metglas Products, Parsippany, NJ, 1990).
62. *The Wall Street Journal*, Tuesday, October 25, 1988, p. B4.
63. B. Damsky, G. Addis, and D. Sharma, *EPRI J.*, *April/May*, p. 43 (1988).
64. *Transmission and Distribution*, *July*, p. 10 (1986).
65. K. C. Lin, E. E. Zook, and J. W. Crockett, *J. Mater. Eng.*, *12*, 51 (1990).
66. H. R. Hilzinger, *IEEE Trans. Magn.*, *MAG-21*, 2020 (1985).
67. H. Harada, in Ref. 8, p. 1691.
68. H. Warlimont and R. Boll, *J. Magn. Magn. Mater.*, *26*, 97 (1982).
69. F. E. Luborsky, in *Glass . . . Current Issues*, edited by A. F. Wright and J. Dupuy, NATO ASI Series E, No. 92, (Martinus-Nijhoff, Dordrecht, 1985), p. 139.
70. C. H. Smith, *IEEE Trans. Magn.*, *MAG-18*, 1376 (1982).
71. The development of switched-mode power supplies can be traced in the conference records of the IEEE Power Electronics Specialist Conference (since 1970) published by IEEE, New York.
72. K. Harada, TG. Nabeshima, R. Hiramatsu, and I. Norigoe, *IEEE Power Electronics Specialist Conference '84 Record* (IEEE, New York, 1984), p. 382.

73. Available from Power Trends, Inc., West Chicago, IL.
74. K. Inomata, *IEEE Transl. J. Magn. Japan*, *TJMJ-1*, 6972 (1985).
75. T. Yamada, T. Inoue, I. Norigoe, and C. Mullett, in *Proc. IEEE Applied Power Electronics Conf.* (IEEE, New York, 1986), p. 134.
76. D. Y. Chen, *Powerconversion Int.*, *Sept./Oct.*, 52 (1980).
77. J. J. Torre, M. Rosen, and C. H. Smith, in *Proc. 4th Int. Power Conversion Conf.* (Intertec Communications Inc., Oxnard, CA, 1982), p. 278.
78. S. Lupi and M. F. Nunes, *J. Magn. Magn. Mater.*, *83*, 367 (1990).
79. C. E. Mullett and D. M. Nathasingh, *PCIM*, *July*, 32 (1988).
80. R. V. Major, T. M. Jasko, and K. J. Cruickshank, *IEEE Trans. Magn.*, *MAG-20*, 1415 (1984).
81. G. Herzer and H.-R. Hilzinger, *J. Magn. Magn. Mater.*, *62*, 143 (1986).
82. *Magnetic Amplifers, a Rising Star in Naval Electronics*, NAVSHIPS 900.172, Electronics Design and Development Division, Bureau of Ships, Department of the Navy (U.S. Govt. Printing Office, Washington, DC, 1954).
83. D. Graetzer, *IEEE Trans. Magn.*, *MAG-16*, 922 (1980).
84. R. Hiramatsu, K. Havada, and T. Ninomiya, *Powerconversion Int.*, *March/April*, 75 (1980).
85. H. R. Hilzinger, *IEEE Trans. Magn.*, *MAG-21*, 2020 (1985).
86. Y. Yoshizawa, K. Yamauchi, T. Yamane, and H. Sugihara, *J. Appl. Phys.*, *64*, 6047 (1988).
87. G. W. Gawrys, *AT&T Technol.*, *1*(1), 2 (1989).
88. H. Hemphill, *Choosing Pulse Transformers for ISDN Applications* (Publ. 81101E, Schaffner Elektrionik, A. G., Luterbach, Switzerland).
89. Product Bulletin PB-261-1, Vacuumschmelze, GMBH, Hanau, Germany (1989).
90. Z. Luo, L. Guang-Di, and S. Song-Yao, in Ref. 8, p. 1679.
91. Y.-S. Yang, in *Proc. 4th Int. Conf. Physics of Magnetics Materials*, edited by W. Gorzkowski, H. K. Lachowicz, and H. Szymczak (World Scientific, Singapore, 1988), p. 298.
92. G. E. Fish, *J. Appl. Phys.*, *57*, 3569 (1985); *J. Phys.*, Colloq. C6, Suppl. no. 9, *46*, C6-207 (1985).
93. D. Nathasingh, Metglas Products, Parsippany, NJ, unpublished data (1985).
94. M. Milkovic, F. E. Luborsky, D. Chen, and R. E. Tompkins, *IEEE Trans. Magn.*, *MAG-13*, 1224 (1977).
95. A. Gytri and T. M. Undeland, in Ref. 77, p. 341.
96. R. D. Brown, J. R. Cost, and J. T. Stanley, *J. Appl. Phys.*, *55*, 1754 (1984).
97. J. E. Lenz, *Proc. IEEE*, *78*, 973 (1990).
98. K. Mohri, *IEEE Trans. Magn.*, *MAG-20*, 942 (1984); and in Ref. 8, p. 1687.
99. H. Warlimont, *Mat. Sci. Eng.*, *99*, 1 (1988); and in Ref. 8, p. 1599.
100. G. E. Fish, *Proc. IEEE*, *78*, 947 (1990).
101. J. D. Livingston, *Phys. Stat. Sol.* (*a*), *70*, 591 (1982).
102. M. Brouha and J. van der Borst, *J. Appl. Phys.*, *50*, 7594 (1979).
103. C. Modzelewski, H. T. Savage, L. T. Kabacoff, and A. E. Clark, *IEEE Trans. Magn.*, *MAG-17*, 2837 (1981).
104. H. T. Savage and M. L. Spano, *J. Appl. Phys.*, *53*, 8092 (1982).
105. P. M. Anderson III, *J. Appl. Phys.*, *53*, 8101 (1982).
106. S. W. Meeks and J. C. Hill, *J. Appl. Phys.* , *54*, 6584 (1983).
107. K. B. Hathaway and M. L. Spano, *J. Appl. Phys.*, *55*, 1765 (1984).
108. M. Wun-Fogle, A. E. Clark, and K. B. Hathaway, *J. Magn. Magn. Mater.*, *54–57*, 393 (1986).
109. B. S. Berry and W. C. Pritchet, *J. Appl. Phys.*, *47*, 3295 (1976).
110. T. Naohara, A. Inoue, H. Tomioka, T. Masumoto, and K. Shinohara, in Ref. 8, p. 1663.

111. K. Mohri, S. Takeuchi, and T. Fujimoto, *IEEE Trans. Magn.*, *MAG-17*, 3370 (1981).
112. J. R. Wiegand, U.S. Patent 3,820,090 (1974).
113. D. C. Webb, D. W. Forester, A. K. Ganguly, and C. Vittoria, *IEEE Trans. Magn.*, *MAG-15*, 1410 (1979).
114. Y. Makino, in Ref. 8, p. 1699.
115. O. V. Nielsen, B. Hernando, J. R. Petersen, and F. Primdahl, *J. Magn. Magn. Mater.*, *83*, 405 (1990).
116. D. Son, *IEEE Trans. Magn.*, *25*, 3420 (1989).
117. K. Mohri, T. Kondo, and J. Yamasaki, in Ref. 8, p. 1659.
118. K. P. Koo and G. H. Sigel, Jr., *Optics Lett.*, *7*, 334 (1982); *Opt. Lett.*, *9*, 257 (1986); K. P. Koo, F. Bucholt, A. Dandridge, and A. B. Tveten, *IEEE Trans. Magn.*, *MAG-22*, 141 (1986).
119. C. J. Nielsen, *Opt. Eng.*, *25*, 1261 (1986).
120. A. Murakami, K. Hosaka, and M. Fukushima, *J. Appl. Phys.*, *64*, 6062 (1988).
121. K. Mohri and E. Sudoh, *IEEE Trans. Magn.*, *MAG-15*, 1806 (1979); *MAG-17*, 1317 (1981).
122. T. Meydan, M. B. Blundell, and K. J. Overshott, *IEEE Trans. Magn.*, *MAG-17*, 3376 (1981).
123. E. E. Mitchell, R. DeMoyer, and J. Vranish, *IEEE Trans. Ind. Elec.*, *IE-33*, 166 (1986).
124. K. Harada, I. Sasada,T. Kawajiri, and M. Inoue, *IEEE Trans. Magn.*, *MAG-18*, 1767 (1982); I. Sasada, A. Hiroike, and K. Harada, *IEEE Trans. Magn.*, *MAG-20*, 951 (1984).
125. K. Shirae and A. Honda, *IEEE Trans. Magn.*, *MAG-17*, 3096 (1981).
126. A. Murakami, K. Hosaka, M. Fukushima, M. Maeda, and N. Tsuya, *IEEE Trans. Magn.*, *24*, 1758 (1988); A. Murakami, K. Hosaka, M. Fukushima, M. Maeda, and N. Tsuya, *IEEE Trans. Magn.*, *25*, 2739 (1989).
127. J. A. Gregor and G. J. Sellers, U.S. Patent RE 32,427 (1987).
128. P. M. Anderson, G. R. Bretts, and J. E. Kearney, U.S. Patents 4,510,489 and 4,510,490 (1985).
129. L. A. Davis and V. R. V. Ramanan, *Key Eng. Mater.*, *13–15*, 733 (1987).
130. L. G. Ferguson and L. H. Charlot, Jr., U.S. Patent 4,654,641 (1987).
131. F. E. Warner, in *Energy Efficient Steels*, edited by A. R. Marder and E. T. Stephenson (TMS-AIME, Pittsburgh, PA, 1977), p. 1.
132. W. R. Mischler, G. M. Rosenburry, P. G. Frishmann, and R. E. Tompkins, IEEE PES Winter Meeting, Atlanta, GA, 1981, paper 81 WM 189-0.
133. H. E. Jordan and E. J. Woods, in *Motor-Con April 1984 Proc.* (Intertec Communications, Oxnard, CA, 1984), p. 37.
134. K. J. Strnat, *Proc. IEEE*, *78*, 923 (1990).
135. G. J. Sellers, *Proc. 1977 IEEE Int. Symp. Electromagnetic Compatibility*, Seattle, WA, (IEEE, New York, 1977), p. 129.
136. G. J. Sellers, J. P. Dismukes, Y. Shiau, J. J. Krstansky, R. E. Sharp, and J. W. Kincaid, Jr., *Proc. 1978 IEEE Int. Symp. Electromagnetic Compatibility*, Atlanta, GA, (IEEE, New York, 1978), p. 152; Y. Shiau, J. Bridges, and G. J. Sellers, ibid., p. 155.
137. L. Borek, *ELEKTRONIK*, *4/82*, 43 (1982).
138. J. Szczydglowski, *J. Magn. Magn. Mater.*, *83*, 393 (1990).
139. VAC Publication FS-M 9, Vacuumschmelze GMBH, Hanau, Germany (1989), p. 24.
140. Delta Research Co., Tokyo, Japan.
141. E. Y. Chu, G. Hofmann, H. Kent, and T. Bernhardt, *IEEE Conf. Record 15th Power Modulator Symp.*, Baltimore, MD, (IEEE, New York, 1982), p. 32.
142. A. Faltens M. Firth, D. Keefe, and S Rosenblum, *IEEE Trans. Nucl. Sci.*, *NS-30*, 3669 (1983).

143. M. Stockton, E. L. Neau, and J. P. VanDevender, *J. Appl. Phys.*, *53*, 2765 (1982).
144. T. J. Pacala, I. S. MacDermid, and J. B. Laudenschlager, *Appl. Phys. Lett.*, *44*, 658 (1984).
145. H. Hatanaka and M. Obara, in *Proc. 7th IEEE Int'l Pulsed Power Conf.*, edited by B. H. Bernstein and J. P. Shannon (IEEE, New York, 1989), p. 671.
146. G. Hinz, in *Proc. 7th Int. PCI Conf.*, Geneva, (Intertec Communications, Inc., Oxnard, CA, 1983), p. 177.
147. D. Birx, E. Cook, S. Hawkins, S. Poor, L. Reginato, J. Schmidt, and M. Smith, *IEEE Trans. Nucl. Sci.*, *NS-30*, 2763 (1983).
148. C. H. Smith, *J. Appl. Phys.*, *64*, 6032 (1988).
149. R. M. Jones, *IEEE Trans. Mag.*, *MAG-19*, 2024 (1983).
150. C. H. Smith and D. Nathasingh, *IEEE Conf. Record 16th Power Modulator Symp.*, Arlington, VA, (IEEE, New York, 1984), p. 240.
151. C. H. Smith and L. Barberi, in *Proc. 5th IEEE Int. Pulsed Power Conf.*, edited by P. J. Turchi and M. F. Rose (IEEE, New York, 1985), p. 664.
152. E. P. Lee and J. Hoving, *Fusion Technol*, *15*(2), part 2A, 369 (1989).
153. T. J. Orzechowski, *Energy and Technology Review*, (Lawrence Livermore National Laboratory, 1986), p. 21.
154. K. Yatsui, *Laser Particle Beams*, *7*, 733 (1989).
155. T. Sawa, Y. Hirose, and K. Inomata, *Toshiba Rev.*, *39*, 735 (1984).
156. *Spike Killer* Technical Information Bulletin, Toshiba Corporation, Metal Products Division, Tokyo, Japan, June 1983.
157. M. A. Newton, D. L. Birx, C. W. Ollis, L. L. Reginato, M. E. Smith, and J. A. Watson, *IEEE Conf. Record 18th Power Modulator Symp.*, Hilton Head, SC, (IEEE, New York, 1988), p. 71.
158. K. Ueno, N. Takahashi, and T. Nozawa, *J. Mater. Eng.*, *12*, 11 (1990).
159. H. Warlimont, *Mater. Sci. Eng.*, *99*, 1 (1988).
160. T. Furukawa, *Amer. Metal Market* (1990).
161. G. Y. Chin, S. Sibley, J. C. Betts, T. D. Schlabach, F. E. Warner, and D. L. Martin, *IEEE Trans. Magn.*, *MAG-15*, 1685 (1979).
162. U. Gonser, *J. Non-Cryst. Solids*, *61*, *62*, 1419 (1984).
163. R. M. White, *J. Appl. Phys.*, *57*, 2997 (1985); *Science*, *229*, No. 4708 (1985), p. 11.
164. G. E. Fish, *Mater. Sci. Eng.*, *B3*, 457 (1989).

20

Rapidly Solidified Nd-Fe-B Alloy for Permanent Magnet Applications

M. S. Guthrie*
Delco Remy Magnequench, General Motors Corporation, Anderson, Indiana

I. INTRODUCTION

The permanent magnet business is radically changing since the introduction of new magnetic materials with high magnetization and high intrinsic coercivity. These materials contain one or more rare earth elements and a transition metal. The resulting crystal structure has only one axis of easy magnetization with very high anisotropy energy. This high anisotropy energy, along with suitable microstructure, results in the very high intrinsic coercivity found in the rare earth permanent magnet materials.

Permanent magnets can suffer irreversible losses in their magnetization when subjected to reverse magnetic fields and/or to elevated temperatures. This is usually the environment of a magnet in applications. The two methods for reducing these losses are: increasing the intrinsic coercivity, or increasing the thickness of the magnet parallel to the magnetization direction and therefore reducing the magnitude of the internal demagnetizing (magnetostatic) field. Material properties determine the intrinsic coercivity. These properties cannot be altered by the magnetic circuit designer. The very high coercivities that have been achieved in the rare earth–transition metal materials allow the designer to use far thinner magnets and thus less total magnet volume. Magnet material is primarily priced by weight, so that cost savings are realized if the higher material cost of a more powerful magnet is offset by a reduction in the space occupied by that magnet.

The first rare earth–transition metal magnet commercially available contained samarium and cobalt. Magnets are produced utilizing one of two intermetallic phases: $SmCo_5$ and Sm_2Co_{17}. Cobalt is relatively expensive and available from somewhat unreliable sources. The so-called rare earth elements are not necessarily

**Current affiliation*: Stackpole Carbon Company, Kane, Pennsylvania.

rare. Samarium, however, comprises only a small fraction of the rare earth elements found in typical ores. Thus, the cost of samarium-cobalt magnets has kept them from being used in many applications. Exceptions to this tendency have been situations where the cost of the magnetic material was not significant when compared to that of other components of a device, and where size reduction of the magnetic circuit was more important than the increased cost of the magnet.

It should be possible to reduce the material cost of rare earth–transition metal permanent magnets if less expensive constituent elements could be used. Obvious choices would be substituting iron for cobalt and cerium, lanthanum, neodymium, or praseodymium for samarium. Nature is not so cooperative. Cerium and lanthanum have no electron orbital magnetic moment, thus lessening the probability of developing useful coercivities. Of the intermetallic phases that form for various combinations of light rare earth and transition metals, most have Curie temperatures below 175°C, making them unsuitable for most applications. Many of the expected phases do not form at all for the light rare earths. The situation for the heavy rare earths is better because more phases form and many of their Curie temperatures are above 175°C. However, the magnetic moments of the heavy rare earths couple antiferromagnetically with those of the transition metals, reducing significantly the saturation magnetization of the ferromagnetic phase. The use of heavy rare earths would then result in a feeble permanent magnet.

The discovery of $Nd_2Fe_{14}B$ was significant in several respects. For example, its magnetic strength is significantly greater than that of any other material. The constituent raw materials are much more abundant than those of other high-energy product magnet materials and are available from politically stable sources. Microcrystallites can be formed via readily controlled rapid solidification (RS) processing, thus allowing a more homogeneous and chemically stable magnet to be manufactured in an isotropic or anisotropic form.

II. THE MAGNEQUENCH PROCESS*

A. Raw Materials Preparation

The manufacture of all rapidly solidified Nd-Fe-B materials begins with a metallothermic reduction of rare earth oxides into a rare earth–iron eutectic. This Neochem process begins with praseodymium oxide and/or neodymium oxide. Neodymium or praseodymium fluoride or chloride may also be used. The use of such nonelemental compounds requires fewer separating operations by the rare earth supplier than do pure neodymium or praseodymium, allowing their purchase at a significantly reduced cost.

Rare earth oxide (or halogen) is mixed with calcium chloride, calcium, and iron as shown in Fig. 1 [1]. The oxygen (fluorine or chlorine) migrates from the neodymium to the calcium and the neodymium is left to alloy with the iron.

$$Nd_2O_3 + 3Ca + 2Fe \xrightarrow{CaCl_2} 2NdFe + 3CaO \quad (1)$$

*MAGNEQUENCH is General Motors Corporation's trademark for RS Nd-Fe-B permanent magnet materials.

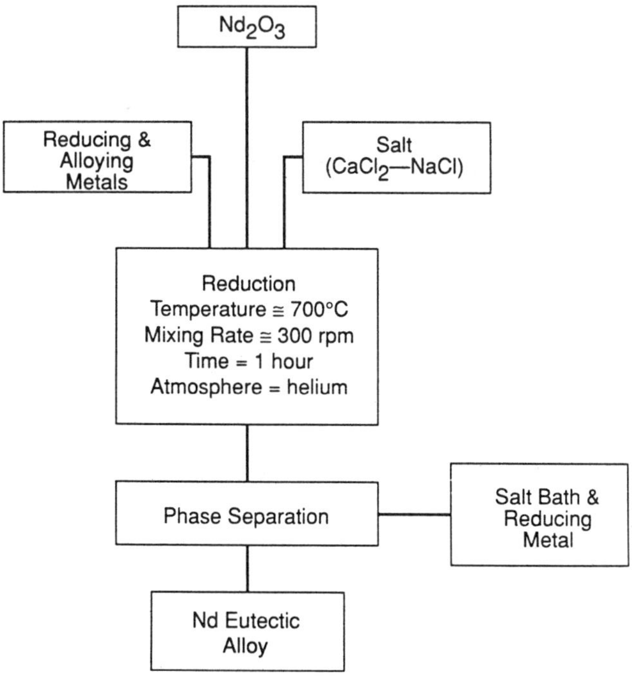

Figure 1 Metallothermic reduction of rare earth oxides.

The calcium chloride is used as a reaction medium. The neodymium-iron eutectic is heavier than the salts and sinks to the bottom of the reaction vessel where it can be removed.

In the alloy furnace, the Nd-Fe eutectic alloy is melted under an inert cover gas and mixed with iron and boron to make the proper alloy. While this alloy consists mostly of $Nd_2Fe_{14}B$, other combinations are used as well.

B. Alloy Chemistry Effects

Praseodymium can be used with, or instead of, neodymium in the preparation of a magnet alloy. The remanence and the coercivities of the resulting magnets are essentially the same. The only real difference when using praseodymium instead of neodymium seems to be the resulting spin reorientation temperatures. The spin reorientation temperature for Nd-Fe-B magnets is 135 K, whereas the spin of the Pr-Fe-B magnet does not reorient.

The hard Φ magnetic phase is 2-14-1 in the neodymium-iron-boron system and comprises most of the magnet. Compared to 2-14-1, the remanence of off-stoichiometric alloys is smaller.

Figure 2 shows the effect of boron on the Nd-Fe system [2]. In $Nd_{0.15}(Fe_{1-y}B_y)_{0.85}$, the structure is mostly Nd_2Fe_{17} phase when $y = 0$. $Nd_{1.1}Fe_4B_4$ is formed as the boron content exceeds $y = 0.05$. This intermetallic phase has no magnetic moment and does not contribute sites for domain wall pinning.

$Nd_2Fe_{14}B$ has a Curie temperature (T_c) of 305°C and magnets based on it have a reversible temperature coefficient of induction of approximately 0.13%/°C

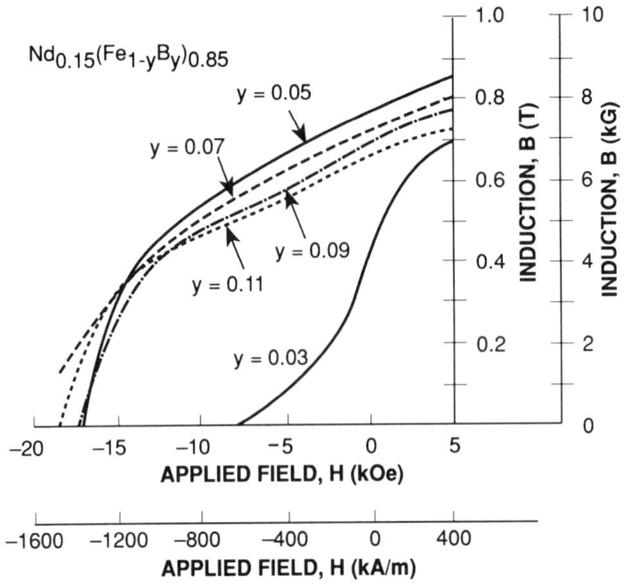

Figure 2 The effect of boron content on magnetic properties.

at 25°C. Since the reversible temperature coefficient of induction is a function of the Curie temperature, and since a lower reversible temperature coefficient of induction is desirable, some alloys of higher T_c are manufactured. For example, in $Nd_2(Fe_1 - {}_xCo_x)_{14}B$ at 25°C, for which $x = 0.05$, $T_c = 360°C$ and the reversible temperature coefficient is 0.10%/°C. When $x = 0.15$, $T_c = 470°C$ and the reversible temperature coefficient is 0.07%/°C. The reversible temperature coefficients of induction and of coercivity both change with temperature. Figure 3 shows the change in the reversible temperature coefficient of induction as the cobalt content changes at 150°C.

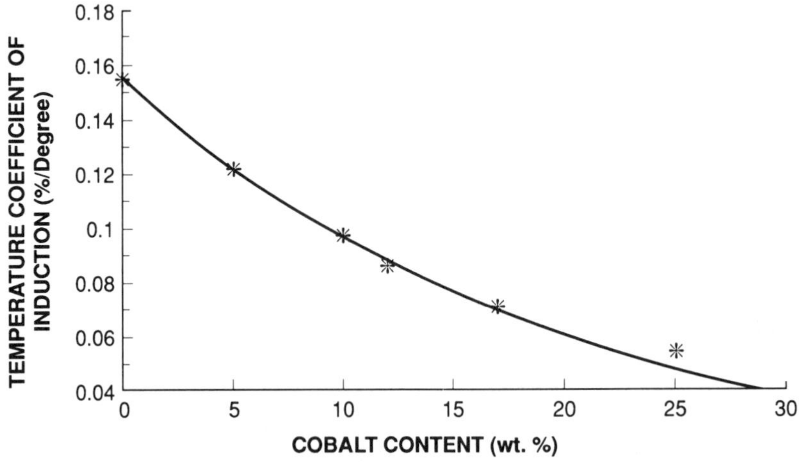

Figure 3 Cobalt content versus temperature coefficient at 150°C.

The substitution of cobalt for iron provides another desirable characteristic. Because the cobalt is less reactive than iron, alloys utilizing some cobalt substitution are less vulnerable to oxidation. The susceptability of Nd-Fe to oxidation has long been a concern to design engineers. The introduction of cobalt also allows the growth of the $Nd(Fe,Co)_4B$ phase, which has neither a magnetic moment nor the capability of acting as pinning sites for the restriction of domain wall motion. This phase constitutes about 3 wt% of the melt-spun ribbon after annealing. Nd-Fe-B magnet coercivity is provided by the Nd_7Fe_3 grain boundary phase. The Φ crystals of the optimally quenched material are small and have the metastable RE_7TM_3 phase completely surrounding them.

In production, RS melt-spun material is overquenched and then annealed. The process parameters necessary for optimal quenching to this final state have too small a window for large-scale manufacturing to be practical. This overquenching and then annealing process results in materials with less than maximum coercivity [3]. Intrinsic coercive fields (H_{ci}) of about 1200 kA/m are typical. H_{ci} is adjusted between 715 to 1435 kA/m daily to produce a variety of products. An H_{ci} of 1675 kA/m is easily achievable; but applications to date have not required this particular material.

Aluminum additions can increase the coercivity of laboratory Nd-Fe-B materials by as much as 50% [4]. However, the effect has not been as great in production. H_{ci} increases of 80–160 kA/m are more common with the addition of aluminum.

Gallium additions to an alloy cause the formation of $(Ga,Nd)_7Fe_3$ phase in the grain boundaries, which provides an additional 320 kA/m in coercivity over the gallium-free material in the hot-formed anisotropic material state (Fig. 4) [5]. This is due to the extra domain wall pinning sites provided by the gallium-containing grain boundary phase.

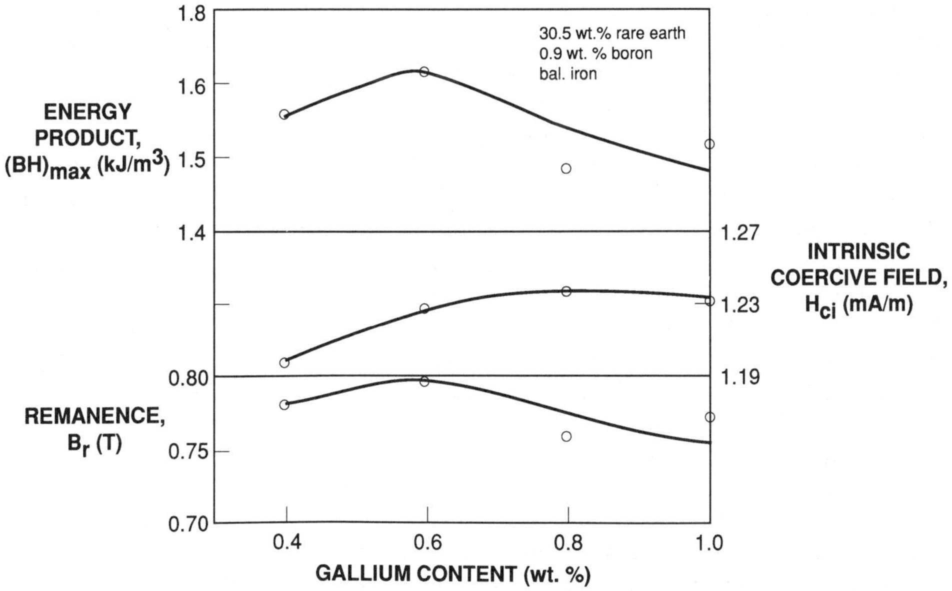

Figure 4 The effect of gallium content on energy product, intrinsic coercive field, and remanent induction.

Heavy rare earth additions such as dysprosium or terbium tend to go into the 2-14-1 matrix phase. These two elements will increase the magnetocrystalline anisotropy and will thereby increase the coercivity of the magnet. Only small quantities of these heavy rare earth additions are required to cause significant increases in coercivity. Even though dysprosium couples antiferromagnetically, it is used extensively in Nd-Fe-B and Sm-Co sintered magnet manufacturing as a coercivity enhancer. The heavy rare earths are not used extensively in melt-spun magnets.

Sm-Ti-Fe alloys have been shown to have coercivities up to 4.6 MA/m (58 kOe) [4]. However, remanence values tend to be below 0.3 T.

The ability to magnetize a permanent magnet is a function of its intrinsic coercivity, H_{ci}. Materials with lower H_{ci} will magnetize to saturation with lesser magnetization fields than materials with higher H_{ci}.

Since the reversible temperature coefficient of intrinsic coercivity is negative in the second quadrant of the hysteresis loop for most permanent magnet materials, the coercivity is decreased as the temperature increases. For isotropic Nd-Fe-B ribbons, this coefficient is $-0.4\%/°C$. A magnet will be more susceptible to a permanent change in induction due to demagnetizing forces at an elevated temperature than would normally be the case.

1. Magnet Types and Their Production

All RS RE-TM-B magnet materials in production today are made by melt spinning, in which a stream of molten alloy is directed onto a rapidly rotating substrate wheel (Chapter 4). On such quenching, 2-14-1 phase forms first, with an intergranular eutectic structure freezing last. Phase size, shape, and uniformity are affected by the quench rate. As quench rate is slowed (via a lower wheel speed for example), precipitated phase size is increased. This will result in a material with low magnetization. If the wheel velocity is greater than optimal (hypercritical quench rate), a smaller, more disordered crystallite size will result in high magnetization but low coercivity, as shown in Figs. 5 and 6 [2]. Amorphous material would have no remanent moment.

In production, the grain boundary phase completely surrounds the small grains at the ribbon-chill wheel surface. As the quench rate lessens toward the opposite side of the ribbon, a lesser percentage of the grain boundaries consist of Nd_7Fe_3 [3]. Some $Nd(Fe,Co)_4B$ is formed.

Most MAGNEQUENCH powder goes to make some type of bonded magnet. Only melt-spun material can be used in the production of bonded Nd-Fe-B magnets. Because the crystallite size is very small (<100 nm diameter), every crystallite on the periphery of the crushed ribbon can be exposed to air, depleted of neodymium, become magnetically soft, yet the magnetic properties of the crushed ribbon will still only be slightly affected. Powder made from sintered RE-TM-B alloy that has been crushed does not have long-term stability. Figure 7 shows the dependence of remanent induction stability on processing technique [6].

In a particle small enough to be used in a bonded magnet which has not been made by a RS process, the surface grains account for a sizeable percentage of the total number of grains in the particle. Thus, when these surface grains become magnetically soft, the magnetic moment of the particle as a whole is lost. Additionally, the rare earth element content of melt-spun RE-TM-B magnets is lower than that of sintered magnets and therefore the affinity of the former for oxygen is reduced.

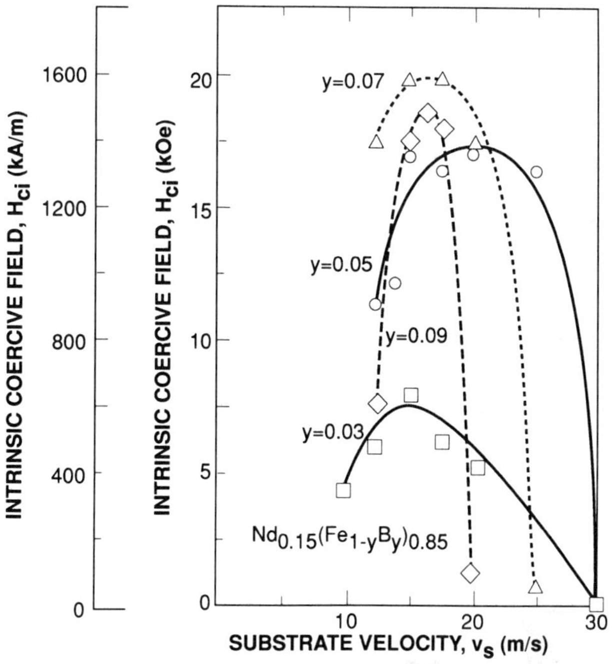

Figure 5 H_{ci} versus quench substrate surface velocity at varying boron content levels.

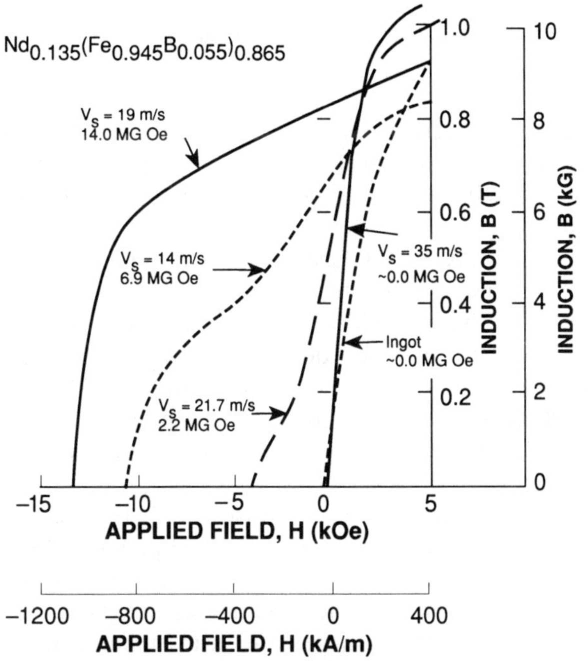

Figure 6 The effect of quench substrate surface velocity on magnetic properties.

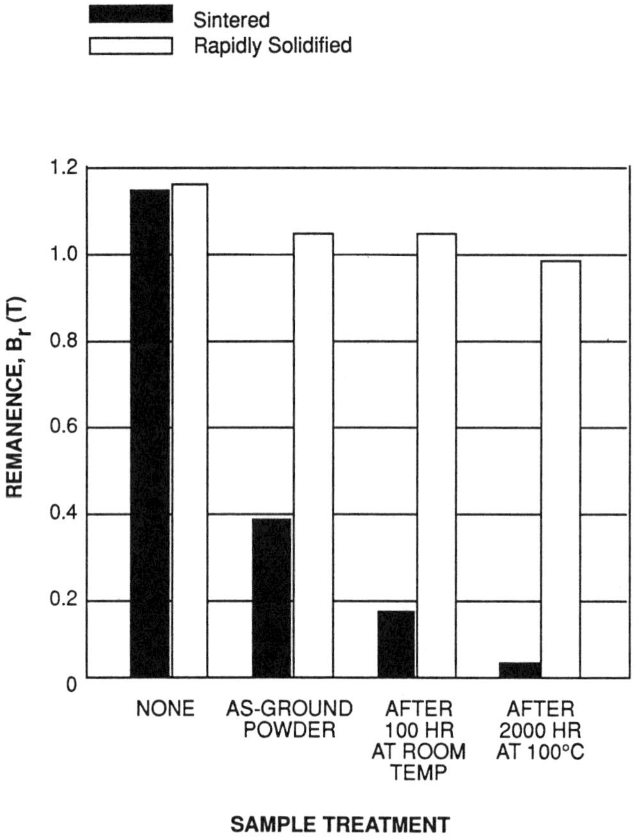

Figure 7 The effect of sample treatment on anisotropic powder remanence.

Isotropic materials can be made by injection molding, compaction molding, or by compression molding annealed powder with a binder. These bonded magnets will have remanent inductions of 0.4 to 0.7 T. B_r is dependent upon the volumetric ratio of magnetic material to binder.

The injection-molding process of MAGNEQUENCH-powder-containing materials is basically the same as that of molded ferrite magnets.

However, the larger and more abrasive RE-TM-B particles make the molding process much more difficult. Much work has been done in developing the right material and method of compounding the annealed, crushed ribbon. Nylon, Ryton, epoxy, polyphenylene sulfide, and other compounds are utilized as bonding media. Parts of small size, with very intricate shapes and tight tolerances are prime candidates for injection-molded magnets.

So-called MQ1 molded magnets are made from a mixture of 2% (by weight) dry epoxy, 0.1% lubricant, and 98% annealed MAGNEQUENCH powder. After thoroughly blending, the powder mix is compaction-molded. The epoxy is subsequently crosslinked in a low-temperature furnace. Since no orienting magnetic field is required, 140° arcs, full rings, and most other shapes are easily manufactured. MQ1 magnets (Fig. 8) have residual inductions of 0.6 to 0.7 T with intrinsic coercive

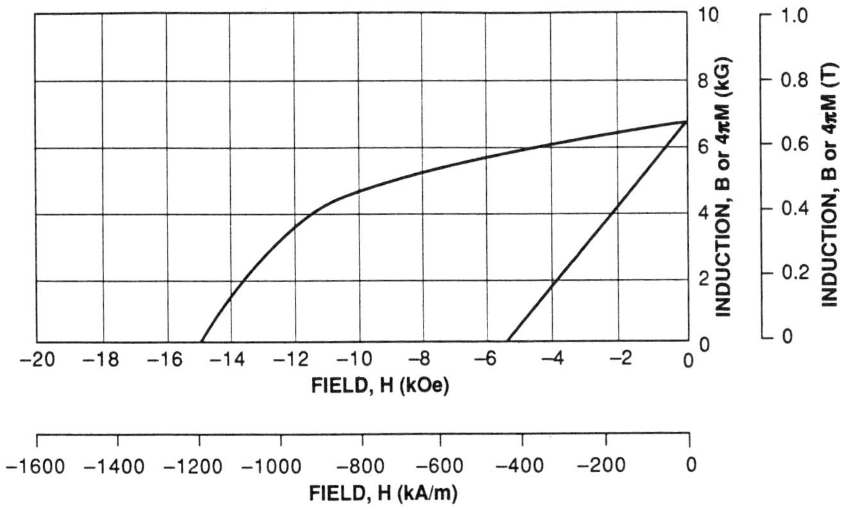

Figure 8 The second quadrant demagnetization curve of an MQ1-A magnet (bonded RS Nd-Fe-B).

fields of 720–1300 kA/m (9.0–16.3 kOe). The magnetic properties are dependent upon the alloy composition of the powder used.

Fully dense isotropic magnets have a similar beginning. The unannealed epoxy-free powder is first compression molded. The resulting low density preforms are then loaded into a 750°C hot die and compacted to full density (~7.6 g/cm^3). This material has a typical residual induction of 0.8 T and is 97% isotropic. Magnetic isotropy aids MQ2 magnets in attaining their very high intrinsic coercivity. The H_{ci} of the MQ2 made from the lower-coercivity "E" alloy is 1400 kA/m (17.6 kOe) (Fig. 9). MQ2 magnets are often used in higher-temperature applications and applications exerting high demagnetization fields.

The full potential of the RE-TM-B crystals cannot be realized until they are all aligned in the same direction. This is done by hot mechanical deformation of MQ2 magnets. During plastic deformation, the constituent crystallites reorient themselves so that the crystallographic *c* axis of the tetragonal $Re_2Tm_{14}B$ phase is normal to the material flow direction during mechanical deformation. The previously equiaxed crystallites are thereby converted to platelets having a typical size of 300 nm diameter by 60 nm thick [2].

The deformation process is very sensitive to alloy composition and process temperature (Fig. 10) [7]. The optimal boron content is defined on the high side by the amount necessary in making the alloy mechanically plastic. A tear may occur if the boron content is too low. If a tear occurs, metal flow during deformation will be lessened and may be in the wrong direction. The lowest boron content allowed is the minimum required for obtaining consistent coercivity.

As the rare earth element content decreases, the necessary cycle time increases and the coercivity decreases. As the rare earth element content becomes larger, magnetization decreases and the affinity for oxidation increases. Excessive tem-

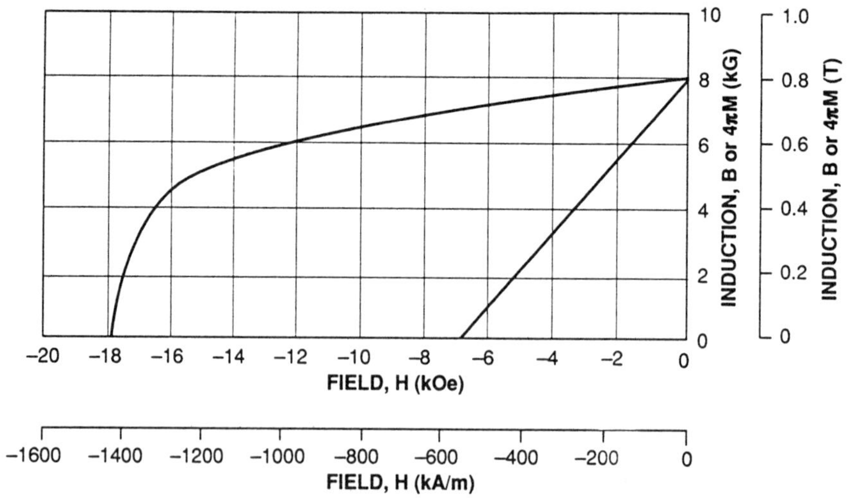

Figure 9 The second quadrant demagnetization curve of an MQ2-E magnet (fully dense, isotropic RS Nd-Fe-B).

peratures in the die cause material to stick to the tools and low temperatures do not allow proper metal flow.

The remanence of a hot-deformed magnet can vary from 0.8 to 1.3 T and up, depending on the amount of deformation (Fig. 11) [6]. The greater the height reduction in the press, the greater the deformation and the greater the anisotropy

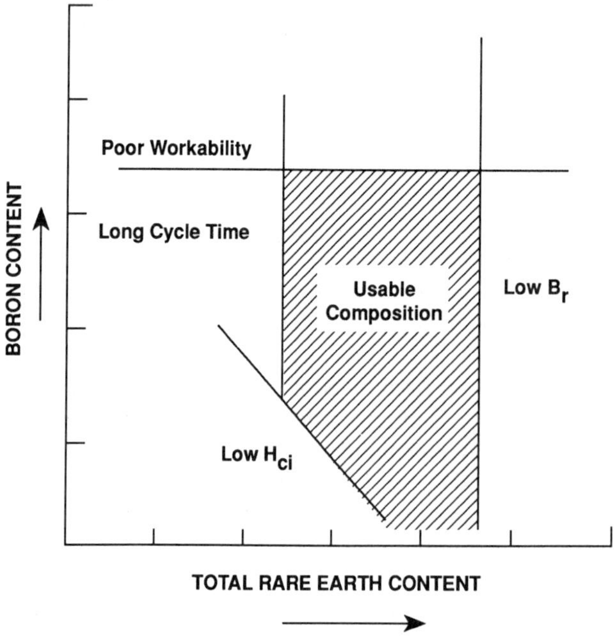

Figure 10 Parameters responsible for defining alloy composition of hot-pressed magnets.

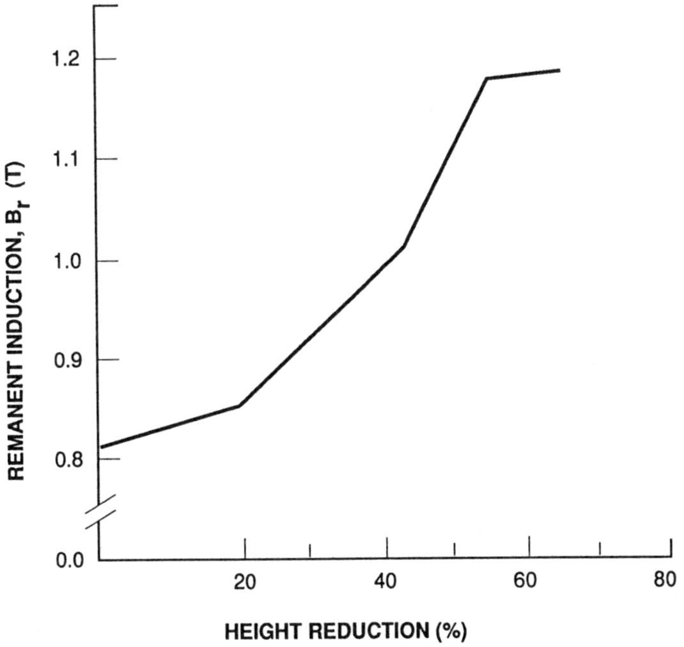

Figure 11 MQ3 magnet remanence versus height reduction by press compaction.

of the resulting material. As the anisotropy increases, B_r increases and H_{ci} decreases.

The coercivity mechanism for thermomechanically aligned material is somewhat different from that of the isotropic material. The magnetic domains in oriented materials extend over many grains. Coercivity is the result of pinning sites at the domain walls by rare-earth-rich second phase particles.

The resulting anisotropic material can have a residual induction of 0.9 to 1.3 T, depending upon the amount of mechanical deformation and alloy composition. Once again, more complex shapes are possible because no orienting field is required. The shapes are only limited to those for which metal flow can be achieved. The development of uniform magnetic properties throughout the magnet requires a uniform amount of metal flow. Conversely, when nonuniform magnetic properties are desired, preforms and die cavities can be designed to provide them accordingly.

MQ3 materials can be physically reduced to a coarse powder and used in bonded magnets. The method of manufacture would be similar to those bonded magnets made from isotropic material, except that an orienting magnetic field is required for alignment of the crystals in most cases. While magnets produced by this process are more shape-limited, they can nonetheless result with residual inductions in excess of 0.8 T.

III. APPLICATIONS AND ADVANTAGES OF HIGH-ENERGY PERMANENT MAGNETS

Magnetic gaskets have been used in refrigerators and freezers for some time. Such gaskets are also utilized on doors to homes. They provide a superior seal and thus increase the barrier to heat and humidity migration. The magnetic force aids in providing a tight, gapless seal. However, now that they are being considered for automotive applications, their relatively low weight is an additional attribute.

Magnetic door gaskets for automobiles can significantly reduce the interior noise level and better maintain the interior environment. The infiltration of dust and fumes can also be effectively restricted. Gaskets utilizing bonded Nd-Fe-B material have been shown to weigh less than nonmagnetic gaskets while concurrently providing a better seal. Other bonded magnetic materials are also available; but the only ones which are economically feasible tend to add as much as 5 lb to the weight of a vehicle.

The largest single market for permanent magnets is the motor industry. There exists a large variety of motors, many of which already use permanent magnets and many more of which are being developed to incorporate permanent magnets. Several factors need to be considered when choosing the best permanent magnet for a motor application. These include motor temperature extremes, speed, torque, current consumption, size, and cost. RS permanent magnet materials have several niches in this market that only they can fill, while also competing directly with other types of permanent magnets that are currently used in other types of motors.

One specific type of dc brushless motor that is extensively used is the permanent magnet micromotor. The majority of these motors are produced in the Asia-Pacific basin. Japan, Korea, and Indonesia are currently producing micromotors at a rate of over 1 billion per year. These are predominantly going into cameras, VCRs, and audio tape drives. Small, expensive, watch motors have been manufactured for some time. Micromotor sales are increasing at a rate of 20% per year.

The reason for this sudden growth is that only recently did a suitable permanent magnet material become available that could be utilized cost-effectively. RS Nd-Fe-B allows small motor magnets to be injection molded in large numbers to tight tolerances quickly and efficiently.

While some small rotary motors contain brushes, most are brushless. Brushes are a more significant problem in micromotors because of the large voltage drop in relation to the total voltage available. Micromotors using moving coils have low inertia and low mechanical time constants, but still have problems associated with brushes. The use of high-energy, low-mass, moving magnets is expanding into brushless micromotors because the requirements of low-inertia, low-power devices are met.

Bonded MAGNEQUENCH materials retain tight dimensional tolerances without the need for grinding. These magnets can be made into most imaginable shapes, including arcs, rings, disks, rotors, stars, etc. High flux densities and high coercivities make RS Nd-Fe-B materials perfect for usage in electric motors.

Bonded RE_2TM_{17} is the only other high energy permanent magnet that can be molded. Bonded 2-17 materials are more limited in their application than bonded Nd-Fe-B due to their method of manufacture, anisotropy, and cost.

Conventional motors contain brushes that carry the electric current to the rotating electromagnetic coils in the rotor. Fig. 12 shows the end view of the magnetic assembly in a two-pole motor. A lesser number of permanent-magnet poles is desired so that the number of brushes, for electric field comutation, can be lessened. This not only simplifies the manufacturing of the brush assembly and the electrical contacts, but also allows larger currents to flow through those larger brushes and

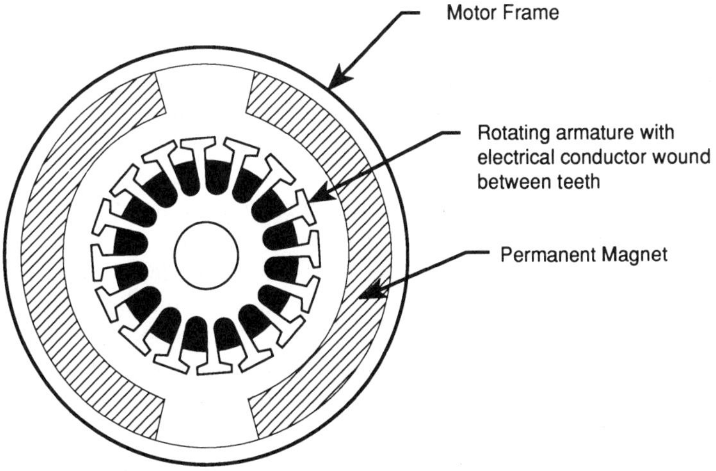

Figure 12 A two-pole brush-containing motor.

contacts. Motors of this kind are used in applications such as slot cars, hand-held drills, vacuum cleaners, automobile engine starters, and lawn mowers.

Most brushless motors have their conductors placed on the outside stator and the permanent magnets are moved to the rotor (Fig. 13). This brushless design is superior in that the voltage drop across the brushes and the shorter motor life due to brush failure are no longer problems. Servo motors and fan motors can take advantage of this energy-saving, compact, silent design which permits little or no leakage flux to escape from the motor. The reduced amount of heat generated and its improved removal also add to the longevity of the brushless motor.

Some applications favor fixed coils in the center of the motor and an outward moving magnet-return path assembly. Turntable and disk and tape drive spindle

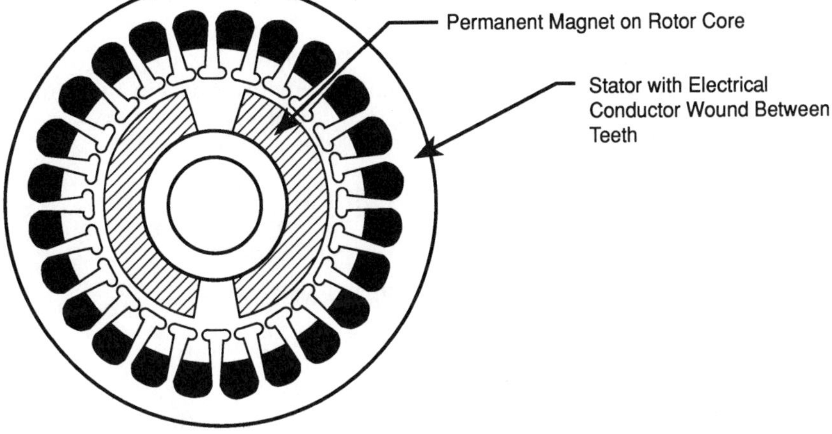

Figure 13 A two-pole brushless motor.

motors can put the electromagnets in the center of the motor, which would otherwise be an unused space.

Electromagnetic interference is significantly reduced when the sparkless commutation of a brushless motor is utilized instead of the brush-type commutation. Brushless motors are required when electric motors are needed in a flammable, or explosive, atmosphere.

Since there is no mechanical device inside a brushless motor to change the direction of current flow in a coil when the magnets move by, the direction of current flow must be reversed at the proper time by some other means. Thus, a special power supply is required. Usually a sensor is used to transmit the position of the rotor (and its magnets) back to the current driver. These special power supplies produce the proper current wave shape and amplitude to drive the motor.

The permeance of a magnet is a direct function of the magnetic circuit. It determines magnet output and resistance to demagnetization and is represented on the demagnetization curve as a B/H ratio. The operating load line (B/H) will be the slope of the straight line from the origin up into the second quadrant of the hysteresis loop. In Fig. 14 B_d = 7200 G and H_d = 4,800 Oe; therefore B/H = 1.5. B/H is not dependent on the magnet material.

$$\frac{B}{H} = \frac{lm}{Am}\frac{Ag}{l_g}\frac{f}{R}$$

l_m = length of the magnet (from pole to pole)

A_m = area of the magnet

A_g = area of the air gap

l_g = length of the air gap

$\frac{f}{R} = \frac{\text{leakage}}{\text{reluctance}}$

In a motor with a short air gap, we can approximate: $B/H \approx l_m/l_g$. An example motor design shows the l_m/l_g = 10.20/0.70 = 14.6. If B/H = 14.6, ceramic 7 material will exhibit a flux density (B_d) of 3350 G. B_d is the usable flux density and is the B coordinate where the loadline intersects the B-H curve. Ceramic 7 is a sintered ferrite having relatively high coercivity and therefore is often chosen for motor applications.

A motor's torque and speed are directly related to magnet flux (Φ). As magnet flux increases, torque increases and speed decreases. For example, in the sample motor (B/H = 14.6):

If the magnet material is ceramic 7:

$$\Phi = B_d \times A_m = 3350 \text{ G} \times 57.66 \text{ cm}^2 = 193{,}161 \text{ Mx}$$

If the magnet material is changed to MQ1:

$$\Phi = B_d \times A_m = 5900 \text{ G} \times 57.66 \text{ cm}^2 = 340{,}194 \text{ Mx}$$

In review:

	Ceramic 7	MQ1	Difference
Total flux	193,000	340,000	76%
Coercivity	4,000	15,000	275%

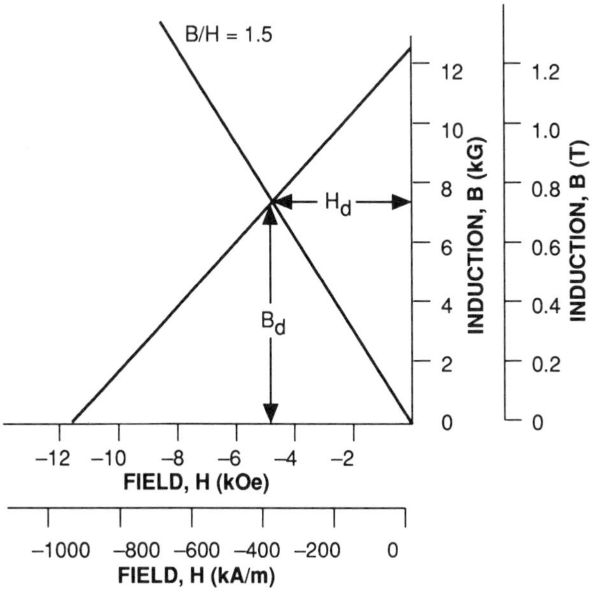

Figure 14 The demagnetization curve of an MQ3 magnet with permeance coefficient.

By using MQ1 permanent magnet material rather than ceramic 7, the motor designer can:

1. Decrease the magnet thickness to 0.90 mm. This will result in a flux level equivalent to the ceramic 7's and will decrease the magnet's weight by 89%.
2. Decrease the magnet axial length by 43%. This will give equivalent flux, decrease magnet and motor weight, and make a significantly shorter motor. A lesser amount of motor steel will also be used.
3. Decrease the current draw of the motor. The original ceramic magnet motor required a $3.50 power control module to handle its necessary electric current. The increase in the permanent magnet flux allows a decrease in the current necessary for equivalent motor performance. The larger magnet could be used, current reduced, and the power module would then be unnecessary.
4. Use a combination of the above. That is, reduce the magnet flux to the point where any further reduction would necessitate the use of a power control module. This is achieved by first reducing magnet thickness (a B/H of 14.7 is inefficient for a high-coercivity material). Next, the axial length of the magnet could be reduced if further flux reductions are possible. These stack length reductions allow the removal of steel and copper from the motor, thus reducing not only motor material cost but also reducing heat (i^2R) losses.

If the room were available to double the number of brushes and contacts in the motor, then conversion from a two-pole motor to a four-pole motor would be desirable. Motor torque and speed are a direct function, and an inverse function respectively, of the number of magnet poles in a motor. If the number of magnet poles in a motor could be doubled, the total magnet flux across the gap could be halved without affecting motor performance.

The previous example compared ceramic 7 to compaction pressed Nd-Fe-B in a motor application. If calculations for an anisotropic magnet made of RS Nd-Fe-B in a four-pole motor were performed, the magnet volume, for the same motor speed and torque, would be reduced by 97% from that of the ceramic 7.

A stepping motor is one in which the rotor can turn in discrete short movements instead of continuously revolving. Step increments vary from 15° down to 0.18° and even smaller. The stepper is a brushless motor which requires a special power source to supply current pulses at the proper rate and polarity of the proper field coils. One of the more common stepping motors is the hybrid. Figure 15 shows the design of a four-phase, 200-step motor with the stepping angle being 1.8°. It consists of a permanent magnet sandwiched between two toothed pole pieces. The coils on either side of the motor will energize when the coils on the other side are off. This enables the motor to step in increments that vary in size from multiple rotor rotations down to half a tooth width.

The computer industry is a large user of stepper motors. Printers, including typewriters, rely on stepper motors for paper advance, ribbon advance, carriage movement, and character wheel motion. X-Y plotters, floppy disk drives, head actuators, and CNC machines are also large users.

Linear stepper motors also exist. They are used in the automobile and in the printing industries. These are very closely related to the permanent magnet rotary stepper motors in their design. They also share the advantage of all stepper motors in that they can immediately change their direction of motion. This capability is very useful in applications such as pen motion on a plotter.

RS permanent magnets can be injection-molded in multicavity dies quickly because they do not need to have an orienting magnetic field present and they do

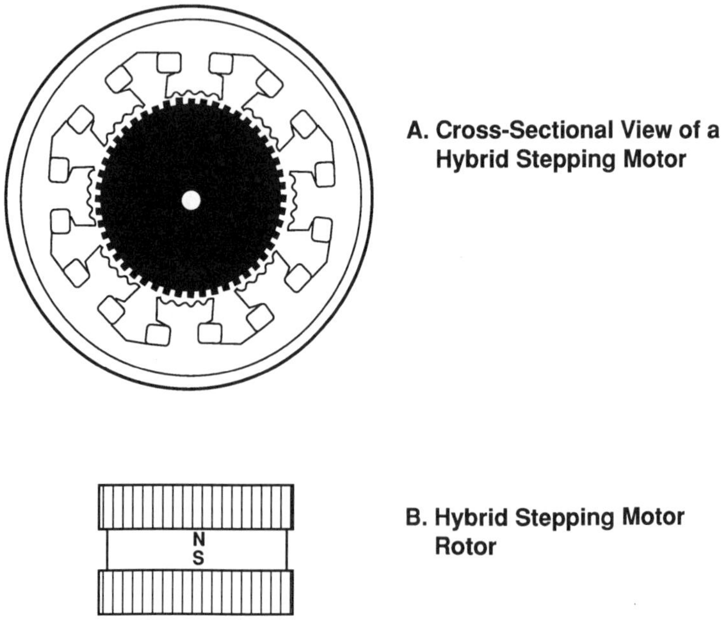

Figure 15 A hybrid stepper motor.

not require demagnetization prior to their removal from the die. Arcs and rings can be pressed from this material, yielding a magnet with high B_r and high H_{ci}. This results in motors with high torque, in a small size, at a relatively low cost. Isotropic materials can be magnetized with a skewed magnetic field, which will allow a motor designer to reduce, or even to eliminate motor cogging. The rotor magnet can be magnetized so that the flux path turns inside the magnet material before it comes back out of the adjacent pole. This results in a higher permeance coefficient and more available flux. Little or no material for the flux path return will be required inside the magnet. This significantly reduces the mass and therefore the response time of the motor.

Anisotropic, radially oriented solid metal rings are also made of RS Nd-Fe-B. These rings provide designers with 360° of MQ3 material. The magnets can be assembled in one piece and magnetized in place.

Moving coil (coreless) motors are constructed with brushes and are attractive because of their very low inertia. In an effort to further decrease the mass of the moving coil, high-induction permanent magnets are used. In the past, Alnico magnets were used extensively. A cast Alnico 5 magnet has a B_r of 1.28 T while a cast Alnico 5-7 has a B_r of 1.35 T. However, both of these materials have very little resistance to demagnetization. Each must work at a permeance coefficient of 18 or greater (this is a B/H of 18 in cgs units and 225 in SI units). Working at this high loadline requires a magnetic circuit with a long magnet length or a short air gap. MQ3 materials are suited for moving coil motors because their flux densities are as high as that of Alnico 5 and their resistances to demagnetization are much higher. MQ3 magnets will not self-demagnetize at temperatures below 180°C, if properly designed.

Printed circuit motor design is similar to that of the conventional coreless motor in that its rotor utilizes an ironless coreless armature (Fig. 16). Whereas most other motor rotors exert force due to a radial magnetic field, the printed circuit motor uses a compact disk-shaped coil in conjunction with a multipole permanent magnet field, the orientations of which are parallel to the coil's axis of rotation. The advantages of RS Nd-Fe-B usage include size reduction, increased resistance to demagnetization, and increase in flux density. Turntables and computer disk drives can take advantage of the increased torque from a flat motor with an MQ magnet. Automotive radiator fan motors benefit from the reduced mass and reduced power consumption. Printed circuit motor design can also be modified by fixing the coils and allowing the permanent magnet to rotate. This will allow the removal of the brushes, add inertia to the rotor, and require the use of a brushless motor power supply.

Permanent magnets are used in a wide variety of applications in conjunction with magnetic field-sensing devices. Magnetic fields can be sensed by Hall-effect devices, magnetoresistive devices, reed switches, and variable-reluctance sensors.

Hall-effect devices are current-carrying materials (usually semiconductors) that develop a voltage drop in the presence of a magnetic field. The conductor is placed such that the magnetic field is perpendicular to the direction of the current flow and the voltage generated is perpendicular to both. Hall-effect devices can be used as field-level detectors or they can be used as switches, merely to indicate the presence of a field. Most gaussmeters used today have Hall-effect devices in their probes to measure the magnitude of the magnetic field present. Many permanent

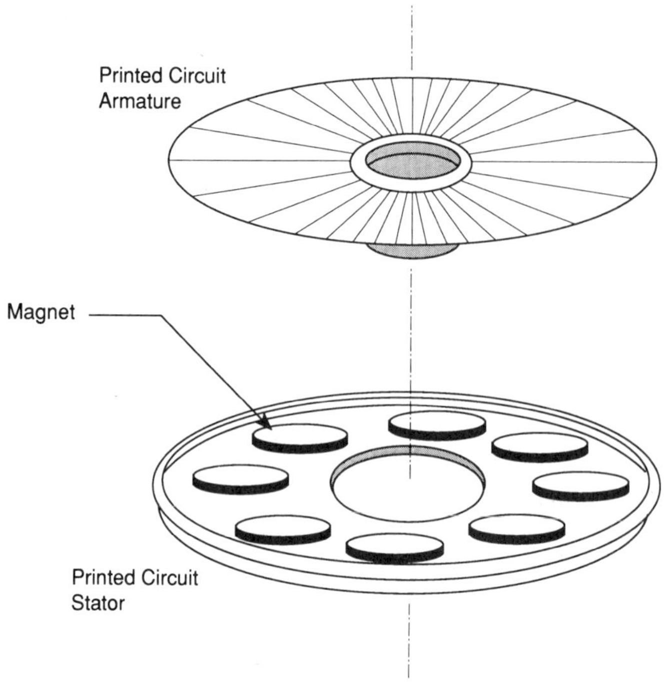

Figure 16 A printed circuit motor.

magnet motors use Hall-effect switches to sense the flux density of the permanent magnet as it goes by and thereby signals the electrical system of the rotor position.

A magnetoresistor is an electrical conductor in which resistance is changed by the presence of a magnetic field. The applications for such devices are somewhat similar to those employing Hall-effect devices. However, a resistance change is sensed instead of a voltage change and only two current leads are necessary instead of four (two to supply the current and two to detect the voltage).

Variable reluctance sensors can detect position or speed. A coil is located in the flux path of a permanent magnet. Voltage is generated when the permeance of the magnet is changed by the changing air gap in the circuit. As the rotational speed of the turning target wheel slows, the sensitivity of the sensor decreases. A stronger magnet enhances this sensitivity. MQ1 and MQ2 magnets are used extensively because they produce large amounts of flux using a small magnet size and at a reasonable cost. The variable-reluctance sensor is frequently used for applications where the speed of a rotating device needs to be monitored. Such applications would include wheel speed monitoring for antilock brake systems and cruise controls, and gear speed monitoring for proper automatic transmission shifting.

Reed switches consist of two "reeds" of electrically conductive ferromagnetic materials in close proximity to one another. When a magnetic field comes near the pair, each conductor is magnetically attracted to the other, as shown in Fig. 17. Reed switches are often used in home security systems. A reed switch will be affixed to the door or window frame, and a permanent magnet will be placed in the door or the window sash. As long as the door or window remain closed, the magnet remains near the switch, and current can pass through the conductors.

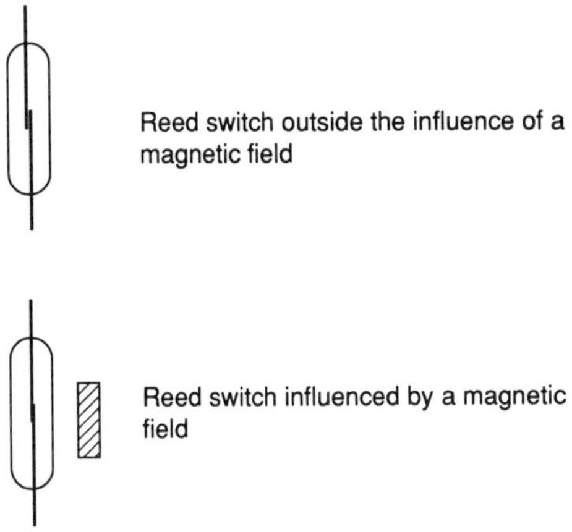

Figure 17 Reed switch activation.

When the aperture is opened and the magnet moves away, the conductors move apart and the electrical circuit is broken, thereby activating the alarm system.

Since the introduction of Alnico 5 permanent magnets in the early 1940s permanent magnets of some type, have been used in the vast majority of loudspeakers. The permanent-magnet-using speakers immediately became competitive with the wound-field electromagnetic speakers, in terms of both performance and cost. The basic design has not changed much in the last 50 years. The moving voice coil in a gap of high flux density has many advantages over other speaker designs. This design also allows the usage of many different types of permanent magnet materials.

Figure 18 shows the initial Alnico 5 speaker circuit. The high-flux-density magnet in the center allows the usage of a material with a relatively small surface area. The flux return path that surrounds the magnet shields the leakage flux from interfering with nearby systems. Systems that may be affected by unwanted stray flux would include computers, a small television's picture tube, or an automobile's dashboard instrumentation. Because Alnico 5 material has a low resistance to demagnetization, its magnets must have relatively long physical length between poles.

In the 1970s, a sudden increase in the price of Alnico raw materials brought loudspeaker manufacturers to redesign their products. Very quickly, every major loudspeaker designer redesigned to use a ferrite magnet. While the residual flux density of ferrite (or ceramic) permanent magnet materials is much lower than that

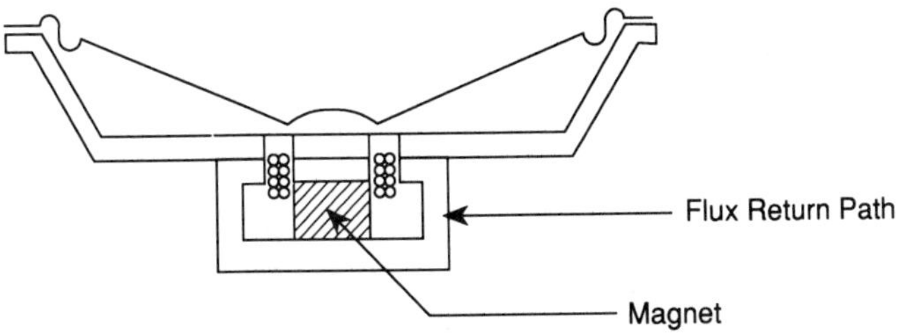

Figure 18 Speaker design utilizing a high-induction, low-coercivity permanent magnet.

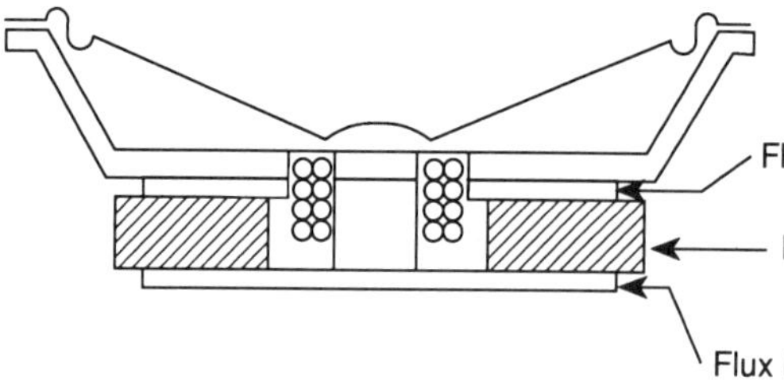

Figure 19 Speaker design utilizing a low-induction permanent magnet.

of Alnico materials, equivalent flux densities could be realized by using the top and bottom plates as flux collectors and focusing all of that flux into the gap.

As shown in Fig. 19, the open sides of the magnet allow flux to leak between the two pole pieces, resulting in speaker efficiencies of about 35% for this type of design. Even with the increased total flux requirement due to increased flux leakage and an increased amount of return iron required to cover the much larger magnet area, and the much larger magnet volume required (because of lower flux density and increased flux leakage), the ceramic design was successful because it was cost effective.

A "best of both worlds" situation has been realized with the new Nd-Fe-B magnet designs. Figure 20 shows a design that looks much like a speaker using a short Alnico 5 magnet. MQ3 has a residual induction value of 1.2 to 1.3 T, about the same as that of Alnico 5. It also has intrinsic coercivity values much greater than those of Alnico 5 or ceramic 5. Therefore, small area magnets with short magnet length can now be used to make loudspeakers. These speakers are characterized by low leakage, low mass, and small size. Low flux leakage allows usage in some applications unsuitable for ceramic magnet speakers. These might include areas near picture tubes, other monitors, and generally electrical lines or beams sensitive to motion or distortion.

The low-mass benefit has created many additional advantages for users of loudspeakers containing MAGNEQUENCH materials. Just as the loudspeaker was the first user of permanent magnet materials, the automobile loudspeaker was the first large volume user of the high-energy MQ3 material. Corporate average fuel economy requirements continue to rise and weight savings are continually sought. Decreased magnet weight and the decreased weight of the flux return path are very much desired, especially in automobiles or aircraft containing more than two speak-

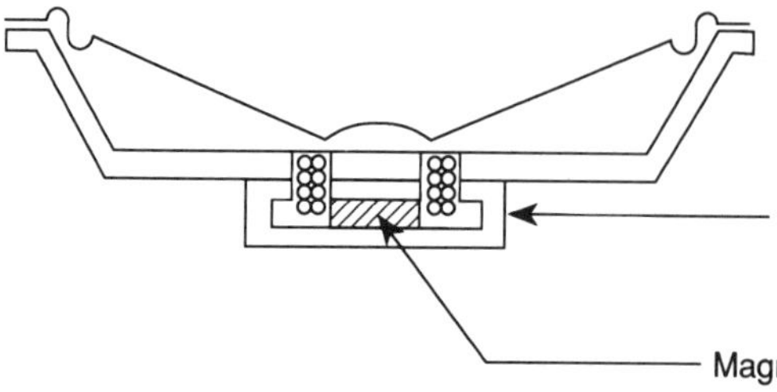

Figure 20 Speaker design utilizing a $RE_2Fe_{14}B$ (high induction, high coercivity) magnet.

ers. The total loudspeaker weight can decrease by as much as 87% [8] when converting from a ferrite ring loudspeaker to a Nd-Fe-B permanent magnet loudspeaker. Low weight is also appreciated by the speaker manufacturer. Shipping charges of a large number of MAGNEQUENCH speaker drivers will be significantly less than that number of ferrite drivers, since the weight of the drivers is significantly reduced.

The greatest benefactors of smaller size and weight loudspeakers will probably be the manufacturers of, assemblers of, and listeners to large sound systems in auditoriums and stadiums since the smaller size allows tighter spacing of horn arc arrays. This results in more precise and even coverage and the need for delay in a horn array is reduced. The load-bearing requirements of these systems can also be greatly reduced.

A microphone and a loudspeaker are designed in much the same way. The only difference is that, instead of optimizing the design for the most accurate acoustic wave generation from a voltage ripple in an electrical circuit, the design for the most accurate voltage ripple generated from an acoustic wave is optimized for microphones. Small size and low mass are important here because the microphone is often carried by someone. A microphone is something the public never wants to see, but cannot do without. Microphones incorporating Nd-Fe-B magnets are currently being produced. These allow the use of smaller mass magnets with larger pole areas. Such changes along with the higher induction have made it possible to increase the flux density in the air gap, providing greater microphone sensitivity. The short, cylindrical shape used is the easiest and fastest shape for MQ3 production. MQ3 materials, with their high residual induction and low price, will enjoy success in this market.

The disk drive actuator market has always been a supporter of state-of-the-art permanent magnet materials. Moving-coil actuators are constructed in a variety of forms and use many different types of coils and permanent magnets. A large user of rapid acceleration, precise positioning actuators is the computer peripheral industry, specifically for use in disk drive actuators which can move masses of 25 g, 100 mm in 10 ms, with an accuracy of ± 0.01 mm.

Moving coil actuators are typically called voice coil actuators, even though most modern designs do not resemble those of a loudspeaker. Since force is directly proportional to flux density, materials that produce high flux density are required for higher forces and shorter access times. In the past, most actuators were constructed in the manner of Figs. 21a or 21b. These designs were used extensively for Alnico and ferrite magnet containing actuators. The pole pieces collected the flux and directed it across the gap.

The disk drive industry was the first to begin using Nd-Fe-B magnets because of customer demand for shorter access time and smaller package size. Magnet cost was not as much a limiting factor as it was in many other applications.

The actuator design shown in Fig. 21c is a linear actuator. A moving coil is positioned around the center flux return path. When current is applied, the forces on the coil produce motion parallel to the axis of the center return path. Because of the shape of the flux path, this design is called an "E-core" actuator. The thickness of the magnet-adjacent return structure sections can be reduced and the leakage flux coming from the end of the actuator can be decreased by placing a flux return path across the front of the actuator (Fig. 21d).

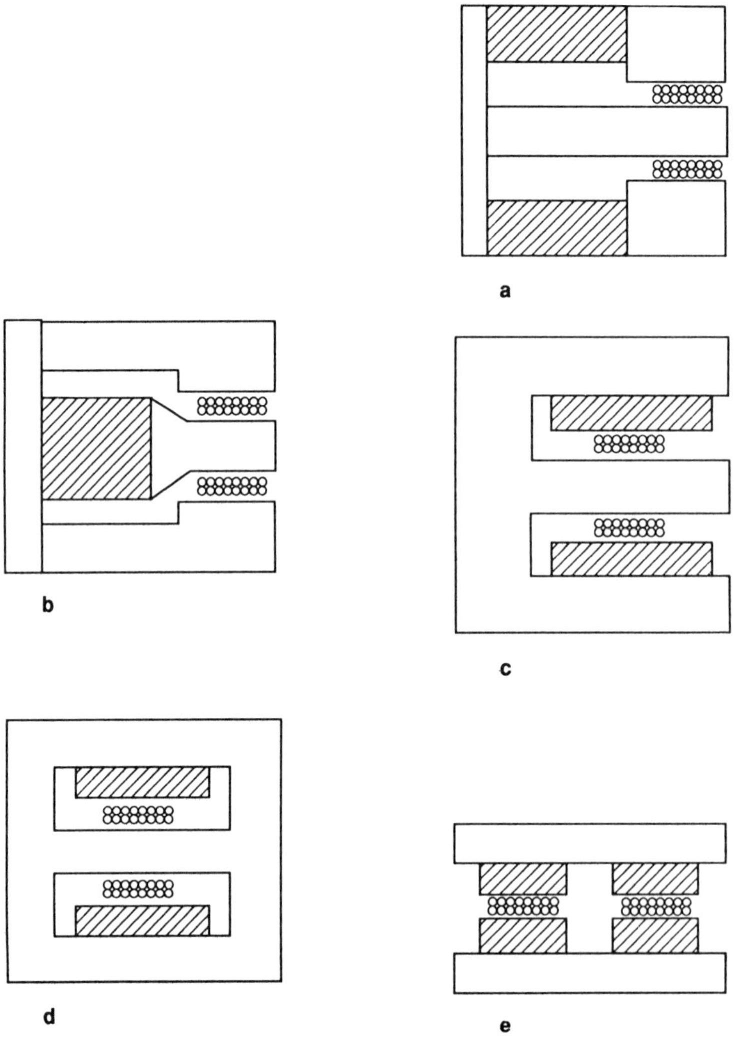

Figure 21 (a) Actuator design utilizing low-induction magnets. (b) Actuator design utilizing a high-induction, low-coercivity magnet. (c) Open-ended E-core actuator utilizing high-induction, high-coercivity magnets. (d) Closed-end E-core actuator utilizing high-induction, high-coercivity magnets. (e) Swing arm actuator utilizing high-coercivity magnets.

The actuator design shown in Fig. 21e utilizes a flat coil in a rotary mechanism. The coil pivots about a point in front of the actuator and the read-write head moves across the memory disk in a direction that is opposite that of coil motion. This design uses one to four magnets.

The MQ3 materials have been shown to be excellent permanent magnet materials for disk drive actuators. Typically, very high residual induction and medium-range coercivity are characteristics of actuator magnet materials. A magnet with higher coercivity is required when larger temperature excursions are prevalent. Disk drive manufacturers who use both MQ3 and sintered Nd-Fe-B magnets have noted the homogeneity of field from the MQ3 materials to be greater. This is because the MQ3 materials can have more homogeneous magnetic properties.

Moving-magnet actuators have been designed for many years and can eliminate problems caused by the rapidly moving coil. Such problems include broken coil leads, coil speed reduction and coil speed nonuniformity. The actuators can also give effective heat transfer to the coils which are mounted against a heat sink. Moving-magnet actuators will be used when the energy-weight product of the permanent magnet equals the energy-weight product of the electromagnetic coil.

The possibility of corrosion and oxidation of Nd-Fe-B materials is troublesome to disk drive manufacturers. This is because small particles of a deteriorating magnet can easily move from the magnet onto the memory disk and cause a head crash and/or erasure of memory. Even if the actuator is not in the sealed environment of the head-disk area, a deteriorating magnet will cause actuator force reductions and nonlinearities. Sintered rare earth magnets are not 100% dense while MQ3 magnets are. Therefore, the former have slightly more surface area that can react with oxygen and moisture. Sintered Nd-Fe-B magnets typically contain 3 wt% more rare earth element than do MQ3 magnets. The majority of this rare earth is neodymium, which is strongly reactive. Thus, reduced surface area and reduced rare earth content of MQ materials contribute in the makeup of a more chemically stable magnet.

Magnetic resonance imaging (MRI) devices have many advantages over other imaging techniques. MRI may be compared to x-ray or ultrasound techniques in that it can create images of a living body's interior without the use of surgical tools or operative procedures. Unlike x-ray imaging, the body to be studied is subjected only to magnetic fields. Nuclear magnetic resonance imaging systems consist of four parts: the bias magnet, the gradient coils, the RF coil, and a receiving antenna. X, Y, and Z gradient coils are used to cause gradients in the magnetic field patterns in these directions. The gradient coils are pulsed so that there is one plane in which a field perturbation does not exist. The receiver scans the magnetic field and constructs an image by the location of the unperturbed areas. Thus, pictures are taken as slices of the object.

When the subject is placed into the magnetic field of the biasing magnet, the nuclear magnetic moments of its constituent atoms are aligned parallel to that applied field. A perturbation in the magnetic field is caused by the RF coils and the dipoles are in turn displaced from the aligned position. When the RF coil field is canceled, the nuclear magnetic moments of the different materials comprising the subject, realign toward the bias magnet field at different rates. The computer, using the receiver's input, can determine the velocity and position of the realigning moments. This data can then be used to record the makeup and distribution of components within a subject.

The bias magnet, which can be either an electromagnet or a permanent magnet structure, is a major problem for MRI systems. The electromagnet structure may be comprised of standard conductor coils, conductor coils with an iron core, or superconducting coils. The resistive conductor coils pass large amounts of electrical energy and are limited to a field strength of about 160 kA/m (2 kOe). Their fields are axial, so that the magnetic field extends far outside the magnet structure. This system results in low image quality. The iron core electromagnets are a little better because they can create a field that is perpendicular to the body inside the magnet. They can shape the field more uniformly and generate a field strength in the 240 kA/m (3 kOe) range. However, power consumption is still high. Superconducting electromagnets generate the highest fields attainable. Their coils are cooled with liquid helium or nitrogen to cryogenic temperatures at which there is no resistance to the flow of electrical current in the coils of the superconductor. Therefore, very high currents can be utilized, resulting in very high magnetic field strengths. Unfortunately, superconducting magnets have high setup and operating costs, produce high magnetic fringe fields, and require special structural support for their large masses and troublesome cryogenic systems.

These devices can weigh as much as 50 to 100 tons. In contrast, the Nd-Fe-B permanent magnet-containing assemblies now on the market are much smaller and lighter (5–45 tons) and are capable of bias fields of 80–400 kA/m (1–5 kOe). MRI devices are usually composed of magnets having different magnetic field alignments, each of which must be precise in order to minimize nonuniformities in the biasing field. Any nonuniformity would make the dipole realignments impossible to accurately compare.

Bonded MQ1 and hot-pressed MQ2 materials are isotropic. Therefore, they can all be made in the same way and each can be magnetized so that its field has the correct orientation in relation to the assembly. This greatly simplifies magnet production. While isotropic MQ magnets are attractive to the MRI industry, they do have the problem of electrical resistivities which are low enough for eddy current production by the changing electromagnetic fields to cause potential problems. Eddy current production in low-resistance materials is a problem because these electrical currents produce transient magnetic fields which in turn cause disturbances in the biasing field.

IV. CONCLUSION

Research on magnetic materials is growing at an ever-increasing rate. Improvements of induction, coercivity, Curie temperature, resistance to corrosion, and ease of magnetization for rare earth–transition metal materials is a goal that is shared worldwide. Improvements in the controllability and cost effectiveness of manufacturing methods are constantly being explored.

Remanence values ranging from 0.2 T up to 1.3 T can currently be consistently exhibited by RS Nd-Fe-B materials varying from isotropic flexible bonded magnet strip to anisotropic fully dense hot-pressed magnets. Coercive fields of 716 kA/m to values in excess of 1.6 MA/m are consistently attainable from materials by suitably varying compositions and methods of manufacture.

Lower-cost raw materials and more controllable manufacturing processes permit the production of affordable magnets having consistent properties. Even though no one magnet material is perfectly suited for every application, the variety of physical and magnetic properties which can be developed by the variety of manufacturing processes from which RS Nd-Fe-B permanent magnets are made, allows them to be the nearest thing to a universal magnet material that has yet been demonstrated.

REFERENCES

1. R. A. Sharma, U.S. Patent No. 4578,242 (1986).
2. J. J. Croat, *J. Mater. Eng.*, *10*, 7, (1988).
3. A. Hutten and P. Haasen, *Acta Metall. Mater.*, *39*, 1 (1991).
4. H. H. Stadelmaier and E.-Th. Henig, *J. Metals*, *32*, (1991).
5. V. Panchanathan and J. J. Croat, *IEEE Trans. Magn.*, *MAG-25*, 4111 (1989).
6. J. J. Croat, IEEE Trans. Magn., *MAG-25*, 3550 (1989).
7. J. W. Herchenroeder, Magnequench Report No. JWH-036 (1991).
8. D. Ratnam, *Machine Design*, June 20, (1991), p. 63.

21

Brazing and Soldering with Rapidly Solidified Alloys

A. Rabinkin and Howard H. Liebermann
AlliedSignal Advanced Materials, Parsippany, New Jersey

I. INTRODUCTION

Metal joining by brazing and soldering is an ancient practice that has become a sophisticated modern technology after having undergone significant evolution and refinement over the years due to the contributions of modern materials science. The ultimate goal of this technology is to join parts into an assembly through metallurgical bonding.This is achieved by placing a relatively low-melting temperature alloy, or "filler metal" (FM) in the clearance (or gap) between the pieces or base materials (BM) to be joined and followed by heating of the assembly until the FM has melted and spread throughout the gap. The molten metal that filled the gap reacts with parts to be brazed and, after solidification, forms an integral solid whole. Figure 1 demonstrates graphically the essence of the brazing/soldering process. Assembly heating can be carried using various means including electromagnetic induction, Joule heating, oven, flame, etc. While joining temperatures above 723 K are arbitrarily associated with brazing rather than with soldering, these processes are essentially similar. In order to distinguish joining materials for both processes, those used at temperatures below 723 K are called solders, whereas those used at temperatures above 723 K are called brazing filler metals. In this chapter we will present some fundamentals of joining processes and how rapidly solidified (RS) FM offers advantageous processability, properties, reliability, and economics.

II. THE ESSENCE OF SOLDERING AND BRAZING PROCESSES

As Fig. 2 illustrates, there are three major stages in any brazing or soldering process. The first stage occurs during heating of an assembled workpiece. At this stage the

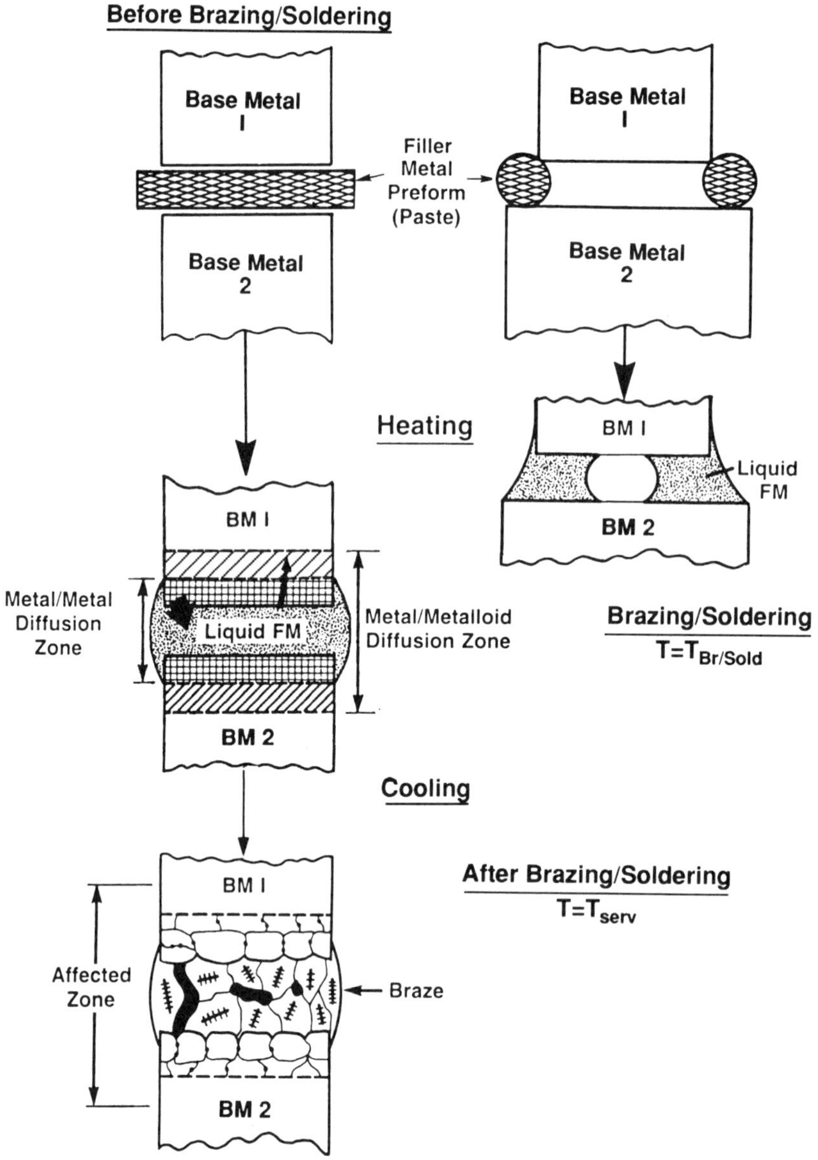

Figure 1 The essence of brazing/soldering processes.

FM melts and flows, filling completely the gap between the parts. Since this gap is usually rather small, on the order of 10 to a maximum of a few hundred micrometers, the degree of penetration of a molten alloy into the gap is determined largely by capillarity effects and, of course, by wetting. Because physical parameters such as part dimensions and the surface tension forces play the major role, this stage may be conditionally called a physical one. The second stage,which normally sets in at a given joining temperature, is characterized by an intensive solid/liquid interaction accompanied by a substantial mass transfer through the interface with

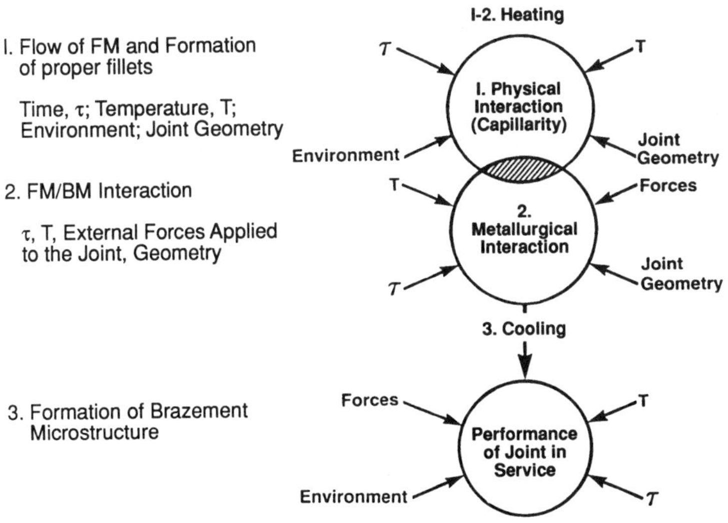

Figure 2 Three major stages in brazing/soldering processes together with external parameters affecting each stage.

strongly uneven rates. Indeed, BM volume immediately adjoining the liquid filler metal dissolves in this stage. At the same time, a small amount of elements from the liquid phase penetrates into the solid BM. Such mass transfer unbalance is due to dramatically different diffusion rates in the solid and liquid phases. Such a redistribution of components in the joint area results in changes of phase composition and sometimes in the onset of crystallization or, in other words, in joint solidification. All the processes of this stage in joining have a distinctly metallurgical character, and therefore the second stage may be properly called a metallurgical one. In soldering, all these processes proceed to a much lesser degree because of a substantial difference in melting temperature of a solder and a BM. Here, dissolution of base metal and mutual diffusion are practically insignificant. However, even in soldering one sometimes needs to protect BM surface from erosion by molten solder. The final stage, which overlaps with the second, is characterized by the formation of the final joint microstructure and progresses vigorously during assembly cooling while the liquid phase is still present. Here, the crystallization process takes place. Subsequent cooling results in only minor changes such as annealing of the joint and partial relief of thermal stresses induced on cooling. This stage has very similar features in both brazing and soldering processes.

In order to better understand specifics of all of these processes it is worth describing some basic phenomena occurring during interaction of liquid and solid metals starting from wetting and flow.

A. Wetting of and Flow Over Solid Metals by Molten FM Alloys

In any brazing/soldering process a molten alloy comes in contact with a surface of solid which may be an alloy, a ceramic or a composite material. For a molten alloy to advance over the solid surface a special relationship has to exist between surface energies of the liquid/gas, solid/gas, and liquid/solid interfaces. The same relationships should, in principle, hold in joining processes where a molten alloy has to fill the gaps existing between surfaces of the parts to be joint. In the simplest case of a liquid drop placed on a clean flat plate (Fig. 3) the liquid behavior (contracting or spreading) and the degree of spreading are first determined by the relationship between these surface energies (surface tensions) expressed by the well-known Young-Laplace equation:

$$\gamma_{sg} = \gamma_{sl} + \gamma_{lg} \cos \theta \tag{1}$$

where s, l, and g stand for solid, liquid, and gas, correspondingly, γ for surface tension, and θ for contact angle. The liquid surface will be curved at the line of contact with the solid only if there is a pressure difference, p, across the surface [1] given by

$$p = \gamma_l\left(\frac{1}{r_1} + \frac{1}{r_2}\right) \tag{2}$$

where r_1 and r_2 are the radii of curvature of the surface.

The ratio $k = (\gamma_{sg} - \gamma_{sl})/\gamma_{lg}$ is sometimes called the *wetting coefficient*. If $k \leq -1$, the solid is not wetted. If $k \geq 1$, the solid is completely wetted. If $-1 < k < 1$, k is equal to cos θ and the solid is only partially wetted. For the elements surface tension values are indicative of their atomic bonds and correlate well with fusion temperatures: the higher the fusion temperature the higher the surface tension. In alloys, the surface tension as a function of composition cannot be generalized and it has a highly specific characteristic depending on the nature of alloy constituents

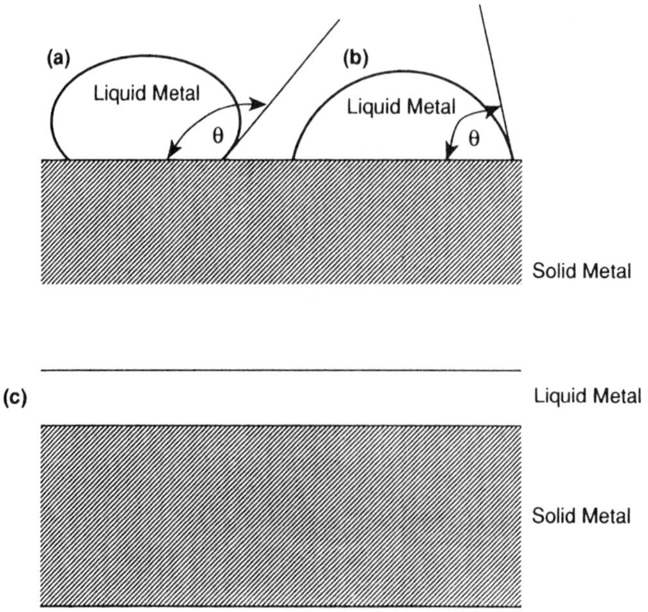

Figure 3 Typical contact angles: (a) $\theta > 90°$, no wetting; (b) $90° > \theta > 0°$, suitable for joint filling; (c) $\theta = 0°$, the most suitable for joint filling [1].

and their atomic interaction [2]. For metals forming nearly ideal solutions in the liquid (solid) state, such as copper and nickel, this characteristic has a nearly linear concentration dependance. Whereas for alloy systems in which intermetallic compounds are formed, such as Zn-Mg alloys, a strong minimum in $\gamma = f(\%)$ in the vicinity of the intermetallic concentration is observed [3]. Concentration effects have particular importance in cases of surface-active additions which tend to segregate at the surface area of liquids (solids). For example, small nonmetallic and metallic additions to liquid iron cause a large reduction in surface tension. Thus, if the oxygen pressure of the atmosphere over molten iron is 6.2×10^{-16} atm, the equilibrium amount of oxygen in solution is 0.01%, resulting in a corresponding 10% reduction of the surface tension of liquid iron [2]. This is the result of the highly surface-active nature of oxygen in iron. Therefore, the wettability of base metals is strongly affected by cleanliness of the heat-treating atmosphere. On one side oxygen as an impurity of the furnace atmosphere can decrease FM surface tension, thus causing better flow; on the other side it reacts with BM, forming oxide films particularly on stainless steels and high-temperature alloys containing active metals such as chromium, titanium, and aluminum. These oxide films have very low surface tension relative to that of the BM itself and therefore can limit or prevent wetting. All transition metals have surface tension higher than 1500 mN/m whereas nontransition metals such as silver, tin, lead, etc., and particularly metalloids such as silicon have surface tension substantially lower than that [4]. The addition of a small amount of oxygen, phosphorus, silicon, boron, etc., to nickel and copper results in a precipitous drop of their surface tension. Therefore, tin- and lead-based alloys have been used as solders whereas copper- and nickel-based alloys with additions of P, Si, and B have been used as brazing filler metals: both groups of alloys in the molten state have surface tension values favorable for strong wetting of all major groups of steels and other BM alloys.

Finally, a complicated character of wetting in real soldering/wetting processes was demonstrated in a detailed study of the spreading of a binary tin-lead alloy over copper in ultrahigh vacuum [5]. It was shown that even in a clean environment, when both molten solder and copper are free of oxygen and oxide films, there is an instant redistribution of solder alloy components in the molten solder drop and at the solid/liquid interface (Fig. 4). Here the surface of advancing liquid is enriched in lead, which is a surface-active element relative to tin and copper because it has

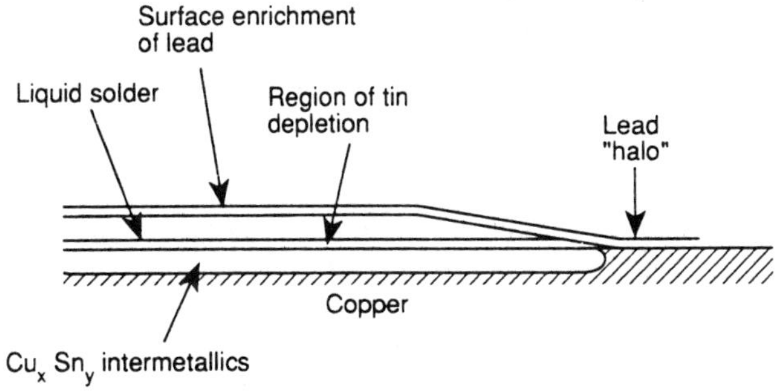

Figure 4 Redistribution of elements and formation of intermetallic compounds during wetting of a copper substrate by liquid Sn-Pb solder [5].

the lowest γ_l of the three alloy constituents. This enrichment continues over the melting front and spreads across the copper forming a monatomic lead "halo" region. This lead halo is formed by a rapid surface diffusion. It precedes the advancing melting front, thus decreasing surface tension of the substrate due to replacing copper having high surface tension with lead. At the same time tin is depleted from the drop layer adjacent to the spreading interface because of formation of Cu_xSn_y intermetallics. Comparison of thermodynamics parameters of the materials involved (Table 1) shows that there are fundamental reasons for this phenomenon caused by the differences in the surface tensions, excess free energies of solution in copper, and heat of the oxide formation. There is no doubt that such an effect should take place to some degree in practically all joining processes where multicomponent alloys are involved, particularly when they contain elements with a high affinity for oxygen and/or having strongly different values of the surface tension.

The advancement of a liquid metal over a clean metallic surface proceeds rather rapidly. In practice, a rather short time is needed for molten FM to fill the gaps in which the capillary effect is operative. Most of brazing/soldering gaps have dimensions satisfying the capillary criteria. For a horizontal gap having thickness, D, the time for liquid advancement, t, over a distance, l, is given by

$$t = \frac{3\eta l^2}{\gamma_l \cos\theta \times D} \tag{3}$$

where η is the viscosity of a liquid alloy [1]. In the case of a vertical joint gap filling is opposed by head pressure of the increasing column height of the molten FM, and t is expressed by a much more complex formula. Here theory does not take into account possible metallurgical interaction between moving liquid and material forming the gap. In fact, the usual filling time is too short to be affected substantially by processes of dissolution of BM and following braze crystallization. According to Milner [1], the filling time by molten tin and copper of a horizontal joint having 125-μm width and 2-cm length and made of tinplate and mild steel, correspondingly, would be about 30 ms. There is only a small difference between rates of filling in vertical and horizontal gaps which one may find in industrial applications. Still in

Table 1 Thermodynamics Data Relevant to Wetting of Copper by Sn-Pb Solder

Property	Cu	Pb	Sn	Remarks
Surface tension, γ_l, mN/m at T_{melt}	1360	468	544	Pb will be absorbed preferentially at the surface of a Sn-Pb drop.
Heat of formation of the highest oxide, kcal/mole	−40	−172	−138.8	Pb will enrich surface of Sn-Sb drop if O_2 is present.
Excess Gibbs energy of solution in solid solution, cal/g-atom (Cu + 0.1 mole fract. of solute at 1400 K)	0	+707	−666	Sn will concentrate at the Cu/Sn-Pb interface forming intermetallics. Pb will be depleted from L/S interface.

practice all metal surfaces are covered by thin oxide films having relatively low surface energies. For example, in the case of nickel, surface tension for the solid phase at 1300°C and the liquid phase at 1500°C is 2100 and 1777 mN/m, respectively [6,4] whereas nickel oxide has $\gamma_s^{NiO} = 400$ mN/m [7]. Therefore, $\gamma_l \cos\theta$ term may be a small if not negative, resulting in a decrease in the driving force for wetting. Recent studies have shown that molten FM spreading may also proceed over an area covered by oxide scale by penetrating beneath it and peeling this oxide film up [8] and/or by partially dissolving the thin oxide film on molten FM in a vacuum. So in many practical cases the liquid front advancement proceeds in small jumplike steps and with much slower spreading rate than in the case of clean liquid/solid interface. This explains the paramount importance of having the surfaces of brazement parts to be clean of grease, oily films, and, particularly, oxide films before and during a joining operation.

Temperature is also one of the important parameters regulating wetting and flow due to its influence on viscosity of liquid metals and surface tension/contact angle. It is well known [9], that there is a fundamental relationship between work of adhesion, W, contact angle, and surface tension of the wetting liquid. Phenomenologically it is as follows:

$$W = \gamma_l(1 + \cos\theta) \tag{4}$$

Both W and γ_l decrease with temperature due to diminished interatomic bonding between atoms in the melt and atoms at solid/liquid interface as well. Therefore, the temperature effect on wetting is a complex function again depending on the nature of materials involved.

III. METALLURGY OF JOINT FORMATION

A. Liquid FM/Solid BM Interaction and Joint Compositional Changes

Let us consider a conventional case in which two BM pieces are separated by a fixed clearance gap, h [10]. The samples are joined using either preplaced foil having the same thickness as the gap, or else the gap is instantly filled by FM originally applied to the periphery of the gap which then melts and flows into the gap on heating. Brazing occurs at brazing temperature, T_{br}, for time τ_{br}. It is convenient to consider the BM as an element or pseudo element A, while the FM is an alloy based on element A, to which a second element B or a combination of various elements is added, forming a multicomponent eutectic-type alloy. The majority of FM alloys is indeed of the eutectic type.

A good practical example in brazing is two pieces of 304 SAE stainless steel (Fe-Ni-Cr-based solid solution) being joined using BNi-3 FM (Ni-based Ni-B-Si ternary eutectic alloy). In this case, iron, nickel, and chromium are considered in combination as metal A, whereas the FM has a composition $A_{1-x}B_x$; silicon and boron are considered in combination as metal B with concentration $x = x_{Si} + x_B$, with x as atom fractions. Figure 5 is a schematic pseudobinary Fe(Ni,Cr)-(Si + B) phase diagram which applies to the 304 SAE/BNi-3 interaction. In reality, this interaction has to be considered using a concentration-temperature cross section of the Fe-Ni-Cr-Si-B phase diagram. One may conveniently trace and understand brazing/soldering processes by following temperature-concentration paths using a

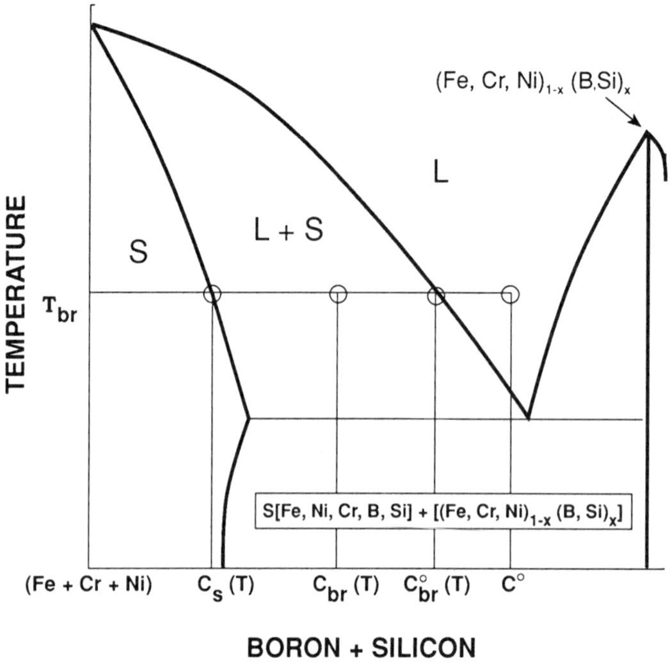

Figure 5 Schematic pseudobinary Fe(Ni,Cr)-(Si + B) phase diagram demonstrating concentration-temperature changes in BNi-3 FM during a brazing cycle.

pseudobinary phase diagram and by considering metallurgical reactions occurring between BM and FM along these paths. The following notation is adopted:

C°	composition of FM
$C^{\circ}_{br}(T)$	composition of a brazement formed at T_{br} and ambient pressure upon saturation by metal A
$C_{br}(T)$	composition of this brazement at the end of the holding time
$C_{s}(T)$	maximum concentration (or solubility) of metal B (Si + B) in the solid metal B at ambient pressure

The BM and FM foil are heated to the brazing temperature, T_{br}, during brazing, and the foil is melted in the T_S–T_L range, where T_S and T_L are the solidus and liquidus temperatures of the alloy, respectively. Thus, the first step in the brazing is as follows:

Heating to T_{br} $\quad A + A_{1-x}B_x \longrightarrow L$

For BNi-3 $\quad S_{C^{\circ}}(\mathrm{Ni} + \mathrm{Si} + \mathrm{B}) \longrightarrow L_{C^{\circ}}(\mathrm{Ni} + \mathrm{Si} + \mathrm{B})$

During the first short period of exposure to temperature, the liquid FM reacts with solid BM until it becomes saturated. No substantial changes in the concentration of solid BM ensues in the brazement at this time. Therefore,

Brazing at $T = T_{br}$ $\quad L_{C^{\circ}} + A \longrightarrow L_{C^{\circ}}$

For BNi-3 $\quad L_{C^{\circ}} + [\mathrm{Fe} + \mathrm{Ni} + \mathrm{Cr}] \longrightarrow L_{C^{\circ}}\ (\mathrm{Fe} + \mathrm{Ni} + \mathrm{Cr} + \mathrm{B} + \mathrm{Si})$

Later, a much slower process starts: namely, elements comprising metal B diffuse into metal A. The rate of diffusion of each individual element comprising B is determined by its diffusional mobility. For example, in the case of BNi-3 boron diffuses much faster than silicon. Therefore, diffusion rapidly depletes the liquid of boron, thus displacing the average brazement composition to $C_{br}(T)$.

B. FM Crystallization and Formation of the Joint

The displacement of concentration from $C^{\circ}_{br}(T)$ to $C_{br}(T)$ results in the crystallization of solid solution phases in the joint as part of a new phase equilibrium. Usually, this crystallization starts at the BM interface and can proceed with the complete disappearance of liquid phase and crystallization of the joint into a single solid solution phase if the holding time is about 12 hr. Normally, holding time is in a 10–30 min range; therefore, the joint contains a substantial amount of the eutectic liquid. As the workpiece cools, the temperature of the braze alloy approaches the eutectic temperature, T_{eu}. Upon cooling below T_{eu}, the remaining liquid, if any, crystallizes at the center of the joint into eutectic mixture of brittle intermetallic and ductile solid solution phases ("central eutectic line"). The volumes of these phases are determined by the position of $C_{br}(T)$ on the corresponding phase diagram. An example of a typical joint microstructure formed as a result of following such a brazing path is shown in Fig. 6 [11]. Here, 409 SAE stainless steel joint is brazed at 1050°C for 10 min using amorphous brazing foil having BNi-3 composition (METGLAS®* MBF-30, for example). The micrograph reveals all characteristic features of joints brazed using eutectic alloys. Even for such a short brazing time, substantial dissolution of the BM has occurred, and boron penetrated some 50 μm into the BM alloy to form $Cr_{1-x}B_x$ boride while silicon is incorporated in the joint phases. The central eutectic zone buildup of nickel borosilicides is clearly seen.

Finally, one has the following phase changes:

Cooling $\quad L_{eu}(T) + S_C \longrightarrow A_{1-x}B_x + A_C$

For BNi-3 $\quad L_{eu} + S[Fe + Ni + Cr + Si + B] \longrightarrow [(Fe,Cr)_{1-x}(B,Si)_x]$
$\qquad + [Fe + Ni + Cr + B + Si]$

The final composition of solid solution phases and borosilicides which are found at ambient temperature depend on how fast the workpiece cools. Usually these compositions are somewhere in between those at T_{eu} and those at equilibrium because the final redistribution of elements in solid phases needs substantial time, which is much longer than even the normal rather slow cooling procedure in vacuum furnaces.

IV. JOINING PROCESS TECHNOLOGY

Cleaning of the BM pieces to be joined and protecting them from oxidation during heating are essential in brazing operations. Chemically active substances (fluxes) are commonly used to provide such cleaning and shielding. However, the use of nonmetallic agents in fluxes is undesirable due to the increased propensity for their

*METGLAS is a registered trademark of AlliedSignal Inc. for rapidly soldified brazing and soldering foils.

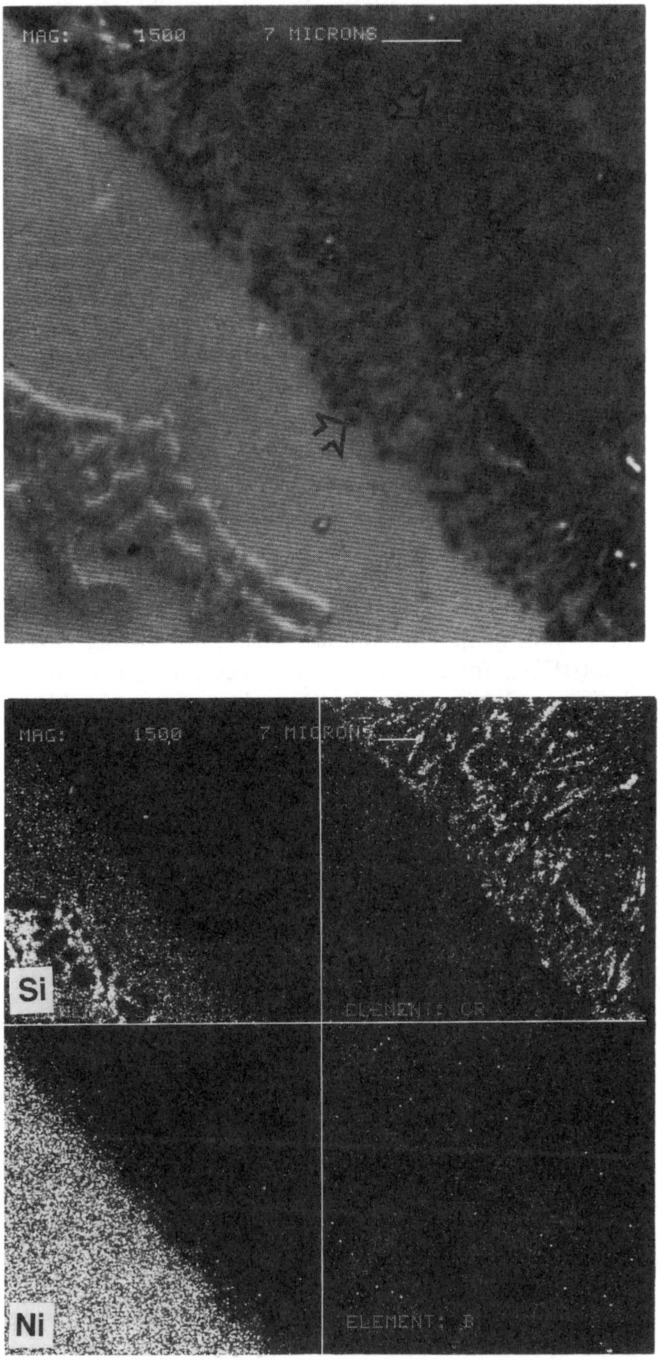

Figure 6 SEM micrograph in back reflection mode and corresponding x-ray maps of the 409 stainless steel/MBF-30 joint brazed at 1055°C for 10 min in −40°C FP 50:50 N_2:H_2 atmosphere. Notice that boron strongly diffuses out of joint forming $Cr_{1-x}B_x$ borides (empty arrows) while silicon is bonded with nickel in the joint [11].

entrapment and the resultant formation of voids in the finished joint. On the other hand, the use of certain active elements added to FM alloys and playing the role of a flux (i.e., makes the FM self-fluxing) is advantageous.

Basic forms of FM include solid preforms and powders, used mostly in the compound form of paste, plastic-bonded tape, and rosin core wire. Conventional paste forms of joining alloys (FM powder + fluxing agent + binder/solvent) are applied at externally accessible locations of the clearance between BM pieces. Such practice requires substantial FM fluid flow during joining to achieve constituent FM phase fusion and binder/solvent extraction. In addition, organic binders decompose when compound FM forms are used in the high-vacuum brazing of parts intended for critical high-temperature service. Such decomposition of binders can result in the formation of soot in the joint, which can act to degrade the performance of expensive vacuum equipment.

A basic requirement for any joint is that its strength and ductility be at least equal that of the BM. In general, the strength of a material increases with decreasing grain size, whereas ductility is affected by the presence of brittle phases, as will be discussed later. Therefore, it is important to limit grain size and the amounts of brittle phases, in addition to limiting porosity in joints. Ideally, the maximum size which the grains in a joint may achieve is equal to half the clearance between the BM pieces. Thus, the use of a smaller clearance during BM joining necessarily limits maximum grain size, promotes higher cooling rates of the FM alloy, and thereby results in a refined joint microstructure. A smaller clearance also promotes improved retention of BM properties because of curtailed BM erosion by the use of a smaller volume of FM. For these reasons, a preplaced self-fluxing thin FM foil used as a preform is superior to powder-containing paste, which requires larger clearances for filling joint cross sections and results in deleterious effects on properties owing to a coarser joint grain size, more fully developed intermetallic compounds, and substantial amounts of contaminants.

The grain and brittle intermetallic size and extent in the parent FM has a direct effect on the strength of the brazed joint. This consideration is particularly important under transient heating/cooling conditions such as torch or belt oven brazing and automatic solder die bonding [12]. It is the rapid dissolution of constituent FM alloy phases which allows the use of higher joining throughput speeds and/or lower joining temperatures. By this reasoning, an amorphous joining material is the epitome of FM microstructural possibility.

V. MAJOR GROUPS OF BRAZE AND SOLDER FILLER METALS

A. Brazing Filler Metals

As stated above, both the choice of specific FM alloy composition and the choice of optimal brazing conditions are mostly determined by joint performance under specific service conditions. Ideally, the composition of an FM alloy must be such that the following four functions are achieved:

1. FM melting temperature lower than that of the corresponding BM.
2. FM surface tension (both solid and liquid states) lower than that of BM to provide driving force for wetting.

3. FM compatible with BM to form good metallic bonding; i.e., FM and BM structure, composition, and properties are similar.
4. FM containing elements causing chemical reduction/decomposition or physical removal of BM oxide film.

Accordingly, four groups of brazing FM alloy compositions have emerged. The first and largest group is eutectic-type alloys having aluminum, nickel, cobalt and copper as a base to which silicon/boron (aluminum- and nickel-based alloys) and phosphorus (copper- and nickel-based alloys) are added. The presence of one or more of these elements in alloys tends to impart the above-mentioned characteristics to the FM. Silicon combined with a small amount of magnesium is successfully used in aluminum-based FM alloys, the surface tension and melting temperature of which are depressed and good bonding promoted while the magnesium addition serves as a fluxing agent. Boron and silicon are used in high-temperature brazing alloys in which the presence of phosphorus (another potential melting temperature depressant and surfactant element) could cause unacceptable joint brittleness. Of the alloy additions which promote self-fluxing of the FM during brazing, boron has the greatest "penetrating power." On the other hand, phosphorus is a beneficial fluxing element for use in copper (low-temperature) brazing, whereas silicon, for example, would cause unacceptable copper joint brittleness. Most of these alloys are brittle because various intermetallic phases precipitate when process by conventional technology. This brittle character has limited the available forms of conventionally produced brazing alloys to powder. Fortunately, it was discovered that the presence of silicon, phosphorus, and boron in many conventional FM alloys having near-eutectic compositions facilitates the conversion of such alloys into a ductile, thin amorphous alloy foil form by using rapid solidification (RS) technology [13], as discussed later. It turns out that, in many cases, a foil form FM is very well suited for use in brazing applications.

The second group of brazing FM alloys is characterized by a phase diagram having a peritectic reaction (copper-tin alloys) or a minimum in the liquidus curve (gold-nickel alloys). These alloys are used mainly in vacuum-brazing applications and, therefore, require no alloying elements playing the role of fluxing agents. The third group of alloys, probably the most widely used, is based on the copper-silver binary eutectic system which is modified by substantial additions of zinc and cadmium (both providing fluxing activity), and minor additions of tin and nickel.

The fourth group of brazing FM alloys consists of eutectic titanium/zirconium-based alloys to which copper and/or nickel are added. The first three groups of conventional brazing FM have been classified by the American Welding Society into eight well-defined classes. The majority of these classes have been processed successfully by rapid solidification technology.

B. Solders

In spite of the wide use and development of solders for millenia, most principal solders are still alloys based on lead- or tin-based alloy systems to which a small amount of silver, zinc, antimony, bismuth, and indium or a combination thereof are added. Many solders are characterized by the segregation of intermetallic phases with relatively high melting temperatures during solidification. These intermetallic phases with low plasticity play a dual role: on one hand, they improve soldered joint strength; on the other hand, they affect negatively the long-term fatigue life expectancy. The latter is of paramount importance for reliable service of electronic devices, particularly silicon dice connected by soldering and continuously subjected to the power and thermal cycling. It is well known that materials with large intermetallic crystals have low resistance to fatigue because of high stress concentration in the vicinity of the crystals. The size of intermetallic crystals formed in joints upon crystallization during transient soldering processes is determined by the initial size of crystals in the starting solder FM and, of course, by dissolution and precipitation processes during the soldering operation. Thus, since RS results in a tremendous refinement of as-cast phase and grain size, subsequent melting/resolidification during rapid automatic die bonding would result in more rapid dissolution of high-temperature phases. This, of course, allows greater throughput and/or lower joining temperatures to be used and, most importantly, higher joint resistance to fatigue.

As strange as it may seem, a national standard classification for solders does not exist in any country. However, there are numerous military and commercial specifications covering individual materials and technological operations.

VI. FILLER METALS BY RAPID SOLIDIFICATION

A. Potential Advantages

One of the first practical applications of RS technology was in the production of ductile amorphous brazing foil from alloys having compositions which previously could only be utilized in powder form or as powder-filled pastes [13–16]. Thus, a "happy marriage" has occurred between one of the oldest and one of the newest metallurgical technologies. As a result, most of FM alloys used in high-temperature brazing are eutectic compositions formed by transition elements, such as nickel, iron, chromium, etc., in combination with metalloids, such as silicon, boron, and phosphorus. In conventional form, i.e., having crystalline structure, all these materials are inherently brittle and cannot be produced in continuous forms such as foil, wire, etc. Therefore, they were available only as powders or its derivatives.

The very presence of metalloids at or near the eutectic concentration promotes RS conversion of such alloys into a ductile amorphous foil. Starting from the discovery of RS potential in manufacturing of flexible brazing foil by Sexton and DeCristofaro [13], practically all developmental work on RS brazing foil has been carried by Metglas Products, Allied-Signal Inc. Known as METGLAS alloy brazing foil, it has found numerous applications during the last 15 years. Later it was also realized that RS may also be successfully applied to the production of solders [12,17]. Such RS solders also bear the METGLAS trade name but have a microcrystalline rather than an amorphous structure. Both product lines, so far, are exclusively produced by Metglas Products, Allied-Signal Inc., which holds practically all basic patents on RS braze and solder FM alloys.

The most important advantage of RS amorphous and microcrystalline FM alloys is their flexibility and ductility. Because a ductile amorphous alloy foil may be used as a preplaced preform, there is no need for large brazement gaps, as those used with pastes, to achieve a complete filling of the braze cross section. Here, amorphous alloy foil, for example, has a particular edge over powder, and polymer-bonded strip forms because of its superior flow characteristics. Indeed, gas-atomized powder has a very large total surface area with subsequently large amounts of surface oxides. These oxides prevent to a certain degree fusion of individual powder particles into a uniform liquid pool. Figure 7 demonstrates this advantage by comparing flow charactertistics of three products having the same composition (BNi-2) but manufactured in the form of RS foil, powder, and polymer-bonded strip [17]. Flow characteristics of the FM forms were measured as a function of the overlap distance. Flowability, defined as S'/S, where S is the cross-sectional area of the fillet where filler metal was preplaced and S' is the cross-sectional area of the fillet formed on the opposite part of the brazement. The RS FM clearly flows more freely than either powder form.

A smaller clearance also promotes improved retention of BM properties because of curtailed BM erosion by the use of a smaller volume of FM. For these reasons, a preplaced self-fluxing thin FM foil used as preform is superior to the use of powder-containing paste which requires larger clearances for filling join cross sections and results in deleterious effects on properties owing to a coarser joint grain size, more fully developed intermetallic compounds and the presence of substantial amounts of contaminants.

As emphasized in Chapter 5 the manufacture of RS products is characterized by high cooling rates (approaching 10^6 K/s) which enable the stabilization of certain alloys into an amorphous solid state, having a spatial distribution of atoms similar to that of liquids. Most METGLAS alloys have such an amorphous structure, with a random, spatially uniform arrangement of the constituent atoms. When rapidly solidified microcrystalline materials are formed, which is the case for *all* RS solder and a few brazing FM alloys, the dimensions of their crystalline grains and phases are usually an order of magnitude smaller than those in material of the very same

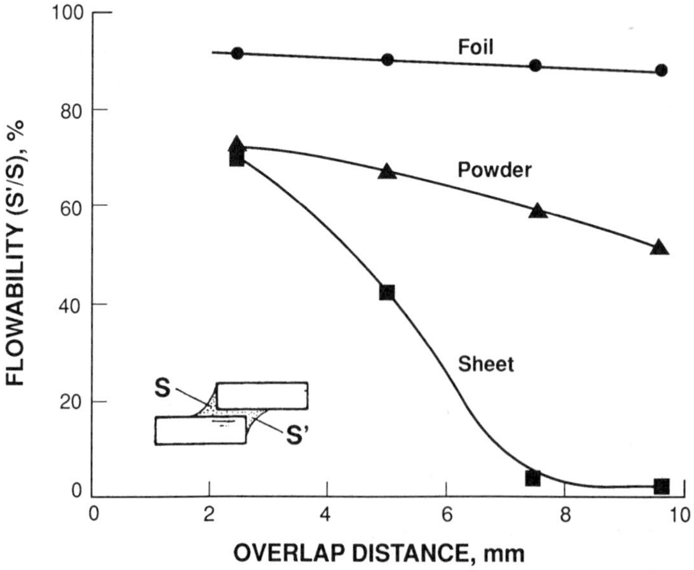

Figure 7 Flowability of BNi-2 RS MBF-20 foil (circles), powder (triangles), and polymer-bonded sheet (squares) [17].

composition but manufactured using conventional process technology. Because RS microcrystalline and, in particular, amorphous materials are compositionally much more uniform, their melting under transient heating occurs over a narrow temperature range. This is a consequence of the shorter distances over which atoms of different elements have to diffuse in order to form a uniform liquid phase. The resulting "instant melting" of RS materials is only one of their important features. From the practical point of view, this provides an opportunity to braze or solder at a lower temperature for a shorter time than in the case of using conventional filler metals. This is particularly important when brazing fine-gauge honeycomb cores, for example, which have to be protected from erosion by molten filler metals during joining. A shorter brazing time is also beneficial in cases wherein BM parts may lose their inherent strength due to annealing during the brazing operation. The joining of cold-deformed stainless steels and of precipitation-hardened superalloys are good examples of where a short brazing time can be critically important.

The absence of contaminating organic solvents bases which powder pastes/tapes contain correspondingly eliminates soot formation and furnace fouling. The low level of gaseous impurities in amorphous alloy foil, due to the specific characteristics of its production technology, is an attractive feature for furnace brazing of critical aircraft engine parts. It has thus, received rapid acceptance for this particular application and maintains a unique niche among various commercial FMs available.

B. Amorphous Alloy Brazing Foil

All conventional FM have been classified by the American Welding Society into eight well-defined classes. These classes can be partially fitted into the four basic groups according to the nature of their parent alloy system. Most of these classifications have been processed successfully by RS technology. Table 2 summarizes the current status of amorphous brazing filler FM which have been produced. It also shows the BM alloys which are compatible with each type of amorphous alloy foil, and the corresponding major areas of foil application as well. The table also lists examples of compositions for alloys which have been cast and tested in small batches. These alloys are undergoing industrial testing and market acceptance.

1. Metal-Metalloid Eutectic Group: Ni-, Cu-, Ni-Pd-, and Al-Based Alloys

a. Nickel-Based Amorphous Alloys. Nickel-based alloy compositions and melting characteristics are shown in Table 3. Their melting troughs determined by DTA are shown in Fig. 8. Compositionally, these alloys originate from ternary Ni-B-Si alloys, the phase diagram of which has a number of ternary eutectics [18], and from a binary $Ni_{81}P_{19}$ eutectic alloy [19]. Ni-B-Si eutectic compositions have a rather wide range of values of their (B + Si) concentrations (12–28 at%), whereas their solidus temperatures, T_S, are located within a range of about 960–1000°C. The $Ni_{81}P_{19}$ binary eutectic has a temperature of 870°C, and the T_s of both alloys containing phosphorus (MBF-60 and Alloy 5) is close to this value (883°C). The majority of these alloys also contain a small amount of iron (3–4.5 wt%) and chromium (3–14 wt%). The former is added to improve brazability, whereas the latter improves the corrosion resistance of joints formed by these alloys. Changes in (B + Si) concentration and in B/Si concentration ratio give to a customer of these alloys a choice of FM alloys with melting temperatures and performance

Table 2 Rapidly Solidified Brazing Filler Metals

No.	Alloy family & type	Metglas Products designation	AWS designation
1	Al-Si, eutectic		BAlSi-2, 3, 4, 5, 6, 7
2	Cu-Sn, peritectic		None
3	Cu-P, eutectic		None
4	Cu-Ag, eutectic		BAg-1, 2, 2a, 3, 4, 8, 13, 18, 19, 20, 21
5	TM-Si-B[a], eutectic (a) (Ni/Fe + Cr)-Si-B	MBF -15, -20, -30, -50, -60, -80	BNi-1, 1a, 2, 3, 6, 7
	(b) (Ni,Pd)-Si-B	MBF -1000 series: 1001, -2, -5, -7, -11	None
	(c) (Co,Cr)-Si-B		BCo-1
6	Cu-Ni-Mn-Si, solid solution		None
7	Cu-(Ti,Zr)-Ni eutectic and peritectic		None

[a]This group includes alloys based on transition metals such as Ni, Fe, and Co.

tailored to their specific needs. Still, there is a limit in the allowable range of B and Si concentration which is determined by potential loss of alloy ductility in the as-cast state [20]. At the same time, there are only two Ni-P-based alloys with rather limited applications from which to choose due to otherwise increased joint brittleness.

Structural state	RS forms	Base metals joined	Major applications current (potential)
Microcrystalline	Foil, powder	Aluminum and aluminum alloys, steel to aluminum and aluminum to berylium	Car radiators, heat exchangers, honeycomb aircraft structures, structural parts
Microcrystalline	Foil	Copper and copper alloys, copper to mild steel, copper to stainless steel	Heat exchangers, structural parts, automotive
Amorphous	Foil, powder	Copper to copper, copper to silver/oxide PM composites	Electrical contacts, bus bars (heat exchangers)
Microcrystalline	Foil, powder	Most ferrous and nonferrous metals, except aluminum and magnesium	Most widely used utility filler metals
Amorphous	Foil, powder	AISI 300 and 400 series steels and nickel and cobalt-base superalloys. Carbon steels, low alloy steels & copper	Aircraft turbine components, automotive, heat exchangers, honeycomb structures (electronic components)
Amorphous	Foil	AISI 300 series stainless steel, cemented carbide, superalloys	Honeycomb structures, cemented carbide/polycrystalline diamond tools, orthodontics, catalytic converters
Amorphous	Foil, powder	Cobalt-based heat- and high-corrosion-resistant superalloys	Aircraft engines (honeycomb marine structure)
Microcrystalline	Foil	Nickel-based heat-resistant alloys, steels	(Honeycomb structures, structural turbine parts)
Amorphous, microcrystalline	Foil	Titanium-zirconium-based alloys	Titanium tubing, aircraft engines, honeycomb aircraft structures, aircraft structural parts, chemical reactors, etc.

All FM alloys of this group react substantially with BM materials because BM atoms have a natural driving force to move, i.e., to be dissolved, into the molten FM alloy. Indeed, FM alloy contains either a large amount of (B + Si) or phosphorus, elements very readily forming a series of intermetallic phases with transition metals. This tendency should be taken into consideration when choosing brazing technology, particularly in the case of fine-gauge parts such as honeycomb structures

Table 3 Nickel-Based Brazing Filler Metals

Alloy designation	AWS & AMS classifications	Nominal chemical compositions, wt% (balance nickel)										Melting characteristics, °C (°F)		Density, g/cm³
		Cr	Fe	Si	C	B	P	Co	Mo	W	Ni	T_S	T_L	
MBF-15		13.0	4.2	4.5	0.03	2.8	—	1.0	—	—	Bal	965 (1,769)	1,103 (2,017)	7.51
MBF-20	AWS BNi2/AMS 4777	7.0	3.0	4.5	0.06	3.2	—	—	—	—	Bal	969 (1,776)	1,024 (1,875)	7.46
MBF-30	AWS BNi3/AMS 4778	—	—	4.5	0.06	3.2	—	—	—	—	Bal	984 (1,803)	1,054 (1,929)	7.94
MBF-50		19.0	—	7.3	0.08	1.5	—	—	—	—	Bal	1,052 (1,926)	1,144 (2,091)	7.49
MBF-60	AWS BNi7	—	—	—	0.10	—	11.0	—	—	—	Bal	883 (1,621)	921 (1,690)	7.91
MBF-80		15.2	—	—	0.06	4.0	—	—	—	—	Bal	1,048 (1,918)	1,091 (1,996)	7.80
Alloy 1	AWS BNi1a/AMS 4776	14.0	4.5	4.5	0.06	3.2	—	—	—	—	Bal	960 (1,760)	1,094 (2,001)	7.44
Alloy 2		11.3	4.0	1.5	0.06	2.2	—	—	—	8.0	Bal	1,022 (1,872)	1,175 (2,147)	8.11
Alloy 3		—	—	7.3	0.06	2.17	—	—	—	—	Bal	987 (1,809)	1,035 (1,895)	7.70
Alloy 4		—	—	5.5	0.06	1.6	—	—	—	—	Bal	985 (1,805)	1,060 (1,939)	
Alloy 5	AWS BNi6	5.3	—	7.3	0.06	1.4	—	—	—	—	Bal	950 (1,742)	1,040 (1,904)	7.72
Alloy 6		14.0	—	—	0.08	—	10.1	—	—	—	Bal	882 (1,620)	924 (1,695)	7.51
Alloy 7		10.0	5.5	—	0.06	3.5	—	23.0	7.0	—	Bal	1,054 (1,929)	1,109 (2,028)	7.98
Alloy 8		—	—	4.0	0.06	2.7	—	20.0	—	—	Bal	988 (1,810)	1,067 (1,953)	7.94

[a]All MBF-XX alloys have completely amorphous structure and are ductile.

which are widely used in the aircraft industry. At the same time, boron and phosphorus diffuse rapidly into BM material, forming a series of brittle borides/phosphides which prefer to segregate at BM alloy grain boundaries, thus creating high joint brittleness.

Overall, the microstructure of joints formed upon brazing with the FM type described above consists of a eutectic-type mixture of solid solution Ni-Fe-Cr-based phases with complex Fe, Ni, and Cr borides/silicides and phosphides. The microstructure described in [21] is a very good example of the microstructure typically formed by this type of FM. Such a microstructure is strong but has a low ductility.

b. Copper-Based Alloys. Copper-based alloys can have a eutectic composition originating in the Cu-P binary system (see Table 4 and Fig. 9). The phosphorus concentration of Alloy 8 is 6.5 wt%, slightly lower than that of the binary Cu-P eutectic (8.3 wt%). However, the addition of 9.3 wt% tin decreases the solidus temperature of this alloy to a level which is substantially lower than that of the binary Cu-P eutectic composition [22]. Alloying with tin and the addition of nickel improve alloy corrosion resistance and stability against oxidation during torch brazing, which is the most typical method of brazing.

The interaction of Alloy 8 with BM material, which are mostly pure copper and copper-based alloys, is similar to that of MBF-XX alloys. Again, a eutectic mixture is formed upon joint crystallization consisting of the copper-based solid solution and copper phosphide phases. The joint formed is weaker than in MBF-XX alloys, but it has higher ductility [23].

c. Nickel-Palladium- and Cobalt-Based Alloys. Nickel-palladium- and cobalt-based alloys are analogous to Ni-B-Si-based FM alloys and, to some degree, are their extension (Table 5 and Fig. 10). In these alloys, palladium is a major constituent along with nickel, thereby augmenting alloy properties, conveying superior performance in corrosion resistance, and, specifically, decreasing melting temperature [24–30].* Two Co-based alloys, in addition to Co, B, and Si, contain substantial amounts of chromium and tungsten (Table 6). Their melting temperatures are the highest among these brazing alloys. Joints made with these FM alloys have high strength even at very high service temperature [31]. At the same time, the presence of chromium and tungsten in large quantities makes their joints particularly heat-resistant. Both Ni-Pd- and Co-based alloys react with BM materials during brazing and develop microstructures in a way similar to Ni-based alloys (see [25–29]).

d. Aluminum-Based Alloys. Aluminum-based alloys are based on binary aluminum-silicon eutectic composition (Table 7). All of these alloys contain a small amount of iron, copper, and magnesium. In some cases the amount of zinc and copper is increased to about 11 wt% in order to improve melting characteristics and strength, correspondingly. These alloys can readily be rapidly solidified into a ductile ribbon form with microcrystalline structure, but their practical application is delayed for economical reasons.

2. *Peritectic Group Alloy*

Alloy 9 is the only commercial copper-based peritectic alloy which so far was produced in ductile ribbon form by RS and which had a microcrystalline structure

*The Pd-Si system has the deepest eutectic trough among all binary transition metal-metalloid alloy systems [30]. Because of its high cost, however, the binary Pd-Si alloys are not commercially viable as a parent system.

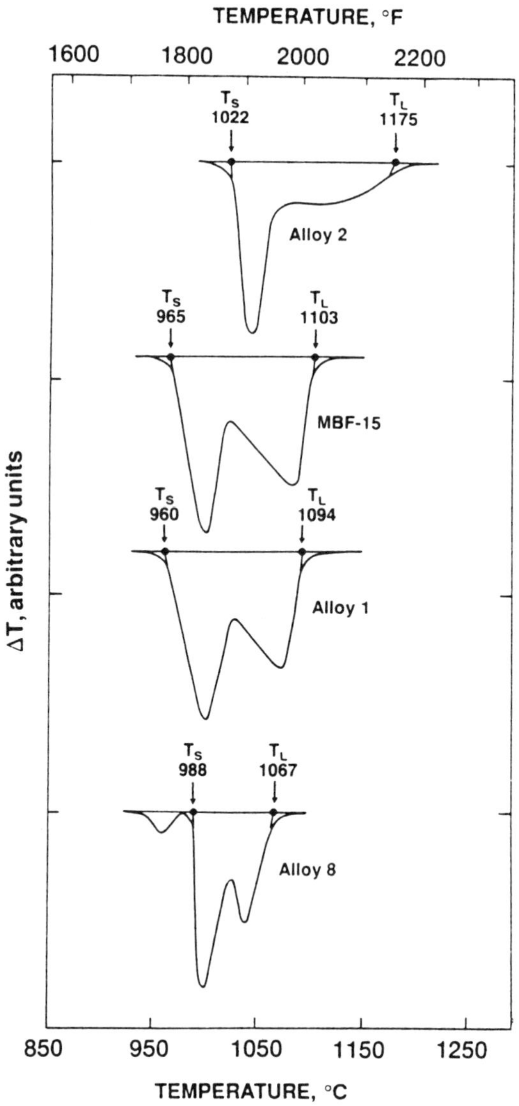

Figure 8 DTA melting troughs of nickel-based brazing alloys.

(see Table 4 and Fig. 9). Its melting characteristics are unique among all FM because of its wide solidus-liquidus range (770–925°C). Nevertheless, this alloy still found an application niche. The interaction of Alloy 9 with various steel BM materials is rather limited, and, therefore, the joint microstructure is determined predominantly by the crystallization of the Alloy 9 itself. This microstructure consists of a Cu(Sn) solid solution matrix phase in which various Cu-Sn electron compound phases are dispersed. The strength of a brazed Alloy 9 joint is higher than that of the Alloy 8 joint, but is more brittle.

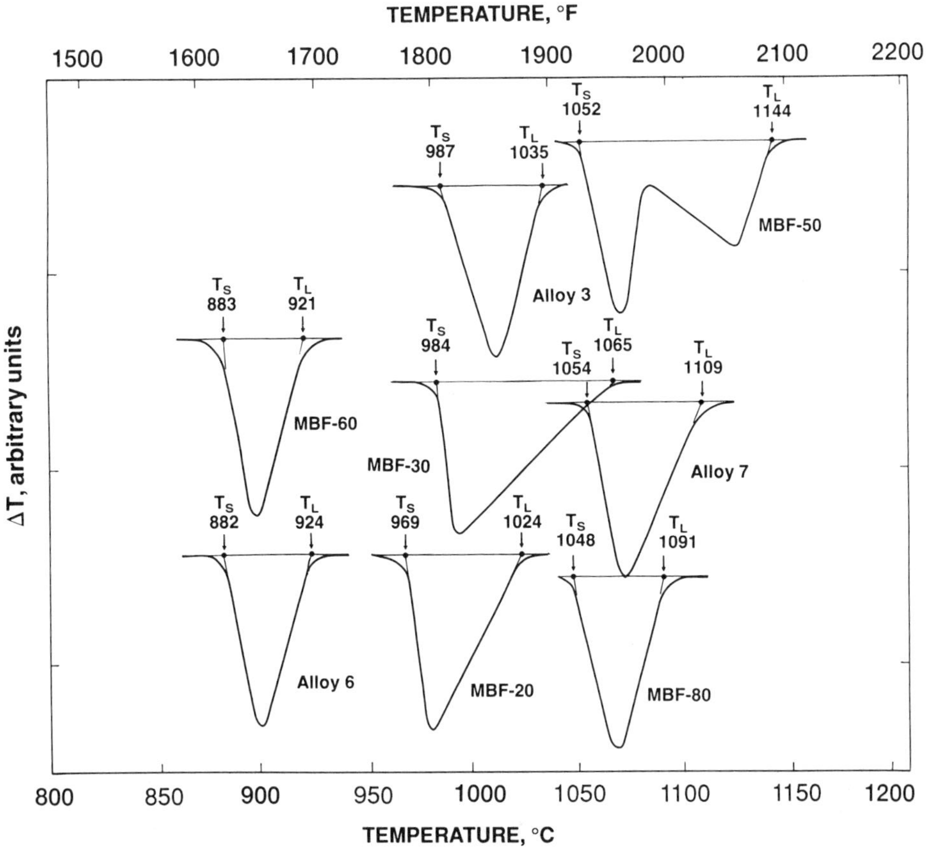

Figure 8 Continued.

3. *Eutectic Copper-Silver-Based Group Alloy*

These alloys are based on the binary Ag-Cu eutectic composition to which substantial amounts of zinc, cadmium, nickel, and tin are added alone or in combination (Table 8). In spite of their wide application in conventional forms and promising results in laboratory studies [32,33], these materials, as in aluminum-based alloys, have not been produced by RS on an industrial scale for economic reasons. Interestingly, practically all existing BAg series compositions with widely varied concentrations of all components can be produced by RS in both ductile and brittle

Table 4 Copper-Based Brazing Filler Metals

	Nominal chemical composition, wt%				Density, g/cm³	Melting characteristics, °C (°F)		Structure	Ductility
	Cu	Ni	Sn	P		T_S	T_L		
Alloy 9	78.5	5.7	9.3	6.5	7.94	595 (1,103)	650 (1,202)	Amorphous	Ductile
Alloy 10	88.0	—	20.0	—	8.58	770 (1,418)	925 (1,697)	Microcrystalline	Ductile

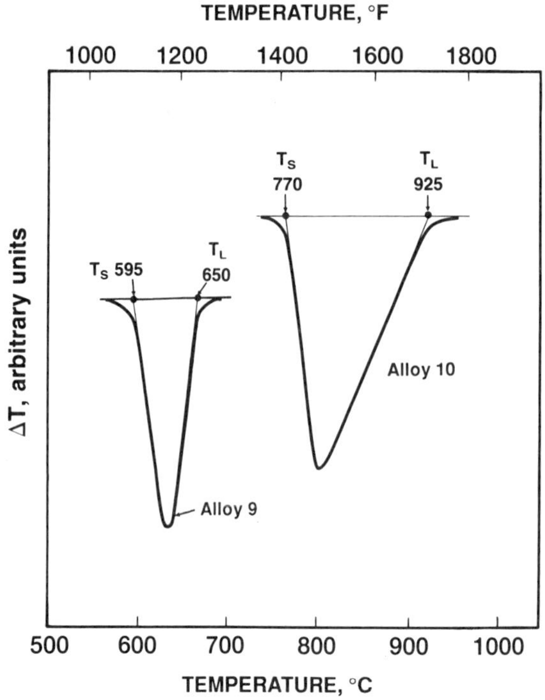

Figure 9 DTA melting troughs of copper-based brazing alloys.

forms. The latter is unusual for RS materials, particularly those having solid solution structure of all phases even in conventional form and which normally retain ductility after all kinds of metallurgical treatment. As indicated in Table 8, casting ribbon thicker than 50–100 μm produces an extremely fine cuboidal phase precipitating in the core of the "thick" ribbon upon relatively slow cooling, resulting in ribbon embrittlement. Here the cooling rate is too slow to prevent the phase from precipitating and the appearance of the associated high stresses. Subsequent annealing results in the growth of the precipitated phase and, eventually, in the appearance of the conventional microstructure having high ductility. Powder produced from pulverized brittle ribbon has shown better performance than the gas-atomized powder. The same is true for RS thin ductile ribbon, which has microcrystalline microstructure and, as a result, better melting characteristics. This, in turn, yields an improved final joint microstructure and better mechanical properties.

4. *Metal-Metal Eutectic Group*: *Ti/Zr-Based Alloys*

This group contains Zr-Ni binary, Ti/Zr-Ni and Ti-Cu-Ni ternary, and Ti-Zr-Cu-Ni quaternary alloys (Table 9 and Fig. 11). All of these alloys are good glass formers upon RS and have rather low melting temperature [34,35]. Copper and nickel are added to binary Ti-Zr compositions to decrease melting temperature and, most importantly, to facilitate conversion of these alloys into the amorphous state. Rapid solidification of these alloys and the testing of brazements produced using them have only recently begun in spite of their potential in titanium brazing, which has been well known for years [36]. The reasons for the two-decade delay in the

Table 5 Nickel-Palladium-based Brazing Filler Metals

Alloy designation	Nominal chemical compositions, wt%								Melting characteristics, °C (°F)		Density, g/cm³
	Cr	Fe	Si	B	Pd	Co	Mo	Ni	T_S	T_L	
MBF-1001	8.6	4.6	—	2.7	35.3	—	—	48.7	945 (1,733)	996 (1,825)	8.80
MBF-1002	8.6	0.9	—	2.7	32.4	—	—	44.6	923 (1,693)	990 (1,814)	8.85
MBF-1005	—	—	6.1	—	46.7	—	—	47.2	810 (1,490)	851 (1,564)	9.93
MBF-1007	—	—	8.1	—	53.8	—	—	38.1	820 (1,508)	870 (1,598)	9.09
MBF-1011	—	—	5.0	—	45.5	5.0	4.5	40.0	847 (1,557)	895 (1,643)	9.11
Alloy 11			8.8	—	41.2	—	—	50.0	714 (1,317)	938 (1,720)	9.60
Alloy 12	—	—	5.1	—	88.3	—	—	6.6	772 (1,442)	820 (1,508)	11.44
Alloy 13	—	21.2	6.2	—	57.6	—	—	15.0	960 (1,760)	1,020 (1,868)	10.13
Alloy 14	—	—	5.0	—	46.8	5.0	—	43.0	820 (1,508)	870 (1,598)	10.00
Alloy 15	—	—	3.8	—	45.8	10.0	4.5	35.2	837 (1,539)	874 (1,605)	8.88
Alloy 16	10.5	—	—	2.4	30.0	—	—	57.1	941 (1,725)	977 (1,790)	9.49

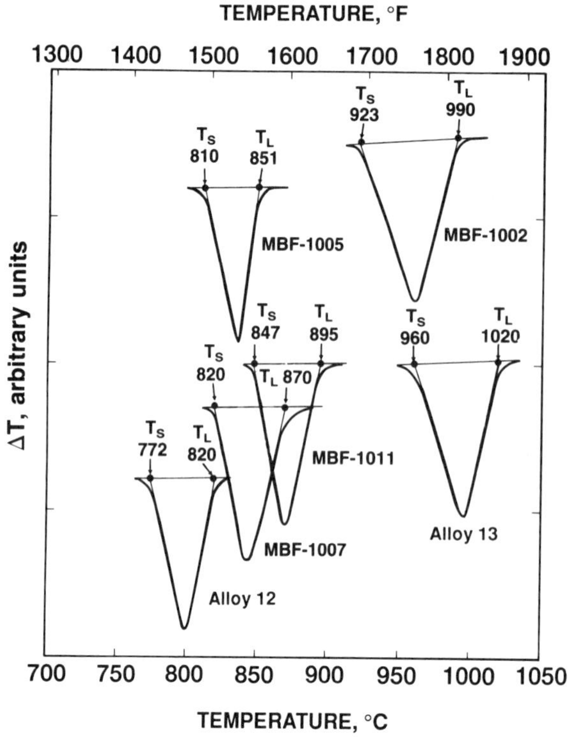

Figure 10 DTA melting troughs of nickel-palladium-based brazing alloys.

production of good-quality ribbon have been serious technical problems involving the containment of active liquid metals in casting crucibles. These problems were partially overcome by Metglas Product [35].

Because all titanium-zirconium-based BM materials used with these alloys contain neither copper nor nickel, the elements readily forming brittle intermetallic phases with Ti/Zr, there is a strong interaction between BM and the FM alloys. On the other hand, this results in a good wetting and bonding between BM and these FM.

The joint microstructure formed upon brazing consists of Ti-Zr-based solid solution phases with high ductility and $[Ti(Zr)]_x[Cu(Ni)]_y$ intermetallic phases. The latter can either be dispersed evenly inside of the matrix when the amount of the molten FM is relatively small, or it can form the central eutectic line in joints having a wide gap. When properly designed, brazed joints may have a strength approaching or even equal to that of the BM [37].

C. Rapidly Solidified Solders

Table 10 shows composition, melting characteristics, and specific gravity of 11 RS solders. These solders can be divided on three groups: tin-based alloys (MSF-101, MSF-111, and Alloy 28), to which silver and/or antimony are added to improve the joint strength; lead-based alloys (MSF-104, MSF-105, MSF-108, Alloy 29, and Alloy 31), to which small amounts of tin, indium, and silver are added alone or in

Table 6 Cobalt-Based Brazing Filler Metals

Alloy designation	Nominal chemical composition, wt%									Density g/cm³	Melting characteristics, °C/°F	
	Co	Cr	Ni	Fe	W	Si	B	C	Other		T_S	T_L
Alloy 17	70.4	21.0	—	—	4.5	1.6	2.4	0.07	—	8.61	1,100 (2,012)	1,200 (2,192)
Alloy 18	57.1	27.3	2.9	2.9	32.9	1.0	2.5	1.0	1.4	8.41	1,130 (2,066)	1,150 (2,102)

Table 7 Aluminum-Based Brazing Filler Metals

AWS classification	Nominal chemical compositions, wt% (balance aluminum)							Melting characteristics, °C (°F)	
	Si	Fe	Cu	Mn	Mg	Zn	Other	T_S	T_L
BAlSi-7	9.0–11.0	0.8	0.25	0.10	1.0–2.0	0.20	—	554 (1,030)	596 (1,105)
BAlSi-4	11.0–13.0	0.8	0.30	0.15	0.10	0.20	—	577 (1,070)	582 (1,080)
BAlSi-11	9.0–11.0	0.8	0.25	0.10	1.0–2.0	0.20	Bi = .02–.20	555 (1,030)	541 (1,105)
BAlSi-3	9.3–10.7	0.8	3.3–4.7	0.15	0.15	0.20	Cr = 0.15	522 (970)	585 (1,085)
—	9.3–10.7	0.8	3.3–4.7	0.07	0.07	9.3–10.7	—	515 (960)	560 (1,040)
BAlSi-8	11.0–13.0	0.8	0.25	0.10	1.0–2.0	0.20	—	559 (1,038)	579 (1,075)
—	11.0–13.0	0.8	0.25	0.10	3.0–4.0	0.20	—	559 (1,038)	579 (1,075)
—	11.0–13.0	0.8	0.25	0.10	2.0–3.0	0.20	RE = 1.0–2.0	559 (1,038)	579 (1,075)

Table 8 Silver-Based RS Brazing Filler Metals

Sample no.	AWS designation	Composition, wt%					Ductility		
		Ag	Cu	Zn	Cd	X	R/S as-cast 20–50 μm thin strip	20–50 μm strip after annealing 200°C, 64 hr	R/S as-cast 50–200 μm thick strip
1	BAg-1	45	15	16	24	—	Brittle	Ductile	Ductile
2	BAg-2	35	26	21	18	—	Brittle	Ductile	Ductile
3	BAg-2a	30	27	23	20	—	Brittle	Ductile	Brittle
4	BAg-3	50	15.5	15.5	16	Ni3.0	Brittle	Ductile	Ductile
5	BAg-4	40	30	28	—	Ni2.0	Brittle	Ductile	Ductile
6	BAg-8	72	28	—	—	—	Brittle	Ductile	Ductile
7	BAg-13	54	40	5	—	Ni1.0	Brittle	Ductile	Ductile
8	BAg-18	60	30	—	—	Sn10.0	Brittle	Ductile	Ductile
9	BAg-19	92.5	7.5	—	—	—	Ductile	Ductile	Ductile
10	BAg-20	30	38	32	—	—	Brittle	Ductile	Ductile
11	BAg-21	63	28.5	—	—	Ni2.5 Sn6.0	Ductile	Ductile	Ductile

Table 9 Titanium/Zirconium-Based Brazing Filler Metals

Alloy designation	Alloy formula, at%	Alloy composition, wt%				Density, g/cm³	Melting characteristics, °C (°F)		Structure	Ductility
		Ti	Zr	Cu	Ni		T_S	T_L		
19	$Zr_{76}Ni_{24}$	—	83.1	—	16.9	6.80	982 (1,788)	986 (1,806)	Amorphous	Ductile
20	$Ti_{48.9}Zr_{25.7}Cu_{14.7}Ni_{10.7}$	37.5	37.5	15.0	10.0	5.92	839 (1,542)	843 (1,548)	Amorphous	Semiduct.
21	$Ti_{74.8}Cu_{12.1}Ni_{13.1}$	70.0	—	15.0	15.0	5.40	902 (1,654)	932 (1,708)	Amorphous + ~10%β-Ti	Semiduct.
22	$Ti_{33.0}Zr_{17.3}Cu_{49.7}$	25.0	25.0	50.0	—	6.67	842 (1,546)	848 (1,558)	Amorphous	Ductile
23	$Ti_{48.8}Zr_{12.8}Cu_{18.4}Ni_{20}$	40.0	20.0	20.0	20.0	6.08	848 (1,558)	856 (1,573)	Amorphous	Ductile
24	$Ti_{65.5}Cu_{12.3}Ni_{22.2}$	60.0	—	25.0	15.0	5.61	901 (1,654)	914 (1,677)	Amorphous	Ductile
25	$Ti_{59.7}Cu_{15.5}Ni_{24.8}$	54.0	—	18.5	27.5	5.83	910 (1,670)	919 (1,686)	Amorphous	Ductile
26	$Ti_{65.5}Cu_{16.6}Ni_{17.9}$	60.0	—	20.0	20.0	5.61	915 (1,679)	936 (1,717)	Amorphous	Ductile
27	$Ti_{20}Zr_{60}Ni_{20}$	12.6	72.0	—	15.4	6.40	792 (1,457)	855 (1,571)	Amorphous	Ductile

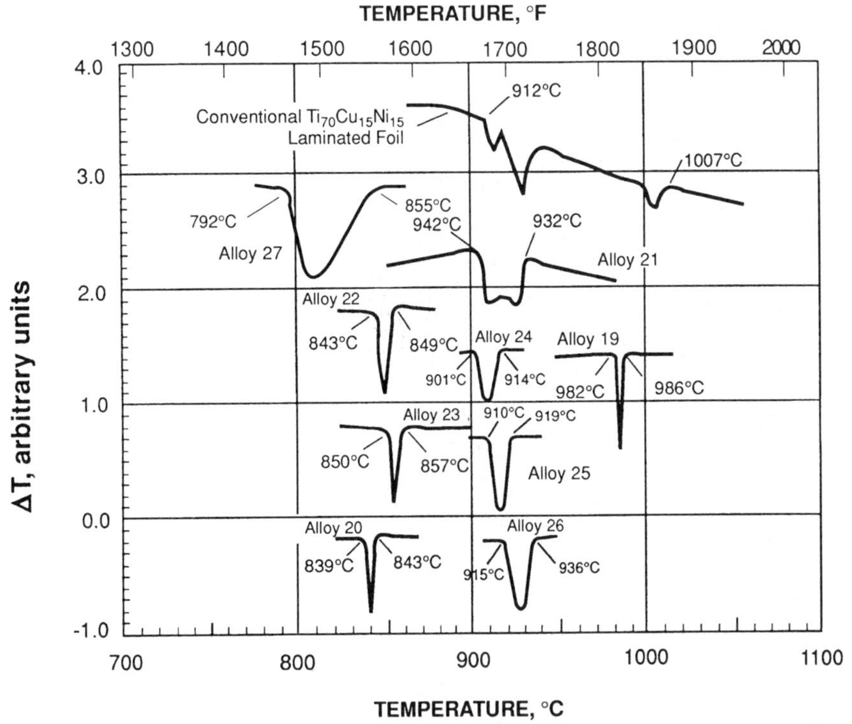

Figure 11 DTA melting troughs of titanium-zirconium-based brazing alloys.

Table 10 RS Solder Alloys

Alloy designation	Composition, wt%	Melting characteristics, °C T_S	T_L	Density, g/cm^2
MSF-101	Sn-25Ag-10Sb	231	236	7.88
MSF-103	Pb-5In-2.5Ag	300	310	11.02
MSF-104	Pb-5Sn	310	314	11.06
MSF-105	Pb-15Sn	183	288	10.48
MSF-107	Pb-5Sn-2.5Ag	287	296	11.02
MSF-108	Pb-10Sn-2Ag	267	299	10.97
MSF-111	Sn-40Pb	182	190	8.50
MSF-113	Pb-25Sn	183	304	9.96
MSF-114	Pb-10Sn	183	269	10.74
MSF-118	Sn-36Pb-2Ag	—	—	8.44
Alloy 28	Sn-5Ag	221	240	7.39
Alloy 29	Pb-1.5Ag-1Sn	309	309	11.28
Alloy 30	Sn-10Sb	235	242	7.25
Alloy 31	Pb-5In-2.5Ag-1.5Sn	296	297	11.02

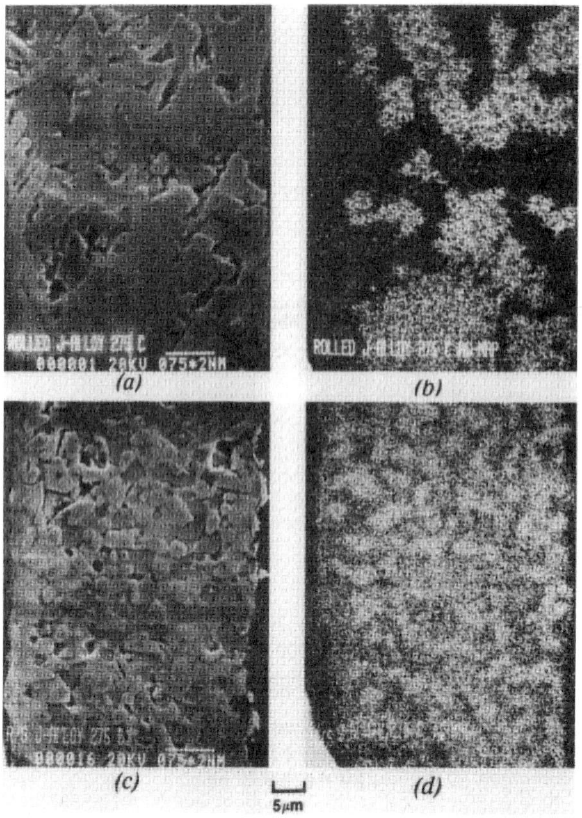

Figure 12 SEM microstructures and x-ray silver dot maps of J-alloy joints: (a,b) conventional foil joint; (c,d) RS MSF-101 foil joint [12].

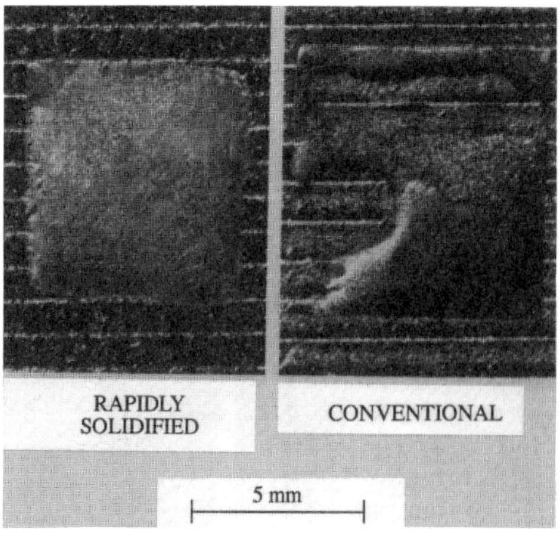

Figure 13 Improved wetting of METGLAS rapidly solidified solder over that of conventional solder.

Table 11 Shear Strength of Some RS and Conventional Solders

	Shear strength	
Alloy	MPa	(lb/in^2)
Sn-25Ag-10Sb (J-alloy)		
Conventional	358 ± 26	(5,200 ± 380)
RS	607 ± 48	(8,800 ± 690)
Pb-5In-25Ag		
Conventional	123 ± 30	(1,800 ± 440)
RS	317 ± 41	(4,600 ± 590)
Sn-3.5Ag		
Conventional	607 ± 48	(8,800 ± 700)
RS	1,103 ± 85	(1,600 ± 1240)

combinations; tin-lead binary alloys (Alloy 30) with concentration close to that of the eutectic of the tin-lead binary system. Addition of silver results in a higher joint strength and in less erosion of BM [12]. Indium improves wetting and resistance to oxidation. It also decreases melting temperature, thus allowing the possibility of varying solder melting characteristics. As can be seen in Table 10, the melting temperature of alloys varies over a wide range (182–310°C), providing a convenient choice of solders with different melting characteristics and properties.

Conventional rolled solders have a coarse microstructure with grain sizes up to 20 μm. Such a coarse microstructure results in a poor flow and attachment of the solder to the BM surface. This phenomenon is particularly detrimental in packaging of electronic devices, which demand an extremely uniform joint microstructure to sustain the high fatigue to which they are subject due to frequent thermal and power cycling [38–40]. All RS solders typically have a grain size of only a few micrometers. Their microstructure consists of a fine mixture of tin-lead-based solid solution phase and intermetallic compounds such as Ag_3Sn (Fig. 12) [12,17]. The use of RS foil which is clean and free from lubricants used in the rolling procedure guarantees the uniform melting of the solder and its well-defined reflowing (Fig. 13). The absence of lubricants also contributes to much less dewetting observed in joints produced with RS foil, because dewetting is the result of gas evolution (due to thermal breakdown or organics) from the parts to be joined [38]. The joint microstructure formed upon soldering with RS solder foil results in a very long service life before any failure occurs under cycling [40] and in a high joint shear strength (Table 11) [39].

VII. APPLICATION OF RS BRAZING AND SOLDER FM

A. Brazing

The application of RS FM was started in those cases where it had uncontested advantage over FM powers with similar chemistry: namely, their flexibility. Many FM powders having BNi series compositions have mostly been used in the aircraft industry, specifically in brazing turbine engine shroud segments, struts, stators, etc.

Preplacement of preforms accurately stamped from RS foil between parts to be brazed results in accurate joints having optimal volume. This, of course, eliminates excessive amount of FM in the work pieces which is always present when brazing is made with powder. Such weight minimization is very beneficial and of ultimate importance to the aircraft manufacturer, who yearns to reduce excessive weight. An additional advantage of displacing the use of powder is the elimination of environmental problems associated with both heavy-metal powder deposition in plant facilities and release of harmful organic vapors evolving during furnace baking during powder usage. Since the late seventies METGLAS brazing foil has been replacing many powder products and is now used in the production of various parts of jet engines and aircraft frames by GE, Pratt & Whitney, Rolls Royce, SNECMA, and others. Table 12 shows an MBF specification cross-reference demonstrating the broad range of materials and applications in which amorphous alloy foil has replaced powders. Ni-based MBF series foil is used in vacuum brazing of such superalloys as Hastalloy X, Inconel 625 and 718, Nimonic 75, IN100 and 300, and 400 SAE series stainless steels. These alloys are also quite compatible with high-chromium-containing FeCrAlY alloys and 17-4 PH precipitation-hardening stainless steel having a substantial amount of aluminum and covered by a dense, strong alumina film for protection from oxidation during service. In fact, high diffusion activity of MBF series alloys and the presence of boron and silicon provides an opportunity to wet well a variety of different alloys and composites including carbon-fiber-reinforced superalloys. When considering a list of nickel-based alloys for choosing an optimal material for a given application, one should take into account the following general features associated with alloy composition. First of all, the higher the combined silicon and boron concentration the better the wettability, although the stronger will be the BM erosion and, at the same time, the higher the amount of intermetallic phases formed. The latter may adversely affect joint strength. In cases where boron is detrimental to the joint strength it is better to choose between Alloy 4 and MBF-50 alloys, having $B < 2$ wt%. To avoid erosion of thin stock parts of a honeycomb structure during brazing, it is recommended to choose Alloy 4 as brazing foil since it contains only 1.6% B. Chromium concentration can be one of the criteria when designing brazements serving under highly corrosive conditions. Some MBF-XX alloys may contain up to 19% Cr, but chromium usually decreases FM flowability.

Nickel-palladium-based MBF series alloys were first developed to replace more expensive gold-based FM and, so far, have been used in the brazing of materials which were previously brazed with the latter. Polycrystalline diamond blanks furnished with cemented carbides [29], special steels used in orthodontic application, and vacuum-deposited nickel and copper films on ceramic parts brazed in vacuum [28] are some of the materials with which these alloys have successfully been used. Most of these alloys have melting temperatures substantially lower than that of MBF-XX series FM and, in this respect, they can be used for brazing under temperature conditions which MBF-XX cannot satisfy. At the same time nickel- palladium-based alloys have very high strength. One of the reasons for their high strength is a smaller amount of brittle intermetallic phases forming in the brazement because of a smaller concentration of metalloids than in simple nickel-based alloys. In fact, the presence of palladium decreases substantially minimal concentration of metalloids necessary for glass formation in this class of FM alloys [30].

Table 12 MBF Specification Cross Reference for Aircraft Turbine Engines

MBF alloy	AWS Spec. no.	AMS Spec. no	Pratt & Whitney Aircraft	General Electric	AiResearch (Garrett) EMS54752	Rolls-Royce MSRR 9500	SNECMA DMR 35	Textron Lycoming
15			PWA 996 CPW 494		−XIII	−705		M3876
20	BNi-2	4777		B50TF204	−II	−97	.302	
30	BNi-3	4778		B50TF205	−I	−114	.304	
50				B50TF217[a]		−722[b]		
60	BNi-6		PWA 36100		−XI			
80				B50TF207	−VIII	−719	.307	

[a]Alternative to B50TF81 powder (AMS4782).
[b]Honeycomb brazing.

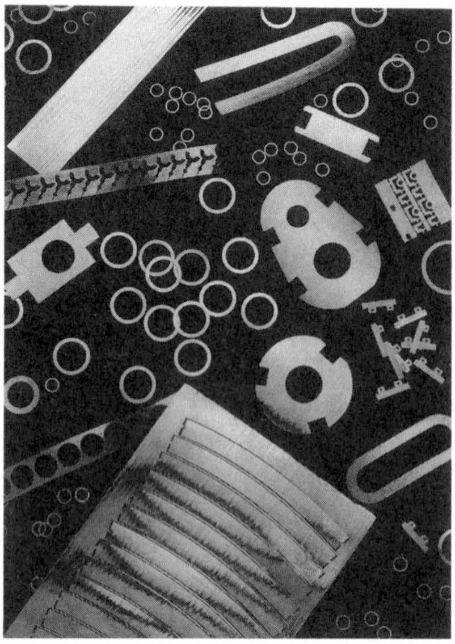

Figure 14 Various preforms stamped from METGLAS amorphous brazing foil.

Copper-based alloys are used in brazing of copper, 300 SAE series stainless steels, silver-based alloys, PM Ag-CdO parts, molybdenum, etc. [10,23,41,42]. Good results were obtained when using titanium-zirconium-based alloys in brazing of CP titanium and Ti-6Al-4V, the workhorse of the aerospace industry. Promising results were obtained in testing titanium/zirconium alloys with Ti-6Al-2Sn-4Zr-2Mo, Ti 1100, Ti-15V-3Al-3Cr-3Sn, and Beta 215 titanium alloys, and γ and α_2 titanium aluminides [43,44]. One rather exotic form of filler metal is a sandwich-

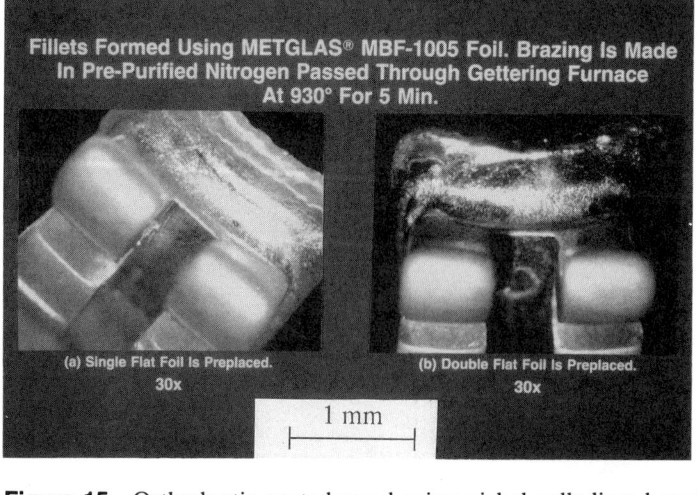

Figure 15 Orthodontic posts brazed using nickel-palladium-based MBF-1005 alloys: (a) stages of METGLAS foil preplacement; (b) a brazed post having a fine braze.

like strip produced from copper-based amorphous alloys hot-clad over a copper core [45]. Such a duplex clad strip may find applications in various heat exchangers.

Amorphous brazing foils, in spite of their rather high microhardness, may be stamped into a variety of intricate brazing preforms, as illustrated in Fig. 14. These preforms may be used in joining parts with circular cross section, miniature parts such as orthodontic posts (Fig. 15), etc.

B. Soldering

So far, all RS solders are mostly used in one major application, namely die bonding in silicon semiconductor power and other devices to metallic (i.e., plated copper) lead frames. The operation of such power devices involves frequent power switching and is accompanied by thermal cycling of the soldered assembly. As a result, severe stresses are generated in the joint because of large differences in the thermal expansion coefficients between FM and BM. The best alloy to withstand fatigue fracture is Sn-25Ag-10Sb, called J-alloy. This alloy has an optimal combination of strength and hardness and is particularly advantageous in RS form (MSF-101, for example). To date, it makes up about 80% of all RS solder production and is mostly used in automated die bonding. MSF-104 is the second most used RS alloy and is applied in devices working at high temperatures.

C. Brazing Application Examples and Advantages

1. Aircraft Frame Structures and Jet Engines

To date, the largest amorphous foil application area is jet engine acoustic shields which absorb engine noise while supporting engines in their attachment to aircraft wings. Such shields are sophisticated honeycomb sandwich-like structures consisting of a cylindrical core enclosed in and brazed to outer and inner skins or face sheets (Fig. 16) [46]. The central core consists of a plurality of cells which have four- and six-sided polygonal cross sections made of fine-gauge heat-resistant alloy. One of the ways of preplacing brazing FM foil into this structure was developed by Rohr Industries, which replaced previously used powder with METGLAS MBF-20 foil. In the Rohr method, foil is attached to the face sheet surfaces. During brazing the foil melts and fills gaps existing between parts. Upon solidification the lightweight, rigid structure is formed in one operational step. Another way to preplace brazing foil was developed by Alloy Spot Welders, Inc., which used advantageous wettability of amorphous brazing foil. Namely, they incorporate the foil only in the core, placing it directly between corrugated heat-resistant alloy as depicted in Fig. 17. In contrast to the Rohr method, there is no FM placed between the face sheets and the core before brazing. Upon melting, the filler metal is drawn to all gaps to be brazed by capillary forces including core/face sheet interfaces. Both methods eliminate the major problem existing when powder filler methods are used, namely clogging of perforation in the inner skin needed for sound attenuation. Additional advantages of METGLAS foil application are controlled amount of braze alloy, high reliability and integrity of structural configuration, and improved strength/stiffness-to-weight ratio because of reduced amount of filler metal. Figure 18 shows remarkable examples of engine parts manufactured using this foil [46].

Open-faced honeycomb structures used in air seals for gas turbine engines is an another example in which amorphous foil has analogous advantage (Fig. 19). Many

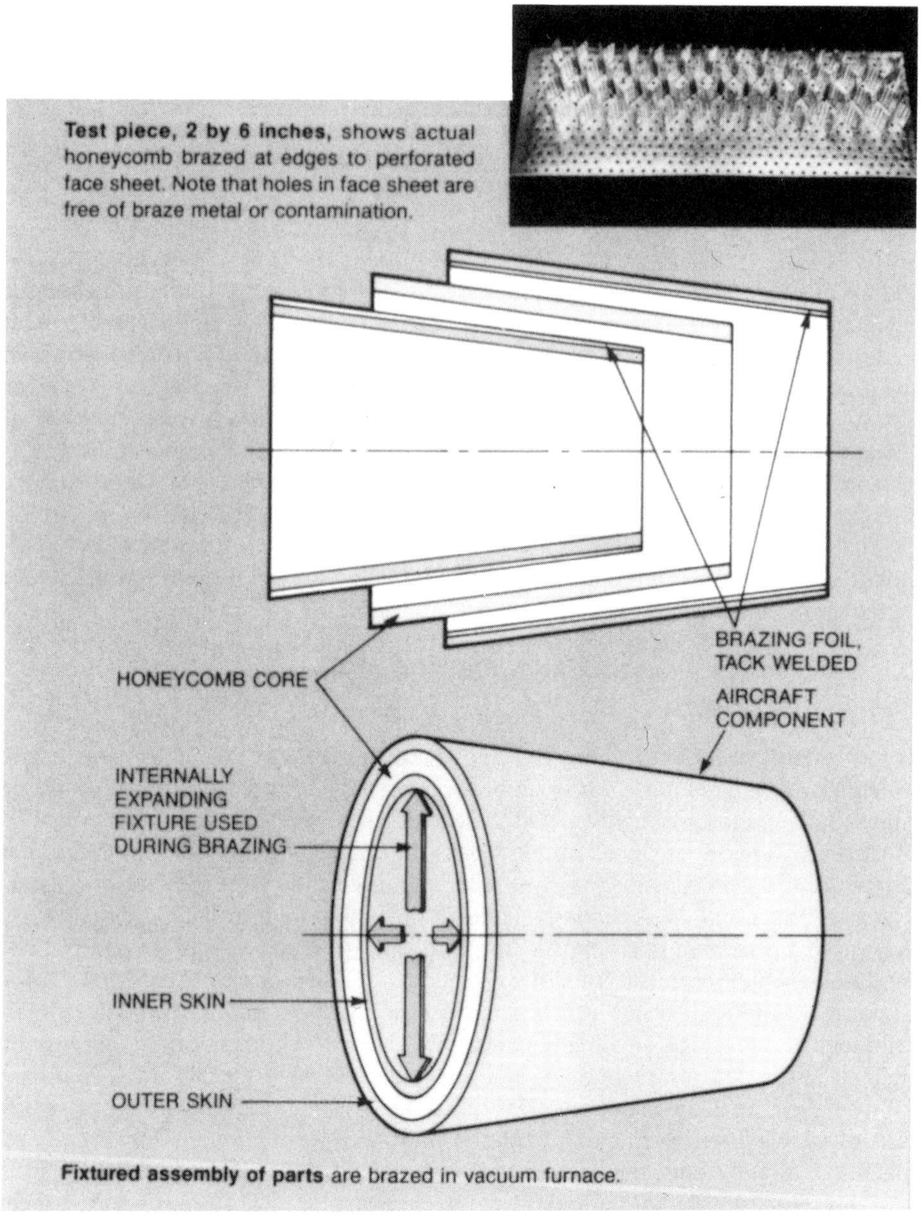
Test piece, 2 by 6 inches, shows actual honeycomb brazed at edges to perforated face sheet. Note that holes in face sheet are free of braze metal or contamination.
BRAZING FOIL, TACK WELDED
HONEYCOMB CORE
AIRCRAFT COMPONENT
INTERNALLY EXPANDING FIXTURE USED DURING BRAZING
INNER SKIN
OUTER SKIN
Fixtured assembly of parts are brazed in vacuum furnace.

Figure 16 Fixtured assembly of parts brazed using Rohr technology of METGLAS brazing foil placement.

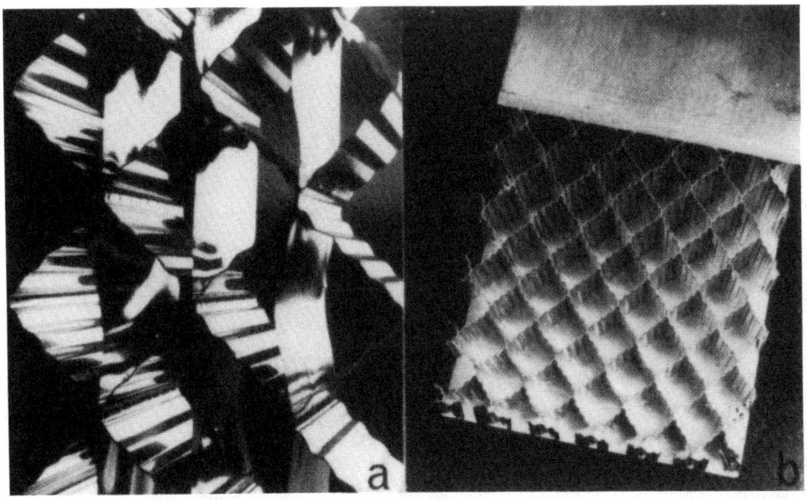

Figure 17 RS METGLAS brazing foil permits constructing closed-cell, brazed, honeycomb structures. Foil is sandwiched between the nodes of a honeycomb core (a), face sheets are placed on the core (b), and the entire assembly is brazed in a vacuum furnace. During brazing, the filler melts and flows to the face sheet/core interface, forming a uniform fillet. The structure is used for acoustic tail pipes on Boeing 727 aircraft.

rotating and stationary turbine engine parts subjected to high loads at high temperature in highly corrosive conditions have also been joined using METGLAS foil. MBF-20 foil is used in the manufacture of Inconel 718 turbine diffusers, 347 SAE stainless steel turbine pump midframes and centerbodies.

Among potential aerospace applications in the near future, one should consider the use of titanium-zirconium-based alloys in brazing a new generation of lightweight turbines in which blades will be made of Ti-6Al-4V alloy (first low-compression/temperature turbine stages) and titanium aluminides (last high-compression/temperature stages). These alloys have also been under testing in brazing of very large compound vanes of a new turbine air compressor. Summing up amorphous alloy applications in this extremely important area, it is impossible to envision present and future modern aerospace industry manufacture without amorphous brazing foil, which only a few years ago did not exist at all.

2. *Heat Exchangers*

Substantial progress has been made in implementing amorphous alloy foil FM in heat exchanger manufacture. For nickel-based alloys this process is already well into production, whereas copper- and titanium-zirconium-based alloys have been successfully tested in experimental samples. For example, large heat exchangers for power generation in electric power plants, power plants of military ships, etc., which have rather elaborate and long systems of finned tubes, were produced using a narrow (about 3 mm (0.125 in)) MBF-20 strip. This strip is preplaced by wrapping between Incolloy 800 tubing and crimped Incolloy 800 or copper fins (Fig. 20) [47]. Because brazing foil may be spooled in very long lengths it is possible to automate the process of fixturing tube, brazing foil, and fin. A few very large units were

Figure 18 Typical engine parts fabricated with METGLAS brazing foil. (1) Exhaust plug for P&W JT9D-7R4 jet engine; (2) tailpipe for Douglas DC-9-50 aircraft; (3) exhaust plug for GE CF6-80 jet engine; and (4) outer nozzle ring for Rolls-Royce RB-211 jet engine [46].

produced so far, and production on a very large scale is expected soon. Plate-shaped heat exchangers for high-pressure application in the aerospace industry, computer cooling systems, etc., were produced using various stainless steels as base metals and MBF-20 brazing FM foil. Copper radiators were also brazed using Alloy 9 copper-phosphorus foil. The most exotic heat exchanger was recently made for cooling blood in life-support systems in which blood circulates during surgery. Here there is a need to preserve the blood cells from destruction due to repeating shocklike contacts with braze fillets located inside the tubing of this system. Fillets formed using powder form of brazing filler metals have a very rough surface and, therefore, cannot be used. Again high corrosive resistance, needed to preserve the

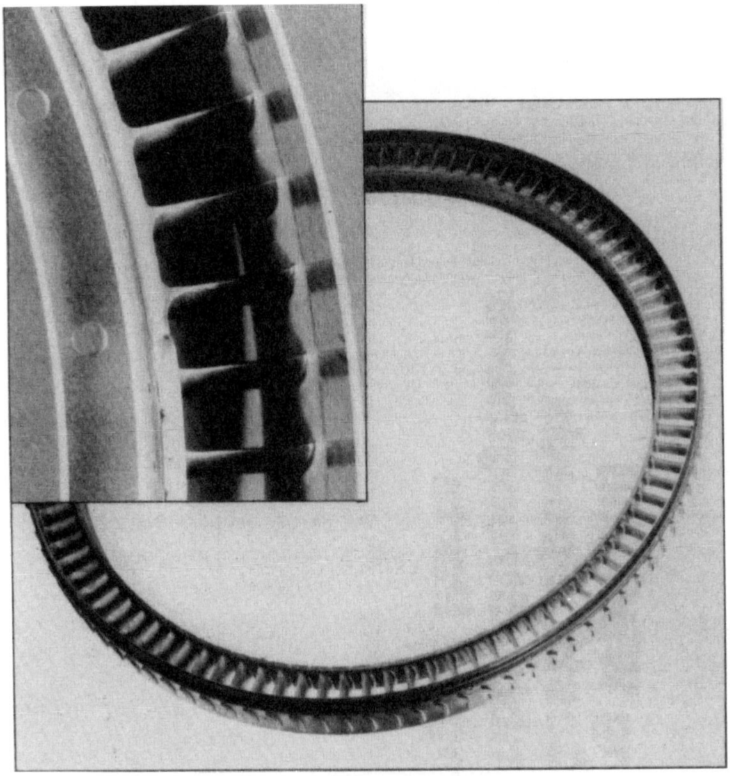

Figure 19 Typical Pratt & Whitney aircraft compressor vane and shroud assembly brazed using METGLAS MBF-50 brazing filler foil [17].

blood from any undesirable chemical reaction, and smooth surface of fillets formed helped to solve this unusual problem.

3. Joining Heat-Resistant Steels and Alloys

METGLAS alloy foils have been tested successfully and applied to brazing a series of heat-resistant steels and alloys such as 302, 304, and 409 SAE stainless steels and FeCrAlY steel. All these materials are used predominantly in honeycomb structures for automotive applications such as monolithic catalytic converters. They all have a strong surface of heat-resistant oxide film built mostly of chromium oxide and containing titanium and aluminum/yttrium as additions.

MBF-20 and -50 wet well these materials due to dissolution of the oxide film but without detrimental erosion of BM which is usually used in a thin-gauge form. Figure 21 shows a group of catalytic converters brazed using MBF-50 foil. The foil may be used for joining the core with the converter case and for joining concentric foil layers of the core. In this case low erosion and good wetting are accompanied by high brazement strength.

High-chromium 409 SAE ferritic steel for high-temperature automotive applications can be brazed with MBF-20 and -30, resulting in very strong joints [11]. The joint strength in this case may exceed that of the BM. Interestingly, good

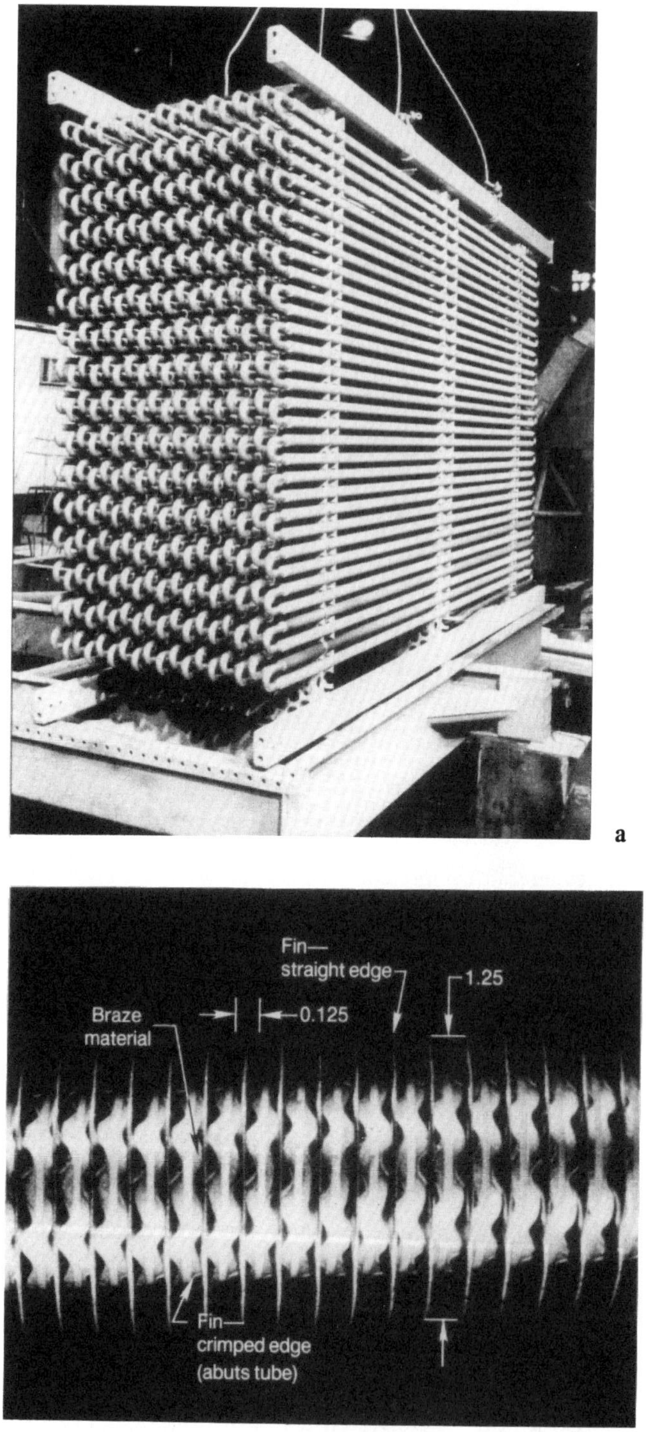

a

b

Figure 20 Finned-tube module for Racer heat-recovery system manufactured using METGLAS MBF-20 foil: (a) general view of the module which stands 10 ft high and contains 576 tubes; (b) enlarged view of finned tube. Overall diameter is 1.25 in. Crimps in fin material flex to allow spiral wrapping of the fin strip. Strip between crimps and tube is 0.125-in-wide MBF-20 50-μm-thick brazing foil [47].

Figure 21 Automotive metallic catalytic converters manufactured by Behr GmbH & Co. (Germany) using METGLAS MBF-50 foil.

wetting and fillet formation may be achieved when brazing is carried out in controlled atmosphere furnaces instead of in much more expensive vacuum furnaces. It was proved that the major condition to achieve good FM flow and wetting is to use furnace gases with a low concentration of water vapor [11]. For example, not only argon but even nitrogen/hydrogen mixtures work well if the frost point of this gas mixture is below −60°C.

4. *Joining Cemented Carbides*

MBF-1011 nickel-palladium-based FM alloy having low silicon concentration has proved to be a good replacement for gold-based alloys heretofore used for brazing cemented carbides [26,29]. Specifically, expensive gold-based alloys are often used for fastening sandwich-like cemented carbide/polycrystalline tools to the cemented carbide studs of oil well drills. The brazed tools are heated up to 400°C during a prolonged drilling cycle. Not only is high strength needed in these joints, but so are high fatigue life and strength stability. MBF-1011 alloy having cobalt and molybdenum additions completely melts below 900°C, forming a fine eutectic structure (Fig. 22). Thus, the brazing temperature does not harm the polycrystalline diamond, which is very sensitive to heating above 900°C. At the same time, the joint eutectic structure is more stable and resistant to recrystallization, which would decrease brazement strength at the service temperature in the case of gold-based alloys. Cobalt and molybdenum additions suppress the deleterious concentration-driven leaching of cobalt binder out of the BM. This leaching causes porosity in the joint vicinity. Therefore, the strength of a MBF-1011 joint, particularly when tested in the shear mode, is very high. MBF-1011 was thoroughly tested in the field

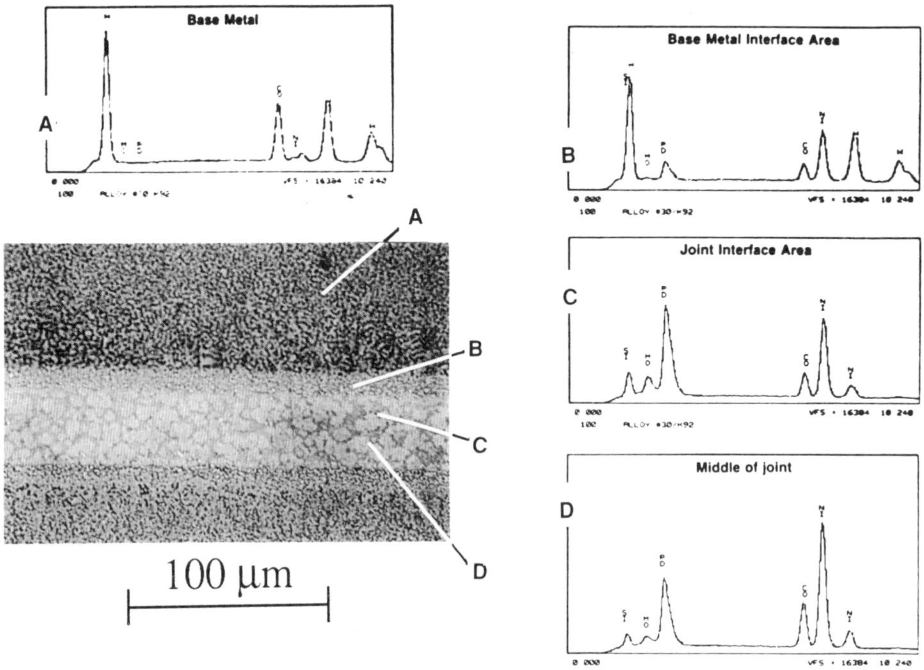

Figure 22 SEM micrograph and elemental distribution the WC/MBF-1011/WC brazement. Palladium and nickel diffuse into the base metal, but molybdenum does not. Cobalt leaching is moderate. Base metal interface does not show porosity [29].

Figure 23 Absence of porosity in joints soldered using METGLAS rapidly solidified solder.

Table 13 Power Cycling Performance of Soft Solders in TO-220 Voltage Regulators

	Number of cycles (in thousands)								
Type of solder	10	20	30	40	50	60	70	80	90
Conventionally cast J-alloy (control)	0/25	2/25	5/23	2/18	0/16	2/16	0/14	0/14	0/14
Rapidly solidified J-alloy	0/25	0/25	0/25	0/25	0/25	0/25	0/25	3/25	0/22

and is now used as a standard material for brazing polycrystalline diamond tools by some companies.

D. Soldering Application

MSF-101 (Sn-25Ag-10Sb) has been used in soldering of TO-220 voltage regulators [39], in which a silicon die was attached to a nickel-silver-plated copper leadframe. Comparison of joints made from both conventional and RS MSF-101 J-alloy showed that the latter melts and flows much more uniformly and consistently than conventional solder. Furthermore, a rough granular surface together with bubbles and excessive voids formed due to evaporation of trapped organics was observed on joints made of conventional solder (Fig. 23). Power cycling data given in Table 13 shows that units soldered with conventional J-alloy begin to fail between 10,000 and 20,000 power cycles, whereas the first failure of the units brazed with RS MSF-101 appears between 70,000 and 80,000 cycles. The remaining units made with RS FM did not show any further failures up to 90,000 cycles. The fine-grain structure formed in joints die-attached with MSF-101 retards crack formation. The thermal fatigue resistance was improved due to the absence of voids usually causing crack propagation. In spite of the very low price of each piece of RS solder needed to join a single electronic device, the replacement of conventional solders with an RS type results in a substantial savings because the number of defective devices after soldering is substantially smaller. Indeed, each silicon die costs hundreds of times

Table 14 Normalized Areal Spread (N.A.S.) of Some RS and Conventional Solders

Alloy	NAS	Temperature, °C
Sn-25Ag-10Sb (J-alloy)		
Conventional	2.14 ± 0.34	350
RS	5.58 ± 0.71	
Pb-5In-2.5Ag		
Conventional	1.50 ± 0.16	375
RS	3.49 ± 0.21	
Sn-3.5Ag		
Conventional	4.59 ± 0.25	285
RS	5.26 ± 0.67	

more than a solder preform, and repair of a faulted device after improper soldering would be uneconomical.

Better wetting and mechanical strength of solder joints manufactured with RS solders was also observed in [38,39]. Tables 11 and 14 demonstrate improved shear strength and normalized areal spread, correspondingly, of joints made using Sn-25Ag-10Sb J-alloy, Pb-5In-2.5Ag, and Sn-3.5Ag alloys as well. The same advantageous factors as in [12,38] contributed to better performance of all three alloys tested elsewhere [39].

VIII. CONCLUSION

As one can see from this chapter, amorphous and RS filler metals constitute one of the three major groups of alloys which were strongly advanced by successful industrial utilization of RS technology. Two other groups are, of course, magnetic and structural high-strength alloys. To date, RS filler metals are considered as standard materials in many critical applications. Especially with recent advances in METGLAS brazing alloy foil production to tonnage quantities, continuing development of new compositions and growth of the list of successful case histories indicates that the potential of amorphous and microcrystalline RS filler metal is far from being exhausted and without a doubt will continue to grow.

ACKNOWLEDGMENT

Thanks are extended to Dr. N. DeCristofaro, who brought us to Metglas Products to work on amorphous metals, and our former and present colleagues D. Bose, A. Datta, K. S. Tan, S. Pounds, and M. Lambert, whose work contributed greatly to the development of amorphous and microcrystalline brazing and soldering materials.

REFERENCES

1. D. R. Milner, *British Weld. J.*, *5*, 90 (1958).
2. B. C. Allen, in *Liquid Metals, Chemistry and Physics*, edited by S. Z. Beer (Marcel Dekker, New York, 1972), p. 172.
3. E. Pelzel, *Metals Progress*, *58*, 252 (1950).
4. C. H. P. Lupis, *Chemical Thermodynamics of Materials* (North-Holland, Amsterdam, 1983), p. 380.
5. G. Smith and C. Lea, *Surf. Interface Anal.*, *9*, 145 (1986).
6. A. Miedema, *Z. Metall.*, *69*, N5, 287 (1978).
7. A. Appen, in *High Temperature Inorganic Coatings* (WNT Warsaw, Warsaw, 1970), p. 49.
8. W. Gale and R. Wallach, *Welding and Cutting*, *N7*, 21 (1991).
9. A. Dupre, in *Theorie Mechanic de la Chaleur* (Gauthier-Villars, Paris, 1869), p. 369.
10. A. Rabinkin and S. Pounds, *Welding J*, *67* (5), 33 (1988).
11. A. Rabinkin and P. Murzyn, presented at the 18th International AWS Brazing Conference, Chicago, IL, 1987 (unpublished).
12. D. Bose and K.-S. Tan, *Solid State Technol.*, *N4*, 166 (1986).
13. P. Sexton and N. DeCristofaro, U.S. Patent No. 4,148,973 (1981).

14. N. DeCristofaro and A. Datta, *Rapidly Solidified Crystalline Alloys*, edited by S. K. Das, B. J. Kear, and C. M. Adam (TMS-AIME, Warrendale, PA, 1985), p. 263.
15. N. DeCristofaro and C. Henschel, *Welding. J.*, *57*, 33 (1978).
16. H. H. Liebermann and A. Rabinkin, in *Encyclopedia of Material Science and Engineering*, first suppl., edited by R. Cahn (Pergamon Press, 1989). p. 59.
17. N. DeCristofaro and D. Bose, in *Rapidly Solidified Materials*: *Properties and Processing*, edited by P. W. Lee and J. H. Moll (ASM, Metals Park, OH, 1986), p. 415.
18. S. Omori, Y. Hashimoto, K. Shoji, K. Hidaka, and Y. Kohira, *Powder Powder Metall. (Japanese)*, *18*, 316 (1972).
19. *Binary Alloy Phase Diagrams*, 2nd ed., edited by T. Massalski (ASM, Metals Park, OH, 1990), p. 2859.
20. I. W. Donald and H. A. Davies, *J. Mater. Sci.*, *15*, 2939 (1980).
21. D. Bose, A. Datta, and N. DeCristofaro, *Welding J*, *60* (10), 29, (1981).
22. In Ref. 19, p. 1449.
23. A. Datta, A. Rabinkin, and D. Bose, *Welding J.*, *63* (10), 14 (1984).
24. N. DeCristofaro, U.S. Patent No. 4,405,391 (1983).
25. D. Bose, A. Datta, A. Rabinkin, and N. DeCristofaro, *Welding J.*, *65* (1), 235 (1986).
26. A. Rabinkin, U.S. Patent No. 4,802,933 (1989).
27. A. Rabinkin, U.S. Patent Nos. 4,746,379 (1988) and 4,928,872 (1990).
28. C. Henschel and W. Frick, *Welding J.*, *64* (10), 45 (1985).
29. A. Rabinkin and S. Pounds, *Refr. Metals Hard Mater.*, *8* (4), 224 (1989).
30. A. Rabinkin, *Mater. Sci. Eng.*, *A124*, 251 (1990).
31. N. DeCristofaro, S. Levinson, and P. Sexton, U.S. Patent No. 4,260,666 (1981).
32. A. Rabinkin, U.S. Patent No. 4,842,955 (1989).
33. A. Rabinkin, F. Reidinger, J. Marti, and L. Bendersky, in *Rapidly Quenched Materials*, Part 1 (Elsevier, New York, 1991), p. 255.
34. L. Tanner and R. Ray, U.S. Patents No. 4,135,924 (1979), No. 4,126,499 (1978), No. 4,148,669 (1979), No. 4,113,478 (1989), and 4,171,992 (1979).
35. A. Rabinkin, H. Liebermann, S. Pounds, T. Taylor, F. Reidinger, and Siu-Ching Lui, *Scripta Metall. Mater.*, *25*, 399 (1991).
36. C. E. Smeltzer and A. N. Hammer, *Titanium Braze Systems for High Temperature Applications*, Technical Report AFML-TR-76-145 (1976).
37. A. Rabinkin and S. Pounds, presented at the *Advanced Aerospace Materials and Processes Conference*, *AEROMAT*, *91* ASM, Long Beach, CA 1991 (unpublished).
38. J. DeVore, *J. Metals*, *N7*, 51 (1984).
39. R. Bowen and D. Peterson, *IEEE Trans.*, *CHMT-10*, N3, 341 (1987).
40. S. Pinamaneni and D. Solomon, *IEEE Trans. Components, Hybrids, Technol.*, *CHMT-3*, N4, 416 (1986).
41. A. Rabinkin, *Welding J*, *67* (10), 29 (1988).
42. A. Rabinkin, *Mater. Lett.*, *2* (6A&B), 487 (1984).
43. S. Huges, in Ref. 37.
44. T. Scoles, in Ref. 37.
45. A. Rabinkin and E. Norin, Metglas Products (1989, private communication).
46. F. Yeaple, *Weld Design Fabric.*, *59*, N6, 20 (1986).
47. *Brazing Foil Joints Fins to Boiler Tubes*, Editorial Note from *Welding Design and Fabrication*, *59*, N6, 25 (1986).

22

Mechanical, Chemical, and Electrical Applications of Rapidly Solidified Alloys

C. Suryanarayana and F. H. Froes
Institute for Materials and Advanced Processes,
University of Idaho, Moscow, Idaho.

I. INTRODUCTION

Rapid solidification (RS) of alloys has been shown to lead to useful constitutional and microstructural effects [1]. RS allows large departures from equilibrium constitution (the identity and compositional ranges of phases formed) resulting in large extensions of solid solubility limits, formation of nonequilibrium or metastable crystalline and quasicrystalline intermediate phases, and amorphous alloys. Retention of disordered crystalline structures in normally ordered materials such as intermetallic compounds has also been reported. Additionally, RS results in changes in the morphology and dimensions of the microstructure (the size and location of the phases present) in the direction of a more uniform and finer microstructure with a large reduction in solute segregation effects. The microstructural features which are refined include grain size, dendrite arm spacing, constituent particles, and precipitates and dispersoids. Each of these attributes of RS can, either individually or in combination, contribute to a better performance of the material produced. It should be mentioned, however, that because of the "batch" nature of the RS techniques in the early years, it was difficult to envision large-scale applications for these exotic products. Small-scale exploratory investigations to apply these materials in an inductance thermometer [2], as homogeneous standards for electron probe [3] and neutron activation analysis [4], and to monitor atmospheric pollution by sulfur dioxide [5] were carried out with some success. Some attention was also paid to potential nuclear [6] and automotive [7] applications.

The introduction of RS techniques to produce wires, tapes, ribbons and powders continuously and with a uniform cross section, the observation of the ductile nature of the amorphous alloy ribbons, and the development of innovative consolidation techniques to retain the desirable attributes of RS in the compacts have given a great fillip to this activity during the 1970s and 1980s. The commercial applications of RS began to be explored, first for metallic glasses [8,9], and, by the mid-1970s, for crystalline materials [10,11]. There are now many comprehensive books [1,12–17], reviews [18–27], and conference proceedings [28–46] which discuss the theoretical and applications-oriented facets of RS. It should be emphasized that generally the applications do not depend on a single property of the RS alloy, but most often on a combination of several desirable properties.

RS amorphous and crystalline alloys find extensive applications as magnetic, brazing, and soldering materials which have been covered in the earlier chapters. The present chapter will deal with applications based on the mechanical, chemical, and electrical properties of RS alloys.

II. MECHANICAL APPLICATIONS

The strength of an alloy can be improved by increasing the strength of the matrix (solid solution strengthening), by introducing second-phase dislocation barriers (precipitation and dispersion strengthening) or by increasing the boundary area (boundary strengthening). RS alloys show enhanced solid solubilities, formation of metastable crystalline and amorphous phases, and grain refinement. Thus, the mechanical strength of the RS alloys can be significantly higher than the ingot-metallurgy (IM) processed alloys. Further, amorphous alloys possess very high strength levels approaching those of dislocation-free single crystals or hard-drawn piano wire. For example, $Fe_{80}B_{20}$ has a tensile yield strength of 3.6 GPa and good ductility in bending or compression, but not in tension [47]. Excellent fatigue resistance of the amorphous alloys has also been recently reported [48]. However, a major disadvantage of amorphous alloys is their poor fracture behavior. This is a consequence of the lack of a mechanism equivalent to that of work hardening in crystalline alloys to disperse slip and so extend the time required for nucleation of fatigue cracks. The variety of potential applications which have been proposed for rapidly solidified alloys because of their strength is shown in Table 1.

Between the amorphous and microcrystalline alloys, applications based on mechanical strength appear to be greater in number for the microcrystalline variety. In fact, it has been shown [49] that some amorphous alloys subjected to a low-temperature devitrification to produce fine-grained microcrystalline material have a better combination of properties than the present amorphous alloys.

Table 1 Potential Applications of RS Alloys Based on Strength

Control cables	Pressure vessels
Flywheels	Mechanical transmission belts
Torque transmission tubes	Cutting tools
Armor penetrators	Dial pointer electrical instruments
Reinforced rubber tires	Spring components
Metal-matrix composites	Reinforced plastics
Extrusion dies	Abrasive wheels
Casting dies	

A. Amorphous Alloys

The high strength and isotropic two-dimensional form of amorphous alloy ribbon suggests applications in planar composites. Indeed, high quality plates have been made of half-inch-wide ribbon in 0°–90° crossply with an epoxy matrix [50]. These specimens have shown a remarkable degree of in-plane isotropy and strength.

Amorphous alloy ribbons have been used for reinforcement of mortar [51], polymers [52,53], and metallic matrix materials [54]. It has been shown that these reinforcements are useful when the amorphous alloy ribbons are in a prestressed condition. The high surface-to-volume ratio of ribbon provides a sufficient adhesion force to avoid relaxation even at the relatively high stress levels which can be carried by these materials. A consequence is the considerable improvement of the ability of the epoxy resin to carry tensile loads. Specific work on fracture has shown an increase in strength of about three orders of magnitude with only 1 vol% of fiber.

By embedding melt-spun aluminum flake into a plastic, a heat dissipator with performance characteristics close to those of aluminum can be achieved [55]. This allows molding to be used in the production of equipment housings at little additional cost. Applications for such materials include food processing equipment, cookware, grinding wheels, egg poachers, hot plates, and electronic housings. The use of RS aluminum flake in a plastic dramatically increases conduction, especially when the aluminum forms an internally connected network [55,56].

Aluminum flake embedded in asphalt gives a superior roof coating because its enhanced reflectivity reduces the interior temperature in sunny climates by as much as 20°C, thus reducing cooling costs. Improved weatherability and increased life also result [55–58].

Thin foils of a variety of metals can be produced by a variety of techniques. One possible application for steel foils is in composite steel foil/paper containers for pressure packaging of products such as coffee, beer, and soft drinks [27]. Other applications involve scrubbing pads and the scrubbing edges of sponge mops where the ability to produce a very sharp ribbon edge and corrosion resistance are attractive [27]. It has also been proposed that crystalline ribbon produced from alloys such as cadmium-copper or nickel-base alloys could be used in heating blankets and wall-type radiant heaters where high resistivity and flex life are desirable.

The commercial benefits of introducing melt-extracted microcrystalline steel fiber into concrete to increase tensile and impact strength and resistance to wear and spalling have been pointed out [59]. Reinforcement for rubber automotive tires is yet another potential application which is still receiving attention [9]. Even though directly cast wire is strong and resistant to corrosion in addition to exhibiting good adherence to rubber, casting speeds of about 300 ft/min are not fast enough to compete with conventional casting and wire-drawing methods, which has prevented market penetration of the direct-cast material.

Because of ease of fabrication from tape, applications such as flywheels also have been proposed [60]. For this application, amorphous alloys are the only soft magnetic materials with sufficient mechanical strength to allow magnetic driving. The use of amorphous alloys in composite flywheels, either of the flat-disk type or the tape-wound type, has been essentially rejected because of the unacceptably limited fatigue properties. While the amorphous alloys are ideally suited for flywheels in many ways, especially as applied to electric vehicles where energy stored per unit volume is particularly important, flywheel applications require high cycle capability by definition.

Other commercial applications include amorphous tungsten-based alloys for kinetic energy armor penetrators [61], as well as dial pointers in electrical instruments [61].

Industrial applications based on the strength of the amorphous alloys have been slow in developing. The principal reason may be fabrication difficulties; for example, the very high hardness of the alloy, causing rapid stamping die wear [25]. Further, the use of such composites in cylindrical pressure vessels has been hampered because of the difficulty in closing off the ends of the chamber. Use could result if applications emerge wherein the electrical and magnetic characteristics of the amorphous alloys are desirable in conjunction with mechanical properties.

B. Microcrystalline Alloys

RS processes for the development of advanced microcrystalline alloys have a number of advantages compared to conventional IM processes, major ones of which include the great flexibility in choice of composition and the high degree of control of microstructural features. The successful development of these alloys is based on fine grain and subgrain size, reduction in constituent particle size, and homogeneous dispersion of intermetallic phases. In addition, RS processing allows additions of certain beneficial alloying elements to an extent beyond that possible by IM. The potential applications of RS microcrystalline alloys will be discussed under two headings: high-strength structural materials and wear-resistant materials.

1. High-Strength Structural Materials

RS processing has been applied extensively to aluminum alloys, much more so than to other metals and, as such, the level of scientific understanding and technological exploitation is now quite mature [62,63]. This fortuitous situation arises because only eight solute elements have solid solubilities greater than 1 at% in aluminum under equilibrium conditions, thus affording great potential to develop new and improved alloys through RS by increasing the solid solubility limits and forming metastable intermediate phases. Further, the development of powder metallurgy (PM) aluminum alloys through consolidation of atomized prealloyed powders has advanced this area further. Because of the cost premium which is associated with some RS products compared with conventional IM counterparts, the major user of these new aluminum alloys is likely to be the performance-driven aerospace industry. Here, high-strength 2XXX and 7XXX series alloys are widely used (50% of the 100–200 thousand ton/year market goes to aerospace [64]) with a typical product mix such as that shown in Table 2. To be cost-effective (based on weight savings), RS alloys should be about 30% stronger, much tougher, and more resistant to fatigue than the conventional 7075 alloy. Table 3 lists the room temperature mechanical properties of these alloys obtained after process optimization. It is apparent that considerably higher strengths can be achieved through RS processing. New compositions also have been developed showing no evidence of pit blistering or exfoliation for strength levels of ~700 MPa, at which wrought 7XXX series

Table 2 Aluminum Product Mix in a Typical Transport Aircraft

Product form	% Total weight
Extrusions	28
Sheet	20
Forgings	18
Plate	16
Bar/forged bar	13
Others	5

Source: Ref. 64.

alloys show extensive exfoliation. These compositions also showed stress corrosion and fatigue resistance superior to those of wrought 7075 [65]. Forgings of near-net shape parts for aircraft structural applications have been made for evaluation from hot-pressed billets of promising compositions up to 1500 kg in size. These alloys contain a small addition of cobalt to control grain size and offer a good balance of properties. One of these, alloy 7090, is specified for a 50-kg forged landing gear door actuator link for the Boeing 757 aircraft, which will save 19% in weight over IM 7175-T73 [66,67]. The alloy CW 67 is being considered as a suitable replacement material for the existing airframe components which are the subject of costly maintenance problems. Other applications include extruded forward C-141B longerons (Warner Robbins Air Logistics Center) [67], and web-rib closed die forgings [68]. The combination of high strength, fracture toughness, and improved corrosion resistance is a significant materials design feature and thus RS alloys possessing this combination of properties have an excellent chance for success.

Improvements in elastic stiffness of aluminum alloys without loss of other properties are of particular interest to airframe designers, especially if combined with

Table 3 Room Temperature Mechanical Properties of RS Aluminum Alloys

Alloy	Temper	0.2% YS (MPa)	UTS (MPa)	%El.	K_{Ic} (MPa$\sqrt{m}$)
IM 7075	T6	510	572	13	—
PM 7075 (+1Ni + 1Fe)	extruded	634	717	9	—
PM 7090	T7 extruded	586	627	10	26
PM 7091	T7 extruded	545	593	12	44
PM CW 67	T7 extruded	580	614	12	47

7075: Al–5.6Zn–2.5Mg–1.6Cu–0.23Cr
7090: Al–8.0Zn–2.5Mg–1.0Cu–1.5Co
7091: Al–6.5Zn–2.5Mg–1.5Cu–0.4Co
CW67: Al–9.0Zn–2.5Mg–1.5Cu–0.14Zr–0.1Ni
Source: Ref. 62.

reduced density. The addition of 1 wt% lithium decreases the density of aluminum by about 3% and increases its elastic modulus by about 6%, resulting in high specific modulus values. Some of the problems encountered in IM Al-Li alloys (severe segregation problems, poor ductility, and fracture toughness) can easily be overcome in the RS alloys. Furthermore, additions of zirconium, copper and magnesium have been found to be beneficial in significantly improving the mechanical properties of Al-Li alloys [11]. These alloys also have superior fatigue crack growth resistance and excellent corrosion and stress corrosion cracking resistance compared with ingot metallurgy 2XXX or 7XXX series alloys [69]. However, alloys with higher weight saving potentials do not appear to be feasible with lithium additions alone and thus substantial additions of beryllium have been made to provide weight savings in the 15–25% range over conventional alloys. Such alloys should lead to applications in areas such as the upper wing skin of a fighter airplane. For wide-bodied subsonic transport aircraft like Boeing 747, the weight saving by using RS Al-Li-base alloys is estimated to result in a reduction of approximately U.S. $5 billion in fuel costs over the anticipated lifetime (about 30 years) of each aircraft. But, the rapid developments in IM Al-Li alloys cast a shadow of doubt on whether RS Al-Li alloys will indeed find applications in the aerospace industry.

RS aluminum alloys meant for elevated temperature applications use insoluble transition and rare earth elements to produce up to 30% volume fraction of stable dispersoids. The prominent alloys developed so far are based on Al-Fe with additions of Ce, Mo, or V + Si, the mechanical properties of which are listed in Table 4 [70–72]. These aluminum alloys with their high temperature strength (on a specific strength basis, these alloys match the performance of titanium alloys up to 200–230°C) are initially destined for a wide range of aerospace applications: gas turbine engines (low-temperature fan and compressor cases, vanes, and blades) [73], missile applications (fins, winglets, rocket motor cases) and airframe structural applications where the improvements in both temperature capability and elastic modulus can be utilized. The RS aluminum alloys can compete with titanium in applications such as impellers [74], rib web parts and in an advanced Delta wing fighter designed to fly at Mach 3 [75]. In the latter study, use of the elevated temperature aluminum alloy rather than titanium allowed a projected 15% reduction in total aircraft weight. Here, components such as upper- and lower-wing

Table 4 Mechanical Properties of Elevated Temperature RS Aluminum Alloys

Alloy	Temp. (°C)	0.2% YS (MPa)	UTS (MPa)	%El.	K_{Ic} (MPa$\sqrt{m}$)	Ref
Al–7.1Fe–6.1Ce	24	524	567	5.7	26	70
(extruded and rolled)	150	403	427	5.0		
	230	278	304	5.0		
Al–8.5Fe–1.3V–1.7Si	24	414	462	12.9	31	71
	150	345	379	7.2		
	230	310	338	8.2		
Al–8Fe–2Mo	24	415	512	3		72
	150	374				
	230	331				

surfaces were considered to be viable candidate applications for RS aluminum alloys.

Another interesting use of RS for aluminum alloys is as feedstock in scrap recycling [76]. This is attractive because much less energy is required to recycle than to extract aluminum from its ores. Further, RS allows refinement of the second-phase constituent particles resulting in attractive mechanical properties. RS processing of a charge of aluminum-based automobile scrap containing 9.6 wt% Si and 0.8 wt% Fe yielded tensile properties intermediate between those of wrought 6063-T6 and 2024-T4 [77]. The composition in question would not otherwise have been acceptable even for castings, let alone as a basis for wrought ingot processing.

RS has been shown to be beneficial to both conventional titanium alloys, such as Ti-6Al-4V, and to novel alloys which can only be produced using RS, such as rare earth or metalloid-containing terminal and intermetallic alloys [78]. The grain size refinement and change in the α-phase morphology produced by RS of Ti-6Al-4V can lead to applications where improved fatigue behavior is required [79]. Many RS nonconventional alloys have enhanced mechanical properties and could find use in high-strength applications (for example, in competition with high-strength steels in landing gears and with low-density alloys for use in both sheet and forging applications). Additionally, dispersoids stable at temperatures well in excess of 550°C in both terminal and intermetallic alloys should lead to applications in advanced engines and supersonic or hypersonic vehicles [78].

Magnesium is the structural metal with the lowest density. Even though improved combinations of properties were reported in magnesium alloys consolidated by extrusion from prealloyed atomized powder [80], low strength, limited ductility, and poor corrosion resistance have precluded the use of these materials in demanding applications. However, recent investigations have indicated that modified magnesium alloys processed through RS display improved strengths and corrosion resistance and have great potential for increased application in automotive and aerospace industries [81].

RS of copper-based alloys results in improved high-temperature alloys having good conductivity, leading to marine, rocket thrust chamber, and plain bearing applications [82]. Potential applications include seawater piping (e.g., in the ballast/cooling systems of deep submergence submarines) and acoustic damping components used to control noise emitted by naval machinery. However, the initial results in both applications are disappointing. It should be possible to adopt new alloying strategies and consolidation techniques to achieve the desired goals. Another application may include Cu-Zr-Cr alloys with good conductivity and low-cycle fatigue behavior (RS giving fine dispersoids) for rocket thrust chambers [83]. Further, Cu-Pb alloys produced by RS (containing more Pb) could replace more expensive conventional bronze bearings (>80% Cu) because of the lower cost of Pb and the fine structure produced by RS.

Nickel-based superalloys in the form of sheets, discs, and blades are used in the hottest sections of gas turbine engines. RS has been used to avoid segregation effects and thereby improve strength in addition to forming a near-net-shape product. RS-processed alloys based on Ni-17 wt% Al with 9–11 wt% molybdenum or chromium show creep-strength temperature capability increases up to 85°C with the molybdenum addition and a 10-fold increase in oxidation resistance with the chromium addition, compared with the strongest unidirectionally solidified alloy

(MAR-M200 +Hf) in current use as a blade material [84]. Such RS alloy compositions are not usable in cast or wrought ingot form because of unacceptable segregation and related effects. Turbine blades have also been fabricated using the RS-processed alloys, resulting in improved cooling efficiency [85]. This, in combination with the higher-temperature capability, should permit higher turbine inlet temperatures, giving greater thrust-to-weight capability or alternatively longer engine life at current inlet temperatures.

2. *Wear-Resistant Materials*

The wear resistance of a commercial high-speed tool steel is due to the presence of uniformly distributed hard, brittle eutectic carbides in a tough matrix. Since the complex solidification phenomenon in tool steels leads to excessive segregation of carbon and alloying elements, it is extremely difficult to achieve uniform distribution of fine carbides even after 98% reduction by hot working. RS processing by extrusion or by hot isostatic pressing of canned atomized alloy powder permits the fabrication of tool steels with higher alloying contents, improved machinability, and greater dimensional stability [86–88]. Thc fine dendrite arm spacing of a few microns and absence of coarse carbides in the atomized powder ensures a relatively uniform distribution of fine carbides during processing and in service.

RS processing of tool steels at high rates of solidification produces homogeneous microstructures with complete solid solubility of alloying elements, which subsequently result in high hardness and wear resistance. Subsequent tempering increases the hardness of these materials even further. RS steels do not soften and have a hardness peak at 615°C, in contrast to commercial tool steels which begin to soften at 300°C and display a secondary hardening peak at 525°C. The high hardness and wear resistance of RS tool steels are expected to improve cutting tool life [89].

Ray et al. [49,90] used metalloid-lean (5–13 at%) amorphous alloys as precursor material to produce a fine-grained (~0.25 μm) crystalline matrix stabilized by ~25 vol% of fine borides or carbides. Such materials can find applications as hot-work die tools requiring hot strength and hardness up to 650°C, oxidation resistance to 700°C and good impact resistance up to 540°C. Trials on Ni-Mo-B based alloys demonstrated considerable extensions in cutting tool life compared with that of conventional M42 tool steel [91], more than doubling of the die life of a mold gate for aluminum die casting (>125,000 shots as compared with 50,000 for H-13 steel) [92], and tripling of the die life of a copper extrusion die compared with conventional high-cobalt Rexalloy or Stellite 6 [24]. The dies made using RS alloy also require less maintenance since they hold their cutting edges longer and possess a greater usable life. Commercial production of RS tool steels is already here, but market penetration of the powder tool steels has yet to fully develop because of high costs [93]. The higher cost is, however, increasingly offset by the enhanced performance and the longer life of tool steels made from RS powders.

Laser treatment to incorporate carbides and borides directly into a thin molten zone can also increase the wear resistance by at least an order of magnitude, making the treatment attractive for improving life of components such as automotive engine valve seats or chainsaw teeth [94].

It has recently been shown that RS of high-silicon Al-Si alloys suppresses the precipitation of large silicon crystals and improves the strength and wear resistance [95]. Further, the addition of Fe improves the heat resistance and fatigue strength.

Potential applications for such alloys include compressor parts (including vanes, rotors, housings, swash plates, shoe disks, and connecting rods), automobile engine parts (including pistons, connecting rods, rocker arms, cylinder liners, shift forks, oil pumps, and bearings) and peripheral components, transmission parts (including synchronizer rings and hubs), and brake disks. Potential applications in the electronics industry include VTR cylinders, gears, shafts, and spindles [95].

III. CHEMICAL APPLICATIONS

Both amorphous and microcrystalline RS alloys have interesting and useful chemical properties. Both corrosion resistance and catalytic properties have been studied for the amorphous alloys, while only the corrosion behavior of microcrystalline alloys has been studied. These and the hydrogen storage applications of RS alloys will now be discussed.

A. Corrosion-Resistant Applications

Remarkable improvements in the resistance to corrosion were reported in 1974 for an RS amorphous Fe-Cr-P-C alloy [96]. Since then this aspect of RS alloys has been reviewed periodically [97–100]. The simultaneous presence of chromium and phosphorus results in the formation of a highly protective hydrated chromium oxyhydroxide film which has been found responsible for the high corrosion resistance of amorphous alloys, and has been explained as due to (a) the single-phase amorphous nature of the material, (b) the chemical homogeneity and the isotropic structure, and (c) the absence of grain boundaries and other imperfections in the material. Improved corrosion resistance combined with high strength has also been claimed for titanium-boride-dispersed microcrystalline austenitic stainless steels [101], and nickel-based alloys [102]. In the latter case, chromium was added for oxidation resistance and molybdenum for pitting resistance. The resistance to oxidation [103] and to corrosion [104] of stainless steels of conventional composition has been reported to be improved for alloys in the RS condition due to a refinement of inclusions and matrix grain size. High levels of aluminum (normally prohibited since they decrease the ductility in ingots) in RS ferritic steels also have been shown to improve the corrosion resistance [105].

The attractive combination of high strength and better corrosion and oxidation resistance of the RS alloys suggests a variety of marine, chemical, and biomedical uses. Table 5 lists the projected applications of corrosion-resistant amorphous alloys [27]. One application in chemical processing is for filters in high-field, high-gradient filtering systems to remove paramagnetic iron oxides from clays, oils, and waters that are otherwise discolored by them [25]. Such a filter made of an iron-based amorphous alloy (having a high magnetic induction) has been claimed to increase the rate of collection of submicron size magnetic particles by a factor of three compared with a conventional ferritic stainless steel filter [106,107]. Amorphous Ni-P alloys have the potential for applications as chemical filters of liquid sodium used in fast breeder reactors [9].

Another potential application for amorphous alloys is as a long-life razor blade or surgical scalpel. While the preparation of high-grade steel for razor blades involves many involved production stages, melt spinning of metallic glass ribbon

Table 5 Projected Applications of Corrosion-Resistant RS Alloys

Marine/biomedical	Magnetic filters
Naval aircraft cables	Surgical scalpels
Torpedo tubes	Long-life razor blades
Chemical filters	Hardface coatings
Reaction vessels	Electric razor
Electrodes	Watch case
Scalpels	Improved stainless steels
Suture clips	Improved copper-based alloys
Oil filters	Improved dental amalgams

is a single-stage, economical process. These glasses can also be honed to superior levels of sharpness because of the absence of microstructure. Antifriction Teflon coatings can also be easily applied to these edges. However, RS ribbons are too thin to be used directly as cutting tools and so they must be attached to, or embedded in, backing materials which increases the manufacturing costs [9]. This factor alone has prevented their commercial exploitation. Evaluation of metallic glasses for electric razors has also been carried out [61].

Microcrystalline alloys have been suggested for applications based on their excellent corrosion resistance. The generally poor corrosion resistance of commercial Ag-Sn-Cu alloy dental amalgams can be improved by RS processing with increased copper contents [108]. Further, the fine grain size of these materials increases the strength and ductility [109]. Extrusions of RS-processed Al-6 to 8 wt% Fe alloy with additions of chromium, manganese, and magnesium are reported [110] to exhibit corrosion resistance in seawater exceeding that of conventional Al-Mn and Al-Mg wrought ingot alloys. Combined with superior hot strength, thermal conductivity and low cost, compared with established Cu-Ni alloys, this suggests applications as condenser and heat-exchanger tubing in desalination plants. A high-strength saltwater-resistant RS Cu-based alloy has the potential for marine use [107]. Similarly, for marine applications, RS Fe-Al bronzes have better corrosion resistance in both Cl-free and Cl-containing environments as compared to sand-cast alloys of the same composition [111].

In spite of the above potential applications for RS ribbon materials, a significant barrier to applications on a significant scale in this area has been the lack of an economic and effective method for applying them to the surfaces of less-corrosion-resistant materials (they cannot be easily welded), or for making the amorphous alloys in bulk form. Laser processing has been shown to overcome the former barrier [100]. Benefits include restoration of protection against intergranular attack to sensitized type 304 stainless steel [112] and increased resistance to attack by chloride solutions of type 614 aluminum bronze [113], both effects attributed to the homogenizing effect of the RS treatment. Amorphous Ni-Cr-P-B alloy layers on mild steel, produced by laser fusion of a prior deposit of the required composition, had a corrosion resistance comparable to that of melt-spun alloy [114].

In addition to the above, corrosion-resistant RS amorphous alloys can be used for decorative purposes. In fact, tie pin frames made from an amorphous Co-Ni-Si-B alloy are being marketed in Japan.

B. Catalytic Applications

Catalysis, directly or indirectly, affects approximately 20% of the U.S. gross national product, with total catalyst use in the U.S. at approximately five billion pounds per year [115]. The high activities of practical solid catalysts are often considered to result from the several imperfections in the solid state such as grain boundaries, dislocations, impurities, etc. Therefore, RS alloys, especially the amorphous variety, have the potential for use in catalytic applications because of their high reactivity and large surface area per unit mass (foils).

Yokoyama et al. [116] demonstrated that Fe-Ni-based amorphous alloys containing P and/or B are two orders of magnitude more effective in catalyzing the hydrogenation of carbon monoxide to yield hydrocarbons (methane, ethylene, ethane, propylene, and propane) than their crystalline counterparts. This enhanced activity was maintained only at temperatures below some critical temperature which is considerably lower than the crystallization temperature of the amorphous alloy. However, the metastable phases formed during the devitrification of the amorphous alloy also showed increased activity, approximately an order of magnitude higher than the alloy under equilibrium conditions (Table 6). Reports also are available on palladium-based amorphous alloys for the oxidation of methanol, sodium formate, and formalin with activities similar to or higher than platinized platinum, suggesting application as an electrode material in methanol-air fuel cells [117]. $Pd_{35}Zr_{65}$ amorphous alloy was also shown to be a highly active methanation catalyst for the reaction of carbon monoxide and hydrogen to produce methane [118]. Other catalytic possibilities include $Pd_{35}Zr_{65}$ as a cathode for hydrogen evolution in water electrolysis equipment [119], $Ni_{62}B_{38}$ for hydrogenation of ethane (the catalytic activity was increased by pulverizing the amorphous ribbon, which also incidentally removes the stable oxide layer on the surface of the ribbon) [120], and Pd-Si and Pd-Ge for the dissociation of hydrogen [121].

Pd-based amorphous alloys also show particularly high electrocatalytic activity for chlorine evolution in hot concentrated NaCl solutions, while retaining high corrosion resistance, suggesting application as a corrosion-resistant, energy-saving anode in the electrolytic soda process [122]. Less-expensive nickel-base amorphous alloys have also been shown to be useful as electrodes for seawater electrolysis [123] and fuel cells [124]. Such enhanced catalytic behavior is not confined to the amorphous alloys among RS materials. Activated microcrystalline Al-30 to 50 at% Ni RS powders were reported to be more effective for catalyzing hydrogenation of a series of organic compounds than were the same alloys in conventional cast form [124].

Table 6 Catalytic Properties of $Ni_{60}Fe_{20}P_{20}$ Ribbons

	Amorphous	Metastable microcrystalline	Stable crystalline
Relative conversion of CO at 260°C	210	20	1
Selectivity to C_2H_4, C_2H_6, C_3H_6, C_3H_8 (%)	38	19	13

Source: Ref. 116.

The catalytic activity and selectivity of a number of Pd-, Fe-, and Zr-based alloys in their amorphous and crystalline states were studied in the hydrogenation of 1-hexene and phenylethylene, dehydrogenation of methylcyclohexane, and dehydrocyclization of *n*-heptane. Most alloys were more active in the amorphous state than in the crystalline state, although the selectivities were found to be almost identical in both the states [125].

C. Hydrogen Storage Applications

Metal hydrides find applications in hydrogen storage systems, hydrogen pumps, and heat pumps. Traditional crystalline materials used for hydrogen storage are intermetallic compounds (e.g., $LaNi_5$), which are brittle and easily disintegrate into very small particles during hydrogen sorption cycles. However, RS unhydrided alloys are strong and ductile and thus are not likely to disintegrate. In an early investigation, it was shown [126] that under similar conditions of temperature and pressure, amorphous alloys have greater hydrogen absorption capacity than their crystalline counterparts. Subsequent work [127] has shown that no plateau exists in the pressure-composition isotherms of amorphous Ni-Zr alloys, in contrast with the crystalline counterpart, indicating that no hydride coexists with a saturated solid solution of hydrogen. It has also been shown [128] that amorphous TiCu absorbs 35% more hydrogen than crystalline TiCu and that even though the maximum hydrogen absorption capacity is determined by the electronic structure of the compound, the crystal structure, i.e., the type and size of interstitial sites in the lattice, may not always permit maximum absorption to take place readily. A critical comparison of the hydrogen sorption behavior in the crystalline and amorphous alloys has recently been presented [129]. Amorphous alloys offer a larger capacity, wide composition range, and higher ductility than crystalline materials for hydrogen storage. Potential applications include fuel for vehicles, heat pumps, and air conditioners [130].

IV. ELECTRICAL APPLICATIONS

RS alloys show high electrical resistivity, a small and often negative temperature coefficient of resistivity, superconductivity in some special cases, and resistance to irradiation damage [131]. The combination of these properties is even more attractive for the amorphous alloys and thus they can compete well with crystalline resistance alloys such as Nichrome. The potential applications of RS alloys based on their electrical properties are listed in Table 7.

Table 7 Potential Applications of RS Alloys Based on Electrical Properties

Precision resistance	Superconductors
Heating elements (low temperature)	Flashbulbs
Low-temperature thermometers	Delay lines
Radiation resistance applications	Sensors
Electron sources	Timing devices
Electrical fuses	

The high electrical resistivity of RS alloys could lead to applications as heating element materials (instead of alloys such as Nichrome or constantan) in products requiring low-temperature heating elements (i.e., TVs, radios, refrigerators, toasters, etc.) [8]. An amorphous alloy resistance thermometer was developed that can measure temperatures from 1.5 to 300 K [132]. The thermometer, made up of the RS $Cr_7Pd_{73}Si_{20}$ amorphous alloy, has the special feature that its sensitivity increases with decreasing temperature. It has a thermo emf of 48 μV/K at 20 K and 60 μV/K at 10 K with a 2-mA current. This can substitute platinum-base thermometers which lose their sensitivity at low temperatures. The amorphous alloy $V_1Ti_{49}Be_{40}Zr_{10}$, which possesses the most negative temperature coefficient of resistivity, is even more attractive for this purpose [133]. The temperature coefficient of resistivity can be altered by changing the alloy chemistry and can also be made zero, when they will be useful as resistance standards. It has also been suggested that such amorphous alloys can be used as magnetoresistance sensors for detecting the presence of magnetic fields, particularly in bubble memory devices [134]. (Variations in a magnetic field give rise to small changes in resistivity. If the resistivity also changes due to temperature, the signals of interest will be obscured. Thus, amorphous alloys with a zero temperature coefficient of resistivity are ideal for this application). Poor workability characteristics have limited the chromium and aluminum levels to 15 and 16 at%, respectively, in conventional wrought resistor alloys based on the Fe-Cr-Al system. Higher resistivity with zero temperature coefficient has been obtained in RS microcrystalline alloys with as much as 30 at% Cr and 15 at% Al [135], which also possess high strength and good ductility. Other potential applications include electron sources in electron-optical equipment, because of uniformity of field emission [136], and electrical fuses, a result of the highly reproducible glass transition temperature where rapid softening leads to circuit interruption [137]. Production of glass-sheathed small-diameter copper conductors for electrical circuits and composites [138], dispersion of hard second-phase particles of low solubility in high conductivity metals to increase their strength without impairing their conductivity [83,139], and incorporation of 8 vol% melt-extracted aluminum fibers into polymers to impart sufficient conductivity for them to carry house currents and to render them effective as electromagnetic shielding materials [140] are other possible applications.

A further potential application is in superconductor use (although the recent developments of high-temperature superconductivity may have reduced the chances of this opportunity materializing), in which the essential requirements are high values of the critical temperature, critical magnetic field, and critical current density. It has been shown that partial crystallization of amorphous alloys produces a fine dispersion of crystalline particles in an amorphous alloy matrix which can act as flux pinning centers and improve the critical current density and transition temperature, while retaining good ductility [141]. The good resistance of amorphous alloys to irradiation could lead to superconductor applications in fusion technology [142], where conventional superconductors exhibit a rapid decline in critical temperature and can also change from crystalline to the amorphous state.

Other applications based on the electrical properties of RS alloys include timing devices (mechanical vibrator) for which Pd-Si and Ni-Si-B amorphous alloys exhibit desirable nonferromagnetic behavior, and a Young's modulus which can be controlled to almost a zero temperature coefficient [143], delay lines which make use of the delay between sonic and electromagnetic waves [143], and flash bulbs, where attributes include shreddability and explosion combustion behavior [144].

V. CONCLUSIONS

A number of potential applications of RS alloys based on their mechanical, chemical, and electrical behavior have been reviewed. These have resulted due to the concerted and worldwide efforts to utilize the excellent combination of high strength, both at room and elevated temperatures, improved stiffness and high corrosion resistance of these microcrystalline and amorphous alloys. Innovative consolidation techniques to retain the attributes of RS and the ability to form near-net shapes through the PM route have enhanced the potential further.

In spite of these tremendous advantages, large-scale applications for the RS products have been slow in developing due to several reasons discussed earlier. Application of RS processing to conventional alloys has yielded only limited improvement in properties and so the strategy should now include novel compositions and the ability to produce materials at lower overall cost than can be achieved with conventional processing. There is little doubt that the RS technology will continue to make a significant impact and will be a major part of the material technology of the future.

VI. ACKNOWLEDGMENT

This review was prepared when one of the authors (C.S.) held a National Research Council–U.S.Air Force Materials Laboratory Senior Research Associateship.

REFERENCES

1. T. R. Anantharaman and C. Suryanarayana, in *Rapidly Solidified Metals: A Technological Overview* (Trans Tech Publications, Switzerland, 1987).
2. R. H. Willens, E. Buehler, and E. A. Nesbitt, *Rev. Sci. Instrum.*, *39*, 194 (1968).
3. J. I. Goldstein, F. J. Majeske, and H. Yakowitz, in *Advances in X-ray Analysis*, edited by J. B. Newkirk and G. P. Mallett (Plenum, New York, 1967), *10*, p. 431.
4. R. C. Dorward, *J. Nucl. Mater.*, *27*, 235 (1968).
5. C. Jansen, B. C. Giessen, and N. J. Grant, *J. Metals*, *20*(12), 10 (1968).
6. G. Beghi, R. Matera, and G. Piatti, *J. Nucl. Mater.*, *26*, 219 (1968).
7. C. F. Dixon and H. M. Skelly, *Int. J. Powder Metall.*, *1*, 28 (1965).
8. J. J. Gilman, *Phys. Today*, *28*(5), 46 (1975).
9. J. J. Gilman, in *Industrial Materials Science and Engineering*, edited by L. E. Murr (Marcel Dekker, New York, 1984), p. 1.
10. J. R. Pickens, *J. Mater. Sci.*, *16*, 1437 (1981).
11. T. E. Tietz and I. G. Palmer, in *Advances in Powder Technology*, edited by G. Y. Chin (ASM, Metals Park, OH, 1981), p. 189.
12. *Glassy Metals I*, edited by H. J. Guntherodt and H. Beck (Springer-Verlag, Berlin, 1981).
13. *Ultrarapid Quenching of Liquid Alloys*, edited by H. Herman, Vol. 20 of Treatise on Materials Science and Technology (Academic, New York, 1981).
14. H. Jones, *Rapid Solidification of Metals and Alloys*, Monograph No. 8 (Institution of Metallurgists, London, 1982).

15. *Amorphous Metallic Alloys*, edited by F. E. Luborsky (Butterworths, London, 1983).
16. *Glassy Metals II*, edited by H. Beck and H. J. Guntherodt (Springer-Verlag, Berlin, 1983).
17. *Metallic Glasses: Production, Properties and Applications*, edited by T. R. Anantharaman (Trans Tech Publications, Switzerland, 1984).
18. T. R. Anantharaman and C. Suryanarayana, *J. Mater. Sci.*, *6*, 1111 (1971).
19. H. Jones, *Rep. Prog. Phys.*, *36*, 1425 (1973).
20. P. Duwez, *Ann. Rev. Mater. Sci.*, *6*, 83 (1976).
21. H. S. Chen, *Rep. Prog. Phys.*, *43*, 353 (1980).
22. L. A. Davis and R. Hasegawa, in *Metallurgical Treatises*, edited by J. K. Tien and J. F. Elliott (TMS-AIME, Warrendale, PA, 1981), p. 301.
23. R. W. Cahn, in *Physical Metallurgy*, 3rd ed., edited by R. W. Cahn and P. Haasen (Elsevier, Amsterdam, 1983), p. 1779.
24. H. Jones, *J. Mater. Sci.*, *19*, 1043 (1984).
25. J. J. Gilman, *Phil. Trans. Roy. Soc. London*, *A322*, 425 (1987).
26. F. H. Froes, W. E. Quist, and S. K. Das, *Metal Powder Rep.*, *43*, 392 (1988).
27. F. H. Froes and R. S. Carbonara, *J. Metals*, *40*(2), 20 (1988).
28. *Metallic Glasses*, edited by J. J. Gilman and H. J. Leamy (ASM, Metals Park, OH, 1978).
29. *Rapidly Quenched Metals*, edited by N. J. Grant and B. C. Giessen (M.I.T. Press, Cambridge, MA, 1976).
30. *Rapidly Quenched Metals III*, edited by B. Cantor (The Metals Society, London, 1978).
31. *Rapidly Quenched Metals IV*, edited by T. Masumoto and K. Suzuki (The Japan Inst. Metals, Sendai, 1982).
32. *Rapidly Quenched Metals V*, edited by S. Steeb and H. Warlimont (Elsevier, Amsterdam, 1985).
33. *Rapidly Quenched Metals 6*, edited by R. W. Cochrane and J. O. Strom-Olsen (Elsevier, London and New York, 1988).
34. *Rapidly Solidified Materials*, edited by P. W. Lee and R. S. Carbonara (ASM, Metals Park, OH, 1986).
35. *Rapidly Solidified Materials: Properties and Processing*, edited by P. W. Lee and J. H. Moll (ASM, Metals Park, OH, 1988).
36. *Processing of Structural Metals by Rapid Solidification*, edited by F. H. Froes and S. J. Savage (ASM, Metals Park, OH, 1987).
37. *Titanium: Rapid Solidification Technology*, edited by F. H. Froes and D. Eylon (TMS, Warrendale, PA, 1986).
38. *Rapid Solidification Processing: Principles and Technologies*, edited by R. Mehrabian, B. H. Kear, and M. Cohen (Claitor's, Baton Rouge, LA, 1978).
39. *Rapid Solidification Processing: Principles and Technologies II*, edited by R. Mehrabian, B. H. Kear, and M. Cohen (Claitor's, Baton Rouge, LA, 1980).
40. *Rapid Solidification Processing: Principles and Technologies III*, edited by R. Mehrabian (National Bureau of Standards, Gaithersburg, MD, 1983).
41. *Rapid Solidification Processing: Principles and Technologies IV*, edited by R. Mehrabian and P. A. Parrish (Claitor's, Baton Rouge, LA, 1988).
42. *Rapidly Solidified Amorphous and Crystalline Alloys*, edited by B. H. Kear, B. C. Giessen, and M. Cohen (Elsevier, New York, 1982).
43. *Rapidly Solidified Metastable Materials*, edited by B. H. Kear and B. C. Giessen (Elsevier, New York, 1984).
44. *Rapidly Solidified Alloys and Their Mechanical and Magnetic Properties*, edited by B. C. Giessen, D. E. Polk, and A. I. Taub (Materials Research Society, Pittsburgh, PA, 1986).

45. *Rapidly Solidified Crystalline Alloys*, edited by S. K. Das, B. H. Kear, and C. M. Adam (TMS, Warrendale, PA, 1985).
46. *Mechanical Behavior of Rapidly Solidified Materials*, edited by S. M. L. Sastry and B. A. MacDonald (TMS, Warrendale, PA, 1986).
47. A. I. Taub, in Ref. 32, p. 1611.
48. M. Hagiwara, A. Inoue, and T. Masumoto, in Ref. 32, p. 1779.
49. R. Ray, *J. Mater. Sci.*, *16*, 2924, 2927 (1981); *Mater. Sci. Eng.*, *52*, 85 (1982); *Metal Prog.*, *121*(7), 29 (1982); *Int. J. Powder Metall.*, *18*, 209 (1982).
50. Y. T. Yeow, *J. Comp. Mater.*, *Suppl. 14* (1980).
51. A. S. Argon, G. W. Hawkins, and H. Y. Kuo, *J. Mater. Sci.*, *14*, 1707 (1979).
52. A. Fels, E. Hornbogen, and K. Friedrich, *J. Mater. Sci. Lett.*, *3*, 639 (1984).
53. W. M. S. B. W. Kadir, C. Hayzelden, and B. Cantor, *J. Mater. Sci.*, *15*, 2663 (1980).
54. S. Y. Cytron, *J. Mater. Sci. Lett.*, *1*, 211 (1982).
55. A. L. Holbrook, in *Progress in Powder Metallurgy*, edited by H. I. Sanderow, W. L. Giebelhausen, and K. M. Kulkarni (Metal Powder Industries Federation, Princeton, NJ, 1985) *41*, 679.
56. R. M. Simon, in Ref. 55, p. 87.
57. Sun-Gard Reflective Roof System Brochure, Transmet Corp., Columbus, OH.
58. UV 2000 Aluminized Roofing Brochure, Transmet Corp., Columbus, OH.
59. R. B. Pond, R. E. Maringer, and C. E. Mobley, in *New Trends in Materials Processing* (ASM, Metals Park, OH, 1976), p. 128.
60. L. A. Davis, N. J. DeCristofaro, and C. H. Smith, in *Metallic Glasses: Science and Technology*, Vol. II, edited by C. Hargitai, I. Bakonyi, and T. Kemeny (Central Res. Inst. Phys., Budapest, Hungary, 1981), p. 1.
61. R. S. Carbonara et al., Battelle MCIC Report 81–45 (1981).
62. C. Suryanarayana, F. H. Froes, S. Krishnamurthy, and Y-W. Kim, *Int. J. Powder Metall.*, *26*, 117 (1990).
63. C. M. Adam and R. E. Lewis, in Ref. 45, p. 157.
64. E. Lavernia, B. Poggiali, I. Servi, J. Clark, F. Katrak, and N. J. Grant, *J. Metals*, *37*(11), 38 (1985).
65. J. P. Lyle and W. S. Cebulak, *Metall. Trans. A*, *6A*, 685 (1985).
66. S. Ashley, *Amer. Metal Market Metal Working News*, *90*(100), 9 (1982).
67. P. R. Bridenbaugh, W. S. Cebulak, F. R. Billman, and G. H. Hildeman, *Light Metal Age*, *43*, 18 (1985).
68. G. J. Hildeman, L. C. Labarre, A. Hafeez, and L. M. Angers, in *High Strength Powder Metallurgy Aluminum Alloys*, edited by G. J. Hildeman and M. J. Koczak (TMS-AIME, Warrendale, PA, 1986), p. 25.
69. S. K. Das and L. A. Davis, *Mater. Sci. Eng.*, *98*, 1 (1988).
70. Y-W. Kim, in *Dispersion-Strengthened Aluminum Alloys*, edited by Y-W. Kim and W. M. Griffith (TMS, Warrendale, PA, 1988), p. 157.
71. P. S. Gilman, M. S. Zedalis, J. M. Peltier, and S. K. Das, *Indust. Heat.*, *56*(2), 30 (1989).
72. F. H. Froes, Y-W. Kim, and F. Hehmann, *J. Metals*, *39*(8), 14 (1987).
73. B. L. Koff, *Aviation Week Space Technol.*, *123*, 45 (April 22, 1985).
74. P. P. Millan, *J. Metals*, *35*(3), 76 (1983).
75. I. F. Sakato and S. L. Langenbeck, in *Proc. SAE Aerospace Conf.* (Long Beach, CA, 1983), p. 1.
76. W. H. Kool, A. Malinowski, H. Kleinjan, and F. J. Kievits, in Ref. 34, p. 383.
77. L. Katgerman, H. Kleinjan, R. W. E. Kropf, and W. G. Zalm, in *Proc. 2nd Int. Symp. Materials and Energy from Refuse*, edited by A. Buekens (KVIV, Antwerp, 1981), p. 57.
78. C. Suryanarayana, F. H. Froes, and R. G. Rowe, *Int. Mater. Rev.*, *36*, 85 (1991).

79. T. F. Broderick, A. G. Jackson, H. Jones, and F. H. Froes, *Metall. Trans. A*, *16A*, 1951 (1985).
80. R. S. Busk and T. E. Leontis, *Trans. AIME*, *188*, 297 (1950).
81. S. K. Das and C. F. Chang, in Ref. 45, p. 137.
82. I. E. Anderson and B. B. Rath, in Ref. 45, p. 219.
83. V. K. Sarin and N. J. Grant, *Metall. Trans.*, *3*, 875 (1972); *Powder Metall. Int.*, *11*, 153 (1979).
84. R. J. Patterson II, A. R. Cox, and E. C. van Reuth, *J. Metals*, *32*(9), 34 (1980).
85. R. J. Patterson II, A. R. Cox, T. D. Tillman, and E. C. van Reuth, in Ref. 39, p. 416.
86. A. Kasak and E. J. Dulis, *Powder Metall.*, *21*, 114 (1978).
87. C. Wick, *Manuf. Eng.*, *85*(3), 52 (1980).
88. H. Takigawa, H. Manto, N. Kawaii, and K. Homma, *Powder Metall.*, *24*, 196 (1981).
89. J. J. Rayment and B. Cantor, *Metall. Trans. A*, *12A*, 155 (1981).
90. R. Ray, V. Panchanathan, and S. Isserow, *J. Metals*, *35*(6), 30 (1983).
91. D. Raybould, *Metal Powder Rep.*, *39*, 282 (1984).
92. Anon, *Precision Metal*, *40*(3), 19 (1982).
93. F. E. Katrak, I. S. Servi, J. R. H. Black, P. Ammann, and N. J. Grant, in Ref. 55, p. 109.
94. J. C. Bittence, *Mater. Eng.*, *95*(4), 57 (1982).
95. N. Amano, Y. Odani, Y. Takeda, and K. Akechi, *Metal Powder Rep.*, *44*, 186 (1989).
96. M. Naka, K. Hashimoto, and T. Masumoto, *J. Jpn. Inst. Metals*, *38*, 835 (1974).
97. Y. Waseda and K. T. Aust, *J. Mater. Sci.*, *16*, 2337 (1981).
98. K. Hashimoto, in Ref. 32, p. 1449.
99. M. D. Archer, C. C. Corke, and B. H. Harji, *Electrochim. Acta*, *32*, 13 (1987).
100. K. Asami, A. Kawashima, and K. Hashimoto, *Mater. Sci. Eng.*, *99*, 475 (1988).
101. R. Ray and T. A. Mozhi, *Metal Powder Rep.*, *44*, 129 (1989).
102. R. Ray, *Metal Powder Rep.*, *39*, 287 (1984).
103. G. J. Yurek, D. Eisen, and A. Garratt-Reed, *Metall. Trans. A*, *13A*, 473 (1982).
104. T. Tsuru, S. X. Zhang, and R. M. Latanision, in Ref. 31, p. 1437.
105. E. R. Slaughter and S. K. Das, in Ref. 39, p. 354.
106. T. Masumoto and K. Hashimoto, *J. Phys.*, *41* (*Suppl. 8*), C8-894 (1980).
107. A. Kawashima, in Ref. 32, p. 1671.
108. L. B. Johnson, H. Otani, M. J. Neary, J. W. Hauck, T. M. Heaven and R. J. Regenitter, *J. Biomech. Mater. Res.*, *8*, 3 (1974).
109. J. V. Wood and S. C. King, *J. Mater. Sci.*, *13*, 1119 (1978).
110. G. Faninger, D. Merz, and H. Winter, in Ref. 29, p. 483.
111. L. Collins, in Ref. 34, p. 435.
112. T. R. Anthony and H. E. Cline, *J. Appl. Phys.*, *49*, 1248 (1980).
113. C. W. Draper, R. E. Woods, and L. S. Meyer, *Corrosion*, *36*, 405 (1980).
114. H. Yoshioka, K. Asami, A. Kawashima, and K. Hashimoto, *Corrosion Sci.*, *27*, 981 (1987).
115. D. L. Cocke, *J. Metals*, *38*(2), 70 (1986).
116. A. Yokoyama, H. Komiyama, H. Inoue, T. Masumoto, and H. M. Kimura, *J. Catal.*, *68*, 355 (1981); *Scripta Metall.*, *15*, 365 (1981); in *Structure and Properties of Amorphous Metals II*, edited by T. Masumoto and T. Imura, Suppl. to Sci. Rep. Res. Inst. Tohoku Univ., *A28*, 217 (1980).
117. A. Kawashima and K. Hashimoto, in Ref. 31, p. 1427.
118. A. Yokoyama, H. Komiyama, H. Inoue, T. Masumoto, and H. M. Kimura, *J. Non-Cryst. Solids*, *61-62*, 619 (1984).
119. M. Enyo et al., *Electrochim. Acta*, *28*, 1573 (1983).

120. H. Yamashita, T. Kaminade, T. Funabiki, and S. Yoshida, *J. Mater. Sci. Lett.*, *4*, 1241 (1985).
121. W. E. Brower, M. S. Matyjaszczyk, T. L. Pettit, and G. V. Smith, *Nature*, *301*, 497 (1983).
122. M. Hara, K. Hashimoto, and T. Masumoto, in Ref. 31, p. 1423.
123. N. Kumagai, Y. Samata, A. Kawashima, K. Asami, and K. Hashimoto, *J. Appl. Electrochem.*, *17*, 347 (1987).
124. C. S. Brooks, F. D. Lemkey, and G. S. Golden, in Ref. 42, p. 397.
125. S. S. Mahmoud, D. A. Forsyth, and B. C. Giessen, in Ref. 44, p. 131.
126. A. J. Maeland, in *Hydrides for Energy Storage*, edited by A. F. Andresen and A. J. Maeland (Pergamon, Oxford, UK, 1978), p. 447.
127. F. H. M. Spit, J. W. Drijver, and S. Radelaar, *Scripta Metall.*, *14*, 1071 (1980).
128. A. J. Maeland, L. E. Tanner, and G. G. Libowitz, *J. Less-Common Metals*, *74*, 279 (1980).
129. A. J. Maeland, in Ref. 32, p. 1507.
130. F. H. M. Spit, K. Blok, E. Hendriks, G. Winkels, W. Turkenberg, J. W. Drijver, and S. Radelaar, in Ref. 31, p. 1635.
131. C. Suryanarayana, in Ref. 17, p. 249.
132. C. C. Tsuei and P. Duwez, *J. Phys. E: Sci. Instrum.*, *4*, 466 (1971).
133. R. Hasegawa and L. E. Tanner, *J. Appl. Phys.*, *48*, 3211 (1977).
134. P. Chaudhari and D. Turnbull, *Science*, *199*, 11 (1978).
135. T. Naohara, A. Inoue, T. Masumoto, and K. Kumada, *Metall. Trans. A*, *13A*, 337 (1982).
136. H. Heinrich, T. Haag, and G. Geiger, *J. Phys. D: Appl. Phys.*, *11*, 2439 (1978).
137. H. Warlimont, *Phys. Technol.*, *11*, 28 (1980).
138. G. Manfre, G. Servi, and C. Ruffino, *J. Mater. Sci.*, *9*, 74 (1974).
139. H. C. Chia and A. Starke, *Wire J.*, *12*(6), 66 (1979).
140. R. E. Maringer and C. E. Mobley, in Ref. 30, Vol. I, p. 49.
141. A. Inoue, A. Hoshi, C. Suryanarayana, and T. Masumoto, *Scripta Metall.*, *14*, 1077 (1980); *J. Appl. Phys.*, *52*, 4711 (1981).
142. W. L. Johnson, in Ref. 12, p. 191.
143. M. Kikuchi, *Sci. Rep. Res. Inst. Tohoku Univ.*, *A31*, 99 (1983).
144. H. C. M. Van der Nieuwenhuizen and K. H. J. Buschow, *Humana*, 99 (1982).

23

Rapid Solidification: Origins, Present, and Future

H. Jones
University of Sheffield, Sheffield, England

I. INTRODUCTION

Attempting to chart the future of any subject involves a combination of judgments concerning the likely course of present overt and underlying trends within the subject area and the possible influence of relevant external factors. These range from the likely strength of competition for development from other subject areas to the prevailing influence of economic and political considerations. Thus any predictions made have large associated elements of speculation and uncertainty. They are necessary, nevertheless, as one element in helping to select possible future pathways for development of a subject.

Rapid solidification has a much longer history as a technology than it has as a science. Shot-making by allowing molten material to spherodize and solidify during free-fall into a collector dates back at least to the 17th century [1] and the product evidently had immediate applications which persist to this day [2]. This was the forerunner of subsequent developments of atomization [3] and spray deposition [4] of melts into rapidly solidified powders and deposited preforms, currently commanding wide interest. Thin strip [5], filament [6], and ribbon [7] casting technology dates back to the 19th century, only to be rediscovered in the 1950s, 1960s and 1970s and again is the object of worldwide current interest and development. Rapid melting and solidification at a surface, as a development of fusion-welding technology, again dates back to the 19th century [8], though the use of electron, laser, or plasma beams to ensure the highest solidification front velocities (up to tens of meters per second) is a relatively recent phenomenon.

The scientific potential of rapid solidification remained more or less hidden until Pol Duwez reported in 1960 [9,10] his seminal discoveries of complete solid solubility extension, production of a new metastable crystalline intermediate phase and formation of a metallic glass in the Ag-Cu, Ag-Ge, and Au-Si systems, respectively. The rapid solidification experiments which led to these discoveries were motivated [11] by curiosity concerning perceived exceptions to the Hume-Rothery rules for phase formation in these systems. The results simultaneously fulfilled and (in the case of the metallic glass) exceeded expectation and launched a new field of scientific endeavour in metallurgy and materials science. It is notable that the initial emphasis on the rapid *quenching* from the melt induced by the gun and derived techniques initiated by Duwez (which interestingly is retained by R. W. Cahn in his excellent introductory chapter to this volume) persisted in literature published on this topic throughout the 1960s [12]. This term connotates the *external* conditions to which the melt is subjected during solidification under such conditions. The term "rapid solidification," in contrast, refers to the *internal* conditions applicable, which were almost literally a closed book until quite recently. With characteristic foresight, Pol Duwez, first featured the term "rapid solidification" in the title of a short early review article eventually published in 1965 [13]. It was used correspondingly by Esslinger [14] in the titles of two papers published in 1966, by Beghi et al [15], by Kumar and Sinha [16], and by Suryanarayana and Anantharaman [17] in 1968, and by Pond and Maddin [18], by Baker and (J. W.) Cahn [19], and by the present author [20] in 1969. These are the only such instances on record during the 1960s amid a plethora of references to rapid cooling or quenching.

The term "rapid solidification" does not, in principle, exclude techniques of rapidly forming nonequilibrium solid starting other than from the molten state. Thus condensation from vapour, deposition from plasmas, and formation by chemical reaction, implantation, or interdiffusion are not excluded. Rapid solidification processing certainly includes all steps subsequent to the initial "solidification" step that influence the product of that critical step, i.e., consolidation, heat treatment, etc., as well as the structure and properties of the product at all stages of processing. This perceived breadth of scope of the subject is particularly important when considering prospects for further development. This is done in Sections II and III. This enables the place of rapid solidification in the future of materials science and engineering to be assessed (Section IV).

II. THE FUTURE OF RAPID SOLIDIFICATION PROCESSING

A. Atomization and Spray Forming

The gun technique introduced by Duwez [9,10,21] was subsequently shown [22] to generate a cloud of minute droplets (mean size ~1 μm) and propel them toward the chill substrate at high velocity (~Mach 1) where they spread and coalesce rapidly to form an unevenly thin splat. The whole process was complete in a few milliseconds and the product contained a very wide range of nonequilibrium microstructures dependent on the local conditions of rapid solidification as well as any subsequent thermal excursions associated with continuing deposition events. The small average spray droplet size produced in this technique tends to isolate nucleant particles and so promotes high undercooling prior to solidification. The high velocity of travel to the substrate also promotes high undercooling in the arriving droplets by maximizing rate of heat extraction from the droplets by the surrounding gas. This high velocity also promotes maximum spreading [23] of the droplets on impact with the chill substrate which ensures that solidification times will be minimized by the combination of a short distance of advance of the solidification front to complete solidification with even more effective heat extraction by the chill substrate.

Many recent developments in melt atomization as a route to RS have been directed towards enhancing heat transfer (use of water or helium rather than air, nitrogen, or argon) and reducing average droplet size from >100 μm to ~10 μm [24], giving notable increases in proportions of derived nonequilibrium microstructure in resulting powder [25,26]. Such measures can, however, result in an increased content of surface films, for example, oxides so that a compromise may need to be reached in each particular situation. The only processes that appear to be able to produce high yields of submicron-sized particulate (to match the Duwez gun technique) are electrohydrodynamic atomization [27] and spark machining [28]. Rates of production are, however, so small as to limit these techniques, for the majority of purposes, to application as research tools only and they have already been shown to be highly effective in this regard.

The burgeoning current interest in melt spray deposition [29] as a means of generating nearly fully dense macrosegregation-free nondendritic microstructures in complex alloys [30] with or without the introduction of ceramic dispersoids [31], is stimulating a substantial research effort on both spray [32] and deposit [33,34] characteristics and their dependence on processing variables. The incident melt spray is considered [35] to comprise a mixture of superheated, rapidly solidified and melt-undercooled droplets at the instant of deposition (as surely was the case, on a smaller scale and over a shorter time interval, in the original gun technique). Optimization of the spray constitution at the instant of deposition, as well as the rate of deposition in relation to the substrate conditions, is crucial for successful running of this process on an industrial scale and presents a considerable challenge both to the process modeler and to the process controller.

B. Strip, Filament, and Ribbon Casting

The chill-block melt-spinning process in its standard form [36] using RF melting in quartz and pressurized expulsion through a cylindrical nozzle, to give melt streaming and ribbon formation (~50 μm thick) on a rotating chill block, has become ubiquitous to research on rapidly solidified (RS) materials throughout the world. Introduction of slot-shaped [37] nozzle(s) in closer proximity to the chill block in order to produce wide thin strip for magnetic or joining applications, or for pulverization [38] into a feedstock more suitable for subsequent consolidation, presents a notable challenge for process control. There have been a number of studies recently [39–45] to provide experimental data that should be used as the basis of realistic process models. Understanding of the role of factors such as wheel wettability and thermomechanical history following ribbon release, is still lacking and provides ample scope for further basic research on the process. The availability of large-scale production facilities [46–48] based on this technology augurs well for the future of RS materials processed by this route.

The corresponding wire-casting process which, in its most recent form, replaces the rotation chill block with a rotating bath of liquid quenchant [48,49], is at a relatively early stage of development. Just as the chill-block technology can be adapted to generate a discontinuous product, such as flake, the wire casting technology can be employed to generate powder rather than continuous filament. The process, in fact, appears to have a relatively narrow operating window for production of continuous filament [51] and further development may be necessary to match its capability with actual requirements for certain applications.

Nozzle blockage is a long-recognized operational hazard in both conventional melt spinning and wire casting. Melt extraction [52], which uses rotating chill disks to generate continuous or discontinuous filament from the surface of a pendant or contained melt, avoids flow of melt through a nozzle as a critical feature and has been successful commercially for a number of years now [53,54]. Its variant for continuous thin strip production, melt overflow [55,56] is now under active development as a technology complementary to chill-block melt-spinning/planar flow casting. Results of any systematic study of these processes have yet to be reported.

The casting of strip or filament between two moving chill surfaces, such as twin rollers [57] (after Bessemer) or a wheel and belt [58] (after Lyman) has received relatively little attention from the rapid solidification community, presumably because processes involving a single moving substrate present fewer operational problems. Requirements for a good, or at least matched, surface finishes on *both* surfaces of the product is changing this position. The need [59] to be able to produce in tonnage quantities good-quality steel strip over a wide range of alloy composition in thicknesses down to 1 mm is a powerful spur to solving these operational problems. There is already [60–62] a substantial research effort worldwide on this aspect of RS technology which will undoubtedly impact on related processing options.

C. Surface-Melt Processing

The use of a directed energy source, such as an electrical discharge, electron beam, laser beam, or plasma beam to achieve rapid incremental melting in contact with an effective heat sink, followed by localized rapid solidification when the power source is switched off or moved away, received only scant attention as a rapid solidification technique prior to the 1980s. Much of the stimulus came from the application of laser treatment to redistribute dopants injected into silicon by ion implantation [63,64]. The rapid solidification that follows this remelting treatment results in large nonequilibrium extensions of solid solubility and even elemental amorphous silicon can result at front velocities estimated to approach or exceed 20 m/s [65,66]. A seminal paper by Boettinger et al. [67] demonstrated convincingly the use of a scanning electron beam to produce rapid solidification in Ag-Cu alloys at predetermined solidification front velocities as high as 4 m/s, enabling the kinetic conditions for extension of solid solubility in this classic system to be quantified for the first time. The coupling between beam-scanning velocity and solidification front velocity in such a technique provides the only available method of achieving quasi-steady-state rapid solidification at front velocities exceeding 10–50 mm/s and is already proving to be crucial in testing the predictions [68–73] of theories of growth under such conditions. Practical applications include: *surface remelting* [74] to reconstitute, homogenize or refine the surface so as to secure or regenerate special properties at the surface; *surface alloying* [75] in which preplaced or injected alloying additions are incorporated into the surface to produce required properties there; *particle injection* [76] in which a dispersion of second phase is incorporated (without being remelted) into the remelt zone at the surface; and *surface cladding* [77] in which a metal or alloy is fused on to the underlying surface with a minimum of dilution by the underlying material. The most extreme conditions attained in this class (possible of all) of RS techniques have been achieved by remelting of an assembly of alternate elemental layers (compositionally modulated films) totaling less than 1 μm in thickness on the surface. Cooling rates during solidification have been estimated to reach 10^{10} [78,79] and 10^{13} [80] K/s for nanosecond and picosecond pulses respectively under such conditions resulting in, for example, the extension [78–81] of glass-forming ranges determined previously for CBMS conditions. The unique power of this class of techniques to generate extreme products of rapid solidification reproducibly under known and controlled conditions combined with its capability to produce surface regions of controlled thickness, composition, structure, and properties ensure their future both as fundamental research tools and as part of the armory of precision production processes for many years to come.

III. THE FUTURE OF RAPIDLY SOLIDIFIED MATERIALS

Rapidly solidified materials are conveniently divided into two types: those with compositions already established in connection with existing process routes such as conventional mold casting, conventional powder metallurgy (blending, pressing, and sintering), or wrought ingot processing; and those with essentially new compositions that cannot readily be processed by existing alternative process routes. The former category includes PM tool steels [82], nickel-based superalloys [83], some 2000 [84] and 7000 [85] series high-strength aluminum alloys and Al-Si based alloys [86], some amorphous brazing alloy compositions [87], melt-extracted stainless steel reinforcing staples [52–54,88], strip-cast steels [59–62], Nd-Fe-B permanent magnet alloys [46,89], and aluminum fillers for electromagnetic shielding [90]. Interest in such RS materials stems mainly from their improved response to processing via the rapid solidification route, or subsequently, resulting in both cost savings and greater dependability in application. The latter category includes amorphous soft magnetic alloys for power and electronic applications [91], microcrystalline Fe–6 wt% Si alloys for soft magnetic applications [92], novel high-strength aluminum-transition metal-based alloys for high-temperature applications [93–95], improved die and wear-resistant materials with high volume fractions of submicron borides and carbides in a steel or superalloy matrix [96–98], and novel high-strength magnesium alloys with enhanced deformability or corrosion resistance [99].

Issues to be addressed in order to ensure more widespread application of both types of material include:

1. Control of melt delivery, contact conditions, and surface quality in thin strip and sheet production for direct application
2. Clean conditions during the preparation and pretreatment of RS particulate for consolidation into fully dense products
3. Minimization of temperature excursions during such processing that lead to undesirable changes in microstructure

4. Achievement of sufficient interparticle shear to ensure effective metal-to-metal bonding across large sections of consolidated RS particulate
5. Maintainance of uniform conditions during production of spray-formed materials with and without injected ceramic particulates
6. Reductions in quantity of or establishment of viable applications for over-spray powder generated as a by-product of spray forming

In a status report published in 1987, Haour and Boswell [100] concluded that RS materials were tending to follow the normal pattern for commercialization of new materials. Products destined for commodity and consumer applications (brazing foils, reinforcing staples, aluminum fillers for EMI shielding, amorphous metals for instrument transformer cores and audio heads, PM tribological alloys, Nd-Fe-B permanent alloys) were undergoing rapid commercialisation and market penetration. Industrially related products were starting to reach commercial maturity, while products intended for energy-related and aerospace sectors (high-strength aluminum alloys, Al-Si cylinder liners, amorphous metals for power distribution transformer cores, P/M nickel-based superalloys) were only just beginning to be commercialized, but, once introduced, could be expected to remain in use for a lengthy period of time. A different study by Katrak et al. [101] reported in 1986, concluded that more rapid commercialization (of RS powders) would result from

1. Production and consolidation of high performance RS powders at high volumes (at least 1000 T/y per plant)
2. Production of relatively simple consolidated shapes (billets, disk, gear stock, tube and forging preforms)
3. Pricing of consolidated shapes at no more than about $1.50 per pound above raw material costs

They estimate that firm market sizes of $40–$350 million per year could result by the year 2000 for tool steels, stainless steels, superalloys and aluminum alloys, with speculative markets as large as $700 million per year, by adopting this strategy. A number of plants for making RS material already satisfy at least the production part of criterion (1) [46,48,102–105]. At least one preform facility satisfies (1) and (2) [104]. High-volume, low-cost products of rapid solidification are evidently already available and finding applications in the market place. Lower-volume, higher-cost applications will take longer to establish though a number of these are in place. Future growth will depend on a continued sustaining of effort on both fronts.

IV. CONCLUSIONS

Rapid solidification technology (RST) has developed, in the three decades since Duwez first drew attention to its potential, from a laboratory tool into a range of production processes several of which have reached an industrial scale of operation. It has been responsible for establishing two new classes of material, the metallic glasses and quasicrystalline solids, new processing routes for established materials and a variety of new high-performance materials. Priorities for future research activity include obtaining a better understanding of how operational variables affect the performance of rapid solidification processes and, how, in turn, these variables give rise to the characteristic microstructures and properties of the resulting material. It is significant that most current initiatives in materials science invariably recognise an implicit or explicit role for RST, whether it be continuous production of thin strip by twin roll casting, development of new magnetic materials, production and processing of intermetallic or metal matrix composites or formulation and processing of novel ceramics. The technology can be expected to remain part of the fabric of materials processing and to sustain many aspects of its development for a number of years to come.

REFERENCES

1. Robert Hooke, *Micrographia*, (1665) p. 22.
2. D. R. Blaskett and D. Boxall, in *Lead and its Alloys* (Ellis Horwood, Chichester, 1990), p. 96.
3. W. Marriott, British patent No. 3,322 (1873).
4. M. U. Schoop, British patent No. 5,712 (1910).
5. H. Bessemer, British patent No. 1,317 (1846).
6. E. Small, U.S. patent 262,625 (1882).
7. N. R. Lyman, U.S. patent 315,045 (1885).
8. de Meritens (1881) quoted by D. Romans and D. Simons, in *Welding Processes and Technology* (Pitman, London, 1968).
9. P. Duwez, R. H. Willens, and W. Klement, *J. Appl. Phys.*, *31*, 1136, 1137, 1500 (1960).
10. W. Klement, R. H. Willens, and P. Duwez, *Nature*, 187, 869 (1960).
11. P. Duwez, *Acta Tech. Belgica Metall.*, *15*, 157 (1975), and in *Glassy Metals I*, edited by H. J. Güntherodt and H. Beck (Springer-Verlag, Berlin, 1981), p. 19.
12. H. Jones and C. Suryanarayana, *J. Mater. Sci.*, *8*, 705–753 (1973).
13. P. Duwez, in *Energetics in Metallurgical Phenomena*, Vol. I, edited by W. M. Mueller (Gordon and Breach, New York, 1965), p. 420.
14. P. Esslinger, *Z. Metall.*, *57*, 12–19, 109–113 (1966).
15. G. Beghi, R. Matera, and G. Piatti, *La Metall. Ital*, *60*, 444–448 (1968).
16. R. Kumar and A. N. Sinha, *Trans. Indian Inst. Metall.*, *21*(1), 9–12 (1968).
17. C. Suryanarayana and T. R. Anantharaman, *Trans. Indian Inst. Metall.*, *21*(3), 67 (1968).
18. R. Pond and R. Maddin, *Trans. Metall. Soc. AIME*, *245*, 2475 (1969).
19. J. C. Baker and J. W. Cahn, *Acta Metall.*, *17*, 575 (1969).
20. H. Jones, *Mater. Sci. Eng.*, *5*, 1 (1969).
21. P. Duwez and R. H. Willens, *Trans. Metall. Soc. AIME*, *227*, 362 (1962).
22. P. Predecki, A. W. Mullendore, and N. J. Grant, *Trans. Metall. Soc. AIME*, *223*, 1581 (1965).
23. H. Jones, *J. Phys. D: Appl. Phys.*, *4*, 1657–60 (1971).
24. M. J. Couper and R. F. Singer, in *Rapidly Quenched Metals*, edited by S. Steeb and H. Warlimont (Elsevier, New York, 1985), p. 1737.
25. A. J. Drehmann and D. Turnbull, *Scripta Metall.*, *15*, 543 (1981).
26. C. G. Levi and R. Mehrabian, *Metall. Trans.*, *13A*, 13 (1982).
27. J. F. Mahoney, *Industr. Res. Dev.*, *24*, 150 (1982).
28. A. E. Berkowitz and J. L. Walter, *Mater. Sci. Eng.*, *55*, 275 (1982); J. L. Walter, *Powder Metall.*, *31*, 267 (1988).
29. A. R. E. Singer, *J. Inst. Metall.*, *100*, 185 (1972).
30. R. G. Brooks, A. G. Leatham, G. R. Dunstan, and C. Moore, in *Powder Metallurgy Superalloys: Aerospace Materials for the 1980s* (MPR, Shrewsbury, 1980), p. 8.
31. A. R. E. Singer and S. Ozbek, *Powder Metall.*, *28*, 72 (1985).

32. *Characterization and Diagnostics of Ceramic and Metal Particulate Processing*, edited by E. J. Lavernia, H. Henein, and I. Anderson (TMS-AIME, Warrendale, PA, 1989), p. 1.
33. D. Schaeffler, A. Lawley, and D. Apelian, in *1989 Advances in Powder Metallurgy*, Vol. II, edited by T. G. Gasbarre and W. F. Jandeska, Jr (MPIF/APMI, Princeton, NJ, 1989), p. 161.
34. F. Akhlaghi, J. Beech, and H. Jones, in *Proc P/M90*, Vol. III (The Institute of Metals, London, 1990), p. 85.
35. E. Guttierrez-Miravete, E. J. Lavernia, G. M. Trapaga, and J. Szekely, *Int. J. Rapid Solidification*, *4*, 125 (1988).
36. H. H. Liebermann and C. D. Graham, Jr, *IEEE Trans. Magn.* MAG-*12*(6), 921 (1976).
37. M. C. Narasimhan, U.S. patent 4,142,571 (1979).
38. S. Pelletier, C. Gelinas, and R. Angers, *Int. J. Powder Metall.*, *26*, 51 (1990).
39. D. H. Warrington, H. A. Davies, and N. Shohoji, in *Rapidly Quenched Metals*, Vol. I, edited by T. Masumoto and K. Suzuki (The Japan Institute of Metals, 1982), p. 69; M. J. Tenwick and H. A. Davies, in Ref. 24, Vol. I, p. 67.
40. A. G. Gillen and B. Cantor, *Acta Metall.*, *33*, 1813 (1985); B. P. Bewlay and B. Cantor, *Int. J. Rapid Solidification*, *2*, 107 (1986).
41. E. Vogt and G. Frommeyer, in Ref. 24, Vol. I, p. 63; in *Rapidly Solidified Materials*, edited by P. W. Lee and R. S. Carbonara (ASM-I, 1985), p. 291; E. Vogt, *Int. J. Rapid Solidification*, *3*, 131 (1987).
42. K. Takeshita and P. H. Shingu, *Trans. Japan Inst. Metals*, *27*, 454 (1986).
43. D. Pang, J. Wang, B. Ding, Q. Song, and Z. Yang, *Mater. Lett.*, *5*, 439 (1987).
44. G. Stephani, H. Mühlbach, H. Fiedler, and G. Richter, *Mater. Sci. Eng.*, *98*, 29 (1988).
45. P. Cremer and J. Bigot, *Mater. Sci. Eng.*, *98*, 95 (1988).
46. Anon, *Metal Powd. Rep.*, *42*, 448 (1987).
47. S. K. Das, *Int. J. Powd. Metall.*, *24*, 175 (1988).
48. Anon, *Adv. Mater. Proc.*, *135*(6), 12 (1989).
49. I. Ohnaka, T. Fukusako, and T. Ohmichi, *J. Japan Inst. Metals*, *45*, 751 (1981); I. Ohnaka, *Int. J. Rapid Solidification*, *1*, 219 (1984/5).
50. R. V. Raman, A. N. Patel, and R. S. Carbonara, *Prog. Powd. Metall.*, *38*, 99 (1982); *Met. Powd. Rep.*, *39*, 106 (1984).
51. J. Liu, L. Arnberg, N. Bäckström, and S. Savage, *Mater. Sci. Eng.*, *98*, 21 (1988).
52. R. E. Maringer and C. E. Mobley, *J. Vac. Sci. Technol*, *11*, 1067 (1974); U.S. patent 3,812,901 (1974).
53. J. Edgington, in *Fibre Reinforced Materials* (Institute of Civil Engrs., London, 1977), p. 129.
54. Anon, *J. Inst. Refract. Eng.*, *Winter*, 17 (1987); *Metal Soc. World*, *3*(11), 5 (1984).
55. T. Gaspar, L. E. Hackman, Y. Sahai, W. A. T. Clark, and J. V. Wood, in *Rapidly Solidified Alloys and their Mechanical and Magnetic Properties*, edited by B. C. Giessen, D. E. Polk, and A. I. Taub (Materials Research Society, Pittsburgh, PA, 1985), p. 23.
56. L. E. Hackman and T. Gaspar, *Indust. Heating*, *Jan.*, 36 (1986).
57. H. S. Chen and C. E. Miller, *Rev. Sci. Instrum.*, *41*, 1237 (1970).
58. M. Sakata and T. Ishibachi, in *Rapidly Quenched Metals*, edited by T. Masumoto and K. Suzuki (The Japan Institute of Metals, Sendai, 1982), p. 141.
59. M. Cygler and M. Wolf, *Iron Steel Maker*, *Aug.*, 27 (1986).
60. G. J. McManus, *Iron Age*, *Aug.*, 14 (1987).
61. *Casting of Near Net Shape Products*, edited by Y. Sahai, J. E. Battles, R. S. Carbonara, and C. E. Mobley (TMS, Warrendale, PA, 1988).
62. T. Yamauchi, T. Nakanori, M. Hasegawa, T. Yabuki, and N. Ohnishi, *Trans. ISI Japan*, *28*, 23 (1988).

63. C. W. White, J. Narayan, and R. T. Young, *Science*, *204*, 461 (1979).
64. R. T. Young and R. F. Ward, *Ann. Rev. Mater. Sci.*, *12*, 323 (1982).
65. J. M. Poate, in *Laser and Electron Beam Interactions with Solids*, edited by B. R. Appleton and G. K. Celler (Elsevier, New York, 1982), p. 125.
66. A. G. Cullis, N. G. Chew, H. C. Webber, and D. J. Smith, *J. Cryst. Growth*, *68*, 624 (1984).
67. W. J. Boettinger, D. Shechtman, R. J. Schaefer, and F. S. Biancaniello, *Metall. Trans.*, *15A*, 55 (1984).
68. W. W. Mullins and R. F. Sekerka, *J. Appl. Phys.*, *35*, 444 (1964).
69. G. H. Gilmer, *Mater. Sci. Eng.*, *65*, 15 (1984).
70. R. Trivedi, *J. Cryst. Growth*, *73*, 289 (1985).
71. W. Kurz, B. Giovanola, and R. Trivedi, *Acta Metall.*, *34*, 823 (1986).
72. J. Lipton, W. Kurz, and R. Trivedi, *Acta Medtall.*, *35*, 957 (1987).
73. R. Trivedi, J. Lipton, and W. Kurz, *Acta Metall.*, *35*, 965 (1987).
74. R. Trivedi, P. Magnin, and W. Kurz, *Acta Metall.*, *35*, 971 (1987).
75. C. W. Draper and J. M. Poate, *Int. Metall. Rev.*, *30*, 85 (1985).
76. K. P. Cooper and J. D. Ayers, *Surf. Eng.*, *1*, 263 (1985).
77. J. Powell and W. M. Steen, U.K. Pat. Appl. 2,090,873A (1982).
78. M. Von Allmen, S. S. Lau, M. Maenpaa, and B. Y. Tsaur, *Appl. Phys. Lett.*, *36*, 205 (1980).
79. M. Von Allmen, E. Huber, A. Blatter, and K. Affolter, *Int. J. Rapid Solidification*, *1*, 15 (1984).
80. N. Bloembergen, in *Laser-Solid Interactions and Laser Processing*, edited by S. D. Ferris et al. (AIP, New York, 1979), p. 1.
81. C-J Lin and F. Spaepen, *Appl. Phys. Lett.*, *41*, 721 (1982).
82. R. W. Bratt, *Metall. Powd. Rep.*, *38*, 475 (1983).
83. W. B. Kent, in P/M in *Aerospace and Defense Technologies*, Vol. 1 (MPIF, Princeton, NJ, 1990), p. 141.
84. O. R. Singleton and P. M. Royster, *J. Metals*, *40*(11), 40 (1988).
85. G. J. Hildeman, L. C. Labarre, A. Hafeez, and L. M. Angers, in *High Strength P/M Al Alloys II*, edited by G. J. Hildeman and M. J. Koczak (TMS, Warrendale, PA, 1986), p. 25; *Metall. Powd. Rep.*, *41*, 775 (1986).
86. N. Amano, Y. Odani, Y. Takeda, and K. Akecki, *Metall. Powd. Rep.*, *44*(3), 186 (1989).
87. N. J. DeCristofaro and A. Datta, in *Rapidly Solidified Crystalline Alloys*, edited by S. K. Das et al. (TMS, Warrendale, PA, 1985), p. 263.
88. L. E. Hackman, *Ceram. Eng. Sci.*, *2*, 849 (1981).
89. J. Ormerod, *Metals Mater.*, *4*, 478 (1988); *Int. J. Powd. Metall.*, *25*, 197 (1989); *Powd. Metall.*, *32*, 244 (1989).

Materials Index

a = amorphous
c = crystalline
l = liquid
v = vapor

subscripts = atom percent
premultipliers = weight percent

Ag:c, 142

Ag-based brazing filler metals:c, 717

Ag-Cu:c, 1, 4, 55, 209
Ag-Cu:s, 1, 54, 57, 59, 60, 758
Ag-7.5Cu:c, 717
Ag-15Cu-16Zn-24Cd:c, 717
Ag-15.5Cu-15.5Zn-16Cd-3Ni:c, 717
Ag-26Cu-21Zn-18Cd:c, 717
Ag-27Cu-23Zn-20Cd:c, 717
Ag-28Cu:c, 717
Ag-28.5Cu-2.5Ni-6Sn:c, 717
Ag-30Cu-10Sn:c, 717
Ag-38Cu-32Zn:c, 717
Ag-40Cu-5Zn-1Ni:c, 717
Ag-40Cu-30Zn-28Cd-2Ni:c, 717

Ag-Ge:c, 6, 11
Ag-Ge:a,c, 755

Ag-Sn-Cu:c, 746

Al:c, 22
Al alloys:a, 415
Al alloys:c, 347, 414, 740

Al-based brazing filler metals:a,c, 716

Al-Be:c, 30

$Al_{87}Co_{8.7}Ce_{4.3}$:a, 414
$Al_{87}Co_{8.7}Y_{4.3}$:a, 414

Al-Cr-Zr:c, 216, 348

Al-4Cu:c, 329
Al-4.5Cu:c, 42

Al-Fe:c, 65
Al-Fe-X:c, 349
Al-Fe-V-Si:c, 350
Al-7.1Fe-6.1Ce:c, 742
Al-8Fe:c,1, 65
Al-8Fe-2Mo:c, 742

Al-8.5Fe-1.3V-1.7Si:c, 742
$Al_{87}Fe_{6.7}Ce_{6.3}$:a, 414
$Al_{87}Fe_{6.7}Gd_{6.3}$:a, 414
$Al_{87}Fe_{7.7}Gd_{5.3}$:a, 414
$Al_{87}Fe_{8.7}Gd_{4.3}$:a, 414
$Al_{90}Fe_{5}Ce_{5}$:a, 414

Al-Ge:c, 317, 319, 334
$Al_{50}Ge_{50}$:a,c, 321

Al-Hf:a, 165, 316
Al-4Hf:c, 316
$Al_{50}Hf_{50}$:a, 185, 187

Al-2.7Li-1Cu-0.5Mg-0.5Zr:c, 358
$Al_{5.5}Li_{3.3}Cu$:c, 245

Al-Mn:a, c, 321
$Al_{89}Mn_{11}$:c, 322

Al-Ni-Fe:c, 59
Al-3.7Ni-1.5Fe:c, 60
$Al_{87}Ni_{8.7}Ce_{4.3}$:a, 414
$Al_{87}Ni_{8.7}Y_{4.3}$:a, 414

$Al_{87}Rh_{8.7}Ce_{4.3}$:a, 414

$Al_{65}Ru_{15}Cu_{20}$:c, 248

Al-Si:c, 131, 207, 210, 215, 216, 223
Al-10Si-0.8Fe-0.25Cu-0.1Mn-1.5Mg-0.2Zn-0.11Bi:c, 716
Al-10Si-0.8Fe-0.25Cu-0.1Mn-1.5Mg-0.2Zn:c, 716
Al-10Si-0.8Fe-4Cu-0.07Mn-0.07Mg-10Zn:c, 716
Al-10Si-0.8Fe-4Cu-0.15Mn-0.15Mg-0.2Zn-0.15Cr:c, 716
Al-12Si-0.8Fe-0.25Cu-0.10Mn-1.5Mg-0.2Zn:c, 716
Al-12Si-0.8Fe-0.25Cu-0.10Mn-2.5Mg-0.2Zn:c, 716
Al-12Si-0.8Fe-0.25Cu-0.10Mn-3.5Mg-0.2Zn:c, 716
Al-12Si-0.8Fe-0.3Cu-0.15Mn-0.1Mg-0.2Zn:c, 716

$Al_{x}Ta_{(100-x)}$:a,c, 602

Al-5.6Zn-2.5Mg-1.6Cu-0.23Cr:c, 741
Al-6.5Zn-2.5Mg-1.5Cu-0.4Co:c, 741
Al-8Zn-2.5Mg-1Cu-1.5Co:c, 741
Al-9Zn-2.5Mg-1.5Cu-0.14Zr-0.1Ni:c, 741

Au-Cu:a,c, 755
$Au_{53.2}Pb_{27.5}Sb_{19.2}$:a, 39

Au-Si:a, 6
Au-Si:a,c, 755

B_4C:c, 132

B_2O_3:a, 331
BAg alloys:c, 706
BAg-1:c, 717
BAg-2:c, 717
BAg-2a:c, 717
BAg-3:c, 717
BAg-4:c, 717
BAg-8:c, 717
BAg-13:c, 717
BAg-18:c, 717
BAg-19:c, 717
BAg-20:c, 717
BAg-21:c, 717

BAlSi alloys:c, 706
BAlSi-3:c, 716
BAlSi-4:c, 716
BAlSi-7:c, 716
BAlSi-8:c, 716
BAlSi-11:c, 716

Be:c, 29

BNi alloys:a,c, 706
BNi-2:c, 704
BNi-3:c, 697, 699

Bi:c,l, 28, 37
$Bi_{84}Sb_{16}$:a, 465

Brazing filler metals:a,c, 701

$CeFe_2H_x$:a, 159

Co:c, 29

Co-based brazing filler metals, 715

Co-B:a, 318, 324
Co-B-Si:a, 146
$Co_{(100-x)}B_x$:a, 382, 400
$Co_{70}B_{20}P_{10}$:a, 518, 525
$Co_{72.5}B_{11}Si_{16.5}$:a, 571
$Co_{72.5}B_{15}Si_{12.5}$:a, 146
$Co_{73}B_{12}Si_{15}$:a, 410
$Co_{75}B_{10}Si_{15}$:a, 397, 571
$Co_{75}B_{25}$:a, 571, 599
$Co_{75}B_{15}Si_{10}$:a, 121
$Co_{75}B_{20}Si_5$:a, 571
$Co_{77.5}B_9Si_{13.5}$:a, 571
$Co_{78}B_{12}Si_{10}$:a, 410
$Co_{81}B_{19}$:a, 323

Co-21Cr-4.5W-1.6Si-2.4B-0.07C:a, 715
Co-27.3Cr-2.9Ni-2.9Fe-32.9W-1Si-2.5B-1C:a, 715
$Co_{70}Cr_{10}P_{20}$:a, 593
$Co_{70.43}Cr_{21}W_{4.5}B_{2.4}Si_{1.6}C_{0.07}$:a, 404

$(Co_{(1-x)}Fe_x)_{75}B_{15}Si_{10}$:a, 413
$(Co_xFe_{(1-x)})_{83}Si_{17}$:a, 410
$(Co_{0.5}Fe_{0.5})_{75}B_{(1-x)}Si_{x25}$:a, 410
$(Co_{0.75}Fe_{0.25})_{(100-x)}(Si_{0.4}B_{0.6})_x$:a, 413
$Co_{37.5}Fe_{37.5}(B_{(1-x)}Si_x)_{25}$:a, 410
$Co_{45}Fe_{23}Cr_{10}B_{13}Si_9$:a, 392, 406
$Co_{58}Fe_5Ni_{10}B_{16}Si_{11}$:a, 412, 419, 574
$Co_{65.6}Fe_{4.7}Ni_{3.4}Nb_{2.2}Cr_{1.1}B_{13.8}Si_{9.1}$:a, 385
$Co_{68}Fe_{4.5}B_{15}Si_{12.5}$:a, 150
$Co_{70}Fe_5B_{10}Si_{15}$:a, 403, 409
$Co_{70.3}Fe_{4.7}B_{10}Si_{15}$:a, 184
$Co_{70.4}Fe_{4.6}B_{10}Si_{15}$:a, 388, 574, 581
$Co_{70.5}Fe_{4.5}B_{15}Si_{10}$:a, 130
$Co_{72.15}Fe_{5.85}Mo_2B_{15}Si_5$:a, 405
$Co_{74}Fe_6B_{20}$:a, 410
$Co_{76.05}Fe_{1.95}Mn_4B_{12}Si_6$:a, 405

$Co_{(78-x)}Nb_xB_{22}$:a, 513
$Co_{(94-x)}Nb_xB_6$:a, 513

$(Co_xNi_{(1-x)})_{75}B_{10}Si_{15}$:a, 573

$Co_xP_{(100-x)}$:a, 518
$Co_{81}P_{19}$:a, 505

$Co_xSn_{(1-x)}$:a, 505

$Co_{(80-x)}V_xB_{20}$:a, 513

Co_5Y:a, 419

Co-Zr:a, 417

Cr:c, 29

Cr-Cu:c, 224

Cr_2O_3:c, 224

Cr-NiAl:c, 60

Cu:c,l, 29, 94

Cu alloys:a,c, 370

Cu-Cr:c, 225

Cu-Nb-Sn:a, 165

Cu-Ni:c,l, 88
$(Cu_xNi_{(1-x)})Zr_2$:a, 297
$Cu_{69}Ni_{30}B_1$:c,l, 98
$Cu_{70}Ni_{30}$:c,l, 97

Cu-Pd-Ti:a, 165

Cu-Sn:c,l, 695
$Cu_{75}Sn_{25}$:a, 184
$Cu_{78}Sn_9Ni_6P_7$:a, 384

$Cu_xTa_{(100-x)}$:a, 603

$Cu_{50}Ti_{50}$:a, 387
$Cu_{67}Ti_{33}$:a, 384
Cu-Zr:a, 330
Cu-Zr-Cr:c, 743
$Cu_{30}Zr_{70}$:a, 330
$Cu_{48}Zr_{52}$:a, 113
$Cu_{50}Zr_{50}$:a, 122
$Cu_{59}Zr_{41}$:a, 385, 386
$Cu_{60}Zr_{40}$:a, 185, 186, 416, 418
$Cu_{67}Ti_{33}$:a, 384

DELTAMAX:c, 639, 640

Fe:c, 29

Fe alloys:c, 368

Fe-16Al:c, 621
$Fe_{67.5}Al_{22.5}Zr_{10}$:a, 410
$Fe_{79.3}Al_{1.1}Cu_{0.7}B_{12.6}Si_{3.8}C_{2.5}$:a, 410
$Fe_{78}Au_2B_{20}$:a, 323

Fe-B:a, 201, 214, 165
Fe-B-Si:a, 146
$Fe_xB_{10}Si_{(90-x)}$:a, 382
$Fe_xB_{13}Si_{(87-x)}$:a, 402
$Fe_{75}B_{(25-y)}Si_y$:a, 382
$Fe_{80}B_{(20-x)}G_x$:a, 511, 519
$Fe_{81}B_{(19-x)}Si_x$:a, 413
$Fe_{82}B_{(18-x)}Si_x$:a, 383
$Fe_{(90-x)}B_xSi_{10}$:a, 398, 399
$Fe_{(100-x)}B_x$:a, 382, 400
$Fe_{70}B_{10}Si_{20}$:a, 382
$Fe_{75}B_{15}Si_{10}$:a, 115, 121, 146, 148, 406, 415
$Fe_{75}B_{19}P_6$:a, 324
$Fe_{75}B_{25}$:a, 324, 331
$Fe_{77.5}B_{15}Si_{7.5}$:a, 578
$Fe_{78}B_{12}Si_{10}$:a, 404
$Fe_{78}B_{13}Si_9$:a, 116, 185, 186, 187, 398, 405, 419
$Fe_{78.5}Ti_1B_{12}Si_{8.5}$:a, 402
$Fe_{79.3}B_{16.4}Si_4C_{0.3}$:a, 398
$Fe_{79.5}Si,B_{20.5}$:a, 401
$Fe_{80}B_{10}P_{10}$:a, 519
$Fe_{80}B_{16}Si_2C_2$:a, 401
$Fe_{80}B_{20}$:a, 272, 321, 324, 385, 411, 560, 738
$Fe_{80.5}B_{15}Si_4C_{0.5}$:a, 397
$Fe_{81}B_{11}Si_8$:a, 402
$Fe_{81}B_{13.5}Si_{3.5}C_2$, 184, 391, 407, 408, 419, 584
$Fe_{81.5}B_{14.5}Si_4$:a, 123, 185, 186, 277, 397
$Fe_{81.5}B_{14.5}Si_{3.9}Te_{0.1}$:a, 397
$Fe_{81.9}B_{12.6}Mn_{0.2}Si_{3.4}C_{1.9}$:a, 410
$Fe_{82}B,Si_{18}$:a, 401
$Fe_{82}B_9Si_9$:a, 401
$Fe_{82}B_{12}Si_6$:a, 323, 388
$Fe_{82}B_{15}Si_3$:a, 388
$Fe_{82}B_{18}$:a, 383, 388, 404
$Fe_{83}B_{17}$:a, 416
$Fe_{85}B_{15}$:a, 389,411
$Fe_{86}B_{14}$:a, 323,404

Fe-C:c, 209, 210
Fe-C-Si:c, 142, 202

Fe-48Co:c, 621, 653
Fe-Co-V:c, 639
$(Fe_{(1-x)}Co_x)_{77}B_{13}Si_{10}$:a, 382
$(Fe_{(1-x)}Co_x)_{83}B_{17}$:a, 410
$(Fe_xCo_{(1-x)})_{75}B_{10}Si_{15}$:a, 567, 572
$(Fe_xCo_{(1-x)})_{80}B_{20}$:a, 573
$Fe_{67}Co_{18}B_{14}Si_1$:a, 386, 387, 395, 408

Fe-Cr:c, 596
Fe-Cr-C:c, 208, 213
Fe-Cr-P-C:a, 745
Fe-Cr-Y-Al:c, 722, 729
Fe-13Cr-11Ni:c, 110
Fe-15Cr-10Ni:c, 110
Fe-19Cr-2Mo:c, 593
Fe-24Cr-14Ni:c, 110
Fe-30Cr-2Mo:c, 593
$Fe_{(75-x)}Cr_xB_{15}Si_{10}$:a, 406
$Fe_{65}Cr_8B_{27}$:a, 412
$Fe_{65}Cr_{10}B_{15}Si_{10}$:a, 402
$Fe_{67}Cr_8B_{15}Si_{10}$:a, 406
$Fe_{68}Cr_{10}B_{13}Si_9$:a, 402
$Fe_{70}Cr_{10}P_{13}C_7$:a, 593, 594
$Fe_{70}Cr_5B_{15}Si_{10}$:a, 406

$Fe_{71.3}Cr_{10}Mo_9P_8C_{1.7}$:a, 185, 187
$Fe_{72.5}Cr_{10}B_{10}Si_{7.5}$:a, 402
$Fe_{75}Cr_3Mo_2P_{13}C_7$:a, 593

$Fe_{73.4}Cu_1Nb_{3.1}B_{9.1}Si_{13.4}$:a, c334, 554, 630
$Fe_{79.3}Cu_{0.7}Al_{1.1}B_{12.6}Si_{3.8}C_{2.5}$:a, 410

$Fe_{74.7}Mo_{8.3}B_{12}Si_5$:a, 400
$Fe_{78}Mo_2B_{20}$:a, 324

Fe-Nd-B:c, 325

Fe-80Ni:c, 621
Fe-Ni-B:a, 319, 323, 332
Fe-Ni-Cr-P-B:a, 142
Fe-Ni-Cr-B-Si:a, c, l, 697
Fe(Ni, Cr)-(Si+B):l, c, a, 697
$(Fe_{(1-x)}Ni_x)_{80}B_{10}Si_{10}$:a, 412
$(Fe_{(1-x)}Ni_x)_{77}B_{13}Si_{10}$:a, 382
$(Fe_xNi_{(1-x)})_{75}P_{16}B_6Al_3$:a, 523
$(Fe_xNi_{(1-x)})_{80}B_{20}$:a, 573
$(Fe_xNi_{(1-x)})_{80}P_{14}B_6$:a, 516
$(Fe_xNi_{(1-x)})Zr_2$:a, 297
$(Fe_{0.5}Ni_{0.5})_{(100-x)}B_x$:a, 400
$(Fe_{0.75}M_{0.25})_{83}B_{17}$:a, 404
$(Fe_{50}Ni_{50})_{(1-x)}B_x$:a, 400
$Fe_{19}Ni_{59}B_{14}Si_8$:a, 121
$Fe_{19.5}Ni_{58.5}B_{14}Si_8$:a, 122
$Fe_{27}Ni_{53}P_{14}B_6$:a, 274, 275
$Fe_{32}Ni_{36}Cr_{14}P_{12}B_6$:a, 257
$Fe_{38}Ni_{38}Al_3P_{14}C_6$:a, 142
$Fe_{39}Ni_{38}P_{14}B_6Al_3$:a, 381
$Fe_{39}Ni_{39}B_{22}$:a, 332, 333
$Fe_{39}Ni_{39}Mo_4B_{12}Si_6$:a, 400
$Fe_{40}Ni_{32}Cr_8P_{12}B_8$:a, 400
$Fe_{40}Ni_{38}Mo_4B_{18}$:a, 386, 403
$Fe_{40}Ni_{40}Mo_3B_5Si_{12}$:a, 236
$Fe_{40}Ni_{40}B_{12}Si_8$:a, 394, 397
$Fe_{40}Ni_{40}P_{14}B_6$:a, 250, 381, 391, 393, 403, 409, 480, 563, 576, 617, 651
$Fe_{40}Ni_{40}B_{20}$:a, 123, 185, 274, 292, 324, 383, 390, 397, 400, 408, 414, 516
$Fe_{41}Ni_{41}B_{18}$:a, 406
$Fe_{42}Ni_{42}B_{16}$:a, 406
$Fe_{55}Ni_{45}$:c, 121

$Fe_{62}Ni_{22}B_{16}$:a, 323
$Fe_{65}Ni_{10}B_{25}$:a, 331
$Fe_{66}Ni_{10}B_{24}$:a, 319, 327
$Fe_{74}Ni_7Si, B_{19}$:a, 401
$Fe_{77}Ni_4Si, B_{19}$:a, 401
$Fe_{85}Ni_{15}$:c, 64

Fe-P-C:a, 146
$Fe_{(100-x)}P_x$:a, 382
$Fe_{(100-x)}P_xSi_y$:a, 411
$Fe_{77}P_{15}C_8$:a, 114
$Fe_{77.5}P_{12.5}C_{10}$:a, 406
$Fe_{80}P_{12.5}C_{7.5}$:a, 518
$Fe_{80}P_{13}C_7$:a, 415, 617
$Fe_{80}P_{20}$:a, 404
$Fe_{86}P_{14}$:a, 331

Fe-3Si:c, 621, 653
Fe-3.2Si:c, 626, 628, 632, 641, 644, 650, 657
Fe-6.5Si:c, 619, 622, 623, 628, 629, 639, 650, 657

$Fe_xSn_{(1-x)}$:a, 511

$(Fe_{(1-x)}T_x)_{90}Zr_{10}$:a, 527

$Fe_{60}Ti_{40}$:a, 418

$Fe_{76.8}V_{6.2}B_{12}Si_5$:a, 400

Fe-Zr:a, 417
$Fe_{(100-x)}Zr_x$:a, 383, 526
Fe_2Zr:a, 418
$Fe_{60}Zr_{40}$:a, 183, 184
$Fe_{78}Zr_2B_{20}$:a, 323
$Fe_{90}Zr_{10}$:a, 526

Ferrites:c, 638, 639, 640, 653

FINEMENT:c, 333, 630

$Ga_{95}Ag_5$:a, 465

$GdNi_2H_{4.35}$:a, 159

$Ge_{(100-x)}Au_x$:a, 465

$H_{5.5}Zr_3Rh$:a, 419
H_3Ni_2Ce:a, c, 419
H_3Ni_2Gd:a, c, 419
$H_{5.5}Zr_3Rh$, 419

$Hf_{35}Ni_{65}$:a, 185

Hg:l, v, 37

In:c, 1 37

InSb:c, 122

Ir:c, 604, 605

MAGNEQUENCH:c, 666

MARKOMET alloy 1064:a, 176, 184

METGLAS alloy 2605:a, 392, 411
METGLAS alloy 2605A:a, 392
METGLAS alloy 2605CO:a, 625, 651, 653
METGLAS alloy 2605S-2:a, 331, 398, 405, 419, 626, 644
METGLAS alloy 2605S-3A:a, 625, 643
METGLAS alloy 2605SC:a, 184, 323, 391, 407, 419, 580, 584, 585, 625, 645, 647, 653
METGLAS alloy 2605SM:a, 625, 643
METGLAS alloy 2605TCA:a, 625, 628, 639, 640, 645
METGLAS alloy 2705M:a, 405, 625, 653
METGLAS alloy 2705MN:a, 405, 625
METGLAS alloy 2714A:a, 625, 639, 640, 653
METGLAS alloy 2826:a, 184, 331, 391, 392, 617, 618, 624, 651
METGLAS alloy 2826A:a, 257, 383, 384, 390, 385
METGLAS alloy 2826B:a, 389
METGLAS alloy 2826MB:a, 403, 625
METGLAS alloy MBF-1000 series:a, 706
METGLAS alloy MBF-1001:a, 713, 725
METGLAS alloy MBF-1002:a, 713, 714
METGLAS alloy MBF-1005:a, 713, 714
METGLAS alloy MBF-1007:a, 713, 714, 724
METGLAS alloy MBF-1011:a, 713, 714, 731

METGLAS alloy MBF-15:a, 708, 710
METGLAS alloy MBF-20:a, 391, 397, 420, 704
METGLAS alloy MBF-20:a, 395, 711, 725, 727, 729, 730
METGLAS alloy MBF-30:a, 699, 711
METGLAS alloy MBF-50:a, 708, 711, 729, 731
METGLAS alloy MBF-60:a, 708, 711
METGLAS alloy MBF-80:a, 708, 711

METGLAS alloy MSF-101:c, 714, 719
METGLAS alloy MSF-103:c, 719
METGLAS alloy MSF-104:c, 714, 719
METGLAS alloy MSF-105:c, 714, 719
METGLAS alloy MSF-107:c, 719
METGLAS alloy MSF-108:c, 714, 719
METGLAS alloy MSF-111:c, 714, 719
METGLAS alloy MSF-114:c, 719
METGLAS alloy MSF-118:c, 719

METSHIELD:a, 618, 651

MQ1-A molded magnets:c, 673, 679
MQ2-E molded magnets:c, 674
MQ3 molded magnets:c, 675, 681

Mg:c, 29
Mg alloys:c, 340
Mg-Zn-Al-Si:c, 342
Mg-Zn-Al-Al-RE:c, 342
$Mg_{37.5}Zn_{37.5}Al_{25}$:a, 465
$Mg_{70}Zn_{30}$:a, 341, 466
$Mg_{70}Zn_{(30-x)}Ga_x$:a, 465
$Mg_{74}Zn_{24}N$:a, 341

Microcrystalline alloys:c, 627

$(Mn_xNi_{(1-x)})_{75}P_{16}Al_3B_6$:a, 240

$Mn_{62}Ti_{38}$:a, 418

Mn-Zn-Fe-O:c, 639

MoB_2:c, 125

$Mo_{52}Ge_{48}$:a, 470
$Mo_{79}Ge_{21}$:a, 471

$Mo_{30}Re_{70}$:a, 470

$(Mo_{(1-x)}Ru_x)_{80}P_{20}$:a, 467

Nanocrystalline alloys:c, 555, 565, 574, 623, 639

Nb:c, 29

Nb alloys:c, 371

$Nb_{75}Ge_{25}$:a, 388, 467

Nb-Ni:a, 165
$Nb_{55}Ni_{45}$:a, 417
$Nb_{59}Ni_{41}$:a, 409
$Nb_{60}Ni_{40}$:a, 417
Nb-Si:c, 51
$Nb_xSi_{(100-x)}$:a, 479, 481
$Nb_{75}Si_{25}$:a, c, 51

$Nb_{75}Sn_{25}$:a, c, 418

$Nd_2Fe_{14}B$:c, 666
Nd_2Fe_{17}:c, 667

Nd-Fe-B:c, 666
$Nd(Fe, Co)_4B$:c, 669
$Nd_{15}(Fe_{(1-y)}B_y)_{85}$:c, 667, 671
$Nd_{13.5}Fe_{0.945}B_{0.05586.5}$:c, 671

Nd-Fe-B-Al:c, 669

Ni:c, 1 29, 95

Ni alloys:c, 369

Ni-based brazing filler metals:a, 708

$Ni_3(Al, Nb)$:c, 304

Ni-17Al-10Mo:c, 743
$Ni_{76.4}Al_2P_{11.8}B_{9.8}$:a, 129

Ni-B:a, 330
Ni-B-Si:a, 146
Ni-B-Si:a, c, 705
$Ni_xB_{(92-x)}Si_8$:a, 400
$Ni_{62}B_{38}$:a, 747
$Ni_{66}B_{34}$:a, 330, 332
$Ni_{70}B_{20}Si_{10}$:a, 184, 407, 415
$Ni_{72}B_{20}Si_8$:a, 397
$Ni_{75}B_{17}Si_8$:a, 129, 184
$Ni_{75}B_{17}Si_8$/WC:a, c, 135, 388, 389
$Ni_{77.5}B_{15}Si_{7.5}$:a, 183, 185, 407, 415
$Ni_{78}B_{12}Si_{10}$/WC:a, 126, 389, 391, 406, 407
$Ni_{78}B_{14}Si_8$:a, 386, 403, 407, 408, 411
$Ni_{78}B_{20}Si_8$:a, 397
$Ni_{80}B_6Si_{14}$:a, 404
$Ni_{80}B_{10}Si_{10}$:a, 404, 405
$Ni_{80}B_{15}Si_5$:a, 404, 405
$Ni_{81.5}B_{14.5}Si_4$:a, 397, 401

Ni-20Co-4Si-2.7B:a, 708
$Ni_{40}Co_{20}Cr_{12}Mo_6Fe_6B_{16}$:a, 400

Ni-Cr-P-B:a, 746
Ni-5.3Cr-7.3Si-1.4B:a, 708
Ni-10Cr-5.5Fe-7Mo-23Co-3.5B:a, 708
Ni-11.3Cr-4.0Fe-8W-1.5Si-2.2B:a, 708
Ni-14Cr-4.5Fe-4.5Si-3.2B:a, 708
Ni-14Cr-10.1P:a, 708
$Ni_{65}Cr_{15}P_{16}B_4$:a, c, 601
$Ni_{69.2}Cr_{6.6}Fe_{2.6}B_{13.7}Si_{7.9}$:a, 391, 395, 397, 420
$Ni_{70}Cr_{10}P_{20}$:a, 593
$Ni_{80.8}Cr_{15.2}B_4$:a, 404

$Ni_{79}Dy_{21}$:a, 505

Ni-Fe:c, 651, 653
Ni-Fe-Mo:c, 639
$Ni_{36}Fe_{32}Cr_{14}P_{12}B_6$:a, 383, 384, 390, 385
$Ni_{39}Fe_{38}Al_3P_{14}B_6$:a, 381
$Ni_{40}Fe_{30}Cr_{10}B_{20}$:a, 404
$Ni_{49}Fe_{29}P_{14}B_6Si_2$:a, 389
$Ni_{60}Fe_{20}P_{20}$:a, 747

Ni-Mo-B:c, 744
$Ni_{55.8}Mo_{25.7}Cr_{9.7}B_{8.8}$:a, 176, 184

Ni-Nb:a, 417
$Ni_xNb_{40}Pt_{(100-x)}$:a, 610
$Ni_{60}Nb_{(40-x)}Al_x$:a, 398
$Ni_{37}Nb_{40}Pt_3$:a, 611
$Ni_{38}Nb_{40}Ir_2$:a, 608
$Ni_{60}Nb_{40}$:a, 417

Ni-P:a, 3, 384
$Ni_{(100-x)}P_x$:a, 384
$Ni_{75}P_{16}B_6Al_3$:a, 480
$Ni_{79}P_{13}B_8$:a, 517
$Ni_{80}P_{14}B_6$:a, 517
$Ni_{81}P_{19}$:a, 705
$Ni_{84}P_{16}$:a, 386
$Ni_{89}P_{11}$:a, 184

NiO:c, 87

Ni-Pd-based brazing filler metals:a, 713

Ni-30Pd-10.5Cr-2.4B:a, 713
Ni-41.2Pd-8.8Si:a, 713
Ni-45.8Pd-10Co-4.5Mo-3.8Si:a, 713
Ni-46.8Pd-5Co-5Si:a, 713
Ni-57.6Pd-21.2Fe-6.2Si:a, 713
Ni-88.3Pd-5.1Si:a, 713
$Ni_{40}Pd_{40}P_{18.5}Si_{1.5}$:a, 387
$Ni_{40}Pd_{40}P_{19}Si_1$:a, 275, 394, 395
$Ni_{40}Pd_{40}P_{20}$:a, 288, 386
$Ni_{60}Pd_{20}P_{20}$:a, 386, 387

Ni-5.5Si-1.6B:a, 708
Ni-7.3Si-2.17B:a, 708

Ni-Sn:a, 165
Ni-Sn:c, 1 95
Ni-5Sn:c, 196
Ni-25Sn:c, 196
$Ni_{(100-x)}Sn_x$:a, 418

Ni-Ti:a, 417
$Ni_xTi_{(100-x)}$:a, c, 417
$Ni_{50}Ti_{50}$:a, 418

Ni-Zr:a, 160, 285, 286, 417, 612
$Ni_{50}Zr_{50}$:a, 160, 418, 419
$Ni_{60}Zr_{40}$:a, 420
$Ni_{65}Zr_{35}$:a, 183, 184
$Ni_{68}Zr_{32}$:a, 164

Pb alloys:c, 714

Pb-1.5Ag-1Sn:c, 719

Pb-Bi:c, 46
Pb-Bi:s, l, 38
$Pb_{70}Bi_{30}$:a, 465

$Pb_{90}Cu_{10}$:a, 466

Pb-5In-2.5Ag:c, 719, 721, 733
Pb-5In-2.5Ag-1.5Sn:c, 719

Pb-Sb:s, 1 40
$Pb_{25}Sb_{25}Au_{50}$:a, 470

Pb-Sn:l, c, 34, 40, 696
Pb-5Sn:c, 719
Pb-10Sn:c, 719
Pb-10Sn-2Ag:c, 719
Pb-15Sn:c, 719
Pb-25Sn:c, 719

Pd:c, 604, 605

$Pd_{57}Au_{43}$:a, 478

$Pd_{73}Cr_7Si_{20}$:a, 749

Pd-Ge:a, 747
$Pd_{80}Ge_8Si_{12}$:a, 387

$Pd_{39}Ir_{40}Ti_2P_{19}$:a, 604
$Pd_{41}Ir_{30}Ru_{10}P_{19}$:a, 605
$Pd_{41}Ir_{30}Ti_{10}P_{19}$:a, 605
$Pd_{41}Ir_{40}P_{19}$:a, 604, 605
$Pd_{45}Ir_{30}Ti_5Pt_1P_{19}$:a, 605
$Pd_{45}Ir_{30}Ti_5Rh_1P_{19}$:a, 605
$Pd_{45}Ir_{30}Ti_5Ru_1P_{19}$:a, 605
$Pd_{46}Ir_{30}Pt_5P_{19}$a, 604
$Pd_{46}Ir_{30}Rh_5P_{19}$:a, 604
$Pd_{46}Ir_{30}Ti_5P_{19}$:a, 605

$Pd_{46}Ni_{36}P_{18}$:a, 421
$Pd_{48}Ni_{32}P_{20}$:a, 385
$Pd_{77.5}Ni_6Si_{16.5}$:a, 393

$Pd_{41}Pt_{40}P_{19}$:a, 604, 605
$Pd_{51}Pt_{30}P_{19}$:a, 604

$Pd_{46}Rh_{25}Ti_5Ru_5P_{10}Si_9$:a, 607
$Pd_{46}Rh_{25}Ti_5Pt_5P_{10}Si_9$:a, 607
$Pd_{46}Rh_{25}Ti_{10}P_{10}Si_9$:a, 607
$Pd_{61}Rh_{20}P_{19}$:a, 604
$Pd_{56}Rh_{25}P_{10}Si_9$:a, 607, 609
$Pd_{51}Rh_{30}P_{19}$:a, 604, 605

Pd-Si:a, 747
Pd-Si-Cu:a, 60
$Pd_{75}Si_{25}$:a, 418
$Pd_{77.5}Si_{16.5}Cu_6$:a, 131, 184, 385, 389, 411, 492, 495
$Pd_{77.5}Si_{16.5}Ni_6$:a, 387, 390
$Pd_{78}Si_{16}Cu_6$:a, 184, 390
$Pd_{80}Si_{20}$:a, 381, 388, 413, 418
$Pd_{80}Si_{12}Ge_8$:a, 387
$Pd_{80}Si_{18}V_2$:a, 413
$Pd_{82}Si_{18}$:a, 87, 187, 292, 381, 411, 413
$Pd_{83}Si_{17}$:a, 390
$Pd_{83.5}Si_{16.5}$:a, 122

$Pd_{77.5}Si_{16.5}Ag_6$:a, 387

Pd-Ti:a, 165
$Pd_{(100-x)}Ti_x$:a, 417

$Pd_{58.8}U_{20.6}Si_{20.6}$:a, 238, 239

$Pd_{80}V_2Si_{18}$:a, 413

$Pd_{81}V_1Si_{18}$:a, 413

Pd-Zr:a, 612

$Pd_{35}Zr_{65}$:a, 747

POWERCORE strip:a, 626, 636

$Pr_{(100-x)}Fe_x$:a, 534

Pr-Fe-B, 667

Pt:c, 604, 605
$Pt_{58.4}Ni_{14.6}P_{27}$:a, 234, 385

RE_7TM_3:c, 669

RE-TM-B magnet alloys:c, 667, 673

REXALLOY:c, 744

RS brazing filler metals:a, c, 706

Rh:c, 604, 605
Ru:c, 604, 605

Ru_2O/Ti:c, 605

$Se_{(100-x)}Sb_x$:a, 410

SENDUST:c, 619, 623, 627, 648

SiC:c, 132

Sm_2Co_{17}:c, 665
$SmCo_5$:c, 665

Sn:c, 1 37

Sn alloys:c, 714

Sn-3.5Ag:c, 721, 733
Sn-5Ag:c, 719
Sn-25Ag-10Sb:c, 719, 721, 733

$Sn_xCu_{(100-x)}$:a, 465

Sn-36Pb-2Ag:c, 719
Sn-40Pb:c, 719

Sn-10Sb:c, 719

Solder filler metals:c, 702, 719

Steel, AISI 9310:c, 181
Steel, cast iron, 207, 210, 215, 217, 223, 226
Steel, SAE 300 series stainless:c, 724
Steel, SAE 304 stainless:c, 697
Steel, SAE 400 series stainless:c, 719, 729
Steel, SAE 409 stainless:c, 700
Steel, stainless:c, 593

STELLITE:c, 744

SUPERMALLOY:c, 639, 640

SUPERMENDUR:c, 639, 653

$(T_{(1-x)}M_x)B_{10}P_{10}$:a, 511

$Te_{85}Ge_{15}$:a, 411

Ti:c, 29

TiC:c, 132

Ti alloys:c, 215, 358
Ti alloys(beta):c, 366
Ti alloys(RE-containing):c, 364
Ti alloys(Ti-containing):c, 363

Ti aluminides:c, 366

Ti-Al:c, 47
Ti-6Al-4V:c, 724, 727, 743
$Ti_{60}Al_{40}$:c, 47

$Ti_{35}Be_{50}Zr_{15}$:a, 421
$Ti_{49}Be_{40}Zr_{10}V_1$:a, 749

$Ti_{50}Be_{40}Zr_{10}$:a, 421

Ti-Cu-Ni:a, 712
$Ti_{(100-x)}Cu_x$:a, 417
$Ti_{50}Cu_{50}$:a, 185
$Ti_{57}Cu_{43}$:a, 418
$Ti_{59.7}Cu_{15.5}Ni_{24.8}$:a, 718
$Ti_{65.5}Cu_{12.3}Ni_{22.2}$:a, 718
$Ti_{65.5}Cu_{16.6}Ni_{17.9}$:a, 718
$Ti_{70}Cu_{15}Ni_{15}$:c, 719
$Ti_{74.8}Cu_{12.1}Ni_{13.1}$:a, 718

$Ti_{68}Ni_{32}$:a, 417

$Ti_xPd_{(100-x)}$:a, c, 417

Ti-X:c, 207, 215

Ti/Zr-Ni:a, 712

Ti/Zr-based brazing filler metals:a, c, 718

Ti-Zr-Cu-Ni:a, 712
$Ti_{20}Zr_{60}Ni_{20}$:a, 718
$Ti_{33}Zr_{17.3}Cu_{49.7}$:a+c, 718
$Ti_{48.8}Zr_{12.8}Cu_{18.4}Ni_{20}$:a, 718
$Ti_{48.9}Zr_{25.7}Cu_{14.7}Ni_{10.7}$:a, 718

VITROVAC alloy 0040:a, 383, 390, 397, 400, 408, 411, 413
VITROVAC alloy 0080:a, 403, 411
VITROVAC alloy 4040:a, 400
VITROVAC alloy 6025:a, 407, 411
VITROVAC alloy 6030:a, 407
VITROVAC alloy 7505:a, 407

WC:c, 126

$W_{67}Ni_{23}B_4C_6$:a, 416

Wear-resisting alloys:c, 744

Zn:c, 29, 142

Zr:c, 29
$Zr_xCu_{(100-x)}$:a, 467
Zr_2Fe:a, c, 418
Zr_2O_3:c, 130
Zr_3Rh:a, 159
$Zr_{60}Ni_{40}$:a, 418
$Zr_{70}Cu_{30}$:a, 496
$Zr_{70}Ni_{30}$:a, 394
$Zr_{70}Pd_{30}$:a, 487
$Zr_{76}Fe_{24}$:a, 389
$Zr_{76}Ni_{24}$:a, 718

Subject Index

AES model, 385, 414
Aging behavior
  microstructural, 294, 414
Amorphous alloys
  Co-rich, 561
  Fe-rich, 561
  history of formation, 408
  transition metal-metalloid, 166
  transition metal-transition metal, 161, 164, 165
Annealing
  conductivity and, 288, 297
  defects, 381
  density and hardness and, 278
  embrittlement, 397
  enthalpy and specific heat, 280
  hydrogen solubility and, 281
  length contraction during, 249
  magnetic field after, 564
  magnetic properties after, 235, 260
  models and mechanisms during, 258
[Annealing]
  positron lifetime, 235, 236, 258, 260
  relaxation, 408
  stress, 564
Antiphase domains, 57
Applications
  acoustic tail pipes, 727
  actuators, 685
  aircraft frame, 725, 741
  amorphous alloys, 624, 739
  armor penetrators, 738, 740
  article surveillance systems, 649
  automotive parts, 224, 737, 744
  bearings, 743
  brazing alloy, 721
  catalytic, 747
  catalytic converters, metallic, 731
  chemical, 745–748
  contact materials, 223
  cookware, 739
  corrosion-resistant, 738, 744

[Applications]
digital telecommunications, 642
disk drive actuators, 685
electric razors, 746
electrical, 748–750
elevated temperature, 742
EMI filters, 642
flashbulbs, 748, 750
flywheels, 738, 739
food processing equipment, 739
free-electron lasers, 654
fuses, 748
gas turbine engines, 742
gaskets, 675
grinding wheels, 739
ground fault interrupters, 643
heat exchangers, 727
high frequency magnetic cores, 620
honeycomb structures, 727
hydrogen storage, 748
importance of combined properties, 622
inductors, output chokes, 640
jet engine, 725
loudspeakers, 683
magnetic amplifiers (magamps), 641
magnetic field sensors, 587
magnetic pulse compression, 620, 656
magnetic resonance imaging (MRI), 687
magnetic separation, 745
magnetic shielding, 651
mechanical performance, 738–745
metal-matrix composites, 738
microcrystalline, 740
motors, 650, 676–681
brushless, 676
printed circuit, 682
stepping, 680
MRI devices, 687
nuclear, 737
orthodontics, 724
permanent magnet, 665
preforms for metal joining, 724
pressure vessels, 738
price trends, soft magnetic alloys, 619
pulsed-power, 652

[Applications]
razor blades, 745
relationship to properties, 622
rocket thrust chamber, 743
scalpels, 746
seawater piping, 743
semiconductor die-bonding, 733
soft magnetic material, 617, 620
solder alloy, 721
superconductors, 748
switch protection, 655
switched-mode power supplies, 620, 637
tire reinforcement, 739
torque sensor, 584–586
transducers, 583, 585, 646
transformers
current measurement, 644
design considerations, 632
distribution, 631
instrumentation, 645
line-frequency, 630
loss evaluation, 631
power, 636
pulse, 654
vehicles, supersonic and hypersonic, 743

Base Metal
joining, 692
Biot number, 68
Bistability, 564
Boundary Layer
formation, 105

Camshaft, 213, 222, 226
Catalytic
potency, 80, 87
sites, 36
Chaos, 584, 587
Chemical properties, 215
anodes for NaCl solution electolysis, 604
anodes for seawater electrolysis, 605
catalysis, 609
catalyst reactions, other, 612
chemical homogeneity and, 594

[Chemical properties]
chemical potential, 32
corrosion resistance, origin, 593
corrosion-resistant alloys, 591
corrosion-resistant coatings, 600
electrodes for electrolysis, 603
fuel cell reactions, 610
microcystalline alloys, 595
oxidation, Nd-Fe-B magnets, 670, 687
oxidation resistance, 745
passive film formation, 593
pitting corrosion, 595, 745
structural relaxation and, 599
Cleaning, 223
Coating, 199
laser and electron beam, 600
sputtering, 602
Composites
continuously reinforced, 133
discretely reinforced, 125, 132, 388
incremental casting, 123
laminated RS, 120
laminated with copper, 391
magnetic properties, 130
particle-reinforced, 125
RS processing of, 119
yield strength, 128
Young's modulus, 127
Conductivity
electrical, 224, 281, 413, 474, 497
annealing effects, 163, 413
critical, 438
electron correlation effects, 438, 452, 455
magnetic fields in, 444, 448
negative temperature coefficient, 433, 447
pressurization and, 413
quantum interference effects, 436, 442, 450, 455
refractory metallic glass, 450
simple amorphous alloy, 440
temperature dependence of, 431, 441, 444, 448, 450
rare earth-containing glasses, 447
solid state amorphization reaction, 163
thermal, 201, 205, 488
Consolidation/compaction
amorphous alloys, 112
dynamic, 115, 175
powders, 157, 171
process, 114, 184
quasistatic, 185
temperature-pressure-time path, 173
warm consolidation, 186
warm pressing, 186
Cooling
constitutional, 49
continuous, 40
convective, 48
hyper, 19, 37
Newtonian, 48
rate of, 44, 47
Cooling rate, 201, 209, 210, 212
calibration of, 12
dendrite arm spacing and, 211
Coordination number, 233, 253, 256
Crack, formation, 203, 216
Creep
continuous heating, 393, 394, 423
Nabarro-Herring equation, 408
preannealing effects, 394
negative, 395
test, 115
Crystallization, 36
eutectoid, 308, 324
growth rate, 309, 324
heating, during, 393
kinetics, 314
metallic glasses, 322
morphology, 323
nucleation, 311
polymorphous (partitionless), 169, 324, 418
primary, 323
surface, 329, 332
effects of treatments, 333
surface-induced, 330
temperature, on continuous heating, 394
thermodynamics, 307
Curie temperature, 256, 257, 260
change with annealing, 272
reversibility, 275

Cylinder
liner, wall, 213, 215, 222

Debye-Waller factor, 441
Decomposition, 201, 209, 211
eutectoid, 319
mechanochemical spinodal, 396
quasicrystals, 321
Diagrams
deformation map, 173, 174
interface condition, 54
phase, *see* Phase diagrams
solidification, 201, 211
Diffusion
amorphous alloys, 161, 282, 411
atomic mobility, 559
atomic size, dependence on, 161
coefficients, 285
crystalline alloys, 161
diffusivity, thermal, 201
kinetics, 33
mechanisms in amorphous alloys, 283
relation to viscosity, 291
relaxation effects and, 282
peed, 53
temperature dependence, liquid, 60
tracer, 162, 286
zone, 692
Dislocations
amorphous metal, 421
Burgers type, 381
Somigliana type, 381
Volterra type, 381
Disorder, 57
Dispersoids, 59
Distribution function
differential (DDF), 238, 239, 247, 248
pair (PDF), 233
radial (RDF), 233
Drop tube, 47
Dugdale model, 380

Elastic properties
atomistic calculations, 384
composition dependence, 381
magnetization effect, 382
stiffness, 741
[Elastic properties]
temperature and annealing effect, 383
Energy
free, 160
surface, 29, 694
Enthalpy
activation, 201
amorphous alloys, 279
specific heat and, 411
Equilibrium
dynamical, 84
global, 24
hierarchy, 24
local, 24, 52
metastable, 24, 32
T_0 curves, 25, 30, 32, 69, 201
Expansion
thermal, 203, 491

Fictive temperature, 272, 414
Filler metal
Ag-based, 717
Al-based, 709, 716
amorphous alloy, 705
crystallization, 699
Cu-based, 709, 711
Cu-Si-based, 711
Ni-based, 705, 708
Ni-Pd and Co-based, 709, 713, 715
potential advantages via RS, 703
RS types, 706
solders, 714
Ti/Zr-based, 712, 718
Fluctuations
compositional, 563
local magnetic easy axes, 554, 555
local electrical fields, 554
topological, 563
Fourier transform, 568
Free energy
parallel tangent construction, 39, 160
Free volume, 497, 564
model for annealing, 380
Fulcher-Vogel equation, 394

Gibbs-Thomson effects, 39, 54

Glass
formation (ability), 34, 36, 51, 168, 201, 203, 209, 214, 263
formation and critical thickness, 146
formation, and composition, 168, 170, 263
Glass transition
structural relaxation and, 260
temperature, 36, 113, 270, 393
thermal fluctuation of stresses and, 261
Grain size, 202, 554
Growth
collision-limited, 54, 55
competitive, 61
dendritic, 62, 90
eutectic, 60
facetted, 27
kinetics, 52
linear, 309
maximum rate of, 61
metastable phase, 27
parabolic, 310
partitionless, 92
rate, temperature dependence of, 324
stability criterion, 91, 202, 210

Hall effect
anomalous, 446
rare earth-containing glasses, 447
refractory glass coefficient, 454
simple amorphous alloys, 442
temperature dependence of, 454
Heat
balance of, 69
flow, 20
fusion of, 18
Heat capacity, 467, 487
liquid, 36
Heat transfer
coefficient, 19, 21, 22, 69, 106
convective, 197, 205
Heating, 196, 197, 209, 217, 229
continuous, 423
rate, 196, 197, 198
Incubation time, 172
Inoculants, 215
Interaction time, 197, 198, 209, 216, 217, 221
Interface
attachment coefficient for, 21
contact angle, 694, 497
response functions of, 52
solid-liquid, 21, 52
stability of, 57
tension, liquid-vapor, 397
wetting and flow at, 693, 720, 733
Intermetallic
growth of, 57
primary, 70
Internal friction (damping), 384, 385, 496
cold working effect, 387
defects and, 422
H peak, 388
magnetic field effect, 388
temperature and annealing effects, 385
Vibrating reed technique, 383
Isoconfigurational
states, 271
flow, 380

Jackson-Hunt, 60
Joining technology, 691
cemented carbide, 731
filling, 694
formation, 699
heat-resistant steels, alloys, 729
microstructure, 720, 732
process technology, 699
shear strength, 721

Kinetics, 314
Kissinger plots
crystallization temperature and heating rate, 398
Curie temperature and heating rate, 412
embrittlement temperature and heating rate, 397
X-ray peak and heating rate, 397
Kolgogorov-Avrami equation, 410
Kondo effect, 444

Laser, 196, 197, 198
absorption, 198, 199, 203, 205, 209, 217
beam, 195, 197, 221
guiding, 203
shaping, 203, 204
commercially available, 219
CO_2, 196, 204, 209, 221, 229
CVL, 198, 209, 217, 221
cw, 198, 209, 217
energy, 198, 217
excimer, 197, 203, 217, 221, 223
focus, 217
focusability, 198, 199
free-electron, 654
gas alloying, 215
glazing, 215
hardening, 205
intensity, 196, 197, 199, 200, 203, 221
metallurgical aspects, 200, 744
Nd: YAG, 198, 199, 203, 209, 221, 224, 229
parameters, 205, 210
plasma formation, 199, 219, 224
power (flux density), 195, 203, 209, 217, 221, 229
process control, 204, 217
pulse, 221
duration, 198, 204, 209, 217
energy, 198, 217
lateral, 203, 204
spatial, 204
temporal, 204, 204
pulsed, 198, 204, 209, 217
quality control, 204
reflectance, 199
repetition, 217
short pulse, 217
stability, pulse-to-pulse, 204, 221
surface alloying, 198, 206, 229
TEA, 198, 209, 217, 221
technical data, 199
treatment of surfaces, 11, 221
visible, 203
wavelength, 198, 199, 200, 203, 205, 217
Leadframes, 223
Low temperature properties, 461
acoustic properties, 492
electron-electron interaction, 478
electron localization, interaction effects, 474
low energy excitations, 462, 482, 498
superconductivity, 462
thermal properties, 486
tunneling model, 483
tunneling, scattering and resistivity, 497
tunneling, states, 462, 483
tunneling, states and relaxation, 497
tunneling, two-level system, 483
ultrasonic properties, 492

Magnetic properties
anisotropy, 411, 554
anelastic, 570
compositional dependence, 566
due to dipolar interactions, 542
energy, magnetocrystalline, 621
field annealing-induced, 564, 568, 573
local, 555
macroscopic, 558, 559, 565
random, 555, 559
single-ion, 561
stress annealing-induced, 570
two-ion, 561
uniaxial, 412
asperomagnet, 509
average moment per TM atom, 513
coercive field, annealing effect, 399, 411, 412
coordination bonding model, 513
correlation length
easy axes, 554
magnetic moments, 554, 556, 559
structural, 555, 557
critical behavior, 524
Curie temperature, 398, 409, 411, 518, 560, 561, 564
deformation map, amorphous alloys, 173
demagnetization curves, 673
density of states, 506

[Magnetic properties]
domains
magnetization rotation, 562
and shear bands, 391, 422
structure, 564, 565
wall motion, 562
early TM-late TM system, 526
energy product, 669
engineering properties, 553
exchange interaction, 506
exchange
correlation length, 555, 556, 558
electron-electron correlation, 554
exciting power, 627
Friedel virtual bound state model, 513
frustration, 508
hysteresis
approach to saturation, 565
coercive force, 562, 563
losses, 560
nucleation field, 565
reentrant flux, 565
remanence, 565
soft materials, 560, 562
hyperfine interactions, 516
magnetic moment, 511
per atom, 561
magnetic moments and exchange, 511
magnetization, compositional dependence, 561
magnetization process, general aspects, 559, 562
magnetization
saturation, 560
temperature dependence, 521, 561
magnetoelastic anisotropy, 561, 562, 570
magnetoelastic behavior, 380
magnetoelastic constants, 583, 584
magnetoelastic coupling, 584
magnetoelastic effects, 582, 647
magnetostriction
amorphous alloy, 559, 572, 574
annealing dependence, 572
and applied stress, 575
and relaxation, 572
compositional dependence, 573

[Magnetic properties]
critical exponents, 583
measurement, 559, 580
pair ordering, 566
saturation, 560
stress dependence, 572
temperature dependence, 574, 581
Mateucci effect, 564, 574, 577, 578
metal-metalloid system, 511
molecular orbital states, 514
Mossbauer spectroscopy, 517
order, magnetic, 508
overall magnetic anisotropy, 541
power loss, 627
POWERCORE® strip, 626
random anisotropy, 509
random axial anisotropy model, 538
RE-TM alloys, 527
RKKY interatction, 507
RS crystalline alloys, 627
sinle-ion anisotropy, 537
sperimagnet, 509
speromagnet, 509
spin glass, 557
spin waves, 512
stress-field-induced, 570
stress-induced, 569
susceptibility, 412, 562, 582
3d- and 4f-based amorphous alloys, 557
3d metal alloy ferromagnetism, 559
Wiedemann effect, inverse, 564, 576–578
Mechanical properties
amorphous alloy, 215, 223, 224, 379
anelastic deformation, strain, 235, 260, 379, 414, 570
boron, 400
cesium toughening, 401
creep and viscous flow, 380, 393
deformation map, 173, 174, 380
density, 280
dislocations, 401
DBTT, 398
ductility, 195
embrittlement annealing, 397

[Mechanical properties]
hydrogen, 401
neutron irradiation and, 397
phosphorus, 402
Sb, Se, Te, 401
eutectic toughening, 402
fatigue, 213, 214, 221, 405, 421, 738, 743
annealing effects, 406
environment effect, 406
overload effect of, 423
fracture, 403
hardness, 202, 212, 215, 216, 224, 226, 277
modulus, atomistic calculations, 384, 385
negative creep, 405
point defects, 407, 420
shear bands, *see* Shear bands
shear strength, joint, 721
strain at fracture, amorphous alloy, 395
stress corrosion cracking, 742
structural relaxation effects, 564
surface-treated alloys, 281
twinning, 212
wear, 213, 214, 215, 216, 219, 221, 246, 407
wear resistance, 744, 745
yield stress, amorphous alloy, 280
Young's modulus, amorphous alloy, 274, 280, 383, 384, 566, 583, 584
Melt
depth, 203, 212
Melt surface tension, *see* Interface
Melt-spinning, 7, 12
free flight, 8
Melting
containerless, 82
electromagnetic levitation, 81
surface, 18, 57, 595
Metallic glasses, *see* Amorphous Alloys
Metastability
state, 79, 559, 565, 574
crystalline alloys, 472, 498
Microsegregation, 58
Microstructure, 206, 214, 216, 226, 229
[Microstructure]
cellular, 211
dendritic, 211, 215
dimension, 202
eutectic, 30, 34, 37, 60, 70, 210, 215, 229
inhomogeneous, 198, 304
joint, 720
metastable, 206
morphology, 201
solidification, 209, 210
two-zone, 71
Momentum traps, 181
Mooij laws, 434
Morphology, 201, 202, 208, 210, 211, 212
amorphous, 221
crystalline, 221
dendritic, 202, 225
structure, 221
Mott model, 433, 447
Mullins-Sekerka, 58, 62

Nonequilibrium, 24, 25, 52
Nucleation, 172, 201, 210, 211, 229, 311
catalytic sites, 35
competitive, 44
heterogeneous, 20, 27, 35, 44, 68, 81, 85, 332, 172
homogeneous, 22, 44, 171
kinetic theory of, 82
kinetics, 35, 40, 46
multiple, 49
powder, 41
rate, 35
rate parameters, 36
surface, 35, 87
transient, 36
undercooled liquid, 50

Ostwald ripening, 310

Partition coefficient, 52, 55
Peclet number, 57, 63
Penrose tiling, 243
Phase
hexatic, 259

[Phase]
icosahedral, 87
metastable, 29, 45, 160, 171, 189, 209
primary, 61
selection of, 421
separation, amorphous, 26, 45, 49, 52
stable, 209
supersaturated solid solution, 160, 171
Phase diagrams
calculation of, 31
metastable, 26, 66, 160
monotectic reaction, 30
Phase separation
embrittlement and, 421
Phase transformations, 303
examples, 315
metastable phases, 318
Poisson statistics, 49
Powder, 10, 42
Al-8Fe, 65
consolidation, 157
growth rate, 70
mechanically alloyed, 165
microstructure, 165
nanocrystalline alloy, 189
size distribution of, 41
solidification, 65
undercooling, 20
Precipitation, 316
thermodynamics, 305
Processes
atomization, 18, 20, 65, 756
background, 1
brazing, 691
bulk sample, 164, 413
chemical reaction, 8, 104, 420
chill block melt-spinning, 416
comparison of, 171
drop tube, 47
dynamic consolidation, 173, 175
electrodeposition, 3, 413
energetic particle, 1, 416
history, 11
hydrogenization, 419
laser treatment, 419
MAGNEQUENCH, 666
mechanical alloying, 164, 417, 423
[Processes]
melt extraction, 18
melt quenching, 414
melt-spinning, 4, 17, 18, 20, 21, 51, 757
metal joining, 691
Nd-Fe-B magnet
compaction-molded, 672, 675
fully dense, 673
injection molded, 672, 675
nomenclature, 7
planar flow casting, 10, 18
rapid solidification, 103, 158, 168
rare earth oxides, reduction, 666
shock wave, 175
soldering, 691
solid state reaction, 159, 170
spark erosion, 11
spray forming, 11, 124, 756
twin-roller, 111
vapor phase, 2
wire, 10, 111, 139
Properties, 195, 210
acoustic, 492
biocompatibility, 223
chemical, *see* Chemical properties
combination of, importance, 622, 745
corrosion, 205, 221, 223, 224, 226, 229
dynamic, 215
elastic, *see* Elastic properties
low temperature, *see* Low temperature properties
magnetic, *see* Magnetic properties
mechanical, *see* Mechanical properties
roughness, 205, 226
self-lubricating, 229
strength, 215, 223
thermal, 196, 486
transport, *see* Transport properties
tribological, 205, 216, 223, 224, 225

Quasicrystals, 472, 298
decomposition, 321
model, 242
structure, 242, 245

Quenching, 196, 197, 199, 200, 201
origins of, 3
rapid, 199, 205, 209, 210, 215
rate, 196, 200, 206, 208, 210, 214, 219, 229
self, 221, 229
techniques, 196
Quenching, liquid state
applications, 13
history, 1
motivation, 4, 7
origins, 4
Quenching, vapor phase, 2
Quenching, solid state, 1

Recalescence, 17, 50, 67
time, 95
Recrystallization, 215, 221
Reflectometry
differential, 204, 217
Relaxation
AES model, 294
and production conditions, 276
and structural defects, 297
kinetics, 293
mechanisms, 294
memory effect, 272
modeling, 293
monotonic, 273
reversible, 274
stress, 392, 560
negative, 395
structural, 235, 260, 269, 559, 564
and free volume model, 380
and glass transition, 260
and property changes, 273, 278, 279
and short range order, 565
and tunneling states, 497
and viscosity measurements, 393
times, log-normal distribution, 414
Resistivity, *see* Conductivity
Ribbon
formation, 104
temperature distribution, 110

Scattering
differential anomalous, 238, 239, 247
EXAFS, 237, 568
high energy, monochromatic, 237
high resolution, 237
neutron, 239
spin-polarized, 241
small angle X-ray, 399, 402
synchrotron radiation, 235
time-of-flight method, 235
X-ray, 232
energy dispersive, 234
Scheil
analysis of solute, 59
Segregation, 202, 211, 221
Shear bands, 390
etching and annealing of, 391
fracture of, 391
free volume model, 380
magnetic domains around, 391
reverse shear of, 391
velocity of, 391
Short range order
bond-orientational order (BOA), 258
compositional (CSRO), 240, 255, 276, 278, 295, 296
CSRO and Curie temperature, 256
topological (TSRO), 276, 278, 295
Shot peening, 390
Simulation, 197
Solid solubility
extension, 3
Solid solution, 201, 211, 213, 214, 224
supersaturated, 201, 202, 209, 210
Solidification, 202, 210, 211, 229
cellular, 58, 201, 210
dendritic, 93, 201, 202, 209
directional, 71
epitaxial, 201, 202, 229
eutectic, 211
fundamentals, 17
glassy, 221, 229
kinetics, 201
metastable, 229
microsegregation, 58

[Solidification]
morphology, 206
partition coefficient, 53
partitionless, 25, 32, 38, 50, 53, 55, 69, 169
planar, 201, 210
rapid, 196, 202, 213, 229
rate, 202, 210, 211
solute diffusion, 57
solute trapping, 53
texture, 203
time-temperature curve, 201, 202, 210, 221
topotaxial, 201
velocity, 201, 202, 211, 212, 229
Solidus
retrograde, 33
Solubility
extension of, 33, 34, 360
hydrogen, 413
Solute
profiles of, 59
redistribution of, 57, 62, 72
trapping, 53
Spacing
dendrite arm, 202, 211, 212
lamellar, 202
Specific heat, *see* Heat Capacity
Stokes-Einstein relation, 380, 395
Stress
atomic level, 254, 261, 381, 420
internal, 555, 562
long range, 559, 560
shear, 577
thermal, 203, 219, 221
torsion, 579, 580
Structure, 201
amorphous alloys, 249, 269, 379, 553
theories, 252
anisotropy, 6
atomic, 231
atomic pair distribution function, 240
"bamboo," 112
bond orientational anisotropy (BOA), 81
correlation length, 555, 557
[Structure]
defects
amorphous alloys, 6
magnetic domain pinning centers, 562
structural, 562, 565
experimental analysis methods, 231
factor, 233, 432
lattice periodicity, 553
martensite, 221
measurements, 282
microcellular, 65
microcrystalline, 51, 740
multigrain, 49
quasicrystals, 242, 245
two-zone (powder), 65
Superconductivity, 462
BCS-type, 463
localization effects on, 469
quasicrystalline and metastable RS alloys, 472
refractory metallic glasses, 450
RS high T_c superconductors, 473, 499
simple amorphous alloys, 464
transition amorphous metals, 466
upper critical field and localization, 469
Supercooling, *see* Undercooling
Surface
layers, 195
treatment laser, 207, 758

Temperature
free energy diagram, 305
gradient, 196, 197, 201, 209, 211, 229
surface, 197, 204, 209
Thermodynamics, 23
free energy diagrams, amorphous alloys, 160
phase transformations, 304
T_0 curves, 169
Transformation, 197, 219, 303
allotropic, 27
continuous cooling, 49, 560
coupled zone, 61, 66
eutectoid, 308, 318

[Transformation]
martensitic, 556
peritectoid decomposition, 308, 321
polymorphous, 307, 320, 325
primary crystallization, 308, 555
quenching, 553
spinodal decomposition, 305, 307, 322
surface-induced, 327
T-T-T curves, 45, 61, 114, 116, 171
Transport properties, 431
d-band conduction, 434
Debye temperature, 433
density of states (DOS), minimum in, 442
electron wave function, 436
Faber-Ziman theory, 432
LCAO method, 435
localization, 436
LT metal-metalloid alloys, 444
low temperature, *see* Low temperature properties
nonmagnetic early-late TM alloys, 449
phonon-assisted tunneling, 435, 484
quantum interference, correlation effects, 436
random tunneling proecesses, 435
RE-TM alloys, 447
resistivity, temperature coefficient, 431
s-d scattering, 433, 447
simple amorphous alloys, 440
Ziman theory, 431

Undercooling
bath, 62
constitutional, 58
control of, 19
curvature, 62
initial, 21, 22, 64, 71
levitation, 87
liquid, 19
maximum, 41
nucleation and, 80, 86
solidification and, 79
solute, 62
thermal, 62

Valve, 215, 222, 226

Viscosity
amorphous alloy, 113, 187, 269, 280, 287, 291
annealing time dependence, 275, 290
diffusion relation, 291
heating rate effect, 394
preannealing effect, 394
temperature dependence of, 394
Viscous flow
amorphous alloys, 114, 287

Wetting coefficient, 694
Wire
cold-drawn, 392, 406
composition range, 149
confluent coolant method, 142
conveyor belt method, 152
fundamental characteristics, 153
glass-coated melt-spinning, 141
in-rotating-water method, 143
magnetization of, 563
mm-scale, 164, 368, 385, 389
necessary conditions, 145
production techniques, 139
Taylor process, 7

Yielding
criteria, 388
deformation and stress annealing, 389
dispersion hardening, 388
free volume model, 388
shear bands, 390
Tresca criterion, 388
von Mises criterion, 380